Infinite Series

$$\sum_{n=m}^{\infty} cr^n = \frac{cr^m}{1-r} \quad \text{(geometric series)}$$

$$e^x = \sum_{n=0}^{\infty} \frac{x^n}{n!}$$

$$\sin x = \sum_{n=0}^{\infty} \frac{(-1)^n}{(2n+1)!} x^{2n+1}$$

$$\cos x = \sum_{n=0}^{\infty} \frac{(-1)^n}{(2n)!} x^{2n}$$

$$(1+x)^s = \sum_{n=0}^{\infty} \binom{s}{n} x^n \quad \text{for } -1 < x < 1 \quad \text{(binomial series)}$$

$$f(x) = \sum_{n=0}^{\infty} \frac{f^{(n)}(a)}{n!} (x-a)^n \quad \text{(Taylor series)}$$

$$r_n(x) = \frac{f^{(n+1)}(t_x)}{(n+1)!} (x-a)^{n+1} \quad \text{(nth Taylor remainder)}$$

Geometric Formulas

 area: $\frac{1}{2} bh$

 area: bh

 area: πr^2

circumference: $2\pi r$

 area: $\frac{\theta}{2} r^2$

 volume: abc

surface area: $2ab + 2ac + 2bc$

 volume: $\frac{4}{3}\pi r^3$

surface area: $4\pi r^2$

 volume: $\pi a^2 h$

surface area: $2\pi ah$

 volume: $\frac{1}{3}\pi a^2 h$

surface area: $\pi a \sqrt{a^2 + h^2}$

$$|PQ| = \sqrt{(x_2 - x_1)^2 + (y_2 - y_1)^2} \quad \text{(distance formula)}$$

$$y - y_1 = m(x - x_1) \quad \text{(point–slope equation of a line)}$$

$$y = mx + b \quad \text{(slope–intercept equation of a line)}$$

$$y - y_1 = \frac{y_2 - y_1}{x_2 - x_1}(x - x_1) \quad \text{(two-point equation of a line)}$$

CALCULUS

WITH ANALYTIC GEOMETRY

ALTERNATE EDITION

CALCULUS

WITH ANALYTIC GEOMETRY
ALTERNATE EDITION

ROBERT ELLIS

University of Maryland
at College Park

DENNY GULICK

University of Maryland
at College Park

HARCOURT BRACE JOVANOVICH, PUBLISHERS
and its subsidiary, Academic Press

San Diego   New York   Chicago   Austin   Washington, D.C.
London   Sydney   Tokyo   Toronto

To Rosemarie and Mark
To Frances, David, Barbara, and Sharon

ISBN: 0-15-505700-6

Library of Congress Catalog Card Number: 87-81139

Printed in the United States of America

Technical Art by Vantage Art, Inc.

Cover credit: "In Equilibrium" 1987, by Masoud Yasami. Courtesy of Elaine Horwitch Galleries, Scottsdale, Arizona

Photo Credits

p. 32: HBJ Photo
57: NASA
64: Courtesy of Educational Development Center, Newton, MA
200: Courtesy of French Cultural Services
216: NASA
306: The Oriental Institute, University of Chicago
307: NASA
345: National Park Service
393: UPI/Bettmann News Photos
400: The Cleveland Museum of Natural History
403: NASA
416: The Convention and Tourist Board of Greater St. Louis
507: From *Sounds of Music* by Charles Taylor, BBC London
585: NASA
586: The Port Authority of New York and New Jersey
595: National Park Service
667: Reproduced by permission of the publisher of *Physics*, 2nd ed., by the Physical Science Study Committee, © 1965, p. 100, D.C. Heath and Company
685: UPI/Bettmann News Photos
705: Stephen J. Potter/Stock, Boston
714: UPI/Bettmann News Photos
771: Swiss National Tourist Office
960: UPI/Bettmann News Photos

PREFACE

The Alternate Edition of *Calculus with Analytic Geometry* contains all the topics that normally constitute a course in calculus of one and several variables. It is suitable for sequences taught in three semesters or in four or five quarters. In the three-semester case, the first semester will usually include the introductory chapter (Chapter 1), the three chapters on limits and derivatives (Chapters 2–4), and the initial chapter on integrals (Chapter 5). The second semester would then include the rest of the discussion of integration (Chapters 6–8) and some combination of the chapters on series (Chapter 9), conic sections (Chapter 10), and the introduction to vectors and vector-valued functions (Chapters 11 and 12). The third semester would include the remainder of those chapters, along with the material on calculus of several variables (Chapters 13 and 14) and Chapter 15, which includes the theorems of Green and Stokes as well as the Divergence Theorem. Finally, an optional chapter on differential equations can be covered in whole or in part during the second or third semester.

The major difference between this Alternate Edition and the Third Edition (which will continue to be available) is that in the Alternate Edition the chapter on applications of integration precedes, rather than follows, techniques of integration. This change was made partly in response to suggestions from reviewers and users of previous editions, and partly in recognition of the advent of symbolic integration, now available with calculators as well as computers.

Other noteworthy changes in this edition:

- The concepts of tangent lines and velocity are discussed at the outset of Chapter 2, in order to document the need for, and provide an immediate application of, the notion of limit. This alteration necessitated extensive rewriting of Sections 2.1 and 2.2 and affected the examples and exercises throughout Chapter 2.
- In order to make the basic limit theorems more accessible to the reader, these theorems are treated in two groups, with the Sum, Constant Multiple, Difference, Product, and Quotient Rules appearing in Section 2.3 and the Squeezing Theorem and Substitution Rule in Section 2.4.
- A new section on integration using tables has been added to the book and appears as Section 8.5.
- The length of a curve described in polar coordinates is considered in Section 6.10, along with areas of regions described in polar coordinates.

Although we have been careful in selecting the order in which the topics appear in this edition, there is flexibility in the choice of topics and the order in which they are introduced. Chapter 1 (which includes a section on trigonometry, so that trigonometric functions can serve as examples throughout the book) is preliminary and can be covered quickly if the student's preparation is sufficient. With a little care, techniques of integration (Chapter 8) can be discussed before applications of the integral (Chapter 6). Sequences and series (Chapter 9) can be studied any time after Chapter 8, conic sections (Chapter 10) any time after Chapter 4, and differential equations (Chapter 16) any time after Chapter 8.

Whenever possible, we use geometric and intuitive motivation to introduce concepts and results, so that students may readily absorb the carefully worded definitions and theorems that follow. The topical development, in which we employ numerous worked examples and almost 900 illustrations, aims for clarity and precision without overburdening the reader with formalism. In keeping with this goal, we have proved most theorems of first-year calculus in the main body of the text but have placed the more difficult proofs in the Appendix. In the chapters on calculus of several variables we have proved selected theorems that aid comprehension of the material.

Exercises appear both at the ends of sections and, for review, at the end of each chapter. Each set begins with a full complement of routine exercises to provide practice in using the ideas and methods presented in the text. These are followed by applied problems and by other exercises of a more challenging nature (identified with an asterisk). To supplement the usual problems from physics and engineering, we have included many from business, economics, biology, chemistry, and other disciplines, as well as a smaller number of exercises suitable for solution on a calculator (indicated by the symbol 🖸). In addition, Chapters 3–15 each end with a collection of cumulative review exercises, which are intended to reinforce the main ideas of the previous chapters. In the interest of accuracy every exercise has been completely worked by each of the authors. Answers to odd-numbered exercises (except those requiring longer explanations) appear at the back of the book.

Throughout the book, statements of definitions, theorems, lemmas, and corollaries, as well as important formulas, are highlighted with tints for easy identification. Numbering is consecutive throughout each chapter for definitions and theorems, and consecutive within each section for examples and formulas. We use the symbol ■ to signal the end of a proof and □ for the end of the solution to an example.

Lists of Key Terms and Expressions, Key Formulas, and Key Theorems appear at the end of each chapter. On the endpapers we have assembled important formulas and results that the student will want to have handy, both for course review and for reference in later studies. Pronunciation of difficult terms and names is shown in footnotes on the pages where they first appear.

We are very grateful to many people who have helped us in a variety of ways as we prepared the various editions of this book. Our thanks go to reviewers Daniel D. Anderson (*University of Iowa*), Raymond J. Cannon, Jr. (*Baylor University*), Douglas Crawford (*College of San Mateo*), Arthur Crummer (*University of Florida*), Robert M. Dieffenbach (*Miami University, Ohio*), J. R. Dorroh (*Louisiana State University*), Daniel Drucker (*Wayne State University*), Bruce Edwards (*University of Florida*), Murray Eisenberg (*University of Massachusetts*), Robert Forward, M.D. (*Grand Forks, North Dakota*), Charles H. Franke (*Seton Hall University*), Robert Gold (*Ohio State University at Columbus*), Jack Goldberg (*University of Michigan*), Stuart Goldenberg (*California Polytechnic University, San Luis Obispo*), Robert B. Hughes (*Boise State University*), Richard Koch (*University of Oregon*), J. D. Konhauser (*Macalester College*), Theodore Laetsch (*University of Arizona*), Peter Lindstrom (*Genessee Community College*), David J. Lutzer (*Miami University, Ohio*), Hugh B. Maynard (*University of Texas at San Antonio*), Peter Nyikos (*University of South Carolina*), Jack Robertson (*Washington State University*), M. M. Subramaniam (*Pennsylvania State University, Delaware Campus*), John Thorpe (*State University of New York at Stonybrook*), Mark S. Ubelhor (*Scott Community College*), Abraham Weinstein (*Nassau Community College*), and Paul Zorn (*St. Olaf College*). Many of our colleagues at the University of Maryland have made contributions to the original writing of this book and to the revisions; we wish to express our appreciation to William Adams, Stuart Antman, Douglas Arnold, Joseph Auslander, Kenneth Berg, Ellen Correl, Jerome Dancis, Gertrude Ehrlich, Craig Evans, Seymour Goldberg, Jacob Goldhaber, Paul Green, Frances Gulick, Bert Hubbard, James Hummel, Nelson Markley, James Owings, Jonathan Rosenberg, Karl Stellmacher, C. Robert Warner, Peter Wolfe, James Yorke, and Mishael Zedek. In addition, we are grateful for comments and suggestions from Bruce L. Aborn (*Bentley College*), Steven Agronsky (*California Polytechnic University, San Luis Obispo*), Robert Baer (*Miami University*), David W. Bange (*University of Wisconsin, LaCrosse*), Don Blevins (*Trinity College*), Thomas T. Bowman (*University of Florida*), Art Bukowski (*University of Alaska at Anchorage*), Martin Buntinas (*Loyola University*), Lawrence O. Cannon (*Utah State University*), Ray Cannon (*Stetson University*), Elizabeth B. Chang (*Hood College*), F. Lee Cook (*University of Alabama, Huntsville*), Craig Cordes (*Louisiana State University*), Brad Crain (*Portland State University*), Hall Crannell (*Catholic University of America*), John S. Cross (*University of Northern Iowa*), Randall Dahlberg (*Seton Hall University*), Leroy Damewood (*Eastern Oregon State College*), Lynn K. Davis (*University of Cincinnati*), Loyal Farmer (*Cameron University*), Gerald Farrell (*California Polytechnic University, San Luis Obispo*), Bill Finch (*University of Florida*), Gregory D. Foley (*North Harris County College*), Robert Fontenot (*Whitman College*), Juan A. Gatica (*University of Iowa*), Donald Gray (*Iowa Western Community College*), Harvey C. Greenwald (*California Polytechnic University, San Luis Obispo*), Charles Groetsch (*University of Cincinnati*), Edwin Halfar (*University of Nebraska*), Leona Henry (*Mercy College*), Stephen R. Hilding (*Gustavus Adolphus College*), Tim Hodges (*University of Cincinnati*), Dean W. Hooner (*Alfred University*), Brindell Horelick (*University of Maryland, Baltimore County*), Shirley Huffman (*Southwest Missouri State University*), Ronald Infante (*Seton Hall University*), Cassius T. Ionescu Tulcea (*Northwestern University*), Bernice Kastner (*Montgomery College*), Dan Kemp (*South Dakota State University*), John T. Kemper (*College of St. Thomas*), Frank Kost (*State University of New York at Oneonta*), Charles Lanski (*University of Southern California*), David Lehmann (*Southwest Missouri State*), Verlyn Lindell (*Augustana College*), Lowell Lynde (*University of Arkansas*), Danny W. McCarthy (*Tulane University*), Jim McKinney (*California State Polytechnic University, Pomona*), Jerome H. Manheim (*California State University, Long Beach*), Bill Marion (*Valparaiso University*), Frank Mathis (*Baylor University*), John Moriarty (*University of Cincinnati*), Roger H. Moritz (*Alfred University*), Kent Morrison (*California Polytechnic State University, San Luis Obispo*), James M. Nare (*University of Tennessee, Chattanooga*), Michael J. Nowak (*United States International University, San Diego*), Jim Osterburg (*University of Cincinnati*), Nancy Jim Poxon (*California State University, Sacramento*), Lolan Redden (*Cumberland College*), Robert Reisel (*Loyola University*), Wayne Rich (*Utah State University*), Joyce Riseberg (*Montgomery College*),

Norlin Rober (*Marshalltown Community College*), Larry P. Runyan (*Shoreline Community College*), John Saccoman (*Seton Hall University*), Helen Salzberg, (*Rhode Island College*), Rollin T. Sandberg (*California State University, Fullerton*), Franklin E. Schroeck (*Florida Atlantic University*), Karen J. Schroeder (*Bentley College*), John Shupert (*Shawnee State Community College*), Kermit Sigmon (*University of Florida*), Madelyn Smith (*Gloucester County College*), James M. Sobota (*University of Wisconsin, La Crosse*), Ken Solem (*Gloucester County College*) Lavinia Spilman (*University of Alabama*), Robert Steward (*Rhode Island College*), Alexander P. Stone (*University of New Mexico*), Keith Stroyan (*University of Iowa*), Virginia Taylor (*University of Lowell*), Terry R. Tiballi (*North Harris County College*), George Van Zwalenberg (*Calvin College*), David S. Watkins (*Utah State University*), Philip M. Whitman (*Rhode Island College*), Robert C. Williams (*Alfred University*), Stephen J. Willson (*Iowa State University*), and Elmar Zemgalis (*Highline Community College*).

Finally, we are grateful to the staff of Harcourt Brace Jovanovich, Inc., for its assistance in the preparation of all editions of this text. For the Alternate Edition, special thanks go to mathematics editor Richard Wallis, for his continuing support and encouragement.

Robert Ellis • *Denny Gulick*

TO THE READER

When you begin to study calculus, you will find that you have encountered many of its concepts and techniques before. Calculus makes extensive use of plane geometry and algebra, two branches of mathematics with which you are already familiar. However, added to these is a third ingredient, which may be new to you: the notion of limit and of limiting processes. From the idea of limit arise the two principal concepts that form the nucleus of calculus; these are the derivative and the integral.

The derivative can be thought of as a rate of change, and this interpretation has many applications. For example, we may use the derivative to find the velocity of an object, such as a rocket, or to determine the maximum and minimum values of a function. In fact, the derivative provides so much information about the behavior of functions that it greatly simplifies graphing them. Because of its broad applicability, the derivative is as important in such disciplines as physics, engineering, economics, and biology as it is in pure mathematics.

The definition of the integral is motivated by the familiar notion of area. Although the methods of plane geometry enable us to calculate the areas of polygons, they do not provide ways of finding the areas of plane regions whose boundaries are curves other than circles. By means of the integral we can find the areas of many such regions. We will also use it to calculate volumes, centers of gravity, lengths of curves, work, and hydrostatic force.

The derivative and the integral have found many diverse uses. The following list, taken from the examples and exercises in this book, illustrates the variety of fields in which these powerful concepts are employed.

	Section
Windpipe pressure during a cough	4.5
Cost of insulating an attic floor	4.5
Blood resistance in vascular branching	4.5
Surface area of a cell in a beehive	4.5
Buffon's needle problem	5.4
Volume of the great pyramid of Cheops	6.1
Force of water on an earth-filled dam	6.8
Growth of a paramecium population	7.3
Magnitude of an earthquake	7.4
Dating of a lunar rock sample	7.5
Amount of an anesthetic needed during an operation	7.5
Mass of binary stars	8.1
Pareto's Law of distribution of income	8.7
Harmonics of a stringed instrument	9.2
Multiplier effect in economics	9.4
Location of the source of a sound	10.3
Rated speed of a banked curve	12.5
Kepler's Laws of planetary motion	12.7
Escape velocity from the earth's gravitational field	12.7
Analysis of a rainbow	13.3
Effect of taxation on production of a commodity	13.3
Law of conservation of energy	15.3
Electric field produced by a charged telephone wire	15.6
Terminal speed of a falling object	16.4
Motion of a spring	16.7

The concepts basic to calculus can be traced, in uncrystallized form, to the time of the ancient Greeks. However, it was only in the sixteenth and early seventeenth centuries that mathematicians developed refined techniques for determining tangents to curves and areas of plane regions. These mathematicians and their ingenious techniques set the stage for Isaac Newton (1642–1727) and Gottfried Leibniz (1646–1716), who are usually credited with the "invention" of calculus because they codified the techniques of calculus and put them into a general setting; moreover, they recognized the importance of the fact that finding derivatives and finding integrals are inverse processes.

During the next 150 years calculus matured bit by bit, and by the middle of the nineteenth century it had become, mathematically, much as we know it today. Thus the definitions and theorems presented in this book were all known a century ago. What is newer is the great diversity of applications, with which we will try to acquaint you throughout the book.

Robert Ellis • Denny Gulick

CONTENTS

CALCULUS

WITH ANALYTIC GEOMETRY

ALTERNATE EDITION

1
FUNCTIONS

In this chapter we will review the basic properties of real numbers, introduce the concept of function, and discuss different types of functions. If you are already familiar with most of the definitions and concepts given, we suggest that you read Chapter 1 quickly and proceed to Chapter 2.

1.1
THE REAL NUMBERS

Real numbers, their properties, and their relationships are basic to calculus. Therefore we begin with a description of some important properties of real numbers.

Types of Real Numbers and the Real Number Line

The best known real numbers are the **integers**:

$$0, \pm 1, \pm 2, \pm 3, \ldots$$

From the integers we derive the **rational numbers**. These are the real numbers that can be written in the form p/q, where p and q are integers and $q \neq 0$. Thus $\frac{48}{37}$, -17, and -1.41 (which is equal to $-\frac{141}{100}$) are rational numbers. Any real number that is not rational is called an **irrational number**. Examples of irrational numbers are π and $\sqrt{2}$. (See Exercise 84 at the end of this section for a proof that $\sqrt{2}$ is irrational.)

There is an order $<$ on the real numbers. If $a \neq b$, then either $a < b$ or $a > b$. For example, $5 < 7$ and $-1 > -2$. If a is less than or equal to b, we write $a \leq b$. If

a is greater than or equal to b, we write $a \geq b$. For example, $x^2 \geq 0$ for any real number x. We say that a is **positive** if $a > 0$ and **negative** if $a < 0$. If $a \geq 0$, we say that a is **nonnegative**.

The real numbers can be represented as points on a horizontal line in such a way that if $a < b$, then the point on the line corresponding to the number a lies to the left of the point on the line corresponding to the number b (Figure 1.1).

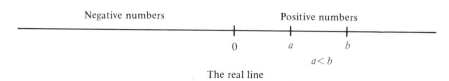

Negative numbers   Positive numbers

0   a b

$a < b$

The real line

FIGURE 1.1

Such a line is called the **real number line**, or **real line**. We think of the real numbers as points on the real line, and *vice versa*. Thus we say that the negative numbers lie to the left of 0 and the positive numbers lie to the right of 0.

Intervals   Certain sets of real numbers, called **intervals**, appear with great frequency in calculus. They can be grouped into nine categories:

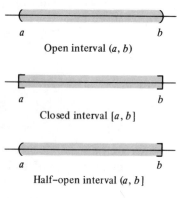

Open interval (a, b)

Closed interval $[a, b]$

Half-open interval $(a, b]$

Half-open interval $[a, b)$

FIGURE 1.2

Name	Notation	Description
Open interval	(a, b)	all x such that $a < x < b$
Closed interval	$[a, b]$	all x such that $a \leq x \leq b$
Half-open interval	$(a, b]$	all x such that $a < x \leq b$
Half-open interval	$[a, b)$	all x such that $a \leq x < b$
Open interval	(a, ∞)	all x such that $a < x$
Open interval	$(-\infty, a)$	all x such that $x < a$
Closed interval	$[a, \infty)$	all x such that $a \leq x$
Closed interval	$(-\infty, a]$	all x such that $x \leq a$
The real line	$(-\infty, \infty)$	all real numbers

Intervals of the form (a, b), $[a, b]$, $(a, b]$, and $[a, b)$ are **bounded intervals**, and a and b are the **endpoints** of each of these intervals. Figure 1.2 shows the four types of bounded intervals. Intervals of the form (a, ∞), $(-\infty, a)$, $[a, \infty)$, $(-\infty, a]$, and $(-\infty, \infty)$ are **unbounded intervals**, and a is the **endpoint** of each of the first four of these intervals. A number that is in an interval but is not an endpoint of the interval is called an **interior point** of the interval.

Caution: The symbols ∞ and $-\infty$ used above are called "infinity" and "minus infinity," respectively. They do not represent numbers.

Notice that (b, b), $(b, b]$, and $[b, b)$ contain no numbers. More generally, (a, b), $(a, b]$, and $[a, b)$ contain no numbers if $b \leq a$. Whenever we write (a, b), $(a, b]$, or

$[a, b)$, we make the implicit assumption that $a < b$. Likewise, we write $[a, b]$ only when $a \leq b$.

Inequalities and Their Properties

Statements such as $a < b$, $a \leq b$, $a > b$, and $a \geq b$ are called **inequalities**. We list several basic laws for inequalities. In what follows $a, b, c,$ and d are assumed to be real numbers.

Trichotomy: Either $a < b$, or $a > b$, or $a = b$,
    and only one of these holds for any given a and b.   (1)

Transitivity: If $a < b$ and $b < c$, then $a < c$.   (2)

Additivity: If $a < b$ and $c < d$, then $a + c < b + d$.   (3)

Positive multiplicativity: If $a < b$ and $c > 0$, then $ac < bc$.   (4)

Negative multiplicativity: If $a < b$ and $c < 0$, then $ac > bc$.   (5)

Replacing $<$ by $\leq$ and $>$ by $\geq$ in laws (2)–(5) yields four new laws for inequalities, which we will also find useful.

The word "trichotomy" in (1) means a threefold division. The trichotomy law states that any two numbers a and b are related in exactly one of the three ways listed in (1). For example, given the two numbers 3.1416 and π, we have either $3.1416 < \pi$, $3.1416 = \pi$, or $3.1416 > \pi$. (The last is actually correct.)

For simplicity of notation, two inequalities are sometimes combined. For example, if $a \leq b$ and $b \leq c$, then we can write $a \leq b \leq c$.

Caution: The multiplication laws, (4) and (5), must be carefully observed. To illustrate their use, we will present several examples. In each, the problem is to solve an inequality, which means to find all real numbers that satisfy the inequality.

Example 1 Solve the inequality $1/x < 3$.

Solution First we observe that 0 cannot be a solution because division by 0 is impossible. Next we multiply through by x to eliminate x from the denominator. For positive x, (4) yields $1 < 3x$ or $\frac{1}{3} < x$. Thus the numbers in $(\frac{1}{3}, \infty)$ constitute one part of the solution of the given inequality. For negative x, (5) yields $1 > 3x$, or $\frac{1}{3} > x$. Since the last inequality is satisfied by all $x < 0$, a second part of the solution consists of all x in $(-\infty, 0)$. Therefore the complete solution consists of all numbers in the interval $(-\infty, 0)$ and all numbers in the interval $(\frac{1}{3}, \infty)$. $\square$

When the solution of an inequality forms only one interval, we will write only that interval as the solution. However, if the solution of an inequality consists of more than one interval, we will refer to the solution as the **union** of these intervals. Thus the solution of the inequality $1/x < 3$ in Example 1 consists of the union of the intervals $(-\infty, 0)$ and $(\frac{1}{3}, \infty)$.

Example 2   Solve the inequality $-1 < -2x + 3 \leq 2$.

Solution   The given inequality is equivalent to the following pair of inequalities:

$$-1 < -2x + 3 \quad \text{and} \quad -2x + 3 \leq 2$$

First subtracting 3 throughout, we find that

$$-4 < -2x \quad \text{and} \quad -2x \leq -1$$

Then dividing throughout by -2 and reversing the inequality signs, we obtain

$$2 > x \quad \text{and} \quad x \geq \frac{1}{2}$$

Thus the solution consists of all numbers x satisfying $\frac{1}{2} \leq x < 2$, that is, the interval $[\frac{1}{2}, 2)$.   □

Since we performed the same algebraic manipulation on both inequalities $-1 < -2x + 3$ and $-2x + 3 \leq 2$, we could have solved the original inequality without splitting it up:

$$-1 < -2x + 3 \leq 2$$
$$-4 < -2x \leq -1$$
$$2 > x \geq \tfrac{1}{2}$$

In solving most inequalities, we will need to find values of x for which a certain expression in x is positive (or negative). We will need to be careful to observe the negative multiplicativity rule (5) when we multiply negative numbers.

Example 3   Solve the inequality

$$\frac{(x - 1)(x - 3)}{x + 2} > 0$$

Solution   First we draw a diagram that shows the signs of the factors $x - 1$ and $x - 3$ of the numerator and $x + 2$ of the denominator.

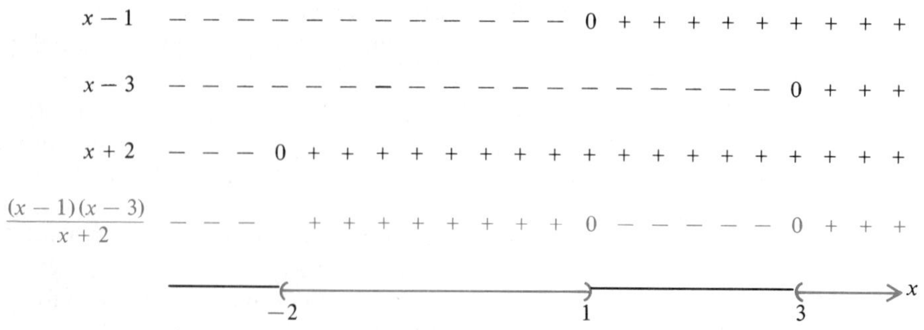

FIGURE 1.3

Then we deduce the sign of $(x - 1)(x - 3)/(x + 2)$ for various values of x, and determine where it is positive. From Figure 1.3 we see that the solution of the given inequality is the union of the intervals $(-2, 1)$ and $(3, \infty)$. $\square$

If we had wished to solve the inequality $\dfrac{(x - 1)(x - 3)}{x + 2} \le 0$, we would have used the same diagram, but at the end we would have selected the union of those intervals on which $(x - 1)(x - 3)/(x + 2)$ is nonpositive, namely $(-\infty, -2)$ and $[1, 3]$.

In Section 2.7 we will discuss a second method of solving inequalities that uses results from calculus.

Absolute Value

The **distance** between a and b on the real line is either $a - b$ or $b - a$, whichever is nonnegative (Figure 1.4). Likewise, the distance between 0 and b is either $b - 0 = b$ or $0 - b = -b$, whichever is nonnegative. The distance between b and 0 is the basis for the definition of the absolute value of b.

FIGURE 1.4

DEFINITION 1.1

The **absolute value** of any real number b is b if $b \ge 0$ and is $-b$ if $b < 0$. The absolute value of b is denoted $|b|$. Thus

$$|b| = \begin{cases} -b & \text{for } b < 0 \\ b & \text{for } b \ge 0 \end{cases}$$

For example, $|6| = 6$, $|0| = 0$, $|-5| = -(-5) = 5$, and $|8 - 17| = |-9| = -(-9) = 9$. Notice that $|b|$ is the larger of b and $-b$, whichever is nonnegative. Geometrically, $|b|$ is the distance between 0 and b. More generally, $|a - b|$ is the distance between the numbers a and b.

We will use the following properties of absolute value:

$$|-a| = |a| \quad \text{and} \quad |a - b| = |b - a| \tag{6}$$

$$|ab| = |a||b| \quad \text{and} \quad |b^2| = |b|^2 \tag{7}$$

$$-|b| \le b \le |b| \tag{8}$$

$$|a + b| \le |a| + |b| \tag{9}$$

$$|a - b| \ge ||a| - |b|| \tag{10}$$

Except for (9) and (10), these properties follow directly from Definition 1.1. We will verify (9) and leave (10) as an exercise. To verify (9), we first use (8):

$$-a \le |a| \quad \text{and} \quad a \le |a|$$
$$-b \le |b| \quad \text{and} \quad b \le |b|$$

Adding these inequalities vertically yields

$$-(a + b) = -a - b \le |a| + |b| \quad \text{and} \quad a + b \le |a| + |b|$$

Since $|a + b|$ is the larger of $a + b$ and $-(a + b)$, it follows that

$$|a + b| \le |a| + |b|$$

which verifies (9).

Next we show that if $b > 0$, then

$$|x| < b \quad \text{if and only if} \quad -b < x < b \tag{11}$$

To verify (11) we notice that $|x| < b$ means that

$$\text{if } x \ge 0, \quad \text{then} \quad x < b$$

and

$$\text{if } x < 0, \quad \text{then} \quad -x < b, \text{ or equivalently, } -b < x$$

From (11) we see that the solution of the inequality $|x| < b$ is the open interval $(-b, b)$. Statements analogous to (11) pertain to inequalities of the form $|x| \le b$, $|x| > b$, and $|x| \ge b$.

Example 4   Solve the inequality $|x - 1| < 3$.

Solution   By (11) with x replaced by $x - 1$ and b replaced by 3, the inequality is equivalent to

$$-3 < x - 1 < 3$$

or equivalently, $-2 < x < 4$. Thus the solution is $(-2, 4)$.   □

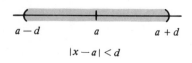

$|x - a| < d$

FIGURE 1.5

Geometrically, $|x - 1| < 3$ means that the distance between x and 1 is less than 3. More generally, $|x - a| < d$ means the distance between x and a is less than d. Thus $|x - a| < d$ if and only if x lies in the interval $(a - d, a + d)$ (Figure 1.5). Algebraically,

$$|x - a| < d \quad \text{if and only if} \quad a - d < x < a + d$$

Example 5   Find all values of x such that $0 < |x - a| < d$, where a is any number and d is any positive number.

Solution The double inequality $0 < |x - a| < d$ means that

$$0 < |x - a| \quad \text{and} \quad |x - a| < d$$

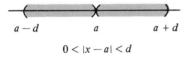

$$0 < |x - a| < d$$

FIGURE 1.6

From $0 < |x - a|$ we know that $x \neq a$. By our comments above, the values of x satisfying $|x - a| < d$ lie in the interval $(a - d, a + d)$. These two observations give the complete solution, which is the union of the intervals $(a - d, a)$ and $(a, a + d)$ (Figure 1.6). □

This concludes our discussion of real numbers, inequalities, and absolute values. The concepts and rules we have given will play an important part in our study of calculus.

EXERCISES 1.1

In Exercises 1–4 determine whether $a < b$ or $a > b$.

1. $a = \frac{4}{9}, b = \frac{7}{16}$
2. $a = -\frac{1}{7}, b = -0.142857$
3. $a = \pi^2, b = 9.8$
4. $a = (3.2)^2, b = 10$

5. Use the fact that $(\sqrt{2})^2 = 2$ to determine whether $\sqrt{2} < 1.41$, $\sqrt{2} = 1.41$, or $\sqrt{2} > 1.41$.

6. Use the fact that $(\sqrt{11})^2 = 11$ to determine whether $\sqrt{11} < 3.3$, $\sqrt{11} = 3.3$, or $\sqrt{11} > 3.3$.

In Exercises 7–14 state whether the interval is open, half-open, or closed and whether it is bounded or unbounded. Then sketch the interval on the real line.

7. $[-4, 5]$
8. $(-2, -1)$
9. $(-\infty, 3)$
10. $[\frac{3}{2}, \frac{5}{2})$
11. $[0, \infty)$
12. $(5, 7)$
13. $(-\infty, -1]$
14. $[-\frac{1}{2}, \frac{1}{2}]$

In Exercises 15–18 write the union of the two intervals as a single interval.

15. $(-3, 2)$ and $[1, 4)$
16. $(-\infty, 0]$ and $[0, 3)$
17. $(1, 3)$ and $(2, \infty)$
18. $(-\infty, \frac{1}{2}]$ and $(0, \infty)$

In Exercises 19–38 solve the inequality.

19. $-6x - 2 > 5$
20. $4 - 3x \geq 7$
21. $-1 \leq 2x - 3 < 4$
22. $-0.1 < 3x + 4 < 0.1$
23. $(x - 1)(x + \frac{1}{2}) \geq 0$
24. $(x-1)(x-2)(x-3) \leq 0$
25. $x(x - \frac{2}{3})(x + \frac{1}{3}) < 0$
26. $\dfrac{x}{(x - 1)(x + 2)} > 0$
27. $\dfrac{(2x - 1)^2}{(x + 1)(x + 3)} \geq 0$
28. $\dfrac{(2x - 3)(4x + 1)}{x - 2} \leq 0$
29. $4x^3 - 6x^2 \leq 0$
30. $3x^2 - 2x - 1 \geq 0$
31. $8x - \dfrac{1}{x^2} > 0$
32. $8x + \dfrac{1}{x^2} < 0$

33. $\dfrac{4x(x^2 - 6)}{x^2 - 4} < 0$
34. $\dfrac{2x(x^2 - 3)}{(x^2 + 1)^3} \geq 0$
35. $\dfrac{t^2 + t - 2}{(t^2 - 1)^3} \geq 0$
36. $\dfrac{t^2 - 2t - 3}{t^2 - 8t + 15} > 0$
37. $\dfrac{2 - x}{\sqrt{9 - 6x}} > 0$
38. $\dfrac{2x^2 - 1}{(1 - x^2)^{1/2}} < 0$

In Exercises 39–42 solve the inequality.

39. $\dfrac{1}{x + 1} > \dfrac{3}{2}$ (*Hint:* Write the inequality as $1/(x + 1) - 3/2 > 0$. Then rewrite the left side as a single fraction.)

40. $\dfrac{1}{3 - x} < -2$
41. $\dfrac{x + 1}{x - 1} \leq \dfrac{1}{2}$
42. $\dfrac{2 - 5x}{3 - 4x} \geq -2$

In Exercises 43–46 evaluate the expression.

43. $-|-3|$
44. $|-\sqrt{2}|^2$
45. $|-5| + |5|$
46. $|-5| - |5|$

In Exercises 47–58 solve the equation.

47. $|x| = 1$
48. $|x| = \pi$
49. $|x - 1| = 2$
50. $|2x - \frac{1}{2}| = \frac{1}{2}$
51. $|6x + 5| = 0$
52. $|3 - 4x| = 2$
53. $|x| = |x|^2$
54. $|x| = |1 - x|$
55. $|x + 1|^2 + 3|x + 1| - 4 = 0$
56. $|x - 2|^2 - |x - 2| = 6$
57. $|x + 4| = |x - 4|$
58. $|x - 1| = |2x + 1|$

In Exercises 59–70 solve the inequality.

59. $|x - 2| < 1$
60. $|x - 4| < 0.1$
61. $|x + 1| < 0.01$
62. $|x + \frac{1}{2}| \leq 2$

63. $|x + 3| \geq 3$

64. $|x - 0.3| > 1.5$

65. $|2x + 1| \geq 1$

66. $|3x - 5| \leq 2$

67. $|2x - \frac{1}{3}| > \frac{2}{3}$

68. $0 < |x - 1| < 0.5$

69. $-1 < |4 - 2x| < 1$

70. $|x - a| < d$

71. Show that $[b, b]$ contains exactly one number.

72. Prove that $[a, b]$ and $[c, d]$ contain precisely the same numbers if and only if $a = c$ and $b = d$.

73. If $x^2 \leq 25$, is it necessarily true that $x \leq 5$? Explain.

74. If $x^3 > 125$, is it necessarily true that $x > 5$? Explain.

75. Is $1/x < x$ for all nonzero x? Explain.

76.   a. Show that $x < x^2$ for $x < 0$ or $x > 1$.
      b. Show that $x^2 < x$ for $0 < x < 1$.

77. Use the definition of absolute value to prove the following.
    a. $|ab| = |a||b|$
    b. $-|b| \leq b \leq |b|$
    c. $|a - b| = |b - a|$

78.   a. Use property (9) of absolute value to prove that

$$|a - b| \geq |a| - |b|$$

for all real numbers a and b. (*Hint:* Show that $|c| \geq |c + b| - |b|$ for all c, and replace c by $a - b$.)

   b. Use (6) and part (a) to prove that for all real numbers a and b,

$$|a - b| \geq |b| - |a|$$

   c. Use parts (a) and (b) to prove that

$$|a - b| \geq ||a| - |b||$$

for all real numbers a and b.

*79. Show that $|a + b| = |a| + |b|$ if and only if $ab \geq 0$ (which means that $a = 0$, $b = 0$, or a and b have the same sign).

80. Prove that if $a < b$, then $a < (a + b)/2 < b$. How is $(a + b)/2$ related to a and b on the real line? The number $(a + b)/2$ is called the **arithmetic mean** of a and b.

*81. Prove that if $0 < a < b$, then $a < \sqrt{ab} < (a + b)/2$. The number $\sqrt{ab}$ is called the **geometric mean** of a and b. (*Hint:* $(\sqrt{b/2} - \sqrt{a/2})^2 > 0$.)

*82. Let $0 < a < b$, and let h be defined by

$$\frac{1}{h} = \frac{1}{2}\left(\frac{1}{a} + \frac{1}{b}\right)$$

Show that $a < h < b$. The number h is called the **harmonic mean** of a and b.

*83. Let $0 < a < b$. Show that

$$\sqrt{b} - \sqrt{a} < \sqrt{b - a}$$

*84. Prove that $\sqrt{2}$ is irrational. (*Hint:* Assume that $\sqrt{2} = p/q$, where p and q are integers such that at most one of them is divisible by 2. It can be shown that a square integer is divisible by 2 only if it is also divisible by 4. Use this fact to show first that p is divisible by 2 and then that q is also divisible by 2. This contradicts the assumption.)

*85. Prove that $\sqrt{3}$ is irrational. (*Hint:* Use the method of Exercise 84.)

86. A rectangle R has length x and width y.
    a. Write an inequality that expresses the condition that the area of R is less than 10.
    b. Write an inequality that expresses the condition that the perimeter of R is at least 47.

87. Show that a square has the largest area of all rectangles having a given perimeter. (*Hint:* Let P be the perimeter. Show that if a rectangle of perimeter P has adjacent sides a and b, with $0 < a \leq b$, then the area of the rectangle is ab and that of the square is $[(a + b)/2]^2$. Now use Exercise 81.)

88. Show that if a square and a circle have equal perimeters, then the circle has a larger area than the square. (*Hint:* Show first that a circle of perimeter P has area $P^2/4\pi$.)

89. Using the results of Exercises 87 and 88, show that a circle has an area larger than any rectangle of equal perimeter.

1.2
POINTS AND LINES IN THE PLANE

In calculus we very often encounter curves lying in a given plane. Using the correspondence between numbers and the points on a line, we will identify points in the plane with pairs of numbers. This will enable us to describe curves in the plane by means of equations.

The Plane   We construct the plane by drawing two real lines that are perpendicular to one another and intersect at their zero points. We call their point of intersection the

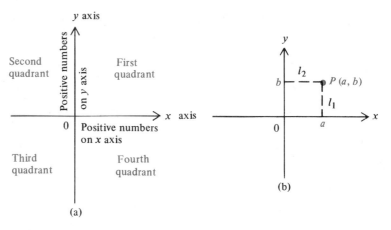

FIGURE 1.7

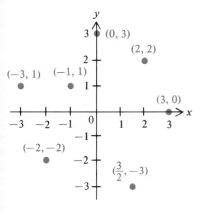

FIGURE 1.8

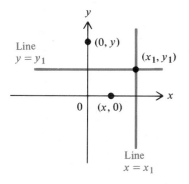

FIGURE 1.9

origin, denoted 0. It is customary to have one of the lines horizontal, with the positive numbers located to the right of 0, and to call it the **x axis**. The other line is usually called the **y axis**, with the positive numbers lying above 0 (Figure 1.7(a)). These axes divide the plane into four quadrants, as shown in Figure 1.7(a).

Now let P be any point in the plane. Draw lines l_1 and l_2 through P perpendicular to the two axes (Figure 1.7(b)). Then the number corresponding to the point on the x axis that lies on l_1 is the **x coordinate** of P, and the number corresponding to the point on the y axis that lies on l_2 is the **y coordinate** of P. Call these numbers a and b, respectively. Then we associate P with the ordered pair (a, b) of numbers. In this way every point in the plane is associated with one and only one ordered pair of numbers, and every ordered pair of real numbers is associated with one and only one point in the plane. Consequently we identify points in the plane with ordered pairs of real numbers. If P is identified with (a, b), we sometimes write $P(a, b)$ for P. Notice that the origin, which is the point whose x and y coordinates are both 0, is not the same as the number 0. Figure 1.8 shows several points in the plane, along with their coordinates.

Observe that the x axis consists of all points of the form $(x, 0)$. Similarly, the y axis consists of all points of the form $(0, y)$. The line perpendicular to the x axis passing through a given point (x_1, y_1) contains all points of the form (x_1, y), and we describe this line by the equation $x = x_1$. The line perpendicular to the y axis passing through the point (x_1, y_1) contains all points of the form (x, y_1), and its equation is $y = y_1$ (Figure 1.9).

Two points in the plane are the same if and only if they have the same x coordinates and the same y coordinates. Thus (a, b) and (c, d) are the same point if and only if $a = c$ and $b = d$.

The Distance Between Two Points

To find the distance between any two points $P(x_1, y_1)$ and $Q(x_2, y_2)$ in the plane, we use the Pythagorean Theorem. The point $R(x_2, y_1)$ in Figure 1.10 has the same x coordinate as Q and the same y coordinate as P. Triangle PQR therefore has a

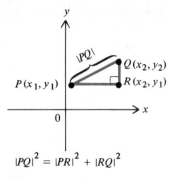

$$|PQ|^2 = |PR|^2 + |RQ|^2$$

FIGURE 1.10

right angle at R. By the Pythagorean Theorem the distance between P and Q is given by the formula

> ### DISTANCE FORMULA
> $$|PQ| = \sqrt{(x_2 - x_1)^2 + (y_2 - y_1)^2}$$

Example 1   Let $P = (3, 4)$ and $Q = (-2, 1)$. Find the distance between P and Q (Figure 1.11).

Solution   By the distance formula,

$$|PQ| = \sqrt{(-2-3)^2 + (1-4)^2} = \sqrt{25+9} = \sqrt{34} \quad \square$$

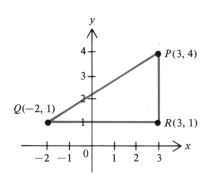

FIGURE 1.11

Lines in the Plane   Identification of points in the plane with ordered pairs of real numbers allows us to use algebra to describe geometric objects such as lines. When we speak of an **equation of a line** l, we mean an equation such as

$$y = 2x + 5$$

having the property that a point P lies on l if and only if the equation is satisfied by the coordinates of P. We will give three methods of obtaining an equation of a line. Since any vertical line is perpendicular to the x axis and hence has an equation of the form $x = x_1$ for some number x_1, we will only consider equations of nonvertical lines in the following discussion.

Let l be any nonvertical line in the plane and $P(x_1, y_1)$ and $Q(x_2, y_2)$ any two distinct fixed points on it (Figure 1.12). Since l is nonvertical, $x_1 \neq x_2$. Let $R(x, y)$ be any third point on the line, and let S and T be located as in Figure 1.12. Then the triangles PTR and PSQ are similar, so that

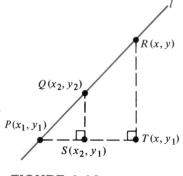

FIGURE 1.12

$$\frac{y - y_1}{x - x_1} = \frac{y_2 - y_1}{x_2 - x_1}$$

Consequently x and y are related by the equation

TWO-POINT EQUATION

$$y - y_1 = \frac{y_2 - y_1}{x_2 - x_1}(x - x_1)$$

(1)

A two-point equation of a line is determined by any two given points on the line.

Example 2   Find a two-point equation of the line that passes through the points (2, 3) and (4, 7) (Figure 1.13).

Solution   Substituting for x_1, y_1, x_2, and y_2 in (1), we obtain

$$y - 3 = \frac{7 - 3}{4 - 2}(x - 2) = 2(x - 2)$$

or $y = 2x - 1$. $\square$

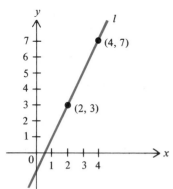

FIGURE 1.13

Let l be a given nonvertical line. It follows from geometry that the fraction

$$\frac{y_2 - y_1}{x_2 - x_1}$$

in (1) is independent of the points (x_1, y_1) and (x_2, y_2) we choose on the line (see Figure 1.14). This fraction is the **slope** of the line l. Letting

$$m = \frac{y_2 - y_1}{x_2 - x_1}$$

(2)

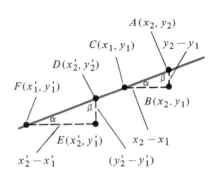

$$\frac{y_2 - y_1}{x_2 - x_1} = \frac{y_2' - y_1'}{x_2' - x_1'}$$

FIGURE 1.14

and substituting in (1), we have

> **POINT–SLOPE EQUATION**
>
> $$y - y_1 = m(x - x_1) \qquad (3)$$

A point–slope equation of a line is determined by the slope and any given point on the line.

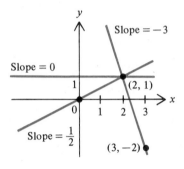

Slope $= -3$

Slope $= 0$

$(2, 1)$

Slope $= \frac{1}{2}$

$(3, -2)$

FIGURE 1.15

Example 3   Find a point–slope equation of the line passing through $(2, 1)$ with the given slope, and sketch the line.

a. slope 0   b. slope $\frac{1}{2}$ c. slope -3

Solution   In each case we use (3), with $x_1 = 2$ and $y_1 = 1$. For (a) we have $m = 0$, so that (3) becomes

$$y - 1 = 0(x - 2) = 0$$

The line is horizontal; it is sketched in Figure 1.15. For (b) we have $m = \frac{1}{2}$, so that (3) becomes

$$y - 1 = \frac{1}{2}(x - 2)$$

To sketch the line, we need a second point on it. Now if we let $x = 0$ in the equation, we obtain

$$y - 1 = \frac{1}{2}(0 - 2) = -1$$

so that $y = 0$. Thus the line passes through both $(0, 0)$ and $(2, 1)$. It is also sketched in Figure 1.15. For part (c), $m = -3$, so that (3) becomes

$$y - 1 = -3(x - 2)$$

Since the slope is -3, or $-3/1$, moving 1 unit to the right and 3 units downward gives a second point $(3, -2)$ on the line. The line also appears in Figure 1.15.   ☐

Figure 1.15 shows that the line with slope 0 is horizontal. More generally, from (3) we see that any line having slope 0 is horizontal. Such a line has an equation of the form

$$y - y_1 = 0, \quad \text{or} \quad y = y_1$$

In drawing the line in part (c) above we noticed that the slope was -3, so an increase of 1 unit in the value of x on the line caused a change of -3 units in the value of y. More generally, if the slope of a given line is m, then a change of 1 unit

|m| large and m positive

$m = 1$

|m| small and m positive

$m = 0$

|m| small and m negative

0

$m = -1$

|m| large and m negative

FIGURE 1.16

in the value of x (say, from x_1 to $x_1 + 1$) causes a change of m units in the value of y (from y_1 to $y_1 + m$). Consequently if (x_1, y_1) is on the line, so is $(x_1 + 1, y_1 + m)$.

We observe that if $m > 0$, then a line with slope m slants upward from left to right, and if $m < 0$, then the line slants downward from left to right. The larger $|m|$ is, the steeper the line is (Figure 1.16).

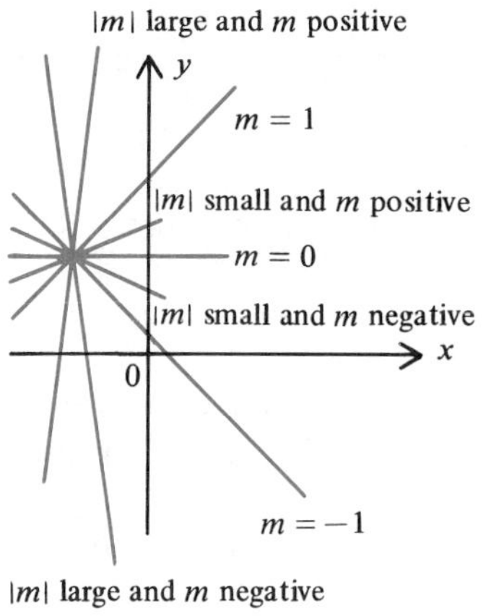

To describe a third type of equation for a line l, recall that we have assumed from the outset that l is *not* vertical. Thus l must cross the y axis. The y coordinate of the point of intersection of l and the y axis is called the **y intercept** of l. Rewriting (3) as an equation for y, we have

$$y = mx + y_1 - mx_1$$

Let $b = y_1 - mx_1$; then we obtain

> **SLOPE–INTERCEPT EQUATION**
> $$y = mx + b$$

(4)

From (4) it follows that the point $(0, b)$ lies on l. But $(0, b)$ also lies on the y axis. Hence b is the y intercept of l. The slope–intercept equation of a line is determined by the slope and y intercept of the line.

Example 4   Find the slope–intercept equation of the line with slope 3 and y intercept -1. Sketch the line.

Solution   By (4) the equation is

$$y = 3x - 1$$

Since $(0, -1)$ is a point on the line and the slope is 3, another point on the line is $(0 + 1, -1 + 3)$, or $(1, 2)$. This enables us to graph the line (Figure 1.17).   □

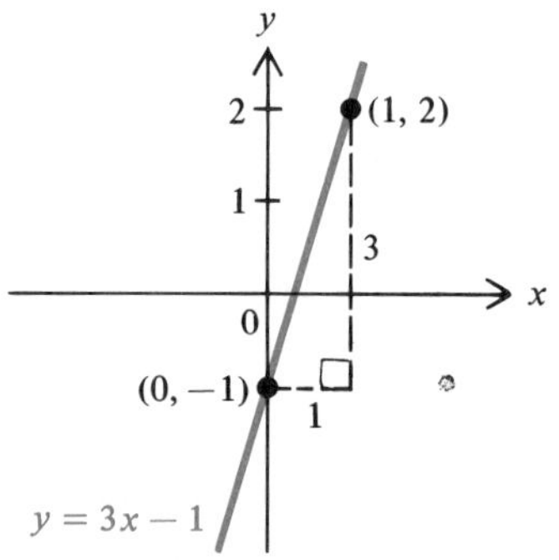

$y = 3x - 1$

FIGURE 1.17

Finally, we mention that any of these types of equation may be used for a given line. The choice is a matter of convenience.

Parallel and Perpendicular Lines

We can determine when two lines are parallel or perpendicular to each other by considering their slopes.

First we discuss parallel lines. Let l_1 and l_2 be two nonvertical lines whose equations are

$$y = m_1 x + b_1 \quad \text{and} \quad y = m_2 x + b_2$$

respectively. Recall from geometry that two lines are *not* parallel if and only if they have precisely one point in common. This fact permits us to characterize parallel (nonvertical) lines.

THEOREM 1.2

> Let l_1 and l_2 be nonvertical lines with slopes m_1 and m_2. Then l_1 and l_2 are parallel if and only if $m_1 = m_2$.

Proof If $m_1 = m_2$, then either l_1 and l_2 do not intersect, in which case they are parallel, or they intersect at a point (x_0, y_0), in which case

$$m_1 x_0 + b_1 = y_0 = m_2 x_0 + b_2$$

Since $m_1 = m_2$, we have $b_1 = b_2$. Thus since the slope–intercept equation uniquely determines a line, l_1 and l_2 are identical and therefore parallel. Hence, if $m_1 = m_2$, then l_1 and l_2 are parallel.

Now suppose $m_1 \neq m_2$. Observe that (x, y) is on both lines if and only if

$$y = m_1 x + b_1 = m_2 x + b_2$$

Since $m_1 \neq m_2$, the equation $m_1 x + b_1 = m_2 x + b_2$ has the unique solution

$$x_0 = \frac{b_2 - b_1}{m_1 - m_2}$$

Thus $(x_0, m_1 x_0 + b_1)$ is a point on both lines, and no other point lies on both lines. Hence the lines are not parallel. ∎

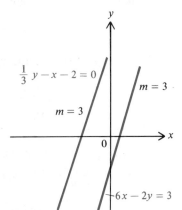

FIGURE 1.18

Example 5   Show that the lines l_1 and l_2 described by the equations

$$6x - 2y = 3 \quad \text{and} \quad \frac{1}{3}y - x - 2 = 0$$

are parallel.

Solution   We rewrite the equations for l_1 and l_2 as slope–intercept equations, obtaining

$$y = 3x - \frac{3}{2} \quad \text{and} \quad y = 3x + 6$$

respectively. Therefore $m_1 = m_2 = 3$, and by Theorem 1.2 the lines l_1 and l_2 are parallel (Figure 1.18). □

Example 6   Show that the lines l_1 and l_2 described by the equations

$$x + 2y = 1 \quad \text{and} \quad x - y = 2$$

are not parallel. Find their point of intersection.

Solution   We find that $m_1 = -\frac{1}{2}$ and $m_2 = 1$. Hence the lines are not parallel. To find their point of intersection, we solve the equations for x and y. Subtracting the second from the first, we obtain $3y = -1$, so that $y = -\frac{1}{3}$. Then the second equation becomes $x - (-\frac{1}{3}) = 2$, or $x = \frac{5}{3}$. This yields $(\frac{5}{3}, -\frac{1}{3})$ as the point of intersection of l_1 and l_2. □

Now we present a characterization of perpendicular (nonvertical) lines.

THEOREM 1.3

Let l_1 and l_2 be nonvertical lines with slopes m_1 and m_2. Then l_1 and l_2 are perpendicular if and only if $m_1 m_2 = -1$.

Proof   Suppose l_1 is perpendicular to l_2, and l_1 and l_2 intersect at $P(x_0, y_0)$. Let $Q(x_1, y_1)$ and $R(x_2, y_2)$ be points distinct from P on the lines l_1 and l_2, respectively (Figure 1.19(a)). Since l_1 is perpendicular to l_2, triangle PRS and triangle QPT are similar. Thus

$$m_1 = \frac{y_1 - y_0}{x_1 - x_0} = \frac{x_0 - x_2}{y_2 - y_0} = \frac{1}{\dfrac{y_2 - y_0}{x_0 - x_2}} = -\frac{1}{\dfrac{y_2 - y_0}{x_2 - x_0}} = -\frac{1}{m_2}$$

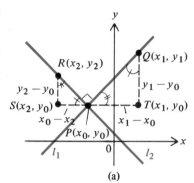

(a)

Therefore $m_1 m_2 = -1$. Conversely, suppose $m_1 m_2 = -1$. Let l_3 be perpendicular to l_1, and let l_3 have slope m_3, so that $m_1 m_3 = -1$. This implies that $m_2 = m_3$. Then by Theorem 1.2, l_2 and l_3 are parallel, so l_1 is perpendicular to l_2. ∎

Example 7   Show that the lines l_1 and l_2 with equations

$$2x - 8y = 3 \quad \text{and} \quad y = 5 - 4x$$

are perpendicular.

Solution   The slopes of l_1 and l_2 are

$$m_1 = \frac{2}{8} = \frac{1}{4} \quad \text{and} \quad m_2 = -4$$

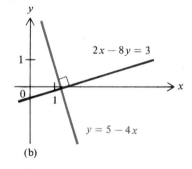

(b)

FIGURE 1.19

Since $m_1 m_2 = -1$, l_1 and l_2 are perpendicular by Theorem 1.3 (Figure 1.19(b)). □

Slope is one of the central concepts of calculus, and we will return to it in Chapters 3 and 4.

EXERCISES 1.2

1. Draw a set of coordinates and plot the following points.
   a. $(2, 1)$ b. $(-1, 3)$ c. $(4, 0)$
   d. $(0, -\frac{3}{2})$ e. $(1, -1)$ f. $(-2, -2)$
   g. $(0, 0)$ h. $(3, -\frac{1}{3})$

2. Let (a, b) be any point in the second quadrant. Describe the locations of the following points.
   a. $(-a, b)$ b. $(a, -b)$ c. $(-a, -b)$

In Exercises 3–12 determine the distance between the pair of points.

3. $(3, 0)$ and $(-2, 0)$ 4. $(0, 0)$ and $(3, 4)$

5. $(2, 1)$ and $(6, -3)$ 6. $(-1, -3)$ and $(-2, 2)$

7. $(6, 5)$ and $(-3, -4)$

8. $(\sqrt{6}, \sqrt{3})$ and $(3\sqrt{6}, -\sqrt{3})$

9. $(\sqrt{2}, 1)$ and $(\sqrt{3}, 2)$

10. (a, b) and (b, a)

11. (a, a) and (b, b)

12. $(a + e, b + e)$ and $(c + e, d + e)$

In Exercises 13–24 find an equation of the line described. Then sketch the line.

13. The line through $(0, 0)$ and $(3, -2)$

14. The line through $(3, 4)$ and $(1, 3)$

15. The line through $(-2, 4)$ and $(-1, 3)$

16. The line through $(-\frac{3}{2}, -\frac{1}{2})$ and $(\frac{1}{2}, 2)$

17. The line through $(2, -1)$ with slope 3

18. The line through $(-3, -2)$ with slope 1

19. The line through $(\frac{1}{2}, \frac{1}{2})$ with slope -1

20. The line through $(0, \pi)$ with slope 0

21. The line with slope -1 and y intercept 0

22. The line with slope $\frac{1}{2}$ and y intercept -1

23. The line with slope 3 and y intercept -3

24. The line with slope -2 and y intercept $\frac{5}{2}$

In Exercises 25–32 determine the slope m and y intercept b of the line with the given equation. Then sketch the line.

25. $y = -x$ 26. $y = 2x - 3$

27. $y = -\frac{1}{3}x - 4$ 28. $y + 3 = x$

29. $y - 1 = 2(x + 3)$ 30. $x - \frac{1}{2}y = 2$

31. $2x + y - 4 = 0$ 32. $\frac{2}{3}y - \frac{1}{3}x = 2$

In Exercises 33–42 decide which pairs of lines are parallel, which are perpendicular, and which are neither. For any pair that is not parallel, find the point of intersection.

33. $y - 2x = 3$ and $y + \frac{1}{2}x = -1$

34. $y = 3 - 2x$ and $3x + \frac{3}{2}y - 4 = 0$

35. $x - y = -1$ and $x = y$

36. $x + y = -1$ and $x = y$

37. $2x + 3y = -1$ and $2y + 2 = 3(x - 1)$

38. $2x + 3y = -1$ and $3x + 2y = 2$

39. $x = 2$ and $x = -5$

40. $x = -1$ and $y = 4$

41. $3y + 6x = 1$ and $y - 3 = -2x$

42. $x - 2y = 8$ and $2x - y = -8$

In Exercises 43–48 find an equation of the line that is parallel to the given line l and passes through the given point P.

43. $l: y = 3x - 1$; $P = (2, -1)$

44. $l: y = -\frac{1}{2}x + 4$; $P = (-1, 0)$

45. $l: x + y = 1$; $P = (0, 0)$

46. $l: 3y + 2x = 5$; $P = (-1, -3)$

47. $l: -3y + 2x = 8$; $P = (2, 1)$

48. $l: 5x - 2y - 1 = 0$; $P = (3, 3)$

In Exercises 49–54 find an equation of the line that is perpendicular to the given line l and passes through the given point P.

49. $l: y = 2x + 1$; $P = (-1, -3)$

50. $l: y = -\frac{1}{3}x - 2$; $P = (0, 0)$

51. $l: 2x + 3y - 6 = 0$; $P = (2, 3)$

52. $l: 3x - y = 0$; $P = (1, 3)$

53. $l: y - 1 = 2(x - 3)$; $P = (4, -5)$

54. $l: y + 4 = -\frac{3}{5}(x - \frac{1}{2})$; $P = (-1, \frac{1}{2})$

55. The **x intercept** of a nonhorizontal line is the x coordinate of the point of intersection of the line and the x axis. Find the x intercept a, if any, of each of the following lines.
    a. $y = -x - 4$ b. $y = 2x - \sqrt{2}$
    c. $x = -3 - 4y$ d. $x - y = 5$
    e. $x = -2$ f. $y = -7$

56. The **two-intercept equation** of a line that is neither horizontal nor vertical and does not pass through the origin has the form $x/a + y/b = 1$. Show that the x and y intercepts of this line are a and b, respectively.

In Exercises 57–60 find the two-intercept equation of the line with the given equation.

57. $y = 4x - 2$ 58. $y - 5 = -10(x + 1)$

59. $2x - 3y = -1$ 60. $\frac{1}{2}y + \frac{3}{2}x + 5 = 0$

In Exercises 61–70 sketch the region in the plane satisfying the given conditions.

61. $x > 0$ 62. $y \le 0$

63. $y > x$ (*Hint:* Consider   64. $x < -y$
    first the line $y = x$.)

65. $x < 0$ and $y < 0$ 66. $x > 0$ and $y < 0$

67. $x < 2$ and $y > 4$ 68. $x \le 3$ and $y \le 2$

69. $x \ge -1$ and $y \ge \frac{1}{2}$ 70. $x > 2$ and $y < 1$

71. a. Describe the region consisting of all (x, y) for which $(x, y) = (x, -y)$.
    b. Describe the region consisting of all (x, y) for which $(x, y) = (-x, y)$.

72. Let $(2, 1), (-3, -2)$, and (a, b) form a triangle. Show that the collection of points (a, b) for which the triangle is isosceles contains a line (with one point deleted). Find an equation of that line.

73. Show that the triangle with vertices $(-1, 2), (\sqrt{3} - 1, 3)$, and $(-1, 4)$ is equilateral.

74. Show that the midpoints of the sides of any rectangle are the vertices of a rhombus (a quadrilateral with all sides of equal length). (*Hint:* Let the vertices of the rectangle be $(0, 0), (a, 0), (0, b)$, and (a, b).)

*75. Show that in any triangle the sum of the squares of the lengths of the medians is equal to three fourths the sum of the squares of the lengths of the sides. (*Hint:* Pick the vertices of the triangle judiciously.)

*76. Show that the sum of the squares of the lengths of the sides of a parallelogram is equal to the sum of the squares of the lengths of the diagonals.

1.3
FUNCTIONS

Scientists and mathematicians often consider correspondences between two sets of numbers. For example, each temperature in degrees Celsius corresponds to a temperature in degrees Fahrenheit. There is also a correspondence between the radius of a circle and the circle's area. In mathematics such correspondences are called functions.

DEFINITION 1.4

> A **function** consists of a domain and a rule. The **domain** is a set of real numbers. The **rule** assigns to each number in the domain one and only one number.

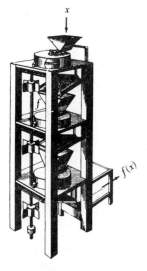

FIGURE 1.20

Functions are normally denoted f, g, or h, and the numbers in the domain are usually denoted by $x, t, a, b,$ or c. The value assigned by a function f to a member x of its domain is written $f(x)$ and is read "f of x" or "the value of f at x." The collection of values (numbers) $f(x)$ that a given function assigns to the members of its domain is called the **range** of f.

We can think of a function as a machine that takes the members of the domain and applies the rule to each to produce the members of the range. As Figure 1.20 suggests, any member x of the domain goes into the machine, is acted on by the function f, and is transformed into a member $f(x)$ of the range.

Caution: We stress two key points in the definition:
1. A function must make an assignment to *each* number in the domain. For example, if the domain of a function f is $(-\infty, \infty)$, then f must make an assignment to each real number.
2. A function can assign *only one* number to any given number in the domain. Thus f cannot assign both 5 and 6 to a single number in its domain.

Examples of Functions

We give several examples of functions. Let f be the function whose domain consists of all real numbers and whose rule assigns to any real. x the number $x^2 - 1$. Then we write

$$f(x) = x^2 - 1 \quad \text{for all } x$$

Next, if the domain of a function g consists of all real numbers except 3 and if

g assigns $(x - 1)/(x - 3)$ to each such number x, then g is described by

$$g(x) = \frac{x - 1}{x - 3} \quad \text{for } x \neq 3$$

Finally, the function h whose domain consists of all numbers greater than or equal to -273.15 and assigns to each such number x the number $\frac{9}{5}x + 32$ can be written

$$h(x) = \frac{9}{5}x + 32 \quad \text{for } x \geq -273.15$$

When the rule of a function is described by one formula or equation, we normally specify the numbers in the domain after the rule. If the domain consists of all real numbers for which the formula or equation is meaningful, then we may omit mention of the domain. Thus we may write

$$f(x) = x^2 - 1 \quad \text{and} \quad g(x) = \frac{x - 1}{x - 3}$$

without specifying the domains of f and g. However, it is necessary to give the domain of h if

$$h(x) = \frac{9}{5}x + 32 \quad \text{for } x \geq -273.15$$

because the rule for h is meaningful for numbers less than -273.15.

When there can be no misinterpretation, we will sometimes let the rule of a function stand for the function itself. For instance, we can replace f and g defined above by $x^2 - 1$ and $(x - 1)/(x - 3)$, respectively.

Sometimes two or more formulas may be needed to define a function. For example,

$$f(x) = \begin{cases} 1 & \text{for } x < 0 \\ x & \text{for } 0 \leq x \leq 2 \\ x^2 & \text{for } 2 < x < 3 \\ x^3 & \text{for } x \geq 3 \end{cases}$$

defines a single function.

The functions we commonly encounter in the physical world can often be described by one or more formulas. Let us consider some examples.

Example 1 Describe the function f that associates with each temperature in degrees Celsius the corresponding temperature in degrees Fahrenheit.

Solution Let x be the temperature in degrees Celsius. Then the temperature in degrees Fahrenheit is $f(x)$. We find it by applying the formula

$$f(x) = \frac{9}{5}x + 32$$

We restrict x to values not less than absolute zero, which for practical purposes is $-273.15°C$, so the domain of f is $[-273.15, \infty)$. Therefore the function f is given by

$$f(x) = \frac{9}{5}x + 32 \quad \text{for } x \geq -273.15 \quad \square$$

Notice that the function in Example 1 is the function h discussed above. What are the values assigned by this function to the numbers 0, 20, and 100? Replacing x by 0, we find that

$$f(0) = \frac{9}{5}(0) + 32 = 32$$

which is associated with the freezing point of water. Similarly,

$$f(20) = \frac{9}{5}(20) + 32 = 68$$

an appropriate room temperature for a person studying diligently. Finally,

$$f(100) = \frac{9}{5}(100) + 32 = 212$$

which corresponds to the boiling point of water at sea level.

In the next example we will employ a formula from physics that will be discussed further in Chapters 2, 3, and 5. If an object is acted on only by the force of gravity, and if at time $t = 0$ the height of the object is h_0 feet above the ground and the velocity is v_0 feet per second, then the object's height $h(t)$ in feet above the ground at time t (seconds) is given by

$$h(t) = -16t^2 + v_0 t + h_0 \tag{1}$$

We call h_0 the **initial height** and v_0 the **initial velocity** of the object. We emphasize that the formula in (1) is valid only as long as the object is subject only to the force of gravity. Implicit in (1) is the fact that if the object is moving upward at $t = 0$, then $v_0 > 0$, whereas if the object is moving downward at $t = 0$, then $v_0 < 0$.

Example 2 Suppose a person on a balcony 160 feet above ground throws a ball upward with an initial velocity of 48 feet per second. Find a formula for the height of the ball until it hits the ground, and determine how long it takes for the ball to hit the ground.

Solution We measure time so that $t = 0$ at the instant the ball is thrown. By assumption $h_0 = 160$ and $v_0 = 48$, and by (1) the height of the ball is given by

$$h(t) = -16t^2 + 48t + 160$$

until the ball strikes the ground. To determine when the ball strikes the ground, we find the values of t for which $h(t) = 0$:

$$-16t^2 + 48t + 160 = 0$$
$$-16(t^2 - 3t - 10) = 0$$
$$-16(t - 5)(t + 2) = 0$$

$$t = -2 \quad \text{or} \quad t = 5$$

Since the ball was thrown at time $t = 0$, it follows that it hits the ground at time $t = 5$. Thus it takes 5 seconds for the ball to hit the ground.   □

Example 3   Substance A undergoes a chemical reaction to become substance B. The amount of A present initially is 3 grams. The rate $f(x)$ at which x grams of A are turned into B is proportional to the product of x and $3 - x$. Express $f(x)$ in terms of x.

Solution   Since $f(x)$ is proportional to $x(3 - x)$, there is a number $c \neq 0$ such that

$$f(x) = cx(3 - x)$$

There are initially 3 grams of substance A, so this relation holds only for $0 \leq x \leq 3$. Consequently

$$f(x) = cx(3 - x) \quad \text{for } 0 \leq x \leq 3 \quad □$$

We now discuss some general classes of functions that will be helpful to us later.

Polynomials and Rational Functions   First we consider the polynomial functions, which are especially amenable to the methods of calculus. Examples of polynomial functions are

$$f(x) = 2x^3 - 4x - 1 \qquad f(x) = \frac{1}{5}x$$

$$f(x) = \pi x^2 + 13 \qquad f(x) = 17$$

In general, a **polynomial** (or **polynomial function**) is any function f of the form

$$f(x) = c_n x^n + c_{n-1} x^{n-1} + \cdots + c_1 x + c_0$$

where $c_n, c_{n-1}, \ldots, c_1$, and c_0 are real numbers with $c_n \neq 0$, and where n is a nonnegative integer (called the **degree** of the polynomial). Zero-degree polynomials are of the form

$$f(x) = c_0 \quad \text{with } c_0 \neq 0$$

and are called **constant functions**. Thus $f(x) = 17$ provides an example of a constant function. By convention no degree is assigned to the constant polynomial 0. First-degree polynomials are of the form

$$f(x) = c_1 x + c_0 \quad \text{with } c_1 \neq 0$$

and are called **linear functions**. The particular linear function defined by $f(x) = x$ is called the **identity function**.

Quotients of polynomials form a second class of functions, called **rational functions**. Examples are

$$f(x) = \frac{1}{x} \qquad\qquad f(x) = x^2 + \sqrt{3}x$$

$$f(x) = \frac{x^4 + 3x^3 + 2x - \pi}{3x^3 - 4x + 1} \qquad f(x) = \frac{x^2 + 3x - 2}{x - 3}$$

Notice that a polynomial is a rational function whose denominator is the constant function 1. For example, the polynomial $f(x) = x^2 + \sqrt{3}x$ can be rewritten

$$f(x) = x^2 + \sqrt{3}x = \frac{x^2 + \sqrt{3}x}{1}$$

Example 4   Let

$$f(x) = \frac{x^2 + x - 2}{x^2 + 5x - 6}$$

Find the domain of f.

Solution   Since $x^2 + 5x - 6 = (x - 1)(x + 6)$, the denominator is 0 for $x = 1$ and $x = -6$. Thus the domain of f consists of all numbers except 1 and -6.   ☐

> **Caution:** It is also possible to factor the numerator in Example 4, which yields $x^2 + x - 2 = (x - 1)(x + 2)$. Thus
>
> $$f(x) = \frac{(x - 1)(x + 2)}{(x - 1)(x + 6)} = \frac{x + 2}{x + 6}$$
>
> Although the expression $(x + 2)/(x + 6)$ is meaningful for $x = 1$, the number 1 is not in the domain of f. Thus the domain of a function must be determined from the original description of the function.

> **Power functions** are a special class of rational functions. They have the form
>
> $$f(x) = x^n$$
>
> where n is an integer. Examples are x^2, x^5, x^{-1}, and x^{-3}. Recall that
>
> $$x^0 = 1, x^{-1} = \frac{1}{x}, x^{-2} = \frac{1}{x^2}, x^{-3} = \frac{1}{x^3}, \ldots, \text{ and } x^{-n} = \frac{1}{x^n}$$
>
> for any integer n. The domain of x^n consists of all real numbers if $n \geq 0$. If $n < 0$, then the domain contains all real numbers except 0, since division by 0 is not defined.

Root Functions   We begin with the square root function. By definition the square root function assigns to each nonnegative number x the nonnegative number y such that $y^2 = x$. We denote y by $\sqrt{x}$ or $x^{1/2}$. We emphasize that $\sqrt{x}$ is defined only for $x \geq 0$ and that $\sqrt{x} \geq 0$ for all $x \geq 0$. Consequently we may write $\sqrt{14}$, $\sqrt{\frac{1}{2}}$, and $\sqrt{0}$, but $\sqrt{-5}$ has no meaning, since there is no number whose square is -5. Moreover, $\sqrt{4} = 2$, not ± 2. In Chapter 2 we will prove that a square root does in fact exist for each $x \geq 0$. Thus we can define the **square root function** by

$$f(x) = \sqrt{x} \quad \text{for } x \geq 0$$

Since $|x|^2 = x^2$ by the definition of absolute value, there is an intimate relation between square root and absolute value:

$$|x| = \sqrt{x^2}$$

Next we define the **cube root function**. It assigns to any number x the unique number y such that $y^3 = x$. We denote y by $\sqrt[3]{x}$ or $x^{1/3}$. In contrast to the square root function, the cube root function has as its domain all real numbers, including negative numbers. For example, $\sqrt[3]{-1} = -1$, $\sqrt[3]{-8} = -2$, and $\sqrt[3]{-\frac{1}{27}} = -\frac{1}{3}$.

More generally, we can define the nth root function for any positive integer n. If n is odd, then for any real number x the nth root $\sqrt[n]{x}$ is the number y such that $y^n = x$. (Later we will show that for every x there is a unique number y with this property.) If n is even, then for any *nonnegative* number x the nth root $\sqrt[n]{x}$ is the *nonnegative* number y such that $y^n = x$. Although $(-y)^n = x$ also when n is even, we exclude negative values for $\sqrt[n]{x}$ when n is even. Thus $(-2)^4 = 2^4 = 16$, but $\sqrt[4]{16} = 2$. In this way $\sqrt[n]{x}$ is unique, whether n is odd or even, so that the nth root function really is a function.

Another notation for $\sqrt[n]{x}$ is $x^{1/n}$. Therefore

$$\sqrt[n]{x} \text{ is defined } \begin{cases} \text{for any real number } x \text{ if } n \text{ is odd} \\ \text{for any nonnegative number } x \text{ if } n \text{ is even} \end{cases}$$

$$\sqrt[n]{x} = x^{1/n} = y \quad \text{if and only if} \quad y^n = x \text{ (with } y \geq 0 \text{ if } n \text{ is even)}$$

It follows that

$$\sqrt[5]{-32} = -2, \quad \sqrt[3]{\frac{27}{8}} = \frac{3}{2}, \quad (16)^{1/4} = 2 \quad \text{and} \quad \sqrt[6]{(-2)^6} = 2$$

In contrast, the expressions $\sqrt[4]{-1}$ and $\sqrt[6]{-\frac{43}{55}}$ have no meaning, since -1 and $-\frac{43}{55}$ are negative and hence not in the domains of the respective root functions. For every positive integer n we also have

$$\sqrt[n]{1} = 1 \quad \text{and} \quad \sqrt[n]{0} = 0$$

Equality of Functions   We say that two functions f and g are **equal**, or the same, if f and g have the same domain and $f(x) = g(x)$ for each x in the common domain. Thus if

$$f(x) = x^2$$

$$g(x) = x^2 \quad \text{for } x \geq 1$$

then f and g are distinct functions, because their domains are different. But if

$$f(x) = x^2 \quad \text{for } x \geq -10$$

$$g(x) = (x - 1)^2 + 2x - 1 \quad \text{for } x \geq -10$$

$$h(y) = y^2 \quad \text{for } y \geq -10$$

then f, g, and h are the same function, because their domains are identical and their rules all assign the same number to each number in the domain. To summarize: If two functions have the same domain and assign the same value to each number in their domain, then the two functions are equal.

Historical Comment   The notion of function was the result of a long development of mathematical thought. Mathematicians from antiquity to the Middle Ages had at best a vague idea of the concept of function. As calculus developed during the latter part of the seventeenth century, the need to make the notion of function precise gradually became apparent. The word "function," which derives from the Latin word for "perform," seems to have been used first by Gottfried Leibniz* (1646–1716).

The Swiss mathematician Leonhard Euler† (1707–1783) was the first to adopt the expression $f(x)$ for the value of a function. He systematically classified many collections of functions. Nevertheless, it was not until the middle of the nineteenth century that the formulation given in Definition 1.4 emerged. This was the work of P. G. Lejeune-Dirichlet‡ (1805–1859). Although there are more formal, set-theoretic versions of the definition of function, the form in Definition 1.4 will suffice for us.

---

* **Leibniz**: Pronounced "*Libe*-nits."
† **Euler**: Pronounced "*Oi*-ler."
‡ **Dirichlet**: Pronounced "Di-ri-*shlay*."

EXERCISES 1.3

In Exercises 1–10 find the numerical value of the function at the given values of a.

1. $f(x) = \sqrt{3}; a = \sqrt{5}, \pi$

2. $f(x) = 2x^2 - 3; a = 1, -2$

3. $f(x) = 1 - x + x^3; a = 0, -1$

4. $f(x) = 1/x; a = 2, \frac{1}{2}$

5. $g(x) = 1/(2x^2); a = \sqrt{2}$

6. $g(x) = \sqrt{x}; a = 4, \frac{1}{25}$

7. $g(t) = \sqrt[3]{t}; a = 27, -\frac{1}{8}$

8. $g(t) = |2 - t|; a = 6$

9. $f(x) = \dfrac{x - 1}{x^2 + 4}; a = 2$

10. $f(x) = \dfrac{3x^2 - 4x - 1}{2x^2 + 5x - 3}; a = -1$

In Exercises 11–32 find the domain of the function.

11. $f(x) = x^3 - 4x + 1$

12. $f(x) = x^6 - \sqrt{2}x^3 - \pi$

13. $k(x) = 1 + x^3 \quad \text{for } -2 \leq x \leq 8$

14. $f(x) = 2x - 3x^5 \quad \text{for } x < 4$

15. $f(x) = \sqrt{x + 2}$

16. $f(x) = \sqrt{2 - 3x}$

17. $f(x) = \sqrt{x^2 + 4}$

18. $f(t) = \sqrt{4 - 9t^2}$

19. $f(t) = \sqrt{3 - \dfrac{1}{t^2}}$

20. $f(t) = \dfrac{t}{\sqrt{t + 5}}$

21. $f(t) = \sqrt[3]{1 - t^2}$

22. $g(x) = (x - 6)^{1/4}$

23. $g(x) = \dfrac{2}{x - 1}$

24. $g(x) = \dfrac{3x - 1}{x - 3}$

25. $g(w) = \dfrac{2w - 8}{w^2 - 16}$

26. $g(w) = \dfrac{w - 1}{w^2 - w - 6}$

27. $k(x) = \dfrac{2x - 3}{x^2 + 4}$

28. $k(x) = \dfrac{1}{x + 1} - \dfrac{2}{x - 1}$

29. $f(x) = \begin{cases} 2x & \text{for } -4 \le x \le -1 \\ 3 & \text{for } 0 < x < 6 \end{cases}$

30. $f(x) = \begin{cases} x^2 + 1 & \text{for } x \le 2 \\ x^2 - 1 & \text{for } x > \sqrt{5} \end{cases}$

*31. $f(x) = \sqrt{1 - \sqrt{9 - x^2}}$

*32. $f(x) = \sqrt{4 - \sqrt{1 + 9x^2}}$

In Exercises 33–38 determine the range of the function.

33. $f(x) = -1$

34. $f(x) = 3x - 2$

35. $f(x) = 3x - 2 \quad \text{for } x < 4$

36. $f(x) = \sqrt{1 - x^2}$

37. $f(x) = \dfrac{1}{x - 1}$

*38. $f(x) = \dfrac{x^2 - 1}{x^2 + 1}$

39. Determine which of the following define a function. Explain your reason for any that do not define a function.
    a. The domain consists of the number -2, which is assigned the number π.
    b. The domain consists of the number -2, which is assigned the numbers -2 and π.
    c. $f(x) = \pm\sqrt{x}$
    d. $f(x) = \pm\sqrt{x^2 + 1}$
    e. $g(x) = \begin{cases} x - 1 & \text{for } x < 0 \\ 12x - 6 & \text{for } x > 0 \end{cases}$
    f. $g(x) = \begin{cases} 2 - x^4 & \text{for } x < 0 \\ x^2 & \text{for } x > 1 \end{cases}$
    g. $g(x) = \begin{cases} 4x + 1 & \text{for } x \le 2 \\ 2x^3 - 7 & \text{for } x \ge 2 \end{cases}$
    h. $g(x) = \begin{cases} 2 - 3x^3 & \text{for } x \le 1 \\ 3x^4 - 3 & \text{for } x \ge 1 \end{cases}$
    i. $f(t) = \begin{cases} t^2 & \text{for } t \text{ rational} \\ t & \text{for } t \text{ irrational} \end{cases}$
    *j. $f(t) = \begin{cases} t^2 & \text{for } t^2 \text{ rational} \\ t & \text{for } t \text{ irrational} \end{cases}$

40. In each of the following, determine whether f and g are the same.
    a. $f(x) = 1 - x^2;\ g(x) = 1 - x^2 \quad \text{for } -1 < x < 1$
    b. $f(x) = \sqrt{x} \quad \text{for } x \ge 0;\ g(x) = \sqrt{x}$
    c. $f(x) = \sqrt{x^2};\ g(x) = |x|$
    d. $f(x) = \dfrac{x^3 - 4x}{x^3 - 4x};\ g(x) = 1$
    e. $f(x) = \dfrac{x - 1}{x^2 - 1};\ g(x) = \dfrac{1}{x + 1}$
    f. $f(x) = \dfrac{x^2 - 5x + 6}{x + 2};\ g(x) = x - 3 \quad \text{for } x \ne -2$

41. Which of the following functions are the same?
    a. $f_1(x) = \sqrt{1 - 6x + 9x^2}$

b. $f_2(x) = 1 - 3x$
c. $f_3(t) = 1 - 3t$
d. $f_4(w) = 1 - 3w \quad \text{for } w \ge 0$
e. $f_5(t) = |1 - 3t|$
f. $f_6(x) = \dfrac{(1 - 3x)^2}{1 - 3x}$

42. Let
$$f(x) = \sqrt{x^2 + 1} - 1 \quad \text{and} \quad g(x) = \dfrac{x^2}{1 + \sqrt{x^2 + 1}}$$
    a. Find the domains of f and g.
    b. Show that $f = g$.

43. Let
$$f(x) = x - \sqrt{x^2 - 1} \quad \text{and} \quad g(x) = \dfrac{1}{x + \sqrt{x^2 - 1}}$$
    a. Find the domains of f and g.
    b. Show that $f = g$.

44. What is the largest possible domain of a function f whose rule is given by $f(x) = \sqrt{x(x - 1)}$?

45. Write a formula for a "negative square root function f," which associates with each nonnegative number x a nonpositive number whose square is x.

46. Let $f(x) = ax^2 + bx + c$, where $a \ne 0$. A **zero** of f is a number x such that $f(x) = 0$. Show that the zeros of f are the numbers x given by the **quadratic formula**:
$$x = \dfrac{-b + \sqrt{b^2 - 4ac}}{2a} \quad \text{or} \quad x = \dfrac{-b - \sqrt{b^2 - 4ac}}{2a}$$
    (*Hint:* Prove that $ax^2 + bx + c = 0$ if and only if
$$\left(x + \dfrac{b}{2a}\right)^2 = \dfrac{b^2 - 4ac}{4a^2}$$
    Then solve for x.) Note that such a zero exists only if $b^2 - 4ac \ge 0$, and there are two zeros if $b^2 - 4ac > 0$, whereas there is only one zero if $b^2 - 4ac = 0$.

In Exercises 47–54 use the result of Exercise 46 to determine the zeros, if any, of the function.

47. $f(x) = x^2 + 6x$

48. $f(x) = 8x^2 - 8x + 2$

49. $f(x) = x^2 - 4x - 5$

50. $f(x) = x^2 - 4x + 5$

51. $g(x) = 3x^2 + 2x + 1$

52. $g(x) = 3x^2 + 2x - 1$

53. $k(t) = 2t^2 + 7t + 3$

54. $k(t) = 6t^2 - 3t - 2$

55. Let $f(x) = ax^2 + bx + c$ and $g(x) = ax^2 + bx - c$. Prove that if f has no zeros, then g has two zeros.

56. Find a formula for the function f that assigns to each x greater than -1 the number obtained by squaring x, then subtracting $2x$, and finally adding $\sqrt{2}$.

57. Find a formula for the function f that assigns to each nonnegative x the number obtained by dividing x by 5,

then taking the cube root of the quotient, and then multiplying the result by the product of $\frac{1}{2}$ and x^2.

58. Find a formula for the function A that expresses the area of a circle in terms of the radius.

59. Find a formula for the function A that expresses the area of an equilateral triangle in terms of the length of one of its sides.

60. Find a formula for the function V that expresses the volume of a cube in terms of the length of an edge.

61. A tank has the form of a right circular cylinder with hemispherical ends. Its volume is 100 cubic inches. Find the length L of the cylinder in terms of the radius of the hemisphere.

62. The Tee-rific Company produces 100,000 golf tees daily and sells them for 5¢ apiece. Assume that the total cost of producing one tee is 2¢. Find the company's profit P (in cents) in terms of the number of working days.

63. A ball is dropped from the top of a tower 784 feet tall. Find the distance the ball has traveled after 3 seconds, and determine how long it takes for the ball to hit the ground.

64. A ball is thrown downward from the roof of a building 96 feet high with a velocity of 16 feet per second. Find the height of the ball after $\frac{1}{2}$ second and after 1 second. When does the ball reach the ground?

65. Starting at noon, A flies 2400 miles from New York to San Francisco at a velocity of 400 miles per hour. B starts the same trip at 2:00 P.M. the same day with a velocity of 800 miles per hour. Express the distance D between A and B at any instant between noon and 5:00 P.M. in terms of the time in hours elapsed after noon.

66. Two cars depart from the same location at the same time. One travels north at 40 miles per hour and the other travels east at 50 miles per hour. Find a formula for the function D that expresses in terms of t the distance between the cars t hours after departure.

67. In the study of the response to acetylcholine by a frog's heart, the formula

$$R(x) = \frac{x}{c + dx}$$

arises, where x denotes the concentration of the drug and c and d are positive constants.
a. Find $R(0)$ and $R(2)$. What is the physical significance of $R(0)$?
b. Find a formula that expresses the concentration x in terms of $R(x)$.

1.4
GRAPHS

The most concise way of describing a function is by means of formulas like

$$f(x) = \frac{x^2}{x + 1}$$

In calculus, however, it is often useful and instructive to sketch a picture of a function. In fact, one of the major achievements of calculus is that it enables us to draw accurate pictures of many functions.

Graph of a Function   The pictorial representation of a function is called a graph.

DEFINITION 1.5

Let f be a function. Then the set of all points $(x, f(x))$ such that x is in the domain of f is called the **graph** of f.

In many cases, by studying the formula that defines a function we can easily draw a satisfactory sketch of its graph.

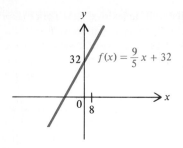

FIGURE 1.21

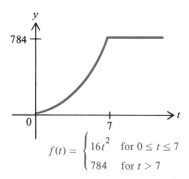

FIGURE 1.22

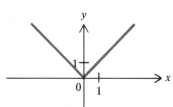

The absolute value function:
$$f(x) = |x| = \begin{cases} -x & \text{for } x < 0 \\ x & \text{for } x \geq 0 \end{cases}$$

FIGURE 1.23

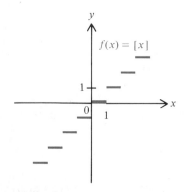

FIGURE 1.24

Example 1   Let $f(x) = \frac{9}{5}x + 32$. Sketch the graph of f.

Solution   By Definition 1.5, the graph of f consists of the set of all points (x, y) in the plane for which $y = \frac{9}{5}x + 32$. This is a line with slope $\frac{9}{5}$ and y intercept 32. The graph is shown in Figure 1.21.   □

If f is any linear function, then $f(x) = mx + b$. As we saw in Section 1.2, the graph of f is a line with slope m and y intercept b.

Sketch the graph of the distance function defined by
$$f(t) = \begin{cases} 16t^2 & \text{for } 0 \leq t \leq 7 \\ 784 & \text{for } t > 7 \end{cases}$$

Solution   We first tabulate several points $(t, f(t))$ on the graph of f:

t	0	$\frac{1}{2}$	1	2	3	4	5	6	7	10
$f(t)$	0	4	16	64	144	256	400	576	784	784

Since the values of $f(t)$ shown here are much larger than those of t, we use a smaller scale on the y axis than on the t axis. Plotting the points in the table and connecting them with a curve, we obtain the graph shown in Figure 1.22.   □

Example 3   Sketch the graph of the **absolute value function**:
$$f(x) = |x|$$

Solution   By the definition of absolute value we have
$$f(x) = |x| = \begin{cases} -x & \text{for } x < 0 \\ x & \text{for } x \geq 0 \end{cases}$$

Therefore the part of the graph to the right of the y axis coincides with the line $y = x$ and the part to the left of the y axis coincides with the line $y = -x$. Thus the graph is as shown in Figure 1.23.   □

Example 4   The **greatest integer**, or **staircase**, **function** is defined by
$$f(x) = [x]$$
where the symbol $[x]$ denotes the greatest integer not larger than x. Sketch the graph of f.

Solution   The values of f are integers. Moreover, if n is any integer and if $n \leq x < n + 1$, then $[x] = n$. Therefore $f(x) = n$ for all x in $[n, n + 1)$, where n is any integer. The graph of f is sketched in Figure 1.24.   □

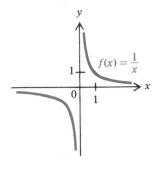

FIGURE 1.25

Example 5   Let $f(x) = 1/x$. Sketch the graph of f.

Solution   We determine several points $(x, f(x))$ on the graph of f:

x	$\frac{1}{4}$	$\frac{1}{2}$	1	2	4	$-\frac{1}{4}$	$-\frac{1}{2}$	-1	-2	-4
$f(x)$	4	2	1	$\frac{1}{2}$	$\frac{1}{4}$	-4	-2	-1	$-\frac{1}{2}$	$-\frac{1}{4}$

Note that for $x > 0$, $1/x$ decreases as x increases. For $x < 0$, $1/x$ increases as x decreases. Note also that $x = 0$ is not in the domain of f. The graph of f is shown in Figure 1.25. $\square$

In the examples above we determined many points and used other information to sketch the graphs. However, this method of analyzing the graphs of functions will be replaced later by very general and powerful methods of calculus which allow us to draw graphs without plotting numerous points.

Until now we have started with a formula for a function and drawn its graph. Can we reverse the process? Suppose we are given a certain curve in the plane. Under what conditions is it necessarily the graph of a function? Since a function can assign only one value to each number in its domain, a curve can represent the graph of a function provided that no more than one point of the curve lies on any given vertical line (Figure 1.26(a)). In that case, we can describe the function as follows. The domain of the function consists of all values of x that are the first coordinates of points on the curve, and the rule assigns to each such x the *unique* value y such that (x, y) is on the curve. Thus if each vertical line contains no more than one point on the curve, each x is assigned a unique value $f(x)$, as required by the definition of a function. The curve shown in Figure 1.26(b) cannot be the graph of a function, since there are vertical lines containing more than one point on the curve.

Finally, we observe that two functions are equal precisely when their graphs are identical.

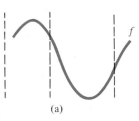

(a)

Each vertical line can cross the graph of a function at most once.

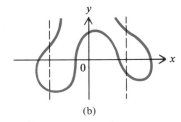

(b)

Some vertical lines cross the curve more than once.

FIGURE 1.26

Dependent and Independent Variables

We generally use the expression (x, y) to denote a point in the plane. We can describe the graph of a function f as the set of all points (x, y) such that x is in the domain of f and $y = f(x)$. For this reason we sometimes specify a function by writing y in terms of x. For instance, we may write

$$y = \tfrac{9}{5}x + 32 \quad \text{instead of} \quad f(x) = \tfrac{9}{5}x + 32$$

$$y = x^2 \qquad\qquad \text{instead of} \quad f(x) = x^2$$

$$y = \frac{1}{x} \qquad\qquad \text{instead of} \quad f(x) = \frac{1}{x}$$

We call x an **independent variable**. Since the value of y depends on that of x, y is a **dependent variable**. We will sometimes say that y is a variable depending on x. We will use both notations, $f(x)$ and y, for the value assigned to a number x in the domain of a function.

Graph of an Equation

There are many equations in x and y that are not of the form $y = f(x)$. Examples of such equations are

$$x^2 + y^2 = 4 \quad \text{and} \quad x = y^2$$

The **graph of an equation** in x and y is the collection of points (x, y) in the plane that satisfy the equation.

Example 6 Sketch the graph of the equation $x^2 + y^2 = 4$.

Solution The graph consists of all points (x, y) that satisfy $x^2 + y^2 = 4$. Recall that the distance r between any point (x, y) in the plane and the origin, $(0, 0)$, is given by

$$r = \sqrt{(x - 0)^2 + (y - 0)^2} = \sqrt{x^2 + y^2}$$

Hence any point (x, y) at a distance r from the origin satisfies the equation

$$x^2 + y^2 = r^2$$

and so the collection of points satisfying $x^2 + y^2 = 4$ consists of those points whose distance from the origin is 2. Thus the graph of $x^2 + y^2 = 4$ is a circle of radius 2 centered at the origin (Figure 1.27). □

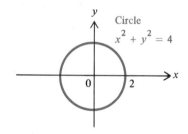

FIGURE 1.27

Example 7 Sketch the graph of the equation $x = y^2$.

Solution We first find a few points satisfying the equation:

x	0	$\tfrac{1}{2}$	1	2	4
y	0	$\pm\sqrt{2}/2$	± 1	$\pm\sqrt{2}$	± 2

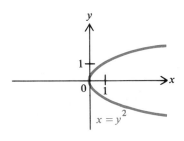

FIGURE 1.28

Since $x = y^2$ and $y^2 \geq 0$ for any y, x cannot be negative. For each positive value of x there are two values of y, namely, $\sqrt{x}$ and $-\sqrt{x}$. As x increases, so must $|y|$. The graph is sketched in Figure 1.28. ☐

Example 8   Sketch the graph of the equation $x^2 + y^4 = -1$.

Solution   Since $x^2 \geq 0$ and $y^4 \geq 0$ for any x and y, there are no numbers x and y satisfying the equation. Consequently there are no points on the graph. ☐

Gaining sufficient knowledge to sketch graphs of many different types of functions and equations is one of the aims of calculus. We will continue to sketch and study graphs in the remainder of this book.

EXERCISES 1.4

In Exercises 1–24 sketch the graph of the function.

1. $f(x) = \frac{1}{2}x + 1$

2. $f(x) = 1 - 3x$ for $-1 \leq x \leq 2$

3. $f(x) = x^2$

4. $f(x) = x^2 + 2$

5. $f(x) = x^2 - 1$ for $-2 \leq x \leq 2$

6. $f(x) = x^3$

7. $y = \sqrt{x}$

8. $y = \sqrt{4 - x^2}$ (*Hint:* See Example 6.)

9. $y = \sqrt{1 - x^2}$

10. $y = \dfrac{x^2 - 16}{x - 4}$ 11. $y = \dfrac{1 - x^2}{x + 1}$

12. $g(x) = |x|$ for $-2 \leq x \leq 3$

13. $g(x) = x|x|$

14. $g(x) = |x + 1|$ (*Hint:* Consider the cases $x + 1 \geq 0$ and $x + 1 < 0$ separately.)

15. $g(x) = |x - 2|$ 16. $f(t) = |t| + t$

17. $f(t) = |t| - t$ 18. $f(t) = \dfrac{|t|}{t}$

19. $f(x) = -[x]$ 20. $f(x) = [-x]$

21. $f(x) = x - [x]$

22. $f(x) = \begin{cases} x^2 & \text{for } 0 \leq x \leq 2 \\ 4 & \text{for } x > 2 \end{cases}$

23. $f(x) = \begin{cases} x & \text{for } x < 0 \\ 2x & \text{for } x \geq 0 \end{cases}$

24. $f(x) = \begin{cases} x^2 & \text{for } x < 0 \\ -x & \text{for } x \geq 0 \end{cases}$

In Exercises 25–27 sketch the graph of the function.

25. The function f defined by
$$f(x) = \begin{cases} 3 & \text{for } x \neq 1 \\ 5 & \text{for } x = 1 \end{cases}$$
A function whose graph resembles that of f can be called a "hiccup" function.

26. The **sign**, or **signum**, function, defined by
$$f(x) = \begin{cases} -1 & \text{for } x < 0 \\ 0 & \text{for } x = 0 \\ 1 & \text{for } x > 0 \end{cases}$$

27. The **diving board** function, defined by
$$f(x) = \begin{cases} 0 & \text{for } x < 0 \\ 1 & \text{for } x \geq 0 \end{cases}$$

In Exercises 28–40 sketch the graph of the equation. In each case determine whether the graph is that of a function.

28. $x^2 + y^2 = 1$ 29. $x^2 + y^2 = 9$

30. $x^2 + y^2 = 9$ for $x \geq 0$ 31. $x^2 + y^2 = 4$ for $y \leq 0$

32. $x^2 + y^2 = 0$ 33. $x = \frac{1}{2}y^2$

34. $y = x^2$ for $x \leq 0$ 35. $x^2 = y^2$

36. $x^3 = 8y^3$ 37. $xy = 0$

38. $|x| = |y|$ 39. $|x| + |y| = 1$

40. $|x| + |y| = 0$

41. Which of the graphs in Figure 1.29 are graphs of functions?

42. Can a horizontal line pass through more than one point on the graph of a function? Explain.

*43. Postage for domestic first-class letters is 22¢ for the first ounce or part ounce, and 17¢ for each additional ounce or part ounce. For $x > 0$, let $P(x)$ denote the postage for a letter weighing x ounces, and assume that the domain of P is $(0, \infty)$. Express $P(x)$ in terms of the greatest integer function.

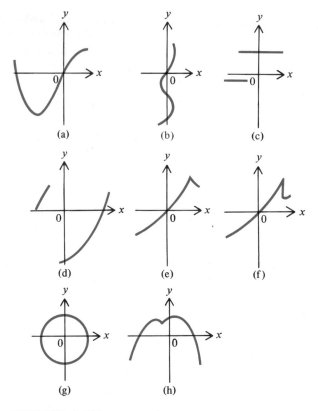

(a)   (b)   (c)

(d)   (e)   (f)

(g)   (h)

FIGURE 1.29

1.5
AIDS TO GRAPHING

There are certain aids to graphing that do not depend on calculus. Among these are the location of coordinate intercepts, symmetry properties, and translations.

Intercepts The intercepts of a graph determine the points where the graph meets the axes. We use the term **x intercept** for the x coordinate of any point where the graph meets the x axis. Similarly, the **y intercept** is the y coordinate of any point where the graph meets the y axis (Figure 1.30). (This definition of y intercept coincides with the original definition given in Section 1.2 for lines.) In theory it is very simple to locate the x and y intercepts from a given equation. More specifically, to find the x intercepts, set $y = 0$ in the equation and solve for x. To find the y intercepts, set $x = 0$ in the equation and solve for y.

Example 1 Find the x and y intercepts of the graph of the equation

$$y = x^2 - 2x - 3$$

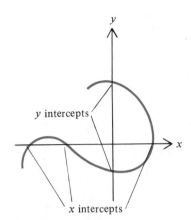

y intercepts

x intercepts

FIGURE 1.30

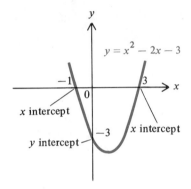

$y = x^2 - 2x - 3$

x intercept

x intercept

y intercept

FIGURE 1.31

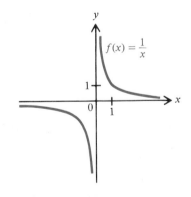

$f(x) = \dfrac{1}{x}$

FIGURE 1.32

Solution   To find the y intercepts, we let $x = 0$ and solve for y. This gives us -3 as the y intercept. To find the x intercepts, we let $y = 0$ and solve for x:

$$0 = x^2 - 2x - 3 = (x - 3)(x + 1)$$

It follows that -1 and 3 are the two x intercepts of the graph (Figure 1.31).   □

The graph of a function f has a y intercept only if 0 is in the domain of f, and in that case the y intercept is $f(0)$. Since $f(0)$ is unique, there can be at most one y intercept of the graph of a function. For example, if

$$f(x) = 2x^3 - 4x^2 + \sqrt{2}$$

then the y intercept is $\sqrt{2}$, since $f(0) = \sqrt{2}$. However, if

$$g(x) = \sqrt{x - 1}$$

then there is no y intercept because 0 is not in the domain of g.

In contrast, the graph of a function may have many x intercepts. As Example 1 shows, if $f(x) = x^2 - 2x - 3$, then the graph of f has two x intercepts, -1 and 3. It is also possible for the graph of a function to have no intercepts, as in the next example.

Example 2   Let $f(x) = 1/x$. Find the intercepts of the graph of f.

Solution   Since 0 is not in the domain of f, there is no y intercept. Moreover, since there is no x such that $0 = 1/x$, there is no x intercept. The graph of f, which is shown in Figure 1.32, has no intercepts at all.   □

It is not always simple, or even possible, to find the intercepts of a given graph. For example, if

$$f(x) = x^5 + x^3 + x^2 - 1$$

then finding any x intercepts is not easy, because this would require solving for x in

$$x^5 + x^3 + x^2 - 1 = 0$$

Later we will show that an x intercept does exist, and we will approximate its numerical value.

Symmetry   Symmetry is a form of balance. For example, part of the beauty of the Taj Mahal lies in the symmetry visible in the building and its surroundings. In mathematics we are interested in symmetry in graphs, especially symmetry with respect to the x and y axes and the origin. There are very simple criteria for determining the existence of such symmetry.

Taj Mahal, Agra, India

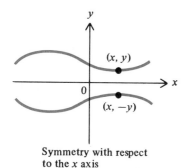

Symmetry with respect
to the x axis

FIGURE 1.33

We say that a graph is **symmetric with respect to the x axis** provided that whenever (x, y) is on the graph, $(x, -y)$ is also on the graph (Figure 1.33). The points above and below the x axis can be considered reflections of each other in the x axis, which we regard as a mirror.

Example 3   Show that the graph of the equation $x = y^2$ is symmetric with respect to the x axis.

Solution   We must show that if (x, y) is on the graph, then $(x, -y)$ is also on the graph. Thus we must show that

$$\text{if } x = y^2, \quad \text{then} \quad x = (-y)^2$$

But this is clear, since

$$(-y)^2 = y^2$$

Figure 1.34 supports our proof.   □

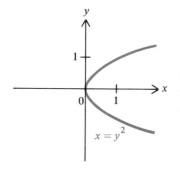

$x = y^2$

FIGURE 1.34

If only *even* powers of y appear in an equation, then the graph of the equation is automatically symmetric with respect to the x axis, since for any integer n, $(-y)^{2n} = y^{2n}$. By this test, the graphs of the following equations are all symmetric with respect to the x axis:

$$x^3y^4 - xy^8 + y^2 = 13$$
$$x^3 + y^4 = y^6 - 1$$
$$x^4y^{-4} + x^5y^2 + x^6 = 9$$

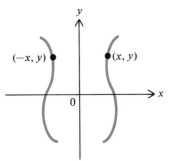

Symmetry with respect
to the y axis

FIGURE 1.35

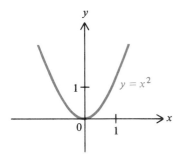

FIGURE 1.36

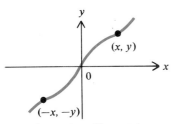

Symmetry with respect
to the origin

FIGURE 1.37

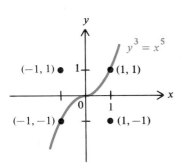

FIGURE 1.38

A graph is **symmetric with respect to the y axis** provided that whenever (x, y) is on the graph, $(-x, y)$ is also on the graph (Figure 1.35). The points to the left and to the right of the y axis can be considered reflections of each other in the y axis, which acts as a mirror.

Example 4   Show that the graph of the equation $y = x^2$ is symmetric with respect to the y axis.

Solution   We must show that

$$\text{if } y = x^2, \quad \text{then} \quad y = (-x)^2$$

But this is clear because

$$(-x)^2 = x^2$$

Therefore the graph is symmetric with respect to the y axis (Figure 1.36).   □

It is noteworthy that the graph of a nonzero function cannot be symmetric with respect to the x axis, because two distinct points on the graph cannot be on the same vertical line. However, many functions have graphs that are symmetric with respect to the y axis. Such functions are called **even functions**. In order for a function to be even, $-x$ must be in the domain of f whenever x is, and the relation $f(-x) = f(x)$ must hold. Any function in which only even powers of x occur is even. Thus the following functions are even:

$$f(x) = 3x^4 - 2x^2 + 5 \quad \text{and} \quad g(x) = \frac{x^2 - 1}{x^2 + 1}$$

Finally, a graph is **symmetric with respect to the origin** if $(-x, -y)$ is on the graph whenever (x, y) is (Figure 1.37). This means that each point on the graph is matched by another point on the graph on the other side of the origin.

Example 5   Show that the graph of the equation $y^3 = x^5$ is symmetric with respect to the origin but not with respect to either axis.

Solution   Since $(-y)^3 = (-x)^5$ whenever $y^3 = x^5$, the graph is symmetric with respect to the origin. Next we observe that $(1, 1)$ is on the graph, but neither $(1, -1)$ nor $(-1, 1)$ is on it. Therefore the graph is not symmetric with respect to either axis (Figure 1.38).   □

A function whose graph is symmetric with respect to the origin is called an **odd function**. In order for a function f to be odd, $-x$ must be in the domain of f whenever x is, and the relation $f(-x) = -f(x)$ must hold. Some examples of odd functions are given by

$$f(x) = x^3 \quad \text{and} \quad g(x) = x(x^2 + 1)$$

Knowledge of symmetry properties reduces the work of sketching graphs, because we can use information about part of the graph to draw the remaining parts.

Translations

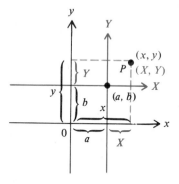

FIGURE 1.39

Graphing an equation is frequently made easier by changing from one set of axes to another. We will call the original set of axes the x and y axes and the new set the X and Y axes. We assume that the X and Y axes are obtained by moving the x and y axes in such a way that each is parallel to its original position (Figure 1.39). Under this condition we say that the X and Y axes are obtained from the x and y axes by a **translation**.

Any point P in the plane has coordinates (x, y) with respect to the x and y axes, and coordinates (X, Y) with respect to the X and Y axes. We would like to determine how these two sets of coordinates are related. Suppose the origin of the XY coordinate system has coordinates (a, b) in the xy coordinate system (Figure 1.39). By inspection we see that

$$x = X + a \quad \text{and} \quad y = Y + b$$

Solving for X and Y, we obtain

$$X = x - a \quad \text{and} \quad Y = y - b \tag{1}$$

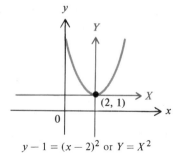

$y - 1 = (x - 2)^2$ or $Y = X^2$

FIGURE 1.40

Example 6   Sketch the graph of the equation

$$y - 1 = (x - 2)^2$$

Solution   If we take $a = 2$ and $b = 1$ in (1), then

$$X = x - 2 \quad \text{and} \quad Y = y - 1$$

The equation to be graphed may then be rewritten

$$Y = X^2$$

This is easily graphed in the XY coordinate system, whose origin is $(a, b) = (2, 1)$ in the xy coordinate system. The graph is shown in Figure 1.40. □

Example 7   Sketch the graph of the equation

$$y = x^2 + 2x + 2$$

Solution   We rewrite the right side as follows:

$$x^2 + 2x + 2 = (x^2 + 2x + 1) + 1 = (x + 1)^2 + 1$$

Thus the original equation becomes

$$y = (x + 1)^2 + 1, \quad \text{or} \quad y - 1 = (x + 1)^2$$

If we let

$$X = x + 1 \quad \text{and} \quad Y = y - 1$$

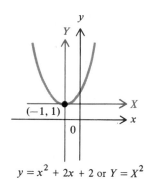

$y = x^2 + 2x + 2$ or $Y = X^2$

FIGURE 1.41

then as in Example 6 we obtain

$$Y = X^2$$

but here the origin of the XY coordinate system is $(a, b) = (-1, 1)$. The graph is shown in Figure 1.41. □

In Example 7 we replaced $x^2 + 2x + 2$ by $(x + 1)^2 + 1$, a step known as **completing the square**. This can be done for any expression of the form $ax^2 + bx + c$, where $a \neq 0$. First we rewrite the quadratic expression $ax^2 + bx + c$ in the following way by factoring out the leading coefficient a:

$$ax^2 + bx + c = a\left(x^2 + \frac{b}{a}x\right) + c \tag{2}$$

Then we take $b/2a$, which is one half the coefficient of x in the parentheses, and square it. The result is added to and subtracted from the expression in the parentheses. Thus the right side of (2) becomes

$$a\left[x^2 + \frac{b}{a}x + \left(\frac{b}{2a}\right)^2 - \left(\frac{b}{2a}\right)^2\right] + c \tag{3}$$

Since the first three terms in the brackets form a square, we separate them from the remainder of the expression, obtaining

$$a\left[x^2 + \frac{b}{a}x + \left(\frac{b}{2a}\right)^2\right] - a\left(\frac{b}{2a}\right)^2 + c = a\left(x + \frac{b}{2a}\right)^2 - \frac{b^2}{4a} + c \tag{4}$$

Letting $X = x + b/2a$, we have

$$ax^2 + bx + c = aX^2 - \frac{b^2}{4a} + c \tag{5}$$

The point of completing the square is that there is no X term in the right side of (5). This simplifies graphing the equation. Some other examples of completing the square are

$$x^2 - 2x = (x - 1)^2 - 1$$

$$3x^2 - x + 1 = 3\left(x - \frac{1}{6}\right)^2 + \frac{11}{12}$$

$$-2x^2 + x + 2 = -2\left(x - \frac{1}{4}\right)^2 + \frac{17}{8}$$

In Example 8 it will be necessary to complete the square twice.

Example 8 Show that the graph of

$$x^2 - 4x + y^2 + 6y = -4$$

is a circle.

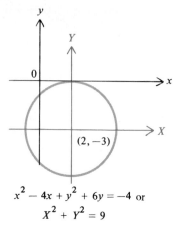

$x^2 - 4x + y^2 + 6y = -4$ or

$X^2 + Y^2 = 9$

FIGURE 1.42

Solution We complete the square for terms in x and for terms in y:

$$x^2 - 4x = x^2 - 4x + 4 - 4 = (x - 2)^2 - 4$$

$$y^2 + 6y = y^2 + 6y + 9 - 9 = (y + 3)^2 - 9$$

Then we substitute these expressions into the original equation, obtaining

$$(x - 2)^2 - 4 + (y + 3)^2 - 9 = -4$$

or

$$(x - 2)^2 + (y + 3)^2 = 9$$

Setting

$$X = x - 2 \quad \text{and} \quad Y = y + 3$$

we obtain

$$X^2 + Y^2 = 9 \tag{6}$$

The last is an equation of a circle of radius 3 centered at the origin of the XY coordinate system. Since that origin has coordinates $(2, -3)$ in the xy system, the graph of the original equation is as shown in Figure 1.42. □

Observe that it is the presence of the first-degree terms $-4x$ and $6y$ that made it difficult to graph the original equation in Example 8. Completing the square and translating gives an equation in X and Y with no first-degree terms, which is easier to graph.

EXERCISES 1.5

In Exercises 1–16 determine all intercepts of the graph of the equation. Then decide whether the graph is symmetric with respect to the x axis, the y axis, or the origin.

1. $x = 3y^2 - 2$
2. $4x^2 + y^2 = 12$
3. $x^2 - y^2 = 1$
4. $x^2 = y^{15} - y^9$
5. $x^4 = 3y^3$
6. $x^4 = 3y^3 + 4$
7. $x^2y^4 - 2x^4 = 1$
8. $2y^2x - 4x^2 = xy^4$
9. $y = x - \dfrac{1}{x}$
10. $y = \dfrac{x}{1 + x^2}$
11. $y = [x]$
12. $|x - 3| = |y + 5|$
13. $y = \sqrt{9 - x^2}$
14. $y^2 = x\sqrt{1 + x^2}$
15. $\sqrt{x} + \sqrt{y} = 1$
16. $y^2 = \dfrac{x^2 + 1}{x^2 - 1}$

In Exercises 17–26 sketch the graph. List the intercepts and describe the symmetry (if any) of the graph.

17. $y = \frac{1}{3}x$
18. $2x = -y^2$
19. $y = x^2 - 3$
20. $|x| = 2$
21. $|y| = 1$
22. $4x^2 + 4y^2 = 9$
23. $x = \sqrt{4 - y^2}$
24. $y = \sqrt{25 - x^2}$
25. $|y| = |3x|$
26. $|2y - 3| = |x + 4|$

In Exercises 27–38 sketch the graph of the given equation with the help of a suitable translation. Show both the x and y axes, and the X and Y axes.

27. $(x - 1)^2 + (y - 3)^2 = 4$
28. $(x + 2)^2 + (y + 4)^2 = \frac{1}{4}$
29. $x^2 - 2x + y^2 = 3$
30. $x^2 + y^2 + 4y = -1$
31. $x^2 + y - 3 = 0$
32. $x^2 - 4x + y = 5$
33. $x^2 + y^2 + 4x - 6y + 13 = 0$
34. $x^2 - 6x + y^2 - y = -9$
35. $y + 4 = \dfrac{1}{x + 2}$
36. $y = |x - 4|$
37. $x - 2 = |y - 2|$
38. $y = \sqrt{x + 3}$

39. Determine which of the following functions are even, which are odd, and which are neither.
    a. $f(x) = -x$
    b. $f(x) = 5x^2 - 3$
    c. $f(x) = x^3 + 1$
    d. $f(x) = (x - 2)^2$
    e. $f(x) = (x^2 + 3)^3$
    f. $y = x(x^2 + 1)^2$
    g. $y = \dfrac{x}{x^2 + 4}$
    h. $y = |x|$
    i. $y = \dfrac{|x|}{x}$

40. Let $f(x) = x^2$ and $g(x) = f(x + 3)$. Using a suitable translation, sketch the graph of g.

41. Let $f(x) = |x|$ and $g(x) = f(x - 2)$. Sketch the graph of g.

42. Let f be a function, and let $g(x) = f(x) - 3$. What is the relationship between the graphs of f and g?

43. Suppose that f is a function and that $g(x) = f(x) + d$ for all x in the domain of f, where d is a constant. What is the relationship between the graphs of f and g?

44. Let $f(x) = |x - 1| + |x + 1|$.
    a. Determine whether the graph of f is symmetric with respect to either axis or the origin.
    b. Sketch the graph of f. (*Hint:* Find alternative expressions for $f(x)$ in the three cases $x < -1$, $-1 \le x \le 1$, and $x > 1$.)

45. Suppose the graph of an equation is symmetric with respect to both axes. Prove that it is symmetric with respect to the origin. Is the converse true?

46. Suppose the graph of an equation is symmetric with respect to the y axis and the origin. Is it necessarily symmetric with respect to the x axis? Explain.

47. Suppose $c > 0$. If $f(c - x) = f(c + x)$ for all x, what property of symmetry does the graph of f have?

48. Show that the points twice as far from the point $(2, -3)$ as from the point $(-1, 0)$ form a circle, and find the center and radius of that circle.

1.6
COMBINING FUNCTIONS

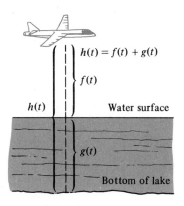

$h(t) = f(t) + g(t)$

$f(t)$

$h(t)$ Water surface

$g(t)$

Bottom of lake

FIGURE 1.43

Many functions arise as combinations of other functions. For example, suppose an airplane is flying over a large lake. Then at any time t, the height $h(t)$ of the airplane above the bottom of the lake is the sum of the height $f(t)$ of the airplane above the surface of the lake and the depth $g(t)$ of the lake directly below the airplane (Figure 1.43). In other words,

$$h(t) = f(t) + g(t)$$

Certain properties of the sum h of f and g can be determined directly from the properties of f and g. Similarly, certain properties of the difference, the product, and the quotient of two functions can be determined directly from the properties of those functions.

Sums, Differences, Products, and Quotients of Functions

Let f and g be functions. We define the **sum** $f + g$, the **difference** $f - g$, and the **product** fg to be the functions whose domains consist of all numbers in the domains of *both* f and g and whose rules are given by

$$(f + g)(x) = f(x) + g(x)$$
$$(f - g)(x) = f(x) - g(x)$$
$$(fg)(x) = f(x)g(x)$$

In each case the domain is the expected one, consisting of those values of x for which both $f(x)$ and $g(x)$ are defined. Because division by 0 is excluded, we give the definition of the quotient of two functions separately. The **quotient** f/g is the function whose domain consists of all numbers x in the domains of both f and g for which $g(x) \ne 0$ and whose rule is given by

$$\left(\frac{f}{g}\right)(x) = \frac{f(x)}{g(x)}$$

Example 1   Let $f(x) = 1/x$ and $g(x) = \sqrt{x}$. Find the domain and rule of $f + g$.

Solution   The domain of f consists of all nonzero numbers, and the domain of g consists of all nonnegative numbers. The only numbers in both domains are the positive numbers, which constitute the domain of $f + g$. For the rule we have

$$(f + g)(x) = f(x) + g(x) = \frac{1}{x} + \sqrt{x} \quad \text{for } x > 0 \quad \square$$

Example 2   Let $f(x) = \sqrt{4 - x^2}$ and $g(x) = \sqrt{x - 1}$. Find the domain and rule of fg.

Solution   The domain of f is the interval $[-2, 2]$, and the domain of g is the interval $[1, \infty)$. Therefore the domain of fg is the set of numbers in both $[-2, 2]$ and $[1, \infty)$. This is just the interval $[1, 2]$. The rule is given by

$$(fg)(x) = f(x)g(x) = \sqrt{4 - x^2}\sqrt{x - 1}$$
$$= \sqrt{(4 - x^2)(x - 1)} \quad \text{for } 1 \le x \le 2 \quad \square$$

Caution: Example 2 illustrates a surprising fact about combinations of functions. Although we found that the domain of fg is the interval $[1, 2]$, the expression $\sqrt{(4 - x^2)(x - 1)}$ is also meaningful for x in $(-\infty, -2]$. This is true because

$$(4 - x^2)(x - 1) \ge 0 \quad \text{for } x \le -2$$

Hence we must be careful to determine the domain of fg from the separate domains of f and g rather than from the rule for fg. Similar comments hold for the domains of $f + g, f - g$, and f/g.

Example 3   Let $f(x) = x + 5$ and $g(x) = 4x^2 - 1$. Find the domain and rule of f/g.

Solution   Since the domains of f and g are all real numbers and since $g(x) = 0$ for $x = -\frac{1}{2}$ and for $x = \frac{1}{2}$, it follows that the domain of f/g consists of all real numbers except $-\frac{1}{2}$ and $\frac{1}{2}$. The rule of f/g is given by

$$\left(\frac{f}{g}\right)(x) = \frac{f(x)}{g(x)} = \frac{x + 5}{4x^2 - 1} \quad \text{for } x \ne \frac{1}{2} \text{ and } x \ne -\frac{1}{2} \quad \square$$

We can add or multiply more than two functions. For example, if f, g, and h are functions, then for all x common to the domains of f, g, and h we have

$$(f + g + h)(x) = f(x) + g(x) + h(x)$$

and
$$(fgh)(x) = f(x)g(x)h(x)$$

In a similar way we can add or multiply more than three functions.

A special case of the product occurs when one of the functions is a constant function:

$$g(x) = c \quad \text{for all } x$$

For any function f, the domain of the product cf is the same as the domain of f. For example,

$$(2f)(x) = 2f(x) \quad \text{and} \quad (\pi f)(x) = \pi f(x) \quad \text{for all } x \text{ in the domain of } f$$

These very simple products occur often in calculus.

Let f and g be functions and c a nonzero number such that

$$g = cf$$

Then we say that the values of g are **proportional** to the values of f. For instance, if $g(r)$ denotes the area of a circle of radius r and if $f(r) = r^2$, then $g(r)$ is proportional to $f(r)$, or $g(r) = cf(r)$, where $c = \pi$. In other words, the area of a circle is proportional to the square of the radius.

Finally, any polynomial

$$g(x) = c_n x^n + c_{n-1} x^{n-1} + \cdots + c_1 x + c_0$$

can be thought of as a combination of constant functions and the identity function $f(x) = x$. Indeed

$$g(x) = c_n (f(x))^n + c_{n-1}(f(x))^{n-1} + \cdots + c_1 f(x) + c_0$$

so that $\qquad g = c_n f^n + c_{n-1} f^{n-1} + \cdots + c_1 f + c_0$

This fact makes polynomials particularly amenable to the methods of calculus.

Composition of Functions

Another way of combining functions that occurs frequently in calculus is illustrated by the formula

$$h(x) = \sqrt{x + 10}$$

To obtain $h(x)$, we first add 10 to x and then take the square root of the result. If we let

$$f(x) = x + 10 \quad \text{and} \quad g(x) = \sqrt{x}$$

then $\qquad h(x) = \sqrt{x + 10} = g(x + 10) = g(f(x))$

Thus $h(x)$ can be obtained by applying f to x and then applying g to the resulting value $f(x)$.

In general, if f and g are two functions, then the **composition** $g \circ f$ of f and g is defined as the function whose rule is

$$(g \circ f)(x) = g(f(x)) \tag{1}$$

and whose domain consists of all numbers x in the domain of f for which the number $f(x)$ is in the domain of g. (This is just the set of all numbers x for which the right side of (1) is defined.) The expression $g \circ f$ is read "g of f," "g composed

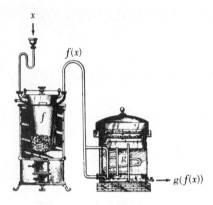

x

$f(x)$

f

g

$g(f(x))$

FIGURE 1.44

with f," or "g circle f." The function $g \circ f$ is the result of performing f and then performing g. If we think of a function as a machine that manufactures numbers in its range out of numbers in its domain, then the composite function $g \circ f$ can be represented as shown in Figure 1.44.

Example 4   Let $f(x) = \sqrt{x - 1}$ and $g(x) = 1/x$. Determine the functions $g \circ f$ and $f \circ g$, and then find $g(f(5))$ and $f(g(\frac{1}{4}))$.

Solution   The rule of $g \circ f$ is given by

$$g(f(x)) = g(\sqrt{x - 1}) = \frac{1}{\sqrt{x - 1}}$$

The domain of f is $[1, \infty)$, so the domain of $g \circ f$ consists of those numbers x in $[1, \infty)$ for which $\sqrt{x - 1}$ is in the domain of g, that is, for which $\sqrt{x - 1} \neq 0$. Since this excludes $x = 1$, the domain of $g \circ f$ is the interval $(1, \infty)$.

The rule for $f \circ g$ is given by

$$f(g(x)) = f\left(\frac{1}{x}\right) = \sqrt{\frac{1}{x} - 1}$$

The domain of g is the set of nonzero numbers, so the domain of $f \circ g$ consists of those nonzero numbers x such that $1/x$ is in the domain $[1, \infty)$ of f. Since $1/x$ is in $[1, \infty)$ whenever x is in $(0, 1]$, the domain of $f \circ g$ is $(0, 1]$.

Finally, we calculate that

$$g(f(5)) = \frac{1}{\sqrt{5 - 1}} = \frac{1}{2} \quad \text{and} \quad f\left(g\left(\frac{1}{4}\right)\right) = \sqrt{\frac{1}{\frac{1}{4}} - 1} = \sqrt{3} \quad \square$$

Example 4 shows that the functions $g \circ f$ and $f \circ g$ may not be equal. In fact, they are usually unequal. In Example 4 not only are the rules for $g \circ f$ and $f \circ g$ different, but the domains of $g \circ f$ and $f \circ g$ do not even have any numbers in common. The difference between $g \circ f$ and $f \circ g$ is much the same as the difference between first taking an examination and then seeing the correct answers, and first seeing the correct answers and then taking the examination.

As an application of composition of functions, we now define a composite function that will be useful later. This is the function $x^{m/n}$, where m and n are integers having no common integer factors other than -1 and 1 and where $n > 0$. Let

$$f(x) = x^m \quad \text{and} \quad g(x) = x^{1/n}$$

Then we define $x^{m/n}$ by

$$x^{m/n} = (g \circ f)(x) = (x^m)^{1/n} \qquad \text{domain:} \begin{cases} \text{all } x \text{ if } m > 0, n \text{ odd} \\ \text{all } x \geq 0 \text{ if } m > 0, n \text{ even} \\ \text{all } x \neq 0 \text{ if } m < 0, n \text{ odd} \\ \text{all } x > 0 \text{ if } m < 0, n \text{ even} \end{cases}$$

Thus the function x^r has been defined for every rational number r.

Example 5   Let $h(x) = (1 - 3x)^{2/3}$. Write h as the composite $g \circ f$ of two functions f and g.

Solution   If $f(x) = 1 - 3x$, then

$$h(x) = (f(x))^{2/3}$$

Hence if we let $g(x) = x^{2/3}$, then

$$h(x) = (f(x))^{2/3} = g(f(x))$$

Consequently $h = g \circ f$. (There are other ways of expressing h as the composite of two functions, but the way we have chosen is a convenient one.)   □

EXERCISES 1.6

In Exercises 1–8 let $f(x) = 2x^2 + x - 4$ and $g(x) = 3 - x^2$. Find the specified values.

1. $(f + g)(-1)$ 2. $(f - g)(2)$ 3. $(fg)(\frac{1}{2})$

4. $(f/g)(-3)$ 5. $f(g(1))$ 6. $g(f(0))$

7. $\dfrac{f(x) - f(2)}{x - 2}$ 8. $\dfrac{g(a) - g(-1)}{a + 1}$

In Exercises 9–14 let $f(x) = \dfrac{x - 1}{x^2 + 1}$ and $g(x) = x^{1/4}$. Find the specified values.

9. $(f + g)(16)$ 10. $(f - g)(1)$ 11. $(fg)(9)$

12. $(f/g)(\frac{1}{4})$ 13. $f(g(1))$ 14. $g(f(1))$

In Exercises 15–22 find the domains and rules of $f + g$, fg, and f/g.

15. $f(x) = 2x + 1$ and $g(x) = 3 - x$

16. $f(x) = x - 2$ and $g(x) = x^2 - 2$

17. $f(x) = \dfrac{2}{x - 1}$ and $g(x) = x - 1$

18. $f(x) = \dfrac{x + 2}{x - 3}$ and $g(x) = \dfrac{x + 3}{x^2 - 4}$

19. $f(t) = t^{3/4}$ and $g(t) = t^2 + 3$

20. $f(t) = \sqrt{1 - t^2}$ and $g(t) = \sqrt{2 + t - t^2}$

21. $f(t) = t^{2/3}$ and $g(t) = t^{3/5} - 1$

22. $f(x) = |x|$ and $g(x) = \sqrt{x^2 - 1}$

In Exercises 23–32 find the domain and rule of $g \circ f$ and $f \circ g$.

23. $f(x) = 1 - x$ and $g(x) = 2x + 5$

24. $f(x) = x^2 + 2x + 3$ and $g(x) = x - 1$

25. $f(x) = x^2$ and $g(x) = \sqrt{x}$

26. $f(x) = x^6$ and $g(x) = x^{3/4}$

27. $f(x) = \sqrt{x}$ and $g(x) = x^2 - 5x + 6$

28. $f(x) = \dfrac{1}{x}$ and $g(x) = x^2 - 3x - 10$

29. $f(x) = \dfrac{1}{x - 1}$ and $g(x) = \dfrac{1}{x + 1}$

30. $f(x) = \dfrac{3}{x + 2}$ and $g(x) = \dfrac{-1}{\frac{1}{3}x + 1}$

31. $f(x) = \sqrt{9 - x^2}$ and $g(x) = [x]$

32. $f(x) = \sqrt{x^2 + 3}$ and $g(x) = \sqrt{x^2 - 4}$

In Exercises 33–40 write h as the composite $g \circ f$ of two functions f and g (neither of which is equal to h).

33. $h(x) = \sqrt{x - 3}$ 34. $h(x) = \sqrt{1 - x^2}$

35. $h(x) = (3x^2 - 5\sqrt{x})^{1/3}$ 36. $h(x) = \left(x + \dfrac{1}{x}\right)^{5/2}$

37. $h(x) = \dfrac{1}{(x + 3)^2 + 1}$ 38. $h(x) = \dfrac{1}{(x^3 - 2x^2)^5}$

39. $h(x) = \sqrt{\sqrt{x} - 1}$ 40. $h(x) = \sqrt{[x] - 1}$

41. Find g if $f(x) = |x|$ and $(f + g)(x) = |x| - |x - 2|$.

42. Find g if $f(x) = (x^2 - 4)/(x + 3)$ and $(fg)(x) = 1$, for $x \neq 2, -2$, and -3.

43. Suppose f is defined on $[0, 4]$ and $g(x) = f(x + 3)$. What is the domain of g?

44. Suppose f is defined on $[a, b]$ and $g(x) = f(x + c)$ for a fixed c. What is the domain of g?

45. For which functions f is there a function g such that $f = g^2$?

46. For which functions f is there a function g such that $f = g^3$?

47. For which functions f is there a function g such that $f = 1/g$?

48. Let f and g be even functions.
    a. Show that $f + g$ is an even function.
    b. Show that fg is an even function.

49. Let f be even and g odd. Show that fg is an odd function.

50. Let $f(x) = 1/x$. Show that $f(f(x)) = x$ for $x \neq 0$.

51. Let a be a real number and $f(x) = a - x$. Show that $f(f(x)) = x$ for all x.

52. Let $f(x) = 1/(1 - x)$. Show that $f(f(f(x))) = x$ for all x different from 0 and 1.

53. Let $f(x) = ax + b$, where a and b are constants, and let p be any real number. Show that if

$$g(x) = f(x + p) - f(x)$$

then g is a constant function.

54. Let c be any number, and define a function f by $f(t) = ct$. Show that $f(x + y) = f(x) + f(y)$ for all numbers x and y.

55. The revenue function R for a certain product is given by

$$R(x) = 5x^2 - \frac{x^4}{10}$$

The cost function C is given by

$$C(x) = 4x^2 - 24x + 38$$

The profit function P is defined as the difference $R - C$. Find the equation that describes P. Then find $P(1)$ and $P(2)$, and show that it is possible to lose money, and also possible to make a profit.

56. Recall that the volume $V(r)$ of a spherical balloon of radius r is given by the formula

$$V(r) = \frac{4}{3}\pi r^3 \quad \text{for } r \geq 0$$

Suppose the radius is given by $r(t) = 3\sqrt{t}$. Write a formula for the volume in terms of t.

57. A sphere with surface area s has a radius $r(s)$ given by

$$r(s) = \frac{1}{2}\sqrt{\frac{s}{\pi}}$$

a. Using the formula in Exercise 56, find a formula for the volume of a sphere in terms of its surface area.
b. Determine the volume corresponding to a surface area of 6.

58. According to Newton's Law of Gravitation, if two bodies are a distance r apart, then the gravitational force $F(r)$ exerted by one body on the other is given by

$$F(r) = \frac{k}{r^2} \quad \text{for } r > 0$$

where k is a positive constant. Suppose that as a function of time t, the distance between the two bodies is given by

$$r(t) = 4000\left(\frac{1 + t}{1 + t^2}\right) \quad \text{for } t \geq 0$$

Find a formula for the force in terms of time.

1.7
TRIGONOMETRIC FUNCTIONS

You are doubtless already acquainted with the trigonometric functions. These functions arise in geometry, but they are also applicable to the study of sound, the motion of a pendulum, and many other phenomena involving rotation or oscillation. For this reason, they are important in the study of calculus.

Radian Measure   We will first discuss the measurement of angles. Recall that there are 360 degrees in a circle (Figure 1.45). However, a more convenient unit of angle measurement for calculus is the **radian**, which is chosen so that a circle contains 2π radians. Thus 2π radians = 360 degrees, so that 1 radian is equal to $360/2\pi$ degrees, or $180/\pi$ degrees (approximately 57.2958 degrees). It is easy to convert radians to

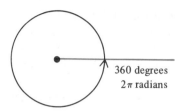

FIGURE 1.45

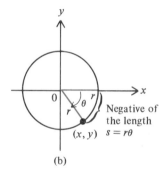

(a)

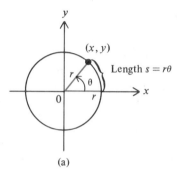

(b)

FIGURE 1.46

degrees or degrees to radians:

If $f(x)$ is the number of degrees in x radians, then $f(x) = \dfrac{180}{\pi}x.$

If $g(x)$ is the number of radians in x degrees, then $g(x) = \dfrac{\pi}{180}x.$

The following is a table of conversions for some common angles.

Degrees	0°	30°	45°	60°	90°	120°	135°	150°	180°	270°	360°
Radians	0	$\dfrac{\pi}{6}$	$\dfrac{\pi}{4}$	$\dfrac{\pi}{3}$	$\dfrac{\pi}{2}$	$\dfrac{2\pi}{3}$	$\dfrac{3\pi}{4}$	$\dfrac{5\pi}{6}$	π	$\dfrac{3\pi}{2}$	2π

In this book we will use radians to measure angles except where otherwise indicated.

Let us draw a circle centered at the origin, with arbitrary radius $r > 0$. Next, we place an angle of θ radians so that its vertex is at the origin and its initial side is on the positive x axis. The angle opens counterclockwise if $\theta \geq 0$ and clockwise if $\theta < 0$ (Figure 1.46(a) and (b)). We allow θ to be greater than 2π. For example, an angle of 3π radians can be obtained by rotating a line through one full revolution (2π radians) and an extra half-revolution. Thus an angle of 3π radians has the same initial and terminal sides as an angle of π radians. The same is true of $-\tfrac{7}{3}\pi$ and $-\tfrac{1}{3}\pi$ radians, since $-\tfrac{7}{3}\pi = -\tfrac{1}{3}\pi - 2\pi$. Since the circle contains 2π radians and its circumference is $2\pi r$, an angle of 1 radian intercepts an arc of length r on the circle. Thus if $\theta > 0$, an angle of θ radians intercepts an arc of length θr on the circle. Calling this arc length s, we have

$$s = r\theta \qquad (1)$$

For $\theta < 0$ formula (1) holds if we think of s as the negative of the length of arc intercepted on the circle (Figure 1.46(b)).

Definitions of the Trigonometric Functions

We define the sine and cosine functions with reference to a circle of radius r centered at the origin and an angle of θ radians, positioned as shown in Figure 1.47. The terminal side of the angle intersects the circle at a unique point (x, y). We define the **sine function** and **cosine function** by

$$\sin\theta = \frac{y}{r} \quad \text{and} \quad \cos\theta = \frac{x}{r} \qquad (2)$$

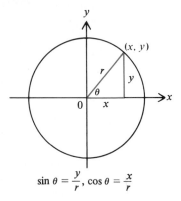

$$\sin\theta = \frac{y}{r}, \cos\theta = \frac{x}{r}$$

FIGURE 1.47

The properties of similar triangles imply that $\sin\theta$ and $\cos\theta$ depend only on θ, not on the value of r. If $r = 1$, then $\sin\theta = y$, which is the length of a halfchord.* The domains of the sine and cosine functions are $(-\infty, \infty)$. Notice that in the expression $\sin\theta$, θ represents a *number*. Thus we write $\sin 1$ instead of $\sin(1$ radian).

Since an angle of θ radians and one of $\theta + 2\pi$ radians have the same terminal side, we can write

$$\sin\theta = \sin(\theta + 2\pi) \quad \text{and} \quad \cos\theta = \cos(\theta + 2\pi)$$

Thus the sine and cosine functions are periodic; they both have period 2π. Consequently, for any integer n and any number θ,

$$\sin\theta = \sin(\theta + 2n\pi) \quad \text{and} \quad \cos\theta = \cos(\theta + 2n\pi) \tag{3}$$

For a circle of radius r centered at the origin, the distance between the origin and a point (x, y) on the circle is r. Using the distance formula, we have

$$x^2 + y^2 = r^2 \tag{4}$$

Substituting for x and y from (2) gives us

$$r^2\cos^2\theta + r^2\sin^2\theta = r^2$$

which yields the famous Pythagorean Identity

$$\sin^2\theta + \cos^2\theta = 1 \tag{5}$$

The table below lists some values of the sine and the cosine. We leave verification of these values to you.

θ	0	$\dfrac{\pi}{6}$	$\dfrac{\pi}{4}$	$\dfrac{\pi}{3}$	$\dfrac{\pi}{2}$	$\dfrac{2\pi}{3}$	$\dfrac{3\pi}{4}$	$\dfrac{5\pi}{6}$	π
$\sin\theta$	0	$\dfrac{1}{2}$	$\dfrac{\sqrt{2}}{2}$	$\dfrac{\sqrt{3}}{2}$	1	$\dfrac{\sqrt{3}}{2}$	$\dfrac{\sqrt{2}}{2}$	$\dfrac{1}{2}$	0
$\cos\theta$	1	$\dfrac{\sqrt{3}}{2}$	$\dfrac{\sqrt{2}}{2}$	$\dfrac{1}{2}$	0	$-\dfrac{1}{2}$	$-\dfrac{\sqrt{2}}{2}$	$-\dfrac{\sqrt{3}}{2}$	-1

Since we normally use x to represent points in the domain of a function, we will usually follow that convention for the sine and cosine functions and replace θ

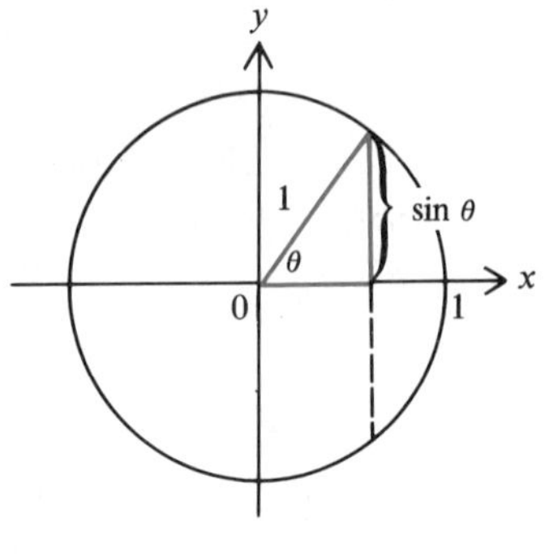

FIGURE 1.48

* This fact seems to have influenced the etymology of the word "sine." In the fifth century A.D. the Indian mathematician Aryabbatta called the sine of an angle by the name "ardha-jya," meaning "half-chord" (Figure 1.48). Soon the word condensed to "jya," or "chord." The Arabs translated the word into their vowelless written language as "jb." Later, as the Arabs once again began adding vowels to their language, "jb" somehow became "jaib," which already meant "bosom." Around the twelfth century, when mathematics was translated into Latin, the Arabic "jaib" was translated into "sinus," the Latin equivalent for "bosom," from which our word "sine" comes.

by x. Thus (5) becomes

$$\sin^2 x + \cos^2 x = 1 \tag{6}$$

By plotting a few points on the graphs of the sine and cosine functions and utilizing (3), we obtain the graphs of these functions, shown in Figure 1.49(a) and (b).

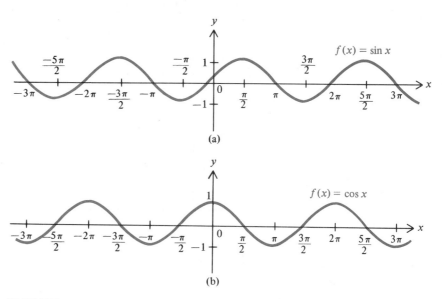

(a)

(b)

FIGURE 1.49

There are four other basic trigonometric functions, which are defined in terms of the sine and the cosine. These are the **tangent**, the **cotangent**, the **secant**, and the **cosecant** functions. Their definitions are

$$\tan x = \frac{\sin x}{\cos x} \quad \text{for } x \neq n\pi + \frac{\pi}{2}, n \text{ any integer}$$

$$\cot x = \frac{\cos x}{\sin x} \quad \text{for } x \neq n\pi, n \text{ any integer}$$

$$\sec x = \frac{1}{\cos x} \quad \text{for } x \neq n\pi + \frac{\pi}{2}, n \text{ any integer}$$

$$\csc x = \frac{1}{\sin x} \quad \text{for } x \neq n\pi, n \text{ any integer}$$

The values of these functions can be quickly computed from the corresponding values of the sine and cosine. Their graphs appear in Figure 1.50(a)–(d).

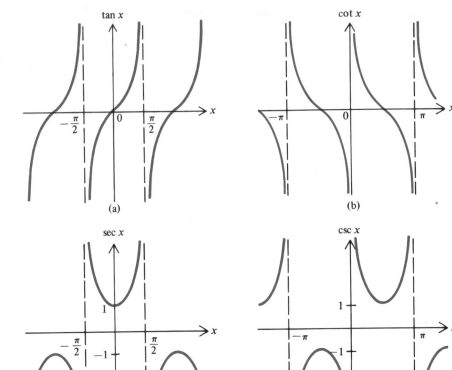

(a)

(b)

(c)

(d)

FIGURE 1.50

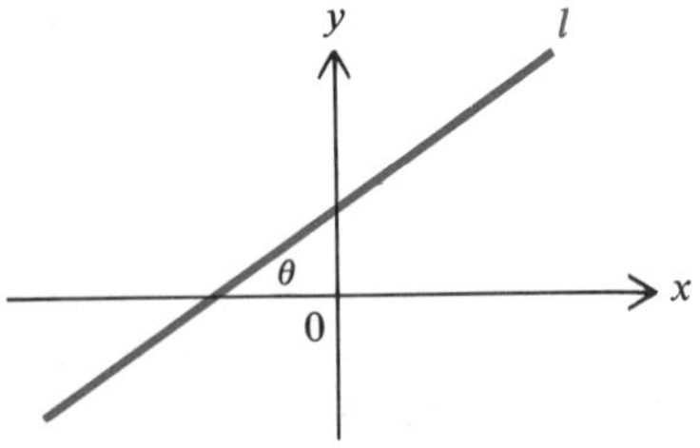

FIGURE 1.51

Observe that if θ is the angle between the positive x axis and any nonvertical line l (Figure 1.51), then

$$\tan \theta = \text{slope of } l \tag{7}$$

This fact relates slopes of lines and the trigonometric functions.

Trigonometric Identities There are many equations, called **trigonometric identities**, that describe relationships among the various trigonometric functions. The most important ones are

$$\sin^2 x + \cos^2 x = 1 \tag{8}$$

$$\sin(x \pm y) = \sin x \cos y \pm \cos x \sin y \tag{9}$$

$$\cos(x \pm y) = \cos x \cos y \mp \sin x \sin y \tag{10}$$

$$\sin(-x) = -\sin x \tag{11}$$

$$\cos(-x) = \cos x \tag{12}$$

Other trigonometric identities can be derived from these. Below we list several of the more useful ones, with n denoting an arbitrary integer.

$$\sec^2 x = 1 + \tan^2 x$$

$$\csc^2 x = 1 + \cot^2 x$$

$$\sin(x + 2n\pi) = \sin x$$

$$\cos(x + 2n\pi) = \cos x$$

$$\sin\left(\frac{\pi}{2} - x\right) = \sin\left(\frac{\pi}{2} + x\right) = \cos x$$

$$\cos\left(\frac{\pi}{2} - x\right) = -\cos\left(\frac{\pi}{2} + x\right) = \sin x$$

$$\sin 2x = 2\sin x \cos x$$

$$\cos 2x = \cos^2 x - \sin^2 x$$

$$\sin^2 x = \frac{1 - \cos 2x}{2}$$

$$\cos^2 x = \frac{1 + \cos 2x}{2}$$

$$\tan(x \pm y) = \frac{\tan x \pm \tan y}{1 \mp \tan x \tan y}$$

The Angle Between Two Lines

Suppose l_1 and l_2 are two intersecting lines. Then we define the **angle from l_1 to l_2** to be the angle θ through which l_1 must be rotated counterclockwise about the point of intersection in order to coincide with l_2 (see θ in Figure 1.52). Thus $0 \leq \theta < \pi$. By using trigonometric identities, we can express θ in terms of the slopes of l_1 and l_2.

THEOREM 1.6

Let l_1 and l_2 be two nonvertical lines that are not perpendicular, with slopes m_1 and m_2, respectively. The tangent of the angle θ from l_1 to l_2 is given by

$$\tan\theta = \frac{m_2 - m_1}{1 + m_1 m_2}$$

Proof Let θ_1 and θ_2 be angles with initial sides along the positive x axis and terminal sides along l_1 and l_2, respectively, chosen so that $0 \leq \theta_1 < \pi$ and $\theta_2 \geq \theta_1$ (Figure 1.52). Then $\theta = \theta_2 - \theta_1$. Since $m_1 = \tan\theta_1$ and $m_2 = \tan\theta_2$,* and since $m_1 m_2 \neq -1$ (why?), we have

$$\tan\theta = \tan(\theta_2 - \theta_1) = \frac{\tan\theta_2 - \tan\theta_1}{1 + \tan\theta_1 \tan\theta_2} = \frac{m_2 - m_1}{1 + m_1 m_2} \quad \blacksquare$$

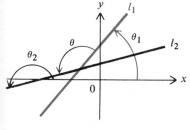

FIGURE 1.52

* If $\theta_2 \geq \pi$ (as in Figure 1.52), then $\tan\theta_2 = \tan(\theta_2 - \pi) = m_2$.

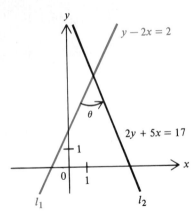

FIGURE 1.53

Example 1   Let the equations of l_1 and l_2 be

$$y - 2x = 2 \quad \text{and} \quad 2y + 5x = 17$$

Find the tangent of the angle θ from l_1 to l_2 (Figure 1.53).

Solution   From the equations of l_1 and l_2 we find that $m_1 = 2$ and $m_2 = -\frac{5}{2}$. Therefore

$$\tan \theta = \frac{m_2 - m_1}{1 + m_1 m_2} = \frac{-\frac{5}{2} - 2}{1 + 2(-\frac{5}{2})} = \frac{-\frac{9}{2}}{-4} = \frac{9}{8} \quad \square$$

The trigonometric functions have many important applications. They will also be valuable to us in illustrating numerous features of calculus.

EXERCISES 1.7

1. Convert to radians.
   a. 210°   b. 315°   c. −405°
   d. 1080°   e. 1°

2. Convert the following radian measures to degrees.
   a. $\dfrac{\pi}{8}$ b. $-\dfrac{3\pi}{10}$ c. $\dfrac{13\pi}{6}$

   d. $-\dfrac{2\pi}{5}$ e. -8π

3. Find the following values.
   a. $\sin \dfrac{11\pi}{6}$ b. $\sin\left(-\dfrac{2\pi}{3}\right)$ c. $\cos \dfrac{5\pi}{4}$

   d. $\cos\left(-\dfrac{7\pi}{6}\right)$ e. $\tan \dfrac{4\pi}{3}$ f. $\tan\left(-\dfrac{\pi}{4}\right)$

   g. $\cot \dfrac{\pi}{6}$ h. $\cot\left(-\dfrac{17\pi}{3}\right)$ i. $\sec 3\pi$

   j. $\sec\left(-\dfrac{\pi}{3}\right)$ k. $\csc \dfrac{\pi}{2}$ l. $\csc\left(-\dfrac{5\pi}{3}\right)$

4. For each of the following intervals, state which of the six trigonometric functions have positive values throughout the interval.
   a. $\left(0, \dfrac{\pi}{2}\right)$ b. $\left(\dfrac{\pi}{2}, \pi\right)$ c. $\left(\pi, \dfrac{3\pi}{2}\right)$

   d. $\left(\dfrac{3\pi}{2}, 2\pi\right)$ e. $(3\pi, 4\pi)$

In Exercises 5–8 find the values of the remaining four trigonometric functions under the given conditions.

5. $\sin x = \frac{4}{5}$ and $\cos x = -\frac{3}{5}$
6. $\cos x = \frac{1}{3}$ and $\tan x = 2\sqrt{2}$
7. $\tan x = -2$ and $\sec x = \sqrt{5}$
8. $\csc x = -\frac{1}{4}\sqrt{65}$ and $\cot x = \frac{7}{4}$

In Exercises 9–14 solve the equation for x in $[0, 2\pi)$.

9. $\sin x = -\frac{1}{2}$ 10. $\tan x = \sqrt{3}$

11. $\sec x = -\dfrac{2\sqrt{3}}{3}$ 12. $\sin 2x = \sin x$

13. $\cos 2x = \cos x$ 14. $\sin x + \tan x = 0$

In Exercises 15–23 solve the inequality for x in $[0, 2\pi)$.

15. $\sin x > -\frac{1}{2}$ 16. $\cos x \le 0$

17. $\tan x \ge 1$ 18. $\cot x < \sqrt{3}$

19. $\sec x < -1$ 20. $\dfrac{2\sqrt{3}}{3} \le \csc x \le 2$

21. $|\cos x| < \frac{1}{2}\sqrt{3}$ 22. $\sin x \le \cos x$

23. $\cot x \ge \tan x$

24. Decide whether each of the following functions is even or odd.
   a. $\sin x$ b. $\cos x$ c. $\tan x$
   d. $\cot x$ e. $\sec x$ f. $\csc x$

In Exercises 25–34 sketch the graph of the function. Indicate any intercepts and symmetry, and determine whether the function is even, odd, or neither.

25. $\sin\left(x - \dfrac{\pi}{2}\right)$ 26. $\sin\left(\dfrac{\pi}{4} - x\right)$

27. $\cos(\pi - x)$ 28. $\cos x - 2$

29. $\tan\left(x + \dfrac{\pi}{2}\right)$ 30. $|\cot x|$

31. $\sec(2\pi - x)$

32. $2 - \csc x$

33. $\sin 2x$

34. $\cos\dfrac{x - \pi}{2}$

In Exercises 35–40 use the identities given in this section to compute the given value.

35. $\sin\dfrac{7\pi}{12}$

36. $\cos\dfrac{5\pi}{12}$

$$\left(Hint: \dfrac{7\pi}{12} = \dfrac{\pi}{3} + \dfrac{\pi}{4}\right)$$

37. $\sin\dfrac{11\pi}{12}$

38. $\cos\dfrac{\pi}{12}$

39. $\tan\dfrac{5\pi}{12}$

40. $\sec\dfrac{7\pi}{12}$

41. Solve the equation $2\sin^2 x + \sin x - 1 = 0$.

42. Solve the equation $4\cos^2 x - 4\sqrt{3}\cos x + 3 = 0$.

43. Using (8), prove that $1 + \tan^2 x = \sec^2 x$.

44. Using (8), prove that $1 + \cot^2 x = \csc^2 x$.

45. Using identities (9)–(12), prove that the following identities hold:

 a. $\sin(\pi - x) = \sin x$

 b. $\sin\left(\dfrac{3\pi}{2} - x\right) = -\cos x$

 c. $\cos(\pi - x) = -\cos x$

 d. $\cos\left(\dfrac{3\pi}{2} - x\right) = -\sin x$

46. a. Using (8), show that $\cos x = \sqrt{1 - \sin^2 x}$ for $0 \le x \le \pi/2$ and for $3\pi/2 \le x \le 2\pi$. Show also that $\cos x = -\sqrt{1 - \sin^2 x}$ for $\pi/2 \le x \le 3\pi/2$.

 b. Using (8), show that $\sin x = \sqrt{1 - \cos^2 x}$ for $0 \le x \le \pi$. Show also that $\sin x = -\sqrt{1 - \cos^2 x}$ for $\pi \le x \le 2\pi$.

In Exercises 47–48 find the tangent of the angle θ from the first line to the second line.

47. $y = 4x - 2;\; 3y = -2x + 7$

48. $y = 3x + 9;\; 4y - 11x = 6$

49. For a function f, if there is a smallest positive number a such that $f(x + a) = f(x)$ for all x in the domain of f, then a is called the **period** of f. Find the periods of the following functions.

 a. $\tan x$ b. $\cot x$ c. $\sec x$

 d. $\csc x$ e. $\sin 3x$ f. $\cos\left(-2x + \dfrac{\pi}{2}\right)$

 g. $4\tan x$ h. $|\sin x|$

50. Through how many complete revolutions does a bicycle wheel with radius one foot turn when the bicycle travels one mile?

51. A beacon is located 2 miles from a straight coastline, as depicted in Figure 1.54. Express the distance between the beacon and the illuminated point x on the coastline as a function of the angle θ in Figure 1.54.

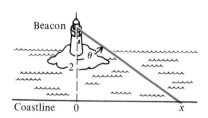

FIGURE 1.54

52. Consider a circular racetrack with a radius of 600 feet. Suppose a straight, narrow path connects two points A and B, diametrically opposite one another, on the track. If a horse is 800 feet around the track from point A, how far from the path is the horse?

53. Prove that the sine of an angle inscribed in a circle of unit diameter is the length of the chord of the subtended arc. (*Hint:* First assume that one side of the angle is a diameter and use the fact that the resulting triangle is a right triangle (Figure 1.55). Then use the fact that all inscribed angles with the same subtended arc are equal.)

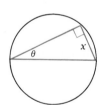

FIGURE 1.55

54. A ladybug is standing on the floor, and an 8-foot giant walks toward her. When the giant is 8 feet away, how big an angle does the giant occupy in the ladybug's field of vision? What happens as the giant comes closer? (This suggests why it is sometimes better to sit far back in a theater with empty rows directly in front of you than to sit in the second row directly behind someone.)

55. A motorist drives down a road toward the intersection with a heavily traveled boulevard (Figure 1.56). A house stands 40 feet from the center of the road and 60 feet from the center of the boulevard. Assuming that the motorist drives down the center of the road, determine how far the motorist will be from the center of the intersection when the angle θ shown in Figure 1.56 is $\pi/6$.

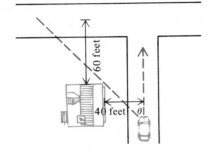

FIGURE 1.56

the function that expresses the volume of the trough in terms of the angle θ.

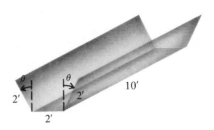

FIGURE 1.57

56. A sack of sand weighing 50 pounds is being dragged by a man whose arm makes an angle of x radians with the ground. If the coefficient of friction is $\mu > 0$, then the minimum force F necessary to drag the sack is given by

$$F(x) = \frac{50\mu}{\mu \sin x + \cos x}$$

   a. If the arm makes an angle of $\pi/4$ radians with the ground, what is the minimum force necessary to drag the load?

   b. If the arm makes an angle of $\pi/3$ radians with the ground, what is the minimum force necessary to drag the load?

   c. What happens if the arm makes an angle of $\pi/2$ radians with the ground?

57. A rectangular sheet of metal 6 feet wide and 10 feet long is bent as in Figure 1.57 to form a trough. Find a formula for

58. The angular displacement $f(t)$ of a pendulum bob is given by

$$f(t) = a \cos 2\pi\omega t$$

where ω is the frequency and a the maximum displacement. Draw the graph of this function for $\omega = 1$ and $a = 2$.

*59. a. Show that the perimeter $p_n(r)$ of a regular polygon of n sides inscribed in a circle of radius r is given by

$$p_n(r) = 2nr \sin \frac{\pi}{n}$$

   c b. Using the result of part (a), find the radius of the smallest circle that can circumscribe the Pentagon Building, each of whose outer walls is 921 feet long.

Key Terms and Expressions

Absolute value
Function
Domain, rule, and range of a function
Polynomial function; rational function
Graph of a function; graph of an equation
Independent variable; dependent variable
x intercept; y intercept

Symmetry with respect to the x axis; with respect to the y axis; with respect to the origin
Even function; odd function
Translation of axes
Sum, product, and quotient of two functions
Composition of two functions
Radian

Key Formulas

$|ab| = |a||b|$
$|a + b| \leq |a| + |b|$
$|a - b| \geq ||a| - |b||$

$m_1 = m_2$ for parallel lines
$m_1 m_2 = -1$ for perpendicular lines
$\sin^2 x + \cos^2 x = 1$

REVIEW EXERCISES

In Exercises 1–6 solve for x.

1. $\dfrac{2x + 3}{(x - 4)^3} > 0$

2. $\dfrac{x}{x - 1} + \dfrac{x + 2}{x - 4} \le 0$

3. $\dfrac{2}{3 - x} \ge 4$

4. $|x + 1| = |2x - 3|$

5. $|4 - 6x| < \frac{1}{2}$

6. $\left|\dfrac{x + 1}{x - 1}\right| \le 1$

7. Find the value of $(|a| + a)/2$ for
   a. $a \ge 0$
   b. $a < 0$

8. Show that for all x and y we have $2xy \le x^2 + y^2$.

In Exercises 9–11 find an equation of the line described. Then sketch the line.

9. The line through $(1, -3)$ and $(0, 1)$

10. The line through $(-2, 2)$ with slope $-\frac{1}{2}$

11. The line with slope $-\frac{1}{3}$ and y intercept 6

12. Find the distance between the points $(-4, -1)$ and $(-2, \frac{1}{2})$.

13. Determine whether the lines $y = 2x - 3$ and $x + 2y = 5$ are parallel, perpendicular, or neither.

14. Find the point of intersection of the lines $x + 3y = 4$ and $2x - 3y = 5$.

15. Let l be the line with equation $2x + 4y = 1$, and let $P = (1, -2)$.
    a. Find an equation of the line parallel to l and passing through P.
    b. Find an equation of the line perpendicular to l and passing through P.

16. Show that the lines $2x + 3y = 1, 2x + 3y = 6, 3x - 2y = 2$, and $3x - 2y = 6$ form a rectangle.

17. Show that the lines $y = x$, $y = -x + 2$, $y = -x + 10$, and $y = x + 8$ form a square.

18. Are the functions x^2 and $\sqrt{x^4}$ equal? Explain.

In Exercises 19–20 find the domain of the function.

19. $f(x) = \dfrac{x^2 + 3x - 4}{x(x^2 + 4x + 3)}$

20. $f(t) = \sqrt{t^3 - 4t}$

In Exercises 21–24 sketch the graph of the function.

21. $f(x) = 2|x - 3|$

22. $f(x) = \sqrt{2 - x^2}$

23. $g(x) = \begin{cases} x^3 & \text{for } x \le 1 \\ -|x| & \text{for } x > 1 \end{cases}$

24. $y = \dfrac{1}{x + 2}$

In Exercises 25–28 sketch the graph of the equation. List the intercepts and describe the symmetry (if any) of the graph.

25. $x + y = 1$

26. $x + y^2 = 1$

27. $2x^2 + 2y^2 = 6$

28. $y = \frac{1}{2}(\sin x + |\sin x|)$

In Exercises 29–32 sketch the graph of the given equation with the help of a suitable translation. Show both the x and y axes, and the X and Y axes.

29. $x^2 + 6x + y + 4 = 0$

30. $x^2 - 4x + y^2 = 5$

31. $x - 2 = \dfrac{1}{y + 3}$

32. $x^2 - x + y^2 - 4y = \frac{19}{4}$

33. Let
$$f(x) = \frac{x + 2}{x^2 - 4x + 3} \quad \text{and} \quad g(x) = \frac{x + 1}{x^2 - 2x - 3}$$
Find the domains and rules of $f - g$ and f/g.

34. Find the domains and rules of $g \circ f$ and $f \circ g$, where
$$f(x) = \frac{x^2 - 1}{2x} \quad \text{and} \quad g(x) = \sqrt{x + 1}$$

35. Let $f(x) = \sqrt{x + 1}$, $g(x) = \sqrt{x + 2}$, and $h(x) = \sqrt{(x + 1)(x + 2)}$.
    a. Find the domains of f, g, and h.
    b. Find the domain of fg, and compare it with the domain of h.

36. Let $f(x) = \sqrt{1 - x^2}$. Show that $f(f(x)) = x$ for $0 \le x \le 1$.

37. Let $f(x) = x/(x - 1)$.
    a. Show that $f(f(x)) = x$ for $x \ne 1$.
    b. Use part (a) to show that $f(f(f(f(x)))) = x$ for $x \ne 1$.

38. Let
$$f(x) = x + \sqrt{x^2 + 1} \quad \text{and} \quad g(x) = x + \sqrt{x^2 - 1}$$
Show that
$$g(\sqrt{x^2 + 1}) = f(x) \quad \text{for} \quad x \ge 0$$
$$f(\sqrt{x^2 - 1}) = g(x) \quad \text{for} \quad x \ge 1$$

39. Let
$$f(x) = \frac{ax + b}{cx - a}$$
Show that if $a^2 + bc \ne 0$, then $f(f(x)) = x$ for all x in the domain of f.

40. Let $h(x) = \dfrac{1}{\sqrt{x^2 + 4}}$. Write h as the composite of two functions f and g (neither of which is equal to h).

41. Find the values of the remaining four trigonometric functions if $\sin x = -\frac{2}{3}$ and $\tan x = -\dfrac{2\sqrt{5}}{5}$.

42. Solve the equation $|\sin x| = |\cos x|$ for x in $[0, 2\pi)$.

43. Solve the inequality $\cos x < -\frac{1}{2}$ for x in $[0, 2\pi)$.

44. Solve the inequality $|\sin x| \ge |\cos x|$ for x in $[0, 2\pi)$.

In Exercises 45–48 sketch the graph of the function. Indicate any intercepts and symmetry, and determine whether the function is even, odd, or neither.

45. $1 - \sin x$ 46. $\cos |x|$

47. $\tan (x - \pi/4)$ 48. $\sin (\pi x/2)$

49. Let $f(x) = x^2 - \cos x$. Determine whether f is even, odd, or neither.

50. Given the partial graph of f in Figure 1.58, complete the graph of f, assuming that
    a. f is an even function
    b. f is an odd function
    c. the graph of f is symmetric with respect to the y axis
    d. the graph of f is symmetric with respect to the origin

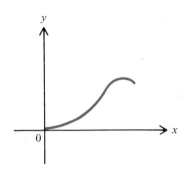

FIGURE 1.58

*51. In each of the following parts, find numbers a and b such that for all x, the following equations hold.
    a. $\sin x + \sqrt{3} \cos x = a \sin (x + b)$
    b. $\sin x + \sqrt{3} \cos x = a \cos (x + b)$

52. Prove that $\sin x \le \tan x \le 2 \sin x$ for $0 \le x \le \pi/3$. (*Hint:* What are the largest and the smallest values of $\cos x$ for $0 \le x \le \pi/3$?)

53. Show that if a square and an equilateral triangle have the same perimeter, then the square has the larger area. Is this true for an arbitrary equilateral triangle and arbitrary rectangle of equal perimeter? Explain.

54. A line with positive slope m contains the point $(-4, 0)$. This line, the negative x axis, and the positive y axis form a triangle. Find a formula for the function A that expresses the area of the triangle in terms of m.

55. Three vertices of a parallelogram are located at $(0, 1)$, $(2, 0)$, and $(3, 2)$. Determine the 3 possible locations of the fourth vertex.

56. Suppose that you drive from A to D, passing through either B or C (Figure 1.59). If you travel at a constant speed of 60 kilometers per hour and the trip lasts 1 hour, do you pass through B?

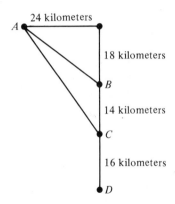

FIGURE 1.59

2

LIMITS AND CONTINUITY

In Chapter 2 we study the notions of limit and continuity. These concepts are fundamental to the main subjects of calculus: the derivative and the integral. Although the topics of limit and continuity are rather theoretical in nature, we will present them as concretely and pictorially as possible. In addition we will apply them to the study of lines tangent to various curves and also to the notion of velocity of an object.

2.1
INFORMAL DISCUSSION OF LIMIT

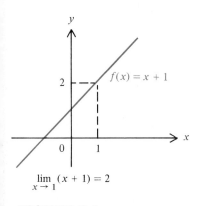

FIGURE 2.1

When we say that "L is the limit of $f(x)$ as x approaches a," we mean, roughly speaking, that $f(x)$ gets close to L as x gets close to a. We express this idea symbolically by the notation

$$\lim_{x \to a} f(x) = L$$

The values of some limits are easy to guess. For example, if x gets close to 1, then $x + 1$ gets close to 2. This suggests that

$$\lim_{x \to 1} (x + 1) = 2 \qquad (1)$$

(See Figure 2.1). In the same manner, as x approaches 4, $\sqrt{x}$ approaches 2 and therefore $\sqrt{x} + 3$ approaches 5. This means that the reciprocal of $\sqrt{x} + 3$

53

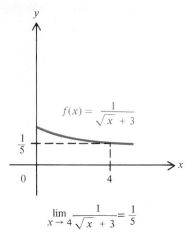

$$f(x) = \frac{1}{\sqrt{x} + 3}$$

$$\lim_{x \to 4} \frac{1}{\sqrt{x} + 3} = \frac{1}{5}$$

FIGURE 2.2

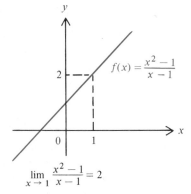

$$f(x) = \frac{x^2 - 1}{x - 1}$$

$$\lim_{x \to 1} \frac{x^2 - 1}{x - 1} = 2$$

FIGURE 2.3

approaches the reciprocal of 5, so

$$\lim_{x \to 4} \frac{1}{\sqrt{x} + 3} = \frac{1}{5}$$

(See Figure 2.2).

In the preceding cases we were able to find the limits without much difficulty. However, finding certain limits requires some ingenuity. For example, consider

$$\lim_{x \to 1} \frac{x^2 - 1}{x - 1}$$

Since $x^2 - 1$ and $x - 1$ both approach 0 as x approaches 1, it might seem that the limit of the quotient should be 0/0. But 0/0 is undefined. As a result, we must try another approach. Notice that

$$\frac{x^2 - 1}{x - 1} = \frac{(x - 1)(x + 1)}{x - 1} = x + 1 \quad \text{for } x \neq 1$$

But from (1) we know that $\lim_{x \to 1}(x + 1) = 2$. Consequently it seems reasonable that

$$\lim_{x \to 1} \frac{x^2 - 1}{x - 1} = \lim_{x \to 1}(x + 1) = 2 \qquad (2)$$

(See Figure 2.3.)

Observe that the left-hand limit in (2) has the form

$$\lim_{x \to a} \frac{f(x) - f(a)}{x - a} \qquad (3)$$

(where $f(x) = x^2$ and $a = 1$). Such limits are vital to calculus, and appear in a variety of contexts. For the remainder of this section we will describe two different settings in which limits of the form in (3) appear, and in Section 2.2 we will present a formal definition of limit.

Tangent Lines   Since the time of ancient Greece, mathematicians have been interested in tangent lines. Early mathematicians studied tangents to simple curves such as circles and spirals (Figure 2.4). Euclid, who was the most prominent of them, conceived of a

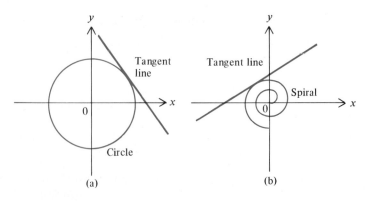

FIGURE 2.4

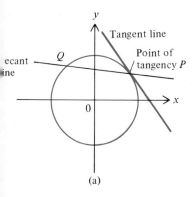

(a)

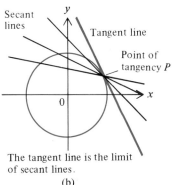

The tangent line is the limit of secant lines.

(b)

FIGURE 2.5

line tangent to a circle as a line that touches the circle at exactly one point. Yet the idea of a tangent line "touching" a curve does not lend itself well to drawing a tangent line, nor does it give a procedure for deriving an equation of a tangent line, which is important in calculus.

The line tangent to a circle at a given point P can be described in another way. If Q is any other point on the circle, then the line joining P and Q is called a **secant line** (Figure 2.5(a)). It turns out that the line tangent to the circle at P is the limit of secant lines through P, in the sense that the slope of the secant line approaches the slope of the tangent line as Q approaches P (Figure 2.5(b)). This property of the tangent line provides a method for finding an equation of the tangent line, even when the curve to which it is tangent is not a circle.

Before we give the general definition of a line tangent to the graph of a function, we first consider the special case $f(x) = x^2$, and select $(1, f(1)) = (1, 1)$ to be the point P at which we wish to define the tangent line (Figure 2.6(a)). If $(x, f(x))$ is any other point on the graph of f, then the slope of the secant line through $(1, 1)$ and $(x, f(x))$ is given by

$$\frac{f(x) - f(1)}{x - 1} = \frac{x^2 - 1}{x - 1} = \frac{(x - 1)(x + 1)}{x - 1} = x + 1 \qquad (4)$$

As x approaches 1, the slope $x + 1$ of the secant line approaches 2, and the secant line seems to slant more and more the way the graph of f does near $(1, 1)$ (Figure 2.6(b)).

More generally, if f is any function, then the slope of the secant line through $(a, f(a))$ and any other point $(x, f(x))$ on the graph of f is

$$\frac{f(x) - f(a)}{x - a}$$

(Figure 2.7(a)). In case the limit of $(f(x) - f(a))/(x - a)$ exists as x approaches a (Figure 2.7(b)), we will define the line tangent to the graph of f at $(a, f(a))$ to be the line through $(a, f(a))$ with slope

$$\lim_{x \to a} \frac{f(x) - f(a)}{x - a} \qquad (5)$$

It might seem that the limit in (5) could always be evaluated by the methods used in (4): just rewrite the expression (5), cancel terms in the numerator and

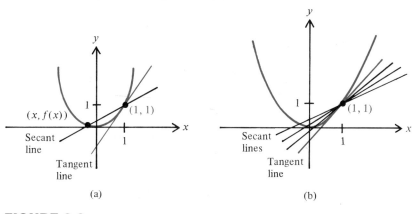

(a)   (b)

FIGURE 2.6

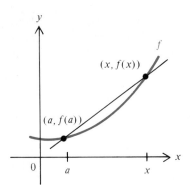

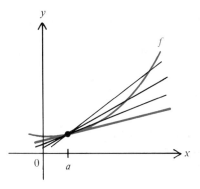

FIGURE 2.7

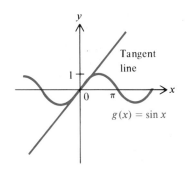

FIGURE 2.8

denominator, and evaluate. But that is not always possible. For example, suppose $f(x) = \sin x$, and let us consider the line tangent to the graph of f at $(0,0)$ (Figure 2.8). By (5) the tangent line in question has slope

$$\lim_{x \to 0} \frac{\sin x - \sin 0}{x - 0} = \lim_{x \to 0} \frac{\sin x}{x} \qquad (6)$$

Here there is no way of canceling terms in the numerator and denominator. Since $\sin x$ approaches 0 as x approaches 0, the quotient $(\sin x)/x$ might appear to approach $0/0$. But as we mentioned before, $0/0$ is undefined, so if the limit in (6) exists, we must find it by a different technique. Having no easy way of rewriting $(\sin x)/x$ to obtain the limit, we use a calculator to find the values of $(\sin x)/x$ for certain values of x close to 0:

x	$\dfrac{\pi}{4}$	$\dfrac{\pi}{30}$	$\dfrac{\pi}{180}$	$\dfrac{\pi}{9000}$
$\dfrac{\sin x}{x}$	0.900316	0.998173	0.999949	0.99999998

Because $\sin(-x) = -\sin x$, it follows that

$$\frac{\sin(-x)}{-x} = \frac{\sin x}{x} \quad \text{for } x \neq 0$$

Thus the chart would show the same values, respectively, for $x = -\pi/4$, $-\pi/30$, $-\pi/180$ and $-\pi/9000$. From the chart one might guess that $(\sin x)/x$ approaches 1 as x approaches 0, that is,

$$\lim_{x \to 0} \frac{\sin x}{x} = 1 \qquad (7)$$

Moreover, if we plot the points we know on the graph of f and then fill in between them, we obtain the curve shown in Figure 2.9(a), which also suggests that the limit should be 1. However, our limiting value of 1 is only a guess. It is more in doubt than the other limits we have found, since we calculated $(\sin x)/x$

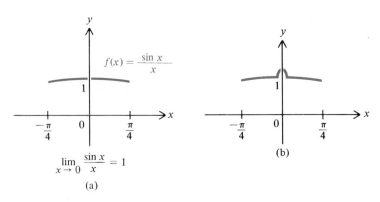

FIGURE 2.9

for only a few values of x. The true graph could possibly be the one shown in Figure 2.9(b), for example, rather than that in Figure 2.9(a). This would suggest a different limit. With the definition of limit that we will give in Section 2.2 we will be able to determine definitively that $\lim_{x \to 0} (\sin x)/x$ exists and equals 1.

Velocity
The limit in (3) also appears in the study of velocity. Suppose a spacecraft is launched from earth and travels vertically upward. If the rocket travels 36 miles during a 2-minute time interval, then the average velocity during that interval is 18 miles per minute. In general, during any time interval the average velocity of the rocket is defined by

$$\text{average velocity} = \frac{\text{distance traveled}}{\text{elapsed time}} \tag{8}$$

Instead of finding average velocity we would like to be able to calculate the velocity at a particular moment (sometimes called instantaneous velocity). This is the number that can be read from a speedometer. One way to calculate the velocity is to consider it as the limit of average velocities. In particular, let $f(t)$ be the height (in miles) of the rocket t minutes after launch. If t is a little larger than 2, then the distance traveled in the interval from 2 to t is $f(t) - f(2)$, and the elapsed time is $t - 2$. Consequently the average velocity during the time interval from 2 to t is given by

$$\frac{f(t) - f(2)}{t - 2}$$

The closer t is to 2, the closer we would expect the average velocity to be to the velocity at 2. It is therefore natural to define the velocity of the rocket at time 2 to be

$$\lim_{t \to 2} \frac{f(t) - f(2)}{t - 2}$$

More generally, the velocity $v(t_0)$ at time t_0 of an object traveling in a straight line with position $f(t)$ at time t is given by

$$v(t_0) = \lim_{t \to t_0} \frac{f(t) - f(t_0)}{t - t_0} \tag{9}$$

To illustrate (9), suppose a spacecraft is headed vertically upward, and that $f(t)$ is its height (in miles) t minutes after launch, with

$$f(t) = t^2 \quad \text{for } 0 \le t \le 2$$

Then the velocity $v(1)$ of the spacecraft at time $t = 1$ is given by

$$v(1) = \lim_{t \to 1} \frac{f(t) - f(1)}{t - 1} = \lim_{t \to 1} \frac{t^2 - 1}{t - 1}$$

This 2-stage Saturn IB rocket launched a 45,900-pound unmanned Apollo spacecraft on a 300-mile-high suborbital flight down the Atlantic Test Range.

Since

$$\lim_{t \to 1} \frac{t^2 - 1}{t - 1} = 2$$

from (2) with x replaced by t, it follows that $v(1) = 2$ (miles per minute).

In Chapter 3 we will discuss further applications that use the limit $\lim_{x \to a} (f(x) - f(a))/(x - a)$. But now it is time for us to formulate a precise definition of limit. This will be done in the next section.

EXERCISES 2.1

In Exercises 1–8, guess the value of the limit.

1. $\lim_{x \to -1} (x + 4)$

2. $\lim_{x \to 5} (-2x + 7)$

3. $\lim_{h \to 0} (2h + h^2)$

4. $\lim_{h \to 0} \left(1 - \frac{h^2}{2}\right)$

5. $\lim_{x \to 2} \dfrac{2x - 5}{4x + 3}$

6. $\lim_{x \to 1/2} \dfrac{3x - 2}{4x - 1}$

7. $\lim_{x \to 3} \dfrac{\sqrt{x + 1}}{2x - 1}$

8. $\lim_{x \to -1/2} \dfrac{1}{2}\sqrt{\dfrac{1}{x} + 6}$

In Exercises 9–18 use the methods of this section to guess the value of the limit.

9. $\lim_{x \to -2} \dfrac{x^2 - 4}{x + 2}$

10. $\lim_{x \to 1} \dfrac{x^2 + 4x - 5}{x - 1}$

11. $\lim_{x \to -5} \dfrac{x^2 + 4x - 5}{x + 5}$

12. $\lim_{x \to \pi} \dfrac{x^2 - 2x + 1}{(x - 1)^2}$

13. $\lim_{x \to 1} \dfrac{x^3 - 1}{x - 1}$

14. $\lim_{x \to -2} \dfrac{x^3 + 8}{x + 2}$

15. $\lim_{x \to 0} 3(\sin^2 x + \cos^2 x)$

16. $\lim_{x \to \pi} \dfrac{\sin 2x}{\sin x \cos x}$

17. $\lim_{x \to -1} -\dfrac{|x|}{x}$

18. $\lim_{x \to -6} \dfrac{x^2}{|x|}$

▣ In Exercises 19–22 evaluate the function at 0.1, 0.01, and 0.001, and at −0.1, −0.01, and −0.001. Then guess the value of $\lim_{x \to 0} f(x)$.

19. $f(x) = \dfrac{\tan x}{x}$

20. $f(x) = \dfrac{\sin 2x}{x}$

21. $f(x) = \dfrac{1 - \cos x}{x}$

22. $f(x) = \dfrac{1 - \cos x}{x^2}$

In Exercises 23–26 find an expression for the quotient $(f(x) - f(a))/(x - a)$, and then guess the slope of the line tangent to the graph of f at $(a, f(a))$.

23. $f(x) = x^2; a = 3$

24. $f(x) = x^2 + 2x; a = -2$

25. $f(x) = 2x^2 + 1; a = 4$

26. $f(x) = x^3 + 1; a = -2$

27. Suppose $\lim_{x \to 1} f(x) = 2$. Which of the graphs in Figure 2.10 could be the graph of f?

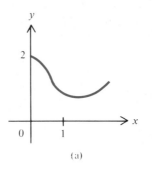
(a)

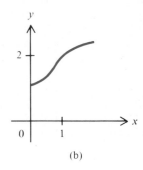
(b)

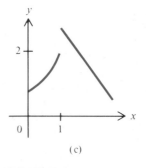
(c)

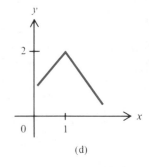
(d)

FIGURE 2.10

28. Suppose $\lim_{x \to a} f(x) = 3$ and $\lim_{x \to a} g(x) = 5$.
   a. Guess the value of $\lim_{x \to a} (f(x) + g(x))$.
   b. Guess the value of $\lim_{x \to a} 2f(x)$.

29. Suppose $\lim_{x \to a} f(x) = -2$ and $\lim_{x \to a} g(x) = 3$.
    a. Guess the value of $\lim_{x \to a} f(x)g(x)$.

    b. Guess the value of $\lim_{x \to a} \dfrac{f(x)}{g(x)}$.

30. Suppose the height of a spacecraft at time t is given by $f(t) = t^2$ for $0 \le t \le 2$. Guess the velocity of the spacecraft after $1/2$ minute.

31. Suppose the height of a spacecraft at time t is given by $f(t) = 2t^2 + 1$ for $0 \le t \le 3$. Find an expression for the spacecraft's average velocity during the time interval between 2 and t (for $t \ne 2$), and then guess its velocity at time 2.

2.2
DEFINITION OF LIMIT

In Section 2.1 we discussed limits informally. In some cases we were able to deduce limits easily. However, when we tried to ascertain whether

$$\lim_{x \to 0} \frac{\sin x}{x} \qquad (1)$$

exists we were reduced to calculating $(\sin x)/x$ for several values of x near 0. Using those calculations, we guessed that the limit exists and guessed its value. The uncertainty about the limit in (1) leads us to seek a formal definition of limit.

In formulating the precise definition of $\lim_{x \to a} f(x)$ we will want to allow f to be undefined at a. After all, we suggested in Section 2.1 that

$$\lim_{x \to 1} \frac{x^2 - 1}{x - 1}$$

exists and equals 2 even though $(x^2 - 1)/(x - 1)$ is not defined at 1.

Next, if f happens to be defined at a, we would like the definition of $\lim_{x \to a} f(x)$ to be independent of the value of $f(a)$. To illustrate this thought, consider the function f defined by

$$f(x) = \begin{cases} 1 \text{ for } x \ne 0 \\ 2 \text{ for } x = 0 \end{cases}$$

Then f is called a "hiccup" function because of the appearance of its graph (Figure 2.11). Notice that $f(0) = 2$. But if f were not defined at 0, it would have only the single value 1, which would suggest that

$$\lim_{x \to 0} f(x) = 1$$

As a result, we would like a definition of $\lim_{x \to a} f(x)$ that is independent of the value of $f(a)$.

Finally, if $\lim_{x \to a} f(x)$ exists, we would like the limit to be unique. The limits discussed so far have this property. For contrast, consider the function defined by

$$f(x) = \begin{cases} 0 \text{ for } x < 0 \\ 1 \text{ for } x \ge 0 \end{cases}$$

which is graphed in Figure 2.12. (This is sometimes called a diving board function.) Does f have a limit as x approaches 0? Notice that in any interval

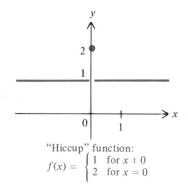

"Hiccup" function:
$f(x) = \begin{cases} 1 & \text{for } x \ne 0 \\ 2 & \text{for } x = 0 \end{cases}$

FIGURE 2.11

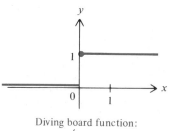

Diving board function:
$f(x) = \begin{cases} 0 & \text{for } x < 0 \\ 1 & \text{for } x \ge 0 \end{cases}$

FIGURE 2.12

about 0, say $(-1/1000, 1/1000)$, the function assumes both values 0 and 1. Therefore 0 and 1 would be equally reasonable candidates for the limit. Since we wish f to have a single limit at 0, if any at all, it appears that f should have no limit at 0.

We are now ready to begin formulating a precise definition of limit, from which we can decide whether a given function has a limit at a particular point, and what the limit must be if it does exist.

The Formal Definition of Limit

We have said that L is the limit of $f(x)$ as x approaches a if $f(x)$ gets close to L as x gets close to a. But precisely what does it mean to say that $f(x)$ gets close to L or that x gets close to a? We begin to answer this question by reinterpreting $\lim_{x \to a} f(x) = L$:

> No matter how close we wish $f(x)$ to be to L, if x is close enough to a (but distinct from a), then $f(x)$ must be at least as close to L as we wish. (2)

We demand that x be chosen distinct from a so that the value of f at a, if it exists, has no influence on the existence or value of the limit.

Figure 2.13 illustrates (2) in two cases. Notice that for $f(x)$ to be within 1 of L (as in Figure 2.13(a)), it suffices to take x in the indicated interval I about a. But if $f(x)$ is to be within $\frac{1}{2}$ of L (as in Figure 2.13(b)), the corresponding interval J about a is smaller. In general, the closer $f(x)$ is to be to L, the closer x must be to a.

To make (2) unambiguous, we first let the Greek letter ε (epsilon*) represent how close we wish $f(x)$ to be to L. Then we can rewrite (2):

> For any $\varepsilon > 0$, if x is any number sufficiently close to a but distinct from a, then the distance between $f(x)$ and L is less than ε. (3)

Once we have chosen $\varepsilon > 0$, we let the Greek letter δ (delta) represent how close x must be to a to ensure that the distance between $f(x)$ and L is less than ε. We can now rewrite (3):

> For any $\varepsilon > 0$, there is a $\delta > 0$ such that
>
> if the distance between x and a is less than δ and if $x \neq a$, then the distance between $f(x)$ and L is less than ε. (4)

Figure 2.14 represents (4) pictorially. The number ε is arbitrary. The number δ is then chosen so as to fulfill the condition of (4); that is, whenever the distance between x and a is less than δ and $x \neq a$, the point $(x, f(x))$ lies in the shaded rectangle shown in Figure 2.14.

In order to put (4) in final form, we observe that the distance between $f(x)$ and L is less than ε when $|f(x) - L| < \varepsilon$. Moreover, $x \neq a$ and the distance

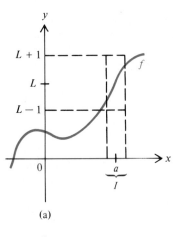

(a)

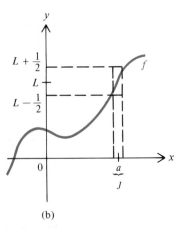

(b)

FIGURE 2.13

* **Epsilon**: Pronounced "*ep*-si-lon."

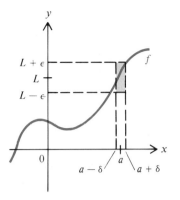

FIGURE 2.14

between x and a is less than δ when $0 < |x - a| < \delta$. Thus we can rewrite (4):

For any $\varepsilon > 0$, there is a $\delta > 0$ such that

$$\text{if } 0 < |x - a| < \delta, \text{ then } |f(x) - L| < \varepsilon \qquad (5)$$

In order for $f(x)$ in (5) to be meaningful, it is necessary that x be in the domain of f whenever $0 < |x - a| < \delta$. For this reason, in defining the limit of $f(x)$ as x approaches a we will require that the domain of f contain all the points in some open interval about a, with the possible exception of a itself.

We now give the formal definition of limit.

DEFINITION 2.1
DEFINITION OF LIMIT

> Let f be a function defined at each point of some open interval containing a, except possibly at a itself. Then a number L is the **limit of $f(x)$ as x approaches a** (or is the **limit of f at a**) if for every number $\varepsilon > 0$ there is a number $\delta > 0$ such that
>
> $$\text{if } 0 < |x - a| < \delta, \quad \text{then} \quad |f(x) - L| < \varepsilon$$

If L is the limit of $f(x)$ as x approaches a, then we write

$$\lim_{x \to a} f(x) = L$$

If such an L can be found, we say that the **limit of f at a exists**, or that f has a limit at a, or that $\lim_{x \to a} f(x)$ exists.

It follows from Definition 2.1 that a function can have at most one limit L at a (see the Appendix). That justifies calling L *the* limit of f at a.

Definition 2.1 gives a precise meaning to the statement "L is the limit of $f(x)$ as x approaches a." Although Definition 2.1 is not needed to convince ourselves of the values of limits such as those appearing in Section 2.1, nevertheless with this definition it is possible to *prove* that such limits are correct, and that the diving board function shown in Figure 2.12 has no limit at 0.

To show how Definition 2.1 can be used to prove that limits exist, we will use it in Examples 1 and 2 to verify two basic limits.

Example 1   Suppose a and c are any numbers. Show that

$$\lim_{x \to a} c = c$$

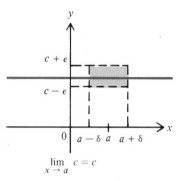

$$\lim_{x \to a} c = c$$

FIGURE 2.15

Solution   By $\lim_{x \to a} c$ we mean $\lim_{x \to a} f(x)$, where $f(x) = c$ for all x. Let $\varepsilon > 0$. We must find a number $\delta > 0$ such that

$$\text{if } 0 < |x - a| < \delta, \quad \text{then} \quad |f(x) - c| < \varepsilon$$

But for *any* positive number δ,

$$\text{if } 0 < |x - a| < \delta, \quad \text{then} \quad |f(x) - c| = |c - c| = 0 < \varepsilon$$

(See Figure 2.15.) Hence in this case we may choose δ to be any positive number. $\square$

From Example 1 it follows that

$$\lim_{x \to 3} 1 = 1, \quad \lim_{x \to -\sqrt{2}} \frac{\pi}{17} = \frac{\pi}{17}, \quad \text{and} \quad \lim_{x \to -2} (-\pi) = -\pi$$

Since the limit of a function f at a does not depend on the value (if any) of f at a, it also follows from Example 1 that if f is the "hiccup" function defined by

$$f(x) = \begin{cases} 1 & \text{for } x \neq 0 \\ 2 & \text{for } x = 0 \end{cases}$$

then

$$\lim_{x \to 0} f(x) = \lim_{x \to 0} 1 = 1$$

(See Figure 2.11.)

Example 2   Show that

$$\lim_{x \to a} x = a$$

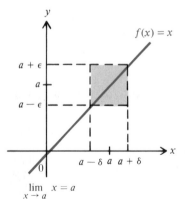

$$\lim_{x \to a} x = a$$

FIGURE 2.16

Solution   Here $f(x) = x$ for all x. Let $\varepsilon > 0$. We must find a number $\delta > 0$ such that

$$\text{if } 0 < |x - a| < \delta, \quad \text{then} \quad |x - a| < \varepsilon$$

Here we can choose δ to be ε because

$$\text{if } 0 < |x - a| < \delta = \varepsilon, \quad \text{then} \quad |x - a| < \varepsilon$$

(See Figure 2.16.) $\square$

From Example 2 we see that

$$\lim_{x \to 4\pi} x = 4\pi \quad \text{and} \quad \lim_{x \to -\sqrt{2}} x = -\sqrt{2}$$

A slight alteration in the solution would show that for any fixed numbers a, b, and c

$$\lim_{x \to a} (bx + c) = ba + c \tag{6}$$

$$\lim_{x \to a} |x| = |a| \tag{7}$$

(see Exercises 28 and 29). It follows, for example, that

$$\lim_{x \to 3} (-4x + 5) = -4 \cdot 3 + 5 = -7$$

and

$$\lim_{x \to -3} |x| = |-3| = 3$$

Now that the notion of limit has been defined in Definition 2.1, we define the **line tangent to the graph of f at $(a, f(a))$** to be the line through $(a, f(a))$ with slope m_a given by

$$m_a = \lim_{x \to a} \frac{f(x) - f(a)}{x - a} \tag{8}$$

provided that the limit exists. In that case,

$$y - f(a) = m_a(x - a)$$

is an equation of the line tangent to the graph of f at $(a, f(a))$.

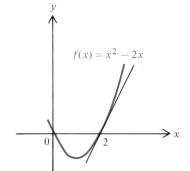

$f(x) = x^2 - 2x$

FIGURE 2.17

Example 3 Let $f(x) = x^2 - 2x$. Find an equation of the line tangent to the graph of f at $(2, 0)$ (Figure 2.17).

Solution By (8) the slope of the line tangent at $(2, 0)$ is

$$\lim_{x \to 2} \frac{f(x) - f(2)}{x - 2}$$

provided that the limit exists. However

$$\lim_{x \to 2} \frac{f(x) - f(2)}{x - 2} = \lim_{x \to 2} \frac{(x^2 - 2x) - 0}{x - 2} = \lim_{x \to 2} \frac{x(x - 2)}{x - 2} = \lim_{x \to 2} x$$

By Example 2, $\lim_{x \to 2} x = 2$. Therefore

$$\lim_{x \to 2} \frac{f(x) - f(2)}{x - 2} = 2$$

so that the slope of the line tangent at $(2, 0)$ is 2. Since the tangent line passes through $(2, 0)$, an equation of the desired tangent line is

$$y - 0 = 2(x - 2), \quad \text{or simply} \quad y = 2(x - 2) \quad \square$$

Now consider an object traveling in a straight line with position $f(t)$ at time t. The **velocity** $v(t_0)$ at time t_0 is defined by the formula

$$v(t_0) = \lim_{t \to t_0} \frac{f(t) - f(t_0)}{t - t_0} \tag{9}$$

provided that the limit exists.

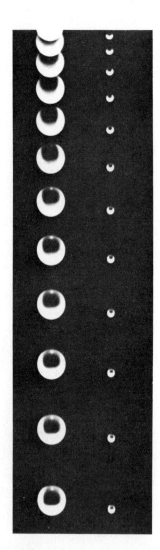

Multiple-flash photograph showing two objects of different sizes and weights dropped at the same moment.

If we have a reasonable formula for the position $f(t)$ of an object at time t, then we can use (9) to find the velocity $v(t_0)$ at a given time t_0. This is illustrated in the next example. Recall from (1) in Section 1.3 that if an object is acted on only by the force of gravity, and if distances are measured in feet and time in seconds, then the object's height $h(t)$ at time t is given by

$$h(t) = -16t^2 + v_0 t + h_0 \tag{10}$$

where h_0 is the initial height and v_0 is the initial velocity.

Example 4   Galileo is reported to have dropped two iron balls from an upper balcony of the Leaning Tower of Pisa, approximately 144 feet above ground. If we neglect air resistance, determine how fast the balls would have been traveling after 2 seconds in flight.

Solution   By our assumptions, $v_0 = 0$ since the balls began with zero velocity, and $h_0 = 144$ because the balls began their journey 144 feet above ground. Thus (10) becomes $h(t) = -16t^2 + 144$. Then by (9) with $t_0 = 2$, along with (6),

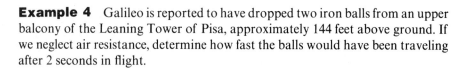

$$v(2) = \lim_{t \to 2} \frac{h(t) - h(2)}{t - 2} = \lim_{t \to 2} \frac{(-16t^2 + 144) - (-16 \cdot 2^2 + 144)}{t - 2}$$

$$= \lim_{t \to 2} \frac{-16(t^2 - 2^2)}{t - 2} = \lim_{t \to 2} \frac{-16(t - 2)(t + 2)}{t - 2} = \lim_{t \to 2} -16(t + 2)$$

$$= \lim_{t \to 2} (-16t - 32) = -16 \cdot 2 - 32 = -64$$

Thus the balls were traveling downward at the rate of 64 feet per second after 2 seconds. $\square$

The fact that both iron balls—one weighing 10 pounds and the other weighing 1 pound—hit the ground simultaneously supported Galileo's hypothesis that the velocity of an object moving under the sole influence of gravity is independent of the weight of the object.

The answer in Example 4 is negative because the height decreases when the object descends. In general, an object traveling upward has a positive velocity, and an object traveling downward has a negative velocity. When we are interested only in the magnitude of the velocity, we talk of the speed of an object. **Speed** is defined as the absolute value of velocity. Thus an object moving with a velocity of -10 feet per second has a speed of $|-10|$ feet per second, or 10 feet per second. After two seconds the speed of the balls in Example 4 is therefore 64 feet per second.

Historical Commentary

The notion of limit is rather complicated. In fact, mathematicians talked about limits for centuries before they were able to define the concept clearly. Even the ancient Greeks had some feeling for limits; for example, Archimedes found an approximation of the value of 2π as the "limit" of the perimeters of regular polygons inscribed in a circle of radius by letting the number of sides grow without bound. Through the Middle Ages and into the Renaissance, mathematicians used various types of limits to compute areas. The seventeenth-century giants of calculus, Isaac Newton (1642–1727) and Gottfried Leibniz (1646–1716), had a good intuitive understanding of limits and even computed very complicated limits. However, neither they nor anyone before them actually defined the concept.

The merely intuitive quality of the idea of limit hampered progress in the development of calculus for a century after Newton and Leibniz. In 1754 the French mathematician Jean-le-Rond d'Alembert* (1717–1783) suggested that the logical basis of calculus would eventually reside in the concept of limit. In time, the great French mathematician Augustin-Louis Cauchy** (1789–1857) molded calculus into a form much like the discipline we have today. However, the definition contained in his 1821 treatise *Cours d'Analyse* still relied on the intuitive notion of a variable approaching a fixed value:

> When the successive values attributed to a variable approach indefinitely close to a fixed value so as to end by differing from it by as little as one wishes, this latter value is called the limit of all the others.[†]

By today's standards this definition is imprecise. Nevertheless, it was a large step toward the formulation proposed in the 1860s by the great German mathematician Karl Weierstrass[††] (1815–1897). It is his definition of limit that we presented in Definition 2.1 and that is universally accepted by mathematicians.

---

* **d'Alembert**: Pronounced "dah-lem-*bair*."
** **Cauchy**: Pronounced "*Co*-shee."
[†] Cauchy, Augustin-Louis, *Oeuvres Complètes*, 2nd Series, Vol. III (Paris, 1882, 1932), p. 19.
[††] **Weierstrass**: Pronounced "*Vire*-shtrahss."

EXERCISES 2.2

In Exercises 1–10 use the results of this section to evaluate the given limit.

1. $\lim\limits_{x \to 2} (-5)$

2. $\lim\limits_{x \to -\sqrt{3}} \dfrac{\sqrt{2}}{3}$

3. $\lim\limits_{x \to 1/2} x$

4. $\lim\limits_{x \to \pi} x$

5. $\lim\limits_{x \to -1} (2x + 5)$

6. $\lim\limits_{x \to 0} \left(-4x - \dfrac{1}{3}\right)$

7. $\lim\limits_{x \to 3/2} |x|$

8. $\lim\limits_{x \to -0.1} |x|$

9. $\lim\limits_{x \to 0} f(x)$, where $f(x) = \begin{cases} 2x & \text{for } x \neq 0 \\ \frac{1}{3} & \text{for } x = 0 \end{cases}$

10. $\lim\limits_{x \to 1/2} f(x)$, where $f(x) = \begin{cases} -\frac{3}{2}x + \frac{1}{4} & \text{for } x \neq \frac{1}{2} \\ 0 & \text{for } x = \frac{1}{2} \end{cases}$

In Exercises 11–20 find an equation of the line l tangent to the graph of f at the given point.

11. $f(x) = \pi;\quad (3, \pi)$

12. $f(x) = -\frac{1}{2}x + 2;\quad (8, -2)$

13. $f(x) = -x^2 + 2;\quad (0, 2)$

14. $f(x) = x^2 - 5;\quad (-1, -4)$

15. $f(x) = 4x^2 + \frac{1}{2};\quad (\frac{1}{2}, \frac{3}{2})$

16. $f(x) = -\frac{1}{2}x^2 + 3;\quad (2, 1)$

17. $f(x) = x^2 + 3x;\quad (1, 4)$

18. $f(x) = -x^2 + 4x;\quad (-1, -5)$

19. $f(x) = 5x^2 - 2x;\quad (\frac{1}{5}, -\frac{1}{5})$

20. $f(x) = -3x^2 - 5x;\quad (-2, -2)$

In Exercises 21–26 assume that $f(t)$ represents the position of an object at time t. Find the velocity $v(t_0)$ of the object at time t_0.

21. $f(t) = 2t + 1; t_0 = 2$

22. $f(t) = -3t + \frac{1}{5}; t_0 = 0$

23. $f(t) = -16t^2; t_0 = 4$

24. $f(t) = -16t^2 + 100; t_0 = 1$

25. $f(t) = -16t^2 + 8t + 54; t_0 = 2$

26. $f(t) = -16t^2 - 2t + 5; t_0 = \frac{1}{2}$

27. Which of the graphs in Figure 2.18 appear to be graphs of functions with a limit at 2?

28. Show that $\lim\limits_{x \to a} |x| = |a|$. (*Hint:* $\big||x| - |a|\big| \le |x - a|$.)

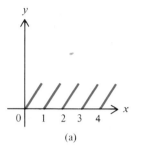
(a)

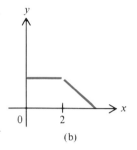
(b)

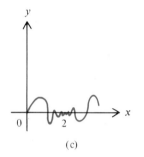
(c)

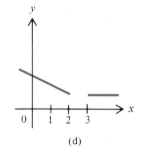
(d)

FIGURE 2.18

*29. Let a, b, and c be any real numbers. Show that

$$\lim_{x \to a} (bx + c) = ba + c$$

(*Hint:* Consider the cases $b = 0$ and $b \neq 0$ separately. For the case $b \neq 0$, let $\delta = \varepsilon/|b|$.)

30. Show that the statements $\lim_{x \to a} f(x) = L$ and $\lim_{x \to a} |f(x) - L| = 0$ are equivalent.

31. Suppose $\lim_{x \to 5} f(x) = 4$. Show that $\lim_{x \to 5} f(x) \neq 2$. (*Hint:* Let $\varepsilon = 1$ and choose $\delta > 0$ such that if $0 < |x - 5| < \delta$, then $|f(x) - 4| < 1$. What does this imply about $|f(x) - 2|$?)

32. Let f be defined at each point of some open interval containing a, except possibly at a itself. Consider the statement

> For every $\delta > 0$ there is an $\varepsilon > 0$ such that if $0 < |x - a| < \delta$, then $|f(x) - L| < \varepsilon$

Is this equivalent to the statement "$\lim_{x \to a} f(x) = L$"? Explain.

*33. Suppose f and g have limits at a, and $f(x) = g(x)$ for all x in some open interval about a except possibly a itself.

Show that

$$\lim_{x \to a} f(x) = \lim_{x \to a} g(x)$$

(*Hint:* Take δ so small that $f(x) = g(x)$ for each x satisfying $0 < |x - a| < \delta$.)

34. Can the graph of a function f have the same tangent line at $(-3, f(-3))$ and at $(2, f(2))$? Give an example showing that it can happen, or explain why it cannot happen.

35. Find the velocity of the balls in Example 4 after they dropped for 1 second.

36. Suppose we try to calculate the velocity of the balls in Example 4 after 4 seconds. Why is the velocity *not* -128 feet per second? (*Hint:* First determine when the balls hit the ground.)

37. Suppose Galileo had thrown the balls downward with a speed of 16 feet per second.

a. Determine the velocity of the balls after 2 seconds in flight.
b. Determine the speed of the balls after 2 seconds in flight.

38. Suppose a stone is thrown upward with an initial velocity of 16 feet per second from a bridge 48 feet above a river. Determine the speed of the stone after 2 seconds.

39. Suppose a rock is thrown downward into a deep hole with an initial velocity of -16 feet per second. Determine the velocity of the rock after 3 seconds.

40. Suppose a ball is thrown from a balcony 112 feet above ground, and assume that after one second the velocity of the ball is -32 feet per second.

a. Determine the ball's initial velocity.
b. Would the answer to part (a) be the same if the balcony were higher? Explain.

2.3
LIMIT THEOREMS

Limits are extremely important throughout calculus. One could evaluate limits by applying the definition of limit directly, but that would be tedious. Fortunately such a procedure will usually not be necessary because of the powerful rules for finding limits which we will present in this and the next section.

The following theorem gives five basic rules for finding limits of combinations of functions, and is proved in the Appendix.

THEOREM 2.2
LIMIT THEOREM

If $\lim_{x \to a} f(x)$ and $\lim_{x \to a} g(x)$ exist, then $\lim_{x \to a} [f(x) + g(x)]$, $\lim_{x \to a} cf(x)$, $\lim_{x \to a} [f(x) - g(x)]$, and $\lim_{x \to a} [f(x)g(x)]$ exist, and

Sum Rule: $\lim_{x \to a} [f(x) + g(x)] = \lim_{x \to a} f(x) + \lim_{x \to a} g(x)$

Constant Multiple Rule: $\lim_{x \to a} cf(x) = c \lim_{x \to a} f(x)$ for any constant c

Difference Rule: $\lim_{x \to a} [f(x) - g(x)] = \lim_{x \to a} f(x) - \lim_{x \to a} g(x)$

Product Rule: $\lim_{x \to a} [f(x)g(x)] = \lim_{x \to a} f(x) \lim_{x \to a} g(x)$

If $\lim_{x \to a} f(x)$ and $\lim_{x \to a} g(x)$ exist and $\lim_{x \to a} g(x) \neq 0$, then $\lim_{x \to a} (f(x)/g(x))$ exists, and

Quotient Rule: $\lim_{x \to a} \dfrac{f(x)}{g(x)} = \dfrac{\lim_{x \to a} f(x)}{\lim_{x \to a} g(x)}$

In Example 1 we will use the limit rules presented above, along with the following limit that appeared in (6) of Section 2.2:

$$\lim_{x \to a} (bx + c) = ba + c \tag{1}$$

Example 1 Evaluate the following limits.

a. $\lim\limits_{x \to -1} x^2$

b. $\lim\limits_{x \to -1} (\pi x + x^2)$

c. $\lim\limits_{x \to -1} \dfrac{x^2}{x + 3}$

Solution a. Since $\lim_{x \to -1} x = -1$ by (1), it follows from the Product Rule that

$$\lim_{x \to -1} x^2 = (\lim_{x \to -1} x)(\lim_{x \to -1} x) = (-1)(-1) = 1$$

b. Using (1), we find that $\lim_{x \to -1} \pi x = \pi(-1) = -\pi$. Since $\lim_{x \to -1} x^2 = 1$ by part (a), the Sum Rule implies that

$$\lim_{x \to -1} (\pi x + x^2) = \lim_{x \to -1} \pi x + \lim_{x \to -1} x^2 = -\pi + 1$$

c. By (1), $\lim_{x \to -1} (x + 3) = -1 + 3 = 2$. Since $\lim_{x \to -1} x^2 = 1$ by part (a), we conclude from the Quotient Rule that

$$\lim_{x \to -1} \frac{x^2}{x + 3} = \frac{\lim\limits_{x \to -1} x^2}{\lim\limits_{x \to -1} (x + 3)} = \frac{1}{2} \quad \square$$

The limit rules can help in showing that a limit cannot exist, as the next example illustrates.

Example 2 Show that $\lim_{x \to 0} 1/x$ does not exist.

Solution Suppose that $\lim_{x \to 0} 1/x$ exists, and let $L = \lim_{x \to 0} 1/x$. Since $1 = x(1/x)$, the Product Rule and (1) would then imply that

$$1 = \lim_{x \to 0} 1 = \lim_{x \to 0} \left(x \cdot \frac{1}{x} \right) = (\lim_{x \to 0} x)\left(\lim_{x \to 0} \frac{1}{x} \right) = 0 \cdot L = 0$$

which is obviously false. Therefore $\lim_{x \to 0} 1/x$ cannot exist. $\square$

We often encounter sums and products of more than two functions. The results are analogous to those in the Sum and Product Rules: If $\lim_{x \to a} f_1(x)$, $\lim_{x \to a} f_2(x), \ldots, \lim_{x \to a} f_n(x)$ exist, then

$$\lim_{x \to a} [f_1(x) + f_2(x) + \cdots + f_n(x)] = \lim_{x \to a} f_1(x) + \lim_{x \to a} f_2(x) + \cdots + \lim_{x \to a} f_n(x)$$

and

$$\lim_{x \to a} [f_1(x)f_2(x) \cdots f_n(x)] = [\lim_{x \to a} f_1(x)][\lim_{x \to a} f_2(x)] \cdots [\lim_{x \to a} f_n(x)] \qquad (2)$$

One consequence of (2) is that

$$\lim_{x \to a} x^n = \lim_{x \to a} \overbrace{(x \cdot x \cdots x)}^{n \text{ factors}} = \overbrace{(\lim_{x \to a} x)(\lim_{x \to a} x) \cdots (\lim_{x \to a} x)}^{n \text{ factors}} = \overbrace{a \cdot a \cdots a}^{n \text{ factors}} = a^n$$

so that

$$\lim_{x \to a} x^n = a^n \text{ for any positive integer } n \qquad (3)$$

By means of the formulas obtained thus far, we can find the limit of *any* polynomial at *any* number. In fact, if $c_n, c_{n-1}, \ldots, c_1$, and c_0 are arbitrary real numbers and n is an arbitrary nonnegative integer, then

$$\lim_{x \to a} (c_n x^n + c_{n-1} x^{n-1} + \cdots + c_1 x + c_0)$$

$$= c_n \lim_{x \to a} x^n + c_{n-1} \lim_{x \to a} x^{n-1} + \cdots + c_1 \lim_{x \to a} x + \lim_{x \to a} c_0$$

$$= c_n a^n + c_{n-1} a^{n-1} + \cdots + c_1 a + c_0$$

It follows that to find the limit of any polynomial f at any point a, we only need to evaluate $f(a)$. In other words,

$$\lim_{x \to a} f(x) = f(a) \qquad (4)$$

Example 3   Evaluate $\lim_{x \to 2} (4x^3 - 6x^2 - 9x)$.

Solution   We use (4), which yields

$$\lim_{x \to 2} (4x^3 - 6x^2 - 9x) = 4(2)^3 - 6(2)^2 - 9(2) = 32 - 24 - 18 = -10 \quad \square$$

Recall that a rational function is the quotient of two polynomials f and g. By (4) and the Quotient Rule,

$$\lim_{x \to a} \frac{f(x)}{g(x)} = \frac{\lim_{x \to a} f(x)}{\lim_{x \to a} g(x)} = \frac{f(a)}{g(a)} \qquad (5)$$

provided that $g(a) \neq 0$. Therefore to find the limit of any rational function at a point a in its domain, we need only evaluate the rational function at a. In particular, if $f(x) = 1$ and $g(x) = x^n$, then we obtain the formula

$$\lim_{x \to a} \frac{1}{x^n} = \frac{1}{a^n} \quad \text{for } a \neq 0$$

Example 4   Evaluate $\lim_{x \to -1} \dfrac{x^3 + 3x + 1}{x^2 - 3\sqrt{5x}}$.

Solution   By (5) we conclude that

$$\lim_{x \to -1} \frac{x^3 + 3x + 1}{x^2 - 3\sqrt{5x}} = \frac{(-1)^3 + 3(-1) + 1}{(-1)^2 - 3\sqrt{5(-1)}} = \frac{-3}{1 + 3\sqrt{5}} \quad \square$$

Although the Quotient Rule does not guarantee the existence of $\lim_{x \to a} f(x)/g(x)$ when $\lim_{x \to a} g(a) = 0$, sometimes it is still possible to evaluate such limits. That was the case with the limits that appeared in the solutions of Examples 3 and 4 in Section 2.2. The next example provides a slightly more complex illustration of this.

Example 5   Find $\lim_{x \to -2} \dfrac{x^3 + 2x^2 - x - 2}{x^2 - 4}$.

Solution   Since $\lim_{x \to -2}(x^2 - 4) = 0$, we cannot apply the Quotient Rule to this function in its original form. However, since $x^3 + 2x^2 - x - 2 = (x + 2)(x^2 - 1)$ and $x^2 - 4 = (x + 2)(x - 2)$, we have

$$\lim_{x \to -2} \frac{x^3 + 2x^2 - x - 2}{x^2 - 4} = \lim_{x \to -2} \frac{(x + 2)(x^2 - 1)}{(x + 2)(x - 2)} = \lim_{x \to -2} \frac{x^2 - 1}{x - 2}$$

$$= \frac{(-2)^2 - 1}{-2 - 2} = -\frac{3}{4}$$

where the last limit to appear was evaluated by (5). $\square$

There is another limit that will be invaluable to us. If r is any fixed rational number, then

$$\lim_{x \to a} x^r = a^r \tag{6}$$

which is valid for any nonzero number a in the domain of x^r, and is even valid for $a = 0$ if $r = m/n$, where m and n are integers, with n odd and positive. This will be a consequence of results to be proved in Section 7.4, so until then we will assume it. From (6) we know that

$$\lim_{x \to -32} x^{1/5} = (-32)^{1/5} = -2 \quad \text{and} \quad \lim_{x \to 9} x^{3/2} = 9^{3/2} = 27$$

Since $\sqrt[n]{x} = x^{1/n}$, it follows from (6) that

$$\lim_{x \to a} \sqrt[n]{x} = \sqrt[n]{a} \quad \begin{cases} \text{for all } a \text{ if } n \text{ is odd} \\ \text{for all } a > 0 \text{ if } n \text{ is even} \end{cases} \tag{7}$$

In particular,

$$\lim_{x \to a} \sqrt{x} = \sqrt{a} \quad \text{for } a > 0 \tag{8}$$

For example,

$$\lim_{x \to 1/4} \sqrt{x} = \sqrt{\frac{1}{4}} = \frac{1}{2}$$

Example 6   Evaluate $\lim_{x \to 9} \dfrac{x(\sqrt{x} - 3)}{x - 9}$.

Solution   Since $\lim_{x \to 9}(x - 9) = 0$, the Quotient Rule cannot be applied directly. However, if we rationalize the numerator by multiplying numerator and denominator by $\sqrt{x} + 3$, we can cancel terms and then use (8) and the Quotient Rule:

$$\lim_{x \to 9} \frac{x(\sqrt{x} - 3)}{x - 9} = \lim_{x \to 9} \frac{x(\sqrt{x} - 3)(\sqrt{x} + 3)}{(x - 9)(\sqrt{x} + 3)} = \lim_{x \to 9} \frac{x(x - 9)}{(x - 9)(\sqrt{x} + 3)}$$

$$= \lim_{x \to 9} \frac{x}{\sqrt{x} + 3} = \frac{9}{\sqrt{9} + 3} = \frac{3}{2} \quad \square$$

Before concluding this section, we remark that one hypothesis of the Quotient Rule is that $\lim_{x \to a} g(x) \neq 0$, that is, either $\lim_{x \to a} g(x) > 0$ or $\lim_{x \to a} g(x) < 0$. In the Appendix we will prove that

$$\text{if } \lim_{x \to a} g(x) > 0, \text{ then } g(x) > 0 \text{ for all } x \text{ sufficiently close to } a \tag{9}$$

whereas

$$\text{if } \lim_{x \to a} g(x) < 0, \text{ then } g(x) < 0 \text{ for all } x \text{ sufficiently close to } a \tag{10}$$

It follows from (9) and (10) that if $\lim_{x \to a} g(x) \neq 0$, then $g(x) \neq 0$ for all x sufficiently close to a with $x \neq a$.

EXERCISES 2.3

In Exercises 1–18 use the results of this section to evaluate the limit.

1. $\lim\limits_{x \to 16} -\frac{1}{2}\sqrt{x}$

2. $\lim\limits_{x \to -32} -\pi x^{1/5}$

3. $\lim\limits_{x \to -1} (-5x^2 + 2x - \frac{1}{2})$

4. $\lim\limits_{x \to -1/2} (4x^4 - 2x^3 + 3x^2 + 5x - \frac{5}{4})$

5. $\lim\limits_{x \to 4} (3x^2 - 5\sqrt{x} - 6|x|)$

6. $\lim\limits_{x \to 1/27} (9x + 6x^{2/3} - 2x^{1/3} + 2)$

7. $\lim\limits_{x \to \sqrt{2}} (x^2 + 5)(\sqrt{2}x + 1)$

8. $\lim\limits_{x \to -1} (x + 1)(4x^3 - 9x + \frac{3}{4})$

9. $\lim\limits_{y \to 0} (2y + \frac{1}{2})(3y^{2/3} - 9)$

10. $\lim\limits_{y \to 4} y^{5/2}(y^{1/2} + y^{-1/2})$

11. $\lim\limits_{y \to 64} (\sqrt[3]{y} + \sqrt{y})^2$

12. $\lim\limits_{y \to 3} \sqrt{y}(y - 1)^4$

13. $\lim\limits_{x \to 2} -\frac{1}{6x}$

14. $\lim\limits_{x \to -1} \dfrac{x - 1}{x^5}$

15. $\lim\limits_{y \to -2} \dfrac{4y - 1}{5y + 4}$

16. $\lim\limits_{y \to 0} \dfrac{(2y^3 - y^2 + \sqrt{2})(3y^3 - \sqrt{2})}{1 - 2y}$

17. $\lim\limits_{t \to 4} \dfrac{t^{-3/2} + 1}{t^{1/2} - 4}$

18. $\lim\limits_{t \to 0} \dfrac{2t^{1/3} - 4}{-3t^{1/3} + 5}$

In Exercises 19–30 reduce the expression and then evaluate the limit.

19. $\lim\limits_{x \to -1} \dfrac{x^2 - 1}{x + 1}$

20. $\lim\limits_{x \to 3} \dfrac{x^2 - 9}{x - 3}$

21. $\lim\limits_{x \to 1} \dfrac{x^3 - 1}{x - 1}$

22. $\lim\limits_{x \to 2} \dfrac{x^2 - 4}{x^3 - 8}$

23. $\lim\limits_{x \to -2} \dfrac{x^4 - 16}{4 - x^2}$

24. $\lim\limits_{x \to -4} \dfrac{x^2 - 16}{|x| - 4}$

25. $\lim\limits_{x \to 9} \dfrac{x - 9}{\sqrt{x} - 3}$

26. $\lim\limits_{x \to 1/16} \dfrac{x^{1/2} - 1/4}{x^{1/4} - 1/2}$

27. $\lim\limits_{y \to 1/27} \dfrac{y^{2/3} - 1/9}{y^{1/3} - 1/3}$

28. $\lim\limits_{y \to 2} \dfrac{\sqrt{y} - \sqrt{2}}{y^2 - 2y}$

29. $\lim\limits_{y \to 1/2} \dfrac{6y - 3}{y(1 - 2y)}$

30. $\lim\limits_{y \to -1} \dfrac{1 + 1/y}{y + 1}$

In Exercises 31–34 reduce the expression and then evaluate the limit.

31. $\lim\limits_{x \to 0} x\left(1 - \dfrac{1}{x}\right)$

32. $\lim\limits_{x \to -2} \left(\dfrac{x^2}{x + 2} - \dfrac{4}{x + 2}\right)$

33. $\lim\limits_{x \to 0} \dfrac{1 + 1/x}{2 + 1/x}$

34. $\lim\limits_{x \to 0} \dfrac{\sqrt[3]{x} + 1/\sqrt[3]{x}}{\sqrt[3]{x} - 1/\sqrt[3]{x}}$

In Exercises 35–42 find an equation of the line l tangent to the graph of f at the given point.

35. $f(x) = x^2 + 4x + 1$; $(-1, -2)$

36. $f(x) = x^6 - 1$; $(1, 0)$

37. $f(x) = \dfrac{1}{x}$; $\left(2, \dfrac{1}{2}\right)$

38. $f(x) = \dfrac{1}{x}$; $\left(-\dfrac{1}{2}, -2\right)$

39. $f(x) = \dfrac{1}{x + 3}$; $\left(-1, \dfrac{1}{2}\right)$

40. $f(x) = \dfrac{1}{x^2}$; $\left(\dfrac{2}{3}, \dfrac{9}{4}\right)$

41. $f(x) = \sqrt{x}$; $(16, 4)$

*42. $f(x) = x^{1/4}$; $(1, 1)$

43. Verify that the following pairs of functions have a common tangent line at the indicated point.
   a. $f(x) = x^2$ and $g(x) = x^3$ at $(0, 0)$
   b. $f(x) = x^2 + 1$ and $g(x) = -x^2 + 1$ at $(0, 1)$

44. Let $f(x) = x^2 - 2x$. Show that the line $y = 8x - 25$ is tangent to the graph of f. What is the point of tangency?

45. Let $f(x) = 1/x$.
   a. For $a \neq 0$, prove that an equation of the line tangent to the graph of f at $(1, 1/a)$ is
   $$y - \dfrac{1}{a} = \dfrac{-1}{a^2}(x - a)$$
   b. Show that the area A of the triangle formed by the coordinate axes and the tangent line in part (a) is independent of the number a.

46. Suppose $\lim\limits_{x \to -1} f(x) = 4$ and $\lim\limits_{x \to -1} [f(x) - g(x)] = 6$. Show that $\lim\limits_{x \to -1} g(x)$ exists and that $\lim\limits_{x \to -1} g(x) = -2$.

47. Suppose $\lim\limits_{x \to \sqrt{2}} f(x) = 3$ and $\lim\limits_{x \to \sqrt{2}} (fg)(x) = -\sqrt{2}$. Show that $\lim\limits_{x \to \sqrt{2}} g(x)$ exists and that $\lim\limits_{x \to \sqrt{2}} g(x) = -\sqrt{2}/3$.

48. What is wrong with the statement, "The limit of the sum of two functions is the sum of the limits"?

49. Find two functions f and g such that $\lim\limits_{x \to 1} (f + g)(x)$ exists but neither $\lim\limits_{x \to 1} f(x)$ nor $\lim\limits_{x \to 1} g(x)$ exists.

50. Suppose $\lim\limits_{x \to a} f(x)$ exists and $\lim\limits_{x \to a} (f + g)(x)$ does not exist. Prove that $\lim\limits_{x \to a} g(x)$ does not exist.

51. Find two functions f and g such that $\lim\limits_{x \to 2} (fg)(x)$ exists but neither $\lim\limits_{x \to 2} f(x)$ nor $\lim\limits_{x \to 2} g(x)$ exists.

52. Let $f(x) < g(x)$ for all $x \neq a$, and assume that
   $$\lim\limits_{x \to a} f(x) = L \quad \text{and} \quad \lim\limits_{x \to a} g(x) = M$$

Show by means of an example that it is not necessarily true that $L < M$.

53. Recall the formula

$$h(t) = -16t^2 + v_0 t + h_0$$

for the height of an object acted on only by the force of gravity. Find $\lim_{t \to t_0} h(t)$.

54. The effect of acetylcholine on a frog's heart is given by the formula

$$f(x) = \frac{x}{a + bx} \quad \text{for } 0 \le x \le 1$$

where a and b are fixed positive constants, x denotes concentration of acetylcholine, and $f(x)$ is a measure of the degree of response. If $a = 8$ and $b = 6$, find $\lim_{x \to 2/3} f(x)$.

2.4
THE SQUEEZING THEOREM AND SUBSTITUTION RULE

This section is devoted to two important limit rules—the Squeezing Theorem and the Substitution Rule. We will use these theorems to help find limits of trigonometric functions.

The Squeezing Theorem The Squeezing Theorem says in effect that if the graphs of f and h converge at a point P in the plane (Figure 2.19) and if the graph of g is "squeezed" between the graphs of f and h, then the graph of g converges with the graphs of f and h at P. This result, which is sometimes called the Pinching Theorem, will be proved in the Appendix.

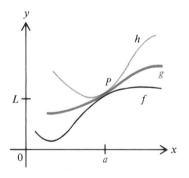

FIGURE 2.19

THEOREM 2.3
SQUEEZING THEOREM

Assume that $f(x) \le g(x) \le h(x)$ for all x in some open interval about a, except possibly a itself. If $\lim_{x \to a} f(x) = \lim_{x \to a} h(x) = L$, then $\lim_{x \to a} g(x)$ exists and $\lim_{x \to a} g(x) = L$.

The Squeezing Theorem will play a critical role in our proof below that $\lim_{x \to a} \sin x = \sin a$ and $\lim_{x \to a} \cos x = \cos a$ for any real number a. In proving

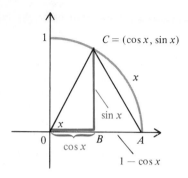

FIGURE 2.20

these formulas we will utilize the following inequalities from geometry (see Figure 2.20):

$$\text{length of segment } BC \leq \text{length of segment } AC$$
$$\leq \text{length of circular arc } AC \tag{1}$$

Now we will show that $\lim_{x \to 0} \sin x = 0$. From (1) it follows that

$$0 \leq \sin x \leq |AC| \leq x \quad \text{for } 0 \leq x \leq \pi/2$$

so that

$$|\sin x| \leq |x| \quad \text{for } 0 \leq x \leq \pi/2$$

Since $\sin(-x) = -\sin x$, and hence $|\sin(-x)| = |\sin x|$, it follows that

$$|\sin x| \leq |x| \quad \text{for } -\pi/2 \leq x \leq 0$$

Consequently

$$|\sin x| \leq |x| \quad \text{for } -\pi/2 \leq x \leq \pi/2$$

or equivalently,

$$-|x| \leq \sin x \leq |x| \quad \text{for } -\pi/2 \leq x \leq \pi/2$$

If we let $f(x) = -|x|$, $g(x) = \sin x$ and $h(x) = |x|$, then since $\lim_{x \to 0} f(x) = 0 = \lim_{x \to 0} h(x)$, the Squeezing Theorem implies that $\lim_{x \to 0} g(x) = 0$, that is,

$$\lim_{x \to 0} \sin x = 0 \tag{2}$$

To show that $\lim_{x \to 0} \cos x = 1$, observe from Figure 2.20 that

$$|\cos x - 1| = |AB| \leq |AC| \leq |x| \quad \text{for } 0 \leq x \leq \pi/2$$

Since $\cos(-x) = \cos x$, it follows that

$$|\cos x - 1| \leq |x| \quad \text{for } -\pi/2 \leq x \leq \pi/2$$

or equivalently,

$$-|x| \leq \cos x - 1 \leq |x| \quad \text{for } -\pi/2 \leq x \leq \pi/2$$

Since $\lim_{x \to 0} -|x| = 0 = \lim_{x \to 0} |x|$, the Squeezing Theorem implies that $\lim_{x \to 0}(\cos x - 1) = 0$. Since $\cos x = (\cos x - 1) + 1$, we can find $\lim_{x \to 0} \cos x$ by using the Sum Rule:

$$\lim_{x \to 0} \cos x = \lim_{x \to 0}[(\cos x - 1) + 1] = \lim_{x \to 0}(\cos x - 1) + \lim_{x \to 0} 1 = 0 + 1 = 1$$

Consequently

$$\lim_{x \to 0} \cos x = 1 \tag{3}$$

Our next goal is to show that $\lim_{x \to a} \sin x = \sin a$ and $\lim_{x \to a} \cos x = \cos a$ for any number a. Because

$$\lim_{x \to a} f(x) = L \text{ if and only if } \lim_{h \to 0} f(a + h) = L \tag{4}$$

(see Exercise 42), we can instead show that

$$\lim_{h \to 0} \sin (a + h) = \sin a \quad \text{and} \quad \lim_{h \to 0} \cos (a + h) = \cos a$$

Example 1 Show that for any number a,

$$\lim_{x \to a} \sin x = \sin a \quad \text{and} \quad \lim_{x \to a} \cos x = \cos a \tag{5}$$

Solution Let a be a fixed number. To prove that $\lim_{h \to 0} \sin (a + h) = \sin a$ and hence that $\lim_{x \to a} \sin x = \sin a$, we use the trigonometric identity

$$\sin (a + h) = \sin a \cos h + \sin h \cos a \tag{6}$$

Since a is fixed, $\sin a$ and $\cos a$ are constants. Applying the Sum and Constant Multiple Rules to (6) and using (2) and (3), we find that

$$\lim_{h \to 0} \sin (a + h) = \lim_{h \to 0} (\sin a \cos h + \sin h \cos a)$$

$$= \sin a \lim_{h \to 0} \cos h + \cos a \lim_{h \to 0} \sin h$$

$$= (\sin a) \cdot 1 + (\cos a) \cdot 0$$

$$= \sin a$$

For the proof that $\lim_{h \to 0} \cos (a + h) = \cos a$ we use the trigonometric identity

$$\cos (a + h) = \cos a \cos h - \sin a \sin h$$

and in a similar way conclude that

$$\lim_{h \to 0} \cos (a + h) = \lim_{h \to 0} (\cos a \cos h - \sin a \sin h)$$

$$= \cos a \lim_{h \to 0} \cos h - \sin a \lim_{h \to 0} \sin h$$

$$= (\cos a) \cdot 1 - (\sin a) \cdot 0$$

$$= \cos a \quad \square$$

The other trigonometric functions have similar properties, as can be verified from (5) by using the limit rules. For example,

$$\lim_{x \to a} \tan x = \lim_{x \to a} \frac{\sin x}{\cos x} = \frac{\lim_{x \to a} \sin x}{\lim_{x \to a} \cos x} = \frac{\sin a}{\cos a} = \tan a \tag{7}$$

for any number a in the domain of the tangent function.

In Example 2 we will verify the limit that appeared in (7) of Section 2.1 and will reappear in Chapter 3.

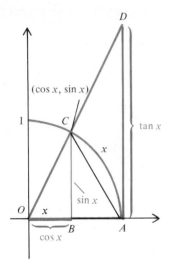

FIGURE 2.21

Example 2   Show that

$$\lim_{x \to 0} \frac{\sin x}{x} = 1$$

Solution   Using Figure 2.21, we obtain the following equations, which are valid for $0 < x < \pi/2$:

$$\text{area of triangle } OAC = \frac{1}{2}|OA||BC| = \frac{1}{2} \cdot 1 \cdot \sin x = \frac{\sin x}{2}$$

$$\text{area of sector } OAC = \frac{x}{2\pi}(\text{area of circle}) = \frac{x}{2\pi}\pi = \frac{x}{2}$$

$$\text{area of triangle } OAD = \frac{1}{2}|OA||AD| = \frac{1}{2} \cdot 1 \cdot \tan x = \frac{1}{2} \cdot \frac{\sin x}{\cos x}$$

It is geometrically clear that

$$\text{area of triangle } OAC \leq \text{area of sector } OAC \leq \text{area of triangle } OAD$$

so that

$$\frac{\sin x}{2} \leq \frac{x}{2} \leq \frac{1}{2} \cdot \frac{\sin x}{\cos x} \quad \text{for } 0 < x < \frac{\pi}{2}$$

Separately, the first and second inequalities yield

$$\frac{\sin x}{x} \leq 1 \quad \text{and} \quad \cos x \leq \frac{\sin x}{x} \tag{8}$$

Combining the inequalities in (8) and using the fact that

$$\cos(-x) = \cos x \quad \text{and} \quad \frac{\sin(-x)}{-x} = \frac{\sin x}{x}$$

we obtain

$$\cos x \leq \frac{\sin x}{x} \leq 1 \quad \text{for } 0 < |x| < \frac{\pi}{2}$$

Since $\lim_{x \to 0} \cos x = 1 = \lim_{x \to 0} 1$, it follows from the Squeezing Theorem that

$$\lim_{x \to 0} \frac{\sin x}{x} = 1 \quad \square$$

There is a companion to the limit appearing in Example 2. We can obtain it from Example 2 by using various limit rules.

Example 3   Show that

$$\lim_{x \to 0} \frac{\cos x - 1}{x} = 0$$

Solution   Notice that $\lim_{x \to 0} x = 0$, so we cannot apply the Quotient Rule directly. However,

$$\lim_{x \to 0} \frac{\cos x - 1}{x} = \lim_{x \to 0} \left(\frac{\cos x - 1}{x} \right) \left(\frac{\cos x + 1}{\cos x + 1} \right) = \lim_{x \to 0} \frac{\cos^2 x - 1}{x(\cos x + 1)}$$

$$= \lim_{x \to 0} \frac{-\sin^2 x}{x(\cos x + 1)} = \lim_{x \to 0} \left(\frac{\sin x}{x} \right) \left(\frac{-\sin x}{\cos x + 1} \right)$$

Now, by Example 2,

$$\lim_{x \to 0} \frac{\sin x}{x} = 1$$

Furthermore

$$\lim_{x \to 0} \frac{-\sin x}{\cos x + 1} = \frac{0}{1 + 1} = 0$$

by the Sum and the Quotient Rules. Thus the Product Rule tells us that

$$\lim_{x \to 0} \frac{\cos x - 1}{x} = \lim_{x \to 0} \frac{\sin x}{x} \lim_{x \to 0} \frac{-\sin x}{\cos x + 1} = 1 \cdot 0 = 0 \quad \square$$

The results of Examples 2 and 3 can be used to evaluate other limits, as Example 4 illustrates.

Example 4   Evaluate $\lim_{x \to 0} \dfrac{\sin x}{x^{2/3}}$.

Solution   From Example 2 we know that

$$\lim_{x \to 0} \frac{\sin x}{x} = 1$$

To use this result, we rewrite $(\sin x)/x^{2/3}$:

$$\frac{\sin x}{x^{2/3}} = \left(\frac{\sin x}{x} \right) x^{1/3}$$

Since $\lim_{x \to 0} x^{1/3} = 0$, it follows from the Product Rule that

$$\lim_{x \to 0} \frac{\sin x}{x^{2/3}} = \lim_{x \to 0} \left[\left(\frac{\sin x}{x} \right) x^{1/3} \right] = \lim_{x \to 0} \frac{\sin x}{x} \lim_{x \to 0} x^{1/3} = 1 \cdot 0 = 0 \quad \square$$

The Substitution Rule   Using the limit rules presented in Section 2.3 and the Squeezing Theorem, we can evaluate limits of rational functions and a variety of trigonometric functions. But as yet we have no convenient method for evaluating limits such as $\lim_{x \to 1} \sqrt{x^3 - 4x^2 + 3x + 2}$. Surely resorting to ε's and δ's has no appeal in this case.

In trying to evaluate the limit above, suppose we first let $y = x^3 - 4x^2 + 3x + 2$ and notice that as x approaches 1, y approaches $(1)^3 - 4(1)^2 + 3(1) + 2 = 2$. It is then tempting to conclude that if we substitute y for $x^3 - 4x^2 + 3x + 2$ and substitute $y \to 2$ for $x \to 1$, then

$$\lim_{x \to 1} \sqrt{x^3 - 4x^2 + 3x + 2} = \lim_{y \to 2} \sqrt{y}$$

Since $\lim_{y \to 2} \sqrt{y} = \sqrt{2}$ by (8) of Section 2.3, it would follow that

$$\lim_{x \to 1} \sqrt{x^3 - 4x^2 + 3x + 2} = \sqrt{2}$$

This is, in fact, a valid conclusion. More generally, if $\lim_{x \to a} f(x) = c$ and if $\lim_{y \to c} g(y)$ exists, then we have the following result:

SUBSTITUTION RULE

$$\lim_{x \to a} g(f(x)) = \lim_{y \to c} g(y) \tag{9}$$

The Substitution Rule is stated and proved in the Appendix. In using (9) to find $\lim_{x \to a} g(f(x))$, we substitute y for $f(x)$, find $c = \lim_{x \to a} y$, and then compute $\lim_{y \to c} g(y)$. Frequently the process is straightforward, as Examples 5 and 6 will show.

Example 5   Evaluate $\lim_{x \to 2} \sqrt{x + \dfrac{1}{x}}$.

Solution   First we let $y = x + 1/x$. Then we notice that

$$\lim_{x \to 2} y = \lim_{x \to 2} \left(x + \frac{1}{x} \right) = 2 + \frac{1}{2} = \frac{5}{2}$$

By the Substitution Rule, we find that

$$\lim_{x \to 2} \sqrt{x + \frac{1}{x}} = \lim_{y \to 5/2} \sqrt{y} = \sqrt{\frac{5}{2}} \quad \square$$

Example 6   Evaluate $\lim_{x \to \pi/3} \cos(x + \pi/6)$.

Solution   Let $y = x + \pi/6$, and notice that

$$\lim_{x \to \pi/3} y = \lim_{x \to \pi/3} \left(x + \frac{\pi}{6} \right) = \frac{\pi}{3} + \frac{\pi}{6} = \frac{\pi}{2}$$

By the Substitution Rule,

$$\lim_{x\to\pi/3}\cos\left(x+\frac{\pi}{6}\right)=\lim_{y\to\pi/2}\cos y=\cos\frac{\pi}{2}=0 \quad \square$$

Sometimes the use of the Substitution Rule is not so immediately transparent. To illustrate this thought we will evaluate a limit that is intimately related to

$$\lim_{x\to 0}\frac{\sin x}{x}=1 \tag{10}$$

Example 7   Evaluate $\lim_{x\to 0}\dfrac{\sin 2x}{x}$.

Solution   Because of the appearance of $2x$ in the numerator, we substitute $y = 2x$ and notice that

$$\lim_{x\to 0} y = \lim_{x\to 0} 2x = 0$$

It follows from the Substitution and Constant Multiple Rules, along with (10), that

$$\lim_{x\to 0}\frac{\sin 2x}{x}=\lim_{y\to 0}\frac{\sin y}{y/2}=\lim_{y\to 0}2\frac{\sin y}{y}=2\lim_{y\to 0}\frac{\sin y}{y}=2\cdot 1=2 \quad \square$$

EXERCISES 2.4

In Exercises 1–6 use the results of this section to evaluate the limit.

1. $\lim_{x\to\pi/3}(\sqrt{3}\sin x - 2x)$

2. $\lim_{x\to 0}\dfrac{x^2-2}{\cos x}$

3. $\lim_{x\to -\pi/3}3x^2\cos x$

4. $\lim_{x\to 0}\dfrac{1-\cos x}{1-x}$

5. $\lim_{y\to 2\pi/3}\dfrac{\pi\sin y\cos y}{y}$

6. $\lim_{y\to 0}\dfrac{\pi\sin y\cos y}{y}$

In Exercises 7–18 use the Substitution Rule to evaluate the limit.

7. $\lim_{x\to 7}(2x-16)^4$

8. $\lim_{x\to 3}\sqrt{3x^3}$

9. $\lim_{x\to\sqrt{5}}(9-x^2)^{-5/2}$

10. $\lim_{x\to 1}\sqrt{\dfrac{2x+1}{2x-1}}$

11. $\lim_{x\to -\pi/6}\cos 2x$

12. $\lim_{x\to -\pi}\sin\left(\dfrac{\pi}{2}-3x\right)$

13. $\lim_{t\to 0}\sin^{4/3}t$

14. $\lim_{t\to\pi/4}\tan^{14}t$

15. $\lim_{t\to\pi/6}\sin^3 t\sec^4 t$

16. $\lim_{t\to -\pi/6}\sqrt[3]{16\cos 2t}$

17. $\lim_{w\to\pi/2}\cos^2(\csc w)$

18. $\lim_{w\to 3\pi/2}\sin\left(\dfrac{\pi}{2}\sin w\right)$

In Exercises 19–33 use the results of this section to evaluate the limit. You may wish to rearrange the given expression before evaluating the limit.

19. $\lim_{x\to 0}\dfrac{\sin 3x}{5x}$

20. $\lim_{x\to 0}\dfrac{\cos 4x-1}{x}$

21. $\lim_{x\to 0}\dfrac{\sin x^{1/3}}{x^{1/3}}$

22. $\lim_{x\to 0}\dfrac{\cos x^2-1}{x^2}$

23. $\lim_{t\to 0}\dfrac{\cos^2 t-1}{t}$

24. $\lim_{t\to 0}\dfrac{\cos t-1}{\sqrt[3]{t}}$

25. $\lim_{y\to 0}\dfrac{\tan y}{y}$

26. $\lim_{y\to 0}\dfrac{\tan\sqrt{2y}}{5y}$

27. $\lim_{y\to 0}\dfrac{\sin y-\tan y}{y}$

28. $\lim_{y\to 0}y\cot y$

29. $\lim\limits_{x \to 0} \dfrac{\sin^2 2x}{8x^2}$

30. $\lim\limits_{x \to 0} \dfrac{\sin^2 x}{1 - \cos x}$

31. $\lim\limits_{x \to \pi} \dfrac{\tan^2 x}{1 + \sec x}$

32. $\lim\limits_{x \to 0} \dfrac{1 - \cos 3x}{\sin 3x}$

33. $\lim\limits_{x \to 0} \dfrac{\cos(x + \pi/2)}{x}$

In Exercises 34–36 let a be a number in the domain of the given function, and verify the formula.

34. $\lim\limits_{x \to a} \cot x = \cot a$

35. $\lim\limits_{x \to a} \sec x = \sec a$

36. $\lim\limits_{x \to a} \csc x = \csc a$

In Exercises 37–40 find an equation of the line l tangent to the graph of the given function at the indicated point.

37. $f(x) = \cos x;\ (0, 1)$

38. $f(x) = \sin 4x;\ (0, 0)$

39. $f(x) = \sqrt{4 - x^2};\ (1, \sqrt{3})$

40. $f(x) = \sqrt{x + x^2};\ (1, \sqrt{2})$

41. Suppose there is a number M such that $|f(x)| \le M|x - a|$ for all $x \ne a$. Show that $\lim\limits_{x \to a} f(x) = 0$.

42. a. Verify (4) by using Definition 2.1. (*Hint:* Let the number x correspond to the number $a + h$, so that $h = x - a$. Then for any $\varepsilon > 0$ the number δ can be chosen to be the same positive number in both limits.)
    b. Verify (4) by using the Substitution Rule.

43. a. Suppose $\lim\limits_{x \to a} f(x) = L$. Using the Substitution Rule and the result of Exercise 28 in Section 2.2, show that $\lim\limits_{x \to a} |f(x)| = |L|$.
    b. Show that if $\lim\limits_{x \to a} |f(x)| = 0$, then $\lim\limits_{x \to a} f(x) = 0$.
    c. Give an example of a function f such that $\lim\limits_{x \to 0} |f(x)| = 1$ but $\lim\limits_{x \to 0} f(x)$ does not exist.

44. Suppose that $\lim\limits_{x \to a} f(x) = 0$ and $|g(x)| \le M$ for a fixed M and for all $x \ne a$. Prove that $\lim\limits_{x \to a} f(x)g(x) = 0$. (*Hint:* Notice that $-M|f(x)| \le f(x)g(x) \le M|f(x)|$.)

45. Using Exercise 44, evaluate the following limits.
    a. $\lim\limits_{x \to 0} x \cos \dfrac{1}{x}$
    b. $\lim\limits_{x \to 0} x \sin \dfrac{1}{x}$

46. Suppose that $f(x) \ge 0$ for all x in the domain of f and $\lim\limits_{x \to a} [f(x)]^2 = L > 0$. Prove that $\lim\limits_{x \to a} f(x) = \sqrt{L}$. (*Hint:* Use the Substitution Rule.)

47. Using the Substitution Rule, prove that if $\lim\limits_{x \to 0} f(x) = L$, then $\lim\limits_{x \to 0} f(x^n) = L$ for any positive integer n.

48. Using Exercise 47, show that $\lim\limits_{x \to 0} \sin x^n = 0$ for any positive integer n.

49. Let $f(x) = \sqrt{1 - x^2}$ and let $0 < a < 1$. The graph of f is a semicircle with radius 1.
    a. Show that the slope of the radial line joining $(0, 0)$ to the point $(a, \sqrt{1 - a^2})$ is $\sqrt{1 - a^2}/a$.
    b. Show that the slope of the line that is tangent at $(a, \sqrt{1 - a^2})$ is $-a/\sqrt{1 - a^2}$. (*Hint:* The tangent line is perpendicular to the radial line.)
    c. If $0 < x < 1$ and $x \ne a$, find the slope m_x of the secant line through $(a, \sqrt{1 - a^2})$ and $(x, \sqrt{1 - x^2})$.
    d. Show that
    $$\lim\limits_{x \to a} m_x = \dfrac{-a}{\sqrt{1 - a^2}}$$
    (This shows that the slope of the tangent line is the limit of the slope of the secant line through $(x, f(x))$ as x approaches a. Thus the definition of tangent line given in Section 2.2 yields the same tangent line for a circle as does the ancient Greek definition.)

*50. A circular hoop of radius 1 foot is suspended from the ceiling by a string 4 feet long (Figure 2.22). Suppose a flea begins to walk counterclockwise from the bottom to the top on the outside of the hoop. At what point on the hoop will the flea first see the point on the ceiling to which the string is attached? (*Hint:* Find the tangent line that passes through the point $(0, 5)$.)

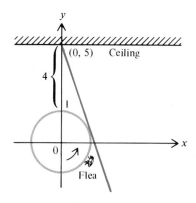

FIGURE 2.22

2.5
ONE-SIDED LIMITS

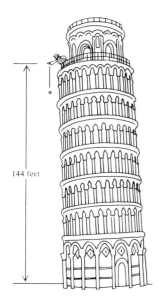

FIGURE 2.23

In Section 2.2 we considered the velocity of an iron ball dropped by Galileo from the Leaning Tower of Pisa (Figure 2.23). Until the ball hit the ground its position $h(t)$ at time t was given by

$$h(t) = -16t^2 + 144 = -16(t^2 - 9)$$

Now we would like to find the velocity of the ball as it struck the ground. Since $h(3) = 0$, the ball hit the ground 3 seconds after it was dropped. After that the ball remained on the ground, so obviously the position of the ball for $t > 3$ will not help us calculate its velocity on impact. However, for $t < 3$ the average velocity of the ball during the time interval $[t, 3]$ was

$$\frac{h(3) - h(t)}{3 - t}, \text{ or equivalently, } \frac{h(t) - h(3)}{t - 3} \tag{1}$$

If we keep $t < 3$ and let t approach 3, the average velocity in (1) should approach the velocity of the ball as it struck the ground. Thus the velocity should be a kind of "half limit," or "left-hand limit" (because t is required to be less than and hence to the left of 3 as it approaches 3). In defining left-hand limits, we will simply replace the inequality $0 < |x - a| < \delta$ in Definition 2.1 by the inequality $-\delta < x - a < 0$, which requires x to be the left of a.

DEFINITION 2.4

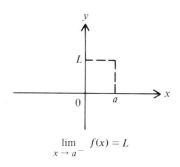

$$\lim_{x \to a^-} f(x) = L$$

FIGURE 2.24

Let f be defined on some open interval (c, a). A number L is the **limit of $f(x)$ as x approaches a from the left** (or the **left-hand limit of f at a**) if for every $\varepsilon > 0$ there is a number $\delta > 0$ such that

$$\text{if } -\delta < x - a < 0, \quad \text{then} \quad |f(x) - L| < \varepsilon$$

In this case we write

$$\lim_{x \to a^-} f(x) = L$$

and say that the **left-hand limit of f at a exists** (Figure 2.24).

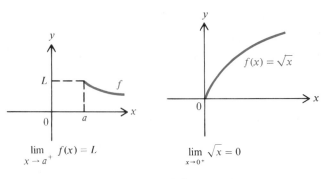

FIGURE 2.25   **FIGURE 2.26**

Right-hand limits are treated in a completely analogous way. Suppose that f is defined on some interval (a, c) to the right of a. A number L is the **limit of $f(x)$ as x approaches a from the right** (or the **right-hand limit of f at a**) if for every $\varepsilon > 0$ there is a number $\delta > 0$ such that

$$\text{if } 0 < x - a < \delta, \quad \text{then} \quad |f(x) - L| < \varepsilon$$

In this case we write $\lim_{x \to a^+} f(x) = L$ and say that the **right-hand limit of f at a exists** (Figure 2.25).

It follows from the preceding definitions that for the function h described above we have

$$\lim_{t \to 3^-} \frac{h(t) - h(3)}{t - 3} = \lim_{t \to 3^-} \frac{-16(t^2 - 9) - 0}{t - 3} = \lim_{t \to 3^-} -16(t + 3) = -96$$

so that as the ball hit the ground its velocity was -96 feet per second.

One can also show that $\lim_{x \to 0^+} \sqrt{x} = 0$, but that neither $\lim_{x \to 0^-} \sqrt{x}$ nor $\lim_{x \to 0} \sqrt{x}$ exists (because $\sqrt{x}$ is not defined to the left of 0 (Figure 2.26).

Right-hand limits and left-hand limits are called **one-sided limits**. Ordinary limits are called **two-sided limits**. All the rules given in Sections 2.3–2.4 for finding limits of combinations of functions, as well as the Squeezing Theorem and the Substitution Rule, remain valid for one-sided limits.

Example 1   Find $\lim_{x \to 0^+} \dfrac{x^{3/2} - 3x + 2}{x - 2\sqrt{x} + 1}$.

Solution   Since $x^{3/2} = (\sqrt{x})^3$ and $\lim_{x \to 0^+} \sqrt{x} = 0$, it follows from the version of the Product Rule for one-sided limits that $\lim_{x \to 0^+} x^{3/2} = 0$. Consequently

$$\lim_{x \to 0^+} \frac{x^{3/2} - 3x + 2}{x - 2\sqrt{x} + 1} = \frac{0 - 0 + 2}{0 - 0 + 1} = 2 \quad \square$$

Care must be used in applying the Substitution Rule to one-sided limits.

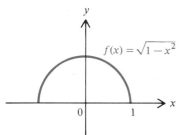

FIGURE 2.27

Example 2 Find $\lim_{x\to 1^-}\sqrt{1-x^2}$.

Solution Let y be the expression $1-x^2$. Since $1-x^2$ approaches 0 from the right as x approaches 1 from the left, it follows that y approaches 0 from the right as x approaches 1 from the left, so we have

$$\lim_{x\to 1^-}\sqrt{1-x^2} = \lim_{y\to 0^+}\sqrt{y} = 0$$

(Figure 2.27).   □

Although $\lim_{x\to 1^-}\sqrt{1-x^2}$ exists by Example 2, $\lim_{x\to 1^+}\sqrt{1-x^2}$ does not exist because $1-x^2$ is negative whenever x lies to the right of 1, and the square root of a negative number is not defined.

The next theorem, which is proved in the Appendix, reveals the relationship between the existence of a two-sided limit at a and the one-sided limits at a.

THEOREM 2.5

Let f be defined on an open interval about a, except possibly at a itself. Then $\lim_{x\to a} f(x)$ exists if and only if both one-sided limits, $\lim_{x\to a^+} f(x)$ and $\lim_{x\to a^-} f(x)$, exist and

$$\lim_{x\to a^+} f(x) = \lim_{x\to a^-} f(x)$$

In that case,

$$\lim_{x\to a} f(x) = \lim_{x\to a^+} f(x) = \lim_{x\to a^-} f(x)$$

When a function is defined by a formula involving more than one equation, one-sided limits are often useful in showing that the function has a limit at a given point.

Example 3 Let

$$f(x) = \begin{cases} 2x+1 & \text{for } x < 2 \\ x+3 & \text{for } x > 2 \end{cases}$$

Find $\lim_{x\to 2} f(x)$.

Solution Since $\lim_{x\to 2}(2x+1) = 5$ and $\lim_{x\to 2}(x+3) = 5$, Theorem 2.5 assures us that

$$\lim_{x\to 2^-} f(x) = \lim_{x\to 2^-}(2x+1) = 5 \quad\text{and}\quad \lim_{x\to 2^+} f(x) = \lim_{x\to 2^+}(x+3) = 5$$

A second application of Theorem 2.5 now yields

$$\lim_{x\to 2} f(x) = 5$$

$$f(x) = \begin{cases} 2x+1 & \text{for } x < 2 \\ x+3 & \text{for } x > 2 \end{cases}$$

FIGURE 2.28

The function is graphed in Figure 2.28.   □

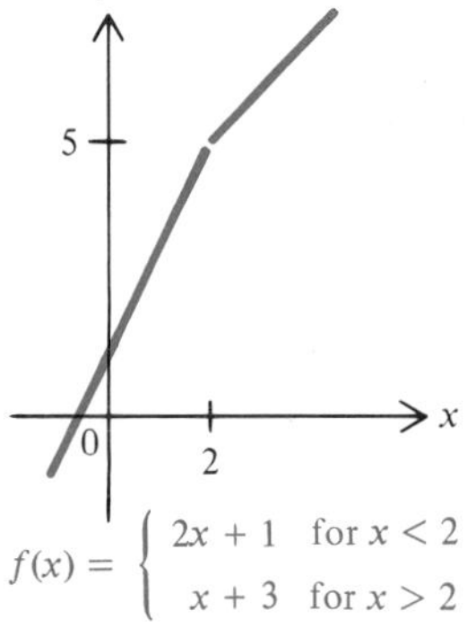

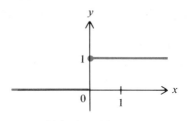

Diving board function:

$$f(x) = \begin{cases} 0 & \text{for } x < 0 \\ 1 & \text{for } x \geq 0 \end{cases}$$

FIGURE 2.29

Example 4 Let f be the diving board function, defined by

$$f(x) = \begin{cases} 0 & \text{for } x < 0 \\ 1 & \text{for } x \geq 0 \end{cases}$$

Find the left-hand and right-hand limits of f at 0. Show that f has no two-sided limit at 0.

Solution Since $\lim_{x \to 0^+} 1 = 1$, we know that $\lim_{x \to 0^+} f(x) = 1$ (see Figure 2.29). Similarly, since $\lim_{x \to 0^-} 0 = 0$, we know that $\lim_{x \to 0^-} f(x) = 0$. Thus both one-sided limits exist, and they are unequal. Thus f has no two-sided limit at 0 by Theorem 2.5. □

Infinite Limits and Vertical Asymptotes

If $\lim_{x \to a^+} f(x)$ does not exist, it may happen that as x approaches a from the right, the value of $f(x)$ becomes indefinitely large or becomes negative and indefinitely large in absolute value (Figure 2.30(a) and (b)). The value of $f(x)$ may behave similarly when the left-hand limit at a does not exist (Figure 2.30(c) and (d)). We introduce a special terminology to cover such cases.

DEFINITION 2.6

Let f be defined on some open interval (a, c).
a. If for every number N there is some $\delta > 0$ such that

$$\text{if } 0 < x - a < \delta, \quad \text{then} \quad f(x) > N$$

then we say that the **limit of $f(x)$ as x approaches a from the right** is ∞. In that case we write $\lim_{x \to a^+} f(x) = \infty$ (Figure 2.30(a)).
b. If for every number N there is some $\delta > 0$ such that

$$\text{if } 0 < x - a < \delta, \quad \text{then} \quad f(x) < N$$

then we say that the **limit of $f(x)$ as x approaches a from the right** is $-\infty$. In that case we write $\lim_{x \to a^+} f(x) = -\infty$ (Figure 2.30(b)).
c. In either case (a) or (b) the vertical line $x = a$ is called a **vertical asymptote of the graph of f**, and we say that f has an **infinite right-hand limit at a**.

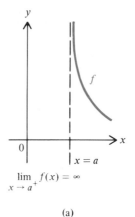

$$\lim_{x \to a^+} f(x) = \infty$$

(a)

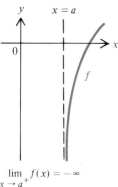

$$\lim_{x \to a^+} f(x) = -\infty$$

(b)

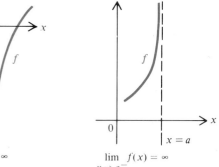

$$\lim_{x \to a^-} f(x) = \infty$$

(c)

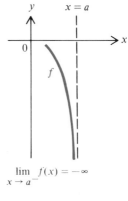

$$\lim_{x \to a^-} f(x) = -\infty$$

(d)

FIGURE 2.30

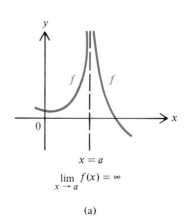

Vertical asymptote

$f(x) = \dfrac{1}{x}$

FIGURE 2.31

There are analogous definitions for the limits

$$\lim_{x \to a^-} f(x) = \infty \quad \text{and} \quad \lim_{x \to a^-} f(x) = -\infty$$

Definition 2.6 implies that $\lim_{x \to 0^+} 1/x = \infty$ and that $\lim_{x \to 0^-} 1/x = -\infty$ (Figure 2.31). This means, in particular, that the line $x = 0$ is a vertical asymptote of the graph of $1/x$. Similarly, if a is any number, then

$$\lim_{x \to a^+} \frac{1}{x - a} = \infty \quad \text{and} \quad \lim_{x \to a^-} \frac{1}{x - a} = -\infty$$

Now suppose that

$$\lim_{x \to a^+} f(x) = \lim_{x \to a^-} f(x) = \infty$$

Then we write $\lim_{x \to a} f(x) = \infty$ and say that the **limit of $f(x)$ as x approaches a is ∞** and that **f has an infinite limit at a** (Figure 2.32(a)). Similar comments hold if ∞ is replaced by $-\infty$ (Figure 2.32(b)). One can show that if $f(x) = 1/x^2$, then $\lim_{x \to 0} f(x) = \infty$ (Figure 2.32(c)), so that $1/x^2$ has an infinite (two-sided) limit at 0. By contrast, if $g(x) = 1/x$, then because $\lim_{x \to 0^-} 1/x = -\infty$ and $\lim_{x \to 0^+} 1/x = \infty$, the two one-sided limits are distinct, so that g does not have an infinite (two-sided) limit at 0.

Caution: If f has an infinite limit at a, then f does not have a limit at a in the sense of Definition 2.1.

The line $x = a$ is a vertical asymptote whenever $\lim_{x \to a^+} f(x) = \infty$ or $-\infty$, and whenever $\lim_{x \to a^-} f(x) = \infty$ or $-\infty$. Pictorially, if $x = a$ is a vertical asymptote of the graph of f, then the graph of f is nearly vertical near the line $x = a$, the points $(x, f(x))$ on the graph of f getting arbitrarily far from the x axis and approaching the line $x = a$ as x approaches a from one or possibly both sides (see Figures 2.30 and 2.32(a),(b)).

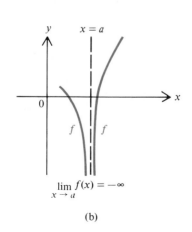

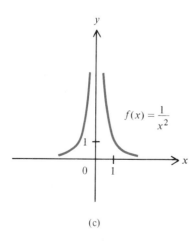

$x = a$

$\lim_{x \to a} f(x) = \infty$

(a)

$\lim_{x \to a} f(x) = -\infty$

(b)

$f(x) = \dfrac{1}{x^2}$

(c)

FIGURE 2.32

Example 5   Let

$$f(x) = \frac{x+2}{x^2-1}$$

Find all vertical asymptotes of the graph of f.

Solution   If a is any number other than 1 or -1, then $a^2 - 1 \neq 0$, so that

$$\lim_{x \to a} \frac{x+2}{x^2-1} = \frac{a+2}{a^2-1}$$

which is a number. Hence the only possible vertical asymptotes are the lines $x = 1$ and $x = -1$. Since

$$\frac{x+2}{x^2-1} = \frac{x+2}{x+1} \cdot \frac{1}{x-1}$$

it follows that

$$\lim_{x \to 1^+} \frac{x+2}{x^2-1} = \lim_{x \to 1^+} \frac{x+2}{x+1} \cdot \frac{1}{x-1} = \infty$$

and

$$\lim_{x \to -1^-} \frac{x+2}{x^2-1} = \lim_{x \to -1^-} \frac{x+2}{x-1} \cdot \frac{1}{x+1} = \infty$$

Thus $x = 1$ and $x = -1$ are indeed vertical asymptotes of the graph of f. Alternatively we could show that $\lim_{x \to 1^-} f(x) = -\infty$ and $\lim_{x \to -1^+} f(x) = -\infty$, which imply that $x = 1$ and $x = -1$ are vertical asymptotes. The graph of f appears in Figure 2.33.   □

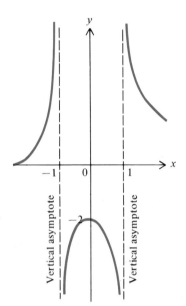

FIGURE 2.33

In general, if f is any rational function with $f(x) = p(x)/q(x)$ (where p and q are polynomials), then the line $x = a$ is a vertical asymptote of the graph of f if $p(a) \neq 0$ and $q(a) = 0$. For instance, if

$$f(x) = \frac{(x-2)(x+3)}{x(x-\pi)(x-4)(x+1)^2}$$

then the graph of f has vertical asymptotes $x = 0$, $x = \pi$, $x = 4$, and $x = -1$.

Caution: It is not always possible to determine the vertical asymptotes of the graph of a rational function by merely finding out where the denominator vanishes. For example, if

$$f(x) = \frac{x+1}{x^2-1}$$

then the denominator equals 0 for $x = -1$ as well as $x = 1$. However,

$$\frac{x+1}{x^2-1} = \frac{x+1}{(x+1)(x-1)} = \frac{1}{x-1}$$

so $\lim_{x \to -1} f(x) = -1/2$. Consequently $x = -1$ is *not* a vertical asymptote of the graph of f, and the only vertical asymptote is the line $x = 1$.

Vertical Tangent Lines   Let $f(x) = x^{1/3}$. From Figure 2.34 it would appear that the vertical line $x = 0$ is tangent to the graph of f at $(0, 0)$. However, since

$$\frac{f(x) - f(0)}{x - 0} = \frac{x^{1/3} - 0}{x - 0} = \frac{1}{x^{2/3}} \tag{2}$$

it follows that

$$\lim_{x \to 0} \frac{f(x) - f(0)}{x - 0}$$

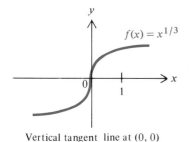

$f(x) = x^{1/3}$

Vertical tangent line at $(0, 0)$

FIGURE 2.34

does not exist (in the sense of Definition 2.1), so that there is no line tangent to the graph of f according to the definition of tangent line given in Section 2.2. Nevertheless, the ratio in (2) has a two-sided infinite limit at 0:

$$\lim_{x \to 0} \frac{f(x) - f(0)}{x - 0} = \lim_{x \to 0} \frac{1}{x^{2/3}} = \infty$$

This leads us to make the following definition.

DEFINITION 2.7

Let f be continuous at a. If

$$\lim_{x \to a} \frac{f(x) - f(a)}{x - a} = \infty \quad \text{or} \quad \lim_{x \to a} \frac{f(x) - f(a)}{x - a} = -\infty$$

then we say that the graph of f has a **vertical tangent line at $(a, f(a))$**. In that case the vertical line $x = a$ is called the **line tangent to the graph of f at a**.

By Definition 2.7 the graph of the function $x^{1/3}$ has a vertical tangent at $(0, 0)$. The same is true of $x^{1/n}$, for any positive odd integer n (see Exercise 78).

EXERCISES 2.5

In Exercises 1–10 determine the one-sided limit.

1. $\lim\limits_{x \to -2^+} (x^3 + 3x - 5)$

2. $\lim\limits_{x \to \pi/3^-} \cos\left(x + \dfrac{\pi}{6}\right)$

3. $\lim\limits_{x \to 2^-} \dfrac{x^2 - 4}{x - 2}$

4. $\lim\limits_{x \to 2^+} \dfrac{x^2 - 4}{x - 2}$

5. $\lim\limits_{x \to 1^+} \dfrac{x^2 + 3x - 4}{x^2 - 1}$

6. $\lim\limits_{x \to -2^-} \dfrac{x^2 - 3x - 10}{x^2 - 9}$

7. $\lim\limits_{t \to 5^+} \dfrac{|t - 5|}{5 - t}$

8. $\lim\limits_{t \to -3^+} \sqrt{t + 3}$

9. $\lim\limits_{t \to 2^-} (t - \sqrt{4 - 2t})$

10. $\lim\limits_{t \to -3^-} \sqrt[4]{-3 - t}$

In Exercises 11–20 determine the infinite limit.

11. $\lim\limits_{x \to 0^-} \dfrac{4}{x^3}$

12. $\lim\limits_{x \to 0^+} \dfrac{4}{x^3}$

13. $\lim\limits_{x \to 0} \dfrac{-1}{x^2}$

14. $\lim\limits_{x \to 0^+} \dfrac{2}{x^{1/4}}$

15. $\lim\limits_{y \to 2^-} \dfrac{3}{y - 2}$

16. $\lim\limits_{y \to 2^+} \dfrac{3}{y - 2}$

17. $\lim\limits_{y \to -1^-} \dfrac{\pi}{y + 1}$

18. $\lim\limits_{z \to 3} \dfrac{2}{(z - 3)^2}$

19. $\lim\limits_{z \to \pi/2^-} \tan z$

20. $\lim\limits_{z \to -\pi/2^+} \sec z$

In Exercises 21–52 decide which of the given one-sided or two-sided limits exist as numbers, which as ∞, which as $-\infty$, and which do not exist. Where the limit is a number, evaluate it.

21. $\lim\limits_{x \to 3/2^+} \sqrt{3 - 2x}$

22. $\lim\limits_{x \to 3/2^-} \sqrt{3 - 2x}$

23. $\lim\limits_{x \to 5^+} \dfrac{1}{x\sqrt{x - 5}}$

24. $\lim\limits_{x \to 1^+} (\sqrt{x^2 - x} + x)$

25. $\lim\limits_{x \to 0^+} \sqrt{x} \cos 2x$

26. $\lim\limits_{x \to 0^-} \sqrt{-x^2 \sin 3x}$

27. $\lim\limits_{x \to 0} \sqrt{\dfrac{1}{x^2}}$

28. $\lim\limits_{x \to 0} \dfrac{x}{|x|}$

29. $\lim\limits_{x \to -1/2^-} \dfrac{4x - 7}{x + \frac{1}{2}}$

30. $\lim\limits_{x \to -1^+} \dfrac{x^2 + 5x + 4}{x + 1}$

31. $\lim\limits_{x \to 2^-} \dfrac{x + 1}{2(x^2 - 4)}$

32. $\lim\limits_{x \to 1^+} \dfrac{x^2 - 3x + 2}{x^2 - 2x + 1}$

33. $\lim\limits_{x \to 5} \dfrac{x^2 - 3}{(x - 5)^2}$

34. $\lim\limits_{x \to -1} \dfrac{x^3 + 3x^2 + 3x + 1}{(x + 1)^2}$

35. $\lim\limits_{y \to 3^-} \dfrac{-1}{\sqrt{3 - y}}$

36. $\lim\limits_{y \to -5^-} \dfrac{1}{\sqrt{-y - 5}}$

37. $\lim\limits_{y \to 1^-} \dfrac{\sqrt{1 - y^2}}{y - 1}$

38. $\lim\limits_{y \to -3^+} \dfrac{\sqrt{9 - y^2}}{y + 3}$

39. $\lim\limits_{y \to 0^-} \dfrac{5 + \sqrt{1 + y^2}}{\sqrt{y}}$

40. $\lim\limits_{t \to \pi/2} \sec^2 t$

41. $\lim\limits_{t \to 0^-} \csc 2t$

42. $\lim\limits_{t \to 0^-} \dfrac{\sin t}{t^2}$

*43. $\lim\limits_{t \to 0} \dfrac{1 - \cos t}{t^2}$

44. $\lim\limits_{x \to 0^-} \dfrac{\sqrt{1 + x} - \sqrt{1 - x}}{x}$

45. $\lim\limits_{x \to 3^+} \left(\dfrac{1}{x - 3} - \dfrac{6}{x^2 - 9} \right)$

46. $\lim\limits_{x \to 3^-} \left(\dfrac{1}{x - 3} - \dfrac{3}{x^2 - 9} \right)$

47. $\lim\limits_{h \to 0^+} \left(\dfrac{1}{h} - \dfrac{1}{\sqrt{h}} \right)$

48. $\lim\limits_{x \to 0} f(x)$, where $f(x) = \begin{cases} 2x - 4 & \text{for } x < 0 \\ -(x + 2)^2 & \text{for } x \geq 0 \end{cases}$

49. $\lim\limits_{x \to -2} f(x)$, where $f(x) = \begin{cases} -1 + 4x & \text{for } x < -2 \\ -9 & \text{for } x > -2 \end{cases}$

50. $\lim\limits_{x \to -1} f(x)$, where $f(x) = \begin{cases} \dfrac{2x + 1}{3x - 1} & \text{for } x < -1 \\ \dfrac{1}{(x - 1)^2} & \text{for } x > -1 \end{cases}$

51. $\lim\limits_{x \to 3^+} f(x)$, where $f(x) = \begin{cases} -\frac{1}{2}x & \text{for } x \leq 3 \\ 3 + \dfrac{1}{x - 3} & \text{for } x > 3 \end{cases}$

52. $\lim\limits_{x \to 0} f(x)$, where $f(x) = \begin{cases} \dfrac{\sin x}{x} & \text{for } x < 0 \\ \dfrac{1 - \cos x}{x} & \text{for } x > 0 \end{cases}$

In Exercises 53–70 find all vertical asymptotes (if any) of the graph of f.

53. $f(x) = \dfrac{1}{x + 4}$

54. $f(x) = \dfrac{x + 2}{x - 3}$

55. $f(x) = \dfrac{2x - 9}{x + \sqrt{2}}$

56. $f(x) = \dfrac{-3x}{x^2 + 9}$

57. $f(x) = \dfrac{x^2 - 1}{x^2 - 4}$

58. $f(x) = \dfrac{\sin x}{x(x^2 - 1)}$

59. $f(x) = \dfrac{-5x + \frac{2}{3}}{(x^2 - 16)(x^2 + 1)}$

60. $f(x) = \dfrac{x}{(x + 1)(x + 2)(x + 3)}$

61. $f(x) = \dfrac{x^2 - 4x - 12}{x^2 - x - 6}$

62. $f(x) = \dfrac{x^2 - 5x + 6}{x^3 - 8}$

63. $f(x) = \dfrac{x^2 + 2x - 15}{x^3 + 7x^2 + 10x}$

64. $f(x) = \dfrac{x + 1}{|x - 1|}$

65. $f(x) = \dfrac{x + 1/x}{x^4 + 1}$

66. $f(x) = \sqrt{\dfrac{x}{x - 1}}$

67. $f(x) = \dfrac{\sin x}{x}$

68. $f(x) = \dfrac{\cos x}{x}$

69. $f(x) = \tan x$

70. $f(x) = \csc x$

In Exercises 71–74 show that f has a vertical tangent line at the given point. Find an equation of the tangent line l.

71. $f(x) = x^{1/5}$; $(0, 0)$

72. $f(x) = (x + 1)^{1/3}$; $(-1, 0)$

73. $f(x) = 1 - 5x^{3/5}$; $(0, 1)$

74. $f(x) = (4 - x)^{3/7}$; $(4, 0)$

75. For each of the six graphs shown in Figure 2.35, decide whether the two-sided limit, the right-hand limit, or the left-hand limit at 2 exists, or whether none of these limits exists.

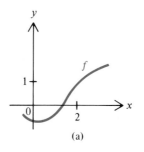

(a)

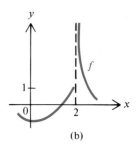

(b)

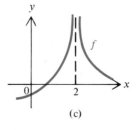

(c)

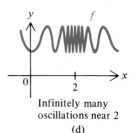

Infinitely many oscillations near 2

(d)

FIGURE 2.35

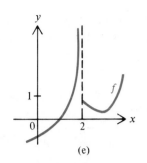

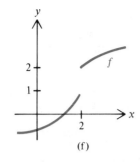

(e)   (f)

FIGURE 2.35 (*cont.*)

76. Show that for each integer n,

$$\lim_{x \to n^-} [x] = n - 1 \quad \text{and} \quad \lim_{x \to n^+} [x] = n$$

77. Suppose $\lim_{x \to a} f(x) = \infty$ and $\lim_{x \to a} g(x) = \infty$. Show by an example that we cannot conclude that

$$\lim_{x \to a} [f(x) - g(x)] = 0$$

78. Let $f(x) = x^{1/n}$, where n is a positive odd integer. Show that the graph of f has a vertical tangent at $(0, 0)$.

79. As in Exercise 54 of Section 2.3, let

$$f(x) = \frac{x}{a + bx} \quad \text{for } 0 \le x \le 1$$

where a and b are positive. Find $\lim_{x \to 1^-} f(x)$.

80. Suppose a rock is dropped into a well 64 feet deep. Determine the velocity of the rock when it hits the bottom of the well.

81. In Exercise 80, suppose the rock were thrown downward with initial velocity -48 feet per second. Determine the velocity of the rock when it crashes into the bottom of the well.

82. Before its booster rocket separates, a rocket's position in miles above the earth is given by $f(t) = -t^4 + 4t^3 + 8t$, where t denotes time in minutes. Assuming that the booster rocket separates after 2 minutes of flight, determine how fast the rocket is rising at that instant.

2.6
CONTINUITY

Let f be a function defined at an arbitrary number a. Then there are three contrasting possibilities for the behavior of f near a:

1. $\lim_{x \to a} f(x)$ does not exist
2. $\lim_{x \to a} f(x)$ exists, but $\lim_{x \to a} f(x) \ne f(a)$
3. $\lim_{x \to a} f(x)$ exists, and $\lim_{x \to a} f(x) = f(a)$

Figure 2.36 reflects these possibilities. Notice that in Figure 2.36(a) and (b) the

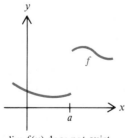

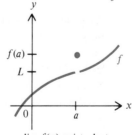

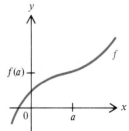

(a)   (b)   (c)

FIGURE 2.36

graphs appear to be broken at a, whereas in Figure 2.36(c) the graph appears to be unbroken at a, with $f(x)$ approaching $f(a)$ as x approaches a. This third type of behavior is of great importance in calculus.

DEFINITION 2.8

A function f is **continuous** at a point a in its domain if

$$\lim_{x \to a} f(x) = f(a)$$

A function f is **discontinuous** at a point a in its domain if f is not continuous at a.

The word "continuous" means much the same in mathematics as in everyday language. Thus we speak of a "continuous hum," in contrast to a hum that is interrupted or suddenly altered in some way.

To show that a function f is continuous at a, we must verify (1) that f is defined at a, (2) that $\lim_{x \to a} f(x)$ exists, and (3) that $\lim_{x \to a} f(x) = f(a)$. Because continuity is defined in terms of limits, we can gain information about continuity from the various limit theorems already stated. For example, earlier we stated that if a is in the domain of the rational function f, then $\lim_{x \to a} f(x) = f(a)$ (see (5) of Section 2.3). Thus any rational function is continuous at every point in its domain. Similarly, each of the six trigonometric functions $\sin x$, $\cos x$, $\tan x$, and so on is continuous at every point in its domain (see Example 1 of Section 2.4 and the remarks following that example).

Example 1   Let $f(x) = (x^2 - 5x + 4)/(x^2 - 9)$. Determine the numbers at which f is continuous.

Solution   Observe that f is a rational function. Since the denominator of f is 0 for $x = 3$ and $x = -3$, f is defined for all x except 3 and -3. Consequently f is continuous at every number except 3 and -3.   □

Example 2   Let $f(x) = (x^2 - 3x + 2)/(x^2 + 5x - 6)$. Determine the numbers at which f is continuous.

Solution   Again f is a rational function. The denominator is 0 for $x = 1$ and $x = -6$, so f is defined for all x except 1 and -6. Therefore f is continuous at every number except 1 and -6.   □

Example 3   Let $f(x) = x^2/(1 + x^2)$. Determine the numbers at which f is continuous.

Solution   Once again f is a rational function, but its denominator $1 + x^2$ is never 0. Thus f is defined for all x, and therefore f is continuous at every real number.   □

Results on the continuity of combinations of functions follow immediately from the corresponding results for limits.

THEOREM 2.9

> Suppose f and g are continuous at a and c is any number. Then $f + g$, cf, and fg are continuous at a. If $g(a) \neq 0$, then f/g is continuous at a.

Proof   By the Sum Rule, we obtain

$$\lim_{x \to a} (f + g)(x) = \lim_{x \to a} [f(x) + g(x)]$$

$$= \lim_{x \to a} f(x) + \lim_{x \to a} g(x) = f(a) + g(a) = (f + g)(a)$$

Consequently $f + g$ is continuous at a. The remaining parts of the proof follow similarly from the limit rules and the definition of continuity.   ∎

Example 4   Let $f(x) = x \sin x + 1$. Show that f is continuous at every number.

Solution   We know that the functions x, $\sin x$, and 1 are continuous at every number. Since f is obtained from these functions by forming first the product $x \sin x$ and then the sum $x \sin x + 1$, we know by Theorem 2.9 that f is continuous at every number.   □

There is a result analogous to Theorem 2.9 for the composite of two functions.

THEOREM 2.10

> If f is continuous at a and g is continuous at $f(a)$, then $g \circ f$ is continuous at a.

Proof   Since f is continuous at a and since g is continuous at $f(a)$, we have

$$\lim_{x \to a} f(x) = f(a) \quad \text{and} \quad \lim_{y \to f(a)} g(y) = g(f(a))$$

If we let $y = f(x)$ and $c = f(a)$ in the Substitution Rule, we find that

$$\lim_{x \to a} g(f(x)) = \lim_{y \to f(a)} g(y) = g(f(a))$$

which means that $g \circ f$ is continuous at a.   ∎

Example 5   Let $h(x) = \sqrt{x - 1}$. Show that h is continuous at 2.

Solution   Let $f(x) = x - 1$ and $g(y) = \sqrt{y}$. Then $h = g \circ f$. We know that f is continuous at 2 and that g is continuous at $f(2) = 1$ since the square root function is continuous at every positive number. It follows from Theorem 2.10 that h is continuous at 2. $\square$

Functions that are continuous at various points in their domains abound in mathematics, as well as in applications. It is often important that a function be continuous wherever possible; in fact, if $\lim_{x \to a} f(x)$ exists but $f(a)$ is either not defined or not equal to $\lim_{x \to a} f(x)$, then mathematicians sometimes redefine f so that it is continuous at a.

Example 6 Let $f(x) = (\sin x)/x$. Define a function f_0 such that f_0 is continuous, and $f_0(x) = f(x)$ for all $x \neq 0$.

Solution Since

$$\lim_{x \to 0} f(x) = \lim_{x \to 0} \frac{\sin x}{x} = 1$$

let

$$f_0(x) = \begin{cases} \dfrac{\sin x}{x} & \text{for } x \neq 0 \\ 1 & \text{for } x = 0 \end{cases}$$

Then f_0 is continuous at 0 since $\lim_{x \to 0} f_0(x) = 1 = f_0(0)$. Furthermore $f_0(x) = f(x)$ for $x \neq 0$, as was required of f_0. $\square$

One-Sided Continuity

In (8) of Section 2.3 we stated that

$$\lim_{x \to a} \sqrt{x} = \sqrt{a} \quad \text{for } a > 0$$

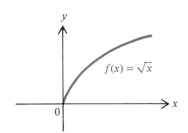

$f(x) = \sqrt{x}$

FIGURE 2.37

This means that the square root function is continuous at every positive number (Figure 2.37). However, as we observed in Section 2.5, $\lim_{x \to 0} \sqrt{x}$ does not exist. As a consequence, under Definition 2.8 the square root function is not continuous at 0. However, it has a right-hand limit at 0. In fact

$$\lim_{x \to 0^+} \sqrt{x} = 0 = \sqrt{0}$$

We express this fact by saying that the square root function is continuous from the right at 0.

DEFINITION 2.11

A function f is **continuous from the right** at a point a in its domain if

$$\lim_{x \to a^+} f(x) = f(a)$$

A function f is **continuous from the left** at a point a in its domain if

$$\lim_{x \to a^-} f(x) = f(a)$$

Since one-sided limits are as easy to determine as two-sided limits, we can ascertain the one-sided continuity of most of the functions we know.

Example 7 Show that the diving board function, defined by

$$f(x) = \begin{cases} 0 & \text{for } x < 0 \\ 1 & \text{for } x \ge 0 \end{cases}$$

is continuous from the right at 0 but not continuous from the left at 0.

Solution Since

$$\lim_{x \to 0^+} f(x) = 1 = f(0)$$

f is continuous from the right at 0. Since

$$\lim_{x \to 0^-} f(x) = 0 \ne f(0)$$

f is not continuous from the left at 0 (Figure 2.38). □

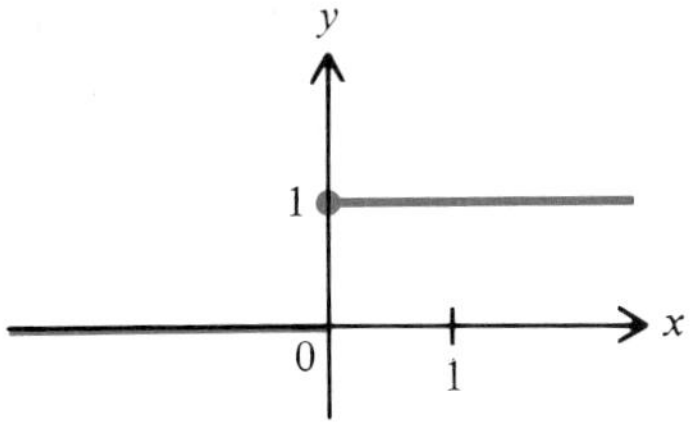

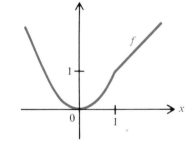

Diving board function:
$$f(x) = \begin{cases} 0 \text{ for } x < 0 \\ 1 \text{ for } x \ge 0 \end{cases}$$

FIGURE 2.38

We observe that a function is continuous at a if and only if it is both continuous from the right and continuous from the left at a. This follows from Theorem 2.5.

Example 8 Let

$$f(x) = \begin{cases} x^2 & \text{for } x \le 1 \\ x & \text{for } x > 1 \end{cases}$$

Show that f is continuous at 1.

Solution Since

$$\lim_{x \to 1^-} f(x) = \lim_{x \to 1^-} x^2 = 1 = f(1)$$

and

$$\lim_{x \to 1^+} f(x) = \lim_{x \to 1^+} x = 1 = f(1)$$

f is continuous from the left and from the right at 1. Therefore f is continuous at 1 (see Figure 2.39). □

FIGURE 2.39

Continuity on Intervals

In the remainder of this book we will almost always be concerned with functions that are continuous on intervals. It is therefore essential that we know precisely what that property means.

DEFINITION 2.12

> **a.** A function is **continuous on an open interval** (a, b), or simply **continuous on** (a, b), if it is continuous at every point in (a, b).
> **b.** A function is **continuous on a closed interval** $[a, b]$, or simply **continuous on** $[a, b]$, if it is continuous at every point in (a, b) and is also continuous from the right at a and continuous from the left at b.

There are analogous definitions of continuity on a half-open interval or on an interval of the form (a, ∞), $[a, \infty)$, $(-\infty, a)$, $(-\infty, a]$, or $(-\infty, \infty)$. For example, the square root function is continuous on $[0, \infty)$, because it is continuous at every point in $(0, \infty)$ and continuous from the right at 0. Moreover, every polynomial is continuous on $(-\infty, \infty)$. In addition, every rational function, every trigonometric function, and every function of the form $x^{m/n}$ is continuous on any open interval throughout which it is defined.

Example 9   Let $f(x) = \sqrt{4 - x^2}$. Show that f is continuous on $[-2, 2]$.

Solution   Since $\sqrt{x}$ is defined for $x \geq 0$, and since $4 - x^2 \geq 0$ if and only if $-2 \leq x \leq 2$, it follows that the domain of f is $[-2, 2]$. Next we observe that

$$\lim_{x \to a} f(x) = \lim_{x \to a} \sqrt{4 - x^2} = \sqrt{4 - a^2} = f(a) \quad \text{if} \quad -2 < a < 2$$

$$\lim_{x \to 2^-} f(x) = \lim_{x \to 2^-} \sqrt{4 - x^2} = \sqrt{4 - 2^2} = 0 = f(2)$$

$$\lim_{x \to -2^+} f(x) = \lim_{x \to -2^+} \sqrt{4 - x^2} = \sqrt{4 - (-2)^2} = 0 = f(-2)$$

Consequently f is continuous on $[-2, 2]$. $\square$

Many of the functions encountered in applications are defined and continuous on closed intervals, as the following five examples illustrate:

1. Degrees Celsius as a function of degrees Fahrenheit:

$$C(x) = \frac{5}{9}(x - 32) \quad \text{for } x \geq -273.15$$

2. The volume of a sphere as a function of its radius:

$$V(r) = \frac{4}{3}\pi r^3 \quad \text{for } r \geq 0$$

3. The speed of a tennis ball tossed up for a serve as a function of time:

$$v(t) = 16 - 32t \quad \text{for } 0 \leq t \leq \tfrac{1}{2}$$

4. The response of an organ to a drug as a function of the strength of the drug:

$$R(x) = \frac{x}{a + bx} \quad \text{for } 0 \leq x \leq 1$$

     where a and b are positive.

5. The cost of production as a function of the amount of production:

$$C(x) = 4\sqrt{x + 3} \quad \text{for } 0 \leq x \leq 50$$

Sometimes, especially in business and economics, functions are idealized so as to be continuous and thus easier to work with. For instance, the cost function in the fifth example might actually be defined only for integer values of x. But for purposes of analysis, the domain of the function may be considered an entire interval, with the formula defining the function extended to all numbers in that interval.

EXERCISES 2.6

In Exercises 1–16 determine whether f is continuous or discontinuous at a. If f is discontinuous at a, determine whether f is continuous from the right at a, continuous from the left at a, or neither.

1. $f(x) = x^2 - 4x + 3; a = 2$

2. $f(x) = \dfrac{1}{2x - 5x^3}; a = -1$

3. $f(x) = \dfrac{x^4 + x^2 - 2}{x^2 - 1}; a = 0$

4. $f(x) = |3x - 1|; a = \frac{1}{3}$

5. $f(x) = \begin{cases} 3 & \text{for } x \neq -1 \\ 0 & \text{for } x = -1 \end{cases} \quad a = -1$

6. $f(x) = \begin{cases} \dfrac{|x - 4|}{x - 4} & \text{for } x \neq 4 \\ 1 & \text{for } x = 4 \end{cases} \quad a = 0, 4$

7. $f(x) = \begin{cases} -1 & \text{for } x < 0 \\ 0 & \text{for } x = 0 \\ 1 & \text{for } x > 0 \end{cases} \quad a = 0$

8. $f(x) = \sqrt{x - 2}; a = 2$

9. $f(t) = \sqrt{1 - t}; a = \frac{1}{2}, 1$

10. $f(t) = \dfrac{\sqrt{t^2 - 4}}{t + 2}; a = 2$

11. $f(t) = t^2 \sqrt{t^2 - t^4}; a = 0, 1$

12. $f(z) = \tan \sqrt[3]{z}; a = 0$

13. $f(z) = \tan \sqrt{z - \pi/2}; a = \pi/2$

14. $f(z) = \sqrt{\cos z}; a = \pi/2$

15. $f(z) = \begin{cases} \dfrac{1}{1 + \tan z} & \text{for } 0 < z < \pi/2 \\ 0 & \text{for } z \geq \pi/2 \end{cases} \quad a = \pi/2$

*16. $f(t) = \sqrt{\sqrt{t + 2} - t}; a = 2$

In Exercises 17–30 explain why f is continuous on the given interval or intervals.

17. $f(x) = \sqrt{2 - \pi x^4 + x^{10}}; (-\infty, \infty)$

18. $f(x) = \dfrac{1 - x}{1 + x^2}; (-\infty, \infty)$

19. $f(x) = \sqrt{x + 3}; [-3, \infty)$

20. $f(x) = \dfrac{\sin x}{x}; (-\infty, 0), (0, \infty)$

21. $f(x) = \begin{cases} \dfrac{\sin x}{x} & \text{for } x \neq 0 \\ 1 & \text{for } x = 0 \end{cases} \quad (-\infty, \infty)$

22. $f(x) = \sqrt{x^2 - x^4}; [-1, 1]$

23. $f(x) = \dfrac{1}{\sqrt{x^2 - x^4}}; (0, 1)$

24. $f(x) = \sqrt{16x^4 - x^2}; (-\infty, -\frac{1}{4}], [\frac{1}{4}, \infty)$

25. $f(x) = \sqrt{\dfrac{1 - x}{3x - 2}}; (\frac{2}{3}, 1]$

26. $f(x) = \dfrac{\sqrt{x - 1}}{\sqrt{2 - x}}; [1, 2)$

27. $f(x) = \tan x; (-\pi/2, \pi/2)$

28. $f(x) = \tan\left(\dfrac{1}{x^2 + 1}\right); (-\infty, \infty)$

29. $f(x) = \cot(1/x); (1/\pi, \infty)$

30. $f(x) = \sec \sqrt{x}; [0, \pi^2/4)$

31. Determine for which of the following functions f we can define $f(a)$ so as to make f continuous at a. For those we can, find the value of $f(a)$ that makes f continuous at a.

a. $f(x) = \dfrac{x^2 - 9}{x - 3}; a = 3$

b. $f(x) = \dfrac{x^2 + 5x + 6}{x + 2}; a = -2$

c. $f(x) = \dfrac{x^2 + 5x + 4}{x - 1}; a = 1$

d. $f(x) = \dfrac{|x|}{x}; a = 0$

e. $f(x) = \dfrac{1 - \cos x}{x}; a = 0$

f. $f(x) = \sec x; a = \pi/2$

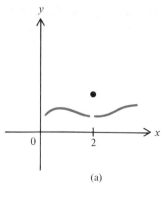

(a)

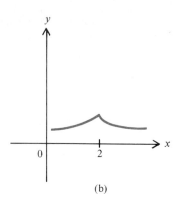

(b)

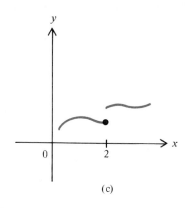

(c)

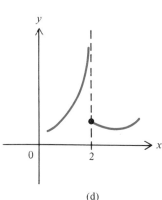

(d)

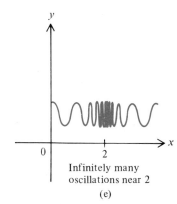

Infinitely many
oscillations near 2

(e)

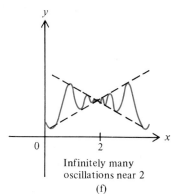

Infinitely many
oscillations near 2

(f)

FIGURE 2.40

32. Determine for which of the following functions f we can redefine $f(a)$ so as to make f continuous at a.

a. $f(x) = \begin{cases} -1 & \text{for } x \neq 2 \\ 1 & \text{for } x = 2 \end{cases} \quad a = 2$

b. $f(x) = \begin{cases} 2x - 3 & \text{for } x < 1 \\ 0 & \text{for } x = 1 \\ 3x - 5 & \text{for } x > 1 \end{cases} \quad a = 1$

c. $f(x) = \begin{cases} x^2 + 1 & \text{for } x < -1 \\ 5 & \text{for } x = -1 \\ x^3 - 6 & \text{for } x > -1 \end{cases} \quad a = -1$

33. Which of the graphs in Figure 2.40 are graphs of functions that are continuous at 2?

34. The greatest integer function f is defined by $f(x) = [x]$, where $[x]$ is the greatest integer not larger than x. Determine where f is discontinuous.

35. Let $g(x) = 3x + 4$ and $h(x) = g(x + 2)$. Using Theorem 2.10, show that h is continuous at -3.

36. Suppose g is continuous at a. Let $h(x) = g(x + b)$ for all x such that $x + b$ is in the domain of g. Show that h is continuous at $a - b$.

37. Prove that if f and g are both continuous at a, then $f - g$ is also continuous at a.

38. Show that the statement "f is continuous at a" is equivalent to the following condition:

For every $\varepsilon > 0$ there is a number $\delta > 0$ such that
if $|x - a| < \delta$, then $|f(x) - f(a)| < \varepsilon$

*39. Let

$$f(x) = \begin{cases} 0 & \text{for } x \text{ rational} \\ 1 & \text{for } x \text{ irrational} \end{cases}$$

Show that f is not continuous at *any* real number. (*Hint:* Every open interval contains both rational and irrational numbers.)

*40. Let

$$f(x) = \begin{cases} 0 & \text{for } x \text{ rational} \\ x & \text{for } x \text{ irrational} \end{cases}$$

Show that f is continuous *only* at 0.

41. Suppose that there is a nonzero number M such that $|f(x) - f(a)| \leq M|x - a|$ for all x in an open interval about a. Show that f is continuous at a.

42. The kinetic energy of a body with mass m is given by

$$K(t) = \tfrac{1}{2}m[v(t)]^2$$

where $v(t)$ is the velocity of the body at time t. Suppose $v(t) = 50/(1 + t^2)$ for all $t \geq 0$. Show that K is continuous on $[0, \infty)$.

43. The speed $S(t)$ of sound in air at t degrees Celsius is given

by

$$S(t) = \frac{c(273.15 + t)}{273.15} \quad \text{for } t \geq -273.15$$

where $c = 1088$, and $S(t)$ is measured in feet per second. Show that S is continuous on $[-273.15, \infty)$.

2.7
THE INTERMEDIATE VALUE THEOREM

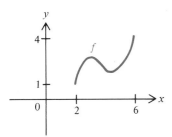

FIGURE 2.41

This final section of Chapter 2 is devoted to an important theorem about continuous functions—the Intermediate Value Theorem. As we will see, the Intermediate Value Theorem can be used to show that every nonnegative number has a square root; it can also be used to determine where continuous functions are positive, where they are negative, and where they are zero.

The Intermediate Value Theorem, which is proved in the Appendix, asserts that the range of a function that is continuous on a closed interval cannot skip any intermediate values. For example, if a function f is continuous throughout the interval $[2, 6]$, and if $f(2) = 1$ and $f(6) = 4$, then every number between 1 and 4 must be in the range of f (see Figure 2.41).

THEOREM 2.13
INTERMEDIATE VALUE THEOREM

Suppose f is continuous on a closed interval $[a, b]$. Let p be any number between $f(a)$ and $f(b)$, so that $f(a) \leq p \leq f(b)$ or $f(b) \leq p \leq f(a)$. Then there exists a number c in $[a, b]$ such that $f(c) = p$.

Pictorially, the Intermediate Value Theorem asserts that the graph of a continuous function cannot jump over a horizontal line (Figure 2.42(a)). In physical terms, you might think of the Intermediate Value Theorem as saying that you cannot cross an infinitely long river without either getting wet or jumping over it (Figure 2.42(b)).

The Intermediate Value Theorem implies that the domain of the square root function consists of *all* nonnegative numbers, as asserted in Chapter 1 when the

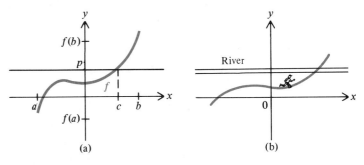

(a)   (b)

FIGURE 2.42

function was defined. To prove the assertion, we select any nonnegative number p and show that there is a number c such that $c \geq 0$ and $c^2 = p$. To that end, we let $f(x) = x^2$ for $x \geq 0$ and observe that f is continuous and

$$f(0) = 0 \leq p \leq p^2 + 2p + 1 = (p + 1)^2 = f(p + 1)$$

Thus $f(0) \leq p \leq f(p + 1)$, so the Intermediate Value Theorem says that there is a number c in $[0, p + 1]$ such that $f(c) = p$, or equivalently, $c^2 = p$. Thus p has a square root. Since p was an arbitrarily chosen nonnegative number, it follows that the square root is defined for every nonnegative number. Similar comments apply to the other root functions.

In the Intermediate Value Theorem the hypothesis that f is continuous is crucial. For example, if

$$f(x) = \begin{cases} 0 & \text{for } -2 \leq x < 0 \\ 1 & \text{for } 0 \leq x \leq 2 \end{cases}$$

then the range of f does not contain all numbers between 0 and 1; in fact, it contains none of them. Of course, f is not continuous on $[-2, 2]$ because it is not continuous at 0, and this is why its graph can "skip" the values between 0 and 1.

Positive and Negative Values of Functions

The consequences of the Intermediate Value Theorem that we will discuss in this and the next subsection involve numbers d of a function f such that $f(d) = 0$. Such a number is called a **zero**, or **root**, of f. For example, let

$$f(x) = x^2 - 4x - 5 = (x - 5)(x + 1)$$

Then $f(x) = 0$ for $x = -1$ and for $x = 5$. Consequently the zeros of f are -1 and 5.

Zeros of functions of the form

$$f(x) = ax^2 + bx + c \qquad \text{where } a \neq 0$$

can be located by means of the quadratic formula (see Exercise 46 in Section 1.3), but for higher-degree polynomials and functions in general there is no simple formula from which we can determine a zero. This is unfortunate, because if the function f happens to be a profit function, then the zeros of f would represent the "break-even" points at which there is neither profit nor loss, and it might be desirable to know these points.

Now let f be continuous on an interval I. If f has both positive and negative values on I, then the Intermediate Value Theorem implies that $f(x) = 0$ for some x in I, that is, f has a zero in I. Equivalently, if f has *no* zero in I, then either $f(x) > 0$ for all x in I or $f(x) < 0$ for all x in I. This fact yields a procedure for discovering the intervals on which a continuous function f is positive, and those on which f is negative.

Example 1   Let $f(x) = (x + 1)^2(x - 2)(x - 3)$. Determine the intervals on which f is positive and those on which f is negative.

Solution   Evidently the zeros of f are $-1, 2,$ and 3, so we will determine the sign of $f(x)$ on each of the intervals $(-\infty, -1), (-1, 2), (2, 3),$ and $(3, \infty)$:

Interval	c in interval	$f(c)$	Sign of $f(x)$ on interval
$(-\infty, -1)$	-2	20	$+$
$(-1, 2)$	0	6	$+$
$(2, 3)$	$\dfrac{5}{2}$	$-\dfrac{49}{16}$	$-$
$(3, \infty)$	4	50	$+$

Consequently f is positive on $(-\infty, -1), (-1, 2)$, and $(3, \infty)$, and is negative on $(2, 3)$. $\square$

The method used in Example 1 applies to a function such as a rational function, even if it is not defined on the whole real line $(-\infty, \infty)$.

Example 2 Let

$$f(x) = \frac{x(x-2)(x^2 + 2x + 4)}{x^3 + 1}$$

Determine the intervals on which f is positive and those on which f is negative.

Solution Since $x^3 + 1 = 0$ if $x = -1$, the domain of f is the union of $(-\infty, -1)$ and $(-1, \infty)$. Next we observe that

$$x^2 + 2x + 4 = (x^2 + 2x + 1) + 3 = (x + 1)^2 + 3 > 0 \qquad \text{for all } x$$

so that $f(x) = 0$ only if $x = 0$ or $x = 2$. Therefore we will determine the sign of $f(x)$ on each of the intervals $(-\infty, -1), (-1, 0), (0, 2)$, and $(2, \infty)$:

Interval	c in interval	$f(c)$	Sign of $f(x)$ on interval
$(-\infty, -1)$	-2	$-\dfrac{32}{7}$	$-$
$(-1, 0)$	$-\dfrac{1}{2}$	$\dfrac{65}{14}$	$+$
$(0, 2)$	1	$-\dfrac{7}{2}$	$-$
$(2, \infty)$	3	$\dfrac{57}{28}$	$+$

Consequently f is positive on $(-1, 0)$ and $(2, \infty)$, and is negative on $(-\infty, -1)$ and $(0, 2)$. $\square$

Finding the intervals of which a function f is positive (or negative) is equivalent to solving the inequality $f(x) > 0$ (or $f(x) < 0$), so one can consider Example 2 as requesting us to determine the solutions of the inequalities

$$\frac{x(x-2)(x^2+2x+4)}{x^3+1} > 0 \quad \text{and} \quad \frac{x(x-2)(x^2+2x+4)}{x^3+1} < 0$$

Thus in addition to the method of solving inequalities discussed in Section 1.1 we have now a second method, which utilizes the Intermediate Value Theorem.

The Bisection Method

Suppose the function g represents the profit from the sale of grain, and is defined by

$$g(x) = x^5 + x^3 + x^2 - 1 \quad \text{for } x \text{ in } [0, 1]$$

where $g(x)$ is in thousands of dollars and x is measured in tons. Since g is continuous, and since $g(0) = -1$ and $g(1) = 2$, the Intermediate Value Theorem assures us that there is a number d in $[0, 1]$ such that $g(d) = 0$. The number d represents the break-even point at which there is neither profit nor loss. In trying to find d we could attempt to factor $g(x)$. But factoring $g(x)$ is futile, so finding the precise solution of the equation $g(x) = 0$ seems impossible. Thus we must be satisfied with approximating d. One method of doing so is called the bisection method.

The bisection method is based on the ancient Roman proverb "divide and conquer," and applies to any function f that is continuous on an interval $[a, b]$ and has the property that $f(a)$ and $f(b)$ have different signs (so either $f(a) < 0$ and $f(b) > 0$, or $f(a) > 0$ and $f(b) < 0$).

Because f is continuous on $[a, b]$, and $f(a)$ and $f(b)$ have different signs, the Intermediate Value Theorem implies that f has a zero in $[a, b]$. Let c be the midpoint of $[a, b]$. If $f(c) = 0$, then c is a zero of f in $[a, b]$, and the search stops. However, if $f(c) \neq 0$, then

either $f(a)$ and $f(c)$ have different signs, so let $I_1 = [a, c]$

or $f(c)$ and $f(b)$ have different signs, so let $I_1 = [c, b]$

Whichever half of $[a, b]$ becomes I_1, we know by the Intermediate Value Theorem that I_1 contains a zero of f, since $f(x)$ has different signs at the endpoints of I_1. Continuing in the same way, we obtain successively smaller subintervals $I_2, I_3, \ldots$, each containing a zero of f. Because the process involves bisecting intervals, it is called the **bisection method** for approximating a zero of f.

The process could be continued indefinitely, so we must have a way of knowing when to stop, that is, when we have found a number c^* sufficiently close to a zero of f. Let $\varepsilon > 0$, and suppose we wish to find a number c^* that approximates a zero d of f with an error less than ε, that is, we wish

$$|c^* - d| < \varepsilon$$

Since each succeeding subinterval is half as long as the preceding subinterval, eventually we obtain a subinterval I_n whose length is less than 2ε. Then the midpoint of I_n can serve as c^*, because d is in I_n and thus

$$|c^* - d| \leq \tfrac{1}{2}(\text{length of } I_n) < \tfrac{1}{2}(2\varepsilon) = \varepsilon$$

Therefore in applying the bisection method we perform a sequence of steps, called an **algorithm**, that stops when we achieve the desired accuracy. The algorithm can be described as follows:

> INPUT: A continuous function f, numbers a and b such that $f(a)$ and $f(b)$ have opposite signs, and an allowable error, or tolerance, ε.
>
> OUTPUT: A number c that approximates a zero of f with error less than ε.
>
> STEP 1: Let the interval I be $[a, b]$.
>
> STEP 2: Let c be the midpoint of I.
>
> STEP 3: Compute the length L of I. If $L \leq 2\varepsilon$, then c is an approximate zero of f with error $< \varepsilon$, so STOP. Otherwise CONTINUE.
>
> STEP 4: Compute $f(c)$. If $f(c) = 0$, then c is a zero of f, so STOP. Otherwise CONTINUE.
>
> STEP 5: If $f(a)$ and $f(c)$ have different signs, replace I by $[a, c]$. Otherwise replace I by $[c, b]$.
>
> STEP 6: Return to Step 2, using the new interval I.

Eventually the process stops (with either Step 3 or Step 4). When the process does stop, the value of c approximates a zero of f with error less than ε.

Example 3   Let $f(x) = x^5 + x^3 + x^2 - 1$. Use the bisection method to find a number in $[0, 1]$ that approximates a zero of f with an error less than $\frac{1}{16}$.

Solution   Since $f(0) = -1$ and $f(1) = 2$, we can begin by letting $a = 0$ and $b = 1$. Using the algorithm, we assemble the following table and the accompanying graph (Figure 2.43):

Interval	Length	Midpoint c	$f(c)$
$[0, 1]$	1	$\dfrac{1}{2}$	$-\dfrac{19}{32}$
$\left[\dfrac{1}{2}, 1\right]$	$\dfrac{1}{2}$	$\dfrac{3}{4}$	$\dfrac{227}{1024}$
$\left[\dfrac{1}{2}, \dfrac{3}{4}\right]$	$\dfrac{1}{4}$	$\dfrac{5}{8}$	$-\dfrac{8843}{32,768}$
$\left[\dfrac{5}{8}, \dfrac{3}{4}\right]$	$\dfrac{1}{8}$		

Since the length of $\left[\frac{5}{8}, \frac{3}{4}\right]$ is $\frac{1}{8}$, and neither $\frac{5}{8}$ nor $\frac{3}{4}$ is a zero of f, the midpoint $\frac{11}{16}$ of $\left[\frac{5}{8}, \frac{3}{4}\right]$ is less than $\frac{1}{16}$ from a zero of f. $\square$

FIGURE 2.43

EXERCISES 2.7

In Exercises 1–8 show that the equation has at least one solution in the given interval.

1. $x^4 - x - 1 = 0$; $[-1, 1]$ (*Hint:* Let $f(x) = x^4 - x - 1$, and apply the Intermediate Value Theorem to f on $[-1, 1]$.)

2. $x^3 - x - 5 = 0$; $[0, 2]$

3. $x^2 + \dfrac{1}{x} = 1$; $[-2, -\frac{1}{2}]$

4. $x^2 + \dfrac{1}{2x} = 2$; $[\frac{1}{4}, 1]$

5. $x^3 + x^2 + x - 2 = 0$; $[-1, 1]$

6. $x^{7/3} + x^{5/3} - 1 = 0$; $[-1, 1]$

7. $\cos x = x$; $\left[0, \dfrac{\pi}{2}\right]$

8. $\sin x = 1 - x$; $\left[0, \dfrac{\pi}{6}\right]$

In Exercises 9–16 solve the inequality for x.

9. $(2 - x)^2(40 - 8x) > 0$

10. $(x + 1)(x - 2)(2x - \frac{1}{4}) \ge 0$

11. $(x^4 + x)(x + 3) \le 0$

12. $\dfrac{x(x - 4)}{2(x - 2)^2} < 0$

13. $\dfrac{(x - 1)(x - 3)}{(2x + 1)(2x - 1)} \ge 0$

14. $\dfrac{6x^2 - 2}{(x^2 - 1)^3} > 0$

15. $\dfrac{-2(x - 1)(x^2 + 2x + 4)}{27 - x^3} < 0$

16. $\dfrac{3x(x^3 - 2)}{(x^3 + 1)^3} \le 0$

In Exercises 17–22 use the bisection method to approximate a zero of f with an error less than $\frac{1}{16}$.

17. $f(x) = x^2 - 2$ (Your answer is an approximation to $\sqrt{2}$.)

18. $f(x) = x^2 - 7$

19. $f(x) = x^3 - 3$

20. $f(x) = x^3 + x - 3$

c 21. $f(x) = \cos x - x$

c 22. $f(x) = \tan x - 2x$ for $0 < x < \pi/2$

In Exercises 23–26 use the bisection method to find an approximation of the given number with an error less than $\frac{1}{16}$.

23. $\sqrt{5}$ 24. $\sqrt{12}$ c 25. $\sqrt{0.7}$ c 26. $\sqrt[3]{10}$

27. Show that every real number has a cube root. (*Hint:* Follow the ideas in the text concerning the existence of square roots.)

28. Show that every real number is the tangent of a number in $(-\pi/2, \pi/2)$.

29. Use the Intermediate Value Theorem to show that among all circles with radius no larger than 10 centimeters, there is one whose area is 200 square centimeters.

30. Use the Intermediate Value Theorem to show that among all right circular cylinders of height 10 meters and radius of the base not exceeding 1 meter, there is one whose volume is 25 cubic meters.

Key Terms and Expressions

Limit
Right-hand limit
Left-hand limit

Tangent line
Velocity
Vertical asymptote

Key Formulas

$$\lim_{x \to a} [f(x) + g(x)] = \lim_{x \to a} f(x) + \lim_{x \to a} g(x)$$

$$\lim_{x \to a} [cf(x)] = c \lim_{x \to a} f(x)$$

$$\lim_{x \to a} [f(x)g(x)] = \lim_{x \to a} f(x) \lim_{x \to a} g(x)$$

$$\lim_{x \to a} \frac{f(x)}{g(x)} = \frac{\lim_{x \to a} f(x)}{\lim_{x \to a} g(x)} \quad \text{if } \lim_{x \to a} g(x) \ne 0$$

$$\lim_{x \to a} g(f(x)) = \lim_{y \to c} g(y), \text{ where } c = \lim_{x \to a} f(x)$$

Key Theorems

Sum Rule   Substitution Rule
Constant Multiple Rule   Squeezing Theorem
Product Rule   Intermediate Value Theorem
Quotient Rule

REVIEW EXERCISES

In Exercises 1–8 evaluate the limit.

1. $\lim\limits_{x \to 2} \dfrac{-4x + 3}{x^2 - 1}$

2. $\lim\limits_{x \to 1/2} \dfrac{2x - 1}{8x^2 - 4x}$

3. $\lim\limits_{x \to 1} x\sqrt{x + 3}$

4. $\lim\limits_{x \to 4} \dfrac{\sqrt{x} - 2}{4 - x}$

5. $\lim\limits_{x \to \pi/4} \sec^3\left(x + \dfrac{3\pi}{4}\right)$

6. $\lim\limits_{x \to 0} \dfrac{\tan^2 x}{x}$

7. $\lim\limits_{v \to 0} \dfrac{5\sin 6v}{4v}$

8. $\lim\limits_{v \to 0} \dfrac{\sin^6 v}{6v^6}$

In Exercises 9–16 determine whether the limit exists as a number, as ∞, or as $-\infty$. If the limit exists, evaluate it.

9. $\lim\limits_{x \to -10} \dfrac{4x}{(x + 10)^2}$

10. $\lim\limits_{x \to 4} \dfrac{2x - 8\sqrt{x} + 8}{\sqrt{x} - 2}$

11. $\lim\limits_{x \to 0^-} \sqrt{9 + \sqrt{-x}}$

12. $\lim\limits_{x \to 5} \dfrac{x^2 - 2x - 15}{x^2 - x - 20}$

13. $\lim\limits_{w \to 0} \dfrac{\sqrt{2 + 3w} - \sqrt{2 - 3w}}{w}$

14. $\lim\limits_{w \to 0^+} \dfrac{\cos\sqrt{w}}{\sin w}$

15. $\lim\limits_{w \to 0^+} \cos\dfrac{1}{w}$

16. $\lim\limits_{x \to 0^+} \dfrac{\sin x}{\sqrt{x}}$

In Exercises 17–22 find all vertical asymptotes (if any) of the graph of f.

17. $f(x) = \dfrac{3 - 2|x|}{x - 4}$

18. $f(x) = \dfrac{(x + 3)^2}{4 - 9x^2}$

19. $f(x) = \dfrac{x^2 + 3x - 4}{x^2 - 5x - 14}$

20. $f(x) = \dfrac{x^2 + x - 2}{x + 2}$

21. $f(x) = \dfrac{(x - 2)^2}{x^2 - 4}$

22. $f(x) = \dfrac{\tan x}{x^2}$

In Exercises 23–26 determine whether f is continuous or discontinuous at a. If f is discontinuous at a, determine whether f is continuous from the right at a, continuous from the left at a, or neither.

23. $f(x) = \sqrt{x^2 - 13};\ a = -\sqrt{13}$

*24. $f(x) = \sqrt{\sqrt{x + 3} - 2x};\ a = 1$

25. $f(x) = \begin{cases} 3x - 4 & \text{for } x < 2 \\ 3x + 4 & \text{for } x \geq 2 \end{cases} \quad a = 2$

26. $f(x) = \begin{cases} \sqrt{2 - x} & \text{for } x < 1 \\ 1 & \text{for } x = 1 \\ 2x^2 - x & \text{for } x > 1 \end{cases} \quad a = 1$

In Exercises 27–30 explain why f is continuous on the given interval.

27. $f(x) = \dfrac{x^2 + 6x - 16}{x - 2};\ (2, \infty)$

28. $f(x) = \sqrt{5 - 2x};\ (-\infty, \tfrac{5}{2}]$

29. $f(x) = \sqrt{\dfrac{3x - \sqrt{2}}{4 - x}};\ \left[\dfrac{\sqrt{2}}{3}, 4\right)$

30. $f(x) = \sec\dfrac{1}{1 + x^2};\ (-\infty, \infty)$

31. Let
$$f(x) = \dfrac{x^2 + 6x - 16}{x - 2}$$
Find the value for $f(2)$ that would make f continuous at 2.

32. Suppose the graph of f is as drawn in Figure 2.44, with domain $[0, 4.5]$.
    a. At which integers does f have a left-hand limit?
    b. At which integers does f have a right-hand limit?
    c. At which integers does f have a limit?
    d. At which integers is f continuous from the left?
    e. At which integers is f continuous from the right?
    f. At which integers is f continuous?

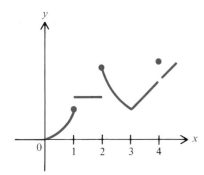

FIGURE 2.44

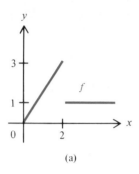

(a)

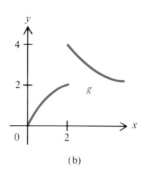

(b)

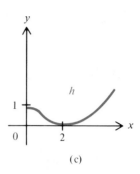

(c)

FIGURE 2.45

33. Consider the functions f, g, and h, whose graphs appear in Figure 2.45. In each of the following, determine whether the limit exists as a number, is ∞ or $-\infty$, or does not exist.
   a. $\lim_{x \to 2} (f(x) + g(x))$
   b. $\lim_{x \to 2} (f(x) - g(x))$
   c. $\lim_{x \to 2} f(x)g(x)$
   d. $\lim_{x \to 2} (f(x) + h(x))$
   e. $\lim_{x \to 2} f(x)h(x)$
   f. $\lim_{x \to 2} \dfrac{f(x)}{h(x)}$

In Exercises 34–35 solve the inequality for x.

34. $\dfrac{(2x - 3)(4 - x)}{(1 + x)^2} < 0$
35. $\dfrac{x + 4}{(x^2 - 4)(1 + x^2)} \geq 0$

ⓒ In Exercises 36–39 show that the equation has at least one solution in the given interval. Then use the bisection method to approximate the solution with an error less than $\frac{1}{16}$.

36. $x^3 + 3x - 3 = 0$; $[-1, 1]$ 37. $\sqrt{x + 1} = x^2$ on $[1, 2]$

38. $\tan x = \cos x$; $[0, \pi/4]$ 39. $\sin^2 x = \cos x$; $[0, \pi/2]$

40. Show that every number in $[-1, 1]$ is the sine of some number in $[-\pi/2, \pi/2]$.

41. What is wrong with the following statement? "The limit of the product of two functions is the product of the limits of the two functions."

42. a. Show that if $\lim_{x \to a} f(x) = L$, then

$$\lim_{x \to -a} f(-x) = L$$

   b. Suppose that f is either an even or an odd function. Using part (a), show that if f is continuous at a, then f is continuous at $-a$.

43. a. Show that if f is continuous at a, then so is $|f|$.
   b. Show that the converse of part (a) is false by letting

$$f(x) = \begin{cases} -1 & \text{for } x \leq 2 \\ 1 & \text{for } x > 2 \end{cases}$$

44. a. Show that if $\lim_{x \to a} f(x) = \infty$, then $\lim_{x \to a} 1/f(x) = 0$.
   b. Show by means of an example that the converse of part (a) is not true. (*Hint:* Choose a function that has both positive and negative values on every open interval about 0.)

45. Verify that the following pairs of functions have parallel tangent lines at the indicated points.
   a. $f(x) = 2 \sin x$ at $(0, 0)$ and $g(x) = -1/x^2$ at $(1, -1)$
   b. $f(x) = x^3$ at $(-1, -1)$ and $g(x) = \frac{3}{8}x^2$ at $(4, 6)$
   c. $f(x) = x^2 + 5x + 1$ at $(-4, -3)$ and $g(x) = -3x - 7$ at $(3, -16)$

46. Verify that the following pairs of functions have perpendicular tangent lines at the indicated points.
   a. $f(x) = \sin x$ at $(0, 0)$ and $g(x) = 1/x$ at $(1, 1)$
   b. $f(x) = \frac{1}{2} \sin x$ at $(0, 0)$ and $g(x) = 1/x^2$ at $(1, 1)$
   c. $f(x) = x - 10$ at $(6, -4)$ and $g(x) = -x + 3$ at $(-5, 8)$

47. Let $f(x) = \sqrt{x}$. Show that the line $4y = x + 4$ is tangent to the graph of f. What is the point of tangency?

48. Suppose a thimble is dropped into a well 64 feet deep. If we use a coordinate system with the positive axis pointing upward, what is its velocity after 1.5 seconds? What is its speed at that moment?

49. A stone dropped from the top of a bridge lands in the river below 6 seconds later.
   a. What distance does the stone travel from bridge to water?
   b. With what speed does the stone hit the water?

3

DERIVATIVES

In Chapter 2 we used limits of the form

$$\lim_{x \to a} \frac{f(x) - f(a)}{x - a} \tag{1}$$

to define tangent lines and the velocity of an object moving in a straight line. However, the limit in (1) also appears in many other contexts, such as marginal revenue and marginal cost in economics, acceleration in physics, and reaction rates in chemistry. Because of the various interpretations of the limit in (1), we will isolate it as an abstract mathematical entity, called a derivative, and study its properties in Chapters 3 and 4.

3.1
THE DERIVATIVE

We are now ready to give the definition of one of the two central concepts of calculus: the derivative.

DEFINITION 3.1
DEFINITION OF
DERIVATIVE

Let a be a number in the domain of a function f. If

$$\lim_{x \to a} \frac{f(x) - f(a)}{x - a}$$

exists, we call this limit the **derivative of f at a** and write it $f'(a)$, so that

$$f'(a) = \lim_{x \to a} \frac{f(x) - f(a)}{x - a} \tag{2}$$

If the limit in (2) exists, we say that f **has a derivative at a**, that f is **differentiable at a**, or that $f'(a)$ **exists**.

The three phrases appearing in Definition 3.1 are all in common use, and we will use them interchangeably. The expression $f'(a)$ is read "the derivative of f at a" or "f prime of a".

By the definition of derivative, $f'(a)$ represents the slope of the line tangent to the graph of f at $(a, f(a))$.

Example 1   Let $f(x) = \frac{1}{4}x^2 + 1$. Find $f'(-1)$ and $f'(3)$, and draw the lines tangent to the graph of f at the corresponding points.

Solution   Using (2), we obtain

$$f'(-1) = \lim_{x \to -1} \frac{(\frac{1}{4}x^2 + 1) - \frac{5}{4}}{x - (-1)} = \lim_{x \to -1} \frac{\frac{1}{4}x^2 - \frac{1}{4}}{x + 1} = \lim_{x \to -1} \frac{\frac{1}{4}(x^2 - 1)}{x + 1}$$

$$= \lim_{x \to -1} \frac{\frac{1}{4}(x + 1)(x - 1)}{x + 1} = \lim_{x \to -1} \frac{1}{4}(x - 1) = -\frac{1}{2}$$

We also obtain

$$f'(3) = \lim_{x \to 3} \frac{(\frac{1}{4}x^2 + 1) - \frac{13}{4}}{x - 3} = \lim_{x \to 3} \frac{\frac{1}{4}x^2 - \frac{9}{4}}{x - 3} = \lim_{x \to 3} \frac{\frac{1}{4}(x^2 - 9)}{x - 3}$$

$$= \lim_{x \to 3} \frac{\frac{1}{4}(x - 3)(x + 3)}{x - 3} = \lim_{x \to 3} \frac{1}{4}(x + 3) = \frac{3}{2}$$

The lines tangent to the graph at the corresponding points are shown in Figure 3.1.   □

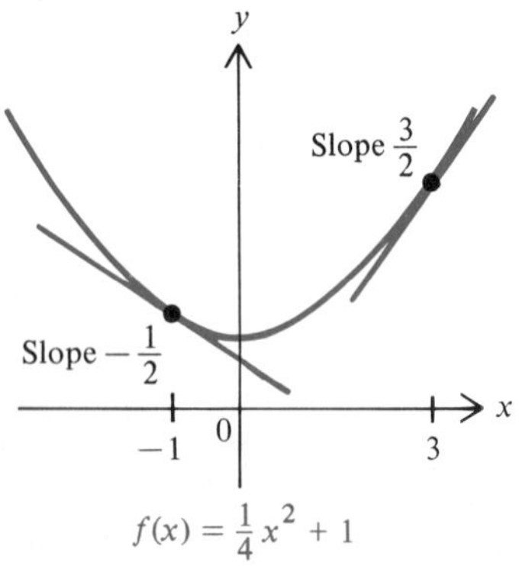

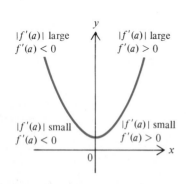

Slope $\frac{3}{2}$

Slope $-\frac{1}{2}$

$f(x) = \frac{1}{4}x^2 + 1$

FIGURE 3.1

Because $f'(a)$ represents the slope of the line tangent to the graph of f at $(a, f(a))$, we sometimes speak of $f'(a)$ as the **slope** of f at a. In Example 1 we found that $f'(-1) = -\frac{1}{2}$ and $f'(3) = \frac{3}{2}$. Thus the slope of f at -1 is $-\frac{1}{2}$ and the slope at 3 is $\frac{3}{2}$ (Figure 3.1).

For an arbitrary function f, if $|f'(a)|$ is large then the graph of f is very steep near $(a, f(a))$, whereas if $|f'(a)|$ is small then the graph of f is nearly horizontal near $(a, f(a))$ (Figure 3.2). Regardless of whether $|f'(a)|$ is large or small, the graph of f rises from left to right if $f'(a) > 0$ and falls from left to right if $f'(a) < 0$. Thus the derivative of a function gives us information that can be very helpful when we wish to graph the function.

At this point let us observe that if $f'(a)$ exists, then the letter x in (2) can be replaced by other letters. For example, we can write

$$f'(a) = \lim_{t \to a} \frac{f(t) - f(a)}{t - a} \tag{3}$$

$|f'(a)|$ large
$f'(a) < 0$

$|f'(a)|$ large
$f'(a) > 0$

$|f'(a)|$ small
$f'(a) < 0$

$|f'(a)|$ small
$f'(a) > 0$

FIGURE 3.2

One might use (3) if t represents time. In particular, suppose an object moves along a straight line on which a coordinate system has been set up, and let $f(t)$ be the position of the object at time t. In Section 2.2 we defined the velocity $v(t_0)$ of the object at time t_0 by the formula

$$v(t_0) = \lim_{t \to t_0} \frac{f(t) - f(t_0)}{t - t_0}$$

In terms of derivatives, this is expressed as

$$v(t_0) = f'(t_0)$$

The notion of derivative is also ideally suited to the fundamental economic concepts of marginal cost and marginal revenue. Suppose a company produces a product such as paint or honey. If little is produced, then the cost per unit of producing additional amounts of the product might be quite high (because the production process is inefficient). As mass production is introduced, the cost per unit of producing additional amounts would probably decline.

To be more specific, let $C(x)$ denote the cost of producing x units of a given product, and consider the cost per unit of producing additional amounts of the product when the production level is a units. If $x > a$, then $C(x) - C(a)$ is the additional cost required to produce the additional quantity $x - a$ of the product. Therefore

$$\frac{C(x) - C(a)}{x - a} = \begin{array}{l}\text{average cost per unit of increasing} \\ \text{production from } a \text{ to } x \text{ units}\end{array}$$

Since

$$\frac{C(x) - C(a)}{x - a} = \frac{C(a) - C(x)}{a - x}$$

the ratio $(C(x) - C(a))/(x - a)$ represents the corresponding average cost when $x < a$. If the ratio approaches a limit as x approaches a, we call that limit the **marginal cost** $m_C(a)$ of producing additional amounts when the production level is a. Thus

$$m_C(a) = \lim_{x \to a} \frac{C(x) - C(a)}{x - a} = C'(a) \tag{4}$$

For instance, if $C(x)$ is the cost in hundreds of dollars of producing x tons of clover honey, then the marginal cost $m_C(2)$ is by definition the cost per ton of producing more honey when the production level is at 2 tons. Intuitively, we would expect this to be very nearly the same as the cost per ton of producing the final microgram of a total of 2 tons of honey.

Example 2   Let $C(x) = 1 + 8x - x^2$ for $0 \le x \le 3$. Find the marginal cost at 2.

Solution   Using (4), we find that

$$m_C(2) = \lim_{x \to 2} \frac{C(x) - C(2)}{x - 2} = \lim_{x \to 2} \frac{(1 + 8x - x^2) - (1 + 8 \cdot 2 - 2^2)}{x - 2}$$

$$= \lim_{x \to 2} \frac{-x^2 + 8x - 12}{x - 2} = \lim_{x \to 2} \frac{-(x - 2)(x - 6)}{x - 2} = \lim_{x \to 2} -(x - 6) = 4 \quad \square$$

Similarly, if $R(x)$ denotes the revenue (money received) when the production level is x, then the **marginal revenue** $m_R(a)$ from producing additional amounts when the production level is a is given by

$$m_R(a) = \lim_{x \to a} \frac{R(x) - R(a)}{x - a} = R'(a) \tag{5}$$

In each of these interpretations—velocity, marginal cost, and marginal revenue—the derivative represents a "rate of change." In fact, if f is any function differentiable at a point a, then we can think of $f'(a)$ as the **rate of change of f at a**.

The Derivative as a Function

Nearly every function we will encounter is differentiable at all numbers, or all but finitely many numbers, in its domain. The function f' that arises when we take the derivative of f at such numbers is called the **derivative of f** (or the **derivative of f with respect to x**). Thus f' is by definition the function whose domain is the collection of numbers at which f is differentiable and whose value at any such number x is given by

$$f'(x) = \lim_{t \to x} \frac{f(t) - f(x)}{t - x} \tag{6}$$

Notice that in (6), x represents any number at which f is differentiable. However, when the limit on the right side of (6) is evaluated, t is the variable and x is regarded as a constant.

Example 3 Let $f(x) = x^2$. Show that $f'(x) = 2x$ for all x.

Solution It follows from (6) that for all x,

$$f'(x) = \lim_{t \to x} \frac{f(t) - f(x)}{t - x} = \lim_{t \to x} \frac{t^2 - x^2}{t - x} = \lim_{t \to x} \frac{(t - x)(t + x)}{t - x}$$

$$= \lim_{t \to x} (t + x) = 2x \quad \square$$

Example 4 Let $f(x) = x^{1/2}$. Show that $f'(x) = \frac{1}{2}x^{-1/2}$ for $x > 0$.

Solution From (6) we find that for $x > 0$,

$$\lim_{t \to x} \frac{f(t) - f(x)}{t - x} = \lim_{t \to x} \frac{t^{1/2} - x^{1/2}}{t - x}$$

$$= \lim_{t \to x} \left(\frac{t^{1/2} - x^{1/2}}{t - x} \cdot \frac{t^{1/2} + x^{1/2}}{t^{1/2} + x^{1/2}} \right)$$

$$= \lim_{t \to x} \frac{t - x}{(t - x)(t^{1/2} + x^{1/2})}$$

$$= \lim_{t \to x} \frac{1}{t^{1/2} + x^{1/2}} = \frac{1}{2x^{1/2}}$$

Therefore

$$f'(x) = \frac{1}{2}x^{-1/2} \quad \text{for } x > 0 \quad \square$$

Many functions we encounter in calculus are derivatives. Indeed, from our comments earlier in the section we find that

velocity is the derivative of the position function: $v(x) = f'(x)$
marginal cost is the derivative of the cost function: $m_C(x) = C'(x)$
marginal revenue is the derivative of the revenue function: $m_R(x) = R'(x)$

Other Notations for the Derivative   The expression $f'(x)$ is neither the only notation for the derivative nor the oldest. Some of the other notations for derivatives are

$$\dot{u}, \quad \frac{dy}{dx}, \quad \frac{d}{dx}f(x)$$

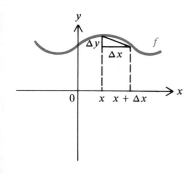

FIGURE 3.3

We will outline the origins of these notations.

Newton developed his calculus in the middle 1660s. He often thought of functions in terms of motion—a reasonable approach, since he was the foremost physicist of the seventeenth century. He referred to his functions as "fluents," and he called his derivatives "fluxions." Later on, he let u denote a function and $\dot{u}$ the derivative of u. The "dot" notation is still in frequent use in certain branches of science today.

At about the same time, Leibniz thought of his functions as relations between variables such as x and y. Motivated by the small triangles that appeared when he attempted to find tangents to curves (Figure 3.3), he adopted the notation $\Delta y/\Delta x$ for the derivative. Here Δy and Δx signified small changes in y and in x, respectively. Today we would write

$$\lim_{\Delta x \to 0} \frac{\Delta y}{\Delta x}$$

emphasizing the limiting process. Later, thinking of the derivative as the quotient of "infinitesimally" small differences in the variables y and x, Leibniz used dy/dx for the derivative of a variable y that depends on x. Thus if $y = x^2$, then using the fact that the derivative of the function x^2 is $2x$ (see Example 3), Leibniz would write

$$\frac{dy}{dx} = 2x \tag{7}$$

A variant form that combines the function notation and variable notation is

$$\frac{d}{dx}f(x)$$

One of the chief advantages of Leibniz's notation is that it allows us to include the value of the function in the expression for the derivative. For example, instead of (7) we can write

$$\frac{d}{dx}(x^2) = 2x$$

This keeps the formula for the function to be differentiated in front of us. For these reasons we will frequently employ the Leibniz notation for the derivative, especially in examples.

> **Caution:** Note carefully that the expression dy/dx is merely the derivative of y as a function of x. Although it has the appearance of a fraction, it is to be regarded as a single entity. We have not assigned separate meanings to dx and dy. Moreover, to indicate the derivative of y at a particular number such as 3 in this notation, we must write something more, such as
>
> $$\frac{dy}{dx}(3) \quad \text{or} \quad \left.\frac{dy}{dx}\right|_{x=3}$$

Each notation has its advantages and its followers. Each continues to be used in textbooks, and each accompanied the parallel growth of mathematics in England and on the Continent.*

At the end of the eighteenth century the French mathematician Joseph Louis Lagrange[†] (1736–1813) began writing the derivative of the function f at x as $f'(x)$, calling f' the function "derived from f."

In Table 3.1 we list four common ways of expressing derivatives and show the relationships between them.

TABLE 3.1

Function	Derivative as a function	Derivative at a point
$f(x) = x^2$	$f'(x) = 2x$	$f'(4) = 8$
$y = x^2$	$\dfrac{dy}{dx} = 2x \quad \text{or} \quad \dfrac{d}{dx}(x^2) = 2x$	$\left.\dfrac{dy}{dx}\right|_{x=4} = 8$
$f(x) = x^2$	$\dfrac{d}{dx} f(x) = 2x \quad \text{or} \quad \dfrac{d}{dx}(x^2) = 2x$	$\left.\dfrac{d}{dx} f(x)\right|_{x=4} = 8$
$u = t^2$	$\dot{u} = 2t$	$\left.\dot{u}\right|_{t=4} = 8$

* A great, and unpleasant, rivalry flared up between the followers of Newton in England and the followers of Leibniz on the Continent. Each party claimed its leader as the inventor of calculus. The English stuck firmly to the dot notation $\dot{u}$, and the Continent retained the Leibniz notation. Mathematical communication across the English Channel dwindled, ultimately slowing mathematical progress in England. Finally, at the start of the nineteenth century, the Analytic Society at Cambridge University sprang up, for the expressed purpose of advancing "the principles of pure d-ism as opposed to the dot-age of the university." Thereafter the Leibniz notation permeated English mathematics, and mathematical communication across the Channel grew.
[†] **Lagrange**: Pronounced "La-*grahnj*."

EXERCISES 3.1

In Exercises 1–12 use Definition 3.1 to find $f'(a)$ for the given value of a.

1. $f(x) = 5; a = 4$
2. $f(x) = -x; a = -3$
3. $f(x) = 2x + 3; a = 1$
4. $f(x) = -4x + 7; a = 0$
5. $f(x) = x^2 - 2; a = -1$
6. $f(x) = x^3; a = 0$
7. $f(x) = 1/x; a = -2$
8. $f(x) = 1/\sqrt{x}; a = 4$
9. $f(x) = |x|; a = \sqrt{2}$
10. $f(x) = |x|; a = -\sqrt{2}$
11. $f(x) = \begin{cases} x^2 & \text{for } x < 2 \\ 4x - 4 & \text{for } x \geq 2 \end{cases}$ $a = 2$
12. $f(x) = \begin{cases} \sqrt{x} & \text{for } x \leq \frac{1}{16} \\ 2x + \frac{1}{8} & \text{for } x > \frac{1}{16} \end{cases}$ $a = \frac{1}{16}$

In Exercises 13–20 use (6) to find the derivative of the function.

13. $f(x) = -\pi$
14. $f(x) = 3x - 7$
15. $f(x) = -5x^2$
16. $f(x) = -5x^2 + x$
17. $g(x) = x^3$
18. $g(x) = 1/x$
19. $k(x) = \dfrac{1}{x^2} - \sqrt{7}$
20. $k(x) = x^{1/3}$
    (*Hint:* $t - x = (t^{1/3} - x^{1/3})(t^{2/3} + t^{1/3}x^{1/3} + x^{2/3})$.)

In Exercises 21–24 find dy/dx.

21. $y = \frac{7}{3}$
22. $y = 1 - \frac{1}{2}x$
23. $y = 3x^2 + 1$
24. $y = 1/x^3$

In Exercises 25–28 find $\dfrac{dy}{dx}\Big|_{x=2}$

25. $y = 0.25$
26. $y = -5x + 9$
27. $y = x^2 - 3$
28. $y = -1/x$

In Exercises 29–36 use Definition 3.1 to determine whether the given function has a derivative at a. Where applicable, find the derivative.

29. $f(x) = x^{1/3}; a = 0$
30. $f(x) = x^{7/3}; a = 0$
31. $f(x) = |x| - x; a = 0$
32. $g(x) = |x - 3|; a = 3$
33. $g(x) = |x + 3|; a = 3$
34. $g(x) = \dfrac{1}{x^2}; a = 0$
35. $k(x) = \begin{cases} -x^2 + 4x & \text{for } x < 0 \\ x^2 - 1 & \text{for } x \geq 0 \end{cases}$ $a = 0$
36. $k(x) = \begin{cases} 3x^2 + 4x & \text{for } x < 0 \\ x^2 + 4x & \text{for } x \geq 0 \end{cases}$ $a = 0$

In Exercises 37–40 find an equation of the line tangent to the graph of f at the given point.

37. $f(x) = x^2; (-2, 4)$
38. $f(x) = 1/x; (-3, -\frac{1}{3})$
39. $f(x) = \sqrt{x}; (4, 2)$
40. $f(x) = \sin x; (0, 0)$

41. Each of the following limits is of the form
$$\lim_{x \to a} [f(x) - f(a)]/(x - a)$$
for an appropriate function f and an appropriate point a. In each case find a formula for f, and calculate $f'(a)$.

   a. $\lim\limits_{x \to 2} \dfrac{x^4 - 16}{x - 2}$
   b. $\lim\limits_{x \to 2} (x^3 + 2x^2 + 4x + 8)$

42. In each case below assume that f is differentiable at a. Describe how each limit is related to $f'(a)$.

   a. $\lim\limits_{t \to a} \dfrac{f(a) - f(t)}{t - a}$
   b. $\lim\limits_{t \to a} \dfrac{f(a) - f(t)}{a - t}$
   c. $\lim\limits_{x \to a} \dfrac{1}{x - a} [f(x) - f(a)]$

43. Assume that $f'(a)$ exists. Express
$$\lim_{h \to 0} \frac{f(a - h) - f(a)}{h}$$
in terms of $f'(a)$. (*Hint:* Show that
$$\lim_{h \to 0} \frac{f(a - h) - f(a)}{h} = -\lim_{h \to 0} \frac{f(a - h) - f(a)}{-h}$$
and then use the Substitution Rule.)

*44. Assume that $f'(a)$ exists. Express
$$\lim_{h \to 0} \frac{f(a + h) - f(a - h)}{h}$$
in terms of $f'(a)$. (*Hint:* Note that $f(a + h) - f(a - h) = f(a + h) - f(a) + f(a) - f(a - h)$. Then use the limit rules and Exercise 43.)

45. a. Suppose f is an even function. If $f'(a) = 2$, find $f'(-a)$. (*Hint:* Draw the graph of such a function f.)
    b. Suppose f is an odd function. If $f'(a) = 2$, find $f'(-a)$. (*Hint:* Draw the graph of such a function f.)

46. a. Suppose f and g are functions such that $f = g$ on an open interval containing a. If $f'(a)$ exists, show that $g'(a)$ exists and that $f'(a) = g'(a)$.
    b. Let $g(x) = |x|$. Find $g'(x)$ for $x > 0$, and then for $x < 0$.

47. Suppose f and g are functions such that $f(a) = g(a)$, and assume that $f'(a)$ exists. Does it follow that $f'(a) = g'(a)$? Explain.

48. Suppose $y = 8x - 23$ is an equation of the line tangent to the graph of a function f at $(5, 17)$. Find $f'(5)$.

49. Suppose f is differentiable at a, and let

$$g(x) = \begin{cases} \dfrac{f(x) - f(a)}{x - a} & \text{if } x \neq a \\ f'(a) & \text{if } x = a \end{cases}$$

Show that g is continuous at a. (*Hint:* Use the definition of continuity.)

50. Suppose a ball is shot upward from 256 feet above the ground, with an initial velocity of 96 feet per second. Find the velocity of the ball after
   a. $\frac{1}{2}$ second   b. 5 seconds

51. In Exercise 50 determine the value of t for which the velocity is 0.

52. Suppose the cost of preparing x barrels of clover honey for sale is $C(x)$ dollars, where

$$C(x) = 400x - (0.1)x^2 \quad \text{for } 0 \leq x \leq 1000$$

Find the marginal cost at 40.

53. Suppose a cost function C is given by $C(x) = 10{,}000 + 3/x$ for $1 \leq x \leq 100$. Find the marginal cost at 50.

54. Suppose a revenue function R is given by

$$R(x) = 40x - x^2 \quad \text{for } 0 \leq x \leq 40$$

Find the marginal revenue at 10.

55. Suppose the revenue resulting from the sale of x barrels of clover honey is $R(x)$ dollars, where

$$R(x) = 450x^{1/2} \quad \text{for } x \geq 0$$

Find the marginal revenue at 16.

56. Suppose it costs $C(x)$ thousand dollars per year to produce x thousand gallons of antifreeze, where

$$C(x) = 3 + 12x - 2x^2 \quad \text{for } 0 \leq x \leq 3$$

   a. Find the marginal cost at 1.
   b. Find the value of a for which $m_C(a) = \frac{1}{2}C(1)$.

57. In Exercise 56 suppose the cost of purchasing the ethylene glycol, a major ingredient of antifreeze, is increased by $500 per thousand gallons. Find the marginal cost at 1,

and compare your answer with the answer to part (a) of Exercise 56.

58. In Exercise 56 suppose the rent of the buildings in which the antifreeze is produced is raised $1000 per year. Find the marginal cost at 1, and compare your answer with the answer to part (a) of Exercise 56.

59. Two ships leave port at the same time. One travels north at 15 knots (that is, 15 nautical miles per hour), and the other west at 20 knots. Show that the distance between the ships increases at a constant rate, and determine the rate of increase.

60. A jet and a Cessna plane each leave the Minneapolis–St. Paul airport at noon. The Cessna travels east at 200 miles per hour, and the jet travels 60° south of east at 400 miles per hour. Show that the distance between the planes increases at a constant rate and determine the rate of increase.

61. Two toy boats start from virtually the same point in a lake and travel in straight lines at an angle of $45°$ with each other (Figure 3.4).
   a. If the speed of each is 2 meters per minute, at what rate does the distance between them increase?
   b. If the distance between them increases at 3 meters a minute and both boats have the same speed, determine their speed.

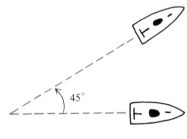

FIGURE 3.4

3.2
DIFFERENTIABLE FUNCTIONS

We begin this section by finding the derivatives of several functions, each of which is differentiable at each number in its domain. Functions having this property are called differentiable functions.

DEFINITION 3.2

> If f is differentiable at each number in its domain, then f is a **differentiable function**.

Example 1   Let $f(x) = c$, where c is a constant. Show that $f'(x) = 0$ for all x.

Solution   By the definition of the derivative we find that

$$f'(x) = \lim_{t \to x} \frac{f(t) - f(x)}{t - x} = \lim_{t \to x} \frac{c - c}{t - x} = \lim_{t \to x} 0 = 0 \quad \text{for all } x \quad \square$$

Example 2   Let $f(x) = x$. Show that $f'(x) = 1$ for all x.

Solution   Again by the definition of the derivative we find that

$$f'(x) = \lim_{t \to x} \frac{f(t) - f(x)}{t - x} = \lim_{t \to x} \frac{t - x}{t - x} = \lim_{t \to x} 1 = 1 \quad \text{for all } x \quad \square$$

We have just now found the derivatives of the functions c and x, and in Section 3.1 we found the derivative of the function x^2. With a little more effort we can find the derivative of x^n, where n is any positive integer. Let $f(x) = x^n$. Since

$$t^n - x^n = (t - x)(t^{n-1} + t^{n-2}x + t^{n-3}x^2 + \cdots + tx^{n-2} + x^{n-1})$$

we have

$$f'(x) = \lim_{t \to x} \frac{t^n - x^n}{t - x} = \lim_{t \to x} \frac{(t - x)(t^{n-1} + t^{n-2}x + t^{n-3}x^2 + \cdots + tx^{n-2} + x^{n-1})}{t - x}$$

$$= \lim_{t \to x} (t^{n-1} + t^{n-2}x + t^{n-3}x^2 + \cdots + tx^{n-2} + x^{n-1})$$

$$= \overbrace{x^{n-1} + x^{n-2} \cdot x + x^{n-3} \cdot x^2 + \cdots + x \cdot x^{n-2} + x^{n-1}}^{n \text{ terms}}$$

$$= nx^{n-1}$$

We have thus derived a basic formula of calculus. In the Leibniz notation the formula is

> $$\frac{d}{dx}(x^n) = nx^{n-1} \quad \text{for any integer } n > 0 \qquad (1)$$

For example, if $f(x) = x^{17}$, then $f'(x) = 17x^{16}$.

Recall from Section 3.1 that $f'(a)$ is sometimes called the rate of change of f at a; in particular, this terminology is used when the function is to be interpreted geometrically or physically. If the function is given by expressing a variable y in terms of a variable x, we call the derivative dy/dx the rate of change of y with respect to x. In the next example we will use (1) to find a rate of change.

Example 3   Find the rate of change of the volume of a cube with respect to its side length.

Solution   The volume V of a cube with side length s is given by

$$V = s^3$$

We are to find dV/ds. From (1) with $n = 3$ it follows that

$$\frac{dV}{ds} = 3s^2$$

Thus the rate of change of the volume with respect to the side length is $3s^2$, that is, 3 times the area of a side.   □

There is an alternative formula for computing the derivative of a function. It arises from the alternative form of the limit (see formula (4) of Section 2.4). We can obtain it from (6) of Section 3.1 by replacing t by $x + h$:

$$f'(x) = \lim_{h \to 0} \frac{f(x + h) - f(x)}{h} \tag{2}$$

Formula (2) is especially convenient when we compute the derivatives of the sine and cosine functions. In our computation we will use the limits derived in Examples 2 and 3 of Section 2.4. These yield

$$\lim_{h \to 0} \frac{\sin h}{h} = 1 \quad \text{and} \quad \lim_{h \to 0} \frac{\cos h - 1}{h} = 0$$

Let $f(x) = \sin x$. Applying formula (2) and the above limits, we obtain

$$f'(x) = \lim_{h \to 0} \frac{\sin(x + h) - \sin x}{h}$$

$$= \lim_{h \to 0} \frac{(\sin x \cos h + \cos x \sin h) - \sin x}{h}$$

$$= \lim_{h \to 0} \left(\frac{\sin x (\cos h - 1)}{h} + \frac{\cos x \sin h}{h} \right)$$

$$= \sin x \lim_{h \to 0} \frac{\cos h - 1}{h} + \cos x \lim_{h \to 0} \frac{\sin h}{h}$$

$$= (\sin x) \cdot 0 + (\cos x) \cdot 1 = \cos x \quad \text{for all } x$$

Consequently

$$\frac{d}{dx} \sin x = \cos x \quad \text{for all } x \tag{3}$$

This establishes that the derivative of the sine function is the cosine function.
Next, let $f(x) = \cos x$. Proceeding as for the sine function, we find that

$$f'(x) = \lim_{h \to 0} \frac{\cos(x + h) - \cos x}{h}$$

$$= \lim_{h \to 0} \frac{(\cos x \cos h - \sin x \sin h) - \cos x}{h}$$

$$= \lim_{h \to 0} \left(\frac{\cos x (\cos h - 1)}{h} - \frac{\sin x \sin h}{h} \right)$$

$$= \cos x \lim_{h \to 0} \frac{\cos h - 1}{h} - \sin x \lim_{h \to 0} \frac{\sin h}{h}$$

$$= (\cos x) \cdot 0 - (\sin x) \cdot 1 = -\sin x \quad \text{for all } x$$

Consequently

$$\frac{d}{dx} \cos x = -\sin x \quad \text{for all } x \tag{4}$$

Therefore the derivative of the cosine function is the negative of the sine function.

The functions c, x^n, $\sin x$, and $\cos x$, which we have discussed in this section, are differentiable. However, there are functions that are not differentiable, as the next example shows.

Example 4   Let $f(x) = |x|$. Show that f is not differentiable by proving that f has a derivative at x if and only if $x \neq 0$.

Solution   First we consider the case $x > 0$. Then $f(x) = x$, and whenever t is close enough to x, we have $t > 0$ and thus $f(t) = t$. Therefore

$$f'(x) = \lim_{t \to x} \frac{f(t) - f(x)}{t - x} = \lim_{t \to x} \frac{t - x}{t - x} = \lim_{t \to x} 1 = 1 \quad \text{for } x > 0$$

Now let $x < 0$; then $f(x) = -x$. Whenever t is close enough to x, we have $t < 0$ and thus $f(t) = -t$. Therefore

$$f'(x) = \lim_{t \to x} \frac{f(t) - f(x)}{t - x} = \lim_{t \to x} \frac{-t - (-x)}{t - x} = \lim_{t \to x} (-1) = -1 \quad \text{for } x < 0$$

Consequently f has a derivative at all x such that $x \neq 0$. Finally we consider the case $x = 0$. Observe that

$$\lim_{t \to 0^+} \frac{f(t) - f(0)}{t - 0} = \lim_{t \to 0^+} \frac{|t|}{t} = \lim_{t \to 0} 1 = 1$$

and

$$\lim_{t \to 0^-} \frac{f(t) - f(0)}{t - 0} = \lim_{t \to 0^-} \frac{|t|}{t} = \lim_{t \to 0^-} -1 = -1$$

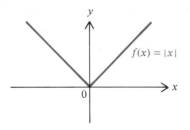

FIGURE 3.5

Since the two one-sided limits are different, we conclude from Theorem 2.5 that

$$\lim_{t \to 0} \frac{f(t) - f(0)}{t - 0}$$

does not exist. Therefore f does not have a derivative at 0. $\quad\square$

Since $f'(0)$ does not exist in Example 4, it follows that the graph of the absolute value function f does not have a tangent line at $(0, 0)$. Notice that the graph of f is bent, or pointed, at $(0, 0)$ (Figure 3.5). More generally, if the graph of a function f is bent, or pointed, at $(a, f(a))$, then f is not differentiable at a, and there is no line tangent to the graph at $(a, f(a))$. Thus differentiability of f at a is associated with smoothness of the graph at $(a, f(a))$.

Differentiability on Intervals

In Example 4 we showed that $|x|$ is not a differentiable function because it is not differentiable at 0. However, it is differentiable at every number in $(-\infty, 0)$ or $(0, \infty)$. If I is an open interval, then we say that a function f is **differentiable on I** if f is differentiable at each point of I. As a result, $|x|$ is differentiable on $(-\infty, 0)$ and on $(0, \infty)$. In Chapter 4 we will be especially interested in functions that are differentiable on open intervals.

If I is a closed interval $[a, b]$ with $a < b$, then we say that f is **differentiable on I** if f is differentiable on (a, b) and if the one-sided limits

$$\lim_{t \to a^+} \frac{f(t) - f(a)}{t - a} \quad \text{and} \quad \lim_{t \to b^-} \frac{f(t) - f(b)}{t - b} \tag{5}$$

both exist. Of course, if f is differentiable at a or at b, then the corresponding one-sided limit in (5) exists. However, there are functions for which the one-sided limits in (5) exist even though f is not differentiable at a or at b. For example, in the solution of Example 4 we saw that

$$\lim_{t \to 0^+} \frac{|t| - 0}{t - 0} = 1 \quad \text{and} \quad \lim_{t \to 0^-} \frac{|t| - 0}{t - 0} = -1$$

Therefore the absolute value function is differentiable on every interval of the form $[-c, 0]$ or $[0, c]$, where $c > 0$, despite the fact that the function is *not* differentiable at 0.

Differentiability on an unbounded interval can be defined similarly. We say that a function f is differentiable on $[a, \infty)$ if f is differentiable at every number in (a, ∞) and if the first one-sided limit in (5) exists. Likewise, f is differentiable on $(-\infty, b]$ if f is differentiable at every number in $(-\infty, b)$ and if the second one-sided limit in (5) exists. For example, the absolute value function is differentiable on $(-\infty, 0]$ and on $[0, \infty)$.

In our final example we study the differentiability of $x^{1/2}$.

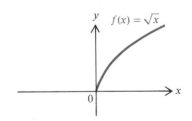

FIGURE 3.6

Example 5 Let $f(x) = x^{1/2}$ (see Figure 3.6). Show that f is differentiable on $(0, \infty)$ but not on $[0, \infty)$.

Solution   By Example 4 of Section 3.1, $f'(x) = \frac{1}{2}x^{-1/2}$ for $x > 0$. Consequently f is differentiable on $(0, \infty)$. For $x = 0$ we have

$$\lim_{t \to 0^+} \frac{f(t) - f(0)}{t - 0} = \lim_{t \to 0^+} \frac{t^{1/2}}{t} = \lim_{t \to 0^+} t^{-1/2} = \infty$$

Since the right-hand limit at 0 does not exist, we conclude that f is not differentiable on $[0, \infty)$.   ☐

EXERCISES 3.2

In Exercises 1–10 use the results of this section to find the derivative of the given function at the given numbers.

1. $f(x) = -2; a = 1$
2. $f(x) = \sqrt{3}; a = -3$
3. $f(x) = x^2; a = \frac{3}{2}, 0$
4. $f(x) = x^3; a = -\frac{1}{4}$
5. $f(x) = x^4; a = \sqrt[3]{2}$
6. $f(x) = x^5; a = -2$
7. $f(x) = x^{10}; a = 1$
8. $f(t) = \sin t; a = \pi/4, \pi/3$
9. $f(t) = \cos t; a = 0, -\pi/3$
10. $f(t) = |t|; a = -\pi, \frac{3}{4}$

In Exercises 11–24 use either (6) of Section 3.1 or (2) of this section to find the derivative of the given function.

11. $f(x) = -2x - 1$
12. $f(x) = 1 - x^2$
13. $f(x) = x^5$
14. $f(x) = x^6$
15. $f(x) = x^2 + x$
16. $f(x) = \dfrac{1}{2x}$
17. $f(x) = \dfrac{x^2 - 1}{x^2 + 1}$
18. $y = 5 \sin x$
19. $y = -3 \cos x$
20. $y = \cos x - \sin x$
21. $y = x^{2/3}$

$\left(\text{Hint: } \dfrac{t^{2/3} - x^{2/3}}{t - x} = \dfrac{(t^{1/3} - x^{1/3})(t^{1/3} + x^{1/3})}{(t^{1/3} - x^{1/3})(t^{2/3} + t^{1/3}x^{1/3} + x^{2/3})}.\right)$

22. $y = x^{3/2}$

$\left(\text{Hint: } \dfrac{t^{3/2} - x^{3/2}}{t - x} = \dfrac{(t^{1/2} - x^{1/2})(t + t^{1/2}x^{1/2} + x)}{(t^{1/2} - x^{1/2})(t^{1/2} + x^{1/2})}.\right)$

*23. $y = \sqrt{x - 1}$ (Hint: Rationalize the numerator of the fraction that arises.)

*24. $y = \sin 2x$

In Exercises 25–30 show that f is differentiable on the given interval.

25. $f(x) = x^2 + x; (-\infty, \infty)$
26. $f(x) = 2x^3 - \sqrt{x}; (0, \infty)$
27. $f(x) = \dfrac{1}{4 - x}; (4, \infty)$
28. $f(x) = \dfrac{1}{4 - x^2}; (-\infty, -2)$

29. $f(x) = |x - 1|; [1, \infty)$
30. $f(x) = |2 + 3x|; (-\infty, -\frac{2}{3}]$
31. Find a in each of the following cases.
   a. $f(x) = -2x^2; f'(a) = 12$
   b. $f(x) = 3x + x^2; f'(a) = 13$
   c. $f(x) = 1/x; f'(a) = -\frac{1}{9}$ (There are two possible values for a.)
   d. $f(x) = \sin x; f'(a) = \sqrt{3}/2$ (There are infinitely many possible values for a.)

32. The height $h(t)$ of an object dropped from an initial height h_0 is given by

$$h(t) = h_0 - 16t^2$$

until the object hits the ground. Find the velocity $v(t)$ during that time interval.

33. Suppose an object moving along the x axis has x coordinate $f(t)$ at time t given by

$$f(t) = -3 \sin t$$

Find a formula for the velocity $v(t)$ at any time t, and in particular find $v(\pi/6)$.

34. Suppose a revenue function R is given by

$$R(x) = \begin{cases} 4x & \text{for } 0 \le x \le 1 \\ 6x - x^2 - 1 & \text{for } 1 < x \le 3 \end{cases}$$

   a. Show that there is a marginal revenue at $x = 1$, and calculate it.
   b. Using part (a), show that the revenue function is differentiable on $[0, 3]$.

35. Let $R(x)$ be the revenue in dollars resulting from the sale of x thousand bushels of wheat, and suppose

$$R(x) = \begin{cases} 1432x & \text{for } 0 < x < 6 \\ 8592 & \text{for } x \ge 6 \end{cases}$$

Show that R is differentiable on $(0, 6)$ and on $(6, \infty)$, but is not differentiable at 6.
(If the government were to set a ceiling on the amount of wheat a single farmer can sell, then the function described above could represent the farmer's revenue.)

36. Show that the rate of change of the circumference of a circle with respect to the radius is constant.

37. Find the rate of change of the area of a circle with respect to its radius.

38. Find the rate of change of the area of a square with respect to the length of a side.

39. a. Find the rate of change of the area of an equilateral triangle with respect to the length of a side.
    b. Let $A(x)$ denote the area of the triangle when the length of a side is x. Find a value of x for which $A'(x) = A(x)$.

40. a. Find the rate of change of the volume of a sphere with respect to the radius.
    b. What relationship does the rate of change bear to the surface area of the sphere?

41. If two bodies are a distance r apart, then the gravitational force $F(r)$ exerted by one body on the other body is given by

$$F(r) = \frac{k}{r^2} \quad \text{for } r > 0$$

where k is a positive constant. Find a formula for the rate of change of the gravitational force with respect to the distance between the two bodies. Show that this rate of change is negative.

42. The rate at which radioactive iodine I^{128} is disintegrating at any given time is proportional to the amount present at that time. If the constant of proportionality is -0.028, write an equation that relates the rate of disintegration at any time t to the amount $A(t)$ present at time t.

43. Assume that the graph of a function f has a tangent line at $(a, f(a))$. The line **normal** to the graph of f at $(a, f(a))$ is the line through $(a, f(a))$ perpendicular to the tangent line. Find equations of the lines normal to the graphs of the following functions at the indicated points on their graphs.
    a. $f(x) = x^2; (-1, 1)$
    b. $f(x) = 2x - 3; (-1, -5)$
    c. $f(x) = \sin x; (0, 0)$

*44. Let $f(x) = x^2$. Show that every normal to the graph of f except the normal passing through the origin intersects the graph of f twice.

3.3
DERIVATIVES OF COMBINATIONS OF FUNCTIONS

Since there are limit rules for sums, differences, products, and quotients of functions, it is natural to ask whether there are corresponding rules for derivatives. There are such rules, but some of the formulas for the derivatives of combinations of functions are quite different from their counterparts for limits. The desirability of such rules can be shown by means of the following example. Suppose the position $f(t)$ of a falling object at any time t is given by

$$f(t) = -16t^2 + 24t + 160$$

This is a fairly complicated formula, and finding the derivative $f'(t)$ directly from the definition would be tedious. But if we knew how to find the derivative of a combination of functions from the derivatives of the individual functions, then obtaining $f'(t)$ would be much simpler.

Before stating and proving the differentiation rules, we establish an elementary result that will play a part in the proofs of Theorem 3.6 and 3.7.

THEOREM 3.3

If f is differentiable at a, then f is continuous at a, that is,

$$\lim_{t \to a} f(t) = f(a)$$

Proof To show that f is continuous at a, it suffices to show that

$$\lim_{t \to a} f(t) = f(a), \quad \text{or equivalently,} \quad \lim_{t \to a} [f(t) - f(a)] = 0$$

Notice that $f'(a)$ exists by hypothesis. Hence by the limit rules we have

$$\lim_{t \to a} [f(t) - f(a)] = \lim_{t \to a} \frac{f(t) - f(a)}{t - a} \cdot (t - a)$$

$$= \lim_{t \to a} \frac{f(t) - f(a)}{t - a} \cdot \lim_{t \to a} (t - a)$$

$$= f'(a) \cdot 0 = 0 \quad \blacksquare$$

The converse of Theorem 3.3 is false: A function can be continuous but not differentiable at a given point. For example, if $f(x) = |x|$, then f is continuous at 0 but is not differentiable at 0 (see Example 4 of Section 3.2).

Derivative of a Sum   For the sum of two differentiable functions we have the following result.

THEOREM 3.4

If f and g are differentiable at a, then $f + g$ is also differentiable at a, and

$$(f + g)'(a) = f'(a) + g'(a)$$

Proof   Using the limit rules, we find that

$$(f + g)'(a) = \lim_{t \to a} \frac{(f + g)(t) - (f + g)(a)}{t - a}$$

$$= \lim_{t \to a} \left(\frac{f(t) - f(a)}{t - a} + \frac{g(t) - g(a)}{t - a} \right)$$

$$= \lim_{t \to a} \frac{f(t) - f(a)}{t - a} + \lim_{t \to a} \frac{g(t) - g(a)}{t - a}$$

$$= f'(a) + g'(a) \quad \blacksquare$$

Theorem 3.4 says that the derivative of the sum of two functions at a given number is equal to the sum of the derivatives, provided that the two functions are differentiable at that number. Alternatively, Theorem 3.4 can be interpreted as saying that

$$(f + g)'(x) = f'(x) + g'(x)$$

for all x at which both f and g are differentiable. This formula is reminiscent of the formula for the limit of a sum of two functions.

Example 1 Let $k(x) = x + \sin x$. Find a formula for $k'(x)$, and then compute $k'(\pi/4)$.

Solution Let $f(x) = x$ and $g(x) = \sin x$, so that $k = f + g$. By Example 2 and (3) of Section 3.2,

$$f'(x) = 1 \quad \text{and} \quad g'(x) = \cos x$$

Therefore Theorem 3.4 implies that

$$k'(x) = f'(x) + g'(x) = 1 + \cos x \quad \text{for all } x$$

Letting $x = \pi/4$, we conclude that

$$k'\left(\frac{\pi}{4}\right) = 1 + \cos\frac{\pi}{4} = 1 + \frac{\sqrt{2}}{2} \quad \square$$

The sum rule for derivatives can also be stated in the Leibniz notation. If u and v are variables that depend on x, then

$$\frac{d}{dx}(u + v) = \frac{du}{dx} + \frac{dv}{dx}$$

Example 2 Find $\dfrac{d}{dx}(\sin x + \cos x)$.

Solution By (3) and (4) of Section 3.2,

$$\frac{d}{dx}\sin x = \cos x \quad \text{and} \quad \frac{d}{dx}\cos x = -\sin x$$

Consequently

$$\frac{d}{dx}(\sin x + \cos x) = \frac{d}{dx}\sin x + \frac{d}{dx}\cos x = \cos x - \sin x \quad \square$$

With practice you should be able to find the derivatives of sums without writing intermediate steps. Thus for Example 2 one would normally just write

$$\frac{d}{dx}(\sin x + \cos x) = \cos x - \sin x$$

Caution: Note that $(f + g)'(a)$ may exist even when $f'(a)$ or $g'(a)$ fails to exist (see Exercise 55). Theorem 3.4 *does not* say that $(f + g)'(a)$ can exist only if $f'(a)$ and $g'(a)$ exist. Rather, it says that if $f'(a)$ and $g'(a)$ both exist, then $(f + g)'(a)$ exists.

We can readily extend Theorem 3.4 to the sum of more than two functions. Indeed, if $f_1, f_2, \ldots, f_n$ are each differentiable at a, so is $f_1 + f_2 + \cdots + f_n$. Moreover,

$$(f_1 + f_2 + \cdots + f_n)'(a) = f'_1(a) + f'_2(a) + \cdots + f'_n(a)$$

Hence if $f(x) = x^2 + x - 12 + \sin x$, then by separating f into the functions $f_1(x) = x^2$, $f_2(x) = x$, $f_3(x) = -12$, and $f_4(x) = \sin x$ and differentiating each of these four functions, we deduce that

$$f'(x) = 2x + 1 + \cos x$$

Derivative of a Constant Multiple of a Function

For a constant multiple of a differentiable function we have the following result.

THEOREM 3.5

If f is differentiable at a, then for any number c the function cf is differentiable at a, and

$$(cf)'(a) = cf'(a)$$

Proof   From the definition of differentiability at a we have

$$(cf)'(a) = \lim_{t \to a} \frac{(cf)(t) - (cf)(a)}{t - a} = \lim_{t \to a} \frac{cf(t) - cf(a)}{t - a}$$

$$= c \lim_{t \to a} \frac{f(t) - f(a)}{t - a} = cf'(a) \quad \blacksquare$$

Theorem 3.5 can be interpreted as saying that

$$(cf)'(x) = cf'(x)$$

for all x at which f is differentiable. It reminds us of the formula for the limit of a constant times a function.

Example 3   Let $k(x) = 4\sqrt{x}$. Find $k'(x)$.

Solution   If $f(x) = \sqrt{x}$, then $k(x) = 4f(x)$. Since $f'(x) = \frac{1}{2}x^{-1/2}$ by Example 4 of Section 3.1, it follows from Theorem 3.5 that

$$k'(x) = 4f'(x) = 4\left(\frac{1}{2}x^{-1/2}\right) = 2x^{-1/2} \quad \square$$

In the Leibniz notation, if u is a variable depending on x and if c is any number, then

$$\frac{d}{dx}(cu) = c\frac{du}{dx}$$

Thus for the derivative of the function in Example 3 we would write

$$\frac{d}{dx}(4\sqrt{x}) = 4\frac{d}{dx}(\sqrt{x}) = 2x^{-1/2}$$

The addition and constant multiple rules for derivatives imply that all polynomials are differentiable functions, and they yield the following formula for the derivative of any polynomial:

$$\frac{d}{dx}(c_n x^n + c_{n-1}x^{n-1} + \cdots + c_1 x + c_0)$$

$$= nc_n x^{n-1} + (n-1)c_{n-1}x^{n-2} + \cdots + 2c_2 x + c_1 \qquad (1)$$

Example 4   Find $\dfrac{d}{dx}\left(3x^8 - \sqrt{2}x^5 + \dfrac{3}{2}x^3 + 20x + 1\right)\Bigg|_{x=-1}$

Solution   First we use (1) to find the derivative of the function at an arbitrary x:

$$\frac{d}{dx}\left(3x^8 - \sqrt{2}x^5 + \frac{3}{2}x^3 + 20x + 1\right) = 3(8x^7) - \sqrt{2}(5x^4) + \frac{3}{2}(3x^2) + 20 + 0$$

$$= 24x^7 - 5\sqrt{2}x^4 + \frac{9}{2}x^2 + 20$$

Then we evaluate the derivative at $x = -1$:

$$\frac{d}{dx}\left(3x^8 - \sqrt{2}x^5 + \frac{3}{2}x^3 + 20x + 1\right)\Bigg|_{x=-1} = 24(-1)^7 - 5\sqrt{2}(-1)^4 + \frac{9}{2}(-1)^2 + 20$$

$$= -24 - 5\sqrt{2} + \frac{9}{2} + 20$$

$$= \frac{1}{2} - 5\sqrt{2} \quad \square$$

The formula in (1) plays a significant role when we wish to find the velocity of an object that is moving vertically above the ground and is acted on only by the force of gravity. Suppose the initial height of the object is h_0 and its initial velocity is v_0. Then as we observed in Section 1.3, until the object hits the ground its height $h(t)$ in feet above the ground after t seconds is given by the formula

$$h(t) = -16t^2 + v_0 t + h_0 \qquad (2)$$

Example 5 Suppose a person 160 feet above the ground tosses a ball vertically upward with an initial velocity of 24 feet per second. Find a formula for the ball's velocity, and determine when the velocity is 0.

Solution Since $h_0 = 160$ and $v_0 = 24$, it follows from (2) that until the ball hits the ground, its height $h(t)$ is given by

$$h(t) = -16t^2 + 24t + 160$$

Since the velocity is the derivative of the height function, it follows from (1) that

$$v(t) = h'(t) = -32t + 24$$

Therefore $v(t) = 0$ when $-32t + 24 = 0$, that is, when $t = \frac{3}{4}$ (seconds). $\square$

In general, if the height of an object is given by

$$h(t) = -16t^2 + v_0 t + h_0$$

then (1) implies that

$$v(t) = h'(t) = -32t + v_0$$

so that the velocity at time t depends on the initial velocity but not the initial position.

As a second consequence of the addition and constant multiple theorems for derivatives, we obtain the difference rule for derivatives:

$$(f - g)'(x) = f'(x) - g'(x) \qquad (3)$$

for all x at which f and g are differentiable.

An important application of the derivative of the difference of two functions occurs in economics. Recall that if R and C are the revenue and cost functions, then the marginal revenue m_R and the marginal cost m_C are given by

$$m_R(x) = R'(x) \quad \text{and} \quad m_C(x) = C'(x)$$

for all x at which R and C are differentiable. Now let P be the profit function, defined by

$$P(x) = R(x) - C(x)$$

for all x in the domains of both R and C. The **marginal profit** m_P is defined by

$$m_P(x) = P'(x)$$

Then (3) tells us that

$$m_P(x) = R'(x) - C'(x) = m_R(x) - m_C(x)$$

for all x at which m_R and m_C are defined.

Example 6 Suppose the revenue and cost functions are given by

$$R(x) = 3x \quad \text{and} \quad C(x) = 4\sqrt{x} \quad \text{for } x > 0$$

Find a formula for the marginal profit function.

Solution From (1) we know that $m_R(x) = 3$, and from Example 3 we know that $m_C(x) = 2x^{-1/2}$. We conclude that

$$m_P(x) = m_R(x) - m_C(x) = 3 - 2x^{-1/2} \quad \square$$

Derivative of a Product   As we have seen, the derivative of a sum of functions is the sum of the derivatives, and the derivative of a difference of functions is the difference of the derivatives. By analogy it is tempting to assume that the derivative of a product of functions is the product of the derivatives. But if that were the case, we would conclude that

$$\frac{d}{dx}(x^2) = \frac{d}{dx}(x \cdot x) \overset{?}{=} 1 \cdot 1 = 1$$

But that is incorrect, because

$$\frac{d}{dx}(x^2) = 2x$$

The correct formula, discovered by Leibniz, is often called the Leibniz rule.

THEOREM 3.6

If f and g are differentiable at a, then fg is also differentiable at a, and

$$(fg)'(a) = f'(a)g(a) + f(a)g'(a)$$

Proof   In addition to using several limit rules, we will employ the device of adding the number 0, in the disguised form $-f(a)g(t) + f(a)g(t)$. We find that

$$(fg)'(a) = \lim_{t \to a} \frac{(fg)(t) - (fg)(a)}{t - a} = \lim_{t \to a} \frac{f(t)g(t) - f(a)g(a)}{t - a}$$

$$= \lim_{t \to a} \frac{f(t)g(t) \overbrace{- f(a)g(t) + f(a)g(t)}^{=0} - f(a)g(a)}{t - a}$$

$$= \lim_{t \to a} \left[\frac{f(t) - f(a)}{t - a} \cdot g(t) \right] + \lim_{t \to a} \left[f(a) \cdot \frac{g(t) - g(a)}{t - a} \right]$$

$$= \lim_{t \to a} \frac{f(t) - f(a)}{t - a} \lim_{t \to a} g(t) + f(a) \lim_{t \to a} \frac{g(t) - g(a)}{t - a}$$

Because g is differentiable, Theorem 3.3 implies that $\lim_{t \to a} g(t) = g(a)$. This allows us to conclude from Definition 3.1 that

$$(fg)'(a) = f'(a)g(a) + f(a)g'(a) \quad \blacksquare$$

Theorem 3.6 can be interpreted as saying that

$$(fg)'(x) = f'(x)g(x) + f(x)g'(x)$$

for all x at which both f and g are differentiable.

Example 7   Let $k(x) = x \sin x$. Find $k'(x)$.

Solution   If $f(x) = x$ and $g(x) = \sin x$, then $k = fg$. Since

$$f'(x) = 1 \quad \text{and} \quad g'(x) = \cos x$$

we have

$$k'(x) = f'(x)g(x) + f(x)g'(x) = \sin x + x \cos x \quad \square$$

In the Leibniz notation, if u and v are variables depending on x, then

$$\frac{d}{dx}(uv) = \frac{du}{dx}v + u\frac{dv}{dx} \tag{4}$$

Example 8   Find $\dfrac{d}{dx}(x^2 \cos x)$.

Solution   Since

$$\frac{d}{dx}(x^2) = 2x \quad \text{and} \quad \frac{d}{dx}\cos x = -\sin x$$

we apply (4) and find that

$$\frac{d}{dx}(x^2 \cos x) = \left[\frac{d}{dx}(x^2)\right]\cos x + x^2\left[\frac{d}{dx}(\cos x)\right]$$

$$= 2x \cos x - x^2 \sin x \quad \square$$

We can extend Theorem 3.6 to the derivative of the product of more than two functions. However, the larger the number of functions, the more complicated the formula becomes. With three functions f, g, and h, the formula is

$$(fgh)'(x) = f'(x)g(x)h(x) + f(x)g'(x)h(x) + f(x)g(x)h'(x) \tag{5}$$

Example 9   Let $k(x) = x^2 \sin x \cos x$. Find $k'(x)$.

Solution   Let $f(x) = x^2$, $g(x) = \sin x$, and $h(x) = \cos x$. Then

$$f'(x) = 2x, \qquad g'(x) = \cos x, \quad \text{and} \quad h'(x) = -\sin x$$

Since $k = fgh$, we use (5) to conclude that

$$k'(x) = 2x \sin x \cos x + x^2 \cos^2 x - x^2 \sin^2 x \quad \square$$

Derivative of a Quotient   We now investigate the derivative of the quotient of two functions.

THEOREM 3.7

If f and g are differentiable at a and $g(a) \neq 0$, then f/g is also differentiable at a, and

$$\left(\frac{f}{g}\right)'(a) = \frac{f'(a)g(a) - f(a)g'(a)}{[g(a)]^2}$$

Proof   Since $g'(a)$ exists by hypothesis, it follows from Theorem 3.3 that g is continuous at a, so that $\lim_{t \to a} g(t) = g(a)$. Because $g(a) \neq 0$ by hypothesis, the comment following (10) of Section 2.3 tells us that $g(t) \neq 0$ for all t in some open interval about a. Therefore f/g is defined throughout some open interval about a, and the following limits exist:

$$\left(\frac{f}{g}\right)'(a) = \lim_{t \to a} \frac{\dfrac{f(t)}{g(t)} - \dfrac{f(a)}{g(a)}}{t - a} = \lim_{t \to a} \frac{f(t)g(a) - f(a)g(t)}{(t - a)g(a)g(t)}$$

$$= \lim_{t \to a} \frac{f(t)g(a) \overbrace{- f(a)g(a) + f(a)g(a)}^{= 0} - f(a)g(t)}{(t - a)g(a)g(t)}$$

$$= \lim_{t \to a} \frac{f(t)g(a) - f(a)g(a)}{(t - a)g(a)g(t)} + \lim_{t \to a} \frac{f(a)g(a) - f(a)g(t)}{(t - a)g(a)g(t)}$$

$$= \lim_{t \to a} \left(\frac{f(t) - f(a)}{t - a} \cdot \frac{g(a)}{g(a)g(t)} \right) - \lim_{t \to a} \left(\frac{f(a)}{g(a)g(t)} \cdot \frac{g(t) - g(a)}{t - a} \right)$$

$$= \frac{f'(a)g(a)}{[g(a)]^2} - \frac{f(a)g'(a)}{[g(a)]^2} = \frac{f'(a)g(a) - f(a)g'(a)}{[g(a)]^2} \quad \blacksquare$$

Theorem 3.7 can be interpreted as saying that

$$\left(\frac{f}{g}\right)'(x) = \frac{f'(x)g(x) - f(x)g'(x)}{[g(x)]^2}$$

for all x at which both f and g are differentiable and $g(x) \neq 0$.

Every rational function is the quotient of two polynomials. Since polynomials are differentiable at every real number, the quotient rule just cited shows that every rational function is differentiable at every number in its domain.

Example 10   Let $k(x) = \dfrac{9x^7}{x^2 + 1}$. Find $k'(x)$.

Solution   If $f(x) = 9x^7$ and $g(x) = x^2 + 1$, then $k = f/g$, $f'(x) = 63x^6$, and $g'(x) = 2x$. Therefore

$$k'(x) = \frac{f'(x)g(x) - f(x)g'(x)}{[g(x)]^2} = \frac{63x^6(x^2 + 1) - (9x^7)(2x)}{(x^2 + 1)^2}$$

$$= \frac{45x^8 + 63x^6}{(x^2 + 1)^2} \quad \square$$

In the Leibniz notation, if u and v are variables depending on x, then

$$\frac{d}{dx}\left(\frac{u}{v}\right) = \frac{\dfrac{du}{dx}v - u\dfrac{dv}{dx}}{v^2} = \frac{v\dfrac{du}{dx} - u\dfrac{dv}{dx}}{v^2}$$

Example 11   Show that $\dfrac{d}{dx}\tan x = \sec^2 x$.

Solution   Since $\tan x = (\sin x)/\cos x$, we conclude that

$$\frac{d}{dx}\tan x = \frac{\cos x\dfrac{d}{dx}\sin x - \sin x\dfrac{d}{dx}\cos x}{\cos^2 x}$$

$$= \frac{\cos x \cos x - \sin x(-\sin x)}{\cos^2 x} = \frac{\cos^2 x + \sin^2 x}{\cos^2 x}$$

$$= \frac{1}{\cos^2 x} = \sec^2 x \quad \square$$

In exactly the same way you can show that

$$\frac{d}{dx}\cot x = -\csc^2 x$$

The formula for the derivative of a quotient becomes more concise when $f(x) = 1$ for all x. In this case the formula is

$$\left(\frac{1}{g}\right)'(x) = \frac{-g'(x)}{[g(x)]^2} \tag{6}$$

or in the Leibniz notation, with $u = 1$,

$$\frac{d}{dx}\left(\frac{1}{v}\right) = \frac{-1}{v^2}\frac{dv}{dx}$$

Example 12 Show that $\dfrac{d}{dx}\sec x = \sec x \tan x$.

Solution From (6) we obtain

$$\frac{d}{dx}\sec x = \frac{d}{dx}\left(\frac{1}{\cos x}\right) = \frac{-(-\sin x)}{\cos^2 x}$$

$$= \frac{1}{\cos x}\cdot\frac{\sin x}{\cos x} = \sec x \tan x \quad \square$$

Similarly,

$$\frac{d}{dx}\csc x = -\csc x \cot x$$

Example 13 Show that $\dfrac{d}{dx}(x^{-n}) = -nx^{-n-1}$, for any positive integer n.

Solution Combining (6) of this section and (1) of Section 3.2, we find that

$$\frac{d}{dx}(x^{-n}) = \frac{d}{dx}\left(\frac{1}{x^n}\right) = \frac{-nx^{n-1}}{(x^n)^2} = \frac{-nx^{n-1}}{x^{2n}} = -nx^{n-1-2n} = -nx^{-n-1} \quad \square$$

For example,

$$\frac{d}{dx}\left(\frac{1}{x}\right) = -\frac{1}{x^2} \quad \text{and} \quad \frac{d}{dx}(x^{-17}) = -17x^{-18}$$

The function x^{-2} appears in Newton's Law of Gravitation and in the formula for the electric force between charges. In addition, x^{-4} appears in the formula for the flow of blood through arteries. Thus functions of the form x^{-n}, where n is positive, arise in the real world.

Example 13, combined with formula (1) of Section 3.2, yields

$$\frac{d}{dx}(x^n) = nx^{n-1} \quad \text{for any nonzero integer } n \tag{7}$$

If we assume that $0 \cdot x^{-1} = 0$, then (7) holds even when $n = 0$. Using the results of the next section, we will be able to show that (7) remains valid when n is replaced by any nonzero rational number.

EXERCISES 3.3

In Exercises 1–32 find the derivative of the given function.

1. $f(x) = -4x^3$

2. $f(x) = x^5 - x^8$

3. $f(x) = 3x^2 - 4x + 2$

4. $f(x) = -2x^3 + 6x^2 - 5x + \sqrt{2}$

5. $f(x) = 4x^4 + 3x^3 + 2x^2 + x$

6. $f(t) = 7t^{-5}$

7. $f(t) = -\dfrac{4}{t^9}$

8. $f(t) = -3t^2 + \dfrac{3}{t^6}$

9. $g(x) = (2x - 3)(x + 5)$

10. $g(x) = (x - 2x^2)^2$

11. $g(x) = \left(1 + \dfrac{1}{x}\right)\left(2 - \dfrac{1}{x}\right)$

12. $g(x) = \left(x - \dfrac{1}{x}\right)\left(x^2 - \dfrac{1}{x^2}\right)$

13. $g(x) = -4x^{-3} + 2\cos x$

14. $g(x) = 3\sin x + 5\cos x$

15. $f(z) = -2z^3 + 4\sec z$

16. $f(z) = z^{-11} + \pi \tan z$

17. $f(z) = z^2 \sin z$

18. $f(x) = \sin x \cos x$

19. $f(x) = \sin^2 x$

20. $f(x) = \sqrt{x}(x^2 + 3\sin x)$

21. $f(x) = -4x \tan x \sec x$

22. $f(x) = (2x + 1)(x - \tan x)$

23. $f(x) = \dfrac{2x + 3}{4x - 1}$

24. $f(x) = \dfrac{-x^2 + 3}{x^2 + 9}$

25. $f(t) = \dfrac{t + 2}{t^2 + 4t + 4}$

26. $f(t) = \dfrac{t^2 - 5t + 4}{t^2 + t - 20}$

27. $f(t) = \dfrac{t^2 + 5t + 4}{t^2 + t - 20}$

28. $f(t) = \dfrac{-2t}{\sin t}$

29. $f(x) = \cot x$ (*Hint:* $\cot x = (\cos x)/\sin x$.)

30. $f(x) = \csc x$

31. $f(y) = \sqrt{y}\sec y$

32. $g(y) = \dfrac{\csc y}{y^2}$

In Exercises 33–43 find dy/dx.

33. $y = (3x + 1)(2x^2 - 5x)$

34. $y = x + \dfrac{1}{x}$

35. $y = x^2 + \dfrac{1}{x^2}$

36. $y = -3x - \dfrac{6}{x + 2}$

37. $y = \dfrac{2x - 1}{5 - 3x}$

38. $y = \dfrac{x^2 + x + 1}{x^2 - x + 1}$

39. $y = \dfrac{x^3 - 1}{x^4 + 1}$

40. $y = 4x \sec x$

41. $y = \csc x \sec x$

42. $y = \dfrac{x^2 + \sqrt{x}}{\sin x \cos x}$

43. $y = \dfrac{x \sin x}{x^2 + 1}$

44. Use trigonometric identities to find dy/dx in each case.
   a. $y = \sin 2x$ b. $y = \sin(-x)$
   c. $y = \cos^2(x/2)$

In Exercises 45–50 find $f'(a)$.

45. $f(x) = -7/x^9$; $a = 1$

46. $f(x) = (3x - \sin x)(x^2 + \cos x)$; $a = 0$

47. $f(x) = \dfrac{1}{\pi}(3x^2 - 4x); a = -2$

48. $f(x) = \dfrac{2x^2 - 1}{x^2 + 1}; a = \sqrt{3}$

49. $f(x) = \dfrac{\sin x}{\sqrt{x}}; a = \dfrac{\pi}{2}$

50. $f(x) = \dfrac{\sec x}{x^2}; a = \pi$

In Exercises 51–53 find an equation of the line tangent to the graph of f at the given point.

51. $f(x) = x^2 - 3x - 4; (2, -6)$

52. $f(x) = \dfrac{x + 1}{x - 1}; (3, 2)$

53. $f(x) = \sin x - \cos x; (\pi/2, 1)$

54. Let $f(x) = 2x^3 - 9x^2 + 12x + 1$. Find the points on the graph of f at which the tangent line is horizontal.

55. Let $f(x) = |x|$ and $g(x) = -|x|$. Find a simple formula for $f + g$. Then show that $(f + g)'(x)$ exists for all x, whereas $f'(x)$ and $g'(x)$ exist only for all nonzero x.

56. Prove that if $f'(a)$ and $(f + g)'(a)$ exist, then $g'(a)$ exists.

57. Suppose $g'(a)$ and $(fg)'(a)$ exist and $g(a) \neq 0$. Prove that $f'(a)$ exists.

58. Suppose that $f'(a)$, $g'(a)$, and $h'(a)$ exist. Prove that $(f + g + h)'(a)$ exists and that

$$(f + g + h)'(a) = f'(a) + g'(a) + h'(a)$$

by using the idea of the proof of Theorem 3.4.

59. Prove that if $f'(a)$, $g'(a)$, and $h'(a)$ exist, then

$$(fgh)'(a) = f'(a)g(a)h(a) + f(a)g'(a)h(a) + f(a)g(a)h'(a)$$

(*Hint:* First, think of fgh as $(fg)h$ and apply Theorem 3.6 to the two functions fg and h. Then rewrite $(fg)'(a)$ wherever it appears, with the help of Theorem 3.6.)

60. Prove the quotient rule (Theorem 3.7) by assuming (6) and then by treating f/g as $f(1/g)$ and using the product rule (Theorem 3.6).

61. a. Find the velocity of the ball in Example 5 when $t = 1$ and when $t = 2$.
    b. Suppose the ball in Example 5 is hurled upward at 48 feet per second. Determine the speed of the ball when $t = 1$.

62. Assume that a ball is thrown vertically upward from the ground with an initial velocity of 64 feet per second. Assuming that the velocity will be 0 when the height of the ball is at its maximum, find the maximum height.

63. Assume that a ball is shot vertically upward from the ground with an initial velocity of 128 feet per second.

a. Assuming that the velocity will be 0 when the height of the ball is at its maximum, find the maximum height.
b. Find the velocity of the ball as it hits the ground again.

64. Assume that the ball in Exercise 63 is shot from 8 feet above the ground. Find the velocity of the ball when it returns to the 8-foot level.

65. If a ball is thrown downward with an initial speed of 16 feet per second (that is, the initial velocity is -16) from the top of a 96-foot building, find a formula for its velocity at time t.

66. If $h(t) = -16t^2 + v_0 t + h_0$ represents the height of an object moving vertically, find a formula for the derivative of the velocity $v(t)$.

67. Suppose that a 5-foot-long chandelier falls from a 35-foot ceiling and that you are 6 feet tall and standing directly under the chandelier (Figure 3.7).
    a. How long do you have to get out of the way?
    b. If you do not duck or get out of the way, how fast will the chandelier be traveling when it hits your head?
    c. If you were only 5 feet tall, how much greater would the speed of the chandelier be when it hit your head?

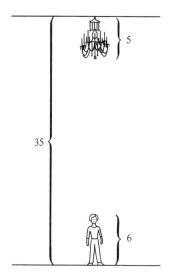

FIGURE 3.7

68. The center of gravity of a long jumper traverses a parabolic path during a jump. Suppose the height $h(x)$ in feet of the jumper's center of gravity during a 26-foot jump satisfies

$$h(x) = \frac{-2}{169}x^2 + \frac{4}{13}x + 3 \quad \text{for } 0 \leq x \leq 26$$

where x is the distance in feet along the ground (Figure 3.8). Show that h is differentiable on $[0, 26]$ and find a formula for $h'(x)$.

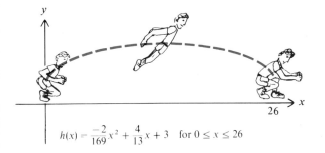

$$h(x) = \frac{-2}{169}x^2 + \frac{4}{13}x + 3 \quad \text{for } 0 \le x \le 26$$

FIGURE 3.8

69. Suppose you are dragging a sack of sand weighing 50 pounds. If your arm makes an angle of x radians with the ground and if the coefficient of friction is $\mu > 0$, then the minimum force $F(x)$ necessary to drag the sack is given by

$$F(x) = \frac{50\mu}{\mu \sin x + \cos x} \quad \text{for } 0 \le x \le \frac{\pi}{2}$$

Show that F is differentiable on $[0, \pi/2]$ and find a formula for $F'(x)$.

70. The response $R(x)$ to a given concentration x of acetylcholine on a frog's heart is given by

$$R(x) = \frac{x}{a + bx} \quad \text{for } 0 \le x \le 1$$

where a and b are fixed positive constants. Show that R is differentiable on $[0, 1]$ and find a formula for R'.

71. The formula

$$f(x) = \frac{100kx^n}{1 + kx^n} \quad \text{for } x > 0$$

where n is a positive integer and k is a constant, appears in the study of the saturation of hemoglobin with oxygen. Find a formula for $f'(x)$.

72. Suppose a yogurt firm finds that its revenue and cost functions are given by

$$R(x) = 15x^{1/2} - x^{3/2} \quad \text{and} \quad C(x) = 3x^{1/2} + 4$$

respectively, for $0 \le x \le 5$. Here x is measured in thousands of gallons and $R(x)$ and $C(x)$ are measured in hundreds of dollars.
 a. Find a formula for the marginal profit m_P, and calculate $m_P(1)$.
 b. Show that $m_P(4) = 0$.

73. Suppose the cost and revenue functions of a gingerbread manufacturing firm are described by

$$C(x) = \frac{x + 3}{\sqrt{x + 1}} \quad \text{and} \quad R(x) = \sqrt{x} \quad \text{for } 1 \le x \le 15$$

Find a value of x for which the profit is 0, and show that for no value of x is the marginal profit $m_P(x) = 0$.

3.4
THE CHAIN RULE

Many of the functions we encounter in mathematics and in applications are composite functions. For example, the function k defined by $k(x) = \sin 3x$, which occurs in the description of the motion of a pendulum, is the composite $g \circ f$, where

$$f(x) = 3x \quad \text{and} \quad g(x) = \sin x$$

If we could discover a general rule for the derivative of a composite function in terms of the derivatives of the component functions, then we would be able to find the derivative of k without resorting to the definition of the derivative (Definition 3.1). We now turn to such a rule. It will be used repeatedly throughout the text.

In order to get an idea of what the formula for $(g \circ f)'(a)$ should be, let f be differentiable at a, g differentiable at $f(a)$, and $f(x) \ne f(a)$ for all x near a. Then by introducing the number 1, written as

$$\frac{f(x) - f(a)}{f(x) - f(a)}$$

we obtain

$$\frac{(g \circ f)(x) - (g \circ f)(a)}{x - a} = \frac{g(f(x)) - g(f(a))}{x - a}$$

$$= \frac{g(f(x)) - g(f(a))}{f(x) - f(a)} \cdot \frac{f(x) - f(a)}{x - a}$$

Using the Product Rule for limits, as well as the Substitution Rule with $y = f(x)$, we conclude that

$$\lim_{x \to a} \frac{(g \circ f)(x) - (g \circ f)(a)}{x - a} = \lim_{x \to a} \left(\frac{g(f(x)) - g(f(a))}{f(x) - f(a)} \cdot \frac{f(x) - f(a)}{x - a} \right)$$

$$= \lim_{x \to a} \frac{g(f(x)) - g(f(a))}{f(x) - f(a)} \lim_{x \to a} \frac{f(x) - f(a)}{x - a}$$

$$= \lim_{y \to f(a)} \frac{g(y) - g(f(a))}{y - f(a)} \lim_{x \to a} \frac{f(x) - f(a)}{x - a}$$

$$= g'(f(a))f'(a)$$

Thus $(g \circ f)'(a) = g'(f(a))f'(a)$. This formula, which is the substance of the following theorem, is valid even if it is not true that $f(x) \neq f(a)$ for all x near a. The complete proof (which is considerably more difficult than the argument just presented) is given in the Appendix.

THEOREM 3.8
THE CHAIN RULE

Let f be differentiable at a, and let g be differentiable at $f(a)$. Then $g \circ f$ is differentiable at a, and

$$(g \circ f)'(a) = g'(f(a))f'(a) \tag{1}$$

The Chain Rule can be interpreted as saying that

$$(g \circ f)'(x) = g'(f(x))f'(x)$$

for all x such that f is differentiable at x and g is differentiable at $f(x)$. The formula in (1) is often written in the form

$$\frac{d}{dx} g(f(x)) = g'(f(x))f'(x) \tag{2}$$

Example 1   Let $k(x) = \sin 3x$. Find a formula for $k'(x)$.

Solution Let $f(x) = 3x$ and $g(x) = \sin x$. Then $k = g \circ f$. Since

$$f'(x) = 3 \quad \text{and} \quad g'(x) = \cos x$$

we conclude that

$$k'(x) = g'(f(x))f'(x) = (\cos 3x)(3) = 3\cos 3x \quad \square$$

Example 2 Find a formula for $\dfrac{d}{dx}\sqrt{1 + x^4}$.

Solution Let $f(x) = 1 + x^4$ and $g(x) = \sqrt{x}$. Then

$$f'(x) = 4x^3 \quad \text{and} \quad g'(x) = \frac{1}{2\sqrt{x}} \quad \text{for } x > 0$$

Therefore

$$\frac{d}{dx}\sqrt{1 + x^4} = g'(f(x))f'(x) = \frac{1}{2\sqrt{f(x)}}(4x^3) = \frac{2x^3}{\sqrt{1 + x^4}} \quad \square$$

When we use the Chain Rule to find a derivative, we need to decide which f and g to choose. In Examples 1 and 2 we chose $f(x)$ to be a suitably simple component of the function to be differentiated, and this determined what $g(x)$ had to be. In practice, the choices of f and g are usually made mentally and not written down. Finally we mention that in finding $(g \circ f)'(x) = g'(f(x))f'(x)$ we compute $g'(x)$ and replace x by $f(x)$ to obtain $g'(f(x))$.

Example 3 Find the derivatives of the following functions.
  a. $k(x) = (x - 2x^3)^{-11}$ b. $y = \cos x^5$

Solution
  a. Using the comments just above, we find that

$$k'(x) = [-11(x - 2x^3)^{-12}](1 - 6x^2) = (-11 + 66x^2)(x - 2x^3)^{-12}$$

  b. Here we have

$$\frac{dy}{dx} = (-\sin x^5)(5x^4) = -5x^4 \sin x^5$$

  or equivalently,

$$\frac{d}{dx}(\cos x^5) = (-\sin x^5)(5x^4) = -5x^4 \sin x^5 \quad \square$$

Observe that when we differentiate a function by using the Chain Rule, we differentiate from the outside inward. Thus we differentiate $(x - 2x^3)^{-11}$ by first differentiating the outer function x^{-11} (at $x - 2x^3$) and then differentiating the inner function $x - 2x^3$. Similarly, we differentiate $\cos x^5$ by first differentiating the outer function $\cos x$ (at x^5) and then differentiating the inner function x^5.

The Chain Rule assumes a very suggestive form in the Leibniz notation. Suppose the functions f and g in the Chain Rule are already given, and let

$$u = f(x) \quad \text{and} \quad y = g(u)$$

Then $y = g(f(x))$, $du/dx = f'(x)$, and $dy/du = g'(u)$. Therefore by (2),

$$\frac{dy}{dx} = \frac{d}{dx} g(f(x)) = g'(f(x))f'(x) = g'(u)f'(x) = \frac{dy}{du}\frac{du}{dx}$$

or more concisely,

$$\frac{dy}{dx} = \frac{dy}{du}\frac{du}{dx} \tag{3}$$

Formula (3) is the Chain Rule in the Leibniz notation.

Caution: Notice that if dy/du and du/dx were quotients rather than expressions for derivatives, we could cancel the du's and make (3) into an identity. But we stress that du has not been defined as an entity, and consequently it is not legitimate to cancel the du's. Nonetheless, the resemblance between (3) and an algebraic identity makes it easy to remember.

Example 4   Let $y = \sin^8 x$. Find dy/dx.

Solution   First we notice that

$$\text{if } u = \sin x, \quad \text{then} \quad y = u^8$$

Then from (3) it follows that

$$\frac{dy}{dx} = \frac{dy}{du}\frac{du}{dx} = (8u^7)(\cos x) = 8\sin^7 x \cos x \quad \square$$

In formula (3), y is represented in two different ways: once as a function of x and once as a function of u. The expression dy/dx is the derivative of y when y is regarded as a function of x. In the same way, dy/du is the derivative of y when y is regarded as a function of u. To show that dy/dx and dy/du may be different, let us consider a simple example. Suppose

$$y = u^2 \quad \text{and} \quad u = \frac{1}{x}$$

Then $y = (1/x)^2 = 1/x^2$, so that

$$\frac{dy}{dx} = \frac{-2}{x^3}, \quad \text{whereas} \quad \frac{dy}{du} = 2u = 2\left(\frac{1}{x}\right) = \frac{2}{x}$$

Therefore

$$\frac{dy}{dx} \neq \frac{dy}{du}$$

and thus when we use variables, it is important to specify the variable with respect to which we differentiate.

Formula (3) is especially useful when y is not given explicitly in terms of x but is given in terms of an intermediate variable, as in the next example.

Example 5 Suppose the radius r of a balloon varies with respect to time according to the equation $r = 1 + 2t$. Find the rate of change of the balloon's volume with respect to time.

Solution If V denotes the volume, then $V = \frac{4}{3}\pi r^3$, while by assumption $r = 1 + 2t$. Therefore (3) tells us that

$$\frac{dV}{dt} = \frac{dV}{dr}\frac{dr}{dt} = 4\pi r^2(2) = 8\pi r^2 = 8\pi(1 + 2t)^2 \quad \square$$

From (7) of Section 3.3 we know that

$$\frac{d}{dx}(x^n) = nx^{n-1} \quad \text{for any nonzero integer } n \tag{4}$$

Now let n be replaced by any nonzero rational number r. Using the Chain Rule, we will show that

$$\frac{d}{dx}(x^r) = rx^{r-1} \tag{5}$$

for all numbers in the domain of x^{r-1}. This result is obviously analogous to the one in (4).

We first consider the case in which $r = 1/m$, where m is a positive integer. Let

$$f(x) = x^{1/m}$$

with domain $(-\infty, \infty)$ if m is an odd integer and domain $[0, \infty)$ if m is an even integer. It will be convenient to use the following algebraic formula:

$$\frac{b - a}{b^m - a^m} = \frac{1}{b^{m-1} + b^{m-2}a + \cdots + ba^{m-2} + a^{m-1}}$$

Letting $t^{1/m} = b$ and $x^{1/m} = a$, we find that

$$\frac{t^{1/m} - x^{1/m}}{(t^{1/m})^m - (x^{1/m})^m} = \frac{1}{t^{(m-1)/m} + t^{(m-2)/m}x^{1/m} + \cdots + t^{1/m}x^{(m-2)/m} + x^{(m-1)/m}}$$

Observe that there are m summands in the denominator on the right, and that as t approaches x each summand approaches $x^{(m-1)/m}$. Therefore

$$\frac{d}{dx}(x^{1/m}) = \lim_{t \to x}\frac{t^{1/m} - x^{1/m}}{t - x} = \lim_{t \to x}\frac{t^{1/m} - x^{1/m}}{(t^{1/m})^m - (x^{1/m})^m} = \frac{1}{mx^{(m-1)/m}} = \frac{1}{m}x^{1/m-1}$$

Thus we have shown that (5) is valid when $r = 1/m$.

If r is any nonzero rational number at all, then we write $r = n/m$, where n is an integer and m is a positive integer, so that

$$x^r = x^{n/m} = (x^n)^{1/m}$$

Then the Chain Rule tells us that

$$\frac{d}{dx}(x^r) = \frac{d}{dx}(x^n)^{1/m} = \left(\frac{1}{m}(x^n)^{(1/m)-1}\right)(nx^{n-1})$$

$$= \frac{n}{m}x^{(n/m)-n}x^{n-1} = \frac{n}{m}x^{(n/m)-1} = rx^{r-1}$$

Thus (5) holds for all nonzero rational r.

Example 6 Find $\dfrac{d}{dx}(x^{-2/5})$.

Solution Applying (5), we find that

$$\frac{d}{dx}(x^{-2/5}) = -\frac{2}{5}x^{(-2/5)-1} = -\frac{2}{5}x^{-7/5} \quad \square$$

Example 7 Let $k(x) = (3x^2 + 1)^{7/4}$. Find $k'(x)$.

Solution Using (5) and the Chain Rule, we find that

$$k'(x) = \frac{7}{4}(3x^2 + 1)^{(7/4)-1}(6x) = \frac{21}{2}x(3x^2 + 1)^{3/4} \quad \square$$

Example 8 Suppose y is a differentiable function of x. Express the derivative of the given function with respect to x in terms of x, y, and dy/dx.
a. y^3 b. $\sin y$ c. $2x^3y^4$

Solution
   a. It might appear that the derivative of y^3 with respect to x would be $3y^2$, but this is false because we must differentiate y^3 with respect to x, not with respect to y. Instead, we use the Chain Rule in the form

$$\frac{dv}{dx} = \frac{dv}{dy}\frac{dy}{dx} \tag{6}$$

where $v = y^3$. We obtain

$$\frac{d}{dx}(y^3) = \left(\frac{d}{dy}(y^3)\right)\frac{dy}{dx} = 3y^2\frac{dy}{dx}$$

b. Using (6) with $v = \sin y$, we have

$$\frac{d}{dx}(\sin y) = \left(\frac{d}{dy}(\sin y)\right)\frac{dy}{dx} = (\cos y)\frac{dy}{dx}$$

c. Since $2x^3y^4$ is the product of the functions $2x^3$ and y^4, we take the derivative of the product, using (6) with $v = y^4$ when we differentiate y^4 with respect to x:

$$\frac{d}{dx}(2x^3y^4) = (6x^2)y^4 + 2x^3\left(\frac{d}{dx}(y^4)\right) = 6x^2y^4 + 2x^3\left(4y^3\frac{dy}{dx}\right)$$

$$= 6x^2y^4 + 8x^3y^3\frac{dy}{dx} \quad \square$$

The Compound Chain Rule

The Chain Rule can be carried a step further. Let

$$k(x) = (h \circ g \circ f)(x) = h(g(f(x)))$$

and let f be differentiable at x, g differentiable at $f(x)$, and h differentiable at $g(f(x))$. Since

$$k(x) = (h \circ g \circ f)(x) = h((g \circ f)(x))$$

a first application of the Chain Rule yields

$$k'(x) = h'(g(f(x)))(g \circ f)'(x)$$

A second application of the Chain Rule, to the term on the right, yields

$$k'(x) = h'(g(f(x)))g'(f(x))f'(x) \tag{7}$$

The derivative of k is thus obtained in a chainlike fashion. In the formula, the derivative of h at the number $g(f(x))$ appears first, then the derivative of g at the number $f(x)$, and finally, the derivative of f at the number x.

Example 9   Let $k(x) = \cos^3 4x$. Find $k'(x)$, and calculate $k'(\pi/6)$.

Solution   Let $f(x) = 4x$, $g(x) = \cos x$, and $h(x) = x^3$. Then $k(x) = h(g(f(x)))$. By (7) we find that

$$k'(x) = (3\cos^2 4x)(-\sin 4x)(4) = -12\cos^2 4x \sin 4x$$

In particular,

$$k'\left(\frac{\pi}{6}\right) = -12\cos^2\frac{2\pi}{3}\sin\frac{2\pi}{3} = -12\left(-\frac{1}{2}\right)^2\left(\frac{\sqrt{3}}{2}\right) = -\frac{3}{2}\sqrt{3} \quad \square$$

In Leibniz notation (7) becomes

$$\frac{du}{dx} = \frac{du}{dv}\frac{dv}{dy}\frac{dy}{dx} \tag{8}$$

where $y = f(x)$, $v = g(y)$, and $u = h(v)$.

Example 10 Suppose y is a function of x. Find $\dfrac{d}{dx}(\sin^3 y)$ in terms of y and $\dfrac{dy}{dx}$.

Solution If we let $v = \sin y$ and $u = v^3$, then

$$\sin^3 y = (\sin y)^3 = v^3 = u$$

so that by (8),

$$\frac{d}{dx}(\sin^3 y) = \frac{du}{dx} = \frac{du}{dv}\frac{dv}{dy}\frac{dy}{dx} = (3v^2)(\cos y)\frac{dy}{dx} = (3\sin^2 y \cos y)\frac{dy}{dx} \quad \square$$

With a little practice you will probably find it easier to apply the Chain Rule from the outside inward, without introducing intermediate variables. For Example 10 one could write

$$\frac{d}{dx}(\sin^3 y) = (3\sin^2 y)(\cos y)\frac{dy}{dx}$$

The Chain Rule can be applied to even longer composites. The procedure is always the same: Differentiate from the outside inward, and multiply the resulting derivatives (evaluated at the appropriate numbers). For example,

$$\frac{d}{dx}[\sin(\cos(\tan^5 x))] = [\cos(\cos(\tan^5 x))][-\sin(\tan^5 x)](5\tan^4 x)(\sec^2 x)$$

Summary of Differentiation Rules

We have now presented all the basic rules of differentiation. Using them, you will be able to differentiate all sorts of functions, including very complicated ones. To have the rules readily available, we list them here:

1. $(f + g)'(x) = f'(x) + g'(x)$

2. $(cf)'(x) = cf'(x)$

3. $(fg)'(x) = f'(x)g(x) + f(x)g'(x)$

4. $\left(\dfrac{f}{g}\right)'(x) = \dfrac{f'(x)g(x) - f(x)g'(x)}{[g(x)]^2}$

5. $(g \circ f)'(x) = \dfrac{d}{dx}g(f(x)) = g'(f(x))f'(x)$

There are still some functions whose derivatives cannot be computed via these rules. However, the derivative of such a function can sometimes be computed directly from the definition. For instance, if $f(x) = x|x|$, then we cannot apply any of the rules to obtain $f'(0)$, because $|x|$ is not differentiable at 0. Nevertheless, using the definition of the derivative, we find that

$$f'(0) = \lim_{x \to 0} \frac{f(x) - f(0)}{x - 0} = \lim_{x \to 0} \frac{x|x| - 0}{x - 0} = \lim_{x \to 0} |x| = 0$$

But the great majority of the differentiable functions you will encounter can be differentiated by rules 1–5.

EXERCISES 3.4

In Exercises 1–34 find the derivative of the function.

1. $f(x) = x^{9/4}$

2. $f(x) = -5x^{-7/3}$

3. $f(x) = x^{2/3} - 7x^{-1/3}$

4. $f(x) = x^{77/6} + x^{-77/6}$

5. $f(x) = (4 - 3x^2)^{400}$

6. $f(x) = (12x^3 - 3x + 4)^{-68}$

7. $f(x) = \left(\dfrac{x-1}{x+1}\right)^3$

8. $f(x) = \sqrt{3x+1}$

9. $f(x) = x\sqrt{2 - 7x^2}$

10. $f(x) = \sqrt{4x - \sqrt{x}}$

11. $f(t) = \sin 5t$

12. $f(t) = 3\cos \pi t^2$

13. $f(t) = \sin^4 t + \cos^4 t$

14. $f(t) = \cos^{-3} t$

15. $g(x) = \tan^6 x$

16. $g(x) = \cot^6 x$

17. $g(x) = \sqrt[3]{1 - \sin x}$

18. $g(x) = (\tan x - \cot x)^{-1/3}$

19. $f(x) = \cos(\sin x)$

20. $f(x) = \tan(\csc x)$

21. $f(x) = \sqrt{x^2 + \dfrac{1}{x^2}}$

22. $f(x) = \dfrac{\sqrt{x^2 - 1}}{x}$

23. $f(x) = \dfrac{1}{x\sqrt{5 - 2x}}$

24. $f(x) = \dfrac{(x^2 + 1)^2}{(x^4 + 1)^4}$

25. $f(x) = x\cos(1/x)$

26. $f(x) = x^2 \sec(1/x^3)$

27. $g(z) = \sqrt{2z - (2z)^{1/3}}$

28. $g(z) = [z^7 + (z^2 - 1)^5]^{-2}$

29. $g(z) = \cos^2(3z^6)$

30. $g(z) = \sqrt{1 + \sin^2 z}$

31. $f(x) = \sqrt[4]{\sec(\tan x)}$

32. $f(x) = \cos(1 + \tan 2x)$

33. $f(x) = \cot^2(2\sqrt{3x + 1})$

34. $f(x) = \sin^2(\cos^2 x)$

In Exercises 35–42 find dy/dx.

35. $y = 3x^{-2/3}$

36. $y = \left(2x + \dfrac{1}{x}\right)^{-6}$

37. $y = -x\sqrt{1 + 3x^2}$

38. $y = \dfrac{1}{(x^8 + 1)^{12} + 1}$

39. $y = \left(\dfrac{1}{x\sin x}\right)^{2/3}$

40. $y = \sin\sqrt{2x + 1}$

41. $y = \tan^3 \frac{1}{2}x$

42. $y = \csc(1 - 3x)^2$

In Exercises 43–52 suppose that y is a differentiable function of x. Express the derivative of the given function with respect to x in terms of x, y, and dy/dx.

43. y^5

44. $y^{-2/3}$

45. $2/y$

46. $\cos y^2$

47. $\sin\sqrt{y}$

48. $\sec\sqrt{y^2 - 1}$

49. $x^3 y^2$

50. $\dfrac{1}{x^2 - xy + y^3}$

51. $\sqrt{x^2 + y^2}$

52. $\cos x^5 y^3$

In Exercises 53–56 find an equation of the line l tangent to the graph of f at the given point.

53. $f(x) = \dfrac{1}{(x+1)^2}$; $(0, 1)$

54. $f(x) = (1 + x^{1/3})^{2/3}$; $(-8, 1)$

55. $f(x) = -2\cos 3x$; $(\pi/3, 2)$

56. $f(x) = \sin\sqrt{x}$; $(\pi^2, 0)$

57. Assume that f and g are differentiable functions, and suppose that the values of f, f', g, and g' at $-3, 0, 1$, and 2 are prescribed as in the table below:

x	$f(x)$	$f'(x)$	$g(x)$	$g'(x)$
-3	2	4	6	11
0	1	2	-3	-7
1	16	-3	4	13
2	-3	2	0	$\sqrt{2}$

Determine $(f \circ g)'(x)$ and $(g \circ f)'(x)$ for as many values of x as the table allows.

58. a. Prove that the derivative of an even function is an odd function.

b. Prove that the derivative of an odd function is an even function.

59. Let n be a positive integer. Assume that we already know that the function $x^{1/n}$ is differentiable but that we do not know a formula for the derivative. Using the equation $x = (x^{1/n})^n$ and the Chain Rule, find a formula for the derivative of $x^{1/n}$.

60. The angular displacement $f(t)$ of a pendulum bob at time t is given by

$$f(t) = a \cos 2\pi\omega t$$

where ω is the frequency and a is the maximum displacement. Find the rate of change of the angular displacement as a function of time. (The rate of change is called the **angular velocity** of the bob.)

61. Suppose the amount $F(t)$ of water (in tons) that flows through a dam from midnight until t hours after midnight is controlled so that

$$F(t) = \frac{336,000}{\pi} \left(1 - \cos\frac{\pi t}{24} \right) \quad \text{for } 0 \leq t \leq 24$$

Show that F is differentiable on $[0, 24]$, and find a formula for the rate of flow of water (in tons per hour).

62. If we ignore air resistance, then the range $R(\theta)$ of a baseball hit at an angle θ with respect to the x axis and with initial velocity v_0 is given by

$$R(\theta) = \frac{v_0^2}{g} \sin 2\theta \quad \text{for } 0 \leq \theta \leq \frac{\pi}{2}$$

where g is the acceleration due to gravity.
a. If $v_0 = 96$ (feet per second) and $g = 32$ (feet per second per second), calculate $R'(\pi/4)$.
b. Determine those values of θ for which $R'(\theta) > 0$.

63. Under certain conditions the percentage efficiency of an internal combustion engine is given by

$$E = 100 \left(1 - \frac{v}{V} \right)^{0.4}$$

where V and v are, respectively, the maximum and minimum volumes of air in each cylinder.
a. If V is kept constant, find the derivative of E with respect to v.
b. If v is kept constant, find the derivative of E with respect to V.

64. If two bodies are a distance r apart, then the gravitational force $F(r)$ exerted by one body on the other is given by

$$F(r) = \frac{k}{r^2} \quad \text{for } r > 0$$

where k is a positive constant. Suppose that as a function

of time, the distance between the two bodies is given by

$$r(t) = 64 + 48t - 16t^2 \quad \text{for } 0 < t < 3$$

a. Find the rate of change of the force with respect to time.
b. Show that $(F \circ r)'(1) = -(F \circ r)'(2)$.

65. Suppose a rocket is launched vertically with a maximum velocity of v_0 miles per second, which is reached moments after takeoff. Suppose also that the velocity $v(r)$ in miles per second when the rocket is a distance of r miles from the center of the earth is given by the formula

$$v(r) = \sqrt{\frac{192,000}{r} + v_0^2 - 48} \quad \text{for } r \geq 4000$$

a. Find a formula for the rate of change in velocity with respect to r.
b. If $v_0 = 8$ miles per second, what is the rate of change of the rocket's velocity with respect to the distance from the center of the earth when that distance is 24,000 miles?

66. The weight $W(t)$ of a tree limb is a function of the age t of the limb. It can be approximated by the formula

$$W(t) = kt^r \quad \text{for } t > 0$$

where r is a rational number between 0 and 1 and k is a positive constant. The **specific rate of growth** is defined to be $W'(t)/W(t)$ for $t > 0$. Find a formula for the specific rate of growth.

67. Recall that the volume V of a spherical balloon is related to the radius r of the balloon by the formula

$$V = \frac{4}{3}\pi r^3$$

Suppose the radius is increasing at the constant rate of 10 inches per minute. Using the Chain Rule, find the rate of change of V with respect to time.

68. The surface area S of the balloon mentioned in Exercise 67 is related to the balloon's radius by the formula

$$S = 4\pi r^2$$

Using the Chain Rule, find the rate of change of the volume with respect to the surface area. (*Hint:* $r = (S/4\pi)^{1/2}$.)

69. If A, x, and h denote the area, length of side, and altitude of an equilateral triangle, respectively, then they are related by the formulas

$$A = \frac{\sqrt{3}}{4}x^2 \quad \text{and} \quad x = \frac{2\sqrt{3}}{3}h$$

Using the Chain Rule, find the rate of change of the area

with respect to the altitude, and determine the rate of change when $h = \sqrt{3}$.

70. Cider is poured into a cylindrical vat 4 feet in diameter and 5 feet tall. After t seconds the cider level is $\frac{1}{3}t$ feet above the base of the vat. Show that the rate of change of the volume with respect to time is constant.

71. The **demand** for a product gives the quantity $D(x)$ that can be sold when the price of one unit of the product is x. The demand almost always has a negative derivative. If

$$D(x) = \sqrt{3 - 2x} \quad \text{for } 0 < x < \frac{3}{2}$$

show that $D'(x) < 0$ for $0 < x < \frac{3}{2}$.

72. Suppose the research department of the Bulb Company determines that the demand for bulbs is given by

$$D(x) = 1000 \left(\frac{6}{x - 16} \right)^{1/3} \quad \text{for } 17 < x < 37$$

where x is in cents.
a. Find $D'(22)$.
b. Show that $D'(x) < 0$ for all x in $(17, 37)$.

3.5
HIGHER DERIVATIVES

If f is a function, then f' is the function that assigns the number $f'(x)$ to each x at which f is differentiable. Since f' is a function, we can carry the process a step further and define $f''(a)$ by the formula

$$f''(a) = \lim_{x \to a} \frac{f'(x) - f'(a)}{x - a} = (f')'(a)$$

whenever this limit exists. We call $f''(a)$ the **second derivative of f at a**. It is often read "f double prime of a." Correspondingly, $f'(a)$ is often called the **first derivative of f at a**. Since $f''(a)$ is merely the derivative of f' at a, finding second derivatives is no more difficult than finding first derivatives.

Example 1   Let $f(x) = \sin x$. Find a formula for $f''(x)$.

Solution   Since $f'(x) = \cos x$, it follows that $f''(x) = -\sin x$.   ☐

Example 2   Let $f(x) = 3x^{1/2}$. Find a formula for $f''(x)$.

Solution   We find that

$$f'(x) = 3\left(\frac{1}{2} x^{-1/2} \right) = \frac{3}{2} x^{-1/2}$$

$$f''(x) = \frac{3}{2}\left(-\frac{1}{2} x^{-3/2} \right) = -\frac{3}{4} x^{-3/2}$$ ☐

For any positive integer $n \geq 3$ we can define the **nth derivative $f^{(n)}(a)$ of f at a** by letting $f^{(n-1)}$ denote the $(n-1)$st derivative and then letting

$$f^{(n)}(a) = \lim_{x \to a} \frac{f^{(n-1)}(x) - f^{(n-1)}(a)}{x - a} = (f^{(n-1)})'(a)$$

if the limit exists. The second derivative, the third derivative, and so on are called **higher derivatives**, to distinguish them from the first derivative. We say that f is **twice differentiable** if $f''(x)$ exists for all x in the domain of f, and f is **n times differentiable** if $f^{(n)}(x)$ exists for all x in the domain of f.

Example 3   Let $f(x) = \cos x$. Show that $f^{(4)}(x) = f(x)$ for all x.

Solution   By taking four successive derivatives we deduce that

$$f'(x) = -\sin x \qquad f^{(3)}(x) = \sin x$$
$$f''(x) = -\cos x \qquad f^{(4)}(x) = \cos x = f(x) \quad \square$$

Example 4   Let $f(x) = x^5 - 3x^4 + 2x - 1$. Find all higher derivatives of f.

Solution   We obtain

$$f'(x) = 5x^4 - 12x^3 + 2 \qquad f^{(4)}(x) = 120x - 72$$
$$f''(x) = 20x^3 - 36x^2 \qquad f^{(5)}(x) = 120$$
$$f^{(3)}(x) = 60x^2 - 72x \qquad f^{(6)}(x) = 0$$

It is apparent also that $f^{(n)}(x) = 0$ for all $n > 6$. $\square$

Notice that for the polynomial f in Example 4, the degree is 5 and $f^{(n)}(x) = 0$ for $n \geq 6$. More generally, the $(n+1)$st and all higher derivatives of any polynomial of degree n are equal to 0. This means that if we differentiate any given polynomial enough times, we will eventually obtain a derivative that is 0.

In the Leibniz notation the second, third, and fourth derivatives are written

$$\frac{d^2 y}{dx^2}, \qquad \frac{d^3 y}{dx^3}, \quad \text{and} \quad \frac{d^4 y}{dx^4}$$

They are read "d squared y, dx squared," and so on.

Example 5   Let $y = x \sin x$. Using the Leibniz notation, find the first three derivatives of y.

Solution   Taking successive derivatives with the help of the product rule, we obtain

$$\frac{dy}{dx} = \sin x + x \cos x$$

$$\frac{d^2y}{dx^2} = \cos x + \cos x - x \sin x = 2\cos x - x \sin x$$

$$\frac{d^3y}{dx^3} = -2\sin x - \sin x - x \cos x = -3\sin x - x \cos x \quad \square$$

Of the higher derivatives, the second derivative is used the most often. It will play a significant role when we analyze graphs of functions in Chapter 4.

Acceleration   Just as the first derivative can be identified with velocity, the second derivative can be identified with acceleration. In conversation we refer to the acceleration of a car or the acceleration of an object due to gravity. In order to define acceleration, we will first define average acceleration. Our discussion will parallel the discussion of velocity in Chapter 2.

Let the motion of an object be in a straight line, and let $v(t)$ be the velocity of the object at time t, as defined in Section 2.2. The **average acceleration** during a time interval from $t = t_0$ to $t = t_1$ is defined to be

$$\frac{\text{difference in velocity}}{\text{time elapsed}} = \frac{(\text{velocity at time } t_1) - (\text{velocity at time } t_0)}{t_1 - t_0}$$

$$= \frac{v(t_1) - v(t_0)}{t_1 - t_0}$$

It seems reasonable to assume that the acceleration at time t_0 should be close to the average acceleration during a time interval having t_0 as one of its endpoints, provided that the length of the interval is small enough. Therefore we define the **acceleration** $a(t_0)$ of the object at time t_0 by the formula

$$a(t_0) = \lim_{t \to t_0} \frac{v(t) - v(t_0)}{t - t_0} = v'(t_0)$$

provided that this limit exists.

If $f(t)$ denotes the position of an object at time t and if the first and second derivatives of f exist at t, then

$$v(t) = f'(t) \quad \text{and} \quad a(t) = v'(t) = f''(t)$$

Hence acceleration is the first derivative of the velocity function and the second derivative of the position function.

Suppose an object moves in a vertical direction and is subject only to the influence of gravity. By (2) in Section 3.3 its height in feet above the ground at time t (in seconds) is given by the formula

$$h(t) = -16t^2 + v_0 t + h_0 \tag{1}$$

where h_0 is the initial height and v_0 is the initial velocity. It is a simple matter to calculate the acceleration of the object.

Example 6 Find the acceleration of an object whose position is described by (1).

Solution Differentiating, we have

$$v(t) = f'(t) = -32t + v_0 \quad \text{and} \quad a(t) = v'(t) = -32 \quad \square$$

A consequence of Example 6 is the fact that any object moving under the sole influence of gravity accelerates at a constant rate, namely -32 feet per second per second.

EXERCISES 3.5

In Exercises 1–18 find $f''(x)$.

1. $f(x) = 5x - 3$

2. $f(x) = x^3 + 3x + 2$

3. $f(x) = -12x^5 + \frac{1}{2}x^4 - \sqrt{1-x}$

4. $f(x) = \dfrac{x+1}{x-1}$

5. $f(x) = \dfrac{2}{(1-4x)^2}$

6. $f(x) = \dfrac{1}{\sqrt{x}}$

7. $f(x) = ax^{-n}$

8. $f(x) = 2x^2 - \dfrac{4000}{x}$

9. $f(x) = \dfrac{1}{x^3 - 1}$

10. $f(x) = \dfrac{x^2}{\sqrt{1-x^2}}$

11. $f(x) = \pi x^{5/2} + \dfrac{\cos x}{x}$

12. $f(x) = \tan x$

13. $f(x) = \sec x$

14. $f(x) = (x^2 + \sin x)^3$

15. $f(x) = x \cot(-4x)$

16. $f(x) = \sqrt{1 + \sin x}$

17. $f(x) = x \tan^3 2x$

18. $f(x) = \sqrt{1 + \sqrt{x}}$

In Exercises 19–30 find d^2y/dx^2.

19. $y = x^{3/2}$

20. $y = \frac{17}{4} - \frac{4}{9}x^2 - 7x^6$

21. $y = (x^4 - \tan x)^3$

22. $y = (1 - x^2)^{3/2}$

23. $y = ax^2 + bx + c$

24. $y = \sqrt{1 + x^4}$

25. $y = \dfrac{1}{3-x}$

26. $y = \dfrac{x}{x^2 - 1}$

27. $y = \csc x$

28. $y = \dfrac{x + \sin x}{x + 2}$

29. $y = \sin x + \cos x$

30. $y = \dfrac{1}{x|x|}$

In Exercises 31–38 find the third derivative of the function.

31. $f(x) = -4x^2 + 5$

32. $f(x) = x^8 - 4x^6 + 3x^4 - 2x^2$

33. $f(x) = \sin x^2$

34. $f(x) = x \cos x$

35. $f(x) = \dfrac{1}{x}$

36. $f(x) = \dfrac{1}{3x^{1/2}}$

37. $f(x) = \dfrac{3x}{4x + 5}$

*38. $f(x) = x^2|x|$

In Exercises 39–46 find d^3y/dx^3.

39. $y = 3x^2$

40. $y = x^{7/2} - 2x^{5/2}$

41. $y = \dfrac{1}{35x^{3/2}}$

42. $y = \csc x$

43. $y = x^2 \sin\dfrac{1}{x}$

44. $y = \dfrac{2}{(1-x)^2}$

45. $y = ax^3 + bx^2 + cx + d$

46. $y = ax^4 + bx^3 + cx^2 + dx + e$

In Exercises 47–50 find the fourth derivative of the function.

47. $f(x) = 3x^8 + \dfrac{3}{4}x^6 - 4x^{3/4} + 2x^{-1}$

48. $f(x) = \sin x - \cos x$

49. $f(x) = \sin \pi x$

50. $f(x) = ax^4 + bx^3 + cx^2 + dx + e$

In Exercises 51–54 find the velocity and acceleration of an object having the given position function.

51. $f(t) = -16t^2 + 3t + 4$

52. $f(t) = -16t^2 - \frac{1}{2}t + 100$

53. $f(t) = 2\sin t - 3\cos t$

54. $f(t) = 3 - 1/t^2$

55. Show that the $(n + 1)$st derivative of any polynomial of degree n is 0. What is the $(n + 2)$nd derivative of such a polynomial?

56. Let $f(x) = \tan x$. Is $f^{(4)}(x) = f(x)$ for $-\pi/2 < x < \pi/2$? Explain your answer.

57. a. If $f''(x) = f(x)$ for all x, what is the relationship between $f'(x)$ and $f^{(3)}(x)$?

b. If $f^{(4)}(x) = f(x)$ for all x, what is the relationship between $f^{(35)}(x)$ and $f^{(31)}(x)$?

*58. Verify the following statements.

a. If $f(x) = \sin x$ and n is a nonnegative integer, then

$$f^{(2n)}(x) = (-1)^n \sin x$$

and $\qquad f^{(2n+1)}(x) = (-1)^n \cos x$

b. If $f(x) = \cos x$ and n is a nonnegative integer, then

$$f^{(2n)}(x) = (-1)^n \cos x$$

and $\qquad f^{(2n+1)}(x) = (-1)^{n+1} \sin x$

59. Let $f(x) = \dfrac{1}{x(x+1)}$. Find a formula for $f^{(n)}(x)$.

$\left(Hint: \dfrac{1}{x(x+1)} = \dfrac{1}{x} - \dfrac{1}{x+1}. \right)$

60. Let $h = fg$. Find $h''(x)$ in terms of f and g and their derivatives.

61. Let $h = g \circ f$. Find $h''(x)$ in terms of f and g and their derivatives.

62. The angular displacement $f(t)$ of a pendulum bob at time t is given by

$$f(t) = a \cos 2\pi \omega t$$

where ω is the frequency and a is the maximum displacement. The first and second derivatives of the angular displacement are the angular velocity and the angular acceleration, respectively, of the bob. Find the angular acceleration of the bob.

*63. A baseball player chasing a fly ball runs in a straight line toward the center field fence. Set up a coordinate system, with feet as units, such that the y axis represents the fence and the player runs along the negative x axis toward the origin. Suppose the player's velocity in feet per second is

$$v(x) = \frac{1}{100}x^2 - \frac{11}{10}x + 25$$

when the player is located at x. What is the acceleration when the player is 1 foot from the fence? (*Hint:* Use the Chain Rule.)

3.6
IMPLICIT DIFFERENTIATION

The differentiable functions we have encountered so far can be described by equations in which y is expressed in terms of x, as in the examples

$$y = \frac{9x^2}{x^2+1} \quad \text{and} \quad y = \tan(\sin x^2) \tag{1}$$

However, suppose y is a differentiable function of x, and that instead of having a formula for y in terms of x, we know only that x and y are related by the equation

$$x^3 + y^3 = 2xy \tag{2}$$

How can we find the derivative of y with respect to x?

Since y is assumed to be a differentiable function of x, the functions $x^3 + y^3$ and $2xy$ are also differentiable functions of x. Thus each side of (2) is a differentiable function of x, and because the two sides are equal, their derivatives must also be equal. Hence we can equate the derivatives of the two sides and then solve for the derivative of y. Let us perform this operation on the equation in (2).

Example 1 Suppose y is a differentiable function of x and satisfies $x^3 + y^3 = 2xy$. Find dy/dx.

Solution First we take the derivatives of the two sides separately. Differentiating $x^3 + y^3$ as the sum of the functions x^3 and y^3, and using the Chain Rule to find the derivative of y^3 with respect to x, we obtain

$$\frac{d}{dx}(x^3 + y^3) = \frac{d}{dx}(x^3) + \frac{d}{dx}(y^3) = 3x^2 + 3y^2\frac{dy}{dx}$$

Differentiating $2xy$ as the product of the functions $2x$ and y yields

$$\frac{d}{dx}(2xy) = \left(\frac{d}{dx}(2x)\right)y + 2x\frac{dy}{dx} = 2y + 2x\frac{dy}{dx}$$

Now we equate the derivatives of the two sides of the original equation, obtaining

$$3x^2 + 3y^2\frac{dy}{dx} = 2y + 2x\frac{dy}{dx}$$

Finally we solve for dy/dx:

$$3y^2\frac{dy}{dx} - 2x\frac{dy}{dx} = 2y - 3x^2$$

$$(3y^2 - 2x)\frac{dy}{dx} = 2y - 3x^2$$

$$\frac{dy}{dx} = \frac{2y - 3x^2}{3y^2 - 2x}$$

provided that the denominator $3y^2 - 2x$ is not 0.   □

In the next example we find a derivative at a specific point.

Example 2   Suppose y is a differentiable function of x satisfying

$$x^3 = y^4 + x^2 \sin y + 1 \qquad (3)$$

Assuming that $y = 0$ for $x = 1$, find $\left.\dfrac{dy}{dx}\right|_{x=1}$

Solution   We first need to find dy/dx, so we differentiate both sides of (3) separately. This yields

$$\frac{d}{dx}(x^3) = 3x^2$$

and

$$\frac{d}{dx}(y^4 + x^2 \sin y + 1) = \frac{d}{dx}(y^4) + \left(\frac{d}{dx}(x^2)\right)\sin y + x^2\frac{d}{dx}(\sin y) + \frac{d}{dx}(1)$$

$$= 4y^3\frac{dy}{dx} + 2x \sin y + x^2(\cos y)\frac{dy}{dx}$$

Equating the derivatives of the two sides of (3), we obtain

$$3x^2 = 4y^3\frac{dy}{dx} + 2x \sin y + x^2(\cos y)\frac{dy}{dx}$$

Now we solve for dy/dx:

$$4y^3\frac{dy}{dx} + x^2(\cos y)\frac{dy}{dx} = 3x^2 - 2x\sin y$$

$$(4y^3 + x^2\cos y)\frac{dy}{dx} = 3x^2 - 2x\sin y$$

$$\frac{dy}{dx} = \frac{3x^2 - 2x\sin y}{4y^3 + x^2\cos y} \tag{4}$$

provided that the denominator is not 0. Now to find $dy/dx|_{x=1}$ under the condition that $y = 0$ for $x = 1$, we substitute $y = 0$ and $x = 1$ in (4) and deduce that

$$\left.\frac{dy}{dx}\right|_{x=1} = \frac{3 - 0}{0 + 1} = 3 \quad \square$$

An equation in which y appears alone on one side of the equality sign and no y appears on the other side is said to **define y explicitly**. Thus each of the equations in (1) defines y explicitly. By contrast, any other kind of equation involving y, in which y is assumed to be a differentiable function of x, is said to **define y implicitly in terms of x**. Thus, provided that such a differentiable function y of x exists, each of the equations in (2) and (3) defines y implicitly.

How can we determine whether there is a differentiable function y that satisfies an equation such as (2) or (3)? Although the answer to this question is not simple, such a function often does exist. In this section and Section 3.7 we will assume without proof that all equations appearing in the text and in the exercises do in fact define one or more differentiable functions y implicitly or explicitly. The process of differentiating both sides of an equation that defines y implicitly in terms of x and then solving for dy/dx is called **implicit differentiation**.

In Examples 1 and 2 the derivative dy/dx was expressed in terms of both x and y. This will usually be the case when we differentiate implicitly. It is usually impossible to find the numerical value of dy/dx at a particular value of x unless the corresponding value of y is known. Indeed, in Example 2 if we had not known that $y = 0$ when $x = 1$, then we would have been unable to determine uniquely the value of y corresponding to $x = 1$. We could have inferred from (3) only that

$$y^4 + \sin y = 0 \tag{5}$$

However, there are two solutions to this equation, as you can convince yourself by drawing the graphs of $\sin y$ and $-y^4$, noting that they cross twice, at 0 and at approximately -0.9496. Without prior information we would not know which solution of (5) to take for y when $x = 1$.

A line tangent to the graph of any function y that satisfies an equation is said to be **tangent to the graph of that equation**.

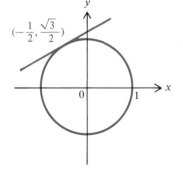

FIGURE 3.9

Example 3   Use implicit differentiation to find an equation of the line tangent to the graph of the equation $x^2 + y^2 = 1$ at the point $(-\frac{1}{2}, \sqrt{3}/2)$ (Figure 3.9).

Solution   Differentiating implicitly the equation $x^2 + y^2 = 1$, we find that

$$2x + 2y\frac{dy}{dx} = 0$$

so that

$$\frac{dy}{dx} = \frac{-x}{y} \quad \text{for } y \neq 0 \tag{6}$$

Therefore the slope of the line tangent to the graph at $(-\frac{1}{2}, \sqrt{3}/2)$ is given by

$$\frac{dy}{dx}\bigg|_{x=-1/2} = \frac{-(-1/2)}{\sqrt{3}/2} = \frac{1}{\sqrt{3}} = \frac{\sqrt{3}}{3}$$

Consequently an equation of the desired tangent line is

$$y - \frac{\sqrt{3}}{2} = \frac{\sqrt{3}}{3}\left[x - \left(-\frac{1}{2}\right)\right] = \frac{\sqrt{3}}{3}\left(x + \frac{1}{2}\right)$$

or more simply,

$$y = \frac{\sqrt{3}}{3}x + \frac{\sqrt{3}}{6} + \frac{\sqrt{3}}{2} = \frac{\sqrt{3}}{3}x + \frac{2}{3}\sqrt{3} \quad \square$$

If y is a function defined implicitly in terms of x, and if d^2y/dx^2 exists, we can differentiate dy/dx to obtain d^2y/dx^2.

Example 4   Let $x^2 + y^2 = 1$. Find d^2y/dx^2 by implicit differentiation.

Solution   We already know from (6) that

$$\frac{dy}{dx} = \frac{-x}{y} \quad \text{for } y \neq 0$$

Taking the derivative of both sides with respect to x and then substituting $-x/y$ for dy/dx, we obtain

$$\frac{d^2y}{dx^2} = \frac{(-1)y - (-x)\dfrac{dy}{dx}}{y^2} = \frac{-y + x\left(-\dfrac{x}{y}\right)}{y^2} = \frac{-(y^2 + x^2)}{y^3} = \frac{-1}{y^3} \quad \square$$

Since it is always possible to solve for dy/dx in terms of y and x, it is also always possible to solve for d^2y/dx^2 in terms of y and x, as we did in Example 4. The principle followed in finding d^2y/dx^2 is the same as the principle followed in finding dy/dx. The only major difference is that the equations involved in finding the second derivative are sometimes grim.

In our final example we will differentiate implicitly, under the assumption that x and y are differentiable functions of a third variable t.

Example 5   Let $x^2 + y^2 = 100$, and assume that x and y are both differentiable functions of t. Find dy/dt in terms of x, y, and dx/dt.

Solution Differentiating both sides of the given equation implicitly with respect to t, we obtain

$$2x\frac{dx}{dt} + 2y\frac{dy}{dt} = 0$$

Therefore

$$2y\frac{dy}{dt} = -2x\frac{dx}{dt}$$

from which we conclude that

$$\frac{dy}{dt} = \frac{-2x\dfrac{dx}{dt}}{2y} = -\frac{x}{y}\frac{dx}{dt}$$

provided that $y \neq 0$. ☐

Were we to desire the value of dy/dt for $x = \frac{1}{2}, y = -\sqrt{3}/2$ and $dx/dt = 5$, we would substitute these values in the equation

$$\frac{dy}{dt} = -\frac{x}{y}\frac{dx}{dt}$$

to obtain

$$\frac{dy}{dt} = -\frac{1/2}{-\sqrt{3}/2}(5) = \frac{5}{3}\sqrt{3}$$

EXERCISES 3.6

In Exercises 1–16 use implicit differentiation to find the derivative of y with respect to x.

1. $3y^2 = -2x^4$

2. $y^2 = \dfrac{x^3}{2 - x}$

3. $y^2 + y = \dfrac{1 + x}{1 - x}$

4. $x^2 = \dfrac{y^2}{y^2 - 1}$

5. $\sec y - \tan x = 0$

6. $\sqrt{x} + \dfrac{1}{\sqrt{y}} = 2$

7. $\dfrac{\sin y}{y^2 + 1} = 3x$

8. $x = \dfrac{1 - \sqrt{y}}{1 + \sqrt{y}}$

9. $x^2 + x^2y^2 + y^3 = 3$

10. $x^2 = \dfrac{x - y^2}{x + y}$

11. $x^2 + y^2 = \dfrac{y^2}{x^2}$

12. $\sqrt{1 + xy} = \dfrac{x}{y} + \dfrac{y}{x}$

13. $\sqrt{xy} + \sqrt{x + 2y} = 4$

14. $2xy = (x^2 + y^3)^{3/2}$

15. $(x^2 + y^2)^{1/2} = 1 + \dfrac{2x}{(x^2 + y^2)^{1/2}}$

16. $(x^2 + y^2)^{1/2} = 1 - \dfrac{x}{(x^2 + y^2)^{1/2}}$

In Exercises 17–28 use implicit differentiation to find the derivative of y with respect to x at the given point.

17. $x^2 + y^2 = y$; $(0, 1)$

18. $x^2 - y^2 = 1$; $(\sqrt{3}, \sqrt{2})$

19. $xy = 2$; $(-2, -1)$

20. $x^4 + xy^3 = 0$; $(-1, 1)$

21. $x^3 + 2xy = 5$; $(1, 2)$

22. $x^2 + 3xy + 2y^2 = 6$; $(-1, -1)$

23. $x^2 + \dfrac{x}{y} = -2; (1, -\frac{1}{3})$

24. $\dfrac{x}{x + 2y} = 1 - y; (1, 0)$

25. $(\sqrt{x} + 1)(\sqrt{y} + 2) = 8; (1, 4)$

26. $(2x + y)^5 = 31 - 1/x; (-1, 4)$

27. $\sin x = \cos y; (\pi/6, \pi/3)$

28. $y^2 \sin 2x = -2y; (\pi/4, -2)$

In Exercises 29–32 find an equation of the line l tangent to the graph of the equation at the given point.

29. $xy^2 = 18; (2, -3)$

30. $x^2 + y^2 = 3y; (-\sqrt{2}, 2)$

31. $\sin(x + y) = 2x; (0, \pi)$

32. $y^2 = \dfrac{x^3}{2 - x}; (1, 1)$

In Exercises 33–36 use implicit differentiation to find d^2y/dx^2.

33. $x^2 - y^4 = 6$

34. $2xy^2 = 4$

35. $x^2 \sin 2y = 1$

36. $x \tan y = y$

In Exercises 37–42 assume that x and y are differentiable functions of t. Find dy/dt in terms of x, y, and dx/dt.

37. $y^2 - x^2 = 4$

38. $x^2 + y^3 = x$

39. $x \sin y = 2$

40. $x^4 y^2 = y$

41. $y = \cos xy^2$

42. $\dfrac{2x + y}{xy^2} = 2$

43. Each equation below describes a single function y. Find dy/dx by implicit differentiation.
    a. $y^5 + y + x = 0$
    b. $y^5 + y^3 + x^3 + y = 0$
    c. $5x = y^3 + \sin y + y$

44. Each of the following equations implicitly describes a single function y that can also be given explicitly. Find dy/dx by implicit differentiation. Then solve the equation explicitly for y. Finally, differentiate again to check the implicit differentiation.

    a. $y^3 = x^2$ b. $\dfrac{8}{y} = x^2 + 4$ c. $y^3 = \dfrac{x^2}{x^2 - 1}$

45. Find the line tangent to the graph of $x^3 + y^3 = 3xy$ at $(\frac{3}{2}, \frac{3}{2})$. Show that the normal at $(\frac{3}{2}, \frac{3}{2})$ passes through the origin.

46. Let l be the line tangent to the astroid $x^{2/3} + y^{2/3} = 4$ at $(2\sqrt{2}, 2\sqrt{2})$. Find the area of the triangle formed by l and the coordinate axes.

47. The **angle** from a curve C_1 to a curve C_2 at a point of intersection (x_0, y_0) is defined as the angle θ from the line l_1 tangent to C_1 at (x_0, y_0) to the line l_2 tangent to C_2 at

(x_0, y_0). The tangent of the angle is given by

$$\tan \theta = \dfrac{m_2 - m_1}{1 + m_1 m_2}$$

where m_1 is the slope of l_1 and m_2 is the slope of l_2. Find the angles from the curve $x^2 + y^2 = 1$ to the curve $(x - 1)^2 + y^2 = 1$ at the two points of intersection.

*48. A circle of radius 1 and center on the y axis is inscribed in the parabola $y = 2x^2$ (Figure 3.10). Determine the points at which the circle and parabola touch. (*Hint:* At such points the circle and parabola have common tangents.)

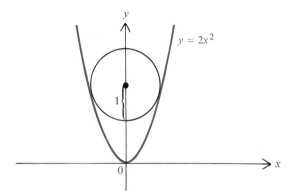

FIGURE 3.10

*49. Consider the lemniscate $(x^2 + y^2)^2 = x^2 - y^2$, shown in Figure 3.11. Find the points on the graph at which the tangents are horizontal. (*Hint:* There are four such points. In finding them you will need to use the equation of the lemniscate twice.)

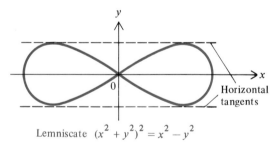

Lemniscate $(x^2 + y^2)^2 = x^2 - y^2$

FIGURE 3.11

3.7
RELATED RATES

When a spherical balloon is inflated, the radius r and the volume V of the balloon are functions of time t. Even if we do not know formulas for r and V as functions of t, at least we know that r and V are related by the equation

$$V = \frac{4}{3}\pi r^3 \tag{1}$$

Differentiating both sides of (1) with respect to t, we find that

$$\frac{dV}{dt} = \frac{dV}{dr}\frac{dr}{dt} = 4\pi r^2 \frac{dr}{dt} \tag{2}$$

The rates dV/dt and dr/dt are related by (2). Therefore we say that dV/dt and dr/dt are **related rates**. If we know, for example, the values of r and dV/dt at a specific time t_0, then we can find the value of dr/dt at t_0 by solving equation (2) for dr/dt.

Example 1 Suppose a spherical balloon is inflated at the rate of 10 cubic inches per minute. How fast is the radius of the balloon increasing when the radius is 5 inches?

Solution   The formula for the volume is given in (1), and by (2),

$$\frac{dV}{dt} = 4\pi r^2 \frac{dr}{dt} \tag{3}$$

Since the volume of the balloon is increasing at 10 cubic inches per minute, we know that $dV/dt = 10$. We wish to find

$$\frac{dr}{dt}\bigg|_{t=t_0}$$

where t_0 is the instant at which $r = 5$. Substituting 10 for dV/dt and 5 for r in (3), we conclude that

$$10 = 4\pi(5)^2 \frac{dr}{dt}\bigg|_{t=t_0}$$

so that

$$\frac{dr}{dt}\bigg|_{t=t_0} = \frac{10}{4\pi(5)^2} = \frac{1}{10\pi}$$

Therefore when the radius is 5 inches, the radius is increasing at the rate of $1/(10\pi)$ inch per minute.   □

Observe that in Example 1 we did not need to know the time t_0 at which $r = 5$. All we needed to know were the values of r and dV/dt at t_0.

Example 2 Suppose that the bigger the balloon is, the harder it is to inflate. In particular, suppose that when the volume V is greater than 10 cubic inches, the balloon is inflated at the rate of $8/V$ cubic inches per minute. How fast is the radius of the balloon increasing when the radius is 2 inches?

Solution Our goal is to find

$$\frac{dr}{dt}\Big|_{t=t_0}$$

where t_0 is the instant at which $r = 2$. On the one hand, (2) implies that

$$\frac{dV}{dt} = 4\pi r^2 \frac{dr}{dt} \tag{4}$$

On the other hand, we have assumed that for $V > 10$,

$$\frac{dV}{dt} = \frac{8}{V} = \frac{8}{\frac{4}{3}\pi r^3} = \frac{6}{\pi r^3} \tag{5}$$

Equating the expressions given in (4) and (5) for dV/dt, we deduce that

$$4\pi r^2 \frac{dr}{dt} = \frac{6}{\pi r^3}$$

or equivalently,

$$\frac{dr}{dt} = \frac{6}{\pi r^3}\frac{1}{4\pi r^2} = \frac{3}{2\pi^2 r^5}$$

For the time t_0 at which $r = 2$ we have

$$\frac{dr}{dt}\Big|_{t=t_0} = \frac{3}{2\pi^2 2^5} = \frac{3}{64\pi^2}$$

Consequently, when the radius is 2 inches, the radius is increasing at the rate of $3/(64\pi^2)$ inches per minute. □

In the next example we will use a diagram to help discover an equation relating the variables.

Example 3 One end of a 13-foot ladder is on the ground, and the other end rests on a vertical wall (Figure 3.12). If the bottom end is drawn away from the wall at 3 feet per second, how fast is the top of the ladder sliding down the wall when the bottom of the ladder is 5 feet from the wall?

Solution At any given instant, let y be the distance between the ground and the top of the ladder, and x the distance between the wall and the bottom of the ladder, as in Figure 3.12. By hypothesis, $dx/dt = 3$, and we must find the value of

$$\frac{dy}{dt}\Big|_{t=t_0}$$

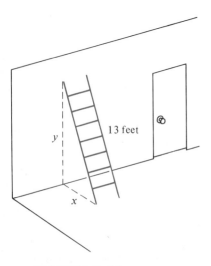

FIGURE 3.12

where t_0 is the time at which $x = 5$. By the Pythagorean Theorem, x and y are related by the equation

$$x^2 + y^2 = 13^2 = 169 \qquad (6)$$

Differentiating the left and right sides of (6) implicitly with respect to t, we obtain

$$2x\frac{dx}{dt} + 2y\frac{dy}{dt} = 0$$

which we solve for dy/dt:

$$2y\frac{dy}{dt} = -2x\frac{dx}{dt}$$

$$\frac{dy}{dt} = -\frac{x}{y}\frac{dx}{dt} \qquad (7)$$

Now at the time t_0 at which the base of the ladder is 5 feet from the wall, we have $x = 5$, so by (6),

$$y^2 = 169 - 5^2 = 144$$

and thus $y = 12$. Substituting $x = 5$, $y = 12$, and $dx/dt = 3$ into (7), we conclude that

$$\left.\frac{dy}{dt}\right|_{t=t_0} = -\frac{5}{12}(3) = -\frac{5}{4}$$

Thus when the bottom of the ladder is 5 feet from the wall, the top is sliding down at the rate of $\frac{5}{4}$ feet per second. $\square$

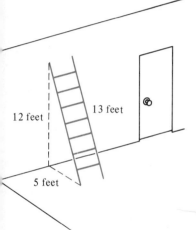

FIGURE 3.13

Caution: In the preceding example, if we had tried to find the rate of slippage from the diagram in Figure 3.13, which represents the situation only at time t_0, we would not have been able to obtain a relationship between the rates dx/dt and dy/dt. (In particular, x does not appear in Figure 3.13.) Thus it is essential to draw a figure that represents the situation at *any* instant, not only at a particular instant.

Example 4   Water is poured into a conical paper cup at the rate of $\frac{2}{3}$ cubic inches per second. If the cup is 6 inches tall and the top of the cup has a radius of 2 inches, how fast does the water level rise when the water is 4 inches deep?

Solution   At any time t let h be the height of the water, V the volume of the water, and r the radius of the top surface of the water (Figure 3.14). Recall that the volume V of a cone is given by

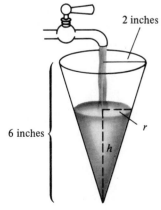

FIGURE 3.14

$$V = \frac{1}{3}\pi r^2 h$$

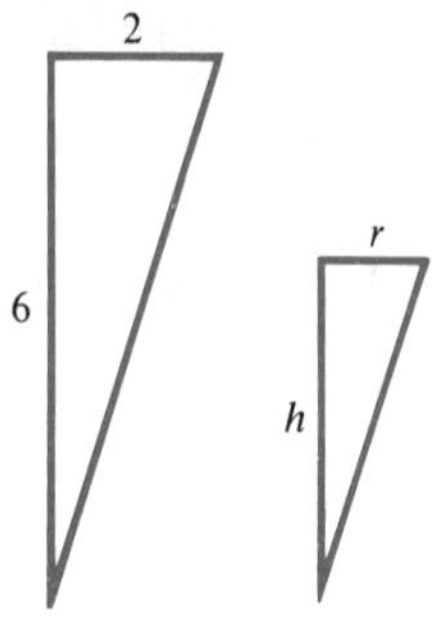

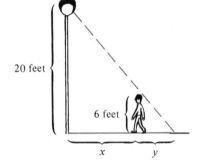

FIGURE 3.15

Since the water enters the cup at the rate of $\frac{2}{3}$ cubic inches per second, we have $dV/dt = \frac{2}{3}$. We wish to determine the value of

$$\left.\frac{dh}{dt}\right|_{t=t_0}$$

where t_0 is the time at which $h = 4$. Using the similar triangles in Figure 3.15, which we have obtained from Figure 3.14, we find that

$$\frac{r}{h} = \frac{2}{6}$$

so that $r = h/3$. As a result we can express V in terms of h alone:

$$V = \frac{1}{3}\pi\left(\frac{h}{3}\right)^2 h = \frac{\pi h^3}{27}$$

Differentiating with respect to t, we find that

$$\frac{dV}{dt} = \frac{dV}{dh}\frac{dh}{dt} = \frac{\pi h^2}{9}\frac{dh}{dt}$$

so that

$$\frac{dh}{dt} = \frac{9}{\pi h^2}\frac{dV}{dt} \tag{8}$$

At time t_0 we have $h = 4$ and $dV/dt = \frac{2}{3}$. Substituting these values into (8) yields

$$\left.\frac{dh}{dt}\right|_{t=t_0} = \frac{9}{\pi(4)^2}\cdot\frac{2}{3} = \frac{3}{8\pi}$$

Thus the water level is rising at the rate of $3/(8\pi)$ inches per second when the water is 4 inches deep. $\square$

Example 5   Pat walks at the rate of 5 feet per second toward a street light whose lamp is 20 feet above the base of the light. If Pat is 6 feet tall, determine the rate of change of the length of Pat's shadow at the moment Pat is 24 feet from the base of the lamppost.

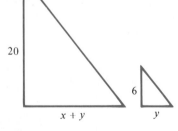

FIGURE 3.16

Solution   At any given time, let x be the distance between Pat and the lamppost and y the length of Pat's shadow (Figure 3.16). We wish to find

$$\left.\frac{dy}{dt}\right|_{t=t_0}$$

for the value of t_0 at which $x = 24$. By the similar triangles appearing in Figure 3.17,

$$\frac{x+y}{20} = \frac{y}{6}$$

so that

$$6x + 6y = 20y, \quad \text{or equivalently,} \quad y = \frac{3}{7}x$$

FIGURE 3.17

Differentiating with respect to t, we obtain

$$\frac{dy}{dt} = \frac{3}{7}\frac{dx}{dt}$$

Since Pat walks at the rate of 5 feet per second toward the lamppost, it follows that $dx/dt = -5$, so that

$$\frac{dy}{dt} = \frac{3}{7}(-5) = -\frac{15}{7}$$

Consequently Pat's shadow shrinks at the rate of $\frac{15}{7}$ feet per second at the moment in question. □

The procedure we have used in solving the related rates examples of this section includes the following steps:

1. Identify and label the different variables. Include the variable whose rate is to be evaluated and those whose rates are given. It may be helpful to sketch a drawing at this stage.
2. Determine an equation connecting the variables from step 1.
3. Differentiate both sides of the equation implicitly and solve for the derivative that will yield the desired rate.
4. Evaluate the derivative by using the given values of the variables and their rates.

This procedure should also help you solve related rates problems.

EXERCISES 3.7

1. Suppose the radius of a spherical balloon is shrinking at $\frac{1}{2}$ inch per minute. How fast is the volume decreasing when the radius is 4 inches?

2. Suppose a snowball remains spherical while it melts, with the radius shrinking at one inch per hour. How fast is the volume of the snowball decreasing when the radius is 2 inches?

3. Suppose the volume of the snowball in Exercise 2 shrinks at the rate of $dV/dt = -2/V$ (cubic inches per hour). How fast is the radius changing when the radius is $\frac{1}{2}$ inch?

4. A spherical balloon is inflated at the rate of 3 cubic inches per minute. How fast is the radius of the balloon increasing when the radius is 6 inches?

5. Suppose a spherical balloon grows in such a way that after t seconds, $V = 4\sqrt{t}$ (cubic inches). How fast is the radius changing after 64 seconds?

6. A spherical balloon is losing air at the rate of 2 cubic inches per minute. How fast is the radius of the balloon shrinking when the radius is 8 inches?

7. Water leaking onto a floor creates a circular pool with an area that increases at the rate of 3 square inches per minute. How fast is the radius of the pool increasing when the radius is 10 inches?

8. A point moves around the circle $x^2 + y^2 = 9$. When the point is at $(-\sqrt{3}, \sqrt{6})$, its x coordinate is increasing at the rate of 20 units per second. How fast is its y coordinate changing at that instant?

9. Suppose the top of the ladder in Example 3 is being pushed up the wall at the rate of 1 foot per second. How fast is the base of the ladder approaching the wall when it is 3 feet from the wall?

10. A ladder 15 feet long leans against a vertical wall. Suppose that when the bottom of the ladder is x feet from the wall, the bottom is being pushed toward the wall at the rate of $\frac{1}{2}x$ feet per second. How fast is the top of the ladder rising at the moment the bottom is 5 feet from the wall?

11. A board 5 feet long slides down a wall. At the instant the bottom end is 4 feet from the wall, the other end is moving down the wall at the rate of 2 feet per second. At that moment,
    a. how fast is the bottom end sliding along the ground?
    b. how fast is the area of the region between the board, ground, and wall changing?

12. Suppose the water in Example 4 is poured in at the rate of $\frac{3}{2}$ cubic inches per second. How fast is the water level rising when the water is 2 inches deep?

13. Suppose that the water level in Example 4 is rising at $\frac{1}{2}$ inch per second. How fast is the water being poured in when the water has a depth of 2 inches?

14. A water trough is 12 feet long, and its cross section is an equilateral triangle with sides 2 feet long. Water is pumped into the trough at a rate of 3 cubic feet per minute. How fast is the water level rising when the depth of the water is $\frac{1}{2}$ foot?

15. A beacon on a lighthouse 1 mile from shore revolves at the rate of 10π radians per minute. Assuming that the shoreline is straight, calculate the speed at which the spotlight is sweeping across the shoreline as it lights up the sand 2 miles from the lighthouse. (*Hint:* In Figure 3.18, x is the coordinate of the point on the shore at which the light shines. Thus dx/dt is the speed of the image of the spotlight moving across the shoreline, and $d\theta/dt = 10\pi$.)

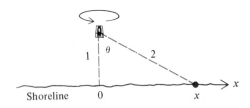

FIGURE 3.18

16. Boyle's Law states that if the temperature of a gas remains constant, then the pressure p and the volume V of the gas satisfy the equation $pV = c$, where c is a constant. If the volume is decreasing at the rate of 10 cubic inches per second, how fast is the pressure increasing when the pressure is 100 pounds per square inch and the volume is 20 cubic inches?

17. A person is pushing a box up the ramp in Figure 3.19 at the rate of 3 feet per second. How fast is the box rising?

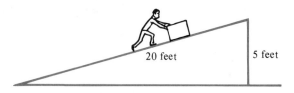

FIGURE 3.19

18. A rope is attached to the bow of a sailboat coming in for the evening. Assume that the rope is drawn in over a pulley 5 feet higher than the bow at the rate of 2 feet per second, as shown in Figure 3.20. How fast is the boat docking when the length of rope from bow to pulley is 13 feet?

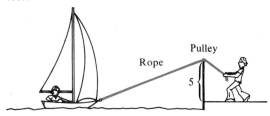

FIGURE 3.20

19. Suppose the rope in Exercise 18 is pulled so that the boat docks at a constant rate of 2 feet per second. How fast is the rope being pulled in when the boat is 12 feet from the dock?

20. As in Exercise 18, assume that the boat is pulled in by a rope attached to the bow passing through a pulley 5 feet above the bow. Assume also that the distance between the bow and the dock decreases as the cube root of the distance; that is, if the distance at time t is y feet, then $dy/dt = -y^{1/3}$ (feet per second). How fast is the length of the rope shrinking when the bow is 8 feet from the dock?

21. A Flying Tiger is making a nose dive along a parabolic path having the equation $y = x^2 + 1$, where x and y are measured in feet. Assume that the sun is directly above the y axis, that the ground is the x axis, and that the distance from the plane to the ground is decreasing at the constant rate of 100 feet per second. How fast is the shadow of the plane moving along the ground when the plane is 2501 feet above the earth's surface? Assume that the sun's rays are vertical.

22. The tortoise and the hare are having their fabled footrace, each moving along a straight line. The tortoise, moving at a constant rate of 10 feet per minute, is 4 feet from the finish line when the hare wakes up 5001 feet from the finish line and darts off after the tortoise. Let x be the distance from the tortoise to the finish line, and suppose the distance y from the hare to the finish line is given by

$$y = 5001 - 2500\sqrt{4 - x}$$

a. How fast is the hare moving when the tortoise is 3 feet from the finish line?

b. Who wins? By how many feet?

23. A baseball diamond is a square with sides 90 feet long. Suppose a baseball player is advancing from second to third base at the rate of 24 feet per second, and an umpire is standing on home plate. Let θ be the angle between the third baseline and the line of sight from the umpire to the runner (Figure 3.21). How fast is θ changing when the runner is 30 feet from third base?

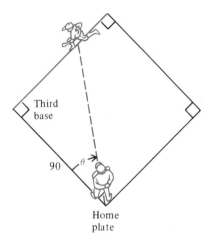

Third base

90 θ

Home plate

FIGURE 3.21

24. Maple and Main Streets are straight and perpendicular to each other. A stationary police car is located on Main Street $\frac{1}{4}$ mile from the intersection of the two streets. A sports car on Maple Street approaches the intersection at the rate of 40 miles per hour. How fast is the distance between the two cars decreasing when the moving car is $\frac{1}{8}$ mile from the intersection?

25. Suppose in Exercise 24 that as the sports car approaches the intersection, the distance between the sports car and the police car decreases at 25 miles per hour. How far from the intersection would the sports car be at the moment when it is traveling 40 miles per hour?

26. A spotlight is on the ground 100 feet from a building that has vertical sides. A person 6 feet tall starts at the spotlight and walks directly toward the building at a rate of 5 feet per second.

a. How fast is the top of the person's shadow moving down the building when the person is 50 feet away from it?

b. How fast is the top of the shadow moving when the person is 25 feet away?

27. A kite 100 feet above the ground is being blown away from the person holding its string, in a direction parallel

to the ground and at the rate of 10 feet per second. At what rate must the string be let out when the length of string already let out is 200 feet?

28. A helicopter flies parallel to the ground at an altitude of $\frac{1}{2}$ mile and at a speed of 2 miles per minute. If the helicopter flies along a straight line that passes directly over the White House, at what rate is the distance between the helicopter and the White House changing 1 minute after the helicopter flies over the White House?

29. When a rocket is two miles high, it is moving vertically upward at a speed of 300 miles per hour. At that instant, how fast is the angle of elevation of the rocket increasing, as seen by an observer on the ground 5 miles from the launching pad?

30. A ferris wheel with radius 25 feet is revolving at the rate of 10 radians per minute. How fast is a passenger rising when the passenger is 15 feet higher than the center of the ferris wheel and is rising?

31. A street light 16 feet high casts a shadow on the ground from a ball that is dropped from a height of 16 feet but 15 feet from the light. How fast is the shadow moving along the ground when the ball is 5 feet from the ground. (*Note:* The distance s from the ball to the ground t seconds after release is given by the equation $s = 16 - 16t^2$.)

*32. Water is released from a conical tank 50 inches tall and 30 inches in radius, and falls into a rectangular tank whose base has an area of 400 square inches (Figure 3.22). The

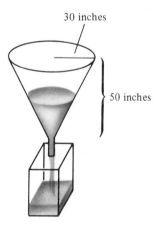

30 inches

50 inches

FIGURE 3.22

rate of release is controlled so that when the height of the water in the conical tank is x inches, the height is decreasing at the rate of $50 - x$ inches per minute. How fast is the water level in the rectangular tank rising when the height of the water in the conical tank is 10 inches? (*Hint:* The total amount of water in the two tanks is constant.)

33. A helicopter 3000 feet high is moving horizontally at the rate of 100 feet per second. It flies directly over a searchlight that rotates so as to always illuminate the helicopter. At how many radians per second is the searchlight rotating when the distance between the helicopter and searchlight is 5000 feet?

34. Suppose a deer is standing 20 feet from a highway on which a car is traveling at a constant rate of v feet per second. Let θ be the angle made by the highway and the line of sight from a passenger to the deer (Figure 3.23). Show that

$$\frac{d\theta}{dt} = \frac{20v}{400 + x^2}$$

(Notice that for x close to 0, $d\theta/dt$ is approximately $v/20$, and thus for the passenger to keep the deer in focus, the passenger's eyes must rotate at the approximate rate of $v/20$ radians per second. This suggests why at large velocities it may be impossible to keep a stationary object near the highway in focus.)

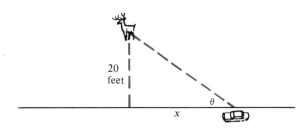

FIGURE 3.23

*35. A 10-foot square sign of negligible thickness revolves about a vertical axis through its center at a rate of 10 revolutions per minute. An observer far away sees it as a rectangle of variable width. How fast is the width changing when the sign appears to be 6 feet wide and is increasing in width? (*Hint:* View the sign from above, and consider the angle it makes with a line pointing toward the observer.)

*36. At night a patrol boat approaches a point on shore along the curve $y = -\frac{1}{2}x^3$, as indicated in Figure 3.24. If the boat moves along the curve so that $dx/dt = -x$, and if its spotlight is pointed straight ahead, determine how fast the illuminated spot on the shore moves when $x = -2$. (*Hint:* You will need to find the x intercept of the line tangent to the curve $y = -\frac{1}{2}x^3$.)

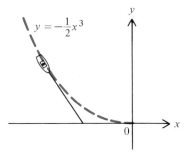

FIGURE 3.24

*37. A deer 5 feet long and 6 feet tall, whose rump is 4 feet above ground as in Figure 3.25, approaches a street light with lamp 20 feet above ground. If the deer proceeds at 3 feet per second, how fast is its shadow changing when the front of the deer is
a. 48 feet from the street light?
b. 24 feet from the street light?
(*Hint:* In parts (a) and (b), determine which yields the shadow, the head or the rump of the deer.)

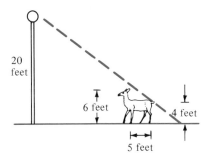

FIGURE 3.25

3.8
TANGENT LINE APPROXIMATIONS AND THE DIFFERENTIAL

As we have seen, tangent lines and derivatives are closely related. In this section we will make use of this relationship to estimate values of functions that are difficult or impossible to obtain exactly.

Tangent Line Approximations

To illustrate the problem of approximating values of functions, let

$$f(x) = \sqrt[3]{x}$$

It follows that

$$f(1.1) = \sqrt[3]{1.1}$$

The expression $\sqrt[3]{1.1}$ is of little practical value as it stands; one would never use $\sqrt[3]{1.1}$ grams of alcohol in a chemical experiment. Instead, one would convert to the decimal expansion $1.03228\ldots$ of $\sqrt[3]{1.1}$ and then use, say, 1.03 grams of alcohol. With a calculator the conversion is easy. But the methods of calculus can often yield very good estimates of numbers such as $\sqrt[3]{1.1}$ without a calculator and without much effort.

To describe a general procedure for approximating values of functions by means of calculus, let f be differentiable at a. By the definition of $f'(a)$, if x approaches a, then

$$\frac{f(x) - f(a)}{x - a}$$

approaches $f'(a)$. Hence for values of x close enough to a, $f(x) - f(a)$ is close to $f'(a)(x - a)$, so that $f(x)$ is close to $f(a) + f'(a)(x - a)$. Thus if f is differentiable at a and if x is close enough to a, then $f(x)$ is approximately equal to $f(a) + f'(a)(x - a)$ (Figure 3.26), and we express this by writing

$$f(x) \approx f(a) + f'(a)(x - a) \tag{1}$$

To emphasize that only values of x close to a are to be considered, we usually replace x by $a + h$, obtaining

$$f(a + h) \approx f(a) + f'(a)h \tag{2}$$

Notice that x is close to a if and only if h is close to 0, so in (2) we only consider values of h close to 0. The approximation given in (2) is most useful when $f(a)$ and $f'(a)$ are easy to compute. To illustrate the use of (2), we will approximate $\sqrt[3]{1.1}$.

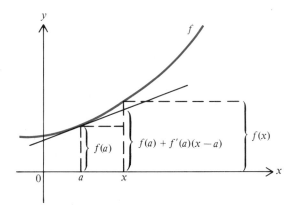

FIGURE 3.26

Example 1 Use (2) to approximate $\sqrt[3]{1.1}$.

Solution Let $f(x) = \sqrt[3]{x}$. Then we seek $f(1.1)$. Notice that 1 is close to 1.1, that $f(1) = 1$, and that $f'(x) = \frac{1}{3}x^{-2/3}$, so that $f'(1) = \frac{1}{3}$. Thus if $a = 1$ and $h = 0.1$ in (2), we find that

$$f(1.1) = f(1 + 0.1) \approx f(1) + f'(1)(0.1) = 1 + \frac{1}{3}(0.1) = \frac{31}{30} \quad \square$$

How accurate is $\frac{31}{30}$ as an approximation of $\sqrt[3]{1.1}$? The true value of $\sqrt[3]{1.1}$, rounded to six digits, is 1.03228. Thus the error introduced by employing (2) is about

$$\frac{31}{30} - 1.03228 \approx 1.03333 - 1.03228 = 0.00105$$

which is a very reasonable degree of precision.

Recall that an equation of the line tangent to the graph of a function f at $(a, f(a))$ is

$$y = f(a) + f'(a)(x - a) \tag{3}$$

Notice that the right sides of (1) and (3) are identical, which means that in using (1) to approximate $f(x)$ we are taking the second coordinate of a point on the tangent line of f (See Figure 3.26 again.). As a result, we call the approximation of $f(x)$ given in (1) or in (2) a **tangent line approximation**.

Example 2 Use a tangent line approximation to estimate the value of $\sqrt{8.5}$.

Solution Let $f(x) = \sqrt{x}$; then $f(9) = \sqrt{9} = 3$, and we seek $f(8.5)$. Observe that

$$f'(x) = \frac{1}{2}x^{-1/2} \quad \text{and} \quad f'(9) = \frac{1}{2}9^{-1/2} = \frac{1}{6}$$

If we let $a = 9$ and $h = -0.5$ in (2), then

$$\sqrt{8.5} = f(8.5) \approx f(9) + f'(9)(-0.5) = 3 + \frac{1}{6}(-0.5) \approx 2.91667 \quad \square$$

Since the true value of $\sqrt{8.5}$ rounded to six digits is 2.91548, our estimate has an error of approximately 0.00119.

The Differential   Let f be a differentiable function. We can replace the number a in (2) by any number x in the domain of f. We then have

$$f(x + h) \approx f(x) + f'(x)h \tag{4}$$

which is equivalent to

$$f(x + h) - f(x) \approx f'(x)h \tag{5}$$

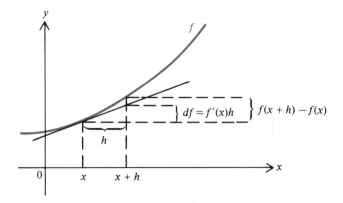

FIGURE 3.27

The number $f(x + h) - f(x)$ on the left-hand side is just a difference of values of f; it is sometimes called the **first difference of f at x with increment h**. You should recognize it as the numerator in the following formula for $f'(x)$:

$$f'(x) = \lim_{h \to 0} \frac{f(x + h) - f(x)}{h}$$

The number $f'(x)h$ appearing on the right-hand side of (5) is usually called the **differential of f (at x with increment h)** and is denoted df. Thus

$$df = f'(x)h$$

Of course, df depends on both x and h, although these variables do not appear in the expression df. The relationship between the differential and the first difference is illustrated in Figure 3.27.

If $g(x) = x$, then we denote dg by dx. Since $g'(x) = 1$, it follows that

$$dx = dg = g'(x)h = h$$

Therefore we may replace the number h by dx when writing the differential df of a function f:

$$df = f'(x)\,dx \tag{6}$$

In the notation of differentials, the approximation in (4) becomes

$$f(x + h) \approx f(x) + df, \quad \text{or} \quad f(x + dx) \approx f(x) + df \tag{7}$$

In the next example we use the differential notation to estimate a value of a trigonometric function.

Example 3 Approximate $\sin \dfrac{7\pi}{36}$ by using differentials.

Solution   We will use the fact that $7\pi/36$ is close to $\pi/6$, because if $f(x) = \sin x$, then $f(\pi/6)$ and $f'(\pi/6)$ are easy to evaluate. Letting $x = \pi/6$ and $dx = \pi/36$, we use (6) to deduce that

$$df = f'(x)\,dx = \left(\cos\frac{\pi}{6}\right)\left(\frac{\pi}{36}\right) = \frac{\sqrt{3}}{2}\frac{\pi}{36} \approx 0.075575$$

Consequently it follows from (7) that

$$\sin\frac{7\pi}{36} \approx f\left(\frac{\pi}{6}\right) + df \approx 0.5 + 0.075575 = 0.575575 \quad \square$$

Once again the estimate is good. Indeed, the actual value of $\sin 7\pi/36$ rounded to six digits is 0.573576, so that the error introduced by the approximation is less than 0.002.

Notice that the method used in Example 3 was the same as that used in Examples 1 and 2, but in differential notation.

The use of differentials shortens certain computations that will appear in Chapters 5–8. If u is a variable depending on x, then by (6),

$$du = u'(x)\,dx = \left(\frac{du}{dx}\right)dx \qquad (8)$$

For example, if $u = x^2$, then

$$du = \left(\frac{du}{dx}\right)dx = \left(\frac{d}{dx}(x^2)\right)dx = 2x\,dx$$

In the same way, if $v = x \sin x$, then

$$dv = \left(\frac{dv}{dx}\right)dx = \left(\frac{d}{dx}(x \sin x)\right)dx = (\sin x + x \cos x)\,dx$$

The differentiation rules can be rephrased in the notation of differentials as follows:

$$d(u + v) = du + dv \qquad d(uv) = v\,du + u\,dv$$

$$d(cu) = c\,du \qquad d\left(\frac{u}{v}\right) = \frac{v\,du - u\,dv}{v^2} \qquad (9)$$

Verification of these formulas is left to the exercises.

The Newton-Raphson Method   In Section 2.6 we discussed the bisection method of finding approximate values for a zero of a function. Now we will describe a second method, called the Newton-Raphson method, after the two mathematicians Isaac Newton and Joseph Raphson. It utilizes lines tangent to the graph of the given function to approximate values for a zero of a function.

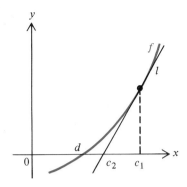

FIGURE 3.28

Let d be a zero of a function f, and assume that f is differentiable on an open interval containing d. We would like to approximate the numerical value of d. We begin the Newton-Raphson method by choosing a number c_1 that we believe is close to d. If by chance $f(c_1) = 0$, then c_1 is a zero of f and we have finished our search. So assume that $f(c_1) \neq 0$. If the line l tangent to the graph of f at $(c_1, f(c_1))$ is not horizontal, we let c_2 be the x intercept of l (Figure 3.28). Ideally c_2 will be closer to d and hence will be a better approximation of d than c_1 is. (This is indeed the case in Figure 3.28.) Now if $f(c_2) = 0$, then c_2 is a zero of f and the process stops. So assume that $f(c_2) \neq 0$. In order to continue the process, we need to find a formula for c_2 in terms of c_1.

To that end we recall that l is tangent to the graph of f at $(c_1, f(c_1))$, so has slope $f'(c_1)$. Therefore an equation of l is

$$y = f(c_1) + f'(c_1)(x - c_1)$$

Since c_2 is the x intercept of l, it follows that c_2 satisfies

$$0 = f(c_1) + f'(c_1)(c_2 - c_1)$$

Solving for c_2, we find that

$$c_2 - c_1 = -\frac{f(c_1)}{f'(c_1)}$$

and thus

$$c_2 = c_1 - \frac{f(c_1)}{f'(c_1)}$$

Repeating the analysis but with c_1 replaced by c_2 and c_2 by c_3, we find a third approximation c_3 (Figure 3.29) given by

$$c_3 = c_2 - \frac{f(c_2)}{f'(c_2)}$$

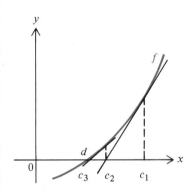

FIGURE 3.29

This procedure may be continued. After we have obtained the approximation c_n, the next approximation c_{n+1} is given by the formula

$$c_{n+1} = c_n - \frac{f(c_n)}{f'(c_n)} \tag{10}$$

The method of approximating a zero of f by means of (10) is called the **Newton-Raphson method**.

Ideally we would like to stop the process when an approximate zero c_n and the zero d of f are within an acceptable tolerance ε of each other, that is, when

$$|c_n - d| < \varepsilon$$

But since we are trying to find the value of d, and hence don't know it, a reasonable alternative is to stop the process when successive approximations c_n and c_{n+1} are within ε of each other, that is, when

$$|c_{n+1} - c_n| < \varepsilon \tag{11}$$

This means, for example, that if we wish to approximate a zero to four decimal places (that is, $\varepsilon = 10^{-4}$), then we would carry out the Newton-Raphson method until we obtain two successive values c_n and c_{n+1} that agree to four decimal places. Then c_{n+1} would be the desired approximate zero.

Now by (10),

$$c_{n+1} - c_n = -\frac{f(c_n)}{f'(c_n)}$$

so that (11) is equivalent to

$$\left|\frac{f(c_n)}{f'(c_n)}\right| < \varepsilon \qquad (12)$$

The algorithm we will use in finding an approximate zero for a given function uses the criterion in (12) to determine when to stop.

INPUT: A differentiable function f, an initial value c, and an allowable error ε.

OUTPUT: A number c that is within ε of the preceding approximation.

STEP 1: Compute $f(c)$. If $f(c) = 0$, then c is a zero of f, so STOP. Otherwise CONTINUE.

STEP 2: Compute $f'(c)$. If $f'(c) = 0$, then STOP because the method cannot be continued. Otherwise CONTINUE.

STEP 3: Compute $f(c)/f'(c)$. If $|f(c)/f'(c)| < \varepsilon$, then $c - (f(c)/f'(c))$ is the desired approximate zero with error ε, so STOP. Otherwise CONTINUE.

STEP 4: Replace c by $c - (f(c)/f'(c))$.

STEP 5: Repeat Steps 1–4 with the new number found in STEP 4, continuing until the process stops with STEP 1, 2, or 3.

Example 4 Let $f(x) = \cos x - x$. Use the Newton-Raphson method until successive approximations are within 10^{-4} of each other.

Solution Notice that $f(0) = 1$ and $f(\pi/2) = -\pi/2$, so by the Intermediate Value Theorem, f has a zero in $[0, \pi/2]$. Since 1 is a convenient number in the interior of the interval, we let 1 be the initial value of c in the Newton-Raphson method. Using the fact that $f'(x) = -\sin x - 1$, we assemble the following table:

c	$f(c)$	$f'(c)$	$\dfrac{f(c)}{f'(c)}$	$c - \dfrac{f(c)}{f'(c)}$
1	$-.459697694$	-1.84147099	$.2496361314$	$.750363868$
$.750363868$	$-.0189230738$	-1.68190495	$.0112509769$	$.739112891$
$.739112891$	$-4.64559998 \times 10^{-5}$	-1.67363254	$2.775758638 \times 10^{-5}$	$.739085133$

Since in the last step, $|f(c)/f'(c)| < 10^{-4}$, it follows that the desired approximate zero is 0.739085133. $\square$

The Newton-Raphson method works very effectively in approximating roots such as square or cube roots, as the next example shows.

Example 5 Approximate $\sqrt{7}$ by using the Newton-Raphson method until successive approximations are within 10^{-8} of each other.

Solution Notice that if $f(x) = x^2 - 7$, then $f'(x) = 2x$ and $f(\sqrt{7}) = 0$. Thus we seek an approximate zero of f. Since 9 is close to 7 and $\sqrt{9} = 3$, we let $c = 3$ and compile the following chart for the Newton-Raphson method:

c	$f(c)$	$f'(c)$	$\dfrac{f(c)}{f'(c)}$	$c - \dfrac{f(c)}{f'(c)}$
3	2	6	.333333333	2.66666667
2.66666667	.111111112	5.33333334	.0208333335	2.64583334
2.64583334	4.3403171×10^{-4}	5.29166667	$8.20217405 \times 10^{-5}$	2.64575131
2.64575131	7.4505806×10^{-9}	5.29150263	$1.40802738 \times 10^{-9}$	2.64575131

Since in the last step, $|f(c)/f'(c)| < 10^{-8}$, it follows that 2.64575131 is the desired approximate value of $\sqrt{7}$. $\square$

How carefully must the initial value of c be chosen in order for the Newton-Raphson method to provide the desired approximate zero in an acceptable number of steps? The answer depends on the function. For example, in Example 5 if we had let $c_1 = 10$ (which is not very close to a zero of f), then it would have taken only three more approximations to come within 10^{-8} of $\sqrt{7}$. However, an unfortunate choice of c_1 may even lead to the failure of the Newton-Raphson method to work at all. Indeed, successive values of c_i might lie outside the domain of f, fail to approach a zero of f, or approach the wrong zero (see Figure 3.30 and Exercises 40, 42, and 43).

Finally we mention that a second criterion for determining when to stop the Newton-Raphson procedure is the following: c_n satisfies

$$|f(c_n)| < \varepsilon$$

Rather than proposing that c_n is genuinely close to a zero, this criterion says that the value of f at c_n is suitably close to zero.

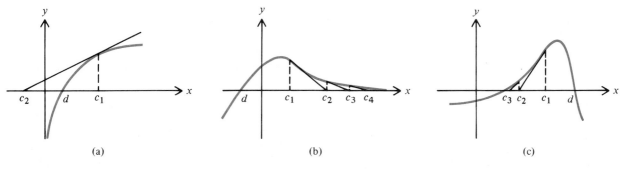

(a)   (b)   (c)

FIGURE 3.30

EXERCISES 3.8

In Exercises 1–12 approximate the given number by using differentials.

1. $\sqrt{101}$ 2. $\sqrt{99.5}$

3. $\sqrt[3]{29}$ 4. $\sqrt[4]{17}$

5. $\sqrt{36.36}$ 6. $(28)^{3/5}$

7. $(28)^{4/3}$ 8. $\dfrac{1}{1 + (1.0175)^2}$

9. $\cos 2\pi/13$ 10. $\cot 11\pi/36$

11. $\sec 4\pi/17$ 12. $\tan 99\pi/100$

In Exercises 13–16 compute df for the given values of a and h.

13. $f(x) = \sqrt{x}; a = 4, h = 0.2$

14. $f(x) = \sqrt[3]{x}; a = 64, h = -0.1$

15. $f(x) = \sqrt{1 + x^3}; a = 2, h = 0.01$

16. $f(x) = \sqrt{1 + x^3}; a = 2, h = -0.001$

In Exercises 17–22 find the differential by using (6) or (8).

17. $f(x) = 5x^3 + 2$ 18. $f(x) = \sin x^2$

19. $f(x) = \sin(\cos x)$ 20. $u = x\sqrt{x - 1}$

21. $u = \sqrt{1 + x^4}$ 22. $u = \dfrac{x^2 + 3}{x^3 - 4}$

23. Prove the following differential rules listed in (9).

   a. $d(u + v) = du + dv$ (*Hint:* Show that the set of equations

   $$d(u + v) = \left(\frac{d}{dx}(u + v)\right)dx = \left(\frac{du}{dx} + \frac{dv}{dx}\right)dx$$

   $$= \frac{du}{dx}dx + \frac{dv}{dx}dx = du + dv$$

   is valid.)

   b. $d(cu) = c\,du$ for any constant c

   c. $d(uv) = v\,du + u\,dv$

   d. $d\left(\dfrac{u}{v}\right) = \dfrac{v\,du - u\,dv}{v^2}$

24. The number $|[f(x + h) - f(x)] - f'(x)h|$ is the error introduced by substituting $df = f'(x)h$ for the true difference $f(x + h) - f(x)$. If f has a second derivative at each number between x and $x + h$ and if

   $$|f''(z)| \le M \quad \text{for all } z \text{ between } x \text{ and } x + h$$

   then it can be shown that

   $$|[f(x + h) - f(x)] - f'(x)h| \le \frac{Mh^2}{2}$$

   Therefore the maximum possible error is $Mh^2/2$. Find the maximum possible error introduced by approximating

the following numbers by means of (4).

   a. $\cos \frac{2}{13}\pi; x = \pi/6$ b. $\sqrt{101}; x = 100$

   c. $\sqrt{99}; x = 100$

In Exercises 25–32 use the Newton-Raphson method to find an approximate solution of the given equation in the given interval. Use the method until successive approximations are within 10^{-6} of each other.

25. $x^3 - 3x - 1 = 0; [1, 2]$ (This equation is related to the study of inscribing a nine-sided polygon in a circle.)

26. $x^3 + x = 1; (-\infty, \infty)$ (*Hint:* Let $f(x) = x^3 + x - 1$.)

27. $x^3 - 2x - 5 = 0; [0, 3]$ (Newton solved this equation himself.)

28. $2x^3 - 5x - 3 = 0; [1, 2]$ 29. $2x^3 - 5x - 3 = 0; (-1, 0)$

30. $x^2 + 4x^6 = 2; [0, 1]$ 31. $\tan x = x; (\pi/2, 3\pi/2)$

32. $x^4 + \sin x = 0; [-2, -\frac{1}{2}]$

In Exercises 33–36 use the Newton-Raphson method to find an approximate value of the given root. Use the method until successive approximations are within 10^{-6} of each other.

33. $\sqrt{15}$ 34. $\sqrt{0.2}$ 35. $\sqrt[3]{9}$ 36. $\sqrt[4]{13}$

37. Let $f(x) = x^4 + 2x^2 - x - \frac{1}{4}$. Use the Newton-Raphson method to find approximations of all zeros of f. Use the method until successive approximations are within 10^{-2} of each other. (*Hint:* f has 2 zeros.)

38. Let $f(x) = x^4 + 2x^3 - x - 1$. Use the Newton-Raphson method to find approximations of all zeros of f. Use the method until successive approximations are within 10^{-2} of each other.

39. Suppose that in approximating a zero of f by the Newton-Raphson method we find that $f(c_n) = 0$ and $f'(c_n) \ne 0$ for some n. What does this imply about c_{n+1}, $c_{n+2}, \ldots$?

40. Let $f(x) = \sqrt{x} - \frac{1}{2}$. In trying to use the Newton-Raphson method to find a zero of f, determine what goes wrong when we choose

   a. $c_1 = 1$ b. $c_1 = 4$

41. Let $f(x) = x^4 + x^2 + 8x - 1$, which has a zero in the interval $[-2, 0]$. Determine what happens when the Newton-Raphson method is used with the initial value of c equal to -1.

42. Let $f(x) = x^3 - x$, and suppose we wish to approximate a zero of f in $(0, 1)$. Determine what happens when the Newton-Raphson method is used with the initial value of c equal to $1/\sqrt{5}$.

43. Let $f(x) = x/(x + 1)^2$. Then 0 is the only zero of f. Determine what happens when the Newton-Raphson

method is used with the initial value of c equal to 2. (*Hint:* Use at least 10 steps in the method.)

c 44. Let $f(x) = 1 + x(x - 1)^6$. Determine what happens when the Newton-Raphson method is used with the initial value of c equal to
   a. 0   b. 1   c. 1.05   d. 1.1

45. Let $N_f(x) = x - \dfrac{f(x)}{f'(x)}$, as might be suggested by (10).

Show that
$$N_f'(x) = \frac{f(x)f''(x)}{[f'(x)]^2}$$

46. A hemispherical dome of radius 20 feet is to be given a coat of paint, $\frac{1}{100}$ inch thick. Use differentials to approximate the volume of paint needed for the job. (*Hint:* Approximate the change in the volume of a hemisphere when the radius increases from 20 feet to $20\frac{1}{1200}$ feet.)

47. Approximate the volume of material in a spherical ball with inner radius 5 inches and outer radius 5.137 inches.

48. One of the most striking results of Einstein's Theory of Relativity is his time dilation concept. If a person P_1 moves with velocity v with respect to an observer P_2, then a watch carried by P_1 appears to be running slower to P_2 than to P_1 by a (dilation) factor of

$$\frac{1}{\sqrt{1 - v^2/c^2}}$$

Here c is the velocity of light, approximately 186,000 miles per second.
   a. If $v = c/2$, what is the dilation factor?
   b. If $v = c/3600$, which is 186,000 miles per *hour*, estimate the dilation factor. (*Hint:* Consider the function $f(x) = 1/\sqrt{1 - x}$ near 0.)

Key Terms and Expressions

Derivative
Function differentiable at a point
Differentiable function;
   function differentiable on an interval
Second derivative; higher derivatives
Chain Rule

Implicit differentiation
Acceleration
Tangent line approximation
Differential
Newton-Raphson method

Key Formulas

$(f + g)'(x) = f'(x) + g'(x)$

$(cf)'(x) = cf'(x)$

$(fg)'(x) = f'(x)g(x) + f(x)g'(x)$

$\left(\dfrac{f}{g}\right)'(x) = \dfrac{f'(x)g(x) - f(x)g'(x)}{(g(x))^2}$

$(g \circ f)'(x) = [g'(f(x))]f'(x)$

$\dfrac{d}{dx}(x^r) = rx^{r-1}$

$\dfrac{d}{dx}(u + v) = \dfrac{du}{dx} + \dfrac{dv}{dx}$

$\dfrac{d}{dx}(cu) = c\dfrac{du}{dx}$

$\dfrac{d}{dx}(uv) = v\dfrac{du}{dx} + u\dfrac{dv}{dx}$

$\dfrac{d}{dx}\left(\dfrac{u}{v}\right) = \dfrac{v\dfrac{du}{dx} - u\dfrac{dv}{dx}}{v^2}$

$\dfrac{dy}{dx} = \dfrac{dy}{du}\dfrac{du}{dx}$

$df = f'(x)\,dx$

REVIEW EXERCISES

In Exercises 1–10 find the derivative of the given function at x.

1. $f(x) = -4x^3 + \dfrac{2}{x^2}$

2. $f(x) = \sqrt{x}(x^2 - 3)^{4/7}$

3. $f(x) = \dfrac{2x - 9}{5x + 7}$

4. $f(x) = \dfrac{4x^2 + 2}{3x - 8}$

5. $g(x) = \dfrac{x}{(2x - 1)^2}$

6. $g(x) = \dfrac{1}{(4 - x^2)^{3/2}}$

7. $f(t) = \cos t \sin 2t$

8. $f(t) = t^2 \sin \dfrac{1}{t}$

9. $f(t) = 5t \tan t + 3 \sec 3t$

10. $f(t) = \tan(\sin t^2)$

In Exercises 11–18 find dy/dx.

11. $y = 4x^3 - \sqrt{3}x + \dfrac{2}{5x}$

12. $y = 3 \sin 2x - \sqrt{x} \cos x$

13. $y = \cot x^3$

14. $y = x^2 \tan^2 x$

15. $y = \dfrac{\sin x}{1 - \sec x}$

16. $y = (3x - 5)^{5/9}$

17. $y = x^3 \sqrt{x^2 - 4}$

18. $y = \dfrac{x^2 - x + 1}{x^2 + x + 1}$

In Exercises 19–26 find an equation of the line l tangent to the graph of f at the given point.

19. $f(x) = 3x^3 - 2x^2 + 4;\ (1, 5)$

20. $f(x) = \dfrac{2x - 1}{5x + 2};\ (0, -\tfrac{1}{2})$

21. $f(x) = x \cos \sqrt{2}x;\ (0, 0)$

22. $f(x) = \sin x - 3 \cos 2x;\ (\pi/6, -1)$

23. $f(x) = x\sqrt{x - 1};\ (5, 10)$

24. $f(x) = (x - 2)^{1/7};\ (2, 0)$

25. $f(x) = \begin{cases} 2 \sin x & \text{for } x < 0 \\ 3x^2 + 2x & \text{for } x \ge 0 \end{cases}$ $(0, 0)$

26. $f(x) = \begin{cases} x^2 \sin(1/x) & \text{for } x \ne 0 \\ 0 & \text{for } x = 0 \end{cases}$ $(0, 0)$

In Exercises 27–32 find the second derivative of f.

27. $f(x) = \tfrac{1}{4}x^{12} - 6x^6 + 12$

28. $f(x) = \cos(3 - x)$

29. $f(t) = (t^2 + 9)^{3/2}$

30. $f(t) = \dfrac{2t + 1}{2t - 1}$

31. $f(x) = \tan^2 x$

32. $f(x) = x \sin x$

In Exercises 33–38 find dy/dx by implicit differentiation.

33. $3y^3 - 4x^2y + xy = -5$

34. $x^2 + y^2 = \dfrac{x^2}{y^2}$

35. $y(\sqrt{x} + 1) = x$

36. $xy = \sqrt{x} + \sqrt{y}$

37. $y^3 + \sin xy^2 = \tfrac{3}{2}$

38. $x \tan x^2 y = y^2 + 1$

In Exercises 39–40 use implicit differentiation to find the derivative at the indicated point.

39. $2x^3 - \sin 4y = x^2y + 2;\ (1, 0)$

40. $x^2 - xy^2 + y^3 = 13;\ (-1, 2)$

In Exercises 41–42 assume that x and y are differentiable functions of t. Find dy/dt in terms of x, y, and dx/dt.

41. $xy = 3$

42. $y = \sin xy^2$

In Exercises 43–46 find df in terms of dx.

43. $f(x) = x^2 \cos x$

44. $f(x) = \dfrac{(3x - 1)^{2/3}}{x}$

45. $f(x) = (x - \sin x)^7$

46. $f(x) = \dfrac{\tan x^2}{x}$

In Exercises 47–50 approximate the given number by using differentials.

47. $1 + \sqrt{10}$

48. $\sqrt[3]{3 + 6(2.01)^2}$

49. $\sec 0.26\pi$

50. $\cot 5\pi/12$

51. Use the Newton-Raphson method to find an approximate solution in $(0, \infty)$ of the equation $x = 2 \sin x$. Use the method until successive approximations are within 10^{-3} of each other.

52. Let $f(x) = x^2 + 2x - 3$ and $g(x) = x^2 - \tfrac{9}{4}x + \tfrac{5}{4}$. Show that the lines tangent to the graphs of f and g at the point of their intersection are perpendicular to one another.

53. Let $f(x) = x^2 + 1$ and $g(x) = x^2 - \cos(\pi/(x^2 + 1))$. Show that the lines tangent to the graphs of f and g at their point of intersection are identical.

54. Let $f(x) = x^2 + 3x$. Find $f'(3)$ from the definition of the derivative.

55. Suppose f is differentiable at a point $a > 0$. Find

$$\lim_{x \to a} \frac{f(x) - f(a)}{x^{1/2} - a^{1/2}}$$

in terms of $f'(a)$. (*Hint:* Eliminate the fractional exponents from the denominator.)

56. Let f be differentiable at 0 and let $g(x) = f(x^2)$. Show that $g'(0) = 0$.

57. Let $y = a \sin x + b \cos x$, where a and b are constant. Show that $d^2y/dx^2 + y = 0$.

58. A triangle is formed by the coordinate axes and the line that passes through the point $(4, 0)$ and is tangent to the graph of $y = 1/x$. Find the area of the triangle.

59. The volume V of a right circular cone of height h and radius r is given by $V = \pi r^2 h/3$.
   a. Find the rate of change of the volume with respect to the radius, assuming that the height is constant.

b. Find the rate of change of the volume with respect to the height, assuming that the radius is constant.

c. Find the rate of change of the height with respect to the volume, assuming that the radius is constant.

60. Water is poured into a conical paper cup so that the height increases at the constant rate of 1 inch per second. If the cup is 6 inches tall and its top has a radius of 4 inches, how fast is the volume of water in the cup increasing when the height is 3 inches?

61. A person 6 feet tall walks at a rate of 150 feet per minute toward a light tower whose searchlight is located 40 feet above the ground. Show that the length of the person's shadow shrinks at a constant rate.

62. A whisperjet travels east from San Francisco toward St. Louis at 500 miles per hour, and a turbojet travels north from New Orleans toward St. Louis at 600 miles per hour. Find the rate of change of the distance between them when the jets are 300 miles apart, and the whisperjet is 100 miles from St. Louis.

63. Suppose it costs $C(x)$ thousand dollars to manufacture x thousand gallons of house paint, where

$$C(x) = 15 + 12x - x^3 \quad \text{for } 0 \le x \le 2$$

Suppose also that the revenue accruing from selling x thousand gallons of the paint is $R(x)$ thousand dollars, where

$$R(x) = \tfrac{45}{4}x$$

Find the value of a for which $m_C(a) = m_R(a)$.

64. A cubical cardboard box measures 3 feet on an outer edge. If the cardboard is $\frac{1}{2}$ inch thick, use differentials to approximate the volume of sand that can be poured into the box.

Cumulative Review, Chapters 1–2

In Exercises 1–4 solve the inequality.

1. $\dfrac{x(x^2 - 3)}{(1 - x)^3} > 0$

2. $\dfrac{6t^2 - 6t + 2}{t^2 - t} \ge 0$

3. $\left| \dfrac{1}{|x|} - 3 \right| < 2$

4. $2 \sin x - 1 \le 0$ for x in $[0, 2\pi)$

5. Let $f(x) = 6x^2 - x - 2$. Determine the values of x for which $f(x) > 0$.

6. Show that $(1 + \sin x + \cos x)^2 - 2(1 + \sin x + \cos x) = \sin 2x$.

7. Let $f(x) = \sqrt{\sqrt{x^2 - 1} - x}$. Find the domain of f.

8. Let $f(x) = \dfrac{x^2}{x - 1}$ and $g(x) = \dfrac{1}{4 - x}$.

   a. Find the domain of $f \circ g$.

   b. Find a formula for $(f \circ g)(x)$.

In Exercises 9–12 determine whether the limit exists as a number, as ∞, or as $-\infty$. If the limit exists, evaluate it.

9. $\displaystyle \lim_{x \to 2} \frac{x^2 - 3x + 2}{x^2 - 5x + 6}$

10. $\displaystyle \lim_{h \to 0} \frac{3 \cos h}{h \csc 2h}$

11. $\displaystyle \lim_{x \to 0} \frac{|x|^3 - x^2}{x^3 + x^2}$

12. $\displaystyle \lim_{x \to 0^+} \frac{\sqrt{1 + x^2} - \sqrt{1 - x^2}}{x^3}$

13. Let $f(x) = \sqrt{2x^2 - 4}$. Find $\displaystyle \lim_{x \to 2} \frac{f(x) - f(2)}{x - 2}$.

14. Let

$$f(x) = \begin{cases} x + 1 & \text{for } x < -1 \\ (x + 1)^2 & \text{for } x \ge -1 \end{cases}$$

Is f continuous at -1? Explain why or why not.

15. Sketch the graph of a function f with all of the following properties.

   a. The domain of f is $[0, 4]$.

   b. f is continuous on $(0, 1)$, $[1, 3]$, and $(3, 4]$.

   c. f is not continuous at 1 or at 3, and is not continuous from the right at 0.

16. Use the bisection method to approximate $\sqrt[4]{15}$ with an error of at most $\frac{1}{16}$.

4

APPLICATIONS
OF THE DERIVATIVE

Chapter 4 will emphasize applications of the derivative to graphing functions. We will learn how to determine where the graph of a differentiable function rises and where it falls; where it has peaks and where it has valleys; where it curves upward and where it curves downward. The concepts we will introduce have applications not only to graphing functions but also to problems in such widely varying areas as engineering, the physical sciences, the biological sciences, and the social sciences.

4.1
MAXIMUM AND MINIMUM VALUES

Suppose the profit function for an oil company is given by

$$f(x) = x - x^3 \quad \text{for } 0 \le x \le 1 \tag{1}$$

where x is the amount of oil produced per year in millions of gallons, $f(x)$ is the profit per year in millions of dollars, and the domain is restricted to $[0, 1]$ because of limitations on plant space and personnel. Naturally, the company wishes to make the largest profit possible, which would correspond to a value of x such that $f(x)$ is the maximum value of f on $[0, 1]$. Does such a value of x exist?

Before we answer this question, let us define precisely what we mean by a maximum value, or a minimum value, of a function.

170

DEFINITION 4.1

A function f **has a maximum value on a set I** if there is a number d in I such that $f(x) \leq f(d)$ for all x in I (Figure 4.1). We call $f(d)$ the **maximum value of f on I**. Similarly, the function f **has a minimum value on I** if there is a number c in I such that $f(x) \geq f(c)$ for all x in I (Figure 4.1). We call $f(c)$ the **minimum value of f on I**. A value of f that is either a maximum value or a minimum value on I is called an **extreme value of f on I**.

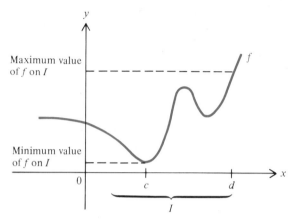

FIGURE 4.1

If the set I is the domain of f, then when it exists, the maximum value of f on I is called the **maximum value** of f. Similarly, when it exists, the minimum value of f on its domain is the **minimum value** of f.

Since $f(x)$ is the y coordinate of the point $(x, f(x))$ on the graph of f, it is evident that $f(d)$ is the maximum value of f on $[a, b]$ if and only if $(d, f(d))$ is the highest point on the graph of f between the lines $x = a$ and $x = b$. Likewise, $f(c)$ is the minimum value of f on $[a, b]$ if and only if $(c, f(c))$ is the lowest point on the graph of f between the lines $x = a$ and $x = b$. Both cases are illustrated in Figure 4.1.

A function f may or may not have extreme values on a set I, depending on f and on I. For example, if $f(x) = x$, then on the interval $[0, 1]$ the function f has the maximum value of 1, and the minimum value of 0. However, on the open interval $(0, 1)$, f has neither a maximum nor a minimum value because f does not assume either 0 or 1 on the open interval $(0, 1)$. In contrast, if

$$f(x) = \begin{cases} x & \text{for } 0 \leq x < 1 \\ \dfrac{1}{2} & \text{for } x = 1 \end{cases}$$

then f has the minimum value of 0, but no maximum value because f does not assume the value 1 (see Figure 4.2).

We can be much more precise about possible extreme values if the function f is continuous and the set I is a closed, bounded interval. The following theorem, which is proved in the Appendix, tells the story.

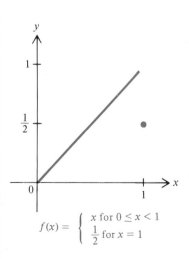

$$f(x) = \begin{cases} x \text{ for } 0 \leq x < 1 \\ \frac{1}{2} \text{ for } x = 1 \end{cases}$$

FIGURE 4.2

THEOREM 4.2

MAXIMUM–MINIMUM
THEOREM

Let f be continuous on a closed, bounded interval $[a, b]$. Then f has a maximum and a minimum value on $[a, b]$.

The function given in (1) by

$$f(x) = x - x^3 \quad \text{for } 0 \leq x \leq 1$$

is continuous on the closed interval $[0, 1]$. Therefore by the Maximum–Minimum Theorem, f has a maximum value on $[0, 1]$. But the theorem does not tell us where in $[0, 1]$ the maximum value occurs, nor does it tell us how to find the maximum value. The next theorem will help in determining such a value.

THEOREM 4.3

Let f be defined on $[a, b]$. If an extreme value of f on $[a, b]$ occurs at a number c in (a, b) at which f has a derivative, then $f'(c) = 0$.

Proof   It suffices to show that f does not assume an extreme value at any number c in (a, b) such that $f'(c)$ exists and $f'(c) \neq 0$. Therefore we assume that $f'(c) \neq 0$. Consider first the case in which $f'(c) > 0$. Since

$$f'(c) = \lim_{x \to c} \frac{f(x) - f(c)}{x - c} > 0$$

we know by (9) of Section 2.3 that for x in some open interval I about c the inequality

$$\frac{f(x) - f(c)}{x - c} > 0$$

holds. For such x, if $x > c$, then

$$f(x) - f(c) = \overbrace{(x - c)}^{\text{positive}} \overbrace{\left(\frac{f(x) - f(c)}{x - c} \right)}^{\text{positive}} > 0$$

Therefore $f(x) > f(c)$, so that f does not have a maximum value at c. However, for x in I and $x < c$, we have

$$f(x) - f(c) = \overbrace{(x - c)}^{\text{negative}} \overbrace{\left(\frac{f(x) - f(c)}{x - c} \right)}^{\text{positive}} < 0$$

Thus $f(x) < f(c)$, so that f does not have a minimum value at c. Hence if $f'(c) > 0$, then f has neither a maximum value nor a minimum value at c. The case in which $f'(c) < 0$ is treated analogously, so we leave it as an exercise. ∎

Let f be defined on $[a, b]$. Theorem 4.3 implies that the only numbers in $[a, b]$ at which f can assume its extreme values are the endpoints a and b and those numbers c in (a, b) such that either $f'(c) = 0$ or f is not differentiable at c. We will call a number c in the domain of f a **critical point** of f if either $f'(c) = 0$ or $f'(c)$ does not exist. Then Theorem 4.3 implies that f can assume an extreme value on $[a, b]$ only at an endpoint or at a critical point in (a, b). This fact provides us with the following general method for determining the extreme values of a continuous function on a closed interval $[a, b]$:

Compute the values of f at all critical points in (a, b) and at the end points a and b. The largest of those values is the maximum value of f on $[a, b]$; the smallest of those values is the minimum value of f on $[a, b]$.

Most of the functions we will encounter are differentiable at all numbers in their domains. For such functions the critical points are just the numbers at which the derivative is 0. If f does not have very many critical points in (a, b), we can apply the method just described, provided that we can evaluate f at the critical points.

Example 1 Let $f(x) = x - x^3$. Find the extreme values of f on $[0, 1]$, and determine at which numbers in $[0, 1]$ they occur.

Solution As we remarked earlier, f has extreme values because it is continuous on $[0, 1]$. Since f is differentiable, the critical points of f are the values of x for which $f'(x) = 0$. But

$$f'(x) = 1 - 3x^2$$

so that $f'(x) = 0$ if $1 - 3x^2 = 0$, that is, if $x = -\frac{1}{3}\sqrt{3}$ or $x = \frac{1}{3}\sqrt{3}$. Since $-\frac{1}{3}\sqrt{3}$ is not in the interval $[0, 1]$, we conclude that an extreme value of f on $[0, 1]$ can occur only at one of the endpoints 0 and 1 or at the critical point $\frac{1}{3}\sqrt{3}$ in $(0, 1)$. To decide which of these give the extreme values, we compute the corresponding values of f:

$$f(0) = 0, \qquad f(1) = 0,$$

$$f\left(\frac{1}{3}\sqrt{3}\right) = \frac{1}{3}\sqrt{3} - \left(\frac{1}{3}\sqrt{3}\right)^3 = \frac{1}{3}\sqrt{3} - \frac{1}{9}\sqrt{3} = \frac{2}{9}\sqrt{3}$$

Consequently the minimum value of f on $[0, 1]$ is 0, and it occurs at 0 and 1. The maximum value of f on $[0, 1]$ is $\frac{2}{9}\sqrt{3}$, and it occurs at $\frac{1}{3}\sqrt{3}$. ☐

By Example 1, the profit function introduced at the beginning of the section has a maximum value of $\frac{2}{9}\sqrt{3}$, which would not have been so easy to guess.

If we alter the interval, the extreme values can be altered. This is illustrated in Example 2, in which we find the extreme values of the function f in Example 1 on a different interval.

Example 2   Let $f(x) = x - x^3$. Find the extreme values of f on $[2, 4]$, and determine at which numbers in $[2, 4]$ they occur.

Solution   We know from the solution of Example 1 that $f'(x) = 0$ only if $x = -\frac{1}{3}\sqrt{3}$ or $x = \frac{1}{3}\sqrt{3}$. Since $\frac{1}{3}\sqrt{3} < 2$, $f'(x) \neq 0$ for all x in $(2, 4)$. Hence the extreme values of f on $[2, 4]$ must occur at the endpoints of the interval. Since

$$f(2) = 2 - 8 = -6 \quad \text{and} \quad f(4) = 4 - 64 = -60$$

we conclude that -6 is the maximum value and occurs at 2, whereas -60 is the minimum value and occurs at 4. $\square$

It is possible for a function to have a critical point that does *not* correspond to an extreme value, as the next example illustrates.

Example 3   Let $f(x) = x^3$. Find the extreme values of f on $[-2, 1]$, and determine at which numbers they occur.

Solution   The function f is differentiable, and $f'(x) = 3x^2$. Therefore $f'(x) = 0$ only for $x = 0$. Since $f(-2) = -8$, $f(0) = 0$, and $f(1) = 1$, it follows that -8 and 1 are the extreme values of f on $[-2, 1]$, and they occur at -2 and 1, respectively (see Figure 4.3). $\square$

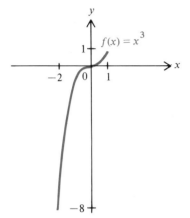

FIGURE 4.3

It may also happen that a function f has an extreme value at a critical point c for which $f'(c)$ does not exist. Indeed, if $f(x) = x^{2/3}$, then $f'(0)$ does not exist, because the limit

$$\lim_{x \to 0} \frac{f(x) - f(0)}{x - 0} = \lim_{x \to 0} \frac{x^{2/3}}{x} = \lim_{x \to 0} x^{-1/3}$$

does not exist. Thus 0 is a critical point of f. However, since $f(x) \geq 0$ for all x and $f(0) = 0$, f has a minimum value at 0 (Figure 4.4). Note also that $f(-1) = f(1) = 1$. This is the maximum value of f on $[-1, 1]$. Thus f attains its maximum value twice on $[-1, 1]$.

We now apply our method of finding extreme values to a problem taken from everyday life.

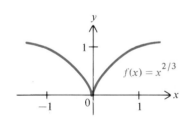

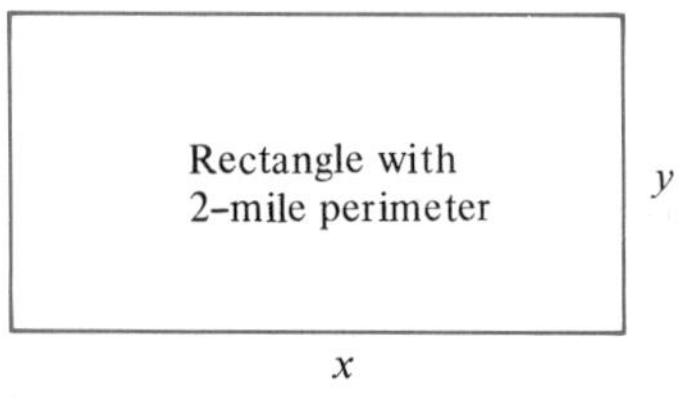

FIGURE 4.4

Example 4   A landowner wishes to use 2 miles of fencing to enclose a rectangular region of maximum area. What should the lengths of the sides be?

Solution   Any rectangular region the landowner could enclose must have a length x and a width y (Figure 4.5). Since the perimeter is to be 2 (miles), we have

$$2x + 2y = 2 \quad \text{for } 0 \leq x \leq 1 \text{ and } 0 \leq y \leq 1$$

Therefore $y = 1 - x$, so the area xy of the rectangle can be written as a

FIGURE 4.5

Rectangle with 2-mile perimeter

function A of x alone:

$$A(x) = x(1 - x) \quad \text{for } 0 \leq x \leq 1$$

The problem has now been reduced to finding the maximum value of A on $[0, 1]$. Since $A'(x) = 1 - 2x$ and thus $A'(x) = 0$ only when $x = \frac{1}{2}$, we know from Theorem 4.3 that A can have its maximum value on $[0, 1]$ only at $0, \frac{1}{2}$, or 1. But $A(0) = 0$, $A(\frac{1}{2}) = \frac{1}{2}(1 - \frac{1}{2}) = \frac{1}{4}$, and $A(1) = 0$. Thus the maximum value of A occurs for $x = \frac{1}{2}$ (mile) (see Figure 4.6); since $y = 1 - x$, the value of y corresponding to $x = \frac{1}{2}$ is also $\frac{1}{2}$ (mile). Consequently the fence should enclose a square region $\frac{1}{2}$ mile on a side. $\square$

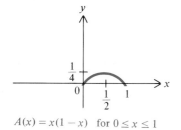

$A(x) = x(1 - x)$ for $0 \leq x \leq 1$

FIGURE 4.6

Relative Extreme Values

In graphing functions we will use the following notion, which is related to the notion of extreme value.

DEFINITION 4.4

> A function f has a **relative maximum value** (respectively, a **relative minimum value**) at c if there is some number $\delta > 0$ such that $f(c)$ is the maximum value (respectively, the minimum value) of f on the interval $[c - \delta, c + \delta]$. A value that is either a relative maximum value or relative minimum value is called a **relative extreme value**.

Relative extreme values correspond to the "hilltops" and "valley bottoms" on the graphs of functions (Figure 4.7).

It follows from Definition 4.4 that if f has a relative extreme value at c, then f is defined on an interval of the form $[c - \delta, c + \delta]$. From Definition 4.4 and Theorem 4.3 it also follows that if f has a relative extreme value at c, then c is a critical point of f. We will use this information in locating relative extreme values in the next example.

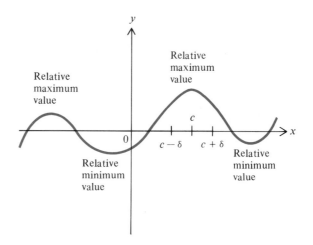

FIGURE 4.7

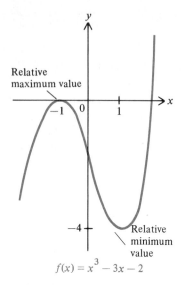

Relative maximum value

$$f(x) = x^3 - 3x - 2$$

FIGURE 4.8

Example 5   Let $f(x) = x^3 - 3x - 2$. Find the relative extreme values of f, and determine at which numbers they occur.

Solution   Since $f'(x)$ exists for all x, relative extreme values can occur only at numbers satisfying $f'(x) = 0$. But

$$f'(x) = 3x^2 - 3 = 3(x^2 - 1) = 3(x + 1)(x - 1)$$

Therefore $f'(x) = 0$ for $x = -1$ and $x = 1$. To show that f has a relative extreme value at 1, we will find an interval about 1 on which f has an extreme value at 1. Consider the interval $[0, 2]$. Calculation shows that $f(0) = -2$, $f(1) = -4$, and $f(2) = 0$. By Theorem 4.3, $f(1)$ is the minimum value of f on $[0, 2]$, and thus $f(1)$ is a relative minimum value of f. Similarly, $f(-2) = -4$, $f(-1) = 0$, and $f(0) = -2$, so that $f(-1)$ is the maximum value of f on $[-2, 0]$. Thus $f(-1)$ is a relative maximum value of f (Figure 4.8).   □

The method used in Example 5 to show that $f(1)$ was a relative extreme value of f may seem a bit puzzling, since it required guessing an interval about 1 on which f attained its minimum value at 1. In Section 4.4 more convenient and reliable methods for finding relative extreme values will be described.

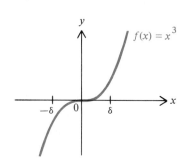

$f(x) = x^3$

FIGURE 4.9

Caution: Notice that f need not have a relative extreme value at c simply because $f'(c) = 0$. To illustrate this, let $f(x) = x^3$. Then $f'(0) = 0$, but $f(0) = 0$ is not an extreme value on *any* interval of the form $[-\delta, \delta]$, where $\delta > 0$. (See Figure 4.9.) In Section 4.6 we will study points at which the derivative is 0 but at which there is no relative extreme value. We also emphasize that if a function is not defined on an interval of the form $[c - \delta, c + \delta]$ about c, then f cannot have a relative extreme value at c—even if f has an extreme value at c. Thus if $f(x) = \sqrt{x}$, then f assumes its minimum value at 0, but f does not have a relative minimum value at 0, since f is not defined on any interval of the form $[-\delta, \delta]$, where $\delta > 0$.

EXERCISES 4.1

In Exercises 1–16 find all critical points (if any) of the given function.

1. $f(x) = x^2 + 4x + 6$
2. $f(x) = 4x^3 - 6x^2 - 9x$
3. $f(x) = 3x^4 + 4x^3 - 12x^2 + 1$
4. $f(x) = x^5 - 5x^3 + 10x - 3$
5. $g(x) = x + \dfrac{1}{x}$
6. $g(x) = x - \dfrac{4}{x^2}$
7. $k(t) = \dfrac{1}{t^2 + 4}$
8. $k(t) = \dfrac{1}{\sqrt{t^2 + 1}}$
9. $k(t) = 3t^{2/5}$
10. $k(t) = 2t - t^{2/3}$
11. $f(x) = \sin x$
12. $f(x) = \cos \sqrt{x}$
13. $f(x) = x + \sin x$
14. $f(x) = \tan x$
15. $f(z) = |z - 2|$
16. $f(z) = z|z + 3|$

In Exercises 17–30 find all extreme values (if any) of the given function on the given interval. Determine at which numbers in the interval these values occur.

17. $f(x) = x^2 - x$; $[0, 2]$
18. $f(x) = x^2 - 5x - 6$; $[\tfrac{5}{2}, 6]$
19. $g(x) = x^4 - 4x$; $[-4, 4]$
20. $g(x) = 1/x$; $(0, 3]$

21. $f(t) = -\dfrac{1}{2t}$; $(0, \infty)$ 22. $f(t) = t^2 - \dfrac{4}{t}$; $[1, 3)$

23. $k(z) = \sqrt{1 + z^2}$; $[-2, 3]$ 24. $k(z) = \sqrt{1 + z}$; $(3, 8]$

25. $f(x) = \sqrt{|x|}$; $(-1, 2)$ 26. $f(x) = x^{2/3}$; $[-8, 8]$

27. $f(x) = -\sin \sqrt[3]{x}$; $[-\pi^3/27, \pi^3/8]$

28. $f(x) = \cos \pi x$; $(\frac{1}{3}, 1]$

29. $f(x) = \tan x/2$; $(-\pi/2, \pi/6)$

30. $f(x) = \csc 3x$; $[\pi/18, \pi/4]$

In Exercises 31–40 find all relative extreme values (if any) of the given function. Determine at which numbers they occur.

31. $f(x) = x^3 - 3x^2$ 32. $f(x) = x^3 - 12x - 1$

33. $f(x) = x^3 + 3x^2 - 9x - 2$ 34. $f(x) = x^4 - 2x^2$

35. $g(x) = x^5 - 20x$ 36. $g(x) = x^5 + 80x + 2$

37. $f(x) = -3 - \sin x$ 38. $f(x) = \cos \frac{1}{2}x$

39. $f(x) = \sec \pi x$ 40. $f(x) = \csc 3x$

In Exercises 41–42 use the bisection method to approximate a critical point of the function in the given interval to within $\frac{1}{16}$.

41. $f(x) = x^4 + 2x^2 - x$; $[0, 1]$

42. $f(x) = x^5 + \frac{9}{2}x^2 + x$; $[-1, 0]$

In Exercises 43–44 use the Newton-Raphson method to approximate a critical point of the function in the given interval until successive approximations are within 10^{-6} of each other.

43. $f(x) = \frac{1}{4}x^4 + x^3 + 2x - 1$; $[-4, -3]$

44. $f(x) = x \cos x$; $[0, \pi/2]$

45. Let $f(x) = \frac{1}{4}x^4 + x^3 - x - 1$. Approximate each of the critical points of f by the Newton-Raphson method until successive approximations are within 10^{-6} of each other.

46. Suppose c and d are not both 0, and let

$$f(x) = \frac{ax + b}{cx + d}$$

Show that f has no critical points unless $ad - bc = 0$, in which case f is a constant function.

47. Assume that f is defined on $[a, b]$ and that $g = -f$. Prove that $f(x_0)$ is the maximum value of f on $[a, b]$ if and only if $g(x_0)$ is the minimum value of g on $[a, b]$.

48. a. Let f be an odd function on $(-\infty, \infty)$, that is, $f(-x) = -f(x)$ for all x. Show that f assumes a relative maximum value at a if and only if f assumes a relative minimum value at $-a$.

    b. What can you say about relative extreme values of an even function f?

49. Prove Theorem 4.3 for the case $f'(c) < 0$. (All that is required is to change certain inequalities in the proof for the case $f'(c) > 0$.)

50. In the proof of Theorem 4.3, where do we use the fact that c is in (a, b)?

*51. Prove Darboux's Theorem: Let f be differentiable on $[a, b]$, and let $f'(a) < m < f'(b)$. Then there exists some number c in (a, b) such that $f'(c) = m$. (*Hint:* Let $g(x) = f(x) - mx$ for $a \le x \le b$. Show that $g'(a) < 0$ and $g'(b) > 0$, and conclude from this that g does not assume its minimum value on $[a, b]$ at either a or b. Now apply the Maximum–Minimum Theorem and Theorem 4.3 to the function g.)

52. Suppose that in Example 4 a stream runs along one side of the landowner's property, so that only three sides must be fenced in. Find the lengths of the sides of the rectangular region having the largest area that can be enclosed.

53. What is the largest possible product of two nonnegative numbers whose sum is 1? Is there a smallest possible product? Explain your answer.

54. A mass connected to a spring moves along the x axis so that its x coordinate at time t is given by

$$x(t) = \sin 2t + \sqrt{3} \cos 2t$$

What is the maximum distance of the mass from the origin?

55. Suppose $R(x)$, $C(x)$, and $P(x)$ denote the revenue, cost, and profit resulting from the manufacture and sale of x units of an item. Recall that

$$P(x) = R(x) - C(x) \quad \text{for } x \ge 0$$

Assume that it is possible to make a maximum profit by manufacturing x_0 units of the item. Show that if R and C are differentiable and $x_0 > 0$, then $R'(x_0) = C'(x_0)$ (that is, the marginal revenue at x_0 equals the marginal cost at x_0).

56. A rectangular sheet of metal 8 inches wide and 100 inches long is folded along the center to form a triangular trough (Figure 4.10). Two extra pieces of metal are attached to the ends of the trough, and the trough is then filled with water.

    a. How deep should the trough be to maximize the capacity of the trough?

    b. What is the maximum capacity?

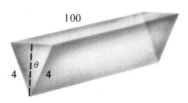

FIGURE 4.10

4.2
THE MEAN VALUE THEOREM

Theorem 4.3 implies that the extreme values of a function f on $[a, b]$ can occur only at the endpoints of $[a, b]$ or at critical points of f in (a, b). The following theorem, named after the seventeenth-century French mathematician Michel Rolle* (1652–1719), implies that if f is continuous on $[a, b]$ and $f(a) = f(b)$, then there always exists at least one critical point of f in (a, b).

THEOREM 4.5
ROLLE'S THEOREM

Let f be continuous on $[a, b]$ and differentiable on (a, b). If $f(a) = f(b)$, then there is a number c in (a, b) such that $f'(c) = 0$.

Proof If f is constant, then its derivative is 0, so that $f'(c) = 0$ for each c in (a, b). If f is not constant, then its maximum and minimum values (which exist by the Maximum–Minimum Theorem) are distinct. Since $f(a) = f(b)$, at least one of these values must occur at a point c in (a, b). By hypothesis, f is differentiable at c, so $f'(c) = 0$ by Theorem 4.3. ∎

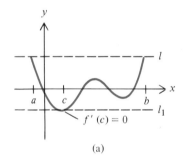

(a)

(b)

FIGURE 4.11

It follows from Rolle's Theorem that if f is continuous on $[a, b]$ and $f(a) = f(b)$, then either there is a number c in (a, b) such that $f'(c)$ does not exist or there is a number c in (a, b) such that $f'(c) = 0$. In either case c is a critical point of f in (a, b).

Notice that the condition $f(a) = f(b)$ cannot be eliminated from Rolle's Theorem. For example, if $f(x) = x$, then $f'(x) = 1$ for all x in any interval (a, b). This implies that $f'(c) \neq 0$ for all c in (a, b).

Rolle's Theorem implies that at some point $(c, f(c))$ on the graph of f, the slope of the tangent line l_1 is 0 (Figure 4.11(a)). But the line l joining $(a, f(a))$ to $(b, f(b))$ is horizontal, because $f(a) = f(b)$. Thus l_1 and l are parallel.

Is there a comparable theorem when $f(a) \neq f(b)$? In other words, must there be a point c in (a, b) such that the line l_1 tangent to the graph of f at $(c, f(c))$ is parallel to the line l joining $(a, f(a))$ and $(b, f(b))$ (Figure 4.11(b))? Since two nonvertical lines are parallel only if their slopes are equal, we can rephrase the question as follows: Does there necessarily exist a point c in (a, b) such that

$$f'(c) = \frac{f(b) - f(a)}{b - a}$$

holds? The Mean Value Theorem supplies the answer.

THEOREM 4.6
MEAN VALUE THEOREM

Let f be continuous on $[a, b]$ and differentiable on (a, b). Then there is a number c in (a, b) such that

$$f'(c) = \frac{f(b) - f(a)}{b - a} \tag{1}$$

*Rolle: Pronounced "Role."

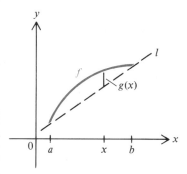

FIGURE 4.12

Proof We introduce an auxiliary function g that allows us to simplify the proof by using Rolle's Theorem. The function g is defined by

$$g(x) = f(x) - \left[f(a) + \frac{f(b) - f(a)}{b - a}(x - a) \right] \quad \text{for } a \le x \le b$$

(See Figure 4.12.) Now g is continuous on $[a, b]$ and differentiable on (a, b), since g is a simple combination of f, constant functions, and a linear function. Substituting in the equation above, we find that

$$g(a) = g(b) = 0$$

so that by Rolle's Theorem there is a number c in (a, b) such that $g'(c) = 0$. However,

$$g'(x) = f'(x) - \frac{f(b) - f(a)}{b - a} \quad \text{for } a < x < b$$

and thus

$$0 = g'(c) = f'(c) - \frac{f(b) - f(a)}{b - a}$$

Solving for $f'(c)$, we obtain

$$f'(c) = \frac{f(b) - f(a)}{b - a} \quad \blacksquare$$

The following example corroborates the Mean Value Theorem for a particular function f and a particular interval $[a, b]$.

Example 1 Let $f(x) = \frac{1}{3}x^3 + 2x$. Find a number c in $(0, 3)$ such that

$$f'(c) = \frac{f(3) - f(0)}{3 - 0}$$

Solution Since

$$\frac{f(3) - f(0)}{3 - 0} = \frac{15 - 0}{3 - 0} = 5$$

we seek a number c in $(0, 3)$ such that $f'(c) = 5$. But

$$f'(x) = x^2 + 2$$

so that c must satisfy

$$c^2 + 2 = 5$$

Therefore $c^2 = 3$, and since c must be in $(0, 3)$, we conclude that $c = \sqrt{3}$. $\square$

When the French mathematician Count Joseph Louis Lagrange first presented the Mean Value Theorem, the terms "average" and "mean" were synonymous. If $f(t)$ denotes the position of an object on the x axis at time t, then the average (or mean) velocity during the interval $[a, b]$ is

$$\frac{f(b) - f(a)}{b - a}$$

Thus by the Mean Value Theorem the mean velocity during an interval $[a, b]$ is equal to the velocity $f'(c)$ at some instant c in (a, b).

Example 2   Pat enters the Pennsylvania turnpike just south of Harrisburg and receives a ticket at toll station 19 with the time 11 A.M. stamped on it. At noon Pat stops at toll station 12 near Crystal Springs, pays the toll for the 80-mile trip on the turnpike, and goes on. A few moments later a police officer arrests Pat for speeding on the turnpike, which has a speed limit of 55 miles per hour. Show that Pat had in fact been speeding somewhere on the pike.

Solution   Let $f(t)$ denote the distance that Pat has traveled during the t hours since 11 A.M. Notice that by assumption,

$$\frac{f(1) - f(0)}{1 - 0} = \frac{80 - 0}{1 - 0} = 80$$

It is natural to suppose that f is a differentiable function. Then the Mean Value Theorem implies that $f'(c) = 80$ for some value of c in $(0, 1)$. Since $f'(c)$ is the velocity of the car at time c, we conclude that at that instant the velocity was 80 miles per hour, and indeed Pat was speeding.   □

EXERCISES 4.2

In Exercises 1–10 find all numbers c in the interval (a, b) for which the line tangent to the graph of f is parallel to the line joining $(a, f(a))$ and $(b, f(b))$.

1. $f(x) = x^2 - 6x; a = 0, b = 4$

2. $f(x) = x - 3x^2; a = -1, b = 3$

3. $f(x) = x^3 - 6x; a = -2, b = 0$

4. $f(x) = x^3 - 6x; a = -2, b = 2$

5. $f(x) = x^3 + 4; a = -2, b = 1$

6. $f(x) = x^3 - 2; a = -3, b = 3$

7. $f(x) = x^3 - 3x^2 + 3x + 1; a = -2, b = 2$

8. $f(x) = -3 + \sqrt{x}; a = 0, b = 1$

9. $f(x) = 1 + x^{1/3}; a = 1, b = 8$

10. $f(x) = 3\left(x + \dfrac{1}{x}\right); a = \frac{1}{3}, b = 3$

11. Let $f(x) = Ax^2 + Bx + C$, where A, B, and C are constants with $A \neq 0$. Show that for any interval $[a, b]$, the number c guaranteed by the Mean Value Theorem is the midpoint of $[a, b]$.

12. Let $f(x) = |x|$. Show that $f(-2) = f(2)$, but there is no number c in $(-2, 2)$ such that $f'(c) = 0$. Does this result contradict Rolle's Theorem? Explain.

13. Suppose $|f'(x)| \leq M$ for $a \leq x \leq b$. Using the Mean Value Theorem, prove that $|f(b) - f(a)| \leq M(b - a)$, so that $f(a) - M(b - a) \leq f(b) \leq f(a) + M(b - a)$.

In Exercises 14–16 use the result of Exercise 13 to determine lower and upper bounds for the given number.

14. $\sqrt{101}$ (*Hint:* Let $f(x) = \sqrt{x}$, $a = 100$ and $b = 101$, and then find bounds for $f(101) = \sqrt{101}$.)

15. $28^{2/3}$ 16. $33^{1/5}$

17. Use the result of Exercise 13 to estimate how far 1.7 can be from the true value of $\sqrt{3}$. (*Hint:* Take $a = 2.89$, $b = 3$.)

18. Use the result of Exercise 13 to estimate how far 2.2 can be from the true value of $\sqrt{5}$.

19. In the aftermath of a car accident it is concluded that one driver slowed to a halt in 9 seconds while skidding 400 feet. If the speed limit was 30 miles per hour, can it be proved that the driver had been speeding? Explain. (*Hint:* 30 miles per hour is equal to 44 feet per second.)

c 20. A racing car accelerated from rest, traveling 2400 feet in 12 seconds. Must the car have been traveling at least 130 miles per hour at some moment during that time interval? Explain. (*Hint:* Convert to miles and hours.)

21. Use Rolle's Theorem to show that there is a solution of the equation $\tan x = 1 - x$ in $(0, 1)$. (*Hint:* Let $f(x) = (x - 1) \sin x$, and find $f(0)$, $f(1)$, and $f'(x)$.)

22. Use Rolle's Theorem to show that there is a solution of the equation $\cot x = x$ in $(0, \pi/2)$. (*Hint:* Let $f(x) = x \cos x$ for x in $[0, \pi/2]$.)

*23. Let $f(x) = x^m (x - 1)^n$, where m and n are positive integers. Show that the number c guaranteed by Rolle's Theorem is unique and that it divides $[0, 1]$ into segments whose lengths have ratio m/n.

24. a. Let g be continuous on $[a, b]$ and differentiable on (a, b). Assume that $g(x) = 0$ for two values of x in (a, b). Show that there is at least one value c in (a, b) such that $g'(c) = 0$.
    b. Let $g(x) = x^4 - 20x^3 - 25x^2 - x + 1$. Use (a) to show that there is some c in $(-1, 1)$ such that $4c^3 - 60c^2 - 50c - 1 = 0$.

*25. Let g be continuous on $[a, b]$ and differentiable on (a, b). Assume that $g(x) = 0$ for three values of x in (a, b). Show that there are at least two values of x in (a, b) such that $g'(x) = 0$.

*26. Suppose f' assumes a value m at most n times. Use the Mean Value Theorem to show that any line with slope m intersects the graph of f at most $n + 1$ times.

27. A number a is a **fixed point** of a function f if $f(a) = a$.
    a. Prove that if $f'(x) < 1$ for every real number x, then f has at most one fixed point.
    b. Let $f(x) = \sin \frac{1}{2}x$. Using part (a), prove that 0 is the only fixed point of f.

4.3
APPLICATIONS OF THE MEAN VALUE THEOREM

Because it is employed so often in proving other theorems, the Mean Value Theorem is one of the most important results in calculus. We will use it now to prove two very different theorems. The first implies that if two functions have identical slopes at each number in an interval, then the functions differ by a constant on that interval.

THEOREM 4.7

a. Let f be continuous on an interval I. If $f'(x) = 0$ for each interior point x of I, then f is constant on I.
b. Let f and g be continuous on an interval I. If $f'(x) = g'(x)$ for each interior point x of I, then $f - g$ is constant on I. In other words, there is a constant C such that $f(x) = g(x) + C$ for all x in I.

Proof   To prove (a), let x and z be arbitrary numbers in I with $x < z$. By the Mean Value Theorem there is a number c in (x, z) such that

$$\frac{f(z) - f(x)}{z - x} = f'(c) \qquad (1)$$

By assumption, $f'(c) = 0$, and thus (1) reduces to

$$f(z) - f(x) = 0$$

Therefore $\qquad\qquad\qquad f(z) = f(x)$

It follows that f assigns the same value at any two points in I, so f is constant on I. To prove (b), notice that

$$(f - g)'(x) = f'(x) - g'(x) = 0$$

so $f - g$ satisfies the conditions of part (a). Consequently $f - g$ is constant on I. In other words, there is a constant C such that $f(x) = g(x) + C$ for all x in I. ∎

If f is a function defined on an interval I, then any function F such that $F'(x) = f(x)$ for each x in I is called an **antiderivative** of f (since f is the derivative of F on I). Thus on any given interval, x^3 is an antiderivative of $3x^2$ and $\sin x$ is an antiderivative of $\cos x$.

By part (b) of Theorem 4.7, once we know a single antiderivative F of a given function, all other antiderivatives can be ascertained by adding constants to F. It follows that on any interval I the only antiderivatives of $3x^2$ are functions of the form $x^3 + C$, and the only antiderivatives of $\cos x$ are functions of the form $\sin x + C$.

Example 1   Find the antiderivatives of the following functions.

    a. $x^2 - 4x$ b. $\sec^2 x$

Solution

    a. Since $\frac{1}{3}x^3$ is an antiderivative of x^2, and $-2x^2$ is an antiderivative of $-4x$, it follows that $\frac{1}{3}x^3 - 2x^2$ is an antiderivative of $x^2 - 4x$. Thus the antiderivatives of $x^2 - 4x$ have the form $\frac{1}{3}x^3 - 2x^2 + C$, where C is any constant.

    b. One antiderivative of $\sec^2 x$ is $\tan x$, so the antiderivatives of $\sec^2 x$ have the form $\tan x + C$, where C is any constant. □

There is precisely one antiderivative that has a specified value at a prescribed point, as the next example illustrates.

Example 2   Let f be such that $f'(x) = \cos x$ and $f(\pi/2) = -1$. Determine the function f.

Solution   Since f and $\sin x$ are both antiderivatives of $\cos x$, by Theorem 4.7(b) there is a constant C such that

$$f(x) = \sin x + C$$

for the appropriate constant C. To determine C, we use the assumption that

$f(\pi/2) = -1$, which yields

$$-1 = f\left(\frac{\pi}{2}\right) = \sin\frac{\pi}{2} + C = 1 + C$$

Therefore $C = -2$, and hence

$$f(x) = \sin x - 2 \quad \square$$

Increasing and Decreasing Functions

The second consequence of the Mean Value Theorem will facilitate the search for extreme values and the analysis of the graph of a function. A function f is said to be **increasing** on an interval I provided that $f(x) \leq f(z)$ whenever x and z are in I and $x < z$. The function f is **strictly increasing** on I provided that $f(x) < f(z)$ whenever x and z are in I and $x < z$. The definitions of **decreasing** and **strictly decreasing** are analogous.

Graphically, a function is strictly increasing on I if its graph slopes upward to the right (Figure 4.13(a)). It is strictly decreasing on I if its graph slopes downward to the right (Figure 4.13(b)). For example if $f(x) = x^2$, then f is strictly decreasing on $(-\infty, 0]$ and strictly increasing on $[0, \infty)$ (Figure 4.14). If $f(x) = x^3$, then f is strictly increasing on $(-\infty, \infty)$ (Figure 4.15).

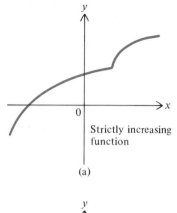

Strictly increasing function

(a)

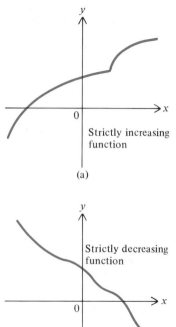

Strictly decreasing function

(b)

FIGURE 4.13

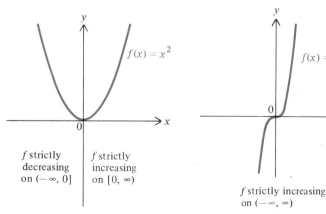

$f(x) = x^2$

f strictly decreasing on $(-\infty, 0]$ | f strictly increasing on $[0, \infty)$

FIGURE 4.14

$f(x) = x^3$

f strictly increasing on $(-\infty, \infty)$

FIGURE 4.15

THEOREM 4.8

Let f be continuous on an interval I and differentiable at each interior point of I.

a. Assume that $f'(x) \geq 0$ at each interior point of I. Then f is increasing on I. Moreover, f is strictly increasing on I if $f'(x) = 0$ for at most a finite number of points in I.

b. Assume that $f'(x) \leq 0$ at each interior point of I. Then f is decreasing on I. Moreover, f is strictly decreasing on I if $f'(x) = 0$ for at most a finite number of points in I.

Proof   We will prove only (a); the proof of (b) is analogous. Let x and z be arbitrary numbers in I, with $x < z$. By assumption f is continuous on $[x, z]$ and differentiable on (x, z). Thus we can apply the Mean Value Theorem to f in order to find a number c in (x, z) such that

$$f'(c) = \frac{f(z) - f(x)}{z - x}$$

Since $f'(c) \geq 0$ and $z - x > 0$, this means that

$$f(z) - f(x) = f'(c)(z - x) \geq 0$$

Therefore $f(z) \geq f(x)$, and thus f is increasing on I. To prove the second statement of (a), let us assume that f is not strictly increasing on I. Since we have shown that f is increasing, this means that there are two distinct points v and w in I such that $v < w$ and $f(v) = f(w)$. This in turn implies that f is constant on $[v, w]$, so that $f'(x) = 0$ for $v < x < w$. In other words, it is *not* true that $f'(x) > 0$ for all x in I except for a finite number of points. Hence if $f'(x) > 0$ except for a finite number of points in I, then f is strictly increasing on I.   ■

The next two examples illustrate the use of Theorem 4.8.

Example 3   Let $f(x) = x^3 + 3x^2 + 3x + 4$. On which intervals is f strictly increasing and on which is it strictly decreasing?

Solution   For all x we have

$$f'(x) = 3x^2 + 6x + 3 = 3(x + 1)^2 \geq 0$$

Thus $f'(x)$ is positive except for $x = -1$. It follows from the last part of Theorem 4.8(a) that f is strictly increasing on $(-\infty, \infty)$ (Figure 4.16).   □

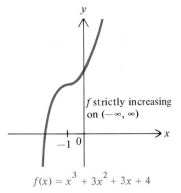

f strictly increasing on $(-\infty, \infty)$

$f(x) = x^3 + 3x^2 + 3x + 4$

FIGURE 4.16

Example 4   Let $f(x) = 2x^3 + 3x^2 - 12x - 3$. On which intervals is f strictly increasing and on which is it strictly decreasing?

Solution   First we find that

$$f'(x) = 6x^2 + 6x - 12 = 6(x + 2)(x - 1).$$

Next we assemble Figure 4.17 to determine the sign of $f'(x)$.

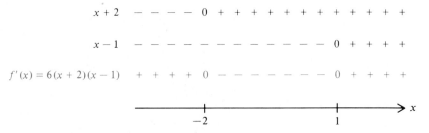

FIGURE 4.17

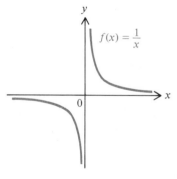

FIGURE 4.18

It follows that $f' \geq 0$ on $(-\infty, -2]$ and $[1, \infty)$, and $f' \leq 0$ on $[-2, 1]$. Moreover, $f'(x) = 0$ only for $x = -2$ and $x = 1$. Therefore Theorem 4.8 implies that f is strictly increasing on $(-\infty, -2]$ and $[1, \infty)$ and is strictly decreasing on $[-2, 1]$. $\square$

Caution: Remember that Theorem 4.8 is stated only for intervals, and thus we can apply it only to intervals. To emphasize this, we observe that if $f(x) = 1/x$, then $f'(x) = -1/x^2 < 0$ for all x in the domain of f. But it is not true that f is strictly decreasing on its domain, because for each x in $(-\infty, 0)$ and each z in $(0, \infty)$, the inequality $f(x) < f(z)$ holds (see Figure 4.18). However, f is strictly decreasing on each of the intervals $(-\infty, 0)$ and $(0, \infty)$.

EXERCISES 4.3

In Exercises 1–10 find all antiderivatives of the given function.

1. 0
2. -2
3. x
4. $3x$
5. $-x^2$
6. x^n for $n \neq -1$
7. $-6x + 5$
8. $4x^2 + 6x - 1$
9. $\sin x$
10. $\cos \pi x$

In Exercises 11–18 determine the function f satisfying the given conditions.

11. $f'(x) = -2, f(0) = 0$
12. $f'(x) = \pi, f(1) = 3\pi$
13. $f'(x) = -x, f(-1) = \sqrt{2}$
14. $f'(x) = \frac{1}{2}x, f(\frac{1}{2}) = -1$
15. $f'(x) = x^2, f(0) = -5$
16. $f'(x) = -\frac{3}{2}x^2, f(-1) = -\frac{1}{2}$
17. $f'(x) = \cos x, f(\pi/3) = 1$
18. $f'(x) = \sec \frac{x}{2} \tan \frac{x}{2}, f(\pi/2) = 2$

In Exercises 19–26 determine all functions f satisfying the given conditions.

19. $f''(x) = 0$ (*Hint:* Use Theorem 4.7 twice.)
20. $f''(x) = 0, f'(-2) = 1$
21. $f''(x) = 0, f'(0) = -1, f(0) = 2$
22. $f''(x) = 0, f'(2) = 3, f(-1) = 1$
23. $f''(x) = \sin x, f'(\pi) = -2, f(0) = 4$
24. $f^{(3)}(x) = 0$ (*Hint:* Use Theorem 4.7 three times.)
25. $f^{(4)}(x) = 0$
26. $f^{(n)}(x) = 0$ for any positive integer n

In Exercises 27–44 find the intervals on which the given function is strictly increasing and those on which it is strictly decreasing.

27. $f(x) = x^2 + x + 1$
28. $f(x) = x^3 - 12x + 4$
29. $f(x) = x^3 - x^2 + x - 1$
30. $f(x) = 4x^3 - 6x^2 - 9x$
31. $f(x) = x^4 - 2x^3 + 1$
32. $f(x) = x^5 + x^3 + 2x - 1$
33. $f(x) = x^5 + x^3 - 2x + 1$
34. $f(x) = x|x|$
35. $g(x) = \sqrt{16 - x^2}$
36. $g(x) = \sqrt{9x^2 - 4}$
37. $g(x) = \dfrac{1}{x + 3}$
38. $g(x) = \dfrac{x - 2}{x - 1}$
39. $k(x) = \dfrac{1}{x^2 + 1}$
40. $k(x) = \dfrac{1}{x^2 - 4}$
41. $f(t) = \tan t$
42. $f(t) = \sin t$
43. $f(t) = 2\cos t - t$
44. $f(t) = \sin t + \cos t$

45. Let F and G be antiderivatives of f and g, respectively, on an interval I.
    a. Prove that $F + G$ is an antiderivative of $f + g$ on I.
    b. Prove that $F - G$ is an antiderivative of $f - g$ on I.
    c. Prove that cF is an antiderivative of cf on I for any constant c.
    d. Is FG necessarily an antiderivative of fg? Explain.
    e. Is F/G necessarily an antiderivative of f/g? Explain.

46. Let f be continuous on $[a, \infty)$.
    a. Show that if $f'(x) \geq 0$ for all x in (a, ∞), then $f(x) \geq f(a)$ for all x in $[a, \infty)$.
    b. Show that if $f'(x) \leq 0$ for all x in (a, ∞), then $f(x) \leq f(a)$ for all x in $[a, \infty)$.

In Exercises 47–50 use Exercise 46 to show that the given inequality holds for the given values of x.

47. $x^4 - 4x \geq -3$ for $x \geq 1$

48. $4x^2 + \dfrac{1}{x} \geq 5$ for $x \geq 1$

49. $\frac{1}{4}x + \dfrac{1}{x} \geq 1$ for $x \geq 2$

50. $\dfrac{1}{x} + \tan\dfrac{1}{x} \leq 1 + \dfrac{\pi}{4}$ for $x \geq \dfrac{4}{\pi}$

51. a. Let $f(x) = x - \sin x$. Show that f is a strictly increasing function.
    b. Use part (a) to show that $\sin x > x$ on $(-\infty, 0)$, and $\sin x < x$ on $(0, \infty)$.

52. Let f and g be continuous on $[a, \infty)$ and differentiable on (a, ∞).
    a. Assume that $f(a) \geq g(a)$ and $f'(x) \geq g'(x)$ for all $x > a$. Prove that $f(x) \geq g(x)$ for any $x \geq a$. (*Hint:* Apply Theorem 4.8 to $f - g$ on $[a, x]$.)
    b. Assume that $f(a) \geq g(a)$ and $f'(x) > g'(x)$ for all $x > a$. Prove that $f(x) > g(x)$ for any $x > a$.

53. Show that $\tan x > x$ for all x in $(0, \pi/2)$.

54. Suppose n is a positive integer greater than 1. Prove that $(1 + x)^n > 1 + nx$ for $x > 0$.

*55. Using Exercises 51(b) and 52(b), prove that $\cos x > 1 - x^2/2$ for all $x > 0$.

*56. Using Exercise 55, show that $\sin x > x - x^3/6$ for all $x > 0$.

57. a. Show that there is exactly one value of x for which $\cos x = 2x$. (*Hint:* Let $f(x) = \cos x - 2x$, and determine how many zeros f could have.)
    c b. Use the Newton-Raphson method to approximate to within 10^{-6} the value of x for which $\cos x = 2x$.

58. Let $f(x) = x^3 + ax^2 + bx + c$, where a, b, and c are constants.
    a. Show that f is strictly increasing on $(-\infty, \infty)$ if $a^2 \leq 3b$.
    b. Assume that $a^2 > 3b$. Show that f is strictly increasing on
    $$(-\infty, (-a - \sqrt{a^2 - 3b})/3]$$
    and on
    $$[(-a + \sqrt{a^2 - 3b})/3, \infty)$$
    and is strictly decreasing on
    $$[(-a - \sqrt{a^2 - 3b})/3, (-a + \sqrt{a^2 - 3b})/3]$$

*59. Suppose f is differentiable on $(-\infty, \infty)$, and for each x and y we have $f(x + y) = f(x) + f(y)$. Show that there is a fixed number c such that $f(x) = cx$ for all x. (*Hint:* Using (2) of Section 3.2, show that $f'(x) = f'(0)$ for all x, and let $c = f'(0)$.)

*60. Use Theorem 4.7(a) to prove that only the constant functions satisfy the inequality
$$|f(x) - f(y)| \leq |x - y|^2$$
for all x and y. Is this true for any integer exponents of $|x - y|$ other than 2? Explain your answer.

61. Show that if a particle has zero velocity for any given period of time, then it stands still during that period.

62. The rate of a certain autocatalytic reaction is given by
$$v(x) = kx(a - x) \quad \text{for } 0 \leq x \leq a$$
where a is the original amount of the substance, x is the amount of the substance produced by the reaction, and k is positive. Determine where the rate of reaction is increasing and where it is decreasing.

63. A weight W is suspended from a ceiling by three strings of length L so that the points of attachment form an equilateral triangle of side S (Figure 4.19). If the weights of the various strings are negligible, then the tension T in the three strings attached to the ceiling is given by
$$T = \frac{W}{3\sqrt{1 - S^2/3L^2}}$$

    a. Show that for fixed values of S and L, T is an increasing function of W.
    b. Show that for fixed values of W and L, T is an increasing function of S.
    c. Show that for fixed values of S and W, T is a decreasing function of L.

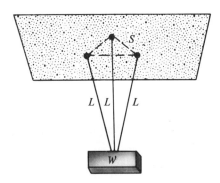

FIGURE 4.19

64. An electrical network consists of a resistor, a capacitor, and an inductor connected in series with a generator producing a voltage V that oscillates between positive and negative according to the equation $V = V_0 \sin \omega t$, where V_0 is the maximum voltage and ω is the angular frequency with $\omega > 0$. The steady-state current also oscillates, with the maximum current $i_{\max}$ given by the

formula

$$i_{max} = \frac{V_0}{\sqrt{\left(\dfrac{1}{\omega C} - \omega L\right)^2 + R^2}}$$

where L, R, and C are positive constants. On what interval is i_{max} an increasing function of ω?

4.4
THE FIRST AND SECOND DERIVATIVE TESTS

Suppose that a function f is differentiable at c. If f has a relative extreme value at c, then we know from Theorem 4.3 that $f'(c) = 0$. Thus in order to locate relative extreme values, we can seek out the values of x for which $f'(x) = 0$. But we still need a method of determining which (if any) of these values of x actually yield relative extreme values of f. In this section we present conditions involving the first and second derivatives of f that guarantee that f has a relative extreme value at c. The conditions will help not only in graphing functions but also in solving applied problems.

Our conditions for relative extreme values are based on the following two observations:

1. If a continuous function f is increasing on the portion of an interval I to the left of c and decreasing on the portion to the right of c, then $f(c)$ is the maximum value of f on I (Figure 4.20). Likewise, if f is decreasing on the portion of I to the left of c and increasing on the portion to the right of c, then $f(c)$ is the minimum value of f on I (Figure 4.21).
2. By Theorem 4.8, a continuous function f is increasing on an interval I if $f'(x) \geq 0$ for all interior points of I; similarly, f is decreasing on an interval I if $f'(x) \leq 0$ for all interior points of I.

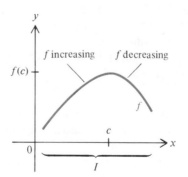

$f(c)$ is the maximum value of f on I.

FIGURE 4.20

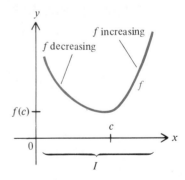

$f(c)$ is the minimum value of f on I.

FIGURE 4.21

The First Derivative Test

To simplify the proof of our next result, we will say that the derivative f' **changes from positive to negative at c** if there exists some number $\delta > 0$ such that $f'(x) > 0$ for all x in $(c - \delta, c)$ and $f'(x) < 0$ for all x in $(c, c + \delta)$. The definition of "f'

changes from negative to positive at c" results from replacing $f'(x) > 0$ by $f'(x) < 0$, and *vice versa*.

Example 1   Let $f(x) = 4x^3 + 9x^2 - 12x + 3$. Determine where f' changes from positive to negative and where it changes from negative to positive.

Solution   First we find the derivative of f:

$$f'(x) = 12x^2 + 18x - 12 = 12\left(x^2 + \frac{3}{2}x - 1\right) = 12(x + 2)\left(x - \frac{1}{2}\right)$$

To determine where f' changes sign, we assemble the chart in Figure 4.22.

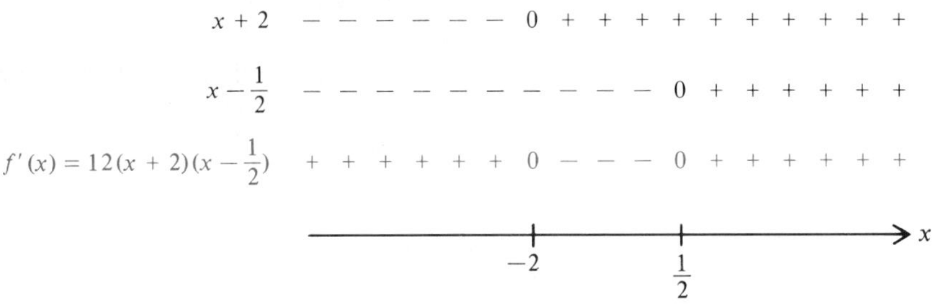

FIGURE 4.22

Consequently f' changes from positive to negative at -2 and from negative to positive at $\frac{1}{2}$. $\square$

Now we are ready to state and prove the following result.

THEOREM 4.9
FIRST DERIVATIVE
TEST

Let f be continuous on an interval I, and let c be in I.
a. If f' changes from positive to negative at c, then f has a relative maximum value at c.
b. If f' changes from negative to positive at c, then f has a relative minimum value at c.

Proof   Because the proofs of the two parts are so similar, we prove only part (a). Since f' changes from positive to negative at c, there is a number $\delta > 0$ such that $f'(x) > 0$ for all x in $(c - \delta, c)$ and $f'(x) < 0$ for all x in $(c, c + \delta)$. By Theorem 4.8, f is strictly increasing on $[c - \delta, c]$ and strictly decreasing on $[c, c + \delta]$. Therefore $f(c)$ is the maximum value of f on $[c - \delta, c + \delta]$, and so f has a relative maximum value at c.   ∎

Example 2   Let $f(x) = 4x^3 + 9x^2 - 12x + 3$. Show that f has a relative maximum value at -2 and a relative minimum value at $\frac{1}{2}$.

Solution   By Example 1 we know that f' changes from positive to negative at -2. Thus the First Derivative Test implies that f has a relative maximum value at

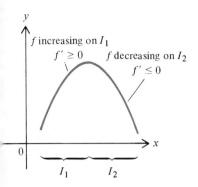

FIGURE 4.23

-2. Similarly, f' changes from negative to positive at $\frac{1}{2}$, so that f has a relative minimum value at $\frac{1}{2}$. $\square$

Together, the First Derivative Test and Theorem 4.8 can help us sketch the graph of a function f. The procedure is as follows. We compute the derivative of f and examine it:

1. From Theorem 4.8, if $f'(x) \geq 0$ for all x in an interval, then f is increasing on that interval, whereas if $f'(x) \leq 0$ for all x in an interval, then f is decreasing on that interval (Figure 4.23).
2. From the First Derivative Test, if f' changes from positive to negative at c, then f has a relative maximum value at c; if f' changes from negative to positive at c, then f has a relative minimum value at c (Figure 4.24).

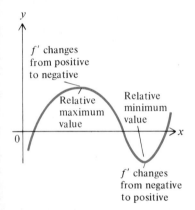

FIGURE 4.24

Example 3   Let $f(x) = \frac{1}{4}x^3 - 3x$. Find the relative extreme values of f, and then sketch the graph of f.

Solution   The derivative is given by

$$f'(x) = \frac{3}{4}x^2 - 3 = \frac{3}{4}(x^2 - 4) = \frac{3}{4}(x + 2)(x - 2)$$

The pertinent information about the sign of $f'(x)$ is summarized in Figure 4.25.

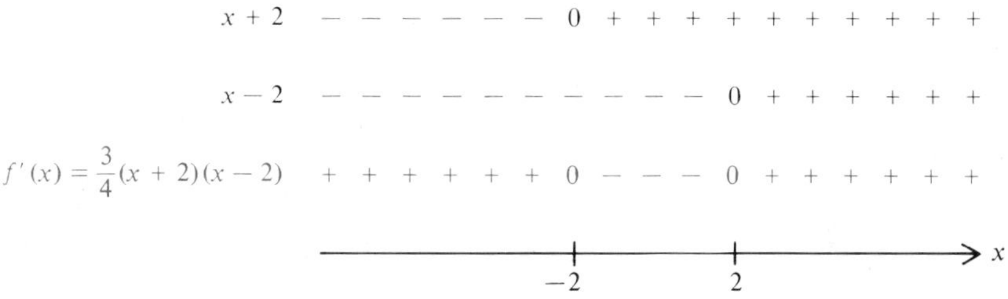

$$x + 2 \quad - - - - - - \ 0 \ + + + + + + + + + +$$

$$x - 2 \quad - - - - - - - - - - \ 0 \ + + + + + +$$

$$f'(x) = \frac{3}{4}(x + 2)(x - 2) \quad + + + + + + \ 0 \ - - - \ 0 \ + + + + + +$$

FIGURE 4.25

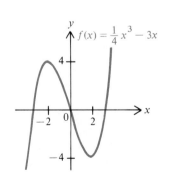

FIGURE 4.26

We conclude that f is strictly increasing on $(-\infty, -2]$ and on $[2, \infty)$ and is strictly decreasing on $[-2, 2]$. Furthermore, f' changes from positive to negative at -2 and from negative to positive at 2. Therefore by the First Derivative Test, $f(-2) = 4$ is a relative maximum value and $f(2) = -4$ a relative minimum value of f. We can now make a fairly accurate sketch of the graph of f, as shown in Figure 4.26. $\square$

Example 4   Let $f(x) = (x - 1)^2(x - 3)^2$. Sketch the graph of f.

Solution   First we find the derivative of f:

$$\begin{aligned} f'(x) &= 2(x - 1)(x - 3)^2 + 2(x - 1)^2(x - 3) \\ &= 2(x - 1)(x - 3)[(x - 3) + (x - 1)] \\ &= 4(x - 1)(x - 2)(x - 3) \end{aligned}$$

Then we prepare Figure 4.27 concerning the sign of $f'(x)$.

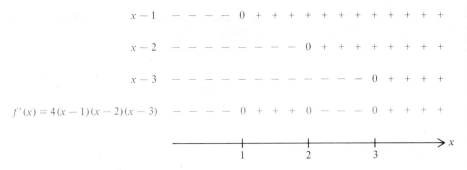

FIGURE 4.27

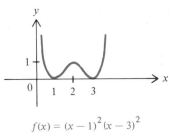

From Figure 4.27 we find that f is strictly increasing on the intervals $[1, 2]$ and $[3, \infty)$ and is strictly decreasing on the intervals $(-\infty, 1]$ and $[2, 3]$. We also find that f' changes from negative to positive at 1 and at 3 and from positive to negative at 2. As a result, the First Derivative Test implies that $f(1) = 0$ and $f(3) = 0$ are relative minimum values and that $f(2) = 1$ is a relative maximum value. This information enables us to sketch the graph of f (Figure 4.28).   □

$$f(x) = (x - 1)^2 (x - 3)^2$$

FIGURE 4.28

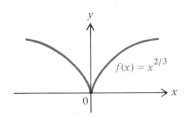

Sometimes we can conclude from the First Derivative Test that a relative extreme value exists at a number c even if $f'(c)$ does not exist. For example, if $f(x) = x^{2/3}$, then

$$f'(x) = \frac{2}{3} x^{-1/3} \quad \text{for } x \neq 0$$

Consequently $f'(x) < 0$ for $x < 0$, and $f'(x) > 0$ for $x > 0$. Thus f' changes from negative to positive at 0. Since f is continuous on $(-\infty, \infty)$, the First Derivative Test tells us that $f(0) = 0$ is a relative minimum value of f (Figure 4.29). Nevertheless, $f'(0)$ does not exist.

FIGURE 4.29

The Second Derivative Test   Now we present our second result on the location of relative extreme values. The result is called the Second Derivative Test because of the importance of the second derivative in the conditions of the theorem.

THEOREM 4.10
SECOND DERIVATIVE TEST

Assume that $f'(c) = 0$.
a. If $f''(c) < 0$, then $f(c)$ is a relative maximum value of f.
b. If $f''(c) > 0$, then $f(c)$ is a relative minimum value of f.

If $f''(c) = 0$, then from this test alone we cannot draw any conclusions about a relative extreme value of f at c.

Proof   We prove (a). By hypothesis,

$$f''(c) = \lim_{x \to c} \frac{f'(x) - f'(c)}{x - c} < 0$$

Since $f'(c) = 0$ by hypothesis, (10) of Section 2.3 implies that for all $x \neq c$ in some interval $(c - \delta, c + \delta)$,

$$\frac{f'(x)}{x - c} = \frac{f'(x) - f'(c)}{x - c} < 0$$

If $c - \delta < x < c$, then $x - c < 0$, so that $f'(x) > 0$. If $c < x < c + \delta$, then $x - c > 0$, so that $f'(x) < 0$. This means that f' changes from positive to negative at c. We conclude by the First Derivative Test that f has a relative maximum value at c. The proof of (b) is analogous to the proof of (a).

   If both $f'(c) = 0$ *and* $f''(c) = 0$, it is possible for f to have a relative minimum value at c, or a relative maximum value at c, or neither (Figure 4.30(a)–(c)), so we cannot conclude anything about a possible relative extreme value of f at c. ■

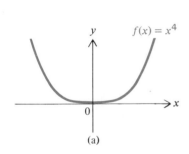

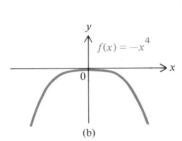

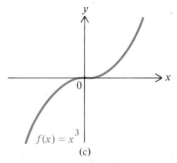

(a)   (b)   (c)

FIGURE 4.30

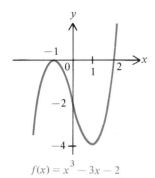

$f(x) = x^3 - 3x - 2$

FIGURE 4.31

Example 5   Let $f(x) = x^3 - 3x - 2$. Using the Second Derivative Test, find the relative extreme values of f.

Solution   By differentiation we obtain

$$f'(x) = 3x^2 - 3 = 3(x - 1)(x + 1) \quad \text{and} \quad f''(x) = 6x$$

Therefore $f'(x) = 0$ when $x = -1$ or $x = 1$. Since

$$f''(-1) = -6 < 0 \quad \text{and} \quad f''(1) = 6 > 0$$

we know from the Second Derivative Test that $f(-1) = 0$ is a relative maximum value of f, whereas $f(1) = -4$ is a relative minimum value of f. These are the only relative extreme values of f. (The graph of f is shown in Figure 4.31.) □

   The Second Derivative Test does not always apply, as the following example shows.

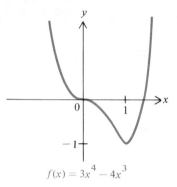

$f(x) = 3x^4 - 4x^3$

FIGURE 4.32

Example 6   Let $f(x) = 3x^4 - 4x^3$. Find the relative extreme values of f, and sketch the graph of f.

Solution   Differentiating, we have

$$f'(x) = 12x^3 - 12x^2 = 12x^2(x - 1) \qquad (1)$$

and
$$f''(x) = 36x^2 - 24x$$

Consequently $f'(x) = 0$ for $x = 0$ and $x = 1$. Since $f''(1) = 12$, the Second Derivative Test tells us that $f(1) = -1$ is a relative minimum value of f. However, since $f''(0) = 0$, the Second Derivative Test cannot be applied to f at 0. But we observe from (1) that $f'(x) < 0$ for all x in $(-\infty, 1)$ except 0, so that f is strictly decreasing on $(-\infty, 1)$. Therefore f cannot have an extreme value at 0. The information we now have leads us to draw the graph in Figure 4.32.   ☐

In Section 4.6 we will return to the function in Example 6 and study its graph further.

EXERCISES 4.4

In Exercises 1–8 determine the values of c at which f' changes from positive to negative, or from negative to positive.

1. $f(x) = x^2 + 6x - 11$
2. $f(x) = x^3 - x^2 - x + 2$
3. $f(x) = 2x^4 - 4x^2 + 3$
4. $f(x) = \dfrac{x}{x^3 - 2}$
5. $f(t) = \dfrac{t^2 - t + 1}{t^2 + t + 1}$
6. $f(t) = \dfrac{1}{\sqrt{t - t^2}}$
7. $f(t) = \sin t + \frac{1}{2}t$
8. $f(t) = \sin t - \cos t$

In Exercises 9–22 use the First Derivative Test to determine the relative extreme values (if any) of the function.

9. $f(x) = -3x^2 + 3x + 7$
10. $f(x) = x^3 - 12x + 2$
11. $f(x) = x^3 + 3x^2 + 4$
12. $f(x) = x^4 - 8x^2 + 1$
13. $g(x) = 4x^2 - \dfrac{1}{x}$
14. $g(x) = \dfrac{1}{x^2 + 1}$
15. $f(x) = \dfrac{x}{16 + x^3}$
16. $f(x) = \sqrt{|x| + 1}$
17. $f(x) = x\sqrt{1 - x^2}$
18. $f(x) = \dfrac{x^2 + x + 1}{x^2 - x + 1}$
19. $k(x) = \cos x + \frac{1}{2}x$
20. $k(x) = \sin x - \dfrac{\sqrt{3}}{2}x$
21. $k(x) = \sin\left(\dfrac{x^2}{1 + x^2}\right)$
22. $k(x) = \dfrac{\cos x}{1 + \sin x}$

In Exercises 23–32 use the Second Derivative Test to determine the relative extreme values of the function.

23. $f(x) = -4x^2 + 3x - 1$
24. $f(x) = x^3 + 6x^2 + 9$
25. $f(x) = x^3 - 3x^2 - 24x + 1$
26. $f(x) = x^4 + \frac{1}{2}x$
27. $f(x) = 3x^4 - 4x^3 - \frac{9}{2}x^2 + \frac{1}{2}$
28. $g(x) = (x^2 + 2)^6$
29. $f(t) = t^2 + \dfrac{1}{t} + 1$
30. $f(t) = t^3 - \dfrac{48}{t^2}$
31. $f(t) = \sin t + \cos t$
32. $f(t) = t + \cos 2t$

In Exercises 33–42 use the First Derivative Test or the Second Derivative Test to determine the relative extreme values, if any, of the function. Then sketch the graph of the function.

33. $f(x) = x^2 + 8x + 12$
34. $f(x) = x^3 - 3x$
35. $f(x) = x^3 + 3x$
36. $f(x) = 2x^3 - 3x^2 - 12x + 5$
37. $f(x) = x^4 + 4x$
38. $f(x) = x^5 - 5x$
39. $f(x) = (x^2 - 1)^2$
40. $f(x) = x^2(x + 3)^2$
41. $f(x) = (x - 2)^2(x + 1)^2$
42. $f(x) = \sqrt{x - x^2}$

In Exercises 43–48 find a point c such that $f'(c) = f''(c) = 0$. Using the derivative of f, show that f does not have a relative extreme value at c.

43. $f(x) = x^5$
44. $f(x) = x^{7/3}$
45. $f(x) = x^5 - x^3$
46. $f(x) = (x - 2)^3$
47. $f(x) = (x - 2)^3(x + 1)$
48. $f(x) = x + \sin x$

49. Prove Theorem 4.10(b).

*50. Find a function f such that $f'(0) = 0$ and $f''(0)$ does not exist, and such that
  a. $f(0)$ is a relative minimum value
  b. $f(0)$ is not a relative extreme value

4.5
APPLICATIONS OF EXTREME VALUES

To solve many applied problems, one needs to find a maximum or minimum value of a suitable function on an interval I. The ideas of Sections 4.1 through 4.4 are particularly well suited to analyzing such problems. As a result, we will devote this section to applied problems. The method we use in solving such a problem will depend on whether the interval is closed and bounded or not.

Extreme Values on a Closed, Bounded Interval

Recall from the Maximum–Minimum Theorem (Section 4.1) that if f is continuous on a closed, bounded interval $[a, b]$, then f assumes a maximum and a minimum value. Moreover, these values can be assumed only at the endpoints a and b of the interval or at critical points in (a, b).

The following two examples illustrate the procedure for solving applied problems involving a closed, bounded interval $[a, b]$.

Example 1   A metal box (without top) is to be constructed from a square sheet of metal that is 10 inches on a side by first cutting square pieces of the same size from the corners of the sheet and then folding up the sides (Figure 4.33). Find the dimensions of the box with the largest volume that can be so constructed.

Solution   Let x be the length (in inches) of the sides of the squares that are cut out (Figure 4.33). Since the original sheet of metal is 10 inches on a side, we have $0 \le x \le 5$. The resulting box has a height of x inches and a base that is $10 - 2x$ inches on a side. This means that the volume V of the box is given by

$$V = x(10 - 2x)^2 = 4x^3 - 40x^2 + 100x \quad \text{for } 0 \le x \le 5$$

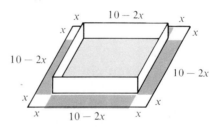

FIGURE 4.33

Thus the problem reduces to finding the maximum value of V on $[0, 5]$. To determine the critical points of V in $(0, 5)$, we first compute $V'(x)$:

$$V'(x) = 12x^2 - 80x + 100 = 4(3x^2 - 20x + 25) = 4(3x - 5)(x - 5)$$

Therefore $V'(x) = 0$ for $x = \frac{5}{3}$ and $x = 5$. Thus the only critical point in $(0, 5)$ is $\frac{5}{3}$. Since

$$V(0) = 0, \qquad V\left(\frac{5}{3}\right) = \frac{5}{3}\left(\frac{20}{3}\right)^2 = \frac{2000}{27}, \quad \text{and} \quad V(5) = 0$$

it follows that the maximum value of V occurs for $x = \frac{5}{3}$. The corresponding value of $10 - 2x$ is $10 - 2(\frac{5}{3}) = \frac{20}{3}$. Consequently the box with the maximum volume has a height of $\frac{5}{3}$ inches and a square base $\frac{20}{3}$ inches on a side.   □

Our next example involves a question that arises in the study of coughing. It is known that coughing is made possible by an increase in the pressure in the lungs, which is accompanied by a decrease in the radius of the windpipe. Suppose we wish to decide whether a smaller radius facilitates or impedes the flow of air through the windpipe.

For simplicity we make the idealized assumption that the windpipe is a cylindrical tube (Figure 4.34). We will also use an equation from physics that relates the rate of flow F (volume per unit time) of a fluid through a cylindrical tube to the radius r of the tube and the pressure difference p at the ends of the tube. The equation is

$$F = \frac{\pi p r^4}{8\eta l} \tag{1}$$

FIGURE 4.34

where l is the length of the tube and η is the viscosity of the fluid. Finally, we will assume that the radius r and pressure difference p are related by the equation

$$p = \frac{r_0 - r}{a} \quad \text{for } \frac{1}{2}r_0 \leq r \leq r_0 \tag{2}$$

where r_0 is the radius when there is no pressure difference, and a is a positive constant. Using (1) and (2), we obtain

$$F = \frac{\pi(r_0 - r)r^4}{8\eta la} = k(r_0 - r)r^4 \quad \text{for } \frac{1}{2}r_0 \leq r \leq r_0 \tag{3}$$

where $k = \pi/8\eta la$ is a constant. We are now ready to use this information in Example 2.

Example 2   Determine the value r of the radius that maximizes the rate of flow F.

Solution   From (3) we find that

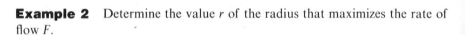

$$F'(r) = k(4r_0 r^3 - 5r^4) = kr^3(4r_0 - 5r)$$

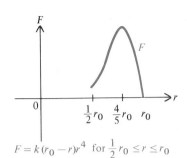

$F = k(r_0 - r)r^4$ for $\frac{1}{2}r_0 \le r \le r_0$

FIGURE 4.35

Since the domain of F is $[\frac{1}{2}r_0, r_0]$, it follows that $F'(r) = 0$ only for $r = \frac{4}{5}r_0$. Since

$$F\left(\frac{1}{2}r_0\right) = k\left(r_0 - \frac{1}{2}r_0\right)\left(\frac{1}{2}r_0\right)^4 = \frac{1}{32}kr_0^5$$

$$F\left(\frac{4}{5}r_0\right) = k\left(r_0 - \frac{4}{5}r_0\right)\left(\frac{4}{5}r_0\right)^4 = \frac{256}{3125}kr_0^5$$

$$F(r_0) = 0$$

we conclude that F assumes its maximum values for $r = \frac{4}{5}r_0$ (Figure 4.35). Therefore contraction of the windpipe might well increase the rate of flow through the windpipe. □

Extreme Values on an Arbitrary Interval

The Maximum–Minimum Theorem applies to closed, bounded intervals. If the interval I is not closed or is not bounded, then a function f that is continuous on I does not automatically have a maximum (or a minimum) value on I. However, as we mentioned in Section 4.4, if f is increasing on the portion of I to the left of c and decreasing on the portion of I to the right of c, then $f(c)$ is the maximum value of f on I (Figure 4.36(a)). Likewise, if f is decreasing on the portion of I to the left of c and increasing on the portion of I to the right of c, then $f(c)$ is the minimum value of f on I (Figure 4.36(b)). Because of the relation between the first derivative and the intervals on which f is increasing or is decreasing, we are led to the following criterion for extreme values, which applies to an arbitrary interval I.

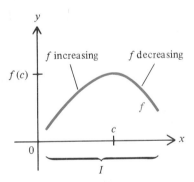

$f(c)$ is the maximum value of f on I

(a)

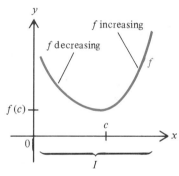

$f(c)$ is the minimum value of f on I

(b)

FIGURE 4.36

THEOREM 4.11

Let f be continuous on an interval I containing c.
a. Assume that for any interior point x of I, $f'(x) \ge 0$ if $x < c$, and $f'(x) \le 0$ if $x > c$. Then $f(c)$ is the maximum value of f on I [Figure 4.37(a)].
b. Assume that for any interior point x of I, $f'(x) \le 0$ if $x < c$, and $f'(x) \ge 0$ if $x > c$. Then $f(c)$ is the minimum value of f on I [Figure 4.37(b)].

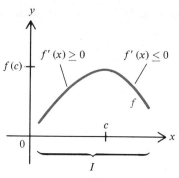

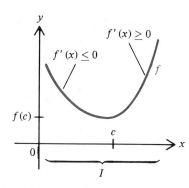

$f(c)$ is the maximum value of f on I $f(c)$ is the minimum value of f on I

(a)   (b)

FIGURE 4.37

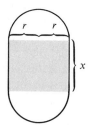

FIGURE 4.38

Example 3   An outdoor track is to be created in the shape shown in Figure 4.38 and is to have a perimeter of 440 yards. Find the dimensions for the track that maximize the area of the rectangular portion of the field enclosed by the track.

Solution   Let r be the radius of the two semicircles (with $r > 0$), and x the length of the rectangular portion of the field. Our goal is to maximize the area A of the rectangular portion. Since the length of the rectangle is x and the width is $2r$, it follows that

$$A = 2rx$$

Observe next that the circumference of the field is $2x + 2\pi r$, so by hypothesis,

$$2x + 2\pi r = 440$$

Therefore $x = 220 - \pi r$ (4)

so that

$$A = 2rx = 2r(220 - \pi r) = 440r - 2\pi r^2 \quad \text{for } 0 < r \le \frac{220}{\pi}$$

To find the maximum value of A, we first take the derivative of A:

$$A'(r) = 440 - 4\pi r = 4(110 - \pi r)$$

It follows that $A'(r) = 0$ for $r = 110/\pi$. Moreover, $A'(r) > 0$ if $0 < r < 110/\pi$, whereas $A'(r) < 0$ if $110/\pi < r \le 220/\pi$. Thus Theorem 4.11 implies that the maximum value of A occurs for $r = 110/\pi$. By (4) the corresponding value of x is given by

$$x = 220 - \pi\left(\frac{110}{\pi}\right) = 110$$

As a result, the area is maximum if the length of the rectangle is 110 yards and the radius of the semicircles is $110/\pi$ yards. $\square$

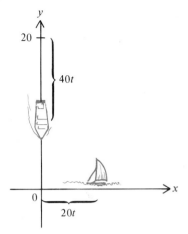

FIGURE 4.39

Example 4   At noon a sailboat is 20 kilometers south of a freighter. The sailboat is traveling east at 20 kilometers per hour, and the freighter is traveling south at 40 kilometers per hour. If visibility is 10 kilometers, could the people on the two ships ever see each other?

Solution   We measure time in hours, and let $t = 0$ correspond to noon. Then at any time $t \geq 0$, the sailboat has traveled $20t$ kilometers and the freighter has traveled $40t$ kilometers. From Figure 4.39 we see that at time t the distance D between the two ships is given by

$$D = \sqrt{(20t)^2 + (20 - 40t)^2} \quad \text{for } t \geq 0$$

The goal is to determine whether $D \leq 10$ for any such value of t, and to accomplish this, we will find the minimum value of D. But notice that the minimum value of D occurs for the same value of t as the minimum value of D^2, the square of the distance. Moreover, the derivative of D^2 is easier to obtain than the derivative of D. So we let $E = D^2$, which means that for $t \geq 0$,

$$E = (20t)^2 + (20 - 40t)^2 = 400[t^2 + (1 - 2t)^2]$$

Then

$$E'(t) = 400[2t + 2(1 - 2t)(-2)] = 400(10t - 4)$$

Therefore $E'(t) = 0$ if

$$10t - 4 = 0$$

which means that

$$t = \frac{2}{5}$$

Since $E'(t) < 0$ for $0 < t < \frac{2}{5}$ and $E'(t) > 0$ for $t > \frac{2}{5}$, it follows from Theorem 4.11 that $E(\frac{2}{5})$ is the minimum value of E, and hence $D(\frac{2}{5})$ is the minimum value of D. Next,

$$D\left(\frac{2}{5}\right) = \sqrt{\left(20 \cdot \frac{2}{5}\right)^2 + \left(20 - 40 \cdot \frac{2}{5}\right)^2} = \sqrt{8^2 + 4^2} = \sqrt{80} < 10$$

so we conclude that the people on the two ships could see each other. In particular, they could see each other $\frac{2}{5}$ of an hour after noon, that is, around 12:24 P.M. $\square$

If $f'(c) = 0$ and the second derivative of f can easily be calculated, it may be simpler to determine what kind of extreme value (if any) f has at c by using the following criterion, which involves the second derivative.

THEOREM 4.12

Let f be continuous on an interval I. Suppose c is in I and $f'(c) = 0$.
a. If $f''(x) \leq 0$ for every interior point x of I, then $f(c)$ is the maximum value of f on I.
b. If $f''(x) \geq 0$ for every interior point x of I, then $f(c)$ is the minimum value of f on I.

Proof   Since the proofs of both parts are similar, we prove only (a). If $f''(x) \leq 0$ for every interior point x of I, then since f'' is the derivative of f', Theorem 4.8 implies that f' is decreasing on I. Since $f'(c) = 0$ by hypothesis, it follows that $f'(x) \geq 0$ for $x < c$ and $f'(x) \leq 0$ for $x > c$. Hence by Theorem 4.11(a), $f(c)$ is the maximum value of f on I. ∎

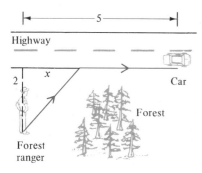

FIGURE 4.40

Example 5   A forest ranger is in a forest 2 miles from a straight road. A car is located 5 miles down the road (Figure 4.40). If the forest ranger can walk 3 miles per hour in the forest and 4 miles per hour along the road, toward what point on the road should the ranger walk in order to minimize the time needed to walk to the car?

Solution   Let x be the distance shown in Figure 4.40. Then by the Pythagorean Theorem the ranger walks $\sqrt{x^2 + 4}$ miles in the forest. The ranger also walks $5 - x$ miles along the road. Using the formula

$$\text{time} = \frac{\text{distance}}{\text{speed}}$$

twice, we find that the total time T the ranger walks is given by

$$T = \overbrace{\frac{\sqrt{x^2 + 4}}{3}}^{\text{time in forest}} + \overbrace{\frac{5 - x}{4}}^{\text{time on road}}$$

Since

$$T'(x) = \frac{2x}{2 \cdot 3\sqrt{x^2 + 4}} - \frac{1}{4} = \frac{x}{3\sqrt{x^2 + 4}} - \frac{1}{4}$$

it follows that $T'(x) = 0$ if

$$3\sqrt{x^2 + 4} = 4x$$

Since x must be nonnegative, when we solve for x we obtain

$$9(x^2 + 4) = 16x^2$$

$$7x^2 = 36$$

$$x = \frac{6}{7}\sqrt{7}$$

Since

$$T''(x) = \frac{3\sqrt{x^2 + 4} - x(3x/\sqrt{x^2 + 4})}{9(x^2 + 4)} = \frac{3(x^2 + 4) - 3x^2}{9(x^2 + 4)^{3/2}} = \frac{4}{3(x^2 + 4)^{3/2}}$$

we find that $T''(x) > 0$ for $x > 0$. By Theorem 4.12 it follows that the minimum value of T occurs for $x = \frac{6}{7}\sqrt{7}$. Thus the total walking time is minimized if the ranger walks toward the point $\frac{6}{7}\sqrt{7}$ miles (approximately 2.3 miles) down the road. □

Since the domain of T in Example 5 is really $[0, 5]$, we could have solved the problem by finding the critical point $\frac{6}{7}\sqrt{7}$ of T on $[0, 5]$ and then using the procedure described in Section 4.1: computing $T(0)$, $T(\frac{6}{7}\sqrt{7})$, and $T(5)$ and determining the smallest of these three values. The answer would have been the same—that $T(\frac{6}{7}\sqrt{7})$ is the minimum value of T on $[0, 5]$. Thus when the interval I is closed and bounded, we can sometimes use more than one method in determining extreme values.

In our final applied problem, we will study the financial effect of placing insulation on a bare attic floor.

Let $C_i(x)$ be the purchase price in dollars of insulation x inches thick. We make the reasonable assumption that

$$C_i(x) = ax \quad \text{for } x > 0$$

where a is a positive constant. If $C_h(x)$ denotes the cost per year of heating the home with this insulation, then we assume further that

$$C_h(x) = \frac{b}{x} \quad \text{for } x > 0$$

where b is also a positive constant.* Over a ten-year period, the total cost $C(x)$ to the homeowner will be

$$C(x) = C_i(x) + 10C_h(x) = ax + \frac{10b}{x} \quad \text{for } x > 0$$

The value of x that yields the minimum cost $C(x)$ over a ten-year period represents the ideal amount of insulation the homeowner should buy. With this in mind we state and solve the following example, in which we assume that the attic floor has an area of 1000 square feet, $a = 50$, and $b = 295$, which would be reasonable in a region like Washington, D.C.

Example 6   Find the value of x that minimizes the cost $C(x)$, where

$$C(x) = 50x + \frac{2950}{x} \quad \text{for } x > 0$$

Solution   Differentiating, we obtain

$$C'(x) = 50 - \frac{2950}{x^2} \quad \text{and} \quad C''(x) = \frac{5900}{x^3}$$

---

* We will tell how to compute a and b for a particular case. First, $a = qA$, where q is the cost in dollars per square foot of insulation 1 inch thick and A is the area of the attic floor in square feet. Second, $b = dcAp$, where d is the number of degree days in the area, c is the conductivity of the insulation, and p is the price per British thermal unit of fuel. If the price in dollars per therm on the gas bill is p_0, then

$$p = \frac{3}{2}p_0\left(\frac{1}{100,000}\right)$$

In Washington, D.C., values for 1987 were approximately as follows: $d = 4200$, $c = 6.5$, $p_0 = 0.72$, and $q = .05$. If the attic has 1000 square feet of floor, then a is approximately 50 and b is approximately 295. You should be able to find out the number of degree days and the price p_0 in your area from your utility company.

It follows that $C'(x) = 0$ for $x = \sqrt{\frac{2950}{50}} = \sqrt{59}$. Moreover, $C''(x) > 0$ for $x > 0$, so from Theorem 4.12 we conclude that the minimum cost $C(x)$ occurs when $x = \sqrt{59}$ (inches), that is, for approximately 7.7 inches of insulation.   □

Our idealized formula in the insulation example does not take into account all possible variables. For instance, it does not reflect installation costs or the fact that some heat is lost through the walls and floor. Nevertheless, the problem as stated and solved is representative of fairly common conditions.

There is a general procedure we have used in solving the applied problems in this section. Below we list the major features of the procedure as a guide for you in solving applied problems involving extreme values.

1. After reading the problem carefully, choose a letter for the quantity to be maximized or minimized, and choose auxiliary variables for the other quantities appearing in the problem.
2. Express the quantity to be maximized or minimized in terms of the auxiliary variables. A diagram is often useful.
3. Choose one auxiliary variable, say x, to serve as master variable, and use the information given in the problem to express all other auxiliary variables in terms of x. Again a diagram may be helpful.
4. Use the results of steps (2) and (3) to express the given quantity to be maximized or minimized in terms of x alone.
5. Use the theory of this chapter to find the desired maximum or minimum value. This usually involves finding a derivative and determining where it is 0, and then either evaluating the given quantity at endpoints and critical points or using Theorem 4.11 or 4.12.

EXERCISES 4.5

1. Find the two positive numbers whose sum is 18 and whose product is as large as possible.

2. Find the two real numbers whose difference is 16 and whose product is as small as possible.

3. A crate open at the top has vertical sides, a square bottom, and a volume of 4 cubic meters. If the crate has the least possible surface area, find its dimensions.

4. Suppose the crate in Exercise 3 has a top. Find the dimensions of the crate with minimum surface area.

5. Show that the entire region enclosed by the outdoor track in Example 3 has maximum area if the track is circular.

6. A Norman window is a window in the shape of a rectangle with a semicircle attached at the top (Figure 4.41). Assuming that the perimeter of the window is 12 feet, find the dimensions that allow the maximum amount of light to enter.

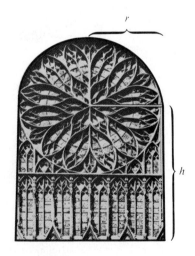

FIGURE 4.41

7. Suppose a window has the shape of a rectangle with an equilateral triangle attached at the top. Assuming that the perimeter of the window is 12 feet, find the dimensions that allow the maximum amount of light to enter.

8. A rectangle is inscribed in a semicircle of radius r, with one side lying on the diameter of the semicircle. Find the maximum possible area of the rectangle.

9. At 3 P.M. an oil tanker traveling west in the ocean at 15 kilometers per hour passes the same point as a luxury liner which arrived at the same spot at 2 P.M. while traveling north at 25 kilometers per hour. At what time were the ships closest together?

10. The coughing problem in Example 2 can be approached from a slightly different point of view. If v denotes the velocity of the air in the windpipe, then F, r, and v are related by the equation

$$F = v(\pi r^2)$$

Consequently by (3),

$$v = \frac{F}{\pi r^2} = \frac{k}{\pi}(r_0 - r)r^2$$

Show that the velocity v is maximized when $r = \frac{2}{3}r_0$. (This shows that constriction of the windpipe during a cough appears to increase the air velocity in the windpipe and facilitate the cough.)

11. In an autocatalytic chemical reaction a substance A is converted into a substance B in such a manner that

$$\frac{dx}{dt} = kx(a - x)$$

where x is the concentration of substance B at time t, a is the initial concentration of substance A, and k is a positive constant. Determine the value of x at which the rate dx/dt of the reaction is maximum.

12. If $C(x)$ is the cost of manufacturing an amount x of a given product and p is the price per unit amount, then the profit $P(x)$ obtaining by selling an amount x is

$$P(x) = px - C(x)$$

(Notice that there is a loss if $P(x)$ is negative.)
   a. If $C(x) = cx$ and $c < p$, is there a maximum profit?
   b. If $C(x) = (x - 1)^2 + 2$, find the maximum profit.

In Exercise 13 we present a mathematical problem that arises in two completely different settings (see Exercises 14 and 15).

13. Let p, q, and r be positive constants with $q < r$, and let

$$f(\theta) = p - q\cot\theta + \frac{r}{\sin\theta} \quad \text{for } 0 < \theta < \frac{\pi}{2}$$

Show that f has a minimum value on $(0, \pi/2)$ at the value of θ for which $\cos\theta = q/r$.

14. This problem derives from the biological study of vascular branching. Assume that a major blood vessel A leads away from the heart (P in Figure 4.42) and that in order for the heart to feed an organ at R, there must be an auxiliary artery somewhere between P and Q. The resistance $\mathcal{R}$ of the blood as it flows along the path PSR is given by

$$\mathcal{R}(\theta) = k\left[\frac{(a - b\cot\theta)}{r_1^4}\right] + k\left(\frac{b}{r_2^4\sin\theta}\right) \quad \text{for } 0 < \theta < \frac{\pi}{2}$$

where k, a, b, r_1, and r_2 are positive constants with $r_1 > r_2$ (see Figure 4.42). Where should the contact at S be made to produce the least resistance? (*Hint:* Using the result of Exercise 13, find the cosine of the angle θ for which $R(\theta)$ is minimized.)

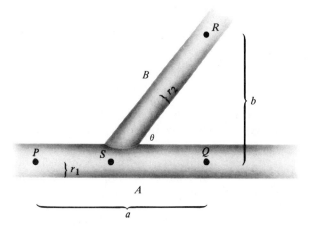

FIGURE 4.42

15. A bee's cell in a hive is a regular hexagonal prism open at the front, with a trihedral apex at the back (the top in Figure 4.43). It can be shown that the surface area of a cell with apex θ is given by

$$S(\theta) = 6ab + \frac{3}{2}b^2\left(-\cot\theta + \frac{\sqrt{3}}{\sin\theta}\right) \quad \text{for } 0 < \theta < \frac{\pi}{2}$$

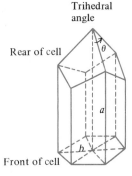

Trihedral angle

Rear of cell

Front of cell

FIGURE 4.43

where a and b are positive constants. Show that the surface area is minimized if $\cos \theta = 1/\sqrt{3}$, so that $\theta \approx 54.7°$. (*Hint:* Use the result of Exercise 13.) Experiments have shown that bee cells have an average angle within 2′ (less than one tenth of one degree) of 54.7°.

16. Suppose we wish to estimate the probability p of rolling a 3 with a loaded die. We roll the die n times and obtain m 3's in a particular order. The probability of this is known to be $p^m(1 - p)^{n-m}$. The **maximum likelihood estimate** of p based on the n rolls is the value of x that maximizes $x^m(1 - x)^{n-m}$ on $[0, 1]$. Show that the maximum likelihood estimate of p is m/n.

17. A company plans to invest $50,000 for the next four years, and initially it buys oil stocks. If it seems profitable to do so, the oil stocks will be sold before the four-year period has lapsed, and the revenue from the sale of the stocks will be placed in tax-free municipal bonds. According to the company's analysis, if the oil stocks are sold after t years, then the net profit $P(t)$ in dollars for the four-year investment is given by

$$P(t) = 2(20 - t)^3 t \quad \text{for } 0 \le t \le 4$$

Determine whether the company should switch from oil stocks to municipal bonds, and if so, after what period of time.

18. If we neglect air resistance, then the range of a ball (or any projectile) shot at an angle θ with respect to the x axis and with an initial velocity v_0 is given by

$$R(\theta) = \left(\frac{v_0^2}{g}\right) \sin 2\theta \quad \text{for } 0 \le \theta \le \frac{\pi}{2}$$

where g is the acceleration due to gravity (32 feet per second per second).
a. Show that the maximum range is attained when $\theta = \pi/4$.
b. If $v_0 = 96$ feet per second and the aim is to snuff out a smouldering cigarette lying on the ground 144 feet away, at what angle should the ball be hit?
c. The maximum height reached by the ball is

$$y_{\max} = \frac{(v_0^2 \sin^2 \theta)}{2g}$$

Why would it be a bad idea to hit the ball so that $y_{\max}$ is maximized?

19. A ring of radius a carries a uniform electric charge Q. The electric field intensity at any point x along the axis of the ring is given by

$$E(x) = \frac{Qx}{(x^2 + a^2)^{3/2}}$$

At what point on the axis is the electric field the greatest?

20. If electric charge is uniformly distributed throughout a circular cylinder (such as a telephone wire) of radius a, then at any point whose distance from the axis of the cylinder is r, the electric field intensity is given by

$$E(r) = \begin{cases} cr & \text{for } 0 \le r \le a \\ ca^2/r & \text{for } r > a \end{cases}$$

where c is a positive constant.
a. Show that $E(r)$ is maximum for $r = a$.
b. Is E differentiable at a? Explain your answer.

21. Find the points on the line $y = 2x - 4$ that are closest to the point $(1, 3)$.

22. Find the points on the parabola $y = x^2 + 2x$ that are closest to the point $(-1, 0)$.

23. Of all the triangles that pass through the point $(1, 1)$ and have two sides lying on the coordinate axes, one has the smallest area. Determine the lengths of its sides.

24. A horse breeder plans to set aside a rectangular region of 1 square kilometer for horses and wishes to build a wooden fence to enclose the region. Since one side of the region will run along a well-traveled highway, the breeder decides to make that side more attractive, using wood that costs three times as much per meter as the wood for the other sides. What dimensions will minimize the cost of the fence?

25. Suppose a landowner wishes to use 3 miles of fencing to enclose an isosceles triangular region of as large an area as possible. What should be the lengths of the sides of the triangle?

26. A manufacturer wishes to produce rectangular containers with square bottoms and tops, each container having a capacity of 250 cubic inches. If the material used for the top and the bottom costs twice as much per square inch as the material for the sides, what dimensions will minimize the cost?

27. A wire of length L is cut into two pieces. One piece is bent to form a square, and the other is bent to form a circle. Determine the minimum possible value for the sum A of the areas of the square and the circle. If the wire is actually cut, is there a maximum value of A?

28. A 12-foot wire is cut into 12 pieces, which are soldered together to form a rectangular frame whose base is twice as long as it is wide (as in Figure 4.44). The frame is then covered with paper.
a. How should the wire be cut if the volume of the frame is to be maximized?
b. How should the wire be cut if the total surface area of the paper is to be maximized?

29. Toward what point on the road should the ranger in Example 5 walk in order to minimize the travel time to the car if the car is located

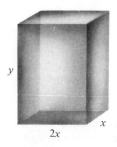

FIGURE 4.44

a. 10 miles down the road?
b. $\frac{1}{2}$ mile down the road?
c. an arbitrary number c of miles down the road?

30. It is known that homing pigeons fly faster over land than over water. Assume that they fly 10 meters per second over land but only 8 meters per second over water. If a pigeon is located at the edge of a straight river 500 meters wide and must fly to its nest, located 1300 meters away on the opposite side of the river (Figure 4.45), what path would minimize its flying time?

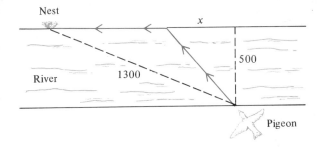

FIGURE 4.45

31. A rectangular printed page is to have margins 2 inches wide at the top and the bottom and margins 1 inch wide on each of the two sides. If the page is to have 35 square inches of printing, determine the minimum possible area of the page itself.

32. Most post offices in the United States have the following limit on the size of a parcel that can be mailed by parcel post: The sum of the length of its longest side and its girth (the largest perimeter of a cross-section perpendicular to the longest side) can be no more than 108 inches.
   a. Find the dimensions of the rectangular parallelepiped with a square base having the largest volume that can be mailed. (There are two cases to be considered, depending on which side is longest.)
   b. Find the dimensions of the right circular cylinder having the largest volume that can be mailed. (Again, there are two cases to consider.)

c. Find the dimensions of the cube having the largest volume that can be mailed.
d. Show that it is possible for a parcel to be mailable and yet have a larger volume than a parcel that is not mailable. (*Hint:* Examine your solutions to (a)–(c).)

33. An isosceles triangle has base 6 and height 12. Find the maximum possible area of a rectangle that can be placed inside the triangle with one side on the base of the triangle.

34. An isosceles triangle is inscribed in a circle of radius r. Find the maximum possible area of the triangle.

35. A cylindrical can with top and bottom has volume V. Find the radius of the can with the smallest possible surface area.

36. A cylinder is inscribed in a sphere with radius R. Find the height of the cylinder with the maximum possible volume.

37. A cylinder is inscribed in a cone of height H and base radius R. Determine the largest possible volume for the cylinder.

38. Find the radius of the cone with given volume V and minimum surface area. (*Hint:* The surface area S of a cone with radius r and height h is given by $S = \pi r \sqrt{r^2 + h^2}$.)

39. Three sides of a trapezoid are of equal length L, and no two are parallel. Find the length of the fourth side that gives the trapezoid maximum area.

40. Find the length of the longest thin, rigid pipe that can be carried from one 10-foot-wide corridor to a similar corridor at right angles to the first. Assume that the pipe has negligible diameter. (*Hint:* Find the length of the shortest line that touches the inside corner of the hallways and extends to the two walls. Use an angle as an auxiliary variable.)

41. After work a person wishes to sit in a long park bounded by two parallel highways 300 meters apart. Suppose one highway is 8 times as noisy as the other. In order to have the quietest repose, how far from the quieter highway should the person sit? (*Hint:* The intensity of noise where the person sits is directly proportional to the intensity of noise at the source and inversely proportional to the square of the distance from the source.)

42. A real estate firm can borrow money at 5% interest per year and can lend the money out. If the amount of money it can lend is inversely proportional to the square of the interest rate at which it lends, what interest rate would maximize the firm's profit per year? (*Hint:* Let x be the loan interest rate. Notice that the profit is the product of the amount borrowed by the firm and the difference between the interest rates at which it lends and borrows.)

43. A company has a daily fixed cost of $5000. If the company produces x units daily, then the daily cost in dollars for labor and materials is $3x$. The daily cost of equipment maintenance is $x^2/2,500,000$. What daily production

minimizes the total daily cost per unit of production? (*Hint:* The cost per unit is the total cost $C(x)$ divided by x.)

44. A company sells 1000 units of a certain product annually, with no seasonal fluctuations in demand. It always reorders the same number x of units, stocks unsold units until no more remain, and then reorders again. If it costs b dollars to stock one unit for one year and there is a fixed cost of c dollars each time the company reorders, how many units should be reordered each time to minimize the total annual cost of reordering and stocking? (*Hint:* The company will have an average inventory of $x/2$ units and must reorder $1000/x$ times per year. Find the annual stocking and reordering costs and minimize their sum.)

*45. A farmer wishes to employ tomato pickers to harvest 62,500 tomatoes. Each picker can harvest 625 tomatoes per hour and is paid $6 per hour. In addition, the farmer must pay a supervisor $10 per hour and pay the union $10 for each picker employed.
   a. How many pickers should the farmer employ to minimize the cost of harvesting the tomatoes?
   b. What is the minimum cost to the farmer?

4.6
CONCAVITY AND INFLECTION POINTS

We now consider other ways the second derivative can help in graphing functions—through the notions of concavity and inflection points.

Concavity In Figure 4.46(a) the tangent lines lie below the graphs, whereas in Figure 4.46(b) the tangent lines lie above the graphs. To distinguish between these two cases, we define the notion of concavity.

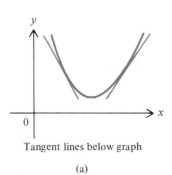

Tangent lines below graph

(a)

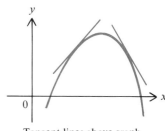

Tangent lines above graph

(b)

FIGURE 4.46

DEFINITION 4.13

   a. Let f be differentiable at c, and let l_c be the line tangent to the graph of f at $(c, f(c))$. The graph of f is **concave upward** at $(c, f(c))$ if there is an open interval I_c about c such that if x is in I_c and $x \neq c$, then $(x, f(x))$ lies above l_c. The graph of f is **concave downward** at $(c, f(c))$ if there is an open interval I_c about c such that if x is in I_c and $x \neq c$, then $(x, f(x))$ lies below l_c.
   b. The graph of a function f is **concave upward** (respectively **concave downward**) **on an open interval I** if it is concave upward (respectively concave downward) at $(x, f(x))$ for each x in I.

We will generally be interested in upward (or downward) concavity on open intervals rather than in concavity at specified points.

Let us assume that $f'(x)$ exists for all x in an open interval I, and let c be fixed in I. Recall that the line l_c tangent to the graph of f at $(c, f(c))$ has the equation

$$y = f(c) + f'(c)(x - c)$$

If we define g on I by

$$g(x) = f(x) - [f(c) + f'(c)(x - c)] = [f(x) - f(c)] - f'(c)(x - c) \qquad (1)$$

then g is very reminiscent of the auxiliary function used in proving the Mean Value Theorem, and is described pictorially in Figure 4.47(a) and (b). If it happens that $g(x) > 0$ for all x in I with $x \neq c$, then the tangent line through $(c, f(c))$ lies below the graph of f (Figure 4.47(a)). If $g(x) < 0$ for all x in I with $x \neq c$, then the tangent line through $(c, f(c))$ lies above the graph of f (Figure 4.47(b)). We will use this fact to find a criterion for concavity involving only values of f''.

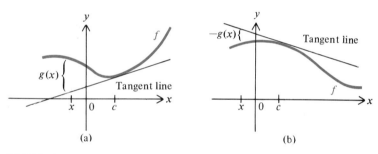

(a)   **(b)**

FIGURE 4.47

THEOREM 4.14

> Assume that f'' exists on an open interval I.
> **a.** If $f''(x) > 0$ for all x in I, then the graph of f is concave upward on I.
> **b.** If $f''(x) < 0$ for all x in I, then the graph of f is concave downward on I.

Proof We prove (a); the proof of (b) is similar and is omitted. Choose an arbitrary c in I. Notice that since $f'' > 0$ on I, f' is strictly increasing on I. For a fixed x in I with $x \neq c$, we can use the Mean Value Theorem to find a number z between x and c such that

$$f(x) - f(c) = f'(z)(x - c) \qquad (2)$$

Substituting (2) into (1) yields

$$g(x) = f'(z)(x - c) - f'(c)(x - c) = [f'(z) - f'(c)](x - c) \qquad (3)$$

Since f' is strictly increasing and z lies between x and c, it follows that the right side of (3) is positive. Therefore $g(x) > 0$ for all x in I with $x \neq c$. Hence l_c lies

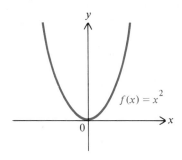

FIGURE 4.48

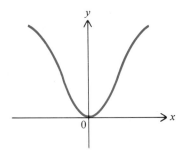

FIGURE 4.49

strictly below the graph of f on I except at the point of tangency. Since c is arbitrary in I, this means that the graph of f is concave upward on I. ■

By using Theorem 4.14, we see that the graph in Figure 4.48 represents $f(x) = x^2$ and that the graph in Figure 4.49 does not. This is true because $f''(x) = 2$ for all x, so that f is concave upward on $(-\infty, \infty)$. More generally, if $f(x) = ax^2 + bx + c$, where $a \neq 0$, then $f''(x) = 2a$. Thus the graph of f is either concave upward on $(-\infty, \infty)$ or concave downward on $(-\infty, \infty)$. The graph of such a function is called a **parabola**, one of the basic types of conic sections, which we will study in Chapter 10.

Next we look at third- and fourth-degree polynomials.

Example 1   Let $f(x) = 3x^4 - 4x^3$. Find the intervals on which the graph of f is concave upward and those on which it is concave downward. Then sketch the graph of f.

Solution   From Example 6 of Section 4.4 we know that

$$f'(x) = 12x^3 - 12x^2 = 12x^2(x - 1)$$

$$f''(x) = 36x^2 - 24x = 36x\left(x - \frac{2}{3}\right)$$

and that $f(1) = -1$ is a relative minimum value of f. Now we determine the sign of $f''(x)$ from Figure 4.50.

$$
\begin{array}{lllllllllllllllll}
x & - & - & - & - & - & 0 & + & + & + & + & + & + & + & + & + \\
x - \dfrac{2}{3} & - & - & - & - & - & - & - & - & - & - & 0 & + & + & + & + & + \\
f''(x) = 36x(x - \dfrac{2}{3}) & + & + & + & + & + & 0 & - & - & - & 0 & + & + & + & + & + & +
\end{array}
$$

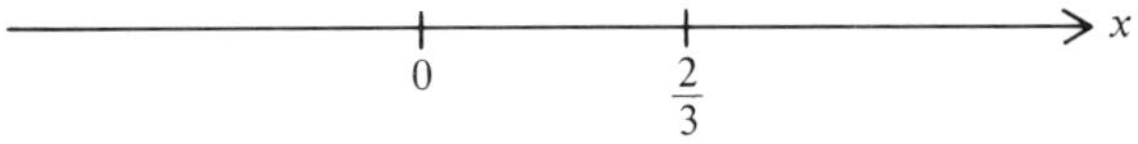

FIGURE 4.50

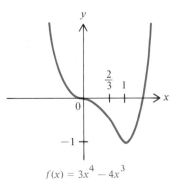

$f(x) = 3x^4 - 4x^3$

FIGURE 4.51

Using the sign of $f''(x)$ along with Theorem 4.14, we deduce that the graph of f is concave upward on $(-\infty, 0)$ and on $(\frac{2}{3}, \infty)$ and is concave downward on $(0, \frac{2}{3})$. From this information we conclude that the graph of f is as shown in Figure 4.51.   □

Example 2   Let $f(x) = 4x^3 - 6x^2 - 9x$. Find the intervals on which the graph of f is concave upward and those on which it is concave downward. Then sketch the graph of f.

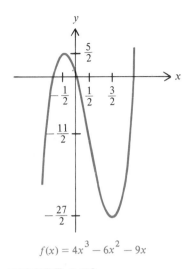

$$f(x) = 4x^3 - 6x^2 - 9x$$

FIGURE 4.52

Solution Differentiating f, we have

$$f'(x) = 12x^2 - 12x - 9 = 12\left(x^2 - x - \frac{3}{4}\right)$$

from which we obtain

$$f'(x) = 12\left(x - \frac{3}{2}\right)\left(x + \frac{1}{2}\right)$$

Next we find that

$$f''(x) = 24x - 12 = 24\left(x - \frac{1}{2}\right)$$

By Theorem 4.14 the graph of f is concave downward on $(-\infty, \frac{1}{2})$ and concave upward on $(\frac{1}{2}, \infty)$. From the first derivative we know that the critical points are $-\frac{1}{2}$ and $\frac{3}{2}$. Since $f''(-\frac{1}{2}) = -24$ and $f''(\frac{3}{2}) = 24$, it follows from the Second Derivative Test that $f(-\frac{1}{2}) = \frac{5}{2}$ is a relative maximum value of f and $f(\frac{3}{2}) = -\frac{27}{2}$ is a relative minimum value of f. We can now draw the graph of f (Figure 4.52). ☐

Inflection Points   In Example 2 the graph changed from concave downward to concave upward at $(\frac{1}{2}, -\frac{11}{2})$ (see Figure 4.52). Points at which the concavity of a graph changes are called inflection points.

DEFINITION 4.15

> Assume that there is a (possibly vertical) line tangent to the graph of f at $(c, f(c))$. Then $(c, f(c))$ is an **inflection point** of the graph of f if there is a number $\delta > 0$ such that the graph of f is concave upward on $(c - \delta, c)$ and concave downward on $(c, c + \delta)$ (or *vice versa*).

Figure 4.53 illustrates the relationship between concavity and inflection points.

Assume that f'' exists and is continuous on an interval containing c. Assume also that the graph of f has an inflection point at $(c, f(c))$, so there is a change of concavity at $(c, f(c))$. Because of the continuity of f'' it is possible to show that $f''(c) = 0$ (see Exercise 41). Thus we are led to the following method of finding inflection points for many functions:

1. Find the values of c for which $f''(c) = 0$.
2. For each value of c found in step 1, determine whether f'' changes sign at c.
3. If f'' changes sign at c, the point $(c, f(c))$ is an inflection point.

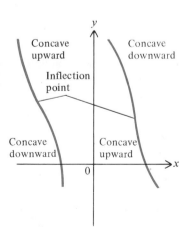

FIGURE 4.53

Example 3 Let $f(x) = x^4 - 6x^2 + 8x + 10$. Find the inflection points of the graph of f.

Solution   The derivatives of f are

$$f'(x) = 4x^3 - 12x + 8$$

$$f''(x) = 12x^2 - 12 = 12(x + 1)(x - 1)$$

Figure 4.54 shows the sign of $f''(x)$.

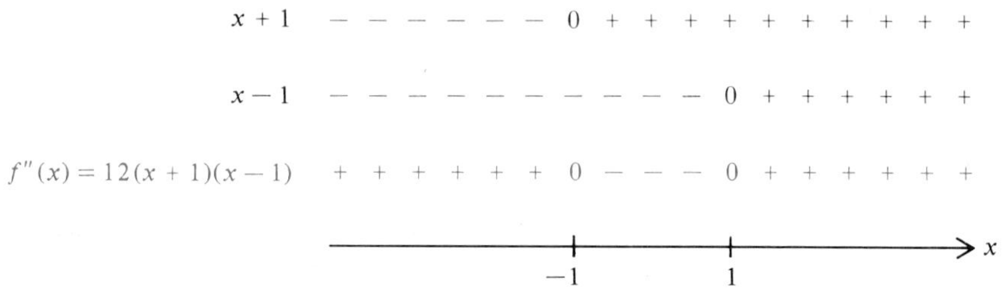

$$x + 1 \quad - \ - \ - \ - \ - \ - \ 0 \ + \ + \ + \ + \ + \ + \ + \ + \ +$$

$$x - 1 \quad - \ - \ - \ - \ - \ - \ - \ - \ - \ - \ 0 \ + \ + \ + \ + \ + \ +$$

$$f''(x) = 12(x + 1)(x - 1) \quad + \ + \ + \ + \ + \ + \ 0 \ - \ - \ - \ 0 \ + \ + \ + \ + \ + \ +$$

```
                                    +           +              → x
                                   -1           1
```

FIGURE 4.54

Observe that $f''(1) = f''(-1) = 0$ and that f'' changes sign at both 1 and -1. It follows that $(1, f(1)) = (1, 13)$ and $(-1, f(-1)) = (-1, -3)$ are inflection points. □

Example 4   Let $f(x) = \sin x$ for $-2\pi \leq x \leq 2\pi$. Find the inflection points of the graph of f, discuss its concavity, and sketch it.

Solution   The derivatives are

$$f'(x) = \cos x \quad \text{and} \quad f''(x) = -\sin x$$

Now $f''(x) = 0$ for $x = -\pi, 0,$ and π. Since f'' changes sign at each of these values of x, it follows that $(-\pi, 0), (0, 0),$ and $(\pi, 0)$ are inflection points. Moreover, the graph of f is concave downward on $(-2\pi, -\pi)$ and $(0, \pi)$ and concave upward on $(-\pi, 0)$ and $(\pi, 2\pi)$. The graph of the sine function appeared in Figure 1.49; we show it again in Figure 4.55, restricted in domain to the interval $[-2\pi, 2\pi]$. □

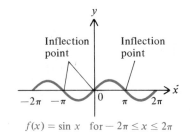

Inflection point   Inflection point

$f(x) = \sin x \quad \text{for} -2\pi \leq x \leq 2\pi$

FIGURE 4.55

Notice that the graph of the sine function in Figure 4.55 has an inflection point wherever it crosses the x axis. Since the sine function (without restricted domain) is periodic with period 2π, its graph therefore has an inflection point wherever it crosses the x axis, that is, at $(n\pi, 0)$ for any integer n. This shows that the graph of a function can have infinitely many inflection points.

Example 5   Let $g(x) = x + \sin x$. Find the inflection points of the graph of g, discuss its concavity, and sketch it.

Solution   Differentiating g, we have

$$g'(x) = 1 + \cos x \quad \text{and} \quad g''(x) = -\sin x$$

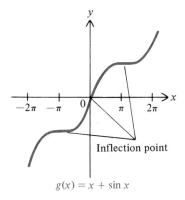

$g(x) = x + \sin x$

FIGURE 4.56

If $f(x) = \sin x$, then $g''(x) = f''(x)$, so g'' changes sign exactly where f'' does. Hence the discussion of the sine function just after Example 4 tells us that $(n\pi, g(n\pi)) = (n\pi, n\pi)$ is an inflection point of the graph of g for any integer n. We also deduce that the graph of g is concave downward on each interval $(2n\pi, (2n + 1)\pi)$ and concave upward on each interval $((2n + 1)\pi, (2n + 2)\pi)$. Does g have any relative extreme values? Note that $g'(x) = 0$ only if $\cos x = -1$, which means $x = (2n + 1)\pi$ for some integer n. Moreover, $g'(x) > 0$ unless $x = (2n + 1)\pi$. It follows from Theorem 4.8 that g is strictly increasing on each bounded closed interval and hence on $(-\infty, \infty)$. Thus g has *no* relative extreme values. The graph of g is drawn in Figure 4.56.   □

Caution: The graph of f need not have an inflection point at $(c, f(c))$ simply because $f''(c) = 0$. Indeed, if $f(x) = x^4$, then $f''(x) = 12x^2 \geq 0$ for all x, so f'' does not change sign at all, and in particular, it does not change sign at 0. Consequently $(0, 0)$ is not an inflection point, even though $f''(0) = 0$ (Figure 4.57).

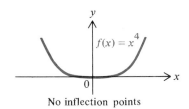

$f(x) = x^4$

No inflection points

FIGURE 4.57

EXERCISES 4.6

In Exercises 1–16 find the intervals on which the graph of the function is concave upward and those on which it is concave downward. Then sketch the graph of the function.

1. $f(x) = -\frac{3}{2}x^2 + x$

2. $f(x) = x^2 + 2x + 4$

3. $f(x) = x^3 + 8$

4. $f(x) = x^3 - 6x^2 + 9x + 2$

5. $g(x) = x^3 - 6x^2 + 12x - 4$

6. $g(x) = x^4 - 6x^2 + 8$

7. $g(x) = x^4 - 4x$ 8. $g(x) = 3x^5 - 5x^3$

9. $f(x) = x + \dfrac{1}{x}$ 10. $f(x) = \dfrac{x}{x^2 + 1}$

11. $f(x) = x\sqrt{x - 1}$ 12. $f(x) = x\sqrt{x^2 - 4}$

13. $f(x) = \sin 2x$ 14. $f(x) = \cos \frac{1}{2}x - 1$

15. $f(x) = \sec x$ 16. $f(x) = \csc \frac{1}{4}x$

In Exercises 17–28 find all inflection points (if any) of the graph of the function. Then sketch the graph of the function.

17. $f(x) = (x + 2)^3$

18. $f(x) = x^3 + 3$

19. $f(x) = x^3 + 3x^2 - 9x - 2$

20. $f(x) = x^3 - \frac{3}{2}x^2 - 6x$

21. $g(x) = 3x^4 + 4x^3$ 22. $g(x) = x^4 - 2x^3$

23. $g(x) = x^9 - 3x^3$ 24. $g(x) = x^{5/3}$

25. $g(x) = \frac{2}{3}x^{2/3} - \frac{3}{5}x^{5/3}$ 26. $g(x) = x\sqrt{1 - x^2}$

27. $f(t) = \tan t$ *28. $f(t) = t - \cos t$

29. In each of the following, draw the graph of a continuous function f having the given properties.
   a. f is increasing and its graph is concave upward on $(-\infty, 0)$, and f is decreasing and its graph is concave downward on $(0, \infty)$.

b. f is decreasing and its graph is concave upward on $(-\infty, 1)$, f is increasing and its graph is concave upward on $(1, 2)$, and f is decreasing and its graph is concave upward on $(2, \infty)$.

c. f is decreasing and its graph is concave upward on $(-\infty, 1)$, f is increasing and its graph is concave upward on $(1, 2)$, and f is increasing and its graph is concave downward on $(2, \infty)$.

d. f is decreasing and its graph is concave downward on $(-\infty, 0)$, f is increasing and its graph is concave downward on $(0, 1)$, f is increasing and its graph is concave upward on $(1, 5)$, and f is decreasing and its graph is concave downward on $(5, \infty)$.

30. Determine which of the graphs in Figure 4.58 could be the graph of a polynomial with the given degree:
   a. second degree   b. third degree
   c. fourth degree   d. fifth degree
   e. sixth degree

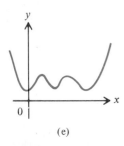

(a)

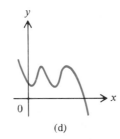

(b)

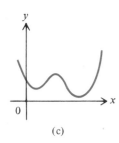

(c)

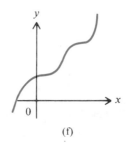

(d)

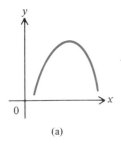

(e)

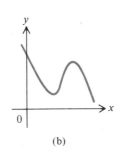

(f)

FIGURE 4.58

31. Recall that a function f is even if $f(-x) = f(x)$ for all x, and f is odd if $f(-x) = -f(x)$ for all x.
   a. If f is even and its graph is concave upward on $(0, \infty)$, what is the concavity on $(-\infty, 0)$?
   b. If f is odd and its graph is concave upward on $(0, \infty)$, what is the concavity on $(-\infty, 0)$?

32. How does the concavity of the graph of a function f compare with the concavity of the graph of $-f$?

*33. Show by giving an example that the graph of the function fg need not be concave upward at $(c, fg(c))$ even if the graph of f is concave upward at $(c, f(c))$ and the graph of g is concave upward at $(c, g(c))$.

34. Let $f(x) = x^{1/3}$. Show that $(0, 0)$ is an inflection point of the graph of f, although neither $f'(0)$ nor $f''(0)$ exists. (Thus it is not absolutely necessary for either the first or the second derivative to exist in order to have an inflection point.)

35. Show that the graph of a second-degree polynomial never has an inflection point.

36. Let $f(x) = ax^3 + bx^2 + cx + d$, with $a \neq 0$. Show that the graph of f has exactly one inflection point, and find it.

37. Let $f(x) = x^5 - cx^3$, where c is a constant. Show that the graph of f has an inflection point at $(0, 0)$.

38. Let n be a positive integer and $f(x) = x^n$. Show that the graph of f has at most one inflection point. Determine those values of n for which the inflection point exists, and find the inflection point.

39. It is known that a polynomial of degree k can have at most k real zeros. Use this fact to determine the maximum number of inflection points of the graph of a polynomial of degree n, where $n \geq 2$.

*40. Let $f(x) = \dfrac{\sin x}{x}$ for $x > 0$.
   a. Show that $f''(x) = g(x)/x^3$, where
   $$g(x) = -x^2 \sin x - 2x \cos x + 2 \sin x$$
   b. Use Theorem 4.8 to show that $g(x) < 0$ for $0 < x < \pi/2$. Use this result to prove that the graph of f is concave downward on $(0, \pi/2)$.
   c. Sketch the graph of f. [*Hint:* Make use of (b) and the fact that as x increases, the graph oscillates between the graphs of $y = 1/x$ and $y = -1/x$. Also use the fact that the graph of f touches the graph of $y = 1/x$ where $\sin x = 1$ and touches the graph of $y = -1/x$ where $\sin x = -1$. Note also that $f'(\pi/2) = -4/\pi^2$ and $f'(x) \neq 0$ for all x in $(0, \pi/2)$, since $\tan x > x$. Thus f decreases on $(0, \pi/2)$.]

41. Show that if f'' exists and is continuous on an open interval containing c and if f has an inflection point at $(c, f(c))$, then $f''(c) = 0$.

4.7

LIMITS AT INFINITY

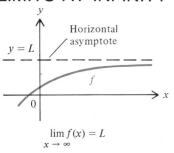

$$\lim_{x \to \infty} f(x) = L$$

FIGURE 4.59

Until now the limits we have encountered have been limits of a function f at a number a: $\lim_{x \to a} f(x)$, $\lim_{x \to a^+} f(x)$, and $\lim_{x \to a^-} f(x)$. Now we will consider the limit of $f(x)$ as x is positive and becomes arbitrarily large, that is, as x approaches ∞. Analogously, we will consider the limit of $f(x)$ as x is negative and $|x|$ becomes arbitrarily large, that is, as x approaches $-\infty$. A knowledge of such limits will aid us in sketching the parts of the graph of f that are far from the y axis.

We can think of the limit of $f(x)$ as x approaches ∞ as a kind of left-hand limit, because x approaches ∞ from the left (Figure 4.59). However, we must find a way of expressing mathematically how "near" a number x is to ∞. To do so, we notice that the larger x is, the nearer it is to ∞. Analogous statements can be made for x approaching $-\infty$. These observations suggest the following definition.

DEFINITION 4.16

a. Let f be defined on an interval (a, ∞). A number L is the **limit of $f(x)$ as x approaches ∞** if for every $\varepsilon > 0$ there is a number M such that

$$\text{if } x > M, \quad \text{then } |f(x) - L| < \varepsilon$$

In this case we write

$$\lim_{x \to \infty} f(x) = L$$

We say that the **limit of $f(x)$ exists as x approaches ∞**, or that f **has a limit at ∞**.

b. Let f be defined on an interval $(-\infty, a)$. A number L is the **limit of $f(x)$ as x approaches $-\infty$** if for every $\varepsilon > 0$ there is a number M such that

$$\text{if } x < M, \quad \text{then } |f(x) - L| < \varepsilon$$

In this case we write

$$\lim_{x \to -\infty} f(x) = L$$

We say that the **limit of $f(x)$ exists as x approaches $-\infty$** or that f **has a limit at $-\infty$**.

c. If either $\lim_{x \to \infty} f(x) = L$ or $\lim_{x \to -\infty} f(x) = L$, then we call the horizontal line $y = L$ a **horizontal asymptote of the graph of f** (Figure 4.59).

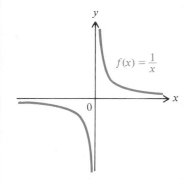

$$f(x) = \frac{1}{x}$$

FIGURE 4.60

The number M in Definition 4.16 corresponds to the number δ in all other definitions of limits given so far. We think of x as "close to" ∞ when $x > M$, just as we say that x is "close to" a when $a - \delta < x < a$. If $\lim_{x \to \infty} f(x) = L$ (or if $\lim_{x \to -\infty} f(x) = L$), then the point $(x, f(x))$ approaches the asymptote $y = L$ as x becomes arbitrarily large positively (or negatively).

From Definition 4.16 it follows that

$$\lim_{x \to \infty} \frac{1}{x} = 0 \quad \text{and} \quad \lim_{x \to -\infty} \frac{1}{x} = 0$$

(Figure 4.60).

The basic limit theorems remain true when a is replaced by ∞ or $-\infty$. Thus the limit as x approaches ∞ or $-\infty$ is unique when it exists. Furthermore, if $\lim_{x \to \infty} f(x)$ and $\lim_{x \to \infty} g(x)$ exist, we have

$$\lim_{x \to \infty} [f(x) + g(x)] = \lim_{x \to \infty} f(x) + \lim_{x \to \infty} g(x)$$

$$\lim_{x \to \infty} f(x)g(x) = \lim_{x \to \infty} f(x) \lim_{x \to \infty} g(x)$$

There are corresponding formulas for limits at $-\infty$.

From the formulas for the limits of a product we conclude that if n is any positive integer, then

$$\lim_{x \to \infty} \frac{1}{x^n} = 0 \quad \text{and} \quad \lim_{x \to -\infty} \frac{1}{x^n} = 0 \tag{1}$$

Indeed,

$$\lim_{x \to \infty} \frac{1}{x^n} = \lim_{x \to \infty} \left(\overbrace{\frac{1}{x} \cdot \frac{1}{x} \cdots \frac{1}{x}}^{n \text{ factors}} \right) = \overbrace{\left(\lim_{x \to \infty} \frac{1}{x} \right) \left(\lim_{x \to \infty} \frac{1}{x} \right) \cdots \left(\lim_{x \to \infty} \frac{1}{x} \right)}^{n \text{ factors}} = 0 \cdot 0 \cdots 0 = 0$$

The limit at $-\infty$ follows analogously.

Example 1 Find the following limits.

a. $\displaystyle \lim_{x \to \infty} \frac{x^2 + 2x - 5}{2x^2 - 6x - 1}$

b. $\displaystyle \lim_{x \to -\infty} \frac{3x + 1}{4 - x^3}$

Solution

a. We divide both numerator and denominator by x^2, which is the highest power of x in the fraction, and then use (1) along with the limit theorems. We obtain

$$\lim_{x \to \infty} \frac{x^2 + 2x - 5}{2x^2 - 6x - 1} = \lim_{x \to \infty} \frac{\dfrac{x^2}{x^2} + \dfrac{2x}{x^2} - \dfrac{5}{x^2}}{\dfrac{2x^2}{x^2} - \dfrac{6x}{x^2} - \dfrac{1}{x^2}} = \lim_{x \to \infty} \frac{1 + \dfrac{2}{x} - \dfrac{5}{x^2}}{2 - \dfrac{6}{x} - \dfrac{1}{x^2}}$$

$$= \frac{1 + 0 - 0}{2 - 0 - 0} = \frac{1}{2}$$

b. As in part (a), we divide by the largest power of x appearing in the fraction, this time x^3:

$$\lim_{x \to -\infty} \frac{3x + 1}{4 - x^3} = \lim_{x \to -\infty} \frac{\dfrac{3x}{x^3} + \dfrac{1}{x^3}}{\dfrac{4}{x^3} - \dfrac{x^3}{x^3}} = \lim_{x \to -\infty} \frac{\dfrac{3}{x^2} + \dfrac{1}{x^3}}{\dfrac{4}{x^3} - 1} = \frac{0 - 0}{0 - 1} = 0 \quad \square$$

Next we evaluate a type of limit that will reappear in Section 4.8.

Example 2 Let $a > 0$. Show that $\lim_{x \to \infty} (\sqrt{x^2 - a^2} - x) = 0$.

Solution First we rewrite the expression in the limit:

$$\sqrt{x^2 - a^2} - x = (\sqrt{x^2 - a^2} - x) \frac{\sqrt{x^2 - a^2} + x}{\sqrt{x^2 - a^2} + x}$$

$$= \frac{x^2 - a^2 - x^2}{\sqrt{x^2 - a^2} + x}$$

$$= \frac{-a^2}{\sqrt{x^2 - a^2} + x}$$

$$= \frac{-a^2/x}{\sqrt{1 - \dfrac{a^2}{x^2}} + 1}$$

Since

$$\lim_{x \to \infty} \frac{-a^2/x}{\sqrt{1 - \dfrac{a^2}{x^2}} + 1} = \frac{0}{\sqrt{1 - 0} + 1} = 0$$

it follows that

$$\lim_{x \to \infty} (\sqrt{x^2 - a^2} - x) = 0 \quad \square$$

Pictorially, the line $y = a$ is a horizontal asymptote of a function f if far from the y axis (either to the left or to the right) the graph of f approaches the line $y = a$ more and more closely. In Example 3 we will determine the asymptotes of a graph.

Example 3 Let

$$f(x) = \frac{2x + 1}{3x - 1}$$

Sketch the graph of f.

Solution We first find the derivatives of f:

$$f'(x) = \frac{2(3x - 1) - 3(2x + 1)}{(3x - 1)^2} = \frac{-5}{(3x - 1)^2}$$

$$f''(x) = \frac{30}{(3x - 1)^3}$$

Consequently f is strictly decreasing on $(-\infty, \frac{1}{3})$ and on $(\frac{1}{3}, \infty)$, and its graph is

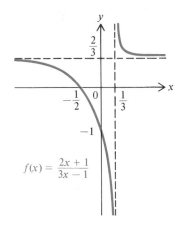

$$f(x) = \frac{2x + 1}{3x - 1}$$

FIGURE 4.61

concave downward on $(-\infty, \frac{1}{3})$ and concave upward on $(\frac{1}{3}, \infty)$. For the asymptotes we observe that

$$\lim_{x \to 1/3^+} f(x) = \lim_{x \to 1/3^+} \frac{2x + 1}{3x - 1} = \infty$$

$$\lim_{x \to 1/3^-} f(x) = \lim_{x \to 1/3^-} \frac{2x + 1}{3x - 1} = -\infty$$

$$\lim_{x \to \infty} f(x) = \lim_{x \to -\infty} f(x) = \frac{2}{3}$$

Therefore the line $x = \frac{1}{3}$ is a vertical asymptote and the line $y = \frac{2}{3}$ is a horizontal asymptote. This information enables us to sketch the graph (Figure 4.61). ☐

Since limits are unique, the graph of a function cannot have more than two horizontal asymptotes, one arising from $\lim_{x \to \infty} f(x)$ and one from $\lim_{x \to -\infty} f(x)$. (See Exercises 29 and 30 for examples in which two distinct horizontal asymptotes actually arise.) Thus for a given function, there may be 0, 1, or 2 horizontal asymptotes. If p and q are polynomial functions and f the rational function defined by $f(x) = p(x)/q(x)$, then we can be more specific about possible horizontal asymptotes of the graph of f:

degree of p < degree of q: one horizontal asymptote, $y = 0$
degree of p = degree of q: one horizontal asymptote, $y = c$, with $c \neq 0$
degree of p > degree of q: no horizontal asymptote

Infinite Limits at Infinity   We conclude this section by examining a type of limit that combines infinite limits with limits at ∞.

DEFINITION 4.17

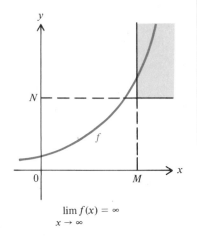

$$\lim_{x \to \infty} f(x) = \infty$$

FIGURE 4.62

Let f be defined on an interval (a, ∞). If for any real number N there is some number M such that

if $x > M$, then $f(x) > N$

we say that the **limit of $f(x)$ as x approaches ∞ is ∞** and write

$$\lim_{x \to \infty} f(x) = \infty$$

(see Figure 4.62). Alternatively, we say that f **has an infinite limit at ∞**.

The definitions of

$$\lim_{x \to \infty} f(x) = -\infty, \qquad \lim_{x \to -\infty} f(x) = \infty, \quad \text{and} \quad \lim_{x \to -\infty} f(x) = -\infty$$

are completely analogous.

Using Definition 4.17, one can show that for any positive integer n,

$$\lim_{x \to \infty} x^n = \infty$$

Furthermore,

$$\lim_{x \to -\infty} x^n = \begin{cases} \infty & \text{for } n \text{ even} \\ -\infty & \text{for } n \text{ odd} \end{cases}$$

It follows that if p is a polynomial of degree at least 1, then

$$\lim_{x \to \infty} p(x) = \infty \quad \text{or} \quad \lim_{x \to \infty} p(x) = -\infty$$

and

$$\lim_{x \to -\infty} p(x) = \infty \quad \text{or} \quad \lim_{x \to -\infty} p(x) = -\infty$$

Which infinite limits occur depends on the coefficient of the term in p with the highest power. For example,

$$\lim_{x \to \infty} (x - x^2) = -\infty \quad \text{and} \quad \lim_{x \to -\infty} (x - x^2) = -\infty$$

whereas

$$\lim_{x \to \infty} (x^3 - 2x + 1) = \infty \quad \text{and} \quad \lim_{x \to -\infty} (x^3 - 2x + 1) = -\infty$$

EXERCISES 4.7

In Exercises 1–16 find the given limit.

1. $\lim\limits_{x \to \infty} \dfrac{2}{x - 3}$

2. $\lim\limits_{x \to \infty} \dfrac{4}{2 - x}$

3. $\lim\limits_{x \to \infty} \dfrac{x}{3x + 2}$

4. $\lim\limits_{x \to \infty} \dfrac{4x^2}{\sqrt{2x - 3}}$

5. $\lim\limits_{x \to \infty} \dfrac{2x^2 + x - 1}{x^2 - x + 4}$

6. $\lim\limits_{t \to \infty} \dfrac{(t - 1)(2t + 1)}{(3t - 2)(t + 4)}$

7. $\lim\limits_{t \to \infty} \dfrac{t}{t^{1/2} + 2t^{-1/2}}$

8. $\lim\limits_{x \to \infty} \dfrac{\sin x}{x}$

9. $\lim\limits_{x \to -\infty} \dfrac{\cos x}{\sqrt{x^2 - 1}}$

10. $\lim\limits_{x \to -\infty} \dfrac{x - \frac{1}{2}}{\frac{1}{2}x + 1}$

11. $\lim\limits_{x \to -\infty} \dfrac{x^2}{4x^3 - 9}$

12. $\lim\limits_{x \to -\infty} \dfrac{2 - 3x - 4x^2}{3x^2 + 6x + 10}$

13. $\lim\limits_{x \to \infty} (x - \sqrt{4x^2 - 1})$

14. $\lim\limits_{x \to \infty} \left(3\sqrt{\dfrac{x^2}{4} - 1} - \dfrac{3}{2}x \right)$

15. $\lim\limits_{x \to \infty} \tan \dfrac{1}{x}$

*16. $\lim\limits_{x \to -\infty} x \tan \dfrac{1}{x}$

In Exercises 17–28 find the horizontal asymptote of the graph of the function. Then sketch the graph of the function.

17. $f(x) = \dfrac{1}{x - 2}$

18. $f(x) = -\dfrac{1}{x + 3}$

19. $f(x) = \dfrac{3x}{2x - 4}$

20. $f(x) = \dfrac{4}{(1 - x)^2}$

21. $f(x) = \dfrac{2x(x + 3)}{9 - x^2}$

22. $f(x) = \dfrac{\sqrt{x + 2}}{(x + 2)^2}$

23. $f(x) = \dfrac{x + 2}{x - 1}$

24. $f(x) = \dfrac{2x - 3}{4 - 6x}$

25. $f(x) = \sqrt{\dfrac{3 - x}{4 - x}}$

26. $f(x) = \sqrt{\dfrac{1 - 3x}{-2 - x}}$

27. $f(x) = \dfrac{1}{x^2 - 4}$

28. $f(x) = \dfrac{-2}{x^2 - 9}$

29. Let

$$f(x) = \dfrac{x|x|}{x^2 + 1}$$

Show that the graph of f has two horizontal asymptotes, and determine them.

30. Let

$$f(x) = \dfrac{\sqrt{2 + 4x^2}}{x}$$

Show that the graph of f has two horizontal asymptotes, and determine them.

31. Suppose $\lim_{x \to -\infty} f(x)/g(x) = 1$ and $\lim_{x \to -\infty} g(x) = \infty$. Show that $\lim_{x \to -\infty} f(x) = \infty$.

*32. Suppose that there is a number M such that $f(x) \geq 0$ for all $x \geq M$, and assume that $\lim_{x \to \infty} [f(x)]^2 = L$. Show that $\lim_{x \to \infty} f(x) = \sqrt{L}$.

33. According to Weiss's law of excitation of tissue, the strength S of an electric current is related to the time t the current takes to excite tissue by the formula

$$S(t) = \frac{a}{t} + b \quad \text{for } t > 0$$

where a and b are constants. Then the limit $\lim_{t \to \infty} S(t)$ is the threshold strength of current below which the tissue will never be excited. Find $\lim_{t \to \infty} S(t)$.

34. Suppose a rocket is launched vertically upward and its velocity $v(r)$ in miles per second at a distance of r miles from the center of the earth is given by the formula

$$v(r) = \sqrt{\frac{192{,}000}{r} + v_0^2 - 48} \quad \text{for } r \geq 4000$$

where v_0 is constant and represents the velocity of the rocket at burnout.
a. If $v_0 = 8$ (miles per second), what is the limit of the velocity as the distance grows without bound?
b. Find the number $v_0 > 0$ for which $\lim_{r \to \infty} v(r) = 0$. This number v_0 is called the **escape velocity** of the rocket. For any smaller value of v_0 the rocket will return to earth.

35. Suppose a rocket is launched from the surface of the earth. The work required to project the rocket from the surface (3960 miles from the center of the earth) to x miles above the surface is given by the formula

$$W(x) = GMm\left(\frac{1}{3960} - \frac{1}{x}\right)$$

where G, M, and m are constants. Find the work required to send the rocket from here to the end of the universe.

The launch of *Voyager 1*, which achieved escape from the earth to travel to the outer planets of the solar system.

36. Suppose the current $I(t)$ flowing in an electrical circuit at time t is given by

$$I(t) = \frac{100}{1 + t^2} + 3 \sin \frac{30t}{\pi} \quad \text{for } t \geq 0$$

Show that

$$\lim_{x \to \infty} \left(I(t) - 3 \sin \frac{30t}{\pi} \right) = 0$$

Thus for large values of t, $I(t)$ is very nearly equal to $3 \sin (30t/\pi)$. The expression $3 \sin (30t/\pi)$ is called the steady state current, and the expression $100/(1 + t^2)$ is the transient current (since it is significant only for small values of t).

4.8
GRAPHING

As we have seen throughout Chapter 4, a knowledge of derivatives helps greatly in sketching the graphs of functions. In this section we collect the methods we have encountered for sketching graphs, and we illustrate their uses.

Table 4.1 lists the items that are most important in graphing a function f.

TABLE 4.1

Property	Test
f has y intercept c	$f(0) = c$
f has x intercept c	$f(c) = 0$
Graph of f is symmetric with respect to the $\begin{cases} y \text{ axis} \\ \text{origin} \end{cases}$	$\begin{aligned} f(-x) &= f(x) \\ f(-x) &= -f(x) \end{aligned}$
f has a relative maximum value at c	$\begin{cases} f'(c) = 0 \text{ and } f' \text{ changes from positive to negative at } c \\ f'(c) = 0 \text{ and } f''(c) < 0 \end{cases}$
f has a relative minimum value at c	$\begin{cases} f'(c) = 0 \text{ and } f' \text{ changes from negative to positive at } c \\ f'(c) = 0 \text{ and } f''(c) > 0 \end{cases}$
f is strictly increasing on an open interval I	$f'(x) > 0$ for all except finitely many x in I
f is strictly decreasing on an open interval I	$f'(x) < 0$ for all except finitely many x in I
Graph of f is concave upward on an open interval I	$f''(x) > 0$ for all x in I
Graph of f is concave downward on an open interval I	$f''(x) < 0$ for all x in I
$(c, f(c))$ is an inflection point of the graph of f	f'' changes sign at c (and usually $f''(c) = 0$)
f has a vertical asymptote $x = c$	$\lim\limits_{x \to c^+} f(x) = \pm\infty$ or $\lim\limits_{x \to c^-} f(x) = \pm\infty$
f has a horizontal asymptote $y = d$	$\lim\limits_{x \to \infty} f(x) = d$ or $\lim\limits_{x \to -\infty} f(x) = d$

Example 1 Let

$$f(x) = \frac{2}{1 + x^2}$$

Sketch the graph of f, noting all relevant properties listed in Table 4.1.

Solution Since $f(0) = 2$, the y intercept is 2. However, there are no x intercepts because $f(x) > 0$ for all x. The fact that $f(-x) = f(x)$ implies that the graph of f is symmetric with respect to the y axis. For the derivatives of f we have

$$f'(x) = \frac{-4x}{(1 + x^2)^2}$$

and

$$f''(x) = \frac{-4(1 + x^2)^2 + 4x(2)(1 + x^2)(2x)}{(1 + x^2)^4} = \frac{4(3x^2 - 1)}{(1 + x^2)^3}$$

Since $f'(x) > 0$ for $x < 0$ and $f'(x) < 0$ for $x > 0$, it follows that f is strictly increasing on $(-\infty, 0]$ and strictly decreasing on $[0, \infty)$, so that $f(0) = 2$ is the maximum value of f. Next we display the sign of $f''(x)$ in Figure 4.63.

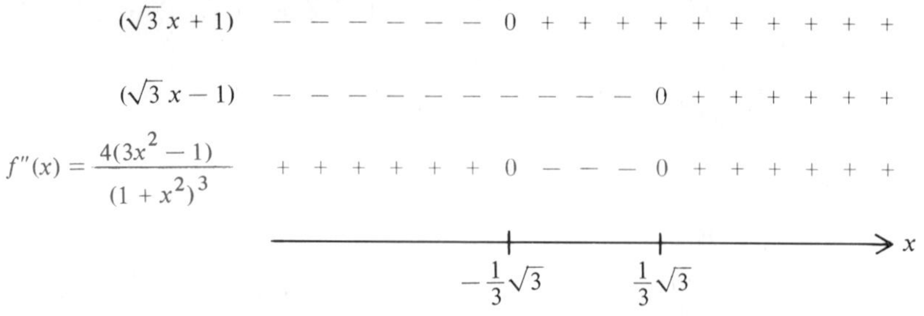

$$(\sqrt{3}\, x + 1) \quad - - - - - - \; 0 \; + + + + + + + + + +$$

$$(\sqrt{3}\, x - 1) \quad - - - - - - - - - - - \; 0 \; + + + + + +$$

$$f''(x) = \frac{4(3x^2 - 1)}{(1 + x^2)^3} \quad + + + + + + \; 0 \; - - - \; 0 \; + + + + + +$$

FIGURE 4.63

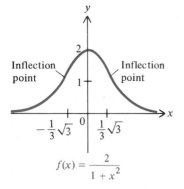

FIGURE 4.64

From the figure we see that the graph of f is concave upward on $(-\infty, -\frac{1}{3}\sqrt{3})$ and on $(\frac{1}{3}\sqrt{3}, \infty)$ and is concave downward on $(-\frac{1}{3}\sqrt{3}, \frac{1}{3}\sqrt{3})$, with inflection points at $(-\frac{1}{3}\sqrt{3}, \frac{3}{2})$ and $(\frac{1}{3}\sqrt{3}, \frac{3}{2})$. Finally, we notice that

$$\lim_{x \to \infty} \frac{2}{1 + x^2} = \lim_{x \to -\infty} \frac{2}{1 + x^2} = 0$$

which means that the x axis is a horizontal asymptote of the graph of f. We are now ready to sketch the graph of f, shown in Figure 4.64.   □

Example 2   Let

$$f(x) = \frac{x}{(x + 1)^2}$$

Sketch the graph of f, noting all relevant properties listed in Table 4.1.

Solution   First, $f(x) = 0$ precisely when $x = 0$, so that the x intercept and y intercept are both 0. Second, there is no symmetry with respect to either the y axis or the origin, because

$$f(-x) = \frac{-x}{((-x) + 1)^2}$$

and consequently $f(-x)$ is different from $f(x)$ and from $-f(x)$. Next we calculate the first and second derivatives of f:

$$f'(x) = \frac{(x + 1)^2 - 2x(x + 1)}{(x + 1)^4} = \frac{1 - x}{(x + 1)^3}$$

and

$$f''(x) = \frac{(-1)(x + 1)^3 - (1 - x)3(x + 1)^2}{(x + 1)^6} = \frac{2(x - 2)}{(x + 1)^4}$$

The signs of $f'(x)$ and $f''(x)$ are shown in Figure 4.65.

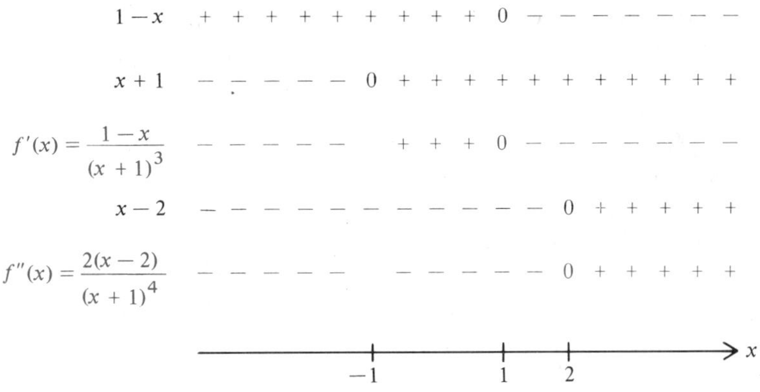

$$1 - x \quad + \ + \ + \ + \ + \ + \ + \ + \ + \ 0 \ - \ - \ - \ - \ - \ -$$

$$x + 1 \quad - \ - \ - \ - \ - \ 0 \ + \ + \ + \ + \ + \ + \ + \ + \ + \ +$$

$$f'(x) = \frac{1 - x}{(x + 1)^3} \quad - \ - \ - \ - \ - \quad + \ + \ + \ 0 \ - \ - \ - \ - \ - \ - \ -$$

$$x - 2 \quad - \ - \ - \ - \ - \ - \ - \ - \ - \ - \ - \ 0 \ + \ + \ + \ + \ +$$

$$f''(x) = \frac{2(x - 2)}{(x + 1)^4} \quad - \ - \ - \ - \ - \quad - \ - \ - \ - \ - \ 0 \ + \ + \ + \ + \ +$$

$$\xrightarrow{\hspace{2cm}} x$$
$$\qquad\qquad -1 \qquad\quad 1 \quad 2$$

FIGURE 4.65

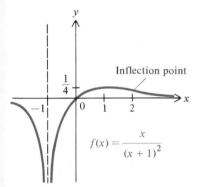

FIGURE 4.66

From the sign of $f'(x)$ it follows that f is strictly increasing on $(-1, 1]$ and is strictly decreasing on $(-\infty, -1)$ and on $[1, \infty)$. Thus $f(1) = \frac{1}{4}$ is a relative maximum value of f. From the sign of $f''(x)$ we deduce that the graph of f is concave downward on $(-\infty, -1)$ and on $(-1, 2)$ and is concave upward on $(2, \infty)$, with an inflection point at $(2, \frac{2}{9})$. Finally, we determine the asymptotes. Since

$$\lim_{x \to \infty} \frac{x}{(x + 1)^2} = \lim_{x \to -\infty} \frac{x}{(x + 1)^2} = 0$$

the line $y = 0$ is a horizontal asymptote. Since

$$\lim_{x \to -1} \frac{x}{(x + 1)^2} = -\infty$$

the line $x = -1$ is a vertical asymptote. Now we are ready to sketch the graph of f, shown in Figure 4.66. ☐

We have found it valuable to graph a function in order to understand its various characteristics. If two functions are graphed on the same coordinate plane, their graphs may form the boundary of a plane region (Figure 4.67). In Section 5.8 we will be interested in regions of this type. To draw such a region, we need only draw the graphs of the two functions and note where the graphs cross each other. We illustrate this procedure in Example 3.

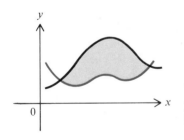

FIGURE 4.67

Example 3   Sketch the graphs of f and g, where

$$f(x) = \frac{2}{1 + x^2} \quad \text{and} \quad g(x) = x^2$$

Shade the region they enclose.

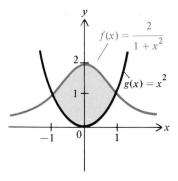

FIGURE 4.68

Solution We graphed f in Example 1, and we know the graph of g from Figure 4.14. To find where the two graphs intersect, we solve the equation $f(x) = g(x)$ for x, that is, we solve

$$\frac{2}{1 + x^2} = x^2$$

We find that

$$x^4 + x^2 - 2 = 0, \quad \text{or} \quad (x^2 + 2)(x^2 - 1) = 0$$

Thus $x = 1$ or $x = -1$. Therefore the graphs of f and g intersect at $(-1, 1)$ and $(1, 1)$. The region is shown in Figure 4.68. □

Ellipses and Hyperbolas

Graphs of certain equations can be analyzed with the methods developed in this chapter. Among the easiest graphs to handle are ellipses and hyperbolas. We give an example of each here, and we will return to the topic of such curves in Chapter 10.

Example 4 Sketch the graph of

$$\frac{x^2}{4} + \frac{y^2}{9} = 1$$

Solution Since both x and y appear to the second power in the equation, a point (x, y) is on the graph if and only if $(-x, y)$ and $(x, -y)$ are on the graph. This means that the graph is symmetric with respect to both the x axis and the y axis. Hence if we draw the part of the graph that lies above the x axis, we can use symmetry to complete the sketch. However, if (x, y) lies on or above the x axis, then $y \geq 0$, so that by solving the given equation for y we find that

$$y = 3 \sqrt{1 - \frac{x^2}{4}} \tag{1}$$

Thus our problem has been reduced to graphing the function described in (1). The domain of this function is $[-2, 2]$, and the range is $[0, 3]$. The x intercepts are -2 and 2, and the y intercept is 3. The derivatives are

$$\frac{dy}{dx} = \frac{-3x}{4\sqrt{1 - x^2/4}}$$

and

$$\frac{d^2y}{dx^2} = \frac{-12\sqrt{1 - x^2/4} - \dfrac{3x^2}{\sqrt{1 - x^2/4}}}{16(1 - x^2/4)} = \frac{-3}{4(1 - x^2/4)^{3/2}}$$

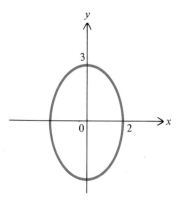

Ellipse $\dfrac{x^2}{4} + \dfrac{y^2}{9} = 1$

FIGURE 4.69

We conclude that y increases on $[-2, 0]$ and decreases on $[0, 2]$ and that the graph is concave downward on $(-2, 2)$. With this information we can sketch the top half and then the complete graph of the given equation (Figure 4.69). □

The graph of an equation

$$\frac{x^2}{a^2} + \frac{y^2}{b^2} = 1 \quad \text{where } a > 0 \text{ and } b > 0$$

is called an **ellipse**. If $a > b$, then the ellipse is as shown in Figure 4.70(a); if $a < b$, then the ellipse is as shown in Figure 4.70(b). For $a = b$, the ellipse is a circle of radius a centered at the origin; see Figure 4.70(c).

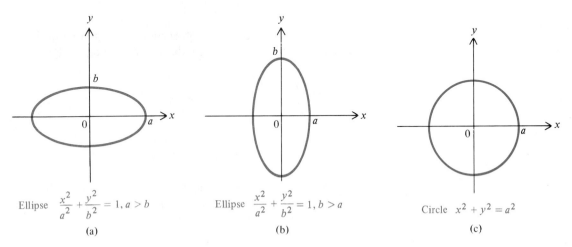

Ellipse $\dfrac{x^2}{a^2} + \dfrac{y^2}{b^2} = 1, a > b$ Ellipse $\dfrac{x^2}{a^2} + \dfrac{y^2}{b^2} = 1, b > a$ Circle $x^2 + y^2 = a^2$

(a)   (b)   (c)

FIGURE 4.70

Example 5   Sketch the graph of

$$\frac{x^2}{4} - \frac{y^2}{9} = 1$$

Solution   As in Example 4, the graph is symmetric with respect to both the x axis and the y axis. For $y \geq 0$ we obtain

$$y = 3\sqrt{\frac{x^2}{4} - 1} \tag{2}$$

The domain of the function in (2) consists of $(-\infty, -2]$ and $[2, \infty)$. The x intercepts are -2 and 2, and there is no y intercept. Furthermore

$$\frac{dy}{dx} = \frac{3x}{4\sqrt{x^2/4 - 1}}$$

and

$$\frac{d^2y}{dx^2} = \frac{12\sqrt{x^2/4 - 1} - \dfrac{3x^2}{\sqrt{x^2/4 - 1}}}{16(x^2/4 - 1)} = \frac{-3}{4(x^2/4 - 1)^{3/2}}$$

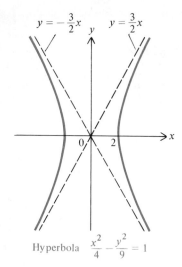

Hyperbola $\dfrac{x^2}{4} - \dfrac{y^2}{9} = 1$

FIGURE 4.71

We conclude that on the interval $(-\infty, -2)$, y is decreasing and the graph is concave downward. Similarly, on the interval $(2, \infty)$, y is increasing and the graph is concave downward. From (2) we also observe that

$$\lim_{x \to \infty} y = \lim_{x \to \infty} 3\sqrt{\frac{x^2}{4} - 1} = \infty$$

In fact, as x becomes larger, y approaches $\frac{3}{2}x$; that is,

$$\lim_{x \to \infty} \left(3\sqrt{\frac{x^2}{4} - 1} - \frac{3}{2}x \right) = 0$$

(Cf. Example 3 and Exercise 14 of Section 4.7.) Combining this information, we sketch the top half and then the complete graph of the equation (Figure 4.71).   □

The graph of

$$\frac{x^2}{a^2} - \frac{y^2}{b^2} = 1 \quad \text{or} \quad \frac{y^2}{b^2} - \frac{x^2}{a^2} = 1$$

is called a **hyperbola**. The graph of the equation on the left is as shown in Figure 4.72(a), and that of the equation on the right is as shown in Figure 4.72(b). Points (x, y) on the hyperbolas approach the lines

$$y = \frac{b}{a}x \quad \text{and} \quad y = \frac{-b}{a}x$$

as x increases without bound. These lines are the **asymptotes** of the hyperbola.

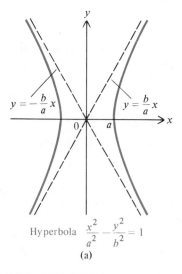

Hyperbola $\dfrac{x^2}{a^2} - \dfrac{y^2}{b^2} = 1$

(a)

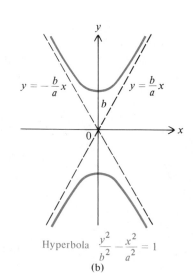

Hyperbola $\dfrac{y^2}{b^2} - \dfrac{x^2}{a^2} = 1$

(b)

FIGURE 4.72

EXERCISES 4.8

In Exercises 1–38 sketch the graph of the given function, noting all relevant properties listed in Table 4.1.

1. $f(x) = x^3 + 3x + 2$

2. $f(x) = x^3 - 8x^2 + 16x - 3$

3. $f(x) = x^4 + 8x^3 + 36x^2 - 3$

4. $f(x) = x^4 - 6x^2$

5. $g(x) = x + \dfrac{4}{x}$

6. $g(x) = 2 + \dfrac{1}{x^2}$

7. $g(x) = \dfrac{1}{x} - \dfrac{1}{x^2}$

8. $g(x) = \dfrac{-x}{x+1}$

9. $k(x) = \dfrac{1+x}{1-x}$

10. $k(x) = \dfrac{x+2}{x-3}$

11. $k(x) = \dfrac{4x}{1+4x^2}$

12. $k(x) = \dfrac{x^2}{1+x^2}$

13. $f(t) = \dfrac{1}{t^2 - 1}$

14. $f(t) = \dfrac{t}{t^2 - 1}$

15. $f(t) = \dfrac{t^2}{t^2 - 1}$

16. $f(t) = \dfrac{1}{t^2 - t}$

17. $f(z) = \dfrac{2z^2}{(z+1)^2}$

18. $f(z) = \dfrac{-z^2}{(2-z)^2}$

19. $f(x) = \dfrac{x^3}{1-x^2}$

20. $f(x) = \dfrac{(x+1)^2}{x-1}$

21. $f(x) = \dfrac{1}{x^3 - 1}$

22. $f(x) = x\sqrt{1+x}$

23. $f(x) = \sqrt{x-1} + \sqrt{x+1}$

24. $f(x) = (1 - x^2)^{3/2}$

25. $f(x) = \dfrac{x^2}{\sqrt{1-x^2}}$

26. $f(x) = x\sqrt{x^2 - 1}$

27. $f(x) = x\sqrt{1+x^2}$

28. $f(x) = \dfrac{x}{\sqrt{1-x}}$

29. $f(x) = \dfrac{\sqrt{x}}{1+\sqrt{x}}$

30. $f(x) = \dfrac{1}{\sqrt{x}} + \sqrt{x}$

31. $f(x) = 4x - 3x^{4/3}$

32. $f(x) = (x+2)(x-1)^{1/3}$

33. $g(x) = |\sin x|$

34. $g(x) = \sin x + \cos x$

35. $g(x) = \sqrt{3}\sin x + \cos x$

36. $g(x) = \dfrac{x}{2} + \sin x$

37. $g(t) = \sin^2 t$

38. $g(t) = \tan^2 t$

In Exercises 39–42 sketch the graph of the function, using the Newton-Raphson method where necessary to find approximate zeros, critical points, and inflection points.

39. $f(x) = x^4 + 4x^3 + 4x^2 - 2$

40. $f(x) = 2x^4 + x^3 + x$

41. $f(x) = \dfrac{x-1}{x(x+1)}$

42. $f(x) = x^5 + 2x^3 - x^2$

In Exercises 43–48 graph each pair of functions. Shade the region(s) the graphs enclose.

43. $f(x) = x^2, g(x) = x$

44. $f(x) = x^2 + 4$ and $g(x) = 12 - x^2$

45. $f(x) = x^3 + x$ and $g(x) = 3x^2 - x$ (*Hint:* The region has two parts.)

46. $f(x) = x^3 + x^2 + 1$ and $g(x) = x^3 + x + 1$

47. $g(x) = 2x/\sqrt{1+x^2}$ and $k(x) = x/\sqrt{1-x^2}$ (*Hint:* The region has two parts.)

48. $g(t) = t - \cos t$ for $-\pi/4 \le t \le 7\pi/4$ and $k(t) = t + \sin t$ for $-\pi/4 \le t \le 7\pi/4$

In Exercises 49–62 sketch the graph of the given equation.

49. $\dfrac{x^2}{9} + \dfrac{y^2}{4} = 1$

50. $\dfrac{x^2}{16} + \dfrac{y^2}{25} = 1$

51. $\dfrac{x^2}{25} + \dfrac{y^2}{16} = 1$

52. $x^2 + y^2 = 1$

53. $x^2 + y^2 = 9$

54. $x^2 + y^2 = 2$

55. $25x^2 + y^2 = 25$

56. $6x^2 + 24y^2 = 96$

57. $\dfrac{x^2}{9} - \dfrac{y^2}{4} = 1$

58. $\dfrac{x^2}{16} - \dfrac{y^2}{36} = 1$

59. $\dfrac{y^2}{16} - \dfrac{x^2}{36} = 1$

60. $\dfrac{y^2}{9} - x^2 = 1$

61. $y^2 - x^2 = 1$

62. $16x^2 - 4y^2 = 64$

In Exercises 63–66 first complete the square; then translate axes to facilitate sketching the graph of the equation.

63. $x^2 - 2x + 4y^2 - 3 = 0$

64. $4y^2 + 16y - 25x^2 + 150x - 309 = 0$

65. $8x^2 - 8x + 4y^2 + 4y = 33$

66. $3x^2 + 12x + 5y^2 + 20y + 31 = 0$

Key Terms and Expressions

Maximum value; minimum value
Relative maximum value; relative minimum value
Critical point
Increasing function; decreasing function
Strictly increasing function; strictly decreasing function
Concave upward; concave downward

Inflection point
Limit at infinity
Horizontal asymptote
Infinite limit at infinity
Parabola; ellipse; hyperbola

Key Theorems

Maximum–Minimum Theorem
Rolle's Theorem
Mean Value Theorem

First Derivative Test
Second Derivative Test

REVIEW EXERCISES

In Exercises 1–2 find the critical points of the given function.

1. $f(x) = x^2\sqrt{2-x}$ 2. $f(x) = \cos x^{1/3}$

In Exercises 3–6 find the extreme values of f on the given interval. Determine at which numbers in the interval they are assumed.

3. $f(x) = x^2 + x + 1; [-2, 2]$

4. $f(x) = \dfrac{x}{3+x^2}; [-4, 4]$

5. $f(x) = x - \sqrt{1-x^2}; [-1, 1]$

6. $f(x) = x^{2/3} - x; [-\frac{1}{8}, \frac{1}{8}]$

7. Let

$$f(x) = \begin{cases} x+1 & \text{for } x \le 0 \\ x-1 & \text{for } x > 0 \end{cases}$$

    a. Show that $f(-1) = f(1) = 0$ but that there is no number c in $(-1, 1)$ such that $f'(c) = 0$.

    b. Why does this not contradict Rolle's Theorem?

8. Let f be a function that is continuous on $[0, 2]$ and differentiable on $(0, 2)$. Suppose that $f(0) < f(2)$ but that f is *not* increasing on $[0, 2]$. Use the Mean Value Theorem to prove that f' takes both positive and negative values on $(0, 2)$.

In Exercises 9–12 find all functions satisfying the given conditions.

9. $f'(x) = x^2 - \sin x$

10. $f'(x) = 5, f(0) = -3$

11. $f''(x) = x^2 - 4$

12. $f''(x) = 0, f(0) = 0, f(1) = -1$

In Exercises 13–16 determine the intervals on which f is increasing and those on which f is decreasing.

13. $f(x) = \frac{1}{3}x^3 - x^2 + x - 2$

14. $f(x) = x^{1/3} - x$

15. $f(x) = \sin x - \frac{1}{8}\tan x$

▣ 16. $f(x) = x^4 + x^3 + x^2 + x$

17. Use Theorem 4.8 to prove that $\sqrt{x+3} \ge \sqrt{3} + x/4$ for $0 \le x \le 1$.

*18. Show that $\sin x \ge 2x/\pi$ for $0 \le x \le \pi/2$. (*Hint:* Let $f(x) = \sin x - 2x/\pi$. By examining $f'(x)$, show that there is an x_0 in $[0, \pi/2]$ such that f is increasing on $[0, x_0]$ and decreasing on $[x_0, \pi/2]$.)

In Exercises 19–22 use the First Derivative Test or the Second Derivative Test to determine the relative extreme values of the function.

19. $f(x) = 3x^4 - 10x^3 + 6x^2 + 3$

20. $f(x) = \dfrac{x-1}{x^2+3}$

21. $f(x) = (x+1)^2(x-2)^4$

22. $f(x) = 2\sqrt{x+1} - \sqrt{x-1}$

In Exercises 23–26 determine the intervals on which the graph of f is concave upward and the intervals on which the graph is concave downward.

23. $f(x) = \frac{1}{2}x^4 + x^3 - 6x^2$ 24. $f(x) = \sqrt{x} + 1/x$

25. $f(x) = \dfrac{1}{1+x^4}$ 26. $f(x) = \sin x + \frac{1}{4}\sin 2x$

In Exercises 27–32 sketch the graph of the function, indicating all relevant properties listed in Table 4.1 of Section 4.8.

27. $f(x) = x^3 - 6x - 1$ 28. $f(x) = x^4 + 2x^3 + 1$

29. $f(x) = \dfrac{1}{x^3} - \dfrac{1}{x}$ 30. $f(x) = \dfrac{1}{4}x - \sqrt{x}$

31. $k(x) = \dfrac{x^2 + 1}{x^2 - 4}$ 32. $f(x) = \cos^2 x$

In Exercises 33–34 sketch the graphs of each pair of functions. Shade the regions they enclose.

33. $f(x) = x^2 + 4x$; $g(x) = x - 2$

34. $f(x) = \dfrac{-8}{7 + 5x}$; $g(x) = \dfrac{8}{x^2 - 1}$

In Exercises 35–40 sketch the graph of the equation.

35. $\dfrac{x^2}{4} + \dfrac{y^2}{4} = 1$ 36. $2x^2 + 3y^2 = 6$

37. $\dfrac{y^2}{25} - \dfrac{x^2}{16} = 1$ 38. $x^2 - 9y^2 = 1$

39. $9x^2 - 18x + y^2 + 4y = 3$

40. $3y^2 - 18y - x^2 - 4x = 2$

41. Suppose the distance $D(v)$ a car can travel on one tank of gas at a velocity of v miles per hour is given by

$$D(v) = \frac{\sqrt{3}}{48}(80v^{3/2} - v^{5/2}) \quad \text{for } 0 \le v \le 75$$

What velocity maximizes D (and hence minimizes fuel consumption per mile)?

42. Let $f(x) = x^m(x - 1)^n$, where m and n are integers greater than or equal to 2. Determine the values of m and n for which the relative extreme values of f are $f(0)$, $f(\frac{1}{4})$, and $f(1)$.

43. Find the point on the graph of $y = x^{1/2}$ that is closest to the point $(4, 0)$.

44. A certain type of vitamin capsule is to have the shape of a cylinder of height h with hemispheres of radius r on each end of the cylinder. For a given amount S of surface area, find the value of r that maximizes the volume of the capsule.

45. A toolshed with a square base and a flat roof is to have a volume of 800 cubic feet. If the floor costs $6 per square foot, the roof $2 per square foot, and the sides $5 per square foot, determine the dimensions of the most economical shed.

46. An entrepreneur makes and sells moonshine whiskey in spherical bottles that cost $60\pi r^2$ cents to make, where r is the radius of the bottle in inches. Suppose the revenue on each cubic inch of whiskey is 15 cents and the largest bottle that can be made has a radius of 5 inches.

a. What radius will maximize the profit on each bottle of whiskey?

b. What radius will minimize the profit on each bottle?

47. An airline company offers a round-trip group flight from New York to London. If x people sign up for the flight, the cost of each ticket is to be $1000 - 2x$ dollars. Find the maximum revenue the airline company can receive from the sale of tickets for the flight.

48. A swimming pool whose bottom has area $5000/\pi$ square feet is to be built in the form of a rectangle with semicircles attached at two opposite ends of the rectangle. Give the dimensions of the pool having a minimum perimeter.

49. The time dilation factor in Einstein's Theory of Relativity is a function of the velocity v of a moving object and is given by

$$f(v) = \frac{1}{\sqrt{1 - v^2/c^2}} \quad \text{for } 0 < v < c$$

where c is the velocity of light. Sketch the graph of f.

50. A carpenter wishes to illuminate a certain point P on the floor with a lamp 5 feet away, which may be raised to any level between 0 and 8 feet above the floor (Figure 4.73). How high above the floor should the carpenter raise the lamp so that the intensity of light at P will be a maximum? (*Hint:* The intensity of light at P is proportional to the cosine of the angle θ that the incident light makes with respect to the vertical, and it is inversely proportional to the square of the distance r from P to the light source.)

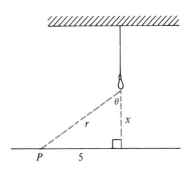

FIGURE 4.73

51. A farmer plans to fence off three rectangular grazing fields along a straight river by erecting one fence parallel to the river and four fences perpendicular to the river. If the total area of the three plots is to be 6400 square feet, what length of fence parallel to the river will minimize the total length of fence required?

52. A small private oil field has a maximum daily yield of 100 barrels. The owner estimates that the daily profit in dollars from a daily production of x barrels is

$$P(x) = 100x - 5x^2 \quad \text{for } 0 \le x \le 100$$

a. What daily profit or loss will result from maximum daily production?
b. What daily production will result in maximum daily profit, and what will that daily profit be?

53. A power company wishes to install a temporary power line. The purchase price of the wire is proportional to the cross-sectional area x of the wire, and the cost due to power loss while the wire is in use is proportional to the reciprocal of x. Thus there are constants a and b such that the total cost $C(x)$ is given by $C(x) = ax + b/x$. Determine the value of the cross-sectional area that minimizes the cost. Show that for such a cross-sectional area the purchase price is equal to the cost due to power loss.

54. During the sixteenth and seventeenth centuries, wine merchants in Linz, Austria, calculated the price of a barrel of wine by first inserting a measuring rod as far as possible into the taphole located in the middle of the side of the barrel (Figure 4.74), and then charging the customer according to the length l of rod inside the barrel. Assuming that the barrel is a circular cylinder, calculate the dimensions of the barrel with largest volume for a given length l of rod.

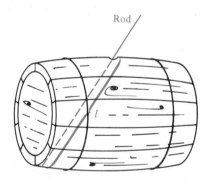

FIGURE 4.74

55. A real estate firm borrows money and then lends the money to its customers at 18 percent interest per year. The amount of money it can borrow is proportional to the square of the interest rate it pays for the money. At what interest rate should the firm borrow money in order to maximize its profit per year from lending?

56. Two houses are being built, one 50 feet from the street and the other 75 feet from the street, as in Figure 4.75. At what point on the edge of the street should a telephone pole be located to minimize the sum of the distances to the two houses?

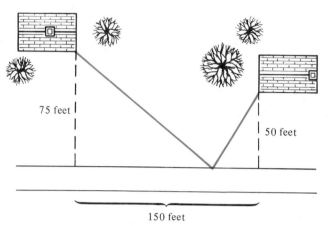

FIGURE 4.75

*57. A cylindrical container with water H feet deep has a spout h feet below the water line. Torricelli's Theorem states that the speed of the water as it flows through the spout is $\sqrt{2gh}$. Determine the value of h that maximizes the distance R as shown in Figure 4.76. Then find the maximum possible value of R. (*Hint:* The distance R equals the product of the velocity $\sqrt{2gh}$ and time t_0, that is, $R = \sqrt{2gh}\, t_0$, where t_0 is the time it takes for a droplet to fall from the spout to the base level of the container. Determine the value of t_0 by using (1) in Section 1.3.)

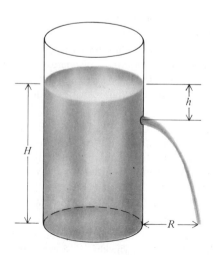

FIGURE 4.76

Cumulative Review, Chapters 1–3

In Exercises 1–4 solve the inequality.

1. $18x(4x^6 - 1) > 0$

2. $||x| - 3| \le 4$

3. $\sqrt{2} \le \sqrt{1 + x^4} < 4$

4. $\sin x \ge 2\cos^2 x - 1$

5. Let $f(x) = \dfrac{1}{\sqrt{x + 3}}$ and $g(x) = \dfrac{1}{2x - 1}$.

   a. Find the domain of $f \circ g$.

   b. Find a formula for $(f \circ g)(x)$.

6. Let $h(x) = 1 + \sin(3x^2 - 2)$. Find functions f and g, neither of which equals h, such that $h = f \circ g$.

In Exercises 7–9 determine whether the limit exists as a number, as ∞, or as $-\infty$. If the limit exists, evaluate it.

7. $\lim\limits_{x \to 4^+} \dfrac{x(x + 4)}{16 - x^2}$

8. $\lim\limits_{x \to -3^-} \dfrac{3|x| + 9}{x^2 - 9}$

9. $\lim\limits_{x \to 0^+} \dfrac{\sin 2x - 2\sqrt{x}\sin x + 4x^2}{x}$

10. Let $f(x) = (x - 4)^2 + 1$. Find the point (a, b) on the graph of f at which the tangent line is perpendicular to the line $y - 2x = 8$.

11. Let

$$f(x) = \begin{cases} (x - 1)^2 & \text{for } x < 1 \\ x^3 - x^2 & \text{for } x \ge 1 \end{cases}$$

   a. Determine whether f is continuous at 1.

   b. Determine whether f is differentiable at 1.

12. Let $f(x) = 2/x^{1/2}$. Use the definition of the derivative to find $f'(4)$.

In Exercises 13–14 find the derivative of f.

13. $f(x) = \dfrac{3x + 1}{\sqrt{x^2 + 2x + 2}}$

14. $f(x) = \sin(\cos(x^3))$

15. Show that

$$\frac{d}{dx}[3(2x + 1)^{5/2} - 5(2x + 1)^{3/2}] = 30x(2x + 1)^{1/2}$$

16. Suppose

$$f(t) = t^2 - \frac{3}{t} + \frac{4}{5}$$

represents the position of an object at time t. Find the acceleration of the object when $t = 2$.

17. Suppose y is a differentiable function such that $xy^2 + 3y - 4x = 17$. Find dy/dx.

18. Suppose a point moves along the right branch of the hyperbola $x^2 - 2y^2 = 1$ in such a way that its y coordinate decreases at the rate of 3 units per second. How fast is its x coordinate increasing when the point is located at $(\sqrt{3}, -1)$?

19. The area of a circular hole in the ground increases at the rate of $2\pi\sqrt{r}$ feet per minute, where r is the radius of the hole in feet. How large is the area of the hole when its radius is increasing at the rate of two feet per minute?

C 20. Let $f(x) = x^4 - x^3 + x - 1$. Use the Newton-Raphson method to approximate a critical point of f until successive approximations are within 10^{-6} of each other.

5

THE INTEGRAL

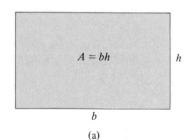

$A = bh$ h

b

(a)

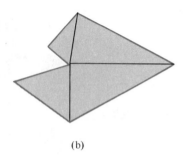

(b)

FIGURE 5.1

Since the time of the ancient Greeks, mathematicians have attempted to calculate areas of plane regions. The most basic plane region is the rectangle, whose area is the product of its base and its height (Figure 5.1(a)). The ancient Greeks used Euclidean geometry to deduce the areas of parallelograms and triangles. They also knew how to compute the area of any polygon by partitioning it into triangles (Figure 5.1(b)).

However, with regions having curved boundaries, such as parabolas, their task was more difficult. Euclidean geometry dealt effectively with lines and planes, but another approach was needed for the areas of other regions. It was Archimedes* (about 287–212 B.C.) who made the first notable advance by his ingenious use of the "method of exhaustion," based on an idea of Eudoxus. With this method, Archimedes found the areas of certain complex regions by inscribing larger and larger polygons of known area in such a region so that it would eventually be "exhausted." The area of the region was then the "limit" of the areas of the inscribed figures. (However, Archimedes had no formal notion of limit.) For example, to find the area of a parabolic region such as R_0 in Figure 5.2(a), Archimedes inscribed an increasing number of triangles, as shown in Figure 5.2(b) and (c). From his calculations he concluded that the area of the region R_0 should be $\frac{4}{3}$ of the area of the single inscribed triangle in Figure 5.2(b).[†] Since it is possible to show that this triangle has area 1, we conclude from Archimedes' result that the area of R_0 should be $\frac{4}{3}$. The definition of area that we will give will also imply that the area of R_0 is $\frac{4}{3}$. The fact that the modern

---

* **Archimedes**: Pronounced "Ar-ki-*mee*-deez."
[†] For further details of his procedure, see Example 5 of Section 9.4.

228

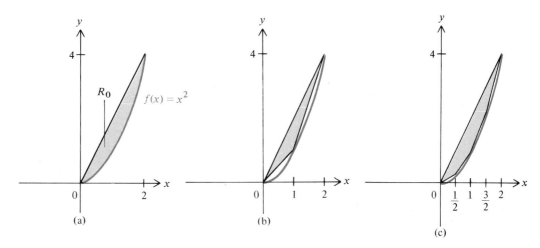

FIGURE 5.2

definition of area stems from Archimedes' method of exhaustion is a tribute to his genius.

In this chapter we will use the problem of computing area to motivate the definition of what we will call the definite integral of a continuous function. Then we will use the definite integral to define the area of a region. Finally, the Fundamental Theorem of Calculus will provide a simple method of computing many definite integrals.

5.1
PREPARATION FOR THE DEFINITE INTEGRAL

As a preview of the method of finding area, we consider the region R_0 shown in Figure 5.2(a). However, it will be more convenient to work with the related region R bounded by the parabola $f(x) = x^2$, the x axis, and the line $x = 2$ (Figure 5.3). Since R and R_0 together comprise a triangle whose area is 4, finding the area of R is equivalent to finding the area of R_0.

Suppose we inscribe rectangles in the region R, as shown in Figure 5.4(a) and (b). Then the sum of the areas of the rectangles is less than the area of R. Similarly, if we circumscribe rectangles about R, as in Figure 5.5(a) and (b), then the sum of the areas of the rectangles is greater than the area of R. Of course, we can find the area of each rectangle as the product of its base and height.

The crucial observation to make about this process is that as the bases of the rectangles become smaller and smaller, the sum of the areas of the rectangles appears to approach the area of R. This suggests that the area of R should be defined as the limit (in a sense to be clarified later) of the sum of the areas of inscribed or circumscribed rectangles. Our definition of area will be based on this idea.

Our assertions thus far about the area of R have rested on three basic properties we expect area to possess:

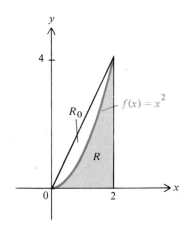

FIGURE 5.3

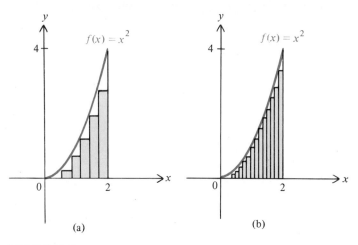

FIGURE 5.4

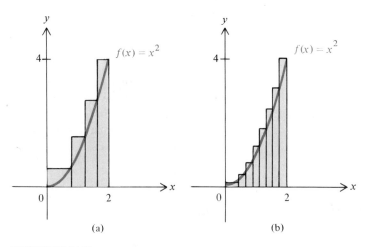

FIGURE 5.5

The Rectangle Property: The area of a rectangle is the product of its base and height.

The Addition Property: The area of a region composed of several smaller regions that overlap in at most a line segment is the sum of the areas of the smaller regions.

The Comparison Property: The area of a region that contains a second region is at least as large as the area of the second region.

You should understand where each of these properties was employed in the preceding discussion. They will play a major role in the definition we will give of area.

Partitions   Now let us consider any region R bounded by the graph of a nonnegative function f that is continuous on an interval $[a, b]$, by the x axis, and by the lines $x = a$ and $x = b$, where $a < b$ (Figure 5.6(a)). Using the three basic properties of area listed above, we set out to define the area of the region R.

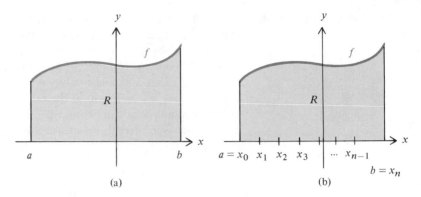

FIGURE 5.6

For any positive integer n we divide $[a, b]$ into subintervals by introducing points of subdivision $x_0, x_1, \ldots, x_n$ (Figure 5.6(b)).

DEFINITION 5.1

> A **partition** of $[a, b]$ is a finite set $\mathscr{P}$ of points $x_0, x_1, \ldots, x_n$ such that $a = x_0 < x_1 < \cdots < x_n = b$. We describe $\mathscr{P}$ by writing
>
> $$\mathscr{P} = \{x_0, x_1, \ldots, x_n\}$$

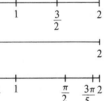

FIGURE 5.7

By definition, any partition of $[a, b]$ must contain both a and b. Except for a and b, the number of points and their placement in $[a, b]$ is arbitrary. For example, $\{0, \frac{1}{2}, 1, \frac{3}{2}, 2\}$ is a partition of $[0, 2]$, as are $\{0, 2\}$ and $\{0, \frac{1}{2}, \pi/4, 1, \pi/2, 3\pi/5, 2\}$ (Figure 5.7). However, $\{\frac{1}{4}, \frac{2}{5}, \frac{1}{2}, \frac{7}{8}, 2\}$ is not a partition of $[0, 2]$ because it does not include 0.

The n subintervals into which a partition $\mathscr{P} = \{x_0, x_1, \ldots, x_n\}$ divides $[a, b]$ are $[x_0, x_1], [x_1, x_2], \ldots, [x_{n-1}, x_n]$. Their lengths are $x_1 - x_0, x_2 - x_1, \ldots, x_n - x_{n-1}$, respectively. We denote the length $x_k - x_{k-1}$ of the kth subinterval $[x_{k-1}, x_k]$ by Δx_k. Thus

$$\Delta x_k = x_k - x_{k-1}$$

In particular, for the partition $\{0, \frac{1}{4}, \frac{1}{2}, 1, 2\}$ of $[0, 2]$, we have

$$x_0 = 0, \qquad x_1 = \tfrac{1}{4}, \qquad x_2 = \tfrac{1}{2}, \qquad x_3 = 1, \qquad x_4 = 2$$

and

$$\Delta x_1 = \tfrac{1}{4} - 0 = \tfrac{1}{4}, \qquad \Delta x_2 = \tfrac{1}{2} - \tfrac{1}{4} = \tfrac{1}{4},$$
$$\Delta x_3 = 1 - \tfrac{1}{2} = \tfrac{1}{2}, \qquad \Delta x_4 = 2 - 1 = 1$$

The length $b - a$ of $[a, b]$ can be written in terms of the lengths $\Delta x_1, \Delta x_2, \ldots, \Delta x_n$ of the subintervals:

$$b - a = \Delta x_1 + \Delta x_2 + \cdots + \Delta x_n$$

Lower and Upper Sums

Having chosen a partition $\mathscr{P}$ of $[a, b]$, we inscribe over each subinterval derived from $\mathscr{P}$ the largest rectangle that lies inside the region R, as was done in Figure 5.4(a) and (b). Since we are assuming that f is continuous on $[a, b]$, we know from the Maximum–Minimum Theorem (Section 4.1) that for each k between 1 and n there exists a smallest value m_k of f on the kth subinterval $[x_{k-1}, x_k]$. If we choose m_k as the height of the kth rectangle R_k, then R_k will be the largest (tallest) rectangle that can be inscribed in R over $[x_{k-1}, x_k]$ (Figure 5.8). Doing this for each subinterval, we create n inscribed rectangles $R_1, R_2, \ldots, R_n$, all lying inside the region R.

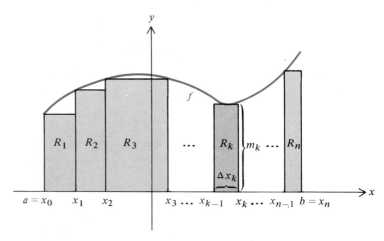

FIGURE 5.8

For each k between 1 and n the rectangle R_k has base $[x_{k-1}, x_k]$ with length Δx_k and has height m_k. Hence the area of R_k is the product $m_k \Delta x_k$. Just as in the case of the parabolic region in Figure 5.4(a) and (b), the sum

$$m_1 \Delta x_1 + m_2 \Delta x_2 + \cdots + m_n \Delta x_n$$

of the areas of all the rectangles should be no larger than the area of R. We denote this sum $L_f(\mathscr{P})$ and call it the **lower sum** of f associated with the partition $\mathscr{P}$. Thus

$$L_f(\mathscr{P}) = m_1 \Delta x_1 + m_2 \Delta x_2 + \cdots + m_n \Delta x_n \tag{1}$$

No matter how we define the area of R, this area must be *at least as large* as the lower sum $L_f(\mathscr{P})$ associated with *any* partition $\mathscr{P}$ of $[a, b]$.

Example 1   Let $f(x) = x^2$ for $0 \le x \le 2$. Find $L_f(\mathscr{P})$ and $L_f(\mathscr{P}')$ for the partitions

$$\mathscr{P} = \{0, \tfrac{1}{2}, 1, \tfrac{3}{2}, 2\} \quad \text{and} \quad \mathscr{P}' = \{0, \tfrac{1}{4}, \tfrac{1}{2}, 1, \tfrac{3}{2}, 2\}$$

Solution   The subintervals associated with the partition $\mathscr{P}$ are $[0, \tfrac{1}{2}]$, $[\tfrac{1}{2}, 1]$, $[1, \tfrac{3}{2}]$, and $[\tfrac{3}{2}, 2]$. Computing the minimum value of f on each of these subintervals, we find that

$$m_1 = f(0) = 0, \qquad m_2 = f(\tfrac{1}{2}) = \tfrac{1}{4}, \qquad m_3 = f(1) = 1, \qquad m_4 = f(\tfrac{3}{2}) = \tfrac{9}{4}$$

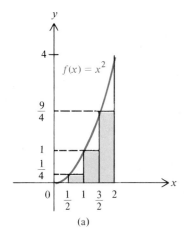

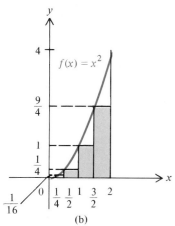

FIGURE 5.9

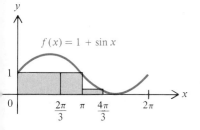

FIGURE 5.10

Moreover,

$$\Delta x_1 = \tfrac{1}{2} - 0 = \tfrac{1}{2}, \qquad \Delta x_2 = 1 - \tfrac{1}{2} = \tfrac{1}{2},$$
$$\Delta x_3 = \tfrac{3}{2} - 1 = \tfrac{1}{2}, \qquad \Delta x_4 = 2 - \tfrac{3}{2} = \tfrac{1}{2}$$

(Figure 5.9(a)). Therefore by (1),

$$L_f(\mathscr{P}) = 0 \cdot \tfrac{1}{2} + \tfrac{1}{4} \cdot \tfrac{1}{2} + 1 \cdot \tfrac{1}{2} + \tfrac{9}{4} \cdot \tfrac{1}{2} = \tfrac{7}{4}$$

The subintervals associated with the partition $\mathscr{P}'$ are $[0, \tfrac{1}{4}]$, $[\tfrac{1}{4}, \tfrac{1}{2}]$, $[\tfrac{1}{2}, 1]$, $[1, \tfrac{3}{2}]$, and $[\tfrac{3}{2}, 2]$. Computing the minimum value of f on each of these subintervals, we find that

$$m_1 = f(0) = 0, \qquad m_2 = f(\tfrac{1}{4}) = \tfrac{1}{16}, \qquad m_3 = f(\tfrac{1}{2}) = \tfrac{1}{4},$$
$$m_4 = f(1) = 1, \qquad m_5 = f(\tfrac{3}{2}) = \tfrac{9}{4}$$

Moreover,

$$\Delta x_1 = \tfrac{1}{4} - 0 = \tfrac{1}{4}, \qquad \Delta x_2 = \tfrac{1}{2} - \tfrac{1}{4} = \tfrac{1}{4}, \qquad \Delta x_3 = 1 - \tfrac{1}{2} = \tfrac{1}{2},$$
$$\Delta x_4 = \tfrac{3}{2} - 1 = \tfrac{1}{2}, \qquad \Delta x_5 = 2 - \tfrac{3}{2} = \tfrac{1}{2}$$

(Figure 5.9(b)). Therefore by (1),

$$L_f(\mathscr{P}') = 0 \cdot \tfrac{1}{4} + \tfrac{1}{16} \cdot \tfrac{1}{4} + \tfrac{1}{4} \cdot \tfrac{1}{2} + 1 \cdot \tfrac{1}{2} + \tfrac{9}{4} \cdot \tfrac{1}{2} = \tfrac{113}{64} \quad \square$$

Example 2 Let $f(x) = 1 + \sin x$ for $0 \le x \le 2\pi$. Find $L_f(\mathscr{P})$ for the partition

$$\mathscr{P} = \left\{0, \frac{2\pi}{3}, \pi, \frac{4\pi}{3}, 2\pi\right\}$$

Solution Notice that the graph of f is the graph of $\sin x$ shifted vertically upward 1 unit (Figure 5.10). Calculating the minimum value of f on each of the subintervals $[0, 2\pi/3]$, $[2\pi/3, \pi]$, $[\pi, 4\pi/3]$, and $[4\pi/3, 2\pi]$ associated with $\mathscr{P}$, we find that

$$m_1 = f(0) = 1, \qquad m_2 = f(\pi) = 1,$$
$$m_3 = f\left(\frac{4\pi}{3}\right) = 1 - \frac{\sqrt{3}}{2}, \qquad m_4 = f\left(\frac{3\pi}{2}\right) = 0$$

Moreover,

$$\Delta x_1 = \frac{2\pi}{3}, \qquad \Delta x_2 = \frac{\pi}{3}, \qquad \Delta x_3 = \frac{\pi}{3}, \qquad \Delta x_4 = \frac{2\pi}{3}$$

From (1) it now follows that

$$L_f(\mathscr{P}) = m_1 \Delta x_1 + m_2 \Delta x_2 + m_3 \Delta x_3 + m_4 \Delta x_4$$
$$= 1\left(\frac{2\pi}{3}\right) + 1\left(\frac{\pi}{3}\right) + \left(1 - \frac{\sqrt{3}}{2}\right)\left(\frac{\pi}{3}\right) + 0\left(\frac{2\pi}{3}\right)$$
$$= \frac{4\pi}{3} - \frac{\pi\sqrt{3}}{6} \quad \square$$

Caution: Observe that m_k may be the value of f at the left endpoint of $[x_{k-1}, x_k]$ (as is m_1 in Example 2), the right endpoint (as are m_2 and m_3 in Example 2), or a point somewhere inside $[x_{k-1}, x_k]$ (as is m_4 in Example 2). The common feature is that each is the minimum value of f on the corresponding subinterval of $[0, 2\pi]$.

By a procedure similar to the one that involves inscribing rectangles to compute a lower sum, we can also circumscribe rectangles and compute an upper sum. Let

$$\mathscr{P} = \{x_0, x_1, \ldots x_n\}$$

be a given partition of $[a, b]$, and let f be continuous and nonnegative on $[a, b]$. Then the Maximum–Minimum Theorem implies that for each k between 1 and n there exists a largest value M_k of f on the kth subinterval $[x_{k-1}, x_k]$ (Figure 5.11). Consequently if we let M_k be the height of the kth rectangle R_k, then R_k will be the smallest possible rectangle circumscribing the appropriate portion of R (Figure 5.11). The area of R_k is $M_k \Delta x_k$, and the sum

$$M_1 \Delta x_1 + M_2 \Delta x_2 + \cdots + M_n \Delta x_n$$

of the areas of the circumscribed rectangles should be no smaller than the area of R. We denote this sum $U_f(\mathscr{P})$ and call it the **upper sum** of f associated with the partition $\mathscr{P}$. Thus

$$U_f(\mathscr{P}) = M_1 \Delta x_1 + M_2 \Delta x_2 + \cdots + M_n \Delta x_n \qquad (2)$$

No matter how we define the area of R, this area must be *no larger* than $U_f(\mathscr{P})$ for *any* partition $\mathscr{P}$ of $[a, b]$.

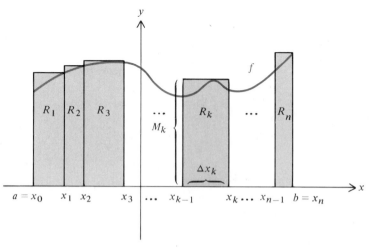

FIGURE 5.11

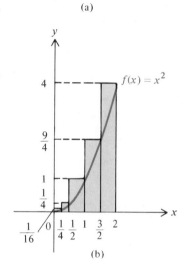

(a)

(b)

FIGURE 5.12

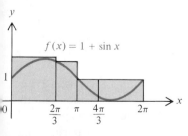

FIGURE 5.13

Example 3 Let $f(x) = x^2$ for $0 \le x \le 2$. Find $U_f(\mathscr{P})$ and $U_f(\mathscr{P}')$ for the partitions

$$\mathscr{P} = \{0, \tfrac{1}{2}, 1, \tfrac{3}{2}, 2\} \quad \text{and} \quad \mathscr{P}' = \{0, \tfrac{1}{4}, \tfrac{1}{2}, 1, \tfrac{3}{2}, 2\}$$

Solution For the partition $\mathscr{P}$ we have

$$M_1 = f(\tfrac{1}{2}) = \tfrac{1}{4}, \qquad M_2 = f(1) = 1, \qquad M_3 = f(\tfrac{3}{2}) = \tfrac{9}{4}, \qquad M_4 = f(2) = 4$$

Since $\mathscr{P}$ is as in Example 1, we know that

$$\Delta x_1 = \Delta x_2 = \Delta x_3 = \Delta x_4 = \tfrac{1}{2}$$

(Figure 5.12(a)). Thus by (2),

$$U_f(\mathscr{P}) = \tfrac{1}{4} \cdot \tfrac{1}{2} + 1 \cdot \tfrac{1}{2} + \tfrac{9}{4} \cdot \tfrac{1}{2} + 4 \cdot \tfrac{1}{2} = \tfrac{15}{4}$$

To calculate $U_f(\mathscr{P}')$, we first find that

$$M_1 = f(\tfrac{1}{4}) = \tfrac{1}{16}, \qquad M_2 = f(\tfrac{1}{2}) = \tfrac{1}{4}, \qquad M_3 = f(1) = 1,$$
$$M_4 = f(\tfrac{3}{2}) = \tfrac{9}{4}, \qquad M_5 = f(2) = 4$$

As in Example 1, for $\mathscr{P}'$ we find that

$$\Delta x_1 = \tfrac{1}{4}, \qquad \Delta x_2 = \tfrac{1}{4}, \qquad \Delta x_3 = \tfrac{1}{2}, \qquad \Delta x_4 = \tfrac{1}{2}, \qquad \Delta x_5 = \tfrac{1}{2}$$

(Figure 5.12(b)). Therefore by (2),

$$U_f(\mathscr{P}') = \tfrac{1}{16} \cdot \tfrac{1}{4} + \tfrac{1}{4} \cdot \tfrac{1}{4} + 1 \cdot \tfrac{1}{2} + \tfrac{9}{4} \cdot \tfrac{1}{2} + 4 \cdot \tfrac{1}{2} = \tfrac{237}{64} \quad \square$$

Example 4 Let $f(x) = 1 + \sin x$ for $0 \le x \le 2\pi$. Find $U_f(\mathscr{P})$ for the partition

$$\mathscr{P} = \left\{0, \frac{2\pi}{3}, \pi, \frac{4\pi}{3}, 2\pi\right\}$$

Solution The subintervals associated with $\mathscr{P}$ were obtained in Example 2. Calculating the maximum value of f on each of these subintervals, we have

$$M_1 = f\left(\frac{\pi}{2}\right) = 2, \qquad M_2 = f\left(\frac{2\pi}{3}\right) = 1 + \frac{\sqrt{3}}{2}$$
$$M_3 = f(\pi) = 1, \qquad M_4 = f(2\pi) = 1$$

(Figure 5.13). Since

$$\Delta x_1 = \frac{2\pi}{3}, \qquad \Delta x_2 = \frac{\pi}{3}, \qquad \Delta x_3 = \frac{\pi}{3}, \qquad \Delta x_4 = \frac{2\pi}{3}$$

we conclude by (2) that

$$U_f(\mathscr{P}) = 2\left(\frac{2\pi}{3}\right) + \left(1 + \frac{\sqrt{3}}{2}\right)\left(\frac{\pi}{3}\right) + 1\left(\frac{\pi}{3}\right) + 1\left(\frac{2\pi}{3}\right)$$
$$= \frac{8\pi}{3} + \frac{\pi\sqrt{3}}{6} \quad \square$$

If f is any continuous nonnegative function on $[a, b]$, and if $\mathscr{P} = \{x_0, x_1, x_2, \ldots, x_n\}$ is an arbitrary partition of $[a, b]$, then m_k and M_k are the minimum and maximum values of f on $[x_{k-1}, x_k]$ for $k = 1, 2, \ldots, n$, so that

$$m_k \leq M_k$$

Therefore

$$L_f(\mathscr{P}) \leq U_f(\mathscr{P}) \tag{3}$$

Moreover, the way we have defined $L_f(\mathscr{P})$ and $U_f(\mathscr{P})$ implies that no matter what partition $\mathscr{P}$ we choose, the area of R should be a number between $L_f(\mathscr{P})$ and $U_f(\mathscr{P})$. That is,

$$L_f(\mathscr{P}) \leq \text{area of } R \leq U_f(\mathscr{P}) \tag{4}$$

Normally $L_f(\mathscr{P})$ and $U_f(\mathscr{P})$ will be quite different numbers. For example, Examples 1 and 3 show that if $f(x) = x^2$ on $[0, 2]$ and $\mathscr{P} = \{0, \frac{1}{2}, 1, \frac{3}{2}, 2\}$, then

$$L_f(\mathscr{P}) = \tfrac{7}{4} \quad \text{and} \quad U_f(\mathscr{P}) = \tfrac{15}{4}$$

Hence we usually cannot determine a unique value for the area of R by using (4) with only one partition $\mathscr{P}$. Instead we must use the lower sums and the upper sums of many different partitions. Fortunately, a simple relationship that holds between certain lower sums and certain upper sums simplifies the task.

In order to see how lower and upper sums are related to each other, we insert one extra point c into $\mathscr{P}$, where c may be any point in $[a, b]$ not already in $\mathscr{P}$. For example, if we insert the point $c = \frac{1}{4}$ into the partition $\mathscr{P} = \{0, \frac{1}{2}, 1, \frac{3}{2}, 2\}$, we obtain the partition $\mathscr{P}' = \{0, \frac{1}{4}, \frac{1}{2}, 1, \frac{3}{2}, 2\}$. Since $\frac{7}{4} < \frac{113}{64}$ and $\frac{237}{64} < \frac{15}{4}$, our calculations in Examples 1 and 3 show that

$$L_f(\mathscr{P}') > L_f(\mathscr{P}) \quad \text{and} \quad U_f(\mathscr{P}') < U_f(\mathscr{P})$$

Thus inserting the additional point c into $\mathscr{P}$ increased the lower sum and decreased the upper sum. As we will show next, this is what usually happens. Suppose c lies in the subinterval $[x_{k-1}, x_k]$ determined by $\mathscr{P}$ (Figure 5.14(a)). Then the effect of inserting the point c is to replace the inscribed rectangle over $[x_{k-1}, x_k]$ by two inscribed rectangles, one over $[x_{k-1}, c]$ and the other over $[c, x_k]$ (Figure 5.14(a)). The illustration shows that the new inscribed rectangles together contain the original rectangle. Consequently their combined area is at least as large as the area of the original rectangle inscribed over $[x_{k-1}, x_k]$. We conclude that inserting the point c into the partition $\mathscr{P}$ causes the lower sum to become larger (or possibly to remain the same). A similar argument shows that the upper sum $U_f(\mathscr{P})$ becomes smaller (or possibly remains the same) when the point c is inserted into $\mathscr{P}$ (Figure 5.14(b)).

Inserting several points into the partition $\mathscr{P}$ has the same effect as inserting them one at a time. Thus we can conclude from the foregoing discussion that the lower sum tends to increase and the upper sum tends to decrease as points are inserted into $\mathscr{P}$. Combining this with (3), we find that if $\mathscr{P}$ and $\mathscr{P}'$ are partitions of $[a, b]$ such that $\mathscr{P}'$ contains all the points of $\mathscr{P}$ and possibly other points in $[a, b]$,

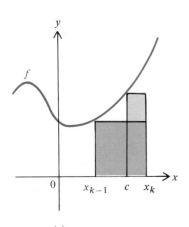

(a)

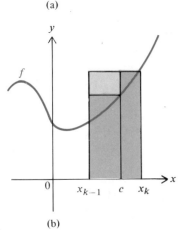

(b)

FIGURE 5.14

then

$$L_f(\mathscr{P}) \le \overset{(3)}{L_f(\mathscr{P}')} \le U_f(\mathscr{P}') \le U_f(\mathscr{P}) \tag{5}$$

Thus the more points we insert into a partition $\mathscr{P}$, the closer the numbers $L_f(\mathscr{P})$ and $U_f(\mathscr{P})$ are to one another—and, by (4), the more restricted we are in choosing a number to serve as the area of R. In the Appendix we prove that the requirement that $L_f(\mathscr{P}) \le$ area of $R \le U_f(\mathscr{P})$ for all partitions $\mathscr{P}$ of $[a, b]$ limits the area of R to exactly one number, and thus we will define the area of R to be this number.

In Section 5.2 we will remove the requirement that f be nonnegative on $[a, b]$. This will lay the foundation for the definition of the definite integral and for a formal definition of area.

EXERCISES 5.1

In Exercises 1–12 compute $L_f(\mathscr{P})$ and $U_f(\mathscr{P})$.

1. $f(x) = 7$; $\mathscr{P} = \{-3, -\frac{5}{2}, -\frac{3}{2}, 0\}$

2. $f(x) = x + 2$; $\mathscr{P} = \{-1, 1, 2\}$

3. $f(x) = x + 2$; $\mathscr{P} = \{-1, -\frac{1}{2}, 0, \frac{1}{2}, 1, \frac{3}{2}, 2\}$

4. $f(x) = x^2$; $\mathscr{P} = \{-1, -\frac{1}{2}, 0, \frac{1}{2}, 1, \frac{3}{2}, 2\}$

5. $f(x) = x^4$; $\mathscr{P} = \{-1, -\frac{1}{2}, 0, \frac{1}{2}, 1, \frac{3}{2}, 2\}$

6. $f(x) = -\dfrac{1}{x}$; $\mathscr{P} = \{-4, -3, -2, -1\}$

7. $f(x) = \sin x$; $\mathscr{P} = \{0, \pi/4, \pi/2\}$

8. $f(x) = \sin x$; $\mathscr{P} = \{0, \pi/6, \pi/4, \pi/3, \pi/2\}$

9. $f(x) = \cos x$; $\mathscr{P} = \{-\pi/3, -\pi/6, 0, \pi/6, \pi/3\}$

10. $f(x) = \sin x + \cos x$; $\mathscr{P} = \{0, \pi/4, \pi/2\}$

11. $f(x) = x^3 + 3x + 3$; $\mathscr{P} = \{0, 1, 2\}$

12. $f(x) = \sin(x + \pi/3) + 1$; $\mathscr{P} = \{0, \pi/3, 2\pi/3\}$

In Exercises 13–16 use a calculator to approximate $L_f(\mathscr{P})$ and $U_f(\mathscr{P})$.

13. $f(x) = \sqrt{x}$; $\mathscr{P} = \{0, \frac{1}{9}, \frac{2}{9}, \ldots, \frac{8}{9}, 1\}$

14. $f(x) = \sqrt{1 - x^2}$; $\mathscr{P} = \{-1, -\frac{3}{4}, -\frac{1}{2}, -\frac{1}{4}, 0, \frac{1}{4}, \frac{1}{2}, \frac{3}{4}, 1\}$

15. $f(x) = \sqrt{1 + x^4}$; $\mathscr{P} = \{-1, -\frac{2}{3}, -\frac{1}{2}, 0, \frac{1}{4}, \frac{1}{2}, \frac{2}{3}, 1\}$

16. $f(x) = x \sin x$; $\mathscr{P} = \left\{0, \dfrac{\pi}{6}, \dfrac{\pi}{4}, \dfrac{\pi}{3}, \dfrac{\pi}{2}\right\}$

In Exercises 17–21 compute $L_f(\mathscr{P})$, $L_f(\mathscr{P}')$, $U_f(\mathscr{P})$, and $U_f(\mathscr{P}')$. Do your answers corroborate formula (5) of this section?

17. $f(x) = 1 + x$; $\mathscr{P} = \{-1, 0, 2\}$; $\mathscr{P}' = \{-1, 0, 1, 2\}$

18. $f(x) = x^2$; $\mathscr{P} = \{0, \frac{1}{2}, 1\}$; $\mathscr{P}' = \{0, \frac{1}{4}, \frac{1}{2}, \frac{3}{4}, 1\}$

19. $f(x) = \sin x$; $\mathscr{P} = \{0, \pi/2, \pi\}$; $\mathscr{P}' = \{0, \pi/4, \pi/2, 3\pi/4, \pi\}$

20. $f(x) = \cos x$; $\mathscr{P} = \{0, \pi/3, \pi/2\}$; $\mathscr{P}' = \{0, \pi/6, \pi/3, \pi/2\}$

21. $f(x) = x + \sin x$; $\mathscr{P} = \{0, \pi/2, \pi\}$; $\mathscr{P}' = \{0, \pi/4, \pi/2, 3\pi/4, \pi\}$

22. Find lower and upper sums for the area of the semicircle defined by $y = \sqrt{4 - x^2}$, with the following partitions:

   a. $\mathscr{P} = \{-2, 0, 2\}$
   b. $\mathscr{P}' = \{-2, 0, 1, 2\}$
   c. $\mathscr{P}'' = \{-2, -1, 0, 1, 2\}$

23. Let $f(x) = 1/x^2$ for $1 \le x \le 3$, and let $\mathscr{P} = \{1, \frac{3}{2}, 2, \frac{5}{2}, 3\}$. By computing $L_f(\mathscr{P})$, approximate the area of the region between the graph of f and the x axis from $x = 1$ to $x = 3$.

24. Why is it impossible to find a function f and a partition $\mathscr{P}$ such that $L_f(\mathscr{P}) = 3$ and $U_f(\mathscr{P}) = 2$?

25. Let $f(x) = x$ for $0 \le x \le 7$. Consider the partitions $\mathscr{P} = \{0, 1, 6, 7\}$, $\mathscr{P}' = \{0, 4, 7\}$, and $\mathscr{P}'' = \{0, (7 - \sqrt{5})/2, 7\}$.

   a. Show that $L_f(\mathscr{P}) \ne L_f(\mathscr{P}')$.
   b. Show that $L_f(\mathscr{P}) = L_f(\mathscr{P}'')$. (It is unusual for different partitions to give rise to equal lower sums.)

26. Let $f(x) = x$ for $0 \le x \le 7$, and let $\mathscr{P} = \{0, c, 7\}$, where $0 < c < 7$. Show that $L_f(\mathscr{P}) \ne U_f(\mathscr{P})$.

27. Let $\mathscr{P}$ be a partition of $[0, 1]$, and let

$$f(x) = \begin{cases} 1 & \text{for } x = 0 \\ \dfrac{1}{x} & \text{for } 0 < x \le 1 \end{cases}$$

Try to find $U_f(\mathscr{P})$. What difficulty do you encounter? Does the same problem arise in trying to find $L_f(\mathscr{P})$?

28. Assume that f and g are continuous and nonnegative on $[a, b]$ and that $f(x) \le g(x)$ for $a \le x \le b$. Show that for

any partition $\mathscr{P}$ of $[a, b]$ the inequalities

$$L_f(\mathscr{P}) \le L_g(\mathscr{P}) \quad \text{and} \quad U_f(\mathscr{P}) \le U_g(\mathscr{P})$$

hold.

29. Let f be continuous and nonnegative on $[a, b]$, and suppose there is a partition $\mathscr{P}$ such that $L_f(\mathscr{P}) = U_f(\mathscr{P})$. Show that f is a constant function. (*Hint:* Sketch the graph of f with the inscribed and circumscribed rectangles for $\mathscr{P}$.)

30. Suppose that f is increasing on $[a, b]$ and $\mathscr{P} = \{x_0, x_1, x_2, \ldots, x_n\}$ is a partition of $[a, b]$ such that

$$\Delta x_k = \frac{b - a}{n} \quad \text{for } 1 \le k \le n$$

Show that

$$U_f(\mathscr{P}) - L_f(\mathscr{P}) = [f(b) - f(a)]\left(\frac{b - a}{n}\right)$$

5.2
THE DEFINITE INTEGRAL

We now drop the assumption made in Section 5.1 that f is nonnegative on $[a, b]$. We assume only that f is continuous on $[a, b]$, and for the present that $a < b$. We still define the lower and upper sums of f for a partition $\mathscr{P}$ of $[a, b]$ by

$$L_f(\mathscr{P}) = m_1 \Delta x_1 + m_2 \Delta x_2 + \cdots + m_n \Delta x_n$$

and

$$U_f(\mathscr{P}) = M_1 \Delta x_1 + M_2 \Delta x_2 + \cdots + M_n \Delta x_n$$

where for any integer k between 1 and n, m_k and M_k are the minimum and maximum values of f on the kth subinterval. As in Section 5.1, we can prove that

$$L_f(\mathscr{P}) \le L_f(\mathscr{P}') \le U_f(\mathscr{P}') \le U_f(\mathscr{P})$$

for any two partitions $\mathscr{P}$ and $\mathscr{P}'$ of $[a, b]$ such that $\mathscr{P}'$ contains all the points of $\mathscr{P}$. In the Appendix we prove that there is a *unique* number greater than or equal to all the lower sums and less than or equal to all the upper sums. This unique number is called the definite integral of f from a to b.

DEFINITION 5.2

Let f be continuous on $[a, b]$. The **definite integral of f from a to b** is the unique number I satisfying

$$L_f(\mathscr{P}) \le I \le U_f(\mathscr{P})$$

for every partition $\mathscr{P}$ of $[a, b]$. This integral is denoted by

$$\int_a^b f(x)\, dx$$

The symbol $\int$ is called an **integral sign**, the numbers a and b are called the **limits of integration**, and the function f appearing in the integral is called the **integrand**.

The definite integral $\int_a^b f(x)\,dx$ is a number depending only on f, a, and b. The variable x appearing in the integral is a "dummy variable"; it may be replaced by any other variable, such as t or u, not already in use. This means that

$$\int_a^b f(x)\,dx = \int_a^b f(t)\,dt = \int_a^b f(u)\,du$$

For the specific case in which $f(x) = x^2$, $a = 0$, and $b = 3$, we have

$$\int_0^3 x^2\,dx = \int_0^3 t^2\,dt = \int_0^3 u^2\,du$$

Caution: We emphasize that the expression dx has no independent meaning in $\int_a^b f(x)\,dx$, although it arose originally from the differential. This expression will play a role later when we develop special methods for computing definite integrals.

We now formally define area in terms of the definite integral.

DEFINITION 5.3

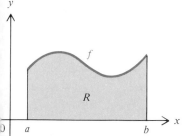

The region between the graph of f and the x axis on $[a, b]$

FIGURE 5.15

Let f be continuous and nonnegative on $[a, b]$, and let R be the region bounded above by the graph of f, below by the x axis, and on the left and right by the lines $x = a$ and $x = b$. Then R is called **the region between the graph of f and the x axis on $[a, b]$** (see Figure 5.15), and the **area** of R is defined to be

$$\int_a^b f(x)\,dx$$

It is sometimes possible to calculate $\int_a^b f(x)\,dx$ from the definition of the integral by calculating formulas for the lower and upper sums, as we do in Examples 1 and 2.

Example 1   Let $f(x) = c$ for $a \le x \le b$. Show that $\int_a^b c\,dx = c(b - a)$.

Solution   Because f assumes only the value c, it follows that for any partition $\mathscr{P} = \{x_0, x_1, \ldots, x_n\}$ of $[a, b]$ and for any k between 1 and n, we have $m_k = c = M_k$. Consequently

$$L_f(\mathscr{P}) = U_f(\mathscr{P}) = c\,\Delta x_1 + c\,\Delta x_2 + \cdots + c\,\Delta x_n$$
$$= c(\Delta x_1 + \Delta x_2 + \cdots + \Delta x_n) = c(b - a)$$

Therefore by Definition 5.2, $\int_a^b c\,dx = c(b - a)$. $\square$

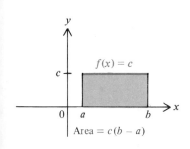

Area $= c(b - a)$

FIGURE 5.16

If $c \ge 0$, the area of the corresponding region is also $c(b - a)$, by Definition 5.3 (see Figure 5.16).

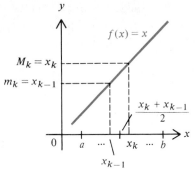

$M_k = x_k$

$m_k = x_{k-1}$

$f(x) = x$

$\dfrac{x_k + x_{k-1}}{2}$

$0 \quad a \quad \cdots \quad x_k \quad \cdots \quad b$

x_{k-1}

FIGURE 5.17

Example 2   Show that

$$\int_a^b x \, dx = \frac{1}{2}(b^2 - a^2)$$

Solution   Let $\mathscr{P} = \{x_0, x_1, \ldots, x_n\}$ be any partition of $[a, b]$. Then for any k between 1 and n we have

$$m_k = x_{k-1} \quad \text{and} \quad M_k = x_k$$

and

$$x_{k-1} < \frac{1}{2}(x_k + x_{k-1}) < x_k$$

(Figure 5.17). Using this information, we find that

$$
\begin{aligned}
L_f(\mathscr{P}) &= m_1 \Delta x_1 + m_2 \Delta x_2 + \cdots + m_n \Delta x_n \\
&= x_0(x_1 - x_0) + x_1(x_2 - x_1) + \cdots + x_{n-1}(x_n - x_{n-1}) \\
&< \frac{1}{2}(x_1 + x_0)(x_1 - x_0) + \frac{1}{2}(x_2 + x_1)(x_2 - x_1) + \cdots \\
&\quad + \frac{1}{2}(x_n + x_{n-1})(x_n - x_{n-1}) \\
&= \frac{1}{2}(x_1^2 - x_0^2) + \frac{1}{2}(x_2^2 - x_1^2) + \cdots + \frac{1}{2}(x_n^2 - x_{n-1}^2) \\
&= \frac{1}{2}(x_n^2 - x_0^2) \\
&= \frac{1}{2}(b^2 - a^2)
\end{aligned}
$$

and that

$$
\begin{aligned}
U_f(\mathscr{P}) &= M_1 \Delta x_1 + M_2 \Delta x_2 + \cdots + M_n \Delta x_n \\
&= x_1(x_1 - x_0) + x_2(x_2 - x_1) + \cdots + x_n(x_n - x_{n-1}) \\
&> \frac{1}{2}(x_1 + x_0)(x_1 - x_0) + \frac{1}{2}(x_2 + x_1)(x_2 - x_1) + \cdots \\
&\quad + \frac{1}{2}(x_n + x_{n-1})(x_n - x_{n-1}) \\
&= \frac{1}{2}(x_1^2 - x_0^2) + \frac{1}{2}(x_2^2 - x_1^2) + \cdots + \frac{1}{2}(x_n^2 - x_{n-1}^2) \\
&= \frac{1}{2}(x_n^2 - x_0^2) \\
&= \frac{1}{2}(b^2 - a^2)
\end{aligned}
$$

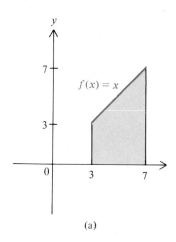

(a)

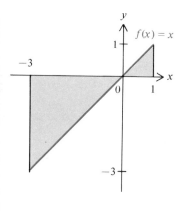

(b)

FIGURE 5.18

From these calculations we conclude that

$$L_f(\mathscr{P}) \leq \frac{1}{2}(b^2 - a^2) \leq U_f(\mathscr{P})$$

for any partition $\mathscr{P}$ of $[a, b]$. Therefore Definition 5.2 tells us that

$$\int_a^b x \, dx = \frac{1}{2}(b^2 - a^2) \quad \square$$

In particular, if $a = 3$ and $b = 7$, we have

$$\int_3^7 x \, dx = \frac{1}{2}(7^2 - 3^2) = 20$$

From this and Definition 5.3 it follows that the area of the trapezoidal region shown in Figure 5.18(a) is 20. In contrast, if $a = -3$ and $b = 1$, we have

$$\int_{-3}^1 x \, dx = \frac{1}{2}[1^2 - (-3)^2] = -4$$

which clearly is not the area of the region shaded in Figure 5.18(b), because area is never negative.

Only slight modifications are necessary in order to prove that

$$\int_a^b -x \, dx = -\frac{1}{2}(b^2 - a^2) = -\int_a^b x \, dx \tag{1}$$

This will also follow directly from Theorem 5.18 in Section 5.5.

One can use much the same general technique as in the solution of Example 2 to show that

$$\int_a^b x^2 \, dx = \frac{1}{3}(b^3 - a^3) \tag{2}$$

(see Exercise 53). In particular, if $a = 0$ and $b = 2$, then (2) tells us that

$$\int_0^2 x^2 \, dx = \frac{1}{3}(2^3 - 0^3) = \frac{8}{3}$$

This implies that the area of the parabolic region R in Figure 5.3, reproduced here as Figure 5.19, is $\frac{8}{3}$. Since the area of triangle composed of R and R_0 is 4, it follows that the area of the region R_0 is $4 - \frac{8}{3} = \frac{4}{3}$. This is compatible with the results obtained by Archimedes more than 2000 years ago.

Evaluating integrals such as $\int_a^b x^2 \, dx$ by using lower sums and upper sums is tedious. To find an integral such as $\int_0^{\pi/2} \cos x \, dx$ by this method would be even more difficult. Fortunately, we will derive a method (in Section 5.4) for computing definite integrals that does not involve lower and upper sums.

In Definition 5.2, which defined $\int_a^b f(x) \, dx$, we assumed explicitly that $a < b$. However, for theoretical purposes and for later applications it will be convenient to give meaning to $\int_a^a f(x) \, dx$ and to $\int_b^a f(x) \, dx$ when $a < b$.

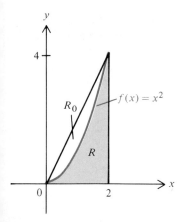

FIGURE 5.19

DEFINITION 5.4

Let f be continuous on $[a, b]$. Then

$$\int_a^a f(x)\, dx = 0 \quad \text{and} \quad \int_b^a f(x)\, dx = -\int_a^b f(x)\, dx$$

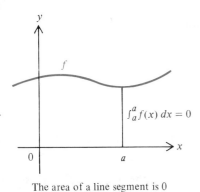

The area of a line segment is 0

FIGURE 5.20

The definition $\int_a^a f(x)\, dx = 0$ implies, for example, that a line segment has area 0 (Figure 5.20). It also implies that

$$\int_\pi^\pi \sin x^2\, dx = 0$$

Example 3 Evaluate $\displaystyle\int_4^1 x^2\, dx$.

Solution By Definition 5.4 we have

$$\int_4^1 x^2\, dx = -\int_1^4 x^2\, dx$$

and by (2) we know that

$$\int_1^4 x^2\, dx = \frac{1}{3}(4^3 - 1^3) = 21$$

Therefore we conclude that

$$\int_4^1 x^2\, dx = -21 \quad \square$$

Riemann Sums   Recall from Definition 5.2 that the definite integral $\int_a^b f(x)\, dx$ is the unique number such that

$$L_f(\mathscr{P}) \leq \int_a^b f(x)\, dx \leq U_f(\mathscr{P})$$

for all partitions $\mathscr{P} = \{x_0, x_1, x_2, \ldots, x_n\}$ of $[a, b]$. Another formulation of the definite integral that does not depend on the minimum value m_k and the maximum value M_k of f on $[x_{k-1}, x_k]$ has widespread use in applications. To prepare for this second formulation, which will appear in connection with Theorem 5.6, we let t_k be an arbitrary number in $[x_{k-1}, x_k]$. Then

$$m_k \leq f(t_k) \leq M_k$$

so that

$$m_k\, \Delta x_k \leq f(t_k)\, \Delta x_k \leq M_k\, \Delta x_k \quad \text{for } k = 1, 2, \ldots, n \tag{3}$$

It follows that

$$L_f(\mathscr{P}) = m_1\, \Delta x_1 + m_2\, \Delta x_2 + \cdots + m_n\, \Delta x_n$$

$$\leq f(t_1)\, \Delta x_1 + f(t_2)\, \Delta x_2 + \cdots + f(t_n)\, \Delta x_n$$

$$\leq M_1\, \Delta x_1 + M_2\, \Delta x_2 + \cdots + M_n\, \Delta x_n$$

$$= U_f(\mathscr{P})$$

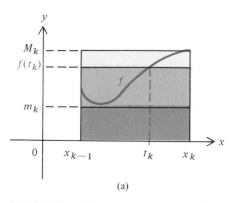

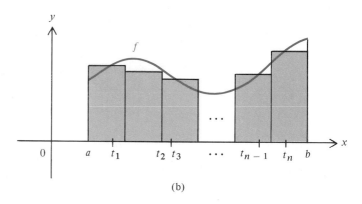

FIGURE 5.21

More succinctly,

$$L_f(\mathscr{P}) \le f(t_1)\Delta x_1 + f(t_2)\Delta x_2 + \cdots + f(t_n)\Delta x_n \le U_f(\mathscr{P}) \qquad (4)$$

If f is nonnegative on $[a, b]$, then the inequalities in (3) and (4) may be interpreted in terms of area. Recall that for $1 \le k \le n$, $m_k \Delta x_k$ and $M_k \Delta x_k$ represent the areas of the inscribed and circumscribed rectangles, respectively, on $[x_{k-1}, x_k]$. The number $f(t_k)\Delta x_k$ in (3) represents the area of the rectangle with base Δx_k and height $f(t_k)$. Equation (3) states that the area of the latter rectangle lies between the areas of the inscribed and circumscribed rectangles (Figure 5.21(a)). The sum $f(t_1)\Delta x_1 + \cdots + f(t_n)\Delta x_n$ in (4) represents the sum of the areas of the rectangles appearing in Figure 5.21(b). Whether f is nonnegative on $[a, b]$ or not, equation (4) states that the sum $f(t_1)\Delta x_1 + \cdots + f(t_n)\Delta x_n$ lies between the lower sum $L_f(\mathscr{P})$ and the upper sum $U_f(\mathscr{P})$. Since $L_f(\mathscr{P})$ and $U_f(\mathscr{P})$ can approximate $\int_a^b f(x)\,dx$ as closely as we please, the same is true of the sum $f(t_1)\Delta x_1 + \cdots + f(t_n)\Delta x_n$, which is called a Riemann* sum.

DEFINITION 5.5

> Let f be continuous on $[a, b]$, and let $\mathscr{P} = \{x_0, x_1, \ldots, x_n\}$ be any partition of $[a, b]$. For each k between 1 and n, let t_k be an arbitrary number in $[x_{k-1}, x_k]$. Then the sum
>
> $$f(t_1)\Delta x_1 + f(t_2)\Delta x_2 + \cdots + f(t_n)\Delta x_n$$
>
> is called a **Riemann sum** for f on $[a, b]$ and is denoted $\Sigma_{k=1}^n f(t_k)\Delta x_k$ (Σ is the Greek letter sigma). Thus
>
> $$\sum_{k=1}^n f(t_k)\Delta x_k = f(t_1)\Delta x_1 + f(t_2)\Delta x_2 + \cdots + f(t_n)\Delta x_n$$

Riemann sums are named for the nineteenth-century German mathematician Georg Bernhard Riemann (1826–1866), who clarified the concept of the integral while employing such sums. The first formal definition of the integral is attributed to him.

If f is continuous on $[a, b]$, then the Maximum–Minimum Theorem implies that each of m_k and M_k has the form $f(t_k)$ for an appropriate t_k in $[x_{k-1}, x_k]$, so

---

* **Riemann**: Pronounced "*Ree*-mahn."

that $L_f(\mathscr{P})$ and $U_f(\mathscr{P})$ are Riemann sums of f on $[a, b]$. However, there are many other Riemann sums, as the next example indicates.

Example 4 Let $f(x) = x^2$ for $-1 \le x \le 2$, and let $\mathscr{P} = \{-1, 0, 1, \frac{3}{2}, 2\}$. Find the Riemann sum for each of the following choices of t_k.

a. $t_k = x_{k-1}$, the left endpoint of the subinterval (Figure 5.22(a)).
b. $t_k = x_k$, the right endpoint of the subinterval (Figure 5.22(b)).
c. $t_k = (x_{k-1} + x_k)/2$, the midpoint of the subinterval (Figure 5.22(c)).

Solution The Riemann sum for (a) is

$$\sum_{k=1}^{4} f(t_k)\,\Delta x_k = f(-1)\cdot 1 + f(0)\cdot 1 + f(1)\cdot \tfrac{1}{2} + f(\tfrac{3}{2})\cdot \tfrac{1}{2}$$
$$= 1\cdot 1 + 0\cdot 1 + 1\cdot \tfrac{1}{2} + \tfrac{9}{4}\cdot \tfrac{1}{2} = \tfrac{21}{8}$$

The Riemann sum for (b) is

$$\sum_{k=1}^{4} f(t_k)\,\Delta x_k = f(0)\cdot 1 + f(1)\cdot 1 + f(\tfrac{3}{2})\cdot \tfrac{1}{2} + f(2)\cdot \tfrac{1}{2}$$
$$= 0\cdot 1 + 1\cdot 1 + \tfrac{9}{4}\cdot \tfrac{1}{2} + 4\cdot \tfrac{1}{2} = \tfrac{33}{8}$$

The Riemann sum for (c) is

$$\sum_{k=1}^{4} f(t_k)\,\Delta x_k = f(-\tfrac{1}{2})\cdot 1 + f(\tfrac{1}{2})\cdot 1 + f(\tfrac{5}{4})\cdot \tfrac{1}{2} + f(\tfrac{7}{4})\cdot \tfrac{1}{2}$$
$$= \tfrac{1}{4}\cdot 1 + \tfrac{1}{4}\cdot 1 + \tfrac{25}{16}\cdot \tfrac{1}{2} + \tfrac{49}{16}\cdot \tfrac{1}{2} = \tfrac{45}{16} \quad \square$$

Because Riemann sums corresponding to various choices of $t_1, t_2, \ldots, t_n$ can be different from one another (as Example 4 illustrates), we give names to certain Riemann sums. If t_k is the left endpoint of the subinterval $[x_{k-1}, x_k]$ for $1 \le k \le n$, as in Example 4(a), then we call the associated Riemann sum a **left sum**. There are analogous definitions for **right sum** and **midpoint sum**. The Riemann sum in Example 4(b) is a right sum, and the Riemann sum in Example 4(c) is a midpoint sum. Notice, however, that since $m_k \le f(t_k) \le M_k$ for all k, it follows that all Riemann sums must lie between the lower sum $L_f(\mathscr{P})$ and the upper sum $U_f(\mathscr{P})$.

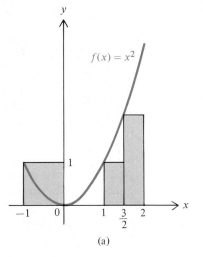

(a)

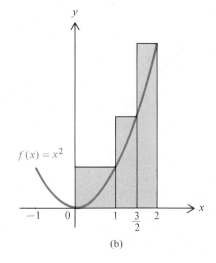

(b)

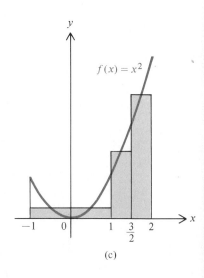

(c)

FIGURE 5.22

An important feature of a Riemann sum $\sum_{k=1}^{n} f(t_k)\Delta x_k$ is that it approximates the definite integral $\int_a^b f(x)\,dx$. For this reason we write

$$\int_a^b f(x)\,dx \approx \sum_{k=1}^{n} f(t_k)\Delta x_k$$

Just as important is the fact that we can approximate $\int_a^b f(x)\,dx$ by a Riemann sum $\sum_{k=1}^{n} f(t_k)\Delta x_k$ as closely as we wish by making the lengths of all the subintervals of the associated partition small enough. The next theorem states this in precise terms; its proof appears in the Appendix.

THEOREM 5.6

Let f be continuous on $[a, b]$. For any $\varepsilon > 0$ there is a number $\delta > 0$ such that the following statement holds: If $\mathscr{P} = \{x_0, x_1, \ldots, x_n\}$ is any partition of $[a, b]$ each of whose subintervals has length less than δ and if $x_{k-1} \le t_k \le x_k$ for each k between 1 and n, then the associated Riemann sum $\sum_{k=1}^{n} f(t_k)\Delta x_k$ satisfies

$$\left| \int_a^b f(x)\,dx - \sum_{k=1}^{n} f(t_k)\Delta x_k \right| < \varepsilon \qquad (4)$$

We sometimes express Theorem 5.6 by saying that the integral $\int_a^b f(x)\,dx$ is the limit of the Riemann sums as the lengths of all the subintervals of $[a, b]$ approach 0. More succinctly, we write

$$\int_a^b f(x)\,dx = \lim_{\|\mathscr{P}\| \to 0} \sum_{k=1}^{n} f(t_k)\Delta x_k \qquad (5)$$

where $\|\mathscr{P}\|$ denotes the largest of the lengths of the subintervals associated with $\mathscr{P}$ and is called the **norm** of $\mathscr{P}$. The right-hand side of (5) is sometimes used as the definition of $\int_a^b f(x)\,dx$. We emphasize that in the remainder of the book, when an equation appears in the form of (5), it is shorthand for the statement in Theorem 5.6.

In the next example we will approximate $\int_1^2 (1/x)\,dx$ by several Riemann sums. When we calculate a Riemann sum, we compute the terms in the sum individually and then add them. Since it is usually convenient to use decimal expansions when performing operations such as addition and multiplication, it is frequently necessary to round off. This induces what is called a **round-off error**. For example, if we were to use 3.14159 for π, then the round-off error would be approximately 0.000003. We will round our final answers to six digits. Although the topic of round-off error is important in its own right, we will not pursue it further in this book.

Example 5   Let $\mathscr{P}_1$, $\mathscr{P}_2$, and $\mathscr{P}_3$ be partitions that divide the interval $[1, 2]$ into 3, 5, and 10 subintervals of equal length, respectively. Approximate

$$\int_1^2 \frac{1}{x}\,dx$$

by the corresponding left sums.

Solution Let $f(x) = 1/x$. For the left sum corresponding to $\mathscr{P}_1$, $\Delta x_k = \frac{1}{3}$, so that

$$\sum_{k=1}^{3} f(x_{k-1})\,\Delta x_k = f(1) \cdot \frac{1}{3} + f\left(\frac{4}{3}\right) \cdot \frac{1}{3} + f\left(\frac{5}{3}\right) \cdot \frac{1}{3}$$

$$= \frac{1}{3}\left(1 + \frac{3}{4} + \frac{3}{5}\right) \approx 0.783333$$

For the left sum corresponding to $\mathscr{P}_2$, $\Delta x_k = \frac{1}{5}$, and therefore

$$\sum_{k=1}^{5} f(x_{k-1})\,\Delta x_k = f(1) \cdot \frac{1}{5} + f\left(\frac{6}{5}\right) \cdot \frac{1}{5} + f\left(\frac{7}{5}\right) \cdot \frac{1}{5} + f\left(\frac{8}{5}\right) \cdot \frac{1}{5} + f\left(\frac{9}{5}\right) \cdot \frac{1}{5}$$

$$= \frac{1}{5}\left(1 + \frac{5}{6} + \frac{5}{7} + \frac{5}{8} + \frac{5}{9}\right) \approx 0.745635$$

For the left sum corresponding to $\mathscr{P}_3$, $\Delta x_k = \frac{1}{10}$, and thus

$$\sum_{k=1}^{10} f(x_{k-1})\,\Delta x_k = f(1) \cdot \frac{1}{10} + f\left(\frac{11}{10}\right) \cdot \frac{1}{10} + f\left(\frac{12}{10}\right) \cdot \frac{1}{10} + \cdots + f\left(\frac{19}{10}\right) \cdot \frac{1}{10}$$

$$= \frac{1}{10}\left(1 + \frac{10}{11} + \frac{10}{12} + \cdots + \frac{10}{19}\right)$$

$$\approx 0.718771 \quad \square$$

Later we will be able to show that $\int_1^2 (1/x)\,dx$ is, accurate to six places, 0.693147. It follows that the accuracy of the approximations in Example 5 improves as the lengths of the subintervals decreases. From Theorem 5.6 we know that we could approximate $\int_1^2 (1/x)\,dx$ as accurately as we might wish if we use a partition with short enough subintervals.

EXERCISES 5.2

In Exercises 1–6 approximate $\int_a^b f(x)\,dx$ by computing $L_f(\mathscr{P})$ and $U_f(\mathscr{P})$.

1. $\displaystyle\int_{-1}^{3} 2x\,dx; \mathscr{P} = \{-1, 0, 1, 2, 3\}$

2. $\displaystyle\int_{-2}^{1} x^2\,dx; \mathscr{P} = \{-2, -1, 0, 1\}$

3. $\displaystyle\int_{-1}^{3} |x|\,dx; \mathscr{P} = \{-1, -\frac{1}{2}, 0, \frac{1}{2}, 1, \frac{3}{2}, 2, \frac{5}{2}, 3\}$

4. $\displaystyle\int_{-\pi/4}^{\pi/4} -\cos x\,dx; \mathscr{P} = \{-\pi/4, 0, \pi/4\}$

5. $\displaystyle\int_{-\pi/4}^{\pi/4} 3\sin x\,dx; \mathscr{P} = \{-\pi/4, 0, \pi/4\}$

6. $\displaystyle\int_{0}^{2} |x - 1|\,dx; \mathscr{P} = \{0, \frac{1}{2}, 1, \frac{3}{2}, 2\}$

In Exercises 7–21 compute the definite integrals by using the definitions and results of this section.

7. $\displaystyle\int_{-2}^{3} 4\,dx$

8. $\displaystyle\int_{0}^{5} 2.1\,dx$

9. $\displaystyle\int_{-1}^{1} -\frac{1}{3}\,dx$

10. $\displaystyle\int_{-1}^{-1} \sqrt{2}\,dx$

11. $\displaystyle\int_{2}^{0} \pi\,dx$

12. $\displaystyle\int_{0}^{3} x\,dx$

13. $\displaystyle\int_{-3}^{3} x\,dx$

14. $\displaystyle\int_{-4}^{-2} x\,dx$

15. $\displaystyle\int_{1/3}^{-1/3} x\,dx$

16. $\displaystyle\int_{2.7}^{2.9} -x\,dx$

17. $\displaystyle\int_{-1/2}^{-7/2} -x\,dx$

18. $\displaystyle\int_{1}^{2} x^2\,dx$

19. $\displaystyle\int_{-5}^{-5} x^2\,dx$ 20. $\displaystyle\int_{-1}^{4} x^2\,dx$ 21. $\displaystyle\int_{2}^{-3} x^2\,dx$

22. Explain why none of the following fit into the mold of Definition 5.2. In each case, is the definition of $U_f(\mathcal{P})$ meaningful?

    a. $\displaystyle\int_{0}^{1} \frac{1}{x}\,dx$ b. $\displaystyle\int_{0}^{\infty} x\,dx$

    c. $\displaystyle\int_{0}^{\pi} \tan x\,dx$ d. $\displaystyle\int_{-1}^{2} \sqrt{x}\,dx$

In Exercises 23–28 find the area A of the region between the graph of f and the x axis on the given interval.

23. $f(x) = \frac{5}{2}; [-2, 3]$ 24. $f(x) = \sqrt{3}; [\sqrt{3}, 3\sqrt{3}]$

25. $f(x) = x; [1, 4]$ 26. $f(x) = -x; [-3, -1]$

27. $f(x) = x^2; [-3, -1]$ 28. $f(x) = x^2; [\sqrt{3}, 2\sqrt{3}]$

In Exercises 29–34 approximate the integral by Riemann sums with the indicated partitions, using first the left sum, then the right sum, and finally the midpoint sum.

29. $\displaystyle\int_{1}^{3} (x^2 - x)\,dx; \mathcal{P} = \{1, 2, 3\}$

30. $\displaystyle\int_{0}^{4} (x - 3)\,dx; \mathcal{P} = \{0, 1, 2, 3, 4\}$

31. $\displaystyle\int_{0}^{2} \sin \pi x\,dx; \mathcal{P} = \left\{0, \frac{1}{2}, 1, 2\right\}$

32. $\displaystyle\int_{-\pi}^{0} \cos x\,dx; \mathcal{P} = \left\{-\pi, -\frac{2\pi}{3}, -\frac{\pi}{3}, 0\right\}$

33. $\displaystyle\int_{1}^{5} \frac{1}{x}\,dx; \mathcal{P} = \{1, 2, 3, 4, 5\}$

34. $\displaystyle\int_{-2}^{-1} \frac{1}{2x + 1}\,dx; \mathcal{P} = \left\{-2, -\frac{3}{2}, -\frac{5}{4}, -1\right\}$

In Exercises 35–38 approximate the integral by Riemann sums with the indicated partition, by using first the left sum and then the right sum.

35. $\displaystyle\int_{\pi/4}^{3\pi/4} \frac{\sin x}{x}\,dx; \mathcal{P} = \left\{\frac{\pi}{4}, \frac{\pi}{2}, \frac{3\pi}{4}\right\}$

36. $\displaystyle\int_{0}^{\pi} \frac{1}{1 + \sin x}\,dx; \mathcal{P} = \left\{0, \frac{\pi}{4}, \frac{\pi}{2}, \frac{3\pi}{4}, \pi\right\}$

37. $\displaystyle\int_{0}^{1} \sin \pi x^2\,dx; \mathcal{P} = \left\{0, \frac{1}{2}, \frac{\sqrt{2}}{2}, \frac{\sqrt{3}}{2}, 1\right\}$

C 38. $\displaystyle\int_{0}^{\pi/2} \sqrt{1 + \cos x}\,dx; \mathcal{P} = \left\{0, \frac{\pi}{6}, \frac{\pi}{4}, \frac{\pi}{3}, \frac{\pi}{2}\right\}$

In Exercises 39–43 approximate the area A of the region between the graph of f and the x axis on $[a, b]$ by using the left sum with the indicated partition.

39. $f(x) = 2x^2 + 3x, a = 0, b = 1, \mathcal{P} = \{0, \frac{1}{4}, \frac{1}{2}, \frac{3}{4}, 1\}$

40. $f(x) = \sqrt{x}, a = 1, b = 4, \mathcal{P} = \{1, 2, 3, 4\}$

41. $f(x) = x/(x + 1), a = 0, b = 2, \mathcal{P} = \{0, \frac{1}{2}, 1, 2\}$

C 42. $f(x) = x^2, a = 1, b = 3, \mathcal{P}$ divides $[1, 3]$ into 10 subintervals of equal length.

C 43. $f(x) = x^2, a = 1, b = 3, \mathcal{P}$ divides $[1, 3]$ into 100 subintervals of equal length.

44. Let f and g be continuous on $[a, b]$, and suppose that $f(x) \leq g(x)$ for $a \leq x \leq b$.

    a. By referring to the definitions of lower and upper sums, show that for any partition $\mathcal{P}$ of $[a, b]$,

    $$L_f(\mathcal{P}) \leq L_g(\mathcal{P}) \quad \text{and} \quad U_f(\mathcal{P}) \leq U_g(\mathcal{P})$$

    (Notice the similarity to Exercise 28 of Section 5.1.)

    b. Show that $\int_a^b f(x)\,dx \leq \int_a^b g(x)\,dx$ by supplying the reasons for the statements labeled 1, 2, and 3 in the following proof by contradiction: Suppose $\int_a^b f(x)\,dx > \int_a^b g(x)\,dx$. Then for any partition $\mathcal{P}$ of $[a, b]$ we have

    $$L_g(\mathcal{P}) \overset{①}{\leq} \int_a^b g(x)\,dx < \int_a^b f(x)\,dx \overset{②}{\leq} U_f(\mathcal{P}) \overset{③}{\leq} U_g(\mathcal{P})$$

    Therefore $\int_a^b f(x)\,dx$ and $\int_a^b g(x)\,dx$ are distinct numbers that lie between every lower sum $L_g(\mathcal{P})$ and upper sum $U_g(\mathcal{P})$. This contradicts the uniqueness of such a number (see Definition 5.2). Therefore

    $$\int_a^b f(x)\,dx \leq \int_a^b g(x)\,dx$$

In Exercises 45–49, use Exercise 44(b) to verify the given inequalities.

45. $\displaystyle\int_{0}^{1} x^6\,dx \leq \int_{0}^{1} x\,dx$

46. $\displaystyle\int_{1}^{2} x\,dx \leq \int_{1}^{2} x^6\,dx$

47. $\displaystyle\int_{1}^{2} \frac{1}{x^6}\,dx \leq \int_{1}^{2} \frac{1}{x}\,dx$

48. $\displaystyle\int_{0}^{\pi/2} \sin x\,dx \leq \int_{0}^{\pi/2} x\,dx$

49. $\displaystyle\int_{0}^{\pi/4} \sin x\,dx \leq \int_{0}^{\pi/4} \cos x\,dx$

*50. Using Examples 1 and 2 as guides, prove that

$$\int_a^b (x + 4)\,dx = \frac{1}{2}(b^2 - a^2) + 4(b - a)$$

51. a. Using the method of Example 2, show that for any constant c and any numbers a and b with $a < b$, we have

$$\int_a^b cx\,dx = \frac{c}{2}(b^2 - a^2)$$

   (*Hint:* Consider the cases $c > 0$, $c = 0$, and $c < 0$ separately.)

   b. Using part (a), along with Definition 5.4, show that for any constant c and any numbers a and b, we have

$$\int_a^b cx\,dx = \frac{c}{2}(b^2 - a^2)$$

52. Using the result of Exercise 51, evaluate the following integrals.

   a. $\displaystyle\int_1^3 4x\,dx$ b. $\displaystyle\int_2^6 -\frac{1}{2}x\,dx$ c. $\displaystyle\int_{-5}^5 \sqrt{3}x\,dx$

   d. $\displaystyle\int_{1/\pi}^0 \pi x\,dx$

*53. a. Let $0 \le a \le b$. Show that

$$\int_a^b x^2\,dx = \frac{1}{3}(b^3 - a^3)$$

   (*Hint:* Use the fact that if $x_{k-1} < x_k$, then

$$x_{k-1}^2 < \frac{1}{3}(x_k^2 + x_k x_{k-1} + x_{k-1}^2) < x_k^2$$

   along with the method of Example 2.)

   b. Use the result of part (a), along with Definition 5.4, to show that for any nonnegative numbers a and b, we have

$$\int_a^b x^2\,dx = \frac{1}{3}(b^3 - a^3)$$

54. Eventually we will show that for any numbers a and b, we have

$$\int_a^b x^3\,dx = \frac{1}{4}(b^4 - a^4)$$

Use this formula to evaluate the following integrals.

   a. $\displaystyle\int_0^2 x^3\,dx$ b. $\displaystyle\int_{-1}^1 x^3\,dx$ c. $\displaystyle\int_2^{-3} x^3\,dx$

55. Using the results of Examples 1 and 2, together with Exercises 53 and 54, guess a formula for

$$\int_a^b x^n\,dx$$

that would hold for any positive integer n.

56. Suppose $f(x) = mx + c$ for $a \le x \le b$. Let $\mathscr{P}$ be any partition that divides $[a, b]$ into equal subintervals. Show that the midpoint sum gives the exact value of $\int_a^b f(x)\,dx$.

57. Let f be any odd function, and let $\mathscr{P}$ be any partition that divides $[-a, a]$ into subintervals of equal length. Describe a simple method of finding a Riemann sum for $\int_{-a}^a f(x)\,dx$ so that

$$\int_{-a}^a f(x)\,dx = \sum_{k=1}^n f(t_k)\Delta x_k$$

   (*Hint:* Recall that $f(-x) = -f(x)$ for $-a \le x \le a$.)

58. Suppose f is continuous on $[1, 3]$ and has values appearing in the following table:

x	1	1.5	1.7	2.1	2.5	3
$f(x)$	2	1	.5	.2	0	.1

Approximate the area A between the graph of f and the x axis on $[1, 3]$ by using the information given in the table and an appropriate

   a. left sum   b. right sum

5.3
SPECIAL PROPERTIES OF THE DEFINITE INTEGRAL

In this section we return to the three basic properties of area from which our definition of integral was derived, and we present them in terms of integrals. The theorems of this section will be used repeatedly throughout the remainder of the book.

Integral Forms of the Three Basic Properties   The first property to appear is an integral form of the Rectangle Property, which is related to Example 1 of Section 5.2.

THEOREM 5.7
RECTANGLE
PROPERTY

For any numbers a, b, and c,

$$\int_a^b c \, dx = c(b - a)$$

Proof If $a < b$, then the result follows directly from Example 1 of Section 5.2. If $a = b$, then by Definition 5.4,

$$\int_a^b c \, dx = \int_a^a c \, dx = 0 = c(b - a)$$

Finally, if $a > b$, then by combining Definition 5.4 and Example 1 of Section 5.2, we obtain

$$\int_a^b c \, dx = -\int_b^a c \, dx = -c(a - b) = c(b - a) \quad \blacksquare$$

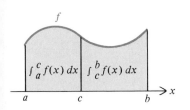

FIGURE 5.23

For the Addition Property we give a version that in special cases can be interpreted geometrically to mean that the area of a region composed of two smaller regions overlapping only in a line segment is the sum of the areas of the two regions (Figure 5.23).

THEOREM 5.8
ADDITION PROPERTY

Let f be continuous on an interval containing a, b, and c. Then

$$\int_a^b f(x) \, dx = \int_a^c f(x) \, dx + \int_c^b f(x) \, dx$$

Proof First we assume that $a < c < b$. By definition $\int_a^b f(x) \, dx$ is the *unique* number such that

$$L_f(\mathscr{P}) \le \int_a^b f(x) \, dx \le U_f(\mathscr{P})$$

for any partition $\mathscr{P}$ of $[a, b]$. Therefore it suffices to show that

$$L_f(\mathscr{P}) \le \int_a^c f(x) \, dx + \int_c^b f(x) \, dx \le U_f(\mathscr{P}) \quad \text{for any } \mathscr{P} \tag{1}$$

So let $\mathscr{P}$ be an arbitrary partition of $[a, b]$, and let $\mathscr{P}'$ contain all the points of $\mathscr{P}$ and also the point c. By the inequalities in (5) of Section 5.1, we have

$$L_f(\mathscr{P}) \le L_f(\mathscr{P}') \le U_f(\mathscr{P}') \le U_f(\mathscr{P}) \tag{2}$$

Now if $\mathscr{P}_1$ is the collection of points in $\mathscr{P}'$ that are also in $[a, c]$, and if $\mathscr{P}_2$ is the collection of points in $\mathscr{P}'$ and in $[c, b]$ (Figure 5.24), then

$$L_f(\mathscr{P}') = L_f(\mathscr{P}_1) + L_f(\mathscr{P}_2) \quad \text{and} \quad U_f(\mathscr{P}') = U_f(\mathscr{P}_1) + U_f(\mathscr{P}_2) \tag{3}$$

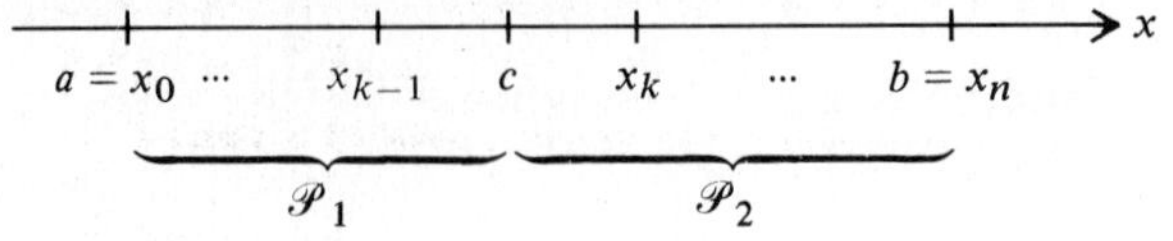

FIGURE 5.24

Moreover, by the definition of $\int_a^c f(x)\,dx$ and $\int_c^b f(x)\,dx$, we have

$$L_f(\mathscr{P}_1) \le \int_a^c f(x)\,dx \le U_f(\mathscr{P}_1) \quad \text{and} \quad L_f(\mathscr{P}_2) \le \int_c^b f(x)\,dx \le U_f(\mathscr{P}_2) \qquad (4)$$

Adding the two sets of inequalities in (4) and applying the equations of (3), we obtain

$$\begin{aligned} L_f(\mathscr{P}') &= L_f(\mathscr{P}_1) + L_f(\mathscr{P}_2) \\ &\le \int_a^c f(x)\,dx + \int_c^b f(x)\,dx \le U_f(\mathscr{P}_1) + U_f(\mathscr{P}_2) = U_f(\mathscr{P}') \end{aligned}$$

Therefore by (2) we conclude that

$$L_f(\mathscr{P}) \le L_f(\mathscr{P}') \le \int_a^c f(x)\,dx + \int_c^b f(x)\,dx \le U_f(\mathscr{P}') \le U_f(\mathscr{P})$$

for any partition $\mathscr{P}$. This is what we needed to prove under the assumption that $a < c < b$.

For any other ordering of a, b, and c the result follows from the first part of the proof and from Definition 5.4. We carry out the proof for the case $b < c < a$ and leave other cases to Exercise 33. If $b < c < a$, then

$$\begin{aligned} \int_a^b f(x)\,dx &\overset{\substack{\text{Definition}\\5.4}}{=} -\int_b^a f(x)\,dx \overset{\substack{\text{first part}\\\text{of proof}}}{=} -\left(\int_b^c f(x)\,dx + \int_c^a f(x)\,dx\right) \\ &= -\int_b^c f(x)\,dx - \int_c^a f(x)\,dx = -\int_c^a f(x)\,dx - \int_b^c f(x)\,dx \\ &\overset{\substack{\text{Definition}\\5.4}}{=} \int_a^c f(x)\,dx + \int_c^b f(x)\,dx \qquad \blacksquare \end{aligned}$$

Example 1 Evaluate $\displaystyle\int_1^3 f(x)\,dx$, where

$$f(x) = \begin{cases} x & \text{for } 1 \le x \le 2 \\ 2 & \text{for } 2 \le x \le 3 \end{cases}$$

(See Figure 5.25.)

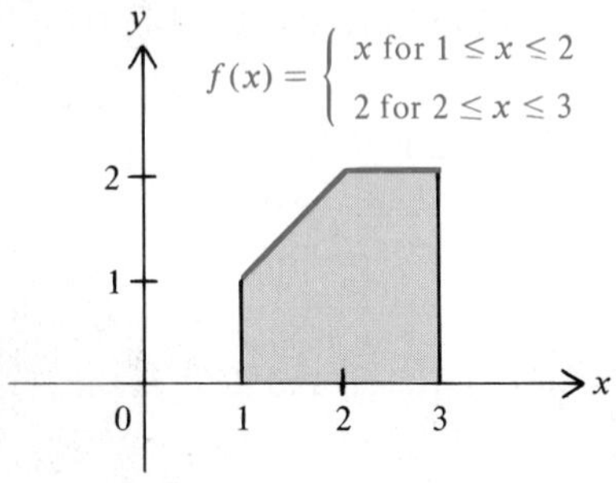

FIGURE 5.25

Solution Notice that f is continuous on $[1, 3]$. By the Addition Property,

$$\int_1^3 f(x)\,dx = \int_1^2 f(x)\,dx + \int_2^3 f(x)\,dx$$

Now by Examples 1 and 2 of Section 5.2,

$$\int_1^2 f(x)\,dx = \int_1^2 x\,dx = \frac{1}{2}(2^2 - 1^2) = \frac{3}{2}$$

and

$$\int_2^3 f(x)\,dx = \int_2^3 2\,dx = 2(3 - 2) = 2$$

We conclude that

$$\int_1^3 f(x)\,dx = \int_1^2 f(x)\,dx + \int_2^3 f(x)\,dx = \frac{3}{2} + 2 = \frac{7}{2} \quad \square$$

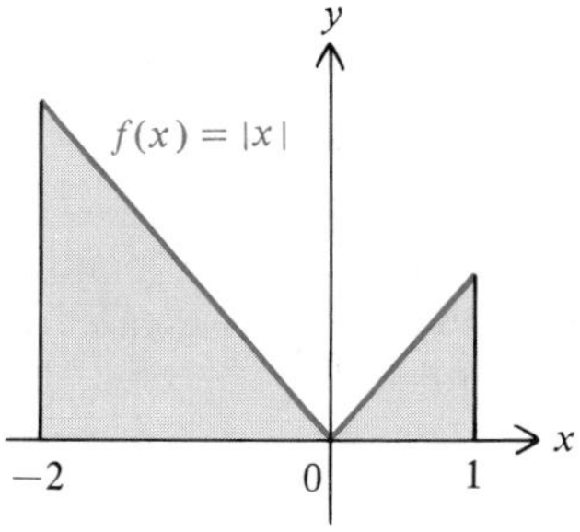

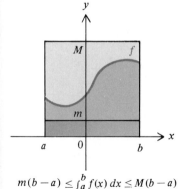

$f(x) = |x|$

FIGURE 5.26

Example 2 Evaluate $\displaystyle\int_1^{-2} |x|\,dx$ (see Figure 5.26).

Solution With the help of the Addition Property and Definition 5.4, we find that

$$\int_1^{-2} |x|\,dx = \int_1^0 |x|\,dx + \int_0^{-2} |x|\,dx = \int_1^0 x\,dx + \int_0^{-2} -x\,dx$$

$$= -\int_0^1 x\,dx + \int_{-2}^0 x\,dx = -\frac{1}{2}(1^2 - 0^2) + \frac{1}{2}[0^2 - (-2)^2]$$

$$= -\frac{1}{2} - 2 = -\frac{5}{2} \quad \square$$

The third property of area is the Comparison Property. We will need only a special case of the Comparison Property. In geometric terms it implies that the area of any region is at least as large as that of any inscribed rectangle and no larger than that of any circumscribed rectangle (Figure 5.27).

$m(b-a) \le \int_a^b f(x)\,dx \le M(b-a)$

FIGURE 5.27

THEOREM 5.9
COMPARISON
PROPERTY

Let f be continuous on $[a, b]$, and suppose $m \le f(x) \le M$ for all x in $[a, b]$. Then

$$m(b - a) \le \int_a^b f(x)\,dx \le M(b - a)$$

Proof Let m_1 and M_1 be the minimum and the maximum values of f on $[a, b]$, which exist because f is continuous on $[a, b]$. Then $m \le m_1 \le M_1 \le M$. Let $\mathscr{P}$ be the trivial partition of $[a, b]$ consisting only of the points a and b (so that $x_0 = a$ and $x_n = x_1 = b$). Then

$$L_f(\mathscr{P}) = m_1(b - a) \quad \text{and} \quad U_f(\mathscr{P}) = M_1(b - a)$$

Since $\int_a^b f(x)\,dx$ lies between $L_f(\mathcal{P})$ and $U_f(\mathcal{P})$, it follows that

$$m(b-a) \le m_1(b-a) = L_f(\mathcal{P})$$

$$\le \int_a^b f(x)\,dx \le U_f(\mathcal{P}) = M_1(b-a) \le M(b-a) \quad \blacksquare$$

The Comparison Property is frequently used to estimate the value of an integral that cannot be computed exactly or easily. A number less than or equal to a given integral is called a **lower bound** for the integral, and a number greater than or equal to the integral is an **upper bound** for the integral.

Example 3   Using the Comparison Property, find lower and upper bounds for

$$\int_{-1}^1 \sqrt{1+x^4}\,dx$$

Solution   We have

$$1 \le \sqrt{1+x^4} \le \sqrt{2} \quad \text{for } -1 \le x \le 1$$

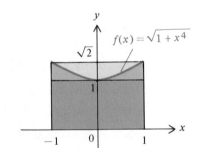

(See Figure 5.28.) By the Comparison Property it follows that

$$2 = 1[1-(-1)] \le \int_{-1}^1 \sqrt{1+x^4}\,dx \le \sqrt{2}[1-(-1)] = 2\sqrt{2}$$

so that

$$2 \le \int_{-1}^1 \sqrt{1+x^4}\,dx \le 2\sqrt{2}$$

FIGURE 5.28

Thus 2 is a lower bound and $2\sqrt{2}$ is an upper bound for the integral. $\square$

In Example 4 of Section 5.5 we will obtain a better, that is, smaller upper bound for $\int_{-1}^1 \sqrt{1+x^4}\,dx$.

Consequences of the Comparison Property

Our first consequence of the Comparison Property tells us that the integral of a nonnegative function is nonnegative.

COROLLARY 5.10

Let f be nonnegative and continuous on $[a, b]$. Then

$$\int_a^b f(x)\,dx \ge 0$$

Proof By hypothesis, $f(x) \geq 0$ for $a \leq x \leq b$. Therefore it is permissible to let $m = 0$ in Theorem 5.9. This implies that

$$\int_a^b f(x)\,dx \geq 0(b-a) = 0 \quad \blacksquare$$

From Corollary 5.10 it follows that the area of any region covered by Definition 5.3 is a nonnegative number. A second consequence of the Comparison Property is an integral form of the Mean Value Theorem.

THEOREM 5.11
MEAN VALUE THEOREM
FOR INTEGRALS

Let f be continuous on $[a, b]$. Then there is a number c in $[a, b]$ such that

$$\int_a^b f(x)\,dx = f(c)(b-a)$$

Proof If $a = b$, then the result is obvious. Thus assume that $a < b$, and let m and M be the minimum and the maximum values of f on $[a, b]$. By the Comparison Property we know that

$$m(b-a) \leq \int_a^b f(x)\,dx \leq M(b-a)$$

Since $a < b$, this means that

$$m \leq \frac{\int_a^b f(x)\,dx}{b-a} \leq M$$

Since f is continuous on $[a, b]$, the Intermediate Value Theorem asserts that there is a number c in $[a, b]$ such that

$$\frac{\int_a^b f(x)\,dx}{b-a} = f(c)$$

that is,

$$\int_a^b f(x)\,dx = f(c)(b-a) \quad \blacksquare$$

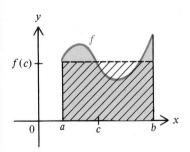

FIGURE 5.29

It follows from the Mean Value Theorem for Integrals that if f is nonnegative and continuous on $[a, b]$, then the area of the region bounded above by the graph of f is the same as the area of a rectangle whose height is $f(c)$ for some properly chosen c in $[a, b]$ (Figure 5.29).

The value $\int_a^b f(x)\,dx/(b-a)$ is called the **average** (or **mean**) **value** of f on the interval $[a, b]$. Later we will see that if f represents velocity, then the average value of velocity on $[a, b]$ is the same as the average velocity defined in Section 2.1. Corresponding statements apply to average cost and average revenue. (See Exercises 62 and 63 of Section 5.4.)

Example 4   Let $f(x) = x^2$ for $0 \leq x \leq 2$. Find the average value of f on $[0, 2]$.

Solution   By definition the average value is

$$\frac{1}{2-0} \int_0^2 x^2 \, dx$$

By (2) of Section 5.2,

$$\frac{1}{2-0} \int_0^2 x^2 \, dx = \frac{1}{2}\left[\frac{1}{3}(2^3 - 0^3)\right] = \frac{4}{3} \quad \square$$

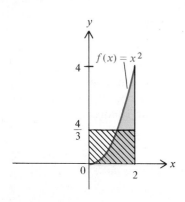

FIGURE 5.30

It follows from Example 4 that the area of the region between the graph of f and the x axis on $[0, 2]$ is the same as the area of the rectangle with base 2 and height $\frac{4}{3}$, which is the average value of f (Figure 5.30).

EXERCISES 5.3

In Exercises 1–6 use the Rectangle Property to evaluate the integral.

1. $\displaystyle\int_3^5 7 \, dx$
2. $\displaystyle\int_{-1}^3 \frac{1}{2} \, dx$
3. $\displaystyle\int_{17}^{100} 1 \, dr$

4. $\displaystyle\int_{-1}^2 -3 \, dx$
5. $\displaystyle\int_2^{-1} -10 \, du$
6. $\displaystyle\int_1^{-1} 5 \, dx$

In Exercises 7–10 evaluate the integrals to corroborate the Addition Property for integrals. Use the results of Section 5.2 in your calculations.

7. $\displaystyle\int_0^1 x \, dx + \int_1^2 x \, dx = \int_0^2 x \, dx$

8. $\displaystyle\int_3^4 x^2 \, dx + \int_4^3 x^2 \, dx = \int_3^3 x^2 \, dx$

9. $\displaystyle\int_1^0 y^2 \, dy + \int_0^2 y^2 \, dy = \int_1^2 y^2 \, dy$

10. $\displaystyle\int_{-2}^{-3} -y \, dy + \int_{-3}^{-6} -y \, dy = \int_{-2}^{-6} -y \, dy$

In Exercises 11–14 let f be a continuous function on $(-\infty, \infty)$. Use the Addition Property to find the values of a and b that make the equation true.

11. $\displaystyle\int_0^2 f(x) \, dx + \int_3^0 f(x) \, dx = \int_a^b f(x) \, dx$

12. $\displaystyle\int_{1/2}^{-1/2} f(x) \, dx + \int_{-1}^{1/2} f(x) \, dx = \int_a^b f(x) \, dx$

13. $\displaystyle\int_a^b f(t) \, dt - \int_5^3 f(t) \, dt = \int_3^1 f(t) \, dt$

14. $\displaystyle\int_\pi^{2\pi} f(t) \, dt - \int_a^b f(t) \, dt = \int_{3\pi}^{2\pi} f(t) \, dt$

In Exercises 15–24 find the maximum and minimum values of the given function on the given interval. Then use the Comparison Property to find upper and lower bounds for the area of the region between the graph of the function and the x axis on the given interval.

15. $f(x) = 13; [-3, 0]$
16. $f(x) = x; [0, 2]$
17. $f(x) = x^2; [-1, 3]$
18. $f(x) = x^3; [0, 4]$

19. $f(x) = \dfrac{1}{x}; [2, 3]$
20. $g(x) = \sin x; [\pi/4, \pi/2]$

21. $g(x) = \cos x; [\pi/4, \pi/3]$

22. $g(x) = \sin x + \cos x; [0, \pi/2]$

23. $h(t) = \tan t; [0, \pi/3]$

24. $g(x) = x^3 - 3x + 3; [0, 2]$

25. Let

$$f(x) = \begin{cases} -x & \text{for } x < 0 \\ x^2 & \text{for } x \geq 0 \end{cases}$$

Find the area A of the region between the graph of f and the x axis on $[-1, 1]$.

26. Let

$$f(x) = \begin{cases} 1 & \text{for } 0 \leq x \leq 1 \\ x^2 & \text{for } x > 1 \end{cases}$$

Find the area A of the region between the graph of f and the x axis on $[0, 4]$.

27. Let $f(x) = c$ for $a \le x \le b$. Show that the average value of f on $[a, b]$ is c.

28. Let $f(x) = x$ for $a \le x \le b$. Show that the average value of f on $[a, b]$ is $(a + b)/2$.

*29. Let $0 \le a < b$ and let $f(x) = x^2$.
  a. Show that the average value of f on $[a, b]$ is $\frac{1}{3}(a^2 + ab + b^2)$.
  b. Show that there is a number c in $[a, b]$ such that
  $$c^2 = \frac{1}{3}(a^2 + ab + b^2)$$

30. Show that $\int_0^1 x^7 \, dx > 0$. (*Hint:* First use the Addition Property with $a = 0$, $b = 1$, and $c = \frac{1}{2}$. Then apply the Comparison Property to each of the resulting integrals.)

31. Show that $\int_0^{\pi/2} \sin x \, dx > 0$.

32. Let f be continuous on $[a, b]$, where $a < b$.
  a. Show that $\int_a^b f(x) \, dx > 0$ if $f(x) > 0$ for $a \le x \le b$. (*Hint:* Since f is continuous, it assumes a minimum value m on $[a, b]$. Clearly $m > 0$.)
  b. Show that $\int_{-\pi/6}^{\pi/6} (\cos x - x^2) \, dx > 0$.

33. Prove the Addition Property for the cases $c < a < b$ and $b < a < c$.

34. Show that the Comparison Property is a consequence of the Mean Value Theorem for Integrals.

5.4
THE FUNDAMENTAL THEOREM OF CALCULUS

The purpose of this section is to develop a general method for evaluating $\int_a^b f(x) \, dx$ that does not necessitate computing various sums. The method will allow us to evaluate many (but not all) of the integrals that arise in applications. We will use the fact (mentioned in Section 5.2 and proved in the Appendix) that if f is continuous on $[a, b]$, then $\int_a^b f(x) \, dx$ is defined. For the present we will assume that $a < b$.

In order to use x as a variable in our discussion, we will substitute $\int_a^b f(t) \, dt$ for $\int_a^b f(x) \, dx$. Recall that $\int_a^b f(t) \, dt$ is by definition a number. If f is continuous on $[a, b]$ and c is any fixed number in $[a, b]$, then for every x in $[a, b]$, f is necessarily continuous on the closed interval with endpoints c and x. Consequently we can associate with any such x the number $\int_c^x f(t) \, dt$ to obtain a function G defined by the equation

$$G(x) = \int_c^x f(t) \, dt \quad \text{for } a \le x \le b \tag{1}$$

For example, if $f(x) = x$ for $0 \le x \le 10$ and if we choose $c = 1$, then

$$G(0) = \int_1^0 t \, dt = -\int_0^1 t \, dt = -\frac{1}{2}(1^2 - 0^2) = -\frac{1}{2}$$

$$G(1) = \int_1^1 t \, dt = 0$$

$$G(2) = \int_1^2 t \, dt = \frac{1}{2}(2^2 - 1^2) = \frac{3}{2}$$

In fact, for any x between 0 and 10 we have

$$G(x) = \int_1^x t \, dt = \frac{1}{2}(x^2 - 1)$$

Notice that in this example, $G' = f$ on $[0, 10]$. We will prove more generally that

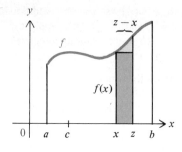

FIGURE 5.31

if f is continuous on $[a, b]$ and G is defined as in (1), then $G' = f$, so that G is an antiderivative of f.

To see geometrically why we can expect to have $G' = f$, let us suppose that f is continuous and nonnegative on $[a, b]$, and let $a \leq c < x < z < b$, as in Figure 5.31. Since $f \geq 0$, it follows from the definition of G in (1) that

$G(z) =$ area of the region between the graph of f and the x axis on $[c, z]$

$G(x) =$ area of the region between the graph of f and the x axis on $[c, x]$

Therefore $G(z) - G(x)$ is the area of the entire shaded region in Figure 5.31, and if z is close to x, this area appears to be close to the area $f(x)(z - x)$ of the darkly shaded rectangle in the figure. We deduce that if $z > x$ and z is close to x, then $G(z) - G(x)$ should be close to $f(x)(z - x)$, and hence that

$$\frac{G(z) - G(x)}{z - x} \quad \text{should be close to } f(x)$$

A similar analysis would show that this approximation also holds if $z < x$ and z is close to x. We conclude intuitively that

$$\lim_{z \to x} \frac{G(z) - G(x)}{z - x} = f(x)$$

that is,

$$G'(x) = f(x)$$

We now state the result in the preceding paragraph formally and prove it by means of the Mean Value Theorem for Integrals.

THEOREM 5.12

> Let f be continuous on $[a, b]$, and let $a \leq c \leq b$. Define G by the equation
>
> $$G(x) = \int_c^x f(t)\,dt \quad \text{for } a \leq x \leq b$$
>
> Then G is differentiable on $[a, b]$, and
>
> $$G'(x) = f(x) \quad \text{for } a \leq x \leq b$$

Proof Fix x in (a, b), and let $a \leq z \leq b$ with $z \neq x$. Then the Addition Property tells us that

$$\frac{G(z) - G(x)}{z - x} = \frac{\int_c^z f(t)\,dt - \int_c^x f(t)\,dt}{z - x} = \frac{\int_x^z f(t)\,dt}{z - x}$$

By the Mean Value Theorem for Integrals, there is a number $c(z)$ between x and z such that

$$\frac{\int_x^z f(t)\,dt}{z - x} = f(c(z))$$

Since $c(z)$ lies between x and z, we know that $\lim_{z \to x} c(z) = x$. Since f is continuous at x by assumption, it follows from the Substitution Rule with

$y = c(z)$ that

$$\lim_{z \to x} \frac{G(z) - G(x)}{z - x} = \lim_{z \to x} f(c(z)) = \lim_{y \to x} f(y) = f(x)$$

Thus $G'(x) = f(x)$. If $x = a$ or $x = b$, then $G'(x)$ is just the appropriate one-sided limit, and the foregoing argument can be altered to apply. ■

We have assumed throughout this discussion that f is defined on the closed interval $[a, b]$. However, the only fact about $[a, b]$ used in proving Theorem 5.12 is that for any points x and z in $[a, b]$ with $x < z$, the interval $[x, z]$ is entirely contained in $[a, b]$. But this remains true if $[a, b]$ is replaced by any other interval I, even an unbounded interval like (a, ∞) or $[a, \infty)$. Therefore we have actually proved the following more general result, which follows from Theorem 5.12 if we replace $[a, b]$ by I and substitute a for c in the formula.

COROLLARY 5.13

Let f be continuous on an interval I (containing more than one point), and let a be any point in I. Define G on I by

$$G(x) = \int_a^x f(t)\,dt \quad \text{for } x \text{ in } I$$

Then G is differentiable on I, and

$$G'(x) = f(x) \quad \text{for all } x \text{ in } I \tag{2}$$

Example 1 Let $G(x) = \int_1^x (1/t)\,dt$ for $x > 0$. Find $G'(x)$.

Solution By Corollary 5.13, with $I = (0, \infty)$ and $a = 1$, we have

$$G'(x) = \frac{1}{x} \quad \text{for } x > 0 \quad \square$$

Example 2 Let $G(x) = \int_0^x t \sin t^3\,dt$ for all x. Find $G'(x)$.

Solution Again by Corollary 5.13,

$$G'(x) = x \sin x^3 \quad \square$$

Next, suppose g is differentiable, and let

$$F(x) = \int_a^{g(x)} f(t)\,dt$$

In order to find $F'(x)$, we let

$$G(x) = \int_a^x f(t)\,dt$$

and notice that

$$F(x) = \int_a^{g(x)} f(t)\,dt = G(g(x))$$

It follows from the Chain Rule and (2) that

$$F'(x) = [G'(g(x))]g'(x) = [f(g(x))]g'(x)$$

Example 3 Let $F(x) = \int_0^{x^2} t \sin t^3\, dt$ for all x. Find $F'(x)$.

Solution If we let $G(x) = \int_0^x t \sin t^3\, dt$ and $g(x) = x^2$, then by the preceding comments,

$$F'(x) = [G'(g(x))]g'(x) = [x^2 \sin(x^2)^3](2x) = 2x^3 \sin x^6 \quad \square$$

At this point let us recall from Theorem 4.7 that all antiderivatives of a function f differ by a constant. For example, the antiderivatives of $2x$ all have the form $x^2 + C$, where C is a constant. For easy reference we list in Table 5.1 a few of the basic functions, along with their antiderivatives. To verify any entry in the column of antiderivatives, simply differentiate it and observe that its derivative is the same as the corresponding entry in the left column.

TABLE 5.1

Function	Antiderivative
c (c a constant)	$cx + C$ (C any constant)
x	$\dfrac{1}{2}x^2 + C$
x^2	$\dfrac{1}{3}x^3 + C$
$px + q$ (p and q constants)	$\dfrac{1}{2}px^2 + qx + C$
x^r (r rational, $r \neq -1$)	$\dfrac{1}{r+1}x^{r+1} + C$
$\sin x$	$-\cos x + C$
$\cos x$	$\sin x + C$

Now we are ready for the most important theorem in calculus.

THEOREM 5.14
FUNDAMENTAL
THEOREM OF CALCULUS

Let f be continuous on $[a, b]$.
a. Then f has an antiderivative on $[a, b]$.
b. If F is any antiderivative of f on $[a, b]$, then

$$\int_a^b f(t)\,dt = F(b) - F(a)$$

Proof To prove (a), we let

$$G(x) = \int_a^x f(t)\,dt \quad \text{for } a \le x \le b$$

It follows from Corollary 5.13 with $I = [a,b]$ that $G' = f$, so that G is an antiderivative of f. To prove (b), we let G be the antiderivative found in part (a). Then

$$G(a) = \int_a^a f(t)\,dt = 0 \quad \text{and} \quad G(b) = \int_a^b f(t)\,dt$$

Now if F is any antiderivative of f, then by Theorem 4.7 we know that $F = G + C$ for some constant C. Consequently

$$\int_a^b f(t)\,dt = G(b) = G(b) - G(a) = [F(b) - C] - [F(a) - C]$$
$$= F(b) - F(a) \quad \blacksquare$$

Example 4 Evaluate $\int_0^2 x^2\,dx$.

Solution From Table 5.1 we know that if $F(x) = \frac{1}{3}x^3$ then F is an antiderivative of x^2, so that by the Fundamental Theorem,

$$\int_0^2 x^2\,dx = F(2) - F(0) = \frac{1}{3}\cdot 2^3 - \frac{1}{3}\cdot 0^3 = \frac{8}{3} \quad \square$$

It is usually simplest to dispense with the symbol F when evaluating integrals. Instead we write, for example,

$$\int_0^2 x^2\,dx = \frac{x^3}{3}\bigg|_0^2 = \frac{1}{3}\cdot 2^3 - \frac{1}{3}\cdot 0^3 = \frac{8}{3}$$

where the expression $|_0^2$ indicates the numbers at which the antiderivative is to be evaluated, in this case 0 and 2.

Example 5 Evaluate $\int_1^4 x^{1/2}\,dx$.

Solution Since $\frac{2}{3}x^{3/2}$ is an antiderivative of $x^{1/2}$, the Fundamental Theorem asserts that

$$\int_1^4 x^{1/2}\,dx = \frac{2}{3}x^{3/2}\bigg|_1^4 = \frac{2}{3}(4^{3/2}) - \frac{2}{3}(1^{3/2}) = \frac{16}{3} - \frac{2}{3} = \frac{14}{3} \quad \square$$

Example 6   Evaluate $\displaystyle\int_0^{\pi/2} \cos x\,dx$.

Solution   Because $\sin x$ is an antiderivative of $\cos x$, we know from the Fundamental Theorem that

$$\int_0^{\pi/2} \cos x\,dx = \sin x\Big|_0^{\pi/2} = \sin\frac{\pi}{2} - \sin 0 = 1 - 0 = 1 \quad \square$$

Example 7   Evaluate $\displaystyle\int_{-1}^{1} (1 + 2x)\,dx$.

Solution   Using Table 5.1 or some trial and error, we find that $x + x^2$ is an antiderivative of $1 + 2x$. Thus

$$\int_{-1}^{1} (1 + 2x)\,dx = (x + x^2)\Big|_{-1}^{1} = (1 + 1^2) - [-1 + (-1)^2] = 2 \quad \square$$

We can extend the Fundamental Theorem of Calculus to the case in which the lower limit of integration is greater than the upper limit.

COROLLARY 5.15

Let f be continuous on $[a, b]$. Then for any antiderivative F of f,

$$\int_b^a f(t)\,dt = F(a) - F(b)$$

Proof   By the Fundamental Theorem,

$$\int_a^b f(t)\,dt = F(b) - F(a)$$

Therefore by Definition 5.4,

$$\int_b^a f(t)\,dt = -\int_a^b f(t)\,dt = -[F(b) - F(a)] = F(a) - F(b) \quad \blacksquare$$

As a consequence of the Fundamental Theorem and Corollary 5.15, once we know an antiderivative F of f, we can compute $\int_a^b f(t)\,dt$ by the formula

$$\int_a^b f(t)\,dt = F(b) - F(a) \tag{3}$$

whether $a < b$ or $a > b$.

Example 8 Find $\int_2^1 2t^3\,dt$.

Solution Since $\dfrac{d}{dt}\left(\dfrac{1}{2}t^4\right) = 2t^3$, we deduce from (3) that

$$\int_2^1 2t^3\,dt = \frac{1}{2}t^4\Big|_2^1 = \frac{1}{2}\cdot 1^4 - \frac{1}{2}\cdot 2^4 = \frac{1}{2} - 8 = -\frac{15}{2} \quad \square$$

Differentiation and Integration as Inverse Processes

Formula (2) can be restated as

$$\frac{d}{dx}\int_a^x f(t)\,dt = f(x) \tag{4}$$

That is, if we start with a continuous function f, integrate it to obtain $\int_a^x f(t)\,dt$, and then differentiate, the result is the original function f. Thus the differentiation has nullified the integration. On the other hand, if we start with a function F having a continuous derivative, first differentiate, and then integrate, we obtain $\int_a^x F'(t)\,dt$. But by the Fundamental Theorem of Calculus,

$$\int_a^x F'(t)\,dt = F(x) - F(a) \tag{5}$$

so we obtain the original function F altered by at most a constant. This time the integration has essentially nullified the differentiation. Thus the two basic processes of calculus, differentiation and integration, are inverses of each other.

Furthermore, whenever we know the derivative F' of a function F, (5) gives us an integration formula. For example, we know already that

$$\frac{d}{dx}\tan x = \sec^2 x$$

Therefore (5) tells us that

$$\int_{\pi/4}^x \sec^2 t\,dt = \tan x - \tan\frac{\pi}{4} = \tan x - 1$$

Formula (5) has numerous applications. For example, in economics the marginal revenue function m_R is by definition the derivative of the total revenue function R. Thus by (5),

$$R(x) - R(a) = \int_a^x R'(t)\,dt = \int_a^x m_R(t)\,dt \tag{6}$$

Likewise, the marginal cost function m_C is by definition the derivative of the total cost function C. Consequently by (5),

$$C(x) - C(a) = \int_a^x C'(t)\,dt = \int_a^x m_C(t)\,dt \tag{7}$$

In physics, the velocity of a particle moving along a straight line is the derivative of the position function. If we use t for the independent variable representing time, f for the position, v for velocity, and s for the variable of integration, we obtain, by suitable substitution in (5),

$$f(t) - f(t_0) = \int_{t_0}^t v(s)\,ds \tag{8}$$

In (8) the number t_0 is arbitrary, and it plays the same role as a does in (5). In applications t_0 is usually a special instant of time. When t_0 is the moment at which motion begins, it is called the **initial time**.

The acceleration a of a particle is the derivative of the velocity. Hence we obtain

$$v(t) - v(t_0) = \int_{t_0}^t a(s)\,ds \tag{9}$$

Near the surface of the earth, the acceleration due to gravity is essentially constant, approximately -32 feet per second per second. Assuming that an object is under the sole influence of gravity, we can derive the formula

$$h(t) = -16t^2 + v_0 t + h_0 \tag{10}$$

for the height of the object at time t. Formula (10) was first presented (without proof) in Section 1.3.

Example 9   Suppose an object undergoes a constant acceleration of -32 feet per second per second. Assume that at time $t = 0$ its initial height is h_0 and initial velocity is v_0. Show that the height $h(t)$ of the object at any time $t > 0$ is given by (10).

Solution   Using (9) with $t_0 = 0$ and $a(s) = -32$, we find that

$$v(t) - v_0 = v(t) - v(0) = \int_0^t -32\,ds = -32s\Big|_0^t = -32t$$

Thus
$$v(t) = v_0 - 32t$$

From this equation, and from (8) with $t_0 = 0$ and f replaced by h, we find that

$$h(t) - h_0 = h(t) - h(0) = \int_0^t (v_0 - 32s)\,ds = (v_0 s - 16s^2)\Big|_0^t = v_0 t - 16t^2$$

from which (10) follows immediately. $\square$

Because differentiation and integration arose from apparently unrelated problems (such as tangents and areas), it was only after mathematicians had worked for centuries with derivatives and integrals separately that Isaac Barrow (1630–1677), who was Newton's teacher, discovered and proved the Fundamental Theorem. His proof was completely geometric, and his terminology far different from ours. Beginning with the work of Newton and Leibniz, the theorem grew in importance, eventually becoming the cornerstone for the study of integration.

EXERCISES 5.4

In Exercises 1–10 find the derivative of each function.

1. $F(x) = \int_0^x t(1 + t^3)^{29}\, dt$

2. $F(x) = \int_3^x \frac{1}{(t + t^3)^{16}}\, dt$

3. $F(y) = \int_y^2 \frac{1}{t^3}\, dt$ $\left(Hint: \int_y^2 f(t)\, dt = -\int_2^y f(t)\, dt. \right)$

4. $F(t) = \int_t^0 x \sin x\, dx$

5. $F(x) = \int_0^{x^2} t \sin t\, dt$

6. $F(x) = \int_1^{-x} t \cos t^4\, dt$

7. $G(y) = \int_y^{y^2} (1 + t^2)^{1/2}\, dt$

$\left(Hint: \int_y^{y^2} f(t)\, dt = \int_y^0 f(t)\, dt + \int_0^{y^2} f(t)\, dt. \right)$

8. $F(x) = \int_{x^2}^{x^3} (1 + t^2)^{1/2}\, dt$

9. $F(x) = \dfrac{d}{dx}\displaystyle\int_0^{4x} (1 + t^2)^{4/5}\, dt$

10. $G(y) = \dfrac{d}{dy}\displaystyle\int_{\sin y}^{2y} \cos t\, dt$

In Exercises 11–36 use (3) to evaluate the integral.

11. $\int_0^1 4\, dx$

12. $\int_1^{12} 0\, dx$

13. $\int_1^3 -y\, dy$

14. $\int_5^2 -4t\, dt$

15. $\int_1^{-3} 3u\, du$

16. $\int_{-b}^b x^5\, dx$, b a constant

17. $\int_0^1 x^{100}\, dx$

18. $\int_0^2 u^{1/2}\, du$

19. $\int_{-1}^1 u^{1/3}\, du$

20. $\int_{16}^2 x^{5/4}\, dx$

21. $\int_1^4 x^{-7/9}\, dx$

22. $\int_0^1 x^{12/5}\, dx$

23. $\int_{-1.5}^{2\pi} (5 - x)\, dx$

24. $\int_0^3 \left(\frac{1}{2}x - 4\right) dx$

25. $\int_{-4}^{-1} (5x + 14)\, dx$

26. $\int_0^{\pi/6} \cos x\, dx$

27. $\int_{-\pi}^{\pi/3} \cos x\, dx$

28. $\int_{\pi/3}^{\pi/4} \sin t\, dt$

29. $\int_{\pi/3}^{-\pi/4} \sin t\, dt$

30. $\int_2^3 \frac{1}{x^3}\, dx$

31. $\int_1^2 \frac{1}{y^4}\, dy$

32. $\int_{-1}^{-2} \left(x - \frac{5}{x^3}\right) dx$

33. $\int_{\pi/6}^{\pi/2} \csc^2 t\, dt$

34. $\int_0^{\pi/4} \sec x \tan x\, dx$

35. $\int_0^{\pi/2} \left(\frac{d}{dx}\sin^5 x\right) dx$

36. $\int_{-1}^1 \left(\frac{d}{dx}\sqrt{1 + x^4}\right) dx$

In Exercises 37–46 compute the area A of the region between the graph of f and the x axis on the given interval.

37. $f(x) = x^4;\ [-1, 1]$

38. $f(x) = \dfrac{1}{x^2};\ [-2, -1]$

39. $f(x) = \sin x;\ [0, 2\pi/3]$

40. $f(x) = \cos x;\ [-\pi/2, \pi/3]$

41. $f(x) = x^{1/2};\ [1, 4]$

42. $f(x) = x^{1/3};\ [1, 8]$

43. $f(x) = \sec^2 x;\ [0, \pi/4]$

44. $f(x) = \csc x \cot x;\ [\pi/4, \pi/2]$

*45. $f(x) = \cos^2 x$; $[0, \pi/2]$ (*Hint:* What is the derivative of $x/2 + (\sin 2x)/4$?)

*46. $f(x) = \sec x \tan^3 x$; $[0, \pi/3]$ (*Hint:* What is the derivative of $(\sec^3 x)/3 - \sec x$?)

47. Evaluate $\int_0^x f(t)\, dt$ for each of the following functions. By differentiating the resulting function, verify formula (4) of this section.
   a. $f(x) = x$ b. $f(x) = -2x^2$
   c. $f(x) = -\sin x$ d. $f(x) = 10x^4$

48. In each of the following, verify formula (5) by first differentiating F, then integrating F' from a to x, and finally comparing that result with $F(x)$.
   a. $F(x) = x + 2$; $a = 1$ b. $F(x) = x^3$; $a = 1$
   c. $F(x) = x^4$; $a = -1$

49. Find the number I satisfying
$$(x_0^2 + 4x_0)\Delta x_1 + \cdots + (x_{n-1}^2 + 4x_{n-1})\Delta x_n \le I$$
$$\le (x_1^2 + 4x_1)\Delta x_1 + \cdots + (x_n^2 + 4x_n)\Delta x_n$$
for every partition $\mathscr{P} = \{x_0, x_1, \ldots, x_n\}$ of $[1, 2]$.

50. Find the number I satisfying
$$(\cos x_1 - \sin x_1)\Delta x_1 + \cdots + (\cos x_n - \sin x_n)\Delta x_n \le I$$
$$\le (\cos x_0 - \sin x_0)\Delta x_1 + \cdots + (\cos x_{n-1} - \sin x_{n-1})\Delta x_n$$
for every partition $\mathscr{P} = \{x_0, x_1, \ldots, x_n\}$ of $[0, \pi/2]$.

51. The velocities (in miles per hour) of a satellite at various times (in seconds) after takeoff are shown in the following chart:

t	0	1	2	3	4	5
v	0	5	15	50	200	500

Using the right endpoints of the subintervals, approximate the distance traveled by the satellite during the first 5 seconds.

52. Suppose the revenue from the first pound of soap sold is $3 and the marginal revenue is given by
$$m_R(x) = 4 - 0.02x \quad \text{for } 1 \le x \le 400$$
   a. What is the revenue accruing from a sale of 30 pounds of soap? (*Hint:* Use formula (6).)
   b. What is the revenue at the point at which the marginal revenue is 0?

53. Assume that the cost of producing the first two pounds of soap is $10.98 and that the marginal cost is given by
$$m_C(x) = 3 - 0.1x \quad \text{for } 0 \le x \le 30$$
Find the total cost involved in producing 30 pounds of soap. (*Hint:* Use formula (7).)

54. A stone is dropped from the top of the Washington Monument, 555 feet above ground, with initial velocity 0. Ignoring air resistance and assuming that the acceleration due to gravity is -32 feet per second per second,
   a. determine the velocity of the stone at any time t.
   b. determine the position of the stone at any time t.
   c. determine how long after release the stone hits the ground.

55. Suppose the velocity of a car, which starts from the origin at $t = 0$ and moves along the x axis, is given by
$$v(t) = 10t - t^2 \quad \text{for } 0 \le t \le 10$$
Find the position of the car
   a. at any time t, with $0 \le t \le 10$.
   b. when its acceleration is 0.

56. The velocity of a bob moving along the x axis on a spring varies with time according to the equation
$$v(t) = 2\sin t + 3\cos t$$
At $t = 0$ the position of the bob is 1. Express the position of the bob as a function of time.

57. The flow of water through a dam is controlled so that the rate $F'(t)$ of flow in tons per hour is given by the equation
$$F'(t) = 14{,}000 \sin\frac{\pi t}{24} \quad \text{for } 0 \le t \le 24$$
How many tons of water flow through the dam per day? (*Hint:* Use formula (5) and the fact that
$$\frac{d}{dt}\left(-\frac{336{,}000}{\pi}\cos\frac{\pi t}{24}\right) = F'(t)$$
for $0 \le t \le 24$.)

58. To make a balloon rise faster, the balloonist drops a sandbag from the bottom of the balloon.
   a. If the sandbag is ejected at an elevation of 528 feet and takes 6 seconds to hit the ground, what was the speed of the balloon at the time of ejection?
   b. If the balloon rises at 4 feet per second and is 992 feet above the ground when the sandbag is ejected, how long will it take the sandbag to hit the ground?

59. A train is cruising at 60 miles per hour when suddenly the engineer notices a cow on the track ahead of the train. The engineer applies the brakes, causing a constant deceleration in the train. Two minutes later the train grinds to a halt, barely touching the cow, which is too frightened to move. How far back was the train when the brakes were applied?

60. Radium disintegrates at a variable rate. Let $R(t)$ be the rate at which radium disintegrates at time t. Express the total amount lost between times t_1 and t_2 as an integral.

61. It can be shown that the volume V of a sphere of radius r is given by the integral

$$\int_{-r}^{r} \pi(r^2 - x^2)\,dx$$

Show that $V = 4\pi r^3/3$.

62. a. For $a \le t \le b$ let $v(t)$ be the velocity of an object at time t. Using the Fundamental Theorem, show that

$$\frac{\int_a^b v(t)\,dt}{b - a}$$

is the average velocity of the object as defined in Section 2.1.

b. Suppose a ball thrown from the top of a cliff has the velocity

$$v(t) = -20 - 32t \quad \text{for } 0 \le t \le 3$$

Find the ball's average velocity during its flight.

63. a. Let $C(x)$ be the cost of producing x units of an object for $a \le x \le b$. Using the Fundamental Theorem, show that

$$\frac{\int_a^b m_C(x)\,dx}{b - a}$$

is the average cost between the ath and bth units produced, as defined in Section 3.1.

b. Suppose an umbrella manufacturing company has a marginal cost function given by

$$m_C(x) = \frac{1}{x^{1/2}} \quad \text{for } 1 \le x \le 4$$

where x represents thousands of umbrellas produced and $m_C(x)$ is measured in thousands of dollars. Find the average cost between the one-thousandth umbrella produced and the four-thousandth umbrella produced.

64. Suppose the voltage in an electrical circuit varies with time according to the formula

$$V(t) = 110 \sin t \quad \text{for } 0 \le t \le \pi$$

Find the average voltage in the circuit.

65. A cylindrical tank 100 feet high and 100 feet in diameter is full of water. The work W (in foot-pounds) required to pump all the water out of the tank is given by

$$W = (2500\pi)(62.5)\int_0^{100} (100 - y)\,dy$$

Compute the work required.

66. A tile manufacturing company plans to produce enamel tiles with the design and color scheme shown in Figure 5.32. Will more white enamel be required than gray?

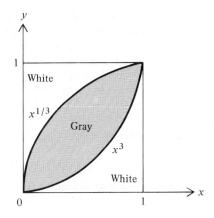

FIGURE 5.32

*67. In a famous eighteenth-century problem known as **Buffon's needle problem**, a needle 1 inch long is dropped onto a hardwood floor whose boards are 2 inches wide. We are asked to determine the probability that the needle will come to rest lying across two boards. As in Figure 5.33(a), we let P be the southernmost point of the

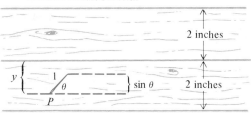

(a)

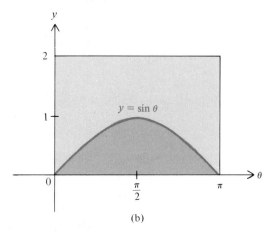

(b)

FIGURE 5.33

needle (or the left-hand endpoint of the needle if the needle lies horizontally). Let y denote the distance from P to the next crack northward, so that $0 \le y < 2$. Let θ denote the positive angle the needle makes with the horizontal to the right of P, with $0 \le \theta < \pi$. Observe that the needle crosses the crack *only* when the pair (θ, y) satisfies $y \le \sin \theta$, that is, when (θ, y) lies in the shaded area in Figure 5.33(b). Since the total set of possibilities for the needle can be identified with the rectangular region with $0 \le y < 2$ and $0 \le \theta < \pi$, the proportion of times that the needle crosses the crack is the fraction

$$\frac{\text{area of shaded region}}{\text{area of rectangle}}$$

It is reasonable to take this proportion to be the probability sought. Calculate this proportion.

68. Rework Exercise 67 with a needle 2 inches long and boards 2 inches wide.

69. Rework Exercise 67 with a needle 2 inches long and boards 5 inches wide.

5.5
INDEFINITE INTEGRALS AND INTEGRATION RULES

When we evaluate a definite integral $\int_a^b f(x)\,dx$ using the Fundamental Theorem of Calculus, the basic problem is to find an antiderivative of f. In this section we present some elementary rules that will help us find antiderivatives of combinations of functions. Recall that if F is an antiderivative of f, then for any constant C the function $F + C$ is also an antiderivative of f, since

$$(F + C)' = F' + C' = f$$

Consequently every function f that has an antiderivative has infinitely many of them—one for each choice of C. By Theorem 4.7 the functions of the form $F + C$, where C is an arbitrary constant, are the *only* antiderivatives of f. As a result, mathematicians group the antiderivatives of f together and call them the indefinite integral of f.

DEFINITION 5.16

> Let f have an antiderivative on an interval I. The collection of antiderivatives of f is called the **indefinite integral** of f on I and is denoted by $\int f(x)\,dx$.

Mathematicians use the notation

$$\int f(x)\,dx = F(x) + C$$

when finding indefinite integrals. For example,

$$\int x^2\,dx = \frac{1}{3}x^3 + C$$

This equation means that $\frac{1}{3}x^3$ is a function whose derivative is x^2 and that the only other functions having derivative x^2 are of the form $\frac{1}{3}x^3 + C$, where C is a constant.

In the notation of indefinite integrals, Table 5.1 in Section 5.4 yields Table 5.2.

TABLE 5.2

$$\int c\,dx = cx + C \quad (c \text{ a constant})$$

$$\int x\,dx = \frac{1}{2}x^2 + C$$

$$\int x^2\,dx = \frac{1}{3}x^3 + C$$

$$\int (px + q)\,dx = \frac{1}{2}px^2 + qx + C \quad (p \text{ and } q \text{ constants})$$

$$\int x^r\,dx = \frac{1}{r+1}x^{r+1} + C \quad (r \text{ rational}, r \neq -1)$$

$$\int \sin x\,dx = -\cos x + C$$

$$\int \cos x\,dx = \sin x + C$$

We now show how to obtain the indefinite integrals of certain combinations of functions from the indefinite integrals of the individual functions.

THEOREM 5.17

If f and g have antiderivatives on an interval I, then

$$\int [f(x) + g(x)]\,dx = \int f(x)\,dx + \int g(x)\,dx$$

Proof This formula means that the sum of an antiderivative of f and an antiderivative of g is an antiderivative of $f + g$. To prove this, let F and G be antiderivatives of f and g, respectively. Then

$$(F + G)' = F' + G' = f + g$$

Hence $F + G$ is an antiderivative of $f + g$. ∎

If f and g are continuous on $[a, b]$, then by the Fundamental Theorem we have the corresponding equation for definite integrals:

$$\int_a^b [f(x) + g(x)]\,dx = \int_a^b f(x)\,dx + \int_a^b g(x)\,dx \qquad (1)$$

Moreover, Theorem 5.17 can be extended to the sum of more than two functions (see Exercise 55).

THEOREM 5.18

If f has an antiderivative on an interval I, and c is a real number, then

$$\int cf(x)\,dx = c \int f(x)\,dx$$

Proof   This formula means that c times an antiderivative of f is an antiderivative of cf. But if F is an antiderivative of f, then

$$(cF)' = c(F') = cf$$

so that cF is an antiderivative of cf.   ■

If f is continuous on $[a, b]$, then the Fundamental Theorem of Calculus implies that the corresponding formula for definite integrals holds:

$$\int_a^b cf(x)\,dx = c \int_a^b f(x)\,dx \tag{2}$$

From Theorems 5.17 and 5.18 we immediately have the following corollary, whose proof is left as an exercise (see Exercise 54).

COROLLARY 5.19

If f and g have antiderivatives on an interval I, then

$$\int [f(x) - g(x)]\,dx = \int f(x)\,dx - \int g(x)\,dx$$

If f and g are continuous on $[a, b]$, then the Fundamental Theorem yields

$$\int_a^b [f(x) - g(x)]\,dx = \int_a^b f(x)\,dx - \int_a^b g(x)\,dx \tag{3}$$

Example 1   Evaluate the indefinite integral $\displaystyle\int (2x - 3\cos x)\,dx$.

Solution   Using first Corollary 5.19 and then Theorem 5.18, we obtain

$$\int (2x - 3\cos x)\,dx = \int 2x\,dx - \int 3\cos x\,dx$$

$$= 2\int x\,dx - 3\int \cos x\,dx = 2\left(\frac{1}{2}x^2\right) - 3\sin x + C$$

$$= x^2 - 3\sin x + C \quad \square$$

Example 2   Evaluate the definite integral $\int_0^1 (4x^2 + 5x^3)\,dx$.

Solution   First method: We break the integral into its components and then integrate. Using (1) and (2), we obtain

$$\int_0^1 (4x^2 + 5x^3)\,dx \overset{(1)}{=} \int_0^1 4x^2\,dx + \int_0^1 5x^3\,dx$$

$$\overset{(2)}{=} 4\int_0^1 x^2\,dx + 5\int_0^1 x^3\,dx$$

$$= \left(4 \cdot \frac{1}{3}x^3\right)\bigg|_0^1 + \left(5 \cdot \frac{1}{4}x^4\right)\bigg|_0^1$$

$$= \frac{4}{3} + \frac{5}{4} = \frac{31}{12}$$

Second method: Here we first find an antiderivative of $4x^2 + 5x^3$ and then evaluate it:

$$\int_0^1 (4x^2 + 5x^3)\,dx = \left(\frac{4}{3}x^3 + \frac{5}{4}x^4\right)\bigg|_0^1 = \frac{4}{3} + \frac{5}{4} = \frac{31}{12} \quad \square$$

Repeated applications of Theorems 5.17 and 5.18 yield the indefinite integral of a general polynomial:

$$\int [c_n x^n + c_{n-1} x^{n-1} + \cdots + c_1 x + c_0]\,dx$$

$$= \frac{c_n}{n+1} x^{n+1} + \frac{c_{n-1}}{n} x^n + \cdots + \frac{c_1}{2} x^2 + c_0 x + C$$

Thus to find the integral of a polynomial

$$c_n x^n + c_{n-1} x^{n-1} + \cdots + c_1 x + c_0$$

just integrate the terms $c_n x^n$, $c_{n-1} x^{n-1}, \ldots, c_1 x$ and c_0 one by one and add the results.

Example 3   Evaluate $\displaystyle\int_1^2 (x^4 - 3x^2 + 4x - 2)\, dx$.

Solution   Integrating the terms one by one, we find that

$$\int_1^2 (x^4 - 3x^2 + 4x - 2)\, dx = \left[\frac{1}{5}x^5 - 3\left(\frac{1}{3}x^3\right) + 4\left(\frac{1}{2}x^2\right) - 2x\right]\Big|_1^2$$

$$= \left(\frac{1}{5}\cdot 32 - 8 + 8 - 4\right) - \left(\frac{1}{5} - 1 + 2 - 2\right)$$

$$= \frac{12}{5} - \left(-\frac{4}{5}\right) = \frac{16}{5} \quad \square$$

Formula (3) yields a simple proof of the following result, which we could also have proved, with more difficulty, in Section 5.3.

COROLLARY 5.20
GENERAL
COMPARISON
PROPERTY

Let f and g be continuous on $[a, b]$, with $g(x) \le f(x)$ for $a \le x \le b$. Then

$$\int_a^b g(x)\, dx \le \int_a^b f(x)\, dx$$

Proof   From the hypotheses it follows that $f(x) - g(x) \ge 0$ for $a \le x \le b$. Then by (3) and Corollary 5.10 with $f - g$ replacing f,

$$\int_a^b f(x)\, dx - \int_a^b g(x)\, dx \overset{(3)}{=} \int_a^b [f(x) - g(x)]\, dx \overset{\substack{\text{Corollary} \\ 5.10}}{\ge} 0$$

Consequently

$$\int_a^b g(x)\, dx \le \int_a^b f(x)\, dx \quad \blacksquare$$

In effect, Corollary 5.20 says that the larger the function, the larger the definite integral. Of course if $b < a$, then all the inequalities must be reversed (see Exercise 53).

Another consequence of Corollary 5.20 is that if $h(x) \le g(x) \le f(x)$ for $a \le x \le b$, then

$$\int_a^b h(x)\, dx \le \int_a^b g(x)\, dx \le \int_a^b f(x)\, dx \tag{4}$$

These inequalities can lead to good lower and upper bounds for integrals that are difficult or impossible to find directly.

Example 4   Show that $2 \le \int_{-1}^{1} \sqrt{1 + x^4}\, dx \le \frac{8}{3}$.

Solution   We have

$$1 \le 1 + x^4 \le 1 + 2x^2 + x^4 = (1 + x^2)^2$$

which implies that

$$1 \le \sqrt{1 + x^4} \le \sqrt{(1 + x^2)^2} = 1 + x^2$$

From (4) we conclude that

$$2 = \int_{-1}^{1} 1\, dx \le \int_{-1}^{1} \sqrt{1 + x^4}\, dx \le \int_{-1}^{1} (1 + x^2)\, dx = \left(x + \frac{x^3}{3} \right)\Big|_{-1}^{1} = \frac{8}{3} \quad \square$$

Since $\frac{8}{3} < 2\sqrt{2}$, Example 4 yields a smaller upper bound for the integral $\int_{-1}^{1} \sqrt{1 + x^4}\, dx$ than the one found in Example 3 of Section 5.3. An even smaller upper bound will be obtained in Section 8.6.

Since $-|f(x)| \le f(x) \le |f(x)|$ for $a \le x \le b$, we obtain as a special case of (4) the inequalities

$$-\int_{a}^{b} |f(x)|\, dx = \int_{a}^{b} -|f(x)|\, dx \le \int_{a}^{b} f(x)\, dx \le \int_{a}^{b} |f(x)|\, dx$$

Therefore, by the definition of the absolute value,

$$\left| \int_{a}^{b} f(x)\, dx \right| \le \int_{a}^{b} |f(x)|\, dx \tag{5}$$

EXERCISES 5.5

In Exercises 1–12 evaluate the indefinite integral.

1. $\int (2x - 7)\, dx$

2. $\int (2x^2 - 7x^3 + 4x^4)\, dx$

3. $\int (2x^{1/3} - 3x^{3/4} + x^{2/5})\, dx$

4. $\int (x^{3/2} + 4x^{1/2} - \pi)\, dx$

5. $\int \left(t^5 - \frac{1}{t^4} \right) dt$

6. $\int \left(\sqrt{y} + \frac{1}{\sqrt{y}} \right) dy$

7. $\int (2\cos x - 5x)\, dx$

8. $\int (\theta^2 + \sec^2 \theta)\, d\theta$

9. $\int (3\csc^2 x - x)\, dx$

10. $\int \left(\frac{1}{y^3} - \frac{1}{y^2} + 2y \right) dy$

11. $\int (2t + 1)^2\, dt$ (*Hint:* Expand the binomial.)

12. $\int \left(t + \frac{1}{t} \right)^2 dt$

In Exercises 13–32 evaluate the definite integral.

13. $\int_{-1}^{2} (3x - 4)\, dx$

14. $\int_{1}^{4} \left(\sqrt{x} + \frac{1}{2\sqrt{x}} \right) dx$

15. $\int_{\pi/4}^{\pi/2} (-7\sin x + 3\cos x)\, dx$

16. $\int_0^\pi (\sin x - 8x^2)\, dx$

17. $\int_{-\pi/4}^{-\pi/2} \left(3x - \dfrac{1}{x^2} + \sin x\right) dx$

18. $\int_{-1/4}^{1/4} (4t - 3)^2\, dt$

19. $\int_{\pi/3}^{\pi/4} (3\sec^2\theta + 4\csc^2\theta)\, d\theta$

20. $\int_1^2 \left(t^2 - \dfrac{1}{t^2}\right)^2 dt$

21. $\int_1^{1/3} (3t + 2)^3\, dt$

22. $\int_{-\pi/4}^0 (\sec\theta)(\tan\theta + \sec\theta)\, d\theta$
    (Hint: First multiply out the product.)

23. $\int_{\pi/2}^\pi \left(\pi\sin x - 2x + \dfrac{5}{x^2} + 2\pi\right) dx$

24. $\int_1^0 x(2x + 5)\, dx$

25. $\int_{-1}^1 (2x + 5)(2x - 5)\, dx$

26. $\int_2^1 (x + 3)^2(x + 1)\, dx$ 27. $\int_4^7 |x - 5|\, dx$

28. $\int_1^0 |2x - 1|\, dx$ 29. $\int_{-3}^4 |-5x + 2|\, dx$

30. $\int_4^6 f(x)\, dx$, where

$$f(x) = \begin{cases} 2x & \text{for } 4 \le x \le 5 \\ 20 - 2x & \text{for } 5 < x \le 6 \end{cases}$$

31. $\int_0^{\pi/2} f(x)\, dx$, where

$$f(x) = \begin{cases} \sec^2 x & \text{for } 0 \le x \le \pi/4 \\ \csc^2 x & \text{for } \pi/4 < x \le \pi/2 \end{cases}$$

32. $\int_{-2}^2 f(x)\, dx$, where

$$f(x) = \begin{cases} |x - 2| & \text{for } -2 \le x \le 1 \\ |x| & \text{for } 1 < x \le 2 \end{cases}$$

In Exercises 33–38 differentiate the function F. Then give F in terms of an indefinite integral. For example, if $F(x) = \cos x^2$, then since $d(\cos x^2)/dx = -2x\sin x^2$, we obtain

$$\int -2x\sin x^2\, dx = \cos x^2 + C$$

33. $F(x) = (1 + x^2)^{10}$ 34. $F(x) = \frac{1}{2}(1 + x)^2$

35. $F(x) = x\sin x - \cos x$ 36. $F(x) = \tan(2x + 1)$

37. $F(x) = 3\sin^7 x$ 38. $F(x) = x\sin x\cos x$

In Exercises 39–44 find the area A of the region between the graph of f and the x axis on the given interval.

39. $f(x) = 3x^2 + 4;\ [-1, 1]$

40. $f(x) = \frac{1}{2}x^3 + 3x;\ [1, 2]$

41. $f(x) = 3\sqrt{x} - 1/\sqrt{x};\ [1, 4]$

42. $f(x) = 8x^{1/3} - x^{-1/3};\ [1, 8]$

43. $f(x) = 2\sin x + 3\cos x;\ [\pi/4, \pi/2]$

44. $f(x) = |x + 1|;\ [-2, 0]$

45. Use the inequalities $0 \le \sin x \le x$ for $0 \le x \le 1$ to show that

   a. $0 \le \displaystyle\int_0^1 \sin x^2\, dx \le \frac{1}{3}$

   b. $0 \le \displaystyle\int_0^{\pi/6} \sin^{3/2} x\, dx \le \frac{2}{5}(\pi/6)^{5/2}$

46. Use the inequalities $0 \le \sin x \le x$ for $0 \le x \le \frac{1}{2}$ to show that

$$0 \le \int_0^{1/2} x\sin x\, dx \le \tfrac{1}{24}$$

47. Let $f(x) = 1 - x$. Show that $|\int_0^2 f(x)\, dx| < \int_0^2 |f(x)|\, dx$. (This inequality shows that the inequality sign in (5) cannot in general be replaced by an equals sign.)

48. Let f be continuous on $[a, b]$, and let $|f(x)| \le M$ for $a \le x \le b$. Prove that

$$\left|\int_a^b f(x)\, dx\right| \le M(b - a)$$

49. Use Exercise 48 to find an upper bound for

$$\left|\int_{-\pi/3}^{-\pi/4} \tan x\, dx\right|$$

50. Show that

$$\lim_{\varepsilon \to 0^+} \frac{1}{\varepsilon} \int_0^\varepsilon x\sin x\, dx = 0$$

(Hint: $0 \le \sin x \le x$ for $x \ge 0$.)

*51. Suppose f has a bounded derivative on $[a, b]$, so that there is a number M such that $|f'(x)| \le M$ for $a \le x \le b$. Assume that $f(a) = 0$.

   a. Using the Mean Value Theorem, show that

$$|f(x)| \le M(x - a) \quad \text{for } a \le x \le b$$

   b. Using the result of (a) and formula (5), show that

$$\left|\int_a^b f(x)\, dx\right| \le \frac{M}{2}(b - a)^2$$

52. Using Exercise 51(b), find upper bounds for the absolute values of the following integrals.

    a. $\int_{1/2}^{1} \left(\frac{1}{x} - 1\right) dx$

    *b. $\int_{0}^{\pi/6} \sin^{3/2} x \, dx$ (*Hint:* Show first that $d(\sin^{3/2} x)/dx \leq \frac{3}{4}\sqrt{2}$ for $0 \leq x \leq \pi/6$.)

53. Suppose that $b < a$ and that f and g are continuous, with $g(x) \leq f(x)$ for $b \leq x \leq a$. Use Corollary 5.20 to prove that

$$\int_{a}^{b} g(x) \, dx \geq \int_{a}^{b} f(x) \, dx$$

54. Prove Corollary 5.19.

55. Prove that if $f_1, f_2, \ldots, f_n$ have antiderivatives on an interval I, then

$$\int [f_1(x) + f_2(x) + \cdots + f_n(x)] \, dx$$

$$= \int f_1(x) \, dx + \int f_2(x) \, dx + \cdots + \int f_n(x) \, dx$$

*56. Find an upper bound for the area $\pi/4$ of the quarter of the unit circle in the first quadrant by calculating the area of the region formed by the tangent to the circle $x^2 + y^2 = 1$ at $(\sqrt{3}/2, \frac{1}{2})$, the x and y axes, and the line $x = 1$.

*57. a. Use the method of Exercise 56 to find an upper bound for the area $\pi/2$ of the quarter of the ellipse

$$\frac{x^2}{4} + y^2 = 1$$

in the first quadrant. (*Hint:* Consider the tangent line at $(\sqrt{2}, 1/\sqrt{2})$.)

   b. Find a lower bound for the area $\pi/2$ by constructing a triangle inside the ellipse.

58. Let a, b, c, and d be positive numbers with $a < b$. The region R bounded by the graph of a function of the form $f(x) = cx + d$, the x axis, and the lines $x = a$ and $x = b$ is a right trapezoid. Show that the area A of R is given by

$$\left(\frac{f(a) + f(b)}{2}\right)(b - a)$$

*59. Let f and g be continuous on $[a, b]$. Show that

$$\left(\int_{a}^{b} f(x)g(x) \, dx\right)^2 \leq \int_{a}^{b} [f(x)]^2 \, dx \int_{a}^{b} [g(x)]^2 \, dx$$

(*Hint:* Let $p(r) = \int_{a}^{b} [f(x) + rg(x)]^2 \, dx$ for all real numbers r. Show that $p(r)$ has the form $Ar^2 + Br + C$, where A, B, and C are constants. Then show that $p(r) \geq 0$ for all r and deduce that $B^2 - 4AC \leq 0$, which yields the desired inequality.)

60. Let r denote the radius of a cylindrical artery of length l and x denote the distance of a given blood cell from the center of a cross section of the artery. The volume V per unit time of the flow of blood through the artery is given by the formula

$$V = \int_{0}^{r} \frac{k}{l} x(r^2 - x^2) \, dx$$

where k is a constant depending on the difference in pressure at the two ends of the artery and on the viscosity of the blood. Calculate V.

5.6
INTEGRATION BY SUBSTITUTION

In the preceding section we transformed addition and constant multiple theorems for derivatives into corresponding theorems for integrals (see Theorems 5.17 and 5.18). Presently we will transform the Chain Rule in the form of

$$\frac{d}{dx} G(f(x)) = G'(f(x))f'(x)$$

into a theorem for integrals. The result we will obtain will be as useful in integration as the Chain Rule is in differentiation, and it will allow us to express many integrals, such as

$$\int \sin^4 x \cos x \, dx \quad \text{and} \quad \int x\sqrt{2x + 1} \, dx$$

in terms of functions familiar to us.

THEOREM 5.21

Let f and g be functions, with both $g \circ f$ and f' continuous on an interval I. If G is an antiderivative of g, then

$$\int g(f(x))f'(x)\,dx = G(f(x)) + C \qquad\qquad (1)$$

Proof Since G is an antiderivative of g, we have $G'(x) = g(x)$. Therefore the Chain Rule implies that

$$\frac{d}{dx}G(f(x)) = G'(f(x))f'(x) = g(f(x))f'(x)$$

In terms of indefinite integrals this becomes

$$\int g(f(x))f'(x)\,dx = G(f(x)) + C$$

which is (1). ■

In the process of applying (1) it is usually convenient to substitute u for $f(x)$ and du for $f'(x)\,dx$. Thus we obtain

$$\int g(\overset{u}{\overbrace{f(x)}})\overset{du}{\overbrace{f'(x)\,dx}} = \int g(u)\,du = G(u) + C = G(f(x)) + C$$

which shows clearly the integration of g to obtain G. For this reason, evaluating an integral by means of (1) is called **integration by substitution**. We illustrate the method in the examples that follow.

Example 1 Find $\displaystyle\int 3x^2(x^3 + 5)^9\,dx$.

Solution We let

$$u = x^3 + 5, \quad \text{so that} \quad du = 3x^2\,dx$$

Then

$$\int 3x^2(x^3 + 5)^9\,dx = \int \overset{u^9}{\overbrace{(x^3 + 5)^9}}\,\overset{du}{\overbrace{(3x^2)\,dx}} = \int u^9\,du = \frac{1}{10}u^{10} + C$$

$$= \frac{1}{10}(x^3 + 5)^{10} + C \quad \square$$

Notice that after the integration was performed we resubstituted $x^3 + 5$ for u, so the answer would be expressed in terms of the original variable x. The variable u is only a temporary convenience.

We observe that it would also be possible to find $\int 3x^2(x^3 + 5)^9\, dx$ by expanding the polynomial $3x^2(x^3 + 5)^9$ and integrating term by term. However, integrating by substitution is much more efficient.

Example 2   Find $\int \sin^4 x \cos x\, dx$.

Solution   We let

$$u = \sin x, \quad \text{so that} \quad du = \cos x\, dx$$

Then

$$\int \sin^4 x \cos x\, dx = \int \overbrace{(\sin x)^4}^{u^4}\overbrace{\cos x\, dx}^{du} = \int u^4\, du = \frac{1}{5}u^5 + C = \frac{1}{5}\sin^5 x + C \quad \square$$

The substitution of u for $x^3 + 5$ worked well in Example 1 because $du = 3x^2\, dx$ and because $3x^2\, dx$ appeared in the original integral. Similarly, the substitution of u for $\sin x$ worked well in Example 2 because $du = \cos x\, dx$ and because $\cos x\, dx$ appeared in the original integral. The method of substitution can still be applied if merely a constant multiple of du appears in the original integral.

Example 3   Find $\int \frac{1}{2}\cos 2x\, dx$.

Solution   We let

$$u = 2x, \quad \text{so that} \quad du = 2\, dx, \quad \text{and thus} \quad dx = \tfrac{1}{2}\, du$$

Then

$$\int \frac{1}{2}\cos 2x\, dx = \int \frac{1}{2}(\cos \overbrace{2x}^{u})\overbrace{dx}^{\frac{1}{2}du} = \int \frac{1}{2}(\cos u)\frac{1}{2}\, du$$

$$= \frac{1}{4}\int \cos u\, du = \frac{1}{4}\sin u + C = \frac{1}{4}\sin 2x + C \quad \square$$

We can use the result of Example 3 to evaluate $\int \cos^2 x\, dx$. Indeed, by using the trigonometric identity

$$\cos^2 x = \tfrac{1}{2} + \tfrac{1}{2}\cos 2x$$

along with Example 3 and the integration rules of Section 5.5, we find that

$$\int \cos^2 x\, dx = \int \left(\frac{1}{2} + \frac{1}{2}\cos 2x\right) dx = \int \frac{1}{2}\, dx + \int \frac{1}{2}\cos 2x\, dx$$

$$= \frac{1}{2}x + \frac{1}{4}\sin 2x + C$$

Therefore

$$\int \cos^2 x \, dx = \frac{1}{2}x + \frac{1}{4}\sin 2x + C \tag{2}$$

By a similar argument,

$$\int \sin^2 x \, dx = \frac{1}{2}x - \frac{1}{4}\sin 2x + C \tag{3}$$

The integrals $\int \sin^2 x \, dx$ and $\int \cos^2 x \, dx$ will occur from time to time throughout the remainder of this book.

Example 4   Find $\int t \sin t^2 \, dt$.

Solution   We let

$$u = t^2, \quad \text{so that} \quad du = 2t \, dt$$

Then

$$\int t \sin t^2 \, dt = \int \overbrace{(\sin t^2)}^{\sin u} \overbrace{t \, dt}^{\frac{1}{2}du} = \int (\sin u)\frac{1}{2} \, du$$

$$= -\frac{1}{2}\cos u + C = -\frac{1}{2}\cos t^2 + C \quad \square$$

Occasionally it is convenient to solve for x (or some expression involving x) in terms of u in order to complete the substitution in the original integral.

Example 5   Find $\int x\sqrt{2x + 1} \, dx$.

Solution   To simplify the expression $\sqrt{2x + 1}$, we let

$$u = 2x + 1, \quad \text{so that} \quad du = 2 \, dx$$

Then

$$\int x\sqrt{2x + 1} \, dx = \int x \overbrace{\sqrt{2x + 1}}^{u^{1/2}} \overbrace{dx}^{\frac{1}{2}du}$$

Thus we still need to find x in terms of u. From the equation $u = 2x + 1$ we deduce that

$$x = \frac{1}{2}(u - 1)$$

Therefore

$$\int x\sqrt{2x + 1}\, dx = \int \overbrace{x}^{\frac{1}{2}(u-1)}\ \overbrace{\sqrt{2x + 1}}^{u^{1/2}}\ \overbrace{dx}^{\frac{1}{2}du}$$

$$= \int \frac{1}{2}(u - 1)u^{1/2} \cdot \frac{1}{2}\, du = \frac{1}{4}\int (u^{3/2} - u^{1/2})\, du$$

$$= \frac{1}{4}\left(\frac{2}{5}u^{5/2} - \frac{2}{3}u^{3/2}\right) + C$$

$$= \frac{1}{10}(2x + 1)^{5/2} - \frac{1}{6}(2x + 1)^{3/2} + C \quad \square$$

Frequently there is more than one substitution that will work. For instance, in Example 5 we could have let $u = \sqrt{2x + 1}$. Then

$$x = \frac{1}{2}(u^2 - 1) \quad \text{and} \quad du = \frac{1}{\sqrt{2x + 1}}\, dx = \frac{1}{u}\, dx, \quad \text{so} \quad u\, du = dx$$

As a result,

$$\int x\sqrt{2x + 1}\, dx = \int \overbrace{x}^{\frac{1}{2}(u^2-1)}\ \overbrace{\sqrt{2x + 1}}^{u}\ \overbrace{dx}^{u\,du}$$

$$= \int \frac{1}{2}(u^2 - 1)u^2\, du$$

$$= \int \left(\frac{1}{2}u^4 - \frac{1}{2}u^2\right) du = \frac{1}{10}u^5 - \frac{1}{6}u^3 + C$$

$$= \frac{1}{10}(2x + 1)^{5/2} - \frac{1}{6}(2x + 1)^{3/2} + C$$

Even though we used a different substitution, the final answer remains the same.

Example 6   Find $\int x^5\sqrt{x^2 - 1}\, dx$.

Solution   In order to simplify the integrand, we let

$$u = \sqrt{x^2 - 1}, \quad \text{so that} \quad du = \frac{x}{\sqrt{x^2 - 1}}\, dx, \quad \text{and thus} \quad u\, du = x\, dx$$

Then we factor out an x from x^5 so that $x\,dx$, which equals $u\,du$, appears in the integral:

$$\int x^5\sqrt{x^2-1}\,dx = \int x^4\overbrace{\sqrt{x^2-1}}^{u}\overbrace{x\,dx}^{u\,du}$$

Now we need to write x^4 in terms of u:

$$u = \sqrt{x^2-1}, \quad\text{so}\quad u^2 = x^2-1$$

Thus
$$x^2 = u^2+1, \quad\text{so}\quad x^4 = (u^2+1)^2$$

Therefore

$$\int x^5\sqrt{x^2-1}\,dx = \int \overbrace{x^4}^{(u^2+1)^2}\ \overbrace{\sqrt{x^2-1}}^{u}\ \overbrace{x\,dx}^{u\,du} = \int (u^2+1)^2\, u\cdot u\,du$$

$$= \int (u^6+2u^4+u^2)\,du = \tfrac{1}{7}u^7 + \tfrac{2}{5}u^5 + \tfrac{1}{3}u^3 + C$$

$$= \tfrac{1}{7}(\sqrt{x^2-1})^7 + \tfrac{2}{5}(\sqrt{x^2-1})^5 + \tfrac{1}{3}(\sqrt{x^2-1})^3 + C \quad \square$$

Substitution with Definite Integrals

Suppose we wish to evaluate a definite integral of the form $\int_a^b g(f(x))f'(x)\,dx$. Using (1) and the Fundamental Theorem of Calculus, we find that

$$\int_a^b g(f(x))f'(x)\,dx = G(f(x))\Big|_a^b = G(f(b)) - G(f(a)) \tag{4}$$

However, since G is an antiderivative of g, we also have

$$\int_{f(a)}^{f(b)} g(u)\,du = G(u)\Big|_{f(a)}^{f(b)} = G(f(b)) - G(f(a)) \tag{5}$$

From (4) and (5) it follows that

$$\int_a^b g(f(x))f'(x)\,dx = \int_{f(a)}^{f(b)} g(u)\,du \tag{6}$$

Thus we now have two methods of evaluating a definite integral of the form $\int_a^b g(f(x))f'(x)\,dx$. One way is to find the indefinite integral $\int g(f(x))f'(x)\,dx$ by substitution, and then evaluate it between the limits a and b, as in (4). The other way is to use (6), which involves using substitution, but with the limits of integration changed before we integrate. Formula (6) is called the **change of variable formula** in integration.

We will illustrate both methods of evaluating a definite integral by substitution in our final example.

Example 7 Evaluate $\displaystyle\int_0^1 \frac{x^5}{(x^6 + 1)^3}\,dx$.

Solution For the first method we find the indefinite integral $\int x^5/(x^6 + 1)^3\,dx$ and then evaluate it between 0 and 1. To achieve this, we let

$$u = x^6 + 1, \quad \text{so that} \quad du = 6x^5\,dx$$

Then

$$\int \frac{x^5}{(x^6 + 1)^3}\,dx = \int \overbrace{\frac{1}{(x^6 + 1)^3}}^{1/u^3}\overbrace{x^5\,dx}^{\frac{1}{6}du} = \int \frac{1}{u^3}\cdot\frac{1}{6}\,du$$

$$= \frac{1}{6}\left(-\frac{1}{2}\cdot\frac{1}{u^2}\right) + C = -\frac{1}{12}\cdot\frac{1}{(x^6 + 1)^2} + C$$

Therefore

$$\int_0^1 \frac{x^5}{(x^6 + 1)^3}\,dx = -\frac{1}{12}\cdot\frac{1}{(x^6 + 1)^2}\Big|_0^1 = -\frac{1}{12}\cdot\frac{1}{4} - \left(-\frac{1}{12}\right) = \frac{1}{16}$$

For the second method we make the same substitution as before but accompany it with a change in limits of integration. Since $u = x^6 + 1$, it follows that

$$\text{if } x = 0 \text{ then } u = 1, \quad \text{and} \quad \text{if } x = 1 \text{ then } u = 2$$

Consequently

$$\int_0^1 \frac{x^5}{(x^6 + 1)^3}\,dx = \int_1^2 \frac{1}{u^3}\cdot\frac{1}{6}\,du = \frac{1}{6}\left(-\frac{1}{2}\cdot\frac{1}{u^2}\right)\Big|_1^2$$

$$= \frac{1}{6}\left(-\frac{1}{8}\right) - \frac{1}{6}\left(-\frac{1}{2}\right) = \frac{1}{16} \quad \square$$

EXERCISES 5.6

In Exercises 1–14 evaluate the integral by making the indicated substitution.

1. $\displaystyle\int \sqrt{4x - 5}\,dx;\ u = 4x - 5$

2. $\displaystyle\int (1 - 5x^2)^{2/3}(10x)\,dx;\ u = 1 - 5x^2$

3. $\displaystyle\int \cos \pi x\,dx;\ u = \pi x$

4. $\displaystyle\int 3\sin(-2x)\,dx;\ u = -2x$

5. $\displaystyle\int x\cos x^2\,dx;\ u = x^2$

6. $\displaystyle\int \sin^3 t\cos t\,dt;\ u = \sin t$

7. $\displaystyle\int \cos^{-4} t\sin t\,dt;\ u = \cos t$

8. $\displaystyle\int \frac{2t-3}{(t^2-3t+1)^2}\,dt;\ u=t^2-3t+1$

9. $\displaystyle\int \frac{2t-3}{(t^2-3t+1)^{7/2}}\,dt;\ u=t^2-3t+1$

10. $\displaystyle\int x\sqrt{x-1}\,dx;\ v=x-1$

11. $\displaystyle\int (x-1)\sqrt{x+1}\,dx;\ v=x+1$

12. $\displaystyle\int x^2\sqrt{x+3}\,dx;\ v=x+3$

13. $\displaystyle\int \sec x\tan x\sqrt{3+\sec x}\,dx;\ u=3+\sec x$

14. $\displaystyle\int \frac{\sqrt[3]{x}}{(\sqrt[3]{x}+1)^5}\,dx;\ u=\sqrt[3]{x}+1$

In Exercises 15–36 evaluate the integral.

15. $\displaystyle\int 3x^2(x^3+1)^{12}\,dx$

16. $\displaystyle\int x^3(2-5x^4)^7\,dx$

17. $\displaystyle\int (2x+3)(x^2+3x+4)^5\,dx$

18. $\displaystyle\int \left(1-\frac{1}{x^2}\right)\left(x+\frac{1}{x}\right)^{-3}\,dx$

19. $\displaystyle\int \sqrt{3x+7}\,dx$

20. $\displaystyle\int \sqrt{4-2x}\,dx$

21. $\displaystyle\int (1+4x)\sqrt{1+2x+4x^2}\,dx$

22. $\displaystyle\int \cos 7x\,dx$

23. $\displaystyle\int_{-1}^{3} \sin \pi x\,dx$

24. $\displaystyle\int_{0}^{1} t^9 \sin t^{10}\,dt$

25. $\displaystyle\int \sin^6 t\cos t\,dt$

26. $\displaystyle\int \cos^{-3} t\sin t\,dt$

27. $\displaystyle\int \sqrt{\sin 2z}\cos 2z\,dz$

28. $\displaystyle\int \sin 3z\sqrt{1-\cos 3z}\,dz$

29. $\displaystyle\int_{0}^{\pi/4} \frac{\sin z}{\cos^2 z}\,dz$

30. $\displaystyle\int_{\pi/2}^{\pi/6} \frac{\cos z}{\sin^3 z}\,dz$

31. $\displaystyle\int \frac{1}{\sqrt{z}}\sec^2\sqrt{z}\,dz$

32. $\displaystyle\int \frac{1}{z^2}\csc^2\frac{1}{z}\,dz$

33. $\displaystyle\int w\left(\sqrt{w^2+1}+\frac{1}{\sqrt{w^2+1}}\right)dw$

34. $\displaystyle\int_{4}^{1} \frac{\sqrt{1+\sqrt{x}}}{\sqrt{x}}\,dx$

35. $\displaystyle\int_{1}^{8} x^{-2/3}\sqrt{1+4x^{1/3}}\,dx$

36. $\displaystyle\int_{-1}^{0} w(\sqrt{1-w^2}+\sin\pi w^2)\,dw$

In Exercises 37–44 evaluate the integral.

37. $\displaystyle\int x\sqrt{x+2}\,dx$

38. $\displaystyle\int \frac{x}{\sqrt{x+3}}\,dx$

39. $\displaystyle\int_{1}^{3} 4x\sqrt{6-2x}\,dx$

40. $\displaystyle\int x^2\sqrt{x+4}\,dx$

41. $\displaystyle\int t^2\sqrt{1-8t}\,dt$

42. $\displaystyle\int t^3\sqrt{3+t^2}\,dt$

43. $\displaystyle\int_{-1}^{2} \frac{t^2}{\sqrt{t+2}}\,dt$

44. $\displaystyle\int_{0}^{1} \frac{\sqrt{x}}{\sqrt{1+\sqrt{x}}}\,dx$

In Exercises 45–50 find the area A of the region between the graph of f and the x axis on the given interval.

45. $f(x)=\sqrt{x+1};\ [0,3]$

46. $f(x)=\sin \pi x;\ [0,1]$

47. $f(x)=\dfrac{x}{(x^2+1)^2};\ [1,2]$

48. $f(x)=x\sqrt{x^2-9};\ [3,5]$

49. $f(x)=\dfrac{1}{x^2}\left(1+\dfrac{1}{x}\right)^{1/2};\ \left[\dfrac{1}{8},\dfrac{1}{3}\right]$

50. $f(x)=-x^{1/3}(1+x^{4/3})^{1/3};\ [-1,0]$

51. Let k be any integer.
  a. Verify that $\int_a^{a+k\pi}\sin^2 x\,dx=k\pi/2$.
  b. Verify that $\int_a^{a+k\pi}\cos^2 x\,dx=k\pi/2$.

52. Let $a>0$, and let f be continuous on $[0,a]$.
  a. Making the substitution $u=a-x$, show that

$$\int_0^a \frac{f(x)}{f(x)+f(a-x)}\,dx=\int_0^a \frac{f(a-u)}{f(u)+f(a-u)}\,du$$

  b. Use part (a) to show that

$$\int_0^a \frac{f(x)}{f(x)+f(a-x)}\,dx=\frac{a}{2}\quad \text{(The answer is independent of } f!)$$

c. Use part (b) to evaluate

$$\int_0^1 \frac{x^4}{x^4 + (1-x)^4}\,dx$$

53. Let f be continuous, and suppose $\int f(x)\,dx = F(x) + C$.
   a. Prove that $\int f(ax+b)\,dx = [F(ax+b)]/a + C$ for any $a \neq 0$ and any b.
   b. Use (a) to evaluate $\int \sin(ax+b)\,dx$.
   c. Use (a) to evaluate $\int (ax+b)^n\,dx$, for any integer $n \neq 0, -1$.

54. Show that $\int_{ca}^{cb}(1/x)\,dx = \int_a^b(1/x)\,ax$ for any positive numbers a, b, and c. (*Hint:* Let $u = x/c$.)

55. Suppose f is continuous on $[-a, 0]$. Prove that

$$\int_{-a}^0 f(x)\,dx = \int_0^a f(-x)\,dx$$

56. Suppose f is continuous on $[-a, a]$.
   a. Use Exercise 55 to show that $\int_{-a}^a f(x)\,dx = \int_0^a [f(x) + f(-x)]\,dx$.
   b. Show that $\int_{-a}^a f(x)\,dx = 0$ if f is odd. (*Hint:* If f is odd, then $f(-x) = -f(x)$.)
   c. Show that $\int_{-a}^a f(x)\,dx = 2\int_0^a f(x)\,dx$ if f is even. (*Hint:* If f is even, then $f(-x) = f(x)$.)

57. Use Exercise 56 to evaluate the following integrals.
   a. $\int_{-\pi/3}^{\pi/3} \sin x\,dx$ b. $\int_{-\pi/4}^{\pi/4} \cos t\,dt$

58. Find the fallacy in the following argument: Since $(1+x^2)^{-1} > 0$ we have $\int_{-1}^1 (1+x^2)^{-1}\,dx > 0$. However, by substituting $u = 1/x$ we obtain

$$\int_{-1}^1 (1+x^2)^{-1}\,dx = \int_{-1}^1 \left(1 + \frac{1}{u^2}\right)^{-1}\left(-\frac{1}{u^2}\right)du$$

$$= -\int_{-1}^1 (1+u^2)^{-1}\,du$$

$$= -\int_{-1}^1 (1+x^2)^{-1}\,dx$$

which implies that $\int_{-1}^1 (1+x^2)^{-1}\,dx = 0$.

59. Suppose an object with mass m_1 is located at a on the x axis and another object with mass m_2 is to the left of the first object on the x axis. Newton's Law of Gravitation implies that the work W required to move the second object from $x_2 < a$ to $x_1 < a$ is given by

$$W = \int_{x_1}^{x_2} \frac{Gm_1m_2}{(x-a)^2}\,dx$$

where G is a constant. Show that

$$W = \frac{Gm_1m_2(x_2 - x_1)}{(x_1 - a)(x_2 - a)}$$

5.7
THE LOGARITHM AS AN INTEGRAL

Corollary 5.13 tells us that if f is continuous on an interval I, there is an antiderivative G of f on I, given by the formula

$$G(x) = \int_a^x f(t)\,dt \quad \text{for } x \text{ in } I$$

where a is any number in I. This is true, whether or not one can find a simple formula for G.

Now consider the special function given by

$$f(x) = \frac{1}{x} = x^{-1}$$

Although f is continuous, none of its antiderivatives has been expressed by a simple formula. Indeed, -1 is the only rational power r for which we cannot write

$$\int x^r\,dx = \frac{x^{r+1}}{r+1} + C$$

(See Table 5.2 in Section 5.5.) Nevertheless, the Fundamental Theorem guarantees that on any interval not containing 0 there is an antiderivative of $1/x$.

Before we continue, we mention that the distinctive property of logarithms is that the logarithm of a product of two positive numbers is the sum of the logarithms of the numbers. Presently we will define the natural logarithm to be a specific antiderivative of $1/x$, and we will then show that the natural logarithm has the property of converting products to sums, just as the common logarithm does. In Section 7.4 we will prove that the natural logarithm is a multiple of the common logarithm.

DEFINITION 5.22

The **natural logarithm** is the function defined on the interval $(0, \infty)$ by

$$\ln x = \int_1^x \frac{1}{t}\, dt$$

The "ln" in the definition stands for the Latin *logarithmus naturalis*. An alternative expression for $\ln x$ is $\log x$. For $x > 1$, $\ln x$ may be considered as the area of the region shaded in Figure 5.34.

Since

$$\int_1^1 \frac{1}{t}\, dt = 0$$

it follows immediately from the definition of the natural logarithm that

$$\ln 1 = 0 \tag{1}$$

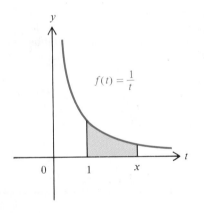

$f(t) = \dfrac{1}{t}$

FIGURE 5.34

Next we observe from Corollary 5.13 that

$$\frac{d}{dx}\left(\int_1^x \frac{1}{t}\, dt\right) = \frac{1}{x}$$

so $\ln x$ is a differentiable function with

$$\frac{d}{dx}\ln x = \frac{1}{x} \tag{2}$$

Since $1/x > 0$ for $x > 0$, we also conclude that $\ln x$ is strictly increasing. The indefinite integral form of (2) is

$$\int \frac{1}{x}\, dx = \ln x + C \tag{3}$$

Example 1 Evaluate $\displaystyle\int_2^6 \frac{1}{x}\, dx$ in terms of logarithms.

Solution By (3),

$$\int_2^6 \frac{1}{x}\,dx = \ln x \Big|_2^6 = \ln 6 - \ln 2 \quad \square$$

Example 2 Show that $\ln 4 > 1$.

Solution By Definition 5.22, Theorem 5.8, and the Comparison Property,

$$\ln 4 = \int_1^4 \frac{1}{t}\,dt = \int_1^2 \frac{1}{t}\,dt + \int_2^3 \frac{1}{t}\,dt + \int_3^4 \frac{1}{t}\,dt$$

$$\geq \frac{1}{2}(2-1) + \frac{1}{3}(3-2) + \frac{1}{4}(4-3) \geq \frac{1}{2} + \frac{1}{3} + \frac{1}{4} > 1$$

(Figure 5.35). $\square$

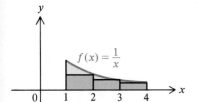

FIGURE 5.35

Since $\ln 1 = 0$ and $\ln 4 > 1$, and since $\ln x$ is differentiable and hence continuous, it follows from the Intermediate Value Theorem (Theorem 2.13) that there is a number, called e, between 1 and 4 such that

$$\ln e = 1$$

Since the function $\ln x$ is strictly increasing, e is unique. It is the value of x for which the area of the shaded region in Figure 5.34 is 1. It can be shown that e is irrational and that it has the nonterminating decimal expansion

$$e = 2.71828182845904523536\ldots$$

The symbol e was first adopted for this number by the great Swiss mathematician Leonhard Euler. It has come to occupy a special place both in mathematics and in applications, and it will reappear in Chapter 7 in full force.

We now prove that the natural logarithm of the product of two positive numbers is the sum of their natural logarithms.

THEOREM 5.23
LAW OF LOGARITHMS

For all $b > 0$ and $c > 0$,

$$\ln bc = \ln b + \ln c$$

Proof Fix $b > 0$. For any $x > 0$ let

$$g(x) = \ln bx$$

The Chain Rule yields

$$g'(x) = \left(\frac{1}{bx}\right)b = \frac{1}{x}$$

Therefore g and $\ln x$ have the same derivative. By Theorem 4.7 they differ by a constant C, that is,

$$\ln bx = \ln x + C$$

Substituting $x = 1$ in this equation and noting that $\ln 1 = 0$, we obtain

$$\ln b = \ln 1 + C = C$$

As a result,

$$\ln bx = \ln x + \ln b$$

When $x = c$, this is equivalent to the equation that was to be verified.   ■

The following three properties of logarithms also play a role in calculus:

$$\ln b^r = r \ln b \quad \text{for } b > 0 \text{ and } r \text{ rational} \tag{4}$$

$$\ln \frac{1}{b} = -\ln b \quad \text{for } b > 0 \tag{5}$$

$$\ln \frac{b}{c} = \ln b - \ln c \quad \text{for } b, c > 0 \tag{6}$$

Although (4)–(6) can be proved from the Law of Logarithms, it is also possible to prove each of them using derivatives, just as we proved Theorem 5.23 (see Exercises 57–59).

Using (6), we can simplify the answer obtained in Example 1. Indeed (6) implies that

$$\ln 6 - \ln 2 = \ln \frac{6}{2} = \ln 3$$

Thus

$$\int_2^6 \frac{1}{x} dx = \ln 3$$

Now we will analyze the graph of the natural logarithm function. First, $\ln 1 = 0$ by (1). Next, by (2),

$$\frac{d}{dx} \ln x = \frac{1}{x} > 0$$

from which we obtain

$$\frac{d^2}{dx^2} \ln x = -\frac{1}{x^2} < 0$$

Therefore, as we noted earlier, $\ln x$ is strictly increasing, and moreover, the graph

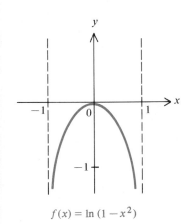

$f(x) = \ln x$

FIGURE 5.36

of $\ln x$ is concave downward on $(0, \infty)$. One can also prove that

$$\lim_{x \to \infty} \ln x = \infty \quad \text{and} \quad \lim_{x \to 0^+} \ln x = -\infty$$

(see Exercise 60). With this information we are able to draw the graph of $\ln x$ (Figure 5.36).

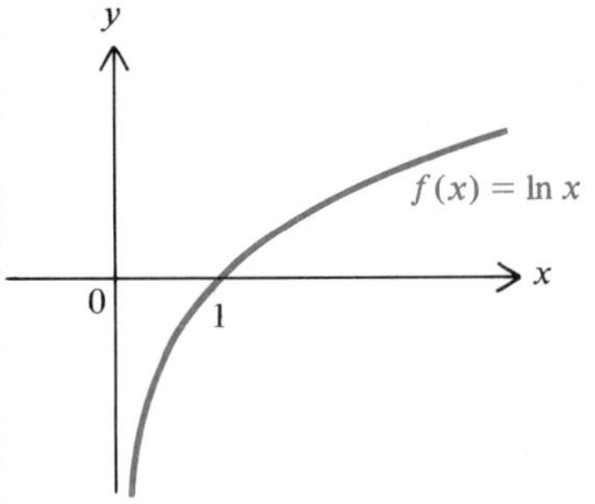

If f is any positive differentiable function, then we can apply the Chain Rule to conclude that

$$\frac{d}{dx} \ln f(x) = \frac{1}{f(x)} f'(x) = \frac{f'(x)}{f(x)} \tag{7}$$

Example 3 Find

$$\frac{d}{dx} \ln (x^2 + x)^{1/3}$$

Solution By using first (4) and then (7) we find that

$$\frac{d}{dx} \ln (x^2 + x)^{1/3} \overset{(4)}{=} \frac{d}{dx} \left[\frac{1}{3} \ln (x^2 + x) \right] \overset{(7)}{=} \frac{1}{3} \cdot \frac{1}{x^2 + x} \cdot \frac{d}{dx} (x^2 + x)$$

$$= \frac{1}{3} \cdot \frac{1}{x^2 + x} (2x + 1) = \frac{2x + 1}{3(x^2 + x)} \quad \square$$

Next we will sketch the graph of a function defined in terms of logarithms.

Example 4 Let $f(x) = \ln (1 - x^2)$. Sketch the graph of f.

Solution The domain of f consists of all x for which $1 - x^2 > 0$, that is, all x in $(-1, 1)$. Since

$$f'(x) = \frac{-2x}{1 - x^2}$$

f has a critical point at 0. In addition,

$$f''(x) = \frac{(-2)(1 - x^2) - (-2x)(-2x)}{(1 - x^2)^2} = \frac{-2(1 + x^2)}{(1 - x^2)^2}$$

which is negative for all x in $(-1, 1)$. Thus $f(0) = 0$ is the maximum value of f on $(-1, 1)$, and the graph is concave downward on $(-1, 1)$. Finally,

$$\lim_{x \to 1^-} f(x) = \lim_{x \to -1^+} f(x) = -\infty$$

We can now sketch the graph (Figure 5.37). $\square$

$f(x) = \ln (1 - x^2)$

FIGURE 5.37

Since the domain of the function $\ln x$ is $(0, \infty)$, the domain of $\ln(-x)$ is $(-\infty, 0)$. Hence by the Chain Rule we have

$$\frac{d}{dx} \ln(-x) = \frac{1}{-x} (-1) = \frac{1}{x}$$

Thus the function $\ln|x|$, which is equal to $\ln(-x)$ on $(-\infty, 0)$, and equal to $\ln x$ on $(0, \infty)$, is an antiderivative of $1/x$ on $(-\infty, 0)$ and on $(0, \infty)$. This gives us the formula

$$\int \frac{1}{x} dx = \ln|x| + C \tag{8}$$

Example 5 Express $\displaystyle\int_{-8}^{-7} \frac{1}{x} dx$ in terms of logarithms.

Solution From (8) we deduce that

$$\int_{-8}^{-7} \frac{1}{x} dx = \ln|x| \Big|_{-8}^{-7} = \ln|-7| - \ln|-8| = \ln 7 - \ln 8 = \ln\frac{7}{8} \quad \square$$

Next we will evaluate

$$\int \frac{f'(x)}{f(x)} dx$$

by substitution. To accomplish that, we let

$$u = f(x), \quad \text{so that} \quad du = f'(x)\,dx$$

Then by (8),

$$\int \frac{f'(x)}{f(x)} dx = \int \overbrace{\frac{1}{f(x)}}^{1/u} \overbrace{f'(x)\,dx}^{du} = \int \frac{1}{u} du \overset{(8)}{=} \ln|u| + C = \ln|f(x)| + C$$

which yields the formula

$$\int \frac{f'(x)}{f(x)} dx = \ln|f(x)| + C \tag{9}$$

It is not necessary to memorize (9) if you remember that when you see an integral of the form in (9), you should think about substituting $u = f(x)$.

Example 6 Find $\displaystyle\int \frac{x^4}{x^5 + 1} dx$.

Solution Let

$$u = x^5 + 1, \quad \text{so that} \quad du = 5x^4\,dx$$

Then

$$\int \frac{x^4}{x^5+1}\,dx = \int \overbrace{\frac{1}{x^5+1}}^{1/u}\overbrace{x^4\,dx}^{\frac{1}{5}\,du} = \int \frac{1}{u}\cdot\frac{1}{5}\,du$$

$$= \frac{1}{5}\ln|u| + C = \frac{1}{5}\ln|x^5+1| + C \quad \square$$

Formula (9) can be used to find the integrals of the four remaining basic trigonometric functions that we have not yet found: $\tan x$, $\sec x$, $\cot x$, and $\csc x$.
To find $\int \tan x\,dx$, we first recall that

$$\tan x = \frac{\sin x}{\cos x}$$

Since the numerator is essentially the derivative of the denominator, we can apply (9) by letting

$$u = \cos x, \quad \text{so that} \quad du = -\sin x\,dx$$

We find that

$$\int \tan x\,dx = \int \overbrace{\frac{1}{\cos x}}^{1/u}\overbrace{\sin x\,dx}^{(-1)\,du} = \int \frac{1}{u}(-1)\,du = -\ln|u| + C = -\ln|\cos x| + C$$

Therefore

$$\int \tan x\,dx = -\ln|\cos x| + C$$

The evaluation of $\int \cot x\,dx$ is analogous. However, the evaluations of $\int \sec x\,dx$ and $\int \csc x\,dx$ are not so transparent. For $\int \sec x\,dx$ recall that

$$\frac{d}{dx}\tan x = \sec^2 x \quad \text{and} \quad \frac{d}{dx}\sec x = \sec x \tan x$$

Thus

$$\frac{d}{dx}(\sec x + \tan x) = \sec x \tan x + \sec^2 x = (\sec x)(\sec x + \tan x)$$

Since

$$\int \sec x\,dx = \int \frac{(\sec x)(\sec x + \tan x)}{\sec x + \tan x}\,dx$$

the calculations above tell us that the numerator of the latter integral is the derivative of the denominator, so we can once again use (9). We let

$$u = \sec x + \tan x, \quad \text{so that} \quad du = (\sec x)(\sec x + \tan x)\,dx$$

Then

$$\int \sec x \, dx = \int \overbrace{\frac{1}{\sec x + \tan x}}^{1/u} \overbrace{[\sec x \, (\tan x + \sec x)] \, dx}^{du}$$

$$= \int \frac{1}{u} du = \ln |u| + C = \ln |\sec x + \tan x| + C$$

Consequently

$$\int \sec x \, dx = \ln |\sec x + \tan x| + C \tag{10}$$

The evaluation of $\int \csc x \, dx$ is similar.

We will return to the study of logarithms in Section 7.4, where we will establish relationships between the natural logarithm and other logarithms.

EXERCISES 5.7

In Exercises 1–6 evaluate the integral. Express your answer in terms of logarithms.

1. $\int_1^3 \frac{1}{x} \, dx$ 2. $\int_2^8 \frac{1}{x} \, dx$ 3. $\int_{1/9}^{1/4} \frac{-1}{3x} \, dx$

4. $\int_{-2}^{-1} \frac{1}{t} \, dt$ 5. $\int_{-4}^{-12} \frac{2}{t} \, dt$ 6. $\int_{-1/16}^{-1/8} \frac{1}{t} \, dt$

In Exercises 7–18 find the domain and the derivative of the function.

7. $f(x) = \ln(x + 1)$ 8. $f(x) = \ln(3x - 2)$

9. $g(x) = x \ln(x^2 - 1)$ 10. $k(t) = \ln(t^2 + 4)^3$

11. $f(x) = \ln \sqrt{\frac{x - 3}{x - 2}}$ 12. $f(t) = t^2 (\ln t)^3$

13. $f(x) = \frac{\ln x}{x - 1}$ 14. $f(x) = \frac{x^2}{\ln x}$

15. $f(t) = \sin(\ln t)$ 16. $g(u) = \ln(\sin u)$

17. $f(x) = \ln(\ln x)$ 18. $f(x) = \ln(x + \sqrt{x^2 - 1})$

In Exercises 19–20 find dy/dx by implicit differentiation.

19. $x \ln(y^2 + x) = 1 + 5y$ 20. $y \ln \frac{y}{x} = \sin y^2$

In Exercises 21–26 find the domain, intercepts, relative extreme values, inflection points, concavity, and asymptotes for the given function. Then draw its graph.

21. $f(x) = \ln |x|$ 22. $f(x) = \ln(x - 2)$

23. $g(x) = \ln(2 + \sin x)$ 24. $f(x) = \ln(1 + x^2)$

25. $f(x) = (\ln x)^2$ 26. $k(t) = \ln \left| \frac{t + 1}{t - 1} \right|$

C 27. Let $f(x) = x^{-2} - \ln x$. Use the Newton-Raphson method to approximate a zero of f to within 10^{-6}.

C 28. Let $f(x) = x \ln x$. Use the Newton-Raphson method to approximate a relative extreme value of f to within 10^{-6}.

In Exercises 29–45 evaluate the integral.

29. $\int \frac{1}{x - 1} \, dx$ 30. $\int \frac{2}{1 - 4x} \, dx$

31. $\int \frac{x}{x^2 + 4} \, dx$ 32. $\int \frac{x^2}{1 - x^3} \, dx$

33. $\int \frac{x^3}{x^4 - 4} \, dx$ 34. $\int_0^{\pi/6} \frac{\cos x}{1 - \sin x} \, dx$

35. $\int_0^{\pi/3} \frac{\sin x}{1 - 3 \cos x} \, dx$ 36. $\int \frac{1}{x(1 + \ln x)} \, dx$

37. $\int_1^4 \frac{1}{\sqrt{x}(1 + \sqrt{x})} \, dx$ 38. $\int_{-1}^0 \frac{x + 2}{x^2 + 4x - 1} \, dx$

39. $\int \frac{\ln z}{z} \, dz$ 40. $\int \frac{(\ln z)^5}{z} \, dz$

41. $\int \frac{\ln(\ln t)}{t \ln t} \, dt$ 42. $\int \frac{\tan \sqrt{t}}{\sqrt{t}} \, dt$

43. $\int \cot t \, dt$

44. $\int \dfrac{1}{1 + x^{1/3}} \, dx$ (*Hint:* Substitute $u = 1 + x^{1/3}$.)

*45. $\int \dfrac{x}{1 + x \tan x} \, dx$ (*Hint:* Substitute $u = x \sin x + \cos x$.)

*46. Evaluate $\int \csc x \, dx$. (*Hint:* Pattern the solution after the evaluation of $\int \sec x \, dx$ in the text.)

In Exercises 47–52 find the area A of the region between the graph of f and the x axis on the given interval.

47. $f(x) = \dfrac{1}{x};\ [e, e^2]$

48. $f(x) = \dfrac{1}{x + 1};\ [0, \sqrt{2} - 1]$

49. $f(x) = \dfrac{x}{2 - x^2};\ [-2, -\sqrt{3}]$

50. $f(x) = \tan 2x;\ [0, \pi/6]$

51. $f(x) = \sec^3 x \tan x;\ \left[\dfrac{\pi}{4}, \dfrac{\pi}{3}\right]$

*52. $f(x) = (1 + \tan x)^2;\ [\pi/4, \pi/3]$

53. Assume that $\ln a = 2.3$ and $\ln b = 4.7$. Find $\ln(a^3 b^2)$.

54. Assume that $\ln a = 0.6$ and that $\ln b = -1.9$. Find $\ln(a^{-3} b^{11})$.

55. a. Show that
$$\int_1^3 \frac{1}{t} \, dt = \int_1^{5/4} \frac{1}{t} \, dt + \int_{5/4}^{6/4} \frac{1}{t} \, dt + \cdots + \int_{11/4}^3 \frac{1}{t} \, dt$$

© b. Use part (a) to show that
$$\ln 3 > \frac{1}{4}\left(\frac{4}{5} + \frac{4}{6} + \cdots + \frac{4}{12}\right) > 1$$

   c. Use part (b) to show that $e < 3$.

56. Show that
$$\ln(x + \sqrt{x^2 - 1}) = -\ln(x - \sqrt{x^2 - 1})$$

   (*Hint:* Multiply $x + \sqrt{x^2 - 1}$ by
$$1 = (x - \sqrt{x^2 - 1})/(x - \sqrt{x^2 - 1})$$
   and then simplify the result. Alternatively, differentiate and then use Theorem 4.7.)

57. Prove that $\ln b^r = r \ln b$ for $b > 0$ and r rational by showing that the functions $\ln x^r$ and $r \ln x$ have the same derivative and the same value at 1.

58. Prove that $\ln(1/b) = -\ln b$ for $b > 0$
   a. by using (4)
   b. by showing that the functions $\ln(1/x)$ and $-\ln x$ have the same derivative and the same value at 1.

59. Prove that $\ln(b/c) = \ln b - \ln c$ for $b, c > 0$
   a. by using (5) and the Law of Logarithms
   b. by showing that the functions $\ln(x/c)$ and $\ln x - \ln c$ have the same derivative and the same value at c.

*60. Use Exercises 57–59 to show that
$$\lim_{x \to \infty} \ln x = \infty \quad \text{and} \quad \lim_{x \to 0^+} \ln x = -\infty$$

61. Find an equation of the line tangent to the curve $y = 1/t$ at $(1, 1)$. Then show that for small $h > 0$, the area below the tangent line on $[1, 1 + h]$ is $h - h^2/2$, and for $h < 0$ and $|h|$ small, the area below the tangent line on $[1 + h, 1]$ is $-(h - h^2/2)$. (Notice that the first area is approximately $\int_1^{1+h} (1/t) \, dt$ and the second area is approximately $\int_{1+h}^1 (1/t) \, dt = -\int_1^{1+h} (1/t) \, dt$. Together these yield the approximation
$$\ln(1 + h) \approx h - \tfrac{1}{2} h^2$$
when $|h|$ is small.)

62. Use the result of Exercise 61 to approximate the following.
   a. $\ln 1.1$ b. $\ln 0.9$
   c. $\ln 1.02$ d. $\ln 0.98$

63. a. Use the inequality $1/t \le 1/\sqrt{t}$ for $t \ge 1$ to show that
$$\ln x = \int_1^x \frac{1}{t} \, dt \le 2(\sqrt{x} - 1) \quad \text{for } x \ge 1$$

   b. Use part (a) to show that
$$\lim_{x \to \infty} \frac{\ln x}{x} = 0$$

   c. Use part (b) and (5) to show that $\lim_{x \to 0^+} x \ln x = 0$.

64. Sketch the graph of $x \ln x$. Use Exercise 63(c) and the fact that $\ln x = -1$ for $x \approx 0.37$.

65. Recall that $\ln e = 1$, and let
$$f(x) = \frac{\ln x}{x}$$

   a. Show that f is strictly increasing on $(0, e]$ and strictly decreasing on $[e, \infty)$.
   b. Find the maximum value of f on $(0, \infty)$.
   c. Sketch the graph of f. Use the second derivative of f, Exercise 63, and the fact that $\ln x = 3/2$ for $x \approx 4.5$.

66. Let a and b be positive rational numbers.
   a. Show that the inequality $a^b > b^a$ is equivalent to $(\ln a)/a > (\ln b)/b$.
   b. Using Exercise 65(a), show that $a^b > b^a$ if $b < a < e$ or $e < a < b$.

67. Use Exercise 51(b) of Section 5.5 to find lower and upper bounds for the integral $\int_1^2 \ln x \, dx$. (The exact value is $(2 \ln 2) - 1$.)

68. Let f be any differentiable function that is not identically 0. Show that

$$\frac{d}{dx}\ln|f(x)| = \frac{f'(x)}{f(x)}$$

 (*Hint:* Use (9).)

69. Let $f_1, f_2, \ldots, f_n$ be differentiable functions, and let $f = f_1 f_2 \cdots f_n$. Assume that f is not identically 0.
 a. Show that

$$\ln|f(x)| = \ln|f_1(x)| + \ln|f_2(x)| + \cdots + \ln|f_n(x)|$$

 b. Use (a) and Exercise 68 to show that

$$\frac{f'(x)}{f(x)} = \frac{f_1'(x)}{f_1(x)} + \frac{f_2'(x)}{f_2(x)} + \cdots + \frac{f_n'(x)}{f_n(x)}$$

 and hence that

$$f'(x) = f(x)\left[\frac{f_1'(x)}{f_1(x)} + \frac{f_2'(x)}{f_2(x)} + \cdots + \frac{f_n'(x)}{f_n(x)}\right]$$

 This method of finding the derivative of a complicated product of functions is called **logarithmic differentiation**.

70. Using logarithmic differentiation, find the derivatives of the following functions.
 a. $f(x) = (x+1)(2x+3)(1-4x)$
 b. $f(x) = \sqrt[3]{(x+3)^2(x-1)}$

*71. Suppose f is defined on $(0, \infty)$, is differentiable at 1, and satisfies

$$f(xy) = f(x) + f(y) \qquad (11)$$

 for all x and y in $(0, \infty)$.
 a. Prove that $f(1) = 0$.
 b. Show that

$$f\left(\frac{y}{x}\right) = f(y) - f(x)$$

 for all x and y in $(0, \infty)$.
 c. Deduce from (a) and (b) that if $0 < |h| < x$, then

$$\frac{f(x+h) - f(x)}{h} = \frac{f\left(\dfrac{x+h}{x}\right)}{h}$$

$$= \frac{1}{x}\frac{f(1+h/x) - f(1)}{h/x}$$

d. From (c) and the hypothesis that f is differentiable at 1, conclude that f' exists on $(0, \infty)$ and that

$$f'(x) = \frac{1}{x}f'(1) \quad \text{for } x > 0$$

 (It follows from (a) and (d) and the definition of $\ln x$ that if f is a differentiable function satisfying (11), then

$$f(x) = f'(1)\ln x \quad \text{for } x > 0)$$

72. A beanbag factory has a marginal revenue function

$$m_R(x) = \frac{2}{x+1}$$

 where x denotes thousands of beanbags sold and $m_R(x)$ denotes dollars received per beanbag.
 a. Determine the total revenue function R. (*Hint:* Use (6) of Section 5.4 and the fact that $R(0) = 0$. What is the derivative of $\ln(x+1)$?)
 b. Demonstrate that R is a reasonable total revenue function by showing that R is increasing and concave downward on the interval $(0, \infty)$.

C 73. The equation

$$\ln w = 4.4974 + 3.135\ln s$$

 has been used to relate the weight w (in kilograms) to the sitting height s (in meters) of people. Using this equation, find the weight of a person whose sitting height is
 a. 1 meter
 b. $\frac{1}{2}$ meter

C 74. Let y represent the weight in ounces of a baby and x the age in months. It has been conjectured that y and x are related by the equation

$$\ln y - \ln(341.5 - y) = c(x - 1.66) \quad \text{for } 0 \le x \le 9$$

 where c is a positive constant.
 a. Show that dy/dx, the rate of weight increase with respect to age of the baby, is a positive function.
 b. Find the age x_0 at which dy/dx is maximum, and find the corresponding weight. According to the equation, it is at this age that the baby gains weight fastest.

5.8
ANOTHER LOOK AT AREA

In Section 5.2 we defined the area of a region of the type shown in Figure 5.38 to be $\int_a^b f(x)\,dx$. However, this definition does not apply to regions like that shown in Figure 5.39(a) (which might, for instance, represent the surface of a lake). Using

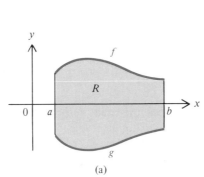

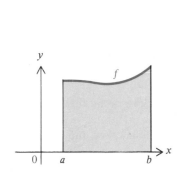

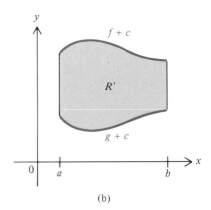

FIGURE 5.38   **FIGURE 5.39**

(a)   (b)

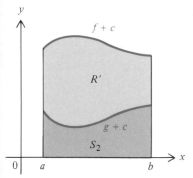

FIGURE 5.40

further properties that we naturally expect area to have, we will now extend the earlier definition of area to include regions whose lower boundaries are not necessarily horizontal.

Let f and g be continuous on an interval $[a, b]$, and assume that $f(x) \geq g(x)$ for $a \leq x \leq b$. We will define the area of the region R that is bounded above by the graph of f, below by the graph of g, and on the sides by the lines $x = a$ and $x = b$ (Figure 5.39(a)). We call R the **region between the graphs of f and g on $[a, b]$**.

Since the area of a region should not be affected by shifting the region vertically, the area A of the region R in Figure 5.39(a) should be the same as the area of the region R' in Figure 5.39(b), which lies above the x axis and is bounded by the graphs of $f + c$ and $g + c$ for an appropriate constant c. However, the area of the region S_1 (which is bounded above by the graph of $f + c$ and below by the x axis) should be the sum of the areas of R' and S_2 in Figure 5.40. Now by Definition 5.3, we have formulas for the areas of S_1 and S_2. Therefore we would expect that

$$A = \overbrace{\int_a^b [f(x) + c]\,dx}^{\text{area of } S_1} - \overbrace{\int_a^b [g(x) + c]\,dx}^{\text{area of } S_2} = \int_a^b [f(x) - g(x)]\,dx$$

Thus we are led to the following definition of the area A:

DEFINITION 5.24

Let f and g be continuous on $[a, b]$, with $f(x) \geq g(x)$ for $a \leq x \leq b$. The area A of the region R between the graphs of f and g on $[a, b]$ is given by

$$A = \int_a^b [f(x) - g(x)]\,dx \qquad (1)$$

Notice that for $a \leq x \leq b$ the integrand $f(x) - g(x)$ represents the height of R at x.

Example 1   Let $f(x) = \cos x$ and $g(x) = \sin x$. Find the area A of the region between the graphs of f and g on $[0, \pi/4]$ (Figure 5.41).

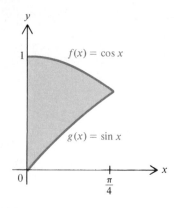

FIGURE 5.41

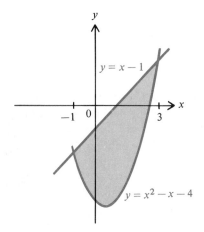

FIGURE 5.42

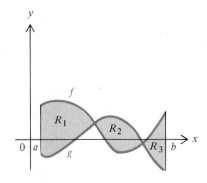

FIGURE 5.43

Solution   Since $\cos x \geq \sin x$ on $[0, \pi/4]$, it follows from (1) that

$$A = \int_0^{\pi/4} [f(x) - g(x)]\,dx = \int_0^{\pi/4} (\cos x - \sin x)\,dx$$

$$= (\sin x + \cos x)\Big|_0^{\pi/4} = \left(\frac{1}{2}\sqrt{2} + \frac{1}{2}\sqrt{2}\right) - (0 + 1) = \sqrt{2} - 1 \quad \square$$

If we seek the area of the region between two graphs that cross exactly twice but we are not given the interval over which to integrate, we first determine where the graphs cross and which graph lies above the other, and then integrate over the corresponding (bounded) interval.

Example 2   Find the area A of the region between the graphs of $y = x^2 - x - 4$ and $y = x - 1$ (Figure 5.42).

Solution   First we determine the x coordinates of the points at which the two curves intersect:

$$x^2 - x - 4 = x - 1$$
$$x^2 - 2x - 3 = 0$$
$$(x + 1)(x - 3) = 0$$

$$x = -1 \quad \text{or} \quad x = 3$$

Thus the region whose area we seek lies between the graphs of $y = x^2 - x - 4$ and $y = x - 1$ on $[-1, 3]$. Since

$$x - 1 \geq x^2 - x - 4 \quad \text{for } -1 \leq x \leq 3$$

(see Figure 5.42), it follows that the height of the region at x is $[(x - 1) - (x^2 - x - 4)]$, so by (1),

$$A = \int_{-1}^{3} [(x - 1) - (x^2 - x - 4)]\,dx = \int_{-1}^{3} (-x^2 + 2x + 3)\,dx$$

$$= \left(-\frac{1}{3}x^3 + x^2 + 3x\right)\Big|_{-1}^{3} = (-9 + 9 + 9) - \left(\frac{1}{3} + 1 - 3\right)$$

$$= \frac{32}{3} \quad \square$$

Now suppose the graphs of f and g cross at one or more points in (a, b). Then the region R between the graphs of f and g on $[a, b]$ is composed of several regions $R_1, R_2, \ldots$, each of the type whose area we have already defined (Figure 5.43). We naturally define the total area A of R to be the sum of the areas of those regions. In order to calculate A, that is, the area of the region between the graphs of f and g on $[a, b]$, we first determine those subintervals on which $f - g \geq 0$ and those on which $f - g \leq 0$. Then we integrate over those subintervals separately to obtain the areas of the corresponding subregions, and add these areas to obtain the total area of A. Example 3 illustrates this technique.

Example 3   Let $f(x) = \sin x$ and $g(x) = \cos x$. Find the area A of the region between the graphs of f and g on $[0, 2\pi]$ (Figure 5.44).

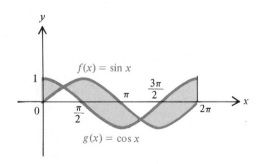

FIGURE 5.44

Solution   In order to determine where $\sin x \geq \cos x$ and where $\sin x \leq \cos x$, we first find the values of x in $[0, 2\pi]$ for which $\sin x = \cos x$, which is equivalent to solving $\tan x = 1$ for x. But $\tan x = 1$ for $x = \pi/4$ and $x = 5\pi/4$. From this you can verify that $\sin x \geq \cos x$ on $[\pi/4, 5\pi/4]$ and $\cos x \geq \sin x$ on $[0, \pi/4]$ and $[5\pi/4, 2\pi]$. Therefore

$$A = \int_0^{\pi/4} (\cos x - \sin x)\,dx + \int_{\pi/4}^{5\pi/4} (\sin x - \cos x)\,dx + \int_{5\pi/4}^{2\pi} (\cos x - \sin x)\,dx$$

$$= (\sin x + \cos x)\Big|_0^{\pi/4} + (-\cos x - \sin x)\Big|_{\pi/4}^{5\pi/4} + (\sin x + \cos x)\Big|_{5\pi/4}^{2\pi}$$

$$= (\sqrt{2} - 1) + (2\sqrt{2}) + (1 + \sqrt{2}) = 4\sqrt{2} \quad \square$$

The region in the next example may at first appear not to be of the form covered by our discussion, but in fact it is.

Example 4   Find the area A of the region R between the parabola $x = \frac{1}{2}y^2$ and the line $y = 2x - 2$ (Figure 5.45).

Solution   First we determine the x coordinates of the points at which the parabola and the line intersect:

$$x = \frac{1}{2}y^2 = \frac{1}{2}(2x - 2)^2 = 2x^2 - 4x + 2$$

which means that

$$2x^2 - 5x + 2 = 0$$

or

$$(2x - 1)(x - 2) = 0$$

Consequently $x = \frac{1}{2}$ or $x = 2$. Next, observe from Figure 5.45 that R may be broken up into two parts: the part over $[0, \frac{1}{2}]$ on the x axis and the part over $[\frac{1}{2}, 2]$

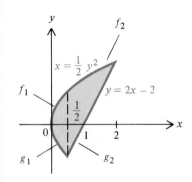

FIGURE 5.45

on the x axis. The part of R over $[0, \frac{1}{2}]$ is between the graphs of f_1 and g_1, where

$$f_1(x) = \sqrt{2x} \quad \text{and} \quad g_1(x) = -\sqrt{2x}$$

The part of R over $[\frac{1}{2}, 2]$ is between the graphs of f_2 and g_2, where

$$f_2(x) = \sqrt{2x} \quad \text{and} \quad g_2(x) = 2x - 2$$

Therefore

$$A = \int_0^{1/2} (f_1(x) - g_1(x))\,dx + \int_{1/2}^{2} (f_2(x) - g_2(x))\,dx$$

$$= \int_0^{1/2} [\sqrt{2x} - (-\sqrt{2x})]\,dx + \int_{1/2}^{2} [\sqrt{2x} - (2x - 2)]\,dx$$

$$= \int_0^{1/2} 2\sqrt{2x}\,dx + \int_{1/2}^{2} (\sqrt{2x} - 2x + 2)\,dx$$

$$= \frac{4\sqrt{2}}{3} x^{3/2} \Big|_0^{1/2} + \left(\frac{2\sqrt{2}}{3} x^{3/2} - x^2 + 2x \right) \Big|_{1/2}^{2}$$

$$= \left(\frac{2}{3} - 0 \right) + \left[\left(\frac{8}{3} - 4 + 4 \right) - \left(\frac{1}{3} - \frac{1}{4} + 1 \right) \right] = \frac{9}{4} \quad \square$$

Reversing the Roles of x and y   Instead of considering a region R as the region between the graphs of two functions of x, it is sometimes convenient to consider R as the region between the graphs of two functions of y (Figure 5.46). Then the area is computed by integrating along the y axis, instead of along the x axis.

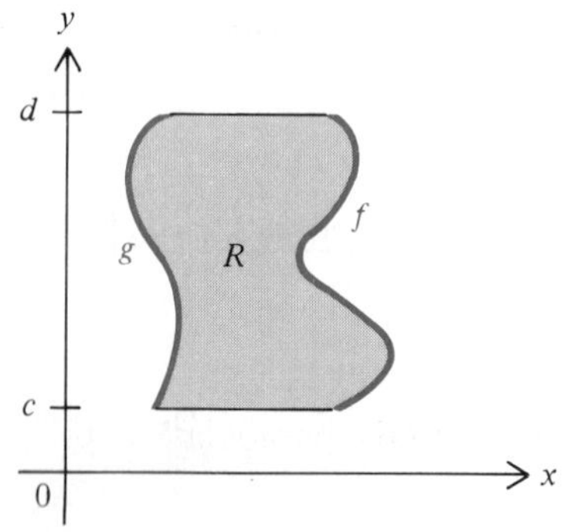

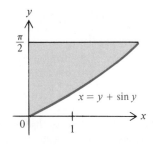

FIGURE 5.46

Example 5   Let R be the region between the y axis and the graph of the equation $x = y + \sin y$ on $[0, \pi/2]$ (Figure 5.47). Find the area A of R.

Solution   Since $y + \sin y \geq 0$ for $0 \leq y \leq \pi/2$, it follows that

$$A = \int_0^{\pi/2} (y + \sin y)\,dy = \left(\frac{1}{2} y^2 - \cos y \right) \Big|_0^{\pi/2}$$

$$= \frac{1}{2} \left(\frac{\pi}{2} \right)^2 - (-1) = \frac{\pi^2}{8} + 1 \quad \square$$

In Example 5 we have no way of describing y in terms of x; therefore it would be impossible to use (1) to determine the area of the region.

If a region can be described both in terms of x and in terms of y, the area is the same, whether we integrate with respect to x or with respect to y. To support this claim, we now return to the region described in Example 4 and find its area by integrating along the y axis.

Example 6   Find the area A of the region R between the parabola $x = \frac{1}{2} y^2$ and the line $y = 2x - 2$ (Figure 5.48).

$x = y + \sin y$

FIGURE 5.47

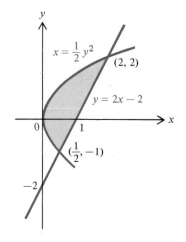

Solution First we determine the y coordinates of points at which the parabola and the line intersect:

$$\frac{1}{2}y^2 = x = \frac{1}{2}(y + 2)$$

$$y^2 - y - 2 = 0$$

$$(y + 1)(y - 2) = 0$$

$$y = -1 \quad \text{or} \quad y = 2$$

Since $\frac{1}{2}(y + 2) \geq \frac{1}{2}y^2$ for $-1 \leq y \leq 2$, it follows that

$$A = \int_{-1}^{2}\left[\frac{1}{2}(y + 2) - \frac{1}{2}y^2\right]dy = \frac{1}{2}\int_{-1}^{2}(y + 2 - y^2)\,dy$$

$$= \frac{1}{2}\left(\frac{1}{2}y^2 + 2y - \frac{1}{3}y^3\right)\Big|_{-1}^{2}$$

$$= \frac{1}{2}\left[\left(2 + 4 - \frac{8}{3}\right) - \left(\frac{1}{2} - 2 + \frac{1}{3}\right)\right] = \frac{9}{4}$$

(the same value we found for the area in Example 4). ☐

FIGURE 5.48

Cavalieri's Principle We conclude this section with a discussion of Cavalieri's Principle. Let f and g be continuous on an interval $[a, b]$, and let A be the area of the region between the graphs of f and g. Using the fact that

$$|f(x) - g(x)| = \begin{cases} f(x) - g(x) & \text{if} \quad f(x) \geq g(x) \\ g(x) - f(x) & \text{if} \quad g(x) \geq f(x) \end{cases}$$

we obtain the formula

$$A = \int_{a}^{b}|f(x) - g(x)|\,dx \tag{2}$$

which holds regardless of the relationship between f and g.

Now suppose f is any continuous function defined on $[a, b]$, and let $g = f + c$, where c is a fixed positive constant. Then the area of the region between the graphs of f and g on $[a, b]$ is

$$\int_{a}^{b}|f(x) - g(x)|\,dx = \int_{a}^{b}c\,dx = c(b - a)$$

In other words, if the distance between the lower and upper boundaries of a region is constant, then the region's area is the product of its length and height, just as in the case of rectangles (Figure 5.49(a)).

The result just obtained is a special case of Cavalieri's Principle. This principle was stated in 1635 by the Milanese mathematician Bonaventura Cavalieri (1598–1647), who used it to help derive formulas for areas of plane regions. In effect Cavalieri's Principle says that if R_1 and R_2 are two regions like those in Figure 5.49(b), with side boundaries $x = a$ and $x = b$, and if the height of R_1 (measured perpendicular to the x axis) is the same as the height of R_2 at every point in $[a, b]$ (Figure 5.49(b)), then R_1 and R_2 have the same area. Cavalieri came

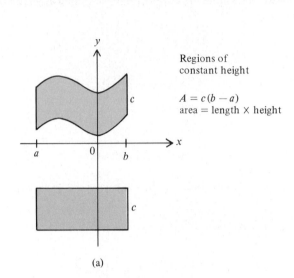

Regions of
constant height

$A = c(b - a)$
area = length × height

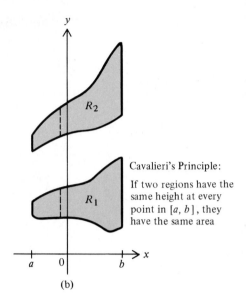

Cavalieri's Principle:

If two regions have the
same height at every
point in $[a, b]$, they
have the same area

(a)   (b)

FIGURE 5.49

to this conclusion several decades before Newton and Leibniz introduced the
general concept of integral. However, by using (2), we can easily prove Cavalieri's
Principle for the case in which the upper boundaries of R_1 and R_2 are the graphs
of continuous functions f_1 and f_2 and the lower boundaries are the graphs of
continuous functions g_1 and g_2. In that case the fact that R_1 and R_2 have identical
height at every point in $[a, b]$ means that

$$|f_1(x) - g_1(x)| = |f_2(x) - g_2(x)| \quad \text{for } a \le x \le b$$

Consequently

$$\int_a^b |f_1(x) - g_1(x)|\, dx = \int_a^b |f_2(x) - g_2(x)|\, dx$$

which means, by (2), that the area of R_1 and the area of R_2 are the same.

EXERCISES 5.8

In Exercises 1–8 find the area A of the region between the
graph of f and the x axis on the given interval.

1. $f(x) = x^2 + 2x$; $[-1, 3]$

2. $f(x) = x^3 - x$; $[-2, 1]$

3. $f(x) = \cos x - \sin x$; $[0, \pi/3]$

4. $f(x) = \tan x$; $[-\pi/4, \pi/6]$

5. $f(x) = x\sqrt{1 - x^2}$; $[-1, 1]$

6. $f(x) = \dfrac{x}{\sqrt{1 + x^2}}$; $[-1, \sqrt{7}]$

7. $f(x) = \dfrac{x}{x^2 - 1}$; $[-\frac{1}{2}, \frac{1}{3}]$

8. $f(x) = \dfrac{\ln x}{x}$; $[\frac{1}{2}, 2]$

In Exercises 9–18 find the area A of the region between the
graphs of the functions on the given interval.

9. $f(x) = x^2$, $g(x) = x^3$; $[-2, 1]$

10. $f(x) = 1/x$, $g(x) = 1/x^2$; $[\frac{1}{2}, 2]$

11. $g(x) = x^2 + 4x$, $k(x) = x - 2$; $[-3, 0]$

12. $g(x) = 2x^3 + 2x^2$, $k(x) = 2x^3 - 2x$; $[-4, 2]$

13. $f(x) = \sec^2 x$, $g(x) = \sec x \tan x$; $[-\pi/3, \pi/6]$

14. $f(x) = \sin 2x$, $g(x) = 2 \cot x$; $[\pi/3, 2\pi/3]$

15. $g(x) = \sin^2 x$, $k(x) = \tan x$; $[-\pi/4, \pi/4]$

16. $g(x) = x^3 + 3x + \cos^2 x$, $k(x) = 4x^2 - \sin^2 x + 1$; $[-1, 2]$

17. $f(x) = x\sqrt{2x + 3}$, $g(x) = x^2$; $[-1, 3]$

18. $f(x) = x(x^2 + 1)^5$, $g(x) = x^2(x^3 + 1)^5$; $[-1, 1]$

In Exercises 19–24 the graphs of f and g enclose a region. Determine the area A of that region.

19. $f(x) = x^3$, $g(x) = x^{1/3}$

20. $f(x) = x^2 + 3$, $g(x) = 12 - x^2$

21. $f(x) = x^2 + 1$, $g(x) = 2x + 9$

22. $f(x) = x^3 + x$, $g(x) = 3x^2 - x$

23. $f(x) = x^3 + 1$, $g(x) = (x + 1)^2$

24. $f(x) = 2 - \sqrt{x}$, $g(x) = \dfrac{\sqrt{x} + 1}{2\sqrt{x}}$

In Exercises 25–28 find the area A of the region between the graphs of the given equations.

25. $y^2 = 6x$ and $x^2 = 6y$

26. $y^2 = x$ and $y = x - 2$

27. $y^2 = 2x - 5$ and $y = x - 4$

28. $y^2 = 3x$ and $y = x^2 - 2x$ (*Hint:* The curves intersect at the points $(0, 0)$ and $(3, 3)$.)

In Exercises 29–30 the graphs of the three equations enclose a region. Determine the area A of that region.

29. $y = x + 2$, $y = -3x + 6$, $y = (2 - x)/3$

30. $y = \frac{3}{2}x$ for $x \geq 0$; $y = -\frac{3}{2}x$ for $x \leq 0$; $y = -x^2 - \frac{3}{2}x + 4$

In Exercises 31–33 find the area A of the region between the graphs of the given equations.

31. $x = y^2 - y$ and $x = y - y^2$

32. $x = 0$ and $x = \cos y$ for $-\pi/2 \leq y \leq 3\pi/2$

33. $x = y^2$ and $x = 6 - y - y^2$

34. Let f, g, and h be continuous on $[a, b]$. Using (2), show that the area A of the region between the graphs of f and g on $[a, b]$ is the same as the area of the region between the graphs of $f + h$ and $g + h$ on $[a, b]$.

5.9
WHO INVENTED CALCULUS?

It is often said that Newton and Leibniz invented calculus. From the vantage point of our knowledge of the derivative and the integral—as well as of the Fundamental Theorem of Calculus, which unites them—we may ask why these two great mathematicians are credited with founding calculus.

Long before Newton and Leibniz, mathematicians knew how to calculate tangents to various curves. Using horizontal tangents, they produced a general procedure for determining maxima and minima of curves. As regards areas, even Archimedes was able to find areas of regions bounded by a few of the more common curves. By the early seventeenth century, the foremost mathematicians had developed very refined ways of evaluating areas and computing what we now call integrals.

Even the Fundamental Theorem did not originate with Newton or Leibniz. Although he did not have our concept of the derivative and the integral, Isaac Barrow, Newton's teacher at Cambridge, stated and proved a geometric equivalent of the Fundamental Theorem. The proof can be found in his *Lectiones Opticae et Geometricae*.

What was left of sufficient importance to justify calling Newton and Leibniz the inventors of calculus? There remained the need for a systematic method of differentiating and integrating general classes of functions, as well as a need for a symbolism that would lend itself to formal, rather than geometrically intuitive, proofs. Newton and Leibniz recognized that the Fundamental Theorem could be used to systematically find integrals of functions they encountered. They also discovered rules for the derivatives and the integrals of combinations of functions

(see Sections 3.3 and 5.5). Newton conceived his idea first, during the middle 1660s, whereas Leibniz began his work during the early 1670s. However, Leibniz published his ideas first, and his symbolism and notation were far superior to those of Newton. In fact, Leibniz's notation survived and is in common use today.

Although we do not minimize the profound implications of Newton's and Leibniz's work, their ideas did rely heavily on previous results. Great as the achievements of Newton and Leibniz were, their ideas constituted only a link between the earlier conception of calculus and the modern approach to the subject. The following passage by L. C. Karpinski is pertinent.

> There is a strong temptation on the part of professional mathematicians and scientists to seek always to ascribe great discoveries and inventions to single individuals. Such ascription serves a didactic end in centering attention upon certain fundamental aspects of the subjects, much as the history of events is conveniently divided into epochs for purposes of exposition. There is in such attributions and division, however, the serious danger that too great a significance will be attached to them. Rarely—perhaps never—is a single mathematician or scientist entitled to receive the full credit for an "innovation," nor does any one age deserve to be called the "renaissance" of an aspect of culture. Back of any discovery or invention there is invariably to be found an evolutionary development of ideas making its geniture possible. The history of the calculus furnishes a remarkably apt illustration of this fact.*

* Karpinski, L. C., "Is There Progress in Mathematical Discovery and Did the Greeks Have Analytic Geometry?" *Isis* **XXVII** (1937), 46–52.

Key Terms and Expressions

The Rectangle Property
The Addition Property
The Comparison Property
Partition
Lower sum; upper sum
Riemann sum
Definite integral

Integrand; limit of integration; integral sign
Area
Average (or mean) value of a function on an interval
Indefinite integral
Method of substitution
Change of variables
Natural logarithm

Key Formulas

$$\int_a^b c\,dx = c(b-a)$$

$$\int_a^b f(x)\,dx = \int_a^c f(x)\,dx + \int_c^b f(x)\,dx$$

$$\int_a^b f(x)\,dx = F(b) - F(a)$$

$$\int [f(x) + g(x)]\,dx = \int f(x)\,dx + \int g(x)\,dx$$

$$\int cf(x)\,dx = c\int f(x)\,dx$$

If $f(x) \le g(x) \le h(x)$ for $a \le x \le b$, then

$$\int_a^b f(x)\,dx \le \int_a^b g(x)\,dx \le \int_a^b h(x)\,dx$$

$$\int_a^b g(f(x))f'(x)\,dx = \int_{f(a)}^{f(b)} g(u)\,du$$

If $f(x) \ge g(x)$ on $[a, b]$, then $A = \int_a^b [f(x) - g(x)]\,dx$

$$\ln x = \int_1^x \frac{1}{t}\,dt \quad \text{and} \quad \frac{d}{dx}\ln x = \frac{1}{x}$$

$\ln bc = \ln b + \ln c$

$\ln b^r = r\ln b$ for r rational

Key Theorems

Fundamental Theorem of Calculus
Mean Value Theorem for Integrals

Law of Logarithms

REVIEW EXERCISES

1. Let $f(x) = 1/x$ and $\mathscr{P} = \{1, \frac{3}{2}, 2, \frac{5}{2}, 3\}$. Compute $L_f(\mathscr{P})$ and $U_f(\mathscr{P})$.

2. Let $f(x) = x\sin x$ and $\mathscr{P} = \{0, \pi/3, 2\pi/3, \pi, 3\pi/2\}$. Compute the left sum, right sum, and midpoint sum.

In Exercises 3–12 find the indefinite integral.

3. $\displaystyle\int (x^{3/5} - 8x^{5/3})\,dx$

4. $\displaystyle\int (3\cos x - 2\sin x)\,dx$

5. $\displaystyle\int (x^3 - 3x + 2 - 2/x)\,dx$

6. $\displaystyle\int (4 - x)^9\,dx$

7. $\displaystyle\int \frac{1 + \sqrt{x+1}}{\sqrt{x+1}}\,dx$

8. $\displaystyle\int \frac{1}{x^2}\sin\frac{1}{x}\,dx$

9. $\displaystyle\int \cos^3 3t\sin 3t\,dt$

10. $\displaystyle\int \frac{\tan(\ln t)}{t}\,dt$

11. $\displaystyle\int \sqrt{1 + \sqrt{x}}\,dx$

*12. $\displaystyle\int \frac{\tan^2 x}{x - \tan x}\,dx$

In Exercises 13–24 evaluate the definite integral.

13. $\displaystyle\int_{-1}^{-2} \left(x^{2/3} - \frac{5}{x^3}\right)dx$

14. $\displaystyle\int_1^2 \frac{x^2 + 2x + 3}{x}\,dx$

15. $\displaystyle\int_{-5\pi/3}^{\pi/4} 5\cos x\,dx$

16. $\displaystyle\int_0^\pi (\sqrt{x} - 3\sin x)\,dx$

17. $\displaystyle\int_{-8}^{-2} \frac{-1}{5u}\,du$

18. $\displaystyle\int_{-3}^2 (u + |u|)\,du$

19. $\displaystyle\int_0^2 (x^2 + 3)(x^3 + 9x + 1)^{1/3}\,dx$

20. $\displaystyle\int_0^1 t^5\sqrt{1 - t^2}\,dt$

21. $\displaystyle\int_{-\pi/4}^{\pi/2} \frac{\cos t}{1 + \sin t}\,dt$

22. $\displaystyle\int_2^5 \frac{x}{\sqrt{x-1}}\,dx$

23. $\displaystyle\int_{1/26}^{1/7} \frac{1}{x^2}\left(\frac{x+1}{x}\right)^{1/3}dx$

24. $\displaystyle\int_{-1}^{\pi/2} f(x)\,dx$, where $f(x) = \begin{cases} x^3 - 2x^2 & \text{for } -1 \le x < 0 \\ \sin x & \text{for } x \ge 0 \end{cases}$

In Exercises 25–28 find the area A of the region between the graph of f and the x axis on the given interval.

25. $f(x) = \frac{7}{4}x^2\sqrt{x} + \frac{1}{\sqrt{x}}$; $[2, 4]$

26. $f(x) = x + 2\sin x$; $[-\pi/2, \pi]$

27. $f(x) = x^2 - 4x + 3$; $[0, 4]$

28. $f(x) = \begin{cases} x - 3 & \text{for } x \le 3 \\ x^2 - 9 & \text{for } x > 3 \end{cases}$ $[1, 4]$

In Exercises 29–31 find the area A of the region between the graphs of the given functions.

29. $f(x) = 2x^5 + 5x^4$; $g(x) = 2x^5 + 20x^2$

30. $y = \dfrac{x^2 + 2}{\sqrt{x+1}}$; $y = \dfrac{3x + 2}{\sqrt{x+1}}$

31. $x = 2y^3 + y^2 + 5y - 7$; $x = y^3 + 4y^2 + 3y - 7$

32. Let $f(x) = x + 2\sin x$. Find the average value of f on the interval $[0, \pi]$.

In Exercises 33–34 sketch the graph of f, giving all pertinent information.

33. $f(x) = \ln(x^2 - 4)$

34. $f(x) = \ln(2 - x - x^2)$

In Exercises 35–42 find the derivative of the given function.

35. $F(x) = \displaystyle\int_0^x t\sqrt{1+t^5}\,dt$ 36. $G(y) = \displaystyle\int_{2y}^{\sin y} t\sin^2 t\,dt$

37. $F(x) = \displaystyle\int_1^{\ln x} \frac{1}{t}\,dt$ 38. $f(x) = \ln(\tan x + \sec x)$

39. $f(x) = x\sin(\ln x) - x\cos(\ln x)$

40. $f(x) = \ln x + \ln 1/x$

41. $f(x) = \ln(\ln\cos x + 2\ln\sec x)$

42. $f(x) = \ln(\ln(\ln x))$

43. In each part, determine which of the integrals can be easily evaluated by means of a substitution of the kind used in this section. For any that can, evaluate the integral.

a. i. $\displaystyle\int \sqrt{x^2+6}\,dx$ ii. $\displaystyle\int x\sqrt{x^2+6}\,dx$

   iii. $\displaystyle\int x^{1/2}\sqrt{x^2+6}\,dx$

b. i. $\displaystyle\int \sin\sqrt{x}\,dx$ ii. $\displaystyle\int \sqrt{x}\sin\sqrt{x}\,dx$

   iii. $\displaystyle\int \frac{1}{\sqrt{x}}\sin\sqrt{x}\,dx$

c. i. $\displaystyle\int \ln(x+1)\,dx$ ii. $\displaystyle\int (x+1)\ln(x+1)\,dx$

   iii. $\displaystyle\int \frac{\ln(x+1)}{x+1}\,dx$

44. By using the Comparison Property show that the following inequalities are valid.

a. $\dfrac{1}{2} \le \displaystyle\int_1^4 \frac{1}{2+x}\,dx \le 1$ b. $\dfrac{3}{5} \le \displaystyle\int_1^3 \frac{t}{1+t^2}\,dt \le 1$

45. Using the fact that $x^4 \le 1 + x^4 \le 2x^4$ for $x \ge 1$, show that

$$\frac{26}{3} \le \int_1^3 \sqrt{1+x^4}\,dx \le \frac{26}{3}\sqrt{2}$$

46. Suppose f is a continuous function satisfying $f(0) = 3$ and $|f(x) - f(y)| \le |x - y|$ for all x and y in $[0, 1]$. Find upper and lower bounds for $\int_0^1 f(x)\,dx$.

47. a. Prove that $\int_1^x (1/t)\,dt \le \int_1^x 1\,dt$ for $x \ge 1$.
   b. Using part (a), prove that $0 \le \ln x \le x - 1$ for $x \ge 1$.
   c. Using part (b), find upper and lower bounds for $\int_1^2 x\ln x\,dx$.

48. a. Let r be a rational number greater than -1. Using the fact that $1/t \le t^r$ for $t \ge 1$, show that

$$\ln x \le \frac{1}{r+1}(x^{r+1} - 1)\quad\text{for } x \ge 1$$

b. Let s be a rational number less than -1. Using ideas similar to those in part (a), show that

$$\ln x \ge \frac{1}{s+1}(x^{s+1} - 1)\quad\text{for } x \ge 1$$

 C c. Use a calculator to approximate $\ln 2$ by using the results of parts (a) and (b), with $x = 2$, and by taking r and s closer and closer to -1.

49. A car moving initially at 44 feet per second (that is, 30 miles per hour) decelerates at the constant rate of 4 feet per second per second. Determine how far the car travels before coming to a stop.

50. Two stones are dropped from the same point, 1 second apart. Find the distance between the two stones at any time after the second stone is dropped, and before the first touches the ground. (*Hint:* Find the position of each stone. Each stone accelerates at the rate of -32 feet per second per second.)

51. Some psychologists believe that a numerical measure of a child's ability to learn during the first four years of life is approximately described by the function

$$f(x) = \frac{5}{3x\ln x - 5x + 10}\quad\text{for } \frac{1}{2} \le x \le 4$$

where x is the age in years. At what age can such a child learn best?

52. From noon to 2 P.M. the temperature T increases with time t in such a way that

$$\frac{dT}{dt} = t^2 + 2t\quad\text{for } 0 \le t \le 2$$

If $T(0) = 60$, find $T(2)$.

53. Suppose a circuit has constant resistance R and variable current i given by

$$i = 110\sin 120\pi t$$

The rate of heat production in the circuit is $i^2 R$. Find the average rate of heat production during the time interval $[0, \frac{1}{60}]$. (*Hint:* See the definition of average value in Section 5.3.)

*54. The following problem appeared in Goursat's *Cours d'Analyse*, one of the most influential early calculus books. Suppose $p(x)$ is a polynomial of degree seven, such that $(x - 1)^4$ is a factor of $p(x) + 1$ and $(x + 1)^4$ is a factor of $p(x) - 1$. Find $p(x)$. (*Hint:* The given information implies that $(x - 1)^3$ and $(x + 1)^3$ are factors of $p'(x)$, so that $p'(x) = B(x - 1)^3(x + 1)^3$ for some constant B.)

55. Suppose a rocket moves in outer space (where no gravitational forces prevail) with initial speed v_0 and initial mass m_0. If gas is ejected from the rocket with speed u_0, then the Law of Conservation of Momentum from

physics implies that the speed v of the rocket increases as the mass decreases in such a way that

$$\frac{dv}{dm} = -\frac{u_0}{m}$$

a. Show that $v = v_0 + u_0 \ln \dfrac{m_0}{m}$.

b. For chemical fuels, 3000 meters per second would be a large value for u_0. Assuming that $u_0 = 3000$, determine the ratio m/m_0 that would result in a change of velocity of 3×10^7 meters per second, which is one-tenth the speed of light. (From the fact that the answer is essentially zero, it should be clear that chemical fuels are not suitable for interstellar travel.)

Cumulative Review, Chapters 1–4

1. Solve the inequality $-\dfrac{(2 + \sin x)}{(x-3)^2}\left(\dfrac{2-x}{4-x}\right)^{-1} < 0$.

2. Let $f(x) = \dfrac{2x+1}{3x+2}$ and $g(x) = \dfrac{2x-1}{2-3x}$.
   a. Find the domains of $f \circ g$ and $g \circ f$.
   b. Show that $(f \circ g)(x) = (g \circ f)(x)$ for all x in the domains of both f and g.
   c. Are $f \circ g$ and $g \circ f$ equal? Explain why or why not.

In Exercises 3–5 find the limit.

3. $\lim\limits_{x \to \pi/2^-} \dfrac{\cos x}{\sin x - 1}$

4. $\lim\limits_{x \to \infty} \dfrac{2x^3 + 3x^2 - 2}{-5x^3 + x - 9}$

5. $\lim\limits_{x \to \pi/4} \dfrac{\tan x - 1}{x - \pi/4}$

6. Let
$$f(x) = \sqrt{\frac{x^2(1+x)}{1-x}} \quad \text{for} \quad -1 \le x < 1$$

Determine whether or not f is
   a. continuous at 0   b. differentiable at 0

7. Let $f(x) = \cot(1/(x^2 + 1))$. Find $f'(x)$.

8. Find the slope of the line tangent to the circle $x^2 + 2x + y^2 - 4y = 0$ at the point on the circle lying on the positive y axis.

9. Let $f(x) = 2x^3 - 9x^2 + 14x - 6$. Show that f is a strictly increasing function.

10. Show that the equation $x^{15} + x^9 + 100x - 9 = 0$ has exactly one real solution. Do not try to find the solution.

11. Let
$$f(x) = cx - \frac{1}{x^3} - \frac{1}{2} \quad \text{for some constant } c \ne 0$$

Determine the value of c for which the equations $f(x) = 0$ and $f'(x) = 0$ have the same root, and find that root.

12. Suppose that f is differentiable and $f'(x) = 2f(x)$ for all x. Show that $f''(x)$ exists for all x, and express $f''(x)$ in terms of $f(x)$.

13. Sketch the graph of a function f defined on $[-2, \infty)$ with all the following properties.
   a. f'' is continuous on $(-2, \infty)$.
   b. f is strictly increasing on $(-2, -1)$ and its graph is concave downward on $(-2, -1)$.
   c. f is strictly decreasing on $(-1, 0)$ and its graph is concave downward on $(-1, 0)$.
   d. f is strictly decreasing on $(0, 1)$ and its graph is concave upward on $(0, 1)$.
   e. The graph of f has an inflection point at the point $(2, 3)$.
   f. $y = 4$ is a horizontal asymptote of the graph of f.

In Exercises 14–15 sketch the graph of f, noting all relevant properties.

14. $f(x) = \dfrac{(x+2)^2}{x-1}$

15. $f(x) = 1 - \dfrac{1}{x} + \dfrac{1}{x^2}$

16. Suppose a spherical balloon is being deflated in such a way that its surface area decreases at the rate of one square inch per second. How fast is the radius of the balloon decreasing when the surface area in square inches equals its volume in cubic inches?

17. Petunia throws a flower pot from her attic window 128 feet above the ground, with initial downward speed of 32 feet per second.
   a. What is the velocity of the pot after 1 second?
   b. How long does it take for the pot to hit the ground?

18. Suppose the acceleration a of an object is related to its velocity v in such a way that $a^3 = 7v^2 + v$. How fast is the acceleration increasing with respect to time when the velocity is 1?

19. A certain toll road averages 24,000 cars per day when the toll is 3 dollars per car. A survey concludes that 300 fewer cars would use the toll road for each 5 cent increase in the toll. What toll would maximize the total revenue?

20. Suppose one leg of a given right triangle is 6 inches long and the other leg is 2 inches long. Determine the dimensions of the rectangle with maximum area that can be inscribed in the triangle, assuming that two sides of the rectangle lie along the legs of the triangle.

6

APPLICATIONS OF THE INTEGRAL

Chapter 6 is devoted to applications of the integral. The mathematical applications will include volume, length of a curve, and surface area. The physical applications will include the computation of the work done by a force, moments of a plane region, and hydrostatic force on a plane surface. In order to present the various applications in a unified manner, we will base our presentations on the fact, stated in Theorem 5.6, that if f is a function continuous on an interval $[a, b]$, then $\int_a^b f(x)\,dx$ is the limit of the Riemann sums for f on $[a, b]$ as the lengths of the subintervals derived from the partitions of $[a, b]$ approach 0.

We will describe briefly our procedure for introducing the applications of the integral. For each application our goal will be to find a formula for a quantity I (such as the volume of a solid region or the work done by a force). We will proceed in the following way.

1. Using properties that we expect the quantity I to have, we will find a function f continuous on an interval $[a, b]$ with the property that for each partition $\mathscr{P}$ of $[a, b]$ there are numbers $t_1, t_2, \ldots, t_n$ in the n subintervals derived from $\mathscr{P}$ such that I should be approximately equal to the Riemann sum

$$\sum_{k=1}^{n} f(t_k)\,\Delta x_k$$

In each case it will be reasonable to expect that $\sum_{k=1}^{n} f(t_k)\,\Delta x_k$ should approach I as the norm of the partition $\mathscr{P}$ tends to 0. As explained in Section 5.2, this idea is expressed by writing

$$I = \lim_{\|\mathscr{P}\| \to 0} \sum_{k=1}^{n} f(t_k)\,\Delta x_k$$

302

2. We will conclude that

$$I = \int_a^b f(x)\,dx$$

In each case, step 1 is the crucial one and is the justification for the formula in step 2.

6.1
VOLUMES: THE CROSS-SECTIONAL METHOD

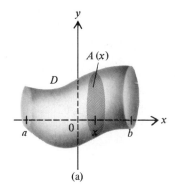

(a)

$A(x) = \pi r^2$

$V = \pi r^2 h$

Right circular cylinder

(b)

FIGURE 6.1

In our first application we will use the integral to find a formula for the volume of a solid region in space whose cross-sectional areas can be computed. The formula will enable us to find the volumes of such well known solids as spheres, cylinders, pyramids, and cones, as well as the volumes of solids that are less well known. Other formulas for volume will be presented in Section 6.2. In Chapter 14 we will give a general definition of volume from which one could obtain all the formulas that will appear in Sections 6.1 and 6.2.

Consider a solid region D having the following description. For every x in an interval $[a, b]$, the plane perpendicular to the x axis at x intersects D in a plane region having area $A(x)$ (Figure 6.1(a)). Our goal is to define the volume of D.

If the cross-sectional area is constant, that is,

$$A(x) = A_0 \quad \text{for } a \leq x \leq b$$

then we define the volume to be $A_0(b - a)$, the product of the constant cross-sectional area A_0 and the length of the interval $[a, b]$. This is consistent with the formula for the volume of a rectangular parallelepiped.

For another example in which the cross-sectional area is constant, consider a right circular cylinder of radius r and height h (Figure 6.1(b)). Here,

$$A(x) = \pi r^2 \quad \text{for } 0 \leq x \leq h$$

Consequently

$$V = \pi r^2 h$$

Now suppose the cross-sectional area A of the three-dimensional region D is a function that is continuous, but not necessarily constant, on $[a, b]$. Let $\mathscr{P} = \{x_0, x_1, \ldots, x_n\}$ be a partition of $[a, b]$. For each k between 1 and n, let t_k be an arbitrary number in the subinterval $[x_{k-1}, x_k]$. If Δx_k is small, the volume ΔV_k of the part of D between x_{k-1} and x_k is approximately equal to the product of the cross-sectional area $A(t_k)$ and the length Δx_k (Figure 6.2). Thus

$$\Delta V_k \approx (\text{cross-sectional area}) \times (\text{length}) = A(t_k)\,\Delta x_k$$

Since the volume V of D is the sum of $\Delta V_1, \Delta V_2, \ldots, \Delta V_n$, it follows that V should be approximately

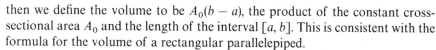

$$\sum_{k=1}^{n} \overbrace{A(t_k)}^{\substack{\text{cross-}\\\text{sectional}\\\text{area}}} \overbrace{\Delta x_k}^{\text{thickness}}$$

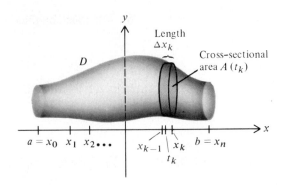

FIGURE 6.2

which is a Riemann sum for A on $[a, b]$. Therefore it seems plausible that

$$V = \lim_{\|\mathscr{P}\| \to 0} \sum_{k=1}^{n} A(t_k) \Delta x_k = \int_a^b A(x)\,dx$$

Thus if a solid region D has cross-sectional area $A(x)$ for $a \le x \le b$, and if A is continuous on $[a, b]$, then we define the volume V of D by the formula

$$V = \int_a^b A(x)\,dx \tag{1}$$

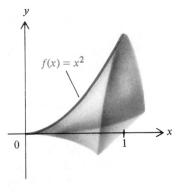

FIGURE 6.3

Example 1 Find the volume V of the solid D in Figure 6.3, whose cross section at x is semicircular with radius x^2, for $0 \le x \le 1$.

Solution Recall that the area of a semicircular region with radius r is $\frac{1}{2}\pi r^2$. Thus the cross-sectional area $A(x)$ at x is given by

$$A(x) = \frac{1}{2}\pi(x^2)^2 = \frac{1}{2}\pi x^4$$

Therefore we conclude from (1) that

$$V = \int_0^1 A(x)\,dx = \int_0^1 \frac{1}{2}\pi x^4\,dx = \frac{1}{10}\pi x^5 \bigg|_0^1 = \frac{\pi}{10} \quad \square$$

We can just as well reverse the roles of x and y and integrate with respect to y by the corresponding formula

$$V = \int_c^d A(y)\,dy \tag{2}$$

We will use (2) to calculate the volume of the pyramid in Figure 6.4(a).

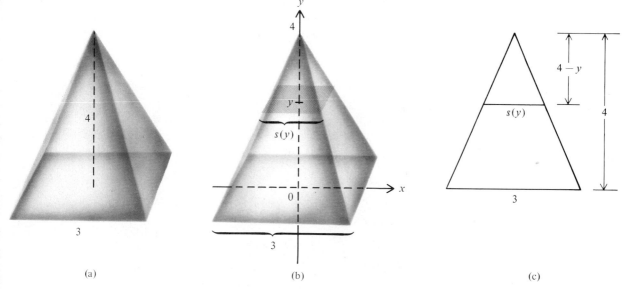

(a)   (b)   (c)

FIGURE 6.4

Example 2 Suppose a pyramid is 4 units tall and has a square base, 3 units on a side. Find the volume V of the pyramid.

Solution Let us place the y axis along the axis of the pyramid, the origin lying at the base of the pyramid, as in Figure 6.4(b). Notice that the pyramid extends from 0 to 4 on the y axis, so the limits of integration will be 0 and 4. Next we observe that the cross section at any y in $[0, 4]$ is a square. From the similar triangles in Figure 6.4(c) we see that the side length $s(y)$ of the square at y satisfies

$$\frac{s(y)}{3} = \frac{4-y}{4}, \quad \text{that is,} \quad s(y) = \frac{3}{4}(4-y)$$

Therefore the cross-sectional area at y is given by

$$A(y) = [s(y)]^2 = \frac{9}{16}(16 - 8y + y^2)$$

It follows from (2) that

$$V = \int_0^4 \frac{9}{16}(16 - 8y + y^2)\,dy = \frac{9}{16}\left(16y - 4y^2 + \frac{1}{3}y^3\right)\bigg|_0^4$$

$$= \frac{9}{16}\left(64 - 64 + \frac{64}{3}\right) = 12 \quad \square$$

More generally, if the pyramid has height h and a square base with side length a, then by substituting h for 4 and a for 3 in Example 2, we would find that

$$V = \frac{1}{3}a^2 h$$

(see Exercise 37). For example, the great pyramid of Cheops was originally (approximately) 482 feet tall and 754 feet square at the base. Therefore its volume was approximately $\frac{1}{3}(754)^2(482)$ cubic feet, that is, 91,341,571 cubic feet.

The Great Pyramid of Cheops

The Disc Method

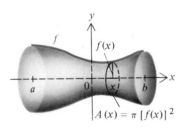

FIGURE 6.5

When the graph of a continuous, nonnegative function f on an interval $[a, b]$ is revolved about the x axis, it generates a solid region having circular cross sections, that is, cross sections that are circular discs (Figure 6.5). Since the radius of the cross section at x is $f(x)$, it follows that

$$A(x) = \pi[f(x)]^2$$

Thus from (1) we obtain a formula for the volume V of the solid that is generated:

$$V = \int_a^b \pi[f(x)]^2 \, dx \tag{3}$$

Because the cross sections are discs, the use of (3) to compute volume is called the **disc method**.

The formulas for the volumes of many well known figures follow readily from (3).

Example 3   Find the volume V of a sphere of radius r.

Solution A sphere is generated by revolving a semicircle about its diameter (Figure 6.6). If we let

$$f(x) = \sqrt{r^2 - x^2} \quad \text{for } -r \le x \le r$$

then from (3) we obtain

$$V = \int_{-r}^{r} \pi(\sqrt{r^2 - x^2})^2 \, dx = \pi \int_{-r}^{r} (r^2 - x^2) \, dx$$

$$= \pi\left(r^2 x - \frac{1}{3}x^3\right)\Big|_{-r}^{r} = \frac{4}{3}\pi r^3 \quad \square$$

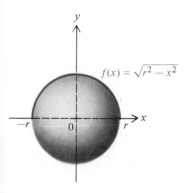

FIGURE 6.6

The formula for the volume of a sphere was known to Archimedes about 250 B.C.

Our next example concerns a frustum of a cone, as shown in Figure 6.7. A two-point equation of the straight line along the upper edge of the frustum is

$$\frac{y - b}{x - 0} = \frac{a - b}{h - 0}$$

or

$$y = b + \frac{a - b}{h} x \tag{4}$$

The frustum is generated by revolving the graph of this equation on $[a, b]$ about the x axis.

Example 4 Find the volume V of the frustum of the cone shown in Figure 6.7.

Solution Since y in (4) represents the radius of the cross section of the frustum at x, it follows from (3) that

$$V = \int_0^h \pi\left(b + \frac{a - b}{h} x\right)^2 dx$$

$$= \frac{\pi h}{3(a - b)}\left(b + \frac{a - b}{h} x\right)^3\Big|_0^h = \frac{\pi h}{3(a - b)}(a^3 - b^3) \quad \square$$

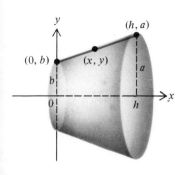

The Apollo 17 Command Module. Such modules are often in the shape of a frustum of a cone.

If $b = 0$ in Example 4, then the resulting region is a complete cone, and the volume reduces to the well-known formula

$$V = \frac{1}{3}\pi a^2 h$$

FIGURE 6.7

The Washer Method Finally, we present a formula for the volume of the solid region generated by revolving a more general plane region about the x axis. Let f and g be functions

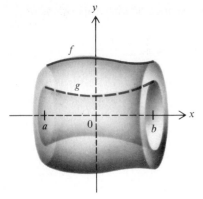

FIGURE 6.8

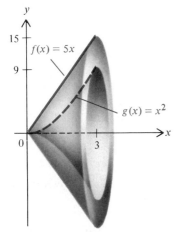

FIGURE 6.9

such that

$$0 \le g(x) \le f(x) \quad \text{for } a \le x \le b$$

Then the plane region between the graph of f and the x axis on $[a, b]$ is composed of two regions: the region between the graphs of f and g on $[a, b]$ and the region between the graph of g and the x axis on $[a, b]$. Therefore the volume of the solid generated by revolving the entire region about the x axis should be equal to the sum of the volumes of the solids generated by revolving the two component regions about the x axis (Figure 6.8). Accordingly, the volume V of the solid generated by revolving the region between the graphs of f and g on $[a, b]$ is given by

$$V = \int_a^b \pi[(f(x))^2 - (g(x))^2]\,dx \qquad (5)$$

When g is different from 0, the method of finding volumes by (5) is sometimes called the **washer method** because the cross sections resemble washers.

Example 5   Let $f(x) = 5x$ and $g(x) = x^2$, and let R be the region between the graphs of f and g on $[0, 3]$. Find the volume V of the solid obtained by revolving R about the x axis (Figure 6.9).

Solution   Since $5x \ge x^2$ for $0 \le x \le 3$, (5) implies that

$$V = \int_0^3 \pi(25x^2 - x^4)\,dx = \pi\left(\frac{25x^3}{3} - \frac{x^5}{5}\right)\Bigg|_0^3$$

$$= \pi\left(225 - \frac{243}{5}\right) = \frac{882\pi}{5} \quad \square$$

EXERCISES 6.1

In Exercises 1–12 let R be the region between the graph of the function and the x axis on the given interval. Find the volume V of the solid obtained by revolving R about the x axis.

1. $f(x) = x^{3/2}$; $[0, 1]$
2. $f(x) = 1 + x^2$; $[-1, 2]$
3. $f(x) = -1/x$; $[-3, -2]$
4. $g(x) = \sqrt{3 - x^2}$; $[0, \sqrt{3}]$
5. $g(x) = \sqrt{\cos x}$; $[0, \pi/6]$
6. $f(x) = \sqrt{1 + \sin^2 x}$; $[-\pi/2, \pi/2]$
7. $g(x) = \sec x$; $[-\pi/4, 0]$
8. $f(x) = (\ln x)/\sqrt{x}$; $[1, e]$
9. $f(x) = x(x^3 + 1)^{1/4}$; $[1, 2]$
10. $f(x) = \dfrac{\sqrt{x}}{1 + x^2}$; $[0, 1]$
11. $f(x) = \tan x$; $[-\pi/6, \pi/3]$
12. $f(x) = \sqrt{x}(1 - x)^{1/4}$; $[0, 1]$

In Exercises 13–16 let R be the region between the graph of f and the y axis on the given interval. Find the volume V of the solid obtained by revolving R about the y axis.

13. $f(y) = \cos y$; $[0, \pi/4]$ 14. $f(y) = y(3 - y)$; $[1, 3]$

15. $f(y) = \sqrt{1 + y^3}$; $[1, 2]$

16. $f(y) = \sqrt{\sin y \cos y}\,(1 + \cos^2 y)^{1/3}$; $[0, \pi/2]$

In Exercises 17–20 let R be the region between the graphs of f and g on the given interval. Find the volume V of the solid obtained by revolving R about the x axis.

17. $f(x) = \sqrt{x + 1}$, $g(x) = \sqrt{x - 1}$; $[1, 3]$

18. $f(x) = x + 1$, $g(x) = x - 1$; $[1, 4]$

19. $f(x) = \cos x + \sin x$, $g(x) = \cos x - \sin x$; $[0, \pi/4]$

*20. $f(x) = 2x - x^2$, $g(x) = x^2 - 2x$; $[0, 1]$ (*Hint:* The washer method will not work. Why?)

In Exercises 21–24 find the volume V of the solid generated by revolving about the x axis the region between the graphs of the given equations.

21. $y = \frac{1}{2}x^2 + 3$ and $y = 12 - \frac{1}{2}x^2$

22. $y = x^{1/2}$ and $y = 2x^{1/4}$

23. $y = 5x$ and $y = x^2 + 2x + 2$

*24. $y = x^3 + 2$ and $y = x^2 + 2x + 2$

In Exercises 25–30 find the volume V of the solid with the given information about its cross sections.

25. The base of the solid is an isosceles right triangle whose legs L_1 and L_2 are each 4 units long. Any cross section perpendicular to L_1 is semicircular.

26. The base of the solid is a square centered at the origin, with side length 6. The area of each cross section perpendicular to an edge of the base equals the distance from the cross section to the origin.

27. The solid has a circular base with radius 1, and the cross sections perpendicular to a fixed diameter of the base are squares. (*Hint:* Center the base at the origin.)

28. The base of a solid is a circle with radius 2, and the cross sections perpendicular to a fixed diameter of the base are equilateral triangles.

29. The base is an equilateral triangle each side of which has length 10. The cross sections perpendicular to a given altitude of the triangle are squares.

30. The base is an equilateral triangle with altitude 10. The cross sections perpendicular to a given altitude of the triangle are semicircles.

31. Suppose f is continuous on $[a, b]$ and the graph of f lies above the line $y = c$. Write down a formula for the volume V of the solid obtained by revolving about the line $y = c$ the region between the graph of f and the line $y = c$ on $[a, b]$.

32. Use the result of Exercise 31 to find the volume V of the solid obtained by revolving about the line $y = -1$ the region between the graph of the equation $y = \sqrt{x + 1}$ and the line $y = -1$ on the interval $[0, 1]$.

33. Find the volume V of the solid obtained by revolving about the line $y = 1$ the region between the graph of the equation $y = \sin x$ and the x axis on the interval $[0, \pi]$. (*Hint:* Pattern your solution after Exercise 31.)

34. Suppose f and g are continuous on $[a, b]$, and let c be such that $c \le g(x) \le f(x)$ for $a \le x \le b$. Write down a formula for the volume V of the solid obtained by revolving about the line $y = c$ the region between the graphs of f and g on $[a, b]$.

35. Use the result of Exercise 34 to find the volume V of the solid obtained by revolving about the line $y = 1$ the region between the graphs of $y = x^2 - x + 1$ and $y = 2x^2 - 4x + 3$.

36. Use the result of Exercise 34 to find the volume V of the solid obtained by revolving about the line $y = 1$ the region between the graphs of $y = x$ and $y = x^2 - 2x + 2$.

37. Show that if a right pyramid has square cross sections, then its volume V is given by

$$V = \frac{1}{3}a^2h$$

where h is the height and a is the length of a side of the base. (*Hint:* Follow the idea used in working Example 2.)

38. Suppose the great pyramid of Cheops had been built with equilateral triangular cross sections instead of square cross sections but had had the same height of 482 feet and base 754 feet on a side. What percentage of the original volume would have resulted?

39. The base of a right triangular pyramid is an equilateral triangle of side a, and the pyramid's height is h. Find the volume V of the pyramid.

40. Let a sphere of radius r be sliced at a distance h from its center. Show that the volume V of the smaller piece cut off is given by

$$V = \frac{\pi(r - h)^2}{3}(2r + h)$$

41. As viewed from above, a swimming pool has the shape of the ellipse

$$\frac{x^2}{400} + \frac{y^2}{100} = 1$$

The cross sections of the pool perpendicular to the ground and parallel to the y axis are squares. If the units are feet, determine the volume V of the pool.

42. Find the volume V of the solid generated when the ellipse

$$\frac{x^2}{a^2} + \frac{y^2}{b^2} = 1$$

is revolved about the x axis. With $a = 100$ and $b = 25$, your answer could be the volume of gas needed to fill a dirigible having these dimensions.

43. The ivory stones used in the Oriental game of *go* have approximately the shape of the solid generated by revolving a certain ellipse about the x axis. If the length of a *go* stone is 2 centimeters and its height is 1 centimeter, what is its volume? (*Hint:* Let $a = \frac{1}{2}$ and $b = 1$ in Exercise 42.)

44. Suppose a ring is obtained by revolving about the x axis the region between the curves $y = 4 - x^2$ and $y = 1$ for $-\sqrt{3} \le x \le \sqrt{3}$. Determine the volume V of the ring.

45. A soda glass has the shape of the surface generated by revolving the graph of $y = 6x^2$ for $0 \le x \le 1$ about the y axis. Soda is extracted from the glass through a straw at the rate of $\frac{1}{2}$ cubic inch per second. How fast is the depth of soda decreasing when the depth is $\frac{3}{2}$ inches?

46. A wooden golf tee has approximately the dimensions of the solid obtained by revolving about the x axis the region bounded by the graphs of f and g, where

$$f(x) = \begin{cases} \frac{1}{2}x & \text{for } 0 \le x \le \frac{1}{2} \\ \frac{1}{4} & \text{for } \frac{1}{2} \le x \le \frac{7}{2} \\ \frac{1}{4}[1 + (x - \frac{7}{2})^2] & \text{for } \frac{7}{2} \le x \le \frac{9}{2} \\ \frac{1}{2} & \text{for } \frac{9}{2} \le x \le 5 \end{cases}$$

and

$$g(x) = \begin{cases} 0 & \text{for } 0 \le x \le \frac{9}{2} \\ x - \frac{9}{2} & \text{for } \frac{9}{2} \le x \le 5 \end{cases}$$

(Figure 6.10). Here x, $f(x)$, and $g(x)$ are measured in centimeters. Determine how much wood goes into a golf tee.

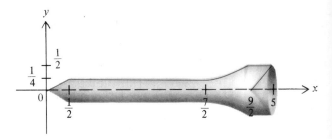

FIGURE 6.10

47. Cavalieri's Principle for volume states that if two solids have the same cross-sectional area at each x between a and b, then the two solids have the same volume. Prove Cavalieri's Principle for volume by using (1).

6.2
VOLUMES: THE SHELL METHOD

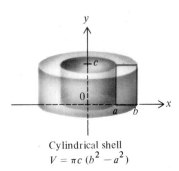

Cylindrical shell
$V = \pi c\,(b^2 - a^2)$

FIGURE 6.11

In Section 6.1 we obtained a formula for the volume of the solid generated by revolving about the x axis the region between the graphs of f and g on $[a, b]$. We can also revolve such a region about the y axis and find a corresponding formula for the volume of the solid so generated. That is the topic of the present section.

To begin, let us determine the volume V of a cylindrical shell obtained by revolving a rectangle about the y axis (Figure 6.11). Suppose the rectangle is bounded by the x axis, the line $y = c$, and the lines $x = a$ and $x = b$, where $b \ge a \ge 0$ and $c \ge 0$; then since the volume of the cylindrical shell is the difference of the volumes of the outer and the inner cylinders, it follows that

$$V = \underbrace{\pi b^2 c}_{\substack{\text{volume of} \\ \text{outer cylinder}}} - \underbrace{\pi a^2 c}_{\substack{\text{volume of} \\ \text{inner cylinder}}} = \pi c(b^2 - a^2) \tag{1}$$

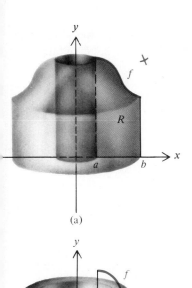

(a)

(b)

FIGURE 6.12

If we replace a by x_{k-1} and b by x_k, then we obtain

$$V = \pi c(x_k^2 - x_{k-1}^2) = \pi c(x_k + x_{k-1})(x_k - x_{k-1}) \tag{2}$$

This formula will be useful when partitions are introduced.

Now let f be a continuous nonnegative function on $[a, b]$, with $a \geq 0$. We wish to define the volume V of the solid region in Figure 6.12(a), obtained by revolving about the y axis the region R between the graph of f and the x axis on $[a, b]$. Let $\mathscr{P} = \{x_0, x_1, \ldots, x_n\}$ be any partition of $[a, b]$. For each k between 1 and n, let t_k be the midpoint $(x_k + x_{k-1})/2$ of the subinterval $[x_{k-1}, x_k]$. If Δx_k is small, the volume ΔV_k of the portion of the solid between the revolved lines $x = x_{k-1}$ and $x = x_k$ is approximately equal to the volume of the corresponding cylindrical shell with height $f(t_k)$ (Figure 6.12(b)). By (2), with $f(t_k)$ replacing c, this means that

$$\Delta V_k \approx \pi f(t_k)(x_k + x_{k-1})(x_k - x_{k-1}) = 2\pi t_k f(t_k)\Delta x_k$$

Therefore the volume V of the solid, which is the sum of $\Delta V_1, \Delta V_2, \ldots, \Delta V_n$, should be approximately

$$\sum_{k=1}^{n} \overbrace{2\pi t_k}^{\text{circumference}} \overbrace{f(t_k)}^{\text{height}} \overbrace{\Delta x_k}^{\text{thickness}}$$

which is a Riemann sum for $2\pi x f$ on $[a, b]$. As a result,

$$V = \lim_{\|\mathscr{P}\| \to 0} \sum_{k=1}^{n} 2\pi t_k f(t_k)\Delta x_k = \int_a^b 2\pi x f(x)\, dx$$

Thus we are led to the following definition of volume:

$$V = \int_a^b 2\pi x f(x)\, dx \tag{3}$$

The emphasis on cylindrical shells justifies the name **shell method** for this way of computing volumes.

Example 1 Let $f(x) = 1 - (x - 2)^2$ for $1 \leq x \leq 3$, and let R be the region between the graph of f and the x axis on $[1, 3]$. Find the volume V of the solid obtained by revolving R about the y axis (Figure 6.13).

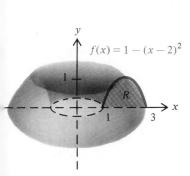

FIGURE 6.13

Solution You can think of the given solid as a cake. Since the height of the solid at any x in $[1, 3]$ is $1 - (x - 2)^2$, (3) implies that

$$V = \int_1^3 2\pi x[1 - (x - 2)^2]\,dx = 2\pi \int_1^3 (-x^3 + 4x^2 - 3x)\,dx$$

$$= 2\pi\left(-\frac{1}{4}x^4 + \frac{4}{3}x^3 - \frac{3}{2}x^2\right)\Bigg|_1^3$$

$$= 2\pi\left[\left(-\frac{81}{4} + 36 - \frac{27}{2}\right) - \left(-\frac{1}{4} + \frac{4}{3} - \frac{3}{2}\right)\right] = \frac{16}{3}\pi \quad \square$$

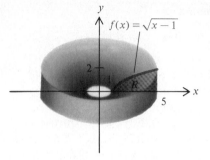

FIGURE 6.14

Example 2   Let $f(x) = \sqrt{x-1}$, and let R be the region between the graph of f and the x axis on $[1, 5]$. Find the volume V of the solid obtained by revolving R about the y axis (Figure 6.14).

Solution   Since the height of the shell at any x in $[1, 5]$ is $\sqrt{x-1}$, it follows from (3) that

$$V = \int_1^5 2\pi x \sqrt{x-1}\, dx$$

To evaluate the integral, we make the substitution

$$u = x - 1, \quad \text{so that } du = dx$$

Now if $x = 1$ then $u = 0$, and if $x = 5$ then $u = 4$. Therefore

$$V = \int_1^5 2\pi x \sqrt{x-1}\, dx = 2\pi \int_0^4 (u+1)\sqrt{u}\, du = 2\pi \int_0^4 (u^{3/2} + u^{1/2})\, du$$

$$= 2\pi \left(\frac{2}{5} u^{5/2} + \frac{2}{3} u^{3/2} \right) \Big|_0^4 = 2\pi \left(\frac{64}{5} + \frac{16}{3} \right) = \frac{544}{15}\pi \quad \square$$

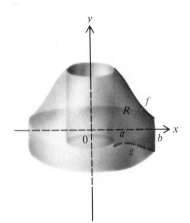

FIGURE 6.15

Finally, we relax the conditions on the upper and lower boundaries of the plane region to be revolved about the y axis. In particular, let f and g be continuous on $[a, b]$, with $a \geq 0$, and suppose that

$$g(x) \leq f(x) \quad \text{for } a \leq x \leq b$$

Then let R be the region between the graphs of f and g on $[a, b]$ (Figure 6.15). The volume V of the solid obtained by revolving R about the y axis is given by

$$V = \int_a^b 2\pi x [f(x) - g(x)]\, dx \tag{4}$$

We will use (4) to find the volume of a solid rather like a single-scoop ice cream cone.

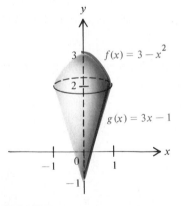

FIGURE 6.16

Example 3   Let $f(x) = 3 - x^2$ and $g(x) = 3x - 1$, and let R be the region between the graphs of f and g on $[0, 1]$. Find the volume V of the solid generated by revolving R about the y axis (Figure 6.16).

Solution   Since $3 - x^2 \geq 3x - 1$ for $0 \leq x \leq 1$, it follows that the height of the solid at any x in $[0, 3]$ is $(3 - x^2) - (3x - 1)$. Thus by (4),

$$V = \int_0^1 2\pi x [(3 - x^2) - (3x - 1)]\, dx$$

$$= 2\pi \int_0^1 (-x^3 - 3x^2 + 4x)\, dx = 2\pi \left(-\frac{1}{4} x^4 - x^3 + 2x^2 \right) \Big|_0^1$$

$$= \frac{3\pi}{2} \quad \square$$

The volumes of certain solids of revolution can be evaluated either by the washer method or by the shell method. As you would expect, the result is the same, whichever method is used. We support this claim with the following example.

Example 4 Let R be the region between the graphs of the equations $y = x^2$ and $y = 2x$ on $[0, 2]$. Find the volume V of the solid generated by revolving R about the x axis (Figure 6.17) by
    a. the washer method   b. the shell method

Solution
    a. For the washer method we integrate with respect to x, using (5) of Section 6.1. Since $2x \geq x^2$ for $0 \leq x \leq 2$, we obtain

$$V = \int_0^2 \pi[(2x)^2 - (x^2)^2]\,dx = \pi \int_0^2 (4x^2 - x^4)\,dx$$

$$= \pi\left(\frac{4}{3}x^3 - \frac{1}{5}x^5\right)\Big|_0^2 = \pi\left(\frac{32}{3} - \frac{32}{5}\right)$$

$$= \frac{64\pi}{15}$$

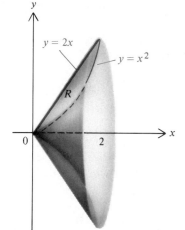

y

$y = 2x$

 $y = x^2$

R

0 2 x

FIGURE 6.17

    b. For the shell method we integrate with respect to y, using (4) of this section with x replaced by y. In order to integrate with respect to y we need to consider the two equations $y = 2x$ and $y = x^2$ as defining functions of y. We obtain the equations

$$x = \tfrac{1}{2}y \quad \text{and} \quad x = \sqrt{y}$$

To determine the limits of integration, we find the y coordinates of the points at which the line $x = \tfrac{1}{2}y$ and the parabola $x = \sqrt{y}$ intersect:

$$\tfrac{1}{2}y = x = \sqrt{y}$$

$$\tfrac{1}{4}y^2 = y$$

$$y^2 - 4y = 0$$

$$y = 0 \quad \text{or} \quad y = 4$$

Since $\sqrt{y} \geq \tfrac{1}{2}y$ for $0 \leq y \leq 4$, we conclude from (4) that

$$V = \int_0^4 2\pi y\left(\sqrt{y} - \frac{1}{2}y\right)dy = 2\pi \int_0^4 \left(y^{3/2} - \frac{1}{2}y^2\right)dy$$

$$= 2\pi\left(\frac{2}{5}y^{5/2} - \frac{1}{6}y^3\right)\Big|_0^4 = 2\pi\left(\frac{64}{5} - \frac{64}{6}\right)$$

$$= \frac{64\pi}{15} \quad \square$$

EXERCISES 6.2

In Exercises 1–8 let R be the region between the graph of the function and the x axis on the given interval. Find the volume V of the solid generated by revolving R about the y axis.

1. $f(x) = \dfrac{4}{x^3}; [1,3]$

2. $f(x) = 2x - \dfrac{3}{x}; [2,3]$

3. $f(x) = (x - 1)^2; [0,2]$

4. $f(x) = \sqrt{x^2 + 1}; [0,\sqrt{3}]$

5. $f(x) = \sin x^2; [\sqrt{\pi}/2, \sqrt{\pi}]$

6. $g(x) = \dfrac{1}{\sqrt{1 - x^2}}; [0, \sqrt{3}/2]$

7. $g(x) = x^2\sqrt{x^2 - 1}; [1,2]$

*8. $g(x) = \sqrt{1 + \sqrt{x}}; [0,4]$

In Exercises 9–12 let R be the region between the graph of f and the y axis on the given interval. By interchanging the roles of x and y in (3), find the volume V of the solid generated by revolving R about the x axis.

9. $f(y) = \dfrac{\ln y}{y^2}; [1,2]$

10. $f(y) = y^2\sqrt{1 + y^4}; [0,1]$

11. $f(y) = \dfrac{y^4}{\sqrt{1 - y^6}}; \left[0, \dfrac{1}{2}\sqrt{2}\right]$

12. $f(y) = \dfrac{y - 1}{2y^3 - 3y^2}; [2,3]$

In Exercises 13–16 let R be the region between the graphs of f and g on the given interval. Find the volume V of the solid obtained by revolving R about the y axis.

13. $f(x) = 1, g(x) = x - 2; [1,3]$

14. $f(x) = \sqrt{x^2 + 1}, g(x) = \sqrt{x^2 - 1}; [1,2]$

15. $f(x) = \cos x^2, g(x) = \sin x^2; [0, \dfrac{1}{2}\sqrt{\pi}]$

16. $f(x) = \sqrt{1 - x^2}, g(x) = -2 + \sqrt{1 + x^2}; [0,1]$

In Exercises 17–18 let R be the region between the graphs of f and g on the given interval. Find the volume V of the solid obtained by revolving R about the x axis.

17. $f(y) = y^2 + 1, g(y) = y\sqrt{1 + y^3}; [0,1]$

18. $f(y) = \dfrac{1}{(y + 2)^2}, g(y) = \dfrac{1}{y + 2}; [0,2]$

(*Hint:* To evaluate the integral, make the substitution $u = y + 2$.)

In Exercises 19–21 find the volume V of the solid generated by revolving the region between the graphs of the equations about the y axis.

19. $y = 2x$ and $y = x^2$

20. $y^2 = x$ and $y = 3x$

21. $y = |x - 2|$ and $y = \dfrac{1}{2}(x - 2)^2 + \dfrac{1}{2}$

22. Let f be continuous and nonnegative on $[a, b]$, and assume that $c \le a$. Let R be the region between the graph of f and the x axis on $[a, b]$. Find a formula for the volume V of the solid obtained by revolving R about the line $x = c$.

23. Use the result of Exercise 22 to find the volume V of the solid obtained by revolving about the line $x = -1$ the region between the graph of $y = x^4$ and the x axis on $[0, 1]$.

24. Find the volume V of the solid obtained by revolving about the line $x = 2\pi$ the region between the graph of $y = (2\pi - x)^{3/2}$ and the x axis on $[0, \pi]$. (*Hint:* Pattern your solution after Exercise 22.)

25. Let f and g be continuous on $[a, b]$, with $g(x) \le f(x)$ for $a \le x \le b$. Let R be the region between the graphs of f and g on $[a, b]$, and assume that $c \le a$. Find a formula for the volume V of the solid obtained by revolving R about the line $x = c$.

26. Use the result of Exercise 25 to find the volume V of the solid obtained by revolving about the line $x = -1$ the region between the graphs of $y = 2x$ and $y = -2x^2 + 4x$.

27. Use the result of Exercise 25 to find the volume V of the solid obtained by revolving about the line $x = -5$ the region between the graphs of $y = x^2 + 4$ and $y = 2x^2 + x + 2$.

28. Using (4), derive a formula for the volume V of a sphere of radius a.

29. Using (3), derive a formula for the volume V of a circular cone having radius a and height h.

*30. A hole of radius $\dfrac{1}{2}$ is drilled vertically through the center of a sphere of radius 2. Find the volume V removed.

6.3
LENGTHS OF CURVES

The ancient Greeks estimated the circumference of a circle by inscribing a polygon of n sides and then computing the perimeter of the polygon. They surmised that the larger n was, the better the perimeter of the polygon approximated the actual circumference of the circle (Figure 6.18). We will use this basic idea to define and compute the lengths of many curves.

FIGURE 6.18

Let f have a continuous derivative on $[a, b]$. If f is linear, that is, if the graph of f is a line segment, then the length $\mathscr{L}$ of the graph is the distance between $(a, f(a))$ and $(b, f(b))$, so that

$$\mathscr{L} = \sqrt{(b - a)^2 + [f(b) - f(a)]^2}$$

(Figure 6.19(a)).

When f is not necessarily linear, we let $\mathscr{P} = \{x_0, x_1, \ldots, x_n\}$ be any partition of $[a, b]$, and approximate the graph of f by a polygonal line L whose vertices are $(x_0, f(x_0)), (x_1, f(x_1)), \ldots, (x_n, f(x_n))$ (Figure 6.19(b)). Let $\Delta \mathscr{L}_k$ be the length of the portion of the graph of f joining $(x_{k-1}, f(x_{k-1}))$ and $(x_k, f(x_k))$. If Δx_k is small,

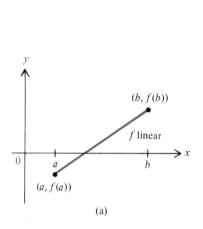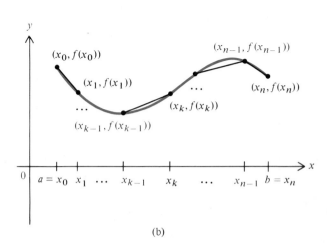

FIGURE 6.19

$\Delta\mathscr{L}_k$ is approximately equal to the length of the line segment joining $(x_{k-1}, f(x_{k-1}))$ and $(x_k, f(x_k))$. In other words,

$$\Delta\mathscr{L}_k \approx \sqrt{(x_k - x_{k-1})^2 + [f(x_k) - f(x_{k-1})]^2} \tag{1}$$

The Mean Value Theorem, applied to f on the interval $[x_{k-1}, x_k]$, implies that

$$f(x_k) - f(x_{k-1}) = f'(t_k)(x_k - x_{k-1})$$

for some t_k in (x_{k-1}, x_k). Therefore (1) can be rewritten

$$\Delta\mathscr{L}_k \approx \sqrt{(x_k - x_{k-1})^2 + [f'(t_k)(x_k - x_{k-1})]^2} = \sqrt{1 + [f'(t_k)]^2}(x_k - x_{k-1})$$

Therefore the total length $\mathscr{L}$ of the graph of f, which is the sum of the lengths $\Delta\mathscr{L}_1, \Delta\mathscr{L}_2, \ldots, \Delta\mathscr{L}_n$, should be approximately

$$\sum_{k=1}^{n} \sqrt{1 + [f'(t_k)]^2}\, \Delta x_k$$

itself a Riemann sum for $\sqrt{1 + (f')^2}$ on $[a, b]$. Therefore it seems plausible that

$$\mathscr{L} = \lim_{\|\mathscr{P}\| \to 0} \sum_{k=1}^{n} \sqrt{1 + [f'(t_k)]^2}\, \Delta x_k = \int_a^b \sqrt{1 + [f'(x)]^2}\, dx$$

This leads us to make the following definition.

DEFINITION 6.1

Let f have a continuous derivative on $[a, b]$. Then the **length** $\mathscr{L}$ of the graph of f on $[a, b]$ is defined by

$$\mathscr{L} = \int_a^b \sqrt{1 + [f'(x)]^2}\, dx \tag{2}$$

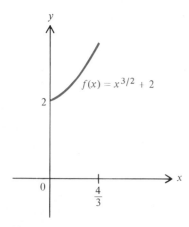

FIGURE 6.20

Example 1   Let $f(x) = x^{3/2} + 2$ for $0 \le x \le \frac{4}{3}$. Find the length $\mathscr{L}$ of the graph of f (Figure 6.20).

Solution   Since $f'(x) = \frac{3}{2}x^{1/2}$, we know from (2) that

$$\mathscr{L} = \int_0^{4/3} \sqrt{1 + \left(\frac{3}{2}x^{1/2}\right)^2}\, dx = \int_0^{4/3} \sqrt{1 + \frac{9}{4}x}\, dx$$

To evaluate the integral, we let

$$u = 1 + \frac{9}{4}x, \quad \text{so that} \quad du = \frac{9}{4}dx$$

If $x = 0$ then $u = 1$, and if $x = \frac{4}{3}$ then $u = 1 + \frac{9}{4} \cdot \frac{4}{3} = 4$. Therefore

$$\mathscr{L} = \int_0^{4/3} \overbrace{\sqrt{1 + \frac{9}{4}x}}^{\sqrt{u}}\ \overbrace{dx}^{\frac{4}{9}du} = \int_1^4 \sqrt{u}\,\frac{4}{9}\,du = \frac{4}{9}\left(\frac{2}{3}u^{3/2}\right)\Big|_1^4 = \frac{56}{27} \quad \square$$

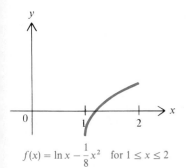

$f(x) = \ln x - \dfrac{1}{8}x^2$ for $1 \le x \le 2$

FIGURE 6.21

Example 2 Let

$$f(x) = \ln x - \frac{1}{8}x^2 \quad \text{for } 1 \le x \le 2$$

Find the length $\mathscr{L}$ of the graph of f (Figure 6.21).

Solution Since $f'(x) = 1/x - x/4$, it follows from (2) that

$$\mathscr{L} = \int_1^2 \sqrt{1 + \left(\frac{1}{x} - \frac{x}{4}\right)^2}\, dx = \int_1^2 \sqrt{1 + \left(\frac{1}{x^2} - \frac{1}{2} + \frac{x^2}{16}\right)}\, dx$$

$$= \int_1^2 \sqrt{\frac{1}{x^2} + \frac{1}{2} + \frac{x^2}{16}}\, dx = \int_1^2 \sqrt{\left(\frac{1}{x} + \frac{x}{4}\right)^2}\, dx$$

$$= \int_1^2 \left(\frac{1}{x} + \frac{x}{4}\right) dx = \left(\ln x + \frac{x^2}{8}\right)\Big|_1^2$$

$$= \ln 2 + \frac{3}{8} \quad \square$$

Notice in the solution of Example 2 that by adding 1 to $(f'(x))^2$ we obtained the perfect square $(1/x + x/4)^2$ inside the radical sign. This enabled us to eliminate the radical sign so that we could perform the integration.

Frequently the integral that arises in using (2) to find the length of a curve cannot be readily evaluated. In those cases one can find approximate values for the length of a graph by using Riemann sums or other more accurate methods to be considered in Section 8.6.

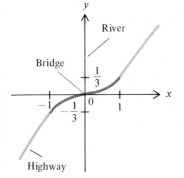

$f(x) = \dfrac{1}{3}x^3$ for $-1 \le x \le 1$

FIGURE 6.22

Example 3 A highway runs northeast across a river, which is represented by the y axis in Figure 6.22. To achieve a crossing at right angles, a curve is introduced into the highway. If

$$f(x) = \frac{1}{3}x^3 \quad \text{for } -1 \le x \le 1$$

then this curve is represented by the graph of f. Suppose the center of the curve is to be painted with a continuous white stripe. Find the approximate length $\mathscr{L}$ of the stripe by dividing the interval $[-1, 1]$ into 4 subintervals of equal length and using the corresponding midpoint sum.

Solution Since $f'(x) = x^2$, we know from (2) that

$$\mathscr{L} = \int_{-1}^1 \sqrt{1 + (x^2)^2}\, dx = \int_{-1}^1 \sqrt{1 + x^4}\, dx$$

In order to find the midpoint sum we let $g(x) = \sqrt{1 + x^4}$, and take ι_k to be the midpoint of the kth subinterval of $[-1, 1]$, for $k = 1, 2, 3$, and 4. We obtain

$$\sum_{k=1}^{4} g(t_k)\Delta x_k = g(t_1)\frac{1}{2} + g(t_2)\frac{1}{2} + g(t_3)\frac{1}{2} + g(t_4)\frac{1}{2}$$

$$= \frac{1}{2}\left(\sqrt{1 + \left(-\frac{3}{4}\right)^4} + \sqrt{1 + \left(-\frac{1}{4}\right)^4} + \sqrt{1 + \left(\frac{1}{4}\right)^4} + \sqrt{1 + \left(\frac{3}{4}\right)^4}\right)$$

$$\approx 2.14930$$

Thus $\mathscr{L} \approx 2.14930$. $\square$

In Section 8.6 we will be able to show that $\mathscr{L} \approx 2.17911$ (accurate to within 0.01).

EXERCISES 6.3

In Exercises 1–16 find the length $\mathscr{L}$ of the graph of the given function.

1. $f(x) = 2x + 3$ for $1 \le x \le 5$

2. $f(x) = \frac{2}{3}x^{3/2}$ for $1 \le x \le 4$

3. $f(x) = x^2 - \frac{1}{8}\ln x$ for $2 \le x \le 3$

4. $f(x) = (2x - 1)^{3/2}$ for $\frac{1}{2} \le x \le 4$

5. $f(x) = \frac{1}{3}(x^2 + 2)^{3/2}$ for $0 \le x \le 3$

6. $g(x) = \frac{2}{3}x^{3/2} - \frac{1}{2}x^{1/2}$ for $1 \le x \le 4$

7. $g(x) = \frac{2}{5}x^{5/4} - \frac{2}{3}x^{3/4}$ for $1 \le x \le 16$

8. $k(x) = x^3 + \frac{1}{12x}$ for $1 \le x \le 3$

9. $k(x) = x^4 + \frac{1}{32x^2}$ for $1 \le x \le 2$

10. $f(x) = \ln(\cos x)$ for $0 \le x \le \pi/4$

11. $f(x) = \ln(x^2 - 1)$ for $2 \le x \le 5$
   $\left(Hint: \frac{x^2 + 1}{x^2 - 1} = 1 + \frac{1}{x - 1} - \frac{1}{x + 1}\right)$

12. $f(x) = \ln(1 + x^2) - \frac{1}{8}\left(\frac{x^2}{2} + \ln x\right)$ for $1 \le x \le 2$

13. $f(x) = \ln(1 + x^3) + \frac{1}{12}\left(\frac{1}{x} - \frac{x^2}{2}\right)$ for $1 \le x \le 2$

14. $f(x) = -\frac{1}{4}\sin x + \ln(\sec x + \tan x)$ for $\pi/4 \le x \le \pi/3$

*15. $f(x) = \tan x - \frac{1}{8}(x + \frac{1}{2}\sin 2x)$ for $0 \le x \le \pi/4$

*16. $f(x) = \ln x$ for $\sqrt{3} \le x \le \sqrt{8}$
   $\left(Hint:$ Make the substitution $u = \sqrt{x^2 + 1}$ and note that
   $\frac{u^2}{u^2 - 1} = 1 + \frac{1}{2}\frac{1}{u - 1} - \frac{1}{2}\frac{1}{u + 1}.\right)$

In Exercises 17–22 let f be a function defined on the given interval and having the indicated derivative f'. (Such a function f exists by the Fundamental Theorem of Calculus.) Find the length $\mathscr{L}$ of the graph of f.

17. $f'(x) = \sqrt{x^2 - 1}$; $[2, 3]$

18. $f'(x) = \sqrt{x^n - 1}$; $[2, 4]$; n a positive integer

19. $f'(x) = \sqrt{\sqrt{x} - 1}$; $[25, 100]$

20. $f'(x) = \sqrt{x^{1/n} - 1}$; $[1, 2^n]$; n a positive integer

21. $f'(x) = \tan x$; $[0, \pi/4]$

22. $f'(x) = \sqrt{\tan^2 x - 1}$; $[\frac{2}{3}\pi, \frac{3}{4}\pi]$

23. Use Riemann sums with $n = 4$ and the midpoint sum to estimate the length $\mathscr{L}$ of the elliptical arc given by

$$f(x) = \frac{3}{2}\sqrt{4 - x^2}$$ for $-1 \le x \le 1$

24. The graph of $x^{2/3} + y^{2/3} = 16$ is an **astroid** (Figure 6.23). Find the length $\mathscr{L}$ of the portion of the astroid in the first quadrant over the interval $[1, 8]$.

*25. Suppose a person is traversing the graph of $y = x^{2/3}$ from the point $(0, 0)$ to the point $(1, 1)$. Find the halfway point on the curve. (*Hint:* If $y = x^{2/3}$, then $x = y^{3/2}$. Integrate along the y axis.)

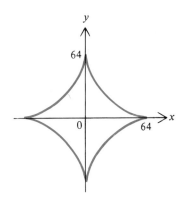

FIGURE 6.23

6.4
LENGTHS OF CURVES DEFINED PARAMETRICALLY

Until now in the book each curve we have discussed has been represented as the graph of a function or an equation. However, consider the cycloid, which is the curve traced out by a point on a circle as the circle rolls along a line (Figure 6.24). Although it is possible to represent the cycloid as the graph of a function, there is another way of representing curves, called a parametric representation, that is more convenient for computing, among other things, the length of one arch of the cycloid. (See Example 4.) This section is devoted to parametric representations of curves and the lengths of such curves.

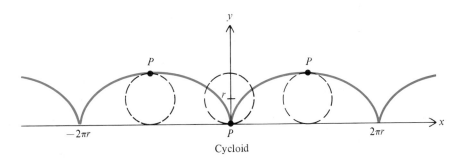

Cycloid

FIGURE 6.24

Suppose a particle traverses a curve C in the plane. Then the x and y coordinates of the point (x, y) on C are functions of time t. We will assume that there are continuous functions f and g on an interval I such that C consists of all points (x, y) such that

$$x = f(t) \quad \text{and} \quad y = g(t) \quad \text{for } t \text{ in } I \tag{1}$$

In such a case the equations in (1) are called **parametric equations** of C, and we say that C **is parametrized by** the equations in (1), with t a **parameter** of C. For

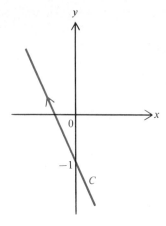

FIGURE 6.25

$a \le t \le b$, we will let

$$P(t) = (f(t), g(t))$$

Example 1   Sketch the curve C parametrized by

$$x = 1 - 2t \quad \text{and} \quad y = -3 + 4t \quad \text{for all } t$$

and indicate the direction $P(t)$ moves along C as t increases.

Solution   We first find a single equation relating x and y that does not involve t. To that end, we solve for t in the first parametric equation to obtain $t = \frac{1}{2} - \frac{1}{2}x$. Next we substitute for t in the second parametric equation:

$$y = -3 + 4\left(\frac{1}{2} - \frac{1}{2}x\right)$$

$$y = -1 - 2x \tag{2}$$

Thus a point (x, y) on C lies on the straight line given by (2), which has slope -2 and y intercept -1 (Figure 6.25). Notice from the parametric equations that $P(0) = (1, -3)$ and $P(1) = (-1, 1)$. Therefore as t increases, $P(t)$ moves upward and to the left on C, as Figure 6.25 indicates.   □

More generally, any set of parametric equations of the form

$$x = a + bt \quad \text{and} \quad y = c + dt \quad \text{with } b \ne 0 \text{ or } d \ne 0 \tag{3}$$

represents a straight line.

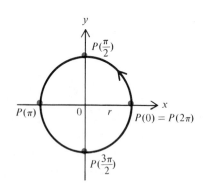

FIGURE 6.26

Example 2   Describe the curve C with parametric equations

$$x = r\cos t \quad \text{and} \quad y = r\sin t \quad \text{for } 0 \le t \le 2\pi$$

Solution   We will show that for $0 \le t \le 2\pi$, $P(t)$ is on the circle $x^2 + y^2 = r^2$. To that end, we square both sides of each of the given equations and add:

$$x^2 + y^2 = (r\cos t)^2 + (r\sin t)^2 = r^2(\cos^2 t + \sin^2 t) = r^2$$

Therefore the points satisfying the given equations also satisfy the equation $x^2 + y^2 = r^2$, and thus the points lie on the designated circle. As t increases, the point $P(t)$ traverses the circle, with $P(0) = (r, 0)$, $P(\pi/2) = (0, r)$, $P(\pi) = (-r, 0)$, $P(3\pi/2) = (0, -r)$, and finally, $P(2\pi) = (r, 0)$. It follows that C is the circle $x^2 + y^2 = r^2$, traversed exactly once and in the counterclockwise direction (Figure 6.26).   □

The parametric equations

$$x = r\cos t \quad \text{and} \quad y = r\sin t \quad \text{for } 0 \le t \le 4\pi$$

represent the same circle as the one in Example 2, but this time the circle is traced out twice in the counterclockwise direction, once for $0 \le t \le 2\pi$ and the second time for $2\pi \le t \le 4\pi$.

Now consider the cycloid mentioned at the outset of this section. If a circle of radius r rolls along the x axis in the positive direction, the curve traced out by a point P on the circle is a cycloid (see Figure 6.24 again). In Section 12.1 we will show that if P is located at the origin at time $t = 0$, then its coordinates x and y are given by

$$x = r(t - \sin t) \quad \text{and} \quad y = r(1 - \cos t) \quad \text{for all } t$$

Notice that y is a periodic function of t with period 2π.

Finally, we observe that if g is any function, then the curve C parametrized by

$$x = t \quad \text{and} \quad y = g(t)$$

is the graph of g. Thus the graph of any function can be parametrized.

Length of a Curve Given Parametrically

Suppose that a curve C is given parametrically by

$$x = f(t) \quad \text{and} \quad y = g(t) \quad \text{for } a \le t \le b$$

where f and g have continuous derivatives on $[a, b]$. Our derivation of the formula for the length $\mathscr{L}$ of C will parallel the derivation of the formula given in Definition 6.1. Let $\mathscr{P} = \{t_0, t_1, \ldots, t_n\}$ be any partition of $[a, b]$, and for $1 \le k \le n$ let $(x_k, y_k) = (f(t_k), g(t_k))$ be the corresponding point on C (Figure 6.27). Let $\Delta \mathscr{L}_k$ be the length of the portion of the curve joining (x_{k-1}, y_{k-1}) and (x_k, y_k). If Δx_k is small, $\Delta \mathscr{L}_k$ is approximately equal to the length of the line segment joining (x_{k-1}, y_{k-1}) and (x_k, y_k). In other words,

$$\Delta \mathscr{L}_k \approx \sqrt{(x_k - x_{k-1})^2 + (y_k - y_{k-1})^2}$$
$$= \sqrt{[f(t_k) - f(t_{k-1})]^2 + [g(t_k) - g(t_{k-1})]^2}$$

By the Mean Value Theorem there are numbers t_k' and t_k'' in $[t_{k-1}, t_k]$ such that

$$f(t_k) - f(t_{k-1}) = f'(t_k')\Delta t_k \quad \text{and} \quad g(t_k) - g(t_{k-1}) = g'(t_k'')\Delta t_k$$

Therefore the total length $\mathscr{L}$ of the graph of the curve, which is the sum of the

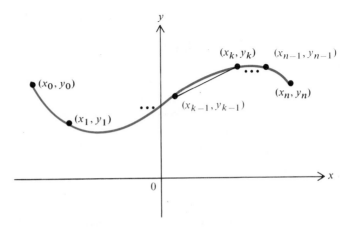

FIGURE 6.27

lengths $\Delta \mathcal{L}_1, \Delta \mathcal{L}_2, \ldots, \Delta \mathcal{L}_n$, should be approximately

$$\sum_{k=1}^{n} \sqrt{[f'(t_k')]^2 + [g'(t_k'')]^2} \, \Delta t_k$$

Although this sum is not a Riemann sum (because t_k' and t_k'' may be distinct numbers), it is closely related to a Riemann sum, and moreover, it can be shown that

$$\lim_{\|\mathcal{P}\| \to 0} \sum_{k=1}^{n} \sqrt{[f'(t_k')]^2 + [g'(t_k'')]^2} \, \Delta t_k = \int_a^b \sqrt{[f'(t)]^2 + [g'(t)]^2} \, dt$$

Consequently it is reasonable to define the length $\mathcal{L}$ of the curve C by the formula

$$\mathcal{L} = \int_a^b \sqrt{[f'(t)]^2 + [g'(t)]^2} \, dt \tag{4}$$

If $x = f(t)$ and $y = g(t)$, then in Leibniz notation (4) becomes

$$\mathcal{L} = \int_a^b \sqrt{\left(\frac{dx}{dt}\right)^2 + \left(\frac{dy}{dt}\right)^2} \, dt \tag{5}$$

Notice that if C is parametrized by

$$x = t \quad \text{and} \quad y = g(t) \quad \text{for } a \leq t \leq b$$

then (4) becomes

$$\mathcal{L} = \int_a^b \sqrt{1 + [g'(t)]^2} \, dt$$

which is equivalent to (2) of Section 6.3 but with g substituted for f. Thus (4) is a more general formula for length than (2) of Section 6.3.

Example 3   Find the circumference $\mathcal{L}$ of the circle of radius r centered at the origin, given parametrically by

$$x = r \cos t \quad \text{and} \quad y = r \sin t \quad \text{for } 0 \leq t \leq 2\pi$$

Solution   Since

$$\frac{dx}{dt} = -r \sin t \quad \text{and} \quad \frac{dy}{dt} = r \cos t$$

it follows from (5) that

$$\mathcal{L} = \int_0^{2\pi} \sqrt{(-r \sin t)^2 + (r \cos t)^2} \, dt = \int_0^{2\pi} \sqrt{r^2 \sin^2 t + r^2 \cos^2 t} \, dt$$

$$= \int_0^{2\pi} \sqrt{r^2} \, dt = r \int_0^{2\pi} 1 \, dt = rt \Big|_0^{2\pi} = 2\pi r$$

Thus the circumference of the circle of radius r is $2\pi r$, a fact familiar from plane geometry. $\square$

In 1658 the British architect and mathematician Christopher Wren calculated the length of an arch of a cycloid, as we will do in the next example.

Example 4 Find the length $\mathscr{L}$ of one arch of the cycloid given parametrically by

$$x = r(t - \sin t) \quad \text{and} \quad y = r(1 - \cos t) \quad \text{for } 0 \le t \le 2\pi$$

(Figure 6.28).

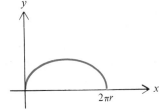

FIGURE 6.28

Solution First we notice that

$$\frac{dx}{dt} = r - r\cos t \quad \text{and} \quad \frac{dy}{dt} = r\sin t$$

Therefore by (5) we have

$$\mathscr{L} = \int_0^{2\pi} \sqrt{(r - r\cos t)^2 + (r\sin t)^2}\, dt$$

$$= \int_0^{2\pi} \sqrt{r^2 - 2r^2\cos t + r^2\cos^2 t + r^2\sin^2 t}\, dt$$

$$= \int_0^{2\pi} \sqrt{2r^2 - 2r^2\cos t}\, dt$$

$$= r\int_0^{2\pi} \sqrt{2(1 - \cos t)}\, dt$$

By the half-angle formula for $\sin t/2$,

$$\frac{1 - \cos t}{2} = \sin^2\frac{t}{2}, \quad \text{so that} \quad \sqrt{2(1 - \cos t)} = \sqrt{4\sin^2\frac{t}{2}}$$

Since $\sin t/2 \ge 0$ for $0 \le t \le 2\pi$, we conclude that

$$\mathscr{L} = r\int_0^{2\pi} \sqrt{4\sin^2\frac{t}{2}}\, dt = 2r\int_0^{2\pi} \sin\frac{t}{2}\, dt$$

$$= -4r\cos\frac{t}{2}\Big|_0^{2\pi} = 8r \quad \square$$

EXERCISES 6.4

In Exercises 1–16 sketch the curve C whose parametric equations are given, and indicate the direction $P(t)$ moves as t increases.

1. $x = 2t + 1$ and $y = 4t - 1$ for all t

2. $x = -2 + 3t$ and $y = 2 - 3t$ for all t

3. $x = 5 - t$ and $y = -4$ for $t \ge 0$

4. $x = 3$ and $y = -1 - t$ for $0 \le t \le 1$

5. $x = 2\cos t$ and $y = 2\sin t$ for $0 \le t \le \pi/2$

6. $x = 3\sin t$ and $y = 3\cos t$ for $-\pi/2 \le t \le \pi/2$

7. $x = 2 - \cos t$ and $y = -1 - \sin t$ for $0 \le t \le 2\pi$

8. $x = -1 + \frac{3}{2}\sin t$ and $y = \frac{1}{2} - \frac{3}{2}\cos t$ for $-\pi \le t \le 3\pi$

9. $x = t$ and $y = 1 - t^2$ for all t

10. $x = t$ and $y = \sqrt{1-t^2}$ for $-1 \le t \le 1$

11. $x = t^3$ and $y = t^2$ for all t

12. $x = \sqrt{t}$ and $y = \sqrt{1-t}$ for $0 \le t \le 1$

13. $x = \tan t$ and $y = \sec t$ for $-\pi/2 < t < \pi/2$

14. $x = \dfrac{1 - \cos 2t}{2}$ and $y = \sin t$ for $-\pi/2 \le t \le 0$

   (*Hint:* Use the half-angle formula to rewrite x.)

15. $x = \dfrac{2t}{1+t^2}$ and $y = \dfrac{1-t^2}{1+t^2}$ for all t

   (*Hint:* Square both sides of each equation.)

16. $x = \dfrac{t^2 + 1}{t^2 - 1}$ and $y = \dfrac{2t}{t^2 - 1}$ for $t \ne -1, 1$

   (*Hint:* Square both sides of each equation.)

In Exercises 17–22 find the length $\mathscr{L}$ of the curve described parametrically.

17. $x = 3t$ and $y = 2t^{3/2}$ for $0 \le t \le 3$

18. $x = 1 - t^2$ and $y = 1 + t^3$ for $0 \le t \le 1$

19. $x = \dfrac{1}{4}t^4 + 1$ and $y = \dfrac{1}{6}t^6 - 1$ for $0 \le t \le 1$

20. $x = \sin t + \cos t$ and $y = \sin t - \cos t$ for $-\pi/3 \le t \le \pi/6$

21. $x = \sin t - t \cos t$ and $y = t \sin t + \cos t$ for $0 \le t \le \pi/2$

*22. $x = \dfrac{2}{3}t^{3/2}$ and $y = \dfrac{4}{9}t^{9/4}$ for $0 \le t \le 4$

23. Let $r > 0$. The equations

$$x = r \cos^3 t \quad \text{and} \quad y = r \sin^3 t \quad \text{for } 0 \le t \le 2\pi$$

parametrize an astroid (Figure 6.29). Determine the length of the astroid. (*Hint:* Remember that square roots are nonnegative.)

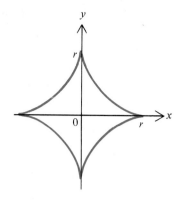

FIGURE 6.29

6.5
SURFACE AREA

A higher-dimensional version of the length of a curve is the surface area of a surface. It has been known since antiquity that the surface area S of a cube of side s is given by

$$S = 6s^2$$

(Figure 6.30), and the surface area S of a cylinder of radius r and height h is given by

$$S = 2\pi r h$$

(Figure 6.31). However, our analysis of the surface areas of other surfaces will be based on the surface area of a frustum of a cone. If the frustum has slant height l and radii r_1 and r_2, as in Figure 6.32, then the surface area S is given by

$$S = 2\pi \left(\frac{r_1 + r_2}{2} \right) l = \pi(r_1 + r_2)l \qquad (1)$$

We will prove (1) in Section 14.3. It can also be derived from the formula for the surface area of a cone. (See Exercise 18.)

More generally, suppose f is defined on $[a, b]$, and $f(x) \ge 0$ for $a \le x \le b$. We will derive a formula for the surface area S of the surface obtained by revolving the graph of f about the x axis (Figure 6.33). In order to be able to use

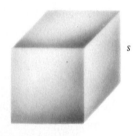

FIGURE 6.30

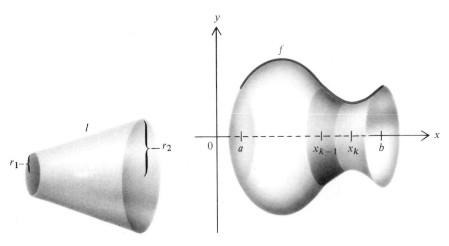

FIGURE 6.31   **FIGURE 6.32**   **FIGURE 6.33**

the length of the graph of f on $[a, b]$ in our calculations of surface area, we will also assume that f is continuously differentiable on $[a, b]$. Now let $\mathscr{P} = \{x_0, x_1, \ldots, x_n\}$ be a partition of $[a, b]$, and let ΔS_k be the area of the portion of the surface between the revolved lines $x = x_{k-1}$ and $x = x_k$. If Δx_k is small, the surface area ΔS_k is approximately equal to the surface area of the frustum of the corresponding cone, that is, the frustum whose slant height is equal to the length of the line between $(x_{k-1}, f(x_{k-1}))$ and $(x_k, f(x_k))$, and the radii of whose ends are $f(x_{k-1})$ and $f(x_k)$ (see Figure 6.33). From (1) this means that

$$\Delta S_k \approx \pi[f(x_{k-1}) + f(x_k)]\sqrt{(x_k - x_{k-1})^2 + [f(x_k) - f(x_{k-1})]^2} \qquad (2)$$

As in our discussion of lengths of curves in Section 6.3, the Mean Value Theorem can be applied to show that for some t_k in $[x_{k-1}, x_k]$,

$$\sqrt{(x_k - x_{k-1})^2 + [f(x_k) - f(x_{k-1})]^2} = \sqrt{1 + [f'(t_k)]^2}\,\Delta x_k \qquad (3)$$

Since Δx_k is assumed to be small, x_{k-1} and x_k are close together, with t_k between them. Since f' and hence f are continuous on $[x_{k-1}, x_k]$, it follows that $f(x_{k-1}) + f(x_k)$ should be approximately $f(t_k) + f(t_k)$, that is, $2f(t_k)$. So substituting from (3) into (2), we deduce that

$$\Delta S_k \approx \pi[2f(t_k)]\sqrt{1 + [f'(t_k)]^2}\,\Delta x_k$$

Consequently the surface area S of the complete surface, which equals the sum of $\Delta S_1, \Delta S_2, \ldots, \Delta S_n$, should be approximately

$$\sum_{k=1}^{n} 2\pi f(t_k)\sqrt{1 + [f'(t_k)]^2}\,\Delta x_k$$

which is a Riemann sum for $2\pi f\sqrt{1 + (f')^2}$ on $[a, b]$. Therefore we are led to define the surface area S by

$$S = \lim_{\|\mathscr{P}\| \to 0} \sum_{k=1}^{n} 2\pi f(t_k)\sqrt{1 + [f'(t_k)]^2}\,\Delta x_k = \int_a^b 2\pi f(x)\sqrt{1 + [f'(x)]^2}\,dx$$

DEFINITION 6.2

Let f be nonnegative and continuously differentiable on $[a, b]$. The **surface area** of the surface obtained by revolving the graph of f about the x axis is defined by

$$S = \int_a^b 2\pi f(x)\sqrt{1 + [f'(x)]^2}\, dx \qquad (4)$$

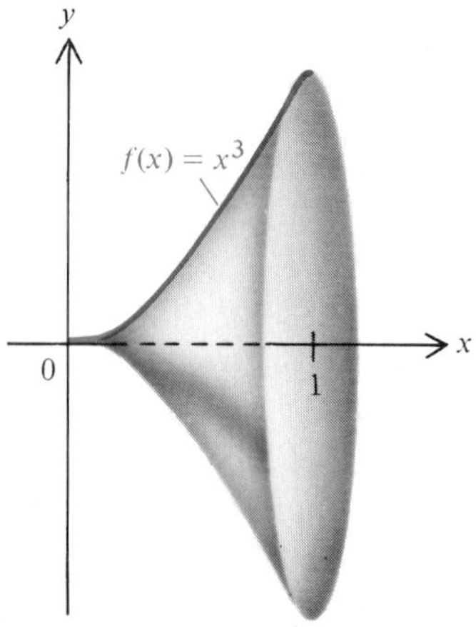

FIGURE 6.34

Example 1 Let $f(x) = x^3$ for $0 \le x \le 1$. Find the surface area S of the surface obtained by revolving the graph of f about the x axis (Figure 6.34).

Solution Since $f'(x) = 3x^2$, it follows from (4) that

$$S = \int_0^1 2\pi x^3 \sqrt{1 + (3x^2)^2}\, dx = 2\pi \int_0^1 x^3 \sqrt{1 + 9x^4}\, dx$$

To evaluate the integral, we let

$$u = 1 + 9x^4, \quad \text{so that} \quad du = 36x^3\, dx$$

Now if $x = 0$ then $u = 1$, and if $x = 1$ then $u = 10$. Therefore

$$S = 2\pi \int_0^1 x^3 \sqrt{1 + 9x^4}\, dx = 2\pi \int_0^1 \overbrace{\sqrt{1 + 9x^4}}^{\sqrt{u}}\, \overbrace{x^3\, dx}^{\frac{1}{36}du}$$

$$= 2\pi \int_1^{10} \sqrt{u}\, \frac{1}{36}\, du = \frac{\pi}{18}\left(\frac{2}{3} u^{3/2}\right)\Big|_1^{10}$$

$$= \frac{\pi}{27}(10^{3/2} - 1) \quad \square$$

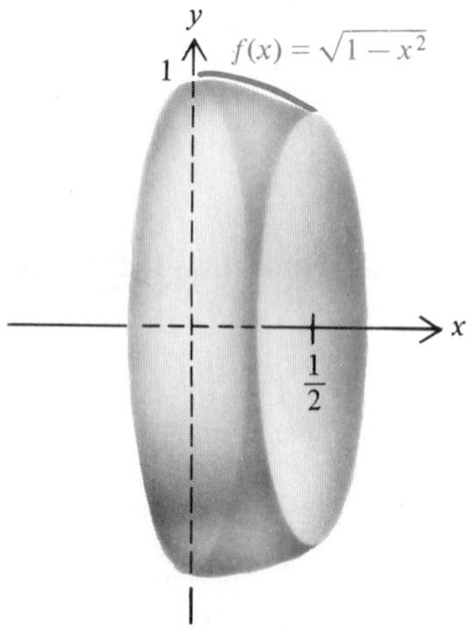

Example 2 Let $f(x) = \sqrt{1 - x^2}$ for $0 \le x \le \frac{1}{2}$. Find the surface area S of the portion of the sphere obtained by revolving the graph of f about the x axis (Figure 6.35).

Solution Since $f'(x) = -x/\sqrt{1 - x^2}$, it follows from (4) that

$$S = \int_0^{1/2} 2\pi \sqrt{1 - x^2}\sqrt{1 + \left(\frac{-x}{\sqrt{1 - x^2}}\right)^2}\, dx$$

$$= 2\pi \int_0^{1/2} \sqrt{1 - x^2}\sqrt{1 + \frac{x^2}{1 - x^2}}\, dx$$

$$= 2\pi \int_0^{1/2} \sqrt{1 - x^2}\sqrt{\frac{1}{1 - x^2}}\, dx = 2\pi \int_0^{1/2} 1\, dx = 2\pi x\Big|_0^{1/2} = \pi \quad \square$$

FIGURE 6.35

Had we considered the portion of the sphere for which $a \le x \le b$, where $-1 < a < b < 1$ (Figure 6.36), similar calculations would show that

$$S = 2\pi(b - a)$$

Thus, rather surprisingly, the surface area depends only on the length of the interval $[a, b]$ and not on its location within $(-1, 1)$ (see Exercise 19).

If the curve that is revolved about the x axis to yield the surface is given parametrically by

$$x = f(t) \quad \text{and} \quad y = g(t) \quad \text{for } a \le t \le b$$

where g is nonnegative on $[a, b]$, then we can obtain a companion formula for the surface area S:

$$S = \int_a^b 2\pi g(t) \sqrt{[f'(t)]^2 + [g'(t)]^2} \, dt \qquad (5)$$

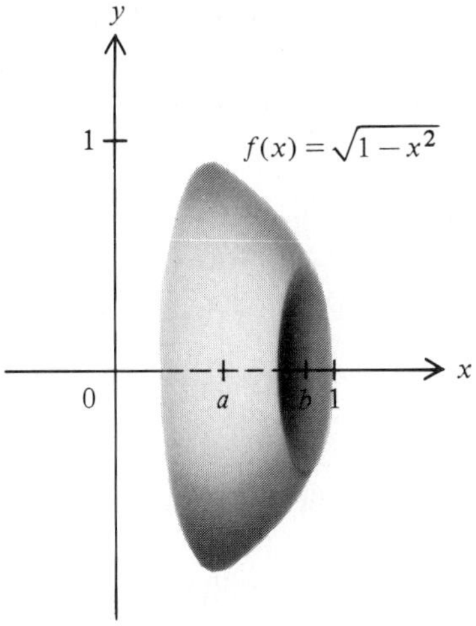

$f(x) = \sqrt{1 - x^2}$

FIGURE 6.36

Example 3 Show that the surface area S of a sphere of radius r is $4\pi r^2$.

Solution Suppose the center of the sphere is at the origin, so that if C is the semicircle defined by

$$x = r \cos t \quad \text{and} \quad y = r \sin t \quad \text{for } 0 \le t \le \pi$$

then the sphere is the surface obtained by revolving C about the x axis (Figure 6.37). It follows from (5) that

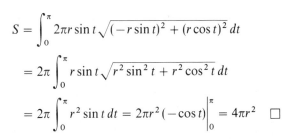

$$S = \int_0^\pi 2\pi r \sin t \sqrt{(-r \sin t)^2 + (r \cos t)^2} \, dt$$

$$= 2\pi \int_0^\pi r \sin t \sqrt{r^2 \sin^2 t + r^2 \cos^2 t} \, dt$$

$$= 2\pi \int_0^\pi r^2 \sin t \, dt = 2\pi r^2 (-\cos t) \Big|_0^\pi = 4\pi r^2 \quad \Box$$

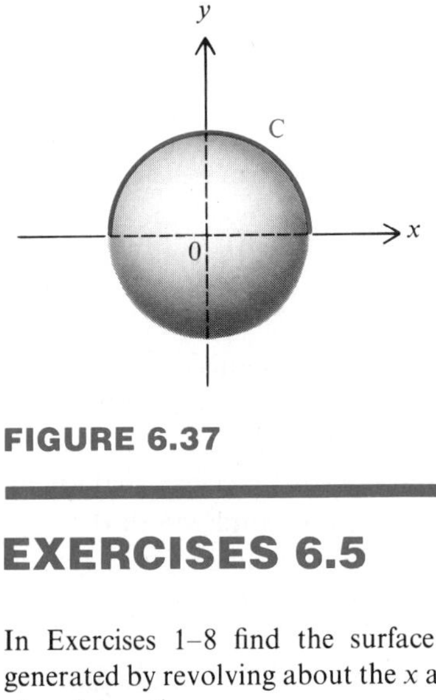

FIGURE 6.37

EXERCISES 6.5

In Exercises 1–8 find the surface area S of the surface generated by revolving about the x axis the graph of f on the given interval.

1. $f(x) = \sqrt{4 - x^2}; \left[-\dfrac{1}{2}, \dfrac{3}{2} \right]$

2. $f(x) = \sqrt{x}; [2, 6]$

3. $f(x) = \sqrt{3 - x}; [-1, 1]$

4. $f(x) = 2\sqrt{1 - x}; [-1, 0]$

5. $f(x) = \dfrac{1}{3} x^3; [0, \sqrt{2}]$

6. $f(x) = \dfrac{1}{2} x^3 + \dfrac{1}{6x}; [1/\sqrt{2}, 1]$

7. $f(x) = \dfrac{2}{3} x^{3/2} - \dfrac{1}{2} x^{1/2}; [1, 2]$

8. $f(x) = \dfrac{1}{4} x^4 + \dfrac{1}{8x^2}; [1, \sqrt{2}]$

In Exercises 9–13 find the surface area S of the surface generated by revolving about the x axis the curve with the given parametric representation.

9. $x = \dfrac{1}{2} t^2$ and $y = t$ for $\sqrt{3} \le t \le 2\sqrt{2}$

10. $x = \dfrac{2}{3}(1 - t)^{3/2}$ and $y = \dfrac{2}{3}(1 + t)^{3/2}$ for $-\dfrac{3}{4} \le t \le 0$

11. $x = \dfrac{1}{3}t^3$ and $y = 2t^{3/2}$ for $1 \leq t \leq 3$

12. $x = \sin^2 t$ and $y = \sin t \cos t$ for $0 \leq t \leq \pi/2$

13. $x = \cos^2 t$ and $y = \sin^2 t$ for $0 \leq t \leq \pi/2$

14. Let $0 \leq a < r$. The portion of the sphere of radius r obtained by revolving the graph of

$$y = \sqrt{r^2 - x^2} \quad \text{for } -a \leq x \leq a$$

about the x axis is called a **zone** of the sphere. Determine its surface area S.

15. Let $r > 0$. The curve C parametrized by

$$x = r\cos^3 t \quad \text{and} \quad y = r\sin^3 t \quad \text{for } 0 \leq t \leq 2\pi$$

is an astroid (see Exercise 23 of Section 6.4). Determine the surface area S of the surface generated by revolving C about the x axis. (*Hint:* S is equal to twice the surface area of the part for which $0 \leq t \leq \pi/2$.)

16. Let C be the curve parametrized by

$$x = \cos^n t \quad \text{and} \quad y = \sin^n t \quad \text{for } 0 \leq t \leq \dfrac{\pi}{2}$$

where n is an integer with $n \geq 2$. Let S be the surface area of the surface generated by revolving C about the x axis. Determine the positive integers n for which we can readily

carry out the integration given in (5) to obtain the surface area S.

17. Let r and h be fixed positive numbers. Consider the curve C parametrized by

$$x = h\cos t \quad \text{and} \quad y = r \quad \text{for } 0 \leq t \leq \dfrac{\pi}{2}$$

and let S be the surface area of the surface generated by revolving C about the x axis.

a. Determine S by using only geometry.

b. Determine S by using (5).

18. Derive (1) from the formula $S = \pi r l$ for the surface area of a cone with slant height l and radius r. (*Hint:* The surface area of the frustum is the difference of the surface areas of two cones, one with radius r_2 and one with radius r_1.)

19. Assume that $[a, b]$ is a closed interval contained in the open interval $(-1, 1)$. Let $f(x) = \sqrt{1 - x^2}$, and let S be the surface area of the region obtained by revolving the graph of f on $[a, b]$ about the x axis (refer to Figure 6.36). Show that

$$S = 2\pi(b - a)$$

(Thus the surface area depends only on the length of the interval $[a, b]$ and not on its location within $(-1, 1)$.)

6.6
WORK

Next we consider the physical concept of work. To introduce this concept, let us imagine a person pushing a wheelbarrow across a yard, from point a to point b. If a constant force c is exerted all the way across the yard, then the amount of work done is defined to be $c(b - a)$, that is, the force times the distance traveled by the wheelbarrow (Figure 6.38). (The product $c(b - a)$ should remind you of the area of a rectangle.) This is a reasonable definition, since we would expect the work done to increase if the person pushes the wheelbarrow farther or with a greater force.

How should the work be defined when the force is variable? The integral will enable us to do this. In order to relate integrals to work, we make certain assumptions. First, we assume for the present that the object on which the force acts moves in a straight line, so that we may think of it as moving along the x axis from a point a to a point b. Second, we assume that at each point x between a and b a certain force $f(x)$ is exerted on the object and that f is continuous on $[a, b]$. Furthermore, we adopt the convention that force is positive if exerted in the direction of the positive x axis and negative if exerted in the direction of the negative x axis. If force is measured in pounds and distance in feet, then work is measured in foot-pounds. In the cgs system of units, force is measured in dynes, distance in centimeters, and work in ergs. In the mks system, force is measured in newtons, distance in meters, and work in joules.

work = force × distance = $c(b - a)$

FIGURE 6.38

As we have seen, if the force f exerted on an object moving from a to b is constant, say $f(x) = c$, then the work W done is $W = c(b - a)$. When f is not necessarily constant, we let $\mathscr{P} = \{x_0, x_1, \ldots, x_n\}$ be any partition of $[a, b]$, and for each k between 1 and n we let t_k be an arbitrary point in the subinterval $[x_{k-1}, x_k]$. If Δx_k is small, then as the object moves from x_{k-1} to x_k, the amount of work ΔW_k done by the force on the object is approximately $f(t_k)\Delta x_k$. It is also reasonable to expect the work W done by the force when the object moves from a to b to be the sum of $\Delta W_1, \Delta W_2, \ldots, \Delta W_n$, which represent the work done on the object as it travels over the successive subintervals of $[a, b]$. Therefore W should be approximately equal to

$$\sum_{k=1}^{n} \overbrace{f(t_k)}^{\text{force}} \overbrace{\Delta x_k}^{\text{distance}}$$

which is a Riemann sum for f on $[a, b]$. Consequently W should be defined by

$$W = \lim_{\|\mathscr{P}\| \to 0} \sum_{k=1}^{n} f(t_k)\Delta x_k = \int_a^b f(x)\,dx$$

DEFINITION 6.3

Suppose an object moves from a to b under the influence of a force f, where f is a continuous function on $[a, b]$. Then the **work** W done by the force on the object is defined by

$$W = \int_a^b f(x)\,dx \tag{1}$$

It is implicit in Definition 6.3 that work is positive if the force acts in the direction of motion of the object and negative if the force opposes the motion of the object.

In our first example we return to the wheelbarrow and compute the work done on it as it moves across the yard, under the condition that it leaks all the while.

Example 1   Suppose we push a leaking wheelbarrow 100 feet and (because of the leak) exert a decreasing force given by

$$f(x) = 60\left(1 - \frac{x^2}{20{,}000}\right) \quad \text{for } 0 \leq x \leq 100$$

Find the work W done by the force on the wheelbarrow.

Solution   Since the force at any x in $[0, 100]$ is $60(1 - x^2/20{,}000)$, it follows from (1) that

$$W = \int_0^{100} 60\left(1 - \frac{x^2}{20{,}000}\right)dx = 60\left(x - \frac{x^3}{60{,}000}\right)\Bigg|_0^{100}$$

$$= 5000 \text{ (foot-pounds)} \quad \square$$

Hooke's Law

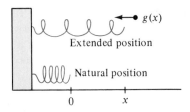

Extended position

Natural position

0   x

FIGURE 6.39

As another example we consider the work done when a spring is stretched. **Hooke's Law** (stated in 1676 by the English geometer Robert Hooke) says that the elastic force $g(x)$ exerted by a spring which has been extended x units beyond its natural length is proportional to x (Figure 6.39). The force is negative, since it opposes the expansion of the spring and hence is exerted in the direction of the negative x axis (Figure 6.39). Thus there is a positive constant k such that

$$g(x) = -kx$$

and it is known that this formula is fairly reliable for reasonably small values of x.

The force $f(x)$ needed in order to hold a spring extended x units beyond its natural length is opposite to $g(x)$ and equal in magnitude, which means that

$$f(x) = -g(x) = kx$$

It follows from (1) that the work W required in order to stretch a spring from a units extended to b units is given by the formula

$$W = \int_a^b f(x)\,dx = \int_a^b kx\,dx \tag{2}$$

Since the force that produces the work opposes the force exerted by the spring, we say that W is the work done *against* the force of the spring.

Example 2   Suppose the work required to stretch a certain spring 1 foot beyond its natural length is 10 foot-pounds. Find the work W required to stretch it from 1 foot beyond its natural length to 3 feet beyond.

Solution   Our assumption implies that

$$10 = \int_0^1 kx\,dx \tag{3}$$

and we desire the value of W, given by

$$W = \int_1^3 kx\,dx$$

From (3) we can determine k:

$$10 = \int_0^1 kx\,dx = \frac{1}{2}kx^2\Big|_0^1 = \frac{1}{2}k(1-0) = \frac{1}{2}k$$

so that $k = 20$. Therefore

$$W = \int_1^3 kx\,dx = \int_1^3 20x\,dx = 10x^2\Big|_1^3 = 10(9-1) = 80 \text{ (foot-pounds)} \quad \square$$

Work Required to Empty a Tank

In our final illustration we will find a formula for the work required to pump water from a tank. We will use the properties of work that led us to Definition 6.3, along with one additional property, which we will prove in Chapter 15: The *same* amount of work is required to move an object against the force of gravity from a point P to a point Q, no matter what path from P to Q the object travels. Thus we may assume that the object moves along a vertical line, and in that case we must exert a force at least as great as the weight of the object to balance the force of gravity.

We wish to compute the amount of work done in pumping water out of a tank and up to a certain level. Suppose first that we lift a solid region of water l feet, and suppose the region has volume V cubic feet (Figure 6.40). By definition the work W done by the lifting force is the product of the weight of the water and the distance l the water is moved. Since the water weighs 62.5 pounds per cubic foot, the weight of the water is $62.5V$, so that

$$W = (\text{weight}) \times (\text{distance}) = 62.5Vl$$

(If the liquid in the tank were not water, the number 62.5 would be replaced by the weight of 1 cubic foot of the other liquid.)

Now suppose the x axis is chosen to be vertical, with the positive numbers above the origin, placed at a convenient reference point for measuring depth, such as the top of the tank. Assume that the water to be pumped extends from a to b on the x axis (Figure 6.41), and let l be the level to which the water is to be pumped. Furthermore, for each x in $[a, b]$ let $A(x)$ be the cross-sectional area of the tank at x, and assume that A is continuous on $[a, b]$. Let $\mathscr{P} = \{x_0, x_1, \ldots, x_n\}$ be any partition of $[a, b]$. For each k between 1 and n, let t_k be any number in $[x_{k-1}, x_k]$. If Δx_k is small, the volume of the water between the levels x_{k-1} and x_k is approximately $A(t_k)\Delta x_k$. Since each particle of water in that portion must travel a distance of approximately $l - t_k$, our earlier discussion implies that the work ΔW_k required to lift the water contained between x_{k-1} and x_k is approximately the product of the weight $62.5A(t_k)\Delta x_k$ and the distance $l - t_k$ traveled, that is, approximately

$$62.5A(t_k)\Delta x_k(l - t_k)$$

As a result, the work W required to lift all the water between the levels $x = a$ and

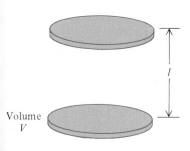

FIGURE 6.40

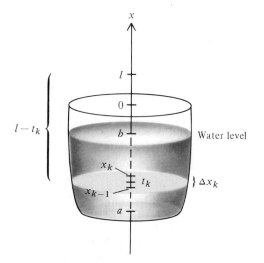

FIGURE 6.41

$x = b$, which is the sum of $\Delta W_1, \Delta W_2, \ldots, \Delta W_n$, should be approximately

$$\sum_{k=1}^{n} 62.5 A(t_k) \Delta x_k (l - t_k)$$

which is a Riemann sum for $62.5A(x)(l - x)$. Therefore the work W done in lifting the water is given by

$$W = \lim_{\|\mathscr{P}\| \to 0} \sum_{k=1}^{n} 62.5(l - t_k)A(t_k)\Delta x_k = \int_a^b 62.5(l - x)A(x)\,dx \qquad (4)$$

Example 3 A swimming pool has the shape of a right circular cylinder with radius 10 feet and depth 8 feet. Assume that the pool contains water to a depth of 5 feet. Find the work W required to pump all but 1 foot of water to the top of the pool.

Solution We place the origin at the bottom of the pool (Figure 6.42). Then $l = 8$, so a particle of water x feet from the bottom is to be raised $8 - x$ feet, and moreover, the water to be pumped out extends from 1 to 5 on the x axis. Furthermore, the cross-sectional area $A(x)$ is constant, with

$$A(x) = \pi(10)^2 = 100\pi$$

Thus it follows from (4) that

$$W = \int_1^5 62.5(8 - x)(100\pi)\,dx = 6250\pi \int_1^5 (8 - x)\,dx$$

$$= 6250\pi \left(8x - \frac{1}{2}x^2 \right)\Big|_1^5 = 125{,}000\pi \text{ (foot-pounds)} \quad \square$$

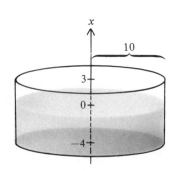

FIGURE 6.42

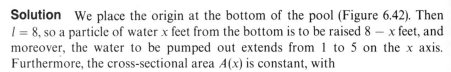

Placement of the origin in pumping problems is a matter of convenience. Were we to place the origin at water level in Example 3 (Figure 6.43), we would still have $A(x) = 100\pi$. However, l would be 3, a particle of water corresponding to x would be raised $3 - x$ feet, and the water to be pumped would extend from -4 to 0 on the x axis. Therefore we would find that

$$W = \int_{-4}^0 62.5(3 - x)(100\pi)\,dx = 6250\pi \int_{-4}^0 (3 - x)\,dx$$

$$= 6250\pi \left(3x - \frac{1}{2}x^2 \right)\Big|_{-4}^0 = 125{,}000\pi \text{ (foot-pounds)}$$

FIGURE 6.43

as we found in Example 3.

Example 4 A hemispherical tank with radius 10 feet is filled with water (Figure 6.44). Find the amount of work W required to pump all the water in the tank to 6 feet above the top of the tank.

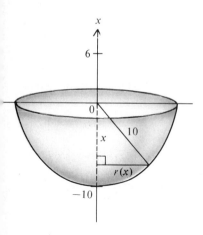

FIGURE 6.44

Solution This time it is convenient to place the origin at the top of the tank. Then $l = 6$, a particle of water corresponding to x is to be raised $6 - x$ feet, and the water to be pumped out extends from -10 to 0 on the x axis. From Figure 6.44 we see that the radius $r(x)$ of the cross section at x satisfies the equation $[r(x)]^2 = 100 - x^2$. Therefore

$$A(x) = \pi[r(x)]^2 = \pi(100 - x^2)$$

Thus it follows from (4) that

$$W = \int_{-10}^{0} 62.5(6 - x)\pi(100 - x^2)\,dx = (62.5)\pi \int_{-10}^{0} (600 - 100x - 6x^2 + x^3)\,dx$$

$$= 62.5\pi\left(600x - 50x^2 - 2x^3 + \frac{1}{4}x^4\right)\Bigg|_{-10}^{0}$$

$$= 62.5\pi(6000 + 5000 - 2000 - 2500)$$

$$= 406{,}250\pi \text{ (foot-pounds)} \quad \square$$

EXERCISES 6.6

1. Determine the work W done on the wheelbarrow of Example 1 if it is pushed only 60 feet.

2. An elevator in the Empire State Building weighs 1600 pounds. Find the work W required to raise the elevator from ground level to the 102nd story, some 1200 feet above ground level.

3. A 10-pound bag of groceries is to be carried up a flight of stairs 8 feet tall. Find the work W done on the bag.

4. Suppose a 120-pound person carries the bag in Exercise 3 up the stairs. Find the total work W done by the person in walking up the stairs with the bag.

5. A sailboat is stationary in the middle of a lake until a strong gust of wind blows it along a straight line. Suppose the force exerted on the sails by the wind when the boat is x miles from its starting point is

$$f(x) = 10^4 \sin x \quad \text{for } 0 \le x \le \pi$$

Find the work W done on the sails by the gust of wind.

6. Suppose a person pushes a thumbtack $\frac{1}{2}$ centimeter long into a bulletin board and the force (in dynes) exerted when the depth of the thumbtack in the bulletin board is x centimeters is given by

$$f(x) = 10{,}000(1 + 2x)^2 \quad \text{for } 0 \le x \le \frac{1}{2}$$

Find the work W done in pushing the thumbtack all the way into the board.

7. A bottle of wine has a cork $\frac{1}{6}$ foot long. A person uncorking the bottle exerts a force to overcome the force of friction between the cork and the bottle. Suppose the force exerted is given by

$$f(x) = 5\left(\frac{1}{6} - x\right) \quad \text{for } 0 \le x \le \frac{1}{6}$$

where x represents the length in feet of the cork extending from the bottle. Determine the work W done in removing the cork.

8. When a certain spring is expanded 1 foot from its natural position and held fixed, the force necessary to hold it is 30 pounds. Find the work W required to stretch the spring an additional foot.

9. Find the work W required to stretch the spring described in Exercise 8 from 3 to 6 feet.

10. If 6 foot-pounds of work are required to compress a spring 1 foot from its natural length, find the work W necessary to compress the spring 1 extra foot. (*Hint:* Hooke's law is also valid for compressing springs.)

11. If 6 foot-pounds of work are required to compress a spring from its natural length of 10 feet to a length of 9 feet, find the work W necessary to stretch the spring from its natural length to a length of 12 feet.

12. Find the work W necessary to pump all the water out of the swimming pool in Example 3.

13. Determine the work W necessary to pump all the water in the pool in Example 3 to 3 feet above the top of the pool.

14. A tank has the shape of the surface generated by revolving the parabolic segment $y = \frac{1}{2}x^2$ for $0 \le x \le 4$ about the y axis. If the tank is full of a fluid weighing 80 pounds per cubic foot, find the work W required to pump the contents of the tank to a level 4 feet above the top of the tank. (*Hint:* Integrate along the y axis.)

15. A tank in the shape of an inverted cone 12 feet tall and 3 feet in radius is full of water. Calculate the work W required to pump all the water over the edge of the tank.

16. Suppose that in Exercise 15 just half the water is pumped out the top edge of the tank and the remaining water is pumped to a level 3 feet above the top of the tank. Compute the total work W done.

17. Suppose a large gasoline tank has the shape of a half-cylinder 4 feet in radius and 10 feet long (Figure 6.45). If the tank is full, find the work W necessary to pump all the gasoline to the top of the tank. Assume the gasoline weighs 42 pounds per cubic foot.

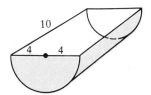

FIGURE 6.45

18. A stationary proton located at the origin of the x axis exerts an attractive force on an electron located at a point x on the negative x axis. The force is given by

$$f(x) = \frac{a}{x^2}$$

where $a = 2.3 \times 10^{-28}$, x is measured in meters, and $f(x)$ is measured in newtons. Determine the work W done on the electron by this force when the electron moves from $x = -10^{-9}$ to $x = -10^{-10}$.

19. A ball weighing 0.2 pounds is thrown vertically upward from a height of 6 feet. Its height after t seconds is given by

$$h(t) = 6 + 8t - 16t^2$$

until the ball strikes the ground again.
   a. Find the work W done on the ball by gravity while the ball descends from its maximum height to the ground.
   b. Find the work W done on the ball on its descent if it is caught 6 feet above the ground.

   c. Find the work W done on the ball from the time it starts its ascent until it is caught 6 feet above the ground.

20. A rocket lifts off the pad at Cape Canaveral. According to Newton's Law of Gravitation, the force of gravity on the rocket is given by

$$f(x) = \frac{-GMm}{x^2}$$

where M is the mass of the earth, m is the mass of the rocket, G is a universal constant, and x is the distance between the rocket and the center of the earth. Take the radius of the earth to be 4000 miles, so that $x \ge 4000$ miles.
   a. Find the work W done against gravity when the rocket rises 1000 miles. Express your answer in terms of G, M, and m.
   b. Find the work W done against gravity when the rocket rises b miles.
   c. Find the limit of the work found in (b) as b approaches ∞, and determine whether it is possible, with a finite amount of work, to send the rocket arbitrarily far away.

21. Suppose a particle with a mass of m units moves along the x axis with position $x(t)$ at time t. According to Newton's second law of motion, the force $F(x)$ acting on the particle and the acceleration d^2x/dt^2 of the particle are related by the equation

$$F(x) = m\frac{d^2x}{dt^2}$$

At time t the kinetic energy of the particle is $m(v(t))^2/2$, where $v(t)$ denotes the velocity dx/dt. Show that when the particle moves from a to b on the x axis during the time interval from t_a to t_b, the work W done by the force is given by the formula

$$W = \frac{1}{2}m[v(t_b)]^2 - \frac{1}{2}m[v(t_a)]^2$$

(This implies that the work done by the force equals the change in kinetic energy.)

22. A bucket of cement weighing 200 pounds is hoisted by means of a windlass from the ground to the tenth story of an office building, 80 feet above the ground.
   a. If the weight of the rope used is negligible, find the work W required to make the lift.
   b. Assume that a chain weighing 1 pound per foot is used in (a), instead of the lightweight rope. Find the work W required to make the lift. (*Hint:* As the bucket is raised, the length of chain that must be lifted decreases.)

23. A container weighing 1 pound is lifted vertically at the rate of 2 feet per second. Water is leaking out of the

container at the rate of $\frac{1}{2}$ pound per second. If the initial weight of water and container is 20 pounds, find the work W done in raising the container 10 feet. (*Hint:* For $0 \le x \le 10$, determine the total weight of water and container when the container has been raised x feet.)

24. A bucket containing water is raised vertically at the rate of 2 feet per second. Water is leaking out of the container at the rate of $\frac{1}{2}$ pound per second. If the bucket weighs 1 pound and initially contains 20 pounds of water, determine the amount of work W required to raise the bucket until it is empty.

*25. A building demolisher consists of a 2000-pound ball attached to a crane by a 100-foot chain weighing 3 pounds per foot. At night the chain is wound up and the ball is secured to a point 100 feet high. Find the work W done by gravity on the ball and the chain when the ball is lowered from its nighttime position to its daytime position at ground level.

*26. The great pyramid of Cheops had square cross sections and was (approximately) 482 feet high and 754 feet square at the base. If the rock used to build the pyramid weighed 150 pounds per cubic foot, find the work W required to lift the rock into place as the pyramid was built.

*27. A thin steel plate in the shape of an isosceles triangle with base 6 feet and height 4 feet is lying flat on the ground. Suppose the weight of any part of the plate equals its area. How much work is required in order to raise the plate to a vertical position, assuming the base remains in contact with the ground? (*Hint:* Let the x axis be perpendicular to the base of the triangle, with the origin on the base. Consider any partition $\mathscr{P}$ of $[a, b]$, and approximate the work by a Riemann sum.)

6.7
MOMENTS AND CENTERS OF GRAVITY

FIGURE 6.46

Children playing on a seesaw quickly learn that a heavier child has more effect on the rotation of the seesaw than does a lighter child and that a lighter child can balance a heavier one by moving farther away from the axis of rotation (Figure 6.46). In this section we define a quantity called "moment," which measures the tendency of a mass to produce rotation. We will use moments to define a point called the center of gravity of a set of points in the plane.

Moments of Point Masses

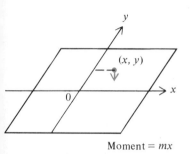

Moment = mx

FIGURE 6.47

Let us first consider the idealized situation in which an object of positive mass m is concentrated at a point (x, y) in the plane. Such an object is called a **point mass**. The **moment** of the point mass **about the y axis** is defined to be mx; we may think of mx as a measure of the tendency of the point mass to rotate about the y axis (Figure 6.47). The larger x or m is, the larger the magnitude of the moment. Thus our definition of moment is consistent with the observation that it is easier to rotate a seesaw about its axis the heavier or farther from the axis one is.

Next, suppose there are several point masses whose masses are $m_1, m_2, \ldots, m_n$, located at the respective points $(x_1, y_1), (x_2, y_2), \ldots, (x_n, y_n)$ in the plane (Figure 6.48). (In the event that these point masses lie on a line, we could associate them with n small children sitting on a seesaw; see Figure 6.49.) We define the **moment** $\mathscr{M}_y$ of the collection of point masses **about the y axis** to be the sum of the moments of the individual point masses about the y axis:

$$\mathscr{M}_y = m_1 x_1 + m_2 x_2 + \cdots + m_n x_n \tag{1}$$

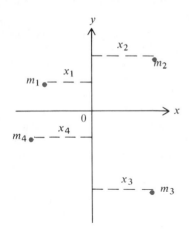

FIGURE 6.48

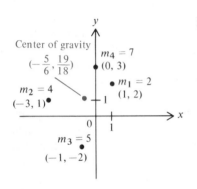

y axis

FIGURE 6.49

If we think of $\mathscr{M}_y$ as a measure of the tendency of the collection of masses to produce a rotation about the *y* axis, then there will be *no* tendency for rotation if $\mathscr{M}_y = 0$. In this case the collection of point masses is said to be in **equilibrium** with respect to the *y* axis. When several children are placed on a seesaw so that equilibrium exists, they can easily rotate the seesaw by pushing against the ground with their feet.

Analogously, we can define the **moment** $\mathscr{M}_x$ of the point masses $m_1, m_2, \ldots,$ m_n **about the x axis** by setting

$$\mathscr{M}_x = m_1 y_1 + m_2 y_2 + \cdots + m_n y_n \tag{2}$$

Then the point masses are in equilibrium with respect to rotation about the *x* axis if $\mathscr{M}_x = 0$.

Now let $m = m_1 + m_2 + \cdots + m_n$ be the combined mass of the point masses just considered, and let us seek a point $(\bar{x}, \bar{y})$ with the property that if we place a point mass with mass *m* at $(\bar{x}, \bar{y})$, then its moments about the *x* and *y* axes will be $\mathscr{M}_x$ and $\mathscr{M}_y$, respectively. The moment of the single point mass about the *y* axis is $m\bar{x}$, and its moment about the *x* axis is $m\bar{y}$; hence by (1) and (2) we need $\bar{x}$ and $\bar{y}$ such that

$$m\bar{x} = \mathscr{M}_y = m_1 x_1 + m_2 x_2 + \cdots + m_n x_n$$

and

$$m\bar{y} = \mathscr{M}_x = m_1 y_1 + m_2 y_2 + \cdots + m_n y_n$$

These conditions on $\bar{x}$ and $\bar{y}$ tell us that

$$\bar{x} = \frac{m_1 x_1 + m_2 x_2 + \cdots + m_n x_n}{m} = \frac{\mathscr{M}_y}{m}$$

$$\bar{y} = \frac{m_1 y_1 + m_2 y_2 + \cdots + m_n y_n}{m} = \frac{\mathscr{M}_x}{m}$$

$$\tag{3}$$

The point $(\bar{x}, \bar{y})$ is called the **center of gravity**, or **centroid**, of the given collection of point masses.

Example 1   Find the two moments and the center of gravity of objects with masses 2, 4, 5, and 7 located at the points $(1, 2)$, $(-3, 1)$, $(-1, -2)$, and $(0, 3)$, respectively.

Solution   By (1) and (2), the moments are given by

$$\mathscr{M}_y = 2(1) + 4(-3) + 5(-1) + 7(0) = -15$$

and

$$\mathscr{M}_x = 2(2) + 4(1) + 5(-2) + 7(3) = 19$$

Since $m = 2 + 4 + 5 + 7 = 18$, it follows from (3) that

$$\bar{x} = \frac{\mathscr{M}_y}{m} = \frac{-15}{18} = \frac{-5}{6} \quad \text{and} \quad \bar{y} = \frac{\mathscr{M}_x}{m} = \frac{19}{18}$$

Therefore the center of gravity of the collection is the point $(-\frac{5}{6}, \frac{19}{18})$ (Figure 6.50). □

FIGURE 6.50

Moments of Plane Regions About the y Axis

We assume from now on that mass is distributed uniformly throughout a plane region R instead of being concentrated at several points. This might be the case if R represented a thin metal plate, or **lamina**. We assume that R is the region between the graphs of two continuous functions f and g on an interval $[a, b]$, where

$$g(x) \leq f(x) \quad \text{for } a \leq x \leq b$$

(Figure 6.51). We also assume that the mass of any subregion of R is equal to its area.

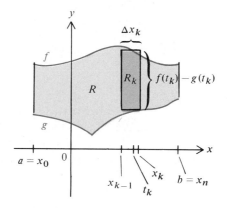

FIGURE 6.51

Let $\mathscr{P} = \{x_0, x_1, \ldots, x_n\}$ be any partition of $[a, b]$. For k between 1 and n, let t_k be an arbitrary number in the subinterval $[x_{k-1}, x_k]$, and let R_k be the portion of R lying between the lines $x = x_{k-1}$ and $x = x_k$. If Δx_k is small, the area ΔA_k of R_k is approximately $[f(t_k) - g(t_k)] \Delta x_k$ (Figure 6.51). Moreover, the moment $\Delta \mathscr{M}_k$ of R_k should be approximately equal to the moment that would result if the entire mass of R_k were concentrated on the line $x = t_k$, that is,

$$\Delta \mathscr{M}_k \approx (\text{distance to } y \text{ axis}) \times (\text{area})$$
$$= t_k \Delta A_k = t_k [f(t_k) - g(t_k)] \Delta x_k$$

Therefore the moment $\mathscr{M}_y$ of the whole region R, which is the sum of the moments $\Delta \mathscr{M}_1, \Delta \mathscr{M}_2, \ldots, \Delta \mathscr{M}_n$, should be approximately

$$\sum_{k=1}^{n} \overbrace{t_k}^{\substack{\text{distance} \\ \text{to } y \text{ axis}}} \overbrace{[f(t_k) - g(t_k)]}^{\text{height}} \overbrace{\Delta x_k}^{\text{width}}$$

which is a Riemann sum for $x(f - g)$ on $[a, b]$. Hence we are led to define $\mathscr{M}_y$ by

$$\mathscr{M}_y = \lim_{\|\mathscr{P}\| \to 0} \sum_{k=1}^{n} t_k [f(t_k) - g(t_k)] \Delta x_k = \int_a^b x[f(x) - g(x)] \, dx \qquad (4)$$

A noteworthy special case occurs when $g = 0$, for then (4) reduces to

$$\mathscr{M}_y = \int_a^b x f(x) \, dx$$

The integral on the right-hand side of this equation makes sense regardless of geometric or physical considerations, and it is normally called the **moment of** f (in contrast to the moment of a region). Moments of functions play a central role in statistics and probability.

Example 2   Find the moment about the y axis of the rectangle R bounded by the lines $x = a$, $x = b$, $y = 0$, and $y = c$.

Solution   Here we have $f(x) = c$ and $g(x) = 0$ (Figure 6.52). Therefore the height of the rectangle at any x in $[a, b]$ is c, so

$$\mathcal{M}_y = \int_a^b x(c - 0)\,dx = \frac{c(b^2 - a^2)}{2} = c(b - a)\left(\frac{a + b}{2}\right) \quad \square$$

Notice that $c(b - a)$ is the area of R and $(a + b)/2$ is the distance from the center of R to the y axis. Notice also that this is the same moment that would result from having the entire mass concentrated at a point on the line $x = (a + b)/2$, midway between the lines $x = a$ and $x = b$.

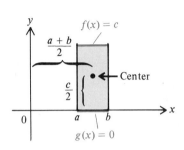

FIGURE 6.52

Moments of Plane Regions About the x Axis

We now turn to the moment of a region about the x axis. First, consider the region R between the graph of a nonnegative continuous function f and the x axis on $[a, b]$. Assume temporarily that f is constant, say

$$f(x) = c \quad \text{for } a \leq x \leq b$$

Then R is a rectangle (see Figure 6.52). We know by analogy with Example 2 that the moment $\mathcal{M}_x$ of R about the x axis should be defined by

$$\mathcal{M}_x = \text{area of } R \times \text{distance from center of } R \text{ to } x \text{ axis}$$

$$= [c(b - a)]\left(\frac{c}{2}\right) = \frac{1}{2}c^2(b - a) \tag{5}$$

If f is not necessarily constant, then as before we let $\mathscr{P} = \{x_0, x_1, \ldots, x_n\}$ be any partition of $[a, b]$. For each k between 1 and n, we let R_k be the portion of R lying between the lines $x = x_{k-1}$ and $x = x_k$. If Δx_k is small and t_k is an arbitrary number in the subinterval $[x_{k-1}, x_k]$, then the moment $\Delta \mathcal{M}_k$ of R_k about the x axis should be approximately equal to the moment about the x axis of a rectangle with height $f(t_k)$ and base $[x_{k-1}, x_k]$ along the x axis (Figure 6.53). By (5) this means that

$$\Delta \mathcal{M}_k \approx \frac{1}{2}[f(t_k)]^2(x_k - x_{k-1}) = \frac{1}{2}[f(t_k)]^2\,\Delta x_k$$

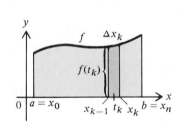

FIGURE 6.53

Since the moment $\mathcal{M}_x$ of R should be the sum of $\Delta \mathcal{M}_1, \Delta \mathcal{M}_2, \ldots, \Delta \mathcal{M}_n$, it follows that $\mathcal{M}_x$ should be approximately equal to

$$\sum_{k=1}^n \frac{1}{2}[f(t_k)]^2\,\Delta x_k$$

and this is a Riemann sum for $\frac{1}{2}f^2$ on $[a, b]$. Accordingly, we are led to define $\mathcal{M}_x$ by

$$\mathcal{M}_x = \lim_{||\mathcal{P}|| \to 0} \sum_{k=1}^{n} \frac{1}{2}[f(t_k)]^2 \, \Delta x_k = \int_a^b \frac{1}{2}[f(x)]^2 \, dx$$

We could expand this argument to show that the moment about the x axis of the region between the graphs of two continuous functions f and g on an interval $[a, b]$ should be defined by

$$\mathcal{M}_x = \int_a^b \frac{1}{2}\{[f(x)]^2 - [g(x)]^2\} \, dx$$

In analogy with (3), we define the center of gravity of R to be the point $(\bar{x}, \bar{y})$ whose coordinates are given by

$$\bar{x} = \frac{\mathcal{M}_y}{A} \quad \text{and} \quad \bar{y} = \frac{\mathcal{M}_x}{A}$$

We incorporate our definitions of the moments and center of gravity of a region R into a single definition.

DEFINITION 6.4

Let f and g be continuous on $[a, b]$, with

$$g(x) \le f(x) \quad \text{for } a \le x \le b$$

and let R be the region between the graphs of f and g on $[a, b]$. Then the **moment $\mathcal{M}_x$ of R about the x axis** is given by

$$\mathcal{M}_x = \int_a^b \frac{1}{2}\{[f(x)]^2 - [g(x)]^2\} \, dx$$

and the **moment $\mathcal{M}_y$ of R about the y axis** is given by

$$\mathcal{M}_y = \int_a^b x[f(x) - g(x)] \, dx$$

If R has positive area A, then the **center of gravity** (or **center of mass**, or **centroid**) of R is the point $(\bar{x}, \bar{y})$ defined by

$$\bar{x} = \frac{\mathcal{M}_y}{A} \quad \text{and} \quad \bar{y} = \frac{\mathcal{M}_x}{A}$$

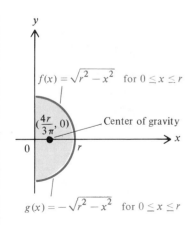

$f(x) = \sqrt{r^2 - x^2}$ for $0 \le x \le r$

$\left(\dfrac{4r}{3\pi}, 0\right)$ —— Center of gravity

$g(x) = -\sqrt{r^2 - x^2}$ for $0 \le x \le r$

FIGURE 6.54

Example 3 Let R be the semicircular region bounded by the y axis and the graphs of f and g, where

$$f(x) = \sqrt{r^2 - x^2} \quad \text{and} \quad g(x) = -\sqrt{r^2 - x^2} \quad \text{for } 0 \le x \le r$$

(Figure 6.54). Find the moments and the center of gravity of R.

Solution   Definition 6.4 tells us directly that

$$M_x = \int_0^r \frac{1}{2}[(r^2 - x^2) - (r^2 - x^2)]\,dx = \frac{1}{2}\int_0^r 0\,dx = 0$$

and

$$M_y = \int_0^r x[\sqrt{r^2 - x^2} - (-\sqrt{r^2 - x^2})]\,dx = 2\int_0^r x\sqrt{r^2 - x^2}\,dx$$

$$= \frac{-2}{3}(r^2 - x^2)^{3/2}\Big|_0^r = \frac{2}{3}r^3$$

Since the area A of R is $\frac{1}{2}\pi r^2$, we have

$$\bar{x} = \frac{M_y}{A} = \frac{\frac{2}{3}r^3}{\frac{1}{2}\pi r^2} = \frac{4r}{3\pi}$$

and

$$\bar{y} = \frac{M_x}{A} = \frac{0}{\frac{1}{2}\pi r^2} = 0$$

Consequently the center of gravity $(4r/3\pi, 0)$ lies on the x axis, approximately $0.42r$ from the origin (Figure 6.54). $\square$

Notice that in Example 3, R was symmetric with respect to the x axis, and the center of gravity turned out to lie on the x axis. More generally, it is possible to show the following:

1. If R is symmetric with respect to the line $x = c$, then the center of gravity of R lies on the line $x = c$; that is, $\bar{x} = c$ (Figure 6.55(a)).

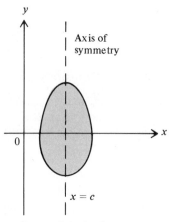

Center of gravity lies
on the axis of symmetry

(a)

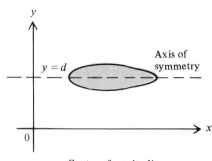

Center of gravity lies
on the axis of symmetry

(b)

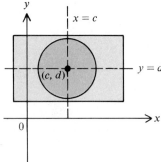

Center of gravity of circle
or rectangle lies at the center

(c)

FIGURE 6.55

2. If R is symmetric with respect to the line $y = d$, then the center of gravity of R lies on the line $y = d$; that is, $\bar{y} = d$ (Figure 6.55(b)).

Therefore symmetry can reduce the effort required to locate the center of gravity of a region.

Example 4 Show that the center of gravity of a circle or a rectangle is its center.

Solution If (c, d) is the center of the region in question, then the region is obviously symmetric with respect to the lines $x = c$ and $y = d$ (Figure 6.55(c)). By the observation above, the center of gravity of the region is (c, d). ◻

The Theorem of Pappus and Guldin

Assume that f and g are continuous on an interval $[a, b]$, with $a \geq 0$ and

$$g(x) \leq f(x) \quad \text{for } a \leq x \leq b$$

Let R be the region between the graphs of f and g on $[a, b]$, with area A. If we compare formula (4) in Section 6.2 for the volume V of the solid obtained by revolving R about the y axis with the formula in Definition 6.4 for the moment $\mathcal{M}_y$ about the y axis, we see that

$$V = 2\pi \mathcal{M}_y$$

But since $\mathcal{M}_y = \bar{x}A$ we may also rewrite this volume as

$$V = (2\pi\bar{x})A$$

Notice that $\bar{x}$ is the distance from the center of gravity of R to the y axis, and the y axis is by assumption the axis of revolution for R. A corresponding result holds when R is revolved about the x axis. These results were first anticipated by the mathematician Pappus of Alexandria about 300 A.D., but they are often credited as well to the Swiss mathematician Paul Guldin, who lived some 1300 years later.

THEOREM 6.5

THEOREM OF PAPPUS AND GULDIN

Let R be a plane region lying completely to one side of a line l. Then the volume V of the solid region generated by revolving R about l is given by

$$V = (2\pi b)A$$

where A is the area of R and b is the distance from the center of gravity of R to the line l.

Theorem 6.5 implies that the volume of the solid generated by revolving a plane region about a line outside the region is equal to the product of the area of the plane region and the distance the center of gravity travels as the region is revolved once about the line. This result can be used to compute the volume of a solid when the area and center of gravity of the plane region are known.

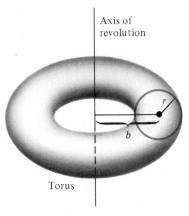

Axis of revolution

Torus

FIGURE 6.56

Example 5   Find the volume V of the **torus** obtained by revolving a circle of radius r about a line lying a distance $b \geq r$ from the center of the circle (Figure 6.56).

Solution   Since the center of gravity of the circle is its center, Theorem 6.5 implies that

$$V = (2\pi b)(\pi r^2) = 2\pi^2 r^2 b \quad \square$$

A torus has the shape of a doughnut. If the axis of revolution is the y axis, then the torus in Figure 6.56 can be considered to be the solid obtained by revolving the circle $(x - b)^2 + y^2 = r^2$ about the y axis. You might try to find the volume of this solid by the shell method of Section 6.2. (See Exercise 31 of this section.)

EXERCISES 6.7

1. A child weighing 30 kilograms sits on a seesaw 1.5 meters to the left of the axis of revolution. Another child, weighing 20 kilograms, is sitting on the right-hand side of the seesaw 2 meters from the axis of revolution. Will the right side of the seesaw rise, or will it fall?

2. Two children weighing 15 and 20 kilograms are sitting on opposite sides of a seesaw, both 2 meters from the axis of revolution. Where on the seesaw should a 10-kilogram toddler sit in order to achieve equilibrium?

3. Suppose the 10-kilogram toddler in Exercise 2 sits down on the seesaw first, 1 meter to the left of the axis of revolution. If the two larger children wish to sit on opposite sides of the axis, equidistant from it, find the location of these children that will ensure equilibrium.

4. A mobile consists of three weights attached to a square piece of cardboard of negligible weight, to be suspended from the ceiling by a string. Assume that the weights have mass 50, 30, and 20 grams, respectively. Suppose that a coordinate system with origin at the center of the square has been set up and that the respective points at which the weights are attached to the square are $(3, -1)$, $(4, 2)$, and $(-1, 1)$ in this coordinate system. Determine the center of gravity of the mobile; that is, find the point on the cardboard at which the string must be attached for the mobile to be balanced.

In Exercises 5–12 calculate the center of gravity of the region R between the graphs of f and g on the given interval.

5. $f(x) = x$, $g(x) = -2$; $[0, 2]$

6. $f(x) = 2x - 1$, $g(x) = x - 2$; $[2, 5]$

7. $f(x) = 2 - x$, $g(x) = -(2 - x)$; $[0, 2]$

8. $f(x) = 3x$, $g(x) = x^2$; $[0, 1]$

9. $f(x) = (x + 1)^2$, $g(x) = (x - 1)^2$; $[1, 2]$

10. $f(x) = x$, $g(x) = 1/x$; $[1, 2]$

11. $f(x) = x + 1$, $g(x) = \sqrt{x + 1}$; $[0, 3]$

12. $f(x) = \dfrac{1}{\sqrt{x - 1}}$, $g(x) = \dfrac{1}{\sqrt{x + 1}}$; $[2, 5]$

In Exercises 13–14 find the center of gravity of the region R between the graphs of the given equations.

13. $y^2 = 2x - 5$ and $y = x - 4$

14. $y = x + 2$, $y = -3x + 6$ and $y = (2 - x)/3$

In Exercises 15–16 find the center of gravity of the region R between the graphs of f and g. Use symmetry wherever applicable.

15. $f(x) = 11 - x^2$ and $g(x) = x^2 + 3$

16. $f(x) = 2 - x^2$ and $g(x) = |x|$

In Exercises 17–22 use the symmetry of the region R to determine the center of gravity of R.

17. R is bounded by the parallelogram with vertices at $(0, 0)$, $(0, 2)$, $(1, 1)$, and $(1, 3)$.

18. R is bounded by the hexagon with vertices at $(0, 0)$, $(0, 6)$, $(1, 1)$, $(1, 5)$, $(-1, 1)$, and $(-1, 5)$.

19. R is bounded by the ellipse

$$\frac{(x - 3)^2}{a^2} + \frac{(y + 1)^2}{b^2} = 1$$

20. R is bounded by the astroid $x^{2/3} + y^{2/3} = 1$.

21. R is bounded by the curves $y = x^2$ and $y = 2 - x^2$

22. R is bounded by the curve $y = 1/x$ and the lines $y = 1 + x$ and $y = -1 + x$.

Exercises 23–26 involve some regions that are important in engineering and architecture. Find the center of gravity of each region.

23. The region is the triangle shown in Figure 6.57(a). (You should recognize the center of gravity as a point familiar from geometry.)

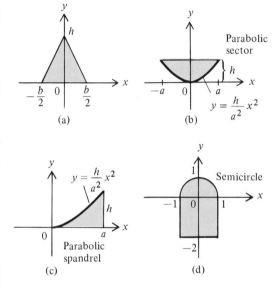

(a)

(b)

(c)

(d)

FIGURE 6.57

24. The region is the **parabolic sector** shown in Figure 6.57(b).

25. The region is the **parabolic spandrel** shown in Figure 6.57(c).

26. The region is the fingerlike shape shown in Figure 6.57(d).

27. Use the Theorem of Pappus and Guldin to obtain an alternate derivation of the center of gravity of the semicircular region R in Example 3.

28. Use the Theorem of Pappus and Guldin and the formula for the volume of a cone to find the center of gravity of a right triangle with legs of length a and c.

29. Let a square $ABCD$ have side length 2 and diagonal AC. Use the Theorem of Pappus and Guldin to find the volume V of the solid obtained by revolving the square about the line that passes through B and is parallel to AC.

30. An equilateral trapezoid has sides with length 8, 4, $2\sqrt{2}$, and $2\sqrt{2}$. Use the Theorem of Pappus and Guldin to find the center of gravity of the trapezoid.

31. Use the shell method to set up an integral for the volume V of the torus in Figure 6.56. (The integral can be evaluated by the methods of Section 8.3; see Exercise 53 of Section 8.3.)

32. Suppose f is an even function defined on $[-a, a]$, and let R be the region between the graph of f and the x axis on $[-a, a]$. Show that the moment of R about the y axis is 0 and hence that the center of gravity of R lies on the y axis.

33. Suppose f is an odd function defined on $[-a, a]$ and is nonnegative on $[0, a]$. Let R be the region between the graph of f and the x axis on $[-a, a]$. Show that the center of gravity of R is the origin.

34. Let R be a plane region with center of gravity $P = (\bar{x}, \bar{y})$. Suppose R is contained in a circle with center at P. Show that the volume of the solid which results from revolving R about a line tangent to the circle is the same for any such line.

35. Sometimes a region R can be divided into regions R_1, $R_2, \dots, R_n$, each of whose center of gravity is easily computed. Then the center of gravity $(\bar{x}, \bar{y})$ of R may be determined from the centers of gravity $(\bar{x}_1, \bar{y}_1)$, $(\bar{x}_2, \bar{y}_2), \dots, (\bar{x}_n, \bar{y}_n)$ of $R_1, R_2, \dots, R_n$ by using the fact that the moment of R about either the x axis or the y axis is the sum of the moments of $R_1, R_2, \dots, R_n$ about that axis. This yields

$$\bar{x} \times \text{area of } R = (\bar{x}_1 \times \text{area of } R_1) + (\bar{x}_2 \times \text{area of } R_2) \\ + \cdots + (\bar{x}_n \times \text{area of } R_n)$$

and

$$\bar{y} \times \text{area of } R = (\bar{y}_1 \times \text{area of } R_1) + (\bar{y}_2 \times \text{area of } R_2) \\ + \cdots + (\bar{y}_n \times \text{area of } R_n)$$

Using these equations, determine the centers of gravity of the regions shown in

a. Figure 6.58(a)
   (*Hint:* The center of gravity lies outside the region.)
b. Figure 6.58(b)
c. Figure 6.58(c)
d. Figure 6.58(d)
e. Figure 6.58(e)

Suppose a curve C is parametrized by

$$x = f(t) \quad \text{and} \quad y = g(t) \quad \text{for } a \le t \le b$$

where f and g are continuously differentiable on $[a, b]$. The **centroid** of C is the point $(\bar{x}, \bar{y}) = (\mathcal{M}_y/\mathcal{L}, \mathcal{M}_x/\mathcal{L})$, where $\mathcal{L}$ is the length of C and

$$\mathcal{M}_x = \int_a^b g(t) \sqrt{(f'(t))^2 + (g'(t))^2} \, dt$$

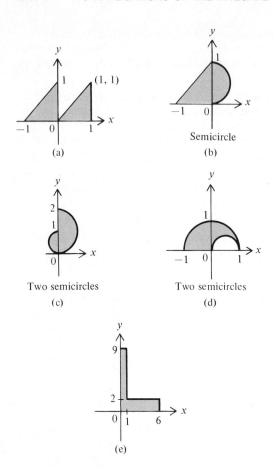

Semicircle
(b)

(a)

Two semicircles
(c)

Two semicircles
(d)

(e)

FIGURE 6.58

and $\quad \mathcal{M}_y = \displaystyle\int_a^b f(t)\sqrt{(f'(t))^2 + (g'(t))^2}\,dt$

The centroid of a wire that is uniformly dense and has the shape of C is also $(\bar{x}, \bar{y})$.

36. Suppose a wire has uniform density and has the shape of the quarter circle C parametrized by

$$x = r\cos t \quad \text{and} \quad y = r\sin t \quad \text{for } 0 \le t \le \pi/2$$

Find the centroid of the wire, and show that the centroid does not lie on the wire.

Suppose a curve C is parametrized by

$$x = f(t) \quad \text{and} \quad y = g(t) \quad \text{for } a \le t \le b$$

where f and g are continuously differentiable on $[a, b]$. The x coordinate of the **centroid** of the surface obtained by revolving C about the x axis is $\bar{x} = \mathcal{M}_y/S$, where S is the surface area of the surface and

$$\mathcal{M}_y = \int_a^b 2\pi f(t)g(t)\sqrt{(f'(t))^2 + (g'(t))^2}\,dt$$

37. Let C be the quarter circle parametrized by

$$x = r\cos t \quad \text{and} \quad y = r\sin t \quad \text{for } 0 \le t \le \pi/2$$

Find the centroid of the surface obtained by revolving C about the x axis.

6.8
HYDROSTATIC FORCE

We devote this section to hydrostatic force, the force exerted by stationary water on a surface such as a plate, a wall, or a dam.

Hydrostatic Pressure   A diver under water experiences pressure due to the weight of the water above. Moreover, the deeper the diver descends, the greater the pressure is. Let us analyze this phenomenon and define it formally.

Consider a horizontal plate A square feet in area at a depth of x feet below the surface of the water (Figure 6.59). The water directly above the plate exerts a force F equal to its weight on the plate. Since the volume of the water directly above the plate is xA cubic feet and water weighs 62.5 pounds per cubic foot, the force F is given by the formula

$$F = 62.5xA$$

A diver under water experiences hydrostatic pressure from all directions.

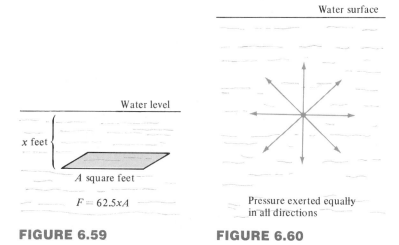

FIGURE 6.59   **FIGURE 6.60**

Dividing by A, we find that

$$\frac{F}{A} = 62.5x \tag{1}$$

The quantity F/A is called the **hydrostatic pressure** on the plate and is measured in pounds per square foot.

The hydrostatic pressure on a horizontal plate depends only on the depth of the plate, not on the area of the plate. As a result, we talk about the hydrostatic pressure at a point under water, meaning the hydrostatic pressure on any horizontal plate containing that point. Hydrostatic pressure is exerted not only downward but in all directions. In fact, Pascal's Principle states that the hydrostatic pressure at a point in water is exerted equally in all directions (Figure 6.60). If the point is x feet below the surface of the water, then the magnitude of the pressure at that point is $62.5x$. Therefore, the deeper the plate is submerged, the greater the hydrostatic pressure on it is—a fact well known to divers.

We are now ready to analyze the hydrostatic force on a vertical plate or surface that is either totally or partially submerged in water.

Hydrostatic Force on a Vertical Plate

The analysis we make of the hydrostatic force on a vertical plate applies equally well to a vertical wall or dam in water. You may think of any of these in the following discussion.

First we place the x axis vertically, with the positive direction upward. Let $x = c$ represent the water level, and let the portion of the plate submerged in water extend from a to b on the x axis (Figure 6.61(a) and (b)). Notice that in Figure 6.61(a) the top of the plate is higher than the water level, so that $c = b$, whereas in Figure 6.61(b) the top of the plate is lower than the water level, so that $c > b$. Next we let $w(x)$ be the width of the plate at x and assume that w is continuous on $[a, b]$.

Let $\mathscr{P} = \{x_0, x_1, \ldots, x_n\}$ be any partition of $[a, b]$. For each k between 1 and n, let t_k be an arbitrary number in the subinterval $[x_{k-1}, x_k]$. If Δx_k is small, the area of the portion S_k of the plate lying between x_{k-1} and x_k on the x axis is approximately $w(t_k) \, \Delta x_k$ (Figure 6.61(b)). In addition, from (1) we know that the

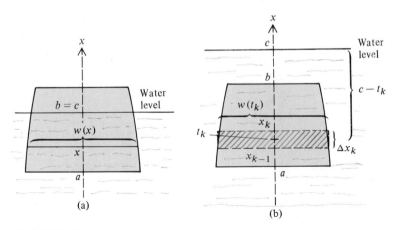

FIGURE 6.61

pressure at any point on S_k is approximately $(6.25)(c - t_k)$. Therefore the force ΔF_k on S_k is approximately the product of the pressure $(62.5)(c - t_k)$ on S_k and its area $w(t_k) \Delta x_k$. Thus

$$\Delta F_k \approx (\text{pressure}) \times (\text{area}) \approx (62.5)(c - t_k)w(t_k)\Delta x_k$$

Since the force F on the plate is the sum of the forces $\Delta F_1, \Delta F_2, \ldots, \Delta F_n$, it follows that F should be approximately

$$\sum_{k=1}^{n} \overbrace{(62.5)(c - t_k)}^{\text{pressure}} \overbrace{w(t_k)\Delta x_k}^{\text{area}}$$

and this is a Riemann sum for the function $(62.5)(c - x)w$ on $[a, b]$. Accordingly, the force F should be defined by

$$F = \lim_{\|\mathscr{P}\| \to 0} \sum_{k=1}^{n} (62.5)(c - t_k)w(t_k)\Delta x_k = \int_a^b (62.5)(c - x)w(x)\,dx$$

DEFINITION 6.6

Assume that a vertical plate is submerged or partially submerged in water with the water level located at $x = c$. Let the submerged portion of the plate extend from a to b on the x axis. Let $w(x)$ be the width of the plate for $a \le x \le b$, and assume that w is continuous on $[a, b]$. Then the **hydrostatic force** F on the plate due to the water is given by

$$F = \int_a^b 62.5(c - x)w(x)\,dx \qquad (2)$$

If water is replaced by some other fluid, such as gasoline or alcohol, then the number 62.5 must be replaced by the weight of one cubic foot of that fluid.

Example 1   An isosceles triangular plate 3 feet tall and 1.5 feet wide is located on a vertical wall of a swimming pool, the bottom vertex 6 feet below water level (Figure 6.62). Find the hydrostatic force F on the plate.

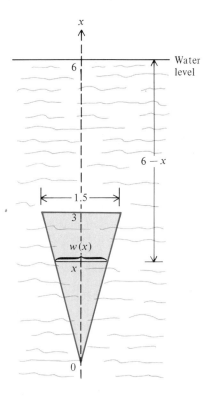

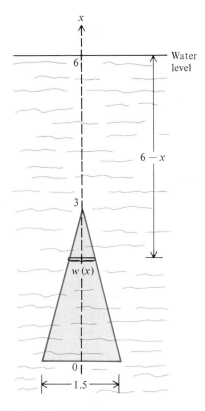

FIGURE 6.62　　　　　　　　**FIGURE 6.63**

Solution　For convenience we place the origin at the bottom vertex of the plate. Then the plate extends from $x = 0$ to $x = 3$, and the depth at a given x is $6 - x$ (Figure 6.62). Moreover, since the width of the triangle is $\frac{1}{2}$ the height, it follows from similar triangles that at any x the width $w(x)$ is given by

$$\frac{w(x)}{1.5} = \frac{x}{3}, \quad \text{so that} \quad w(x) = \frac{1}{2}x$$

Consequently (2) implies that the force F is given by

$$F = \int_0^3 (62.5)(6 - x)\left(\frac{1}{2}x\right) dx = 31.25 \int_0^3 (6x - x^2)\, dx$$

$$= 31.25\left(3x^2 - \frac{x^3}{3}\right)\bigg|_0^3 = (31.25)(18) = 562.5 \,(\text{pounds}) \quad \square$$

　　Had the triangle been placed with the vertex at the top, 3 feet below water level, then we could place the origin again at the bottom of the plate (Figure 6.63), so once again the plate would extend from $x = 0$ to $x = 3$ and the depth at a given x would be $6 - x$. But now at any x the width $w(x)$ would, by similar triangles, satisfy

$$\frac{w(x)}{1.5} = \frac{3 - x}{3} \quad \text{so} \quad w(x) = \frac{3 - x}{2}$$

(Figure 6.63). Therefore the force F would be given by

$$F = \int_0^3 (62.5)(6 - x)\left(\frac{3 - x}{2}\right) dx = 31.25 \int_0^3 (18 - 9x + x^2) dx$$

$$= 31.25\left(18x - \frac{9}{2}x^2 + \frac{x^3}{3}\right)\Big|_0^3 = (31.25)(22.5) = 703.125 \text{ (pounds)}$$

EXERCISES 6.8

1. Suppose the top edge of the plate in Example 1 were at water level. Compute the hydrostatic force F on the plate.

2. Suppose a rectangular dam is 1000 feet wide and 100 feet high. Compute the hydrostatic force F on the dam when the water level is
   a. 100 feet above the bottom.
   b. 50 feet above the bottom.

3. Suppose the end of a water trough has the shape of an equilateral triangle (pointed down) with side length 2 feet. Compute the hydrostatic force F on the end of the trough when the water level is 1 foot above the bottom. (*Hint:* Let the origin be at water level.)

4. A flat semicircular floodlight on a vertical wall of a swimming pool has a radius of 1 foot, with a diameter located at water level. Compute the hydrostatic force F on the submerged part of the floodlight.

5. A pool surrounding a water fountain has flat vertical sides that contain lights having the shape of an equilateral triangle with sides of length 2 feet. The arrangement of the lights is shown in Figure 6.64. Assuming that the water level is 3 feet above the bottoms of the triangles, compute the hydrostatic force F on each type of light. (*Hint:* Let the origin be at water level.)

7. A cubical block of marble 1 foot on a side is submerged in water but does not touch bottom. Suppose two of its faces are parallel to the water surface. Show that the difference between the hydrostatic force exerted on the bottom face and the hydrostatic force on the top face is equal to 62.5 pounds. (This difference is called the **buoyant force**. Archimedes' Principle states that any body submerged in water is buoyed up by a force equal to the weight of the water it displaces.)

*8. a. Using (2), show that the hydrostatic force F equals $62.5hA$, where A is the area of the submerged portion of the plate and h is the depth of the water at the center of gravity of the submerged portion.
   b. Using the results of part (a) and Example 3 of Section 6.7, calculate the force on a vertical semicircular dam with radius 100 feet when the water level is at the top of the dam (Figure 6.65).
   c. Suppose that the water level is at the top of a rectangular dam. Use the result of part (a) to show that the hydrostatic force on the bottom half of the dam is three times that on the top half.

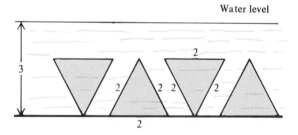

Water level

FIGURE 6.64

6. A cubical block of marble 1 foot on a side rests at the bottom of a swimming pool 6 feet deep. Determine the total hydrostatic force F on the five sides touching the water.

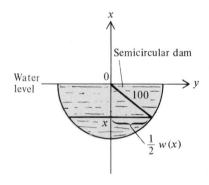

FIGURE 6.65

9. Suppose that in our analysis of hydrostatic pressure we replace the assumption that the wall is vertical by the assumption that it makes an angle θ with the vertical

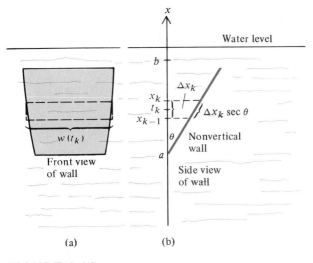

FIGURE 6.66

(a) (b)

(Figure 6.66). Show that the hydrostatic force F on the portion between x_{k-1} and x_k would be

$$62.5(c - t_k)[(\sec \theta)w(t_k)\,\Delta x_k]$$

and that the hydrostatic force F would be given by

$$F = 62.5 \sec \theta \int_a^b (c - x)w(x)\,dx$$

10. Suppose a rectangular earth-filled dam 1000 feet wide makes an angle of $\pi/3$ with the vertical. Using the result of Exercise 9, compute the hydrostatic force F on the dam when the water level is 50 feet above the bottom.

11. Using the method employed to derive formula (2), show that the hydrostatic force F on the sides of a cylindrical tank filled with water and having a radius of 20 feet and a height of 30 feet is given by

$$F = 62.5 \int_0^{30} 40\pi x\,dx$$

12. In an offshore search for oil in water 100 feet deep, drilling is done through a cylindrical pipe having an exterior diameter of 3 feet. Using a formula like the one obtained in Exercise 11, compute the hydrostatic force F on the exterior of the pipe.

6.9
POLAR COORDINATES

Since Chapter 1 we have identified each point in the plane with its coordinates in a Cartesian coordinate system. However, there are other coordinate systems, one of the most prominent being the polar coordinate system. We introduce this system now and will employ it in Section 6.10 to find the areas of certain regions.

The Polar Coordinate System

We begin with a Cartesian coordinate system in the plane. For any point P other than the origin, let r be the distance between P and the origin, and θ an angle having its initial side on the positive x axis and its terminal side on the line segment joining P and the origin (Figure 6.67(a)). Then the pair (r, θ) is called a **set of polar coordinates** for the point P. For each P there are infinitely many possible choices of θ, any two differing by a multiple of 2π. For convenience we also let $(-r, \theta + \pi)$ be a set of polar coordinates for the point P whenever (r, θ) is. (You may think of $(-r, \theta + \pi)$ for $r > 0$ as corresponding to a point reached by moving a distance r in the direction opposite that of the angle $\theta + \pi$ (Figure 6.67(b)).) If (r, θ) is one set of polar coordinates for P, then any other set will be one of the following:

$$(r, \theta + 2n\pi) \quad \text{for } n \text{ any integer}$$
$$(-r, \theta + (2n + 1)\pi) \quad \text{for } n \text{ any integer}$$

(1)

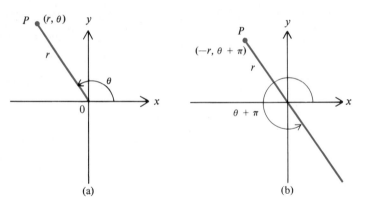

FIGURE 6.67

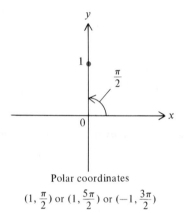

Polar coordinates

$(1, \frac{\pi}{2})$ or $(1, \frac{5\pi}{2})$ or $(-1, \frac{3\pi}{2})$

FIGURE 6.68

For example, the point that lies 1 unit from the origin on the positive y axis has polar coordinates $(1, \pi/2)$; by virtue of (1) it also has, among others, polar coordinates $(1, 5\pi/2)$ and $(-1, \pi/2 + \pi) = (-1, 3\pi/2)$ (Figure 6.68). But despite the unlimited number of sets of polar coordinates for any point P other than the origin, we stress that P has *only one* set of polar coordinates (r, θ) such that $r > 0$ and $0 \le \theta < 2\pi$. Finally, we assign the origin polar coordinates $(0, \theta)$, where θ may be any number. Figure 6.69 shows several points in the plane along with some of their polar coordinates.

Since we use the expression (a, b) to denote both Cartesian and polar coordinates, we will explicitly state which kind of coordinates are involved where there is danger of confusion. Except in this section and Section 6.10, we will assume unless stated otherwise that Cartesian coordinates are to be used.

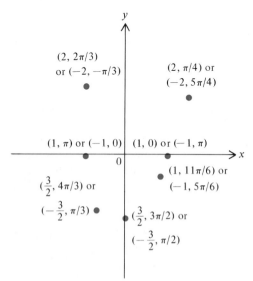

Polar coordinates of some points

FIGURE 6.69

Although most maps use Cartesian coordinates, maps in polar coordinates do exist. In addition, the dance a bee performs to communicate the location of a source of food seems to be related to polar coordinates. The orientation of the bee's body locates the direction of the food, and the intensity of the dance indicates the distance to the source.

Conversion Between Cartesian and Polar Coordinates

Every point in the plane has both Cartesian and polar coordinates. We will now see that it is possible to convert from each type of coordinates to the other. Suppose a point P in the plane has polar coordinates (r, θ) and Cartesian coordinates (x, y). Then from the definition of the sine and the cosine we deduce that

$$x = r\cos\theta \quad \text{and} \quad y = r\sin\theta \tag{2}$$

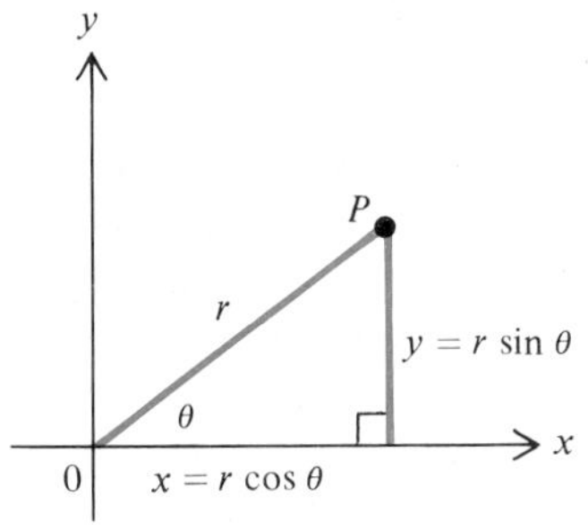

FIGURE 6.70

for all values of r and θ (Figure 6.70). From (2) we see that x and y are uniquely determined by r and θ and we also obtain formulas for r^2 and $\tan\theta$:

$$r^2 = x^2 + y^2 \quad \text{and} \quad \tan\theta = \frac{y}{x} \quad \text{for } x \neq 0 \tag{3}$$

Example 1   Find the Cartesian coordinates of the point P having polar coordinates $(3, 23\pi/6)$.

Solution   From (2) we obtain

$$x = 3\cos\frac{23\pi}{6} = 3\cos\frac{-\pi}{6} = 3\cos\frac{\pi}{6} = \frac{3\sqrt{3}}{2}$$

and

$$y = 3\sin\frac{23\pi}{6} = 3\sin\frac{-\pi}{6} = -3\sin\frac{\pi}{6} = -\frac{3}{2}$$

Therefore $(3\sqrt{3}/2, -\frac{3}{2})$ are the Cartesian coordinates of P.   □

Although we cannot determine r and θ uniquely from (x, y) merely by applying (3), it is possible to determine all sets of polar coordinates for a given point (x, y) in Cartesian coordinates.

Example 2   Find all sets of polar coordinates for the point P having Cartesian coordinates $(-5, 5\sqrt{3})$.

Solution   First we find the polar coordinates (r, θ) for P such that $r > 0$ and $0 \leq \theta < 2\pi$. From (3) we know that

$$r^2 = (-5)^2 + (5\sqrt{3})^2 = 25 + 75 = 100$$

Therefore $r = 10$. We know also by (3) that

$$\tan \theta = \frac{5\sqrt{3}}{-5} = -\sqrt{3} \tag{4}$$

From (4) and the fact that $(-5, 5\sqrt{3})$ lies in the second quadrant, it follows that $\theta = 2\pi/3$. Thus one set of polar coordinates is $(10, 2\pi/3)$. Consequently by (1) any set of polar coordinates of P must be of the form

$$\left(10, \frac{2\pi}{3} + 2n\pi\right) \quad \text{where } n \text{ is an integer}$$

or

$$\left(-10, \frac{5\pi}{3} + 2n\pi\right) \quad \text{where } n \text{ is an integer} \quad \square$$

Caution: Notice that $r = 10$ and $\theta = -\pi/3$ also satisfy the equations in (3) when $x = -5$ and $y = 5\sqrt{3}$. Nevertheless $(10, -\pi/3)$ is not a set of polar coordinates for the point $(-5, 5\sqrt{3})$. The reason is simple: The original point $(-5, 5\sqrt{3})$ lies in the second quadrant, but the point with polar coordinates $(10, -\pi/3)$ lies in the fourth quadrant. Therefore when we convert from Cartesian to polar coordinates, it is not enough merely to choose r and θ to satisfy (3). We must be sure as well that the point with polar coordinates (r, θ) lies in the correct quadrant.

Polar Equations and Graphs

Just as we can graph an equation involving the Cartesian coordinates x and y, we can also graph an equation involving the polar coordinates r and θ. The **polar graph** of such an equation is defined to be the collection of all points in the plane having a set of polar coordinates (r, θ) that satisfies the equation. A **polar equation** of a collection of points in the plane is an equation in r and θ whose polar graph is the given collection of points.

Our next two examples involve polar equations of circles.

Example 3   Find a polar equation of the circle $x^2 + y^2 = a^2$, where $a > 0$ (Figure 6.71(a)).

Solution   Using (3), we find immediately that $r^2 = a^2$, or more simply, $r = a$. $\square$

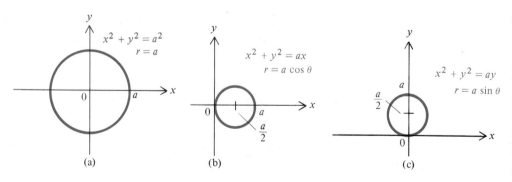

(a)   (b)   (c)

FIGURE 6.71

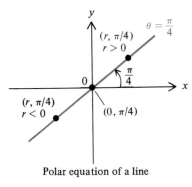

Polar equation of a line

FIGURE 6.72

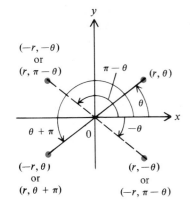

FIGURE 6.73

Example 4   Find a polar equation of the circle $x^2 + y^2 = ax$ (Figure 6.71(b)).

Solution   Using (2) and (3), we find that $r^2 = ar \cos \theta$, or $r = a \cos \theta$.   □

A similar argument shows that a polar equation of the circle $x^2 + y^2 = ay$ is

$$r = a \sin \theta$$

(Figure 6.71(c)).

Example 5   Find a polar equation of the line passing through the origin and making an angle of $\pi/4$ with respect to the positive x axis (Figure 6.72).

Solution   From the definition of polar coordinates we find that any point on the line has polar coordinates $(r, \pi/4)$, where r may be either positive, negative, or 0. Consequently a polar equation of the line is $\theta = \pi/4$.   □

More generally, a polar equation of the line passing through the origin and making an angle θ_0 with the positive x axis is given by $\theta = \theta_0$.

The types of symmetry we have studied with respect to rectangular coordinates play just as significant a role in polar graphs as in graphs in rectangular coordinates. The points and polar coordinates shown in Figure 6.73 suggest some conditions for symmetry of polar graphs. These conditions are listed in Table 6.1.

TABLE 6.1

Symmetry	Conditions for symmetry
With respect to the x axis	If (r, θ) satisfies the equation, so does $(r, -\theta)$ or $(-r, \pi - \theta)$
With respect to the y axis	If (r, θ) satisfies the equation, so does $(-r, -\theta)$ or $(r, \pi - \theta)$
With respect to the origin	If (r, θ) satisfies the equation, so does $(-r, \theta)$ or $(r, \theta + \pi)$

However, it is possible for a graph to be symmetric with respect to the origin or one of the axes without satisfying any of the corresponding conditions (see Exercises 46 and 47).

We apply these criteria for symmetry to facilitate the drawing of graphs. In the examples that follow we will frequently associate the name of a familiar graph with the equation that defines it. For instance, we say "the circle $r = 2 \cos \theta$" instead of using the longer phrase "the circle whose equation is $r = 2 \cos \theta$."

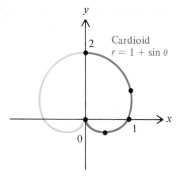

FIGURE 6.74

Example 6   Sketch the cardioid* $r = 1 + \sin\theta$.

Solution   First we observe that $\sin(\pi - \theta) = \sin\theta$, so that from Table 6.1 it is evident that the cardioid is symmetric with respect to the y axis. Thus, since the sine function has period 2π, we need consider only those values of θ between $-\pi/2$ and $\pi/2$; then we can use symmetry to complete the graph. From the chart below we obtain the special points appearing in Figure 6.74.

θ	$-\pi/2$	$-\pi/6$	0	$\pi/6$	$\pi/2$
$r = 1 + \sin\theta$	0	$\frac{1}{2}$	1	$\frac{3}{2}$	2

Next we notice that as θ increases from $-\pi/2$ to $\pi/2$, r increases from 0 to 2. Thus that part of the cardioid to the right of the y axis appears in Figure 6.74 as the portion of the curve in dark color. Using the symmetry with respect to the y axis, we obtain the complete graph, whose appearance justifies the name "cardioid," which means "heart-shaped."   □

Observe that the equation $r = 1 + \sin\theta$ has the same polar graph as the equation $r^2 = r + r\sin\theta$. (The only problem occurs for $r = 0$, and both graphs include the origin, which corresponds to $r = 0$.) However, by (2) and (3) the polar graph of the polar equation $r^2 = r + r\sin\theta$ is the same as the graph of

$$x^2 + y^2 = \sqrt{x^2 + y^2} + y \tag{5}$$

It is clearly easier to graph the cardioid from its polar equation $r = 1 + \sin\theta$ than from (5).

Example 7   Sketch the limaçon† $r = \frac{1}{2} + \cos\theta$.

Solution   Since $\cos(-\theta) = \cos\theta$, it follows that the limaçon is symmetric with respect to the x axis. Consequently we need consider only those values of θ between 0 and π; then we can use symmetry to complete the graph. From the chart below we determine the special points in Figure 6.75.

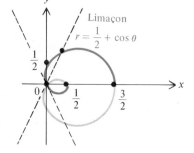

FIGURE 6.75

θ	0	$\pi/3$	$\pi/2$	$2\pi/3$	π
$r = \frac{1}{2} + \cos\theta$	$\frac{3}{2}$	1	$\frac{1}{2}$	0	$-\frac{1}{2}$

Now we notice that as θ increases from 0 to $2\pi/3$, r decreases from $\frac{3}{2}$ to 0. Similarly, as θ increases from $2\pi/3$ to π, r decreases from 0 to $-\frac{1}{2}$; since r is negative, the graph of the portion corresponding to $2\pi/3 \leq \theta \leq \pi$ lies in the fourth quadrant instead of the second quadrant. With this information we sketch

---

* **Cardioid:** Pronounced "*car*-dee-oid."
† **Limaçon:** Pronounced "*lee*-ma-sohn."

the part of the graph corresponding to $0 \leq \theta \leq \pi$ (darkly colored in Figure 6.75) and then use symmetry to sketch the remainder of the graph (lightly colored in Figure 6.75). Does the appearance of the graph justify the name "limaçon," which means "snail"? ☐

Example 8 Sketch the three-leaved rose $r = \cos 3\theta$.

Solution Since $\cos(-3\theta) = \cos 3\theta$, the polar graph is symmetric with respect to the x axis. Therefore we will graph the equation $r = \cos 3\theta$ for $0 \leq \theta \leq \pi$ and then use symmetry to complete the graph. The chart below allows us to plot the special points appearing in Figure 6.76.

θ	0	$\pi/3$	$\pi/2$	$2\pi/3$	π
$r = \cos 3\theta$	1	-1	0	1	-1

Next we notice that as θ increases from 0 to $\pi/3$, r decreases from 1 to -1; since r is negative for $\pi/6 < \theta \leq \pi/3$, the corresponding part of the graph lies in the third quadrant instead of the first quadrant. Similarly, as θ increases from $\pi/3$ to $2\pi/3$, r increases from -1 to 1; since r is negative for $\pi/3 \leq \theta < \pi/2$, the corresponding part of the graph lies in the third quadrant. Finally, as θ increases from $2\pi/3$ to π, r decreases again from 1 to -1; since r is negative for $5\pi/6 < \theta \leq \pi$, the corresponding part of the graph lies in the fourth quadrant. We use this information to sketch the portion of the graph of $r = \cos 3\theta$ corresponding to $0 \leq \theta \leq \pi$, which appears in Figure 6.76. Because of symmetry with respect to the x axis, the remainder of the graph is obtained by reflection through the x axis, which duplicates the curve we already have. ☐

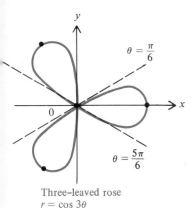

Three-leaved rose
$r = \cos 3\theta$

FIGURE 6.76

EXERCISES 6.9

1. Find the Cartesian coordinates of the following points, which are in polar coordinates.
   a. $(3, \pi/4)$
   b. $(-2, -\pi/6)$
   c. $(3, 7\pi/3)$
   d. $(5, 0)$
   e. $(-2, \pi/2)$
   f. $(-2, 3\pi/2)$
   g. $(4, 3\pi/4)$
   h. $(0, 6\pi/7)$
   i. $(-1, 23\pi/3)$
   j. $(-1, -23\pi/3)$
   k. $(1, 3\pi/2)$
   l. $(3, -5\pi/6)$

2. Find all sets of polar coordinates for each of the following points, which are in Cartesian coordinates.
   a. $(3, 3)$
   b. $(4, -4)$
   c. $(0, 5)$
   d. $(-4, 0)$
   e. $(3, 3\sqrt{3})$
   f. $(-\frac{1}{3}, \sqrt{3}/3)$
   g. $(-3, \sqrt{3})$
   h. $(-2\sqrt{3}, 2)$
   i. $(0, 0)$
   j. $(-5\sqrt{3}, -5)$

In Exercises 3–11 write the equation in polar coordinates. Express the answer in the form $r = f(\theta)$ wherever possible.

3. $2x + 3y = 4$
4. $y^2 = 4x$
5. $x^2 + 9y^2 = 1$
6. $9x^2 + y^2 = 4y$
7. $(x^2 + y^2)^2 = x^2 - y^2$
8. $x^2 + y^2 = 2x(3y^2 - x^2)$
9. $x^2 + y^2 = x(x^2 - 3y^2)$
10. $y^2 = \dfrac{x^3}{2 - x}$
11. $y^2 = \dfrac{x^2(3 - x)}{1 + x}$

In Exercises 12–18 write the polar equation as an equation in Cartesian coordinates.

12. $r = 5$
13. $r = 3\cos\theta$
14. $\tan\theta = 6$
15. $\cot\theta = 3$

16. $r \cot \theta = 3$

17. $r = \sin 2\theta$

18. $r = 2 \sin \theta \tan \theta$

19. Show that the polar graph of the equation $r = 1 + \cos \theta$ is the same as the polar graph of $r^2 = r + r \cos \theta$.

20. Show that the polar graph of the equation $r = 2 + \cos \theta$ is not the same as the polar graph of $r^2 = 2r + r \cos \theta$.

In Exercises 21–39 sketch the polar graph of the equation. Each graph has a familiar form. It may be convenient to convert the equation to rectangular coordinates.

21. $r = 5$

22. $r = -2$

23. $r = 0$

24. $\theta = 3\pi/2$

25. $\theta = -7\pi/6$

26. $|\theta| = \pi/3$

27. $r \sin \theta = 5$

28. $r = \sin \theta$

29. $r = -\frac{3}{2} \cos \theta$

30. $r \sin(\theta - \pi/2) = 3$

31. $r \cos(\theta - \pi/3) = 2$

32. $r \cos(\theta + \pi/4) = -2$

33. $r = 2 \cot \theta \csc \theta$

34. $r = -3 \tan \theta \sec \theta$

35. $r(\sin \theta + \cos \theta) = 1$

36. $r^2(4 \cos^2 \theta + \sin^2 \theta) = 4$

37. $r^2(\cos^2 \theta - 9 \sin^2 \theta) = 9$

38. $r = \dfrac{2}{3 \cos \theta - 2 \sin \theta}$

39. $r = \dfrac{-1}{\cos \theta + 4 \sin \theta}$

In Exercises 40–58 sketch the polar graph of the given equation. Note any symmetries.

40. $r = 3 \sin 2\theta$ (four-leaved rose)

41. $r = 2 \cos 2\theta$ (four-leaved rose)

42. $r = -4 \cos 3\theta$ (three-leaved rose)

43. $r = -4 \sin 3\theta$ (three-leaved rose)

44. $r = -\sin 4\theta$ (eight-leaved rose)

45. $r = 2 \cos 6\theta$ (twelve-leaved rose)

46. $r = \cos \dfrac{\theta}{2}$ (*Hint:* Not all symmetry can be deduced from Table 6.1.)

47. $r = \sin \dfrac{\theta}{2}$ (*Hint:* Not all symmetry can be deduced from Table 6.1.)

48. $r^2 = \sin \theta$

49. $r^2 = 25 \cos \theta$

50. $r^2 = 9 \sin 2\theta$ (lemniscate of Bernoulli)

51. $r^2 = 4 \cos 2\theta$ (lemniscate of Bernoulli)

52. $r = 1 + 2 \sin \theta$ (limaçon of Pascal)

53. $r = 2 - \cos \theta$ (limaçon of Pascal)

54. $r = 3(1 - \sin \theta)$ (cardioid)

55. $r = 3 \tan \theta$ (kappa curve)
    (*Hint:* Find $\lim_{\theta \to \pi/2^-} r \cos \theta$.)

56. $r\theta = 2$ (hyperbolic spiral)

57. $r = 2\theta$ (spiral of Archimedes)

58. $r = \sin \theta + \cos \theta$ (*Hint:* Transform the equation into Cartesian coordinates first.)

59. Show that each of the following pairs of equations has the same polar graph.
    a. $r = 3(\cos \theta + 1)$ and $r = 3(\cos \theta - 1)$ (*Hint:* Show that if (r, θ) satisfies one equation, then $(-r, \theta + \pi)$ satisfies the other.)
    b. $r = 2(\sin \theta + 1)$ and $r = 2(\sin \theta - 1)$
    c. $r = \theta$ and $r = \theta - 2\pi$

60. Show that the polar graphs of the equations $r = \frac{1}{2}(1 + \cos \theta)$ and $r^2 = -\cos \theta$ intersect at $(1, 0)$, even though no single set of polar coordinates for $(1, 0)$ satisfies both equations.

*61. Find a polar equation of the collection of points the product of whose distances from the points $(1, 0)$ and $(-1, 0)$ is 1.

6.10
LENGTH AND AREA IN POLAR COORDINATES

As we saw in Section 6.9, many curves such as circles and cardioids are easier to describe in polar coordinates than in rectangular coordinates. In most cases it is easier to use polar coordinates in order to compute the lengths of such curves and the areas of corresponding regions.

Length in Polar Coordinates   Consider a nonnegative function f defined on $[\alpha, \beta]$, with $0 \le \beta - \alpha \le 2\pi$. The polar graph of f is the set of points (x, y) with polar coordinates (r, θ) satisfying $r = f(\theta)$ and $\alpha \le \theta \le \beta$ (Figure 6.77). Thus by (2) of Section 6.9 the

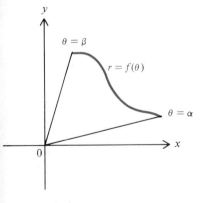

FIGURE 6.77

polar graph has the parametric representation

$$x = f(\theta)\cos\theta \quad \text{and} \quad y = f(\theta)\sin\theta \quad \text{for } \alpha \le \theta \le \beta \tag{1}$$

If f' is continuous on $[\alpha, \beta]$, then we may apply (5) of Section 6.4 to the parametric equations in (1) to find the length $\mathscr{L}$ of C:

$$\mathscr{L} = \int_\alpha^\beta \sqrt{\left(\frac{dx}{d\theta}\right)^2 + \left(\frac{dy}{d\theta}\right)^2}\, d\theta$$

$$= \int_\alpha^\beta \sqrt{[f'(\theta)\cos\theta - f(\theta)\sin\theta]^2 + [f'(\theta)\sin\theta + f(\theta)\cos\theta]^2}\, d\theta$$

$$= \int_\alpha^\beta \sqrt{(f'(\theta))^2(\cos^2\theta + \sin^2\theta) + (f(\theta))^2(\sin^2\theta + \cos^2\theta)}\, d\theta$$

$$= \int_\alpha^\beta \sqrt{(f'(\theta))^2 + (f(\theta))^2}\, d\theta$$

Therefore the length $\mathscr{L}$ of C is given by

$$\mathscr{L} = \int_\alpha^\beta \sqrt{(f'(\theta))^2 + (f(\theta))^2}\, d\theta \tag{2}$$

Example 1   Use (2) to determine the circumference of the circle $r = 5$.

Solution   If $r = 5$, then $f(\theta) = 5$ for $0 \le \theta \le 2\pi$. Then (2) becomes

$$\mathscr{L} = \int_0^{2\pi} \sqrt{0^2 + 5^2}\, d\theta = \int_0^{2\pi} 5\, d\theta = 10\pi \quad \square$$

Example 2   Let C be the cardioid $r = 1 + \cos\theta$ for $0 \le \theta \le 2\pi$. Find the length $\mathscr{L}$ of C.

Solution   Because of the symmetry of C with respect to the x axis, the length $\mathscr{L}$ of C is twice the length of the top half of C. Thus we let $f(\theta) = 1 + \cos\theta$ for $0 \le \theta \le \pi$. It follows from (2) that

$$\mathscr{L} = 2\int_0^\pi \sqrt{(-\sin\theta)^2 + (1 + \cos\theta)^2}\, d\theta = 2\int_0^\pi \sqrt{2 + 2\cos\theta}\, d\theta$$

since C is symmetric with respect to the x axis. By the half-angle formula for $\cos\theta/2$,

$$\frac{1 + \cos\theta}{2} = \cos^2\frac{\theta}{2}$$

Since $\cos\theta/2 \ge 0$ for $0 \le \theta \le \pi$, we find that

$$\sqrt{2 + 2\cos\theta} = \sqrt{4\cos^2\frac{\theta}{2}} = 2\cos\frac{\theta}{2}$$

We conclude that

$$\mathcal{L} = 2 \int_0^\pi \sqrt{2 + 2\cos\theta}\, d\theta = 4 \int_0^\pi \cos\frac{\theta}{2}\, d\theta = 8 \sin\frac{\theta}{2}\Big|_0^\pi = 8 \quad \square$$

Area in Polar Coordinates

When we prepared to find the area of a region in rectangular coordinates (in Chapter 5), we divided the region into vertical strips, approximated the area of each such strip by the area of a rectangle, and added the estimates (Figure 6.78). To find the area of a region in polar coordinates (Figure 6.79(a)) we will divide the region into "pie-shaped" sectors (Figure 6.79(b)), approximate the areas of the sectors, and find their sum.

We begin by determining the area A of a sector S with angle ϕ and radius r (Figure 6.79(c)). Since the area A is $\phi/(2\pi)$ times the area (πr^2) of the circle, A is given by

$$A = \frac{\phi}{2\pi}(\pi r^2) = \frac{1}{2}r^2\phi \tag{3}$$

Now consider a continuous, nonnegative function f defined on $[\alpha, \beta]$, with $0 \le \beta - \alpha \le 2\pi$, and let R be the region consisting of all points in the plane whose polar coordinates satisfy

$$0 \le r \le f(\theta) \quad \text{and} \quad \alpha \le \theta \le \beta$$

Our objective is to define the area A of R. Let $\mathscr{P} = \{\theta_0, \theta_1, \ldots, \theta_n\}$ be any partition of $[\alpha, \beta]$, and for each k between 1 and n, let $\Delta\theta_k = \theta_k - \theta_{k-1}$. If t_k is an arbitrary number in the interval $[\theta_{k-1}, \theta_k]$, and if $\Delta\theta_k$ is small, then the area ΔA_k of the region R_k between the lines $\theta = \theta_{k-1}$ and $\theta = \theta_k$ is approximately equal to the area of a circular sector of angle $\Delta\theta_k$ and radius $f(t_k)$ (Figure 6.80). Thus by (3) with r replaced by $f(t_k)$ and ϕ replaced by $\Delta\theta_k$, ΔA_k is approximately $\frac{1}{2}[f(t_k)]^2 \Delta\theta_k$. Since the area A of R is the sum of the areas $\Delta A_1, \Delta A_2, \ldots, \Delta A_n$, it follows that A should be approximately equal to

$$\sum_{k=1}^n \frac{1}{2}[f(t_k)]^2 \Delta\theta_k$$

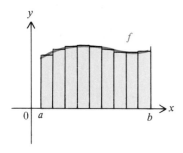

FIGURE 6.78

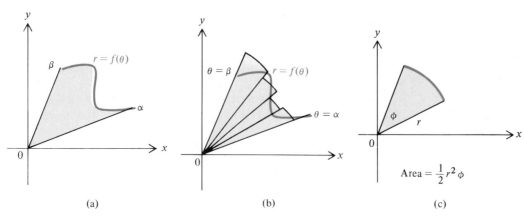

(a)

(b)

(c)

FIGURE 6.79

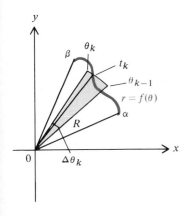

FIGURE 6.80

and this is a Riemann sum for $\frac{1}{2}f^2$ on $[\alpha, \beta]$. Therefore the area A should be given by the formula

$$A = \lim_{\|\mathscr{P}\| \to 0} \sum_{k=1}^{n} \frac{1}{2}[f(t_k)]^2 \, \Delta\theta_k = \int_{\alpha}^{\beta} \frac{1}{2}[f(\theta)]^2 \, d\theta$$

Thus the area A of the region consisting of all points in the plane having polar coordinates (r, θ) that satisfy $0 \le r \le f(\theta)$ and $\alpha \le \theta \le \beta$ is given by

$$A = \int_{\alpha}^{\beta} \frac{1}{2}[f(\theta)]^2 \, d\theta \qquad (4)$$

The areas of some regions have been given both by a formula in rectangular coordinates ((2) of Section 5.8) and by a formula in polar coordinates ((4) above). In such cases the same area results from applying either formula (although we will not prove that here).

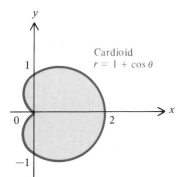

FIGURE 6.81

Example 3 Let R be the region enclosed by the cardioid $r = 1 + \cos\theta$ (Figure 6.81). Find the area A of R.

Solution Let

$$f(\theta) = 1 + \cos\theta$$

If we let $\alpha = 0$ and $\beta = 2\pi$, then by (4), along with (2) of Section 5.6, we have

$$A = \int_{0}^{2\pi} \frac{1}{2}[f(\theta)]^2 \, d\theta = \frac{1}{2}\int_{0}^{2\pi} (1 + \cos\theta)^2 \, d\theta$$

$$= \frac{1}{2}\int_{0}^{2\pi} (1 + 2\cos\theta + \cos^2\theta) \, d\theta$$

$$= \frac{1}{2}\left(\theta + 2\sin\theta + \frac{\theta}{2} + \frac{1}{4}\sin 2\theta\right)\Bigg|_{0}^{2\pi}$$

$$= \frac{3\pi}{2} \qquad \square$$

Finding the area of the cardioid in rectangular coordinates would be a formidable task, for it would involve finding the area of the region enclosed by the graph of $x^2 + y^2 = \sqrt{x^2 + y^2} + x$.

Let f and g defined on $[\alpha, \beta]$, where $0 \le \beta - \alpha \le 2\pi$, and suppose

$$0 \le g(\theta) \le f(\theta) \quad \text{for } \alpha \le \theta \le \beta$$

We assume that the area of the region determined by the graph of f between the lines $\theta = \alpha$ and $\theta = \beta$ is the sum of the areas of the darkly shaded region

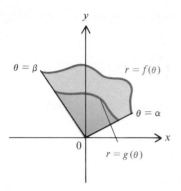

FIGURE 6.82

determined by the graph of g and the lightly shaded region between the graphs of f and g (Figure 6.82). Then it follows from (4) that the area A of the region consisting of all points in the plane having polar coordinates (r, θ) that satisfy

$$g(\theta) \le r \le f(\theta) \quad \text{and} \quad \alpha \le \theta \le \beta$$

is given by

$$A = \int_\alpha^\beta \frac{1}{2}\{[f(\theta)]^2 - [g(\theta)]^2\}\, d\theta \tag{5}$$

If $g = 0$, then the area described in (5) is just that appearing in (4). But g need not be 0, as the next example shows.

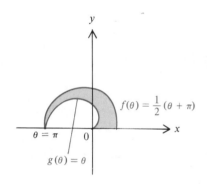

FIGURE 6.83

Example 4   Find the area A of the region shaded in Figure 6.83.

Solution   If we let

$$f(\theta) = \frac{1}{2}(\theta + \pi) \quad \text{and} \quad g(\theta) = \theta$$

then by (5) we know that

$$A = \int_0^\pi \frac{1}{2}\{[f(\theta)]^2 - [g(\theta)]^2\}\, d\theta = \frac{1}{2}\int_0^\pi \left[\frac{1}{4}(\theta + \pi)^2 - \theta^2\right] d\theta$$

$$= \frac{1}{2}\left[\frac{1}{12}(\theta + \pi)^3 - \frac{1}{3}\theta^3\right]\Bigg|_0^\pi = \frac{1}{2}\left[\left(\frac{1}{12}(8\pi^3) - \frac{1}{3}\pi^3\right) - \frac{1}{12}\pi^3\right] = \frac{\pi^3}{8} \quad \square$$

In our final example we must determine the limits of integration before we can set up the integral.

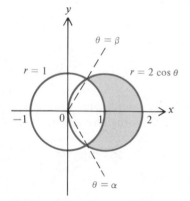

FIGURE 6.84

Example 5   Find the area A of the region inside the circle $r = 2\cos\theta$ and outside the circle $r = 1$ (Figure 6.84).

Solution   The region in question lies between the lines $\theta = \alpha$ and $\theta = \beta$ that pass through the points of intersection of the two circles (Figure 6.84). To find the points of intersection we solve the equation

$$2\cos\theta = 1, \quad \text{or} \quad \cos\theta = \frac{1}{2}$$

We find that $\theta = -\pi/3$ or $\theta = \pi/3$, so that $\alpha = -\pi/3$ and $\beta = \pi/3$. Taking

$$f(\theta) = 2\cos\theta \quad \text{and} \quad g(\theta) = 1$$

in (5) and again using (2) of Section 5.6, we obtain

$$A = \int_{-\pi/3}^{\pi/3} \frac{1}{2}(4\cos^2\theta - 1)\,d\theta = \int_{-\pi/3}^{\pi/3} 2\cos^2\theta\,d\theta - \int_{-\pi/3}^{\pi/3} \frac{1}{2}\,d\theta$$

$$= 2\left(\frac{1}{2}\theta + \frac{1}{4}\sin 2\theta\right)\Bigg|_{-\pi/3}^{\pi/3} - \frac{1}{2}\theta\Bigg|_{-\pi/3}^{\pi/3}$$

$$= \frac{\pi}{3} + \frac{\sqrt{3}}{2} \quad \square$$

EXERCISES 6.10

In Exercises 1–6 find the length $\mathscr{L}$ of the graph of the given equation.

1. $r = 2\cos\theta$

2. $r = \theta^2$ for $0 \le \theta \le 4\sqrt{2}$

3. $r = 2 - 2\cos\theta$

4. $r = \sin^2\dfrac{\theta}{2}$ for $0 \le \theta \le \pi$

5. $r = \sin^2\dfrac{\theta}{2}$ for $-\dfrac{\pi}{2} \le \theta \le \dfrac{3\pi}{2}$

6. $r = \sin^3\dfrac{\theta}{3}$ for $0 \le \theta \le 2\pi$

In Exercises 7–22 find the area A of the region bounded by the graphs of the given equations.

7. $r = 4$

8. $r = a$, where $a > 0$

9. $r = 3\sin\theta$

10. $r = 3\sin\theta$ for $0 \le \theta \le \pi/3$, and the line $\theta = \pi/3$

11. $r = -2\cos\theta$

12. $r = 9\sin 2\theta$ (four-leaved rose)

13. $r = 9\cos 2\theta$ for $\pi/4 \le \theta \le \pi/2$, and the line $\theta = \pi/2$

14. $r = -4\sin 3\theta$ (three-leaved rose)

15. $r = \frac{1}{2}\cos 3\theta$ (three-leaved rose)

16. $r = 6\sin 4\theta$ (eight-leaved rose)

17. $r = 2(1 - \sin\theta)$ (cardioid)

18. $r = 2 + 2\cos\theta$ (cardioid)

19. $r = 4 + 3\cos\theta$ (limaçon)

20. $r^2 = 9\sin 2\theta$ (lemniscate)

21. $r^2 = 25\cos\theta$

22. $r^2 = -\cos\theta$

In Exercises 23–29 find the area A of the region inside the first curve and outside the second curve.

23. $r = 5$ and $r = 1$

24. $r = 5$ and $r = 2(1 + \cos\theta)$

25. $r = 1$ and $r = \sin\theta$

26. $r = 1$ and $r = \cos 2\theta$

27. $r = 1$ and $r^2 = \cos 2\theta$

28. $r = 5(1 + \cos\theta)$ and $r = 2\cos\theta$

29. $r = 2 + \cos\theta$ and $r = -\cos\theta$

In Exercises 30–34 find the area A of the indicated region.

30. The region common to the two circles $r = \cos\theta$ and $r = \sin\theta$.

31. The region common to the circle $r = \cos\theta$ and the cardioid $r = 1 - \cos\theta$.

32. The region inside the circle $r = \cos\theta$ and outside the cardioid $r = 1 - \cos\theta$.

33. The region outside the cardioid $r = 1 + \cos\theta$ and inside the cardioid $r = 1 + \sin\theta$.

* 34. The region outside the small loop and inside the large loop of $r = 1 + 2\cos\theta$.

35. The graph of

$$y^2 = \frac{x^2(1 + x)}{1 - x}$$

is called a **strophoid** (Figure 6.85).
   a. Show that $r = \sec\theta - 2\cos\theta$ for $-\pi/2 < \theta < \pi/2$ is a polar equation of this strophoid.
   b. Find the area A enclosed by the loop of the strophoid.

36. Suppose C is the graph of the equation $r = f(\theta)$ for $\alpha \le \theta \le \beta$, where $0 \le \alpha \le \beta \le \pi$. Assume that f' is continuous on $[\alpha, \beta]$. Let S be the surface area of the surface

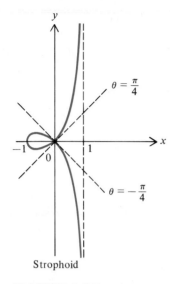

Strophoid

FIGURE 6.85

obtained by revolving C about the x axis. Show that

$$S = \int_{\alpha}^{\beta} 2\pi f(\theta) \sin\theta \sqrt{(f'(\theta))^2 + (f(\theta))^2}\, d\theta \qquad (6)$$

(*Hint:* Use (5) in Section 6.5 and (2) in Section 6.9.)

In Exercises 37–38 use (6) to determine the surface area S of the surface obtained by revolving the graph of the given equation about the x axis.

37. $r = 1 + \cos\theta$ for $0 \leq \theta \leq \pi$ (half a cardioid)

38. $r = \sqrt{2\cos 2\theta}$ for $0 \leq \theta \leq \dfrac{\pi}{4}$ (one-fourth of a lemniscate)

FORMULAS FOR APPLICATIONS OF THE INTEGRAL

Type	Formula
Cross-sectional volume	$V = \int_a^b A(x)\,dx$
Volume of revolution about x axis (disc method)	$V = \begin{cases} \int_a^b \pi[f(x)]^2\,dx \\ \int_a^b \pi\{[f(x)]^2 - [g(x)]^2\}\,dx \end{cases}$
Volume of revolution about y axis (shell method)	$V = \begin{cases} \int_a^b 2\pi x f(x)\,dx \\ \int_a^b 2\pi x[f(x) - g(x)]\,dx \end{cases}$
Length of a curve	$\begin{cases} \mathscr{L} = \int_a^b \sqrt{1 + [f'(x)]^2}\,dx \\ \mathscr{L} = \int_a^b \sqrt{[f'(t)]^2 + [g'(t)]^2}\,dt \end{cases}$
Surface area	$\begin{cases} S = \int_a^b 2\pi f(x)\sqrt{1 + [f'(x)]^2}\,dx \\ S = \int_a^b 2\pi g(t)\sqrt{[f'(t)]^2 + [g'(t)]^2}\,dt \end{cases}$
Work	$W = \int_a^b f(x)\,dx$
Work to pump water from a tank	$W = \int_a^b 62.5(l - x)A(x)\,dx$
Moment about x axis	$\mathscr{M}_x = \int_a^b \frac{1}{2}\{[f(x)]^2 - [g(x)]^2\}\,dx$
Moment about y axis	$\mathscr{M}_y = \int_a^b x[f(x) - g(x)]\,dx$
Center of gravity	$(\bar{x}, \bar{y})\!:\ \bar{x} = \dfrac{\mathscr{M}_y}{A}$ and $\bar{y} = \dfrac{\mathscr{M}_x}{A}$
Hydrostatic force	$F = \int_a^b 62.5(c - x)w(x)\,dx$
Length of curve in polar coordinates	$\mathscr{L} = \int_\alpha^\beta \sqrt{(f'(\theta))^2 + (f(\theta))^2}\,d\theta$
Area of region in polar coordinates	$A = \begin{cases} \int_\alpha^\beta \frac{1}{2}[f(\theta)]^2\,d\theta \\ \int_\alpha^\beta \frac{1}{2}\{[f(\theta)]^2 - [g(\theta)]^2\}\,d\theta \end{cases}$

REVIEW EXERCISES

In Exercises 1–2 find the volume V of the solid obtained by revolving about the x axis the region between the graphs of f and g on the given interval.

1. $f(x) = 1 + \sqrt{x}, g(x) = 1 - x; [0, 1]$

2. $f(x) = \sec x, g(x) = \cos x; [-\pi/4, 0]$

In Exercises 3–4 find the volume V of the solid obtained by revolving about the y axis the region between the graphs of f and g on the given interval.

3. $f(x) = \sin(\pi + x^2), g(x) = \sin x^2; [\sqrt{\pi}, \sqrt{2\pi}]$

4. $f(x) = \sqrt{1 + x}, g(x) = (\ln x)/x^2; [1, 3]$

5. Let $f(x) = x^3$, and let R be the region between the graph of f and the x axis on $[0, 2]$. Then R is called a **cubical spandrel**.
   a. Using the shell method, find the volume V of the solid obtained by revolving R about the y axis.
   b. Find the center of gravity of R.
   c. Rework part (a) using the Theorem of Pappus and Guldin.

6. The graph of $f(x) = x^2$ divides the rectangle bounded by the lines $x = 0, x = 2, y = 0$, and $y = 4$ into a lower region A and an upper region B. Compare the volumes of the solids obtained by revolving A and B
   a. about the x axis
   b. about the y axis

7. Suppose that c is a fixed positive number. Let $f(x) = x + c/x$ and $g(x) = x$, and let R be the region between the graphs of f and g on $[1, 3]$.
   a. Find the volume V_1 of the solid obtained by revolving R about the y axis.
   b. Find the volume V_2 of the solid obtained by revolving R about the x axis.
   c. Use the results of (a) and (b) to show that there is no positive value of c for which $V_1 = V_2$.

8. One way of constructing a ring is to remove a cylinder from the center of a sphere. If the resulting ring is $\frac{1}{4}$ inch wide, show that the volume of the ring is $\pi/384$ cubic inches, regardless of the radius r of the sphere from which the ring is made (which could even be as big as the sun). (*Hint:* Revolve the region R depicted in Figure 6.86 about the x axis, and compute the volume V of the solid generated.)

9. The horizontal cross sections of a cone 10 feet high are regular hexagons. The hexagon at the base has sides 2 feet long. Find the volume V of the cone.

10. A wedge is cut out of a circular cylinder of cheese with a radius a by first making a cut halfway through the cheese

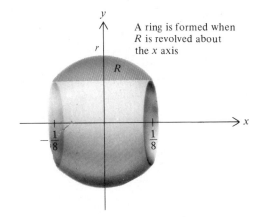

A ring is formed when R is revolved about the x axis

FIGURE 6.86

FIGURE 6.87

perpendicular to the axis of the cylinder and then making another cut halfway through the cheese at an angle θ with respect to the first cut (Figure 6.87). Find the volume V of cheese removed. (*Hint:* The cross sections of the wedge perpendicular to the first cut are rectangles.)

In Exercises 11–12 find the length $\mathscr{L}$ of the graph of f.

11. $f(x) = \ln(\sin x)$ for $\pi/6 \le x \le 5\pi/6$

12. $f(x) = \dfrac{1}{60}x^5 + \dfrac{1}{x^3}$ for $1 \le x \le 2$

In Exercises 13–14 sketch the graph of the curve represented parametrically.

13. $x = 2t + 1$ and $y = 4 - 6t$

14. $x = 3\cos t$ and $y = 3\sin t$ for $-\pi/2 \le t \le \pi/4$

In Exercises 15–16 find the length $\mathcal{L}$ of the curve given parametrically.

15. $x = t^2$ and $y = \dfrac{1}{6}(4t + 1)^{3/2}$ for $1 \le t \le 4$

16. $x = \dfrac{1}{3}\cos^3 t$ and $y = \dfrac{1}{2}\sin^2 t$ for $0 \le t \le \pi/2$

17. Using the right sum with the partition $\mathcal{P} = \{-\pi/3, -\pi/6, 0, \pi/6 \text{ and } \pi/3\}$, estimate the length $\mathcal{L}$ of the graph of $y = \tan x$ for $-\pi/3 \le x \le \pi/3$.

In Exercises 18–19 find the surface area S of the surface obtained by revolving the graph of f about the x axis.

18. $f(x) = \sqrt{2x - 1}$ for $2 \le x \le 8$

19. $f(x) = \dfrac{2}{5}x^{5/2} + \dfrac{1}{2}x^{-1/2}$ for $1 \le x \le 2$

In Exercises 20–21 find the surface area S of the surface obtained by revolving about the x axis the curve with the given parametric representation.

20. $x = \pi + \dfrac{1}{3}t^3$ and $y = \dfrac{1}{2}t^2$ for $1 \le t \le \sqrt{3}$

21. $x = \dfrac{1}{2}\sin^2 t$ and $y = \cos t$ for $0 \le t \le \pi/2$

22. A spherical gasoline tank 40 feet in diameter contains gasoline to a depth of 5 feet in the middle. How many cubic feet of gasoline are in the tank?

23. Suppose 4 foot-pounds of work are required in order to stretch a spring from a natural length of $\frac{1}{2}$ foot to a length of 1 foot. Find the work W necessary to stretch the spring from a length of $\frac{3}{2}$ feet to a length of 2 feet.

24. A cylindrical well 20 feet deep and 3 feet in radius is dug. Assuming that the soil weighs 150 pounds per cubic foot, calculate the work W required to raise the soil to ground level.

25. An underground tank containing 144π cubic feet of water has the shape of a right circular cone with its vertex at the bottom. If the radius of the cone is 10 feet at the top and the height is 20 feet, find the work W required to pump all the water to a point 16 feet above the top of the tank.

26. A chamber in the shape of a right circular cylinder contains a gas, which can be compressed or expanded by a piston (Figure 6.88). If the chamber is maintained at a constant temperature, then according to Boyle's Law the pressure p and the volume V of the gas are related by

$$pV = c$$

where c is a constant. The force F on the piston is the product of its area A and the pressure in the gas (that is, $F = pA$).
a. Show that the force F exerted on the piston by the gas

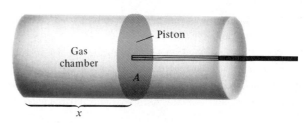

FIGURE 6.88

is c/x when the piston is x units from the left end of the cylinder.
b. Compute the work W done on the piston by the gas when the piston moves from $x = 1$ to $x = 2$.

27. Let $f(x) = 1 + 2x - x^2$ and $g(x) = x^2 - 2x + 1$, and let R be the region between the graphs of f and g on $[0, 2]$. Find the center of gravity of R.

28. Find the center of gravity of the region bounded by the graphs of f and g, where $f(x) = x$ and $g(x) = x^2$.

29. Find the center of gravity of the region bounded by the graphs of the equations $y^2 = 3x$ and $y = x^2 - 2x$. (*Hint:* The curves meet for $x = 0$ and $x = 3$.)

30. Find the center of gravity of the region above the x axis that is bounded by the x axis and the semicircles $r = 3$ and $r = 5$ for $0 \le \theta \le \pi$.

31. Let R be the region between the graphs of $y = (x - 1)^2$ and $y - 2 = -(x - 1)^2$. Using the Theorem of Pappus and Guldin, show that the volume of the solid obtained by revolving R about the x axis equals the volume of the solid obtained by revolving R about the y axis.

32. A vertical dam 100 feet tall has the shape of a trapezoid 200 feet wide at the top and 100 feet wide at the bottom. Find the hydrostatic force F on the dam
a. if the water is 50 feet deep.
b. if the water is 100 feet deep.

33. Show that the hydrostatic force on a vertical surface completely submerged in water is equal to the product of the pressure at the center of gravity of the surface and the area of the surface.

In Exercises 34–37 sketch the polar graph of the equation.

34. $r = \sin 5\theta$ 35. $r = 2\cos\theta - 2$

36. $r = \sqrt{3} - 2\sin\theta$ 37. $r^2 = \frac{1}{4}\cos 2\theta$

38. a. Sketch the polar graphs of $r = 2\sin 2\theta$ and $r = 2\sin\theta$.
b. Find all points of intersection of the graphs found in part (a).

39. Find the length $\mathcal{L}$ of the polar graph of the equation $r = 1 + \sin\theta$ for $0 \le \theta \le \pi$. (*Hint:* In the integral, use the fact that $\sin\theta = \cos(\pi/2 - \theta)$.)

40. Find the area A of the region common to the circles $r = 2\cos\theta$ and $r = \sin\theta + \cos\theta$.

41. Find the area A of the region inside the lemniscate $r^2 = 2\sin 2\theta$ and outside the circle $r = 1$.

42. A water clock is a glass water container with a hole in the bottom through which water can trickle. Such a clock is

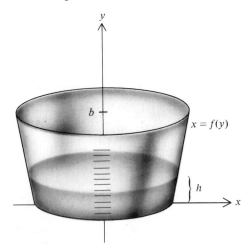

FIGURE 6.89

calibrated for measuring time by placing markings on the exterior of the container corresponding to water levels at equally spaced times (Figure 6.89).

a. Let f be a function of y that is continuous on an interval $[0, b]$, and suppose the container has the shape of the surface obtained by revolving the graph of f on $[0, b]$ about the y axis. If V denotes the volume of water and h the height of the water level above the bottom, then h depends on time t, and V depends on h and t. Show that

$$\frac{dV}{dt} = \pi[f(h)]^2\frac{dh}{dt}$$

b. Suppose A is the area of the hole in the bottom of the container. If the hole is very small, then it is known from physics that the rate dV/dt depends on the height h according to the equation

$$\frac{dV}{dt} = cA\sqrt{h}$$

where c is a (negative) constant. Find a formula for f if dh/dt is to be a constant k. (In such a case, the markings on the exterior would be equally spaced.)

Cumulative Review, Chapters 1–5

In Exercises 1–4 find the limit.

1. $\lim\limits_{x\to 3^+} \dfrac{\sqrt{x^2 - 9}}{x - 3}$

2. $\lim\limits_{x\to 0} \dfrac{\sec x - 1}{x}$

3. $\lim\limits_{x\to 0} \sin x \sin\dfrac{1}{x}$

4. $\lim\limits_{x\to\infty} \dfrac{2x^2 - \sin x}{4 - 3x^2}$

5. Let $f(x) = x\sqrt{\ln x}$. Find $f'(e^4)$.

6. Let $f(x) = \dfrac{1}{2}\ln\left|\dfrac{1 - \cos x}{1 + \cos x}\right|$. Show that $f'(x) = \csc x$.

7. Let $f(x) = (x - 1)\cos x$. Find a formula for $f^{(24)}(x)$.

8. Find an equation of the line tangent to the graph of the equation $xy^3 - x^2 = y^2 + xy + 5$ at the point $(3, 2)$.

9. A shadow in the shape of an equilateral triangle is growing 9 square inches per minute. At what rate does the height of the triangle grow when the area is $\sqrt{3}$ square inches?

In Exercises 10–11 sketch the graph of f, indicating all pertinent information.

10. $f(x) = -x^4 + 3x^2 - 2$

11. $f(x) = x^5 - \dfrac{10}{3}x^3 + 5x$

12. Suppose a shoreline has the shape of the parabola $y = \frac{1}{5}x^2$, where x and y are measured in miles, and that a fog light located at $(0, 2)$ revolves at the rate of $\frac{1}{2}$ radian per second. How fast does the x coordinate of the point of illumination on the shoreline change at the instant the point $(1, \frac{1}{5})$ is illuminated?

13. Suppose the velocity of an object is given by $v(t) = \frac{1}{4}t^2 + \sin t$ for $0 \le t \le \pi/2$. For what value of t in $[0, \pi/2]$ is the acceleration maximal?

14. A cylindrical tank with a volume of 40π cubic meters is to be built. Materials for the sides cost \$10 per square meter and for the top and bottom the materials cost \$25 per square meter. Find the dimensions that will minimize the cost.

15. Find the point(s) on the graph of $y = 2/(1 + x^2)$ that are closest to the origin.

16. Suppose the acceleration of an object is given by

$$a(t) = \frac{1}{t(\ln t + 2)} \quad \text{for } t \geq 1$$

If $v(1) = 5$, find $v(e)$.

17. Let $g(x) = \int_x^{x+\pi} \sin^{2/3} t \, dt$. Show that g is a constant function.

In Exercises 18–21 evaluate the integral.

18. $\displaystyle\int \frac{1}{\sqrt{2x - 3}} \, dx$

19. $\displaystyle\int_0^\pi \frac{\sin x}{2 + \cos x} \, dx$

20. $\displaystyle\int x(1 + \sqrt{x})^{1/3} \, dx$

21. $\displaystyle\int \frac{1}{t} (\ln t)^{5/3} \, dt$

22. a. Determine, if possible, a positive integer n such that $\int_1^n 1/x^{.9} \, dx > 1000$.

b. Determine, if possible, a positive integer n such that $\int_1^n 1/x^{1.1} \, dx > 1000$.

7
INVERSE FUNCTIONS

Any function f associates with each number x in its domain a unique number y in its range. Is it possible to reverse this procedure and associate x with the number y? Although it is not always possible to do so and still obtain a function, in some cases this can be done. The resulting function is called the inverse of f. By applying this procedure to certain familiar functions we obtain many new and useful functions.

In this chapter we begin by discussing general properties of inverses. Then we turn to inverses of the natural logarithm function and of trigonometric functions. The inverse of the natural logarithm function, which we call the exponential function, has many scientific applications, some of which we discuss in this chapter. Finally, we introduce a method for evaluating limits of quotients that is especially suited to the functions defined in this chapter.

7.1
INVERSE FUNCTIONS

Certain pairs of functions are "opposites" of one another. For example, the function f representing the conversion from degrees Celsius to degrees Fahrenheit is opposite to the function g representing the conversion from degrees Fahrenheit to degrees Celsius. Similarly, if f represents the conversion from inches to centimeters, then f is opposite to the function g representing the conversion from centimeters to inches.

In mathematical terms, how are the two functions f and g opposite to each other? The answer is that

$$f(x) = y \quad \text{if and only if} \quad g(y) = x$$

We use the term "inverse" to describe such a relationship.

DEFINITION 7.1

Let f be a function. Then f **has an inverse** provided that there is a function g such that the domain of g is the range of f and such that

$$f(x) = y \quad \text{if and only if} \quad g(y) = x \tag{1}$$

for all x in the domain of f and all y in the range of f.

For a particular function f, it might seem possible that many functions g could satisfy (1). However, any function g satisfying (1) must assign to each y in its domain every x such that $f(x) = y$. Therefore, since g is a function, there can be only one such x, that is, one x such that $f(x) = y$. Thus g is uniquely determined, and we make the following definition.

DEFINITION 7.2

Assume that the function f has an inverse, and let f^{-1} be the unique function having as its domain the range of f and satisfying

$$f(x) = y \quad \text{if and only if} \quad f^{-1}(y) = x \tag{2}$$

for all x in the domain of f and all y in the range of f. Then f^{-1} is the **inverse** of f.

Many functions have inverses. For the present let us show only that the Fahrenheit-to-Celsius function is the inverse of the Celsius-to-Fahrenheit function. To do this, let

$$f(x) = \frac{9}{5}x + 32 \quad \text{for } x \geq -273.15 \tag{3}$$

and

$$g(y) = \frac{5}{9}(y - 32) \quad \text{for } y \geq -459.67 \tag{4}$$

so that f represents the conversion from degrees Celsius to degrees Fahrenheit and g represents the conversion from degrees Fahrenheit to degrees Celsius. If $f(x) = y$, then by (3), $y = \frac{9}{5}x + 32$, and by (4) this implies that

$$g(y) = \frac{5}{9}\left[\left(\frac{9}{5}x + 32\right) - 32\right] = \frac{5}{9} \cdot \frac{9}{5}x = x$$

Similarly, if $g(y) = x$, then (4) tells us that $x = \frac{5}{9}(y - 32)$, and thus (3) implies that

$$f(x) = \frac{9}{5}\left[\frac{5}{9}(y - 32)\right] + 32 = y - 32 + 32 = y$$

Hence the functions f and g defined by (3) and (4) satisfy (1). Since f is linear and increasing and $f(-273.15) = -459.67$, the range of f is $[-459.67, \infty)$, which is the domain of g. Consequently g is the inverse of f. In the same way you can show that the functions representing the conversion from inches to centimeters and from centimeters to inches are inverses of each other (see Exercise 56).

> **Caution:** The number $f^{-1}(x)$ is almost always different from $[f(x)]^{-1} = 1/f(x)$. Thus the function f^{-1} is almost always different from the function $1/f$. For example, if $f(x) = \frac{9}{5}x + 32$, then
>
> $$f^{-1}(x) = \frac{5}{9}(x - 32) \quad \text{whereas} \quad \left(\frac{1}{f}\right)(x) = \frac{1}{\frac{9}{5}x + 32} = \frac{5}{9x + 160}$$
>
> The difference between f^{-1} and $1/f$ is similar to the difference between walking backwards and walking upside down on one's hands.

Properties of Inverses

First, observe from Definition 7.2 that the domain of f^{-1} is the range of f, whereas the range of f^{-1} is the domain of f. From this observation and from (2) we can derive three elementary relationships between f and f^{-1}, which we group together in a theorem.

THEOREM 7.3

> Let f have an inverse. Then f and f^{-1} have the following properties.
> **a.** $(f^{-1})^{-1} = f$
> **b.** $f^{-1}(f(x)) = x$ for all x in the domain of f
> **c.** $f(f^{-1}(y)) = y$ for all y in the range of f

Proof   To prove (a), recall first that the domain and range of f are the range and domain, respectively, of f^{-1}. Furthermore, (2) is equivalent to

$$f^{-1}(y) = x \quad \text{if and only if} \quad f(x) = y$$

for all y in the domain of f^{-1} and all x in the range of f^{-1}. Hence f^{-1} has an inverse, which is evidently f. To prove (b), simply substitute $f(x)$ for y in the equation $f^{-1}(y) = x$ in (2). Similarly, substitution of $f^{-1}(y)$ for x in the equation $f(x) = y$ in (2) establishes (c).   ∎

We would like to be able to tell whether a given function has an inverse. If f has an inverse, then for each y in the range of f there is a single x such that $f(x) = y$; equivalently, two distinct values of x in the domain of f are associated with two distinct values of y in the range of f. Conversely, if any two distinct values of x in the domain of f are assigned two distinct values in the range of f, then for any y in the range of f we can let $g(y) = x$ for the unique x such that $f(x) = y$, thereby defining a function g that satisfies (1). Thus g is the inverse of f. In summary, we have the following criterion for the existence of an inverse of a function.

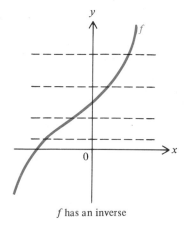

f has an inverse

FIGURE 7.1

A function f has an inverse if and only if for any two numbers x_1 and x_2 in the domain of f,

$$\text{if } x_1 \neq x_2, \quad \text{then} \quad f(x_1) \neq f(x_2) \tag{5}$$

Pictorially the condition in (5) means that a horizontal line can intersect the graph of f at most once (Figure 7.1). Thus if $f(x) = x^2$, then f does not have an inverse, because the horizontal line $y = 1$ intersects the graph of f twice (Figure 7.2).

Although (5) can frequently help in deciding that a function does not have an inverse, it is not so easy to use (5) to determine that a function has an inverse. For that purpose one usually applies the following theorem.

THEOREM 7.4

Every strictly increasing function and every strictly decreasing function has an inverse.

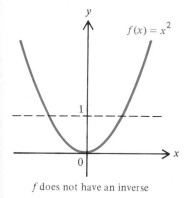

f does not have an inverse

FIGURE 7.2

Proof   Let f be strictly increasing, and suppose that x_1 and x_2 are in the domain of f with $x_1 \neq x_2$. Then either $x_1 < x_2$, in which case $f(x_1) < f(x_2)$ (because f is strictly increasing), or $x_2 < x_1$, in which case $f(x_2) < f(x_1)$ (again because f is strictly increasing). In either case, $f(x_1) \neq f(x_2)$. Therefore f satisfies the criterion in (5) and hence has an inverse. The proof for strictly decreasing functions is similar.   ■

Theorem 7.4 is especially easy to apply to differentiable functions whose domains are intervals. By Theorem 4.8, a function f is strictly increasing (and hence has an inverse) if $f'(x) > 0$ for all x in the interval, or if $f'(x) \geq 0$ for all x in the interval and $f'(x) = 0$ for at most finitely many values of x. Similar comments apply to strictly decreasing functions.

Thus we can conclude in each case below that f has an inverse.

$$f(x) = 3x - 2 \qquad\qquad f'(x) = 3 > 0 \tag{6}$$

$$f(x) = 1 - 2x^3 \qquad\qquad \begin{cases} f'(x) = -6x^2 \leq 0 \\ f'(x) = 0 \text{ only if } x = 0 \end{cases} \tag{7}$$

$$f(x) = x^7 + 8x^3 + 4x - 2 \qquad f'(x) = 7x^6 + 24x^2 + 4 > 0 \tag{8}$$

$$\begin{aligned} f(x) &= \sin x \\ &\text{for } -\pi/2 \leq x \leq \pi/2 \end{aligned} \qquad \begin{aligned} f'(x) &= \cos x > 0 \\ &\text{for } -\pi/2 < x < \pi/2 \end{aligned} \tag{9}$$

$$f(x) = \ln x \qquad\qquad f'(x) = \frac{1}{x} > 0 \quad \text{for } x > 0 \tag{10}$$

Formulas for Inverses   Often the clearest way to describe a function is to write its rule by one or more formulas. For functions such as those described in (6) and (7), there is a method that generates a formula for the inverse:

> 1. Write $y = f(x)$.
> 2. Solve for x in terms of y.
> 3. Write $f^{-1}(y)$ for x in step 2.

Let us see how the method works.

Example 1 Let $f(x) = 3x - 2$. Write a formula for the inverse of f.

Solution Following the steps listed, we obtain

$$y = 3x - 2$$

$$x = \frac{y + 2}{3}$$

Thus

$$f^{-1}(y) = \frac{y + 2}{3}$$

Since we customarily use x as the independent variable, we replace y by x to obtain

$$f^{-1}(x) = \frac{x + 2}{3} \quad \square$$

Example 2 Let $f(x) = 1 - 2x^3$. Write a formula for the inverse of f.

Solution Again following the steps listed, we find that

$$y = 1 - 2x^3$$

$$\frac{y - 1}{-2} = x^3$$

$$x = \sqrt[3]{\frac{1 - y}{2}}$$

Thus

$$f^{-1}(y) = \sqrt[3]{\frac{1 - y}{2}}$$

Converting to the variable x, we have

$$f^{-1}(x) = \sqrt[3]{\frac{1 - x}{2}} \quad \square$$

It is not always possible to find a simple formula for an inverse, as is illustrated by (8)–(10). In such a case we sometimes give the inverse a special name, as we will do in Section 7.6 for the restricted sine function appearing in (9) and in Section 7.3 for the natural logarithm function in (10). For other functions such as the one appearing in (8), the inverse will remain unnamed.

Graphs of Inverses

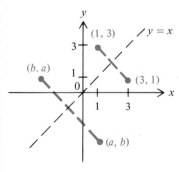

FIGURE 7.3

To find a method of graphing the inverse of f, we begin by noting that if $(1, 3)$ is on the graph of f, then $f(1) = 3$, so that $f^{-1}(3) = 1$. This means that $(3, 1)$ is on the graph of f^{-1}. In general (a, b) is on the graph of f if and only if (b, a) is on the graph of f^{-1}. But $(1, 3)$ and $(3, 1)$ are symmetric with respect to the line $y = x$, as are (a, b) and (b, a) (Figure 7.3). Thus the graph of f^{-1} is obtained by simply reflecting the graph of f through the line $y = x$.

Example 3 For each function f, sketch the graph of f and f^{-1} on the same coordinate system.
a. $f(x) = 3x - 2$ b. $f(x) = 1 - 2x^3$ c. $f(x) = \sin x$ for $-\pi/2 \le x \le \pi/2$

Solution In each case the graph of f^{-1} is obtained by reflecting the graph of f through the line $y = x$. The graphs appear in Figure 7.4(a)–(c).   □

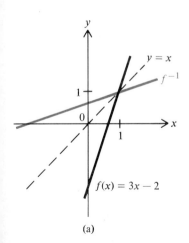

(a)

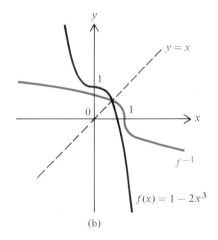

(b)

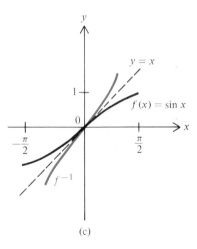

(c)

FIGURE 7.4

EXERCISES 7.1

In Exercises 1–19 determine whether the given function has an inverse. If an inverse exists, give the domain and range of the inverse and graph the function and its inverse.

1. $f(x) = x - 4$
2. $f(x) = 3x + 6$
3. $f(x) = x^5$
4. $f(x) = 5x^7 + 4x^3$
5. $f(x) = -x^8$
6. $f(x) = 4\sqrt[5]{x}$
7. $g(x) = \sqrt[6]{x}$
8. $g(t) = \sqrt{t^7}$
9. $f(t) = \sqrt{4 - t}$
10. $f(t) = \sqrt{1 - t^2}$
11. $f(x) = x + |x|$
12. $f(x) = x + \sin x$
13. $f(x) = x - \sin x$
14. $f(x) = x^2 + \sin x$
15. $f(z) = \tan z$
16. $f(z) = \tan z$ for $-\pi/2 < z < \pi/2$
17. $f(z) = \sec z$ for $0 < z < \pi/2$
18. $g(t) = \ln (3 - t)$
19. $k(x) = \ln x^2$

20. Let f be a periodic function; that is, suppose there is a positive number a such that

$$f(a + x) = f(x)$$

for all x in the domain of f. Show that f does not have an inverse.

In Exercises 21–24, use Exercise 20 to show that the given function does not have an inverse.

21. $f(x) = \sin^3 x$

22. $g(x) = \sin x - \cos x$

23. $f(x) = 2\tan 3x - 3\cos 4x + 17$

24. $g(x) = x - [x]$

In Exercises 25–32 find a formula for the inverse of each function.

25. $f(x) = -4x^3 - 1$

26. $f(x) = -2x^5 + \dfrac{9}{4}$

27. $g(x) = \sqrt{1 + x}$

28. $g(t) = \sqrt{3 - 2t}$

29. $f(x) = \dfrac{1 - 2x}{5x}$

30. $f(x) = \dfrac{x + 1}{x - 1}$

31. $k(t) = \dfrac{t - 1}{t + 1}$

32. $f(x) = \dfrac{3x + 5}{x - 4}$

In Exercises 33–44 find an interval on which f has an inverse. (*Hint:* Find an interval on which $f' > 0$ or on which $f' < 0$.)

33. $f(x) = x^2 - 4$

34. $f(x) = x^2 - 3x + 2$

35. $f(t) = t^4 + 2t^2 + 1$

36. $f(x) = x^3 + 5x + 1$

37. $f(x) = x^3 - 5x + 1$

38. $f(x) = \dfrac{x}{1 + x^2}$

39. $f(u) = \dfrac{1}{1 + u^2}$

40. $f(x) = \cos x$

41. $f(x) = \tan x$

42. $f(x) = \sin^2 x$

43. $f(x) = \csc x$

44. $f(x) = \sec x$

45. Let

$$f(x) = \int_0^x \sqrt{1 + t^4}\, dt \quad \text{for all } x$$

Show that f has an inverse.

46. Let

$$f(x) = \int_0^x \sin^4(t^2)\, dt \quad \text{for all } x$$

Show that f has an inverse.

47. Show that if $f = f^{-1}$, then the domain and range of f are the same.

48. Show that the following functions are their own inverses, and sketch their graphs.

a. $f(x) = x$ b. $f(x) = -x$

c. $f(x) = 1/x$ d. $f(x) = \sqrt{1 - x^2}$ for $0 \le x \le 1$

49. Assume that f has an inverse.

a. Suppose the graph of f lies in the first quadrant. In which quadrant does the graph of f^{-1} lie?

b. Suppose the graph of f lies in the second quadrant. In which quadrant does the graph of f^{-1} lie?

50. Show that if f and g both have inverses, then $g \circ f$ has an inverse and

$$(g \circ f)^{-1} = f^{-1} \circ g^{-1}$$

51. Using (10), along with Exercises 3, 31, and 50, show that the following functions have inverses.

a. $k(x) = \left(\dfrac{x - 1}{x + 1}\right)^5$ c. $k(x) = \dfrac{\ln x - 1}{\ln x + 1}$

b. $k(x) = \ln\left(\dfrac{x - 1}{x + 1}\right)$

52. Assume that g has an inverse, and let $f(x) = -x$. Using Exercise 50, show that

$$(g \circ f)^{-1}(x) = -g^{-1}(x)$$

for all x in the domain of g^{-1}.

53. Assume that f has an inverse, and let a be a fixed number. Let

$$g(x) = f(x + a)$$

for all x such that $x + a$ is in the domain of f. Show that g has an inverse and that $g^{-1}(x) = f^{-1}(x) - a$.

54. Assume that f has an inverse, and let a be a fixed number different from 0. Let

$$g(x) = f(ax)$$

for all x such that ax is in the domain of f. Show that g has an inverse and that $g^{-1}(x) = f^{-1}(x)/a$.

55. a. Can a polynomial of even degree have an inverse? Explain.

b. Can a polynomial of odd degree have an inverse? Explain.

56. Let f be the function representing the conversion from inches to centimeters, and let g be the function representing the conversion from centimeters to inches. Then

$$f(x) = 2.54x \quad \text{for } x \ge 0$$

and

$$g(x) = \dfrac{1}{2.54}x \quad \text{for } x \ge 0$$

Show that f and g are inverses of one another.

7.2
CONTINUITY AND DIFFERENTIABILITY OF INVERSE FUNCTIONS

As we observed in Section 7.1, the function $\ln x$ has an inverse. Although as yet we have no formula describing the inverse of $\ln x$, it is possible to use the continuity and differentiability of $\ln x$ to deduce the continuity and differentiability of the inverse. More generally, one can prove that the inverse of a continuous function is continuous and that the inverse of a differentiable function is differentiable.

Our first theorem concerns the continuity of inverses. Its proof appears in the Appendix.

THEOREM 7.5

Let f be continuous on an interval I, and let the values assigned by f to the points in I form the interval J. If f has an inverse, then f^{-1} is continuous on J.

Before giving the theorem on differentiability of inverses, we temporarily assume that the inverse f^{-1} of a function f is differentiable and analyze the derivative of f^{-1}. To begin with, we recall that the graph of f^{-1} can be obtained from the graph of f by reflecting the graph of f through the line $y = x$. As Figure 7.5 suggests, we would expect f^{-1} to increase (or decrease) rapidly at $f(a)$ if f increases (or decreases) slowly at a, and *vice versa*. This means that $|(f^{-1})'(f(a))|$ should be large if $|f'(a)|$ is small, and small if $|f'(a)|$ is large. Moreover, if $f'(a) = 0$, then it appears that $(f^{-1})'(f(a))$ does not exist, because the line tangent to the graph of f^{-1} is vertical. Now we are ready to state and prove the theorem.

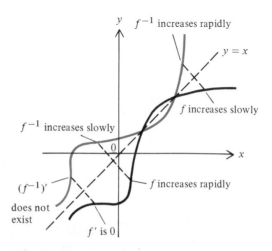

FIGURE 7.5

THEOREM 7.6

Suppose that f has an inverse and is continuous on an open interval I containing a. Assume also that $f'(a)$ exists, $f'(a) \neq 0$, and $f(a) = c$. Then $(f^{-1})'(c)$ exists, and

$$(f^{-1})'(c) = \frac{1}{f'(a)} \qquad (1)$$

Proof Using the fact that $f^{-1}(c) = a$ and the definition of the derivative, we find that

$$(f^{-1})'(c) = \lim_{y \to c} \frac{f^{-1}(y) - f^{-1}(c)}{y - c} = \lim_{y \to c} \frac{f^{-1}(y) - a}{f(f^{-1}(y)) - f(a)} \qquad (2)$$

provided that the latter limit exists. We will simultaneously show that it does exist and find its value. First notice that f^{-1} is continuous at c by Theorem 7.5. Therefore

$$\lim_{y \to c} f^{-1}(y) = f^{-1}(c) = a$$

so that if $x = f^{-1}(y)$, then x approaches a as y approaches c. Moreover, the fact that f^{-1} has an inverse and $f^{-1}(c) = a$ implies that $f^{-1}(y) \neq a$ for $y \neq c$. Consequently (2) and the Substitution Rule for Limits (with x substituting for $f^{-1}(y)$) imply that

$$(f^{-1})'(c) = \lim_{y \to c} \frac{f^{-1}(y) - a}{f(f^{-1}(y)) - f(a)} = \lim_{x \to a} \frac{x - a}{f(x) - f(a)}$$

$$= \frac{1}{\displaystyle\lim_{x \to a} \frac{f(x) - f(a)}{x - a}} = \frac{1}{f'(a)} \qquad \blacksquare$$

Caution: Notice from (1) that $(f^{-1})'(c)$ is obtained by evaluating the derivative of f at a, not at c.

Example 1 Let $f(x) = x^7 + 8x^3 + 4x - 2$. Find $(f^{-1})'(-2)$.

Solution In order to use (1), we must first find the value of a for which $f(a) = -2$. But $f(0) = -2$, so $a = 0$. Since $f'(x) = 7x^6 + 24x^2 + 4$, it follows that $f'(0) = 4$. Thus we conclude from (1) that

$$(f^{-1})'(-2) = \frac{1}{f'(0)} = \frac{1}{4} \qquad \square$$

Had we desired the number $(f^{-1})'(11)$ for the function f in Example 1, we would first have needed to determine the value of a for which $f(a) = 11$. By checking the values of f at a few numbers, we would have found that $f(1) = 11$. Then (1) would have yielded

$$(f^{-1})'(11) = \frac{1}{f'(1)} = \frac{1}{35}$$

Despite the fact that we have used (1) to find $(f^{-1})'(-2)$ and $(f^{-1})'(11)$ when $f(x) = x^7 + 8x^3 + 4x - 2$, it is not so easy to find the numerical value of $(f^{-1})'(c)$ in general. For instance, to find the value of $(f^{-1})'(0)$ we would first need to determine the value of a for which $f(a) = 0$, that is, for which

$$a^7 + 8a^3 + 4a - 2 = 0$$

That task turns out to be impossible to accomplish, and thus we find it impossible to obtain the numerical value of $(f^{-1})'(0)$. More generally, to apply (1) to find the numerical value of $(f^{-1})'(c)$, we must be able to find the value of a from the equation $f(a) = c$.

Formula (1) takes a very simple form in the Leibniz notation. Suppose $y = f(x)$ and f has an inverse, so that $x = f^{-1}(y)$. Then dy/dx is the derivative of f, whereas dx/dy is the derivative of f^{-1}. Using this notation, we may rewrite formula (1) in the easily remembered form

$$\frac{dx}{dy} = \frac{1}{\dfrac{dy}{dx}} \tag{3}$$

Example 2 Let $y = \ln x$. Find dx/dy.

Solution By (10) of Section 7.1 we know that the function y has an inverse. Since

$$\frac{d}{dx} \ln x = \frac{1}{x}$$

we conclude from (3) that

$$\frac{dx}{dy} = \frac{1}{\dfrac{dy}{dx}} = \frac{1}{\dfrac{1}{x}} = x \quad \square$$

Thus the derivative of the inverse of the natural logarithm function is the inverse itself. We will study this function in detail in the next section.

Example 3 Let $y = x^5 + 2x$. Find $\dfrac{dx}{dy}$ and $\dfrac{dx}{dy}\bigg|_{y=-3}$.

Solution   The function y has an inverse because its derivative, $5x^4 + 2$, is a positive function. Using (3), we deduce that

$$\frac{dx}{dy} = \frac{1}{\dfrac{dy}{dx}} = \frac{1}{5x^4 + 2} \tag{4}$$

Since $y = x^5 + 2x$, it follows that $y = -3$ for $x = -1$. We conclude from (4) that

$$\frac{dx}{dy}\bigg|_{y=-3} = \frac{1}{5(-1)^4 + 2} = \frac{1}{7} \quad \square$$

As the solution of Example 3 illustrates, when we wish to evaluate dx/dy at a given value c of y by using (3), we must evaluate the right side of (3) at the value a of x for which $y = c$.

EXERCISES 7.2

In Exercises 1–8 find $(f^{-1})'(c)$.

1. $f(x) = x^3 + 7; c = 6$
2. $f(x) = 5x^5 + 4x^3; c = 9$
3. $f(x) = x + \sin x; c = 0$
4. $f(x) = x + \sqrt{x}; c = 2$
5. $f(x) = 4 \ln x; c = 0$
6. $f(x) = \tan x$ for $-\pi/2 < x < \pi/2; c = \sqrt{3}$
7. $f(t) = 3t - (1/t^3)$ for $t < 0; c = -2$
*8. $f(t) = t \ln t; c = 2e^2$

In Exercises 9–14 find dx/dy.

9. $y = x^9 + 7x$
10. $y = x - 2/x$ for $x < 0$
11. $y = \ln(x^3 + 1)$
12. $y = x + \cos x$
13. $y = \sin x$ for $-\pi/2 < x < \pi/2$
14. $y = \tan x$ for $-\pi/2 < x < \pi/2$
15. Suppose $f(x) = \int_0^x \sqrt{1 + t^4}\, dt$ for all x, and let $c = f(1)$. Find $(f^{-1})'(c)$. (*Hint:* Do not attempt to evaluate $f(1)$.)
16. Let $g(x) = \int_\pi^x \cos t^3\, dt$ for all x, and let $g(\pi^{1/3}) = c$. Find $(g^{-1})'(c)$.
17. Let $y = x^3$. Then

$$\frac{dy}{dx} = 3x^2 \quad \text{and} \quad \frac{dx}{dy} = \frac{1}{3y^{2/3}}$$

Show that equation (3) holds for this function if we

evaluate dy/dx at $x = 2$ and dx/dy at $y = 8$ but that it does not hold if we evaluate dy/dx at $x = 2$ and dx/dy at $y = 2$.

18. Let f be a function with an inverse, and suppose $f'(a) = 0$. Show that f^{-1} is not differentiable at $f(a)$. (*Hint:* Prove by contradiction, using the Chain Rule and differentiating the equation $f^{-1}(f(x)) = x$ implicitly.)

19. Using Exercise 18, show that the following functions are not differentiable at the given value of c.
    a. f^{-1}, where $f(x) = x + \sin x; c = \pi$
    b. f^{-1}, where $f(x) = x^5 + x^3 - 4; c = -4$

20. Let $0 \le a < b$, and let f be nonnegative, strictly increasing, and continuous on $[a, b]$, so that f^{-1} exists. Let R_1 be the lightly shaded region and R_2 the darkly shaded region in Figure 7.6, with respective areas A_1 and A_2.
    a. Show that $A_1 + A_2 = bf(b) - af(a)$.

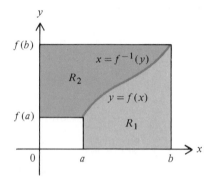

FIGURE 7.6

b. Use (a) to prove that

$$\int_a^b f(x)\,dx + \int_{f(a)}^{f(b)} f^{-1}(y)\,dy = bf(b) - af(a)$$

21. Use Exercise 20(b) to evaluate $\int_0^1 [(x-1)^{1/3} + 1]^{1/2}\,dx$.

22. a. Let $f(x) = -\sqrt{x}$ for $x > 0$. Show that the graphs of f and f^{-1} are concave upward on their respective domains.

    b. Let $f(x) = \sqrt{x}$ for $x > 0$. Show that the graph of f is concave downward on $(0, \infty)$, whereas the graph of f^{-1} is concave upward on $(0, \infty)$.

*23. Assume that f has an inverse, that $f''(f^{-1}(x))$ exists, and that $f'(f^{-1}(x)) \neq 0$. Show that

$$(f^{-1})''(x) = \frac{-f''(f^{-1}(x))}{[f'(f^{-1}(x))]^3}$$

24. Let the domain of f be an open interval I, and suppose f^{-1} exists. Suppose f'' exists on I and $f'(x) \neq 0$ and $f''(x) \neq 0$ for all x in I.

    a. Suppose f is strictly increasing on I. Show that the graph of f^{-1} is concave upward on its domain if the graph of f is concave downward on I, and the graph of f^{-1} is concave downward on its domain if the graph of f is concave upward on I. (*Hint:* Use Exercise 23.)

    b. Suppose f is strictly decreasing on I. Show that the graph of f^{-1} is concave upward on its domain if the graph of f is concave upward on I, and the graph of f^{-1} is concave downward on its domain if the graph of f is concave downward on I.

    c. Suppose that a is in I and that the graph of f has an inflection point at $(a, f(a))$. What can you say about an inflection point for the graph of f^{-1}?

7.3
THE NATURAL EXPONENTIAL FUNCTION

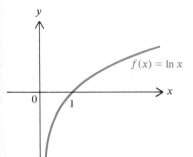

FIGURE 7.7

The first inverse we will study in detail is the inverse of the natural logarithm function $\ln x$. Recall from Section 5.7 that $\ln x$ is defined for all x in $(0, \infty)$ by

$$\ln x = \int_1^x f(t)\,dt$$

The graph of $\ln x$ is shown in Figure 7.7. Noteworthy features of $\ln x$ that we will use below are:

$$\ln 1 = 0 \tag{1}$$

$$\ln bc = \ln b + \ln c \quad \text{for positive } b \text{ and } c \tag{2}$$

$$\frac{d}{dx}\ln x = \frac{1}{x} > 0 \quad \text{for all } x > 0 \tag{3}$$

By (3), $\ln x$ is strictly increasing and therefore has an inverse, which we call the **natural exponential function**. For the present we designate its value at x by $\exp x$ (never by $\ln^{-1} x$). It follows that

$$\exp x = y \quad \text{if and only if} \quad x = \ln y \\ \text{for any } x \quad \text{and} \quad \text{for } y > 0 \tag{4}$$

Thus for any number x, $\exp x$ is the positive number y whose natural logarithm is x. In particular, $\exp 0 = 1$ because $\ln 1 = 0$.

Although $\exp x$ is defined for all numbers x, its values are positive because $\ln x$ is defined only for positive numbers. It follows from Theorem 7.3 that

$$\exp(\ln x) = x \quad \text{for } x > 0 \tag{5}$$

$$\ln(\exp x) = x \quad \text{for all } x \tag{6}$$

Recall from Section 5.7 that e is the unique positive number such that $\ln e = 1$. In order to relate e to the natural exponential function defined above, we recall from Section 1.6 that for each number a, a^r has meaning for any rational number r. In particular, this is true of e^r. We will now give meaning to the expression e^r when r is not rational.

When r is substituted for x and e^r for y, (4) reads

$$\exp r = e^r \quad \text{if and only if} \quad r = \ln e^r$$

But from (4) of Section 5.7,

$$\ln e^r = r \ln e = r \quad \text{for any rational } r$$

Consequently

$$\exp r = e^r \quad \text{for any rational } r \tag{7}$$

Since the domain of the natural exponential function is all real numbers, $\exp r$ is defined for all real numbers, and we are led to define e^r for *all* real numbers r by (7). Thus we obtain the function e^x defined by

$$e^x = \exp x \quad \text{for any real number } x$$

It follows that the function e^x is the natural exponential function, so e^x is the inverse of the natural logarithm function $\ln x$. In the notation of e^x, (4) becomes

$$e^x = y \quad \text{if and only if} \quad x = \ln y$$
$$\text{for any } x \quad \text{and} \quad \text{for } y > 0$$

Similarly, (5) and (6) become

$$e^{\ln x} = x \quad \text{for } x > 0 \tag{8}$$
$$\ln e^x = x \quad \text{for all } x \tag{9}$$

The Law of Logarithms in (2) has a companion involving exponentials, called the Law of Exponents.

THEOREM 7.7
LAW OF EXPONENTS

For all numbers b and c,

$$e^{b+c} = e^b e^c \tag{10}$$

Proof Let $u = e^b$ and $v = e^c$. Then $\ln u = b$ and $\ln v = c$. Consequently the Law of Logarithms implies that

$$\ln uv = \ln u + \ln v = b + c$$

As a result, we conclude from (8) that

$$e^{b+c} = e^{\ln uv} = uv = e^b e^c \quad \blacksquare$$

If $c = -b$, then (10) becomes

$$e^{b-b} = e^b e^{-b}$$

Since $e^{b-b} = e^0 = 1$, it follows that $1 = e^b e^{-b}$, so that

$$e^{-b} = \frac{1}{e^b} \qquad (11)$$

The Derivative of the Function e^x

The function e^x has a very unusual property, which is derived from the fact that the function e^x is the inverse of the function $\ln x$. Let $y = e^x$, so that $x = \ln y$. By (3) of Section 7.2,

$$\frac{d}{dx} e^x = \frac{1}{\dfrac{d}{dy} \ln y} = \frac{1}{\dfrac{1}{y}} = y = e^x$$

Therefore

$$\frac{d}{dx} e^x = e^x \qquad (12)$$

In other words, the natural exponential function *is its own derivative*. Since $e^x > 0$ for all x, we know from (12) that e^x is strictly increasing.

Next we use (12) to deduce that

$$\frac{d^2}{dx^2} e^x = \frac{d}{dx} e^x = e^x$$

This and the fact that $e^x > 0$ imply that the graph of e^x is concave upward on $(-\infty, \infty)$. Since

$$\lim_{x \to \infty} \ln x = \infty \quad \text{and} \quad \lim_{x \to 0^+} \ln x = -\infty$$

we conclude that

$$\lim_{x \to \infty} e^x = \infty \quad \text{and} \quad \lim_{x \to -\infty} e^x = 0 \qquad (13)$$

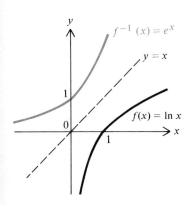

FIGURE 7.8

These results, coupled with the fact that the graph of e^x is the reflection of the graph of $\ln x$ through the line $y = x$, enable us to sketch the graph of e^x (Figure 7.8).

Functions of the form $e^{h(x)}$, where h is a differentiable function, arise frequently. By the Chain Rule,

$$\frac{d}{dx} e^{h(x)} = h'(x) e^{h(x)} \qquad (14)$$

Hence we can find the derivative of $e^{h(x)}$ provided we can find the derivative of h. Let us consider the special case in which $h(x) = kx$.

Example 1 Let $f(x) = e^{kx}$, where k is a constant. Find a formula for $f'(x)$, and then sketch the graph of f.

Solution Using the Chain Rule, we deduce that

$$f'(x) = e^{kx}(k) = ke^{kx}$$

Since $e^{kx} > 0$, it follows that $f'(x)$ is positive, 0, or negative, depending on whether $k > 0$, $k = 0$, or $k < 0$. In addition,

$$f''(x) = k^2 e^{kx}$$

so the graph of f is concave upward unless $k = 0$. Moreover, from (13) we know that if $k > 0$, then

$$\lim_{x \to \infty} e^{kx} = \infty \quad \text{and} \quad \lim_{x \to -\infty} e^{kx} = 0$$

whereas if $k < 0$, then

$$\lim_{x \to \infty} e^{kx} = 0 \quad \text{and} \quad \lim_{x \to -\infty} e^{kx} = \infty$$

With this information we can draw the graph of f—a graph whose shape indeed depends on the value of k (Figure 7.9). □

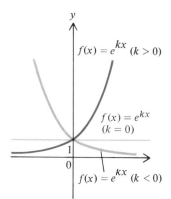

FIGURE 7.9

Functions of the form e^{kx}, or more generally ce^{kx} (where c and k are both constants), are very prevalent in applications, and we will devote Section 7.5 to them.

We end this subsection with a study of a function whose graph is bell-shaped. Closely related functions are fundamental in probability and statistics.

Example 2 Let $f(x) = e^{-x^2/2}$. Find a formula for $f'(x)$, and sketch the graph of f.

Solution Using the Chain Rule, we find that

$$f'(x) = e^{-x^2/2}\left(-\frac{2x}{2}\right) = -xe^{-x^2/2}$$

Since $e^{-x^2/2}$ is positive for any x, it follows that $f'(x) > 0$ for $x < 0$ and $f'(x) < 0$ for $x > 0$. Therefore f is strictly increasing on $(-\infty, 0]$ and strictly decreasing on $[0, \infty)$. In particular, $f(0) = 1$ is the maximum value of f. Next,

$$f''(x) = -e^{-x^2/2} - xe^{-x^2/2}\left(-\frac{2x}{2}\right) = -e^{-x^2/2} + x^2 e^{-x^2/2} = (x^2 - 1)e^{-x^2/2}$$

Again $e^{-x^2/2}$ is positive for any x, and thus the sign of $f''(x)$ is as shown in Figure 7.10.

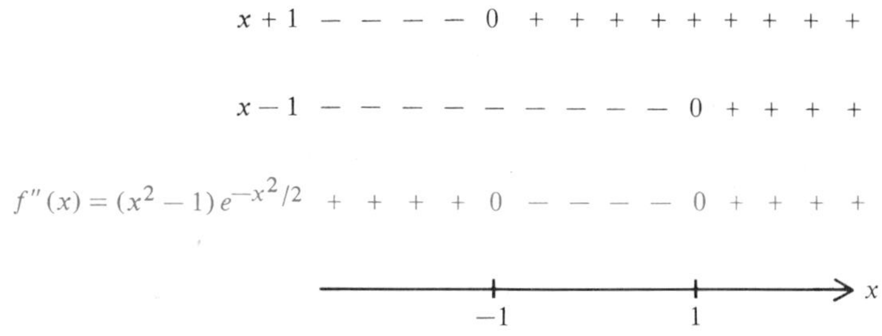

$$x + 1 \quad - \ - \ - \ - \ 0 \ + \ + \ + \ + \ + \ + \ + \ + \ +$$

$$x - 1 \quad - \ - \ - \ - \ - \ - \ - \ - \ - \ 0 \ + \ + \ + \ +$$

$$f''(x) = (x^2 - 1)e^{-x^2/2} \quad + \ + \ + \ + \ 0 \ - \ - \ - \ - \ 0 \ + \ + \ + \ +$$

$$\xrightarrow{\hspace{1cm}\underset{-1}{\mid}\hspace{1.5cm}\underset{1}{\mid}\hspace{1cm}} x$$

FIGURE 7.10

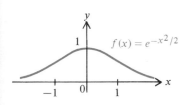

FIGURE 7.11

Consequently the graph of f is concave upward on $(-\infty, -1)$ and on $(1, \infty)$ and is concave downward on $(-1, 1)$ with inflection points at $(-1, f(-1)) = (-1, e^{-1/2})$ and at $(1, f(1)) = (1, e^{-1/2})$. Moreover, from (13) we deduce that

$$\lim_{x \to \infty} f(x) = 0 = \lim_{x \to -\infty} f(x)$$

Finally, we notice that $f(-x) = f(x)$, so the graph of f is symmetric with respect to the y axis. With all this information, we are prepared to sketch the graph of f (Figure 7.11). □

The Integral of e^x

The integral form of

$$\frac{d}{dx} e^x = e^x$$

is

$$\int e^x \, dx = e^x + C$$

Thus the natural exponential function is an indefinite integral of itself.

Example 3   Find $\int e^{-3x} \, dx$.

Solution   Using substitution, we let

$$u = -3x, \quad \text{so that} \quad du = -3 \, dx$$

Therefore

$$\int \underset{e^{-3x}}{e^u} \, \overset{-\frac{1}{3} \, du}{dx} = -\frac{1}{3} \int e^u \, du = -\frac{1}{3} e^u + C = -\frac{1}{3} e^{-3x} + C \quad □$$

Example 4   Evaluate $\int_0^\pi (\sin x)e^{\cos x}\, dx$.

Solution   This time we let

$$u = \cos x, \quad \text{so that} \quad du = -\sin x\, dx$$

For the change in the limits of integration we find that

$$\text{if } x = 0 \text{ then } u = 1, \quad \text{and} \quad \text{if } x = \pi \text{ then } u = -1$$

Consequently

$$\int_0^\pi (\sin x)e^{\cos x}\, dx = \int_0^\pi e^{\cos x}\overset{e^u}{\overbrace{\phantom{e^{\cos x}}}}\underset{\sin x\, dx}{\overbrace{}}^{-du} = -\int_1^{-1} e^u\, du$$

$$= -e^u \Big|_1^{-1} = -e^{-1} - (-e^1)$$

$$= e - e^{-1} \quad \square$$

EXERCISES 7.3

In Exercises 1–6 simplify the expression.

1. $\ln\sqrt{e}$ 2. $\ln e^3$ 3. $\ln e^e$

4. $\ln|\ln(1/e)|$ 5. $e^{\ln 3x}$ 6. $\ln(e^{\ln e})$

In Exercises 7–16 find the derivative of the function.

7. $f(x) = e^{4x}$ 8. $f(x) = e^{-\sqrt{x}}$

9. $f(x) = e^{(x^5)}$ 10. $f(x) = (x+1)e^{-x}$

11. $f(x) = \dfrac{e^x + 1}{e^x - 1}$ 12. $y = e^{\cos z}$

13. $y = \tan e^{3z}$ 14. $y = e^{2z}\ln z$

15. $y = e^{-z}\sin az$ 16. $y = e^{-\ln x}$

In Exercises 17–18 find dy/dx by implicit differentiation.

17. $x^2 e^y = \ln xy$ 18. $e^{-x/y} = y^2 - \sqrt{x}$

In Exercises 19–20 find a formula for the nth derivative of f, where n is any positive integer.

19. $f(x) = e^{4x}$ 20. $f(x) = xe^x$

21. Let $y = xe^{2x}$. Show that $x\dfrac{dy}{dx} = y(1 + 2x)$.

22. Let $y = xe^{-x^2/2}$. Show that $x\dfrac{dy}{dx} = y(1 - x^2)$.

In Exercises 23–28 sketch the graph of the function, noting any relative extreme values, concavity, inflection points, and asymptotes.

23. $f(x) = e^{-x}$ 24. $f(x) = e^x + e^{-x}$

25. $f(x) = e^x \cos\sqrt{3}\,x$ 26. $f(x) = \dfrac{e^x}{1 + e^x}$

27. $f(x) = \ln(e^x + e^{-x})$ 28. $f(x) = \ln(e^x + 1)$

In Exercises 29–45 evaluate the integral.

29. $\displaystyle\int e^{-4x}\, dx$ 30. $\displaystyle\int e^{\sqrt{2x+3}}\, dx$

31. $\displaystyle\int_0^{\sqrt{2}} ye^{y^2}\, dy$ 32. $\displaystyle\int_1^4 \dfrac{e^{\sqrt{y}}}{\sqrt{y}}\, dy$

33. $\displaystyle\int_0^{\pi/2} (\cos y)e^{\sin y}\, dy$ 34. $\displaystyle\int \dfrac{e^{\tan y}}{\cos^2 y}\, dy$

35. $\displaystyle\int \dfrac{e^t}{e^t + 1}\, dt$ 36. $\displaystyle\int_{-1}^1 \dfrac{e^t - e^{-t}}{e^t + e^{-t}}\, dt$

37. $\displaystyle\int \dfrac{e^{2t}}{\sqrt{e^{2t} - 4}}\, dt$ 38. $\displaystyle\int e^y e^{(e^y)}\, dy$

39. $\displaystyle\int \dfrac{e^{-t}\ln(1 + e^{-t})}{1 + e^{-t}}\, dt$ 40. $\displaystyle\int_{-1}^1 \dfrac{1}{e^x}\, dx$

41. $\displaystyle\int \dfrac{1}{e^{-x/2}}\, dx$ 42. $\displaystyle\int \sqrt{e^x}\, dx$

43. $\displaystyle\int \dfrac{1}{1 + e^x}\, dx$ 44. $\displaystyle\int \dfrac{1}{1 + e^{-x}}\, dx$

*45. $\displaystyle\int e^{(x-e^x)}\, dx$

46. Let $f(x) = e^x/(e^{2x} + 1)$. Show that f is an even function.

47. Let $f(x) = 2e^{-3x}$. Determine an equation of the line tangent to the graph of f at $(0, 2)$.

48. Find the point at which the line $y = -4x - 7$ is tangent to the graph of $y = e^{x^2 - 4}$.

In Exercises 49–52 determine at which points the graphs of the given pair of functions intersect.

49. $f(x) = e^x$ and $g(x) = e^{1-x}$ (*Hint:* Take natural logarithms.)

50. $f(x) = e^x$ and $g(x) = e^{-x^2}$

51. $f(x) = e^{3x}$ and $g(x) = 2e^{-x}$

52. $f(x) = 5e^{-2x}$ and $g(x) = 3e^x$

* 53. Let $f(x) = e^x$ and $g(x) = \ln x$. Show that the graphs of f and g do not intersect.

54. a. Show that $e^x \geq 1 + x$ for $x \geq 0$.
    b. Use part (a) to prove that $e^x \neq x$ for all x.
    c. Use part (a) to show that $\lim_{x \to \infty} e^x = \infty$.

In Exercises 55–56 find a formula for the inverse of the function.

55. $f(x) = \dfrac{e^x - 1}{e^x + 1}$

56. $f(x) = \dfrac{3e^x - 2}{e^x + 4}$

In Exercises 57–58 use the Newton-Raphson method to approximate a solution of the equation with an error of at most 10^{-6}.

57. $e^{-x} = x$

58. $e^x = 2 - x$

59. Evaluate $\int_1^e \ln x \, dx$ by using the formula

$$\int_a^b f(x) \, dx + \int_{f(a)}^{f(b)} f^{-1}(y) \, dy = bf(b) - af(a)$$

appearing in Exercise 20 of Section 7.2.

60. Let R be a positive number. Show that

$$\int_0^{\pi/2} e^{-R \sin x} \, dx < \frac{\pi}{2R}$$

which is an inequality known as **Jordan's inequality.** (*Hint:* By Review Exercise 18 in Chapter 4, $\sin x \geq 2x/\pi$ for $0 \leq x \leq \pi/2$.)

In Exercises 61–62 find the area A of the region between the graph of f and the x axis on the given interval.

61. $f(x) = e^{-x}$; $[0, 1]$

62. $f(x) = 2e^{4x}$; $[-\frac{1}{2}, \frac{1}{4}]$

63. Find the area A of the region bounded by the graphs of $y = e^{2x}$ and $y = e^{-2x}$ and the line $x = \frac{1}{2}$.

64. Find the area A of the region in the first quadrant bounded by the curves $y = 3e^x$ and $y = 2 + e^{2x}$.

65. A rectangle with base on the x axis has its two upper vertices on the graph of $y = e^{-x^2}$. Show that if these vertices are inflection points of the graph, then the rectangle has the maximum possible area.

66. Let $f(x) = e^x$, and let R be the region between the graph of f and the x axis on $[1, 2]$. Find the volume V of the solid obtained by revolving R about the x axis.

67. Let $f(x) = e^{(x^2)}$ and $g(x) = e^{-x^2}$. Find the volume V of the solid obtained by revolving about the y axis the region between the graphs of f and g on $[0, 1]$.

68. Find the volume V of the solid obtained by revolving about the line $y = 1$ the region between the graph of the equation $y = e^{-2x}$ and the x axis on the interval $[0, 1]$. (*Hint:* Use Exercise 31 of Section 6.1.)

69. Find the length $\mathcal{L}$ of the curve described parametrically by $x = e^t \sin t$ and $y = e^t \cos t$ for $0 \leq t \leq \pi$.

70. Let $f(x) = e^x + \frac{1}{4}e^{-x}$ for $0 \leq x \leq 1$.
    a. Determine the length $\mathcal{L}$ of the graph of f.
    b. Determine the surface area S of the surface obtained by revolving the graph of f about the x axis.

<u>C</u> 71. By the **Bouguer-Lambert Law**, which is fundamental to the study of photometry, if a beam of light with intensity c at the surface of a lake travels vertically downward, then at a depth of x meters the intensity $f(x)$ of the beam is given by the formula

$$f(x) = ce^{-1.4x}$$

where the constant 1.4 is the so-called **absorption coefficient** of pure water. Use a calculator to calculate the ratio of the intensity 10 meters below the surface of the lake to the intensity at the surface itself; that is, find $f(10)/f(0)$. (From the number you obtain it will be apparent why most flora cannot survive at depths greater than 10 meters.)

72. Let $f(x) = a/(1 + be^{-kax})$ for $x \geq 0$, where a, b and k are positive numbers.
    a. Show that

$$f'(x) = \frac{a^2 bke^{-kax}}{(1 + be^{-kax})^2}$$

    b. Show that

$$f''(x) = \frac{a^3 bk^2 e^{-kax}(-1 + be^{-kax})}{(1 + be^{-kax})^3}$$

    c. Sketch the graph of f, noting the inflection point and asymptotes. This graph is known as the **logistic curve** and was introduced in 1838 by the Belgian mathematician P. F. Verhulst. The logistic curve has played a significant role in population ecology. Normally $f(x)$ denotes the population as a function of time x, under the condition that the population has a stable age structure and a stable source of food, but limited space, which inhibits the size of the population. The number a

to which the population tends with time is called the **carrying capacity** of the medium in which the population lives.

73. In 1934 the biologist G. F. Gause placed 20 paramecia in 5 cubic centimeters of a saline solution with a constant amount of food and measured their growth on a daily basis. He found that the population $f(t)$ at any time t (in days) was approximately

$$f(t) = \frac{449}{1 + e^{5.4094 - 1.0235t}} \quad \text{for } t \geq 0$$

a. Determine the carrying capacity of the medium.
b. Use Exercise 72 to determine the time at which the population is growing fastest. (This corresponds to the inflection point of the graph of f.)

74. The **Gompertz growth curve** is the graph of the function f defined by

$$f(x) = ae^{(-be^{-cx})} \quad \text{for } x \geq 0$$

where a, b, and c are positive constants. Sketch the graph of f, noting any extreme values, inflection points, and asymptotes. The Gompertz growth curve appears in population studies.

75. It has been shown that the concentration y of a drug in the blood t minutes after injection is given by

$$y = \frac{c}{a - b}(e^{-bt} - e^{-at})$$

where a, b, and c are positive constants and $a > b$.
a. Find the maximum value of y.
b. What happens to the values of y as t becomes very large?

76. If an electrical network has a capacitance C, voltage V, and resistance R connected in series, then the charge $Q(t)$ on the capacitor at time $t \geq 0$ is governed by the equation

$$Q(t) = VC + ce^{-t/RC}$$

where c is some constant.
a. If $Q(0) = 0$, find c.
b. Assuming that C, V, and R are positive and that $Q(0) = 0$, sketch the graph of Q.
c. At what time t is the capacitor 90 percent charged? (In other words, for what t is $Q(t) = 0.9VC$?)

7.4
GENERAL EXPONENTIAL AND LOGARITHM FUNCTIONS

Now we study other exponential and logarithm functions that are derived from the natural exponential and logarithm functions.

Let $a > 0$. Then a^r is defined for any rational number r (see Section 1.6). From (8) of Section 7.3 with x replaced by a^r, we know that

$$a^r = e^{\ln(a^r)}$$

Since $\ln a^r = r \ln a$ by (4) of Section 5.7, we deduce that

$$a^r = e^{r \ln a}$$

whenever r is rational. But $e^{r \ln a}$ makes sense for any real number r. Thus we define a^r for any real number r by the equation $a^r = e^{r \ln a}$.

DEFINITION 7.8

For any $a > 0$ and any number r,

$$a^r = e^{r \ln a} \tag{1}$$

In the expression a^r, the number a is the **base** and r is the **exponent** (or **power**).

This may seem a strange way to define a^r, but to justify it, let us observe the following two facts. First, for rational values of r, Definition 7.8 assigns the same value to a^r that the earlier definition in Section 1.6 did. Second, if $a = e$, then by (1),

$$a^r = e^{r \ln e} = e^r$$

so that the definition of a^r in (1) conforms to the original definition of e^r when $a = e$.

To begin our study of a^r, let us notice that

$$a^0 = e^{0 \ln a} = e^0 = 1$$

and

$$a^1 = e^{1 \ln a} = e^{\ln a} = a$$

Thus

$$a^0 = 1 \quad \text{and} \quad a^1 = a \tag{2}$$

Next we take the natural logarithms of both sides of the equation in (1). We find that

$$\ln a^r = \ln e^{r \ln a}$$

By (9) of Section 7.3,

$$\ln e^{r \ln a} = r \ln a$$

so we conclude that

$$\ln a^r = r \ln a \tag{3}$$

which extends (4) of Section 5.7 to arbitrary values of r and which will be used in the following discussion.

From the original Law of Exponents (Theorem 7.7), along with the definition of a^r, we can prove very general laws of exponents.

THEOREM 7.9
GENERAL LAWS OF
EXPONENTS

Let $a > 0$, and let b and c be any numbers. Then the following formulas hold:

a. $a^{b+c} = a^b a^c$

b. $a^{-b} = \dfrac{1}{a^b}$

c. $(a^b)^c = a^{bc}$

Proof To prove (a), we use the definition of a^r, along with the Law of Exponents:

$$a^{b+c} = e^{[(b+c) \ln a]} = e^{(b \ln a + c \ln a)} = e^{b \ln a} e^{c \ln a} = a^b a^c$$

To prove (b), we take $c = -b$ in (a) and refer to (2) to obtain

$$1 \overset{(2)}{=} a^0 = a^{b-b} = a^b a^{-b}$$

Dividing by a^b (which is never 0), we conclude that

$$a^{-b} = \frac{1}{a^b}$$

To prove (c), we use (3), obtaining

$$(a^b)^c = e^{c \ln (a^b)} \overset{(3)}{=} e^{bc \ln a} = a^{bc} \quad \blacksquare$$

From the General Laws of Exponents it follows, for example, that

$$\pi^2 \pi^{-4} = \pi^{2-4} = \pi^{-2}$$

However, the laws apply even when the exponents are not rational. For instance,

$$\pi^{\sqrt{2}} \pi^{\sqrt{5}} = \pi^{\sqrt{2}+\sqrt{5}}$$

Exponential and Power Functions

In (1), if the base a stays fixed and r varies, we obtain the **exponential function** a^x defined by

$$a^x = e^{x \ln a} \tag{4}$$

The derivatives of the function a^x are obtained by differentiating $e^{x \ln a}$ by means of the Chain Rule (see (14) of Section 7.3):

$$\frac{d}{dx} a^x = \frac{d}{dx} (e^{x \ln a}) = (e^{x \ln a})(\ln a) = (\ln a)a^x$$

or in more condensed form,

$$\frac{d}{dx} a^x = (\ln a)a^x \tag{5}$$

Thus the derivative of a^x is merely a constant multiple of a^x. Moreover, since $a^x > 0$ for all x and $\ln a > 0$ only for $a > 1$, (5) implies that a^x is strictly increasing if $a > 1$ and is strictly decreasing if $0 < a < 1$. It also follows from (5) that

$$\frac{d^2}{dx^2} a^x = (\ln a)^2 a^x$$

so that if $a \neq 1$, the graph of a^x is concave upward on $(-\infty, \infty)$. Finally, formula (4) above, along with (13) of Section 7.3, implies that

$$\text{if } a > 1, \quad \text{then} \quad \lim_{x \to \infty} a^x = \infty \text{ and } \lim_{x \to -\infty} a^x = 0$$

whereas

$$\text{if } 0 < a < 1, \quad \text{then} \quad \lim_{x \to \infty} a^x = 0 \text{ and } \lim_{x \to -\infty} a^x = \infty$$

With this information we can draw the graph of a^x. As we notice in Figure 7.12, the appearance of the graph of a^x depends on the value of a.

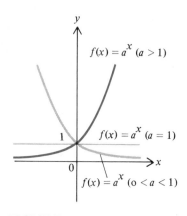

FIGURE 7.12

Example 1 Let $f(x) = x^2 2^x$. Sketch the graph of f, indicating all pertinent information.

Solution The x and y intercepts of f are both 0, since $f(0) = 0$ and $f(x) > 0$ for all $x \neq 0$. Next, by (5) with $a = 2$ we find that

$$f'(x) = (2x)2^x + x^2 (\ln 2)2^x$$

Now $f'(x) = 0$ if

$$(2\dot{x})2^x + x^2 (\ln 2)2^x = 0$$

Since $2^x > 0$ for all x, this equation is equivalent to

$$2x + x^2 \ln 2 = 0$$

which occurs if $x = 0$ or $x = -2/\ln 2$. Furthermore,

$$f''(x) = (2)2^x + (2x)(\ln 2)2^x + (2x)(\ln 2)2^x + x^2 (\ln 2)^2 \, 2^x$$

$$= [2 + 4x \ln 2 + x^2 (\ln 2)^2]2^x$$

By the quadratic formula, $(\ln 2)^2 x^2 + (4 \ln 2)x + 2 = 0$ if

$$x = \frac{-4 \ln 2 \pm \sqrt{16 (\ln 2)^2 - 8 (\ln 2)^2}}{2 (\ln 2)^2} = -\frac{2}{\ln 2} \pm \frac{\sqrt{2}}{\ln 2}$$

so the graph of f is concave upward on $(-\infty, (-2-\sqrt{2})/\ln 2)$ and on $((-2 + \sqrt{2})/\ln 2, \infty)$, and is concave downward on $((-2-\sqrt{2})/\ln 2, (-2 + \sqrt{2})/\ln 2)$. It follows that $f(0) = 0$ is a relative minimum value, $f(-2/\ln 2)$ is a relative maximum value, and that the points $((-2 \pm \sqrt{2})/\ln 2, f((-2 \pm \sqrt{2})/\ln 2))$ are inflection points. With the help of a calculator we find that

$$f\left(\frac{-2}{\ln 2}\right) \approx 1.13, \qquad f\left(\frac{-2 + \sqrt{2}}{\ln 2}\right) \approx .40, \quad \text{and} \quad f\left(\frac{-2 - \sqrt{2}}{\ln 2}\right) \approx .80$$

Using all this information, we obtain the graph in Figure 7.13. (It will follow from Section 7.8 that the x axis is a horizontal asymptote.) □

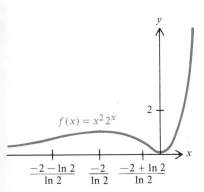

$f(x) = x^2 2^x$

$\dfrac{-2 - \ln 2}{\ln 2} \qquad \dfrac{-2}{\ln 2} \qquad \dfrac{-2 + \ln 2}{\ln 2}$

FIGURE 7.13

If in (1) the base varies and the exponent r is fixed, we obtain the **power function** x^r defined by

$$x^r = e^{r \ln x} \tag{6}$$

The power functions x^r discussed in Sections 1.6 and 3.4 were defined only for rational values of r. By (6) the power functions x^r are now defined for *every* real value of r. In general the domain of the function x^r consists of all $x > 0$ (although for certain values of r such as 2 and 3 the function x^r is defined for all x). To study x^r for arbitrary values of r, however, we will assume that $x > 0$.

The derivative of the function x^r is obtained by differentiating $e^{r \ln x}$ by means of the Chain Rule:

$$\frac{d}{dx} x^r = \frac{d}{dx}(e^{r \ln x}) = e^{r \ln x}\left(\frac{r}{x}\right) = \frac{r}{x} x^r = rx^{r-1}$$

or more simply,

$$\frac{d}{dx} x^r = rx^{r-1} \tag{7}$$

For rational values of r this formula is familiar from (5) of Section 3.4. It follows from (7) that

$$\frac{d^2}{dx^2} x^r = \frac{d}{dx}(rx^{r-1}) = r(r-1)x^{r-2}$$

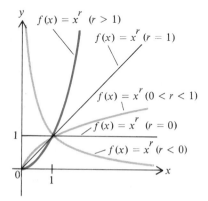

By using the derivatives of x^r, we can draw the graph of x^r. As you might well expect, the appearance of the graph depends on the value of r, as shown in Figure 7.14.

It is also possible for *both* the base and the exponent in a^r to vary; in that case, we obtain a function of the form $f(x)^{g(x)}$. Then by (1),

$$f(x)^{g(x)} = e^{g(x)\ln f(x)} \tag{8}$$

FIGURE 7.14

Any time we must perform an operation (such as differentiation) on a function of the form $f(x)^{g(x)}$, we use (8) to write the function with base e.

Example 2   Let $f(x) = (2x)^{\sin x}$. Find $f'(x)$.

Solution   By (8) we know that

$$f(x) = (2x)^{\sin x} = e^{(\sin x)(\ln 2x)}$$

Using the Chain Rule, we find that

$$f'(x) = e^{(\sin x)(\ln 2x)}\left[(\cos x)(\ln 2x) + \frac{(\sin x)2}{2x}\right]$$

$$= (2x)^{\sin x}\left[(\cos x)(\ln 2x) + \frac{\sin x}{x}\right] \quad \square$$

Now we turn to integrals of a^x and x^r. If we divide both sides of (5) by $\ln a$, we obtain

$$\frac{d}{dx}\left(\frac{1}{\ln a}a^x\right) = a^x$$

In integral form this becomes

$$\int a^x \, dx = \frac{1}{\ln a}a^x + C \tag{9}$$

Example 3   Evaluate $\int_0^3 3^x \, dx$.

Solution   By (9) we have

$$\int_0^3 3^x \, dx = \frac{1}{\ln 3} 3^x \Big|_0^3 = \frac{1}{\ln 3} 3^3 - \frac{1}{\ln 3} 3^0 = \frac{27}{\ln 3} - \frac{1}{\ln 3} = \frac{26}{\ln 3} \quad \square$$

The formula for the integral of x^r is obtained from (7) by first replacing r by $r + 1$ to obtain

$$\frac{d}{dx}(x^{r+1}) = (r + 1)x^r$$

and then dividing both sides by $r + 1$, which produces

$$\frac{d}{dx}\left(\frac{1}{r + 1} x^{r+1}\right) = x^r \quad \text{for } r \neq -1$$

In integral form this becomes

$$\int x^r \, dx = \frac{1}{r + 1} x^{r+1} + C \quad \text{for } r \neq -1 \tag{10}$$

Example 4   Evaluate $\int_1^2 x^{\sqrt{3}} \, dx$.

Solution   By (10),

$$\int_1^2 x^{\sqrt{3}} \, dx = \frac{1}{\sqrt{3} + 1} x^{\sqrt{3}+1} \Big|_1^2 = \frac{1}{\sqrt{3} + 1}\left(2^{\sqrt{3}+1}\right) - \frac{1}{\sqrt{3} + 1}(1^{\sqrt{3}+1})$$

$$= \frac{2(2^{\sqrt{3}}) - 1}{\sqrt{3} + 1} \quad \square$$

Logarithms to Different Bases   In the preceding subsection we proved that a^x is strictly increasing for $a > 1$ and strictly decreasing for $0 < a < 1$. Therefore, provided that $a > 0$ and $a \neq 1$, there is an inverse for a^x, which we denote by $\log_a x$ and call the **logarithm to the base a**. When $a = e$, we obtain the natural logarithm function, which we write as before as $\ln x$; thus

$$\ln x = \log_e x$$

For the special case $a = 10$, $\log_{10} x$ is called the **common logarithm of x** and is usually written $\log x$.

For logarithms to the base a, parts (b) and (c) of Theorem 7.3 become

$$a^{\log_a x} = x \tag{11}$$

and
$$\log_a a^x = x \tag{12}$$

We can show that the logarithm to any base is a constant multiple of the natural logarithm.

THEOREM 7.10

> For any $a > 0$ such that $a \neq 1$ we have
>
> $$\log_a x = \frac{\ln x}{\ln a} \quad \text{for all } x > 0 \tag{13}$$

Proof   The simplest way to verify (13) is to take natural logarithms of both sides of (11) and use (3) with $\log_a x$ replacing r:

$$\ln x = \ln(a^{\log_a x}) = (\log_a x)(\ln a)$$

Since $a \neq 1$ by assumption, it follows that $\ln a \neq 0$, so that

$$\log_a x = \frac{\ln x}{\ln a} \quad \blacksquare$$

From (13) we can easily find the derivative of $\log_a x$:

$$\frac{d}{dx} \log_a x = \frac{d}{dx} \frac{\ln x}{\ln a} = \frac{1}{\ln a} \frac{d}{dx} \ln x = \frac{1}{x \ln a}$$

Formula (13) also makes it possible to calculate $\log_a x$ once we know the natural logarithms of x and a. Since most calculators have keys for natural logarithms, the conversion formula is very useful.

Example 5   Calculate $\log_2 3$ by using natural logarithms.

Solution   By (13),

$$\log_2 3 = \frac{\ln 3}{\ln 2}$$

By calculator we find that

$$\log_2 3 \approx 1.58496 \quad \square$$

Another consequence of (13) is the General Law of Logarithms.

THEOREM 7.11
GENERAL LAW OF
LOGARITHMS

> If a, b, and c are positive, then
>
> $$\log_a bc = \log_a b + \log_a c$$

Proof By virtue of (13) and the original Law of Logarithms (Theorem 5.23), we find that

$$\log_a bc \overset{(13)}{=} \frac{\ln bc}{\ln a} = \frac{\ln b + \ln c}{\ln a} = \frac{\ln b}{\ln a} + \frac{\ln c}{\ln a} \overset{(13)}{=} \log_a b + \log_a c \quad \blacksquare$$

From (3) and (13) it follows that for any r and any positive a and b,

$$\log_a b^r = r \log_a b$$

Finally, (13) enables us to determine the general shape of the graph of $\log_a x$ from the graph of $\ln x$, because $\log_a x$ is just a constant multiple of $\ln x$. Figure 7.15 shows the graphs for $a > 1$ and for $0 < a < 1$.

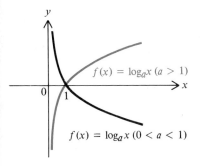

$f(x) = \log_a x \ (a > 1)$

$f(x) = \log_a x \ (0 < a < 1)$

FIGURE 7.15

Earthquakes   Several formulas have been proposed for converting seismographic readings into a unified scale that would represent the magnitude of an earthquake. The scale most commonly used is the Richter scale, devised by the American geologist Charles F. Richter. In order to describe measurement on the Richter scale, we first introduce a reference, or zero-level, earthquake, which by definition is any earthquake whose largest seismic wave would measure 0.001 millimeter

The effect of the Alaska Good Friday earthquake (see Exercise 53) on a main street in Anchorage. Part of the street was left some 20 feet below the rest of it.

(1 micron) on a standard seismograph located 100 kilometers from the epicenter of the earthquake. The **magnitude** M of a given earthquake can be obtained by the formula

$$M = \log \frac{x}{a} = \log x - \log a \qquad (14)$$

where x is the amplitude measured on a standard seismograph of the largest seismic wave of the given earthquake and a is the amplitude on the same seismograph of the largest seismic wave of a zero-level earthquake with the same epicenter. Values of a for various distances from the epicenter have been tabulated, so one only needs to measure the number x and know the distance to the epicenter in order to be able to assess the magnitude of a given earthquake.

Example 6 In August 1979, an earthquake near San Francisco registered approximately 5.9 on the Richter scale. Find the ratio of the amplitude of the largest seismic wave to that of the famous San Francisco earthquake of 1906, which measured 8.4 in magnitude.

Solution Let M_1 and M_2 denote the magnitudes of the 1979 and 1906 earthquakes, respectively, and x_1 and x_2 their respective maximum wave amplitudes. We need to find the value of x_1/x_2. By (14), $5.9 = \log(x_1/a)$, so that $10^{5.9} = x_1/a$, and thus $x_1 = a \cdot 10^{5.9}$. Likewise, $x_2 = a \cdot 10^{8.4}$. Consequently

$$\frac{x_1}{x_2} = \frac{a \cdot 10^{5.9}}{a \cdot 10^{8.4}} = 10^{5.9 - 8.4} = 10^{-2.5} \approx 0.003162 \qquad \square$$

The physical damage caused by an earthquake increases with magnitude, but fortunately the frequency of earthquakes decreases with magnitude. The following table indicates damage and frequency in relation to magnitude.

Magnitude	Result Near the Epicenter	Approximate Number per Year
≥ 8	near total damage	0.2
7.0–7.9	serious damage to buildings	14
6.0–6.9	moderate damage to buildings	185
5.0–5.9	slight damage to buildings	1000
4.0–4.9	felt by most people	2800
3.0–3.9	felt by some people	26,000
2.0–2.9	not felt but recorded	800,000

Theoretically, the magnitude of an earthquake can be any real number. In practice, however, a standard seismograph can only record those earthquakes whose magnitudes exceed 2. On the other end of the scale, no magnitude has ever exceeded 9.0, and only half a dozen times has one been in the neighborhood of 8.5.

EXERCISES 7.4

In Exercises 1–6 simplify the expression.

1. $\log_9 3$
2. $\log_2 \frac{1}{4}$
3. $\log_4 4^x$
4. $\log_{1/4} 2^x$
5. $7^{\log_7 2x}$
6. $5^{2\log_5 x}$

In Exercises 7–22 find the derivative of the function.

7. $f(x) = 5^x$
8. $f(x) = 2^{-x}$
9. $f(x) = 3^{5x-7}$
10. $f(x) = \dfrac{2x}{1 + 2^x}$
11. $g(x) = a^x \sin bx$
12. $g(x) = a^x \ln x$
13. $g(x) = x \log_2 x$
14. $g(x) = \log_3 (x^2 + 4)$
15. $y = t^t$
16. $y = t^{\sin t}$
17. $y = t^{2/t}$
18. $y = \left(1 + \dfrac{1}{t}\right)^t$
19. $f(x) = (\cos x)^{\cos x}$
20. $f(x) = (\ln x)^{\ln x}$
21. $f(x) = (2x)^{\sqrt{2}}$
22. $f(x) = (\cos x)^{\pi x^2}$

In Exercises 23–24 sketch the graph of the function, noting all pertinent information.

23. $f(x) = 2^x - 2^{-x}$
24. $f(x) = \log_2 (1 + x^2)$

In Exercises 25–26 determine at which points the graphs of the given functions intersect.

25. $f(x) = \log_3 x$ and $g(x) = \log_2 x$
26. $f(x) = 3^x$ and $g(x) = 2^{(x^2)}$
    (*Hint:* Take natural logarithms.)

In Exercises 27–34 evaluate the integral.

27. $\displaystyle\int 2^x \, dx$
28. $\displaystyle\int_1^2 10^x \, dx$
29. $\displaystyle\int_{-2}^0 3^{-x} \, dx$
30. $\displaystyle\int_0^1 4^{3x} \, dx$
31. $\displaystyle\int x \cdot 5^{-x^2} \, dx$
32. $\displaystyle\int \dfrac{\log_2 x}{x} \, dx$
33. $\displaystyle\int x^{2\pi} \, dx$
34. $\displaystyle\int (\sin x)^e \cos x \, dx$

35. Find the area A of the region bounded by the graph of $y = 2^{-x} - \frac{1}{4}$ and the two axes.

36. Find the area A of the region between the graphs of $y = 2^x$ and $y = 3^x$ on $[0, 2]$.

37. Find the area A of the region bounded by the line $x = 16$ and the graphs of
$$y = \frac{3}{x} \quad \text{and} \quad y = \frac{\log_2 x}{x}$$

38. Find the volume V of the solid generated by revolving

about the x axis the region between the graph of $y = 2^x$ and the x axis on $[0, 1]$.

C In Exercises 39–42 calculate the logarithm by using Theorem 7.10.

39. $\log_3 5$
40. $\log_{1/2} \frac{1}{3}$
41. $\log_\pi e$
42. $\log_{\sqrt{2}} \sqrt{\pi}$

43. By Exercise 65 of Section 5.7, $(\ln x)/x$ is strictly increasing on $(0, e]$ and strictly decreasing on $[e, \infty)$. Using this fact, prove that $a^b > b^a$ for $0 < b < a \leq e$ and for $e \leq a < b$.

44. Use Exercise 43 to show that
    a. $\pi^e < e^\pi$
    b. $2^{\sqrt{x}} > x$ for $0 < x < 4$

45. Let $f(x) = x \log_a x$. Determine the values of a for which the graph of f is concave upward on $(0, \infty)$ and the values of a for which the graph is concave downward on $(0, \infty)$.

46. Let $f(x) = (2^x + 1)/(2^x - 3)$. Find a formula for $f^{-1}(x)$.

47. Use Theorem 7.9 to prove that
$$a^{b+c+d} = a^b a^c a^d \quad \text{for all } b, c, \text{ and } d \text{ and all } a > 0$$

48. a. Prove the **change of base formula**
$$\log_b x = \log_b a \log_a x$$
    where a, b, and x are positive.
    b. Using (a), find a formula that converts logarithms to the base 7 into logarithms to the base 4.

49. Let $a > 0$ and $b > 0$. Using Exercise 48(a), show that
    a. $\log_b a = \dfrac{1}{\log_a b}$
    b. $\log_{1/a} x = -\log_a x$ for $x > 0$

50. Determine the ratio x/a for the powerful earthquake that occurred in Japan on June 12, 1978, whose magnitude was 7.5.

51. If an earthquake has magnitude 2, find the amplitude of its largest wave 100 kilometers from the epicenter.

52. Suppose the amplitude of the maximal seismic wave is doubled. By how much is the magnitude of the earthquake increased?

53. The magnitude of the Good Friday Alaska earthquake of March 28, 1964, is given sometimes as 8.4 and sometimes as 8.5. What is the ratio of the maximum amplitude of an earthquake of magnitude 8.4 to that of an earthquake of magnitude 8.5?

54. The strongest earthquakes ever recorded occurred off the coast of Ecuador and Colombia in 1906, and in Japan in

1933. Each had a magnitude of 8.9. Find the ratio of the amplitude of the largest wave of such a quake to the corresponding amplitude of a zero-level quake.

Suppose the frequency of a particular sound wave is 1000 hertz (cycles per second). Let I_0 be the intensity of the sound wave at the threshold of audibility, approximately 10^{-16} watts per square centimeter. If x denotes the intensity of the sound wave, then the **noise level** $L(x)$ of the sound wave is given by

$$L(x) = 10 \log \frac{x}{I_0}$$

The units for $L(x)$ are called decibels, in honor of Alexander Graham Bell. Exercises 55–59 concern noise level.

55. The noise level of a whisper is about 20 decibels, and that of ordinary conversation is around 65 decibels. Determine the ratio of the intensity of a whisper to that of conversation.

56. Determine the number of decibels that corresponds to each of the following intensities.
    a. 10^{-11} (refrigerator motor)
    b. 10^{-4} (electric typewriter)
    c. 10^2 (jet airplane engine 100 feet away)

57. What is the difference in the noise levels of two sounds, one of which is 1000 times as intense as the other?

58. What is the ratio of the intensity of a given sound to that of one that is 100 decibels higher?

59. The human ear can just barely distinguish between two sounds if one is 0.6 decibels higher than the other. What is the ratio of the intensity of one sound to that of another sound that is lower than the first and is just barely distinguishable from the first sound?

60. Let $r \neq 0$. Show that the functions x^r and $x^{1/r}$, defined for $x > 0$, are inverses of one another. (Thus even when r is irrational we might call $x^{1/r}$ the rth root of x.)

61. Logarithms were invented by John Napier, who considered two particles, P and Q, traveling along paths described in Figure 7.16, according to the formulas

$$P'(t) = 10^7 - P(t) \quad \text{for } t > 0$$
$$Q(t) = 10^7 t \qquad \text{for } t > 0$$

In this way P travels slower and slower on its finite path as it approaches the point a distance 10^7 from the origin, while Q travels at a constant rate on its infinitely long path. Napier defined his logarithmic function, which we denote by the symbol N, by the equation

$$N(10^7 - P(t)) = Q(t) = 10^7 t \quad \text{for } t > 0 \qquad (15)$$

In this exercise we analyze the function N.
a. By taking derivatives of both sides of (15) with respect to t, show that

$$[N'(10^7 - P(t))][P(t) - 10^7] = 10^7 \quad \text{for } t > 0 \quad (16)$$

b. Letting $x = 10^7 - P(t)$ in (16), show that

$$N'(x) = \frac{-10^7}{x} \quad \text{for } 0 < x < 10^7$$

and therefore that

$$N(x) = -10^7 \ln x + C \quad \text{for } 0 < x \leq 10^7$$

c. Using the fact that $N(10^7) = 0$, prove that

$$N(x) = -10^7 \ln x + 10^7 \ln 10^7 \quad \text{for } 0 < x < 10^7$$

d. Let $b = e^{-1/10^7}$. Show that

$$-10^7 = \log_b e$$

With the help of Exercise 48(a) conclude that

$$-10^7 \ln x = (\log_b e) \ln x = \log_b x$$

e. Using (c) and (d), show that

$$N(x) = \log_b x + C \quad \text{for } 0 < x < 10^7$$

where $C = 10^7 \ln 10^7$. This proves that N is actually the logarithm to the base b, shifted by C.

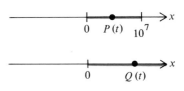

FIGURE 7.16

7.5
EXPONENTIAL GROWTH AND DECAY

Certain strains of algae and bacteria, if given sufficient food and left unmolested, reproduce at a rate proportional to the number present at any given time. As a result, if $f(t)$ represents the number at time t, so that $f'(t)$ represents the rate of growth of the number at time t, then there is a positive constant k such that

$$f'(t) = kf(t) \tag{1}$$

In Example 1 we will consider a specific case of such growth of algae.

Now consider an entirely different physical phenomenon. Carbon has three naturally occurring isotopes: C^{12} and C^{13}, which are stable, and C^{14}, which is unstable and radioactive. In the atmosphere the loss of C^{14} atoms through radioactive decay is offset by the creation of new C^{14} atoms by cosmic radiation. In plants the loss of C^{14} atoms is offset by the intake of C^{14} during photosynthesis, and in living animals the C^{14} is replenished through consumption of vegetation. It turns out that the ratio between the three isotopes of carbon is essentially constant, both in the atmosphere and in living organisms. However, in a dead organism the ratio varies with time, the number of C^{14} atoms decreasing at any given time at a rate proportional to the amount present at that time. Consequently if $f(t)$ denotes the amount of C^{14} in, say, a bone at time t, then f satisfies (1) for an appropriate negative constant k. This is the basis for estimating age by "carbon 14 dating," as we will see in Example 2.

Thus the equation in (1) arises in the description of two quite different physical phenomena. In fact, it arises in the solutions of many diverse physical and mathematical problems. It would therefore be of interest to determine all functions that satisfy (1). Of course, the constant function 0 satisfies (1), but other functions, such as e^{kt}, also satisfy (1). Indeed, if $f(t) = e^{kt}$, then the Chain Rule implies that

$$f'(t) = \frac{d}{dt}(e^{kt}) = ke^{kt} = kf(t)$$

Now let us determine the form of the functions satisfying (1).

Assume that f is a continuous function on $[0, \infty)$ satisfying (1), and let

$$g(t) = e^{-kt}f(t) \tag{2}$$

where k is the constant appearing in (1). Then differentiating and using (1), we find that

$$g'(t) = -ke^{-kt}f(t) + e^{-kt}f'(t) \overset{(1)}{=} -ke^{-kt}f(t) + e^{-kt}kf(t) = 0$$

for all $t > 0$. Then Theorem 4.7 implies that g is a constant function, so that for some number c we have

$$g(t) = c \quad \text{for } t \geq 0$$

Therefore from (2) we see that

$$f(t) = g(t)e^{kt} = ce^{kt} \quad \text{for } t \geq 0 \tag{3}$$

Thus any function f satisfying (1) must be a constant multiple of e^{kt}. Finally, observe that

$$f(0) = ce^{k \cdot 0} = c$$

so we can substitute $f(0)$ for c in (3). This yields

$$f(t) = f(0)e^{kt} \quad \text{for } t \geq 0$$

We summarize our result in a theorem.

THEOREM 7.12

Suppose f is continuous on $[0, \infty)$ and
$$f'(t) = kf(t) \quad \text{for } t > 0$$
Then
$$f(t) = f(0)e^{kt} \quad \text{for } t \geq 0 \tag{4}$$

If a function f satisfies (4), then we say that f **grows exponentially** if $k > 0$ and that f **decays exponentially** if $k < 0$. For future reference we notice that by the same analysis, if f satisfies (1) for all real numbers t, then f also satisfies (4) for all real numbers t.

Example 1   It is known that a certain kind of algae in the Dead Sea can double in population every 2 days. Assuming that the population of algae grows exponentially, beginning now with a population of 1,000,000, determine what the population will be after one week.

Solution   Let $f(t)$ denote the number of algae t days from now, for $t \geq 0$. We wish to find $f(7)$. By hypothesis, $f(0) = 1,000,000$, and combined with (4) this means that
$$f(t) = f(0)e^{kt} = 1,000,000e^{kt} \quad \text{for } t \geq 0 \tag{5}$$
In order to find $f(7)$, we first determine k. By using (5) and the hypothesis that $f(2) = 2,000,000$ we deduce that
$$2,000,000 = f(2) = 1,000,000e^{2k}$$
Thus
$$2 = e^{2k}$$
so that by taking logarithms of both sides, we have
$$\ln 2 = 2k$$
and therefore
$$k = \frac{1}{2}\ln 2$$
Consequently from (5),
$$f(t) = 1,000,000e^{(\ln 2)t/2} \tag{6}$$
The right side of (6) can be simplified by noting that
$$e^{(\ln 2)t/2} = (e^{\ln 2})^{t/2} = 2^{t/2}$$
which means that
$$f(t) = 1,000,000(2^{t/2}) \quad \text{for } t \geq 0$$

To find $f(7)$, we use the last equation with $t = 7$ and find that

$$f(7) = 1,000,000(2^{7/2}) \approx 11,313,709 \quad \square$$

The time it takes for a population to double is called its **doubling time**. Thus the doubling time for the algae in Example 1 is 2 days. More generally, if a population grows exponentially with doubling time d, then the solution of Example 1 can be modified to show that at time t the population is given by

$$f(t) = f(0)2^{t/d} \quad \text{for } t \geq 0$$

(see Exercise 6).

The **half-life** of a radioactive substance is the length of time it takes for half of a given amount of the substance to disintegrate through radiation. By international agreement the half-life of C^{14} is considered to be 5568 years.* This means that 5568 years after an organism dies, it should contain half as much C^{14} as it did when it was alive.

Example 2 Let $f(t)$ denote the amount of C^{14} present in an organism t years after its death. Determine the constant k in equation (4) and the percentage of C^{14} that should remain 1000 years after the death of the organism.

Solution If t denotes the number of years after death, then

$$f(t) = f(0)e^{kt} \quad \text{for } t \geq 0$$

The hypothesis that the half-life of C^{14} is 5568 years means that $f(5568) = \frac{1}{2}f(0)$, so that

$$\frac{1}{2} = \frac{f(5568)}{f(0)} = e^{5568k}$$

Taking logarithms, we obtain

$$5568k = \ln\frac{1}{2} = -\ln 2$$

so that

$$k = \frac{-\ln 2}{5568} \approx -1.24 \times 10^{-4}$$

Therefore

$$f(t) \approx f(0)e^{-1.24 \times 10^{-4}t} \quad \text{for } t \geq 0 \tag{7}$$

To determine the percentage of C^{14} left after 1000 years, we need only calculate $f(1000)/f(0)$:

$$\frac{f(1000)}{f(0)} \approx e^{-1.24 \times 10^{-4} \times 1000} = e^{-0.124} \approx 0.883380$$

Thus approximately 88.338 percent remains after a millennium. $\square$

---

* Recent measurements indicate that the half-life of C^{14} is actually closer to 5730 years.

Example 3   Suppose that of the original amount of C^{14} in a human bone uncovered in Kenya only $\frac{1}{10}$ remains today. How long ago did death occur?

Solution   If t denotes the number of years from death until now, then we know by hypothesis that

$$f(t) = \frac{1}{10} f(0)$$

and by (7) that

$$f(t) \approx f(0)e^{-1.24 \times 10^{-4}t}$$

Consequently

$$e^{-1.24 \times 10^{-4}t} \approx \frac{f(t)}{f(0)} = \frac{1}{10}$$

Taking logarithms, we find that

$$-1.24 \times 10^{-4}t \approx \ln\frac{1}{10} = -\ln 10$$

Solving for t, we obtain

$$t \approx \frac{-\ln 10}{-1.24 \times 10^{-4}} \approx 18{,}569.2$$

Thus death occurred approximately 18,569 years ago.   ☐

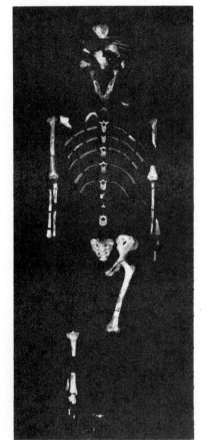

The skeleton of "Lucy," a possible ancestor of the human species discovered in Ethiopia by Donald Johanson. By various dating techniques, including the potassium-argon method, Johanson has concluded that the skeleton is 3.75 million years old.

Carbon dating is effective only for dates between approximately 1500 A.D. and 50,000 B.C. For objects significantly older, such as million-year-old fossils and billion-year-old lunar rocks, the potassium–argon and rubidium–strontium dating methods can be used (see Exercises 27–29).

Finally, we consider the air pressure $f(t)$ at an altitude of t meters above sea level. The higher the altitude is, the lower the air pressure is. Moreover, it turns out that to a high degree of accuracy, the rate at which the air pressure decreases with increasing altitude is proportional to the pressure. Consequently it is reasonable to assume that f satisfies (4), so that there is a negative constant k such that

$$f(t) = f(0)e^{kt} \quad \text{for } t \geq 0$$

If t is measured in meters and $f(t)$ in "atmospheres," with $f(0) = 1$ (atmosphere), then experimental results indicate that

$$k \approx -1.25 \times 10^{-4}$$

so that we take

$$f(t) = e^{-1.25 \times 10^{-4}t} \quad \text{for } t \geq 0 \tag{8}$$

Example 4   The atmospheric pressure outside a jet passenger plane is 0.28 atmospheres. Determine the altitude of the plane.

Solution   If t denotes the altitude of the plane in meters above sea level, then by hypothesis, $f(t) = 0.28$. To find t, we use (8):

$$0.28 = f(t) = e^{-1.25 \times 10^{-4}t}$$

Taking logarithms of both sides, we obtain

$$\ln 0.28 = -1.25 \times 10^{-4}t$$

so that

$$t = \frac{\ln 0.28}{-1.25 \times 10^{-4}} \approx 10,183.7 \text{ (meters)}$$

Consequently the plane is approximately 10,184 meters above sea level.   □

An altitude of 10,184 meters is approximately 33,412 feet, a typical altitude of a plane in transcontinental flight. Since at that altitude the air pressure is about $\frac{1}{4}$ the air pressure at sea level, it is not surprising that the pressure inside the plane would be regulated.

As we have seen, problems concerning exponential growth or decay involve four quantities: $t, k, f(0)$, and $f(t)$. If we know any three of them (or if we know the ratio $f(t)/f(0)$ and either t or k), then we can determine the remaining quantity by using (4). The problem normally revolves about finding that remaining quantity.

The functions arising in our examples concerning exponential growth and decay are determined from experimental data in idealized conditions. As a result, we should expect the formulas we obtain to be only approximately accurate. Moreover, we frequently idealize the functions themselves. After all, the number of bacteria or C^{14} atoms is always an integer, and yet we use a continuous function to represent the number of bacteria at various times. However, these idealizations lead us to very valuable information about different sorts of natural phenomena, and this is the utility of the theory of exponential growth and decay.

EXERCISES 7.5

1. How long would it take for the algae mentioned in Example 1 to
   a. quadruple in number?
   b. triple in number?

2. Suppose that a colony of bacteria is growing exponentially. If 12 hours are required for the number of bacteria to grow from 4000 to 6000, find the doubling time.

3. Experiment has shown that under ideal conditions and a constant temperature of 28.5°C, the population of a certain type of flour beetle doubles in 6 days and 20 hours. Suppose that there are now 1500 such beetles. How long ago were there 1200?

4. If the population of the world is not unduly affected by war, famine, or new technology, then it is reasonable to assume that the population will grow (at least for a long period of time) at an exponential rate. Using this assumption and the census figures which show that

   the world population in 1962 was 3,150,000,000

   the world population in 1978 was 4,238,000,000

   determine what the world population should be in 1990.

5. Suppose the populations of two countries are growing exponentially. Suppose also that one country has a population of 50,000,000 and a doubling time of 20 years, whereas the other has a population of 20,000,000 and a doubling time of 10 years. How long will it be until the two countries have the same population?

6. Suppose the population $f(t)$ of a given species grows exponentially, so that $f(t) = f(0)e^{kt}$ for some positive constant k.
   a. Show that the population doubles during any time interval of duration $(\ln 2)/k$. Thus $(\ln 2)/k$ is the doubling time d.
   b. Show that $f(t) = f(0)2^{t/d}$.

7. Suppose the amount $f(t)$ of a radioactive substance decays exponentially, so that $f(t) = f(0)e^{kt}$ for some negative constant k.
   a. Show that the amount decreases by half in any time interval of duration $-(\ln 2)/k$. Thus $-(\ln 2)/k$ is the half-life h.
   b. Show that $f(t) = f(0)(\frac{1}{2})^{t/h}$.

8. Suppose an unknown radioactive substance produces 4000 counts per minute on a Geiger counter at a certain time, and only 500 counts per minute 4 days later. Assuming that the amount of radioactive substance is proportional to the number of counts per minute, determine the half-life of the radioactive substance.

9. Suppose you have a cache of radium, whose half-life is approximately 1590 years. How long would you have to wait for one tenth of it to disappear?

10. Uranium 238 is a radioactive isotope. Of any given amount, 85.719 percent remains after 1 billion years. Compute the half-life of uranium 238.

11. The so-called "Pittsburgh man," unearthed near the town of Arella, Pennsylvania, shows that civilization existed there from around 12,300 B.C. to 13,000 B.C. Calculate the difference between the percentage of C^{14} lost prior to 2000 A.D. from a bone of someone who died in 12,300 B.C. and that from someone who died in 13,000 B.C.

12. Richard Leakey's "1470" skull, found in Kenya, is reputed to be 1,800,000 years old. Show that the percentage of C^{14} remaining now would be negligible, and hence that in this instance dating by means of C^{14} would be meaningless.

13. Iodine 131, which is used for treating cancer of the thyroid gland, is also used to detect leaks in water pipes. It has a half-life of 8.14 days. Suppose a water company wishes to use 100 mg of iodine 131 to search for a leak and must take delivery 2 days before using it. How much iodine 131 should be purchased?

14. Suppose a small quantity of radon gas, which has a half-life of 3.8 days, is accidentally released into the air in a laboratory. If the resulting radiation level is 50% above the "safe" level, how long should the laboratory remain vacated?

15. Cabins in jet passenger planes are often pressurized to a pressure equivalent to that at 1600 meters above sea level.

Use (8) to determine the ratio of the cabin pressure to the atmospheric pressure at sea level.

16. Mountain climbers normally wear gas masks when they are higher than 7000 meters above sea level. Determine the air pressure at 7000 meters.

17. Halley's Law states that the barometric pressure $p(t)$ in inches of mercury at t miles above sea level is given by

$$p(t) \approx 29.92e^{-0.2t} \quad \text{for } t \geq 0$$

Find the barometric pressure
   a. at sea level
   b. 5 miles above sea level
   c. 10 miles above sea level

18. Assume that the air pressure $p(t)$ in pounds per square foot at t feet above sea level is given by

$$p(t) \approx 2140e^{-0.000035t} \quad \text{for } t \geq 0$$

and that an airplane is losing altitude at the rate of 20 miles per hour. At what rate is the air pressure just outside the plane increasing when the plane is 2 miles above sea level?

19. For an operation a dog is anesthetized with sodium pentobarbitol, which is eliminated exponentially from the blood stream. Assume that of any sodium pentobarbitol in the blood stream, half is eliminated in 5 hours. Assume also that to anesthetize a dog, 20 milligrams of sodium pentobarbitol are required for each kilogram of body weight. What single dose of sodium pentobarbitol would be required to anesthetize a dog weighing 10 kilograms for half an hour?

20. If a sum of S dollars is invested at p percent interest and interest is compounded continuously, then the amount $A(t)$ of money accumulated after t years is given by

$$A(t) = Se^{pt/100} \quad \text{for } t \geq 0$$

In terms of S, how much money will there be after 10 years, if the interest rate is 6 percent?

21. Using the equation of Exercise 20, determine what interest rate would be required in order to double a sum of money within 10 years.

22. If one of your grandparents had put $100 into a savings account 75 years ago at 4 percent interest compounded continuously, how much would be in the account now? Suppose George Washington had done the same 200 years ago, but at 3 percent interest. How much would be in the account now?

23. The **unimolecular reaction theory** says that if a substance is dissolved in a large container of solvent, then the rate of reaction is proportional to the amount of the remaining substance. Suppose a sugar cube 1 cubic inch in volume is

dropped into a jug of iced tea. If there is $\frac{3}{4}$ of a cubic inch of the cube left after 1 minute, then when is there $\frac{1}{2}$ cubic inch left?

24. When an electric condenser discharges electricity to the ground through resistance, the charge on the condenser decreases at a rate proportional to the amount of charge. If the charge was 5×10^{-2} coulombs 4 seconds ago and is 10^{-3} coulombs now, what was the charge 1 second ago?

25. This exercise leads to a different method of obtaining (4). Assume that

$$f'(t) = kf(t) \quad \text{for } t > 0 \qquad (9)$$

that $f(t) > 0$ for all $t > 0$, and that f is continuous on $[0, \infty)$.

a. Divide both sides of (9) by $f(t)$ and integrate both sides of the resulting equation in order to obtain

$$k(t - a) = \int_a^t \frac{f'(s)}{f(s)} ds = \ln \frac{f(t)}{f(a)}$$

where a is any fixed positive number.

b. Now show that

$$f(t) = f(0)e^{kt} \quad \text{for all } t \geq 0$$

(*Hint:* Using (a), equate $e^{k(t-a)}$ and $e^{\ln(f(t)/f(a))}$ and then let a tend to 0 from the right.)

26. Suppose f has the property that

$$f(t) = \int_0^t f(s) ds \quad \text{for all } t \geq 0$$

Show that $f(t) = 0$ for all $t \geq 0$.

27. Suppose a "parent" substance decays exponentially into a "daughter" substance. Then the amount $P(t)$ of the parent remaining after t years is given by

$$P(t) = P(0)e^{-\lambda t}$$

and the amount $D(t)$ of the daughter substance is given by

$$D(t) = P(0) - P(t)$$

Show that

$$t = \frac{1}{\lambda} \ln \left(\frac{D(t)}{P(t)} + 1 \right)$$

(This formula can be used to determine the age of an object; see Exercise 28.)

28. For the decay of rubidium (Rb^{87}) into strontium (Sr^{87}) the value of λ in Exercise 27 is 1.39×10^{-11}.

a. Biotite taken from a sample of granite in the Grand Canyon contained 202 parts per million of rubidium and 3.96 parts per million of strontium.* Determine the age of the sample of granite.

* Dale Nations, *The Record of Geological Time: A Vicarious Trip* (New York: McGraw-Hill, 1975), p. 48.

A lunar breccia found by the Apollo 16 mission.

b. Some lunar breccias, formed by meteors and collected during the Apollo 16 mission, were determined by the rubidium–strontium method to be approximately 4.53 billion years old. Assuming that the breccias contained no strontium when they were formed, determine the present ratio of strontium to rubidium in the breccias.

29. Potassium (K^{40}) decays into two substances, calcium (Ca^{40}) and argon (Ar^{40}). By measuring the amounts of potassium and argon it is possible to determine the age of an object from which they are extracted. The formula, which is more complicated than the one in Exercise 27, is

$$t = (1.885)10^9 \ln \left(9.068 \frac{D(t)}{P(t)} + 1 \right)$$

where $D(t)$ is the amount of argon and $P(t)$ the amount of potassium at time t (in years).

a. A sample of basalt taken from lava in the Grand Canyon contained 1.95×10^{-12} moles per gram of argon and 2.885×10^{-8} moles per gram of potassium.[†] Determine the age of the basalt.

b. A rock from the lunar plains collected during the Apollo 16 mission was determined by the potassium–argon method to be approximately 4.19 billion years old. Assuming that the rock contained no argon when it was formed, determine the present ratio of the amount of argon to the amount of potassium in the lunar rock.

† Dale Nations, *The Record of Geological Time: A Vicarious Trip* (New York: McGraw-Hill, 1975), p. 48.

7.6
THE INVERSE TRIGONOMETRIC FUNCTIONS

In this section we give special names to six inverse functions, one for each of the basic trigonometric functions. The inverse trigonometric functions are used in evaluating integrals, as well as in other applications.

The Arcsine Function   By the criterion in (5) of Section 7.1, the sine function does not have an inverse because it is periodic. However, if we restrict the domain of the sine function to $[-\pi/2, \pi/2]$, then the resulting function is strictly increasing (because its derivative is positive except at $-\pi/2$ and $\pi/2$). Hence the restricted function has an inverse, which we call the **arcsine function**. The domain of the arcsine is $[-1, 1]$, and its range is $[-\pi/2, \pi/2]$. Its value at x is usually written arcsin x or $\sin^{-1} x$. As a consequence,

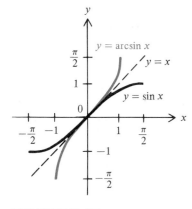

FIGURE 7.17

$$\boxed{\begin{array}{l} \arcsin x = y \quad \text{if and only if} \quad \sin y = x \\ \text{for } -1 \le x \le 1 \quad \text{and} \quad -\pi/2 \le y \le \pi/2 \end{array}}$$

Thus if $-1 \le x \le 1$, then arcsin x is the number y between $-\pi/2$ and $\pi/2$ whose sine is x. For example,

$$\arcsin(-1) = -\frac{\pi}{2}, \quad \arcsin 0 = 0, \quad \arcsin\frac{1}{2} = \frac{\pi}{6}, \quad \arcsin\frac{\sqrt{2}}{2} = \frac{\pi}{4}$$

The graph of the arcsine function appears in Figure 7.17. We also have the equations

$$\boxed{\begin{array}{l} \arcsin(\sin x) = x \quad \text{for } -\pi/2 \le x \le \pi/2 \\ \sin(\arcsin x) = x \quad \text{for } -1 \le x \le 1 \end{array}}$$

Caution: Although arcsin $(\sin x)$ is defined for all real numbers, the equation $\arcsin(\sin x) = x$ is valid only for x in $[-\pi/2, \pi/2]$ because that interval is the range of the arcsine function. Thus care must be exercised when we evaluate arcsin $(\sin x)$ with x outside $[-\pi/2, \pi/2]$.

Example 1   Evaluate $\arcsin\left(\sin\dfrac{5\pi}{6}\right)$.

Solution   First we use the fact that $\sin 5\pi/6 = \frac{1}{2}$, so that

$$\arcsin\left(\sin\frac{5\pi}{6}\right) = \arcsin\frac{1}{2}$$

Since $\sin \pi/6 = \frac{1}{2}$, it follows that $\arcsin \frac{1}{2} = \pi/6$, so that

$$\arcsin\left(\sin\frac{5\pi}{6}\right) = \frac{\pi}{6} \quad \square$$

Sometimes an expression involving both trigonometric functions and their inverses occurs. It may be possible to simplify such an expression by drawing and analyzing a judiciously chosen right triangle.

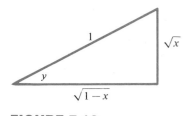

Example 2   Simplify the expression $\sec(\arcsin \sqrt{x})$.

Solution   We will evaluate $\sec(\arcsin \sqrt{x})$ by evaluating $\sec y$ for the value of y in $(-\pi/2, \pi/2)$ such that $\arcsin \sqrt{x} = y$, that is, $\sin y = \sqrt{x}$. Since $\sin y = \sqrt{x} \geq 0$, it follows that $0 \leq y < \pi/2$. Applying the Pythagorean Theorem to the triangle in Figure 7.18, we find $\sec y = 1/\sqrt{1-x}$. Therefore

$$\sec(\arcsin \sqrt{x}) = \sec y = 1/\sqrt{1-x} \quad \square$$

FIGURE 7.18

Next we derive a formula for the derivative of the arcsine function. Toward that end, let $y = \arcsin x$. Then $-\pi/2 < y < \pi/2$ and $x = \sin y$. Thus $\cos y \geq 0$, so that

$$\cos y = \sqrt{1 - \sin^2 y} = \sqrt{1 - x^2}$$

Therefore by (3) of Section 7.2 we find that

$$\frac{d}{dx}\arcsin x = \frac{1}{\dfrac{d}{dy}\sin y} = \frac{1}{\cos y} = \frac{1}{\sqrt{1-x^2}}$$

Consequently

$$\frac{d}{dx}\arcsin x = \frac{1}{\sqrt{1-x^2}} \quad \text{for } -1 < x < 1 \tag{1}$$

Example 3   Let $f(x) = \arcsin x^4$. Find $f'(x)$.

Solution   Using (1), along with the Chain Rule, we find that

$$f'(x) = \frac{1}{\sqrt{1-(x^4)^2}}(4x^3) = \frac{4x^3}{\sqrt{1-x^8}} \quad \square$$

The integral formula corresponding to (1) is

$$\int \frac{1}{\sqrt{1-x^2}}\,dx = \arcsin x + C \tag{2}$$

We have thus expanded the collection of integrals that we can evaluate.

Example 4   Find $\displaystyle\int \frac{1}{\sqrt{9 - x^2}}\,dx$.

Solution   We wish the denominator to have the form $\sqrt{1 - u^2}$ for a suitable u, so we first prepare the integral by factoring out the 9 in the square root:

$$\int \frac{1}{\sqrt{9 - x^2}}\,dx = \int \frac{1}{\sqrt{9(1 - x^2/9)}}\,dx = \frac{1}{3}\int \frac{1}{\sqrt{1 - (x/3)^2}}\,dx$$

Next we make the substitution

$$u = \frac{x}{3}, \quad \text{so that} \quad du = \frac{1}{3}\,dx$$

Therefore by (2),

$$\int \frac{1}{\sqrt{9 - x^2}}\,dx = \frac{1}{3}\int \overbrace{\frac{1}{\sqrt{1 - (x/3)^2}}}^{1/\sqrt{1 - u^2}} \overbrace{dx}^{3\,du} = \frac{1}{3}\int \frac{1}{\sqrt{1 - u^2}}\,3\,du$$

$$\overset{(2)}{=} \arcsin u + C = \arcsin\frac{x}{3} + C \quad \square$$

By the same method as used in Example 4 you can show that

$$\int \frac{1}{\sqrt{a^2 - x^2}}\,dx = \arcsin\frac{x}{a} + C \quad \text{for } a > 0 \tag{3}$$

A second way of proving (3) is to differentiate the right-hand side of (3) with the help of the Chain Rule and (1):

$$\frac{d}{dx}\arcsin\frac{x}{a} = \frac{1}{\sqrt{1 - (x/a)^2}}\left(\frac{1}{a}\right) = \frac{1}{\sqrt{a^2 - x^2}}$$

The result is exactly the integrand in (3).

Example 5   Find $\displaystyle\int \frac{1}{\sqrt{4x - x^2}}\,dx$.

Solution   We wish the denominator to have the form $\sqrt{a^2 - u^2}$ for suitable a and u, so we first rewrite the denominator by completing the square under the radical:

$$\int \frac{1}{\sqrt{4x - x^2}}\,dx = \int \frac{1}{\sqrt{4 - (4 - 4x + x^2)}}\,dx = \int \frac{1}{\sqrt{4 - (x - 2)^2}}\,dx$$

Now we make the substitution

$$u = x - 2, \quad \text{so that} \quad du = dx$$

and then use (3) with $a = 2$:

$$\int \frac{\overbrace{1}^{1/\sqrt{4-u^2}}}{\sqrt{4 - (x-2)^2}} \overset{du}{dx} = \int \frac{1}{\sqrt{4 - u^2}} du \overset{(3)}{=} \arcsin \frac{u}{2} + C = \arcsin \frac{x-2}{2} + C$$

Thus

$$\int \frac{1}{\sqrt{4x - x^2}} dx = \int \frac{1}{\sqrt{4 - (x-2)^2}} dx = \arcsin \frac{x-2}{2} + C \quad \square$$

If $\int_3^{2+\sqrt{2}} 1/\sqrt{4x - x^2}\, dx$ were the integral in Example 5, we would still make the substitution $u = x - 2$, but the limits of integration would require suitable alterations. (If $x = 3$ then $u = 1$, and if $x = 2 + \sqrt{2}$ then $u = \sqrt{2}$.) Therefore

$$\int_3^{2+\sqrt{2}} \frac{1}{\sqrt{4x - x^2}} dx = \int_3^{2+\sqrt{2}} \frac{1}{\sqrt{4 - (x-2)^2}} dx = \int_1^{\sqrt{2}} \frac{1}{\sqrt{4 - u^2}} du$$

$$= \arcsin \frac{u}{2} \bigg|_1^{\sqrt{2}} = \arcsin \frac{\sqrt{2}}{2} - \arcsin \frac{1}{2}$$

$$= \frac{\pi}{4} - \frac{\pi}{6} = \frac{\pi}{12}$$

In conclusion, we mention that although it is customary to choose the interval $[-\pi/2, \pi/2]$ as the range of the arcsine function, we could equally well have used $[\pi/2, 3\pi/2]$, or any other interval on which the sine function has an inverse.

The Arctangent Function

To define an inverse for the tangent function, we restrict the tangent function to $(-\pi/2, \pi/2)$. The resulting inverse function is called the **arctangent function**. Its domain is $(-\infty, \infty)$, and its range is $(-\pi/2, \pi/2)$. We usually write its value at x as $\arctan x$ or $\tan^{-1} x$. As a consequence,

> $\arctan x = y$ if and only if $\tan y = x$
>
> for any x and for $-\pi/2 < y < \pi/2$

Thus for any x, $\arctan x$ is the number y between $-\pi/2$ and $\pi/2$ whose tangent is x. For instance,

$$\arctan 0 = 0, \quad \arctan 1 = \frac{\pi}{4}, \quad \text{and} \quad \arctan(-\sqrt{3}) = -\frac{\pi}{3}$$

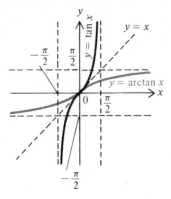

FIGURE 7.19

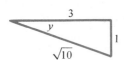

FIGURE 7.20

The graph of the arctangent function is shown in Figure 7.19. We also have the relations

$$\arctan(\tan x) = x \quad \text{for } -\pi/2 < x < \pi/2$$
$$\tan(\arctan x) = x \quad \text{for all } x$$

Example 6   Evaluate $\csc(\arctan(-\tfrac{1}{3}))$.

Solution   We will evaluate $\csc(\arctan(-\tfrac{1}{3}))$ by evaluating $\csc y$ for the value of y in $(-\pi/2, \pi/2)$ such that $\arctan(-\tfrac{1}{3}) = y$, that is, $\tan y = -\tfrac{1}{3}$. Since $\tan y < 0$, it follows that $-\pi/2 < y < 0$, so that $\csc y < 0$. By the Pythagorean Theorem, the hypotenuse of the triangle in Figure 7.20 has length $\sqrt{10}$. Therefore $\csc y = -\sqrt{10}$. As a result,

$$\csc\left(\arctan\left(-\frac{1}{3}\right)\right) = \csc y = -\sqrt{10} \quad \square$$

In order to find the derivative of the arctangent function, we let $y = \arctan x$. Then $x = \tan y$ and

$$\sec^2 y = \tan^2 y + 1 = x^2 + 1$$

Therefore by (3) in Section 7.2,

$$\frac{d}{dx}\arctan x = \frac{1}{\dfrac{d}{dy}\tan y} = \frac{1}{\sec^2 y} = \frac{1}{x^2 + 1}$$

In other words,

$$\frac{d}{dx}\arctan x = \frac{1}{x^2 + 1} \tag{4}$$

Example 7   Find $\dfrac{d}{dx}\arctan e^{3x}$.

Solution   By (4), along with the Chain Rule,

$$\frac{d}{dx}\arctan e^{3x} = \frac{1}{(e^{3x})^2 + 1}(e^{3x} \cdot 3) = \frac{3e^{3x}}{e^{6x} + 1} \quad \square$$

The indefinite integral version of (4) is

$$\int \frac{1}{x^2 + 1}\,dx = \arctan x + C \tag{5}$$

Example 8 Find $\displaystyle\int \frac{1}{x^2 + 16}\,dx$.

Solution We wish the denominator to have the form $u^2 + 1$ for suitable u, so we factor out 16:

$$\int \frac{1}{x^2 + 16}\,dx = \int \frac{1}{16\left(\dfrac{x^2}{16} + 1\right)}\,dx = \frac{1}{16}\int \frac{1}{(x/4)^2 + 1}\,dx$$

Then we let

$$u = \frac{x}{4}, \quad \text{so that} \quad du = \frac{1}{4}\,dx$$

Therefore by (5),

$$\int \frac{1}{x^2 + 16}\,dx = \frac{1}{16}\int \overbrace{\frac{1}{(x/4)^2 + 1}}^{1/(u^2+1)}\overbrace{dx}^{4\,du} = \frac{1}{16}\int \frac{1}{u^2 + 1}\,4\,du$$

$$= \frac{1}{4}\arctan u + C = \frac{1}{4}\arctan\frac{x}{4} + C \quad \square$$

In general, if $a > 0$, then

$$\int \frac{1}{x^2 + a^2}\,dx = \frac{1}{a}\arctan\frac{x}{a} + C \tag{6}$$

This can be shown by letting $u = x/a$ and proceeding exactly as in the solution of Example 8, or by differentiating both sides of (6). Because integrals of the form in (6) occur often (especially in Section 8.4), it is advisable that you memorize (6).

Example 9 Find $\displaystyle\int \frac{1}{2x^2 + 2x + 1}\,dx$.

Solution We wish the denominator to have the form $u^2 + a^2$ for suitable u and a, so we factor out 2 and then complete the square:

$$\int \frac{1}{2x^2 + 2x + 1}\,dx = \frac{1}{2}\int \frac{1}{x^2 + x + \frac{1}{2}}\,dx = \frac{1}{2}\int \frac{1}{(x + \frac{1}{2})^2 + (\frac{1}{2})^2}\,dx$$

Now we make the substitution

$$u = x + \frac{1}{2}, \quad \text{so that} \quad du = dx$$

and conclude from (6), with $a = \frac{1}{2}$, that

$$\int \frac{1}{2x^2 + 2x + 1} \, dx = \frac{1}{2} \int \frac{1}{u^2 + (\frac{1}{2})^2} \, du$$

$$= \frac{\frac{1}{2}}{\frac{1}{2}} \arctan \frac{u}{\frac{1}{2}} + C = \arctan 2\left(x + \frac{1}{2}\right) + C$$

$$= \arctan (2x + 1) + C \quad \square$$

We conclude this subsection with an application of the arctangent function.

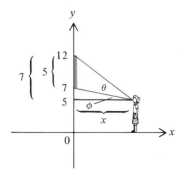

FIGURE 7.21

Example 10 Suppose a large oil painting of water lilies by Claude Monet is on display at the National Gallery in Washington, D.C. In order to accommodate a large number of viewers at one time, the painting is placed with its bottom and top edges 7 and 12 feet above the floor, respectively (Figure 7.21). Assume that a spectator's eyes are 5 feet above the floor, and that the spectator has the best view when the angle θ shown in the illustration is largest. Determine which distance x from the wall will allow the best view of the water lilies.

Solution We seek the maximum value of θ, regarded as a function of x. If ϕ is as in Figure 7.21, then we have

$$\tan (\theta(x) + \phi(x)) = \frac{7}{x}, \quad \text{so that} \quad \theta(x) + \phi(x) = \arctan \frac{7}{x}$$

and

$$\tan \phi(x) = \frac{2}{x}, \quad \text{so that} \quad \phi(x) = \arctan \frac{2}{x}$$

Therefore

$$\theta(x) = [\theta(x) + \phi(x)] - \phi(x) = \arctan \frac{7}{x} - \arctan \frac{2}{x} \quad \text{for } x > 0$$

Consequently by the Chain Rule,

$$\theta'(x) = \frac{-7/x^2}{1 + (7/x)^2} - \frac{-2/x^2}{1 + (2/x)^2}$$

$$= \frac{-7}{x^2 + 49} + \frac{2}{x^2 + 4} = \frac{-5(x^2 - 14)}{(x^2 + 49)(x^2 + 4)}$$

Since $x > 0$, this implies that $\theta'(x) = 0$ only for $x = \sqrt{14}$. Since $\theta'(x) > 0$ for $0 < x < \sqrt{14}$ and $\theta'(x) < 0$ for $x > \sqrt{14}$, it follows from the First Derivative Test that θ achieves its maximum value for $x = \sqrt{14}$. Therefore the spectator has the best view when standing $\sqrt{14} \approx 3.74$ feet from the wall. $\quad \square$

The Remaining Inverse Trigonometric Functions

Now we turn to the inverses of the remaining trigonometric functions. We define the arccosine, arccotangent, arcsecant, and arccosecant functions by the following formulas:

$\arccos x = y$ if and only if $\cos y = x$ for $-1 \le x \le 1$ and for $0 \le y \le \pi$

$\text{arccot}\, x = y$ if and only if $\cot y = x$ for any x and for $0 < y < \pi$

$\text{arcsec}\, x = y$ if and only if $\sec y = x$
    for $|x| \ge 1$ and for $0 \le y < \pi/2$ or $\pi \le y < 3\pi/2$

$\text{arccsc}\, x = y$ if and only if $\csc y = x$
    for $|x| \ge 1$ and for $0 < y \le \pi/2$ or $\pi < y \le 3\pi/2$

The graphs of the four new functions are in Figure 7.22(a)–(d). Of these functions, only the arcsecant function appears with any frequency in the sequel.

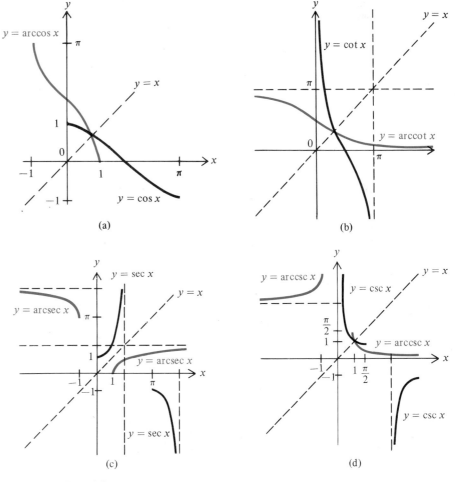

FIGURE 7.22

The same process that we used in finding the derivatives of the arcsine and arctangent functions yields the derivatives of the inverse functions just defined:

$$\frac{d}{dx}\arccos x = \frac{-1}{\sqrt{1-x^2}} \qquad \frac{d}{dx}\operatorname{arcsec} x = \frac{1}{x\sqrt{x^2-1}}$$

$$\frac{d}{dx}\operatorname{arccot} x = \frac{-1}{x^2+1} \qquad \frac{d}{dx}\operatorname{arccsc} x = \frac{-1}{x\sqrt{x^2-1}}$$

Corresponding integration formulas are as follows (where we assume that $a > 0$):

$$\int \frac{1}{\sqrt{a^2-x^2}}\,dx = -\arccos\frac{x}{a} + C \tag{7}$$

$$\int \frac{1}{x^2+a^2}\,dx = -\frac{1}{a}\operatorname{arccot}\frac{x}{a} + C \tag{8}$$

$$\int \frac{1}{x\sqrt{x^2-a^2}}\,dx = \frac{1}{a}\operatorname{arcsec}\frac{x}{a} + C \tag{9}$$

$$\int \frac{1}{x\sqrt{x^2-a^2}}\,dx = -\frac{1}{a}\operatorname{arccsc}\frac{x}{a} + C \tag{10}$$

Notice that we can use either the arcsine or the arccosine function to evaluate the integral in (3) and (7). From now on we will use the arcsine function for this purpose. Similar comments hold for the arctangent and arccotangent functions in (6) and (8), and for the arcsecant and arccosecant functions in (9) and (10). For uniformity we will use only the arctangent and arcsecant functions hereafter.

EXERCISES 7.6

In Exercises 1–14 find the numerical value of the expression.

1. $\arcsin\sqrt{3}/2$
2. $\arcsin 1$
3. $\arccos\sqrt{2}/2$
4. $\arccos(-\frac{1}{2})$
5. $\arctan(-1/\sqrt{3})$
6. $\arctan\sqrt{3}$
7. $\operatorname{arccot}\sqrt{3}$
8. $\operatorname{arccot}(-1)$
9. $\operatorname{arcsec}(-\sqrt{2})$
10. $\operatorname{arcsec} 2$
11. $\operatorname{arccsc} 2\sqrt{3}/3$
12. $\sin^{-1}(-\frac{1}{2})$
13. $\cos^{-1}(-\sqrt{3}/2)$
14. $\tan^{-1} 1$

In Exercises 15–24 find the numerical value of the expression.

15. $\sin(\arcsin(-\frac{1}{2}))$
16. $\sin(\arccos\sqrt{2}/2)$
17. $\cos(\arctan(-1))$
18. $\cos(\operatorname{arccsc} 2\sqrt{3}/3)$
19. $\tan(\operatorname{arcsec}\sqrt{2})$
20. $\sec(\arccos\sqrt{3}/2)$
21. $\csc(\operatorname{arccot}(-\sqrt{3}))$
22. $\arctan(\tan 0)$
23. $\arcsin(\cos\pi/6)$
24. $\operatorname{arccot}(\tan\pi/3)$

In Exercises 25–32 simplify the expression.

25. $\cos(\arcsin x)$
26. $\sin(\operatorname{arcsec} x)$
27. $\sec(\arctan x)$
28. $\tan(\operatorname{arccsc} x/2)$
29. $\cos(\operatorname{arccot} x^2)$
30. $\tan(\arccos\sqrt{1-x})$
*31. $\cos(2\arcsin x)$
*32. $\sin(2\arcsin x)$

In Exercises 33–40 find the derivative of f.

33. $f(x) = \arccos(-3x)$
34. $f(x) = x\arcsin x^2$
35. $f(t) = \arctan\sqrt{t}$
36. $f(x) = \arctan\dfrac{x+1}{x-1}$
37. $f(x) = \operatorname{arccot}\sqrt{1-x^2}$
38. $f(u) = \operatorname{arcsec}(-u)$
39. $f(x) = \operatorname{arcsec}(\ln x)$
40. $f(x) = (\operatorname{arccsc} x)^2$

In Exercises 41–60 evaluate the indefinite integral.

41. $\displaystyle\int \frac{1}{x^2+16}\,dx$
42. $\displaystyle\int \frac{1}{t^2+6}\,dt$

43. $\displaystyle\int \frac{1}{9x^2 + 16}\,dx$

44. $\displaystyle\int \frac{1}{x^2 + 4x + 7}\,dx$

45. $\displaystyle\int \frac{1}{2x^2 + 4x + 6}\,dx$

46. $\displaystyle\int \frac{1}{t^2 - 3t + 3}\,dt$

47. $\displaystyle\int \frac{1}{\sqrt{9 - 4x^2}}\,dx$

48. $\displaystyle\int \frac{1}{\sqrt{4 - 9x^2}}\,dx$

49. $\displaystyle\int \frac{1}{x\sqrt{x^2 - 25}}\,dx$

50. $\displaystyle\int \frac{1}{x\sqrt{4x^4 - 25}}\,dx$

51. $\displaystyle\int \frac{x^3}{\sqrt{1 - x^8}}\,dx$

52. $\displaystyle\int \frac{1}{x \ln x \sqrt{(\ln x)^2 - 1}}\,dx$

53. $\displaystyle\int \frac{e^{-x}}{1 + e^{-2x}}\,dx$

54. $\displaystyle\int \frac{e^{3x}}{\sqrt{1 - e^{6x}}}\,dx$

55. $\displaystyle\int \frac{\arctan 2x}{1 + 4x^2}\,dx$

56. $\displaystyle\int \frac{1}{x[1 + (\ln x)^2]}\,dx$

57. $\displaystyle\int \frac{\cos t}{9 + \sin^2 t}\,dt$

58. $\displaystyle\int \frac{\sec^2 x}{\sqrt{4 - \tan^2 x}}\,dx$

59. $\displaystyle\int \frac{\cos 4x}{\sin 4x \sqrt{16 \sin^2 4x - 4}}\,dx$

60. $\displaystyle\int \frac{x^{n-1}}{1 + x^{2n}}\,dx$, n a positive integer

In Exercises 61–64 evaluate the definite integral.

61. $\displaystyle\int_0^2 \frac{1}{\sqrt{16 - x^2}}\,dx$

62. $\displaystyle\int_{2\sqrt{2}}^{2\sqrt{3}} \frac{1}{\sqrt{16 - t^2}}\,dt$

63. $\displaystyle\int_{-2}^{2\sqrt{3}-2} \frac{1}{u^2 + 4u + 8}\,du$

64. $\displaystyle\int_{4\sqrt{3}/3}^{4} \frac{1}{x\sqrt{x^2 - 4}}\,dx$

65. Find the area A of the region bounded by the y axis and the graphs of
$$y = \frac{1}{x^2 - 2x + 4} \quad \text{and} \quad y = \frac{1}{3}$$

66. Find the area A of the region below the graphs of $y = 1/x$ and $y = 1/\sqrt{1 - x^2}$, above the x axis, and between the lines $x = \frac{1}{2}$ and $x = \frac{1}{2}\sqrt{3}$.

67. Evaluate $\int_0^1 \arcsin x\,dx$ by using the formula
$$\int_a^b f(x)\,dx + \int_{f(a)}^{f(b)} f^{-1}(y)\,dy = bf(b) - af(a)$$
appearing in Exercise 20 of Section 7.2.

*68. Let a, $b > 0$. By making the substitution $u = \tan x$, evaluate
$$\int \frac{1}{a^2 \sin^2 x + b^2 \cos^2 x}\,dx$$

69. Prove that the graph of the arctangent function has an inflection point at $x = 0$. (*Hint:* Find the second deriva-tive of the arctangent function. Then analyze the concav-ity of the graph of the arctangent.)

*70. Prove that
$$\arcsin x + \arcsin y = \arcsin(x\sqrt{1 - y^2} + y\sqrt{1 - x^2})$$
provided that the value of the expression on the left-hand side lies in $[-\pi/2, \pi/2]$. (*Hint:* Use the formula for $\sin(a + b)$, along with Exercise 25.)

71. Prove that
$$\arctan \frac{x}{\sqrt{1 - x^2}} = \arcsin x \quad \text{for } -1 < x < 1$$
(*Hint:* You can prove this by differentiating both sides. You might convince yourself that the formula is correct by drawing a right triangle whose sides have length x, $\sqrt{1 - x^2}$, and 1.)

*72. Prove that
$$\arctan x + \arctan y = \arctan \frac{x + y}{1 - xy} \quad \text{for } xy \neq 1$$
provided that the value of the expression on the left-hand side lies in $(-\pi/2, \pi/2)$.

73. Using Exercise 72, verify the following relations involving the number $\pi/4$. These can be used to estimate π. (We will use (d) for this purpose in Chapter 9.)
   a. $\arctan \frac{1}{2} + \arctan \frac{1}{3} = \pi/4$
   b. $2 \arctan \frac{1}{3} + \arctan \frac{1}{7} = \pi/4$
   c. $\arctan \frac{120}{119} - \arctan \frac{1}{239} = \pi/4$
   *d. $4 \arctan \frac{1}{5} - \arctan \frac{1}{239} = \pi/4$ (*Hint:* Use Exercise 72 twice to find $4 \arctan \frac{1}{5}$, and then use part (c).)

*74. Suppose a, b, and c are numbers satisfying $bc = 1 + a^2$. Show that
$$\arctan \frac{1}{a + b} + \arctan \frac{1}{a + c} = \arctan \frac{1}{a} \quad (11)$$
provided that a, $a + b$, and $a + c$ are not 0 and that the value of the expression on the left-hand side lies in $(-\pi/2, \pi/2)$. This formula was discovered by C. L. Dodgson (Lewis Carroll). The equation in Exercise 73(a) results from (11) if $a = b = 1$ and $c = 2$.

*75. Show that the following identities are valid:
   a. $\arcsin\left(\dfrac{x}{3} - 1\right) = \dfrac{\pi}{2} - 2 \arcsin\sqrt{1 - \dfrac{x}{6}}$
   b. $\arcsin\left(\dfrac{x}{3} - 1\right) = 2 \arcsin\dfrac{\sqrt{x}}{\sqrt{6}} - \dfrac{\pi}{2}$

76. a. Use Theorem 4.7 to show that there is a constant c such that
$$\arcsin x + \arccos x = c \quad \text{for } -1 \leq x \leq 1$$
   b. Determine the value of c in part (a).

77. Let $f(x) = \arctan x + \arctan (1/x)$. Show that f is constant on each of the intervals $(-\infty, 0)$ and $(0, \infty)$. Find the constants.

*78. Let $f(x) = \arctan \dfrac{x + 1}{x - 1}$.

   a. Show that $f'(x) = \dfrac{-1}{x^2 + 1}$ for $x \ne 1$.
   b. Find a formula for f''.
   c. Show that $\lim_{x \to 1^+} f(x) = \pi/2$ and $\lim_{x \to 1^-} f(x) = -\pi/2$.
   d. Sketch the graph of f.
   e. Show that there is no constant C such that $f(x) = -\arctan x + C$ for *all* $x \ne 1$, although the functions f and $-\arctan x$ have the same derivative when $x \ne 1$.
   f. Find constants C_1 and C_2 such that

$$f(x) = -\arctan x + C_1 \quad \text{for } x < 1$$
$$f(x) = -\arctan x + C_2 \quad \text{for } x > 1$$

79. Suppose that the painting in Example 10 is to be hung on the wall in a way that provides the best view to observers standing precisely 10 feet from the wall. Assuming that an observer's eyes are 5 feet from the floor, how high should the picture be hung? Would the answer be the same for an observer standing 15 feet from the wall? Explain your answer.

80. An observer watches a rocket blast off 20,000 feet away. At the moment the rocket appears to make an angle of $\pi/4$ radians with the horizon (Figure 7.23), the observer calculates that the angle of elevation of the rocket is

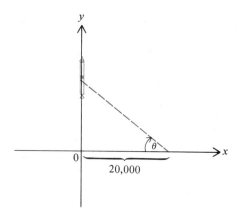

FIGURE 7.23

changing at the rate of $\pi/60$ radians per second and that the rocket seems to move vertically. How fast is the rocket then traveling?

81. Let $f(x) = 1/(1 - x^2)^{1/4}$, and let R be the region between the graph of f and the x axis on $[0, 1/2]$. Find the volume V of the solid obtained by revolving R about the x axis.

82. Let $f(y) = 1/\sqrt{1 - y^4}$, and let R be the region between the graph of f and the y axis on $[0, \frac{1}{2}\sqrt{2}]$. Find the volume V of the solid obtained by revolving R about the x axis.

83. Let $f(x) = \frac{1}{3}x^3 + x - \frac{1}{4}\arctan x$ for $0 \le x \le 1$. Find the length $\mathscr{L}$ of the graph of f.

7.7
HYPERBOLIC FUNCTIONS

In this section we briefly describe the hyperbolic functions, which are defined in terms of the exponential function and have applications in engineering.

The Hyperbolic Functions

The two most important hyperbolic functions are defined as follows:

$$\sinh x = \frac{e^x - e^{-x}}{2} \quad \text{and} \quad \cosh x = \frac{e^x + e^{-x}}{2}$$

We read $\sinh x$ as "hyperbolic sine of x"; mathematicians often give it the shortened pronunciation "shin x" or "sinch x." Analogously, $\cosh x$ is read "hyperbolic cosine of x" or, for short, "cosh x."

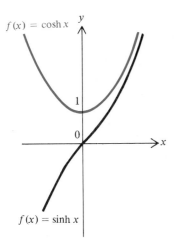

$f(x) = \cosh x$

$f(x) = \sinh x$

FIGURE 7.24

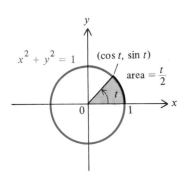

$x^2 + y^2 = 1$

$(\cos t, \sin t)$

area $= \dfrac{t}{2}$

FIGURE 7.25

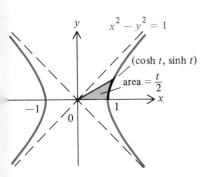

$x^2 - y^2 = 1$

$(\cosh t, \sinh t)$

area $= \dfrac{t}{2}$

FIGURE 7.26

Notice that $\sinh x$ is an odd function with $\sinh 0 = 0$, and that $\cosh x$ is an even function with $\cosh 0 = 1$. Since $0 < e^x < 1$ for $x < 0$, and $e^x > 1$ for $x > 0$, it follows that

$$\sinh x < 0 \quad \text{for } x < 0 \quad \text{and} \quad \sinh x > 0 \quad \text{for } x > 0 \qquad (1)$$

$$\cosh x > 0 \quad \text{for all } x \qquad (2)$$

The derivatives of $\sinh x$ and $\cosh x$ are given by

$$\frac{d}{dx}\sinh x = \frac{e^x + e^{-x}}{2} = \cosh x \qquad (3)$$

$$\frac{d}{dx}\cosh x = \frac{e^x - e^{-x}}{2} = \sinh x \qquad (4)$$

From (2) and (3) we conclude that $\sinh x$ is a strictly increasing function, and from (1) and (4) we conclude that $\cosh x$ is strictly decreasing on $(-\infty, 0]$ and strictly increasing on $[0, \infty)$, so that $\cosh x$ has a minimum value at 0. Evidently

$$\frac{d^2}{dx^2}\sinh x = \sinh x \quad \text{and} \quad \frac{d^2}{dx^2}\cosh x = \cosh x$$

The graphs of these hyperbolic functions are shown in Figure 7.24. Direct calculation shows that

$$\cosh^2 x - \sinh^2 x = \frac{e^{2x} + 2 + e^{-2x}}{4} - \frac{e^{2x} - 2 + e^{-2x}}{4} = 1$$

and this reduces to

$$\boxed{\cosh^2 x - \sinh^2 x = 1} \qquad (5)$$

the fundamental identity relating $\sinh x$ and $\cosh x$. Formula (5) is reminiscent of the fundamental trigonometric identity $\sin^2 x + \cos^2 x = 1$, and formulas (3) and (4) remind us of the formulas for the derivatives of the sine and the cosine. Whereas the point $(\cos t, \sin t)$ lies on the unit circle (Figure 7.25), formula (5) shows that $(\cosh t, \sinh t)$ lies on the hyperbola $x^2 - y^2 = 1$ (Figure 7.26). The unit circle and the hyperbola are related in yet another way. The region shaded in Figure 7.25 has area $t/2$, and the region shaded in Figure 7.26 also has area $t/2$ (see Exercise 47).

If c is a nonzero constant, then the graph of the function

$$y = c \cosh \frac{x}{c}$$

is called a **catenary**, from the Latin word *catena*, meaning chain. A flexible, inelastic chain of uniform density suspended at its two ends will take the shape of an arc of a catenary when it is subjected only to the influence of gravity (Figure 7.27). Thus a cable attached to telephone poles hangs in this shape. It turns out that certain structures are strongest when built in the shape of a catenary. It is for this reason that the Jefferson "Gateway Arch to the West" in

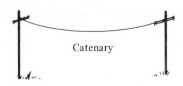

Catenary

FIGURE 7.27

Gateway Arch to the West

FIGURE 7.28

St. Louis was constructed in the shape of a catenary. For this arch the number c is approximately -127.7 (Figure 7.28).

Continuing the analogy of the hyperbolic functions with trigonometric functions, we define the other four hyperbolic functions in terms of $\sinh x$ and $\cosh x$:

$$\tanh x = \frac{\sinh x}{\cosh x} \qquad \operatorname{sech} x = \frac{1}{\cosh x}$$

$$\coth x = \frac{\cosh x}{\sinh x} \qquad \operatorname{csch} x = \frac{1}{\sinh x}$$

The derivatives of these new hyperbolic functions follow readily from the derivatives of $\sinh x$ and $\cosh x$:

$$\frac{d}{dx}\tanh x = \operatorname{sech}^2 x \qquad \frac{d}{dx}\operatorname{sech} x = -\operatorname{sech} x \tanh x$$

$$\frac{d}{dx}\coth x = -\operatorname{csch}^2 x \qquad \frac{d}{dx}\operatorname{csch} x = -\operatorname{csch} x \coth x$$

(see Exercises 17–20). Their graphs appear in Figure 7.29(a)–(d). The six

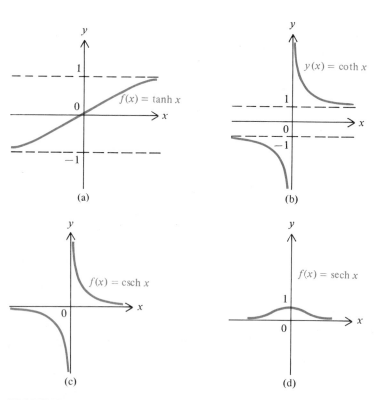

FIGURE 7.29

hyperbolic functions are related by many identities, called **hyperbolic identities**. We list a few of them:

$$\tanh^2 x + \operatorname{sech}^2 x = 1 \qquad \coth^2 x - \operatorname{csch}^2 x = 1$$

$$\sinh(-x) = -\sinh x \qquad \cosh(-x) = \cosh x$$

$$\sinh(x \pm y) = \sinh x \cosh y \pm \cosh x \sinh y$$

$$\cosh(x \pm y) = \cosh x \cosh y \pm \sinh x \sinh y$$

$$\tanh(x \pm y) = \frac{\tanh x \pm \tanh y}{1 \pm \tanh x \tanh y}$$

These identities, which are easily verified from the definitions of the hyperbolic functions, are very similar to the corresponding trigonometric identities listed in Section 1.7. However, in one way the hyperbolic functions are very different from trigonometric functions. Whereas all the trigonometric functions are periodic, *none* of the hyperbolic functions is periodic. The reason is that the hyperbolic functions are defined from the exponential function, which is certainly not periodic.

The Inverse Hyperbolic Sine Function

As you might expect, the hyperbolic functions have inverses, at least when their domains are suitably restricted. Because we will have little need for them later on, we study only the inverse hyperbolic sine function, to give you an idea of how to analyze the inverse hyperbolic functions.

Since $\sinh x$ is a strictly increasing function with $(-\infty, \infty)$ as both domain and range, $\sinh x$ has an inverse with the same domain and range. We denote the value of the inverse at x by $\sinh^{-1} x$:

$$\sinh^{-1} x = y \quad \text{if and only if} \quad \sinh y = x \quad \text{for all } x \text{ and all } y$$

In order to derive a formula for $\sinh^{-1} x$ in terms of x, we note that

$$\sinh y = \frac{e^y - e^{-y}}{2} = \frac{e^{2y} - 1}{2e^y}$$

Hence if $x = \sinh y$, then

$$2e^y x = e^{2y} - 1, \quad \text{or} \quad (e^y)^2 - 2xe^y - 1 = 0$$

Our aim now is to write y as a function of x. Using the quadratic formula and the fact that $e^y > 0$, we deduce that

$$e^y = \frac{2x + \sqrt{4x^2 + 4}}{2} = x + \sqrt{x^2 + 1} \tag{6}$$

We can find y by taking natural logarithms in (6). This yields

$$y = \ln(x + \sqrt{x^2 + 1})$$

Thus our formula for $\sinh^{-1} x$ is

$$\sinh^{-1} x = \ln(x + \sqrt{x^2 + 1}) \quad \text{for all } x \tag{7}$$

Differentiating (7), we obtain

$$\frac{d}{dx}\sinh^{-1}x = \frac{1 + \dfrac{2x}{2\sqrt{x^2+1}}}{x + \sqrt{x^2+1}} = \frac{1}{\sqrt{x^2+1}} \tag{8}$$

The associated indefinite integral is

$$\int \frac{1}{\sqrt{1+x^2}}\,dx = \sinh^{-1}x + C \tag{9}$$

Example 1   Evaluate $\displaystyle\int_{-1}^{3}\frac{1}{\sqrt{1+x^2}}\,dx.$

Solution   From (9) and (7) we obtain

$$\int_{-1}^{3}\frac{1}{\sqrt{1+x^2}}\,dx \overset{(9)}{=} \sinh^{-1}x\Big|_{-1}^{3} = \sinh^{-1}3 - \sinh^{-1}(-1)$$

$$\overset{(7)}{=} \ln(3 + \sqrt{10}) - \ln(-1 + \sqrt{2})$$

$$= \ln\frac{3 + \sqrt{10}}{-1 + \sqrt{2}} \quad \square$$

EXERCISES 7.7

In Exercises 1–12 find the numerical value of the expression.

1. $\sinh 0$
2. $\cosh 0$
3. $\tanh 0$
4. $\tanh 1$
5. $\coth(-1)$
6. $\sinh(\ln 2)$
7. $\sinh(\ln 3)$
8. $\cosh(\ln 3)$
9. $\coth(\ln 4)$
10. $\operatorname{csch}(\ln \pi^2)$
11. $\operatorname{sech}(\ln \sqrt{2})$
12. $\sinh^{-1}\frac{4}{3}$

In Exercises 13–16 simplify the expression.

13. $\sinh(\ln x)$
14. $\cosh(\ln x)$
15. $\tanh(\ln x)$
16. $\sinh^{-1}\dfrac{1-x^2}{2x}$

In Exercises 17–20 establish the formula.

17. $\dfrac{d}{dx}\tanh x = \operatorname{sech}^2 x$

18. $\dfrac{d}{dx}\coth x = -\operatorname{csch}^2 x$

19. $\dfrac{d}{dx}\operatorname{sech} x = -\operatorname{sech} x \tanh x$

20. $\dfrac{d}{dx}\operatorname{csch} x = -\operatorname{csch} x \coth x$

In Exercises 21–26 differentiate the function.

21. $f(x) = \operatorname{sech}\sqrt{x}$
22. $f(x) = \cosh\sqrt{1-x^2}$
23. $f(x) = \sinh^2\sqrt{1-x^2}$
24. $f(x) = e^{\operatorname{csch} x}$
25. $f(x) = \cosh(\arctan e^{2x})$
26. $f(x) = \sinh^{-1}(-3x^2)$

27. Let $y = \dfrac{\sinh^{-1}x}{\sqrt{1+x^2}}$. Show that $(1+x^2)\dfrac{dy}{dx} + xy = 1$.

In Exercises 28–34 evaluate the integral.

28. $\displaystyle\int \operatorname{csch}^2 x\,dx$
29. $\displaystyle\int \operatorname{sech}^2 x\,dx$
30. $\displaystyle\int \tanh x\,dx$
31. $\displaystyle\int \operatorname{sech} x\,dx$
32. $\displaystyle\int e^x \sinh x\,dx$
33. $\displaystyle\int_{5}^{10}\frac{1}{\sqrt{x^2+1}}\,dx$
34. $\displaystyle\int_{0}^{1}\frac{x}{\sqrt{1+x^4}}\,dx$ (*Hint:* Substitute $u = x^2$.)

35. Find the area A of the region between the catenary $y = 4 \cosh x/4$ and the x axis on $[-4, 4]$.

36. Let $f(x) = \cosh x$ for $0 \le x \le \ln 2$. Find the length $\mathscr{L}$ of the graph of f.

37. Suppose we substitute $\sinh x$ for $\cosh x$ in Exercise 36. Why is it more difficult to find the corresponding length? (Do not attempt to evaluate the integral.)

38. Let $f(x) = \cosh x$. Find the surface area S of the surface generated by revolving about the x axis the graph of f on $[0, 1]$.

39. Let $f(x) = 1/\sqrt{1 + x^2}$. Find the center of gravity of the region between the graph of f and the x axis on $[0, 1]$.

40. Show that $\cosh x \ge 1$ for all x by using the fact that $z + 1/z \ge 2$ for all $z > 0$.

41. Show that
$$\cosh x > \frac{e^{|x|}}{2} \quad \text{for all } x$$

42. a. Prove that $e^x = \cosh x + \sinh x$ for all x.
    b. Prove that $e^{-x} = \cosh x - \sinh x$ for all x.

43. a. Prove that
$$(\cosh x + \sinh x)^n = \cosh nx + \sinh nx$$
   for all x. (*Hint:* Use Exercise 42(a).)
    b. Prove that
$$(\cosh x - \sinh x)^n = \cosh nx - \sinh nx$$
   for all x. (*Hint:* Use Exercise 42(b).)

44. Prove that $\sinh 2x = 2 \sinh x \cosh x$.

45. Prove that
$$\cosh 2x = \cosh^2 x + \sinh^2 x = 2 \sinh^2 x + 1$$

*46. Prove that for $t > 0$,
$$\int_1^{\cosh t} \sqrt{x^2 - 1}\, dx = \frac{\sinh 2t}{4} - \frac{t}{2}$$

(*Hint:* Let $x = \cosh u$, and then use Exercise 45.)

*47. With the help of Exercises 44 and 46 show that the shaded region in Figure 7.26 has area $t/2$.

48. Let
$$f(x) = c \cosh \frac{x}{c} = c\left(\frac{e^{x/c} + e^{-x/c}}{2}\right) \quad \text{for all } x \ge 0$$
   where c is a positive constant.
    a. Show that $f(0) = c$.
    b. Show that $f^{-1}(x) = c \ln\left[(x + \sqrt{x^2 - c^2})/c\right]$ for $x \ge c$. (We usually denote f^{-1} by $\cosh^{-1}$ when $c = 1$.)

49. Using (7) and the formula
$$\cosh^{-1} x = \ln(x + \sqrt{x^2 - 1}) \quad \text{for } x \ge 1$$
   from Exercise 48(b), prove that the following formulas are valid.
    a. $\sinh^{-1}\sqrt{x^2 - 1} = \cosh^{-1} x \quad$ for $x \ge 1$
    b. $\cosh^{-1}\sqrt{x^2 + 1} = \sinh^{-1} x \quad$ for $x \ge 0$

50. An electric wire connecting a telephone pole to a house hangs in the shape of a catenary $y = c \cosh x/c$, where the units are feet (Figure 7.30). Find the length $\mathscr{L}$ of the wire. (*Hint:* First find the value of c.)

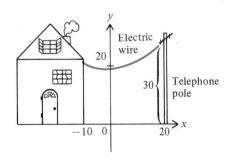

FIGURE 7.30

7.8
L'HÔPITAL'S RULE

In Chapter 2 we stated the Quotient Rule for Limits:

$$\lim_{x \to a} \frac{f(x)}{g(x)} = \frac{\lim_{x \to a} f(x)}{\lim_{x \to a} g(x)}$$

provided that the limits on the right side exist *and* $\lim_{x \to a} g(x) \ne 0$. However, there are many examples of quotient functions f/g that have limits at a even though $\lim_{x \to a} g(x) = 0$. Perhaps the simplest example of such a function is x/x, with $a = 0$:

$$\lim_{x \to 0} \frac{x}{x} = 1$$

A less trivial example is $(\sin x)/x$, again with $a = 0$:

$$\lim_{x \to 0} \frac{\sin x}{x} = 1$$

Nevertheless, except for canceling factors where possible, we have thus far no systematic method for evaluating limits of quotients in which the denominator approaches 0. In this section we will describe such a method.

We will need the following generalization of the Mean Value Theorem.

THEOREM 7.13
GENERALIZED MEAN
VALUE THEOREM

Let f and g be continuous on $[a, b]$ and differentiable on (a, b). If $g'(x) \neq 0$ for $a < x < b$, then there is a number c in (a, b) such that

$$\frac{f(b) - f(a)}{g(b) - g(a)} = \frac{f'(c)}{g'(c)} \tag{1}$$

Proof We introduce a special function much like the one appearing in the proof of the Mean Value Theorem:

$$h(x) = [f(b) - f(a)]g(x) - [g(b) - g(a)]f(x) \quad \text{for } a \leq x \leq b$$

Being a combination of f and g, the function h is continuous on $[a, b]$ and differentiable on (a, b). Thus by the Mean Value Theorem there is a number c in (a, b) such that

$$\frac{h(b) - h(a)}{b - a} = h'(c)$$

But a simple calculation reveals that $h(a) = h(b)$, and thus $h'(c) = 0$. Another calculation shows that

$$h'(x) = [f(b) - f(a)]g'(x) - [g(b) - g(a)]f'(x) \quad \text{for } a < x < b$$

Since $h'(c) = 0$, it follows that

$$[f(b) - f(a)]g'(c) - [g(b) - g(a)]f'(c) = 0 \tag{2}$$

By assumption, $g'(x) \neq 0$ for $a < x < b$. Consequently $g'(c) \neq 0$. It also follows that $g(a) \neq g(b)$ (because if $g(a) = g(b)$, then the Mean Value Theorem would imply that $g'(x) = 0$ for some x in (a, b), contradicting the hypothesis). Since $g'(c) \neq 0$ and $g(a) \neq g(b)$, we can divide both sides of (2) by $[g(b) - g(a)]g'(c)$ to obtain

$$\frac{f(b) - f(a)}{g(b) - g(a)} - \frac{f'(c)}{g'(c)} = 0$$

which is equivalent to (1). ∎

If we let $g(x) = x$ for $a \leq x \leq b$, then (1) reduces to the equation of the Mean Value Theorem, because in this case

$$\frac{f(b) - f(a)}{b - a} = \frac{f(b) - f(a)}{g(b) - g(a)} = \frac{f'(c)}{g'(c)} = \frac{f'(c)}{1} = f'(c)$$

This justifies the name of the Generalized Mean Value Theorem.

The Generalized Mean Value Theorem plays a significant part in proving l'Hôpital's Rule,* an important result named after the seventeenth- and eighteenth-century French mathematician the Marquis de l'Hôpital.† It consists of several variations on the theme of using derivatives to evaluate limits of quotients. We will study the variations separately.

The Indeterminate Form 0/0

If $\lim_{x \to a^+} f(x) = 0 = \lim_{x \to a^+} g(x)$, then we say that $\lim_{x \to a^+} f(x)/g(x)$ has the **indeterminate form 0/0**. The same notion applies if $\lim_{x \to a^+}$ is replaced by $\lim_{x \to b^-}$, $\lim_{x \to c}$, $\lim_{x \to \infty}$, or $\lim_{x \to -\infty}$. The following limits therefore have the indeterminate form 0/0:

$$\lim_{x \to \pi/2^-} \frac{\cos x}{\sin x - 1} \quad \text{and} \quad \lim_{x \to 0} \frac{e^x - x - 1}{x^2}$$

Our first version of l'Hôpital's Rule concerns limits of the indeterminate form 0/0.

THEOREM 7.14 L'HÔPITAL'S RULE

Let L be a real number or ∞ or $-\infty$.

a. Suppose f and g are differentiable on (a, b) and $g'(x) \neq 0$ for $a < x < b$. If

$$\lim_{x \to a^+} f(x) = 0 = \lim_{x \to a^+} g(x) \quad \text{and} \quad \lim_{x \to a^+} \frac{f'(x)}{g'(x)} = L$$

then

$$\lim_{x \to a^+} \frac{f(x)}{g(x)} = L = \lim_{x \to a^+} \frac{f'(x)}{g'(x)}$$

An analogous result holds if $\lim_{x \to a^+}$ is replaced by $\lim_{x \to b^-}$, or by $\lim_{x \to c}$, where c is any number in (a, b). (In the latter case f and g need not be differentiable at c.)

* **L'Hôpital:** Pronounced "*Lo*-pi-tal,"
† L'Hôpital's Rule should actually be called "Bernoulli's Rule," because it appears in correspondence from Johann Bernoulli to l'Hôpital. But l'Hôpital and Bernoulli had made an agreement under which l'Hôpital paid Bernoulli a monthly fee for solutions to certain problems, with the understanding that Bernoulli would tell no one of the arrangement. As a result the rule described in Theorem 7.14 first appeared publicly in l'Hôpital's 1696 treatise *Analyse des infiniment petits pour l'intelligence des lignes courbes*. It was only recently discovered that the rule, its proof, and relevant examples all appeared in a 1694 letter from Bernoulli to l'Hôpital.

b. Suppose f and g are differentiable on (a, ∞) and $g'(x) \neq 0$ for $x > a$. If

$$\lim_{x \to \infty} f(x) = 0 = \lim_{x \to \infty} g(x) \quad \text{and} \quad \lim_{x \to \infty} \frac{f'(x)}{g'(x)} = L$$

then

$$\lim_{x \to \infty} \frac{f(x)}{g(x)} = L = \lim_{x \to \infty} \frac{f'(x)}{g'(x)}$$

An analogous result holds if $\lim_{x \to \infty}$ is replaced by $\lim_{x \to -\infty}$.

Proof   We establish the formula involving the right-hand limits in (a). Define F and G on $[a, b]$ by

$$F(x) = \begin{cases} f(x) & \text{for } a < x < b \\ 0 & \text{for } x = a \end{cases}$$

$$G(x) = \begin{cases} g(x) & \text{for } a < x < b \\ 0 & \text{for } x = a \end{cases}$$

Then

$$\lim_{x \to a^+} F(x) = \lim_{x \to a^+} f(x) = 0 = F(a)$$

so that F is continuous on $[a, b]$. The same is true of G. Moreover, F and G are differentiable on (a, b), since they agree with f and g, respectively, on (a, b). Consequently if x is any number in (a, b), then F and G are continuous on $[a, x]$ and differentiable on (a, x). By the Generalized Mean Value Theorem, this means that there is a number $c(x)$ in (a, x) such that

$$\frac{F(x)}{G(x)} = \frac{F(x) - F(a)}{G(x) - G(a)} = \frac{F'(c(x))}{G'(c(x))}$$

Because $F = f$ and $G = g$ on (a, b), this means that

$$\frac{f(x)}{g(x)} = \frac{f'(c(x))}{g'(c(x))}$$

Since $a < c(x) < x$, we know that

$$\lim_{x \to a^+} c(x) = a$$

so we can use the Substitution Rule with $y = c(x)$ to conclude that

$$\lim_{x \to a^+} \frac{f(x)}{g(x)} = \lim_{x \to a^+} \frac{f'(c(x))}{g'(c(x))} = \lim_{y \to a^+} \frac{f'(y)}{g'(y)} = \lim_{x \to a^+} \frac{f'(x)}{g'(x)} = L$$

This proves the equation involving right-hand limits in (a). The results involving left-hand and two-sided limits are proved analogously. Part (b) is more difficult to prove, and we omit its proof.   ◼

Normally when we use l'Hôpital's Rule, the differentiability of f and g are obvious, as is a suitable interval on which $g(x) \neq 0$ and $g'(x) \neq 0$. Therefore in the examples that follow we will not refer to these two hypotheses of l'Hôpital's Rule.

Example 1 Find $\displaystyle\lim_{x \to 0} \frac{\sin 4x}{\sin 3x}$.

Solution Notice first that

$$\lim_{x \to 0} \sin 4x = 0 = \lim_{x \to 0} \sin 3x$$

As a result we can apply part (a) of l'Hôpital's Rule, which yields

$$\lim_{x \to 0} \frac{\sin 4x}{\sin 3x} = \lim_{x \to 0} \frac{4 \cos 4x}{3 \cos 3x} = \frac{4 \cdot 1}{3 \cdot 1} = \frac{4}{3} \quad \square$$

Example 2 Evaluate $\displaystyle\lim_{x \to \pi/2^-} \frac{\cos x}{\sin x - 1}$.

Solution Since

$$\lim_{x \to \pi/2^-} \cos x = 0 = \lim_{x \to \pi/2^-} (\sin x - 1)$$

we can apply part (a) of l'Hôpital's Rule and obtain

$$\lim_{x \to \pi/2^-} \frac{\cos x}{\sin x - 1} = \lim_{x \to \pi/2^-} \frac{-\sin x}{\cos x} = \lim_{x \to \pi/2^-} (-\tan x) = -\infty \quad \square$$

In the next example we will apply part (b) of l'Hôpital's Rule.

Example 3 Find $\displaystyle\lim_{x \to \infty} \frac{(\pi/2) - \arctan x}{1/x}$.

Solution Since $\displaystyle\lim_{x \to \infty} \arctan x = \pi/2$, we deduce that

$$\lim_{x \to \infty} \left(\frac{\pi}{2} - \arctan x \right) = 0 = \lim_{x \to \infty} \frac{1}{x}$$

Therefore part (b) of l'Hôpital's Rule applies to give us

$$\lim_{x \to \infty} \frac{(\pi/2) - \arctan x}{1/x} = \lim_{x \to \infty} \frac{-1/(1 + x^2)}{-1/x^2} = \lim_{x \to \infty} \frac{x^2}{1 + x^2} = 1 \quad \square$$

In the next example we will determine the limit by applying l'Hôpital's Rule twice in succession.

Example 4 Find $\displaystyle\lim_{x \to 0} \frac{e^x - x - 1}{x^2}$.

Solution Since

$$\lim_{x \to 0} (e^x - x - 1) = 0 = \lim_{x \to 0} x^2$$

l'Hôpital's Rule tells us that

$$\lim_{x \to 0} \frac{e^x - x - 1}{x^2} = \lim_{x \to 0} \frac{e^x - 1}{2x}$$

provided that the latter limit exists. However,

$$\lim_{x \to 0} (e^x - 1) = 0 = \lim_{x \to 0} 2x$$

so that a second application of l'Hôpital's Rule yields

$$\lim_{x \to 0} \frac{e^x - 1}{2x} = \lim_{x \to 0} \frac{e^x}{2} = \frac{1}{2}$$

We conclude that

$$\lim_{x \to 0} \frac{e^x - x - 1}{x^2} = \lim_{x \to 0} \frac{e^x - 1}{2x} = \frac{1}{2} \quad \square$$

The Indeterminate Form ∞ / ∞

Suppose that $\lim_{x \to a^+} f(x) = \infty$ or $-\infty$ and that $\lim_{x \to a^+} g(x) = \infty$ or $-\infty$. Then we say that $\lim_{x \to a^+} f(x)/g(x)$ has the **indeterminate form** ∞ / ∞. The same notion applies if $\lim_{x \to a^+}$ is replaced by $\lim_{x \to b^-}$, $\lim_{x \to c}$, $\lim_{x \to \infty}$, or $\lim_{x \to -\infty}$.

Our second version of l'Hôpital's Rule involves limits with indeterminate form ∞/∞. We give it now, without proof.

THEOREM 7.15
L'HÔPITAL'S RULE

Let L be a real number or ∞ or $-\infty$.
a. Suppose f and g are differentiable on (a, b) and $g'(x) \neq 0$ for $a < x < b$. If

$$\lim_{x \to a^+} f(x) = \infty \text{ or } -\infty, \quad \lim_{x \to a^+} g(x) = \infty \text{ or } -\infty, \text{ and } \lim_{x \to a^+} \frac{f'(x)}{g'(x)} = L$$

then

$$\lim_{x \to a^+} \frac{f(x)}{g(x)} = L = \lim_{x \to a^+} \frac{f'(x)}{g'(x)}$$

An analogous result holds if $\lim_{x \to a^+}$ is replaced by $\lim_{x \to b^-}$, or by $\lim_{x \to c}$, where c is any number in (a, b). (In the latter case, neither f nor g will be differentiable at c.)
b. Suppose f and g are differentiable on (a, ∞) and $g'(x) \neq 0$ for $x > a$. If

$$\lim_{x \to \infty} f(x) = \infty \text{ or } -\infty, \quad \lim_{x \to \infty} g(x) = \infty \text{ or } -\infty, \text{ and } \lim_{x \to \infty} \frac{f'(x)}{g'(x)} = L$$

then

$$\lim_{x \to \infty} \frac{f(x)}{g(x)} = L = \lim_{x \to \infty} \frac{f'(x)}{g'(x)}$$

An analogous result holds if $\lim_{x \to \infty}$ is replaced by $\lim_{x \to -\infty}$.

Example 5 Find $\displaystyle\lim_{x\to 0^+}\frac{1/x}{e^{1/x^2}}$.

Solution Since

$$\lim_{x\to 0^+}\frac{1}{x}=\infty=\lim_{x\to 0^+}e^{1/x^2}$$

it follows from part (a) of l'Hôpital's Rule that

$$\lim_{x\to 0^+}\frac{1/x}{e^{1/x^2}}=\lim_{x\to 0^+}\frac{-1/x^2}{e^{1/x^2}(-2/x^3)}$$

$$=\lim_{x\to 0^+}\frac{xe^{-1/x^2}}{2}=\frac{0\cdot 0}{2}=0 \quad \square$$

Let us see how we can use the limit just found. Suppose

$$f(x)=\begin{cases} e^{-1/x^2} & \text{for } x\neq 0 \\ 0 & \text{for } x=0 \end{cases}$$

and suppose we wish to determine $f'(0)$. By definition,

$$f'(0)=\lim_{x\to 0}\frac{f(x)-f(0)}{x-0}=\lim_{x\to 0}\frac{f(x)-0}{x-0}=\lim_{x\to 0}\frac{e^{-1/x^2}}{x}$$

if the limit exists. It is tempting to use l'Hôpital's Rule on the fraction $e^{-1/x^2}/x$, but unfortunately a straight application of the rule does not help. However, we get results by transforming both numerator and denominator to obtain

$$\lim_{x\to 0}\frac{e^{-1/x^2}}{x}=\lim_{x\to 0}\frac{1/x}{e^{1/x^2}}$$

From Example 5 we know that

$$\lim_{x\to 0^+}\frac{1/x}{e^{1/x^2}}=0$$

and by a similar calculation,

$$\lim_{x\to 0^-}\frac{1/x}{e^{1/x^2}}=0$$

Therefore

$$f'(0)=\lim_{x\to 0}\frac{e^{-1/x^2}}{x}=0$$

In the next example we determine the relationship between e^x and x^2 as x grows without bound.

Example 6 Find $\displaystyle\lim_{x\to\infty}\frac{e^x}{x^2}$.

Solution   Evidently

$$\lim_{x \to \infty} e^x = \infty = \lim_{x \to \infty} x^2$$

so that by part (b) of l'Hôpital's Rule we have

$$\lim_{x \to \infty} \frac{e^x}{x^2} = \lim_{x \to \infty} \frac{e^x}{2x}$$

Since

$$\lim_{x \to \infty} e^x = \infty = \lim_{x \to \infty} 2x$$

another application of part (b) of l'Hôpital's Rule is necessary. We find that

$$\lim_{x \to \infty} \frac{e^x}{2x} = \lim_{x \to \infty} \frac{e^x}{2} = \infty$$

Consequently

$$\lim_{x \to \infty} \frac{e^x}{x^2} = \infty \quad \square$$

In a similar way you can show that

$$\lim_{x \to \infty} \frac{e^x}{x^n} = \infty$$

for any positive integer n. We sometimes express this by saying that as x approaches ∞, e^x goes to infinity faster than any power of x.

Other Indeterminate Forms   Various indeterminate forms, such as $0 \cdot \infty$, 0^0, 1^∞, ∞^0, and $\infty - \infty$, can usually be converted into the indeterminate form $0/0$ or ∞/∞ and then evaluated by one of the versions of l'Hôpital's Rule given in Theorems 7.14 and 7.15.

Example 7   Find $\lim\limits_{x \to 0^+} x \ln x$.

Solution   Since

$$\lim_{x \to 0^+} x = 0 \quad \text{and} \quad \lim_{x \to 0^+} \ln x = -\infty$$

the given limit is of the form $0 \cdot \infty$ (more precisely, $0 \cdot (-\infty)$). However, we can transform it into the indeterminate form ∞/∞ by rewriting it as

$$\lim_{x \to 0^+} \frac{\ln x}{1/x}$$

because

$$\lim_{x \to 0^+} \ln x = -\infty \quad \text{and} \quad \lim_{x \to 0^+} \frac{1}{x} = \infty$$

By l'Hôpital's Rule (Theorem 7.15(a)) we conclude that

$$\lim_{x \to 0^+} \frac{\ln x}{1/x} = \lim_{x \to 0^+} \frac{1/x}{-1/x^2} = \lim_{x \to 0^+} (-x) = 0$$

so that
$$\lim_{x \to 0^+} x \ln x = 0 \qquad \square$$

Occasionally we apply l'Hôpital's Rule to the logarithm of an expression in order to evaluate the limit of the expression.

Example 8 Find $\lim_{x \to 0^+} x^x$.

Solution The limit evidently has the indeterminate form 0^0. By (8) in Section 7.4, $x^x = e^{x \ln x}$, and consequently

$$\lim_{x \to 0^+} x^x = \lim_{x \to 0^+} e^{x \ln x}$$

Since the exponential function is continuous, it follows that

$$\lim_{x \to 0^+} e^{x \ln x} = e^{\lim_{x \to 0^+} (x \ln x)}$$

if the limit on the right side exists. But by Example 7,

$$\lim_{x \to 0^+} x \ln x = 0$$

Consequently

$$\lim_{x \to 0^+} x^x = \lim_{x \to 0^+} e^{x \ln x} = e^0 = 1 \qquad \square$$

Let $f(x) = x^x$. We will use the information from Example 8 to help sketch the graph of f. Since $x^x = e^{x \ln x} > 0$ and the domain of f is $(0, \infty)$, f has no intercepts. Differentiating x^x, we obtain

$$\frac{d}{dx}(x^x) = \frac{d}{dx}(e^{x \ln x}) = e^{x \ln x}(\ln x + 1) = x^x(\ln x + 1) \tag{3}$$

Thus $f'(x) = 0$ if $\ln x + 1 = 0$, that is, if $\ln x = -1$, or $x = e^{-1}$. We use (3) to obtain the second derivative of x^x:

$$\frac{d^2}{dx^2}(x^x) = x^x(\ln x + 1)^2 + x^x\left(\frac{1}{x}\right)$$

$$= x^x\left[(\ln x + 1)^2 + \frac{1}{x}\right] > 0 \quad \text{for all } x > 0$$

It follows that $f(e^{-1}) = (1/e)^{1/e} \approx 0.69$ is the absolute minimum value of f, and the graph is concave upward on $(0, \infty)$. The only major feature of the graph we need to attend to is its appearance near the y axis. This is determined in part by

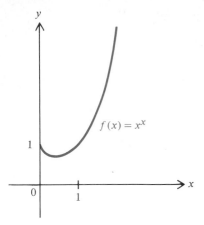

FIGURE 7.31

$\lim_{x \to 0^+} x^x$, which is 1 by Example 8. We now have the information needed to sketch the graph, which appears in Figure 7.31.

The final example of this section yields a formula for e.

Example 9 Show that $\displaystyle\lim_{x \to \infty} \left(1 + \frac{1}{x}\right)^x = e$.

Solution As in Example 8, we first find the limit of the logarithm of the expression on the left. This means finding a number b such that

$$\lim_{x \to \infty} \ln\left(1 + \frac{1}{x}\right)^x = b$$

Our answer for the original limit will then be e^b. We find that

$$\lim_{x \to \infty} \ln\left(1 + \frac{1}{x}\right)^x = \lim_{x \to \infty} x \ln\left(1 + \frac{1}{x}\right) = \lim_{x \to \infty} \frac{\ln\left(1 + 1/x\right)}{1/x}$$

This expression is now prepared for l'Hôpital's Rule (Theorem 7.14(b)), because

$$\lim_{x \to \infty} \ln\left(1 + \frac{1}{x}\right) = 0 = \lim_{x \to \infty} \frac{1}{x}$$

As a result,

$$\lim_{x \to \infty} \frac{\ln\left(1 + 1/x\right)}{1/x} = \lim_{x \to \infty} \frac{\dfrac{1}{(1 + 1/x)}\left(-\dfrac{1}{x^2}\right)}{-1/x^2} = \lim_{x \to \infty} \frac{1}{1 + 1/x} = \frac{1}{1 + 0} = 1$$

Thus $b = 1$, so that

$$\lim_{x \to \infty} \left(1 + \frac{1}{x}\right)^x = e^1 = e \quad \square$$

Thus e could also be defined by means of the limit in Example 9, as is done in some textbooks.

EXERCISES 7.8

In Exercises 1–8 use the version of l'Hôpital's Rule given in Theorem 7.14 to find the limit.

1. $\displaystyle\lim_{x \to a} \frac{x^{16} - a^{16}}{x - a}$

2. $\displaystyle\lim_{x \to 0} \frac{\sin x}{x}$

3. $\displaystyle\lim_{x \to 0} \frac{\cos x - 1}{x}$

4. $\displaystyle\lim_{x \to \pi/2^-} \frac{\cos x}{\sin x - 1}$

5. $\displaystyle\lim_{x \to 0^+} \frac{\sin 8\sqrt{x}}{\sin 5\sqrt{x}}$

6. $\displaystyle\lim_{x \to 1} \frac{\ln x}{x - 1}$

7. $\displaystyle\lim_{x \to \infty} \frac{e^{1/x} - 1}{1/x}$

8. $\displaystyle\lim_{x \to \infty} \frac{\tan 1/x}{1/x}$

In Exercises 9–14 use the version of l'Hôpital's Rule given in Theorem 7.15 to find the limit.

9. $\displaystyle\lim_{x \to \pi/2^-} \frac{\tan x}{\sec x + 1}$

10. $\displaystyle\lim_{x \to 0^+} \frac{x^{-1/2}}{\ln x}$

11. $\displaystyle\lim_{x \to \infty} \frac{x}{\ln x}$

12. $\displaystyle\lim_{x \to \infty} \frac{x}{e^{2x}}$

13. $\displaystyle\lim_{x\to\infty}\frac{\ln(1+x)}{\ln x}$

14. $\displaystyle\lim_{x\to\infty}\frac{\ln x}{\ln(\ln x)}$

In Exercises 15–50 use l'Hôpital's Rule to find the limit.

15. $\displaystyle\lim_{x\to0}\frac{1-\cos x}{\sin x}$

16. $\displaystyle\lim_{x\to0}\frac{\tan 4x}{\tan 2x}$

17. $\displaystyle\lim_{x\to\pi/2^-}\frac{\tan 4x}{\tan 2x}$

18. $\displaystyle\lim_{x\to0^-}\frac{\tan x}{x^2}$

19. $\displaystyle\lim_{x\to0}\frac{1-\cos 2x}{1-\cos 3x}$

20. $\displaystyle\lim_{x\to0^+}\frac{1-\cos\sqrt{x}}{\sin x}$

21. $\displaystyle\lim_{x\to0}\frac{\sqrt{1+x}-\sqrt{1-x}}{x}$

22. $\displaystyle\lim_{x\to0^+}\frac{e^{1/x}}{\ln x}$

23. $\displaystyle\lim_{x\to0^+}\frac{\sin x}{e^{\sqrt{x}}-1}$

24. $\displaystyle\lim_{x\to0}\frac{e^{(x^2)}-1}{e^x-1}$

25. $\displaystyle\lim_{x\to0}\frac{5^x-3^x}{x}$

26. $\displaystyle\lim_{x\to1}\frac{4^x-3^x-1}{x-1}$

27. $\displaystyle\lim_{x\to0^+}x\ln|\ln x|$

28. $\displaystyle\lim_{x\to0}\frac{\arcsin x}{x}$

29. $\displaystyle\lim_{x\to1^-}\frac{(\pi/2)-\arcsin x}{\sqrt{1-x^2}}$

30. $\displaystyle\lim_{x\to0}\frac{x^3}{x-\sin x}$

31. $\displaystyle\lim_{x\to1}\frac{\ln x-x+1}{x^3-3x+2}$

32. $\displaystyle\lim_{x\to0}\frac{\tanh x-\sinh x}{x^2}$

33. $\displaystyle\lim_{x\to0}(\csc x-\cot x)$

34. $\displaystyle\lim_{x\to\pi/2}(\pi^2-4x^2)\tan x$

35. $\displaystyle\lim_{x\to0^+}\sin x\ln(\sin x)$

36. $\displaystyle\lim_{x\to0^+}x^{\sin x}$

37. $\displaystyle\lim_{x\to0^+}\left(\ln\frac{1}{x}\right)^x$

38. $\displaystyle\lim_{x\to1/2^-}(\tan\pi x)^{1-2x}$

39. $\displaystyle\lim_{x\to\pi/2^-}\frac{\ln(\cos x)}{\tan x}$

40. $\displaystyle\lim_{x\to\infty}\frac{e^x}{x^3}$

41. $\displaystyle\lim_{x\to\infty}x\sin\frac{1}{x}$

42. $\displaystyle\lim_{x\to\infty}\frac{\ln x}{x^2}$

43. $\displaystyle\lim_{x\to\infty}\frac{\ln(x^2+1)}{\ln x}$

44. $\displaystyle\lim_{x\to\infty}\frac{\log_4 x}{x}$

45. $\displaystyle\lim_{x\to\infty}\frac{e^{(e^x)}}{e^x}$

46. $\displaystyle\lim_{x\to\infty}\left(1-\frac{1}{x}\right)^x$

47. $\displaystyle\lim_{x\to\infty}\left(1+\frac{1}{x^2}\right)^x$

48. $\displaystyle\lim_{x\to\infty}\frac{1}{x(\pi/2-\arctan x)}$

49. $\displaystyle\lim_{x\to\infty}x^{-1/2}\ln(\ln x)$

50. $\displaystyle\lim_{x\to\infty}x^2\left(1-x\sin\frac{1}{x}\right)$

The limits appearing in Exercises 51–52 appear as the first two illustrations of l'Hôpital's Rule in l'Hôpital's original text. Find these limits, assuming that $a>0$.

51. $\displaystyle\lim_{x\to a}\frac{a^2-ax}{a-\sqrt{ax}}$

52. $\displaystyle\lim_{x\to a}\frac{\sqrt{2a^3x-x^4}-a\sqrt[3]{a^2x}}{a-\sqrt[4]{ax^3}}$.

In Exercises 53–54 sketch the graph of each function, noting any relative extreme values and concavity. Use l'Hôpital's Rule to determine the horizontal asymptotes, if any, and to determine the behavior of the function near the indicated point a.

53. $f(x)=xe^{-x}$

54. $f(x)=x\ln x;\ a=0$

55. Let
$$f(x)=\begin{cases}x(\ln x)^2 & \text{for }x>0\\ 0 & \text{for }x=0\end{cases}$$
Show that $\lim_{x\to0^+}f(x)=f(0)$, meaning that f is continuous from the right at 0.

56. Let
$$f(x)=\begin{cases}\dfrac{\cos x}{x^2-\pi^2/4} & \text{for }x\neq-\dfrac{\pi}{2}\text{ and }\dfrac{\pi}{2}\\[2mm] -\dfrac{1}{\pi} & \text{for }x=-\dfrac{\pi}{2}\text{ or }\dfrac{\pi}{2}\end{cases}$$
a. Show that f is continuous at $-\pi/2$ and $\pi/2$.
b. Show that f is differentiable at $-\pi/2$ and $\pi/2$.

57. a. Show that $\lim_{x\to0^+}x^{(x^x)}=0$.
b. Show that $\lim_{x\to0^+}(x^x)^x=1$.

58. Why is the following "application" of l'Hôpital's Rule invalid?
$$\frac{1}{\pi/2}=\lim_{x\to\pi/2}\frac{\sin x}{x}=\lim_{x\to\pi/2}\frac{\cos x}{1}=\frac{0}{1}=0$$

59. Try to use l'Hôpital's Rule to find $\lim_{x\to\infty}x/\sqrt{x^2+1}$, and see what happens. Then determine the limit by a different method.

60. Try to evaluate $\lim_{x\to0}e^{-1/x^2}/x$ by applying l'Hôpital's Rule directly to the fraction $e^{-1/x^2}/x$, and see what happens.

61. Prove the formulas involving the left-hand limits in Theorem 7.14(a).

62. Suppose $\lim_{x\to a}g(x)=0$. Show that $\lim_{x\to a}f(x)/g(x)$ exists only if $\lim_{x\to a}f(x)=0$.

*63. Show that we can still reach the conclusion of the Generalized Mean Value Theorem if we replace the condition that $g'(x)\neq0$ for $a<x<b$ by the conditions that $g(a)\neq g(b)$ and that $g'(x)\neq0$ for any x such that $f'(x)=0$. (*Hint:* Prove that under the new conditions $g'(c)\neq0$; then use the relevant portions of the proof of the Generalized Mean Value Theorem.)

64. Let $f(x)=x^2$ and $g(x)=x^3$ for $-1\le x\le1$. Show that there is no c in $(-1,1)$ such that
$$\frac{f(1)-f(-1)}{g(1)-g(-1)}=\frac{f'(c)}{g'(c)}$$

Why does this not violate the Generalized Mean Value Theorem?

65. Let $f(x) = c \cosh(x/c)$, where c is a positive constant. Find

$$\lim_{x \to \infty} \frac{f^{-1}(x)}{c \ln 2x}$$

(*Hint:* Use the formula for $f^{-1}(x)$ in Exercise 48 of Section 7.7, along with l'Hôpital's Rule.)

66. Let R be a rectangle in the first quadrant with base on the x axis, one vertex at the origin, and the opposite vertex on the graph of e^{-x}.

   a. Show that for any $\varepsilon > 0$ the area of R is less than ε if the base is sufficiently large.

   b. Show that the rectangle with maximal area has a base of length 1.

67. A right triangle T in the first quadrant has legs on the axes and hypotenuse tangent to the curve $y = e^{-x}$.

   a. Show that for any $\varepsilon > 0$ the area of T is less than ε if the base is sufficiently large.

   b. Show that the area of T is maximal if the base of T has length 2.

Key Terms and Expressions

Inverse function
Exponential function
Exponent; base

Exponential growth; exponential decay
Hyperbolic function

Key Formulas

$f^{-1}(f(x)) = x$ and $f(f^{-1}(x)) = x$

$\dfrac{dx}{dy} = \dfrac{1}{\dfrac{dy}{dx}}$

$(f^{-1})'(c) = \dfrac{1}{f'(a)}, \quad$ where $c = f(a)$

$\displaystyle\int e^x \, dx = e^x + C \quad$ and $\quad \dfrac{d}{dx} e^x = e^x$

$a^r = e^{r \ln a} \quad$ for $a > 0$

$a^{b+c} = a^b a^c, \qquad a^{-b} = \dfrac{1}{a^b}, \quad$ and $\quad (a^b)^c = a^{bc}$

$\log_a bc = \log_a b + \log_a c \quad$ for $a, b, c > 0$

$\log_a b^r = r \log_a b$

$\log_a x = \dfrac{\ln x}{\ln a}$

$\displaystyle\int \frac{1}{\sqrt{1 - x^2}} \, dx = \arcsin x + C$

and $\quad \dfrac{d}{dx} \arcsin x = \dfrac{1}{\sqrt{1 - x^2}}$

$\displaystyle\int \frac{1}{x^2 + 1} \, dx = \arctan x + C$

and $\quad \dfrac{d}{dx} \arctan x = \dfrac{1}{x^2 + 1}$

$\displaystyle\int \frac{1}{x\sqrt{x^2 - 1}} \, dx = \operatorname{arcsec} x + C$

and $\quad \dfrac{d}{dx} \operatorname{arcsec} x = \dfrac{1}{x\sqrt{x^2 - 1}}$

$\cosh^2 x - \sinh^2 x = 1$

$\dfrac{d}{dx} \sinh x = \cosh x$

$\dfrac{d}{dx} \cosh x = \sinh x$

Key Theorems

General Laws of Exponents
General Law of Logarithms

l'Hôpital's Rule

REVIEW EXERCISES

In Exercises 1–6 determine whether the function has an inverse.

1. $f(x) = 3x^3 + 5x^5 - 10$

2. $f(x) = \dfrac{x^2 - 1}{x^2 + 1}$

3. $f(x) = 1 - \dfrac{1}{x}$

4. $f(x) = x - \dfrac{1}{x}$

5. $g(x) = \sin 2x + \cos x$

6. $g(x) = x + \cos x$

In Exercises 7–8 find a formula for f^{-1}.

7. $f(x) = \dfrac{3x - 2}{-x + 1}$

8. $f(x) = \dfrac{x^3}{4 - x^3}$

9. Let f be an even function on $(-a, a)$. Show that f does not have an inverse.

10. Let

$$f(x) = \frac{e^x}{1 + e^x}$$

a. Show that f has an inverse.
b. Find $(f^{-1})'(\frac{1}{2})$.

In Exercises 11–18 find dy/dx.

11. $y = \ln(1 + 2^x)$

12. $y = x^{\cos x}$

13. $y = \arctan \sinh x$

14. $y = x^{(2^x)}$

15. $y = \arcsin(1 - x^2)^{1/3}$

16. $y = \dfrac{\sinh x}{x}$

17. $y = \log_4(\arctan x^2)$

18. $y \sinh^{-1} x + e^y = y^5$

In Exercises 19–34 evaluate the integral.

19. $\displaystyle \int \frac{e^x}{\sqrt{1 + e^x}}\, dx$

20. $\displaystyle \int \frac{e^x}{\sqrt{1 - e^{2x}}}\, dx$

21. $\displaystyle \int \frac{e^x}{\sqrt{1 + e^{2x}}}\, dx$

22. $\displaystyle \int \frac{e^x}{e^x + e^{-x}}\, dx$

23. $\displaystyle \int (\sec x \tan x) e^{\sec x}\, dx$

24. $\displaystyle \int e^x \cosh x\, dx$

25. $\displaystyle \int_0^1 x^2\, 5^{-x^3}\, dx$

26. $\displaystyle \int \frac{\sin x}{\sqrt{1 - 4\cos^2 x}}\, dx$

27. $\displaystyle \int \frac{3}{1 + 4t^2}\, dt$

28. $\displaystyle \int \frac{t^4}{t^{10} + 1}\, dt$

29. $\displaystyle \int_{-5/4}^{5/4} \frac{1}{\sqrt{25 - 4t^2}}\, dt$

30. $\displaystyle \int \frac{e^{1 + \ln x}}{\sqrt{1 + x^2}}\, dx$

31. $\displaystyle \int 2^x \sinh 2^x\, dx$

32. $\displaystyle \int \frac{\sqrt{x}}{\sqrt{1 + x^3}}\, dx$

33. $\displaystyle \int \frac{x}{x^4 + 4x^2 + 10}\, dx$

34. $\displaystyle \int \frac{1}{x\sqrt{1 - (\ln x)^2}}\, dx$

35. Using (3) of Section 7.2 and a hyperbolic identity, give an alternative derivation of the formula

$$\frac{d}{dx} \sinh^{-1} x = \frac{1}{\sqrt{1 + x^2}}$$

36. Verify the following two equations.

$$\cosh 2u - 1 = \frac{1}{2}(e^u - e^{-u})^2$$

$$\cosh 2u + 1 = \frac{1}{2}(e^u + e^{-u})^2$$

37. a. Using derivatives, show that $e^x \geq 1 + x$ for all x.
    b. Show that

$$e^x \leq \frac{1}{1 - x} \quad \text{for } 0 \leq x < 1$$

   (*Hint:* Replace x by $-x$ in part (a).)
    c. Show that

$$\ln \frac{x + 1}{x} \leq \frac{1}{x} \leq \ln \frac{x}{x - 1} \quad \text{for } x > 1$$

   (*Hint:* Replace x by $1/x$ in parts (a) and (b).)
    d. Approximate $\ln 1.05$ by using (c) with $x = 20$ and $x = 21$.

38. Let

$$f(x) = \left(1 + \frac{1}{x}\right)^x$$

   Show that f is strictly increasing on $(0, \infty)$. (*Hint:* Using Exercise 37(c), show that $\ln f(x)$ is strictly increasing.)

39. a. By applying the Mean Value Theorem to $\ln x$ on the interval $[a, b]$, show that

$$\frac{b-a}{b} \leq \ln b - \ln a \leq \frac{b-a}{a} \quad \text{for } 0 < a \leq b$$

b. Using part (a), show that

$$\frac{b-1}{b} \leq \ln b \leq b-1 \quad \text{for } b \geq 1$$

c. Using part (b), approximate $\ln 1.05$.

40. Show that for any integer n,

$$\ln 2^n a = n \ln 2 + \ln a$$

(Hence it is possible to compute the logarithm of any positive number if the logarithms of all numbers between 1 and 2 are known.)

In Exercises 41–48 evaluate each limit by l'Hôpital's Rule.

41. $\displaystyle\lim_{t \to 0} \frac{\sinh at}{t}$

42. $\displaystyle\lim_{x \to 0} \frac{\tan x - x}{\sin x - x}$

43. $\displaystyle\lim_{x \to 0^+} \frac{\ln x}{\ln(\sin x)}$

44. $\displaystyle\lim_{x \to 0^+} \frac{e^{1/x}}{1 - \cot x}$

45. $\displaystyle\lim_{x \to 0^+} \left(\frac{1}{x} - \frac{1}{\arctan x}\right)$

46. $\displaystyle\lim_{x \to \infty} x^{1/2} \sin\frac{1}{x}$

47. $\displaystyle\lim_{x \to \infty} \left(\frac{x+1}{x-1}\right)^x$

48. $\displaystyle\lim_{x \to 1^+} (\ln x)^{x-1}$

49. Let $f(x) = x + \sin x$ and $g(x) = x$. Show that $\lim_{x \to \infty} [f(x)/g(x)]$ exists but $\lim_{x \to \infty} [f'(x)/g'(x)]$ does not exist.

50. Evaluate

$$\lim_{x \to \infty} \left(\frac{2x+1}{2x-1}\right)^x$$

After calculating the limit, evaluate $[(2x+1)/(2x-1)]^x$ for $x = 2$ and compare it with the limit.

In Exercises 51–53 sketch the graph of the function, indicating any relative extreme values, concavity, and asymptotes.

51. $f(x) = (2x)^x$

52. $f(x) = e^{-x} \sin x \quad$ for $x > 0$

53. $f(x) = e^{-2x} - e^{-3x} \quad$ for $x > 0$

54. Let

$$f(x) = \frac{\sqrt{x}}{x-1} e^{\sqrt{x}} \quad \text{and} \quad g(x) = \frac{1}{\sqrt{x}(x-1)} e^{\sqrt{x}}$$

Find the area A of the region between the graphs of f and g on $[4, 9]$. (Hint: To evaluate the integral, substitute $u = \sqrt{x}$.)

55. Let

$$f(x) = \frac{x}{\sqrt{x^2-1}} \quad \text{and} \quad g(x) = \frac{\sqrt{x^2-1}}{x}$$

Find the area A of the region between the graphs of f and g on $[\sqrt{2}, 2]$.

56. Let R be the region between the graphs of $y = e^{-x}$ and $y = e^{-2x}$ on $[0, \ln 2]$. Find the volume V of the solid generated by revolving R about the x axis.

57. Find the length $\mathcal{L}$ of the curve given parametrically by $x = \frac{1}{2}\sinh 2t - t$ and $y = 2\cosh t$ for $0 \leq t \leq 1$. (Hint: $\cosh 2t - 1 = 2\sinh^2 t$.)

58. If an earthquake measures 6.1 on the Richter scale, what will one that is 100 times as intense measure?

59. The formula

$$\log N = 8.73 - 1.15M$$

has been used to approximate the average number N of earthquakes per year with a magnitude of between $M - 0.05$ and $M + 0.05$ (that is, within 0.05 of M) for $M \geq 7$.

a. Determine the magnitude M for which $N = 1$.

🖩 *b. Determine the average number of earthquakes with magnitude between 7.45 and 8.05. (Hint: Take $M = 7.5, 7.6, \ldots, 8.0$ in succession.)

60. It is known that approximately 99.27 percent of all uranium on earth now occurs in the form of the isotope U^{238} and that approximately 0.72 percent occurs in the form of the isotope U^{235}.

a. Assume (as scientists sometimes do) that when the earth was created, there were approximately equal amounts of U^{238} and U^{235}. Using this assumption and the fact that U^{238} has a half-life of 4.5×10^9 years and U^{235} a half-life of 7.1×10^8 years, determine the age of the earth. (Hint: If $f(t)$ is the amount of U^{238} at time t and $g(t)$ is the amount of U^{235} at time t, then by assumption we have $f(0) = g(0)$ and

$$f(a) = \frac{99.27}{0.72} g(a)$$

where a is the present age of the earth. Now use Exercise 7(b) of Section 7.5.)

b. Through other calculations it is estimated that the true age of the earth is between 4×10^9 and 5×10^9 years. Assuming that the age of the earth is 4.5×10^9 years, determine the original ratio $f(0)/g(0)$ of U^{238} to U^{235}.

Cumulative Review, Chapters 1–6

In Exercises 1–2 find the limit.

1. $\lim\limits_{x \to \sqrt{2}} \dfrac{2\sqrt{2} - 2x}{8 - 4x^2}$

2. $\lim\limits_{x \to \infty} \dfrac{(1 + \sin^2(x))}{\sqrt{x}}$

In Exercises 3–4 find $f'(x)$.

3. $f(x) = \dfrac{1 + x^3}{1 - x^3}$

4. $f(x) = \displaystyle\int_{x^2}^{2x+1} \dfrac{1}{\sqrt{t + 2}}\, dt$

5. Let $f(x) = \ln x - \frac{1}{8}x^2$. Show that

$$\sqrt{1 + (f'(x))^2} = \frac{1}{x} + \frac{x}{4}$$

6. Show that the slope of the line tangent to the graph of $2x^{3/2} - 2y^{3/2} = 3xy^{1/2}$ at the point (a, b) is positive if $a > b$, and is negative if $a < b$.

7. Let $f(x) = \dfrac{x - 4}{2x + 6}$. Show that $f(x) \geq \frac{3}{20}$ for all $x \geq 7$ by

   a. solving the inequality $\dfrac{x - 4}{2x + 6} \geq \dfrac{3}{20}$

   b. showing that f is increasing on $[7, \infty)$ and then using this fact.

In Exercises 8–9 sketch the graph of f, indicating all pertinent information.

8. $f(x) = \dfrac{\cos x}{1 - \sin x}$

9. $f(x) = x^4 - 8x^2$

10. Find a positive value of c that makes the parabolas $y = x^2 + 3$ and $x = cy^2$ have the same tangent line at a point, and find the point of tangency.

11. A circular racetrack has a radius of 200 feet. Pat sits in the best seat in the stands, located at a point on the edge of the track. At a certain instant a greyhound is on the track, 200 feet from Pat. The dog is moving around the track with a speed of 50 feet per second. How fast is the distance between the greyhound and Pat changing at that instant? (*Hint:* Use the Law of Cosines.)

12. A rectangular window is to be placed inside a parabolic arch whose height and base are each 36 inches. Determine the largest possible area A of the window.

In Exercises 13–14 evaluate the integral.

13. $\displaystyle\int \dfrac{x}{4 + x^2}\, dx$

14. $\displaystyle\int_0^{\pi/3} \dfrac{1 + \tan^2 x}{\sec x}\, dx$

15. Suppose the velocity of a car (in miles per hour) is given by $v(t) = 40 + 40/(4 + t^2)$ for $0 \leq t \leq 2$, where t is in hours. How far does the car travel in the two hours?

16. Let $f(x) = x^2 + 2x$ and $g(x) = \dfrac{4}{4 + x} - 1$. Find the area A of the region between the graphs of f and g on $[0, 1]$.

17. Find the area A of the region that lies in the first quadrant below the graph of $(y^2/x^4) - x^3 = 8$ and between the lines $x = 1$ and $x = 2$.

18. Find the value of c for which $\int_{1/2}^{2} 2/x^2\, dx = \int_{1/2}^{2} (x^2 + c)\, dx$.

19. Let R be the region between the graphs of $y = 4 - x^2$ and $y = 1 - (|x| - 1)^2$.

    a. Determine the volume V of the double-holed solid obtained by revolving R about the x axis.

    b. Determine the center of gravity of R.

20. Suppose a pyramidal water tank with square base 20 feet on a side is 30 feet tall, and rests on its base. If the tank is full of water, determine the work W required to pump all but the bottom three feet of water to a point two feet above the top of the tank.

21. Find the area A of the region in the first quadrant that lies inside the polar graph of $r = \cos\theta$ and outside the polar graph of $r = 1 - \cos\theta$.

8

TECHNIQUES OF INTEGRATION

This chapter is devoted to techniques, or methods, of integration. Recall that in Section 5.6 we studied the substitution method of integration. By the techniques in Sections 8.1 through 8.4, we will greatly increase the kinds of functions we can integrate. Section 8.5 is concerned with evaluating integrals using a table of integrals, and Section 8.6 is dedicated to finding approximate values of definite integrals whose numerical values are difficult or impossible to calculate. We conclude the chapter by defining the integrals of a new collection of functions.

Below, we summarize the basic integrals that have so far appeared.

$$\int c \, dx = cx + C \qquad\qquad \int x^r \, dx = \frac{1}{r+1} x^{r+1} + C$$

$$\int x \, dx = \frac{1}{2} x^2 + C \qquad\qquad \int \frac{1}{x} \, dx = \ln|x| + C$$

$$\int \sin x \, dx = -\cos x + C \qquad\qquad \int \cot x \, dx = \ln|\sin x| + C$$

$$\int \cos x \, dx = \sin x + C \qquad\qquad \int \sec x \, dx = \ln|\sec x + \tan x| + C$$

$$\int \tan x \, dx = -\ln|\cos x| + C \qquad\qquad \int \csc x \, dx = -\ln|\csc x + \cot x| + C$$

$$\int e^x \, dx = e^x + C \qquad\qquad \int a^x \, dx = \frac{1}{\ln a} a^x + C$$

$$\int \frac{1}{\sqrt{a^2 - x^2}} \, dx = \arcsin \frac{x}{a} + C \qquad\qquad \int \frac{1}{x^2 + a^2} \, dx = \frac{1}{a} \arctan \frac{x}{a} + C$$

8.1
INTEGRATION BY PARTS

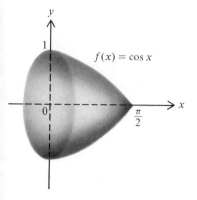

$f(x) = \cos x$

FIGURE 8.1

Suppose we wish to find the volume V of the solid obtained by revolving about the y axis the region between the graph of $y = \cos x$ and the x axis on $[0, \pi/2]$ (Figure 8.1). By the results of Section 6.2 we know that V is given by the formula

$$V = \int_0^{\pi/2} 2\pi x \cos x \, dx \tag{1}$$

However, as yet we have no technique for evaluating this integral. In the present section we will introduce a method for evaluating integrals such as the one in (1). The method is derived from the product rule for differentiation, just as the sum rule for integration (Theorem 5.17) is derived from the sum rule for differentiation.

Let F and G be differentiable functions. From the product rule we know that

$$(FG)'(x) = F'(x)G(x) + F(x)G'(x)$$

or by rearranging,

$$F(x)G'(x) = (FG)'(x) - F'(x)G(x) \tag{2}$$

If F' and G' are continuous, then we can restate (2) in the language of indefinite integrals as

$$\int F(x)G'(x)\,dx = \int [(FG)'(x) - F'(x)G(x)]\,dx$$

$$= \int (FG)'(x)\,dx - \int F'(x)G(x)\,dx$$

or equivalently,

$$\int F(x)G'(x)\,dx = F(x)G(x) - \int F'(x)G(x)\,dx \tag{3}$$

Now suppose we wish to evaluate $\int f(x)\,dx$ but cannot readily do so. If f can be rewritten as the product FG', then (3) tells us that

$$\int f(x)\,dx = \int F(x)G'(x)\,dx$$

$$= F(x)G(x) - \int F'(x)G(x)\,dx \tag{4}$$

If in addition $\int F'(x)G(x)\,dx$ can be readily evaluated, then $\int f(x)\,dx$ can be evaluated by means of (4). This method of evaluating $\int f(x)\,dx$, by "splitting" the integrand f into two parts F and G', is known as **integration by parts**. We summarize this result and its definite integral version as a theorem.

THEOREM 8.1
INTEGRATION BY PARTS

Let F and G be differentiable on $[a, b]$, and assume that F' and G' are continuous on $[a, b]$. Then

$$\int F(x)G'(x)\,dx = F(x)G(x) - \int F'(x)G(x)\,dx \qquad (5)$$

and

$$\int_a^b F(x)G'(x)\,dx = F(x)G(x)\Big|_a^b - \int_a^b F'(x)G(x)\,dx \qquad (6)$$

Formula (5) can be simplified by formally substituting $u = F(x)$ and $v = G(x)$, so that $du = F'(x)\,dx$ and $dv = G'(x)\,dx$. We obtain

$$\int \overset{u}{\overbrace{F(x)}}\overset{dv}{\overbrace{G'(x)\,dx}} = \overset{u}{\overbrace{F(x)}}\,\overset{v}{\overbrace{G(x)}} - \int \overset{v}{\overbrace{G(x)}}\overset{du}{\overbrace{F'(x)\,dx}}$$

or more simply,

$$\int u\,dv = uv - \int v\,du$$

which is shorter and easier to remember than (5). In this notation (4) becomes

$$\int f(x)\,dx = \int u\,dv = uv - \int v\,du \qquad (7)$$

To apply (7), we must choose u and dv so that $f(x)\,dx = u\,dv$. In particular, $f(x)$ is to be split into two parts, one becoming u and the other joining with dx to become dv. In order to obtain $uv - \int v\,du$ from $\int u\,dv$, we must be able to determine du from u and v from dv. As a result, u should be a function that is easy to differentiate, and dv should be chosen so that v can be readily found by integration. Moreover, $\int v\,du$ should be no more difficult, and preferably less difficult, to integrate than $\int u\,dv$. Let us see how all this works in specific examples.

Example 1   Find $\int x\cos x\,dx$.

Solution   The integrand $x\cos x$ can naturally be split into the two parts x and $\cos x$. We let

$$u = x \quad \text{and} \quad dv = \cos x\,dx$$

Then

$$du = dx \quad \text{and} \quad v = \sin x$$

Consequently integration by parts yields

$$\int \overset{u}{\overbrace{x}} \overset{dv}{\overbrace{\cos x \, dx}} = \overset{u}{\overbrace{x}} \overset{v}{\overbrace{\sin x}} - \int \overset{v}{\overbrace{\sin x}} \overset{du}{\overbrace{dx}} = x \sin x + \cos x + C \quad \square$$

It follows from Example 1 that the volume V of the solid of revolution mentioned at the outset of the section is given by

$$V = \int_0^{\pi/2} 2\pi x \cos x \, dx = 2\pi \int_0^{\pi/2} x \cos x \, dx = 2\pi (x \sin x + \cos x) \Big|_0^{\pi/2}$$

$$= 2\pi \left(\frac{\pi}{2} - 1 \right) = \pi(\pi - 2)$$

As we said earlier, it is important that the integral $\int v \, du$ in (7) be no more difficult to integrate than $\int u \, dv$. For example, in calculating $\int x \cos x \, dx$, suppose we had let $u = \cos x$ and $dv = x \, dx$. Then we could write $du = -\sin x \, dx$ and $v = \frac{1}{2}x^2$, but the integration would not be made any easier, because in that case,

$$\int x \cos x \, dx = \overset{v}{\overbrace{\frac{1}{2}x^2}} \overset{u}{\overbrace{\cos x}} - \int \overset{v}{\overbrace{\frac{1}{2}x^2}} \overset{du}{\overbrace{(-\sin x) \, dx}}$$

and the second integral is harder to evaluate than the first.

Example 2   Find $\int 2xe^{3x} \, dx$.

Solution   Since $2xe^{3x}$ splits into $2x$ and e^{3x}, we let

$$u = 2x \quad \text{and} \quad dv = e^{3x} \, dx$$

Then

$$du = 2 \, dx \quad \text{and} \quad v = \frac{1}{3}e^{3x}$$

Therefore

$$\int \overset{u}{\overbrace{2x}} \overset{dv}{\overbrace{e^{3x} \, dx}} = \overset{u}{\overbrace{2x}} \overset{v}{\overbrace{\left(\frac{1}{3}e^{3x} \right)}} - \int \overset{v}{\overbrace{\frac{1}{3}e^{3x}}} \overset{du}{\overbrace{2 \, dx}} = \frac{2}{3}xe^{3x} - \int \frac{2}{3}e^{3x} \, dx$$

$$= \frac{2}{3}xe^{3x} - \frac{2}{9}e^{3x} + C \quad \square$$

Example 3   Find $\int \ln x \, dx$.

Solution   This time we write $\ln x$ as $(\ln x) \cdot 1$, so that $\ln x$ is the product of $\ln x$

and 1. This simple trick helps us find the integral. To see how, we let

$$u = \ln x \quad \text{and} \quad dv = 1 \, dx = dx$$

Then
$$du = \frac{1}{x} \, dx \quad \text{and} \quad v = x$$

Therefore

$$\int \overset{u}{\overbrace{\ln x}} \, \overset{dv}{\overbrace{dx}} = \overset{u}{\overbrace{(\ln x)}} \overset{v}{\overbrace{x}} - \int \overset{v}{\overbrace{x}} \cdot \overset{du}{\overbrace{\frac{1}{x} \, dx}} = x \ln x - \int 1 \, dx$$

$$= x \ln x - x + C \quad \square$$

Notice that in the solution of Example 3, we let $u = \ln x$ and $dv = dx$, which led to the simpler integral $\int v \, du = \int x(1/x) \, dx$. A corresponding choice of u can be effective if the integrand involves $\arctan x$ or $\arcsin x$, as you can see in the exercises.

In the next example we will need two successive applications of integration by parts to evaluate the integral.

Example 4 Find $\displaystyle\int_0^1 x^2 e^{-x} \, dx$.

Solution We start by letting

$$u = x^2 \quad \text{and} \quad dv = e^{-x} \, dx$$

Then
$$du = 2x \, dx \quad \text{and} \quad v = -e^{-x}$$

Therefore

$$\int_0^1 \overset{u}{\overbrace{x^2}} \, \overset{dv}{\overbrace{e^{-x} \, dx}} = \overset{u}{\overbrace{x^2}} \overset{v}{\overbrace{(-e^{-x})}} \bigg|_0^1 - \int_0^1 \overset{v}{\overbrace{(-e^{-x})}} \overset{du}{\overbrace{2x \, dx}} = -e^{-1} + \int_0^1 2xe^{-x} \, dx \quad (8)$$

Next we perform integration by parts on $\int_0^1 2xe^{-x} \, dx$, with

$$u = 2x \quad \text{and} \quad dv = e^{-x} \, dx$$

Then
$$du = 2 \, dx \quad \text{and} \quad v = -e^{-x}$$

We find that

$$\int_0^1 \overset{u}{\overbrace{2x}} \, \overset{dv}{\overbrace{e^{-x} \, dx}} = \overset{u}{\overbrace{(2x)}} \overset{v}{\overbrace{(-e^{-x})}} \bigg|_0^1 - \int_0^1 \overset{v}{\overbrace{(-e^{-x})}} \overset{du}{\overbrace{2 \, dx}} = -2e^{-1} + \int_0^1 2e^{-x} \, dx \quad (9)$$

The last integral in (9) can be evaluated directly, and in fact,

$$\int_0^1 2e^{-x} \, dx = -2e^{-x} \bigg|_0^1 = -2e^{-1} + 2 \quad (10)$$

Now we combine the results of (8)–(10) to conclude that

$$\int_0^1 x^2 e^{-x} dx = -e^{-1} + (-2e^{-1}) + (-2e^{-1} + 2) = -5e^{-1} + 2 \quad \square$$

In the following example we will see that two successive integrations by parts do not quite complete the solution.

Example 5 Find $\int e^{-x} \cos x \, dx$.

Solution Let

$$u = e^{-x} \quad \text{and} \quad dv = \cos x \, dx$$

Then

$$du = -e^{-x} dx \quad \text{and} \quad v = \sin x$$

Therefore

$$\int \overset{u}{\overbrace{e^{-x}}} \overset{dv}{\overbrace{\cos x \, dx}} = \overset{u}{\overbrace{e^{-x}}} \overset{v}{\overbrace{\sin x}} - \int \overset{v}{\overbrace{(\sin x)}} \overset{du}{\overbrace{(-e^{-x}) \, dx}}$$

$$= e^{-x} \sin x + \int e^{-x} \sin x \, dx$$

Next we apply a second integration by parts to $\int e^{-x} \sin x \, dx$ by letting

$$u = e^{-x} \quad \text{and} \quad dv = \sin x \, dx$$

Then

$$du = -e^{-x} dx \quad \text{and} \quad v = -\cos x$$

Thus

$$\int \overset{u}{\overbrace{e^{-x}}} \overset{dv}{\overbrace{\sin x \, dx}} = \overset{u}{\overbrace{e^{-x}}} \overset{v}{\overbrace{(-\cos x)}} - \int \overset{v}{\overbrace{(-\cos x)}} \overset{du}{\overbrace{(-e^{-x}) \, dx}}$$

$$= -e^{-x} \cos x - \int e^{-x} \cos x \, dx$$

Combining the results of the two integrations by parts, we obtain

$$\int e^{-x} \cos x \, dx = e^{-x} \sin x + \int e^{-x} \sin x \, dx$$

$$= e^{-x} \sin x - e^{-x} \cos x - \int e^{-x} \cos x \, dx$$

Notice that the original integral has once again appeared—as the last integral. To complete the solution, we add $\int e^{-x} \cos x \, dx$ to both sides of the equation and obtain

$$2 \int e^{-x} \cos x \, dx = e^{-x} \sin x - e^{-x} \cos x + C_1$$

Consequently

$$\int e^{-x} \cos x \, dx = \frac{1}{2}(e^{-x} \sin x - e^{-x} \cos x) + C \quad \square$$

In conclusion we remark that integration by parts is effective with integrals involving a polynomial and either an exponential, a logarithmic, or a trigonometric function. More specifically, integration by parts is especially well adapted to integrals of the form

$$\int (\text{polynomial}) \sin ax \, dx, \qquad \int (\text{polynomial}) \cos ax \, dx,$$

$$\int (\text{polynomial}) e^{ax} \, dx, \qquad \int (\text{polynomial}) \ln x \, dx$$

In all except $\int (\text{polynomial}) \ln x \, dx$, the most effective choice of u is the polynomial, since the derivatives of a polynomial are simpler than the polynomial (and the antiderivatives of $\sin ax$, $\cos ax$, and e^{ax} are essentially the same). For $\int (\text{polynomial}) \ln x \, dx$ the choice of $u = \ln x$ is effective, and in fact, $\int v \, du$ becomes the integral of a polynomial.

Reduction Formulas for $\int \sin^n x \, dx$ and $\int \cos^n x \, dx$

First we will discuss a formula for $\int \sin^n x \, dx$, for $n \geq 2$. In order to use integration by parts, we notice that $\sin^n x = (\sin^{n-1} x)(\sin x)$, and we let

$$u = \sin^{n-1} x \quad \text{and} \quad dv = \sin x \, dx$$

Then

$$du = (n-1)\sin^{n-2} x \cos x \, dx \quad \text{and} \quad v = -\cos x$$

Therefore

$$\int \sin^n x \, dx = \int \overbrace{\sin^{n-1} x}^{u} \overbrace{\sin x \, dx}^{dv}$$

$$= \overbrace{(\sin^{n-1} x)}^{u} \overbrace{(-\cos x)}^{v} - \int \overbrace{(-\cos x)}^{v} \overbrace{(n-1)\sin^{n-2} x \cos x \, dx}^{du}$$

$$= -\sin^{n-1} x \cos x + (n-1)\int \sin^{n-2} x \cos^2 x \, dx$$

$$= -\sin^{n-1} x \cos x + (n-1)\int \sin^{n-2} x (1 - \sin^2 x) \, dx$$

Thus

$$\int \sin^n x \, dx = -\sin^{n-1} x \cos x + (n-1)\int \sin^{n-2} x \, dx - (n-1)\int \sin^n x \, dx$$

Combining the terms that contain $\int \sin^n x \, dx$ yields

$$n \int \sin^n x \, dx = -\sin^{n-1} x \cos x + (n - 1) \int \sin^{n-2} x \, dx$$

so that we have the following formula for $\int \sin^n x \, dx$:

$$\int \sin^n x \, dx = -\frac{1}{n} \sin^{n-1} x \cos x + \frac{n-1}{n} \int \sin^{n-2} x \, dx \qquad (11)$$

The formula that arises for $\int \cos^n x \, dx$ is established in exactly the same way:

$$\int \cos^n x \, dx = \frac{1}{n} \cos^{n-1} x \sin x + \frac{n-1}{n} \int \cos^{n-2} x \, dx \qquad (12)$$

The formulas in (11) and (12) are called **reduction formulas**, because the exponent is reduced (from n to $n - 2$). By using the formulas repeatedly, we can reduce the power of the sine or the cosine to 0 or 1. We mention that if $n - 2 = 0$, then by convention $\sin^{n-2} x$ and $\cos^{n-2} x$ are both 1.

Using (11) with $n = 2$, we find that

$$\int \sin^2 x \, dx = -\frac{1}{2} \sin x \cos x + \frac{1}{2} \int 1 \, dx$$

$$= -\frac{1}{2} \sin x \cos x + \frac{1}{2} x + C$$

Since $-\frac{1}{2} \sin x \cos x = -\frac{1}{4} \sin 2x$, it follows that

$$\int \sin^2 x \, dx = -\frac{1}{4} \sin 2x + \frac{1}{2} x + C$$

which is equivalent to (3) of Section 5.6, which we obtained by integrating by substitution. Similarly, if $n = 2$ then (12) yields

$$\int \cos^2 x \, dx = \frac{1}{4} \sin 2x + \frac{1}{2} x + C$$

which is equivalent to (2) of Section 5.6.

In Exercises 45 and 46 you are asked to verify that

$$\int \sin^4 x \, dx = -\frac{1}{4} \sin^3 x \cos x - \frac{3}{8} \sin x \cos x + \frac{3}{8} x + C \qquad (13)$$

$$\int \cos^4 x \, dx = \frac{1}{4} \cos^3 x \sin x + \frac{3}{8} \cos x \sin x + \frac{3}{8} x + C \qquad (14)$$

Example 6 Find $\displaystyle\int \cos^5 x\, dx$.

Solution Letting $n = 5$ in the reduction formula for $\int \cos^n x\, dx$, we obtain

$$\int \cos^5 x\, dx = \frac{1}{5}\cos^4 x \sin x + \frac{4}{5}\int \cos^3 x\, dx$$

A second application of the reduction formula, to $\int \cos^3 x\, dx$, yields

$$\int \cos^3 x\, dx = \frac{1}{3}\cos^2 x \sin x + \frac{2}{3}\int \cos x\, dx$$

$$= \frac{1}{3}\cos^2 x \sin x + \frac{2}{3}\sin x + C_1$$

Consequently

$$\int \cos^5 x\, dx = \frac{1}{5}\cos^4 x \sin x + \frac{4}{5}\left(\frac{1}{3}\cos^2 x \sin x + \frac{2}{3}\sin x + C_1\right)$$

$$= \frac{1}{5}\cos^4 x \sin x + \frac{4}{15}\cos^2 x \sin x + \frac{8}{15}\sin x + C \quad \square$$

For the definite integral $\int_0^{\pi/6} \cos^5 x\, dx$, we would have

$$\int_0^{\pi/6} \cos^5 x\, dx = \left.\left(\frac{1}{5}\cos^4 x \sin x + \frac{4}{15}\cos^2 x \sin x + \frac{8}{15}\sin x\right)\right|_0^{\pi/6}$$

$$= \frac{1}{5}\left(\frac{\sqrt{3}}{2}\right)^4 \frac{1}{2} + \frac{4}{15}\left(\frac{\sqrt{3}}{2}\right)^2 \frac{1}{2} + \frac{8}{15}\left(\frac{1}{2}\right)$$

$$= \frac{9}{160} + \frac{12}{120} + \frac{8}{30} = \frac{203}{480}$$

EXERCISES 8.1

In Exercises 1–28 find the indefinite integral.

1. $\displaystyle\int x \sin x\, dx$

2. $\displaystyle\int x \sec^2 x\, dx$

3. $\displaystyle\int x \ln x\, dx$

4. $\displaystyle\int x \ln x^2\, dx$

5. $\displaystyle\int (\ln x)^2\, dx$

6. $\displaystyle\int x^2 \ln x\, dx$

7. $\displaystyle\int x^3 \ln x\, dx$

8. $\displaystyle\int x e^{-x}\, dx$

9. $\displaystyle\int x^2 e^{4x}\, dx$

10. $\displaystyle\int x^2 \sin x\, dx$

11. $\displaystyle\int x^3 \cos x\, dx$

12. $\displaystyle\int e^x \sin x\, dx$

13. $\displaystyle\int e^{3x} \cos 3x\, dx$

14. $\displaystyle\int \frac{\sin x}{e^x}\, dx$

15. $\displaystyle\int t \cdot 2^t\, dt$

16. $\displaystyle\int t \cdot 3^{-t}\, dt$

17. $\displaystyle\int t^2 \cdot 4^t\, dt$

18. $\displaystyle\int \log_6 x\, dx$

19. $\displaystyle\int t \sinh t\, dt$

20. $\displaystyle\int t^2 \cosh t\, dt$

21. $\displaystyle\int \arctan x\, dx$

22. $\displaystyle\int \arcsin 3x\, dx$

23. $\int \arccos(-7x)\,dx$

24. $\int x^n \ln x\,dx$, where n is a positive integer

25. $\int x^n \ln x^m\,dx$, where m and n are positive integers

26. $\int \sin(\ln x)\,dx$ (*Hint:* Let $u = \sin(\ln x)$, and integrate by parts twice.)

27. $\int \cos(\ln x)\,dx$

28. $\int x\ln(x+1)\,dx$ (*Hint:* Take $u = \ln(x+1)$, $dv = x\,dx$, and $v = (x^2 - 1)/2$.)

In Exercises 29–34 evaluate the definite integral.

29. $\int_0^1 xe^{5x}\,dx$

30. $\int_0^{\pi/2} (x + x\sin x)\,dx$

31. $\int_0^\pi t^2 \cos t\,dt$

32. $\int_1^4 x^{3/2}\ln x\,dx$

33. $\int_{-\pi/3}^{\pi/4} x\sec^2 x\,dx$

34. $\int_1^b x^2 (\ln x)^2\,dx$, where $b > 0$

In Exercises 35–42 first make a substitution and then use integration by parts to evaluate the integral.

35. $\int_0^1 \ln(x+1)\,dx$

36. $\int e^x \ln(1 + e^x)\,dx$

37. $\int x\sin ax\,dx$

38. $\int e^{6x}\cos e^{3x}\,dx$

39. $\int \sin x \arctan(\cos x)\,dx$

40. $\int \frac{1}{x}\arcsin(\ln x)\,dx$

41. $\int \cos\sqrt{t}\,dt$

42. $\int \frac{(\ln t)^2}{t^2}\,dt$ (*Hint:* Let $u = \ln t$.)

In Exercises 43–44 evaluate the integral with the help of the reduction formulas in (11) and (12).

43. $\int_0^{\pi/2} \cos^3 \frac{x}{2}\,dx$

44. $\int \sin^5 x\,dx$

45. Verify equation (13).

46. Verify equation (14).

In Exercises 47–50 use integration by parts to establish the reduction formula.

47. $\int \cos^n x\,dx = \frac{1}{n}\cos^{n-1} x\sin x + \frac{n-1}{n}\int \cos^{n-2} x\,dx$, where n is an integer greater than or equal to 2

48. $\int (\ln x)^n\,dx = x(\ln x)^n - n\int (\ln x)^{n-1}\,dx$, where n is a positive integer (*Hint:* Let $u = (\ln x)^n$.)

49. $\int \frac{1}{(x^2 + a^2)^{n+1}}\,dx = \frac{x}{2na^2(x^2 + a^2)^n}$
$+ \frac{2n-1}{2na^2}\int \frac{1}{(x^2 + a^2)^n}\,dx$
$\left(\text{\textit{Hint:} Use integration by parts on } \int \frac{1}{(x^2 + a^2)^n}\,dx \text{ with}\right.$
$u = \frac{1}{(x^2 + a^2)^n}$ and $\left. v = x.\right)$

*50. $\int \sec^n x\,dx = \frac{\sec^{n-2} x\tan x}{n-1} + \frac{n-2}{n-1}\int \sec^{n-2} x\,dx$,
where n is an integer greater than or equal to 2 (*Hint:* Let $u = \sec^{n-2} x$ and $v = \tan x$. After integrating by parts, you will have an integral of the form $\int g(x)\tan^2 x\,dx$. Write this as $\int g(x)(\sec^2 x - 1)\,dx = \int g(x)\sec^2 x\,dx - \int g(x)\,dx$, and solve algebraically for the original integral.)

In Exercises 51–52 use the results of Exercises 48 and 50 to evaluate the integral.

51. $\int (\ln x)^3\,dx$

52. $\int_{\pi/3}^{\pi/4} \sec^5 x\,dx$

53. Verify the following formulas, where $a \neq 0$, $b \neq 0$.
   a. $\int e^{ax}\sin bx\,dx = \frac{e^{ax}}{a^2 + b^2}(a\sin bx - b\cos bx) + C$
   b. $\int e^{ax}\cos bx\,dx = \frac{e^{ax}}{a^2 + b^2}(a\cos bx + b\sin bx) + C$

In Exercises 54–55 find the area A of the region between the graph of f and the x axis on the given interval.

54. $f(x) = \ln x$; $[1, 2]$

55. $f(x) = \arctan x$; $[0, 1]$

56. Find the area A of the region bounded by the graphs of $y = 3x\ln x$ and $y = x^2 \ln x$.

In Exercises 57–59 let R be the region between the graph of f and the x axis on the given interval. Find the volume V of the solid obtained by revolving R about the x axis.

57. $f(x) = \sqrt{x}\,e^x$; $[0, 1]$

58. $f(x) = x^{1/2}\ln x$; $[1, 2]$

59. $f(x) = \sin^{3/2} x$; $[0, \pi]$

60. Find the length $\mathscr{L}$ of the polar graph of the equation $r = \sin^4 \frac{\theta}{4}$ for $0 \leq \theta \leq 2\pi$.

*61. Let n be any positive even integer. Find the length $\mathscr{L}$ of the polar graph of the equation $r = \sin^n \dfrac{\theta}{n}$ for $0 \le \theta \le n\pi/2$.

62. Let $f(x) = x^2 - \dfrac{1}{8}\ln x$. Find the surface area S of the surface generated by revolving about the x axis the graph of f on $[1, 2]$.

63. Suppose that one arch of the cycloid given parametrically by

$$x = r(t - \sin t) \quad \text{and} \quad y = r(1 - \cos t) \quad \text{for } 0 \le t \le 2\pi$$

is revolved around the x axis. Find the surface area S of the surface.

In Exercises 64–65 calculate the center of gravity of the region R between the graphs of f and g on the given interval.

64. $f(x) = \cos x$, $g(x) = \sin x$; $[0, \pi/4]$

65. $f(x) = 1 + \ln x$, $g(x) = 1 - \ln x$; $[1, 2]$

66. Certain binary stars are believed to have the same mass m. Spectroscopic measurements (based on the Doppler shift) yield an "observed mass" m_0. The true mass m is then estimated by means of the formula

$$m = \frac{1}{c} m_0, \quad \text{where} \quad c = \int_0^{\pi/2} \sin^4 x \, dx$$

Determine the number c.

8.2
TRIGONOMETRIC INTEGRALS

Integrals such as

$$\int \sin^5 x \cos^4 x \, dx, \qquad \int \tan^3 x \sec^3 x \, dx, \quad \text{and} \quad \int \sin 5x \cos 3x \, dx$$

are called **trigonometric integrals** because their integrands are combinations of trigonometric functions. This section is devoted to trigonometric integrals, especially those in which the integrands are composed of powers of the basic trigonometric functions.

Integrals of the Form $\int \sin^m x \cos^n x \, dx$

We have already encountered integrals of the form $\int \sin^m x \cos^n x \, dx$. Indeed, if $n = 0$, then the integral becomes $\int \sin^m x \, dx$; likewise, if $m = 0$, then the integral becomes $\int \cos^n x \, dx$; both were analyzed in Section 8.1 by means of reduction formulas.

Next, if $n = 1$, then $\int \sin^m x \cos^n x \, dx$ becomes $\int \sin^m x \cos x \, dx$, which is evaluated by substituting $u = \sin x$. (See Example 2 of Section 5.6.) Similarly, if $m = 1$, then we obtain $\int \sin x \cos^n x \, dx$, which is evaluated by substituting $u = \cos x$.

Now we turn to those integrals of the form $\int \sin^m x \cos^n x \, dx$ in which m and n are positive integers with $m \ge 2$ and $n \ge 2$. Evaluation of the integrals involves substitution. There are 2 cases: m or n is odd (or both are odd), and m and n are both even.

If n is odd, as in $\int \sin^2 x \cos^3 x \, dx$, we factor out $\cos x$ and write the rest of the integrand in terms of $\sin x$, using the Pythagorean Identity $\cos^2 x = 1 - \sin^2 x$.

Example 1 Find $\int \sin^2 x \cos^3 x \, dx$.

Solution As indicated above, we factor out $\cos x$ and rearrange as follows:

$$\int \sin^2 x \cos^3 x \, dx = \int \sin^2 x \cos^2 x \cos x \, dx$$

$$= \int (\sin^2 x)(1 - \sin^2 x) \cos x \, dx$$

Now we evaluate the integral by substituting

$$u = \sin x, \quad \text{so that} \quad du = \cos x \, dx$$

We obtain

$$\int \sin^2 x \cos^3 x \, dx = \int \overbrace{(\sin^2 x)}^{u^2} \overbrace{(1 - \sin^2 x)}^{1-u^2} \overbrace{\cos x \, dx}^{du}$$

$$= \int u^2 (1 - u^2) \, du$$

$$= \int (u^2 - u^4) \, du$$

$$= \frac{1}{3} u^3 - \frac{1}{5} u^5 + C$$

$$= \frac{1}{3} \sin^3 x - \frac{1}{5} \sin^5 x + C \quad \square$$

If m is odd in $\int \sin^m x \cos^n x \, dx$, we follow the same general procedure, but in this case we factor out $\sin x$ and write the rest of the integrand in terms of $\cos x$, using the Pythagorean Identity $\sin^2 x = 1 - \cos^2 x$.

Example 2 Find $\int \sin^5 x \cos^4 x \, dx$.

Solution Using the suggestion made above, we factor out $\sin x$ and rearrange as follows:

$$\int \sin^5 x \cos^4 x \, dx = \int \sin^4 x \sin x \cos^4 x \, dx$$

$$= \int (\sin^2 x)^2 \cos^4 x \sin x \, dx$$

$$= \int (1 - \cos^2 x)^2 \cos^4 x \sin x \, dx$$

Next we substitute

$$u = \cos x, \quad \text{so that} \quad du = -\sin x \, dx$$

Therefore

$$\int \sin^5 x \cos^4 x \, dx = \int \overbrace{(1 - \cos^2 x)^2}^{(1-u^2)^2} \overbrace{\cos^4 x}^{u^4} \overbrace{\sin x \, dx}^{-du}$$

$$= -\int (1 - u^2)^2 u^4 \, du$$

$$= -\int (u^4 - 2u^6 + u^8) \, du$$

$$= -\frac{1}{5} u^5 + \frac{2}{7} u^7 - \frac{1}{9} u^9 + C$$

$$= -\frac{1}{5} \cos^5 x + \frac{2}{7} \cos^7 x - \frac{1}{9} \cos^9 x + C \quad \square$$

If m and n are both even in the integral $\int \sin^m x \cos^n x \, dx$, the evaluation is more complicated. Three trigonometric identities help reduce the exponents m and n:

$$\sin x \cos x = \frac{1}{2} \sin 2x \tag{1}$$

$$\sin^2 x = \frac{1 - \cos 2x}{2} \tag{2}$$

$$\cos^2 x = \frac{1 + \cos 2x}{2} \tag{3}$$

In addition, the following two formulas, derived in Section 5.6, will be helpful:

$$\int \sin^2 x \, dx = \frac{1}{2} x - \frac{1}{4} \sin 2x + C \tag{4}$$

and

$$\int \cos^2 x \, dx = \frac{1}{2} x + \frac{1}{4} \sin 2x + C \tag{5}$$

Example 3 Find $\int \sin^2 x \cos^4 x \, dx$.

Solution   Using (1) and (3) together, we find that

$$\int \sin^2 x \cos^4 x \, dx = \int (\sin^2 x \cos^2 x) \cos^2 x \, dx$$

$$= \int (\sin x \cos x)^2 \cos^2 x \, dx$$

$$\overset{(1),\,(3)}{=} \int \left(\frac{1}{2}\sin 2x\right)^2 \left(\frac{1+\cos 2x}{2}\right) dx$$

$$= \frac{1}{8} \int \sin^2 2x \, dx + \frac{1}{8} \int \sin^2 2x \cos 2x \, dx$$

For the first integral on the right we let

$$u = 2x, \quad \text{so that} \quad du = 2\,dx$$

and for the second integral we let

$$v = \sin 2x, \quad \text{so that} \quad dv = 2\cos 2x \, dx$$

Then with the help of (4) we find that

$$\int \sin^2 x \cos^4 x \, dx = \frac{1}{8} \int \sin^2 \overset{u}{2x} \overset{\frac{1}{2}du}{dx} + \frac{1}{8} \int \underbrace{\sin^2 2x}_{v^2} \underbrace{\cos 2x \, dx}_{\frac{1}{2}dv}$$

$$= \frac{1}{8} \int (\sin^2 u) \frac{1}{2} du + \frac{1}{8} \int v^2 \cdot \frac{1}{2} dv$$

$$\overset{(4)}{=} \frac{1}{16}\left(\frac{1}{2}u - \frac{1}{4}\sin 2u\right) + \frac{1}{16}\left(\frac{1}{3}v^3\right) + C$$

$$= \frac{1}{16}\left(x - \frac{1}{4}\sin 4x\right) + \frac{1}{48}\sin^3 2x + C \quad \square$$

An alternative way to evaluate $\int \sin^m x \cos^n x \, dx$ when m and n are even is to use the identity $\sin^2 x + \cos^2 x = 1$, but this time to transform the integral into integrals of the form $\int \sin^k x \, dx$ or of the form $\int \cos^k x \, dx$, which can be evaluated by the reduction formulas (11) and (12) of Section 8.1. Thus for the integral $\int \sin^4 x \cos^6 x \, dx$ we would have

$$\int \sin^4 x \cos^6 x \, dx = \int (\sin^2 x)^2 \cos^6 x \, dx = \int (1 - \cos^2 x)^2 \cos^6 x \, dx$$

$$= \int (\cos^6 x - 2\cos^8 x + \cos^{10} x) \, dx$$

$$= \int \cos^6 x \, dx - 2 \int \cos^8 x \, dx + \int \cos^{10} x \, dx$$

and to evaluate the integrals on the right we would use (12) of Section 8.1.

Table 8.1 summarizes our analysis of $\int \sin^m x \cos^n x \, dx$.

TABLE 8.1
EVALUATION OF $\int \sin^m x \cos^n x \, dx$ for $m \geq 0$ and $n \geq 0$

	Method	Useful Identities
n odd	substitute $u = \sin x$	$\cos^2 x = 1 - \sin^2 x$
m odd	substitute $u = \cos x$	$\sin^2 x = 1 - \cos^2 x$
m and n even	reduce to smaller powers of m or n	$\begin{cases} \sin x \cos x = \dfrac{1}{2} \sin 2x \\[2mm] \sin^2 x = \dfrac{1 - \cos 2x}{2} \\[2mm] \cos^2 x = \dfrac{1 + \cos 2x}{2} \end{cases}$

Integrals of the Form $\int \tan^m x \sec^n x \, dx$

For many nonnegative integer values of m and n, integrals of the form $\int \tan^m x \sec^n x \, dx$ are handled by substitution. Again, the procedure we use to evaluate the integral depends on the evenness or oddness of m and n.

If n is even and $n > 0$, as in $\int \tan^3 x \sec^4 x \, dx$, we factor out $\sec^2 x$ and write the rest of the integrand in terms of $\tan x$, using the identity $\sec^2 x = 1 + \tan^2 x$.

Example 4   Find $\int \tan^3 x \sec^4 x \, dx$.

Solution   As we said above, we will factor out $\sec^2 x$ and then rearrange:

$$\int \tan^3 x \sec^4 x \, dx = \int \tan^3 x \sec^2 x \sec^2 x \, dx$$

$$= \int \tan^3 x (1 + \tan^2 x) \sec^2 x \, dx$$

Now we make the substitution

$$u = \tan x, \quad \text{so that} \quad du = \sec^2 x \, dx$$

It follows that

$$\int \tan^3 x \sec^4 x \, dx = \int \overbrace{\tan^3 x}^{u^3} \overbrace{(1 + \tan^2 x)}^{1 + u^2} \overbrace{\sec^2 x \, dx}^{du}$$

$$= \int u^3 (1 + u^2) \, du = \int (u^3 + u^5) \, du$$

$$= \frac{1}{4} u^4 + \frac{1}{6} u^6 + C = \frac{1}{4} \tan^4 x + \frac{1}{6} \tan^6 x + C \quad \square$$

Now let us assume that m is odd and $n > 0$ in $\int \tan^m x \sec^n x \, dx$, as it is in $\int \tan^3 x \sec^3 x \, dx$. In this case we factor out $\sec x \tan x$ and write the rest of the integrand in terms of $\sec x$, using the identity $\tan^2 x = \sec^2 x - 1$.

Example 5 Find $\displaystyle\int \tan^3 x \sec^3 x \, dx$.

Solution As indicated above, we first factor out $\sec x \tan x$ and then rearrange:

$$\int \tan^3 x \sec^3 x \, dx = \int (\tan^2 x \sec^2 x) \sec x \tan x \, dx$$

$$= \int [(\sec^2 x - 1) \sec^2 x] \sec x \tan x \, dx$$

At this point we substitute

$$u = \sec x, \quad \text{so that} \quad du = \sec x \tan x \, dx$$

We obtain

$$\int \tan^3 x \sec^3 x \, dx = \int [(\overbrace{\sec^2 x - 1}^{u^2 - 1}) \overbrace{\sec^2 x}^{u^2}] \overbrace{\sec x \tan x \, dx}^{du}$$

$$= \int (u^2 - 1) u^2 \, du$$

$$= \int (u^4 - u^2) \, du$$

$$= \frac{1}{5} u^5 - \frac{1}{3} u^3 + C$$

$$= \frac{1}{5} \sec^5 x - \frac{1}{3} \sec^3 x + C \quad \square$$

As Examples 4 and 5 illustrate, if n is even or m odd, we can readily evaluate $\int \tan^m x \sec^n x \, dx$. The remaining possibility is that n is odd and m is even. By using the identity $\tan^2 x = \sec^2 x - 1$ we can reduce the problem to finding integrals of the form $\int \sec^n x \, dx$, where n is odd. However, these integrals are not so easy to evaluate, even when $n = 1$ or $n = 3$. If $n = 1$, then $\int \sec^n x \, dx$ becomes $\int \sec x \, dx$, which by (10) of Section 5.7 is given by

$$\int \sec x \, dx = \ln |\sec x + \tan x| + C \tag{6}$$

The integration of $\int \sec^3 x \, dx$ is more involved. Using integration by parts, we let

$$u = \sec x \quad \text{and} \quad dv = \sec^2 x \, dx$$

Then

$$du = \sec x \tan x \, dx \quad \text{and} \quad v = \tan x$$

Therefore

$$\int \sec^3 x \, dx = \int \overbrace{\sec x}^{u} \overbrace{\sec^2 x \, dx}^{dv}$$

$$= \overbrace{\sec x}^{u} \overbrace{\tan x}^{v} - \int \overbrace{\tan x}^{v} \overbrace{\sec x \tan x \, dx}^{du}$$

$$= \sec x \tan x - \int \sec x \tan^2 x \, dx$$

$$= \sec x \tan x - \int (\sec x)(\sec^2 x - 1) \, dx$$

$$= \sec x \tan x - \int \sec^3 x \, dx + \int \sec x \, dx$$

By combining both occurrences of $\int \sec^3 x \, dx$, we obtain

$$\int \sec^3 x \, dx = \frac{1}{2} \sec x \tan x + \frac{1}{2} \int \sec x \, dx$$

so that by (6),

$$\int \sec^3 x \, dx = \frac{1}{2} \sec x \tan x + \frac{1}{2} \ln |\sec x + \tan x| + C \qquad (7)$$

Example 6   Find $\int \tan^2 x \sec x \, dx$.

Solution   Since m is even and n is odd, we use the identity $\tan^2 x = \sec^2 x - 1$ to reduce the integrand to terms involving powers of $\sec x$ and then use (6) and (7):

$$\int \tan^2 x \sec x \, dx = \int (\sec^2 x - 1)(\sec x) \, dx$$

$$= \int (\sec^3 x - \sec x) \, dx$$

$$= \int \sec^3 x \, dx - \int \sec x \, dx$$

$$= \frac{1}{2} \sec x \tan x + \frac{1}{2} \ln |\sec x + \tan x|$$

$$\quad - \ln |\sec x + \tan x| + C$$

$$= \frac{1}{2} \sec x \tan x - \frac{1}{2} \ln |\sec x + \tan x| + C \quad \square$$

Our various ways of evaluating integrals of the form $\int \tan^m x \sec^n x\, dx$ are summarized in Table 8.2.

TABLE 8.2
EVALUATION OF $\int \tan^m x \sec^n x\, dx$ for $m \geq 0$ and $n > 0$

	Method	Useful Identity
n even	substitute $u = \tan x$	$\sec^2 x = 1 + \tan^2 x$
m odd	substitute $u = \sec x$	$\tan^2 x = \sec^2 x - 1$
m even, n odd	reduce to powers of $\sec x$ alone	$\tan^2 x = \sec^2 x - 1$

Since $\csc^2 x = 1 + \cot^2 x$, the same techniques allow us to find integrals of the form

$$\int \cot^m x \csc^n x\, dx$$

(See Exercises 28–32.)

Conversion to Sine and Cosine   Trigonometric integrals that are not of the varieties discussed so far can often be simplified by expressing the integrands in terms of sines and cosines and then using the methods already discussed.

Example 7   Find $\displaystyle\int \cos^2 x \tan^5 x\, dx.$

Solution   We write $\tan^5 x$ in terms of sines and cosines and combine like terms:

$$\int \cos^2 x \tan^5 x\, dx = \int \cos^2 x \frac{\sin^5 x}{\cos^5 x}\, dx$$

$$= \int \frac{\sin^4 x}{\cos^3 x} \sin x\, dx$$

$$= \int \frac{(1 - \cos^2 x)^2}{\cos^3 x} \sin x\, dx$$

Substituting

$$u = \cos x, \quad \text{so that} \quad du = -\sin x\, dx$$

gives us

$$\int \cos^2 x \tan^5 x \, dx = \int \frac{\overbrace{(1 - \cos^2 x)^2}^{(1-u^2)^2/u^3}}{\cos^3 x} \overbrace{\sin x \, dx}^{-du}$$

$$= -\int \frac{(1 - u^2)^2}{u^3} \, du = -\int \frac{1 - 2u^2 + u^4}{u^3} \, du$$

$$= \int \left(-\frac{1}{u^3} + \frac{2}{u} - u \right) du$$

$$= \frac{1}{2u^2} + 2 \ln |u| - \frac{u^2}{2} + C$$

$$= \frac{1}{2 \cos^2 x} + 2 \ln |\cos x| - \frac{\cos^2 x}{2} + C \quad \square$$

Integrals of the Form
$\int \mathbf{sin}\ \mathit{ax}\ \mathbf{cos}\ \mathit{bx}\ \mathit{dx}$

Evaluating such integrals depends on the trigonometric identity

$$\sin x \cos y = \frac{1}{2} \sin (x - y) + \frac{1}{2} \sin (x + y)$$

With the appropriate replacements, this identity becomes

$$\sin ax \cos bx = \frac{1}{2} \sin (a - b)x + \frac{1}{2} \sin (a + b)x \qquad (8)$$

Notice that $\frac{1}{2} \sin (a - b)x$ and $\frac{1}{2} \sin (a + b)x$ are easy to integrate by substitution.

Example 8 Find $\displaystyle\int \sin 5x \cos 3x \, dx$.

Solution Using (8) with $a = 5$ and $b = 3$, we find that

$$\int \sin 5x \cos 3x \, dx = \int \left(\frac{1}{2} \sin 2x + \frac{1}{2} \sin 8x \right) dx$$

$$= -\frac{1}{4} \cos 2x - \frac{1}{16} \cos 8x + C \quad \square$$

Integrals of the form

$$\int \sin ax \sin bx \, dx \quad \text{and} \quad \int \cos ax \cos bx \, dx$$

can be found by similar techniques (see Exercises 50–53).

EXERCISES 8.2

In Exercises 1–58 evaluate the integral.

1. $\displaystyle\int \sin^3 x \cos^2 x \, dx$

2. $\displaystyle\int \sin^3 x \cos^3 x \, dx$

3. $\displaystyle\int \sin^3 3x \cos 3x \, dx$

4. $\displaystyle\int_0^{\pi/2} \sin^2 t \cos^5 t \, dt$

5. $\displaystyle\int \frac{1}{x^2} \sin^5 \frac{1}{x} \cos^2 \frac{1}{x} \, dx$

6. $\displaystyle\int \sin^8 6x \cos^3 6x \, dx$

7. $\displaystyle\int \sin^2 y \cos^2 y \, dy$

8. $\displaystyle\int_0^{\pi/2} \sin^4 x \cos^2 x \, dx$

9. $\displaystyle\int \sin^4 x \cos^4 x \, dx$

10. $\displaystyle\int \sin^6 w \cos^4 w \, dw$

11. $\displaystyle\int \sin^{-10} x \cos^3 x \, dx$

12. $\displaystyle\int \sin^{-17} x \cos^5 x \, dx$

13. $\displaystyle\int (1 + \sin^2 x)(1 + \cos^2 x) \, dx$

14. $\displaystyle\int \sin^5 x \cos^{1/2} x \, dx$

15. $\displaystyle\int_0^{\pi/4} \frac{\sin^3 x}{\cos^2 x} \, dx$

16. $\displaystyle\int \frac{\cos^3 x}{\sin^{5/2} x} \, dx$

17. $\displaystyle\int \tan^5 x \sec^2 x \, dx$

18. $\displaystyle\int \tan^3 5x \sec^2 5x \, dx$

19. $\displaystyle\int_0^{\pi/4} \tan^5 t \sec^4 t \, dt$

20. $\displaystyle\int x \tan^3 x^2 \sec^4 x^2 \, dx$

21. $\displaystyle\int_{5\pi/4}^{4\pi/3} \tan^3 x \sec x \, dx$

22. $\displaystyle\int \tan x \sec^3 x \, dx$

23. $\displaystyle\int \frac{1}{\sqrt{x}} \tan^3 \sqrt{x} \sec^3 \sqrt{x} \, dx$

24. $\displaystyle\int \tan^4 (1 - y) \sec^4 (1 - y) \, dy$

25. $\displaystyle\int \tan^3 x \sec^4 x \, dx$

26. $\displaystyle\int_0^{\pi/3} \tan x \sec^{3/2} x \, dx$

27. $\displaystyle\int \tan x \sec^5 x \, dx$

28. $\displaystyle\int \csc^3 x \, dx$

29. $\displaystyle\int \cot^3 x \csc^2 x \, dx$

30. $\displaystyle\int \cot^3 s \csc^4 s \, ds$

31. $\displaystyle\int_{\pi/4}^{\pi/2} \cot^3 x \csc^3 x \, dx$

32. $\displaystyle\int_{\pi/3}^{\pi/4} \cot x \csc^3 x \, dx$

33. $\displaystyle\int \cot x \csc^{-2} x \, dx$

34. $\displaystyle\int \frac{\cot t}{\csc^3 t} \, dt$

35. $\displaystyle\int \frac{\tan x}{\cos^3 x} \, dx$

36. $\displaystyle\int \tan^3 x \csc^2 x \, dx$

37. $\displaystyle\int \frac{\tan^2 x}{\sec^5 x} \, dx$

38. $\displaystyle\int \sin^5 w \cot^3 w \, dw$

39. $\displaystyle\int \frac{\tan x}{\sec^2 x} \, dx$

40. $\displaystyle\int \frac{\sec^4 x}{\tan^5 x} \, dx$

41. $\displaystyle\int \tan^2 x \, dx$ (*Hint:* $\tan^2 x = \sec^2 x - 1$.)

42. $\displaystyle\int \tan^3 x \, dx$

43. $\displaystyle\int \tan^4 x \, dx$

44. $\displaystyle\int \cot^5 x \, dx$

45. $\displaystyle\int \sin 2x \cos 3x \, dx$

46. $\displaystyle\int_0^{2\pi/3} \sin x \cos 2x \, dx$

47. $\displaystyle\int \sin(-4x) \cos(-2x) \, dx$

48. $\displaystyle\int \sin 3x \cos \tfrac{1}{2} x \, dx$

49. $\displaystyle\int \sin \tfrac{1}{2} x \cos \tfrac{2}{3} x \, dx$

50. $\displaystyle\int \sin ax \sin bx \, dx$ (*Hint:* Use the identity $\sin x \sin y = \tfrac{1}{2} \cos(x - y) - \tfrac{1}{2} \cos(x + y)$.)

51. $\displaystyle\int \sin 2x \sin 3x \, dx$

52. $\displaystyle\int \cos ax \cos bx \, dx$ (*Hint:* Use the identity $\cos x \cos y = \tfrac{1}{2} \cos(x + y) + \tfrac{1}{2} \cos(x - y)$.)

53. $\displaystyle\int \cos 5x \cos(-3x) \, dx$

54. $\displaystyle\int \frac{1}{1 + \sin x} \, dx$ (*Hint:* Multiply the integrand by $(1 - \sin x)/(1 - \sin x)$.)

55. $\displaystyle\int_{\pi/4}^{\pi/2} \frac{1}{1 + \cos x} \, dx$

*56. $\displaystyle\int \frac{1}{(1 + \sin x)^2} \, dx$

57. $\displaystyle\int \frac{1 + \cos x}{\sin x} \, dx$

58. $\displaystyle\int \frac{1 + \sin x}{\cos x} \, dx$

59. Let m and n be positive integers. Prove the following.

a. $\displaystyle\int_{-\pi}^{\pi} \sin mx \cos nx \, dx = 0$

b. $\displaystyle\int_{-\pi}^{\pi} \cos mx \cos nx \, dx = \begin{cases} 0 & \text{if } m \neq n \\ \pi & \text{if } m = n \end{cases}$

c. $\displaystyle\int_{-\pi}^{\pi} \sin mx \sin nx \, dx = \begin{cases} 0 & \text{if } m \neq n \\ \pi & \text{if } m = n \end{cases}$

*60. Evaluate

$$\int \frac{\sin x - 5\cos x}{\sin x + \cos x} \, dx$$

by finding numbers a and b such that

$$\sin x - 5\cos x = a(\sin x + \cos x) + b(\cos x - \sin x)$$

In Exercises 61–62 find the area A of the region between the graph of f and the x axis on the given interval.

61. $f(x) = \sin^2 x \cos^3 x; \ [0, \pi/2]$

62. $f(x) = \sec^4 x; \ [-\pi/3, \pi/4]$

63. Find the area A of the portion of the region between the graphs of $y = \tan^3 x$ and $y = \frac{1}{4}\tan x \sec^4 x$ that lies between the lines $x = 0$ and $x = \pi/3$.

64. Let $f(x) = \sin^3(x^2)\cos^2(x^2)$, and let R be the region between the graph of f and the x axis on $[0, \sqrt{\pi}]$. Find the volume V of the solid generated by revolving R about the y axis.

65. Let C be described parametrically by $x = \ln(\sec t + \tan t)$ and $y = \sec t$ for $0 \leq t \leq \pi/4$. Find the surface area S of the surface generated by revolving C about the x axis.

8.3
TRIGONOMETRIC SUBSTITUTIONS

In Section 5.6 we introduced integration by substitution. For an integral of the form $\int f(g(x))g'(x)\,dx$ consider the substitution

$$u = g(x), \quad \text{so that} \quad du = g'(x)\,dx$$

This substitution leads to the integral formula

$$\int f(g(x))g'(x)\,dx = \int f(u)\,du$$

In order to prepare for a different kind of substitution, let

$$u = \arcsin x, \quad \text{so that} \quad du = \frac{1}{\sqrt{1-x^2}}\,dx \tag{1}$$

To simplify the equations in (1), we first notice that $u = \arcsin x$ is equivalent to $x = \sin u$, and by the definition of the arcsine function, $-\pi/2 \leq u \leq \pi/2$. Since $\cos u \geq 0$ for $-\pi/2 \leq u \leq \pi/2$, we deduce that

$$\sqrt{1-x^2} = \sqrt{1-\sin^2 u} = \cos u$$

and thus the second equation in (1) can be rewritten

$$du = \frac{1}{\cos u}\,dx, \quad \text{or equivalently,} \quad dx = \cos u\,du$$

Thus instead of making the substitution appearing in (1), we can equally well substitute

$$x = \sin u, \quad \text{so that} \quad dx = \cos u\,du \tag{2}$$

where $-\pi/2 \leq u \leq \pi/2$.

More generally, in order to evaluate an integral $\int f(x)\,dx$, we can make a substitution of the form

$$x = g(u), \quad \text{so that} \quad dx = g'(u)\,du \tag{3}$$

thus obtaining the integral formula

$$\int f(x)\,dx = \int f(g(u))g'(u)\,du \tag{4}$$

The corresponding definite integral formula is

$$\int_{g(a)}^{g(b)} f(x)\,dx = \int_a^b f(g(u))g'(u)\,du \tag{5}$$

Normally the function g in (3) involves trigonometric functions (as when $x = \sin u$ in (2)); in such cases the substitution in (3) is called a **trigonometric substitution**. Trigonometric substitutions are especially valuable when the integral contains square roots of the form $\sqrt{a^2 - x^2}$, $\sqrt{x^2 + a^2}$, or $\sqrt{x^2 - a^2}$, where $a \geq 0$.

Integrals Containing $\sqrt{a^2 - x^2}$

If we let $x = a \sin u$, with $a > 0$ and $-\pi/2 \leq u \leq \pi/2$, then $a \cos u \geq 0$, so that

$$\sqrt{a^2 - x^2} = \sqrt{a^2 - a^2 \sin^2 u} = \sqrt{a^2(1 - \sin^2 u)} = \sqrt{a^2 \cos^2 u} = a \cos u$$

Thus if an integral contains $\sqrt{a^2 - x^2}$, we can eliminate the square root by substituting

$$x = a \sin u, \quad \text{so that} \quad dx = a \cos u\,du$$

Example 1 Find $\displaystyle\int \frac{1}{x^2 \sqrt{16 - x^2}}\,dx$.

Solution Because $\sqrt{16 - x^2} = \sqrt{4^2 - x^2}$, we substitute

$$x = 4 \sin u, \quad \text{so that} \quad dx = 4 \cos u\,du$$

Then

$$\int \frac{1}{x^2 \sqrt{16 - x^2}}\,dx = \int \frac{1}{16 \sin^2 u \sqrt{16 - 16 \sin^2 u}}(4 \cos u)\,du$$

$$= \int \frac{1}{(16 \sin^2 u)4\sqrt{1 - \sin^2 u}}(4 \cos u)\,du$$

$$= \frac{1}{16} \int \frac{\cos u}{\sin^2 u \cos u}\,du = \frac{1}{16} \int \csc^2 u\,du$$

$$= -\frac{1}{16}\cot u + C$$

In order to write the answer in terms of the original variable x, we draw the triangle in Figure 8.2, in which $x = 4 \sin u$. From the triangle we see that

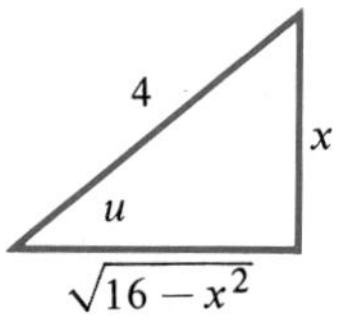

FIGURE 8.2

$$\cot u = \frac{\sqrt{16 - x^2}}{x}$$

Thus

$$\int \frac{1}{x^2\sqrt{16-x^2}}\,dx = -\frac{1}{16}\cot u + C$$

$$= -\frac{\sqrt{16-x^2}}{16x} + C \quad \square$$

Had the integral in Example 1 been

$$\int \frac{x^2}{\sqrt{16-x^2}}\,dx$$

then the same substitution $x = 4\sin u$, along with (4) of Section 8.2, would have yielded

$$\int \frac{x^2}{\sqrt{16-x^2}}\,dx = \int \frac{16\sin^2 u}{\sqrt{16-16\sin^2 u}}(4\cos u)\,du$$

$$= 16\int \sin^2 u\,du$$

$$= 8u - 4\sin 2u + C$$

As always, we must write the answer in terms of the original variable x. Using the double angle formula and referring to Figure 8.2 again, we find that

$$\sin 2u = 2\sin u\cos u = 2\left(\frac{x}{4}\right)\frac{\sqrt{16-x^2}}{4}$$

$$= \frac{x}{8}\sqrt{16-x^2}$$

Since $x = 4\sin u$, it follows that $u = \arcsin(x/4)$, and thus

$$\int \frac{x^2}{\sqrt{16-x^2}}\,dx = 8u - 4\sin 2u + C$$

$$= 8\arcsin\frac{x}{4} - \frac{x}{2}\sqrt{16-x^2} + C$$

Example 2 Evaluate $\displaystyle\int_{-5/2}^{5/2}\sqrt{25-4x^2}\,dx$.

Solution Because $\sqrt{25-4x^2} = \sqrt{5^2-(2x)^2}$, we are led to substitute

$$2x = 5\sin u, \quad \text{so that} \quad x = \frac{5}{2}\sin u, \quad \text{and thus} \quad dx = \frac{5}{2}\cos u\,du$$

For the limits of integration we notice that

$$\text{if } x = -\frac{5}{2} \text{ then } u = -\frac{\pi}{2}, \quad \text{and} \quad \text{if } x = \frac{5}{2} \text{ then } u = \frac{\pi}{2}$$

Therefore with the help of (5) of Section 8.2, we conclude that

$$\int_{-5/2}^{5/2} \sqrt{25 - 4x^2}\, dx = \int_{-5/2}^{5/2} \sqrt{5^2 - (2x)^2}\, dx$$

$$= \int_{-\pi/2}^{\pi/2} \sqrt{5^2 - 5^2 \sin^2 u}\left(\frac{5}{2}\cos u\right) du$$

$$= \int_{-\pi/2}^{\pi/2} 5\sqrt{1 - \sin^2 u}\left(\frac{5}{2}\cos u\right) du = \frac{25}{2}\int_{-\pi/2}^{\pi/2} \cos^2 u\, du$$

$$= \frac{25}{2}\left(\frac{1}{2}u + \frac{1}{4}\sin 2u\right)\Big|_{-\pi/2}^{\pi/2}$$

$$= \frac{25}{2}\left(\frac{\pi}{4} - \left(-\frac{\pi}{4}\right)\right) = \frac{25}{4}\pi \quad \square$$

We can use Example 2 to find the area A of the region enclosed by the ellipse

$$\frac{x^2}{(5/2)^2} + \frac{y^2}{5^2} = 1$$

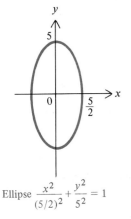

Ellipse $\dfrac{x^2}{(5/2)^2} + \dfrac{y^2}{5^2} = 1$

FIGURE 8.3

(Figure 8.3). This area is twice the area of the region bounded above by the graph of $y = \sqrt{25 - 4x^2}$ on the interval $[-\frac{5}{2}, \frac{5}{2}]$. Using the result of Example 2, we find that

$$A = 2\int_{-5/2}^{5/2} \sqrt{25 - 4x^2}\, dx = 2\left(\frac{25\pi}{4}\right) = \frac{25\pi}{2}$$

More generally, it is possible to show that the area of the region enclosed by the ellipse

$$\frac{x^2}{a^2} + \frac{y^2}{b^2} = 1 \quad \text{with } a > 0 \text{ and } b > 0$$

is $\pi a b$ (see Exercise 50). This result will be employed in the study of planetary orbits in Section 12.7.

Integrals Containing
$\sqrt{x^2 + a^2}$

If we let $x = a\tan u$, with $a > 0$ and $-\pi/2 < u < \pi/2$, then $a\sec u \geq 0$, so that

$$\sqrt{x^2 + a^2} = \sqrt{a^2\tan^2 u + a^2} = \sqrt{a^2(\tan^2 u + 1)} = \sqrt{a^2\sec^2 u} = a\sec u$$

Thus if an integral contains $\sqrt{x^2 + a^2}$, we can eliminate the square root by substituting

$$x = a\tan u, \quad \text{so that} \quad dx = a\sec^2 u\, du$$

Example 3   Find $\displaystyle\int \frac{1}{x^2\sqrt{x^2 + 1}}\, dx$.

Solution   We substitute

$$x = \tan u, \quad \text{so that} \quad dx = \sec^2 u\, du$$

Therefore

$$\int \frac{1}{x^2 \sqrt{x^2+1}} dx = \int \frac{1}{\tan^2 u \sqrt{\tan^2 u + 1}} (\sec^2 u)\, du$$

$$= \int \frac{1}{\tan^2 u \sec u} (\sec^2 u)\, du = \int \frac{\sec u}{\tan^2 u}\, du$$

$$= \int \frac{\cos u}{\sin^2 u}\, du = -\frac{1}{\sin u} + C$$

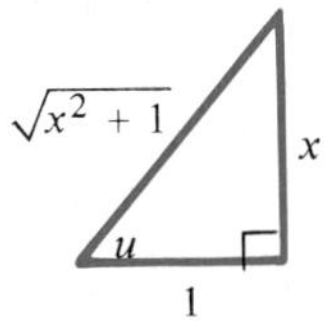

FIGURE 8.4

To give the answer in terms of x, we use the triangle in Figure 8.4, with $x = \tan u$. We find that

$$\int \frac{1}{x^2 \sqrt{x^2+1}} dx = -\frac{1}{\sin u} + C = -\frac{\sqrt{x^2+1}}{x} + C \quad \square$$

Example 4   Find $\displaystyle\int \frac{1}{\sqrt{1+4x^2}}\, dx$.

Solution   Because

$$\sqrt{1+4x^2} = \sqrt{1+(2x)^2}$$

we substitute

$$2x = \tan u, \quad \text{so that} \quad x = \frac{1}{2}\tan u, \quad \text{and thus} \quad dx = \frac{1}{2}\sec^2 u\, du$$

Then

$$\int \frac{1}{\sqrt{1+4x^2}}\, dx = \int \frac{1}{\sqrt{1+(2x)^2}}\, dx = \int \frac{1}{\sqrt{1+\tan^2 u}} \left(\frac{1}{2}\sec^2 u\right) du$$

$$= \int \frac{1}{\sec u} \left(\frac{1}{2}\sec^2 u\right) du$$

$$= \frac{1}{2} \int \sec u\, du$$

Using (6) of Section 8.2 for $\int \sec u\, du$, and then using the fact that $\tan u = 2x$, so that $\sec u = \sqrt{1+\tan^2 u} = \sqrt{1+4x^2}$, we conclude that

$$\int \frac{1}{\sqrt{1+4x^2}}\, dx = \frac{1}{2} \int \sec u\, du = \frac{1}{2} \ln|\sec u + \tan u| + C$$

$$= \frac{1}{2} \ln|\sqrt{1+4x^2} + 2x| + C \quad \square$$

If we were to seek the length $\mathscr{L}$ of the curve described parametrically by

$$x = 1 + \arctan 2t \quad \text{and} \quad y = 1 - \tfrac{1}{2}\ln(1+4t^2) \quad \text{for } 0 \le t \le 1$$

then by (5) of Section 6.4,

$$\mathcal{L} = \int_0^1 \sqrt{\left(\frac{2}{1+4t^2}\right)^2 + \left(\frac{-4t}{1+4t^2}\right)^2}\, dt = \int_0^1 \sqrt{\frac{4}{(1+4t^2)^2} + \frac{16t^2}{(1+4t^2)^2}}\, dt$$

$$= \int_0^1 \sqrt{\frac{4(1+4t^2)}{(1+4t^2)^2}}\, dt = 2\int_0^1 \frac{1}{\sqrt{1+4t^2}}\, dt$$

By the result of Example 4, it would follow that

$$\mathcal{L} = 2\left(\frac{1}{2}\ln|\sqrt{1+4t^2} + 2t|\right)\Big|_0^1 = \ln(\sqrt{5} + 2)$$

Integrals Containing $\sqrt{x^2 - a^2}$

If we let $x = a\sec u$, with $0 \le u < \pi/2$ or $\pi \le u < 3\pi/2$, then $a\tan u \ge 0$, so that

$$\sqrt{x^2 - a^2} = \sqrt{a^2\sec^2 u - a^2} = \sqrt{a^2(\sec^2 u - 1)} = \sqrt{a^2\tan^2 u} = a\tan u$$

Thus if an integral contains $\sqrt{x^2 - a^2}$, then we can eliminate the square root by substituting

$$x = a\sec u, \quad \text{so that} \quad dx = a\sec u\tan u\, du$$

Example 5   Find $\displaystyle\int_{-6}^{-3} \frac{\sqrt{x^2 - 9}}{x}\, dx$.

Solution   The domain of the integrand consists of $(-\infty, -3]$ and $[3, \infty)$, but since the interval over which we must integrate is $[-6, -3]$, we seek an antiderivative whose domain is contained in $(-\infty, -3]$. Since $\sqrt{x^2 - 9} = \sqrt{x^2 - 3^2}$, we let

$$x = 3\sec u, \quad \text{so that} \quad dx = 3\sec u\tan u\, du$$

and notice that $\sqrt{x^2 - 9} = \sqrt{9\sec^2 u - 9} = 3\tan u$. For the limits of integration we observe that

$$\text{if } x = -6 \text{ then } u = \frac{4\pi}{3}, \quad \text{and} \quad \text{if } x = -3 \text{ then } u = \pi$$

Therefore

$$\int_{-6}^{-3} \frac{\sqrt{x^2 - 9}}{x}\, dx = \int_{4\pi/3}^{\pi} \frac{\sqrt{9\sec^2 u - 9}}{3\sec u}(3\sec u\tan u)\, du$$

$$= \int_{4\pi/3}^{\pi} \frac{3\tan u}{3\sec u}(3\sec u\tan u)\, du$$

$$= 3\int_{4\pi/3}^{\pi} \tan^2 u\, du = 3\int_{4\pi/3}^{\pi} (\sec^2 u - 1)\, du$$

$$= 3(\tan u - u)\Big|_{4\pi/3}^{\pi} = 3(\tan\pi - \pi) - 3\left(\tan\frac{4\pi}{3} - \frac{4\pi}{3}\right)$$

$$= \pi - 3\sqrt{3} \quad \square$$

Table 8.3 summarizes the trigonometric substitutions that we have discussed.

TABLE 8.3
TRIGONOMETRIC SUBSTITUTIONS

Expression in Integrand	Substitution
$\sqrt{a^2 - x^2}$	$x = a\sin u$, with $-\pi/2 \leq u \leq \pi/2$ $dx = a\cos u\,du$
$\sqrt{x^2 + a^2}$	$x = a\tan u$, with $-\pi/2 < u < \pi/2$ $dx = a\sec^2 u\,du$
$\sqrt{x^2 - a^2}$	$x = a\sec u$, with $0 \leq u < \pi/2$ or $\pi \leq u < 3\pi/2$ $dx = a\sec u\tan u\,du$

Integrals Containing $\sqrt{bx^2 + cx + d}$

In order to evaluate an integral containing an expression of the form $\sqrt{bx^2 + cx + d}$, where $b \neq 0$ and $c \neq 0$, we first complete the square in $bx^2 + cx + d$ and then make a trigonometric substitution.

Example 6   Find $\displaystyle\int \frac{1}{\sqrt{9x^2 + 6x + 2}}\,dx$.

Solution   First we complete the square in the denominator:

$$\sqrt{9x^2 + 6x + 2} = \sqrt{(3x + 1)^2 + 1}$$

Then we let

$$3x + 1 = \tan u, \quad \text{so that} \quad 3\,dx = \sec^2 u\,du, \quad \text{and thus} \quad dx = \frac{1}{3}\sec^2 u\,du$$

and notice that $\sqrt{(3x + 1)^2 + 1} = \sqrt{\tan^2 u + 1} = \sec u$. Using (6) of Section 8.2, we find that

$$\int \frac{1}{\sqrt{9x^2 + 6x + 2}}\,dx = \int \frac{1}{\sqrt{(3x + 1)^2 + 1}}\,dx$$

$$= \int \frac{1}{\sec u}\left(\frac{1}{3}\sec^2 u\right)du$$

$$= \frac{1}{3}\int \sec u\,du = \frac{1}{3}\ln|\sec u + \tan u| + C$$

Referring to our formulas for $\tan u$ and $\sec u$, we conclude that

$$\int \frac{1}{\sqrt{9x^2 + 6x + 2}}\,dx = \frac{1}{3}\ln|\sqrt{(3x + 1)^2 + 1} + 3x + 1| + C$$

$$= \frac{1}{3}\ln|\sqrt{9x^2 + 6x + 2} + 3x + 1| + C \quad \square$$

EXERCISES 8.3

In Exercises 1–45 evaluate the integral.

1. $\int_0^{1/2} \sqrt{1 - 4x^2}\, dx$

2. $\int_0^4 \sqrt{16 - x^2}\, dx$

3. $\int_{-2}^2 \sqrt{1 - \frac{x^2}{4}}\, dx$

4. $\int \sqrt{1 + (2x - 1)^2}\, dx$

5. $\int \frac{1}{(9 + t^2)^2}\, dt$

6. $\int_{4\sqrt{2}}^8 \frac{1}{(t^2 - 16)^{3/2}}\, dt$

7. $\int \frac{1}{(x^2 + 1)^{3/2}}\, dx$

8. $\int \frac{1}{(2x^2 + 1)^{3/2}}\, dx$

9. $\int_0^1 \frac{1}{(3x^2 + 2)^{5/2}}\, dx$

10. $\int \frac{x^2}{(x^2 - 2)^{5/2}}\, dx$

11. $\int \frac{1}{(9x^2 - 4)^{5/2}}\, dx$

12. $\int \frac{1}{(y^2 - 4)^{3/2}}\, dy$

13. $\int \frac{1}{(3 - x^2)^{3/2}}\, dx$

14. $\int_1^{\sqrt{2}} \frac{1}{\sqrt{2x^2 - 1}}\, dx$

15. $\int_0^{5/4} \frac{1}{\sqrt{25 - 4x^2}}\, dx$

16. $\int \frac{x^2}{\sqrt{1 - x^2}}\, dx$

17. $\int \frac{1}{\sqrt{4x^2 + 4x + 2}}\, dx$

18. $\int_1^{\sqrt{3}} \frac{1}{\sqrt{1 + x^2}}\, dx$

19. $\int \frac{1}{(1 - 2w^2)^{5/2}}\, dw$

20. $\int_{\sqrt{2} - 2}^{(2\sqrt{3}/3) - 2} \frac{1}{(t^2 + 4t + 3)^{3/2}}\, dt$

21. $\int_0^1 \frac{1}{2x^2 - 2x + 1}\, dx$

*22. $\int \sqrt{z^2 - 9}\, dz$

23. $\int \sqrt{x - x^2}\, dx$

24. $\int \frac{\sqrt{25 - z^2}}{z^2}\, dz$

25. $\int \frac{x^2}{\sqrt{9x^2 - 1}}\, dx$

26. $\int_{-4/\sqrt{3}}^{-2\sqrt{2}} \frac{1}{x\sqrt{x^2 - 4}}\, dx$

27. $\int \frac{1}{x\sqrt{x^2 + 4}}\, dx$

28. $\int \frac{1}{x^2\sqrt{9 - 4x^2}}\, dx$

29. $\int \frac{1}{x^2\sqrt{4x^2 - 9}}\, dx$

30. $\int \frac{1}{w^2\sqrt{4w^2 + 9}}\, dw$

31. $\int \frac{x^2}{\sqrt{1 + x^2}}\, dx$

32. $\int \frac{x^2}{(1 + x^2)^{3/2}}\, dx$

33. $\int_2^{2\sqrt{2}} \frac{\sqrt{x^2 - 4}}{x}\, dx$

34. $\int \frac{\sqrt{4 + x^2}}{x^3}\, dx$

35. $\int \sqrt{4 + x^2}\, dx$

36. $\int \frac{\sqrt{x^2 + 1}}{x^4}\, dx$

37. $\int_{3\sqrt{2}}^6 \frac{1}{x^4\sqrt{x^2 - 9}}\, dx$

38. $\int \frac{2x - 3}{\sqrt{4x - x^2 - 3}}\, dx$

39. $\int \frac{x}{\sqrt{2x^2 + 12x + 19}}\, dx$

40. $\int \frac{1}{(w^2 + 2w + 5)^{3/2}}\, dw$

*41. $\int \sqrt{x^2 + 6x + 5}\, dx$

42. $\int \frac{e^{3w}}{\sqrt{1 - e^{2w}}}\, dw$

43. $\int e^w \sqrt{1 + e^{2w}}\, dw$

44. $\int_{\sqrt{2}}^2 \operatorname{arcsec} x\, dx$

45. $\int x \arcsin x\, dx$ (*Hint:* Use integration by parts.)

46. a. Evaluate $\int \left(\frac{1 - x}{1 + x}\right)^{1/2} dx$ by letting $x = \cos 2u$ and recalling that $1 + \cos 2u = 2\cos^2 u$ and $1 - \cos 2u = 2\sin^2 u$.

   *b. Evaluate $\int_1^{\cosh 2} \left(\frac{x - 1}{x + 1}\right)^{1/2} dx$ by letting $x = \cosh 2u$.

In Exercises 47–49 find the area A of the region between the graph of f and the x axis on the given interval.

47. $f(x) = \sqrt{1 - x^2}$; $[0, 1]$

48. $f(x) = \frac{x^2}{\sqrt{1 - x^2}}$; $\left[-\frac{1}{2}, \frac{1}{2}\right]$

49. $f(x) = \sqrt{9 + x^2}$; $[0, 3]$

50. Show that the region enclosed by the ellipse

$$\frac{x^2}{a^2} + \frac{y^2}{b^2} = 1$$

has area πab.

51. The Ellipse in Washington, D.C., which is located in front of the White House, is approximately 1500 feet long and approximately 1280 feet wide. How many square feet of grass should be purchased in order to sod the Ellipse completely? (*Hint:* An equation for the Ellipse is $x^2/750^2 + y^2/640^2 = 1$.)

52. Suppose a washer for a faucet has the shape of the solid generated by revolving about the x axis the figure bounded by the y axis, the lines $x = 1$ and $y = \frac{1}{2}$, and the curve $y = 1 + \sqrt{1 - x^2}$ for $0 \leq x \leq 1$. Determine the volume V of the washer.

53. A torus is a doughnut-shaped region generated by revolving a circle about a line that does not intersect the circle (Figure 8.5).

   a. Determine the volume V of the torus if an equation of

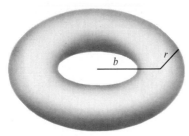

Torus

FIGURE 8.5

the circle is $(x - b)^2 + y^2 = r^2$, the line is the y axis, and $r < b$.

b. Which doughnut should cost more, one with $b = 4$ and $r = 2$, or one with $b = 6$ and $r = 1$?

54. Let $f(x) = x^2 - x$ for $0 \le x \le 1$. Find the length $\mathscr{L}$ of the graph of f.

*55. Let $f(x) = x^2$ for $0 \le x \le 1/2$. Find the length $\mathscr{L}$ of the graph of f.

56. Let $f(x) = \sin x$ for $0 \le x \le \pi$. Find the surface area S of the surface generated by revolving the graph of f about the x axis.

57. The curve C parametrized by

$$x = \sqrt{2} \cos t \quad \text{and} \quad y = 2 \sin t \quad \text{for } 0 \le t \le \pi$$

is half an ellipse, and the surface obtained by revolving C around the x axis is called an **ellipsoid**. Determine the surface area S of the ellipsoid.

58. Let $f(x) = \sqrt{1 - x^2}$ and $g(x) = -(1 + x)$. Calculate the center of gravity of the region R between the graphs of f and g on $[0, 1]$.

59. Suppose a large gasoline tank has the shape of a half-cylinder 4 feet in radius and 10 feet long (Figure 8.6). If the tank is full of gasoline weighing 42 pounds per cubic foot, find the work W necessary to pump all the gasoline through a spout at the top.

60. A flat circular floodlight on a vertical wall of a swimming pool has a radius of 1 foot, and its highest point is 3 feet

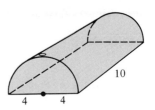

FIGURE 8.6

below the water level. Compute the hydrostatic force F on the light.

*61. In long jumping, the horizontal length of the jump is measured. Suppose the center of gravity of the jumper travels the path described by

$$y = \frac{-2x^2}{169} + \frac{4x}{13} + 3 \quad \text{for } 0 \le x \le 26$$

where x is measured in feet, and suppose the center of gravity starts and finishes 3 feet above the ground (Figure 8.7). This means that the horizontal length of the jump is 26 feet. Show that the actual distance $\mathscr{L}$ the center of gravity travels is approximately 26.4046 feet. (*Hint:* First evaluate the indefinite integral, by substituting $\tan u = \frac{4}{169}x - \frac{4}{13}$ and then using (7) of Section 8.2.)

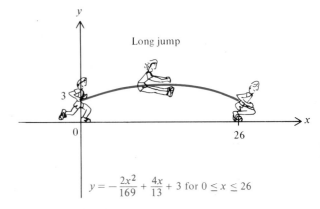

Long jump

$$y = -\frac{2x^2}{169} + \frac{4x}{13} + 3 \text{ for } 0 \le x \le 26$$

FIGURE 8.7

8.4
PARTIAL FRACTIONS

In this section we set up a three-part procedure to follow in integrating rational functions. We will use that procedure to find integrals such as

$$\int \frac{2x + 3}{x^3 + 2x^2 + x} dx \quad \text{and} \quad \int \frac{x^2 + 2x + 7}{x^3 + x^2 - 2} dx$$

The formula

$$\int \frac{1}{x^2 + a^2}\,dx = \frac{1}{a}\arctan\frac{x}{a} + C \tag{1}$$

derived in Section 7.6, will come into play.

Preparing Rational Functions for Integration

Before integrating any rational function, we normally must make three preparations, which we list as (i), (ii), and (iii).

(i) When applicable, divide the numerator of the rational function by its denominator, to ensure that the degree of the numerator of any resulting remainder is *less* than the degree of the denominator.

Example 1 Perform operation (i) on $\dfrac{2x^3}{x^2 + 3}$.

Solution We perform the division much as we would if the numerator and denominator were integers:

$$\begin{array}{r} 2x \\ x^2 + 3\,\overline{\smash{\big)}\,2x^3 } \\ \underline{2x^3 + 6x} \\ -6x \end{array}$$

Therefore

$$\frac{2x^3}{x^2 + 3} = 2x - \frac{6x}{x^2 + 3}$$

Notice that the degree of the numerator in the remainder is 1, while the degree of the denominator is 2. $\square$

It follows from Example 1 that

$$\int \frac{2x^3}{x^2 + 3}\,dx = \int 2x\,dx - \int \frac{6x}{x^2 + 3}\,dx$$

Since the first integral on the right side of the equation is easy to find, the problem of finding the integral on the left is reduced to finding the second integral on the right.

Example 2 Perform operation (i) on $\dfrac{x^2 - 1}{x^2 + 1}$.

Solution In this case the degree of both numerator and denominator is 2, and since the degree of the numerator is *not less than* the degree of the denominator, we must divide:

$$\begin{array}{r} 1 \\ x^2+1\,\overline{\big)\,x^2-1} \\ \underline{x^2+1} \\ -2 \end{array}$$

Therefore

$$\frac{x^2-1}{x^2+1} = 1 - \frac{2}{x^2+1}$$

and the right-hand side complies with (i). □

In the case of

$$\frac{x^2+3}{2x^3+x^2+1}$$

operation (i) would require no work, since the degree of the numerator is already less than the degree of the denominator.

After (i) has been performed, operation (ii) is to be performed on the resulting remainder, which is itself a rational function.

> (ii) Factor the numerator and the denominator of the remainder into linear and quadratic factors, that is, into expressions of the form
>
> $$\text{constants,} \quad (x-a)^r, \quad \text{and} \quad (x^2+bx+c)^s$$
>
> where r and s are positive integers. Factor quadratic factors wherever possible. Finally, reduce the fraction if any common factors appear in both numerator and denominator.

Example 3   Perform operation (ii) on $\dfrac{2x+4}{x^2+3x+2}$.

Solution   First, we factor the numerator and denominator:

$$2x+4 = 2(x+2)$$
$$x^2+3x+2 = (x+2)(x+1)$$

Therefore

$$\frac{2x+4}{x^2+3x+2} = \frac{2(x+2)}{(x+2)(x+1)} = \frac{2}{x+1} \quad □$$

Example 4   Perform operation (ii) on $\dfrac{2x+3}{x^3+2x^2+x}$.

Solution Here we find that

$$2x + 3 = 2(x + \tfrac{3}{2})$$

$$x^3 + 2x^2 + x = x(x^2 + 2x + 1)$$

$$= x(x + 1)^2$$

As a result,

$$\frac{2x + 3}{x^3 + 2x^2 + x} = \frac{2(x + \tfrac{3}{2})}{x(x + 1)^2} \quad \square$$

In Examples 3 and 4 the factorization was easy. In more difficult cases the following lemma can simplify finding factors of a polynomial $P(x)$.

LEMMA 8.2

The expression $x - a$ is a factor of $P(x)$ if and only if $P(a) = 0$.

Proof If $x - a$ is a factor of $P(x)$, then there is a polynomial function R such that

$$P(x) = (x - a)R(x)$$

But then

$$P(a) = (a - a)R(a) = 0$$

To prove the converse, let $P(x)$ be given by

$$P(x) = c_n x^n + c_{n-1} x^{n-1} + \cdots + c_1 x + c_0$$

and assume that $P(a) = 0$. Then

$$P(a) = c_n a^n + c_{n-1} a^{n-1} + \cdots + c_1 a + c_0 = 0$$

so that

$$\begin{aligned} P(x) &= P(x) - P(a) \\ &= c_n(x^n - a^n) + c_{n-1}(x^{n-1} - a^{n-1}) + \cdots + c_1(x - a) \end{aligned} \tag{2}$$

But for any integer $k \geq 2$, we know that

$$x^k - a^k = (x - a)(x^{k-1} + x^{k-2}a + \cdots + xa^{k-2} + a^{k-1})$$

Therefore $x - a$ is a factor of each of the summands appearing on the right side of (2). Thus $x - a$ is a factor of $P(x)$. ∎

Example 5 Perform (ii) on $\dfrac{x^2 + 2x + 7}{x^3 + x^2 - 2}$.

Solution By the quadratic formula, the numerator has no real roots, and hence cannot be factored. To factor the denominator, notice first that

$$x^3 + x^2 - 2 = 0 \quad \text{for} \quad x = 1$$

By Lemma 8.2, $x - 1$ must therefore be a factor of $x^3 + x^2 - 2$. Carrying out the division of $x^3 + x^2 - 2$ by $x - 1$ gives us

$$x^3 + x^2 - 2 = (x - 1)(x^2 + 2x + 2)$$

Again the quadratic formula tells us that $x^2 + 2x + 2$ has no real roots, so cannot be factored. Thus we have

$$\frac{x^2 + 2x + 7}{x^3 + x^2 - 2} = \frac{x^2 + 2x + 7}{(x - 1)(x^2 + 2x + 2)} \tag{3}$$

as the desired way of rewriting the original function.   □

After performing operations (i) and (ii) we obtain a rational function g such that the denominator of $g(x)$ is a product of constants, factors of the form $(x - a)^r$, and factors of the form $(x^2 + bx + c)^s$. The final step in the preparation involves equating $g(x)$ with a sum of terms arising from the factors appearing in the denominator of $g(x)$. For every factor $(x - a)^r$ appearing in the denominator of $g(x)$ we include an expression of the form

$$\frac{A_1}{x - a} + \frac{A_2}{(x - a)^2} + \cdots + \frac{A_r}{(x - a)^r} \tag{4}$$

where $A_1, A_2, \ldots, A_r$ must be determined. For every factor of the form $(x^2 + bx + c)^s$ we include an expression of the form

$$\frac{B_1 x + C_1}{x^2 + bx + c} + \frac{B_2 x + C_2}{(x^2 + bx + c)^2} + \cdots + \frac{B_s x + C_s}{(x^2 + bx + c)^s} \tag{5}$$

where again the numbers $B_1, B_2, \ldots, B_s$, and $C_1, C_2, \ldots, C_s$ must be determined. For instance,

$$\frac{2(x + \frac{3}{2})}{x(x + 1)^2} = \frac{A}{x} + \frac{B}{x + 1} + \frac{C}{(x + 1)^2} \tag{6}$$

and

$$\frac{x^2 + 2x + 7}{(x - 1)(x^2 + 2x + 2)} = \frac{A}{x - 1} + \frac{Bx + C}{x^2 + 2x + 2} \tag{7}$$

where the constants A, B, and C must be determined.

We are now ready to describe the third and final preparation.

> (iii) Rewrite the transformed rational function $g(x)$ as a sum of expressions of the types appearing in (4) and (5).

Because of the fractions that arise from step (iii), this method of transforming rational functions is known as the method of **partial fractions**.

Examples of Integration by Partial Fractions   In the next three examples we will perform (i)–(iii) as needed and then integrate the resulting integrals.

Example 6   Evaluate $\int \dfrac{2x + 3}{x^3 + 2x^2 + x}\, dx.$

Solution   In Example 4 we performed (ii) on the integrand and obtained

$$\frac{2x + 3}{x^3 + 2x^2 + x} = \frac{2(x + \frac{3}{2})}{x(x + 1)^2}$$

Next we perform (iii) by rewriting the right side:

$$\frac{2(x + \frac{3}{2})}{x(x + 1)^2} = \frac{A}{x} + \frac{B}{x + 1} + \frac{C}{(x + 1)^2} \tag{8}$$

which is just (6) again. To calculate the constants A, B, and C, we clear fractions by multiplying both sides of (8) by $x(x + 1)^2$. We obtain

$$2(x + \tfrac{3}{2}) = A(x + 1)^2 + Bx(x + 1) + Cx$$

This equation must hold for all x, so in particular it must hold for $x = 0$ and for $x = -1$. For $x = 0$ we have

$$3 = A(1) + B(0) + C(0), \quad \text{so that} \quad A = 3$$

For $x = -1$ we have

$$2\left(\frac{1}{2}\right) = A(0) + B(0) + C(-1), \quad \text{so that} \quad C = -1$$

Therefore

$$2\left(x + \frac{3}{2}\right) = 3(x + 1)^2 + Bx(x + 1) - x$$

so that by combining like powers of x on the right side of the equation we obtain

$$2\left(x + \frac{3}{2}\right) = (3 + B)x^2 + (5 + B)x + 3$$

This equation can be true for all values of x only if the coefficients of like powers of x on the two sides of the equation are the same. But then

$$3 + B = 0, \quad \text{so that} \quad B = -3$$

As a result, equation (8) becomes

$$\frac{2(x + \frac{3}{2})}{x(x + 1)^2} = \frac{3}{x} - \frac{3}{x + 1} - \frac{1}{(x + 1)^2}$$

Therefore

$$\int \frac{2x + 3}{x^3 + 2x + x}\, dx = \int \frac{2(x + \frac{3}{2})}{x(x + 1)^2}\, dx$$

$$= \int \frac{3}{x} \, dx - \int \frac{3}{x+1} \, dx - \int \frac{1}{(x+1)^2} \, dx$$

$$= 3 \ln |x| - 3 \ln |x+1| + \frac{1}{x+1} + C_1$$

$$= 3 \ln \left| \frac{x}{x+1} \right| + \frac{1}{x+1} + C_1$$

Thus we have evaluated the integral under consideration. □

Example 7   Evaluate $\int \frac{x^2 + 2x + 7}{x^3 + x^2 - 2} \, dx$.

Solution   To achieve (ii) and (iii), we combine (3) and (7):

$$\frac{x^2 + 2x + 7}{x^3 + x^2 - 2} = \frac{A}{x-1} + \frac{Bx + C}{x^2 + 2x + 2} \tag{9}$$

Next, to clear fractions in the right-hand sum of (9), we multiply through by $(x - 1)(x^2 + 2x + 2)$ and obtain

$$x^2 + 2x + 7 = A(x^2 + 2x + 2) + (Bx + C)(x - 1)$$

This equation must hold for all x, so in particular it must hold for $x = 1$. For $x = 1$ we get

$$10 = A(5), \quad \text{so that} \quad A = 2$$

Therefore

$$x^2 + 2x + 7 = 2(x^2 + 2x + 2) + (Bx + C)(x - 1)$$

Next we combine like powers of x:

$$x^2 + 2x + 7 = (2 + B)x^2 + (4 + C - B)x + (4 - C) \tag{10}$$

This can be true for all values of x only if the coefficients of like powers of x on both sides of (10) are equal. Thus

$$1 = 2 + B, \quad \text{so that} \quad B = -1$$
$$7 = 4 - C, \quad \text{so that} \quad C = -3$$

(Since $4 + C - B = 4 - 3 + 1 = 2$, the coefficients of x on both sides of (10) are equal as well.) As a result, we can rewrite (9) as follows:

$$\frac{x^2 + 2x + 7}{x^3 + x^2 - 2} = \frac{2}{x-1} - \frac{x+3}{x^2 + 2x + 2}$$

Thus

$$\int \frac{x^2 + 2x + 7}{x^3 + x^2 - 2} \, dx = \int \frac{2}{x-1} \, dx - \int \frac{x+3}{x^2 + 2x + 2} \, dx$$

To evaluate the right-hand integral, we first complete the square in the denominator to obtain

$$x^2 + 2x + 2 = (x + 1)^2 + 1$$

and then substitute

$$u = x + 1, \quad \text{so that} \quad du = dx \quad \text{and} \quad x + 3 = u + 2$$

Therefore

$$\int \frac{x + 3}{x^2 + 2x + 2} \, dx = \int \frac{x + 3}{(x + 1)^2 + 1} \, dx$$

$$= \int \frac{u + 2}{u^2 + 1} \, du$$

$$= \int \frac{u}{u^2 + 1} \, du + 2 \int \frac{1}{u^2 + 1} \, du$$

$$= \frac{1}{2} \int \frac{2u}{u^2 + 1} \, du + 2 \int \frac{1}{u^2 + 1} \, du$$

$$= \frac{1}{2} \ln (u^2 + 1) + 2 \arctan u + C$$

$$= \frac{1}{2} \ln ((x + 1)^2 + 1) + 2 \arctan (x + 1) + C$$

Consequently

$$\int \frac{x^2 + 2x + 7}{x^3 + x^2 - 2} \, dx = \int \frac{2}{x - 1} \, dx - \int \frac{x + 3}{x^2 + 2x + 2} \, dx$$

$$= 2 \ln |x - 1| + \frac{1}{2} \ln ((x + 1)^2 + 1) + 2 \arctan (x + 1) + C \quad \square$$

Finally, we evaluate an integral in which the quadratic factor $x^2 + 1$ in the denominator is squared.

Example 8 Evaluate $\displaystyle\int \frac{1}{x(x^2 + 1)^2} \, dx$.

Solution To execute operation (iii), we write

$$\frac{1}{x(x^2 + 1)^2} = \frac{A}{x} + \frac{Bx + C}{x^2 + 1} + \frac{Dx + E}{(x^2 + 1)^2}$$

Next we clear fractions:

$$1 = A(x^2 + 1)^2 + (Bx + C)x(x^2 + 1) + (Dx + E)x \tag{11}$$

For $x = 0$ this equation becomes

$$1 = A(1) + C(0) + E(0), \quad \text{so that} \quad A = 1$$

After we replace A by 1, multiply out, and combine like powers, equation (11) becomes

$$1 = (1 + B)x^4 + Cx^3 + (2 + B + D)x^2 + (C + E)x + 1$$

Equating coefficients of like powers of x, we find that

$$1 + B = 0, \quad \text{so that} \quad B = -1$$

$$C = 0$$

$$2 + B + D = 0, \quad \text{so that} \quad D = -1$$

$$C + E = 0, \quad \text{so that} \quad E = 0$$

Therefore

$$\frac{1}{x(x^2 + 1)^2} = \frac{1}{x} - \frac{x}{x^2 + 1} - \frac{x}{(x^2 + 1)^2}$$

so that

$$\int \frac{1}{x(x^2 + 1)^2}\,dx = \int \frac{1}{x}\,dx - \int \frac{x}{x^2 + 1}\,dx - \int \frac{x}{(x^2 + 1)^2}\,dx$$

Substituting

$$u = x^2 + 1, \quad \text{so that} \quad du = 2x\,dx$$

we find that

$$-\int \frac{x}{x^2 + 1}\,dx - \int \frac{x}{(x^2 + 1)^2}\,dx = -\int \frac{1}{u}\cdot\frac{1}{2}\,du - \int \frac{1}{u^2}\cdot\frac{1}{2}\,du$$

$$= -\frac{1}{2}\ln|u| + \frac{1}{2}\cdot\frac{1}{u} + C_1$$

$$= -\frac{1}{2}\ln(x^2 + 1) + \frac{1}{2}\cdot\frac{1}{x^2 + 1} + C_1$$

Consequently

$$\int \frac{1}{x(x^2 + 1)^2}\,dx = \int \frac{1}{x}\,dx - \int \frac{x}{x^2 + 1}\,dx - \int \frac{x}{(x^2 + 1)^2}\,dx$$

$$= \ln|x| - \frac{1}{2}\ln(x^2 + 1) + \frac{1}{2(x^2 + 1)} + C_2 \quad \square$$

When we evaluate the integral of a rational function by partial fractions, the result can always, at least in principle, be given in terms of rational functions, logarithms, and arctangents. You can check that this is true with Examples 6–8.

EXERCISES 8.4

In Exercises 1–29 find the integral.

1. $\displaystyle \int \frac{x}{x+1}\,dx$

2. $\displaystyle \int \frac{x^2}{x^2+1}\,dx$

3. $\displaystyle \int \frac{x^2}{x^2-1}\,dx$

4. $\displaystyle \int \frac{t^2-1}{t^2+1}\,dt$

5. $\displaystyle \int \frac{x^2+4}{x(x-1)^2}\,dx$

6. $\displaystyle \int \frac{2x^3+x^2+12}{x^2-4}\,dx$

7. $\displaystyle \int_3^4 \frac{5}{(x-2)(x+3)}\,dx$

8. $\displaystyle \int \frac{5x}{(x-2)(x+3)}\,dx$

9. $\displaystyle \int \frac{3t}{t^2-8t+15}\,dt$

10. $\displaystyle \int \frac{2}{x^2-x-6}\,dx$

11. $\displaystyle \int_{-1}^0 \frac{x^2+x+1}{x^2+1}\,dx$

12. $\displaystyle \int \frac{2(1-x)}{x(x^2+2x-1)}\,dx$

13. $\displaystyle \int \frac{x^2+x+1}{x^2-1}\,dx$

14. $\displaystyle \int \frac{x^2+2x+1}{x^2-1}\,dx$

15. $\displaystyle \int_0^1 \frac{u-1}{u^2+u+1}\,du$

16. $\displaystyle \int \frac{2x}{(x+1)^2}\,dx$

17. $\displaystyle \int \frac{3x}{(x-2)^2}\,dx$

18. $\displaystyle \int \frac{4x}{(x+1)(x+2)(x+3)}\,dx$

19. $\displaystyle \int \frac{-x}{x(x-1)(x-2)}\,dx$

20. $\displaystyle \int \frac{x+1}{x(x-1)^2}\,dx$

21. $\displaystyle \int \frac{u^3}{(u+1)^2}\,du$

22. $\displaystyle \int_0^1 \frac{x^3}{(x+1)^3}\,dx$

$\left(\text{Hint: } \left(\dfrac{x}{x+1}\right)^3 = \left(1-\dfrac{1}{x+1}\right)^3.\right)$

23. $\displaystyle \int \frac{1}{(1-x^2)^2}\,dx$

$\left(\text{Hint: } \dfrac{1}{(1-x^2)^2} = \left(\dfrac{1}{2(x+1)} - \dfrac{1}{2(x-1)}\right)^2.\right)$

24. $\displaystyle \int \frac{1}{x^4+1}\,dx$

(Hint: $x^4+1 = (x^2+\sqrt{2}x+1)(x^2-\sqrt{2}x+1)$.)

25. $\displaystyle \int \frac{x}{(x+1)^2(x-2)}\,dx$

26. $\displaystyle \int \frac{1}{(x-1)(x^2+1)^2}\,dx$

27. $\displaystyle \int \frac{-x^3+x^2+x+3}{(x+1)(x^2+1)^2}\,dx$

28. $\displaystyle \int \frac{1}{(x^2+1)(x-2)}\,dx$

29. $\displaystyle \int \frac{x^2-1}{x^3+3x+4}\,dx$

(Hint: Find a root of the denominator.)

In Exercises 30–36 find the integral by means of the indicated substitution.

30. $\displaystyle \int \frac{\sqrt{x}+1}{x+1}\,dx;\ u = \sqrt{x}$

31. $\displaystyle \int \frac{1}{x\sqrt{x+1}}\,dx;\ u = \sqrt{x+1}$

32. $\displaystyle \int_0^{\pi/4} \tan^3 x\,dx;\ u = \tan x$

33. $\displaystyle \int \frac{\sqrt{x}}{1+\sqrt[3]{x}}\,dx;\ u = \sqrt[6]{x}$

34. $\displaystyle \int \frac{1}{\sqrt[4]{x}(1+\sqrt{x})}\,dx;\ u = \sqrt[4]{x}$

*35. $\displaystyle \int_{-5/3}^{-1} \sqrt{\frac{x+1}{x-1}}\,dx;\ u = \sqrt{\frac{x+1}{x-1}}$

*36. $\displaystyle \int_{-1/9}^{-1/2} \frac{1}{x}\left(\frac{x}{x+1}\right)^{1/3}dx;\ u = \left(\frac{x}{x+1}\right)^{1/3}$

In Exercises 37–43 evaluate the integral by first using substitution or integration by parts and then using partial fractions.

37. $\displaystyle \int \frac{\sin^2 x \cos x}{\sin^2 x + 1}\,dx$

38. $\displaystyle \int \frac{e^x}{1-e^{2x}}\,dx$

39. $\displaystyle \int \frac{e^x}{1-e^{3x}}\,dx$

40. $\displaystyle \int \frac{e^x}{e^{2x}+3e^x+2}\,dx$

41. $\displaystyle \int x \arctan x\,dx$

42. $\displaystyle \int x^3 \arctan x\,dx$

43. $\displaystyle \int \ln(x^2+1)\,dx$

44. Find the following integrals, which are related.

a. $\int \dfrac{x^2}{(x^2-4)^2}\,dx$

$\left(Hint: \dfrac{x^2}{(x^2-4)^2} = \dfrac{1}{4}\left(\dfrac{1}{x-2} + \dfrac{1}{x+2}\right)^2.\right)$

b. $\int \dfrac{\sqrt{x+4}}{x^2}\,dx$

(*Hint:* Let $u = \sqrt{x+4}$ and use part (a).)

In Exercises 45–47 make the substitution $u = \tan x/2$ and use the equations

$$\sin x = \frac{2u}{1+u^2}, \cos x = \frac{1-u^2}{1+u^2}, \text{ and } dx = \frac{2}{1+u^2}\,du$$

45. Find $\int \dfrac{1}{2+\sin x}\,dx$ 46. Find $\int \dfrac{1}{3-\cos x}\,dx$

47. Find $\int \dfrac{1}{2\cos x + \sin x}\,dx$

48. The reduction formula

$$\int \frac{1}{(x^2+a^2)^{n+1}}\,dx = \frac{x}{2na^2(x^2+a^2)^n}$$
$$+ \frac{2n-1}{2na^2}\int \frac{1}{(x^2+a^2)^n}\,dx$$

which appeared in Exercise 49 of Section 8.1, helps in the evaluation of certain rational functions. Use partial fractions, along with the above formula, to evaluate the integral

$$\int \frac{x^2-x+1}{x(x^2+1)^2}\,dx$$

In Exercises 49–50 find the area A of the region between the graph of f and the x axis on the given interval.

49. $f(x) = \dfrac{x^3}{x^2+1}$; on $[0,3]$

50. $f(x) = \dfrac{4x^2}{(x^2-1)(x^2+1)}$; $\left[0, \dfrac{\sqrt{3}}{3}\right]$

51. Find the area A of the region between the graphs of

$$y = \frac{x^2}{(x-2)(x^2+1)} \quad \text{and} \quad y = \frac{1}{x-3}$$

52. Let $f(x) = \dfrac{1}{2x^3 - x^4}$, and let R be the region between the graph of f and the x axis on $[1/2, 1]$. Find the volume V of the solid generated by revolving R about the y axis.

53. Let $f(y) = \dfrac{1}{(y+2)^2}$ and $g(y) = \dfrac{1}{y+2}$, and let R be the region between the graphs of f and g on $[0,2]$. Find the volume V of the solid obtained by revolving R about the x axis.

54. Let $f(x) = \dfrac{x}{x+2}$, and let R be the region between the graph of f and the x axis on $[0,2]$.
   a. Find the center of gravity of R.
   b. Use (a) and the Theorem of Pappus and Guldin to determine the volume V of the solid generated by revolving R about the x axis.

55. Let $g(x) = \ln(x^2-1)$ for $2 \le x \le 5$. Find the length $\mathscr{L}$ of the graph of f.

56. Find the length $\mathscr{L}$ of the curve described parametrically by $x = \arcsin t$ and $y = \dfrac{1}{2}\ln(1-t^2)$ for $0 < t \le \dfrac{1}{2}$.

57. Suppose that two substances interact chemically and that the original concentrations of the two substances are a and b. Let $x(t)$ be the decrease in the concentrations of the two substances after an elapsed time of t units, and let $r > 0$. Then under certain circumstances these quantities are related by the equation $dx/dt = r(a-x)(b-x)$, which can be rewritten

$$\frac{dt}{dx} = \frac{1}{r(a-x)(b-x)} \tag{12}$$

   a. Use partial fractions on the right side of (12) and then show by integration that

$$t = \frac{1}{r(a-b)}\ln\left|\frac{a-x}{b-x}\right| + C_1$$

   where C_1 is a constant.
   b. Assume that $x < a$ and $x < b$. Use (a) to show that

$$x = \frac{a - bCe^{r(a-b)t}}{1 - Ce^{r(a-b)t}}$$

   where $C = e^{r(b-a)C_1}$.

8.5
INTEGRATION USING THE TABLE OF INTEGRALS

The techniques of integration we have discussed so far—integration by substitution and trigonometric substitution, integration by parts and integration by partial fractions—allow us to evaluate many integrals. Mastery of these techniques gives one the knowledge to proceed confidently when encountering new integrals. Why then does this text include a Table of Integrals containing formulas for over 120 integrals?

One answer is that it is frequently easier to evaluate integrals by applying formulas in the Table of Integrals than by resorting to techniques of integration. However, even when we use the Table of Integrals to help solve integrals, it may be necessary to employ basic techniques of integration, especially substitution.

Before turning to the examples, we observe that the Table of Integrals in the back of the text is organized by type. When we wish to locate a formula for an integral, we first try to find a suitable type from the headings in the table, and then look at the formulas for that type.

Example 1   Use the Table of Integrals to evaluate $\int e^{-2x} \cos 3x \, dx$.

Solution   Because of the occurrence of e^{-2x}, we look at the group labeled "Exponential and Logarithmic Forms," and locate Formula 50 for the integral $\int e^{ax} \cos bx \, dx$. Applying the formula with $a = -2$ and $b = 3$, we find that

$$\int e^{-2x} \cos 3x \, dx = \frac{e^{-2x}}{(-2)^2 + 3^2}(-2 \cos 3x + 3 \sin 3x) + C$$

$$= \frac{e^{-2x}}{13}(-2 \cos 3x + 3 \sin 3x) + C \quad \square$$

Example 2   Use the Table of Integrals to evaluate $\int_0^3 x^2 \sqrt{9 - x^2} \, dx$.

Solution   The integrand of the indefinite integral $\int x^2 \sqrt{9 - x^2} \, dx$ has a square root of the form $\sqrt{a^2 - x^2}$, so we look in the group labeled "Forms Involving $a^2 - x^2$." Formula 68 applies, with $a = 3$:

$$\int x^2 \sqrt{9 - x^2} \, dx = \frac{x}{8}(2x^2 - 9)\sqrt{9 - x^2} + \frac{81}{8} \arcsin \frac{x}{3} + C$$

Therefore

$$\int_0^3 x^2 \sqrt{9 - x^2} \, dx = \left[\frac{x}{8}(2x^2 - 9)\sqrt{9 - x^2} + \frac{81}{8} \arcsin \frac{x}{3} \right]\Big|_0^3$$

$$= \frac{81\pi}{16} - 0 = \frac{81\pi}{16} \quad \square$$

Frequently in order to evaluate an integral by using the Table of Integrals we need first to make a substitution that converts the integrand to a form appearing in the table. This is illustrated in the next two examples.

Example 3   Use the Table of Integrals to evaluate $\int x^2 \sqrt{1 - 9x^2} \, dx$.

Solution   Despite the similarity of the integrands in this example and

Example 2, the present example is not covered directly by the Table because $9x^2$ appears rather than x^2. However, if we substitute $u = 3x$, then $du = 3\,dx$, and the given integral becomes

$$\int \left(\frac{1}{3}u\right)^2 \sqrt{1 - u^2}\, \frac{1}{3}\, du = \int \frac{1}{27} u^2 \sqrt{1 - u^2}\, du = \frac{1}{27} \int u^2 \sqrt{1 - u^2}\, du$$

which we can evaluate by applying Formula 68 with $a = 1$:

$$\frac{1}{27} \int u^2 \sqrt{1 - u^2}\, du = \frac{1}{27}\left[\frac{u}{8}(2u^2 - 1)\sqrt{1 - u^2} + \frac{1}{8}\arcsin u\right] + C$$

Therefore

$$\int x^2 \sqrt{1 - 9x^2}\, dx = \frac{1}{27}\left[\frac{u}{8}(2u^2 - 1)\sqrt{1 - u^2} + \frac{1}{8}\arcsin u\right] + C$$

$$= \frac{1}{216}[3x(18x^2 - 1)\sqrt{1 - 9x^2} + 1\arcsin 3x] + C \quad \square$$

Example 4 Use the Table of Integrals to evaluate $\displaystyle\int \frac{\sin^2 x \cos x}{(2 + 3\sin x)^2}\, dx$.

Solution None of the trigonometric forms listed in the Table of Integrals seems appropriate for this integral. However, because of the double occurrence of $\sin x$ and the single occurrence of $\cos x$ as a factor in the integrand, we substitute $u = \sin x$, so that $du = \cos x\, dx$. Then the integral becomes

$$\int \frac{u^2}{(2 + 3u)^2}\, du$$

which occurs (with x instead of u) in Formula 101. Letting $a = 2$ and $b = 3$ in Formula 101, we obtain

$$\int \frac{\sin^2 x \cos x}{(2 + 3\sin x)^2}\, dx = \int \frac{u^2}{(2 + 3u)^2}\, du = \frac{1}{27}\left(2 + 3u - \frac{4}{2 + 3u} - 4\ln|2 + 3u|\right) + C$$

$$= \frac{1}{27}\left(2 + 3\sin x - \frac{4}{2 + 3\sin x} - 4\ln|2 + 3\sin x|\right) + C \quad \square$$

The Table of Integrals contains reduction formulas which are also effective in evaluating integrals.

Example 5 Use the Table of Integrals to evaluate $\int 32x^4 e^{2x}\, dx$.

Solution In order to modify the given integral to fit the form of Formula 42, we substitute $u = 2x$, so that $du = 2\,dx$. Then $\int 32x^4 e^{2x}\, dx = \int 32(\frac{1}{16}u^4)e^u \frac{1}{2}\, du = \int u^4 e^u\, du$. Applying Formula 42 with x replaced by u, and with n successively equal to 4, 3, 2, and 1, we find that

$$\int 32x^4 e^u\, du = \int u^4 e^u\, du = u^4 e^u - 4\int u^3 e^u\, du$$

$$= u^4 e^u - 4\left(u^3 e^u - 3\int u^2 e^u\, du\right)$$

$$= u^4 e^u - 4u^3 e^u + 12\left(u^2 e^u - 2\int u e^u\, du\right)$$

$$= u^4 e^u - 4u^3 e^u + 12u^2 e^u - 24\left(u e^u - \int e^u\, du\right)$$

$$= u^4 e^u - 4u^3 e^u + 12u^2 e^u - 24u e^u + 24 e^u + C$$

$$= 16x^4 e^{2x} - 32x^3 e^{2x} + 48x^2 e^{2x} - 48x e^{2x} + 24 e^{2x} + C \quad \square$$

Each of the integrals evaluated above by using the Table of Integrals can be resolved by means of the techniques of integration we have already discussed. However, even if an integral can be evaluated by such techniques, it may not be clear what technique or techniques would be fruitful. Tables such as the one by Gröbner and Hofreiter, which includes formulas for over 1000 integrals, can be of great assistance in evaluating more complicated integrals that occasionally arise in applications. For example, it might not be apparent how to proceed in evaluating integrals such as

$$\int x\sqrt{\frac{a+x}{a-x}}\,dx \quad \text{and} \quad \int \sqrt{\frac{x}{x^3-a}}\,dx$$

These are included as Formulas 122 and 123 in the group headed "Miscellaneous Formulas" at the end of the Table of Integrals (see Exercises 17 and 18).

EXERCISES 8.5

In Exercises 1–18 use the Table of Integrals to evaluate the given integral.

1. $\int \sqrt{x^2+9}\,dx$

2. $\int \frac{1}{(16-x^2)^{3/2}}\,dx$

3. $\int_0^1 e^{5x}\sin\frac{1}{2}x\,dx$

4. $\int_1^2 \frac{1}{x(3x-2)}\,dx$

5. $\int \frac{1}{4x^2-9}\,dx$

6. $\int \frac{1}{x\sqrt{2x-4x^2}}\,dx$

7. $\int \frac{\sqrt{10x-\frac14 x^2}}{x}\,dx$

8. $\int \frac{\sqrt{\frac19 x^2-5}}{x^2}\,dx$

9. $\int \frac{e^{\sqrt{x}}\sinh 2\sqrt{x}}{\sqrt{x}}\,dx$

10. $\int_{\pi/6}^{\pi/2} \frac{\cos x}{\sin^2 x}\sqrt{1+\sin^2 x}\,dx$

11. $\int_0^1 e^{2x}\cos e^x\,dx$

12. $\int \frac{x^3}{3-2x^2}\,dx$

13. $\int \sin(\ln x)\,dx$

14. $\int \frac{\sqrt{x}}{(2-3\sqrt{x})^2}\,dx$

15. $\int \sqrt{2\sqrt{x}-x}\,dx$

16. $\int \frac{1}{x\ln x\sqrt{-4+\ln x}}\,dx$

17. $\int x\sqrt{\frac{2+x}{2-x}}\,dx$

18. $\int \sqrt{\frac{x}{x^3+1}}\,dx$

8.6
THE TRAPEZOIDAL RULE AND SIMPSON'S RULE

Despite the various methods that exist for calculating integrals—those we have described as well as many we have not—it is frequently difficult or impossible to express an indefinite integral in terms of familiar functions and thereby compute the exact numerical values of corresponding definite integrals. For example, we are unable to evaluate exactly the integral

$$\int_{-1}^{1} \sqrt{1 + x^4}\, dx \tag{1}$$

Such an integral turns up in the analysis of highways and railroad tracks (see Section 6.3). Although it is impossible to find the exact numerical value of (1), the methods to be described in the present section give very good estimates for this value.

When we approximate the integral $\int_a^b f(x)\, dx$ of a nonnegative function f by a Riemann sum, we are, in effect, replacing the segment of the graph of f on each subinterval by a suitable horizontal line (Figure 8.8); the Riemann sum then equals the sum of the areas of the corresponding rectangles. If we allow the segments of the graph of f to be replaced instead by nonhorizontal lines or parabolas, then with nearly the same effort, we can generally approximate $\int_a^b f(x)\, dx$ with much greater precision than with Riemann sums. It is to these more refined types of approximation that we now turn. For purposes of exposition, we will assume that f is continuous and nonnegative throughout $[a, b]$. Despite this restriction, the formulas we will derive remain valid for functions that are not necessarily nonnegative.

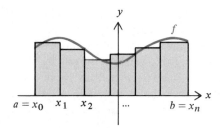

Riemann sum: horizontal lines
replace portions of the graph

FIGURE 8.8

The Trapezoidal Rule

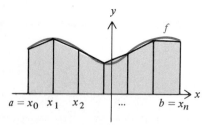

Trapezoidal Rule: possibly nonhorizontal
lines replace portions of the graph

FIGURE 8.9

To describe the first type of approximation, we take a partition $\mathscr{P} = \{x_0, x_1, \ldots, x_n\}$ that divides the interval $[a, b]$ into n subintervals of equal length $(b - a)/n$. We join each pair of points $(x_{k-1}, f(x_{k-1}))$ and $(x_k, f(x_k))$ on the graph of f by a straight line, thereby creating trapezoidal regions (Figure 8.9). Since we assumed that f is nonnegative, the formula for the area of the kth trapezoid is

$$\left(\frac{f(x_{k-1}) + f(x_k)}{2} \right)(x_k - x_{k-1}) = \frac{b - a}{2n} [f(x_{k-1}) + f(x_k)]$$

(see Exercise 58 of Section 5.5). Using this formula for each of the subintervals, we obtain the sum

$$\frac{b - a}{2n} \{ [f(x_0) + f(x_1)] + [f(x_1) + f(x_2)] + \cdots + [f(x_{n-1}) + f(x_n)] \}$$

as an approximation of $\int_a^b f(x)\, dx$. In other words,

$$\int_a^b f(x)\, dx \approx \frac{b - a}{2n} [f(x_0) + 2f(x_1) + 2f(x_2) + \cdots + 2f(x_{n-1}) + f(x_n)] \tag{2}$$

Because of the way it was derived, (2) is called the **Trapezoidal Rule**.

Example 1   Use the Trapezoidal Rule with $n = 3$, $n = 5$, and $n = 10$ to compute approximate values for

$$\ln 2 = \int_1^2 \frac{1}{x}\,dx$$

Solution   By the Trapezoidal Rule with $n = 3$ we have

$$\int_1^2 \frac{1}{x}\,dx \approx \frac{1}{6}\left[f(1) + 2f\!\left(\frac{4}{3}\right) + 2f\!\left(\frac{5}{3}\right) + f(2) \right]$$

$$= \frac{1}{6}\left(1 + \frac{6}{4} + \frac{6}{5} + \frac{1}{2} \right)$$

$$= 0.7$$

For $n = 5$ we have

$$\int_1^2 \frac{1}{x}\,dx \approx \frac{1}{10}\left[f(1) + 2f\!\left(\frac{6}{5}\right) + 2f\!\left(\frac{7}{5}\right) + 2f\!\left(\frac{8}{5}\right) + 2f\!\left(\frac{9}{5}\right) + f(2) \right]$$

$$= \frac{1}{10}\left(1 + \frac{10}{6} + \frac{10}{7} + \frac{10}{8} + \frac{10}{9} + \frac{1}{2} \right)$$

$$\approx 0.695635$$

Finally, for $n = 10$ we have

$$\int_1^2 \frac{1}{x}\,dx \approx \frac{1}{20}\left[f(1) + 2f\!\left(\frac{11}{10}\right) + 2f\!\left(\frac{12}{10}\right) + 2f\!\left(\frac{13}{10}\right) + 2f\!\left(\frac{14}{10}\right) + 2f\!\left(\frac{15}{10}\right) \right.$$

$$\left. + 2f\!\left(\frac{16}{10}\right) + 2f\!\left(\frac{17}{10}\right) + 2f\!\left(\frac{18}{10}\right) + 2f\!\left(\frac{19}{10}\right) + f(2) \right]$$

$$= \frac{1}{20}\left(1 + \frac{20}{11} + \frac{20}{12} + \frac{20}{13} + \frac{20}{14} + \frac{20}{15} + \frac{20}{16} + \frac{20}{17} + \frac{20}{18} + \frac{20}{19} + \frac{1}{2} \right)$$

$$\approx 0.693771 \quad \square$$

Since $\ln 2 \approx 0.693147$, our calculations confirm that larger numbers of subintervals in a partition tend to yield better estimates for $\ln 2$. In fact, for $n = 10$ the estimate is within 10^{-3} of the value of $\ln 2 = \int_1^2 \frac{1}{x}\,dx$.

The value given by the Trapezoidal Rule approximates, but normally is not equal to, the value of the corresponding definite integral. To register the difference between the two values, we define the error incurred in using the Trapezoidal Rule to approximate the integral. More specifically, we define the **nth Trapezoidal Rule error E_n^T** by

$$E_n^T = \left| \int_a^b f(x)\,dx - \frac{b - a}{2n}\left[f(x_0) + 2f(x_1) + \cdots + 2f(x_{n-1}) + f(x_n) \right] \right|$$

Using the approximate value 0.693147 for ln 2, we find from Example 1 that

$$E_3^T \approx |0.693147 - 0.7| = 0.006853$$

$$E_5^T \approx |0.693147 - 0.695635| = 0.002488$$

$$E_{10}^T = |0.693147 - 0.693771| = 0.000624$$

Notice that the error decreases when n increases.

To present a formula for an upper bound of E_n^T, we suppose f has a continuous second derivative on $[a, b]$ and let

$$M = \text{maximum of the numbers } |f''(x)| \quad \text{for } a \leq x \leq b$$

Then it can be proved that the error E_n^T incurred in using the Trapezoidal Rule to approximate $\int_a^b f(x)\,dx$ satisfies

$$E_n^T \leq \frac{(b - a)^3 M}{12n^2} \tag{3}$$

Formula (3) implies that as n increases, the upper bound for E_n^T shrinks.

Example 2 Determine a value of n that guarantees an error of no more than 10^{-3} in the approximation by the Trapezoidal Rule of

$$\ln 2 = \int_1^2 \frac{1}{x}\,dx$$

Solution To use (3), we let $a = 1$, $b = 2$, and $f(x) = 1/x$. We find that $f''(x) = 2/x^3$, so that

$$|f''(x)| = \left|\frac{2}{x^3}\right| \leq 2 \quad \text{for } 1 \leq x \leq 2$$

Thus $M = 2$, so by (3) we conclude that

$$E_n^T \leq \frac{(2 - 1)^3 2}{12n^2} = \frac{2}{12n^2} = \frac{1}{6n^2}$$

To ensure that $E_n^T \leq 10^{-3}$, we choose n so that $1/(6n^2) \leq 10^{-3}$, that is, $n^2 \geq 10^3/6 = 166\frac{2}{3}$. Therefore if $n \geq 13$, then $E_n^T \leq 10^{-3}$. $\square$

If for Example 2 we let $T = $ the value of the Trapezoidal Rule for $n = 13$, then since $E_{13}^T \leq 10^{-3}$, it follows that

$$\left|\int_1^2 \frac{1}{x}\,dx - T\right| \leq 0.001$$

or equivalently,

$$T - 0.001 < \int_1^2 \frac{1}{x}\,dx < T + 0.001 \tag{4}$$

But by the Trapezoidal Rule applied to $\int_1^2 \frac{1}{x}\,dx$ with $n = 13$,

$$T = \frac{1}{26}\left[1 + 2\left(\frac{13}{14}\right) + 2\left(\frac{13}{15}\right) + \cdots + 2\left(\frac{13}{25}\right) + \frac{1}{2}\right] \approx 0.693517$$

(accurate to 6 places), so that

$$0.693516 < T < 0.693518 \tag{5}$$

Combining (4) and (5), we conclude that

$$0.692 < 0.693516 - 0.001 < \int_1^2 \frac{1}{x}\,dx < 0.693518 + 0.001 < 0.695$$

Thus we have lower and upper bounds for $\ln 2 = \int_1^2 1/x\,dx$.

The point of Example 2 is that even without approximating $\int_1^2 \frac{1}{x}\,dx$ by means of the Trapezoidal Rule, we can guarantee that $E_{13}^T \le 10^{-3}$.

Simpson's Rule

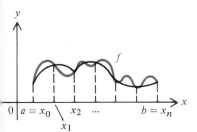

y

$0 \quad a = x_0 \quad x_2 \quad \cdots \quad b = x_n$

x_1

Simpson's Rule: parabolic segments replace portions of the graph

FIGURE 8.10

For the second of our two types of approximation we substitute parabolas for the straight lines used in the Trapezoidal Rule (Figure 8.10). We begin by taking a partition $\mathcal{P} = \{x_0, x_1, \ldots, x_n\}$, but now in addition to assuming that the partition divides the interval $[a, b]$ into n subintervals of equal length

$$h = \frac{b - a}{n}$$

we will assume that n is an *even* number. A tedious computation would verify that the parabola passing through three consecutive points $(x_0, f(x_0)), (x_1, f(x_1))$, and $(x_2, f(x_2))$ on the graph of f is defined by

$$p(x) = f(x_0) + \frac{f(x_1) - f(x_0)}{h}(x - x_0)$$

$$+ \frac{f(x_0) - 2f(x_1) + f(x_2)}{2h^2}(x - x_0)(x - x_1) \tag{6}$$

Since p is only a second-degree polynomial in the variable x, it is of course possible to find

$$\int_{x_0}^{x_2} p(x)\,dx$$

After a somewhat lengthy algebraic manipulation we conclude that

$$\int_{x_0}^{x_2} p(x)\,dx = \frac{h}{3}[f(x_0) + 4f(x_1) + f(x_2)]$$

(see Exercise 37). Since f is by assumption nonnegative, the integral just evaluated represents the area of the region between the graph of p and the x axis on $[x_0, x_2]$.

The same is done for the subintervals $[x_2, x_4], [x_4, x_6], \ldots, [x_{n-2}, x_n]$; each of the indexes $2, 4, 6, \ldots, n - 2, n$ must be even, which explains why we required n to be even. The sum of the areas under the parabolas so obtained serves as an approximation to $\int_a^b f(x)\,dx$. That sum can be shown to be the right-hand member of the formula

$$\int_a^b f(x)\,dx \approx \frac{b-a}{3n}[f(x_0) + 4f(x_1) + 2f(x_2) + 4f(x_3) + \cdots$$
$$+ 2f(x_{n-2}) + 4f(x_{n-1}) + f(x_n)]$$

(7)

This formula is widely known as **Simpson's Rule**, after the Englishman Thomas Simpson (1710–1761).* We emphasize that it holds for all functions that are continuous on $[a, b]$, whether or not they are nonnegative. The beauty of the formula is that it is essentially as easy to use as the Trapezoidal Rule or any Riemann sum, and yet it generally produces far more accurate estimates than do the other kinds of approximation presented. In fact, variants of Simpson's Rule are actually used by computers to approximate definite integrals.

To illustrate the accuracy of Simpson's Rule, we return to the approximation of

$$\int_1^2 \frac{1}{x}\,dx$$

In contrast to our approximations by the Trapezoidal Rule, where we took $n = 3$, $n = 5$, and $n = 10$, with Simpson's Rule we can only take $n = 10$, or any other positive even integer n.

Example 3   Compute an approximate value for

$$\int_1^2 \frac{1}{x}\,dx$$

by using Simpson's Rule with $n = 10$.

Solution   By Simpson's Rule,

$$\int_1^2 \frac{1}{x}\,dx \approx \frac{1}{3(10)}\left[f(1) + 4f\left(\frac{11}{10}\right) + 2f\left(\frac{12}{10}\right) + 4f\left(\frac{13}{10}\right) + 2f\left(\frac{14}{10}\right)\right.$$
$$\left. + 4f\left(\frac{15}{10}\right) + 2f\left(\frac{16}{10}\right) + 4f\left(\frac{17}{10}\right) + 2f\left(\frac{18}{10}\right) + 4f\left(\frac{19}{10}\right) + f(2)\right]$$
$$= \frac{1}{30}\left(1 + \frac{40}{11} + \frac{20}{12} + \frac{40}{13} + \frac{20}{14} + \frac{40}{15} + \frac{20}{16} + \frac{40}{17} + \frac{20}{18} + \frac{40}{19} + \frac{1}{2}\right)$$
$$\approx 0.693150 \quad \square$$

Since $\ln 2 \approx 0.693147$, the estimate is within 10^{-6} of the actual value, and the amount of work needed to find it is reasonable.

* Thomas Simpson, who was trained to be a weaver, was also a first-class, self-taught mathematician.

In our next example we will use Simpson's Rule to help find an approximate value for the length of a graph.

Example 4 A highway runs northeast across a river, which is represented by the y axis in Figure 8.11. To achieve a crossing at right angles, a curve is introduced into the highway. If

$$f(x) = \frac{1}{3}x^3 \quad \text{for } -1 \le x \le 1$$

then this curve is represented by the graph of f. Suppose the center of the curve is to be painted with a continuous white stripe. By using Simpson's Rule with $n = 4$, find the approximate length of the stripe.

Solution Since $f'(x) = x^2$, we know from (2) in Section 6.3 that the legnth $\mathcal{L}$ is given by

$$\mathcal{L} = \int_{-1}^{1} \sqrt{1 + (x^2)^2}\, dx = \int_{-1}^{1} \sqrt{1 + x^4}\, dx$$

Simpson's Rule with $n = 4$ yields

$$\mathcal{L} = \int_{-1}^{1} \sqrt{1 + x^4}\, dx \approx \frac{2}{3(4)}\left[f(-1) + 4f\left(-\frac{1}{2}\right) + 2f(0) + 4f\left(\frac{1}{2}\right) + f(1) \right]$$

$$= \frac{1}{6}\left(\sqrt{2} + 4\sqrt{\frac{17}{16}} + 2\cdot 1 + 4\sqrt{\frac{17}{16}} + \sqrt{2} \right)$$

$$\approx 2.17911 \quad \square$$

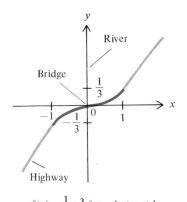

y

River

Bridge

$\frac{1}{3}$

-1 $-\frac{1}{3}$ 0 1 x

Highway

$f(x) = \frac{1}{3}x^3$ for $-1 \le x \le 1$

FIGURE 8.11

The **nth Simpson's Rule error** E_n^S incurred in using Simpson's Rule to approximate the integral is defined by

$$E_n^S = \left| \int_a^b f(x)\, dx - \frac{b-a}{3n}[f(x_0) + 4f(x_1) + 2f(x_2) + \cdots + 4f(x_{n-1}) + f(x_n)] \right|$$

Using the approximate value 0.693147 for ln 2, we find from Example 3 that

$$E_{10}^S \approx |0.693147 - 0.693150| = 0.000003 = 3 \times 10^{-6}$$

which is significantly smaller than E_{10}^T for $f(x) = 1/x$ on $[1, 2]$.

An upper bound for E_n^S is derived from the fourth derivative of f on $[a, b]$. Let us assume that f has a continuous fourth derivative on $[a, b]$, and let

$$M = \text{maximum of the numbers } |f^{(4)}(x)| \quad \text{for } a \le x \le b$$

Then the error E_n^S incurred in using Simpson's Rule to approximate $\int_a^b f(x)\, dx$ satisfies

$$E_n^S \le \frac{(b-a)^5 M}{180 n^4} \tag{8}$$

The upper bound for E_n^S in (8) shrinks in proportion to n^4, and as a result, approximation by Simpson's Rule becomes excellent as the number n of subintervals grows.

Example 5   Determine a value of n that guarantees an error of no more than 10^{-3} in the approximation by Simpson's Rule of

$$\ln 2 = \int_1^2 \frac{1}{x} dx$$

Solution   We use (8) with $a = 1$, $b = 2$, and $f(x) = 1/x$. We find that $f^{(4)}(x) = 24/x^5$, so that

$$|f^{(4)}(x)| = \left|\frac{24}{x^5}\right| \le 24 \quad \text{for } 1 \le x \le 2$$

Thus $M = 24$, so by (8) we conclude that

$$E_n^S \le \frac{(2-1)^5(24)}{180n^4} = \frac{24}{180n^4} = \frac{2}{15n^4}$$

To be certain that $E_n^S \le 10^{-3}$, we choose n so that $2/(15n^4) \le 10^{-3}$, that is, $n^4 \ge \frac{2}{15} \times 10^3 \approx 133.333$. Since $4^4 = 256$ and $3^4 = 81$, it suffices to let $n = 4$, with the result that $E_4^S \le 10^{-3}$.   □

Observe that the value of n obtained from the error estimate for Simpson's Rule is smaller than that obtained for the Trapezoidal Rule. However, to use (8) it is necessary to be able to find not only $f^{(4)}(x)$ but also an upper bound for M. Sometimes, as in Example 5, that poses no problem. Usually, however, it is difficult or impossible to find an upper bound for M that yields a usable upper bound for E_n. In fact, even for the integral $\int_{-1}^1 \sqrt{1 + x^4} \, dx$, it is only with difficulty that we can determine that the error E_4^S of the approximation in Example 4 is less than 10^{-2}.

EXERCISES 8.6

Approximate the integrals in Exercises 1–12 by each of the two rules presented in this section, with the indicated number n of subintervals.

1. $\int_1^3 \frac{1}{x} dx; n = 2$

2. $\int_1^3 \frac{1}{x} dx; n = 4$

3. $\int_1^7 \frac{1}{t} dt; n = 6$

4. $\int_0^4 \frac{1}{1 + x^2} dx; n = 4$

5. $\int_0^2 \sqrt{1 + x^3} \, dx; n = 2$

6. $\int_{1/4}^{1/2} \ln w \, dw; n = 2$

7. $\int_1^5 [(x \ln x) - x] \, dx; n = 4$

8. $\int_1^5 \frac{\ln x}{x} dx; n = 4$

9. $\int_{-1}^1 \sin \pi x^2 \, dx; n = 4$

10. $\int_{\pi/4}^{3\pi/4} \frac{\sin x}{x} dx; n = 2$

11. $\int_0^\pi \frac{1}{1 + \sin x} dx; n = 4$

12. $\int_1^3 \frac{1}{y + y^3} dy; n = 4$

C In Exercises 13–16 use Simpson's Rule with the indicated number n of subintervals to approximate the given integral.

13. $\int_0^1 \sqrt{1 + x^6} \, dx; n = 6$

14. $\int_2^6 \cos \sqrt{x} \, dx; n = 8$

15. $\int_1^3 \ln(x^2+1)\,dx; \; n=10$ 16. $\int_{-1}^2 e^{(x^2)}\,dx; \; n=12$

In Exercises 17–20 find an integer n such that the given inequality is satisfied for the approximation of the integral by the indicated rule.

17. $E_n^T \le 10^{-2}; \; \int_1^2 \ln x \, dx;$ Trapezoidal Rule

18. $E_n^S \le 10^{-2}; \; \int_1^2 \ln x \, dx;$ Simpson's Rule

19. $E_n^S \le 10^{-4}; \; \int_1^5 (x\ln x - x)\,dx;$ Simpson's Rule

20. $E_n^S \le 10^{-4}; \; \int_0^1 e^{-x^2}\,dx;$ Simpson's Rule
    (*Hint:* Choose a reasonable upper bound for M.)

In Exercises 21–24 first find an upper bound for the error introduced by approximating the integral by the indicated rule. Then find lower and upper bounds for the value of the integral.

21. $\int_1^3 \frac{1}{x}\,dx;$ Trapezoidal Rule with $n=4$

22. $\int_{-1}^1 \sin \pi x^2 \, dx;$ Trapezoidal Rule with $n=4$

23. $\int_1^3 \frac{1}{x}\,dx;$ Simpson's Rule with $n=4$

24. $\int_{-1}^1 \sin \pi x^2 \, dx;$ Simpson's Rule with $n=4$
    (*Hint:* Choose a reasonable upper bound for M.)

In Exercises 25–26 approximate the area A of the region between the graph of f and the x axis on the given interval by using Simpson's Rule with $n=4$.

25. $f(x) = \dfrac{1}{1+\cos x}; \; [-\pi/3, \pi/3]$

26. $f(x) = \dfrac{\pi \cos x}{x}; \; [\pi/2, 3\pi/2]$

27. It is known that $\pi = \int_0^1 4/(1+x^2)\,dx$.
    a. Approximate π by using
       i. the Trapezoidal Rule with $n=6$
       ii. Simpson's Rule with $n=4$
    b. Using the fact that $\pi \approx 3.14159$, estimate the errors that arise from the approximations of π in (a).

28. Approximate $\int_0^{\pi/2} \sqrt{1+\cos x}\,dx$ by using Simpson's Rule with $n=2$. Calculate the exact value of the integral by using the relation
    $$\sqrt{1+\cos x} = \sqrt{2}\cos\frac{x}{2}$$
    and estimate the error that arises from the approximation.

29. Find the smallest positive integer n that is guaranteed by (8) to yield an error of no more than 10^{-4} in the approximation of $\ln 8$ by Simpson's Rule by considering $\ln 8$ as
    a. $\int_1^8 \frac{1}{x}\,dx$ b. $3\int_1^2 \frac{1}{x}\,dx$
    (By comparing the answers to (a) and (b), we see that making algebraic changes that shorten the interval of integration can drastically reduce the computations required in using Simpson's Rule to approximate definite integrals.)

30. Let f be continuous on $[a, b]$ and $\mathscr{P}$ a partition of $[a, b]$ with n subintervals of equal length. Also let L_n and R_n be the left and right sums of f on $[a, b]$, and T_n the corresponding value for the Trapezoidal Rule. Show that
    $$T_n = \frac{1}{2}(L_n + R_n)$$
    (Thus T_n is the average of the left and right sums.)

31. Assume that $f(x) \ge 0$ for $a \le x \le b$, and let the Trapezoidal Rule give the value T_n for the approximation of $\int_a^b f(x)\,dx$.
    a. Show that if the graph of f is concave upward on $[a, b]$, then
       $$T_n \ge \int_a^b f(x)\,dx$$
       (*Hint:* Draw a picture, and note the relationship of the trapezoids to the region under the graph of f.)
    b. Show that if the graph of f is concave downward on $[a, b]$, then
       $$T_n \le \int_a^b f(x)\,dx$$

32. Prove that the Trapezoidal Rule gives the exact value for $\int_a^b f(x)\,dx$ if f is linear. (*Hint:* Use (3).)

33. Prove that Simpson's Rule gives the exact value for $\int_a^b f(x)\,dx$ if f is any polynomial of degree at most 3. (*Hint:* Use (8).)

34. Let $f(x) = x^4$ for $0 \le x \le 1$, and let $n=2$. Show that
    $$E_n^S = \frac{M(b-a)^5}{180n^4}$$
    so that the inequality in (8) can actually reduce to an equality. (A similar remark is valid for (3), where the function $f(x) = x^2$ for $0 \le x \le 1$ suffices.)

35. Let $a>0$, and let k be a nonnegative integer. Suppose we wish to approximate $\int_{-a}^a x^k\,dx$ by using Simpson's Rule with $n=2$.

a. Show that if k is odd, then Simpson's Rule gives the exact value of the integral.

b. Show that if $k = 2$, then Simpson's Rule gives the exact value of the integral.

c. Does the conclusion of part (b) hold whenever k is an even positive integer? Explain.

36. Prove that the sum appearing in the Trapezoidal Rule for $\int_a^b f(x)\,dx$ is a genuine Riemann sum for f on $[a, b]$. (*Hint:* Use the Intermediate Value Theorem on each sub-interval $[x_{k-1}, x_k]$ associated with the partition $\mathscr{P} = \{x_0, x_1, \ldots, x_n\}$ to obtain a number c_k in $[x_{k-1}, x_k]$ such that

$$\frac{f(x_{k-1}) + f(x_k)}{2} = f(c_k)$$

*37. Let p be the second-degree polynomial defined by (6). Show that

$$\int_{x_0}^{x_2} p(x)\,dx = \frac{h}{3}[f(x_0) + 4f(x_1) + f(x_2)]$$

☑ 38. Use Simpson's Rule with $n = 8$ to approximate the length $\mathscr{L}$ of a bridge cable in the shape of the curve $y = 16 - \frac{1}{16}x^2$ for $-16 \le x \le 16$.

☑ 39. The equations

$$x = 2\cos t \quad \text{and} \quad y = 3\sin t$$

parametrize an ellipse. Use Simpson's Rule with $n = 8$ to estimate the circumference $\mathscr{L}$ of the ellipse.

☑ 40. The width of a piece of corrugated cardboard is 2 inches, and the cardboard has the shape of the graph of $y = \frac{1}{16}\sin 4\pi x$ for $0 \le x \le 2$ (Figure 8.12). Approximate the arc length $\mathscr{L}$ of the curved portion of the cardboard by Simpson's Rule with $n = 8$.

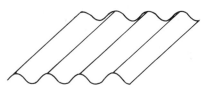

FIGURE 8.12

☑ 41. An egg has the shape of the ellipse generated by revolving about the x axis the half ellipse parametrized by

$$x = 2\cos t \quad \text{and} \quad y = 3\sin t \quad \text{for } 0 \le t \le \pi$$

Use Simpson's Rule with $n = 6$ to estimate the surface area S of the egg.

8.7
IMPROPER INTEGRALS

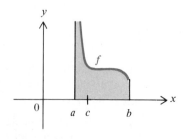

FIGURE 8.13

So far we have defined $\int_a^b f(x)\,dx$ only for f continuous on $[a, b]$. It follows from the Maximum–Minimum Theorem that such a function f is bounded on $[a, b]$ in the sense that for some number M, $|f(x)| \le M$ for all x in $[a, b]$. More generally, if I is any interval, then we say that f is **bounded on I** if there is a constant M such that

$$|f(x)| \le M \quad \text{for all } x \text{ in } I$$

A function not bounded on a given interval I inside its domain is said to be **unbounded on I**. In this section we will consider the geometric question of whether an area can be defined for the region under the graph of a nonnegative function that is unbounded on a bounded interval (Figure 8.13). We would also like to know whether it is possible to define the area of the region under the graph of a nonnegative function on an unbounded interval (Figure 8.14). Although it may

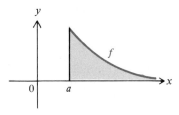

FIGURE 8.14

seem that the areas of such regions must be infinite, we will find that in certain cases they should be defined to be finite. To accomplish this we will define the definite integral when either the integrand or the interval of integration is unbounded. Such integrals are called **improper integrals**, as opposed to integrals of continuous functions over bounded intervals, which are called **proper integrals**. (Of course, improper integrals are as proper an object of study as proper integrals, but this is the usual and traditional terminology.)

Integrals with Unbounded Integrands

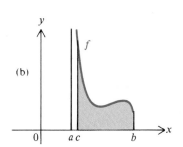

(a)

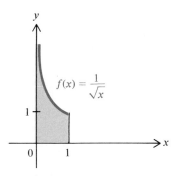

(b)

FIGURE 8.15

To simplify the discussion, we say that f is **unbounded near c** if f is unbounded either on every open interval of the form (c, x) or on every open interval of the form (x, c). For example, $1/x$ and $1/\sqrt{x}$ are unbounded near 0.

We now consider a function f that is continuous at every point in $(a, b]$ and unbounded near a (see Figure 8.13). By assumption f is continuous on the interval $[c, b]$ for any c in (a, b), so that $\int_c^b f(x)\,dx$ is defined for such c. If the one-sided limit

$$\lim_{c \to a^+} \int_c^b f(x)\,dx$$

exists, then we define $\int_a^b f(x)\,dx$ to be that limit and say that the integral $\int_a^b f(x)\,dx$ **converges**. Otherwise, we say that $\int_a^b f(x)\,dx$ **diverges** and do not assign any number to the integral. If f is nonnegative on $(a, b]$, then as c approaches a from the right, the integrals $\int_c^b f(x)\,dx$ represent areas of larger and larger regions (Figure 8.15(a) and (b)). In addition, if $\int_a^b f(x)\,dx$ converges, the **area** of the region between the graph of f and the x axis on $[a, b]$ is defined to be $\int_a^b f(x)\,dx$. By contrast, if f is nonnegative on $(a, b]$ and if $\int_a^b f(x)\,dx$ diverges, then we say that the corresponding region has infinite area and write

$$\int_a^b f(x)\,dx = \infty$$

Example 1 Show that $\displaystyle\int_0^1 \frac{1}{\sqrt{x}}\,dx$ converges and compute its value.

Solution The integral is of the kind just discussed, because $1/\sqrt{x}$ is continuous at every point in $(0, 1]$ and is unbounded near 0. For $0 < c < 1$ we have

$$\int_c^1 \frac{1}{\sqrt{x}}\,dx = 2\sqrt{x}\,\Big|_c^1 = 2(1 - \sqrt{c})$$

and

$$\lim_{c \to 0^+} 2(1 - \sqrt{c}) = 2$$

Consequently $\int_0^1 1/\sqrt{x}\,dx$ converges, and

$$\int_0^1 \frac{1}{\sqrt{x}}\,dx = 2$$

This means that the region shaded in Figure 8.16 has area 2. ☐

FIGURE 8.16

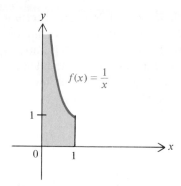

FIGURE 8.17

Example 2 Show that $\int_0^1 \frac{1}{x} dx$ diverges.

Solution The function $1/x$ is continuous on $(0, 1]$ and unbounded near 0, and for $0 < c < 1$ we have

$$\int_c^1 \frac{1}{x} dx = \ln x \Big|_c^1 = \ln 1 - \ln c = -\ln c$$

But

$$\lim_{c \to 0^+} (-\ln c) = \infty$$

Therefore the integral $\int_0^1 1/x \, dx$ diverges, which means that the region shaded in Figure 8.17 has infinite area. □

If f is continuous at every point of $[a, b)$ and is unbounded near b, the integral $\int_a^c f(x) \, dx$ is defined for all c in (a, b). If

$$\lim_{c \to b^-} \int_a^c f(x) \, dx$$

exists, then we define $\int_a^b f(x) \, dx$ to be that limit and say that the integral $\int_a^b f(x) \, dx$ **converges**. Otherwise, we say that $\int_a^b f(x) \, dx$ **diverges**. If f is nonnegative, then $\int_a^b f(x) \, dx$ represents the area of the corresponding region, and is finite or infinite depending on the convergence or divergence of $\int_a^b f(x) \, dx$.

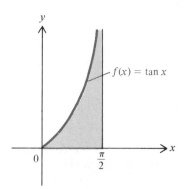

FIGURE 8.18

Example 3 Determine whether $\int_0^{\pi/2} \tan x \, dx$ converges.

Solution The integrand is continuous on $[0, \pi/2)$ and is unbounded near $\pi/2$. For $0 < c < \pi/2$ we have

$$\int_0^c \tan x \, dx = -\ln(\cos x) \Big|_0^c = -\ln(\cos c) + \ln(\cos 0) = -\ln(\cos c)$$

Since

$$\lim_{c \to \pi/2^-} \cos c = 0$$

it follows that

$$\lim_{c \to \pi/2^-} [-\ln(\cos c)] = \infty$$

Consequently the integral $\int_0^{\pi/2} \tan x \, dx$ diverges, and the corresponding region has infinite area (Figure 8.18). □

If f is continuous at every point of (a, b) and is unbounded near both a and b, we say that $\int_a^b f(x) \, dx$ **converges** if for some point d in (a, b) both the integrals $\int_a^d f(x) \, dx$ and $\int_d^b f(x) \, dx$ converge. Otherwise, we say that $\int_a^b f(x) \, dx$ **diverges**.

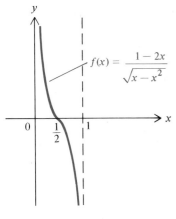

FIGURE 8.19

Example 4 Determine whether $\int_0^1 \dfrac{1-2x}{\sqrt{x-x^2}}\,dx$ converges.

Solution The integrand is unbounded near both the endpoints 0 and 1 and is continuous on $(0, 1)$ (Figure 8.19). Consequently the integral is of the type under consideration. If we let $d = \frac{1}{2}$, then we need to analyze the convergence of

$$\int_0^{1/2} \frac{1-2x}{\sqrt{x-x^2}}\,dx \quad \text{and} \quad \int_{1/2}^1 \frac{1-2x}{\sqrt{x-x^2}}\,dx$$

For $0 < c < \frac{1}{2}$ we have

$$\int_c^{1/2} \frac{1-2x}{\sqrt{x-x^2}}\,dx = 2\sqrt{x-x^2}\,\Big|_c^{1/2} = 2\left(\sqrt{\frac{1}{4}} - \sqrt{c-c^2}\right) = 1 - 2\sqrt{c-c^2}$$

Since

$$\lim_{c \to 0^+} \sqrt{c - c^2} = 0$$

this implies that

$$\int_0^{1/2} \frac{1-2x}{\sqrt{x-x^2}}\,dx = \lim_{c \to 0^+} \int_c^{1/2} \frac{1-2x}{\sqrt{x-x^2}}\,dx = \lim_{c \to 0^+} (1 - 2\sqrt{c-c^2}) = 1$$

A similar computation shows that the second improper integral also converges and that

$$\int_{1/2}^1 \frac{1-2x}{\sqrt{x-x^2}}\,dx = -1$$

Therefore the original integral converges, and

$$\int_0^1 \frac{1-2x}{\sqrt{x-x^2}}\,dx = \int_0^{1/2} \frac{1-2x}{\sqrt{x-x^2}}\,dx + \int_{1/2}^1 \frac{1-2x}{\sqrt{x-x^2}}\,dx = 1 - 1 = 0 \quad \square$$

The last type of unbounded function to be considered in this subsection consists of functions that are continuous at every point of $[a, b]$ except at a point d in (a, b), near which f is unbounded. We say that $\int_a^b f(x)\,dx$ **converges** if both the integrals $\int_a^d f(x)\,dx$ and $\int_d^b f(x)\,dx$ converge, and otherwise we say that $\int_a^b f(x)\,dx$ **diverges**. For example,

$$\int_{-1}^2 \frac{1}{x^{1/3}}\,dx$$

converges (see Exercise 18) and

$$\int_{-1}^2 \left(\frac{1}{x} + \frac{1}{x^2}\right)dx$$

diverges (see Exercise 19). Again, the interpretation of $\int_a^b f(x)\,dx$ as the area of a region applies if f is nonnegative.

Integrals over Unbounded Intervals

Integrals of the form $\int_a^{\infty} f(x)\,dx$ and $\int_{-\infty}^{b} f(x)\,dx$ are also called improper integrals. If f is continuous on $[a, \infty)$, then $\int_a^b f(x)\,dx$ is a proper integral for any $b \geq a$. This fact allows us to say that $\int_a^{\infty} f(x)\,dx$ **converges** if $\lim_{b \to \infty} \int_a^b f(x)\,dx$ exists. In that event,

$$\int_a^{\infty} f(x)\,dx = \lim_{b \to \infty} \int_a^b f(x)\,dx$$

As before, the integral **diverges** otherwise. The improper integral $\int_{-\infty}^{b} f(x)\,dx$ is handled in an analogous way. When $f \geq 0$, the connection between these integrals and areas holds as before.

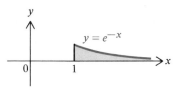

FIGURE 8.20

Example 5 Show that the shaded region in Figure 8.20 has finite area, and compute the area.

Solution We must evaluate $\int_1^{\infty} e^{-x}\,dx$, which is equal to $\lim_{b \to \infty} \int_1^b e^{-x}\,dx$. For $b \geq 1$ we have

$$\int_1^b e^{-x}\,dx = -e^{-x}\Big|_1^b = -e^{-b} + e^{-1}$$

Since $\lim_{b \to \infty} -e^{-b} = 0$, the improper integral converges, and

$$\int_1^{\infty} e^{-x}\,dx = \lim_{b \to \infty} \int_1^b e^{-x}\,dx = \lim_{b \to \infty}(-e^{-b} + e^{-1}) = 0 + e^{-1} = e^{-1}$$

Thus the region in Figure 8.20 has finite area, and the area is e^{-1}. □

By the definitions given in this section, we can investigate the area of an unbounded region by studying the convergence or divergence of an associated improper integral. Appropriate improper integrals can also yield the volume of an unbounded solid and the surface area of an unbounded surface. To illustrate these ideas, we will analyze the volume and the surface area of **Gabriel's horn**, which is the solid obtained by revolving about the x axis the region between the graph of $y = 1/x$ and the x axis on $[1, \infty)$ (Figure 8.21).

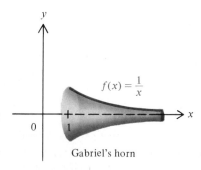

Gabriel's horn

FIGURE 8.21

Example 6 Let $f(x) = 1/x$. Denote by R the region between the graph of f and the x axis on $[1, \infty)$, and let D be the solid generated by revolving R about the x axis.
a. Show that the volume V of D is finite.
b. Show that the surface area S of D is infinite.

Solution
a. We use (3) of Section 6.1, with the limits of integration replaced by 1 and ∞, which makes the integral improper. We obtain the following:

$$V = \int_1^{\infty} \pi \frac{1}{x^2}\,dx = \pi \lim_{b \to \infty} \int_1^b \frac{1}{x^2}\,dx = \pi \lim_{b \to \infty} \left(-\frac{1}{x}\right)\Big|_1^b = \pi \lim_{b \to \infty}\left(1 - \frac{1}{b}\right) = \pi$$

Therefore the volume V is finite.

b. We use (4) of Section 6.5, with the limits of integration replaced by 1 and ∞, so that again an improper integral arises. We find that

$$S = \int_1^\infty 2\pi \frac{1}{x} \sqrt{1 + \left(-\frac{1}{x^2}\right)^2}\, dx = 2\pi \lim_{b \to \infty} \int_1^b \frac{1}{x} \sqrt{1 + \frac{1}{x^4}}\, dx$$

But

$$\frac{1}{x} \sqrt{1 + \frac{1}{x^4}} \geq \frac{1}{x} \quad \text{for } x \geq 1$$

so that

$$\int_1^b \frac{1}{x} \sqrt{1 + \frac{1}{x^4}}\, dx \geq \int_1^b \frac{1}{x}\, dx$$

Since

$$2\pi \lim_{b \to \infty} \int_1^b \frac{1}{x}\, dx = 2\pi \lim_{b \to \infty} \ln b = \infty$$

the surface area S is infinite. $\square$

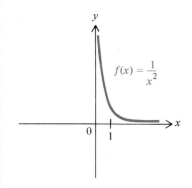

FIGURE 8.22

$f(x) = \dfrac{1}{x^2}$

A surprising consequence of the results of Example 6 is the fact that one could fill the interior of Gabriel's horn with a finite amount of paint (namely, π cubic units of paint), whereas it would take an infinite amount of paint to cover the surface of Gabriel's horn!

Integrals can be improper in more than one way. For example, the integrand in $\int_0^\infty 1/x^2\, dx$ is unbounded near 0, and the interval of integration is also unbounded (see Figure 8.22). In such a case we say that the integral converges if for some $d > 0$, both $\int_0^d 1/x^2\, dx$ and $\int_d^\infty 1/x^2\, dx$ (each of which is improper in only one way) converge. It can be proved that if $\int_0^d f(x)\, dx$ and $\int_d^\infty f(x)\, dx$ both converge for a single d in $(0, \infty)$, then the integrals converge for *all* d in $(0, \infty)$. Therefore to show that $\int_0^\infty f(x)\, dx$ diverges, it is enough to find *one* d in $(0, \infty)$ for which $\int_0^d f(x)\, dx$ or $\int_d^\infty f(x)\, dx$ diverges.

Example 7   Show that $\displaystyle\int_0^\infty \frac{1}{x^2}\, dx$ diverges.

Solution   Let $d = 1$. We will show that $\int_0^1 1/x^2\, dx$, and hence $\int_0^\infty 1/x^2\, dx$, diverges. To determine that $\int_0^1 1/x^2\, dx$ diverges we compute $\int_c^1 1/x^2\, dx$ for any c in $(0, 1)$. Since

$$\int_c^1 \frac{1}{x^2}\, dx = -\frac{1}{x}\bigg|_c^1 = \frac{1}{c} - 1$$

and since $\lim_{c \to 0^+} (1/c - 1) = \infty$, we know that $\int_0^1 1/x^2\, dx$ diverges. This implies that $\int_0^\infty 1/x^2\, dx$ diverges. $\square$

We say that $\int_{-\infty}^{\infty} f(x)\,dx$ **converges** if both $\int_{-\infty}^{d} f(x)\,dx$ and $\int_{d}^{\infty} f(x)\,dx$ converge, for some number d. In that case,

$$\int_{-\infty}^{\infty} f(x)\,dx = \int_{-\infty}^{d} f(x)\,dx + \int_{d}^{\infty} f(x)\,dx$$

Example 8 Determine whether $\displaystyle\int_{-\infty}^{\infty} \frac{1}{x^2 + 1}\,dx$ converges.

Solution The integrand is continuous on $(-\infty, \infty)$. To show that the integral converges, we will show that

$$\int_{-\infty}^{0} \frac{1}{x^2 + 1}\,dx \quad \text{and} \quad \int_{0}^{\infty} \frac{1}{x^2 + 1}\,dx$$

both converge. First we have

$$\int_{-\infty}^{0} \frac{1}{x^2 + 1}\,dx = \lim_{a \to -\infty} \int_{a}^{0} \frac{1}{x^2 + 1}\,dx = \lim_{a \to -\infty} \arctan x \Big|_{a}^{0}$$

$$= \lim_{a \to -\infty} (\arctan 0 - \arctan a) = \frac{\pi}{2}$$

A similar calculation shows that

$$\int_{0}^{\infty} \frac{1}{x^2 + 1}\,dx = \frac{\pi}{2}$$

Therefore the given improper integral converges, and moreover,

$$\int_{-\infty}^{\infty} \frac{1}{x^2 + 1}\,dx = \int_{-\infty}^{0} \frac{1}{x^2 + 1}\,dx + \int_{0}^{\infty} \frac{1}{x^2 + 1}\,dx = \frac{\pi}{2} + \frac{\pi}{2} = \pi \quad \square$$

EXERCISES 8.7

In Exercises 1–24 determine whether the improper integral converges. If it does, determine the value of the integral.

1. $\displaystyle\int_{0}^{1} \frac{1}{x^{0.9}}\,dx$

2. $\displaystyle\int_{0}^{1} \frac{1}{x^{1.1}}\,dx$

3. $\displaystyle\int_{3}^{4} \frac{1}{(t - 4)^2}\,dt$

4. $\displaystyle\int_{1}^{5} \frac{1}{(x - 1)^{1.5}}\,dx$

5. $\displaystyle\int_{3}^{4} \frac{1}{\sqrt[3]{x - 3}}\,dx$

6. $\displaystyle\int_{0}^{2} \frac{x}{\sqrt[3]{x^2 - 2}}\,dx$

7. $\displaystyle\int_{0}^{\pi/2} \sec^2 \theta\,d\theta$

8. $\displaystyle\int_{-\pi/2}^{\pi/4} \tan x\,dx$

9. $\displaystyle\int_{0}^{\pi} \frac{\sin x}{\sqrt{1 + \cos x}}\,dx$

10. $\displaystyle\int_{0}^{2} \frac{\ln x}{x}\,dx$

11. $\displaystyle\int_{1}^{2} \frac{1}{w \ln w}\,dw$

12. $\displaystyle\int_{1}^{2} \frac{1}{w(\ln w)^{1/2}}\,dw$

13. $\displaystyle\int_{0}^{1} \frac{3x^2 - 1}{\sqrt[3]{x^3 - x}}\,dx$

14. $\displaystyle\int_{0}^{1} \frac{3x^2 - 1}{x^3 - x}\,dx$

15. $\displaystyle\int_{0}^{\pi} \csc^2 x\,dx$

16. $\displaystyle\int_{-\pi/2}^{\pi/2} \sec \theta\,d\theta$

17. $\displaystyle\int_{0}^{1} \frac{1}{e^t - e^{-t}}\,dt$

18. $\displaystyle\int_{-1}^{2} \frac{1}{x^{1/3}}\,dx$

19. $\int_{-1}^{2} \left(\frac{1}{x} + \frac{1}{x^2} \right) dx$

20. $\int_{0}^{2} \frac{1}{(x-1)^{4/3}} dx$

21. $\int_{0}^{2} \frac{1}{(x-1)^{1/3}} dx$

22. $\int_{-\pi/2}^{\pi/2} \frac{1}{x^2} \sin \frac{1}{x} dx$

23. $\int_{0}^{1} \frac{1}{\sqrt{1-t^2}} dt$

24. $\int_{0}^{1} \frac{e^t}{\sqrt{e^t - 1}} dt$

In Exercises 25–60 determine whether the improper integral converges. If it does, determine the value of the integral.

25. $\int_{0}^{\infty} \frac{1}{x} dx$

26. $\int_{1}^{\infty} \frac{1}{x^{1.1}} dx$

27. $\int_{0}^{\infty} \frac{1}{(2+x)^{\pi}} dx$

28. $\int_{-\infty}^{0} \sqrt{4-x} \, dx$

29. $\int_{0}^{\infty} \sin y \, dy$

30. $\int_{0}^{\infty} \cos x \, dx$

31. $\int_{0}^{\infty} \frac{1}{(1+x)^3} dx$

32. $\int_{-\infty}^{0} \frac{1}{1-w} dw$

33. $\int_{0}^{\infty} \frac{x}{1+x^2} dx$

34. $\int_{0}^{\infty} \frac{x}{(1+x^2)^4} dx$

35. $\int_{3}^{\infty} \ln x \, dx$

36. $\int_{-\infty}^{-1} \ln(-x) \, dx$

37. $\int_{2}^{\infty} \frac{1}{x(\ln x)^3} dx$

38. $\int_{2}^{\infty} \frac{1}{x \ln x} dx$

39. $\int_{-\infty}^{0} \frac{1}{(x+3)^2} dx$

40. $\int_{2/\pi}^{\infty} \frac{1}{x^2} \sin \frac{1}{x} dx$

41. $\int_{1}^{\infty} \frac{1}{\sqrt{x}(\sqrt{x}+1)} dx$

42. $\int_{-\infty}^{0} e^{4x} dx$

43. $\int_{0}^{\infty} e^{4x} dx$

44. $\int_{0}^{\infty} e^{-ex} dx$

45. $\int_{0}^{\infty} xe^{-x} dx$

46. $\int_{0}^{\infty} e^x \sin x \, dx$

47. $\int_{1}^{\infty} \frac{1}{\sqrt{x^2 - 1}} dx$

48. $\int_{-\infty}^{-2} \frac{1}{x^2 + 4} dx$

49. $\int_{1}^{\infty} \frac{1}{t\sqrt{t^2 - 1}} dt$

50. $\int_{0}^{\infty} \frac{1}{t^2 - 2t + 1} dt$

51. $\int_{-2}^{\infty} \frac{1}{t^2 + 4t + 8} dt$

52. $\int_{0}^{\infty} \frac{x \cos x - \sin x}{x^2} dx$

$\left(\text{Hint: Compute } \frac{d}{dx} \left(\frac{\sin x}{x} \right). \right)$

53. $\int_{-\infty}^{\infty} x \, dx$

54. $\int_{-\infty}^{\infty} x \sin x^2 \, dx$

55. $\int_{-\infty}^{\infty} x \sin x \, dx$

56. $\int_{-\infty}^{\infty} \frac{x^3}{x^4 + 1} dx$

57. $\int_{-\infty}^{\infty} \frac{x^3}{(x^4 + 1)^2} dx$

58. $\int_{-\infty}^{\infty} xe^{-x^2} dx$

59. $\int_{-\infty}^{\infty} \frac{1}{x^2 - 6x + 10} dx$

60. $\int_{-\infty}^{\infty} \frac{x \sin x + 2 \cos x - 2}{x^3} dx$

(*Hint:* Differentiate $(1 - \cos x)/x^2$, and use the fact that $\lim_{x \to 0} (1 - \cos x)/x^2 = \frac{1}{2}$.)

61. a. Show that

$$\frac{1}{x(x+1)} = \frac{1}{x} - \frac{1}{x+1}$$

b. Does the equation

$$\int_{1}^{\infty} \frac{1}{x(x+1)} dx = \int_{1}^{\infty} \frac{1}{x} dx - \int_{1}^{\infty} \frac{1}{x+1} dx$$

hold? Explain why or why not.

62. a. Show that $\int_{0}^{1} 1/x^r \, dx$ converges if $r < 1$ and diverges otherwise.

b. Show that $\int_{1}^{\infty} 1/x^r \, dx$ converges if $r > 1$ and diverges otherwise.

c. Conclude from (a) and (b) that $\int_{0}^{\infty} 1/x^r \, dx$ diverges for any r.

In Exercises 63–68 decide whether the region between the graph of the integrand and the x axis on the interval of integration has finite area. If it does, calculate the area.

63. $\int_{-\infty}^{0} \frac{1}{(x-3)^2} dx$

64. $\int_{-\infty}^{3} \frac{1}{(x-3)^2} dx$

65. $\int_{2}^{\infty} \frac{\ln x}{x} dx$

66. $\int_{0}^{1} \frac{\ln x}{-x} dx$

67. $\int_{2}^{\infty} \frac{1}{\sqrt{x+1}} dx$

68. $\int_{-1}^{1} \frac{1}{\sqrt{x+1}} dx$

69. a. Suppose f and g are continuous on $[a, \infty)$. Show that if $\int_{a}^{\infty} f(x) \, dx$ and $\int_{a}^{\infty} g(x) \, dx$ converge, then $\int_{a}^{\infty} (f(x) + g(x)) \, dx$ converges.

b. Show that if $\int_{a}^{\infty} f(x) \, dx$ converges, then so does $\int_{a}^{\infty} cf(x) \, dx$ for any number c.

70. Let f and g be continuous on $[a, \infty)$, and assume that $0 \le g(x) \le f(x)$ for $x \ge a$. Show that if $\int_{a}^{\infty} g(x) \, dx = \infty$, then $\int_{a}^{\infty} f(x) \, dx = \infty$, and consequently $\int_{a}^{\infty} f(x) \, dx$ diverges.

71. With the help of Exercise 70, show that each of the following integrals diverges.

a. $\int_{1}^{\infty} \frac{1}{1 + x^{1/2}} dx$ (*Hint:* Let $g(x) = \frac{1}{2} x^{-1/2}$.)

b. $\displaystyle\int_0^\infty \frac{1}{\sqrt{2+\sin x}}\,dx$

c. $\displaystyle\int_2^\infty \frac{\ln x}{\sqrt{x^2-1}}\,dx$

d. $\displaystyle\int_2^\infty \frac{1}{(1+x)\ln x}\,dx$

72. Let

$$I_n = \int_0^\infty x^n e^{-x}\,dx$$

for any nonnegative integer n. Using integration by parts, show that

$$I_n = nI_{n-1}$$

for $n \geq 1$. From this show that

$$I_n = n(n-1)(n-2)\cdots 2\cdot 1$$

73. Suppose that $\int_a^\infty f(t)\,dt$ converges, and let $x > a$.
a. Show that $\int_a^\infty f(t)\,dt = \int_a^x f(t)\,dt + \int_x^\infty f(t)\,dt$.
b. Use (a) to show that $\dfrac{d}{dx}\displaystyle\int_x^\infty f(t)\,dt = -f(x)$.

74. Let $f(x) = \dfrac{1}{\sqrt{2\pi}}e^{-x^2/2}$. Then f is the **standard normal density function**, which is fundamental in the study of statistics, and whose graph is the famous bell-shaped curve in Figure 8.23. Our goal is to estimate the area A between the graph of f and the x axis on $(-\infty, \infty)$.
a. Use Simpson's Rule with $n = 8$ to estimate the area A_1 between the graph of f and the x axis on $[-2, 2]$.
b. Use Simpson's Rule with $n = 50$ to estimate the area A_2 between the graph of f and the x axis on $[-5, 5]$.
c. Using the results of parts (a) and (b), guess the value of A.

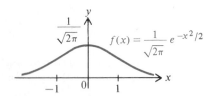

FIGURE 8.23

75. Suppose $f(x) = 1/x^p$ for $x \geq 1$, where p is a fixed positive number. Let V denote the volume of the solid generated by revolving the graph of f about the x axis. Determine those values of p for which V is finite. For such values of p, determine V.

76. Suppose a projectile is launched from earth at time $t = 1$ with a velocity $v(t) = 1/\ln t$ for $t \geq 2$. Show that the

projectile is eventually as far away from earth as one can imagine.

77. The formula

$$V(s) = \int_0^s -\ln(2\sin x)\,dx$$

gives the volume of a certain geometric object. Show that the volume is maximized on $[0, \pi]$ for $s = \pi/6$.

78. If A is the amount of radioactive substance at time 0, then the amount $f(t)$ remaining at any time $t > 0$ (measured in years) is given by

$$f(t) = Ae^{kt}$$

where k is a negative constant. The **mean life** M of an atom in the substance is given by

$$M = -\frac{1}{A}\int_0^\infty tkf(t)\,dt$$

a. Using $k = -1.24 \times 10^{-4}$, find the mean life of a C^{14} atom.
b. Using $k = -4.36 \times 10^{-4}$, find the mean life of a radium atom.

79. Economists calculate the **present sale value** P of land that can be rented for r dollars annually by the formula

$$P = \int_0^\infty re^{-it}\,dt$$

where i is the prevailing interest rate and where e^{-it} is called the **discounting factor**. Show that

$$P = \frac{r}{i}$$

80. Let s denote the annual subsistence level of people in the United States. For $x \geq s$ let $G(x)$ be the number of people with annual income between s and x, and assume that there is a continuous function g such that $G(x) = \int_s^x g(t)\,dt$ for all $x \geq s$. Let P be the total population of the United States, and assume that every individual's income is at least at subsistence level. Pareto's Law of distribution of income in a population asserts that under very general conditions,

$$P - G(x) = \int_x^\infty g(t)\,dt \approx cx^{-1.5} \quad \text{for } x \geq s$$

where c is a constant. Assuming that

$$\int_x^\infty g(t)\,dt = cx^{-1.5} \quad \text{for } x \geq s$$

and using Exercise 73, determine the function g. (Data collected in England, Ireland, Germany, Peru, Italy, and the United States confirm Pareto's Law to a remarkable degree.)

Key Terms and Expressions

Integration by parts
Trapezoidal Rule
Simpson's Rule

Proper integral; improper integral
Convergent improper integral; divergent improper
  integral

Key Formulas

$$\int u\,dv = uv - \int v\,du$$

$$\int_{g(a)}^{g(b)} f(x)\,dx = \int_a^b f(g(u))g'(u)\,du, \quad \text{where } x = g(u)$$

$$\int_a^b f(x)\,dx \approx \frac{b-a}{2n}[f(x_0) + 2f(x_1) + 2f(x_2) + \cdots + 2f(x_{n-1}) + f(x_n)]$$

$$\int_a^b f(x)\,dx \approx \frac{b-a}{3n}[f(x_0) + 4f(x_1) + 2f(x_2) + 4f(x_3) + \cdots + 2f(x_{n-2}) + 4f(x_{n-1}) + f(x_n)], \quad \text{where } n \text{ is even}$$

REVIEW EXERCISES

In Exercises 1–24 evaluate the indefinite integral.

1. $\int \ln(x^2 + 9)\,dx$

2. $\int (\ln x)^4\,dx$

3. $\int x\csc^2 x\,dx$

4. $\int x^3 e^{3x}\,dx$

5. $\int x\cosh x\,dx$

6. $\int x^2 \arcsin x\,dx$

7. $\int x\cos^2 x\,dx$

8. $\int \cos^2 x \sin^4 x\,dx$

9. $\int x^2 \sin x^3 \cos x^3\,dx$

10. $\int \sec^4 x\,dx$

11. $\int \tan^5 x\,dx$

12. $\int \frac{\tan^3 x}{\sec^5 x}\,dx$

13. $\int x^3 \cos x^2\,dx$

14. $\int \sqrt{e^t - 1}\,dt$

15. $\int t^2 \sqrt{1 - 3t}\,dt$

16. $\int \frac{t^7}{(1 - t^4)^3}\,dt$

17. $\int \frac{\cos x}{1 + \cos x}\,dx$

$\left(\textit{Hint:} \text{ Multiply the}\right.$

$\text{integrand by } \dfrac{1 - \cos x}{1 - \cos x}.\Big)$

18. $\int \frac{x^3}{\sqrt{x^8 - 1}}\,dx$

19. $\int \frac{x^2}{(x^2 + 2x + 10)^{5/2}}\,dx$

20. $\int \frac{1}{x^3 - 1}\,dx$

21. $\int \frac{x^4}{(x^2 + 1)^2}\,dx$

22. $\int \frac{1}{x(x^2 + x + 1)}\,dx$

23. $\int \frac{x}{x^2 + 3x - 18}\,dx$

24. $\int \frac{\sqrt{2x + 1}}{x + 1}\,dx$

(*Hint:* Substitute
$u = \sqrt{2x + 1}$ and then
use partial fractions.)

In Exercises 25–52 determine whether the integral is proper or improper. If the integral is improper, determine whether it converges. If it is either a proper integral or a convergent improper integral, evaluate the integral.

25. $\int_{-2}^1 \frac{1}{3x + 4}\,dx$

26. $\int_\pi^{3\pi/2} \frac{\cos x}{1 + \sin x}\,dx$

27. $\int_0^{\pi/4} x\sec x \tan x\,dx$

28. $\int_0^1 \frac{x}{x + 2}\,dx$

29. $\int_1^4 \frac{1}{1 + \sqrt{x}}\,dx$

30. $\int_1^5 \frac{x}{\sqrt{x - 1}}\,dx$

31. $\int_0^{\pi/4} \sin^4 x \cos^3 x\,dx$

32. $\int_0^{\pi/3} \tan^5 x \sec x\,dx$

33. $\int_0^{\pi/4} (\tan^3 x + \tan^5 x)\,dx$

34. $\int_0^1 \frac{x^3}{\sqrt{1 - x^2}}\,dx$

35. $\displaystyle\int_1^{\sqrt{2}} \frac{\sqrt{x^2-1}}{x^2}\,dx$

36. $\displaystyle\int_0^1 \frac{x^2}{1+x^6}\,dx$

37. $\displaystyle\int_0^{\sqrt[3]{2}/2} \frac{\sqrt{x}}{\sqrt{1-x^3}}\,dx$

38. $\displaystyle\int_0^1 x^5\sqrt{1-x^2}\,dx$

(*Hint:* Substitute $u = x^{3/2}$.)

39. $\displaystyle\int_0^{\sqrt{3}} \sqrt{x^2+1}\,dx$

40. $\displaystyle\int_{-1}^1 \frac{x}{x^2+2x+5}\,dx$

41. $\displaystyle\int_{-5}^0 \frac{x}{x^2+4x-5}\,dx$

42. $\displaystyle\int_{-1}^1 \frac{5x^3}{x^2+x-6}\,dx$

43. $\displaystyle\int_0^{\pi/2} \frac{1}{1-\sin x}\,dx$ $\left(Hint:\ \text{Multiply the integrand by}\ \dfrac{1+\sin x}{1+\sin x}.\right)$

44. $\displaystyle\int_0^{\pi/4} \frac{1}{1+\tan x}\,dx$ (*Hint:* Substitute $u = 1 + \tan x$ and then use partial fractions.)

45. $\displaystyle\int_0^1 x\ln x\,dx$

46. $\displaystyle\int_0^\infty x\ln x\,dx$

47. $\displaystyle\int_1^\infty \frac{1}{x(\ln x)^2}\,dx$

48. $\displaystyle\int_3^\infty \frac{1}{1+\ln x}\,dx$

49. $\displaystyle\int_0^\infty e^{-x}\cos x\,dx$

50. $\displaystyle\int_1^\infty x^2 e^{-x^3}\,dx$

51. $\displaystyle\int_1^\infty \frac{1}{x(x^2+1)}\,dx$

52. $\displaystyle\int_0^\infty \frac{1}{x(x^2+4)}\,dx$

53. Use integration by parts to evaluate $\displaystyle\int \frac{\ln x}{x}\,dx$.

54. a. Using the substitution $x = \sin u$, show that
$$\int_0^1 x^m(1-x^2)^n\,dx = \int_0^{\pi/2} \sin^m u\cos^{2n+1}u\,du$$
where m and n are nonnegative integers.

   b. Use part (a) to evaluate
$$\int_0^1 x^3(1-x^2)^{10}\,dx$$

55. a. By making the substitution $u = 1 - x$, show that
$$\int_0^1 x^n(1-x)^m\,dx = \int_0^1 x^m(1-x)^n\,dx$$
for any nonnegative integers m and n.

   b. Use part (a) to evaluate
$$\int_0^1 x^2(1-x)^{10}\,dx$$

56. a. By making the substitution $x = u/(1-u)$ show that
$$\int_0^b \frac{x^{m-1}}{(1+x)^{m+n}}\,dx = \int_0^{b/(1+b)} u^{m-1}(1-u)^{n-1}\,du$$
for any positive integers m and n and any positive number b.

   b. Use part (a) to show that
$$\int_0^\infty \frac{x}{(1+x)^5}\,dx = \int_0^1 u(1-u)^2\,du = \frac{1}{12}$$

*57. Make the substitution $x = 1 - 3\cos^2 u$ in order to evaluate
$$\int_{-2}^{-1/2} \left(\frac{2+x}{1-x}\right)^{1/2}\,dx$$

58. a. Show that the integral
$$I_n = \int_0^1 (1-x^2)^n\,dx$$
satisfies the reduction formula
$$I_n = \frac{2n}{2n+1}I_{n-1}$$
for any positive integer n. (*Hint:* Write $(1-x^2)^n = (1-x^2)^{n-1} - x^2(1-x^2)^{n-1}$, express I_n as a sum of two integrals, and evaluate the second integral by parts.)

   b. Use part (a) to evaluate $\int_0^1 (1-x^2)^4\,dx$.

*59. Use the trigonometric identities
$$\tan\frac{x}{2} = \frac{\sin x}{1+\cos x} \quad\text{and}\quad \sec^2\frac{x}{2} = \frac{2}{1+\cos x}$$
to find
$$\int \frac{\tan\left(\dfrac{\pi}{4}+\dfrac{x}{2}\right)}{\sec^2\dfrac{x}{2}}\,dx$$

60. Approximate $\int_0^1 \sqrt{1-x^2}\,dx$ (which has the value $\pi/4$) by using
   a. the Trapezoidal Rule with $n = 4$
   b. Simpson's Rule with $n = 4$

C 61. Approximate $\int_0^1 \sqrt{1-x^2}\,dx$ by using
   a. the Trapezoidal Rule with $n = 10$
   b. Simpson's Rule with $n = 10$

62. Determine values of n that ensure errors of less than 10^{-4} when $\int_2^{2.5} \sqrt{x^2-1}\,dx$ is approximated by the Trapezoidal Rule and by Simpson's Rule.

C 63. Using a calculator, approximate the integral in Exercise 62 with an error less than 10^{-4}
   a. by the Trapezoidal Rule
   b. by Simpson's Rule

In Exercises 64–65 find the area A of the region between the graph of f and the x axis on the given interval.

64. $f(x) = \cos^3 x$; $[0, \pi/2]$

65. $f(x) = \sqrt{9 - x^2}$; $[-3, 0]$

66. Let $f(x) = \sqrt{1 + x^5}$. Use Simpson's Rule with $n = 4$ to approximate the area of the region between the graph of f and the x axis on $[0, 4]$.

In Exercises 67–71 determine whether the area of the region between the graph of f and the x axis on the given interval is finite or infinite. If the area is finite, determine its numerical value.

67. $f(x) = \dfrac{\cos x}{1 + \sin x}$; $[\pi, 3\pi/2)$

68. $f(x) = \dfrac{\cos x}{\sqrt{1 + \sin x}}$; $[\pi, 3\pi/2)$

69. $f(x) = \dfrac{x^3}{2 + x^4}$; $(-\infty, \infty)$

70. $f(x) = \dfrac{x^3}{\sqrt{2 + x^4}}$; $(-\infty, \infty)$

71. $f(x) = \dfrac{x^3}{(2 + x^4)^2}$; $(-\infty, \infty)$

72. Let $f(x) = x + \sin x$ and $g(x) = x$, and let R be the region between the graphs of f and g on $[0, \pi]$.
   a. Find the volume V of the solid obtained by revolving R about the y axis.
   b. Find the volume V of the solid obtained by revolving R about the x axis.

73. a. Find the center of gravity of the region R in Exercise 72.
   b. Rework part (a) of Exercise 72 using the Theorem of Pappus and Guldin, and the result of part (a) of this exercise.

74. Let $f(x) = \ln x$, and let R denote the region between the graph of f and the x axis on $[1, \sqrt{3}]$.
   a. Determine the perimeter $\mathscr{L}$ of R.
   b. Determine the volume V of the solid generated by revolving R about the y axis.

75. Use Simpson's Rule with $n = 6$ to approximate the length $\mathscr{L}$ of one arch of the sine curve. In other words, if $f(x) = \sin x$ for $0 \le x \le \pi$, approximate the length $\mathscr{L}$ of the graph of f.

76. Find the length $\mathscr{L}$ of the curve described parametrically by $x = t^2$ and $y = 2t + 1$ for $0 \le t \le \sqrt{3}$.

77. Find the surface area S of the surface obtained by revolving about the x axis the curve parametrized by $x = e^t \cos t$ and $y = e^t \sin t$ for $0 \le t \le \pi/2$.

78. Suppose the side of a wine barrel has the shape of a surface obtained by revolving about the x axis the graph of the equation

$$y = \cos x \quad \text{for} \quad -\arcsin\frac{1}{\sqrt{3}} \le x \le \arcsin\frac{1}{\sqrt{3}}$$

Find the total surface area S of the barrel, including the side and both ends.

79. A gasoline tank in the shape of a cylinder 20 feet long and 10 feet in diameter lies on its side with its center 8 feet beneath the surface of the ground.
   a. How much work is done in emptying the full tank out to a level 3 feet above the ground? (Take the weight of gasoline to be 42 pounds per cubic foot.)
   b. Suppose that two hemispheres, each of radius 5 feet, were welded to the ends of the cylinder. How much work would be done in raising a full tank of gasoline to 3 feet above the ground?

80. Find the area A of the region enclosed by the polar graph of $r = \sin^2 \theta$.

Cumulative Review, Chapters 1–7

In Exercises 1–2 evaluate the limit.

1. $\lim\limits_{x \to 0} \dfrac{\sin x - \sin(\sin x)}{x^3}$

2. $\lim\limits_{x \to \infty} \left(1 - \dfrac{3}{x}\right)^x$

3. Find the value(s) of c for which $\lim\limits_{x \to \infty} (\sqrt{cx + 1} - \sqrt{x})$ exists.

In Exercises 4–5 find $f'(x)$.

4. $f(x) = \sqrt{\dfrac{x + 1}{3x - 2}}$

5. $f(x) = \dfrac{e^{2x}}{e^x + 1}$

6. Let $f(x) = \dfrac{1}{1 - x}$. Find a formula for $f^{(5)}(x)$.

7. Suppose that the velocity and acceleration of a given particle are positive and are related by the equation $a^2 - a = 2v^2 - 6v$. Show that $v = 3a$ at the instant that $da/dt = 6\,dv/dt$.

8. A car going 60 feet per second passes a car going 50 feet per second along a straight highway. The distance between the drivers is 10 feet when they are alongside one another. Let θ denote the angle made by one edge of the

highway and the line joining the two drivers. How fast is θ changing when $\theta = \pi/3$?

c 9. Let $f(x) = e^x - 1$. For $x > 0$, let $A(x)$ be the area between the graph of f and the x axis on $[0, x]$. Use the Newton-Raphson method to approximate to within 0.01 the value of x for which $A(x) = 1$.

In Exercises 10–11 sketch the graph of f, indicating all pertinent information.

10. $f(x) = x^{2/3}(x - 2)^{1/3}$ 11. $f(x) = \dfrac{x}{(x - 3)^2}$

12. Find the largest possible area A of an isosceles trapezoid inscribed in a circle of radius a with the diameter of the circle serving as the largest side of the trapezoid.

13. Of all right triangles with hypotenuse 1, which has the smallest ratio of circumference to area? (*Hint:* Express the ratio in terms of one of the angles of the triangle.)

14. A radioactive substance has a half-life of 24 hours. How long will it take for 99% of a given amount to disintegrate?

In Exercises 15–17 evaluate the integral.

15. $\displaystyle\int_0^2 \frac{1}{4 + 3x^2}\, dx$ 16. $\displaystyle\int e^{-3x}(e^{5x} + 1)\, dx$

17. $\displaystyle\int \frac{1}{\sqrt{-x^2 + 6x - 8}}\, dx$

18. Consider the region R bounded by the curves $y = e^{2x}$ and $y = e^{-x}$ and the line $x = \frac{1}{2}$. Find the area A of R.

19. The base of a solid is semicircular with diameter 2. The cross sections perpendicular to the diameter of the base are also semicircular. Find the volume V of the solid.

20. Let $f(x) = \frac{3}{2}x^{2/3}$ for $-27 \le x \le -1$. Find the length $\mathscr{L}$ of the graph of f.

21. Let C be parametrized by $x = \frac{1}{3}t^3$ and $y = \frac{1}{5}t^5$ for $0 \le t \le 1$. Determine the surface area S of the surface generated by revolving C about the x axis.

9

SEQUENCES AND SERIES

This is our last chapter concerning functions whose domains and ranges are sets of real numbers. Early in Chapter 9 we study sequences, which by definition are functions having sets of integers as their domains. We use sequences in the study of series, which are sums of infinite collections of numbers. Series can be used to represent many of the differentiable functions we have already encountered, such as the polynomial, exponential, logarithmic, sine, and cosine functions. A major advantage of the series representation of functions is that it allows us to approximate numbers such as e, π, and $\sqrt{2}$.

9.1
POLYNOMIAL APPROXIMATION AND TAYLOR'S THEOREM

In preceding chapters we have found approximate values of numbers such as e and $\ln 2$. Through Taylor's Theorem, calculus provides a remarkably powerful and general method of estimating the values of certain differentiable functions with any prescribed degree of accuracy. In particular, Taylor's Theorem will enable us to obtain an excellent approximation of numbers like e and $\ln 2$ with a minimum of effort.

Suppose we wish to approximate a value $f(x)$ of a function f and we know the value $f(a)$ at a nearby number a. Because values of polynomials are easy to compute, we will use polynomials to approximate $f(x)$. These polynomials will be manufactured from higher derivatives of f, so we will need to assume that f has such higher derivatives. For simplicity we assume that $a = 0$; later, in Section 9.9, we will let a be any number.

Taylor Polynomials

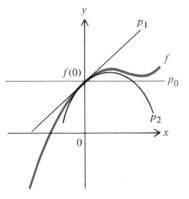

$f(0)$

FIGURE 9.1

The simplest kind of polynomial is constant, and the constant polynomial p_0 that best approximates values of f near 0 is given by

$$p_0(x) = f(0) \tag{1}$$

since both f and p_0 have the same value at 0 (Figure 9.1). In most cases a better approximation is obtained by letting

$$p_1(x) = f(0) + f'(0)x \tag{2}$$

Then p_1 is linear, and at 0 it has both the same value and the same derivative as f (Figure 9.1). Notice that $p_1(x)$ gives the tangent line approximation of $f(x)$ discussed in Section 3.9. Still better accuracy is usually afforded by a polynomial p_2 having the same value and same first and second derivatives at 0 as f. You can check that if we let

$$p_2(x) = f(0) + f'(0)x + \frac{f''(0)}{2}x^2 \tag{3}$$

then p_2 has these properties (Figure 9.1). The polynomial of degree (at most) 3 that would approximate the values of f is defined by

$$p_3(x) = f(0) + f'(0)x + \frac{f''(0)}{2}x^2 + \frac{f^{(3)}(0)}{6}x^3 \tag{4}$$

and you can check that p_3 and f have the same value and the same first, second, and third derivatives at 0 (see Exercise 30). For this reason we expect $p_3(x)$ to be a better estimate of $f(x)$ than $p_2(x)$.

To simplify the formula for the higher-degree polynomials that will approximate the values of f, we define the **factorial** of any nonnegative integer n as follows:

$$n! = n(n-1)(n-2)(n-3)\cdots 3\cdot 2\cdot 1 \quad \text{for } n \geq 1$$
$$0! = 1$$

Thus

$$1! = 1 \qquad\qquad 4! = 4\cdot 3\cdot 2\cdot 1 = 24$$
$$2! = 2\cdot 1 = 2 \qquad 5! = 5\cdot 4\cdot 3\cdot 2\cdot 1 = 120$$
$$3! = 3\cdot 2\cdot 1 = 6 \qquad 6! = 6\cdot 5\cdot 4\cdot 3\cdot 2\cdot 1 = 720$$

An often-used property of factorials that follows directly from the definition is

$$(n+1)! = (n+1)n! \quad \text{for } n \geq 0$$

Now we define a polynomial p_n of degree (at most) n by the formula

$$p_n(x) = f(0) + f'(0)x + \frac{f''(0)}{2!}x^2 + \frac{f^{(3)}(0)}{3!}x^3 + \cdots + \frac{f^{(n)}(0)}{n!}x^n$$

The polynomial p_n is called the **nth Taylor polynomial** of f about 0, after the English mathematician Brook Taylor (1685–1731), the author of an early calculus book (published in 1717). It and f have the same value at 0, as well as the same derivatives at 0, up through the nth derivative. Notice that the Taylor polynomials of f for $n = 0, 1, 2$, and 3 appear in (1), (2), (3), and (4).

Example 1   Let $f(x) = e^x$. Find a formula for the nth Taylor polynomial of f and for $p_n(1)$. Then compute $p_5(1)$.

Solution   In order to use the formula for $p_n(x)$, we calculate $f^{(k)}(0)$ for any nonnegative integer k. But for each such k, $f^{(k)}(x) = e^x$, so that $f^{(k)}(0) = e^0 = 1$. Therefore

$$p_n(x) = f(0) + f'(0)x + \frac{f''(0)}{2!}x^2 + \frac{f^{(3)}(0)}{3!}x^3 + \cdots + \frac{f^{(n)}(0)}{n!}x^n$$

$$= 1 + x + \frac{x^2}{2!} + \frac{x^3}{3!} + \cdots + \frac{x^n}{n!}$$

As a result,

$$p_n(1) = 1 + 1 + \frac{1}{2!} + \frac{1}{3!} + \cdots + \frac{1}{n!}$$

For $n = 5$ we obtain

$$p_5(1) = 1 + 1 + \frac{1}{2!} + \frac{1}{3!} + \frac{1}{4!} + \frac{1}{5!} = \frac{163}{60} \approx 2.71667 \quad \square$$

Since $p_n(x)$ is intended as an approximation to $f(x)$, we should examine how well $p_5(1)$ approximates $f(1) = e$. The value of e is 2.71828 (accurate to six digits), and $p_5(1) \approx 2.71667$, so $p_5(1)$ approximates e with an error of about 0.00161.

Example 2   Let $f(x) = \ln(1 + x)$. Find a formula for the nth Taylor polynomial of f, and then calculate $p_6(1)$.

Solution   First we calculate the derivatives of f:

$$f(x) = \ln(1 + x) \qquad\qquad f(0) = 0$$

$$f'(x) = \frac{1}{1 + x} \qquad\qquad f'(0) = 1$$

$$f''(x) = \frac{-1}{(1 + x)^2} \qquad\qquad f''(0) = -1$$

$$f^{(3)}(x) = \frac{(-1)^2 2!}{(1 + x)^3} \qquad\qquad f^{(3)}(0) = 2!$$

$$f^{(4)}(x) = \frac{(-1)^3 3!}{(1 + x)^4} \qquad\qquad f^{(4)}(0) = -3!$$

In general, for $k \geq 1$,

$$f^{(k)}(x) = \frac{(-1)^{k-1}(k-1)!}{(1+x)^k}, \qquad f^{(k)}(0) = (-1)^{k-1}(k-1)!$$

Consequently

$$p_n(x) = f(0) + f'(0)x + \frac{f''(0)}{2!}x^2 + \frac{f^{(3)}(0)}{3!}x^3 + \cdots + \frac{f^{(n)}(0)}{n!}x^n$$

$$= x - \frac{x^2}{2} + \frac{x^3}{3} - \frac{x^4}{4} + \cdots + \frac{(-1)^{n-1}}{n}x^n$$

We conclude that

$$p_6(1) = 1 - \frac{1}{2} + \frac{1}{3} - \frac{1}{4} + \frac{1}{5} - \frac{1}{6} = \frac{37}{60} \approx 0.616667 \quad \square$$

We expect $p_n(x)$ to approximate $f(x)$. Since the value of $f(1) = \ln 2$ is 0.693147 (accurate to six digits) and since $p_6(1) \approx 0.616667$, we find that $p_6(1)$ approximates ln 2 with an error of about 0.07648.

Example 3   Let $f(x) = 1/(1 - x)$. Find a formula for the nth Taylor polynomial of f, and compute $p_n(2)$.

Solution   The derivatives of f are

$$f(x) = \frac{1}{1-x} \qquad\qquad f(0) = 1$$

$$f'(x) = \frac{1}{(1-x)^2} \qquad\qquad f'(0) = 1$$

$$f''(x) = \frac{2!}{(1-x)^3} \qquad\qquad f''(0) = 2!$$

$$f^{(3)}(x) = \frac{3!}{(1-x)^4} \qquad\qquad f^{(3)}(0) = 3!$$

In general,

$$f^{(k)}(x) = \frac{k!}{(1-x)^{k+1}} \qquad\qquad f^{(k)}(0) = k!$$

As a result,

$$p_n(x) = f(0) + f'(0)x + \frac{f''(0)}{2!}x^2 + \frac{f^{(3)}(0)}{3!}x^3 + \cdots + \frac{f^{(n)}(0)}{n!}x^n$$

$$= 1 + x + x^2 + x^3 + \cdots + x^n$$

In particular,

$$p_n(2) = 1 + 2 + 2^2 + 2^3 + \cdots + 2^n \quad \square$$

Taylor's Theorem

In Examples 1 and 2 the value of the Taylor polynomial provided a reasonable approximation to the corresponding value of the given function. Indeed, we found that if $f(x) = e^x$, then

$$|f(1) - p_5(1)| = |e - p_5(1)| \approx 0.00161$$

and if $f(x) = \ln(1 + x)$, then

$$|f(1) - p_6(1)| = |\ln 2 - p_6(1)| \approx 0.07648$$

By contrast, if $f(x) = 1/(1 - x)$ (as in Example 3), then for $n \geq 1$,

$$|f(2) - p_n(2)| = \left| \frac{1}{1 - 2} - (1 + 2 + 2^2 + \cdots + 2^n) \right|$$

$$= |-1 - (1 + 2 + 2^2 + \cdots + 2^n)|$$

$$\geq 2^n$$

Consequently $p_n(2)$ is not reasonably close to $f(2)$ for *any* value of n. In fact, the larger n becomes, the worse $p_n(2)$ approximates $f(2)$.

If f is a function with higher derivatives at 0, then in order to measure how well $p_n(x)$ approximates $f(x)$, we define the **nth Taylor remainder r_n** of f by the formula

$$r_n(x) = f(x) - p_n(x) = f(x) - \left[f(0) + f'(0)x + \frac{f''(0)}{2!}x^2 + \cdots + \frac{f^{(n)}(0)}{n!}x^n \right]$$

The value of $r_n(x)$ tells us how well $p_n(x)$ approximates $f(x)$. The smaller $r_n(x)$ is, the better $p_n(x)$ approximates $f(x)$.

A theorem due to Brook Taylor provides us with a means of estimating the nth Taylor remainder, and hence the accuracy involved in replacing $f(x)$ by the value of the nth Taylor polynomial at x.

THEOREM 9.1
TAYLOR'S THEOREM

Let n be a nonnegative integer, and suppose $f^{(n+1)}(x)$ exists for each x in an open interval I containing 0. For each $x \neq 0$ in I, there is a number t_x strictly between 0 and x such that

$$f(x) = f(0) + f'(0)x + \cdots + \frac{f^{(n)}(0)}{n!}x^n + \frac{f^{(n+1)}(t_x)}{(n+1)!}x^{n+1} \qquad (5)$$

and

$$r_n(x) = \frac{f^{(n+1)}(t_x)}{(n+1)!}x^{n+1} \qquad (6)$$

(Equation (5) is known as **Taylor's Formula**, and (6) is known as the **Lagrange Remainder Formula**.)

We are not proving Taylor's Theorem here because in Section 9.9 we will state and prove a slightly more general theorem. Notice that if $n = 0$, then formula (5) in Taylor's Theorem becomes

$$f(x) = f(0) + f'(t_x)x$$

or equivalently,

$$f(x) - f(0) = f'(t_x)(x - 0)$$

which is equivalent to the formula in the Mean Value Theorem (with a, b, and c replaced, respectively, by 0, x, and t_x). Observe also that the number t_x in Taylor's Theorem depends on n as well as on x.

We can use Taylor's Theorem to achieve approximations with a prescribed accuracy, as the next two examples show.

Example 4   Approximate the number e with an error less than 0.001.

Solution   Let $f(x) = e^x$. If we approximate $e = e^1 = f(1)$ by $p_n(1)$, then the error introduced will be less than 0.001 if $|f(1) - p_n(1)| = |r_n(1)| < 0.001$. Consequently if we find an n such that $|r_n(1)| < 0.001$, then $p_n(1)$ will be the desired approximation. Now by Taylor's Theorem we know that there is a number t_1 between 0 and 1 such that

$$r_n(1) = \frac{f^{(n+1)}(t_1)}{(n+1)!} 1^{n+1}$$

Since $f^{(n+1)}(t_1) = e^{t_1}$ and $e < 4$ (as we observed in Section 5.7), it follows that

$$e^{t_1} < 4^{t_1} < 4^1 = 4$$

so that

$$r_n(1) = \frac{e^{t_1}}{(n+1)!} < \frac{4}{(n+1)!} \tag{7}$$

By computing the value of $4/(n+1)!$ for $n = 1, 2, \ldots, 6$, we discover that $|r_n(1)| < 0.001$ if $n \geq 6$. Thus $p_6(1)$ is the desired approximation to e. By the general formula for $p_n(1)$ in Example 1,

$$p_6(1) = 1 + 1 + \frac{1}{2!} + \frac{1}{3!} + \frac{1}{4!} + \frac{1}{5!} + \frac{1}{6!} = \frac{1957}{720} \approx 2.71806$$

We conclude that 2.71806 approximates e with an error less than 0.001.   □

From (7) we find that

$$|r_n(1)| = |e - p_n(1)| = \left| e - \left(1 + 1 + \frac{1}{2!} + \frac{1}{3!} + \cdots + \frac{1}{n!} \right) \right| < \frac{4}{(n+1)!} \tag{8}$$

Since $4/(n+1)!$ gets small as n gets large, we could make the difference between e and $p_n(1)$ as small as we please by taking n large enough.

Since $\sqrt{e} = e^{1/2}$, we could also approximate $\sqrt{e}$ by using the same Taylor polynomials of the function e^x. But for $\sqrt{e}$ we would utilize the numbers $p_n(\frac{1}{2})$ instead of using $p_n(1)$.

Example 5   Approximate ln 2 with an error less than 0.1.

Solution   Let $f(x) = \ln(1 + x)$. As in Example 4, we will approximate $f(1)$ by $p_n(1)$ for an appropriate value of n. By the formula for $f^{(k)}(x)$ in Example 2, with k

replaced by $n + 1$, we have

$$f^{(n+1)}(x) = \frac{(-1)^n n!}{(1 + x)^{n+1}}$$

so that by (6), with $x = 1$, there is a number t_1 strictly between 0 and 1 such that

$$r_n(1) = \frac{f^{(n+1)}(t_1)}{(n+1)!} = \frac{(-1)^n n!}{(1 + t_1)^{n+1}(n+1)!} = \frac{(-1)^n}{(1 + t_1)^{n+1}(n+1)}$$

Since $t_1 > 0$, we have

$$|r_n(1)| < \frac{1}{n+1} \tag{9}$$

so that if $n \geq 9$, then $|r_n(1)| < 0.1$. Thus $p_9(1)$ approximates $\ln 2$ with an error less than 0.1. The general formula for $p_n(x)$ appearing in the solution of Example 2 yields

$$p_9(1) = 1 - \frac{1}{2} + \frac{1}{3} - \frac{1}{4} + \frac{1}{5} - \frac{1}{6} + \frac{1}{7} - \frac{1}{8} + \frac{1}{9} \approx 0.745635$$

and this is the desired approximation. $\square$

From (9) we find that

$$|r_n(1)| = |\ln 2 - p_n(1)| = \left| \ln 2 - \left(1 - \frac{1}{2} + \frac{1}{3} - \frac{1}{4} + \cdots + (-1)^{n-1}\frac{1}{n} \right) \right|$$
$$< \frac{1}{n+1} \tag{10}$$

Since $1/(n + 1)$ gets small as n becomes large, we could make the difference between $\ln 2$ and $p_n(1)$ as small as we please by taking n large enough.

In approximating e and $\ln 2$ by values of the Taylor polynomials, we noticed that the error introduced could be made as small as we wished by picking n sufficiently large. But taking larger values of n means adding up more numbers in the nth Taylor polynomial. This suggests the possibility of attaching a meaning to the sum of an infinite number of numbers. In fact this can be done, and we will find that e and $\ln 2$ can be not only approximated but also represented by a sum of an infinite collection of numbers. It is even possible to create entirely new functions through the process of summing infinite collections of numbers. We will devote most of this chapter to these topics.

EXERCISES 9.1

1. Let $f(x) = \sin x$. Sketch the graphs of $f, p_0, p_1, p_2,$ and p_3 in the same coordinate plane.

2. Let $f(x) = \cos x$. Sketch the graphs of $f, p_0, p_1, p_2,$ and p_3 in the same coordinate plane.

In Exercises 3–12 find a formula for an arbitrary Taylor polynomial of f.

3. $f(x) = x^2 - x - 2$

4. $f(x) = x^5 + 3x + 4$

5. $f(x) = \dfrac{1}{1 + x}$

6. $f(x) = \dfrac{1}{1 + 2x}$

7. $f(x) = e^{-x}$

8. $f(x) = e^{3x}$

9. $f(x) = \cosh x$

10. $f(x) = \ln\dfrac{1 + x}{1 - x}$

$\left(Hint:\ \ln\dfrac{1 + x}{1 - x} = \ln(1 + x) - \ln(1 - x).\right)$

11. $f(x) = \sin x$ 12. $f(x) = \cos x$

C In Exercises 13–19 use the results of Exercises 3–12 and the examples in this section to approximate the given number with an error less than the value shown.

13. $e^{1/2}$; 0.001 14. $e^{-1/2}$; 0.0001

15. $\ln 1.1$; 0.00001 16. $\sin \pi/10$; 0.001

17. $\cos(-\pi/3)$; 0.001

18. $\ln 2$; 0.01 (*Hint:* Use Exercise 10 with $x = \frac{1}{3}$.)

19. $\ln \frac{3}{2}$; 0.01 (*Hint:* Use Exercise 10.)

In Exercises 20–27 find the nth Taylor polynomial of f for the given values of n.

20. $f(x) = \sin x^2$; $n = 3$

21. $f(x) = e^{-(x^2)}$; $n = 3$

22. $f(x) = \arcsin x$; $n = 2$

23. $f(x) = \ln(\cos x)$; $n = 2$

24. $f(x) = \tan x$; $n = 3$ and $n = 4$

25. $f(x) = \sec x$; $n = 2$ and $n = 3$

*26. $f(x) = \begin{cases} \dfrac{\sin x}{x} & \text{for } x \neq 0 \\ 1 & \text{for } x = 0 \end{cases}$ $n = 2$

*27. $f(x) = \begin{cases} e^{-1/x^2} & \text{for } x \neq 0 \\ 0 & \text{for } x = 0 \end{cases}$ $n = 2$

28. Find the third Taylor polynomial of the arctangent function, and use it to approximate $\pi/4$.

29. Let $f(x) = \sqrt{1 + x}$.
 a. Find the second Taylor polynomial of f.
 b. Use the solution to part (a) to approximate $\sqrt{2}$.
 c. Use the solution to part (a) to approximate $\sqrt{1.1}$.

30. Assume that $f^{(3)}(0)$ exists. Verify that the third Taylor polynomial of f, defined by (4), has the same value and the same first, second, and third derivatives at 0 as f does.

9.2
SEQUENCES

In Section 9.1 we used the numbers

$$1 + 1 + \frac{1}{2!} + \frac{1}{3!} + \cdots + \frac{1}{n!} \quad \text{for } n \geq 0 \tag{1}$$

to approximate e, and we observed that the approximation can be made as good as we wish by taking n large enough. The numbers in (1) give rise to a function f whose domain is the set of all nonnegative numbers:

$$f(n) = 1 + 1 + \frac{1}{2!} + \frac{1}{3!} + \cdots + \frac{1}{n!} \quad \text{for } n \geq 0$$

Consider also the numbers

$$\left(1 + \frac{0.05}{n}\right)^n \quad \text{for } n \geq 1 \tag{2}$$

If \$1.00 is deposited in an account at 5 percent interest compounded n times per year, then $(1 + 0.05/n)^n$ is the number of dollars in the account at the end of 1 year (see Exercise 45). The larger n is, the larger the amount of money at the end of the year is. The numbers in (2) give rise to a function whose domain is the set of all positive numbers:

$$g(n) = \left(1 + \frac{0.05}{n}\right)^n \quad \text{for } n \geq 1$$

This section is devoted to functions with domains like those of f and g.

DEFINITION 9.2

A **sequence** is a function whose domain is the collection of all integers greater than or equal to a given integer m (usually 0 or 1).

Thus the functions f and g described above are sequences. Other examples of sequences are h and k, where

$$h(n) = \left(\frac{1}{2}\right)^n \quad \text{for } n \geq 0$$

$$k(n) = n^2 \quad \text{for } n \geq 4$$

In general, if

$$f(n) = a_n \quad \text{for } n \geq 1 \tag{3}$$

then the ordered set of numbers $a_1, a_2, \ldots$ completely determines the sequence. As a result we normally suppress the symbol f and just write $\{a_n\}_{n=1}^{\infty}$ for the sequence defined in (3). Similarly if

$$f(n) = a_n \quad \text{for } n \geq m$$

then we would write $\{a_n\}_{n=m}^{\infty}$ for the sequence. In this notation we express our original sequences f, g, h, and k as

$$\left\{1 + 1 + \frac{1}{2!} + \frac{1}{3!} + \cdots + \frac{1}{n!}\right\}_{n=0}^{\infty}, \quad \left\{\left(1 + \frac{0.05}{n}\right)^n\right\}_{n=1}^{\infty}, \quad \left\{\left(\frac{1}{2}\right)^n\right\}_{n=0}^{\infty}, \quad \{n^2\}_{n=4}^{\infty}$$

The initial four terms of the third sequence are

$$a_0 = \left(\frac{1}{2}\right)^0 = 1, \qquad a_1 = \left(\frac{1}{2}\right)^1 = \frac{1}{2},$$

$$a_2 = \left(\frac{1}{2}\right)^2 = \frac{1}{4}, \qquad a_3 = \left(\frac{1}{2}\right)^3 = \frac{1}{8}$$

Similarly, the initial four terms of the fourth sequence are

$$a_4 = 4^2 = 16, \qquad a_5 = 5^2 = 25,$$

$$a_6 = 6^2 = 36, \qquad a_7 = 7^2 = 49$$

The symbol n in $\{a_n\}_{n=m}^{\infty}$ is called an **index**, and m is called the **initial index**. Observe that the symbol used for the index is immaterial. For example, $\{a_n\}_{n=m}^{\infty}$ and $\{a_j\}_{j=m}^{\infty}$ are the same sequence.

Convergent Sequences

For any sequence $\{a_n\}_{n=m}^{\infty}$ we may consider the behavior of a_n as n increases without bound. For example, we already know that if

$$a_n = 1 + 1 + \frac{1}{2!} + \frac{1}{3!} + \cdots + \frac{1}{n!}$$

then a_n can be made as close to e as we wish by taking n sufficiently large.

However, for

$$a_n = \left(1 + \frac{0.05}{n}\right)^n$$

it might not be so clear whether a_n approaches some specific number as n increases without bound. But if a_n represents a bank balance after 1 year with interest compounded n times per year, the behavior of a_n for large n would be of interest to an investor or a bank director. Now we state precisely what it means for the numbers a_n in a sequence to approach a fixed value as n increases without bound.

DEFINITION 9.3

Let $\{a_n\}_{n=m}^{\infty}$ be a sequence. A number L is the **limit** of $\{a_n\}_{n=m}^{\infty}$ if for every $\varepsilon > 0$ there is an integer N such that

$$\text{if } n \geq N, \quad \text{then} \quad |a_n - L| < \varepsilon \tag{4}$$

In this case we write

$$\lim_{n \to \infty} a_n = L$$

If such a number L exists, we say that $\{a_n\}_{n=m}^{\infty}$ **converges** (or **converges to L**), or that $\lim_{n \to \infty} a_n$ **exists**. If such a number L does not exist, we say that $\{a_n\}_{n=m}^{\infty}$ **diverges** or that $\lim_{n \to \infty} a_n$ **does not exist**.

As with limits of functions, the limit of a sequence is unique if it exists.

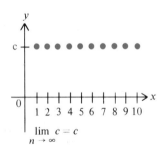

$$\lim_{n \to \infty} c = c$$

FIGURE 9.2

Example 1 Let c be any number, and let $a_n = c$ for $n \geq 1$. Show that $\lim_{n \to \infty} a_n = c$.

Solution For any $\varepsilon > 0$, let $N \geq 1$. Observe that

$$\text{if } n \geq N, \quad \text{then} \quad |a_n - c| = |c - c| = 0 < \varepsilon$$

Consequently $\lim_{n \to \infty} a_n = c$ (Figure 9.2).   □

We express the result of Example 1 by writing $\lim_{n \to \infty} c = c$. Specifically, $\lim_{n \to \infty} 1 = 1$ and $\lim_{n \to \infty} (-5) = -5$.

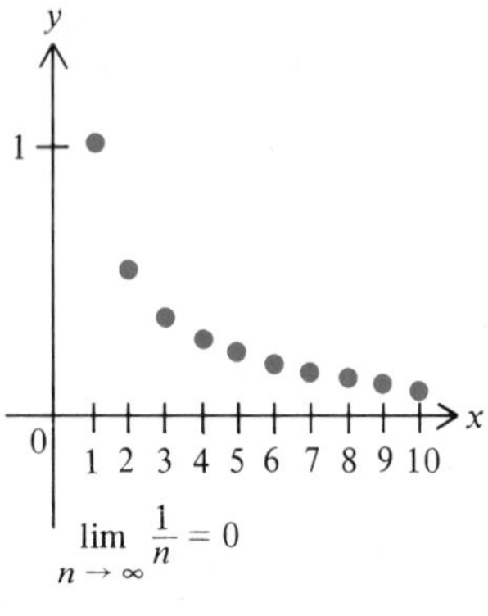

Example 2 Show that $\{1/n\}_{n=1}^{\infty}$ converges and that $\lim_{n \to \infty} 1/n = 0$.

Solution For any $\varepsilon > 0$, let N be an integer greater than $1/\varepsilon$. It follows that

$$\text{if } n \geq N, \quad \text{then} \quad \left|\frac{1}{n} - 0\right| = \frac{1}{n} \leq \frac{1}{N} < \varepsilon$$

$$\lim_{n \to \infty} \frac{1}{n} = 0$$

FIGURE 9.3

This shows that $\lim_{n \to \infty} 1/n = 0$ (Figure 9.3).   □

The sequence $\{1/n\}_{n=1}^{\infty}$ is known as the **harmonic sequence**; this name derives from the fact that $1/n$ is the harmonic mean of $1/(n-1)$ and $1/(n+1)$ (see Exercise 82 of Section 1.1). There is also a musical interpretation for the harmonic sequence. The fundamental tone of a violin string of length l is obtained by vibrating the whole string. The first harmonic (or overtone) is obtained by lightly touching the vibrating string at its midpoint. The second harmonic (or overtone) is obtained by lightly touching the vibrating string at a point one third of the way up the string. In general, the $(n-1)$st harmonic is obtained by lightly touching the vibrating string at a point $1/n$ of the way up the string. If l is taken to be 1, then the lengths that determine the fundamental and the harmonics form the harmonic sequence $\{1/n\}_{n=1}^{\infty}$.

Example 3   Show that $\{(-1)^n\}_{n=1}^{\infty}$ diverges.

Solution   We have

$$(-1)^n = \begin{cases} 1 & \text{for } n \text{ even} \\ -1 & \text{for } n \text{ odd} \end{cases}$$

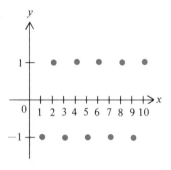

FIGURE 9.4

so the graph is as shown in Figure 9.4. If $L \geq 0$, then for odd values of n we have

$$|(-1)^n - L| = |-1 - L| \geq 1$$

But if $L < 0$, then for even values of n we have

$$|(-1)^n - L| = |1 - L| \geq 1$$

Consequently for any ε such that $0 < \varepsilon \leq 1$, there is no number L satisfying (4). Therefore the given sequence diverges. $\square$

The sequence in Example 3 diverges because the numbers in the sequence alternate between 1 and -1 (and hence do not approach a limit). A sequence can also diverge because the numbers in the sequence grow arbitrarily large in absolute value. An example of such a sequence is $\{n^2\}_{n=0}^{\infty}$. To describe such sequences we make the following definition.

DEFINITION 9.4

Let $\{a_n\}_{n=m}^{\infty}$ be a sequence. If for every number M there is an integer N such that

$$\text{if } n \geq N, \quad \text{then} \quad a_n > M \tag{5}$$

we say that $\{a_n\}_{n=m}^{\infty}$ **diverges to** ∞, and we write

$$\lim_{n \to \infty} a_n = \infty$$

Similarly, if for every number M there is an integer N such that

$$\text{if } n \geq N, \quad \text{then} \quad a_n < M$$

we say that $\{a_n\}_{n=m}^{\infty}$ **diverges to** $-\infty$, and we write

$$\lim_{n \to \infty} a_n = -\infty$$

We could apply Definition 9.4 to prove that $\lim_{n \to \infty} n^2 = \infty$. But referring to Definition 9.4 to show that a sequence diverges, or referring to Definition 9.3 to show that a sequence such as those in Examples 1–3 converges or diverges is tedious. One way to avoid constantly using Definitions 9.3 and 9.4 arises from the fact that the definition of $\lim_{n \to \infty} a_n = L$ is very reminiscent of the definition of $\lim_{x \to \infty} f(x) = L$ (see Definition 4.16), and similarly, that the definition of $\lim_{n \to \infty} a_n = \infty$ (or $-\infty$) is reminiscent of the definition of $\lim_{x \to \infty} f(x) = \infty$ (or $-\infty$). These observations lead us to the following theorem.

THEOREM 9.5

Let $\{a_n\}_{n=m}^{\infty}$ be a sequence, L a number, and f a function defined on $[m, \infty)$ such that $f(n) = a_n$ for $n \geq m$. If $\lim_{x \to \infty} f(x) = L$, then $\{a_n\}_{n=m}^{\infty}$ converges and $\lim_{n \to \infty} a_n = L$. If $\lim_{x \to \infty} f(x) = \infty$ (or $\lim_{x \to \infty} f(x) = -\infty$), then $\{a_n\}_{n=m}^{\infty}$ diverges, and $\lim_{n \to \infty} a_n = \infty$ (or $\lim_{n \to \infty} a_n = -\infty$). Thus

$$\lim_{n \to \infty} a_n = \lim_{x \to \infty} f(x)$$

Proof   First we assume that $\lim_{x \to \infty} f(x) = L$, and let $\varepsilon > 0$. Then there is some integer N such that

$$\text{if } x \geq N, \quad \text{then} \quad |f(x) - L| < \varepsilon$$

This implies that

$$\text{if } n \geq N, \quad \text{then} \quad |a_n - L| = |f(n) - L| < \varepsilon$$

so that by (4), $\lim_{n \to \infty} a_n = L$. If $\lim_{x \to \infty} f(x) = \infty$, then for any M there is an integer N such that

$$\text{if } x \geq N, \quad \text{then} \quad f(x) > M$$

so that

$$\text{if } n \geq N, \quad \text{then} \quad a_n > M$$

But by (5), this means that $\lim_{n \to \infty} a_n = \infty$. Similarly, $\lim_{x \to \infty} f(x) = -\infty$ implies that $\lim_{n \to \infty} a_n = -\infty$. ■

Example 4   Show that $\lim_{n \to \infty} n^2 = \infty$.

Solution   Let

$$f(x) = x^2 \quad \text{for } x \geq 1$$

Then $f(n) = n^2$ for $n \geq 1$. Since $\lim_{x \to \infty} x^2 = \infty$, we conclude from Theorem 9.5 that

$$\lim_{n \to \infty} n^2 = \infty \quad \square$$

Example 5   Show that

$$\lim_{n \to \infty} \frac{1}{n^r} = 0 \quad \text{for } r > 0$$

Solution   Let

$$f(x) = \frac{1}{x^r} \quad \text{for } x \geq 1$$

Then $f(n) = 1/n^r$ for $n \geq 1$, and by our analysis of power functions in Section 7.4, we know that $\lim_{x \to \infty} f(x) = 0$. As a result, Theorem 9.5 implies that

$$\lim_{n \to \infty} \frac{1}{n^r} = 0 \quad \square$$

Example 6   Let r be any number. Show that the sequence $\{r^n\}_{n=1}^{\infty}$ diverges for $|r| > 1$ and for $r = -1$. Show that for all other values of r the sequence converges, with

$$\lim_{n \to \infty} r^n = \begin{cases} 1 & \text{for } r = 1 \\ 0 & \text{for } |r| < 1 \end{cases}$$

Solution First we consider nonnegative values of r. Let

$$f(x) = r^x \quad \text{for } x \geq 1$$

so that $f(n) = r^n$ for $n \geq 1$. It follows from our analysis of exponential functions in Section 7.4 that

$$\lim_{x \to \infty} r^x = \begin{cases} 0 & \text{for } 0 \leq r < 1 \\ 1 & \text{for } r = 1 \\ \infty & \text{for } r > 1 \end{cases}$$

By Theorem 9.5 this means that

$$\lim_{n \to \infty} r^n = \begin{cases} 0 & \text{for } 0 \leq r < 1 \\ 1 & \text{for } r = 1 \\ \infty & \text{for } r > 1 \end{cases} \tag{6}$$

Thus $\{r^n\}_{n=1}^{\infty}$ diverges for $r > 1$ and converges for $0 \leq r \leq 1$. Next we consider negative values of r. If $r = -1$, then $\{r^n\}_{n=1}^{\infty}$ becomes $\{(-1)^n\}_{n=1}^{\infty}$, which diverges by Example 3. If $r \neq -1$, then since $|r^n| = |r|^n$, we know from (6) that

$$\lim_{n \to \infty} |r^n| = \lim_{n \to \infty} |r|^n = \begin{cases} 0 & \text{for } -1 < r < 0 \\ \infty & \text{for } r < -1 \end{cases}$$

It follows that $\lim_{n \to \infty} r^n = 0$ when $-1 < r < 0$ and that $\lim_{n \to \infty} r^n$ does not exist when $r < -1$. This completes the solution of the example. □

For any number r, the sequence $\{r^n\}_{n=1}^{\infty}$ is called a **geometric sequence**, because for each n, r^n is the geometric mean of r^{n-1} and r^{n+1}.

Example 7 Show that

$$\lim_{n \to \infty} \left(1 + \frac{1}{n}\right)^n = e$$

Solution If we let

$$f(x) = \left(1 + \frac{1}{x}\right)^x \quad \text{for } x \geq 1$$

then $f(n) = (1 + 1/n)^n$ for $n \geq 1$. By Example 9 of Section 7.8 we know that

$\lim_{x \to \infty} f(x) = e$. Consequently Theorem 9.5 yields

$$\lim_{n \to \infty} \left(1 + \frac{1}{n} \right)^n = e \quad \square$$

By letting $g(x) = (1 + 0.05/x)^x$ and using the same ideas as in Example 7, we can deduce that $\lim_{x \to \infty} g(x) = e^{0.05}$. Thus

$$\lim_{n \to \infty} \left(1 + \frac{0.05}{n} \right)^n = e^{0.05} \approx 1.05127 \tag{7}$$

(see Exercise 25). This implies that the sequence in (2) converges to $e^{0.05}$. The limit in (7) is the basis of compounding interest "continuously." If a bank offers 5 percent interest compounded continuously, then (7) tells us that during a year $1000 grows to approximately $1051.27.

Example 8   Show that

$$\lim_{n \to \infty} \sqrt[n]{c} = 1 \quad \text{for } c > 0$$

Solution   Notice that

$$\sqrt[n]{c} = c^{1/n} = e^{(1/n)\ln c}$$

Thus we let

$$f(x) = e^{(1/x)\ln c} \quad \text{for } x \geq 1$$

so that f is continuous and $f(n) = e^{(1/n)\ln c} = \sqrt[n]{c}$ for $n \geq 1$. Then, since $\lim_{x \to \infty} (1/x)\ln c = 0$, we have

$$\lim_{x \to \infty} f(x) = \lim_{x \to \infty} e^{(1/x)\ln c} = e^0 = 1$$

It follows from Theorem 9.5 that $\lim_{n \to \infty} \sqrt[n]{c} = 1$. $\square$

Example 9   Show that

$$\lim_{n \to \infty} \sqrt[n]{n} = 1$$

Solution   Imitating the solution of Example 8, we notice that

$$\sqrt[n]{n} = n^{1/n} = e^{(1/n)\ln n}$$

Thus we let

$$f(x) = e^{(1/x)\ln x} \quad \text{for } x \geq 1$$

so that f is continuous and $f(n) = e^{(1/n)\ln n} = \sqrt[n]{n}$ for $n \geq 1$. Since

$$\lim_{x \to \infty} x = \infty = \lim_{x \to \infty} \ln x$$

l'Hôpital's Rule implies that

$$\lim_{x \to \infty} \frac{\ln x}{x} = \lim_{x \to \infty} \frac{1/x}{1} = \lim_{x \to \infty} \frac{1}{x} = 0$$

and thus

$$\lim_{x \to \infty} f(x) = \lim_{x \to \infty} e^{(1/x)\ln x} = e^0 = 1$$

From Theorem 9.5 we conclude that $\lim_{n \to \infty} \sqrt[n]{n} = 1$. $\square$

EXERCISES 9.2

In Exercises 1–4 write the initial four terms of the sequence.

1. $\left\{\dfrac{1}{n}\right\}_{n=3}^{\infty}$ 2. $\left\{\dfrac{1}{3^n}\right\}_{n=0}^{\infty}$ 3. $\left\{\dfrac{k-1}{k+1}\right\}_{k=1}^{\infty}$ 4. $\left\{k - \dfrac{1}{k}\right\}_{k=5}^{\infty}$

In Exercises 5–12 use Definition 9.3 or 9.4 to verify the equation.

5. $\lim_{n \to \infty} (-2) = -2$
6. $\lim_{n \to \infty} \dfrac{1}{n^2} = 0$
7. $\lim_{n \to \infty} \dfrac{3n+1}{n} = 3$
8. $\lim_{n \to \infty} \dfrac{2n-1}{n+1} = 2$
9. $\lim_{n \to \infty} \sqrt{n} = \infty$
10. $\lim_{n \to \infty} (-2n^3) = -\infty$
11. $\lim_{k \to \infty} e^k = \infty$
12. $\lim_{k \to \infty} \dfrac{k^2+1}{k} = \infty$

In Exercises 13–32 evaluate the limit as a number, ∞, or $-\infty$.

13. $\lim_{n \to \infty} \left(\pi + \dfrac{1}{n}\right)$
14. $\lim_{n \to \infty} (\pi - n)$
15. $\lim_{j \to \infty} (0.8)^j$
16. $\lim_{j \to \infty} \dfrac{3^j}{2^j}$
17. $\lim_{n \to \infty} e^{-n}$
18. $\lim_{n \to \infty} e^{1/n}$
19. $\lim_{n \to \infty} \dfrac{n+3}{n^2-2}$
20. $\lim_{n \to \infty} \dfrac{5n^2+1}{4-3n^2}$
21. $\lim_{n \to \infty} \dfrac{2n^2-4}{-n-5}$
22. $\lim_{n \to \infty} \cos\dfrac{\pi}{n}$
23. $\lim_{n \to \infty} n \sin\dfrac{\pi}{n}$
24. $\lim_{n \to \infty} \ln\dfrac{1}{n}$
25. $\lim_{n \to \infty} \left(1 + \dfrac{0.05}{n}\right)^n$
26. $\lim_{k \to \infty} \left(1 + \dfrac{1}{3k}\right)^k$
27. $\lim_{k \to \infty} (1+k)^{1/(2k)}$
28. $\lim_{k \to \infty} \arctan k$
29. $\lim_{k \to \infty} \arcsin\left(\dfrac{1}{\sqrt{2}}\cos\dfrac{1}{k}\right)$
30. $\lim_{k \to \infty} \sqrt[k]{2k}$
31. $\lim_{n \to \infty} \int_{-1/n}^{1/n} e^x \, dx$
32. $\lim_{n \to \infty} \int_{1+1/n}^{2-1/n} \dfrac{1}{x} \, dx$

In Exercises 33–38 determine whether the sequence converges or diverges. If it converges, find its limit.

33. $\{-4n\}_{n=-2}^{\infty}$
34. $\left\{\dfrac{n-1}{n}\right\}_{n=1}^{\infty}$
35. $\left\{\dfrac{1}{n^2-1}\right\}_{n=2}^{\infty}$
36. $\{2^n\}_{n=1}^{\infty}$
37. $\left\{\left(-\dfrac{1}{3}\right)^n\right\}_{n=1}^{\infty}$
38. $\left\{\dfrac{1}{n} - n\right\}_{n=2}^{\infty}$

39. Show that if
$$a_n = \frac{1}{n^2} + \frac{2}{n^2} + \frac{3}{n^2} + \cdots + \frac{n}{n^2}$$
then a_n is a Riemann sum for $\int_0^1 x \, dx$ for each $n \geq 1$. Find $\lim_{n \to \infty} a_n$.

40. Show that if
$$a_n = \frac{1^2}{n^3} + \frac{2^2}{n^3} + \frac{3^2}{n^3} + \cdots + \frac{n^2}{n^3}$$
then a_n is a Riemann sum for $\int_0^1 x^2 \, dx$ for each $n \geq 1$. Find $\lim_{n \to \infty} a_n$.

41. Prove that a convergent sequence has a unique limit.

*42. Prove that if $\lim_{n \to \infty} a_{2n} = \lim_{n \to \infty} a_{2n+1}$, then $\lim_{n \to \infty} a_n$ exists and
$$\lim_{n \to \infty} a_n = \lim_{n \to \infty} a_{2n} = \lim_{n \to \infty} a_{2n+1}$$

43. Suppose the number of bacteria in a culture is growing exponentially, with a doubling time of 10 hours. Suppose also that there are initially 1000 bacteria in the culture. Find a formula for the number a_n of bacteria in the culture after n hours.

44. If \$1000 is deposited in a savings account at an interest rate of r percent per year, then the number of dollars (principal plus interest) in the account after 1 year is $1000(1 + 0.01r)$. Write a formula for the sequence that gives the amount of money in the account after n years for any positive integer n.

45. Suppose P dollars are deposited in a savings account at an interest rate of r percent per year, compounded n times a year.
   a. Show that the amount of money in the account after 1

year is given by

$$R_n = P\left(1 + \frac{0.01r}{n}\right)^n$$

dollars (*Hint:* The amount after $1/n$ years is $P(1 + 0.01r/n)$.)
   b. Find

$$R = \lim_{n \to \infty} P\left(1 + \frac{0.01r}{n}\right)^n$$

This is the amount in the account after 1 year if the interest is compounded "continuously."
   c. If $P = 1000$ and $r = 5$, find the difference between the amounts after 1 year if the interest is compounded continuously and if it is compounded quarterly.

9.3
CONVERGENCE PROPERTIES OF SEQUENCES

In this section we continue our analysis of the convergence and divergence of sequences. Since sequences are functions, we may add, subtract, multiply, and divide sequences just as we do functions. Rules for computing the limits of combinations of sequences are analogous to the rules for limits of combinations of functions. We present these rules now.

Let $\{a_n\}_{n=m}^{\infty}$ and $\{b_n\}_{n=m}^{\infty}$ be convergent sequences. Then the sum $\{a_n + b_n\}_{n=m}^{\infty}$, any scalar multiple $\{ca_n\}_{n=m}^{\infty}$, the product $\{a_nb_n\}_{n=m}^{\infty}$, and (provided $\lim_{n \to \infty} b_n \neq 0$) the quotient $\{a_n/b_n\}_{n=m}^{\infty}$ all converge, with

$$\lim_{n \to \infty} (a_n + b_n) = \lim_{n \to \infty} a_n + \lim_{n \to \infty} b_n \tag{1}$$

$$\lim_{n \to \infty} ca_n = c \lim_{n \to \infty} a_n \tag{2}$$

$$\lim_{n \to \infty} a_nb_n = \lim_{n \to \infty} a_n \lim_{n \to \infty} b_n \tag{3}$$

$$\lim_{n \to \infty} \frac{a_n}{b_n} = \frac{\lim_{n \to \infty} a_n}{\lim_{n \to \infty} b_n} \tag{4}$$

Example 1 Find $\lim_{n \to \infty} \frac{2n - 1}{3n + 5}$.

Solution We divide the numerator and the denominator by n and then use (1)–(4), along with Examples 1 and 2 of Section 9.2. This yields

$$\lim_{n \to \infty} \frac{2n - 1}{3n + 5} = \lim_{n \to \infty} \frac{2 - 1/n}{3 + 5/n} = \frac{\lim_{n \to \infty} 2 - \lim_{n \to \infty} 1/n}{\lim_{n \to \infty} 3 + 5 \lim_{n \to \infty} 1/n} = \frac{2 - 0}{3 + 0} = \frac{2}{3} \quad \square$$

The version of the Squeezing Theorem for sequences is as follows:

If $\lim_{n \to \infty} a_n = \lim_{n \to \infty} b_n$, and if $\{c_n\}_{n=m}^{\infty}$ is any sequence such that $a_n \leq c_n \leq b_n$ for $n \geq m$, then $\{c_n\}_{n=m}^{\infty}$ converges, and moreover,

$$\lim_{n \to \infty} c_n = \lim_{n \to \infty} a_n = \lim_{n \to \infty} b_n \tag{5}$$

Example 2   Show that $\lim_{n \to \infty} \dfrac{\sin^2 n}{n} = 0$.

Solution   We observe that $0 \leq \sin^2 n \leq 1$, so that

$$0 \leq \frac{\sin^2 n}{n} \leq \frac{1}{n}$$

Since $\lim_{n \to \infty} 0 = 0$ and $\lim_{n \to \infty} 1/n = 0$, it follows from the Squeezing Theorem for sequences that

$$\lim_{n \to \infty} \frac{\sin^2 n}{n} = 0 \quad \square$$

Example 3   Show that $\lim_{n \to \infty} \dfrac{\ln n}{n} = 0$.

Solution   First we will show that for $n \geq 1$ we have $0 \leq (\ln n)/n \leq 2/\sqrt{n}$. To that end we observe that for $t \geq 1$, we have $1/t \leq 1/\sqrt{t}$, so that

$$\ln n = \int_1^n \frac{1}{t} \, dt \leq \int_1^n \frac{1}{\sqrt{t}} \, dt = 2\sqrt{t} \Big|_1^n = 2(\sqrt{n} - 1) \leq 2\sqrt{n}$$

Therefore

$$0 \leq \frac{\ln n}{n} \leq \frac{2\sqrt{n}}{n} = \frac{2}{\sqrt{n}} \tag{6}$$

Since $\lim_{n \to \infty} 0 = 0$ and $\lim_{n \to \infty} 2/\sqrt{n} = 0$, it follows from (5) that

$$\lim_{n \to \infty} \frac{\ln n}{n} = 0 \quad \square$$

Example 4   Show that $\lim_{n \to \infty} \left(1 + 1 + \dfrac{1}{2!} + \dfrac{1}{3!} + \cdots + \dfrac{1}{n!} \right) = e$.

Solution   Recall from (8) of Section 9.1 that

$$\left| e - \left(1 + 1 + \frac{1}{2!} + \frac{1}{3!} + \cdots + \frac{1}{n!} \right) \right| < \frac{4}{(n+1)!}$$

Since

$$0 \leq \frac{4}{(n+1)!} < \frac{4}{n}$$

for every integer $n \geq 1$, and since $\lim_{n \to \infty} 0 = 0$ and $\lim_{n \to \infty} 4/n = 0$, it follows from (5) that

$$\lim_{n \to \infty} \left[e - \left(1 + 1 + \frac{1}{2!} + \frac{1}{3!} + \cdots + \frac{1}{n!} \right) \right] = 0$$

or equivalently,

$$\lim_{n \to \infty} \left(1 + 1 + \frac{1}{2!} + \frac{1}{3!} + \cdots + \frac{1}{n!} \right) = e \quad \square$$

Example 5 Show that $\lim_{n \to \infty} \left(1 - \frac{1}{2} + \frac{1}{3} - \frac{1}{4} + \cdots + (-1)^{n-1} \frac{1}{n} \right) = \ln 2$.

Solution From (10) of Section 9.1 we know that

$$\left| \ln 2 - \left(1 - \frac{1}{2} + \frac{1}{3} - \frac{1}{4} + \cdots + (-1)^{n-1} \frac{1}{n} \right) \right| < \frac{1}{n+1} < \frac{1}{n}$$

so that, by the Squeezing Theorem,

$$\lim_{n \to \infty} \left[\ln 2 - \left(1 - \frac{1}{2} + \frac{1}{3} - \frac{1}{4} + \cdots + (-1)^{n-1} \frac{1}{n} \right) \right] = 0$$

Thus

$$\lim_{n \to \infty} \left(1 - \frac{1}{2} + \frac{1}{3} - \frac{1}{4} + \cdots + (-1)^{n-1} \frac{1}{n} \right) = \ln 2 \quad \square$$

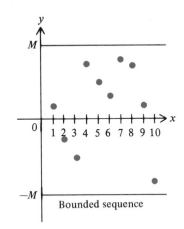

FIGURE 9.5

Bounded sequence

Bounded Sequences In analogy with boundedness for a function (see Section 8.7), we say that a sequence $\{a_n\}_{n=m}^{\infty}$ is **bounded** if there is a number M such that $|a_n| \le M$ for every $n \ge m$ (Figure 9.5). Otherwise, we say that the sequence is **unbounded**. In this terminology the sequences $\{1/n\}_{n=2}^{\infty}$ and $\{(-1)^n\}_{n=1}^{\infty}$ are bounded, whereas the sequence $\{n^2\}_{n=4}^{\infty}$ is unbounded.

Now we show that convergent sequences are always bounded, or equivalently, that unbounded sequences are always divergent.

THEOREM 9.6

> **a.** If $\{a_n\}_{n=m}^{\infty}$ converges, then $\{a_n\}_{n=m}^{\infty}$ is bounded.
> **b.** If $\{a_n\}_{n=m}^{\infty}$ is unbounded, then $\{a_n\}_{n=m}^{\infty}$ diverges.

Proof To prove (a), suppose $\lim_{n \to \infty} a_n = L$, which implies that there is an N such that

$$\text{if } n \ge N, \quad \text{then} \quad |a_n - L| < 1$$

Therefore

$$\text{if } n \ge N, \quad \text{then} \quad |a_n| = |a_n - L + L| \le |a_n - L| + |L| < 1 + |L|$$

Let M be a number larger than $|a_m|, |a_{m+1}|, \ldots, |a_{N-1}|$, and $1 + |L|$. Then $|a_n| \le M$ for $n \ge m$, and thus the sequence is bounded. Since part (b) is logically equivalent to (a), the proof of the theorem is complete. ∎

Theorem 9.6 tells us immediately that the unbounded sequences

$$\{2^n\}_{n=0}^{\infty} \quad \text{and} \quad \{\ln n\}_{n=2}^{\infty}$$

diverge.

Caution: Theorem 9.6 does *not* imply that all bounded sequences converge, and indeed that is not the case. For example, $\{(-1)^n\}_{n=1}^{\infty}$ is bounded, but it diverges, as we proved in Example 3 of Section 9.2.

We say that the sequence $\{a_n\}_{n=m}^{\infty}$ is **increasing** if as a function it is increasing. This is equivalent to saying that $a_n \le a_{n+1}$ for each $n \ge m$ (Figure 9.6(a)). Similarly, $\{a_n\}_{n=m}^{\infty}$ is **decreasing** if $a_n \ge a_{n+1}$ for each $n \ge m$ (Figure 9.6(b)). Thus of the sequences

$$\left\{1 - \frac{1}{n}\right\}_{n=1}^{\infty}, \qquad \left\{1 + 1 + \frac{1}{2!} + \frac{1}{3!} + \cdots + \frac{1}{n!}\right\}_{n=1}^{\infty}, \quad \text{and} \quad \left\{\sin\frac{1}{n}\right\}_{n=1}^{\infty}$$

the first and second are increasing, and the third is decreasing. As with other types of functions, a sequence need not be increasing or decreasing. For example, $\{(-1)^n\}_{n=m}^{\infty}$ is neither increasing nor decreasing, because it oscillates between 1 and -1.

Although a bounded sequence may diverge, a bounded sequence that is either increasing or decreasing must converge. The proof of this assertion is given in the Appendix, but Figure 9.6(c) suggests why we might expect it to be true.

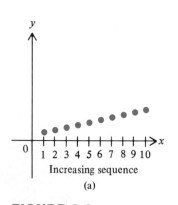

Increasing sequence
(a)

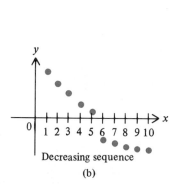

Decreasing sequence
(b)

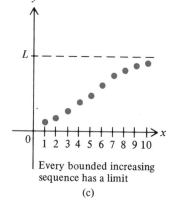

Every bounded increasing sequence has a limit
(c)

FIGURE 9.6

THEOREM 9.7

A bounded sequence $\{a_n\}_{n=m}^{\infty}$ that is either increasing or decreasing converges. If the sequence is increasing, then the limit is the smallest number L such that $a_n \le L$ for $n \ge m$. If the sequence is decreasing, then the limit is the largest number L such that $a_n \ge L$ for $n \ge m$.

Although Theorem 9.7 will be of theoretical interest to us (especially in Section 9.7), it can be used in showing that certain sequences converge.

Example 6 Let

$$a_n = 1 + \frac{1}{1 \cdot 1!} + \frac{1}{2 \cdot 2!} + \frac{1}{3 \cdot 3!} + \cdots + \frac{1}{n \cdot n!} \quad \text{for } n \geq 1$$

Show that $\{a_n\}_{n=1}^{\infty}$ converges.

Solution Since

$$a_{n+1} = a_n + \frac{1}{(n+1)(n+1)!} > a_n$$

the sequence is increasing. To show that it is bounded, we use Example 4 and the observation that

$$0 < a_n = 1 + \frac{1}{1 \cdot 1!} + \frac{1}{2 \cdot 2!} + \frac{1}{3 \cdot 3!} + \cdots + \frac{1}{n \cdot n!}$$

$$< 1 + 1 + \frac{1}{2!} + \frac{1}{3!} + \cdots + \frac{1}{n!} < e$$

Since $\{a_n\}_{n=1}^{\infty}$ is increasing and bounded, Theorem 9.7 implies that the sequence converges. □

If $m_2 > m_1$, then the two sequences $\{a_n\}_{n=m_1}^{\infty}$ and $\{a_n\}_{n=m_2}^{\infty}$ either both converge to the same limit or both diverge. (This is true because we can always take the integer N in Definition 9.3 or 9.4 to be larger than m_2.) Consequently in determining whether a sequence converges or diverges, we may ignore the first few terms and apply any of our results to the portion of the sequence that remains. For example, the sequence

$$-\pi, 10, \pi, \sqrt{2}, 1, \frac{1}{2}, \frac{1}{3}, \frac{1}{4}, \frac{1}{5}, \cdots$$

converges to 0 because if the first four terms are removed, we have the sequence

$$1, \frac{1}{2}, \frac{1}{3}, \frac{1}{4}, \frac{1}{5}, \cdots$$

which we know converges to 0.

It follows from the preceding discussion that if $\{a_n\}_{n=1}^{\infty}$ converges, and if k is any positive integer, then the sequence $\{a_{n+k}\}_{n=1}^{\infty}$ converges, and moreover

$$\lim_{n \to \infty} a_{n+k} = \lim_{n \to \infty} a_n$$

For instance, taking $a_n = 1/n$ and $k = 2$, we find that

$$\lim_{n \to \infty} \frac{1}{n+2} = \lim_{n \to \infty} \frac{1}{n} = 0$$

and taking $a_n = r^n$ and $k = -1$, we find that

$$\lim_{n \to \infty} r^{n-1} = \lim_{n \to \infty} r^n = 0 \quad \text{for } -1 < r < 1$$

EXERCISES 9.3

In Exercises 1–6 determine whether the sequence is bounded or unbounded.

1. $\left\{1 + \dfrac{2}{n}\right\}_{n=1}^{\infty}$

2. $\{e^{1/n}\}_{n=1}^{\infty}$

3. $\left\{n + \dfrac{1}{2n}\right\}_{n=1}^{\infty}$

4. $\{\cos n^2\}_{n=0}^{\infty}$

5. $\{\cosh k\}_{k=10}^{\infty}$

6. $\{\operatorname{sech} k\}_{k=1}^{\infty}$

In Exercises 7–28 find the limit.

7. $\lim\limits_{n \to \infty}\left(2 + \dfrac{1}{n}\right)$

8. $\lim\limits_{n \to \infty}\left(4 - \dfrac{2}{n}\right)$

9. $\lim\limits_{n \to \infty}\left(\dfrac{1}{n} - \dfrac{1}{n+1}\right)$

10. $\lim\limits_{n \to \infty}\left(1 - \dfrac{1}{n}\right)\left(1 + \dfrac{1}{n^2}\right)$

11. $\lim\limits_{n \to \infty} \dfrac{2^{n+1}}{5^{n+2}}$

12. $\lim\limits_{n \to \infty} \dfrac{2^{n-1} + 3}{3^{n+2}}$

13. $\lim\limits_{k \to \infty} \dfrac{k \mid 1}{k}$

14. $\lim\limits_{k \to \infty} \dfrac{2k - 1}{k^2 - 7}$

15. $\lim\limits_{n \to \infty} \dfrac{4n^3 - 5}{6n^3 + 3}$

16. $\lim\limits_{n \to \infty} \sqrt[n]{5}\left(\dfrac{n - 1}{n + 1}\right)$

17. $\lim\limits_{n \to \infty} \sqrt[n]{3n}$

18. $\lim\limits_{n \to \infty} \dfrac{\sqrt{n^2 - 1}}{2n}$

19. $\lim\limits_{n \to \infty} \dfrac{\sqrt{n} + 1}{\sqrt{n} - 1}$

20. $\lim\limits_{n \to \infty} (\sqrt{n + 1} - \sqrt{n})$

21. $\lim\limits_{n \to \infty} \sqrt{n}(\sqrt{n + 1} - \sqrt{n})$

22. $\lim\limits_{n \to \infty} \dfrac{1 - (1 + 1/n)^2}{1 - (1 + 1/n)}$

23. $\lim\limits_{n \to \infty} \dfrac{(-1)^n}{n}$

24. $\lim\limits_{n \to \infty} \dfrac{\sin^2 n}{n}$

25. $\lim\limits_{n \to \infty} \dfrac{1}{ne^n}$

26. $\lim\limits_{n \to \infty} \dfrac{1}{n^2(n^2 + 1)}$

27. $\lim\limits_{j \to \infty} \dfrac{\ln\left(1 + \dfrac{1}{j}\right)}{j}$

28. $\lim\limits_{n \to \infty} \sqrt[n]{\ln n}$ (*Hint*:

$1 \le \ln n \le n$ for $n \ge 3$.)

29. From Section 7.7 we know that

$$\tanh x = \dfrac{\sinh x}{\cosh x} = \dfrac{e^x - e^{-x}}{e^x + e^{-x}} = \dfrac{1 - e^{-2x}}{1 + e^{-2x}}$$

Using this formula, prove that $\lim_{n \to \infty} \tanh n = 1$.

30. a. Assume that $\lim_{n \to \infty} a_n = 0$ and that $\{b_n\}_{n=m}^{\infty}$ is bounded. Show that $\lim_{n \to \infty} a_n b_n = 0$. (*Hint*: Suppose that $|b_n| \le M$ for all n, and show that for any $\varepsilon > 0$, $|a_n|$ is eventually less than ε/M.)

b. Let $\{c_n\}_{n=1}^{\infty}$ be a sequence. Using part (a), show that if $\lim_{n \to \infty} |c_n| = 0$, then $\lim_{n \to \infty} c_n = 0$. (*Hint*: Let $a_n = |c_n|$, and take b_n to be 1 or -1, according to whether $c_n \ge 0$ or $c_n < 0$.)

c. Using part (a), verify the following limits.

   i. $\lim\limits_{n \to \infty} \dfrac{\sin n}{n} = 0$

   ii. $\lim\limits_{n \to \infty} \dfrac{1}{n^2} \ln\left(1 + \dfrac{(-1)^n}{n}\right) = 0$

   iii. $\lim\limits_{n \to \infty} \dfrac{2 + (-1)^n}{e^n} = 0$

   *iv. $\lim\limits_{n \to \infty} \dfrac{2n + (-1)^n}{e^{2n}} = 0$

31. If the sequence $\{a_n\}_{n=1}^{\infty}$ converges and the sequence $\{b_n\}_{n=1}^{\infty}$ diverges, what can you say about the sequence $\{a_n + b_n\}_{n=1}^{\infty}$?

*32. Let $\{a_n\}_{n=1}^{\infty}$ be the sequence $\sqrt{2}$, $\sqrt{2 + \sqrt{2}}$, $\sqrt{2 + \sqrt{2 + \sqrt{2}}}, \ldots$, where in general,

$$a_{n+1} = \sqrt{2 + a_n}$$

a. Show that $\{a_n\}_{n=1}^{\infty}$ is a bounded increasing sequence. (*Hint*: Show that if $a_n \le 2$, then $a_n \le a_{n+1} \le 2$.)

b. Using (a) and Theorem 9.7, show that $\{a_n\}_{n=1}^{\infty}$ converges to a number r.

c. Using (b) and the fact that

$$a_{n+1}^2 = 2 + a_n$$

show that $r = 2$.

*33. Let $\{a_n\}_{n=1}^{\infty}$ be the sequence $\sqrt{2}$, $(\sqrt{2})^{\sqrt{2}}$, $(\sqrt{2})^{(\sqrt{2}^{\sqrt{2}})}, \ldots$ where in general,

$$a_{n+1} = (\sqrt{2})^{a_n}$$

a. Show that $\{a_n\}_{n=1}^{\infty}$ converges to a number $L \le 2$. (*Hint*: Show that $a_n < 2$ for all n, and use Exercise 65(a) of Section 5.7 to show that $\{a_n\}_{n=1}^{\infty}$ is increasing.)

b. Show that $L = 2$. (*Hint*: Show that $L = (\sqrt{2})^L$, and use Exercise 65(a) of Section 5.7 again.)

*34. Let $a_n = 1 + \frac{1}{2} + \cdots + 1/n - \ln n$ for $n \ge 1$.

a. Using the definition of the natural logarithm, show that

$$\ln(n + 1) - \ln n \ge \dfrac{1}{n + 1} \quad \text{for } n \ge 1$$

b. Using (a), show that the sequence $\{a_n\}_{n=1}^{\infty}$ is decreasing.

c. Using the fact that

$$\int_1^n \frac{1}{t}\, dt = \int_1^2 \frac{1}{t}\, dt + \int_2^3 \frac{1}{t}\, dt + \cdots$$
$$+ \int_{n-1}^n \frac{1}{t}\, dt \quad \text{for } n \geq 2$$

show that

$$\ln n \leq 1 + \frac{1}{2} + \cdots + \frac{1}{n-1}$$

and hence that $a_n \geq 0$ for all n.

d. Show that $\{a_n\}_{n=1}^\infty$ converges.

The limit c to which the sequence converges is called the **Euler-Mascheroni constant** and has been calculated to be approximately 0.577216 (accurate to six places).

*35. Suppose a rabbit colony begins with one pair of adult rabbits. Assume that every pair of adult rabbits produces two offspring every month (one male and one female) and that rabbits become adults at the age of 2 months and live forever. The problem is to find how many pairs of adult rabbits there are after n months. If we let a_n be the number of pairs of adult rabbits after n months have passed, then we have $a_1 = 1$, $a_2 = 1$, and

$$a_{n+1} = a_n + a_{n-1} \quad \text{for } n \geq 2 \qquad (7)$$

The sequence $\{a_n\}_{n=1}^\infty$ is called the **Fibonacci sequence**, after the Italian mathematician known as Fibonacci (his real name was Leonardo de Pisa), who in 1202 posed and solved the problem just described. The first 14 terms of the sequence are

$$1, 1, 2, 3, 5, 8, 13, 21, 34, 55, 89, 144, 233, 377$$

Now let

$$b_n = \frac{a_{n+1}}{a_n} \quad \text{for } n \geq 1 \qquad (8)$$

Assuming that $\lim_{n \to \infty} b_n$ exists and that $\lim_{n \to \infty} b_n = b$, show that

$$b = \frac{1 + \sqrt{5}}{2}$$

This means that after many years the number of pairs of adult rabbits in the colony would increase by about 62 percent every month. (*Hint:* By substituting (8) into (7), show that

$$b_n = 1 + \frac{1}{b_{n-1}}$$

Then deduce that $b = 1 + 1/b$ and solve for b.)

9.4
INFINITE SERIES

Of course, we know how to add a finite collection of numbers. But suppose we wish to add the inifinite collection of numbers $\frac{1}{2}, \frac{1}{4}, \frac{1}{8}, \frac{1}{16}, \ldots$. If we begin adding them in the order of their appearance, we obtain successively the sums $\frac{1}{2}, \frac{3}{4}, \frac{7}{8}, \frac{15}{16}, \ldots$, which seem to approach 1. Thus it is reasonable to define the sum

$$\frac{1}{2} + \frac{1}{4} + \frac{1}{8} + \frac{1}{16} + \cdots$$

to be 1. Similarly, since we showed in Example 4 of Section 9.3 that

$$\lim_{n \to \infty} \left(1 + 1 + \frac{1}{2!} + \frac{1}{3!} + \cdots + \frac{1}{n!}\right) = e$$

it is natural to define the sum

$$1 + 1 + \frac{1}{2!} + \frac{1}{3!} + \cdots$$

to be e.

In general, to determine whether it is possible to define the sum of the numbers in a sequence $\{a_n\}_{n=1}^{\infty}$, we simply begin adding the numbers in the order of their indexes $(a_1, a_2, a_3, \ldots)$ and ascertain whether the resulting sums approach a limit. The sums obtained, which are

$$s_1 = a_1$$

$$s_2 = a_1 + a_2$$

$$s_3 = a_1 + a_2 + a_3$$

$$\vdots$$

are called partial sums. If the sequence of partial sums has a limit, we call that limit the sum and denote it $\sum_{n=1}^{\infty} a_n$. But whether the partial sums approach a limit or not, we use the expressions $\sum_{n=1}^{\infty} a_n$ and $a_1 + a_2 + a_3 + \cdots$ to indicate the numbers to be added. Each such expression is called a series. Examples of series are

$$\sum_{n=1}^{\infty} \frac{1}{2^n} = \frac{1}{2} + \frac{1}{4} + \frac{1}{8} + \frac{1}{16} + \cdots$$

and

$$\sum_{n=1}^{\infty} (-1)^n = -1 + 1 - 1 + 1 - 1 + \cdots$$

So far we have discussed partial sums of the sequence $\{a_n\}_{n=1}^{\infty}$ and the sum of the series $\sum_{n=1}^{\infty} a_n$. Corresponding concepts for $\{a_n\}_{n=m}^{\infty}$ (where the initial index m need not be 1) are defined analogously, and appear in Definition 9.8.

DEFINITION 9.8

Let $\{a_n\}_{n=m}^{\infty}$ be a sequence. For each positive integer j, the **jth partial sum** s_j is the sum of the first j terms of the sequence. If $\lim_{j \to \infty} s_j$ exists, we say that the series $\sum_{n=m}^{\infty} a_n$ **converges** and we call $\lim_{j \to \infty} s_j$ the **sum of the series**. Otherwise, we say that the series $\sum_{n=m}^{\infty} a_n$ **diverges**. The numbers $a_m, a_{m+1}, a_{m+2}, \ldots$ are the **terms** of the series $\sum_{n=m}^{\infty} a_n$. Finally, if $\sum_{n=m}^{\infty} a_n$ converges, we also use the expression $\sum_{n=m}^{\infty} a_n$ to denote the sum of the series.

Since the limit of a sequence is unique, it follows from Definition 9.8 that the sum of a convergent series is unique.

Almost all series we will consider are of the form $\sum_{n=1}^{\infty} a_n$ or $\sum_{n=0}^{\infty} a_n$. For $\sum_{n=1}^{\infty} a_n$ the jth partial sum is

$$s_j = a_1 + a_2 + \cdots + a_j$$

and for $\sum_{n=0}^{\infty} a_n$ the jth partial sum is

$$s_j = a_0 + a_1 + \cdots + a_{j-1}$$

Example 1   Show that $\displaystyle\sum_{n=0}^{\infty} \frac{1}{n!} = 1 + 1 + \frac{1}{2!} + \frac{1}{3!} + \cdots = e.$

Solution   In Example 4 of Section 9.3 we showed that

$$\lim_{n \to \infty} \left(1 + 1 + \frac{1}{2!} + \frac{1}{3!} + \cdots + \frac{1}{n!} \right) = e \tag{1}$$

If $a_n = 1/n!$ for $n \geq 0$, then

$$s_j = a_0 + a_1 + a_2 + \cdots + a_{j-1} = 1 + 1 + \frac{1}{2!} + \cdots + \frac{1}{(j-1)!}$$

and (1) implies that $\lim_{j \to \infty} s_j = e$. Thus $\sum_{n=0}^{\infty} 1/n! = e$. $\square$

Example 2   Show that $\sum_{n=1}^{\infty} (-1)^{n-1} \dfrac{1}{n} = 1 - \dfrac{1}{2} + \dfrac{1}{3} - \dfrac{1}{4} + \cdots = \ln 2$.

Solution   In Example 5 of Section 9.3 we proved that

$$\lim_{n \to \infty} \left(1 - \frac{1}{2} + \frac{1}{3} - \frac{1}{4} + \cdots + (-1)^{n-1} \frac{1}{n} \right) = \ln 2 \tag{2}$$

If $a_n = (-1)^{n-1}/n$, for $n \geq 1$, then

$$s_j = a_1 + a_2 + a_3 + \cdots + a_j = 1 - \frac{1}{2} + \frac{1}{3} - \frac{1}{4} + \cdots + (-1)^{j-1} \frac{1}{j}$$

and (2) implies that $\lim_{j \to \infty} s_j = \ln 2$. Thus

$$\sum_{n=1}^{\infty} (-1)^{n-1} \frac{1}{n} = \ln 2 \quad \square$$

Example 3   Show that

$$\sum_{n=1}^{\infty} \frac{1}{n(n+1)} = \frac{1}{1 \cdot 2} + \frac{1}{2 \cdot 3} + \frac{1}{3 \cdot 4} + \cdots$$

converges, and find its sum.

Solution   Observe that

$$\frac{1}{n(n+1)} = \frac{1}{n} - \frac{1}{n+1} \quad \text{for } n \geq 1$$

and hence

$$s_j = \left(1 - \frac{1}{2} \right) + \left(\frac{1}{2} - \frac{1}{3} \right) + \left(\frac{1}{3} - \frac{1}{4} \right) + \cdots + \left(\frac{1}{j-1} - \frac{1}{j} \right) + \left(\frac{1}{j} - \frac{1}{j+1} \right) \tag{3}$$

Thus
$$s_j = 1 - \frac{1}{j+1}$$

The expression in (3) could be simplified because adjacent pairs of numbers cancel each other. Since

$$\lim_{j \to \infty} s_j = \lim_{j \to \infty} \left(1 - \frac{1}{j+1} \right) = 1$$

we have simultaneously proved that the given series converges and found the sum of the series:

$$\sum_{n=1}^{\infty} \frac{1}{n(n+1)} = \lim_{j \to \infty} s_j = 1 \quad \square$$

The series $\sum_{n=1}^{\infty} 1/n(n+1)$ is called a **telescoping series**, because when we write the partial sums as in (3), all but the first and last terms cancel.

Example 4 Show that the series

$$\sum_{n=1}^{\infty} \frac{1}{n} = 1 + \frac{1}{2} + \frac{1}{3} + \cdots$$

diverges.

Solution We will show that

$$s_{2^j} \geq 1 + j\left(\frac{1}{2}\right) \quad \text{for each } j \tag{4}$$

and thus that the sequence of partial sums is unbounded and must diverge. To prove (4), we observe that

$$s_{2^1} = s_2 = 1 + \frac{1}{2}$$

$$s_{2^2} = s_4 = 1 + \frac{1}{2} + \frac{1}{3} + \frac{1}{4} \geq 1 + \frac{1}{2} + \overbrace{\frac{1}{4} + \frac{1}{4}}^{\frac{1}{2}} = 1 + \frac{2}{2}$$

$$s_{2^3} = s_8 = 1 + \frac{1}{2} + \frac{1}{3} + \frac{1}{4} + \frac{1}{5} + \frac{1}{6} + \frac{1}{7} + \frac{1}{8}$$

$$\geq 1 + \frac{1}{2} + \overbrace{\frac{1}{4} + \frac{1}{4}}^{\frac{1}{2}} + \overbrace{\frac{1}{8} + \frac{1}{8} + \frac{1}{8} + \frac{1}{8}}^{\frac{1}{2}} \geq 1 + \frac{3}{2}$$

In general we arrange the terms making up s_{2^j} into several groups and then substitute smaller values for the terms so that each group has sum $\frac{1}{2}$:

$$s_{2^j} = 1 + \frac{1}{2} + \left(\frac{1}{3} + \frac{1}{4} \right) + \left(\frac{1}{5} + \frac{1}{6} + \frac{1}{7} + \frac{1}{8} \right) + \cdots + \left(\frac{1}{2^{j-1}+1} + \cdots + \frac{1}{2^j} \right)$$

$$\geq 1 + \frac{1}{2} + \overbrace{\frac{1}{4} + \frac{1}{4}}^{\frac{1}{2}} + \overbrace{\frac{1}{8} + \frac{1}{8} + \frac{1}{8} + \frac{1}{8}}^{\frac{1}{2}} + \cdots + \overbrace{\frac{1}{2^j} + \cdots + \frac{1}{2^j}}^{\frac{1}{2}}$$

$$= 1 + j\left(\frac{1}{2}\right)$$

Since
$$\lim_{j \to \infty} \left[1 + j\left(\frac{1}{2}\right) \right] = \infty$$

it follows that the sequence $\{s_j\}_{j=1}^{\infty}$ of partial sums is unbounded, as we wished to prove; consequently $\Sigma_{n=1}^{\infty} 1/n$ diverges. $\square$

As we have just seen, the series $\Sigma_{n=1}^{\infty} 1/n$ diverges even though its terms converge to 0. This series is called the **harmonic series**, because its terms form the harmonic sequence. The divergence of the harmonic series does not imply that the jth partial sum increases rapidly as j increases. On the contrary, s_j grows very slowly. Computer calculations have shown that

$$s_j \geq 20 \quad \text{only if} \quad j \geq 272{,}400{,}600$$

and

$$s_j \geq 100 \quad \text{only if} \quad j \geq 1.5 \times 10^{43} \text{ (approximately)}$$

In fact, if j is very large, then s_j is approximately equal to $c + \ln j$, where c is the Euler–Mascheroni constant, approximately equal to 0.577216. (See Exercise 34 of Section 9.3.)

Comparison of $\sum_{n=m_1}^{\infty} a_n$ **and** $\sum_{n=m_2}^{\infty} a_n$

Let us compare the partial sums of the two series $\Sigma_{n=2}^{\infty} a_n$ and $\Sigma_{n=5}^{\infty} a_n$. Notice that for any positive integer $j \geq 5$,

$$\underbrace{\frac{1}{2^2} + \frac{1}{3^2} + \cdots + \frac{1}{j^2}}_{\text{partial sum of } \Sigma_{n=2}^{\infty} 1/n^2} = \left(\frac{1}{2^2} + \frac{1}{3^2} + \frac{1}{4^2}\right) + \underbrace{\left(\frac{1}{5^2} + \frac{1}{6^2} + \cdots + \frac{1}{j^2}\right)}_{\text{partial sum of } \Sigma_{n=5}^{\infty} 1/n^2} \quad (5)$$

From (5) it follows that if $\Sigma_{n=2}^{\infty} 1/n^2$ converges, then so does $\Sigma_{n=5}^{\infty} 1/n^2$, and vice versa. Moreover, if the two series converge, their sums are related by

$$\sum_{n=2}^{\infty} \frac{1}{n^2} = \left(\frac{1}{2^2} + \frac{1}{3^2} + \frac{1}{4^2}\right) + \sum_{n=5}^{\infty} \frac{1}{n^2}$$

A similar argument would show that if $m_2 > m_1$, then the series $\Sigma_{n=m_1}^{\infty} a_n$ converges if and only if the series $\Sigma_{n=m_2}^{\infty} a_n$ converges. In addition, if the two series converge, then

$$\sum_{n=m_1}^{\infty} a_n = (a_{m_1} + a_{m_1+1} + \cdots + a_{m_2-1}) + \sum_{n=m_2}^{\infty} a_n \quad (6)$$

Thus the index m at which the summation of the series $\Sigma_{n=m}^{\infty} a_n$ begins is irrelevant to the convergence of the series—although, as (6) indicates, the actual sum of the series *is* affected by the index m. For that reason, although most of our theorems will be stated for series of the form $\Sigma_{n=1}^{\infty} a_n$ (with initial index 1), all these theorems can be applied to a series $\Sigma_{n=m}^{\infty} a_n$ with arbitrary initial index m.

A Divergence Test   Our first theorem about series will tell us immediately that certain series diverge.

THEOREM 9.9

> **a.** If $\sum_{n=1}^{\infty} a_n$ converges, then $\lim_{n \to \infty} a_n = 0$.
> **b.** If $\lim_{n \to \infty} a_n$ is not 0 (or does not exist), then $\sum_{n=1}^{\infty} a_n$ diverges.

Proof   To prove (a), notice that for $n \geq 1$,

$$a_n = (a_1 + a_2 + \cdots + a_n) - (a_1 + a_2 + \cdots + a_{n-1}) = s_n - s_{n-1} \tag{7}$$

Since $\sum_{n=1}^{\infty} a_n$ converges by assumption, we know that $\lim_{n \to \infty} s_n$ exists. Therefore by the comments at the end of Section 9.3, $\lim_{n \to \infty} s_{n-1}$ also exists, and

$$\lim_{n \to \infty} s_{n-1} = \lim_{n \to \infty} s_n$$

Now we conclude from (7) that

$$\lim_{n \to \infty} a_n = \lim_{n \to \infty} (s_n - s_{n-1}) = \lim_{n \to \infty} s_n - \lim_{n \to \infty} s_{n-1} = 0$$

This proves (a). Since (b) is logically equivalent to (a), the proof of the theorem is complete. ∎

Theorem 9.9(b) is sometimes called the **nth term test** for divergence. This test tells us immediately that the following series diverge:

$$\sum_{n=1}^{\infty} 61, \quad \sum_{n=1}^{\infty} \left(1 + \frac{1}{n}\right), \quad \sum_{n=-4}^{\infty} n, \quad \sum_{n=0}^{\infty} 2^{(n^2)}, \quad \sum_{n=0}^{\infty} (-1)^n, \quad \sum_{n=1}^{\infty} n \sin \frac{1}{n}$$

Geometric Series   A **geometric series** is a series of the form $\sum_{n=m}^{\infty} cr^n$, where r and c are constants and $c \neq 0$. (The name results from the fact that cr^n is the geometric mean of cr^{n-1} and cr^{n+1}.) The convergence of geometric series depends entirely on the choice of r, as we see next.

THEOREM 9.10
GEOMETRIC SERIES
THEOREM

> Let r be any number, and let $c \neq 0$ and $m \geq 0$. Then the geometric series $\sum_{n=m}^{\infty} cr^n$ converges if and only if $|r| < 1$. For $|r| < 1$,
>
> $$\sum_{n=m}^{\infty} cr^n = \frac{cr^m}{1-r} \tag{8}$$

Proof   We consider the cases $|r| \geq 1$ and $|r| < 1$ separately. If $|r| \geq 1$, then $|cr^n| \geq |c|$ for all $n \geq m$, so the terms do not go to 0; thus by Theorem 9.9(b) the series $\sum_{n=m}^{\infty} cr^n$ diverges. If $|r| < 1$, then we use the identity

$$(1 - r)(1 + r + r^2 + \cdots + r^{j-1}) = 1 - r^j$$

which implies that

$$s_j = cr^m + cr^{m+1} + \cdots + cr^{m+j-1} = cr^m(1 + r + r^2 + \cdots + r^{j-1})$$

$$= cr^m \left(\frac{1 - r^j}{1 - r}\right)$$

(a)

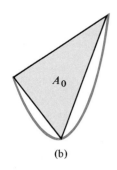

(b)

(c)

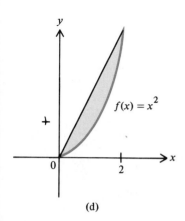

(d)

FIGURE 9.7

Since $\lim_{j \to \infty} r^j = 0$ by Example 6 of Section 9.2, it follows that

$$\lim_{j \to \infty} s_j = \frac{cr^m}{1 - r} \lim_{j \to \infty} (1 - r^j) = \frac{cr^m}{1 - r} \quad \blacksquare$$

The number r is called the **ratio** of the geometric series. By the Geometric Series Theorem, the sum of a convergent geometric series is equal to the first term, cr^m, divided by $1 - r$.

Example 5   Show that $\sum_{n=0}^{\infty} \left(\frac{1}{4}\right)^n = \frac{4}{3}$.

Solution   The given series is a geometric series with ratio $r = \frac{1}{4}$; thus the Geometric Series Theorem asserts that the series converges. Since the first term is $\left(\frac{1}{4}\right)^0 = 1$, it follows that

$$\sum_{n=0}^{\infty} \left(\frac{1}{4}\right)^n = \frac{1}{1 - \frac{1}{4}} = \frac{4}{3} \quad \square$$

The series in Example 5 is historically important. When Archimedes computed the area of a parabolic region such as that drawn in Figure 9.7(a), he began by inscribing an initial triangle with area A_0 (Figure 9.7(b)). Then he inscribed smaller and smaller triangles in the outer portions of the region, as in Figure 9.7(c). He calculated the sum of the areas of the triangles as

$$A = A_0 \left(1 + \frac{1}{4} + \frac{1}{4^2} + \frac{1}{4^3} + \cdots \right) = A_0 \sum_{n=0}^{\infty} \left(\frac{1}{4}\right)^n$$

By Example 5 we know that the sum of the series is $\frac{4}{3}$, so that

$$A = \frac{4}{3} A_0$$

In particular, for the parabolic region described in Figure 9.7(d), it can be shown that $A_0 = 1$, and thus $A = \frac{4}{3}$. This corroborates our results in Section 5.2 (see the comments after formula (2) of Section 5.2). Although Archimedes did not use series in the strict sense, his work involved the germ of this concept.

Example 6   Show that $\sum_{n=2}^{\infty} 14(0.3)^n = 1.8$.

Solution   The given series is a geometric series with ratio $r = 0.3$. Since the first term is $14(0.3)^2$, it follows that

$$\sum_{n=2}^{\infty} 14(0.3)^n = \frac{14(0.3)^2}{1 - 0.3} = 1.8 \quad \square$$

Of course the Geometric Series Theorem also tells us that not all geometric series converge. In fact, $\sum_{n=6}^{\infty} 3(5.4)^n$ diverges, since $r = 5.4 > 1$.

There is an intimate relationship between geometric series and repeating decimals. For example the decimal $0.33333\ldots$ is defined by

$$0.33333\ldots = 3\left(\frac{1}{10}\right) + 3\left(\frac{1}{10}\right)^2 + 3\left(\frac{1}{10}\right)^3 + \cdots = \sum_{n=1}^{\infty} 3\left(\frac{1}{10}\right)^n$$

Similarly

$$0.454545\ldots = 45\left(\frac{1}{100}\right) + 45\left(\frac{1}{100}\right)^2 + 45\left(\frac{1}{100}\right)^3 + \cdots = \sum_{n=1}^{\infty} 45\left(\frac{1}{100}\right)^n$$

Geometric series allow us to express any repeating decimal as a fraction (and hence as a rational number). For example,

$$0.33333\ldots = \sum_{n=1}^{\infty} 3\left(\frac{1}{10}\right)^n = \frac{3(1/10)}{1 - 1/10} = \frac{1}{3}$$

$$0.45454545\ldots = \sum_{n=1}^{\infty} 45\left(\frac{1}{100}\right)^n = \frac{45(1/100)}{1 - 1/100} = \frac{45}{99} = \frac{5}{11}$$

Since this procedure will transform any repeating decimal into a fraction, every repeating decimal is a rational number. (The converse is also true: Every rational number has a repeating decimal expansion.) However, the procedure just demonstrated does not work if the decimal is not repeating. For instance, the number $e = 2.71828\ldots$, which is not a repeating decimal, can be written in the form of a series as

$$2 + 7\left(\frac{1}{10}\right) + 1\left(\frac{1}{10}\right)^2 + 8\left(\frac{1}{10}\right)^3 + 2\left(\frac{1}{10}\right)^4 + 8\left(\frac{1}{10}\right)^5 + \cdots$$

The series converges because its partial sums are increasing and bounded (by, say, 3). In fact, the series converges to e. However, the series is *not* a geometric series.

Combinations of Series

If $\sum_{n=1}^{\infty} a_n$ and $\sum_{n=1}^{\infty} b_n$ are two series, we can add them term by term; we can also multiply all the terms of either series by a single number c. These operations generate two new series, $\sum_{n=1}^{\infty} (a_n + b_n)$ and $\sum_{n=1}^{\infty} ca_n$, whose convergence is guaranteed by the convergence of the original series, as we now prove.

THEOREM 9.11

a. If $\sum_{n=1}^{\infty} a_n$ and $\sum_{n=1}^{\infty} b_n$ converge, then $\sum_{n=1}^{\infty} (a_n + b_n)$ also converges, and

$$\sum_{n=1}^{\infty} (a_n + b_n) = \sum_{n=1}^{\infty} a_n + \sum_{n=1}^{\infty} b_n$$

b. If $\sum_{n=1}^{\infty} a_n$ converges and if c is any number, then $\sum_{n=1}^{\infty} ca_n$ also converges, and

$$\sum_{n=1}^{\infty} ca_n = c \sum_{n=1}^{\infty} a_n$$

Proof To prove (a), let s_j, s'_j, and s''_j be the jth partial sums of $\sum_{n=1}^{\infty} (a_n + b_n)$, $\sum_{n=1}^{\infty} a_n$, and $\sum_{n=1}^{\infty} b_n$, respectively. Then

$$
\begin{aligned}
s_j &= [(a_1 + b_1) + (a_2 + b_2) + \cdots + (a_j + b_j)] \\
&= (a_1 + a_2 + \cdots + a_j) + (b_1 + b_2 + \cdots + b_j) \\
&= s'_j + s''_j
\end{aligned}
$$

As a result,

$$
\sum_{n=1}^{\infty} (a_n + b_n) = \lim_{j \to \infty} s_j = \lim_{j \to \infty} (s'_j + s''_j) = \lim_{j \to \infty} s'_j + \lim_{j \to \infty} s''_j = \sum_{n=1}^{\infty} a_n + \sum_{n=1}^{\infty} b_n
$$

The proof of (b) uses the same ideas, so we leave it as an exercise. ∎

The result of Theorem 9.11(a) can be extended to the sum of three or more series. We also remark that by combining parts (a) and (b) of Theorem 9.11, we can conclude that if $\sum_{n=1}^{\infty} a_n$ and $\sum_{n=1}^{\infty} b_n$ converge, then $\sum_{n=1}^{\infty} (a_n - b_n)$ converges and

$$
\sum_{n=1}^{\infty} (a_n - b_n) = \sum_{n=1}^{\infty} a_n - \sum_{n=1}^{\infty} b_n \tag{9}
$$

Example 7 Show that the series $\sum_{n=1}^{\infty} \left(\dfrac{4}{2^n} - \dfrac{2}{n(n+1)} \right)$ converges, and find its sum.

Solution The Geometric Series Theorem implies that

$$
\sum_{n=1}^{\infty} \frac{4}{2^n} = \frac{4\left(\frac{1}{2}\right)}{1 - \frac{1}{2}} = 4
$$

From Example 3 and Theorem 9.11(b) we know that

$$
\sum_{n=1}^{\infty} \frac{2}{n(n+1)} = 2 \sum_{n=1}^{\infty} \frac{1}{n(n+1)} = 2
$$

Consequently (9) assures us that

$$
\sum_{n=1}^{\infty} \left(\frac{4}{2^n} - \frac{2}{n(n+1)} \right) = 4 - 2 = 2 \quad \square
$$

Changing Indexes It is always possible to regard $\sum_{n=m}^{\infty} a_n$ as a series $\sum_{n=0}^{\infty} b_n$ with initial index 0. We accomplish this by letting

$$
b_0 = a_m
$$
$$
b_1 = a_{m+1}
$$
$$
\vdots
$$
$$
b_n = a_{m+n}
$$

Clearly, the jth partial sum of $\sum_{n=0}^{\infty} b_n$ coincides with the jth partial sum of

$\sum_{n=m}^{\infty} a_n$, for both are the sum of the same j numbers. Therefore $\sum_{n=m}^{\infty} a_n$ and $\sum_{n=0}^{\infty} b_n$ are essentially the same series. In particular, they have the same sum whenever they converge. Since $b_n = a_{m+n}$ for $n \geq 0$, we may also write

$$\sum_{n=m}^{\infty} a_n = \sum_{n=0}^{\infty} a_{m+n} \tag{10}$$

The change of the initial index from m to 0 has been offset by replacing a_n by a_{m+n}. For example,

$$\sum_{n=3}^{\infty} \frac{1}{n!} = \sum_{n=0}^{\infty} \frac{1}{(n+3)!}$$

Caution: It is critical to distinguish between the series $\sum_{n=1}^{\infty} a_n$ and the sequence $\{a_n\}_{n=1}^{\infty}$. If you neglect to maintain this distinction, you are making an error comparable to confusing a book with the pages inside the book.

EXERCISES 9.4

In Exercises 1–5 compute the fourth partial sum of each series.

1. $\sum_{n=1}^{\infty} 1$

2. $\sum_{n=1}^{\infty} n$

3. $\sum_{n=0}^{\infty} \left(\frac{1}{3}\right)^n$

4. $\sum_{n=1}^{\infty} (-1)^n$

5. $\sum_{n=2}^{\infty} \frac{(-1)^n}{n}$

In Exercises 6–14 determine whether the given series must diverge because its terms do not converge to 0.

6. $\sum_{n=1}^{\infty} \left(\frac{-1}{7}\right)^n$

7. $\sum_{n=1}^{\infty} \left(1 + \frac{1}{n}\right)$

8. $\sum_{n=0}^{\infty} \frac{n^2}{n+1}$

9. $\sum_{n=2}^{\infty} (-1)^n \frac{1}{n^2}$

10. $\sum_{n=1}^{\infty} \sin n\pi$

11. $\sum_{n=1}^{\infty} \sin\left(\frac{\pi}{2} - \frac{1}{n}\right)$

12. $\sum_{n=2}^{\infty} \tan\left(\frac{\pi}{2} - \frac{1}{n}\right)$

13. $\sum_{n=1}^{\infty} n \sin \frac{1}{n}$

14. $\sum_{n=1}^{\infty} \left(1 + \frac{1}{n}\right) \ln\left(1 + \frac{1}{n}\right)$

In Exercises 15–22 find a formula for the partial sums of the series. For each series, determine whether the partial sums have a limit. If so, find the sum of the series.

15. $\sum_{n=1}^{\infty} 1$

16. $\sum_{n=1}^{\infty} \left(\frac{1}{4}\right)^n$

17. $\sum_{n=0}^{\infty} (-1)^n$

18. $\sum_{n=3}^{\infty} \frac{1}{n(n-1)}$

19. $\sum_{n=1}^{\infty} \left(\frac{1}{n+1} - \frac{1}{n+2}\right)$

20. $\sum_{n=1}^{\infty} \left(\frac{1}{n^3} - \frac{1}{(n+1)^3}\right)$

21. $\sum_{n=1}^{\infty} [n^3 - (n+1)^3]$

*22. $\sum_{n=1}^{\infty} \frac{1}{n(n+1)(n+2)}$ $\left(Hint: \frac{1}{n(n+1)(n+2)} = \frac{1}{2}\left(\frac{1}{n} - \frac{1}{n+1}\right) - \frac{1}{2}\left(\frac{1}{n+1} - \frac{1}{n+2}\right).\right)$

23. Let $\{a_n\}_{n=1}^{\infty}$ be a sequence. Show that the nth partial sum of the series $\sum_{n=1}^{\infty} (a_n - a_{n+1})$ is $a_1 - a_{n+1}$. Conclude that this series converges if and only if $\lim_{n \to \infty} a_n$ exists, in which case the sum is $a_1 - \lim_{n \to \infty} a_n$. Series of the form just described (for example, those in Exercises 19–21) are telescoping series.

24. Let $\{a_n\}_{n=1}^{\infty}$ be the Fibonacci sequence (see Exercise 35 of Section 9.3). Use the fact that $a_{n-1} = a_{n+1} - a_a$ and the idea of Exercise 23 to show that $a_1 + a_2 + \cdots + a_n = a_{n+2} - 1$.

In Exercises 25–34 determine whether or not the series converges, and if so, find its sum.

25. $\sum_{n=1}^{\infty} 5\left(\frac{4}{7}\right)^n$

26. $\sum_{n=1}^{\infty} \left(\frac{7}{4}\right)^n$

27. $\sum_{n=0}^{\infty} (-1)^n (0.3)^n$

28. $\sum_{n=2}^{\infty} (0.33)^n$

29. $\sum_{n=1}^{\infty} 5\left(\frac{1}{2}\right)^{n+1}$

30. $\sum_{n=0}^{\infty} \frac{5^n}{7^{n+1}}$

31. $\sum_{n=1}^{\infty} \frac{3^{n+3}}{5^{n-1}}$

32. $\sum_{n=0}^{\infty} \frac{2^n + 5^n}{2^n 5^n}$

33. $\sum_{n=1}^{\infty} (-1)^n \frac{3^n}{2^{n+2}}$

34. $\sum_{n=0}^{\infty} (-3)\left(\frac{2}{3}\right)^{2n}$

In Exercises 35–39 express the given series $\Sigma_{n=1}^{\infty} a_n$ in the form $c + \Sigma_{n=4}^{\infty} a_n$.

35. $\frac{1}{7} + \frac{1}{7^2} + \frac{1}{7^3} + \cdots$

36. $1 + \frac{2}{3} + \frac{2^2}{3^2} + \frac{2^3}{3^3} + \cdots$

37. $\sum_{n=1}^{\infty} \frac{1}{n^2 + 1}$

38. $\sum_{n=1}^{\infty} (-1)^n \frac{1}{n}$

39. $\sum_{n=1}^{\infty} \frac{1}{n^2}$

In Exercises 40–47 express the repeating decimal as a fraction.

40. $0.6666666\ldots$

41. $0.72727272\ldots$

42. $0.024242424\ldots$

43. $0.232232232\ldots$

44. $0.453232232232\ldots$

45. $27.56123123123\ldots$

46. $0.00649649649649\ldots$

47. $0.86400000\ldots$

48. a. It is known that $\Sigma_{n=1}^{\infty} 1/n^2 = \pi^2/6$. Use this to find the sum of $\Sigma_{n=3}^{\infty} 1/n^2$.
    b. It is known that $\Sigma_{n=1}^{\infty} 1/n^4 = \pi^4/90$. Use this to find the sum of $\Sigma_{n=1}^{\infty} 1/[\pi^4(n+1)^4]$.

49. It is known that $\Sigma_{n=1}^{\infty} (-1)^{n+1}/n = \ln 2$. Use this to find the sum of $\Sigma_{n=4}^{\infty} (-1)^{n+1}/n$.

50. Show that $\Sigma_{n=0}^{\infty} (-1)^n r^n = 1/(1+r)$ if $-1 < r < 1$.

51. Show that $\Sigma_{n=0}^{\infty} (-1)^n r^{2n} = 1/(1+r^2)$ if $-1 < r < 1$.

52. Prove Theorem 9.11(b).

53. The Swiss mathematician Leonhard Euler used ideas expressed in the Geometric Series Theorem to deduce that if $r > 0$, then

$$\sum_{n=0}^{\infty} \left(\frac{1}{r}\right)^n = \frac{1}{1 - 1/r} = \frac{-r}{1-r} \quad \text{and} \quad \sum_{n=1}^{\infty} r^n = \frac{r}{1-r}$$

Then he concluded that

$$\left(\cdots + \frac{1}{r^4} + \frac{1}{r^3} + \frac{1}{r^2} + \frac{1}{r} + 1\right) + (r + r^2 + \cdots)$$

$$= \sum_{n=0}^{\infty} \left(\frac{1}{r}\right)^n + \sum_{n=1}^{\infty} r^n = \frac{-r}{1-r} + \frac{r}{1-r} = 0$$

Since all the terms in the series on the left side of the equation are positive, this is absurd. Why is this argument invalid?

54. Find the fallacy in the following argument: Let

$$a = 1 + 2 + 4 + 8 + \cdots$$

Then

$$2a = 2 + 4 + 8 + 16 + \cdots$$
$$= (1 + 2 + 4 + 8 + \cdots) - 1 = a - 1$$

Thus

$$a = -1$$

*55. a. Let q be an integer greater than or equal to 2. Show that

$$0 < q! \sum_{n=q+1}^{\infty} \frac{1}{n!} < 1$$

  b. Show that e is irrational. (*Hint:* Suppose e is rational. Then there are positive integers p and q such that $e = p/q$ and $q \geq 2$. Show that

$$p[(q-1)!] = q! \sum_{n=0}^{q} \frac{1}{n!} + q! \sum_{n=q+1}^{\infty} \frac{1}{n!} \quad (11)$$

Using (a), decide which terms in (11) must be integers and which cannot be integers, and conclude that the assumption that e is rational is invalid.)

56. A ball is dropped from a height of 1 meter onto a smooth surface. On each bounce the ball rises to 60 percent of the height it reached on the previous bounce. Find the total distance the ball travels.

[C] 57. a. Determine the smallest value of j such that $\Sigma_{n=1}^{j} 1/n \geq 3$.
    b. Use a calculator to determine the smallest value of j such that $\Sigma_{n=1}^{j} 1/n \geq 4$.

58. Some drugs administered to people are eliminated from the body exponentially with the passage of time. Suppose a dosage c of such a drug is given at intervals of length t. The amount c_n of the drug in the body just after the nth injection (at the beginning of the nth time period) is given by the formula

$$c_n = c + ce^{-kt} + ce^{-2kt} + \cdots + ce^{-(n-1)kt}$$

where k is positive. Show that $\{c_n\}_{n=1}^{\infty}$ forms the sequence of partial sums of a convergent geometric series, and compute the sum of that series. As n increases, the amount of the drug in the body at the beginning of the nth period approaches the sum of the series. The treatment of diseases such as glaucoma is based upon such considerations.

59. A person earns $1500 a month and during each month spends a certain fraction p (with $0 < p < 1$) of all money accumulated during that month and the previous months. Find a formula for the wealth w_n of such a person after n months, and show that the wealth has a limit as n approaches ∞.

60. In normal times a nation spends about 90 percent of its income. This means that out of a given $100 income, approximately $90 would be spent first-hand, then about 90 percent of the $90 already spent would be spent second-hand, and so on. Thus the same $100 is actually worth many times the initial $100 to the general economy. (The effect of the process of spending and respending a certain amount of money is called the **multiplier effect**.) Under the assumption of a 90 percent reutilization of any given income, how much is $100 really worth to the economy?

61. a. Two trains, each traveling 15 miles per hour, approach each other on a straight track. When the trains are 1 mile apart, a bee begins flying back and forth between the trains at 30 miles per hour. Express the distance the bee travels before the trains collide as an infinite series, and find its sum.

  b. Find a simple solution of the bee problem without using series. (*Hint:* Determine how long the bee flies.) (It is said that a similar problem was posed to the great twentieth-century mathematician John von Neumann (1903–1957), who solved it almost instantly in his head. When the poser of the problem suggested that by the quickness of his response, he must have solved the problem the simple way, von Neumann replied that he had actually solved the problem by summing a series.)

*62. One of Zeno's paradoxes purports to prove that Achilles, who runs 10 times faster than the tortoise, cannot overtake the tortoise, who has a 100-yard lead. The argument runs as follows: While Achilles runs the 100 yards, the tortoise runs an additional 10 yards. While Achilles runs that 10 yards, the tortoise runs one additional yard. While Achilles runs that yard, the tortoise runs $\frac{1}{10}$ yard, and so on. Thus Achilles is always behind the tortoise and never catches up. By summing two infinite series, show that Achilles does in fact catch up with the tortoise, and at the same time determine how many yards it takes him to do it.

*63. An indeterminately large number of identical blocks 1 unit long are stacked on top of each other. Show that it is possible for the top block to protrude as far from the bottom block as we wish without the blocks toppling (Figure 9.8). (*Hint:* The center of gravity of the top block must lie over the second block; the center of gravity of the top two blocks must lie over the third block, and so on. Thus the top block can protrude up to $\frac{1}{2}$ unit from the end of the second block, the second block can protrude up to $\frac{1}{4}$ unit from the end of the third block, the third block can protrude up to $\frac{1}{6}$ unit from the end of the fourth block, and so on. Assuming that the center of gravity of the first $(n-1)$ blocks lies over the end of the nth block, show that the nth block can protrude up to $1/2n$ units from the end of the $(n+1)$st block.)

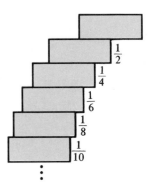

FIGURE 9.8

*64. An ant crawls at the rate of 1 foot per minute along a rubber band, which can be stretched uniformly. Suppose the rubber band is initially 1 yard long and is stretched an additional yard at the end of each minute.

  a. If the ant begins at one end of the band, does it ever reach the other end? If so, how long does it take the ant? (*Hint:* Use Exercise 57(a).)

  b. Suppose the band is originally 20 feet long and is stretched an additional 20 feet at the end of each minute. How long would it take the ant to reach the end? (*Hint:* Use the data following the solution of Example 4.)

(Adapted from Martin Gardner, "Mathematical Games," *Scientific American*, March 1975, p. 112.)

9.5
NONNEGATIVE SERIES: THE INTEGRAL TEST AND THE COMPARISON TESTS

Thus far we have always proved that a series converges by actually computing its value. However, for most convergent series the exact sum is difficult or impossible to find. This is true even of such series as $\sum_{n=1}^{\infty} 1/n^2$ or $\sum_{n=1}^{\infty} 1/n^3$. Yet in such cases it may suffice to know at least that the series converges. In Sections 9.5–9.7 we will formulate tests for determining the convergence or divergence of series. For the present we will restrict our attention to **nonnegative series**, that is, to series whose terms are nonnegative. For simplicity we assume that the initial index is 1. The partial sums $\{s_j\}_{j=1}^{\infty}$ of a nonnegative series $\Sigma_{n=1}^{\infty} a_n$ form an increasing sequence:

$$s_j = a_1 + a_2 + \cdots + a_j \leq a_1 + a_2 + \cdots + a_j + a_{j+1} = s_{j+1} \quad \text{for } j \geq 1$$

Consequently if $\{s_j\}_{j=1}^{\infty}$ is bounded, then $\lim_{j \to \infty} s_j$ exists (by Theorem 9.7), so $\Sigma_{n=1}^{\infty} a_n$ converges. By contrast, if $\{s_j\}_{j=1}^{\infty}$ is unbounded, then $\lim_{j \to \infty} s_j$ cannot exist (by Theorem 9.6(b)), so $\Sigma_{n=1}^{\infty} a_n$ diverges.

In this section we will discuss two types of convergence tests, one that compares a given nonnegative series with an improper integral and one that compares a given nonnegative series with another series.

The Integral Test   The test that involves comparing a nonnegative series with an improper integral is the following.

THEOREM 9.12
INTEGRAL TEST

Let $\{a_n\}_{n=1}^{\infty}$ be a nonnegative sequence, and let f be a continuous, decreasing function defined on $[1, \infty)$ such that

$$f(n) = a_n \quad \text{for } n \geq 1 \qquad (1)$$

Then the series $\Sigma_{n=1}^{\infty} a_n$ converges if and only if the integral $\int_1^{\infty} f(x)\,dx$ converges.

Proof   From (1) and the fact that f is decreasing it follows that

$$0 \leq a_2 + a_3 + \cdots + a_n \leq \int_1^n f(x)\,dx \quad \text{for } n \geq 2 \qquad (2)$$

and

$$\int_1^n f(x)\,dx \leq a_1 + a_2 + \cdots + a_{n-1} \quad \text{for } n \geq 2 \qquad (3)$$

(Figure 9.9). From (2) we conclude that the series $\Sigma_{n=2}^{\infty} a_n$, and hence $\Sigma_{n=1}^{\infty} a_n$, converges if the integral converges, and from (3) we conclude that the integral converges if the series $\Sigma_{n=1}^{\infty} a_n$ converges.   ∎

It follows from (2) and (3) in the proof of the Integral Test that if either $\Sigma_{n=1}^{\infty} a_n$ or $\int_1^{\infty} f(x)\,dx$ converges, then so does the other, and moreover,

$$\sum_{n=2}^{\infty} a_n \leq \int_1^{\infty} f(x)\,dx \leq \sum_{n=1}^{\infty} a_n$$

The Integral Test is tailor-made for **p series**, which are (nonnegative) series of the form $\Sigma_{n=1}^{\infty} 1/n^p$, where p is a fixed number.

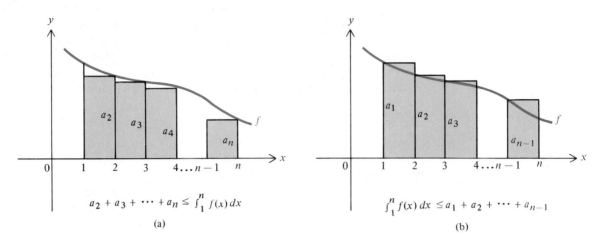

FIGURE 9.9

Example 1 Show that $\sum_{n=1}^{\infty} \dfrac{1}{n^p}$ converges if and only if $p > 1$.

Solution If $p \le 0$, the terms $1/n^p$ do not tend to 0 as n increases, so by Theorem 9.9 the series $\sum_{n=1}^{\infty} 1/n^p$ diverges. Thus from here on in the proof, we assume that $p > 0$. For $p = 1$ the series is the harmonic series, which we know diverges. So we assume that $p \ne 1$. The ideal function f to use in applying the Integral Test is defined by

$$f(x) = \frac{1}{x^p} \quad \text{for } x \ge 1$$

since $f(n) = 1/n^p$ for $n \ge 1$ and since f is continuous and decreasing on $[1, \infty)$. To determine whether $\int_1^{\infty} f(x)\,dx$ converges, we must ascertain whether

$$\lim_{b \to \infty} \int_1^b \frac{1}{x^p}\,dx$$

exists. However, since $p \ne 1$, we find that

$$\int_1^b \frac{1}{x^p}\,dx = \frac{1}{(-p+1)x^{p-1}}\bigg|_1^b = \frac{1}{(-p+1)}\left(\frac{1}{b^{p-1}} - 1\right)$$

Since $\lim_{b \to \infty} (1/b^{p-1})$ exists if $p > 1$ and does not exist if $0 < p < 1$, it follows that $\int_1^{\infty} f(x)\,dx$ converges if $p > 1$ and diverges if $0 < p < 1$. Consequently the Integral Test tells us that $\sum_{n=1}^{\infty} 1/n^p$ converges if $p > 1$ and diverges if $0 < p < 1$. $\square$

From Example 1 we know that $\sum_{n=1}^{\infty} 1/n^{1.001}$ converges and $\sum_{n=1}^{\infty} 1/n^{0.999}$ diverges. We also know that $\sum_{n=1}^{\infty} 1/n^2$ and $\sum_{n=1}^{\infty} 1/n^3$ converge. More generally, $\sum_{n=1}^{\infty} 1/n^p$ converges for any integer $p > 1$. However, the result of Example 1 does not give us the sum of any of these series. In fact, the value of $\sum_{n=1}^{\infty} 1/n^2$ baffled mathematicians until the middle of the eighteenth century, when Leonhard Euler proved that

$$\sum_{n=1}^{\infty} \frac{1}{n^2} = \frac{\pi^2}{6}$$

Today the sum of the p series $\sum_{n=1}^{\infty} 1/n^p$ is known for every positive even integer p but not for a single odd integer greater than 1. Thus, even though $\sum_{n=1}^{\infty} 1/n^3$ is known to converge, its sum is unknown (although it has recently been proved that the sum is irrational).

Since the initial few terms of a series do not affect its convergence, we may sometimes prefer to define the function f for the Integral Test on an interval different from $[1, \infty)$. In the next example it will be helpful to define f on $[2, \infty)$.

Example 2 Show that $\displaystyle\sum_{n=2}^{\infty} \frac{1}{n \ln n}$ diverges.

Solution In this case we let

$$f(x) = \frac{1}{x \ln x} \quad \text{for } x \geq 2$$

Then f is continuous and decreasing on $[2, \infty)$, and

$$f(n) = \frac{1}{n \ln n} \quad \text{for } n \geq 2$$

Next we observe that

$$\int_2^{\infty} f(x)\,dx = \lim_{b \to \infty} \int_2^b \frac{1}{x \ln x}\,dx = \lim_{b \to \infty} \left(\ln (\ln x) \Big|_2^b \right)$$
$$= \lim_{b \to \infty} [\ln (\ln b) - \ln (\ln 2)] = \infty$$

This means that $\int_2^{\infty} f(x)\,dx$ diverges. By the Integral Test, the given series also diverges. $\square$

The Integral Test is most effective when the function f to be used is easily integrated, as was the case in Examples 1 and 2.

The Comparison Tests Our next convergence test involves the comparison of two series.

THEOREM 9.13
COMPARISON TEST

a. If $\sum_{n=1}^{\infty} b_n$ converges and $0 \leq a_n \leq b_n$ for all $n \geq 1$, then $\sum_{n=1}^{\infty} a_n$ converges, and $\sum_{n=1}^{\infty} a_n \leq \sum_{n=1}^{\infty} b_n$.
b. If $\sum_{n=1}^{\infty} b_n$ diverges and $0 \leq b_n \leq a_n$ for all $n \geq 1$, then $\sum_{n=1}^{\infty} a_n$ diverges.

Proof To prove (a), we first let s_j and s'_j be the jth partial sums of $\sum_{n=1}^{\infty} a_n$ and $\sum_{n=1}^{\infty} b_n$, respectively. Since $\sum_{n=1}^{\infty} b_n$ converges by hypothesis, $\lim_{j \to \infty} s'_j$ exists, and we denote this limit by L. Since $0 \le a_n \le b_n$ for all n, it follows that $\{s_j\}_{j=1}^{\infty}$ and $\{s'_j\}_{j=1}^{\infty}$ are increasing and that

$$0 \le s_j \le s'_j \le L \quad \text{for } j \ge 1 \tag{4}$$

But this implies that $\{s_j\}_{j=1}^{\infty}$ is a bounded, increasing sequence, so by Theorem 9.7 we know that $\lim_{j \to \infty} s_j$ exists. Consequently $\Sigma_{n=1}^{\infty} a_n$ converges. Moreover, from (4) it follows that $\Sigma_{n=1}^{\infty} a_n = \lim_{j \to \infty} s_j \le L = \Sigma_{n=1}^{\infty} b_n$. This proves (a). To prove (b), we need only observe that if $\Sigma_{n=1}^{\infty} a_n$ were to converge, then by part (a) the series $\Sigma_{n=1}^{\infty} b_n$ would necessarily converge, contrary to the hypothesis of (b). Therefore $\Sigma_{n=1}^{\infty} a_n$ must diverge.   ■

We use the Comparison Test to deduce that a given nonnegative series converges by comparing it with a known convergent series having larger terms. To show that a given nonnegative series diverges, we compare it with a known divergent series having smaller nonnegative terms.

Example 3   Show that $\displaystyle\sum_{n=1}^{\infty} \frac{1}{2^n + 1}$ converges.

Solution   By the Geometric Series Theorem we know that the geometric series $\Sigma_{n=1}^{\infty} 1/2^n$ converges. Since

$$\frac{1}{2^n + 1} \le \frac{1}{2^n} \quad \text{for } n \ge 1$$

it follows from the Comparison Test that $\Sigma_{n=1}^{\infty} 1/(2^n + 1)$ converges.   □

Example 4   Show that $\displaystyle\sum_{n=1}^{\infty} \frac{1}{2\sqrt{n} - 1}$ diverges.

Solution   Observe that

$$\frac{1}{2\sqrt{n} - 1} \ge \frac{1}{2\sqrt{n}} \quad \text{for } n \ge 1$$

Since $\Sigma_{n=1}^{\infty} 1/\sqrt{n}$ is a p series with $p = \frac{1}{2}$, it diverges. Therefore $\Sigma_{n=1}^{\infty} 1/(2\sqrt{n})$ also diverges. We conclude from the Comparison Test that $\Sigma_{n=1}^{\infty} 1/(2\sqrt{n} - 1)$ diverges.   □

Example 5   Show that $\displaystyle\sum_{n=0}^{\infty} \frac{3\sin^2 n}{n!}$ converges.

Solution   Here we notice that

$$\frac{3\sin^2 n}{n!} \le \frac{3}{n!} \quad \text{for } n \ge 0$$

In Example 1 of Section 9.4 we noted that $\Sigma_{n=0}^{\infty} 1/n!$ converges. This means that

$\Sigma_{n=0}^{\infty} 3/n!$ also converges, so by the Comparison Test the given series converges too. $\square$

Example 6 Show that $\displaystyle\sum_{n=1}^{\infty} \frac{1}{2^n - 1}$ converges.

Solution It might be tempting to compare the given series with the convergent series $\Sigma_{n=1}^{\infty} 1/2^n$. However,

$$\frac{1}{2^n - 1} \geq \frac{1}{2^n} \quad \text{for } n \geq 1$$

and thus it is impossible to determine the convergence or divergence of the given series by comparing it with the series $\Sigma_{n=1}^{\infty} 1/2^n$. But since

$$\frac{1}{2^n - 1} \leq \frac{1}{2^n - 2^{n-1}} = \frac{1}{2^{n-1}} \quad \text{for } n \geq 1 \tag{5}$$

and since $\Sigma_{n=1}^{\infty} 1/2^{n-1}$ is a convergent geometric series, the Comparison Test implies that the given series converges as well. $\square$

In general, when we try to determine whether a given series $\Sigma_{n=1}^{\infty} a_n$ converges or diverges by using the Comparison Test, we select a series $\Sigma_{n=1}^{\infty} b_n$ and find a nonzero number c such that either $\Sigma_{n=1}^{\infty} b_n$ converges and $a_n \leq cb_n$ for all $n \geq 1$, or $\Sigma_{n=1}^{\infty} b_n$ diverges and $a_n \geq cb_n$ for all $n \geq 1$. We did this in Examples 5 and 6. However, finding such a number c is sometimes rather difficult. In such instances the following test may be easier to apply.

THEOREM 9.14
LIMIT COMPARISON TEST

Let $\Sigma_{n=1}^{\infty} a_n$ and $\Sigma_{n=1}^{\infty} b_n$ be nonnegative series. Suppose $\lim_{n \to \infty} a_n/b_n = L$, where L is a positive number.
a. If $\Sigma_{n=1}^{\infty} b_n$ converges, then $\Sigma_{n=1}^{\infty} a_n$ converges.
b. If $\Sigma_{n=1}^{\infty} b_n$ diverges, then $\Sigma_{n=1}^{\infty} a_n$ diverges.

Proof Since $\lim_{n \to \infty} a_n/b_n = L$, there is an integer N such that

$$\frac{1}{2}L \leq \frac{a_n}{b_n} \leq 2L \quad \text{for } n \geq N$$

Consequently $a_n \leq 2Lb_n$ and $a_n \geq \frac{1}{2}Lb_n$ for $n \geq N$. From these inequalities and the Comparison Test it follows that if $\Sigma_{n=1}^{\infty} b_n$ converges, then so in turn do $\Sigma_{n=N}^{\infty} 2Lb_n, \Sigma_{n=N}^{\infty} a_n$, and $\Sigma_{n=1}^{\infty} a_n$. Likewise, if $\Sigma_{n=1}^{\infty} b_n$ diverges, then so in turn do $\Sigma_{n=N}^{\infty} \frac{1}{2}Lb_n, \Sigma_{n=N}^{\infty} a_n$, and $\Sigma_{n=1}^{\infty} a_n$. $\blacksquare$

Example 7 Show that $\displaystyle\sum_{n=1}^{\infty} \frac{4n - 3}{n^3 - 5n - 7}$ converges.

Solution The terms of the series are rational functions of n. To find a series with which to compare the given series, we disregard all but the highest powers of n

appearing in the numerator and denominator, obtaining

$$\frac{4n}{n^3} = \frac{4}{n^2}$$

Therefore we will compare the given series with $\sum_{n=1}^{\infty} 4/n^2$. Since $\sum_{n=1}^{\infty} 1/n^2$ is a convergent p series (with $p = 2$), $\sum_{n=1}^{\infty} 4/n^2$ converges. Because

$$\lim_{n\to\infty} \frac{(4n-3)/(n^3-5n-7)}{4/n^2} = \lim_{n\to\infty} \frac{4n^3-3n^2}{4n^3-20n-28}$$

$$= \lim_{n\to\infty} \frac{4-3/n}{4-20/n^2-28/n^3} = 1$$

part (a) of the Limit Comparison Test implies that the given series converges.   □

Example 8   Show that $\displaystyle\sum_{n=1}^{\infty} \frac{1}{\sqrt[3]{8n^2-5n}}$ diverges.

Solution   We disregard all but the highest power of n in the denominator, obtaining

$$\frac{1}{\sqrt[3]{8n^2}} = \frac{1}{2n^{2/3}}$$

Accordingly, we compare the given series with $\sum_{n=1}^{\infty} 1/2n^{2/3}$. Since $\sum_{n=1}^{\infty} 1/n^{2/3}$ is a divergent p series (with $p = \frac{2}{3}$), $\sum_{n=1}^{\infty} 1/2n^{2/3}$ diverges. Because

$$\lim_{n\to\infty} \frac{1/\sqrt[3]{8n^2-5n}}{1/\sqrt[3]{8n^2}} = \lim_{n\to\infty} \sqrt[3]{\frac{8n^2}{8n^2-5n}} = \lim_{n\to\infty} \sqrt[3]{\frac{8}{8-5/n}} = 1$$

part (b) of the Limit Comparison Test implies that the given series diverges.   □

EXERCISES 9.5

In Exercises 1–28 use the Comparison Test, the Limit Comparison Test, or the Integral Test to determine whether the series converges or diverges.

1. $\displaystyle\sum_{n=1}^{\infty} \frac{1}{(n+1)^2}$

2. $\displaystyle\sum_{n=1}^{\infty} \frac{1}{\sqrt{n}}$

3. $\displaystyle\sum_{n=1}^{\infty} \frac{1}{\sqrt{n^3+1}}$

4. $\displaystyle\sum_{n=2}^{\infty} \frac{1}{\sqrt{n^2-1}}$

5. $\displaystyle\sum_{n=1}^{\infty} \frac{1}{\sqrt{n^2+1}}$

6. $\displaystyle\sum_{n=1}^{\infty} \frac{1}{n+\sqrt{n}}$

7. $\displaystyle\sum_{n=1}^{\infty} \frac{1}{e^{(n^2)}}$ $\left(Hint:\right.$

Compare with $\displaystyle\sum_{n=1}^{\infty} \left(\frac{1}{e}\right)^n.\Big)$

8. $\displaystyle\sum_{n=2}^{\infty} \frac{1}{n\sqrt{\ln n}}$

9. $\displaystyle\sum_{n=3}^{\infty} \frac{1}{(n-1)(n-2)}$

10. $\displaystyle\sum_{n=2}^{\infty} \frac{n}{n^3-n-1}$

11. $\displaystyle\sum_{n=2}^{\infty} \frac{n^2-1}{n^3-n-1}$

12. $\displaystyle\sum_{n=3}^{\infty} \frac{3+\cos n}{n^2-4}$

13. $\displaystyle\sum_{n=1}^{\infty} \frac{n}{7^n}$

14. $\displaystyle\sum_{n=1}^{\infty} \frac{\sqrt{n}}{n^2+3}$

15. $\displaystyle\sum_{n=4}^{\infty} \frac{\sqrt{n}}{n^2-3}$

16. $\displaystyle\sum_{n=1}^{\infty} \frac{\cos^2 n}{n^{3/2}}$

17. $\displaystyle\sum_{n=1}^{\infty} \frac{n}{(n^3+1)^{3/7}}$

18. $\displaystyle\sum_{n=1}^{\infty} \frac{1}{\sqrt[3]{n^2+1}}$

19. $\displaystyle\sum_{n=2}^{\infty} \frac{1}{n\sqrt{n^2-1}}$

20. $\displaystyle\sum_{n=1}^{\infty} ne^{(-n^2)}$

21. $\displaystyle\sum_{n=1}^{\infty} \frac{\arctan n}{n^2 + 1}$

22. $\displaystyle\sum_{n=2}^{\infty} \frac{1}{n(\ln n)^2}$

23. $\displaystyle\sum_{n=1}^{\infty} \frac{\ln n}{n^2}$

(*Hint:* $\ln n \le 2\sqrt{n}$ for $n \ge 1$ by the solution of Example 3 of Section 9.3.)

24. $\displaystyle\sum_{n=2}^{\infty} \frac{1}{(\ln n)^2}$

(*Hint:* See Exercise 23.)

25. $\displaystyle\sum_{n=2}^{\infty} \frac{1}{(\ln n)^n}$

26. $\displaystyle\sum_{n=1}^{\infty} ne^{-n}$

27. $\displaystyle\sum_{n=1}^{\infty} \sin\frac{1}{n}$

28. $\displaystyle\sum_{n=1}^{\infty} \frac{1}{n\sqrt[n]{n}}$

29. Determine those values of p for which the series $\sum_{n=2}^{\infty} 1/[n(\ln n)^p]$ converges.

30. Prove that if $\sum_{n=1}^{\infty} a_n$ is a convergent nonnegative series, then so is $\sum_{n=1}^{\infty} a_n^2$. (*Hint:* Show that $a_n^2 \le a_n$ for sufficiently large n.)

31. Show that if $\sum_{n=1}^{\infty} a_n$ and $\sum_{n=1}^{\infty} b_n$ are convergent nonnegative series, then $\sum_{n=1}^{\infty} a_n b_n$ converges. (*Hint:* Use the fact that $b_n \le 1$ for sufficiently large values of n.)

32. It is known that if $a_n > 0$ for all n and $\lim_{n\to\infty} a_{n+1}/a_n = r$, then $\lim_{n\to\infty} \sqrt[n]{a_n} = r$. Use this fact to evaluate

$$\lim_{n\to\infty} \sqrt[n]{\frac{n^n}{n!}}$$

*33. Using (2) and (3) from the proof of the Integral Test, find

$$\lim_{j\to\infty} \left(\frac{\sum_{n=1}^{j} (1/n)}{\ln j} \right)$$

34. A rocket is launched from earth. On its journey it consumes one fourth of its fuel during the first 100 miles, one ninth of its initial fuel during the second 100 miles, and in general, $1/(n+1)^2$ of its initial fuel during the nth 100 miles. Does the rocket ever use up all its fuel? (*Hint:* $\sum_{n=1}^{\infty} 1/n^2 = \pi^2/6$.)

9.6
NONNEGATIVE SERIES: THE RATIO TEST AND THE ROOT TEST

In this section we present two more convergence tests for nonnegative series: the Ratio Test and the Root Test. These are intrinsic tests in that they involve only the terms of the series being tested; it is not necessary to manufacture another series, an improper integral, or anything else against which to compare the given series. In this sense the Ratio Test and the Root Test are easier to apply than are the Comparison Test and the Integral Test. However, even though no comparison is required in the application of the Ratio Test and the Root Test, in proving the validity of these tests we will compare the given series with a geometric series. Thus both the Ratio Test and the Root Test ultimately depend on the Comparison Test (as well as an understanding of geometric series).

The Ratio Test   The more commonly used of the two tests in the Ratio Test.

THEOREM 9.15
RATIO TEST

Let $\sum_{n=1}^{\infty} a_n$ be a nonnegative series. Assume that $a_n \ne 0$ for all n and that

$$\lim_{n\to\infty} \frac{a_{n+1}}{a_n} = r \quad \text{(possibly } \infty\text{)}$$

a. If $0 \le r < 1$, then $\sum_{n=1}^{\infty} a_n$ converges.
b. If $r > 1$, then $\sum_{n=1}^{\infty} a_n$ diverges.

If $r = 1$, then from this test alone we cannot draw any conclusion about the convergence or divergence of $\sum_{n=1}^{\infty} a_n$.

Proof   First we assume that $0 \leq r < 1$, and we let s be any number such that $r < s < 1$. Since

$$\lim_{n \to \infty} \frac{a_{n+1}}{a_n} = r \quad \text{and} \quad r < s$$

there is an integer N such that for $n \geq N$ we have $a_{n+1}/a_n \leq s$, or equivalently, $a_{n+1} \leq a_n s$. By letting $n = N$ we obtain $a_{N+1} \leq a_N s$, and then by letting $n = N + 1$ we obtain

$$a_{N+2} \leq a_{N+1}s \leq (a_N s)s = a_N s^2$$

In general, for any positive integer n we find that

$$0 \leq a_{N+n} \leq a_{N+n-1}s \leq a_{N+n-2}s^2 \leq \cdots \leq a_{N+1}s^{n-1} \leq a_N s^n \tag{1}$$

Since $0 < s < 1$, the geometric series $\sum_{n=0}^{\infty} a_N s^n$ converges. From (1) and the Comparison Test we conclude that $\sum_{n=0}^{\infty} a_{N+n}$, which is the same series as $\sum_{n=N}^{\infty} a_n$, also converges. From the convergence of $\sum_{n=N}^{\infty} a_n$ we conclude that $\sum_{n=1}^{\infty} a_n$ converges. This proves (a). The proof of (b) is analogous, but with $r > s > 1$, with all the inequalities in the proof of (a) reversed.

To verify the final statement of the theorem, we recall the divergence of $\sum_{n=1}^{\infty} 1/n$ and the convergence of $\sum_{n=1}^{\infty} 1/n^2$, and compute the corresponding ratios:

$$\sum_{n=1}^{\infty} \frac{1}{n} \text{ diverges} \quad \text{and} \quad \lim_{n \to \infty} \frac{1/(n+1)}{1/n} = \lim_{n \to \infty} \frac{n}{n+1} = \lim_{n \to \infty} \frac{1}{1+1/n} = 1$$

$$\sum_{n=1}^{\infty} \frac{1}{n^2} \text{ converges} \quad \text{and} \quad \lim_{n \to \infty} \frac{1/(n+1)^2}{1/n^2} = \lim_{n \to \infty} \frac{n^2}{(n+1)^2} = \lim_{n \to \infty} \frac{1}{(1+1/n)^2} = 1$$

Therefore if $\lim_{n \to \infty} a_{n+1}/a_n = 1$, it is impossible to determine the convergence or divergence of the series merely by the Ratio Test.   ∎

Example 1   Show that $\displaystyle\sum_{n=1}^{\infty} \frac{2^n}{n!}$ converges.

Solution   We find that

$$r = \lim_{n \to \infty} \frac{a_{n+1}}{a_n} = \lim_{n \to \infty} \frac{2^{n+1}/(n+1)!}{2^n/n!} = \lim_{n \to \infty} \frac{2}{n+1} = 0$$

Since $r < 1$, the series converges.   □

Example 2   Show that $\displaystyle\sum_{n=1}^{\infty} \frac{2^n}{n^2}$ diverges.

Solution   A routine computation yields

$$r = \lim_{n \to \infty} \frac{a_{n+1}}{a_n} = \lim_{n \to \infty} \frac{2^{n+1}/(n+1)^2}{2^n/n^2} = \lim_{n \to \infty} 2\left(\frac{n}{n+1}\right)^2 = 2$$

Since $r > 1$, the series diverges.   □

Example 3 Show that $\displaystyle\sum_{n=1}^{\infty} \frac{n!}{n^n}$ converges.

Solution We find that

$$r = \lim_{n \to \infty} \frac{a_{n+1}}{a_n} = \lim_{n \to \infty} \frac{(n+1)!/(n+1)^{n+1}}{n!/n^n} = \lim_{n \to \infty} \frac{(n+1)!}{n!} \cdot \frac{n^n}{(n+1)^{n+1}}$$

$$= \lim_{n \to \infty} (n+1) \frac{n^n}{(n+1)^{n+1}} = \lim_{n \to \infty} \frac{n^n}{(n+1)^n}$$

$$= \lim_{n \to \infty} \frac{1}{[(n+1)/n]^n} = \frac{1}{\lim_{n \to \infty}(1+1/n)^n} = \frac{1}{e}$$

where the last equality results from Example 7 of Section 9.2. Since $r < 1$, the series converges. $\square$

Caution: Do not be misled by the solution to Example 3. We have *not* shown that $\Sigma_{n=1}^{\infty} n!/n^n$ converges to $1/e$—merely that the ratios of adjacent terms in the series converge to $1/e$. Incidentally, mathematicians have no idea what the sum of the series is.

In case $\lim_{n \to \infty} a_{n+1}/a_n$ fails to exist, no conclusion can be drawn from the Ratio Test (see Exercise 24).

The Root Test   A useful companion of the Ratio Test is the Root Test.

THEOREM 9.16
ROOT TEST

Let $\Sigma_{n=1}^{\infty} a_n$ be a nonnegative series, and assume that

$$\lim_{n \to \infty} \sqrt[n]{a_n} = r \quad (\text{possibly } \infty)$$

a. If $0 \leq r < 1$, then $\Sigma_{n=1}^{\infty} a_n$ converges.
b. If $r > 1$, then $\Sigma_{n=1}^{\infty} a_n$ diverges.

If $r = 1$, then from this test alone we cannot draw any conclusion about the convergence or divergence of $\Sigma_{n=1}^{\infty} a_n$.

Proof Assume first that $0 \leq r < 1$, and let s be any number such that $r < s < 1$. Since

$$\lim_{n \to \infty} \sqrt[n]{a_n} = r \quad \text{and} \quad r < s$$

there is an integer N such that for $n \geq N$ we have $\sqrt[n]{a_n} \leq s$, or equivalently, $a_n \leq s^n$. Since $s < 1$, the geometric series $\Sigma_{n=N}^{\infty} s^n$ converges. Then by the Comparison Test $\Sigma_{n=N}^{\infty} a_n$ converges, and hence $\Sigma_{n=1}^{\infty} a_n$ also converges. Thus (a) is proved. The case in which $r > 1$ is proved analogously.

To verify the final statement of the theorem, we recall the divergence of $\sum_{n=1}^{\infty} 1/n$ and the convergence of $\sum_{n=1}^{\infty} 1/n^2$, and compute corresponding roots (with the help of Example 9 of Section 9.2):

$$\sum_{n=1}^{\infty} \frac{1}{n} \text{ diverges } \quad \text{and} \quad \lim_{n \to \infty} \sqrt[n]{\frac{1}{n}} = \lim_{n \to \infty} \frac{1}{\sqrt[n]{n}} = 1$$

$$\sum_{n=1}^{\infty} \frac{1}{n^2} \text{ converges } \quad \text{and} \quad \lim_{n \to \infty} \sqrt[n]{\frac{1}{n^2}} = \lim_{n \to \infty} \frac{1}{(\sqrt[n]{n})^2} = \lim_{n \to \infty} \frac{1}{\sqrt[n]{n}} \lim_{n \to \infty} \frac{1}{\sqrt[n]{n}} = 1$$

Therefore if $\lim_{n \to \infty} \sqrt[n]{a_n} = 1$, it is impossible to draw any conclusion from the Root Test about the convergence of $\sum_{n=1}^{\infty} a_n$. ∎

Example 4   Show that $\sum_{n=1}^{\infty} \left(\frac{\ln n}{1000} \right)^n$ diverges.

Solution   A simple computation shows that

$$r = \lim_{n \to \infty} \left[\left(\frac{\ln n}{1000} \right)^n \right]^{1/n} = \lim_{n \to \infty} \frac{\ln n}{1000} = \infty$$

Therefore the series diverges by the Root Test. □

Example 5   Show that $\sum_{n=1}^{\infty} \frac{n}{2^n}$ converges.

Solution   Taking the nth roots of the terms of the series and using Example 9 of Section 9.2, we find that

$$r = \lim_{n \to \infty} \sqrt[n]{\frac{n}{2^n}} = \lim_{n \to \infty} \frac{\sqrt[n]{n}}{2} = \frac{1}{2}$$

Thus the Root Test implies that the series converges. □

Once again, it is impossible to draw any conclusion from the Root Test if $\lim_{n \to \infty} \sqrt[n]{a_n}$ does not exist (see Exercise 25).

The Ratio Test is likely to be effective when factorials or powers appear in the terms of the series, whereas the Root Test is likely to be effective when powers (and not factorials) appear in the terms of the series. This is one reason why the Ratio Test is more frequently used than the Root Test.

EXERCISES 9.6

In Exercises 1–23 determine whether the series converges or diverges. In some cases you may need to use tests other than the Ratio and Root Tests.

1. $\sum_{n=0}^{\infty} \frac{n!}{2^n}$

2. $\sum_{n=1}^{\infty} \frac{n}{10^n}$

3. $\sum_{n=1}^{\infty} \frac{n! 3^n}{10^n}$

4. $\sum_{n=1}^{\infty} \frac{n!}{2^{(n^2)}}$

5. $\sum_{n=1}^{\infty} \left(\frac{n}{2n+5} \right)^n$

6. $\sum_{n=1}^{\infty} \frac{n!}{(2n)!}$

7. $\displaystyle\sum_{n=1}^{\infty} \frac{(2n)!}{(n!)^2}$

8. $\displaystyle\sum_{n=2}^{\infty} \frac{2^{2n-2}}{(2n-2)!}$

9. $\displaystyle\sum_{n=0}^{\infty} n^{100} e^{-n}$

10. $\displaystyle\sum_{n=1}^{\infty} \frac{(1.1)^n}{n^7}$

11. $\displaystyle\sum_{n=1}^{\infty} \frac{n^{1.7}}{(1.7)^n}$

12. $\displaystyle\sum_{n=1}^{\infty} \frac{\ln n}{e^n}$

13. $\displaystyle\sum_{n=0}^{\infty} \frac{n!}{e^n}$

14. $\displaystyle\sum_{n=1}^{\infty} n\left(\frac{\pi}{4}\right)^n$

15. $\displaystyle\sum_{n=2}^{\infty} \frac{1}{(\ln n)^n}$

16. $\displaystyle\sum_{n=1}^{\infty} \frac{n+5}{n^3}$

17. $\displaystyle\sum_{n=1}^{\infty} \frac{1\cdot3\cdot5\cdots(2n-1)}{2\cdot4\cdot6\cdots(2n)}$

18. $\displaystyle\sum_{n=1}^{\infty} \frac{1\cdot3\cdot5\cdots(2n+1)}{2\cdot5\cdot8\cdots(3n+2)}$

19. $\displaystyle\sum_{n=1}^{\infty} \frac{(2n)!}{n!(2n)^n}$ (*Hint:* Use the ideas in the solution of Example 3.)

20. $\displaystyle\sum_{n=2}^{\infty} \left(\frac{n!}{n^n}\right)^n$ $\left(Hint: \dfrac{n!}{n^n} \le \dfrac{1}{2} \text{ for } n \ge 2.\right)$

*21. $\displaystyle\sum_{n=1}^{\infty} \frac{\sin 1/n!}{\cos 1/n!}$

*22. $\displaystyle\sum_{n=1}^{\infty} \left(\sum_{k=1}^{n} \frac{1}{k}\right)^n$

*23. $\displaystyle\sum_{n=1}^{\infty} a_n$, where $a_n = \begin{cases} 0 & \text{for } n \text{ even} \\ \left(\dfrac{n}{2n+1}\right)^n & \text{for } n \text{ odd} \end{cases}$

24. a. Let $a_{2n} = 1/n^2$ and $a_{2n+1} = 1/(2n+1)^2$. Show that $\lim_{n\to\infty} a_{n+1}/a_n$ does not exist but $\sum_{n=1}^{\infty} a_n$ converges.
    b. Construct a nonnegative series $\sum_{n=1}^{\infty} a_n$ such that $\lim_{n\to\infty} a_{n+1}/a_n$ does not exist but $\sum_{n=1}^{\infty} a_n$ diverges.

25. a. Let $\sum_{n=1}^{\infty} a_n$ be the series in Exercise 23. Show that $\lim_{n\to\infty} \sqrt[n]{a_n}$ does not exist but $\sum_{n=1}^{\infty} a_n$ converges.
    b. Construct a nonnegative series $\sum_{n=1}^{\infty} a_n$ such that $\lim_{n\to\infty} \sqrt[n]{a_n}$ does not exist but $\sum_{n=1}^{\infty} a_n$ diverges.

9.7
ALTERNATING SERIES AND ABSOLUTE CONVERGENCE

Now that we have established criteria for the convergence and divergence of nonnegative series, we are ready to study series involving both positive and negative numbers.

Alternating Series

If the terms in a series are alternately positive and negative, we call the series an **alternating series**. For example, the series

$$\sum_{n=1}^{\infty} (-1)^{n+1} 3^n = 3 - 9 + 27 - 81 + \cdots$$

and

$$\sum_{n=1}^{\infty} (-1)^n \frac{1}{(2n)!} = -\frac{1}{2} + \frac{1}{24} - \frac{1}{720} + \cdots$$

are alternating series. There is a simple convergence test credited to Leibniz that often applies to such a series.

THEOREM 9.17
ALTERNATING SERIES TEST

Let $\{a_n\}_{n=1}^{\infty}$ be a decreasing sequence of positive numbers such that $\lim_{n\to\infty} a_n = 0$. Then the alternating series $\sum_{n=1}^{\infty} (-1)^n a_n$ and $\sum_{n=1}^{\infty} (-1)^{n+1} a_n$ converge. Furthermore, for either series the sum s and the sequence of partial sums $\{s_j\}_{j=1}^{\infty}$ satisfy the inequality

$$|s - s_j| \le a_{j+1} \tag{1}$$

Proof We will prove that $\sum_{n=1}^{\infty} (-1)^{n+1} a_n$ converges and that (1) holds for this series. Because $\sum_{n=1}^{\infty} (-1)^n a_n = \sum_{n=1}^{\infty} -(-1)^{n+1} a_n$, it will then follow that $\sum_{n=1}^{\infty} (-1)^n a_n$ converges and that (1) holds for $\sum_{n=1}^{\infty} (-1)^n a_n$. For our proof

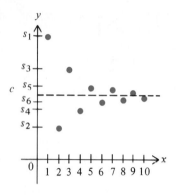

FIGURE 9.10

we let $\{s_j\}_{j=1}^{\infty}$ be the sequence of partial sums of the series $\sum_{n=1}^{\infty}(-1)^{n+1}a_n$. We will first show that $\{s_{2j}\}_{j=1}^{\infty}$ is bounded and increasing, whereas $\{s_{2j+1}\}_{j=0}^{\infty}$ is bounded and decreasing (Figure 9.10). Then we can apply Theorem 9.7 to both sequences and thereby conclude that both sequences converge. We begin by observing that

$$s_{2j+2} - s_{2j} = (a_1 - a_2 + \cdots + a_{2j-1} - a_{2j} + a_{2j+1} - a_{2j+2})$$

$$-(a_1 - a_2 + \cdots + a_{2j-1} - a_{2j}) \tag{2}$$

$$= a_{2j+1} - a_{2j+2} \geq 0$$

since $\{a_n\}_{n=1}^{\infty}$ is decreasing. Thus the sequence $\{s_{2j}\}_{j=1}^{\infty}$ is increasing. Next we notice that

$$s_{2j} = a_1 \overbrace{- a_2 + a_3}^{\leq 0} \overbrace{- a_4 + a_5}^{\leq 0} - \cdots \overbrace{- a_{2j-2} + a_{2j-1}}^{\leq 0} \overbrace{- a_{2j}}^{\leq 0} \tag{3}$$

Now (2) and (3) together imply that $s_2 \leq s_{2j} \leq a_1$, so that $\{s_{2j}\}_{j=1}^{\infty}$ is bounded. A similar argument shows that $\{s_{2j+1}\}_{j=0}^{\infty}$ is decreasing and bounded. From Theorem 9.7 we conclude that $\{s_{2j}\}_{j=1}^{\infty}$ and $\{s_{2j+1}\}_{j=0}^{\infty}$ both converge. Next we let

$$c = \lim_{j \to \infty} s_{2j} \quad \text{and} \quad d = \lim_{j \to \infty} s_{2j+1}$$

Because

$$s_{2j+1} - s_{2j} = (a_1 - a_2 + \cdots - a_{2j} + a_{2j+1}) - (a_1 - a_2 + \cdots - a_{2j}) = a_{2j+1}$$

and because $\lim_{j \to \infty} a_{2j+1} = 0$ by hypothesis, we deduce that

$$d - c = \lim_{j \to \infty} s_{2j+1} - \lim_{j \to \infty} s_{2j} = \lim_{j \to \infty} a_{2j+1} = 0$$

Consequently $d = c$, so that

$$\lim_{j \to \infty} s_{2j+1} = \lim_{j \to \infty} s_{2j}$$

As a result, $\lim_{j \to \infty} s_j$ exists. But that is equivalent to the convergence of the series $\sum_{n=1}^{\infty}(-1)^{n+1}a_n$.

Finally, to verify the inequality in (1), let s be the sum of the series $\sum_{n=1}^{\infty}(-1)^{n+1}a_n$, and observe that $\{s_{2j}\}_{j=1}^{\infty}$ is an increasing sequence with limit s, so that

$$0 \leq s - s_{2j} \tag{4}$$

Similarly, $\{s_{2j+1}\}_{j=0}^{\infty}$ is a decreasing sequence with limit s, so that

$$s - s_{2j} \leq s_{2j+1} - s_{2j} \tag{5}$$

Combining (4) and (5) yields

$$0 \leq s - s_{2j} \leq s_{2j+1} - s_{2j} = a_{2j+1} \tag{6}$$

Similar computations yield

$$0 \leq s_{2j+1} - s \leq s_{2j+1} - s_{2j+2} = a_{2j+2} \tag{7}$$

Combining (6) and (7), we obtain (1). ■

It is usually simple to ascertain whether an alternating series satisfies the hypotheses of the Alternating Series Test. If the hypotheses are satisfied, we can conclude that the series converges.

Example 1   Show that the **alternating harmonic series**

$$\sum_{n=1}^{\infty} (-1)^{n-1}\frac{1}{n} = \sum_{n=1}^{\infty} (-1)^{n+1}\frac{1}{n} = 1 - \frac{1}{2} + \frac{1}{3} - \frac{1}{4} + \cdots$$

converges.

Solution   We have $a_n = 1/n$, so that $\{a_n\}_{n=1}^{\infty}$ is a decreasing, nonnegative sequence such that $\lim_{n \to \infty} a_n = 0$. Therefore the alternating harmonic series satisfies the conditions of the Alternating Series Test and consequently must converge.   □

Example 2   Show that $\sum_{n=2}^{\infty} \frac{(-1)^n}{\ln n}$ converges.

Solution   Since $\{1/\ln n\}_{n=2}^{\infty}$ is a decreasing, nonnegative sequence and $\lim_{n \to \infty} 1/\ln n = 0$, the series satisfies the hypotheses of the Alternating Series Test. Consequently the series must converge.   □

In our next example we will use (1) to approximate the sum of an alternating series.

Example 3   Find an approximation of the series $\sum_{n=1}^{\infty} (-1)^{n+1}\frac{1}{n^3}$ with an error less than 0.001.

Solution   A partial sum s_j will approximate the sum s of the series with an error less than 0.001 if $|s - s_j| < 0.001$. If we find a value of j for which $a_{j+1} < 0.001$, then by (1) we will have $|s - s_j| < 0.001$. Since $a_{j+1} = 1/(j+1)^3$, our goal is to find a value of j for which $1/(j+1)^3 < 0.001$. By taking $j = 10$ we find that

$$\frac{1}{(j+1)^3} = \frac{1}{11^3} < 0.001$$

and the desired approximation is

$$\sum_{n=1}^{\infty} (-1)^{n+1}\frac{1}{n^3} \approx \frac{1}{1^3} - \frac{1}{2^3} + \frac{1}{3^3} - \frac{1}{4^3} + \frac{1}{5^3} - \frac{1}{6^3} + \frac{1}{7^3} - \frac{1}{8^3} + \frac{1}{9^3} - \frac{1}{10^3}$$

$$\approx 0.901116 \quad \square$$

Absolute and Conditional Convergence

Since the convergence tests studied so far do not apply directly to a series $\sum_{n=1}^{\infty} a_n$ that is neither nonnegative nor alternating, the normal procedure with such a series is to study the convergence of $\sum_{n=1}^{\infty} |a_n|$. The latter series is nonnegative and thus amenable to our convergence tests, and as we now prove, convergence of $\sum_{n=1}^{\infty} |a_n|$ implies convergence of $\sum_{n=1}^{\infty} a_n$.

THEOREM 9.18

> If $\Sigma_{n=1}^{\infty} |a_n|$ converges, then $\Sigma_{n=1}^{\infty} a_n$ converges.

Proof Since

$$0 \leq |a_n| - a_n \leq |a_n| + |a_n| = 2|a_n|$$

and since $\Sigma_{n=1}^{\infty} |a_n|$ and hence $\Sigma_{n=1}^{\infty} 2|a_n|$ converge by hypothesis, it follows from the Comparison Test that $\Sigma_{n=1}^{\infty} (|a_n| - a_n)$ converges. Thus the comment preceding (9) of Section 9.4 implies that

$$\sum_{n=1}^{\infty} a_n = \sum_{n=1}^{\infty} [|a_n| - (|a_n| - a_n)]$$

converges. ∎

Using Theorem 9.18, we can conclude that $\Sigma_{n=1}^{\infty} a_n$ converges by merely observing that the associated nonnegative series $\Sigma_{n=1}^{\infty} |a_n|$ converges.

Example 4 Show that the series $\displaystyle\sum_{n=1}^{\infty} \frac{\sin n}{n^3}$ converges.

Solution By calculating a few values of $\sin n$, we find that the series is neither nonnegative nor alternating. Therefore none of the earlier tests applies directly to it. However, since

$$\left| \frac{\sin n}{n^3} \right| \leq \frac{1}{n^3} \quad \text{for } n \geq 1$$

and since $\Sigma_{n=1}^{\infty} 1/n^3$ converges because it is a p series with $p = 3$, we know by the Comparison Test that $\Sigma_{n=1}^{\infty} |(\sin n)/n^3|$ converges. Consequently Theorem 9.18 assures us that the given series also converges. □

> **Caution:** It is wrong to conclude that the convergence of a series $\Sigma_{n=1}^{\infty} a_n$ implies the convergence of the series $\Sigma_{n=1}^{\infty} |a_n|$. A simple example is provided by the (convergent) alternating harmonic series $\Sigma_{n=1}^{\infty} (-1)^{n+1}(1/n)$. Its associated nonnegative series is the harmonic series $\Sigma_{n=1}^{\infty} 1/n$, which we know diverges. Thus when a series $\Sigma_{n=1}^{\infty} a_n$ converges, the series $\Sigma_{n=1}^{\infty} |a_n|$ may or may not converge.

DEFINITION 9.19

> Let $\Sigma_{n=1}^{\infty} a_n$ be a convergent series. If $\Sigma_{n=1}^{\infty} |a_n|$ converges, we say that the series $\Sigma_{n=1}^{\infty} a_n$ **converges absolutely**. If $\Sigma_{n=1}^{\infty} |a_n|$ diverges, we say that the series $\Sigma_{n=1}^{\infty} a_n$ **converges conditionally**.

In this terminology we say that the series $\sum_{n=1}^{\infty} (\sin n)/n^3$ converges absolutely, whereas the series $\sum_{n=1}^{\infty} (-1)^{n+1}(1/n)$ converges conditionally. Of course, all convergent nonnegative series converge absolutely.

The Generalized Convergence Tests

By combining Theorem 9.18 with our tests for nonnegative series, we obtain convergence tests that apply to any series, nonnegative or not.

THEOREM 9.20
GENERALIZED CONVERGENCE TESTS

Let $\sum_{n=1}^{\infty} a_n$ be a series.

a. *Generalized Comparison Test.* If $|a_n| \leq |b_n|$ for $n \geq 1$, and if $\sum_{n=1}^{\infty} |b_n|$ converges, then $\sum_{n=1}^{\infty} a_n$ converges (absolutely).

b. *Generalized Limit Comparison Test.* If $\lim_{n \to \infty} |a_n/b_n| = L$, where L is a positive number, and if $\sum_{n=1}^{\infty} |b_n|$ converges, then $\sum_{n=1}^{\infty} a_n$ converges (absolutely).

c. *Generalized Ratio Test.* Suppose that $a_n \neq 0$ for $n \geq 1$ and that

$$\lim_{n \to \infty} \left| \frac{a_{n+1}}{a_n} \right| = r \quad (\text{possibly } \infty)$$

If $r < 1$, then $\sum_{n=1}^{\infty} a_n$ converges (absolutely). If $r > 1$, then $\sum_{n=1}^{\infty} a_n$ diverges.

If $r = 1$, then from this test alone we cannot draw any conclusion about the convergence of the series.

d. *Generalized Root Test.* Suppose that

$$\lim_{n \to \infty} \sqrt[n]{|a_n|} = r \quad (\text{possibly } \infty)$$

If $r < 1$, then $\sum_{n=1}^{\infty} a_n$ converges (absolutely). If $r > 1$, then $\sum_{n=1}^{\infty} a_n$ diverges.

If $r = 1$, then from this test alone we cannot draw any conclusion about the convergence of the series.

Proof To prove (a) and (b) and the first parts of (c) and (d), we simply apply Theorems 9.13–9.16 to $\sum_{n=1}^{\infty} |a_n|$ and then conclude from Theorem 9.18 that $\sum_{n=1}^{\infty} a_n$ converges. To prove the second parts of (c) and (d), we observe that in each case, $\{a_n\}_{n=1}^{\infty}$ does not converge to 0, and consequently by Theorem 9.9 the series $\sum_{n=1}^{\infty} a_n$ diverges. ∎

Example 5 Show that

$$\sum_{n=1}^{\infty} \frac{x^n}{n} = x + \frac{x^2}{2} + \frac{x^3}{3} + \frac{x^4}{4} + \cdots$$

converges absolutely for $|x| < 1$, converges conditionally for $x = -1$, and diverges for $x = 1$ and for $|x| > 1$.

Solution If $x = 0$, the series obviously converges. If $x \neq 0$, then

$$\lim_{n \to \infty} \left| \frac{x^{n+1}/(n+1)}{x^n/n} \right| = \lim_{n \to \infty} \left| \frac{n}{n+1} x \right| = |x|$$

Therefore the Generalized Ratio Test implies that the given series converges for $|x| < 1$ and diverges for $|x| > 1$. For $x = 1$ the series becomes the harmonic series $\sum_{n=1}^{\infty} 1/n$, which diverges. For $x = -1$ the series becomes

$$\sum_{n=1}^{\infty} \frac{(-1)^n}{n}$$

This is the negative of the alternating harmonic series and consequently converges. Since

$$\sum_{n=1}^{\infty} \left| \frac{(-1)^n}{n} \right| = \sum_{n=1}^{\infty} \frac{1}{n}$$

which diverges, we conclude that $\sum_{n=1}^{\infty} (-1)^n/n$ converges conditionally. $\quad\square$

Example 6   Show that

$$\sum_{n=0}^{\infty} \frac{(-1)^n}{2n+1} x^{2n+1} = x - \frac{x^3}{3} + \frac{x^5}{5} - \frac{x^7}{7} + \cdots$$

converges absolutely for $|x| < 1$, converges conditionally for $|x| = 1$, and diverges for $|x| > 1$.

Solution   If $x = 0$, the series converges. If $x \neq 0$, then we have

$$\lim_{n \to \infty} \left| \frac{\dfrac{(-1)^{n+1}}{2(n+1)+1} x^{2(n+1)+1}}{\dfrac{(-1)^n}{2n+1} x^{2n+1}} \right| = \lim_{n \to \infty} \left| \frac{2n+1}{2n+3} x^2 \right| = |x^2|$$

Consequently the Generalized Ratio Test implies that the series converges absolutely for $|x| < 1$ and diverges for $|x| > 1$. It remains to consider the cases in which $|x| = 1$. For $x = -1$ the series becomes

$$\sum_{n=0}^{\infty} \frac{-(-1)^n}{2n+1} = \sum_{n=0}^{\infty} \frac{(-1)^{n+1}}{2n+1}$$

and this converges by the Alternating Series Test. For $x = 1$ the series reduces to

$$\sum_{n=0}^{\infty} \frac{(-1)^n}{2n+1}$$

which also converges by the Alternating Series Test. It is easy to show by using the Integral Test or the Limit Comparison Test that

$$\sum_{n=0}^{\infty} \frac{1}{2n+1}$$

diverges. Hence the given series converges conditionally for $|x| = 1$. $\quad\square$

If $\{a_n\}_{n=1}^{\infty}$ is a sequence and if

$$\lim_{n \to \infty} \left| \frac{a_{n+1}}{a_n} \right| = r < 1 \quad \text{or} \quad \lim_{n \to \infty} \sqrt[n]{|a_n|} = r < 1$$

then we know from the Generalized Ratio and Root Tests that the series $\Sigma_{n=1}^{\infty} a_n$ converges. It then follows from Theorem 9.9 that

$$\lim_{n \to \infty} a_n = 0$$

This result is useful in itself, and since we will refer to it later, we state it as a corollary.

COROLLARY 9.21

Let $\{a_n\}_{n=1}^{\infty}$ be a sequence. If

$$\lim_{n \to \infty} \left| \frac{a_{n+1}}{a_n} \right| = r < 1 \quad \text{or} \quad \lim_{n \to \infty} \sqrt[n]{|a_n|} = r < 1$$

then
$$\lim_{n \to \infty} a_n = 0$$

Example 7   Show that $\lim_{n \to \infty} x^n/n! = 0$ for all x.

Solution   If $x = 0$, then clearly the limit is 0. If $x \neq 0$, let

$$a_n = \frac{x^n}{n!}$$

Then

$$r = \lim_{n \to \infty} \left| \frac{a_{n+1}}{a_n} \right| = \lim_{n \to \infty} \left| \frac{x^{n+1}/(n+1)!}{x^n/n!} \right| = \lim_{n \to \infty} \left| x \frac{n!}{(n+1)!} \right| = |x| \lim_{n \to \infty} \frac{1}{(n+1)} = 0$$

Since $r < 1$, the result follows from Corollary 9.21.   ☐

This section completes our study of tests for convergence and divergence of series. A table at the end of the chapter summarizes the tests.

EXERCISES 9.7

In Exercises 1–14 determine whether the series converges or diverges.

1. $\sum_{n=1}^{\infty} (-1)^{n+1} \dfrac{1}{2n+1}$

2. $\sum_{n=2}^{\infty} (-1)^n \dfrac{1}{\ln n}$

3. $\sum_{n=1}^{\infty} (-1)^n \dfrac{2n+1}{5n+1}$

4. $\sum_{n=3}^{\infty} \dfrac{\cos n\pi}{\sqrt{n}}$

5. $\sum_{n=1}^{\infty} (-1)^n \dfrac{n+2}{n^2+3n+5}$

6. $\sum_{n=2}^{\infty} (-1)^{n+1} \dfrac{n+1}{4n}$

7. $\sum_{n=1}^{\infty} (-1)^n \dfrac{\ln n}{n}$

8. $\sum_{n=1}^{\infty} (-1)^n \dfrac{(\ln n)^2}{n}$

9. $\sum_{n=1}^{\infty} (-1)^n \dfrac{(\ln n)^p}{n}$
where p is any positive integer

10. $\sum_{n=1}^{\infty} (-1)^{n+1} \dfrac{\sqrt{n}}{2n+1}$

11. $\sum_{n=1}^{\infty} (-1)^{n+1} \dfrac{n!}{100^n}$

12. $\sum_{n=1}^{\infty} (-1)^n \cot\left(\dfrac{\pi}{2} - \dfrac{1}{n}\right)$

13. $\sum_{n=1}^{\infty} (-1)^{n+1} \dfrac{n^2}{2n+1}$

14. $\sum_{n=1}^{\infty} (-1)^{n+1} \dfrac{1}{n^{1/10}}$

15. For which positive values of p does the series $\Sigma_{n=1}^{\infty} (-1)^n (1/n^p)$ converge?

In Exercises 16–19 use (1) to find an upper bound for the error of the sum of the first four terms of the series as an approximation of the sum of the series.

16. $\sum_{n=1}^{\infty} (-1)^{n+1} \frac{1}{n}$

17. $\sum_{n=1}^{\infty} (-1)^n \frac{1}{n^2}$

18. $\sum_{n=1}^{\infty} (-1)^{n+1} \frac{1}{n!}$

19. $\sum_{n=1}^{\infty} (-1)^n \frac{1}{\sqrt{n+4}}$

In Exercises 20–22 approximate the sum of the given series with an error less than 0.01.

20. $\sum_{n=1}^{\infty} (-1)^{n+1} \frac{1}{n^3}$

21. $\sum_{n=2}^{\infty} (-1)^{n+1} \frac{1}{1+n+6n^2}$

22. $\sum_{n=1}^{\infty} (-1)^n \frac{8}{10^n + 1}$

In Exercises 23–33 determine which series diverge, which converge conditionally, and which converge absolutely.

23. $\sum_{n=1}^{\infty} (-1)^{n+1} \frac{1}{3n+4}$

24. $\sum_{n=1}^{\infty} n\left(\frac{4}{5}\right)^n$

25. $\sum_{n=1}^{\infty} (-1)^n \frac{n^n}{n!}$

26. $\sum_{n=3}^{\infty} (-1)^{n+1} \frac{1}{n(n-2)}$

27. $\sum_{n=1}^{\infty} (-1)^{n+1} \frac{1}{n^{1/n}}$

28. $\sum_{n=2}^{\infty} (-1)^n \frac{1}{n(\ln n)}$

29. $\sum_{n=2}^{\infty} (-1)^{n+1} \frac{1}{n(\ln n)^2}$

30. $\sum_{n=1}^{\infty} \frac{\sin n}{n^2 + 1}$

31. $\sum_{n=2}^{\infty} (-1)^{n+1} \frac{1}{(\ln n)^n}$

32. $\sum_{n=2}^{\infty} (-1)^{n+1} \frac{1}{(\ln n)^{1/n}}$

33. $\sum_{n=1}^{\infty} (-1)^n \frac{1 \cdot 3 \cdot 5 \cdots (2n+1)}{2 \cdot 5 \cdot 8 \cdots (3n+2)}$

In Exercises 34–39 use Corollary 9.21 to verify the given limit.

34. $\lim_{n \to \infty} \frac{n!}{n^n} = 0$

35. $\lim_{n \to \infty} \frac{(n+1)^2}{n!} = 0$

36. $\lim_{n \to \infty} \frac{x^{2n}}{n^n} = 0$ for all x

37. $\lim_{n \to \infty} \frac{x^{2n}}{n} = 0$ for $|x| < 1$

38. $\lim_{n \to \infty} \frac{x^n}{2^n} = 0$ for $|x| < 2$

39. $\lim_{n \to \infty} \frac{n! x^n}{n^n} = 0$ for $|x| < e$

40. Show that the series $\sum_{n=1}^{\infty} \frac{n!}{n^n} x^n$ converges for $|x| < e$.

41. Suppose $\sum_{n=1}^{\infty} a_n$ converges and is not necessarily nonnegative. Give an example to show that $\sum_{n=1}^{\infty} a_n^2$ need not converge. (*Hint:* Let $\sum_{n=1}^{\infty} a_n$ be an alternating series whose terms approach 0 very slowly.)

*42. If $\sum_{n=1}^{\infty} a_n$ is absolutely convergent, must $\sum_{n=1}^{\infty} (a_n + a_{n+1})$ be absolutely convergent? Explain your answer.

*43. Let $a \neq 0$, and assume that $\lim_{n \to \infty} a_n = a$ and $a_n \neq 0$ for all n. Show that $\sum_{n=1}^{\infty} |a_{n+1} - a_n|$ converges if and only if

$$\sum_{n=1}^{\infty} \left| \frac{1}{a_{n+1}} - \frac{1}{a_n} \right|$$

converges.

9.8
POWER SERIES

In Examples 5 and 6 of Section 9.7 we determined those values of x for which the series

$$\sum_{n=1}^{\infty} \frac{x^n}{n} \quad \text{and} \quad \sum_{n=0}^{\infty} \frac{(-1)^n}{2n+1} x^{2n+1}$$

converges. Such series are called power series.

DEFINITION 9.22

A series of the form $\sum_{n=0}^{\infty} c_n x^n$ is called a **power series**.

If $c_0 = 0$ we usually write the power series as $\sum_{n=1}^{\infty} c_n x^n$, and if $c_1 = 0$ as well, we normally write the series as $\sum_{n=2}^{\infty} c_n x^n$. In general the initial index of a power series can be any nonnegative number. For the sake of uniformity in writing power series, we will always assume that $x^0 = 1$ for all x, including $x = 0$, so that

$$\sum_{n=0}^{\infty} c_n x^n = c_0 + c_1 x + c_2 x^2 + c_3 x^3 + \cdots$$

If $c_n = 0$ for all $n > N$, then the power series is just a polynomial of degree at most N. For instance,

$$2 - 3x + 4x^2 = \sum_{n=0}^{\infty} c_n x^n$$

where

$$c_0 = 2, \quad c_1 = -3, \quad c_2 = 4, \quad \text{and} \quad c_n = 0 \quad \text{for } n \geq 3$$

Just as we wish to know whether a given series converges or diverges, with a given power series $\sum_{n=0}^{\infty} c_n x^n$ we wish to determine those values of x for which the series converges and those for which it diverges. Notice that every power series automatically converges for $x = 0$; thus every power series defines a function whose domain is the collection of those values of x for which the power series converges. Of course, this domain may consist of 0 alone, as is illustrated by the following example.

Example 1 Show that

$$\sum_{n=0}^{\infty} n! x^n = 1 + x + 2x^2 + 6x^3 + 24x^4 + \cdots$$

converges only for $x = 0$.

Solution If $x \neq 0$, then

$$\lim_{n \to \infty} \left| \frac{(n+1)! x^{n+1}}{n! x^n} \right| = \lim_{n \to \infty} |(n+1)x| = |x| \lim_{n \to \infty} (n+1) = \infty$$

Thus the Generalized Ratio Test implies that the series diverges. □

In contrast, a power series can converge for all numbers x.

Example 2 Show that

$$\sum_{n=0}^{\infty} \frac{x^n}{n!} = 1 + x + \frac{1}{2}x^2 + \frac{1}{6}x^3 + \frac{1}{24}x^4 + \cdots$$

converges for every number x.

Solution If $x \neq 0$, then

$$\lim_{n \to \infty} \left| \frac{x^{n+1}/(n+1)!}{x^n/n!} \right| = \lim_{n \to \infty} \left| x \frac{n!}{(n+1)!} \right| = |x| \lim_{n \to \infty} \frac{1}{(n+1)} = 0$$

Thus the Generalized Ratio Test implies that the series converges for all $x \neq 0$, and hence the series converges for all x. $\square$

Between the two extremes given in Examples 1 and 2 are power series that converge for some nonzero values of x and diverge for other values of x. One of the most prominent such series is $\sum_{n=0}^{\infty} x^n$, which for each nonzero value of x is a geometric series with ratio x. Thus we know from the Geometric Series Theorem that the series diverges for $|x| \geq 1$ and converges for $|x| < 1$, with

$$\sum_{n=0}^{\infty} x^n = \frac{1}{1-x} \tag{1}$$

Notice in particular that this series converges for all x in an open interval about 0. In general, if there is a nonzero value of x for which a power series $\sum_{n=0}^{\infty} c_n x^n$ converges, then the power series converges for all values of x in some open interval about 0, as we now prove.

LEMMA 9.23

a. If $\sum_{n=0}^{\infty} c_n s^n$ converges, then $\sum_{n=0}^{\infty} c_n x^n$ converges absolutely for $|x| < |s|$.
b. If $\sum_{n=0}^{\infty} c_n s^n$ diverges, then $\sum_{n=0}^{\infty} c_n x^n$ diverges for $|x| > |s|$.

Proof   To prove (a), let $|x| < |s|$, and rewrite $\sum_{n=0}^{\infty} c_n x^n$ as follows:

$$\sum_{n=0}^{\infty} c_n x^n = \sum_{n=0}^{\infty} c_n s^n \left(\frac{x}{s}\right)^n$$

Since $\sum_{n=0}^{\infty} c_n s^n$ converges by hypothesis, we know from Theorem 9.9(a) that $\lim_{n \to \infty} c_n s^n = 0$. Consequently there is a positive integer N such that $|c_n s^n| \leq 1$ for all $n \geq N$. This means that

$$|c_n x^n| = |c_n s^n| \left|\frac{x}{s}\right|^n \leq \left|\frac{x}{s}\right|^n \quad \text{for } n \geq N$$

Because $|x| < |s|$, which means that $|x/s| < 1$, we know that the geometric series $\sum_{n=N}^{\infty} |x/s|^n$ converges absolutely. By the Generalized Comparison Test it follows that $\sum_{n=N}^{\infty} c_n x^n$ converges absolutely, and hence $\sum_{n=0}^{\infty} c_n x^n$ also converges absolutely. This proves (a). To prove (b), we assume that $\sum_{n=0}^{\infty} c_n s^n$ diverges, and we only need to observe that if $\sum_{n=0}^{\infty} c_n x^n$ were to converge, then by reversing the roles of s and x in part (a), we would find that $\sum_{n=0}^{\infty} c_n s^n$ would converge after all, which contradicts our assumption. Therefore $\sum_{n=0}^{\infty} c_n x^n$ diverges.   ■

Lemma 9.23 provides much more information about the convergence of a power series than a first reading might reveal. We saw in Examples 1 and 2 that a power series $\sum_{n=0}^{\infty} c_n x^n$ might converge only for $x = 0$ and that it might converge for all x; otherwise, the series must converge for some nonzero value of x and diverge for some other nonzero value of x. Lemma 9.23 implies that in this case there is a number $R > 0$ such that the series converges for $|x| < R$ and diverges for $|x| > R$; this result is fundamental to the study of power series.

THEOREM 9.24

Let $\sum_{n=0}^{\infty} c_n x^n$ be a power series. Then exactly one of the following conditions holds.
a. $\sum_{n=0}^{\infty} c_n x^n$ converges only for $x = 0$.
b. $\sum_{n=0}^{\infty} c_n x^n$ converges for all x.
c. There is a number $R > 0$ such that $\sum_{n=0}^{\infty} c_n x^n$ converges for $|x| < R$ and diverges for $|x| > R$.

We call the number R in part (c) of Theorem 9.24 the **radius of convergence** of $\sum_{n=0}^{\infty} c_n x^n$. If $\sum_{n=0}^{\infty} c_n x^n$ satisfies (a), then we let $R = 0$. If $\sum_{n=0}^{\infty} c_n x^n$ satisfies (b), then we let $R = \infty$. Therefore *every* power series has a radius of convergence R, which is either a nonnegative number or ∞. The collection of values of x for which $\sum_{n=0}^{\infty} c_n x^n$ converges is called the **interval of convergence** of $\sum_{n=0}^{\infty} c_n x^n$. From conditions (a)–(c) it is apparent that the interval of convergence takes one and only one of the following forms:

$$[0,0], \quad (-R, R), \quad [-R, R), \quad (-R, R], \quad [-R, R], \quad (-\infty, \infty)$$

In Table 9.1 we verify that each of these types of intervals actually arises as the interval of convergence of a suitable power series.

The radius of convergence R need not be 0, 1, or ∞. For example, the geometric series

$$\sum_{n=0}^{\infty} \frac{x^n}{2^n} = \sum_{n=0}^{\infty} \left(\frac{x}{2}\right)^n$$

TABLE 9.1

Interval	Example	Reference
$[0,0]$	$\sum_{n=0}^{\infty} n! x^n$	Example 1
$(-1, 1)$	$\sum_{n=0}^{\infty} x^n$	Theorem 9.10 and the comment after Example 2
$[-1, 1)$	$\sum_{n=1}^{\infty} \frac{x^n}{n}$	Example 5 of Section 9.7
$(-1, 1]$	$\sum_{n=1}^{\infty} \frac{(-x)^n}{n}$	The preceding example altered
$[-1, 1]$	$\sum_{n=0}^{\infty} \frac{(-1)^n}{2n+1} x^{2n+1}$	Example 6 of Section 9.7
$(-\infty, \infty)$	$\sum_{n=0}^{\infty} \frac{x^n}{n!}$	Example 2

converges for all x in $(-2, 2)$, because $|x/2| < 1$ for exactly those values of x; thus $R = 2$. In fact, a similar argument shows that any positive number R is the radius of convergence of some power series.

The next example shows the power of Theorem 9.24.

Example 3 Determine the interval of convergence of the power series

$$\sum_{n=0}^{\infty} \frac{x^n}{n^{1/2}}$$

Solution For $x = -1$ the series becomes $\sum_{n=1}^{\infty} (-1)^n(1/n^{1/2})$, which converges by the Alternating Series Test. For $x = 1$ the series becomes $\sum_{n=1}^{\infty} 1/n^{1/2}$, which diverges because it is a p series with $p = \frac{1}{2}$. Since the original series converges for $x = -1$ and diverges for $x = 1$, Theorem 9.24 assures us that the only possible value for the radius of convergence is $R = 1$, and consequently the interval of convergence is $[-1, 1)$. □

In most cases it is not possible to find the interval of convergence as easily as in Example 3. The usual procedure is to use the Generalized Ratio Test or the Generalized Root Test to ascertain the radius of convergence R, and then to test the endpoints $x = R$ and $x = -R$ separately for convergence. We illustrate this procedure now.

Example 4 Determine the interval of convergence of the power series $\sum_{n=1}^{\infty} nx^n$.

Solution For $x \neq 0$ we obtain

$$\lim_{n \to \infty} \left| \frac{(n+1)x^{n+1}}{nx^n} \right| = \lim_{n \to \infty} \left| \frac{n+1}{n} x \right| = |x|$$

Therefore by the Generalized Ratio Test we know that the series converges for $|x| < 1$ and diverges for $|x| > 1$. By Theorem 9.24 the radius of convergence R is 1. But since we can draw no conclusions from the Generalized Ratio Test when $|x| = 1$, we must test the two values 1 and -1 separately. For those values of x we obtain the series $\sum_{n=1}^{\infty} n$ and $\sum_{n=1}^{\infty} (-1)^n n$, both of which diverge, since the terms in the series fail to converge to 0. It follows that the interval of convergence is $(-1, 1)$. □

Differentiation of Power Series

Since $\sum_{n=0}^{\infty} c_n x^n$ is a function, we may ask whether its derivative exists. Perhaps surprisingly, a power series with a nonzero radius of convergence is *always* differentiable; moreover, the derivative is obtained from $\sum_{n=0}^{\infty} c_n x^n$ by differentiating term by term, the way we differentiate polynomials.

THEOREM 9.25

DIFFERENTIATION THEOREM FOR POWER SERIES

Let $\sum_{n=0}^{\infty} c_n x^n$ be a power series with radius of convergence $R > 0$. Then $\sum_{n=1}^{\infty} n c_n x^{n-1}$ has the same radius of convergence, and

$$\frac{d}{dx}\left(\sum_{n=0}^{\infty} c_n x^n\right) = \sum_{n=1}^{\infty} n c_n x^{n-1} = \sum_{n=1}^{\infty} \frac{d}{dx}(c_n x^n) \quad \text{for } |x| < R$$

The proof of the Differentiation Theorem is complicated, and we defer it to the Appendix. Notice, however, that the initial index of

$$\sum_{n=1}^{\infty} n c_n x^{n-1} = \sum_{n=1}^{\infty} \frac{d}{dx}(c_n x^n)$$

is 1, not 0, since the derivative of $c_0 x^0$ is 0.

Example 5 Show that

$$\frac{d}{dx}\left(\sum_{n=0}^{\infty} \frac{x^n}{n!}\right) = \sum_{n=0}^{\infty} \frac{x^n}{n!} \tag{2}$$

Solution We know from Example 2 that the series $\sum_{n=0}^{\infty} x^n/n!$ converges for all x. The Differentiation Theorem tells us that $\sum_{n=1}^{\infty} n x^{n-1}/n!$ converges as well and that

$$\frac{d}{dx}\left(\sum_{n=0}^{\infty} \frac{x^n}{n!}\right) = \sum_{n=1}^{\infty} \frac{n x^{n-1}}{n!} = \sum_{n=1}^{\infty} \frac{x^{n-1}}{(n-1)!} = \sum_{n=0}^{\infty} \frac{x^n}{n!} \quad \square$$

If we define a function f by

$$f(x) = \sum_{n=0}^{\infty} \frac{x^n}{n!} \quad \text{for all } x$$

then (2) may be rewritten

$$f'(x) = f(x) \quad \text{for all } x$$

Since $f(0) = \sum_{n=0}^{\infty} 0^n/n! = 1$, we conclude from the comment following Theorem 7.12 that

$$f(x) = e^x \quad \text{for all } x$$

Therefore

$$e^x = \sum_{n=0}^{\infty} \frac{x^n}{n!} = 1 + x + \frac{x^2}{2!} + \frac{x^3}{3!} + \cdots \tag{3}$$

Thus we have found a formula for e^x in terms of power series, as we suggested at the end of Section 9.1. Other such formulas will occur below.

Formula (3) yields some related formulas:

$$\frac{1}{e} = e^{-1} = \sum_{n=0}^{\infty} \frac{(-1)^n}{n!}$$

$$e^{-x} = \sum_{n=0}^{\infty} \frac{(-x)^n}{n!} = \sum_{n=0}^{\infty} \frac{(-1)^n}{n!} x^n$$

$$e^{(x^2)} = \sum_{n=0}^{\infty} \frac{1}{n!}(x^2)^n = \sum_{n=0}^{\infty} \frac{1}{n!} x^{2n}$$

Caution: Although the Differentiation Theorem says that the radii of convergence of the series $\Sigma_{n=0}^{\infty} c_n x^n$ and $\Sigma_{n=1}^{\infty} n c_n x^{n-1}$ are the same, it does not imply that the *intervals* of convergence are the same. On the contrary, the interval of convergence of $\Sigma_{n=1}^{\infty} x^n/n$ is $[-1, 1)$, whereas the interval of convergence of its derivative $\Sigma_{n=1}^{\infty} x^{n-1}$ is $(-1, 1)$.

The Differentiation Theorem states that a power series $\Sigma_{n=0}^{\infty} c_n x^n$ with a nonzero radius of convergence can be differentiated once. However, because the derivative is itself a power series with the same radius of convergence, the derivative also may be differentiated, and thus the original power series can be differentiated twice. By repeating this process, we conclude that a power series with radius of convergence $R > 0$ has derivatives of all orders on $(-R, R)$. The values of the derivatives of the series at 0 are closely related to the numbers $c_0, c_1, c_2, \ldots$, as we see in the next theorem. For convenience, we will denote $f(0)$ by $f^{(0)}(0)$.

THEOREM 9.26

Suppose a power series $\Sigma_{n=0}^{\infty} c_n x^n$ has radius of convergence $R > 0$. Let

$$f(x) = \sum_{n=0}^{\infty} c_n x^n \quad \text{for } -R < x < R \tag{4}$$

Then f has derivatives of all orders on $(-R, R)$, and

$$f^{(n)}(0) = n! c_n \quad \text{for } n \geq 0 \tag{5}$$

Consequently

$$f(x) = \sum_{n=0}^{\infty} \frac{f^{(n)}(0)}{n!} x^n \quad \text{for } -R < x < R \tag{6}$$

Proof   From the discussion preceding this theorem we know that f has derivatives of all orders on $(-R, R)$. By substituting $x = 0$ into (4) we obtain

$$f^{(0)}(0) = f(0) = c_0 = 0! c_0$$

Next we differentiate both sides of (4), using the Differentiation Theorem:

$$f'(x) = \sum_{n=1}^{\infty} n c_n x^{n-1} \tag{7}$$

By again substituting $x = 0$, we obtain

$$f'(0) = c_1 = 1!c_1$$

Differentiation of both sides of (7) yields

$$f''(x) = \sum_{n=2}^{\infty} n(n-1)c_n x^{n-2}$$

Substituting $x = 0$ once more, we find that

$$f''(0) = 2(1)c_2 = 2!c_2$$

In the same way, performing n differentiations on both sides of (4) and then substituting $x = 0$ into the result yields (5). Finally, if we substitute for c_n from (5) into (4), we obtain (6). ■

COROLLARY 9.27

Let $R > 0$, and suppose $\sum_{n=0}^{\infty} c_n x^n$ and $\sum_{n=0}^{\infty} b_n x^n$ are power series that converge for $-R < x < R$. If

$$\sum_{n=0}^{\infty} c_n x^n = \sum_{n=0}^{\infty} b_n x^n \quad \text{for } -R < x < R$$

then $c_n = b_n$ for each $n \geq 0$.

Proof Let $f(x) = \sum_{n=0}^{\infty} c_n x^n = \sum_{n=0}^{\infty} b_n x^n$ for $-R < x < R$. Then Theorem 9.26 implies that

$$c_n = \frac{f^{(n)}(0)}{n!} = b_n \quad ■$$

In particular, Corollary 9.27 tells us that two polynomials

$$c_n x^n + c_{n-1} x^{n-1} + \cdots + c_1 x + c_0$$

and $\qquad b_m x^m + b_{m-1} x^{m-1} + \cdots + b_1 x + b_0$

which are of course power series having all but a finite number of their coefficients 0, are the same function if and only if

$$m = n \quad \text{and} \quad b_j = c_j \quad \text{for } 1 \leq j \leq n$$

Integration of Power Series

Suppose that $R > 0$ and that $\sum_{n=0}^{\infty} c_n x^n$ converges for $-R < x < R$. Then we know that $\sum_{n=0}^{\infty} c_n x^n$ is differentiable, and hence continuous, on $(-R, R)$. It is therefore possible to integrate the function $\sum_{n=0}^{\infty} c_n x^n$ over any closed interval in $(-R, R)$. The next theorem states that we can carry out the integration term by term, just as we do differentiation. We also defer the proof of this theorem to the Appendix.

THEOREM 9.28
INTEGRATION THEOREM
FOR POWER SERIES

Let $\sum_{n=0}^{\infty} c_n x^n$ be a power series with radius of convergence $R > 0$. Then $\sum_{n=0}^{\infty} (c_n/(n+1)) x^{n+1}$ has the same radius of convergence, and

$$\int_0^x \left(\sum_{n=0}^{\infty} c_n t^n \right) dt = \sum_{n=0}^{\infty} \left(\int_0^x c_n t^n \, dt \right) = \sum_{n=0}^{\infty} \frac{c_n}{n+1} x^{n+1} \quad \text{for } |x| < R$$

The Integration Theorem is an invaluable tool for expressing many well known functions as power series.

Example 6 Show that

$$\ln(1+x) = \sum_{n=0}^{\infty} \frac{(-1)^n}{n+1} x^{n+1} = \sum_{n=1}^{\infty} \frac{(-1)^{n-1}}{n} x^n \quad \text{for } |x| < 1 \qquad (8)$$

Solution From (1) we know that

$$\frac{1}{1-x} = \sum_{n=0}^{\infty} x^n \quad \text{for } |t| < 1$$

Replacing x by $-t$ in this equation, we obtain

$$\frac{1}{1+t} = \sum_{n=0}^{\infty} (-1)^n t^n \quad \text{for } |t| < 1 \qquad (9)$$

Using (9) and the Integration Theorem, we find that

$$\ln(1+x) = \int_0^x \frac{1}{1+t} \, dt = \int_0^x \left(\sum_{n=0}^{\infty} (-1)^n t^n \right) dt$$

$$= \sum_{n=0}^{\infty} \frac{(-1)^n}{n+1} x^{n+1} \quad \text{for } |x| < 1 \quad \square$$

We can write the result of Example 6 in expanded form as follows:

$$\ln(1+x) = x - \frac{x^2}{2} + \frac{x^3}{3} - \frac{x^4}{4} + \cdots$$

By Example 2 of Section 9.4, the series just given for $\ln(1+x)$ holds equally well when $x = 1$. The power series expansion of $\ln(1+x)$ for $-1 < x \leq 1$ is sometimes known as **Mercator's series**, after the Danish mathematician Nicolaus Mercator (about 1620–1687). From Mercator's series one can derive the power series for $\ln[(1+x)/(1-x)]$, which has been used to obtain quick and accurate estimates for logarithms (see Exercise 46 of this section and Exercises 18 and 19 of Section 9.1).

Using (9) and the Integration Theorem, we can express arctan x as a power series.

Example 7 Show that

$$\arctan x = \sum_{n=0}^{\infty} \frac{(-1)^n}{2n+1} x^{2n+1} \quad \text{for } |x| < 1 \tag{10}$$

Solution If $|t| < 1$, then $|t^2| < 1$. Therefore we may substitute t^2 for t in (9) and obtain

$$\frac{1}{1+t^2} = \sum_{n=0}^{\infty} (-1)^n t^{2n} \quad \text{for } |t| < 1$$

Then the Integration Theorem yields

$$\arctan x = \int_0^x \frac{1}{1+t^2} \, dt = \int_0^x \left(\sum_{n=0}^{\infty} (-1)^n t^{2n} \right) dt$$

$$= \sum_{n=0}^{\infty} \frac{(-1)^n}{2n+1} x^{2n+1} \quad \text{for } |x| < 1 \quad \square$$

In expanded form the power series for $\arctan x$ becomes

$$\arctan x = x - \frac{x^3}{3} + \frac{x^5}{5} - \frac{x^7}{7} + \cdots$$

This is called **Gregory's series**, after the Scottish mathematician James Gregory (1638–1675), who achieved (along with, but independently from, Newton) most of the outstanding early results on series.

Although it is not so easy to prove, Gregory's series for $\arctan x$ also holds for $x = 1$. This value of x yields the following expression for $\pi/4$:

$$\frac{\pi}{4} = \arctan 1 = \sum_{n=0}^{\infty} (-1)^n \frac{1}{2n+1} = 1 - \frac{1}{3} + \frac{1}{5} - \frac{1}{7} + \cdots$$

This series, which was independently discovered by Leibniz and is often called the Leibniz series, is an alternating series that converges *very* slowly. It is impractical to use it to estimate $\pi/4$, because a large number of terms would be needed to ensure even moderate accuracy. In fact, to guarantee an error less than 0.0001 would require 5000 terms! However, we can use the power series expansion for $\arctan x$ along with the equation

$$\frac{\pi}{4} = 4 \arctan \frac{1}{5} - \arctan \frac{1}{239} \tag{11}$$

which appeared in Exercise 73 of Section 7.6, to obtain a very good approximation of $\pi/4$ with few calculations.

Example 8 Using (11), approximate $\pi/4$ with an error less than 0.0001. Use the approximation obtained to estimate π with an error less than 0.0004.

Solution   If we approximate $4 \arctan \frac{1}{5}$ and $\arctan \frac{1}{239}$, each with an error less than 0.00005, then the desired approximation of $\pi/4$ will have an error less than 0.0001. The power series for $\arctan x$ given in (10) satisfies the conditions of the Alternating Series Test for $0 \leq x < 1$, so we use (1) of Section 9.7 to find an upper bound for the error introduced by using the first few terms of the series for $4 \arctan \frac{1}{5}$ and for $\arctan \frac{1}{239}$. In approximating $4 \arctan \frac{1}{5}$, we need n large enough to ensure that

$$4 \left| \frac{(-1)^{n+1}}{2(n+1)+1} \left(\frac{1}{5} \right)^{2(n+1)+1} \right| = 4 \left(\frac{1}{2n+3} \right) \left(\frac{1}{5^{2n+3}} \right) < 0.00005$$

which you can check happens if $n \geq 2$. Therefore

$$4 \arctan \frac{1}{5} \approx 4 \left[\frac{1}{5} - \frac{1}{3} \left(\frac{1}{5} \right)^3 + \frac{1}{5} \left(\frac{1}{5} \right)^5 \right] \approx 0.789589$$

with an error less than 0.00005. For $\arctan \frac{1}{239}$, we need n large enough so that

$$\left| \frac{(-1)^{n+1}}{2(n+1)+1} \left(\frac{1}{239} \right)^{2(n+1)+1} \right| = \frac{1}{2n+3} \left(\frac{1}{239^{2n+3}} \right) < 0.00005$$

which happens if $n \geq 0$. Consequently

$$\arctan \frac{1}{239} \approx \frac{1}{239} \approx 0.004184$$

with an error less than 0.00005. Therefore by (11),

$$\frac{\pi}{4} = 4 \arctan \frac{1}{5} - \arctan \frac{1}{239} \approx 0.789589 - 0.004184 = 0.785405 \qquad (12)$$

with an error less than $0.00005 + 0.00005 = 0.0001$. Finally, multiplication of the numbers in (12) by 4 yields

$$\pi \approx 3.14162$$

with an error less than 0.0004.   □

To emphasize the accuracy that can be obtained by means of (11), we mention that in 1706 John Machin employed (11) to calculate the first 100 digits in the decimal expansion of π. More recently formula (11) has been used to compute an approximation of π that is accurate to over 16,000 places. The related formula

$$\pi = 24 \arctan \frac{1}{8} + 8 \arctan \frac{1}{57} + 4 \arctan \frac{1}{239}$$

was used in 1962 by D. Shanks and J. W. Wrench, Jr., to calculate the first 100,000 digits in the decimal expansion of π. It took a computer over 8 hours to perform the task. Recently a high-speed computer has calculated π to some 6,000,000 places.

EXERCISES 9.8

In Exercises 1–20 find the interval of convergence of the given series.

1. $\displaystyle\sum_{n=1}^{\infty} \frac{x^n}{n^2}$

2. $\displaystyle\sum_{n=0}^{\infty} 2^n x^n$

3. $\displaystyle\sum_{n=1}^{\infty} \frac{1}{\sqrt{n}\, 3^n} x^n$

4. $\displaystyle\sum_{n=0}^{\infty} \frac{n}{4^n} x^n$

5. $\displaystyle\sum_{n=0}^{\infty} \frac{(-1)^n}{n+1} x^{2n}$

6. $\displaystyle\sum_{n=1}^{\infty} \frac{2}{3^{n+1} n^2} x^n$

7. $\displaystyle\sum_{n=1}^{\infty} \frac{(-1)^n}{2n-1} x^{n+1}$

8. $\displaystyle\sum_{n=1}^{\infty} \frac{n-1}{n^{2n}} x^n$

9. $\displaystyle\sum_{n=1}^{\infty} \frac{(-1)^n}{n^n} x^n$

10. $\displaystyle\sum_{n=1}^{\infty} \frac{2^n}{n^n} x^n$

11. $\displaystyle\sum_{n=0}^{\infty} \frac{n!}{(2n)!} x^n$

12. $\displaystyle\sum_{n=0}^{\infty} \frac{4^{n+1}}{\pi^{n+2}} x^{n+3}$

13. $\displaystyle\sum_{n=1}^{\infty} \frac{2^n}{n(3^{n+2})} x^{n+1}$

14. $\displaystyle\sum_{n=3}^{\infty} \frac{1}{n^3-4} x^n$

15. $\displaystyle\sum_{n=2}^{\infty} (\ln n) x^n$

16. $\displaystyle\sum_{n=2}^{\infty} \frac{\ln n}{n} x^n$

17. $\displaystyle\sum_{n=2}^{\infty} \frac{\ln n}{n^2} x^n$

18. $\displaystyle\sum_{n=2}^{\infty} \frac{1}{\ln n} x^n$

19. $\displaystyle\sum_{n=0}^{\infty} x^{(n^2)}$

20. $\displaystyle\sum_{n=1}^{\infty} x^{n!}$

In Exercises 21–25 find the radius of convergence of the given series.

21. $\displaystyle\sum_{n=1}^{\infty} \frac{n!}{n^n} x^n$

22. $\displaystyle\sum_{n=1}^{\infty} \frac{n^n}{n!} x^n$

23. $\displaystyle\sum_{n=1}^{\infty} \frac{1^2 \cdot 3^2 \cdot 5^2 \cdots (2n-1)^2}{2^2 \cdot 4^2 \cdot 6^2 \cdots (2n)^2} x^{2n}$

24. $\displaystyle\sum_{n=2}^{\infty} (-1)^{n+1} \frac{1 \cdot 3 \cdot 5 \cdots (2n-3)}{2^n n!} x^n$

25. $\displaystyle\sum_{n=1}^{\infty} \frac{1 \cdot 3 \cdot 5 \cdots (2n-1)}{2^n [1 \cdot 4 \cdot 7 \cdots (3n-2)]} x^n$

In Exercises 26–29, let $f(x)$ be the sum of the series. Find $f'(x)$ and $\int_0^x f(t)\, dt$.

26. $\displaystyle\sum_{n=0}^{\infty} \frac{(-1)^n}{n+1} x^n$

27. $\displaystyle\sum_{n=1}^{\infty} (n+1) x^n$

28. $\displaystyle\sum_{n=0}^{\infty} \frac{1}{n^2+1} x^{n+1}$

29. $\displaystyle\sum_{n=1}^{\infty} \frac{5}{n} x^{(n^2)}$

In Exercises 30–37 approximate the value of the integral with an error less than the given error, by first using the Integration Theorem to express the integral as an infinite series and then approximating the infinite series by an appropriate partial sum.

30. $\displaystyle\int_0^1 \cos\sqrt{x}\, dx;\ 10^{-3}$ $\left(\text{Hint: } \cos x = \displaystyle\sum_{n=0}^{\infty} \frac{(-1)^n}{(2n)!} x^{2n}.\right)$

31. $\displaystyle\int_0^2 \sin x^2\, dx;\ 10^{-2}$ $\left(\text{Hint: } \sin x = \displaystyle\sum_{n=0}^{\infty} \frac{(-1)^n}{(2n+1)!} x^{2n+1}.\right)$

32. $\displaystyle\int_0^1 \cos x^2\, dx;\ 10^{-7}$

33. $\displaystyle\int_0^1 \frac{1-e^{-x}}{x}\, dx;\ 10^{-3}$

34. $\displaystyle\int_0^{1/5} \arctan x\, dx;\ 10^{-5}$

35. $\displaystyle\int_0^{1/2} \frac{x^2}{1+x}\, dx;\ 10^{-3}$

36. $\displaystyle\int_0^1 \frac{x^3}{2+x}\, dx;\ 10^{-3}$

*37. $\displaystyle\int_{-1}^0 e^{x^2}\, dx;\ 10^{-3}$

38. Using the power series expansion for e^x given in (3), show that

a. $\cosh x = \displaystyle\sum_{n=0}^{\infty} \frac{1}{(2n)!} x^{2n}$ for all x

b. $\sinh x = \displaystyle\sum_{n=0}^{\infty} \frac{1}{(2n+1)!} x^{2n+1}$ for all x

39. Using formula (1) and the Differentiation Theorem, show that

$$\sum_{n=1}^{\infty} n x^n = \frac{x}{(1-x)^2}$$

40. Show that both

$$\sum_{n=0}^{\infty} \frac{(-1)^n}{(2n)!} x^{2n} \quad \text{and} \quad \sum_{n=0}^{\infty} \frac{(-1)^n}{(2n+1)!} x^{2n+1}$$

converge for all x.

41. Let

$$f(x) = \sum_{n=0}^{\infty} \frac{(-1)^n}{(2n)!} x^{2n}$$

$$g(x) = \sum_{n=0}^{\infty} \frac{(-1)^n}{(2n+1)!} x^{2n+1}$$

a. Show that $f'(x) = -g(x)$ and $g'(x) = f(x)$.
b. Show that $f''(x) = -f(x)$ and $g''(x) = -g(x)$.
c. What functions do you know that satisfy the properties of (a) and (b)?

42. Find a power series expansion for $(e^x - 1)/x$ and use it to verify that

$$\lim_{x \to 0} \frac{e^x - 1}{x} = 1$$

43. Find a power series expansion for $(e^x - 1 - x)/x^2$ and use it to evaluate

$$\lim_{x \to 0} \frac{e^x - 1 - x}{x^2}$$

44. Using (8), evaluate

$$\lim_{x \to 0} \frac{\ln(1 + x)}{x}$$

45. Express $\ln(1 + x^2)$ as a power series. What is the radius of convergence of that series?

46. Express $\ln[(1 + x)/(1 - x)]$ as a power series. (*Hint:* $\ln[(1 + x)/(1 - x)] = \ln(1 + x) - \ln(1 - x)$.)

47. a. Show that $\ln\dfrac{1}{1 - x} = \displaystyle\sum_{n=1}^{\infty} \frac{x^n}{n}$.

    b. Using (a), show that $\ln 2 = \displaystyle\sum_{n=1}^{\infty} \frac{1}{n2^n}$.

    c. Using the fact that

$$\sum_{n=N}^{\infty} \frac{1}{n2^n} \le \sum_{n=N}^{\infty} \frac{1}{N2^n}$$

    estimate $\ln 2$ with an error less than 0.01.

C 48. Approximate $\arctan \frac{1}{2}$ with an error less than 0.001.

C 49. Suppose that in Example 8 we had used $n = 3$ rather than $n = 2$ in the estimation of $4 \arctan \frac{1}{5}$. Find an upper bound for the error introduced for the estimate of $\pi/4$.

50. a. Show that

$$\frac{\arctan t}{t} = \sum_{n=0}^{\infty} \frac{(-1)^n}{2n + 1} t^{2n}$$

$$= 1 - \frac{t^2}{3} + \frac{t^4}{5} - \frac{t^6}{7} + \cdots \quad \text{for } 0 < |t| < 1$$

    b. Using part (a), conclude that

$$\lim_{t \to 0} (\arctan t)/t = 1$$

51. a. Using (1), with x replaced by $-t^4$, show that

$$\frac{t^2}{1 + t^4} = \sum_{n=0}^{\infty} (-1)^n t^{4n + 2} \quad \text{for } -1 < t < 1$$

    b. Using part (a), express $\int_0^{1/2} t^2/(1 + t^4)\, dt$ as the sum of a power series.

52. The **error function**, which is defined by

$$\text{erf}(x) = \frac{2}{\sqrt{\pi}} \int_0^x e^{-t^2}\, dt \quad \text{for all } x$$

    is prominent in statistics. Estimate $\text{erf}(1)$ with an error less than 0.01.

53. a. Using (3), find a power series expansion for xe^x.

    b. Show that

$$\sum_{n=0}^{\infty} \frac{1}{n!(n + 2)} = 1$$

    (*Hint:* Evaluate $\int_0^1 xe^x\, dx$ and $\int_0^1 (\sum_{n=0}^{\infty} c_n x^n)\, dx$, where $\sum_{n=0}^{\infty} c_n x^n$ is the power series obtained in (a).)

54. Let $R > 0$ be arbitrary. Find an example of a power series $\sum_{n=0}^{\infty} c_n x^n$ whose radius of convergence is R.

55. Show that if the radius of convergence of $\sum_{n=0}^{\infty} c_n x^n$ is R, then

$$\lim_{n \to \infty} \left| \frac{c_{n+1}}{c_n} \right| = \frac{1}{R}$$

    provided that the limit exists and is not 0.

56. Let $f(x) = \sum_{n=0}^{\infty} c_n x^n$ have a nonzero radius of convergence.

    a. Using Corollary 9.27, prove that if f is an even function, then $c_n = 0$ for all odd integers n. (*Hint:* Find the series expansion for $f(-x)$, and equate it to the series expansion for $f(x)$.)

    b. Prove that if f is an odd function, then $c_n = 0$ for all even integers n.

57. A pendulum L meters long swings back and forth, with maximum angle of deflection ϕ. Let $x = \sin \phi/2$, and let $g = 32$ (feet per second per second) be the acceleration due to gravity. Then the period of oscillation T of the pendulum is given by the formula

$$T(x) = 2\pi \sqrt{\frac{L}{g}} \left(1 + \sum_{n=1}^{\infty} \frac{1^2 \cdot 3^2 \cdot 5^2 \cdots (2n - 1)^2}{2^2 \cdot 4^2 \cdot 6^2 \cdots (2n)^2} x^{2n} \right)$$

The series on the right converges for $|x| < 1$ (see Exercise 23). Notice that

$$\left| \frac{T(x)}{2\pi \sqrt{L/g}} - 1 \right| = \sum_{n=1}^{\infty} \frac{1^2 \cdot 3^2 \cdot 5^2 \cdots (2n - 1)^2}{2^2 \cdot 4^2 \cdot 6^2 \cdots (2n)^2} x^{2n} \quad (13)$$

Usually ϕ is very small in pendulum clocks, say, $0 < \phi \le \pi/12$, and consequently

$$0 < x^2 = \sin^2 \phi/2 < 0.018$$

Show that if $x^2 < 0.018$, then the series in (13) is bounded by 0.005. (*Hint:* The coefficient of x^{2n} in the series is less than or equal to $\frac{1}{4}$ for all n.) It is because of (13) that the period of a pendulum of length L is often assumed to be $2\pi \sqrt{L/g}$, rather than $T(x)$ for the correct value of x. We also deduce from (13) that as the maximum angle of deflection of the pendulum shrinks slightly while the clock runs down, the period $T(x)$ stays practically the same, and thus the clock keeps good time.

9.9
TAYLOR SERIES

In the preceding section we showed that

$$e^x = \sum_{n=0}^{\infty} \frac{x^n}{n!} \quad \text{for all } x \tag{1}$$

$$\ln(1+x) = \sum_{n=0}^{\infty} \frac{(-1)^n}{n+1} x^{n+1} \quad \text{for } -1 < x < 1 \tag{2}$$

$$\arctan x = \sum_{n=0}^{\infty} \frac{(-1)^n}{2n+1} x^{2n+1} \quad \text{for } -1 < x < 1 \tag{3}$$

In each case we say that we have a power series representation of the given function. More generally, if f is a function, if I is an open interval containing 0, and if

$$f(x) = \sum_{n=0}^{\infty} c_n x^n \quad \text{for } x \text{ in } I$$

then we say that we have a **power series representation of f on I**. The main advantage of having a power series representation of a function f on I is that the value of f at any point in I is the sum of a convergent series and hence can be approximated by its partial sums. Since the partial sums of a power series are polynomials, values of functions having power series representations are easily approximated and are therefore tractable (with the help of computers, in some cases). In this section we will study functions that have power series representations, and in so doing we will complete the discussion of the approximation of functions by polynomials that we initiated in Section 9.1.

For a given function f we wish to determine whether there is a power series $\sum_{n=0}^{\infty} c_n x^n$ and an open interval I containing 0 such that

$$f(x) = \sum_{n=0}^{\infty} c_n x^n \quad \text{for } x \text{ in } I \tag{4}$$

We can already eliminate many functions such as $|x|$ from consideration, because Theorem 9.26 asserts that if f has such a power series representation, then f must have derivatives of all orders on I. Moreover, formula (5) in Theorem 9.26 tells us that if f has a representation in the form of (4), then $c_n = f^{(n)}(0)/n!$. Thus a function can have only one power series representation, and (4) can be rewritten

$$f(x) = \sum_{n=0}^{\infty} \frac{f^{(n)}(0)}{n!} x^n \quad \text{for } x \text{ in } I \tag{5}$$

DEFINITION 9.29

Suppose that f has derivatives of all orders at 0. Then the **Taylor series of f** is the power series

$$\sum_{n=0}^{\infty} \frac{f^{(n)}(0)}{n!} x^n$$

From the discussion above we know that once we find a power series representation of a function, that power series must be the Taylor series for the function. It is therefore apparent that the series appearing in (1)–(3) are the Taylor series of their respective functions. Moreover, if f is a polynomial, say

$$f(x) = c_0 + c_1 x + c_2 x^2 + \cdots + c_n x^n$$

then f is a power series (with $c_j = 0$ for $j > n$), and consequently f is its own Taylor series.

For a function f that is not necessarily a polynomial, we use the nth Taylor polynomial p_n and the nth Taylor remainder r_n of f defined in Section 9.1 by

$$p_n(x) = f(0) + f'(0)x + \frac{f''(0)}{2!}x^2 + \frac{f^{(3)}(0)}{3!}x^3 + \cdots + \frac{f^{(n)}(0)}{n!}x^n$$

and

$$r_n(x) = f(x) - p_n(x)$$

It follows that the Taylor polynomials of f are the partial sums of the Taylor series of f and that

$$f(x) = \sum_{n=0}^{\infty} \frac{f^{(n)}(0)}{n!} x^n \quad \text{if and only if} \quad \lim_{n \to \infty} r_n(x) = 0 \tag{6}$$

Thus if we wish to show that for a given value of x, the Taylor series of f converges to $f(x)$, we must show that $\lim_{n \to \infty} r_n(x) = 0$. To accomplish this we usually use the Lagrange form of the remainder given in Taylor's Theorem (Theorem 9.1):

$$r_n(x) = \frac{f^{(n+1)}(t_x)}{(n+1)!} x^{n+1} \tag{7}$$

where t_x lies between 0 and x.

Example 1 Show that

$$\sin x = x - \frac{x^3}{3!} + \frac{x^5}{5!} - \frac{x^7}{7!} + \cdots = \sum_{n=0}^{\infty} \frac{(-1)^n}{(2n+1)!} x^{2n+1} \quad \text{for all } x$$

Solution Let $f(x) = \sin x$. The derivatives of f repeat themselves in groups of four:

$$f(x) = \sin x, \qquad f'(x) = \cos x, \qquad f''(x) = -\sin x, \qquad f^{(3)}(x) = -\cos x,$$

$$f^{(4)}(x) = \sin x, \qquad f^{(5)}(x) = \cos x, \qquad f^{(6)}(x) = -\sin x, \qquad f^{(7)}(x) = -\cos x$$

and so on. Thus the derivatives of even order involve $\sin x$, whereas the derivatives of odd order involve $\cos x$. More precisely, for each nonnegative integer k,

$$f^{(2k)}(x) = (-1)^k \sin x \quad \text{and} \quad f^{(2k+1)}(x) = (-1)^k \cos x \qquad (8)$$

Therefore

$$f^{(2k)}(0) = 0 \quad \text{and} \quad f^{(2k+1)}(0) = (-1)^k$$

Since $f^{(2k)}(0) = 0$, all the coefficients of the even powers of x in the Taylor series for $\sin x$ are 0. Thus we suppress the even powers and write the Taylor series of $\sin x$ as

$$x - \frac{x^3}{3!} + \frac{x^5}{5!} - \frac{x^7}{7!} + \cdots = \sum_{n=0}^{\infty} \frac{(-1)^n}{(2n+1)!} x^{2n+1}$$

To show that the Taylor series converges to $\sin x$ for every x, we will show that $\lim_{n \to \infty} r_n(x) = 0$ for all x. From (8) we deduce that, irrespective of the integer n or the numbers x and t_x, we have $|f^{(n+1)}(t_x)| \leq 1$, so that by (7),

$$|r_n(x)| = \left| \frac{f^{(n+1)}(t_x)}{(n+1)!} x^{n+1} \right| \leq \frac{|x|^{n+1}}{(n+1)!} \qquad (9)$$

It follows from Example 7 of Section 9.7 that

$$\lim_{n \to \infty} \frac{|x|^{n+1}}{(n+1)!} = \lim_{n \to \infty} \frac{|x|^n}{n!} = 0$$

Therefore $\lim_{n \to \infty} r_n(x) = 0$ for all x. Consequently (6) implies that the Taylor series of the sine function converges to $\sin x$ for every x. $\square$

From (9) we notice that

$$|r_n(x)| \leq \frac{(\pi/2)^{n+1}}{(n+1)!}$$

for every x in $[-\pi/2, \pi/2]$. Taking $n = 10$, we find that $|r_{10}(x)| \leq 0.000004$ for all x in $[-\pi/2, \pi/2]$. Therefore if we wish to tabulate the values of $\sin x$ on $[-\pi/2, \pi/2]$ accurate to five digits, we need only calculate the values of the tenth Taylor polynomial, which is the same as the ninth Taylor polynomial

$$x - \frac{x^3}{3!} + \frac{x^5}{5!} - \frac{x^7}{7!} + \frac{x^9}{9!}$$

By an analysis similar to that in Example 1, it is possible to show that

$$\cos x = 1 - \frac{x^2}{2!} + \frac{x^4}{4!} - \frac{x^6}{6!} + \cdots = \sum_{n=0}^{\infty} \frac{(-1)^n}{(2n)!} x^{2n}$$

The error estimate given in (9) remains valid for the cosine. Accordingly, it is possible to prepare a table of values of $\cos x$ on $[-\pi/2, \pi/2]$ accurate to five digits by using the tenth Taylor polynomial of $\cos x$:

$$1 - \frac{x^2}{2!} + \frac{x^4}{4!} - \frac{x^6}{6!} + \frac{x^8}{8!} - \frac{x^{10}}{10!}$$

Caution: Do not jump to the conclusion that the Taylor series of every function that is infinitely differentiable at 0 converges to the function on an open interval containing 0. The classic counterexample is given by

$$f(x) = \begin{cases} e^{-1/x^2} & \text{for } x \neq 0 \\ 0 & \text{for } x = 0 \end{cases}$$

By the formula for f, $f(0) = 0$, and an application of l'Hôpital's Rule shows that $f'(0) = 0$ (see the remark following Example 5 of Section 7.8). With a great deal more effort it is possible to show that f has derivatives of all orders at 0 and that

$$f^{(n)}(0) = 0 \quad \text{for all } n \geq 0$$

As a result, the Taylor series of f is 0, which implies that the Taylor series of f does not converge to $f(x)$ for any $x \neq 0$.

Taylor Series About an Arbitrary Point

If a function f is to have a Taylor series

$$\sum_{n=0}^{\infty} \frac{f^{(n)}(0)}{n!} x^n \tag{10}$$

then f must have derivatives at 0 and hence must be defined in an open interval about 0. Since $\ln x$ is not defined in an open interval about 0, $\ln x$ does not have a Taylor series of the form given in (10) for $f(x) = \ln x$. However, in (2) we presented the Taylor series for $\ln (1 + x)$:

$$\ln(1 + x) = \sum_{n=0}^{\infty} \frac{(-1)^n}{n+1} x^{n+1} \quad \text{for } -1 < x < 1 \tag{11}$$

Now if $0 < x < 2$, then $-1 < x - 1 < 1$, so that by (11) we find that

$$\ln x = \ln(1 + (x - 1))$$

$$= \sum_{n=0}^{\infty} \frac{(-1)^n}{n+1} (x - 1)^{n+1} \quad \text{for } 0 < x < 2 \tag{12}$$

The series in (12) is also a power series, but it contains powers of $x - 1$ rather than powers of x, and its interval of convergence is centered at 1 rather than at 0.

In general, we are interested in the possibility of expressing a function as a power series

$$\sum_{n=0}^{\infty} c_n(x - a)^n \tag{13}$$

in powers of $x - a$, where a can be any fixed number. All the results we obtained for ordinary power series have their counterparts for power series of the type given in (13). In particular, if f has derivatives of all orders at a, then we call

$$\sum_{n=0}^{\infty} \frac{f^{(n)}(a)}{n!}(x-a)^n \tag{14}$$

the **Taylor series of f about the number a.*** The **nth Taylor polynomial p_n of f about a** is defined by

$$p_n(x) = f(a) + f'(a)(x-a) + \frac{f''(a)}{2!}(x-a)^2 + \cdots + \frac{f^{(n)}(a)}{n!}(x-a)^n$$

and the **nth Taylor remainder r_n of f about a** is defined by

$$r_n(x) = f(x) - p_n(x)$$

Taylor's Theorem

There is a version of Taylor's Theorem, which originally appeared as Theorem 9.1, for Taylor series about an arbitrary number a. It provides us with a way of determining those values of x for which the Taylor series of a function f converges to $f(x)$. It also allows us to estimate the accuracy obtained in using a Taylor polynomial $p_n(x)$ of a function f to approximate the value of $f(x)$.

THEOREM 9.30
TAYLOR'S THEOREM

Let n be a nonnegative integer, and suppose $f^{(n+1)}(x)$ exists for each x in an open interval I containing a. For each $x \neq a$ in I, there is a number t_x strictly between a and x such that

$$f(x) = f(a) + f'(a)(x-a) + \cdots + \frac{f^{(n)}(a)}{n!}(x-a)^n + \frac{f^{(n+1)}(t_x)}{(n+1)!}(x-a)^{n+1} \tag{15}$$

and

$$r_n(x) = \frac{f^{(n+1)}(t_x)}{(n+1)!}(x-a)^{n+1} \tag{16}$$

(Equation (15) is known as **Taylor's Formula**, and (16) is known as the **Lagrange Remainder Formula**.)

Proof In the proof of the Mean Value Theorem we introduced an auxiliary function g, which allowed us to apply Rolle's Theorem. Here we also introduce an

---

* If $a = 0$, the Taylor series becomes the series $\sum_{n=0}^{\infty} f^{(n)}(0)x^n/n!$, which we have already discussed in detail and which is frequently called a Maclaurin series, after the Scottish mathematician Colin Maclaurin (1698–1746).

auxiliary function g. For any fixed $x \neq a$ in I and any t in I, let

$$g(t) = f(x) - f(t) - f'(t)(x - t) - \frac{f''(t)}{2!}(x - t)^2 - \frac{f^{(3)}(t)}{3!}(x - t)^3 - \cdots$$

$$- \frac{f^{(n)}(t)}{n!}(x - t)^n - r_n(x)\frac{(x - t)^{n+1}}{(x - a)^{n+1}}$$

Then $g(x) = g(a) = 0$, as you can check by substituting and using the definition of $r_n(x)$. Next we find $g'(t)$ (remembering that x is fixed):

$$g'(t) = 0 - f'(t) + [f'(t) - f''(t)(x - t)]$$

$$+ \left(\frac{2f''(t)}{2!}(x - t) - \frac{f^{(3)}(t)}{2!}(x - t)^2\right)$$

$$+ \left(\frac{3f^{(3)}(t)}{3!}(x - t)^2 - \frac{f^{(4)}(t)}{3!}(x - t)^3\right) + \cdots$$

$$+ \left(\frac{nf^{(n)}(t)}{n!}(x - t)^{n-1} - \frac{f^{(n+1)}(t)}{n!}(x - t)^n\right)$$

$$+ (n + 1)r_n(x)\frac{(x - t)^n}{(x - a)^{n+1}}$$

As you see, adjacent pairs of terms cancel each other, leaving

$$g'(t) = \frac{-f^{(n+1)}(t)}{n!}(x - t)^n + (n + 1)r_n(x)\frac{(x - t)^n}{(x - a)^{n+1}}$$

By Rolle's Theorem, there is a number t_x strictly between a and x such that $g'(t_x) = 0$. This means that

$$0 = \frac{-f^{(n+1)}(t_x)}{n!}(x - t_x)^n + (n + 1)r_n(x)\frac{(x - t_x)^n}{(x - a)^{n+1}}$$

Solving for $r_n(x)$, we obtain

$$r_n(x) = \frac{f^{(n+1)}(t_x)}{(n + 1)!}(x - a)^{n+1}$$

This proves (16), and (15) now follows from the definition of $r_n(x)$. ■

The number t_x given in Theorem 9.30 depends on n as well as x.

Example 2   Express the polynomial

$$f(x) = 2x^3 - 9x^2 + 11x - 1 \tag{17}$$

as a polynomial in $(x - 2)$.

Solution   We wish to write $f(x)$ in the form of (14) with $a = 2$, and to do so we must compute the derivatives of f at 2:

$$f(x) = 2x^3 - 9x^2 + 11x - 1 \qquad f(2) = 1$$
$$f'(x) = 6x^2 - 18x + 11 \qquad f'(2) = -1$$
$$f''(x) = 12x - 18 \qquad f''(2) = 6$$
$$f^{(3)}(x) = 12 \qquad f^{(3)}(2) = 12$$
$$f^{(n)}(x) = 0 \quad \text{for } n \geq 4 \qquad f^{(n)}(2) = 0 \quad \text{for } n \geq 4$$

Therefore

$$f(x) = 1 + (-1)(x - 2) + \frac{6}{2!}(x - 2)^2 + \frac{12}{3!}(x - 2)^3$$
$$= 1 - (x - 2) + 3(x - 2)^2 + 2(x - 2)^3 \quad \square$$

Although the form of the polynomial just obtained looks quite different from the polynomial given in (17), both polynomials represent the same function.

Example 3   Find the Taylor series of $\sin x$ about $\pi/6$, and show that it converges to $\sin x$ for all x.

Solution   Let $f(x) = \sin x$. This time we wish to write $f(x)$ in the form of (14) with $a = \pi/6$. The derivatives of f are

$$f(x) = \sin x \qquad f(\pi/6) = \frac{1}{2}$$

$$f'(x) = \cos x \qquad f'(\pi/6) = \frac{\sqrt{3}}{2}$$

$$f''(x) = -\sin x \qquad f''(\pi/6) = \frac{-1}{2}$$

$$f^{(3)}(x) = -\cos x \qquad f^{(3)}(\pi/6) = \frac{-\sqrt{3}}{2}$$

$$f^{(4)}(x) = \sin x \qquad f^{(4)}(\pi/6) = \frac{1}{2}$$

Since $f^{(4)}(x) = f(x)$, we deduce that

$$|f^{(n+1)}(t_x)| \leq 1 \quad \text{for all } x$$

With this information we can apply (16) to obtain

$$|r_n(x)| = \left| \frac{f^{(n+1)}(t_x)}{(n+1)!}(x - \pi/6)^{n+1} \right| \leq \left| \frac{(x - \pi/6)^{n+1}}{(n+1)!} \right| \qquad (18)$$

With a slight modification of the result of Example 7 of Section 9.7, we deduce that

$$\lim_{n \to \infty} \frac{(x - \pi/6)^{n+1}}{(n+1)!} = \lim_{n \to \infty} \frac{(x - \pi/6)^n}{n!} = 0$$

which implies that

$$\lim_{n \to \infty} r_n(x) = 0 \quad \text{for all } x$$

As a result, the Taylor series of f converges to f. Moreover

$$f^{(4n)}\left(\frac{\pi}{6}\right) = \frac{1}{2}, \qquad f^{(4n+1)}\left(\frac{\pi}{6}\right) = \frac{\sqrt{3}}{2},$$

$$f^{(4n+2)}\left(\frac{\pi}{6}\right) = \frac{-1}{2}, \qquad f^{(4n+3)}\left(\frac{\pi}{6}\right) = \frac{-\sqrt{3}}{2}$$

Therefore

$$f(x) = \frac{1}{2} + \frac{\sqrt{3}}{2}\left(x - \frac{\pi}{6}\right) - \frac{1}{2}\frac{1}{2!}\left(x - \frac{\pi}{6}\right)^2 - \frac{\sqrt{3}}{2} \cdot \frac{1}{3!}\left(x - \frac{\pi}{6}\right)^3 + \cdots$$

$$= \frac{1}{2} + \frac{\sqrt{3}}{2}\left(x - \frac{\pi}{6}\right) - \frac{1}{4}\left(x - \frac{\pi}{6}\right)^2 - \frac{\sqrt{3}}{12}\left(x - \frac{\pi}{6}\right)^3 + \cdots$$

In condensed form the series is given by

$$f(x) = \frac{1}{2}\sum_{n=0}^{\infty} (-1)^n \left(\frac{1}{(2n)!}(x - \pi/6)^{2n} + \frac{\sqrt{3}}{(2n+1)!}(x - \pi/6)^{2n+1}\right) \quad \square$$

Example 4 Using the Taylor series obtained in Example 3, approximate $\sin 7\pi/36$ with an error less than 0.000003.

Solution A calculation based on (18) shows that

$$\left| r_n\left(\frac{7\pi}{36}\right) \right| < 0.000003$$

for $n \geq 3$. Therefore the third Taylor polynomial about $\pi/6$, obtained from the solution of Example 3, furnishes the desired approximation:

$$p_3\left(\frac{7\pi}{36}\right) = \frac{1}{2} + \frac{\sqrt{3}}{2}\left(\frac{\pi}{36}\right) - \frac{1}{4}\left(\frac{\pi}{36}\right)^2 - \frac{\sqrt{3}}{12}\left(\frac{\pi}{36}\right)^3 \approx 0.573575 \quad \square$$

If we had estimated $\sin(7\pi/36)$ by using Taylor polynomials about 0, we would have needed many more terms to achieve the accuracy obtained in Example 4. In general, one obtains a better approximation for $f(x)$ by considering the Taylor polynomials about a point a nearer to x.

In this section we have described Taylor series of several kinds of functions: polynomials, e^x, $\sin x$, $\cos x$, $\ln(1 + x)$, $\arctan x$, and others related to these functions. But we caution you that in practice it is usually *very* difficult to write Taylor series for most other functions, unless they are rather closely related to those functions whose Taylor series we already know. As a result, in the exercises we will concentrate on Taylor series of the functions just mentioned and their relatives.

Historical Comment on Series

Power series, which eventually came to be called Taylor series, contributed greatly to the growth of calculus. These series allowed Newton and Gregory and their successors to analyze properties of functions with a single theory and to approximate values of functions easily; as a result, Taylor series became immensely important to mathematicians such as Euler and Lagrange during the eighteenth century. In fact, Lagrange isolated Taylor series as the notion fundamental to calculus. He maintained that all one had to do in order to understand a continuous function f was to find the derivatives $f^{(n)}(a)$ at a given number a of one's choice. Since he, like all other mathematicians at that time, had no clear concept of convergence, he did not realize that not every continuous function has a Taylor series. In time, mathematicians found that Taylor series did not provide the sole key to understanding continuous functions. Moreover, there appeared other kinds of series, such as Fourier series, which are basic to the study of waves. As we have seen, Taylor series can be used to provide estimates of such numbers as e, $\ln 2$, and π; yet the most economical estimates of irrational numbers often come from other series and from other considerations. In spite of the limitations of Taylor series, they have had a major impact on the development of calculus, and they continue to be of theoretical interest.

EXERCISES 9.9

In Exercises 1–8 find the third Taylor polynomial of the given function about a.

1. $f(x) = x^4 - x + 2; a = -1$
2. $f(x) = \sqrt{x}; a = 1$
3. $f(x) = \sqrt{x}; a = 4$
4. $f(x) = \ln x; a = 1$
5. $g(x) = \cos x; a = \pi/4$
6. $f(x) = \arctan x; a = 1$
7. $k(x) = \csc x; a = \pi/2$
8. $k(x) = \tan x; a = \pi/3$

In Exercises 9–15 find the Taylor series of f about a. In each case show that the Taylor series you find converges to f by demonstrating that $\lim_{n \to \infty} r_n(x) = 0$ for all x.

9. $f(x) = 4x^2 - 2x + 1; a = 0, -3$
10. $f(x) = 5x^3 + 4x^2 + 3x + 2; a = 0, 2$
11. $f(x) = \cos x; a = 0$
12. $f(x) = \cos x; a = \pi/3$
13. $f(x) = \sin x; a = \pi/3$
14. $f(x) = \sin x; a = 3\pi/4$
15. $f(x) = e^x; a = -2$

In Exercises 16–20 find the Taylor series of f about the given number a. (Do not be concerned with whether the series converges to the given function.)

16. $f(x) = \dfrac{1}{x}; a = -1$
17. $f(x) = \dfrac{1}{x}; a = 3$
18. $f(x) = \dfrac{1}{x + 2}; a = 0$
19. $f(x) = \ln x; a = 2$
*20. $f(x) = \sqrt{x}; a = 1$
21. Let $\Sigma_{n=0}^{\infty} c_n x^n$ be the Taylor series for f, and let $m \geq 0$. Show that $\Sigma_{n=0}^{\infty} c_n x^{n+m}$ is the Taylor series for the function $x^m f$.

In Exercises 22–36 find the Taylor series of the given function about a. Use the series already obtained in the text or in previous exercises.

22. $f(x) = \sin 2x; a = 0$
23. $f(x) = \cos x^2; a = 0$
24. $f(x) = \ln 3x; a = 1$

25. $f(x) = \ln(1 - 2x); a = 0$

26. $g(x) = \ln \dfrac{1 + x}{1 - x}; a = 0$

27. $f(x) = x \ln(1 + x^2); a = 0$

28. $g(x) = \sinh x; a = 0$ 29. $k(x) = 2^x; a = 0$

30. $f(x) = 10^x; a = 0$ 31. $f(x) = \dfrac{x - 1}{x + 1}; a = 1$

32. $f(x) = \sin^2 x; a = 0$ (*Hint:* Observe that $\sin^2 x = \frac{1}{2}(1 - \cos 2x)$, and then determine the series.)

33. $f(x) = \cos^2 x; a = 0$

34. $f(x) = e^x \cos x; a = 0$ (Write out only the first four terms of the series.)

35. $f(x) = \begin{cases} \dfrac{\sin x}{x} & \text{for } x \neq 0 \\ 1 & \text{for } x = 0 \end{cases}$ $a = 0$

36. $f(x) = \begin{cases} \dfrac{(\sin x) - x}{x^3} & \text{for } x \neq 0 \\ -\frac{1}{6} & \text{for } x = 0 \end{cases}$ $a = 0$

37. Find the Taylor series about 0 of

$$\frac{-3x + 2}{2x^2 - 3x + 1}$$

by using a partial fraction decomposition of the rational function.

*38. Let $f(x) = \tan x$. Using the fact that $f(0) = 0$ and $f'(x) = 1 + [f(x)]^2$, find the sum of the first six terms in the Taylor series of f about 0.

*39. Let $f(x) = e^{(x^2)} \int_0^x e^{-(t^2)} \, dt$.
   a. Show that $f(0) = 0$ and $f'(x) = 2xf(x) + 1$.

b. Find the Taylor series of f about 0. (*Hint:* Note that $f^{(n)}(x) = 2(n - 1)f^{(n-2)}(x) + 2xf^{(n-1)}(x)$ for $n \geq 2$.)

*40. a. Show that if there is a number M such that $|f^{(n)}(x)| \leq M$ for all x and for all $n \geq 0$, then the Taylor series of f about any given point a converges to $f(x)$ for all x.

   b. Let $0 < \delta < 1$. Suppose that for each x in $(a - \delta, a + \delta)$ except a there is a number M_x such that

$$|f^{(n+1)}(t)| \leq \frac{M_x n!}{|x - a|^{n+1}} \quad \text{for all } t \text{ between } x \text{ and } a \text{ and for all } n \geq 0$$

   Show that the Taylor series of f about a converges to $f(x)$ for all x in $(a - \delta, a + \delta)$.

*41. Assume that $\{a_n\}_{n=1}^{\infty}$ is the Fibonacci sequence, defined by $a_1 = a_2 = 1$, and $a_{n+2} = a_{n+1} + a_n$ for $n \geq 1$.
   a. Show that the radius of convergence of $\sum_{n=1}^{\infty} a_n x^n$ is at least $\frac{1}{2}$. (*Hint:* Show that $0 \leq a_n \leq 2a_{n-1}$, so that $0 \leq a_n \leq 2^n$.)
   b. Show that

$$\sum_{n=1}^{\infty} a_n x^n = \frac{x}{1 - x - x^2} \quad \text{for } |x| < \frac{1}{2}$$

   (*Hint:* Show that

$$\sum_{n=1}^{\infty} a_n x^n - x \sum_{n=1}^{\infty} a_n x^n - x^2 \sum_{n=1}^{\infty} a_n x^n = x$$

   for $|x| < \frac{1}{2}$.)

42. Let f be the position function of a car that is stationary for $t \leq 0$ and is in motion for $t > 0$. Show that f does not have a power series representation on any open interval containing 0.

9.10
BINOMIAL SERIES

We conclude Chapter 9 by investigating Taylor series about 0 of functions given by

$$f(x) = (1 + x)^s$$

where s is any fixed number. These functions are called **binomial functions**, because they arise from the binomial, or two-term, expression $1 + x$. They had enormous influence on the early growth of calculus.

We begin by letting $s = \frac{1}{2}$, so that

$$f(x) = (1 + x)^{1/2}$$

In order to derive the Taylor series of f, we first find the derivatives of f:

$$f(x) = (1 + x)^{1/2} \qquad\qquad f(0) = 1$$

$$f'(x) = \frac{1}{2}(1 + x)^{-1/2} \qquad\qquad f'(0) = \frac{1}{2}$$

$$f''(x) = \left(\frac{1}{2}\right)\left(\frac{-1}{2}\right)(1 + x)^{-3/2} \qquad\qquad f''(0) = -\frac{1}{4}$$

$$f^{(3)}(x) = \left(\frac{1}{2}\right)\left(\frac{-1}{2}\right)\left(\frac{-3}{2}\right)(1 + x)^{-5/2} \qquad\qquad f^{(3)}(0) = \frac{3}{8}$$

$$f^{(4)}(x) = \left(\frac{1}{2}\right)\left(\frac{-1}{2}\right)\left(\frac{-3}{2}\right)\left(\frac{-5}{2}\right)(1 + x)^{-7/2} \qquad f^{(4)}(0) = -\frac{15}{16}$$

and in general,

$$f^{(n)}(x) = \left(\frac{1}{2}\right)\left(\frac{-1}{2}\right)\left(\frac{-3}{2}\right)\cdots\left(\frac{-2n + 3}{2}\right)(1 + x)^{(-2n+1)/2} \quad \text{for } n \geq 2$$

$$f^{(n)}(0) = (-1)^{n+1} \frac{1 \cdot 3 \cdot 5 \cdots (2n - 3)}{2^n} \quad \text{for } n \geq 2$$

Consequently the Taylor series of f is

$$1 + \frac{1}{2}x + \sum_{n=2}^{\infty} (-1)^{n+1} \frac{1 \cdot 3 \cdot 5 \cdots (2n - 3)}{2^n n!} x^n \tag{1}$$

which can also be written

$$1 + \frac{1}{2}x + \sum_{n=2}^{\infty} (-1)^{n+1} \frac{1 \cdot 3 \cdot 5 \cdots (2n - 3)}{2 \cdot 4 \cdot 6 \cdots (2n)} x^n$$

By the Ratio Test you can show that the radius of convergence of the series in (1) is 1 (see Exercise 24 of Section 9.8).

More generally, let s be any number and let $f(x) = (1 + x)^s$. Then we define the **binomial coefficient** $\binom{s}{n}$ by the formulas

$$\binom{s}{0} = 1 \quad \text{, and} \quad \binom{s}{n} = \frac{s(s - 1)(s - 2)\cdots(s - (n - 1))}{n!} \quad \text{for } n \geq 1$$

In particular,

$$\binom{s}{1} = s \quad \text{and} \quad \binom{s}{2} = \frac{s(s - 1)}{2}$$

Moreover, if s is a positive integer, then

$$\binom{s}{n} = \frac{s!}{n!(s - n)!}$$

Using binomial coefficients, we find that the Taylor series of f can be written

$$\sum_{n=0}^{\infty} \binom{s}{n} x^n$$

This series is called a **binomial series**. It is possible (although not easy) to show that

$$(1 + x)^s = \sum_{n=0}^{\infty} \binom{s}{n} x^n \quad \text{for} \quad \begin{cases} -1 < x < 1 \text{ if } s \leq -1 \\ -1 < x \leq 1 \text{ if } -1 < s < 0 \\ -1 \leq x \leq 1 \text{ if } s > 0 \text{ and } s \text{ is not an integer} \\ \text{all } x \text{ if } s \text{ is a nonnegative integer} \end{cases} \tag{2}$$

It follows from (2), for example, that

$$(1 + x)^{1/2} = \sum_{n=0}^{\infty} \binom{1/2}{n} x^n$$

$$= 1 + \frac{1}{2} x + \frac{1}{2!} \left(\frac{1}{2}\right)\left(\frac{-1}{2}\right) x^2 + \frac{1}{3!} \left(\frac{1}{2}\right)\left(\frac{-1}{2}\right)\left(\frac{-3}{2}\right) x^3 + \cdots$$

$$(1 - x^2)^{1/2} = \sum_{n=0}^{\infty} \binom{1/2}{n}(-1)^n x^{2n}$$

$$= 1 - \frac{1}{2} x^2 + \frac{1}{2!} \left(\frac{1}{2}\right)\left(\frac{-1}{2}\right) x^4 - \frac{1}{3!} \left(\frac{1}{2}\right)\left(\frac{-1}{2}\right)\left(\frac{-3}{2}\right) x^6 + \cdots \tag{3}$$

$$(1 + x)^{1/3} = \sum_{n=0}^{\infty} \binom{1/3}{n} x^n$$

$$= 1 + \frac{1}{3} x + \frac{1}{2!} \left(\frac{1}{3}\right)\left(\frac{-2}{3}\right) x^2 + \frac{1}{3!} \left(\frac{1}{3}\right)\left(\frac{-2}{3}\right)\left(\frac{-5}{3}\right) x^3 + \cdots \tag{4}$$

Newton and Mercator discovered the binomial series independently during the mid-1660s. At that time Mercator had already gained recognition as a mathematician, but the binomial series was among Newton's initial successes, achieved while he was a student at Cambridge University. Thereafter binomial series played a key role in Newton's method of differentiating and integrating, and his facility with series led him to profound insights into the subject we now call calculus.

Before turning to approximate values of numbers, we will estimate the error $r_N(x)$ introduced by taking the Nth Taylor polynomial of $(1 + x)^s$, for certain values of s. If $|s| \leq 1$ and $n \geq 1$, then

$$\left|\binom{s}{n}\right| \leq \left|\binom{s}{n-1}\right|$$

as you can verify by writing out the binomial coefficients involved and comparing them (see Exercise 19). Then the Nth Taylor remainder $r_N(x)$ satisfies

$$|r_N(x)| \leq \sum_{n=N+1}^{\infty} \left|\binom{s}{n}\right| |x|^n \leq \sum_{n=N+1}^{\infty} \left|\binom{s}{1}\right| |x|^n = |s| \frac{|x|^{N+1}}{1 - |x|} \tag{5}$$

This estimate of the remainder allows us to approximate various roots of numbers by means of the Taylor polynomials of binomial series.

Example 1   Find an approximate value of $\sqrt[3]{28}$ with an error less than 0.0001.

Solution   We begin by writing $\sqrt[3]{28}$ as

$$b\sqrt[3]{1+x} = b(1+x)^{1/3}$$

where b is a convenient number and $|x|$ is small:

$$\sqrt[3]{28} = \sqrt[3]{27+1} = \sqrt[3]{27}\sqrt[3]{1+\tfrac{1}{27}} = 3\sqrt[3]{1+\tfrac{1}{27}}$$

This means that if we find an approximation A to $\sqrt[3]{1+\tfrac{1}{27}}$ with an error less than, say, 0.00003, then by taking $3A$ for $\sqrt[3]{28}$ we will introduce an error of no more than $3(0.00003) < 0.0001$. From (5) with $s = \tfrac{1}{3}$ and $x = \tfrac{1}{27}$ we obtain

$$|r_N(\tfrac{1}{27})| \le \frac{1}{3}\frac{(1/27)^{N+1}}{26/27} = \frac{1}{78(27)^N}$$

so that $|r_N(\tfrac{1}{27})| < 0.00003$ if $N \ge 2$. Consequently the desired approximation is obtained by taking the first three terms of the series in (4), with $x = \tfrac{1}{27}$:

$$\sqrt[3]{28} = 3\sqrt[3]{1+\frac{1}{27}} \approx 3\left[1 + \left(\frac{1}{3}\right)\left(\frac{1}{27}\right) + \left(\frac{1}{2!}\right)\left(\frac{1}{3}\right)\left(\frac{-2}{3}\right)\left(\frac{1}{27}\right)^2\right] \approx 3.03658 \quad \square$$

If instead we had desired an estimate of $\sqrt[6]{60}$, we would have rewritten it

$$\sqrt[6]{60} = \sqrt[6]{64\left(1 - \frac{4}{64}\right)} = \sqrt[6]{64}\,\sqrt[6]{1-\frac{1}{16}} = 2\sqrt[6]{1-\frac{1}{16}}$$

In this case $\sqrt[6]{60}$ has the form $b\sqrt[6]{1+x}$, where $b = 2$ and $x = -\tfrac{1}{16}$.

EXERCISES 9.10

In Exercises 1–6 approximate the number with an error less than 0.001.

1. $\sqrt{1.05}$ 2. $\sqrt[3]{9}$ 3. $\sqrt[4]{83}$

4. $\sqrt[5]{35}$ 5. $\sqrt[6]{65}$ 6. $29^{2/3}$

In Exercises 7–14 find the Taylor series of f about a.

7. $f(x) = \dfrac{1}{\sqrt{1+x}}$; $a = 0$

8. $f(x) = (1+x^2)^{1/3}$; $a = 0$

9. $f(x) = (1+x)^{-8/5}$; $a = 0$

10. $f(x) = \dfrac{1}{\sqrt{1-x^2}}$; $a = 0$

11. $f(x) = \dfrac{x}{\sqrt{1-x^2}}$; $a = 0$

12. $f(x) = (1-x^2)^{5/2}$; $a = 0$

13. $f(x) = \sqrt{1-(x+1)^2}$; $a = -1$

14. $f(x) = \sqrt{2x - x^2}$; $a = 1$

15. Use (3) and the Integration Theorem to show that

$$\sum_{n=0}^{\infty}\binom{1/2}{n}\frac{(-1)^n}{2n+1} = \frac{\pi}{4}$$

16. Using the Integration Theorem and the series derived in Exercise 10, show that for $-1 < x < 1$,

$$\arcsin x = x + \frac{1}{2\cdot 3}x^3 + \frac{1\cdot 3}{2\cdot 4\cdot 5}x^5 + \frac{1\cdot 3\cdot 5}{2\cdot 4\cdot 6\cdot 7}x^7 + \cdots$$

$$= x + \sum_{n=1}^{\infty}\frac{1\cdot 3\cdot 5\cdots(2n-1)}{[2\cdot 4\cdot 6\cdots(2n)](2n+1)}x^{2n+1}$$

17. Newton found the following integrals by first obtaining power series representations for the integrands and then

integrating the power series term by term. In each part below, use Newton's ideas to carry out the integration, and then determine the radius of convergence of the resulting power series. Take $a > 0$.

a. $\displaystyle\int_0^x \sqrt{a^2 + t^2}\, dt$ b. $\displaystyle\int_0^x \sqrt{a^2 - t^2}\, dt$

18. a. Find the numerical values of $\binom{6}{2}$ and $\binom{1/2}{4}$.

b. Show that

$$\sum_{n=0}^{s}\binom{s}{n} = 2^s$$

for any positive integer s. (*Hint:* Use (2).)

c. Show that

$$\sum_{n=0}^{s}\binom{s}{n}(-1)^n = 0$$

for any positive integer s. (*Hint:* Use (2).)

19. Prove that if $|s| \le 1$ and $n \ge 1$, then

$$\left|\binom{s}{n}\right| \le \left|\binom{s}{n-1}\right|$$

20. The first partial sum $1 + x/2$ of the binomial series for $(1 + x)^{1/2}$ is often used as an approximation of $(1 + x)^{1/2}$. Using the Lagrange form of the remainder $r_1(x)$, show that the error introduced is at most $(0.172)x^2$, provided that $|x| < 0.19$.

21. A tunnel 200 miles long connects two points on the earth's surface. Assuming that the earth's radius is 4000 miles and using Exercise 20, approximate the maximum depth of the tunnel.

22. Suppose a cable hangs from the point A to the origin, as in Figure 9.11, and has a uniformly distributed load of p pounds per foot. Suppose q is the tension force at 0. Assume furthermore that the x coordinate of A is x. Then the cable traces out the parabolic curve given by

$$f(t) = \frac{pt^2}{2q} \quad \text{for } 0 \le t \le x$$

so that in particular,

$$A = \left(x, \frac{px^2}{2q}\right)$$

a. Show that the length $\mathscr{L}$ of the cable is given by

$$\mathscr{L} = \int_0^x \sqrt{1 + (f'(t))^2}\, dt$$

$$= \sum_{n=0}^{\infty}\binom{1/2}{n}\frac{1}{2n+1}\frac{p^{2n}}{q^{2n}}x^{2n+1}$$

b. Determine the radius of convergence of the series given in (a).

23. Suppose a suspension bridge has span a and sag b (Figure 9.12). Then it can be shown that the length $\mathscr{L}$ of the supporting cables is given by

$$\mathscr{L} = 2\int_0^{a/2}\left(1 + \frac{64b^2}{a^4}x^2\right)^{1/2} dx$$

Using the binomial series and the Integration Theorem, show that $\mathscr{L}$ is given approximately by

$$\mathscr{L} \approx a\left[1 + \frac{8}{3}\left(\frac{b^2}{a^2}\right) - \frac{32}{5}\left(\frac{b^4}{a^4}\right)\right]$$

and use this formula to approximate the length of a cable on a bridge having a span of 500 feet and a sag of 40 feet.

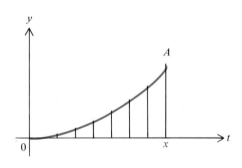

Uniformly loaded cable

FIGURE 9.11

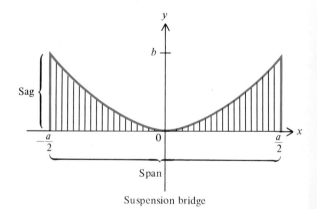

Suspension bridge

FIGURE 9.12

SUMMARY OF CONVERGENCE TESTS FOR SERIES

Test	Statement of test	Comments
nth Term Test	If $\lim\limits_{n \to \infty} a_n \neq 0$, then $\sum\limits_{n=1}^{\infty} a_n$ diverges.	Can show divergence but not convergence.
Comparison Test	If $|a_n| \leq |b_n|$ for $n \geq 1$ and if $\sum\limits_{n=1}^{\infty} |b_n|$ converges, then $\sum\limits_{n=1}^{\infty} a_n$ converges (absolutely). If $a_n \geq b_n \geq 0$ for $n \geq 1$ and if $\sum\limits_{n=1}^{\infty} b_n$ diverges, then $\sum\limits_{n=1}^{\infty} a_n$ diverges.	Usually $\sum\limits_{n=1}^{\infty} b_n$ is a series whose convergence or divergence is well known.
Limit Comparison Test	Let $\lim\limits_{n \to \infty} |a_n/b_n| = L$, where L is a positive number. If $\sum\limits_{n=1}^{\infty} |b_n|$ converges, then $\sum\limits_{n=1}^{\infty} a_n$ converges (absolutely). If $a_n \geq 0$ and $b_n \geq 0$ and if $\sum\limits_{n=1}^{\infty} b_n$ diverges, then $\sum\limits_{n=1}^{\infty} a_n$ diverges.	Usually $\sum\limits_{n=1}^{\infty} b_n$ is a series whose convergence or divergence is well known.
Integral Test	If $f(n) = a_n$ for $n \geq 1$ and if f is continuous, decreasing, and nonnegative, then $\sum\limits_{n=1}^{\infty} a_n$ converges if and only if $\int_1^{\infty} f(x)\,dx$ converges.	Most effective if f is easy to integrate.
Ratio Test	Let $\lim\limits_{n \to \infty} \left| \dfrac{a_{n+1}}{a_n} \right| = r$ (possibly ∞). If $r < 1$, then $\sum\limits_{n=1}^{\infty} a_n$ converges (absolutely). If $r > 1$, then $\sum\limits_{n=1}^{\infty} a_n$ diverges.	Most effective if a_n involves factorials or nth powers. If $r = 1$, test yields no conclusion.
Root Test	Let $\lim\limits_{n \to \infty} \sqrt[n]{|a_n|} = r$ (possibly ∞). If $r < 1$, then $\sum\limits_{n=1}^{\infty} a_n$ converges (absolutely). If $r > 1$, then $\sum\limits_{n=1}^{\infty} a_n$ diverges.	Most effective if a_n involves nth powers. If $r = 1$, test yields no conclusion.
Alternating Series Test	If $\{a_n\}_{n=1}^{\infty}$ is decreasing and nonnegative and if $\lim\limits_{n \to \infty} a_n = 0$, then $\sum\limits_{n=1}^{\infty} (-1)^n a_n$ and $\sum\limits_{n=1}^{\infty} (-1)^{n+1} a_n$ both converge.	

Key Terms and Expressions

Sequence
Convergent sequence; divergent sequence
Increasing sequence; decreasing sequence
Bounded sequence
Series; sum of a series; convergent series; divergent
  series
Partial sum
Geometric series

Nonnegative series
p series
Absolute convergence; conditional convergence
Power series
Radius of convergence; interval of convergence
Taylor series; nth Taylor polynomial; nth Taylor
  remainder
Binomial series; binomial coefficient

Key Theorems

Taylor's Theorem
Differentiation Theorem for Power Series
Integration Theorem for Power Series

Key Formulas

$$\sum_{n=m}^{\infty} cr^n = \frac{cr^m}{1 - r}$$

$$e^x = \sum_{n=0}^{\infty} \frac{x^n}{n!}$$

$$\sin x = \sum_{n=0}^{\infty} \frac{(-1)^n}{(2n+1)!} x^{2n+1}$$

$$\cos x = \sum_{n=0}^{\infty} \frac{(-1)^n}{(2n)!} x^{2n}$$

$$\ln(1 + x) = \sum_{n=0}^{\infty} \frac{(-1)^n}{n+1} x^{n+1}$$

$$\arctan x = \sum_{n=0}^{\infty} \frac{(-1)^n}{2n+1} x^{2n+1}$$

$$(1 + x)^s = \sum_{n=0}^{\infty} \binom{s}{n} x^n$$

$$f(x) = \sum_{n=0}^{\infty} \frac{f^{(n)}(a)}{n!} (x - a)^n$$

$$r_n(x) = \frac{f^{(n+1)}(t_x)}{(n+1)!} (x - a)^{n+1}$$

REVIEW EXERCISES

In Exercises 1–4 find the limit.

1. $\lim\limits_{n \to \infty} \left(1 + \dfrac{e}{n}\right)^n$

2. $\lim\limits_{k \to \infty} \dfrac{(2k)!}{2^k k^k}$

3. $\lim\limits_{n \to \infty} \left(\sqrt{n^2 + n} - \sqrt{n^2 - n}\right)$

4. $\lim\limits_{n \to \infty} \left(\dfrac{1^3}{n^4} + \dfrac{2^3}{n^4} + \cdots + \dfrac{n^3}{n^4}\right)$
   (*Hint:* The sum is a Riemann sum.)

5. Show that
$$\lim_{n \to \infty} \left(\frac{1}{n+1} + \frac{1}{n+2} + \cdots + \frac{1}{2n}\right) = \ln 2$$
   (*Hint:* The sum is a Riemann sum.)

6. Prove that if f is continuous at a, then
$$\lim_{n \to \infty} f\left(a + \frac{1}{n}\right) = f(a)$$

In Exercises 7–17 determine whether the series converges or diverges.

7. $\displaystyle\sum_{n=1}^{\infty} \frac{1}{(n + n\sqrt{n})}$

8. $\displaystyle\sum_{n=1}^{\infty} \frac{(\ln n)^2}{n^3}$

9. $\displaystyle\sum_{n=1}^{\infty} \frac{\sqrt{n}}{n^2 + n + 1}$

10. $\displaystyle\sum_{n=1}^{\infty} n^2 e^{-n/2}$

11. $\displaystyle\sum_{n=2}^{\infty} \frac{6^n}{n^2 (\ln n)^2}$

12. $\displaystyle\sum_{n=2}^{\infty} \frac{(\ln n)^4}{n}$

13. $\displaystyle\sum_{n=1}^{\infty} \frac{3^n}{n^3}$

14. $\displaystyle\sum_{n=1}^{\infty} \left(\frac{n^n}{n!}\right)^n$

15. $\displaystyle\sum_{n=4}^{\infty} (-1)^n \frac{\sqrt{n}}{n-3}$

16. $\displaystyle\sum_{n=0}^{\infty} \frac{(2n)!}{2^n n!}$

17. $\displaystyle\sum_{n=1}^{\infty} \sin^2 \frac{1}{n}$

18. Express the repeating decimal 27.1318318318... as a fraction.

19. Using the fact that

$$\ln 2 = \sum_{n=1}^{\infty} (-1)^{n+1} \frac{1}{n} = 1 - \frac{1}{2} + \frac{1}{3} - \frac{1}{4} + \cdots$$

and grouping adjacent pairs of terms, show that

$$\ln 2 = \sum_{n=0}^{\infty} \frac{1}{(2n + 1)(2n + 2)}$$

$$= \frac{1}{1 \cdot 2} + \frac{1}{3 \cdot 4} + \frac{1}{5 \cdot 6} + \cdots$$

20. Using the fact that

$$\sum_{n=0}^{\infty} (-1)^n \frac{1}{2n + 1} = \pi/4$$

determine the sum of

$$\sum_{n=5}^{\infty} (-1)^n \frac{1}{2n + 1}$$

21. Show that $\displaystyle\sum_{n=0}^{\infty} (-1)^n \frac{2n + 3}{(n + 1)(n + 2)} = 1$.

$$\left(Hint: \frac{2n + 3}{(n + 1)(n + 2)} = \frac{1}{n + 1} + \frac{1}{n + 2}.\right)$$

22. Does the series $\sum_{n=1}^{\infty} (\sqrt{n^2 + 1} - \sqrt{n^2 - 1})/n$ converge?

23. Show that

$$(1 - x) \sum_{n=1}^{\infty} nx^n = \frac{x}{1 - x} \quad \text{for } |x| < 1$$

24. What is wrong with the following argument? Since

$$\sum_{n=1}^{\infty} x^n = \frac{x}{1 - x}$$

if we let $x = -1$, then we obtain

$$\sum_{n=1}^{\infty} (-1)^n = \frac{-1}{2}$$

25. a. Let $\{a_n\}_{n=1}^{\infty}$ and $\{b_n\}_{n=1}^{\infty}$ be two convergent sequences such that $a_n \leq b_n$ for each n. Show that

$$\lim_{n \to \infty} a_n \leq \lim_{n \to \infty} b_n$$

b. Prove that if $\Sigma_{n=1}^{\infty} a_n$ converges absolutely, then

$$\left| \sum_{n=1}^{\infty} a_n \right| \leq \sum_{n=1}^{\infty} |a_n|$$

(*Hint:* Apply part (a) to the partial sums of $\Sigma_{n=1}^{\infty} a_n$ and of $\Sigma_{n=1}^{\infty} |a_n|$, and then to the partial sums of $\Sigma_{n=1}^{\infty} (-a_n)$ and of $\Sigma_{n=1}^{\infty} |a_n|$, and combine the results.)

26. Find the interval of convergence of

$$\sum_{n=2}^{\infty} (-1)^n \frac{(\ln n)^2}{n^2} x^n$$

27. Find the interval of convergence of

$$\sum_{n=0}^{\infty} \frac{3^n}{5^{2n}} x^{3n}$$

28. Find the radius of convergence of

$$\sum_{n=0}^{\infty} \frac{(n!)^2}{(2n)!} x^{2n}$$

29. Find the radius of convergence of

$$\sum_{n=1}^{\infty} \frac{n^{2n}}{(2n)!} x^n$$

30. Find the third Taylor polynomial of $\sec x$ about $\pi/6$.

31. Let $f(x) = \sqrt{1 + x^4}$. Find the second Taylor polynomial of f about 0.

32. Let $f(x) = x^6 - 3x^4 + 2x - 1$.
a. Find the fifth Taylor polynomial of f about 0.
b. Find the fourth Taylor polynomial of f about -1.
c. Find the Taylor series of f about -1.

In Exercises 33–35 find the Taylor series of f about a.

33. $f(x) = \sin x; \ a = \pi/4$

34. $f(x) = \cos 2x; \ a = -\pi/6$

35. $f(x) = \dfrac{x - 1}{x + 1}; a = 0$

In Exercises 36–39 use Taylor polynomials to approximate the number with an error less than 0.001.

36. $\sqrt{95}$ 37. $\sqrt[4]{17}$ 38. $e^{-1/3}$ 39. $\sin \pi/5$

40. Suppose a square is inscribed in a circle of radius 1, and thereafter circles and squares are alternately inscribed in one another, as in Figure 9.13. Determine the area A of the shaded region in Figure 9.13.

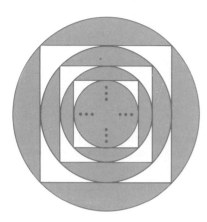

FIGURE 9.13

41. Suppose $\sum_{n=1}^{\infty} a_n$ is a series such that the jth partial sum $s_j = 2 - 2/j$, for each positive integer j.
    a. Find $\sum_{n=1}^{25} a_n$.
    b. Determine whether the series converges or diverges, and if it converges, what its sum is.
    c. Find $\lim_{n \to \infty} a_n$.
    d. Find a formula for a_n in terms of n.

42. The **Bessel function J_0 of order zero** is defined by
$$J_0(x) = \sum_{n=0}^{\infty} (-1)^n \frac{1}{4^n (n!)^2} x^{2n}$$
    a. Show that the series converges for all x.
    b. Show that $xJ_0''(x) + J_0'(x) + xJ_0(x) = 0$ for all x.

43. Using (1) of Section 9.8 with x replaced by t^2, along with the Integration Theorem for Power Series and the formula
$$\tanh^{-1} x = \int_0^x \frac{1}{1 - t^2} dt \quad \text{for } -1 < x < 1$$
    show that
$$\tanh^{-1} x = \sum_{n=0}^{\infty} \frac{1}{2n + 1} x^{2n+1}$$

44. Using (9) of Section 7.7, a binomial series, and the Integration Theorem for Power Series, show that
$$\sinh^{-1} x = \sum_{n=0}^{\infty} \binom{-1/2}{n} \frac{x^{2n+1}}{2n + 1} = x - \frac{1}{2}\left(\frac{1}{3}\right) x^3$$
$$+ \frac{1 \cdot 3}{2 \cdot 4}\left(\frac{1}{5}\right) x^5 - \frac{1 \cdot 3 \cdot 5}{2 \cdot 4 \cdot 6}\left(\frac{1}{7}\right) x^7 + \cdots$$
$$= x + \sum_{n=1}^{\infty} (-1)^n \frac{1 \cdot 3 \cdot 5 \cdots (2n - 1)}{2 \cdot 4 \cdot 6 \cdots (2n)} \frac{1}{2n + 1} x^{2n+1}$$

45. Estimate $\int_0^{1/4} 1/(1 + x^{3/2}) dx$ with an error less than 0.001 by using the Taylor series for $1/(1 - x)$. (*Hint:* Replace x by $-x^{3/2}$ in the Taylor series.)

Cumulative Review, Chapters 1–8

In Exercises 1–2 evaluate the limit.

1. $\displaystyle \lim_{x \to 0^-} \frac{\cos x + \frac{1}{2}x^2 - 1}{x^5}$

2. $\displaystyle \lim_{x \to \infty} \frac{\ln(1 + \sin^2(e^x))}{\sqrt{x}}$

3. Let $f(x) = 3^x/x^3$. Find $\lim_{x \to 0^+} f(x)$.

4. Find $\dfrac{d}{dx}(\arctan \sqrt{x - 1})$.

5. Let $f(x) = \ln(\ln(3x - 4))$.
   a. Find the domain of f. b. Find $f'(x)$.

6. Let $f(x) = x^5 - 3x^3$. For what value(s) of x in $[-1, 1]$ is the slope of the line tangent to the graph of f at $(x, f(x))$ greatest?

7. Let $e^x + e^{2x} - \frac{1}{3}y^3 + y = 10$. Suppose there is a point (a, b) on the graph of the equation at which $\dfrac{dy}{dx}(1 - y^2) = -6$. Find the value of a.

8. Let $f(x) = x^2 e^{-x}$. Sketch the graph of f, indicating all pertinent information.

9. A stone dropped into a still pond sends out a circular ripple whose enclosed area increases at a constant rate of $\frac{1}{2}$ square foot per second. How rapidly is the radius of the ripple increasing when the radius is one foot?

10. A circular cylinder is to be generated by revolving a rectangle with a given perimeter p around one of its sides. Find the largest possible volume of the circular cylinder.

In Exercises 11–13 evaluate the integral.

11. $\displaystyle \int \frac{\cos x}{e^{3x}} dx$

12. $\displaystyle \int_{\pi/3}^{\pi/2} \sin x \cos^2 \frac{x}{2} dx$

13. $\displaystyle \int_{2\sqrt{2}}^{4} \frac{\sqrt{16 - x^2}}{x^3} dx$

14. Find the area A of the portion of the region between the graphs of

$$y = \left(\frac{x}{x-1}\right)^2 \quad \text{and} \quad y = \left(\frac{x}{x+2}\right)^2$$

that lies in the second quadrant.

In Exercises 15–16 determine whether the improper integral converges. If it converges, evaluate the integral.

15. $\displaystyle\int_{-\infty}^{\infty} \frac{e^x}{(1+e^x)^2}\, dx$

16. $\displaystyle\int_{1/e}^{e} \frac{1}{x(\ln x)^3}\, dx$

17. Find the length $\mathscr{L}$ of the curve parametrized by $x = \sec t$ and $y = \ln(\sec t + \tan t)$ for $0 \le t \le \pi/4$.

18. Let R be the region between the graphs of $y = 1 + e^x$ and $y = 1 - e^{-x}$ on $[0, \ln 2]$. Find the center of gravity of R.

19. Find the area A of the region that lies between the polar graphs of $r = \theta$ and $r = \theta^3$ for $0 \le \theta \le 1$.

10

CONIC SECTIONS

Conic sections are among the best known plane figures. They have been of special interest to mathematicians since the time of Apollonius, who during the third century B.C. wrote eight treatises on conic sections and became known to his contemporaries as "the Great Geometer." The conic sections arise when a double right circular cone is cut by a plane. Depending on how the plane cuts the cone, the intersection forms a curve called a parabola, an ellipse, or a hyperbola (Figure 10.1(a)–(c)) or, in "degenerate" cases, a point, a line, or two intersecting lines (Figure 10.1(d)–(f)). Since we have already studied lines in Chapter 1, we will limit our discussion in this chapter to parabolas, ellipses, and hyperbolas.

Our definitions of these three conic sections will be given in terms of points, lines, and distances, rather than in terms of planes and cones. Using the definitions, we will derive equations of the conic sections and show that any second-degree equation $Ax^2 + Bxy + Cy^2 + Dx + Ey + F = 0$ is (except in degenerate cases) an equation of a parabola, an ellipse, or a hyperbola.

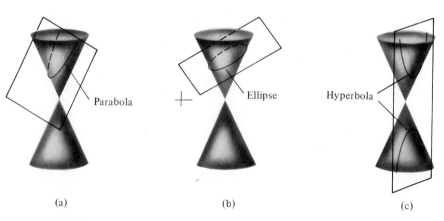

(a)   (b)   (c)

FIGURE 10.1

580

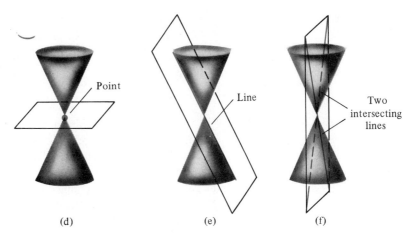

Point

Line

Two
intersecting
lines

(d)   (e)   (f)

FIGURE 10.1 (*continued*)

10.1
THE PARABOLA

What do the following have in common?

1. The path of a golf ball in flight
2. The shape of the reflector in an automobile headlight
3. The shape of the mirror or reflector in certain types of telescopes
4. The shape of a cable on a suspension bridge

Although the shapes in Figure 10.2(a)–(d), which represent these four objects, may look quite different from one another, they all are related to curves called parabolas.

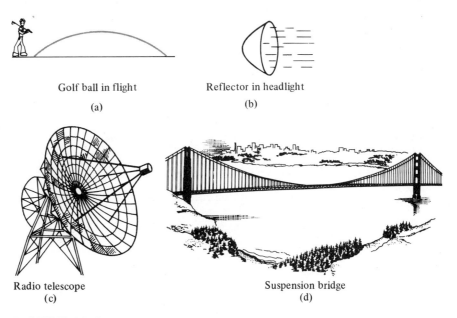

Golf ball in flight
(a)

Reflector in headlight
(b)

Radio telescope
(c)

Suspension bridge
(d)

FIGURE 10.2

DEFINITION 10.1

Let l be a fixed line in the plane and P a fixed point not on l. The set of all points in the plane equidistant from l and P is called a **parabola**. The line l is called the **directrix** and the point P the **focus** of the parabola.

It follows from Definition 10.1 and plane geometry (see Figure 10.3) that a parabola is always symmetric with respect to the line through the focus perpendicular to the directrix. This line is called the **axis** of the parabola. By definition the point on the axis midway between the focus and the directrix lies on the parabola; this point is called the **vertex** of the parabola (Figure 10.3). We say that the parabola is in **standard position** if its vertex is the origin and its axis is either the x axis or the y axis.

Parabolas in Standard Position

Let us now find an equation of a parabola in standard position. If the axis of the parabola is the y axis, then the focus will be a point $(0, c)$ for some $c \neq 0$ and the directrix l will be the line $y = -c$ (Figure 10.4(a) or (b)). By definition a point (x, y) is on the parabola if and only if the distance from (x, y) to l is equal to the distance from (x, y) to $(0, c)$. This means that

$$|y - (-c)| = \sqrt{(x - 0)^2 + (y - c)^2}$$

or

$$\sqrt{(y + c)^2} = \sqrt{x^2 + (y - c)^2}$$

But because the expressions within the square root signs are nonnegative, this is

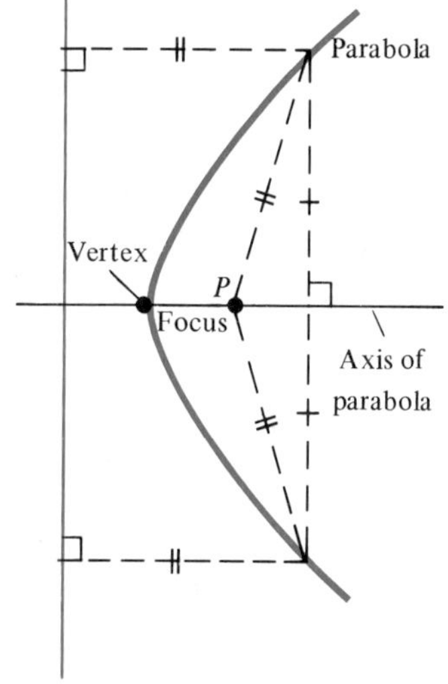

Directrix

Parabola

Vertex

P

Focus

Axis of parabola

A parabola is symmetric with respect to its axis

FIGURE 10.3

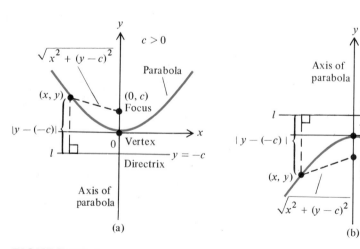

(a)

(b)

FIGURE 10.4

equivalent to the equation

$$(y + c)^2 = x^2 + (y - c)^2$$

or

$$y^2 + 2cy + c^2 = x^2 + y^2 - 2cy + c^2$$

Simplifying, we obtain

$$x^2 = 4cy \qquad (1)$$

If we solve for y in (1), we obtain

$$y = \frac{x^2}{4c}$$

It follows that

$$\frac{dy}{dx} = \frac{x}{2c} \quad \text{and} \quad \frac{d^2y}{dx^2} = \frac{1}{2c}$$

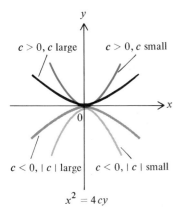

$x^2 = 4cy$

FIGURE 10.5

Therefore the parabola described by (1) is concave upward if $c > 0$ and concave downward if $c < 0$. If $|c|$ is large, the focus and directrix are far apart and the parabola looks wide and blunt. If $|c|$ is small, the focus and directrix are close together and the parabola looks long and slender (Figure 10.5).

Now if we assume that the axis of the parabola is the x axis, so that its focus is a point $(c, 0)$ and its directrix is the line $x = -c$, then a similar derivation (with x and y interchanged) yields the equation

$$y^2 = 4cx$$

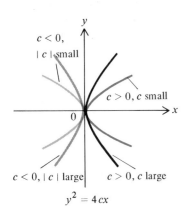

$y^2 = 4cx$

FIGURE 10.6

for the parabola. In this case the parabola is symmetric with respect to the x axis, opening to the right if $c > 0$ and to the left if $c < 0$ (Figure 10.6).

For reference we list the two forms for equations of parabolas in standard position:

$$x^2 = 4cy \quad \begin{cases} \text{focus is } (0, c) \\ \text{directrix is } y = -c \\ \text{symmetry with respect to the } y \text{ axis} \end{cases} \qquad (2)$$

$$y^2 = 4cx \quad \begin{cases} \text{focus is } (c, 0) \\ \text{directrix is } x = -c \\ \text{symmetry with respect to the } x \text{ axis} \end{cases} \qquad (3)$$

Notice also that

$$2|c| = \text{distance from focus to directrix}$$

We can analyze parabolas in standard position by investigating their equations.

Example 1 Find the focus and directrix of the parabola $y^2 = -16x$, and sketch it.

Solution   The given equation is in the form $y^2 = 4cx$, with $c = -4$. Therefore the focus is the point $(-4, 0)$ and the directrix is the line $x = 4$. The parabola is sketched in Figure 10.7.   □

Example 2   A parabola has focus $(0, -3)$ and directrix $y = 3$. Find an equation for the parabola, and sketch it.

Solution   The formula to use here is $x^2 = 4cy$, with $c = -3$. Therefore an equation for the parabola is $x^2 = -12y$. The parabola appears in Figure 10.8.   □

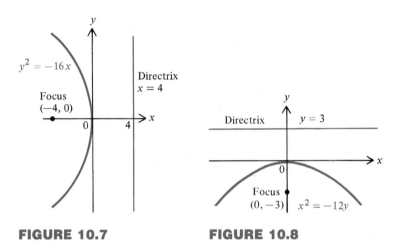

FIGURE 10.7   **FIGURE 10.8**

Translation of Parabolas   If a parabola is not in standard position, we may be able to find an equation for it and sketch it if we first locate the axis and the vertex of the parabola and then translate the coordinate axes (see Section 1.5).

Example 3   Find an equation of the parabola whose focus is $(3, 1)$ and whose directrix is $x = 1$. Sketch the parabola.

Solution   Since the axis of the parabola is the line through the focus $(3, 1)$ that is perpendicular to the directrix $x = 1$, we know that the axis is the line $y = 1$. Because the vertex is midway between the focus and the directrix and is on the axis, the vertex is the point $(2, 1)$. Therefore the parabola is in standard position for the XY coordinate system obtained by translating axes so that the XY origin is the point $(2, 1)$ in the xy coordinate system (Figure 10.9). The focus is then $(1, 0)$ in the XY system, and the directrix is the line $X = -1$. From (3) we know that an equation for the parabola in the XY coordinate system is $Y^2 = 4X$. Consequently we can draw the parabola (Figure 10.9). Since $X = x - 2$ and $Y = y - 1$, an equation of the parabola in the xy coordinate system is

$$(y - 1)^2 = 4(x - 2)$$

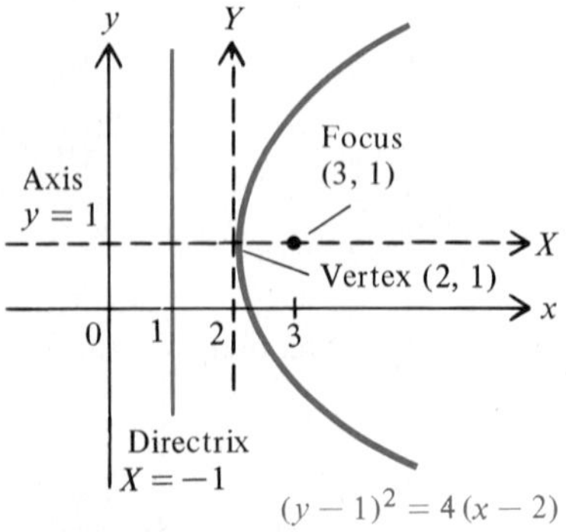

FIGURE 10.9

or

$$y^2 - 2y - 4x + 9 = 0 \quad □$$

If either $A = 0$ or $C = 0$, but not both, then

$$Ax^2 + Cy^2 + Dx + Ey + F = 0$$

is an equation of a parabola (or a degenerate conic section). Completing the square enables us to locate the vertex, the focus, the directrix, and the axis. With this information we can sketch the parabola.

Example 4   Show that $x^2 + 2x - 6y - 17 = 0$ is an equation of a parabola. Sketch the parabola, and determine its focus, directrix, and axis.

Solution   First we complete the square in x by adding 1 to each side of the equation. We obtain

$$x^2 + 2x + 1 - 6y - 17 = 1$$

which condenses to

$$x^2 + 2x + 1 = 6y + 18$$

or

$$(x + 1)^2 = 6(y + 3)$$

In the translated XY system whose origin is $(-1, -3)$, we have $X = x + 1$ and $Y = y + 3$, so that the equation becomes

$$X^2 = 6Y$$

By formula (2) the focus is $(0, \frac{3}{2})$ and the directrix is $Y = -\frac{3}{2}$ in the XY system. Therefore in the xy system the focus is $(-1, -\frac{3}{2})$, the directrix is $y = -\frac{9}{2}$, the vertex is $(-1, -3)$, and the axis is the line $x = -1$. The parabola is shown in Figure 10.10.   □

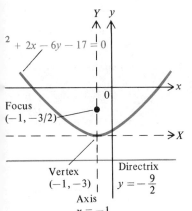

FIGURE 10.10

Applications of Parabolas

There are many theoretical and practical applications of parabolas. We list several of them, leaving details to the exercises.

1. When an object moves only under the influence of the earth's gravity and eventually strikes the earth, its path is normally parabolic (or linear if the motion is vertical). Thus the path of a golf ball, baseball, or football in flight would be parabolic if there were no air resistance.

2. The shapes of certain mirrors can be obtained by revolving a parabola about its axis. Such mirrors, called parabolic mirrors, are found in automobile headlights and reflecting telescopes. In the headlight all light rays emitted from a light source at the focus are reflected from the mirror along lines parallel to the axis of the parabola. In a reflecting telescope all incoming rays from a star, which are essentially parallel to the axis, are reflected by the mirror; they then pass through the focus, where the eyepiece of the telescope should be located. The same principle applies to parabolic radio telescopes.

3. A suspended cable tends to hang in the shape of a parabola if the weight of the cable is small compared to the weight it supports and if the weight

A NASA radio telescope in Goldstone, California

is uniformly distributed along the cable. Bridges that are supported by cables in this fashion are called suspension bridges. All the larger bridges in the world, and most of the more famous ones, such as the Golden Gate Bridge, the George Washington Bridge, the Brooklyn Bridge, and the Delaware Memorial Bridge, are suspension bridges.

The George Washington Bridge

EXERCISES 10.1

In Exercises 1–14 find an equation of the parabola with the given properties.

1. The focus is $(-2, 0)$, and the directrix is $x = 2$.
2. The focus is $(0, \frac{1}{2})$, and the directrix is $y = -\frac{1}{2}$.
3. The focus is $(0, -6)$, and the directrix is $y = 6$.
4. The focus is $(2, -3)$, and the directrix is $y = 3$.
5. The focus is $(0, 0)$, and the directrix is $x = 5$.
6. The focus is $(-5, -2)$, and the directrix is $x = 3$.
7. The vertex is $(1, 0)$, and the directrix is $x = -2$.
8. The vertex is $(0, 2)$, and the directrix is $y = 4$.
9. The focus is $(3, 3)$, and the vertex is $(3, 2)$.
10. The focus is $(-5, -2)$, and the vertex is $(5, -2)$.
11. The parabola is in standard position, is symmetric with respect to the x axis, and passes through the point $(-1, 1)$.
12. The parabola is in standard position, is symmetric with respect to the y axis, and passes through the point $(-1, 2)$.
13. The parabola is the collection of points (x, y) whose distance from $(3, 4)$ is the same as the distance from the line $y = 2$.

*14. The line $x = 4$ is the directrix, the line $y = -2$ is the axis, and the point $(0, -3)$ is on the parabola. (There are two possible solutions.)

In Exercises 15–22 sketch the parabola having the given equation, and locate its focus, vertex, directrix, and axis.

15. $y^2 = 3x$
16. $y^2 = \dfrac{-x}{2}$
17. $x^2 = y$
18. $x^2 = -6y$
19. $(x - 1)^2 = y + 2$
20. $(y + 3)^2 = 4x - 3$
21. $(x - 2)^2 = 3(y - 3)$
22. $4(x + 1) = 2(y + 2)^2$

In Exercises 23–27 find an XY coordinate system in which the parabola with the given equation is in standard position. Then sketch the parabola.

23. $x^2 - 6x - 2y + 1 = 0$
24. $2x^2 + 4x - 5y + 7 = 0$
25. $3y^2 - 5x + 3y - \frac{17}{4} = 0$
26. $-4y^2 - \dfrac{x}{2} + 4y = 1$
27. $4y^2 - \sqrt{2}x + 2y = \dfrac{1}{\sqrt{2}} - 1$

28. Find an equation of the parabola that has a vertical axis, its vertex at $(-1, 3)$, and slope 2 at $x = 1$.

29. Find the point on the parabola $y^2 = 4x$ that is closest to the point $(1, 0)$.

30. Find the point on the parabola $y^2 = 4cx$ that is closest to the focus.

31. The line segment that passes through the focus, is parallel to the directrix, and has its endpoints on the parabola is called the **latus rectum**. Show that if a parabola is in standard position and the focus is c units from the origin, then the length of the latus rectum is $4c$.

32. Find an equation for the line tangent to the parabola $y^2 = 8x$ at $(2, 4)$.

33. Find the number d such that the line $x + y = d$ is tangent to the parabola $x^2 = 2y$, and determine the point of tangency.

*34. Find equations for the two lines that are tangent to the parabola $x^2 = 2y$ and pass through $(-1, -4)$.

35. Suppose a golf ball is driven so that it travels a distance of 600 feet as measured along the ground and reaches an altitude of 200 feet. If the origin represents the tee and if the ball travels in the positive x direction over flat ground, find an equation for the path of the golf ball.

36. This exercise will show that a ray of light emitted from the focus of a parabolic mirror is reflected from the mirror along a line parallel to the axis of the mirror. Let $y^2 = 4cx$, where $c > 0$, represent the parabola shown in Figure 10.11, and let $Q(x_0, y_0)$ be on the parabola. Let l_1 represent an incident ray from the focus $P(c, 0)$ to Q, l_2 a horizontal line emanating from Q, and l_3 the line tangent to the parabola at Q. Assume that $x_0 \neq 0$ and $y_0 > 0$.

a. Show that
$$y - y_0 = \sqrt{\frac{c}{x_0}}(x - x_0)$$
is an equation of l_3.

b. Show that the x intercept of l_3 is $-x_0$.

c. Using the fact that the distance between P and Q is equal to the distance from Q to the directrix, show that $|PQ| = |PR|$.

d. From (c) conclude that $\theta_1 = \theta_2$. (From this result and the Law of Reflection, which states that the angle of incidence of the light ray is equal to the angle of reflection, it follows that every light ray emitted from the focus of the mirror is reflected along a line parallel to the axis of the mirror.)

37. If the diameter of a parabolic mirror is 10 inches and if the mirror is 5 inches deep at the center, how far is the focus from the center of the mirror?

In a suspension bridge the horizontal distance between the supports is called the span, and the vertical distance between the points where the cable is attached to the supports and the lowest point of the cable is called the sag of the cable (Figure 10.12). In Exercises 38–40 assume that the origin is at the lowest point on the cable.

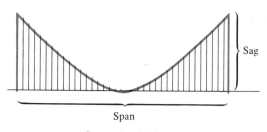

Span

Sag

Suspension bridge

FIGURE 10.12

38. Assume that the span and sag of a suspension bridge are a and b, respectively. Determine an equation of the parabola that represents the hanging cable. (*Hint:* Let the y axis lie along the axis of the parabola.)

39. The George Washington Bridge across the Hudson River from New York to New Jersey has a span of 3500 feet and a sag of 316 feet. Determine an equation of the parabola that represents the cable.

40. Suppose an architect wishes to design a suspension bridge for automobile traffic across the Mississippi River, and needs to make the span a half-mile long. For aesthetic reasons the architect feels that the angle the cable makes with the support should be $30°$. What would the sag be?

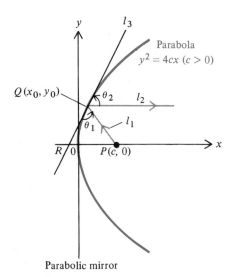

Parabolic mirror

FIGURE 10.11

10.2
THE ELLIPSE

Although ellipses had been studied by the Greeks, they were brought into prominence in the seventeenth century by Johannes Kepler's discovery that planets move around the sun in elliptical orbits. Ellipses can be defined in the following way.

DEFINITION 10.2

Let P_1 and P_2 be two points in the plane, and let k be a number greater than the distance between P_1 and P_2. The set of all points P in the plane such that

$$|P_1 P| + |P_2 P| = k$$

is called an **ellipse**. The points P_1 and P_2 are called the **foci** of the ellipse.

It follows from Definition 10.2 and plane geometry (see Figure 10.13) that an ellipse is symmetric with respect to the line through the two foci of the ellipse. The midpoint of the segment between the two foci is called the **center** of the ellipse (Figure 10.14). We say that an ellipse is in **standard position** if its center lies at the origin and if the foci lie on either the x axis or the y axis. In the event that the foci are the same point, the ellipse is a circle.

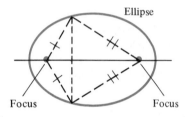

An ellipse is symmetric with respect to the line joining its foci.

FIGURE 10.13

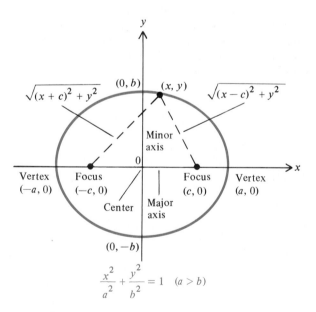

FIGURE 10.14

Ellipses in Standard Position

Let us now find an equation of an ellipse in standard position. If the foci are the points $(-c, 0)$ and $(c, 0)$, with $c \geq 0$ (Figure 10.14), then the distance between the foci is $2c$. For convenience we let $k = 2a$. It follows by hypothesis that $2a =$

$k > 2c$, that is, $a > c$. Then by the definition of distance, a point (x, y) is on the ellipse if and only if

$$\sqrt{(x - (-c))^2 + (y - 0)^2} + \sqrt{(x - c)^2 + (y - 0)^2} = 2a$$

or

$$\sqrt{(x + c)^2 + y^2} = 2a - \sqrt{(x - c)^2 + y^2}$$

Squaring and canceling, we obtain

$$a^2 - cx = a\sqrt{(x - c)^2 + y^2}$$

Squaring again yields

$$a^4 - 2cxa^2 + c^2x^2 = a^2[(x - c)^2 + y^2]$$

which can be rewritten

$$(a^2 - c^2)x^2 + a^2y^2 = a^2(a^2 - c^2) \tag{1}$$

If we now set

$$b = \sqrt{a^2 - c^2} \tag{2}$$

then $0 < b \le a$, and division by a^2b^2 in (1) yields

$$\frac{x^2}{a^2} + \frac{y^2}{b^2} = 1 \tag{3}$$

We have just seen that any point on the ellipse must satisfy equation (3). It is possible to show that any point satisfying equation (3) lies on the ellipse (see Exercise 47). Therefore (3) is an equation of the ellipse whose foci are $(-c, 0)$ and $(c, 0)$ and for which $k = 2a$.

From (3) we infer that the x intercepts of the ellipse are $-a$ and a, while the y intercepts are b and $-b$. We call the points $(-a, 0)$ and $(a, 0)$ the **vertices** of the ellipse, the line segment between $(-a, 0)$ and $(a, 0)$ the **major axis** of the ellipse, and the line segment between $(0, -b)$ and $(0, b)$ the **minor axis** (Figure 10.14).

To confirm that the sketch in Figure 10.14 represents the graph of equation (3), we begin by noting that $-a \le x \le a$ and $-b \le y \le b$ for any point (x, y) on the ellipse. Second, we learn from (3) that the ellipse is symmetric with respect to the x axis, the y axis, and the origin. This means in particular that if we find the portion of the ellipse lying above the x axis, then the rest can be obtained through symmetry. However, by taking $y > 0$ we can transform (3) into the equation

$$y = b\sqrt{1 - \frac{x^2}{a^2}} = \frac{b}{a}\sqrt{a^2 - x^2}$$

Therefore

$$\frac{dy}{dx} = \frac{-bx}{a\sqrt{a^2 - x^2}}$$

and

$$\frac{d^2y}{dx^2} = \frac{-ab}{(a^2 - x^2)^{3/2}}$$

Since the second derivative is negative, the portion of the ellipse for which $y > 0$ is concave downward. This information confirms that the sketch in Figure 10.14 represents the graph of the equation in (3).

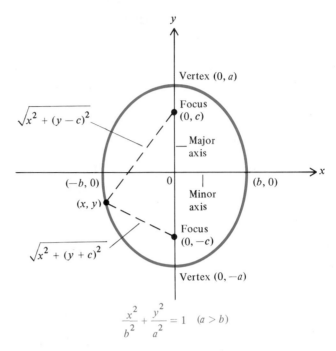

FIGURE 10.15

If the foci of the ellipse are on the y axis, say at $(0, c)$ and $(0, -c)$, where $c \geq 0$, then we can find an equation of the ellipse by simply interchanging x and y in (3). We obtain

$$\frac{x^2}{b^2} + \frac{y^2}{a^2} = 1 \tag{4}$$

The graph of this equation is the ellipse shown in Figure 10.15.

For later reference we list the two forms of equations of ellipses in standard position:

$$\frac{x^2}{a^2} + \frac{y^2}{b^2} = 1 \quad (a \geq b > 0) \quad \begin{cases} \text{foci are } (-c, 0) \text{ and } (c, 0), \\ \quad \text{where } c = \sqrt{a^2 - b^2} \\ \text{vertices are } (-a, 0) \text{ and } (a, 0) \\ \text{symmetry with respect to the} \\ \quad x \text{ axis, } y \text{ axis, and origin} \end{cases} \tag{5}$$

$$\frac{x^2}{b^2} + \frac{y^2}{a^2} = 1 \quad (a \geq b > 0) \quad \begin{cases} \text{foci are } (0, -c) \text{ and } (0, c), \\ \quad \text{where } c = \sqrt{a^2 - b^2} \\ \text{vertices are } (0, -a) \text{ and } (0, a) \\ \text{symmetry with respect to the} \\ \quad x \text{ axis, } y \text{ axis, and origin} \end{cases} \tag{6}$$

The numbers a, b, and c have the following geometric interpretation:

$$2a = \text{length of major axis}$$
$$2b = \text{length of minor axis}$$
$$2c = \text{distance between foci}$$

(7)

Example 1   Sketch the ellipse

$$x^2 + 4y^2 = 8$$

and find its foci.

Solution   Dividing both sides of the equation by 8, we obtain

$$\frac{x^2}{8} + \frac{y^2}{2} = 1$$

Since $8 > 2$, this equation is in the form of (5), with $a = \sqrt{8} = 2\sqrt{2}$ and $b = \sqrt{2}$. Therefore the foci lie on the x axis. To determine their coordinates, we observe that

$$c = \sqrt{a^2 - b^2} = \sqrt{8 - 2} = \sqrt{6}$$

Consequently the foci are the points $(-\sqrt{6}, 0)$ and $(\sqrt{6}, 0)$. The graph of the ellipse and its foci are shown in Figure 10.16.   □

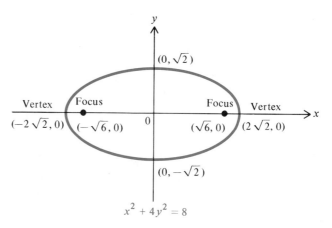

FIGURE 10.16

Example 2   Sketch the ellipse

$$3x^2 + y^2 = 3$$

and locate the foci and vertices.

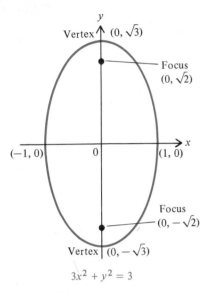

$3x^2 + y^2 = 3$

FIGURE 10.17

Solution The given equation is equivalent to

$$\frac{x^2}{1} + \frac{y^2}{3} = 1$$

This is in the form of (6), with $a = \sqrt{3}$ and $b = 1$. Hence the foci and the major axis lie on the y axis. Since $a = \sqrt{3}$, the vertices are $(0, -\sqrt{3})$ and $(0, \sqrt{3})$, and since

$$c = \sqrt{a^2 - b^2} = \sqrt{3 - 1} = \sqrt{2}$$

the foci are the points $(0, -\sqrt{2})$ and $(0, \sqrt{2})$. The ellipse is shown in Figure 10.17. ☐

If the foci and the length $2a$ of the major axis are known, it is possible to draw the ellipse immediately. Simply place a tack at each focus, tie the ends of a string of length $2a$ to the tacks, and then move a pencil around with its lead touching the string, keeping the string taut. The resulting figure is by definition an ellipse (Figure 10.18).

Translation of Ellipses

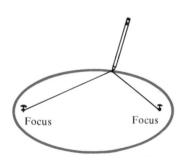

FIGURE 10.18

When an ellipse is not in standard position, we may still be able to obtain an equation for it by determining the location of its center and other pertinent information and then translating the coordinate axes.

Example 3 Find an equation of the ellipse whose foci are $(2, -1)$ and $(2, 7)$ and whose major axis has length 12. Sketch the ellipse.

Solution Since the foci $(2, -1)$ and $(2, 7)$ lie on the major axis, the major axis is on the line $x = 2$. The center of the ellipse is midway between the foci and is therefore the point $(2, 3)$. In the XY system with origin $(2, 3)$, the ellipse is in standard position with the major axis on the Y axis. Therefore an equation of the ellipse has the form

$$\frac{X^2}{b^2} + \frac{Y^2}{a^2} = 1$$

Since the major axis has length 12 by hypothesis, it follows from (7) that $2a = 12$, or $a = 6$. Since the distance between the two foci is 8, we infer from (7) that $2c = 8$, or $c = 4$. Consequently by (6) we have

$$b = \sqrt{a^2 - c^2} = \sqrt{36 - 16} = \sqrt{20}$$

Therefore an equation of the ellipse in the XY system is

$$\frac{X^2}{20} + \frac{Y^2}{36} = 1$$

This means that in the xy system the equation is

$$\frac{(x - 2)^2}{20} + \frac{(y - 3)^2}{36} = 1$$

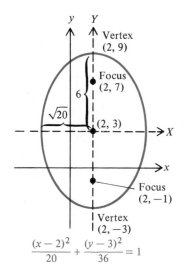

$$\frac{(x-2)^2}{20} + \frac{(y-3)^2}{36} = 1$$

FIGURE 10.19

The ellipse is shown in Figure 10.19. $\square$

Any equation of the form

$$Ax^2 + Cy^2 + Dx + Ey + F = 0$$

in which A and C are either both positive or both negative (so that $AC > 0$), is an equation of an ellipse or a degenerate conic section. (See Exercise 42.) We can find the center, vertices, and foci of the ellipse by completing the squares and translating the coordinate axes.

Example 4   Show that

$$9x^2 - 18x + 4y^2 + 16y = 11$$

is an equation of an ellipse, and sketch the ellipse.

Solution   By completing the squares we find that

$$9(x^2 - 2x + 1) + 4(y^2 + 4y + 4) = 11 + 9 + 16 = 36$$

so that

$$\frac{(x-1)^2}{4} + \frac{(y+2)^2}{9} = 1$$

In the XY system with origin $(1, -2)$, the equation becomes

$$\frac{X^2}{4} + \frac{Y^2}{9} = 1$$

This is an equation of an ellipse whose vertices are $(0, -3)$ and $(0, 3)$ and whose foci are $(0, \sqrt{5})$ and $(0, -\sqrt{5})$ in the XY system. Therefore in the xy system the center of the ellipse is $(1, -2)$, the vertices are $(1, -5)$ and $(1, 1)$, and the foci are $(1, -2 - \sqrt{5})$ and $(1, -2 + \sqrt{5})$. The ellipse is shown in Figure 10.20. $\square$

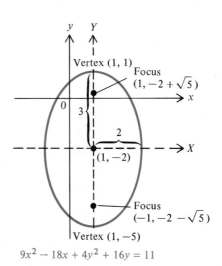

$$9x^2 - 18x + 4y^2 + 16y = 11$$

FIGURE 10.20

Applications of Ellipses   We list a few applications of ellipses, again leaving details to the exercises:

1. According to Kepler's laws of planetary motion, planets move in elliptical orbits with the sun at one focus. Some comets, such as Halley's comet, also move in elliptical orbits around the sun.
2. An electron in an atom moves in an approximately elliptical orbit with the nucleus at one focus.
3. The dome of a whispering gallery is in the shape of an ellipse that has been revolved about its major axis. Sound emanating from one focus bounces from any point on the dome to the other focus. It is said that John C. Calhoun used this phenomenon in the Statuary Hall of the Capitol in Washington, D.C., in order to eavesdrop on his adversaries.
4. Ellipses are aesthetically pleasing and are often used in architecture and in the design of formal gardens.

EXERCISES 10.2

In Exercises 1–17 write an equation of the ellipse having the given properties. Then sketch the ellipse.

1. The foci are $(2, 0)$ and $(-2, 0)$, and the vertices are $(3, 0)$ and $(-3, 0)$.
2. The foci are $(0, 3)$ and $(0, -3)$, and the vertices are $(0, 5)$ and $(0, -5)$.
3. The foci are $(2, 1)$ and $(2, -1)$, and the length of the major axis is 4.
4. The foci are $(2, 3)$ and $(2, -3)$, and the length of the minor axis is 5.
5. The foci are $(4, \pi)$ and $(-4, \pi)$, and the length of the minor axis is 9.
6. The vertices are $(5, 0)$ and $(-5, 0)$, and the ellipse passes through $(3, -4)$.
7. The vertices are $(0, 16)$ and $(0, -16)$, and the ellipse passes through $(2\sqrt{3}, 4)$.
8. The foci are $(1, 0)$ and $(-1, 0)$, and the ellipse passes through $(-1, \frac{3}{2})$.
9. The foci are $(0, 0)$ and $(4, 0)$, and the vertices are $(-1, 0)$ and $(5, 0)$.
10. The foci are $(2, 3)$ and $(2, 7)$, and the vertices are $(2, 0)$ and $(2, 10)$.
11. The center is $(6, 1)$, one focus is $(3, 1)$, and one vertex is $(10, 1)$.
12. The center is $(3, 2)$, one vertex is $(-5, 2)$, and the length of the minor axis is 6.

13. The ellipse passes through $(-1, 1)$ and $(\frac{1}{2}, -2)$ and is in standard position.
14. The ellipse passes through $(2, 1)$ and $(1, \frac{1}{2}\sqrt{7})$ and is in standard position.
15. The major axis extends from $(3, -4)$ to $(3, 4)$, and the ellipse passes through the origin.
16. The minor axis extends from $(3, -2)$ to $(3, 2)$, and the ellipse passes through the origin.
17. The minor axis extends from $(1, 0)$ to $(5, 0)$, and the ellipse passes through $(4, 2\sqrt{3})$.

In Exercises 18–27 find the foci and vertices of the ellipse having the given equation.

18. $x^2 + y^2 = 100$

19. $\dfrac{x^2}{9} + \dfrac{y^2}{25} = 1$

20. $\dfrac{x^2}{36} + y^2 = 1$

21. $\dfrac{x^2}{64} + \dfrac{y^2}{49} = 1$

22. $x^2 = 1 - \dfrac{y^2}{4}$

23. $\dfrac{x^2}{3} + y^2 = 3$

24. $4x^2 + y^2 = 8$

25. $4x^2 + y^2 = 1$

26. $4(x - 1)^2 + y^2 = 1$

27. $25(x + 1)^2 + (y - 3)^2 = 1$

In Exercises 28–34 show that the equation represents an ellipse. Sketch the ellipse, and indicate the foci, center, and vertices.

28. $9x^2 + 4y^2 - 36x + 8y + 4 = 0$
29. $x^2 + 2y^2 - 2x - 4y = 1$

30. $25x^2 + 16y^2 + 150x - 96y = 31$
31. $x^2 - 8x + 2y^2 + 12 = 0$
32. $x^2 + 3y^2 + 12y = 24$
33. $x^2 + 2x + 2y^2 + 4y = 1$
34. $8x^2 + 8x + 2y^2 - 20y - 12 = 0$
35. For what value of c will the ellipse

$$\frac{(x-1)^2}{4} + \frac{(y+2)^2}{9} = c$$

pass through the origin?

36. Suppose the major axis of an ellipse in standard position lies along the y axis and has length $4\sqrt{2}$. Assuming the distance between the foci equals the length of the minor axis, find an equation of the ellipse.

37. Find an equation of the ellipse in standard position that passes through $(0, 2)$ and has slope $1/\sqrt{2}$ at the point (x, y) on the ellipse with $x = -2$ and $y > 0$.

38. Show that an equation of the line tangent to the ellipse $x^2/a^2 + y^2/b^2 = 1$ at the point (x_0, y_0) is

$$\frac{xx_0}{a^2} + \frac{yy_0}{b^2} = 1$$

39. Find two values of d such that the line $2x + y = d$ is tangent to the ellipse $4x^2 + y^2 = 8$. Find the points of tangency.

40. Find an equation for the collection of points such that the distance from each point to $(3, 0)$ is half the distance to the line $x = -3$. Show that your equation is an equation of an ellipse.

41. The line segment that passes through a focus of an ellipse, is perpendicular to the major axis of the ellipse, and has its endpoints on the ellipse is called a **latus rectum.** Show that the ellipse

$$\frac{x^2}{a^2} + \frac{y^2}{b^2} = 1$$

has a latus rectum of length $2b^2/a$.

42. Consider the equation

$$Ax^2 + Cy^2 + Dx + Ey + F = 0 \qquad (8)$$

where A and C are positive, and let

$$r = \frac{D^2}{4A} + \frac{E^2}{4C} - F$$

By completing the squares in (8), show that
a. if $r > 0$, then the graph of the equation is an ellipse.
b. if $r = 0$, then the graph of the equation is a point.
c. if $r < 0$, then the graph of the equation consists of no points.

43. Consider the equation

$$2x^2 + 4y^2 + 8x - 16y + F = 0$$

Use the information from Exercise 42 to find all values of F such that the graph of the equation
a. is an ellipse.
b. is a point.
c. consists of no points at all.

44. Suppose a rectangle with horizontal and vertical sides is to be inscribed in the ellipse

$$\frac{x^2}{a^2} + \frac{y^2}{b^2} = 1$$

What location of the vertices of the rectangle will yield the largest area?

45. The planet Mars travels around the sun in an ellipse whose equation is approximately

$$\frac{x^2}{(228)^2} + \frac{y^2}{(227)^2} = 1$$

where x and y are measured in millions of kilometers. Find the ratio of the length of the major axis to the length of the minor axis.

46. The Ellipse in Washington, D.C., has a major axis approximately 0.285 miles long and a minor axis approximately 0.241 miles long. Determine the distance from a vertex to the closest focus of the ellipse.

The Ellipse in Washington, D.C.

*47. Show that if x and y satisfy (3), then the point (x, y) lies on the ellipse. (*Hint:* Using (2), compute the distances from (x, y) to $(-c, 0)$ and from (x, y) to $(c, 0)$. Notice that both $a^2 - cx$ and $a^2 + cx$ are positive, so that the sum of the two distances is $2a$).

*48. Prove that a ray of light emitted from one focus of an elliptical mirror and reflected by the mirror must pass through the other focus (Figure 10.21).

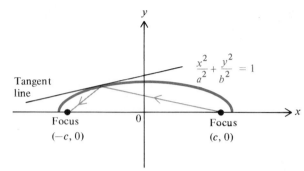

Angle of incidence is equal to angle of reflection

FIGURE 10.21

10.3
THE HYPERBOLA

The last type of nondegenerate conic section is the hyperbola.

DEFINITION 10.3

Let P_1 and P_2 be two distinct points in the plane, and let k be a positive number less than the distance between P_1 and P_2. The set of all points P in the plane such that

$$\big||P_1P| - |P_2P|\big| = k$$

is called a **hyperbola**. The points P_1 and P_2 are called the **foci** of the hyperbola.

Like an ellipse, a hyperbola is symmetric with respect to the line through the two foci (Figure 10.22). This line is the **principal axis** of the hyperbola, and the point midway between the two foci is the **center** of the hyperbola (Figure 10.23).

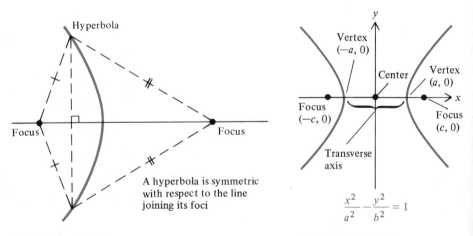

A hyperbola is symmetric with respect to the line joining its foci

FIGURE 10.22

FIGURE 10.23

The points at which the hyperbola meets the principal axis are the **vertices** of the hyperbola, and the line segment they determine is the **transverse axis** of the hyperbola. Finally, we say that a hyperbola is in **standard position** if its center is the origin and its foci are either on the x axis or on the y axis.

Hyperbolas in Standard Position

Suppose the foci of a hyperbola in standard position are located at $(-c, 0)$ and $(c, 0)$, with $c > 0$, so that the distance between the foci is $2c$. If we let $k = 2a$, then by hypothesis $2a = k < 2c$, so that $a < c$. By definition a point (x, y) is on the hyperbola if and only if

$$\left| \sqrt{(x + c)^2 + (y - 0)^2} - \sqrt{(x - c)^2 + (y - 0)^2} \right| = 2a$$

By algebraic manipulations similar to those made for the ellipse, we can transform this equation into the following equation of the hyperbola:

$$\frac{x^2}{a^2} - \frac{y^2}{b^2} = 1 \tag{1}$$

where

$$b = \sqrt{c^2 - a^2}$$

From (1) we see that the hyperbola described above has no y intercepts and that its x intercepts are $-a$ and a. Moreover, $|x| \geq a$ for any point (x, y) on the hyperbola, and the hyperbola is symmetric with respect to the x axis, the y axis, and the origin. For $y > 0$ we find that

$$y = \frac{b}{a}\sqrt{x^2 - a^2} \tag{2}$$

so that

$$\frac{dy}{dx} = \frac{bx}{a\sqrt{x^2 - a^2}}$$

and

$$\frac{d^2y}{dx^2} = \frac{-ab}{(x^2 - a^2)^{3/2}}$$

Therefore the portion of the hyperbola for which $y > 0$ is concave downward. For accurate graphing of the parts of the hyperbola far from the origin, we use (2), along with Example 2 of Section 4.7, to deduce that

$$\lim_{x \to \infty} \left(y - \frac{b}{a}x \right) = \frac{b}{a} \lim_{x \to \infty} \left(\sqrt{x^2 - a^2} - x \right) = 0$$

This means that the hyperbola approaches the line $y = (b/a)x$ as x approaches ∞. Because the hyperbola is symmetric with respect to the x axis, it also approaches the line $y = -(b/a)x$. We call the lines

$$y = \frac{b}{a}x \quad \text{and} \quad y = -\frac{b}{a}x \tag{3}$$

asymptotes of the hyperbola. This information implies that the hyperbola is as sketched in Figure 10.24(a).

Now assume that the foci of a hyperbola are located at the points $(0, -c)$ and $(0, c)$ on the y axis, and again let $k = 2a$ in Definition 10.3. Then the equation of

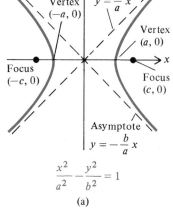

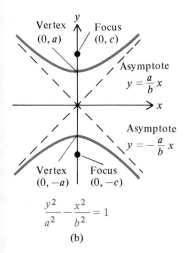

FIGURE 10.24

the hyperbola becomes

$$\frac{y^2}{a^2} - \frac{x^2}{b^2} = 1 \tag{4}$$

where as before

$$b = \sqrt{c^2 - a^2} \tag{5}$$

In this case the asymptotes are the lines $y = (a/b)x$ and $y = (-a/b)x$. The hyperbola is shown in Figure 10.24(b).

For reference we list the two forms of equations of hyperbolas in standard position:

$$\frac{x^2}{a^2} - \frac{y^2}{b^2} = 1 \begin{cases} \text{foci are } (-c, 0) \text{ and } (c, 0), \text{ where } c = \sqrt{a^2 + b^2} \\[4pt] \text{vertices are } (-a, 0) \text{ and } (a, 0) \\[4pt] \text{asymptotes are the lines } y = (b/a)x \text{ and} \\ \quad y = (-b/a)x \\[4pt] \text{symmetry with respect to the } x \text{ axis, } y \text{ axis,} \\ \quad \text{and origin} \end{cases} \tag{6}$$

$$\frac{y^2}{a^2} - \frac{x^2}{b^2} = 1 \begin{cases} \text{foci are } (0, -c) \text{ and } (0, c), \text{ where } c = \sqrt{a^2 + b^2} \\[4pt] \text{vertices are } (0, -a) \text{ and } (0, a) \\[4pt] \text{asymptotes are the lines } y = (a/b)x \text{ and} \\ \quad y = (-a/b)x \\[4pt] \text{symmetry with respect to the } x \text{ axis, } y \text{ axis,} \\ \quad \text{and origin} \end{cases} \tag{7}$$

The numbers a and c have the following geometric interpretation:

$$\begin{aligned} 2a &= \text{length of transverse axis} \\ 2c &= \text{distance between foci} \end{aligned} \tag{8}$$

Example 1   Find the foci and the asymptotes of the hyperbola

$$x^2 - \frac{y^2}{4} = 1$$

and sketch the hyperbola.

Solution   The equation is in the form of (6), with $a = 1$ and $b = 2$. Therefore $c = \sqrt{1 + 4} = \sqrt{5}$. As a result the foci are $(-\sqrt{5}, 0)$ and $(\sqrt{5}, 0)$. Moreover, the

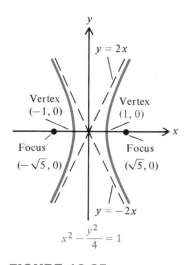

FIGURE 10.25

asymptotes are the lines $y = 2x$ and $y = -2x$. Consequently the hyperbola is as shown in Figure 10.25. ☐

Example 2 Find the foci and the asymptotes of the hyperbola whose equation is

$$9y^2 - 16x^2 = 144$$

and sketch the hyperbola.

Solution Dividing both sides of the equation by 144 yields

$$\frac{y^2}{16} - \frac{x^2}{9} = 1$$

This equation is in the form of (7), with $a = 4$ and $b = 3$. Therefore $c = \sqrt{4^2 + 3^2} = 5$, so that the foci are located at $(0, -5)$ and $(0, 5)$. The asymptotes are the lines $y = \frac{4}{3}x$ and $y = -\frac{4}{3}x$. The graph is shown in Figure 10.26. ☐

Translation of Hyperbolas

If a hyperbola is situated so that a translation of the axes brings it into standard position, then we can readily sketch it.

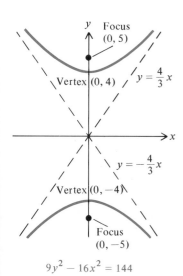

FIGURE 10.26

Example 3 Find an equation of the hyperbola whose foci are $(-2, -1)$ and $(-2, 9)$ and whose transverse axis has length 6. Sketch the hyperbola.

Solution The principal axis is the line $x = -2$, and the center of the hyperbola is $(-2, 4)$, midway between the two foci. Since the distance between the foci is $2c$, it follows that $c = 5$. In the XY system with origin at $(-2, 4)$, the hyperbola is in standard position with foci $(0, -5)$ and $(0, 5)$ and with the transverse axis lying along the Y axis. As a result, an equation of the hyperbola has the form

$$\frac{Y^2}{a^2} - \frac{X^2}{b^2} = 1$$

To determine a we observe that the transverse axis has length 6, so that by (8) we have $2a = 6$, or $a = 3$. It is now simple to determine b from (7):

$$b = \sqrt{c^2 - a^2} = \sqrt{5^2 - 3^2} = 4$$

Consequently the hyperbola has the equation

$$\frac{Y^2}{9} - \frac{X^2}{16} = 1$$

with asymptotes $Y = \frac{3}{4}X$ and $Y = -\frac{3}{4}X$. It follows that in the xy system the equation is

$$\frac{(y - 4)^2}{9} - \frac{(x + 2)^2}{16} = 1$$

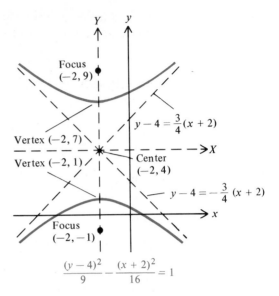

FIGURE 10.27

and the asymptotes are

$$y - 4 = \frac{3}{4}(x + 2) \quad \text{and} \quad y - 4 = -\frac{3}{4}(x + 2)$$

The hyperbola is shown in Figure 10.27. □

Any equation of the form

$$Ax^2 + Cy^2 + Dx + Ey + F = 0$$

with A and C of opposite sign (that is, $AC < 0$) is an equation of a hyperbola or a degenerate conic section. (See Exercise 36.) We can sketch the hyperbola after completing the squares and translating the coordinate axes.

Example 4 Show that

$$16x^2 - y^2 - 32x - 6y = 57$$

is an equation of a hyperbola, and sketch the hyperbola.

Solution By completing the squares we find that

$$16(x^2 - 2x + 1) - (y^2 + 6y + 9) = 57 + 16 - 9 = 64$$

so that

$$\frac{(x - 1)^2}{4} - \frac{(y + 3)^2}{64} = 1$$

Therefore an equation in the XY system with origin at $(1, -3)$ is

$$\frac{X^2}{4} - \frac{Y^2}{64} = 1$$

This equation is in the form of (6), with $a = 2$ and $b = 8$. Consequently its graph is a hyperbola. Since

$$c = \sqrt{2^2 + 8^2} = 2\sqrt{17}$$

the foci are $(2\sqrt{17},\ 0)$ and $(-2\sqrt{17},\ 0)$ in the XY system. Moreover, the asymptotes are $Y = 4X$ and $Y = -4X$. Translating to the xy system, we find that the foci are $(1 + 2\sqrt{17},\ -3)$ and $(1 - 2\sqrt{17},\ -3)$ and that the asymptotes are the lines

$$y + 3 = 4(x - 1) \quad \text{and} \quad y + 3 = -4(x - 1)$$

The hyperbola is sketched in Figure 10.28.   □

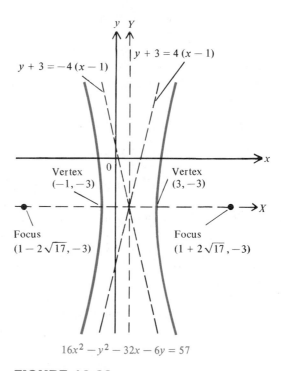

FIGURE 10.28

Applications of Hyperbolas

In conclusion, we mention a few applications of hyperbolas.

1. Comets that do not move in elliptical orbits around the sun almost always move in hyperbolic orbits. (In theory they can also move in parabolic orbits.)
2. Boyle's Law, relating the pressure p and the volume V of a perfect gas at constant temperature, states that $pV = c$, where c is a constant. The graph of p as a function of V is a hyperbola. In Section 10.4 we will demonstrate this for the case $c = 1$.
3. Hyperbolas can be used to locate the source of a sound heard at three different locations (see Exercise 41).

EXERCISES 10.3

In Exercises 1–12 find an equation of the hyperbola having the given properties. Then sketch the hyperbola.

1. The foci are $(9, 0)$ and $(-9, 0)$, and the vertices are $(4, 0)$ and $(-4, 0)$.

2. The foci are $(\sqrt{2}, 0)$ and $(-\sqrt{2}, 0)$, and the vertices are $(1, 0)$ and $(-1, 0)$.

3. The foci are $(0, 1)$ and $(0, -1)$, and the vertices are $(0, \frac{1}{3})$ and $(0, -\frac{1}{3})$.

4. The foci are $(2, 0)$ and $(-2, 0)$, and the length of the transverse axis is 2.

5. The foci are $(0, 5)$ and $(0, -5)$, and the length of the transverse axis is 6.

6. The foci are $(\sqrt{5}, 0)$ and $(-\sqrt{5}, 0)$, and the asymptotes are $y = 2x$ and $y = -2x$.

7. The foci are $(0, 10)$ and $(0, -10)$, and the asymptotes are $y = 3x$ and $y = -3x$.

8. The foci are $(2, 1)$ and $(-4, 1)$, and the asymptotes are $y = \sqrt{3}x + \sqrt{3} + 1$ and $y = -\sqrt{3}x - \sqrt{3} + 1$

9. The foci are $(0, 9)$ and $(0, -1)$, and the asymptotes are $y = \frac{4}{3}x + 4$ and $y = -\frac{4}{3}x + 4$.

10. The vertices are $(4, 0)$ and $(-4, 0)$, and the asymptotes are perpendicular to one another.

*11. The hyperbola passes through $(\sqrt{20}, 8)$ and $(\sqrt{8}, 4)$ and is in standard position.

*12. The hyperbola passes through $(2, -2\sqrt{7})$ and $(1, -2)$ and is in standard position.

In Exercises 13–23 find the vertices, foci, and asymptotes of the given hyperbola.

13. $\dfrac{x^2}{9} - \dfrac{y^2}{16} = 1$

14. $\dfrac{y^2}{16} - \dfrac{x^2}{4} = 1$

15. $\dfrac{x^2}{25} - \dfrac{y^2}{49} = 1$

16. $9x^2 - 16y^2 = 144$

17. $y^2 - 9x^2 = 25$

18. $x^2 - y^2 = 1$

19. $4x^2 - y^2 = 25$

20. $6y^2 - 3x^2 = 4$

21. $\dfrac{(x + 3)^2}{25} - \dfrac{(y + 1)^2}{144} = 1$

22. $\dfrac{(y - 2)^2}{121} - \dfrac{(x + 2)^2}{121} = 1$

23. $(x + \sqrt{2})^2 - \dfrac{(y - 3)^2}{81} = 1$

In Exercises 24–29 show that the equation represents a hyperbola. Sketch the hyperbola and indicate its center, vertices, and asymptotes.

24. $x^2 + 4x - 4y^2 = 12$

25. $x^2 - y^2 + 6x + 12y = 36$

26. $x^2 - 2x - 4y^2 - 12y = -8$

27. $4x^2 - 9y^2 - 8x - 36y = 68$

28. $4x^2 - 16x - 9y^2 - 54y = 101$

29. $x^2 + 2\sqrt{2}x - y^2 + 2\sqrt{2}y = 1$

30. For what value of c does the hyperbola

$$\frac{(x - 1)^2}{4} - \frac{(y + 2)^2}{9} = c$$

pass through the origin?

31. Let $y_0 \neq 0$. Show that an equation of the line tangent to the hyperbola $x^2/a^2 - y^2/b^2 = 1$ at the point (x_0, y_0) is

$$\frac{xx_0}{a^2} - \frac{yy_0}{b^2} = 1$$

32. Find the values of d for which the line $2y - x = d$ is tangent to the hyperbola $6y^2 - 3x^2 = 9$.

33. Show that any line parallel to but distinct from an asymptote of a hyperbola intersects the hyperbola exactly once.

34. A hyperbola for which $a = b$ is called **equilateral**. Show that a hyperbola is equilateral if and only if its asymptotes are perpendicular to one another.

35. The line segment that passes through a focus of a hyperbola, is perpendicular to the principal axis, and has its endpoints on the hyperbola is called a **latus rectum**. Show that if an equation of the hyperbola is

$$\frac{x^2}{a^2} - \frac{y^2}{b^2} = 1$$

then the length of a latus rectum is $2b^2/a$.

36. Consider the equation

$$Ax^2 + Cy^2 + Dx + Ey + F = 0 \quad \text{where } AC < 0 \quad (9)$$

and let

$$r = \frac{D^2}{4A} + \frac{E^2}{4C} - F$$

a. Show that the graph of equation (9) is a hyperbola if $r \neq 0$.

b. Show that the graph of the given equation is a pair of intersecting lines if $r = 0$.

37. Use the results of Exercise 36 to find all values of F such that the graph of the equation

$$3x^2 - 4y^2 + 6x + 8y + F = 0$$

a. is a hyperbola.

b. consists of two intersecting lines.

38. Find an equation for the collection of points such that the distance from each point to (3, 0) is twice the distance to the line $x = -3$. Show that the equation represents a hyperbola.

39. Why is it impossible for a hyperbola to have foci at $(-2, 0)$ and $(2, 0)$ and vertices at $(-3, 0)$ and $(3, 0)$?

40. Thunder is heard by Marian and Jack, who are talking to each other by telephone, 8800 feet apart. Marian hears the thunder 4 seconds before Jack does. Sketch a graph of the locations where the lightning could have struck. Take the speed of sound to be 1100 feet per second. (*Hint:* Suppose Marian is at (4400, 0) and Jack is at $(-4400, 0)$. Use Definition 10.3 and (6).)

*41. In Exercise 40, suppose Bruce is wired into the conversation and is midway between Marian and Jack. If Bruce hears the thunder 1 second after Marian does, determine where the lightning strikes in relation to the three persons involved.

10.4
ROTATION OF AXES

Let us summarize the information we have about conic sections thus far. Except in degenerate cases, the graph of the equation

$$Ax^2 + Cy^2 + Dx + Ey + F = 0 \tag{1}$$

is

1. a parabola if $AC = 0$ but not both A and C are zero.
2. an ellipse if $AC > 0$.
3. a hyperbola if $AC < 0$.

Now we would like to analyze the graph of any second-degree equation of the form

$$Ax^2 + Bxy + Cy^2 + Dx + Ey + F = 0 \tag{2}$$

Since (2) reduces to (1) when $B = 0$, we assume in this section that $B \neq 0$. We will show that in the XY coordinate system obtained by rotating the x and y axes through a suitable angle about the origin, (2) reduces to

$$A'X^2 + C'Y^2 + D'X + E'Y + F' = 0 \tag{3}$$

This equation has the same form as (1) and has already been analyzed thoroughly. The purpose of rotating the x and y axes is to eliminate the xy term in (2).

To gain an understanding of the relationship between such an XY coordinate system and the given xy system, we assume that the XY coordinate system is obtained by rotating the x and y axes through an angle θ about the origin (Figure 10.29). Then any point P in the plane has coordinates (x, y) and (X, Y) in the two systems. If P is not the origin, let ϕ be the angle from the positive X axis to the line segment joining P to the origin, and let r be the length of that line segment (Figure 10.29). Then we have

$$X = r\cos\phi \quad \text{and} \quad Y = r\sin\phi$$

and

$$x = r\cos(\theta + \phi) \quad \text{and} \quad y = r\sin(\theta + \phi)$$

Using the trigonometric identities for $\sin(\theta + \phi)$ and $\cos(\theta + \phi)$, we obtain

$$x = r\cos(\theta + \phi) = r(\cos\theta\cos\phi - \sin\theta\sin\phi)$$
$$= (\cos\theta)(r\cos\phi) - (\sin\theta)(r\sin\phi)$$

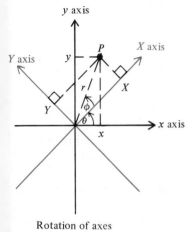

Rotation of axes

FIGURE 10.29

and thus
$$x = (\cos \theta)X - (\sin \theta)Y \tag{4}$$

Similarly
$$y = r \sin(\theta + \phi) = r(\sin \theta \cos \phi + \cos \theta \sin \phi)$$
$$= (\sin \theta)(r \cos \phi) + (\cos \theta)(r \sin \phi)$$

and thus
$$y = (\sin \theta)X + (\cos \theta)Y \tag{5}$$

We can now substitute (4) and (5), respectively, for x and y in (2); we conclude that

$$A[(\cos \theta)X - (\sin \theta)Y]^2 + B[(\cos \theta)X - (\sin \theta)Y][(\sin \theta)X + (\cos \theta)Y]$$
$$+ C[(\sin \theta)X + (\cos \theta)Y]^2 + D[(\cos \theta)X - (\sin \theta)Y]$$
$$+ E[(\sin \theta)X + (\cos \theta)Y] + F = 0$$

Combining like terms (those involving X^2, XY, Y^2, and so on) leads to an equation of the form

$$A'X^2 + [-2A \cos \theta \sin \theta + B(\cos^2 \theta - \sin^2 \theta) + 2C \cos \theta \sin \theta]XY$$
$$+ C'Y^2 + D'X + E'Y + F' = 0 \tag{6}$$

The exact expressions for A', C', D', E', and F' are irrelevant at this moment. What is important is that the equation in (6) will be in the form of (3) if the expression in brackets is 0. But this means that

$$-A \sin 2\theta + B \cos 2\theta + C \sin 2\theta = 0$$

or
$$(A - C) \sin 2\theta = B \cos 2\theta \tag{7}$$

If $A = C$, the equation in (7) is satisfied by $\theta = \pi/4$. If $A \neq C$, then $\cos 2\theta \neq 0$, so we can divide by $(A - C) \cos 2\theta$ and deduce that

$$\tan 2\theta = \frac{\sin 2\theta}{\cos 2\theta} = \frac{B}{A - C}$$

Thus the expression in the brackets in (6) will be 0 if we choose θ so that

$$\tan 2\theta = \frac{B}{A - C} \quad \text{if } A \neq C$$
$$\theta = \frac{\pi}{4} \quad \text{if } A = C \tag{8}$$

The discussion above yields the following procedure for converting an equation in the form of (2) to an equivalent equation in the form of (3):

1. Determine θ between 0 and $\pi/2$ by using (8) with the values of A, B, and C for the given equation (in the form of (2)).
2. Substitute the value of θ found in step 1 into (4) and (5) in order to express x and y in terms of X and Y.

3. Substitute the expressions for x and y found in step 2 into the given equation.
4. Simplify the equation resulting from step 3 in order to obtain an equation in the form of (3).

This procedure is illustrated in the next two examples.

Example 1   Show that the graph of the equation

$$xy = 1$$

is a hyperbola, and find the angle θ between 0 and $\pi/2$ through which the x and y axes must be rotated for the hyperbola to be in standard position in the rotated system. Then sketch the hyperbola.

Solution   For the equation $xy = 1$ we have $A = 0 = C$ in (2), so that $\theta = \pi/4$ in (8). Since $\cos \pi/4 = \sin \pi/4 = \sqrt{2}/2$, formulas (4) and (5) become

$$x = \frac{\sqrt{2}}{2}(X - Y) \quad \text{and} \quad y = \frac{\sqrt{2}}{2}(X + Y)$$

Therefore the equation $xy = 1$ becomes

$$\left(\frac{\sqrt{2}}{2}(X - Y)\right)\left(\frac{\sqrt{2}}{2}(X + Y)\right) = 1$$

or

$$\frac{X^2}{2} - \frac{Y^2}{2} = 1$$

This is an equation of a hyperbola with vertices $(-\sqrt{2}, 0)$ and $(\sqrt{2}, 0)$ in the XY coordinate system. Since the x and y axes were rotated through an angle of $\pi/4$ to obtain the XY system, the hyperbola is as shown in Figure 10.30.   □

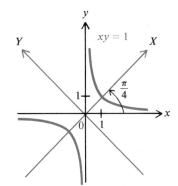

FIGURE 10.30

Example 2   Use rotation of axes to eliminate the xy term in the equation

$$73x^2 - 72xy + 52y^2 + 30x + 40y - 75 = 0 \tag{9}$$

Determine the type of conic section it represents, and sketch the graph of the equation.

Solution   In this example $A = 73$, $B = -72$, and $C = 52$, so that equation (8) becomes

$$\tan 2\theta = \frac{B}{A - C} = -\frac{72}{21} = -\frac{24}{7}$$

Since $\tan 2\theta < 0$ and $0 < \theta < \pi/2$, we have $\pi/2 < 2\theta < \pi$, so that $\cos 2\theta < 0$. From the Pythagorean Theorem we find that the hypotenuse of the triangle in Figure 10.31 is 25. Therefore

$$\cos 2\theta = -\frac{7}{25}$$

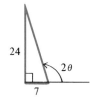

FIGURE 10.31

Since $0 < \theta < \pi/2$, both $\cos \theta$ and $\sin \theta$ are nonnegative. This implies that

$$\cos \theta = \sqrt{\frac{1 + \cos 2\theta}{2}} = \sqrt{\frac{1 - 7/25}{2}} = \frac{3}{5}$$

and

$$\sin \theta = \sqrt{\frac{1 - \cos 2\theta}{2}} = \sqrt{\frac{1 + 7/25}{2}} = \frac{4}{5}$$

From (4) and (5) we find that

$$x = \frac{3X - 4Y}{5} \quad \text{and} \quad y = \frac{4X + 3Y}{5}$$

Now we substitute for x and y into (9) to obtain

$$\frac{73}{25}(3X - 4Y)^2 - \frac{72}{25}(3X - 4Y)(4X + 3Y) + \frac{52}{25}(4X + 3Y)^2$$

$$+ \frac{30}{5}(3X - 4Y) + \frac{40}{5}(4X + 3Y) - 75 = 0$$

Expanding the squared expressions and combining like terms, we derive the equation

$$25X^2 + 100Y^2 + 50X - 75 = 0$$

Completing the square and dividing by 100 yields

$$\frac{(X + 1)^2}{4} + Y^2 = 1$$

We recognize this as an equation of an ellipse. The graph is shown in Figure 10.32, along with both sets of axes. $\quad\square$

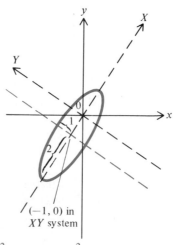

$$73x^2 - 72xy + 52y^2 + 30x + 40y - 75 = 0$$

FIGURE 10.32

Although the exact expressions for A' or C' in (3) are of no theoretical interest to us, it is possible to show (see Exercise 17) that

$$B^2 - 4AC = -4A'C' \qquad (10)$$

The number $4A'C'$ is valuable because, as we infer from the three criteria listed at the beginning of this section, the sign of $A'C'$ (and hence of $4A'C'$) determines the type of conic section involved. Combining (10) with these three criteria, we conclude that if the graph of

$$Ax^2 + Bxy + Cy^2 + Dx + Ey + F = 0 \qquad (11)$$

is not degenerate, then it is

1. a parabola if $B^2 - 4AC = 0$.
2. an ellipse if $B^2 - 4AC < 0$.
3. a hyperbola if $B^2 - 4AC > 0$.

The expression $B^2 - 4AC$ is called the **discriminant** of (11), and indeed it serves to discriminate between the different types of conic section.

EXERCISES 10.4

In Exercises 1–10 remove the xy term by rotation of axes. Then decide what type of conic section is represented by the equation, and sketch its graph.

1. $xy = -4$
2. $x^2 + \sqrt{3}xy = 3$
3. $x^2 - xy + y^2 = 2$
4. $9x^2 - 24xy + 2y^2 - 75 = 0$
5. $145x^2 + 120xy + 180y^2 = 900$
6. $10x^2 - 12xy + 10y^2 - 16\sqrt{2}x + 16\sqrt{2}y = 16$
7. $16x^2 - 24xy + 9y^2 - 5x - 90y + 25 = 0$
8. $16x^2 + 24xy + 9y^2 + 100x - 50y = 0$
9. $2x^2 - 72xy + 23y^2 + 100x - 50y = 0$
10. $2x^2 + 4xy + 2y^2 + 28\sqrt{2}x - 12\sqrt{2}y + 16 = 0$

In Exercises 11–13 use rotation of axes to show that the graph of the given equation is a degenerate conic section.

11. $9x^2 - 24xy + 2y^2 = 0$

12. $145x^2 + 120xy + 180y^2 = 0$
13. $145x^2 + 120xy + 180y^2 = -900$
14. Show that if $B > 0$, then the graph of

$$x^2 + Bxy = F$$

is a hyperbola if $F \neq 0$, and two intersecting lines if $F = 0$.
15. Assume that $B \neq 0$. Describe the graph of

$$Bxy + Dx + Ey + F = 0$$

16. Let R be the region between the line $y = x$ and the parabola $x^2 + 2xy + y^2 - \sqrt{2}x + \sqrt{2}y = 2$. Let D be the solid region obtained by revolving R about the line $y = x$. Find the volume V of D.

*17. Determine formulas for the numbers A' and C' in (6), and show that if θ satisfies (8), then

$$B^2 - 4AC = -4A'C'$$

10.5
A UNIFIED DESCRIPTION OF CONIC SECTIONS

We defined the parabola in terms of a focus and a directrix, whereas we defined the ellipse and the hyperbola in terms of two foci. In this section we will show that ellipses and hyperbolas can also be described in terms of a focus and a directrix.

Such a formulation lends itself to a description of the conic sections in polar coordinates. We will give such a description at the conclusion of the section.

Eccentricity   The key to the unified treatment of the conic sections is the concept of eccentricity. The **eccentricity** e of an ellipse or a hyperbola is defined by

$$e = \frac{\text{distance between the foci}}{\text{distance between the vertices}}$$

For an ellipse the distance between the foci is $2c$ and the distance between the vertices is $2a$. Consequently the formula for e becomes

$$e = \frac{c}{a} \tag{1}$$

From the definition of b in (2) of Section 10.2 we know that

$$c^2 = a^2 - b^2 \tag{2}$$

so that $0 \le e < 1$. From (1) and (2) we see that

$$e^2 = 1 - \frac{b^2}{a^2} \tag{3}$$

Thus the larger e is, the smaller b^2/a^2 is, and hence the more disparate the major and minor axes are (Figure 10.33(a) and (b)). In the extreme case in which $e = 0$ we find that $a = b$, so that the ellipse is a circle.

Now let us assume that the ellipse has eccentricity $e \ne 0$ and is in standard position with one focus P_1 located at $(c, 0)$ and equation

$$\frac{x^2}{a^2} + \frac{y^2}{b^2} = 1$$

(Figure 10.34). If $P(x, y)$ is any point on the ellipse, then from (1)–(3) we deduce that

$$|PP_1| = \sqrt{(x - c)^2 + y^2}$$

$$= \sqrt{x^2 - 2cx + c^2 + b^2\left(1 - \frac{x^2}{a^2}\right)}$$

$$= \sqrt{x^2\left(1 - \frac{b^2}{a^2}\right) - 2cx + c^2 + b^2}$$

$$= \sqrt{e^2 x^2 - 2aex + a^2} = \sqrt{(ex - a)^2} = |ex - a| = e\left|x - \frac{a}{e}\right|$$

If we let l_1 denote the line $x = a/e$, then the distance $|Pl_1|$ from (x, y) to l_1 is

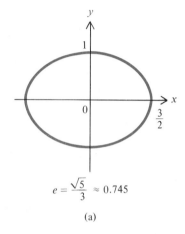

$e = \dfrac{\sqrt{5}}{3} \approx 0.745$

(a)

$e = \dfrac{2\sqrt{2}}{3} \approx 0.943$

(b)

FIGURE 10.33

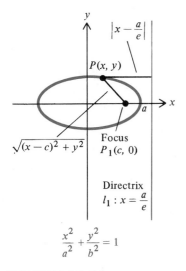

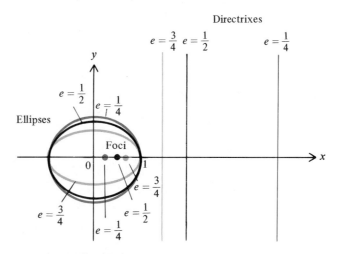

$|x - a/e|$ (Figure 10.34). Consequently

$$|PP_1| = e|Pl_1| \tag{4}$$

We have assumed that $0 < e < 1$. But if we were to let $e = 1$ in (4), then the equation would become $|PP_1| = |Pl_1|$, which is the defining equation for a parabola with directrix l_1. With this in mind, we call the line l_1 given by $x = a/e$ a **directrix** of the ellipse. (A second directrix l_2, defined to be the line $x = -a/e$, arises in a similar way if we let P_2 be the second focus $(-c, 0)$.) Figure 10.35 indicates the relative positions of the focus and the directrix when the eccentricity varies but the major axis remains the same.

FIGURE 10.35

If the conic section is a hyperbola whose foci are $2c$ apart and whose transverse axis has length $2a$, the eccentricity is again given by

$$e = \frac{c}{a}$$

Since $c^2 = a^2 + b^2$, it follows that $e > 1$. In fact,

$$e^2 = 1 + \frac{b^2}{a^2}$$

For a hyperbola in standard position, if P_1 is a focus located at $(c, 0)$ and if $P(x, y)$ is any point on the hyperbola, then calculations similar to those performed above for the ellipse establish that

$$|PP_1| = \sqrt{(x - c)^2 + y^2} = e\left|x - \frac{a}{e}\right|$$

Once again we let l_1 denote the line $x = a/e$ and call l_1 a **directrix** of the hyperbola. We conclude that

$$|PP_1| = e|Pl_1| \tag{5}$$

This is the same formula as in (4). However, now we have $e > 1$, rather than $0 < e < 1$. By the same method you can show that if $P_2 = (-c, 0)$ and if l_2 denotes the line $x = -a/e$ (also called a directrix of the hyperbola), then any point P on the hyperbola satisfies

$$|PP_2| = e|Pl_2|$$

Figure 10.36 shows the relative positions of the focus and the directrix when the eccentricity varies but the transverse axis remains the same.

The formula $|PP_1| = e|Pl_1|$ holds for any noncircular ellipse, parabola, or hyperbola, regardless of its position in the xy plane, because translations and rotations of the coordinate axes do not change distances between fixed points and lines. Thus we already have proved the first half of the following theorem.

THEOREM 10.4

For any noncircular, nondegenerate conic section there exist a line l_1, a point P_1 not on l_1, and a number $e > 0$ such that each point P on the conic section satisfies

$$|PP_1| = e|Pl_1|$$

Conversely, if l_1 is a line, P_1 a point not on l_1, and e is a positive number, then the set of points P in the plane that satisfy the equation

$$|PP_1| = e|Pl_1|$$

is a conic section. Moreover, the conic section is

1. an ellipse if $0 < e < 1$.
2. a parabola if $e = 1$.
3. a hyperbola if $e > 1$.

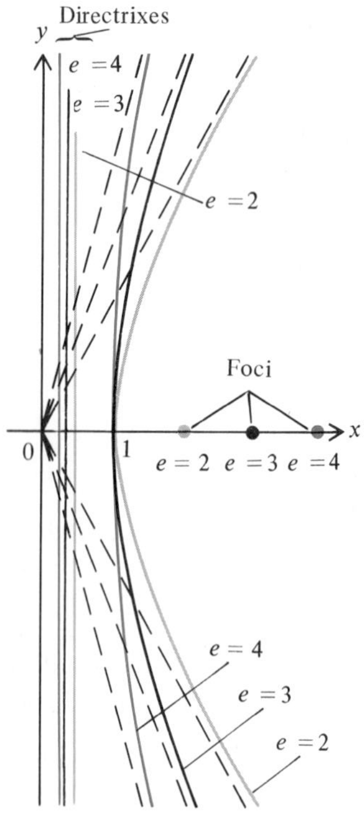

FIGURE 10.36

Proof The first half has already been proved. To simplify the proof of the converse, assume that P_1 is the origin and l_1 is the line $x = d$. If $P(x, y)$ satisfies

$$|PP_1| = e|Pl_1|$$

then

$$\sqrt{x^2 + y^2} = e|d - x|$$

or

$$x^2 + y^2 = e^2(d - x)^2 \tag{6}$$

If $e = 1$, then (6) describes a parabola, as you can verify directly. If $e \neq 1$, then by completing the squares and combining like terms we can transform (6) into

$$(1 - e^2)\left(x + \frac{de^2}{1 - e^2}\right)^2 + y^2 = \frac{d^2 e^2}{1 - e^2} \tag{7}$$

Equation (7) is an equation of an ellipse if $0 < e < 1$ and is an equation of a hyperbola if $e > 1$. Since translations and rotations of conic sections are again conic sections of the same type, we have proved the converse. ∎

When we know the eccentricity of a conic section and the value of a, b, or c, we can find an equation that describes the conic section.

Example 1   The elliptical orbit of Mars has an eccentricity of approximately 0.093, and its major axis has a length of approximately 456 million kilometers. Find an equation for the orbit.

Solution   Let us assume that the center is at the origin and that the major axis lies on the x axis and has length 456. Then $a = 228$, and therefore

$$c = ae = (228)(0.093) = 21.204$$

As a result,

$$b = \sqrt{a^2 - c^2} = \sqrt{(228)^2 - (21.204)^2} \approx 227$$

Consequently the orbit is given (approximately) by the equation

$$\frac{x^2}{(228)^2} + \frac{y^2}{(227)^2} = 1 \quad \square$$

Notice that since the eccentricity of Mars's orbit is very small, the orbit is practically circular. This is true of most of the planets in our own solar system. We consider this topic in more detail in the exercises.

Polar Equations for the Conic Sections

Now that we have described all noncircular conic sections in terms of distances from foci and directrixes, we will derive polar equations for these conic sections.

Let us consider a conic section having eccentricity $e > 0$, focus P_1 located at the origin, and corresponding directrix l_1 perpendicular to the x axis. Suppose l_1 lies to the right of the origin, with equation $x = k$ (Figure 10.37). If $P(x, y)$ is any point on the conic section to the left of l_1 with polar coordinates (r, θ), where $r > 0$, then by Theorem 10.4 we know that

$$|PP_1| = e|Pl_1|$$

Since

$$|PP_1| = r \quad \text{and} \quad |Pl_1| = k - x = k - r \cos \theta$$

it follows that

$$r = |PP_1| = e|Pl_1| = e(k - r \cos \theta)$$

Solving for r, we obtain

$$r = \frac{ek}{1 + e \cos \theta} \tag{8}$$

Equation (8) completely describes the given conic section if it is a parabola or an ellipse, because all the points on the conic section lie to the left of l_1. However, if the conic section is a hyperbola, then it contains points to the right of l_1 as well.

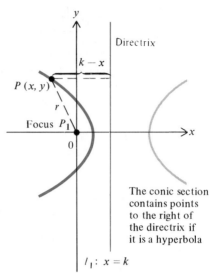

The conic section contains points to the right of the directrix if it is a hyperbola

l_1: $x = k$

Conic section with eccentricity e, focus at the origin, and directrix $x = k$

FIGURE 10.37

We will now show that the points to the right of l_1 also lie on the polar graph of (8). More precisely, we will show that if $P(x, y)$ is on the hyperbola to the right of l_1 and P has polar coordinates (r, θ) with $r > 0$, then the alternative polar coordinates $(-r, \theta + \pi)$ of P satisfy (8). The proof is as follows. For such a point P we have

$$|PP_1| = r$$

and

$$x = r \cos \theta = (-r) \cos (\theta + \pi)$$

so that

$$|Pl_1| = x - k = (-r) \cos (\theta + \pi) - k$$

From Theorem 10.4 we deduce that

$$r = |PP_1| = e|Pl_1| = e[(-r) \cos (\theta + \pi) - k]$$

Solving for $-r$, we obtain

$$-r = \frac{ek}{1 + e \cos (\theta + \pi)}$$

This means in particular that $(-r, \theta + \pi)$ satisfies (8), as asserted.

It is possible to reverse the steps above and show that conversely, every point $P(r, \theta)$ satisfying (8) lies on the associated conic section.

If we had chosen the coordinate system so that the directrix was the line $x = -k$, where $k > 0$, then a similar argument would demonstrate that an equation of the conic section is

$$r = \frac{ek}{1 - e \cos \theta} \tag{9}$$

Likewise, if the directrix were perpendicular to the y axis, having the equation $y = k$, where $k > 0$, then the equation would be

$$r = \frac{ek}{1 + e \sin \theta} \qquad (10)$$

If the directrix were the line $y = -k$ with $k > 0$, then the equation would be

$$r = \frac{ek}{1 - e \sin \theta} \qquad (11)$$

Thus (8)–(11) are equations of conic sections with a focus at the origin and a directrix perpendicular to the x axis or the y axis.

Example 2   Sketch the polar graph of the equation

$$r = \frac{6}{3 + 2 \cos \theta}$$

Solution   To put the equation into the form of one of the equations (8)–(11), we divide both numerator and denominator by 3, obtaining

$$r = \frac{2}{1 + \frac{2}{3} \cos \theta}$$

Comparing this equation with (8), we find that

$$e = \frac{2}{3} \quad \text{and} \quad k = \frac{2}{e} = \frac{2}{2/3} = 3$$

Since $e < 1$, the conic section is an ellipse with one focus at the origin; the associated directrix is the line $x = 3$ (Figure 10.38).   □

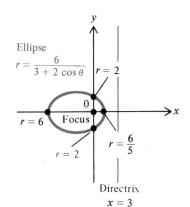

Ellipse

$r = \dfrac{6}{3 + 2 \cos \theta}$ $r = 2$

$r = 6$ Focus

$r = 2$ $r = \dfrac{6}{5}$

Directrix
$x = 3$

FIGURE 10.38

EXERCISES 10.5

In Exercises 1–12 find the eccentricity of the conic section with the given equation.

1. $\dfrac{x^2}{9} + \dfrac{y^2}{25} = 1$

2. $\dfrac{x^2}{64} + \dfrac{y^2}{49} = 1$

3. $\dfrac{x^2}{9} - \dfrac{y^2}{25} = 1$

4. $\dfrac{y^2}{25} - \dfrac{x^2}{9} = 1$

5. $4x^2 + y^2 = 8$

6. $6y^2 - 3x^2 = 4$

7. $4(x - 3)^2 + (y + 3)^2 = 8$

8. $x^2 = -6y$

9. $x^2 + 2y^2 - 2x - 4y = 1$

10. $x^2 + 4x - 4y^2 = 12$

11. $49x^2 - 9y^2 + 98x - 36y = 428$

12. $3y^2 - 5x + 3y - 6 = 0$

In Exercises 13–21 find an equation of the conic section possessing the given properties, and sketch the conic section.

13. The foci are $(9, 0)$ and $(-9, 0)$, and the eccentricity is $\frac{3}{5}$.

14. The foci are $(9, 0)$ and $(-9, 0)$, and the eccentricity is $\frac{5}{3}$.

15. The foci are $(0, 1)$ and $(0, -1)$, and the eccentricity is 2.

16. The vertices are $(5, 0)$ and $(-5, 0)$, and the eccentricity is $\frac{4}{5}$.

17. The vertices are $(0, 0)$ and $(0, 10)$, and the eccentricity is $\frac{13}{5}$.

18. The center is $(-2, 3)$, a vertex is $(-2, 0)$, and the eccentricity is $\frac{1}{2}$.

19. The hyperbola is in standard position, passes through $(\sqrt{20}, 8)$, and has eccentricity $\sqrt{17}$. (There are two such hyperbolas.)

20. The ellipse is in standard position, passes through $(-1, 1)$, and has eccentricity $\sqrt{3}/2$. (There are two such ellipses.)

21. The major axis of the ellipse has length 8 and is on the x axis, a focus is located at the origin, and the eccentricity is $\frac{1}{2}$. (There are two such conic sections.)

In Exercises 22–29 indicate the type of conic section represented by the given equation, and find an equation of a directrix.

22. $r = \dfrac{1}{1 + \frac{1}{2}\sin\theta}$

23. $r = \dfrac{1}{1 - \sin\theta}$

24. $r = \dfrac{1}{1 - 2\cos\theta}$

25. $r = \dfrac{25}{5 - 3\sin\theta}$

26. $r = \dfrac{9}{3 - 5\cos\theta}$

27. $r = \dfrac{1}{2 - 2\cos\theta}$

28. $r = \dfrac{12}{4 - 5\sin\theta}$

29. $r = \dfrac{3}{1 + \sin\theta}$

30. a. Sketch the graphs of
$$r = \frac{1}{2 + \cos\theta} \quad \text{and} \quad r = \frac{1}{1 - \cos\theta}$$
   b. Find the points where the graphs in (a) intersect.

The eccentricities and lengths of the major axes of the planetary orbits are given in the accompanying table.

Planet	Eccentricity	Length of major axis (in millions of kilometers)
Mercury	0.206	116
Venus	0.007	216
Earth	0.017	299
Mars	0.093	456
Jupiter	0.048	1557
Saturn	0.056	2854
Uranus	0.047	5738
Neptune	0.008	8996
Pluto	0.249	11,800

31. Find an approximate equation for the orbit of the earth around the sun.

*32. Assuming that the sun is located at a focus, approximate the minimum distance from the earth to the sun.

*33. From the information given in the table, is it possible that Neptune and Pluto could collide? Explain your answer.

Key Terms and Expressions

Parabola
Ellipse
Hyperbola

Standard position of conic section
Eccentricity
Rotation of axes

Key Formulas

$$x^2 = 4cy \quad \text{and} \quad y^2 = 4cx$$

$$\frac{x^2}{a^2} + \frac{y^2}{b^2} = 1 \quad \text{and} \quad \frac{x^2}{b^2} + \frac{y^2}{a^2} = 1$$

$$\frac{x^2}{a^2} - \frac{y^2}{b^2} = 1 \quad \text{and} \quad \frac{y^2}{a^2} - \frac{x^2}{b^2} = 1$$

$$e = \frac{c}{a}$$

REVIEW EXERCISES

In Exercises 1–12 write an equation for the conic section.

1. The parabola with focus $(-3, 4)$ and directrix $x = 5$.

2. The parabola with focus $(0, 4)$ and vertex $(0, 2)$.

3. The parabola having directrix $y = -10$ and axis $x = -2$ and passing through $(6, -2)$.

4. The ellipse having foci $(\sqrt{3}, 0)$ and $(-\sqrt{3}, 0)$ and passing through $(\sqrt{3}, 2)$.

5. The ellipse having vertices $(0, 2\sqrt{2})$ and $(0, -2\sqrt{2})$ and passing through $(1, \sqrt{6})$.

6. The ellipse with center $(-1, -3)$, focus $(-1, -2)$, and vertex $(-1, -5)$.

7. The ellipse with foci $(1, 1)$ and $(1, -3)$ and major axis of length 8.

8. The hyperbola in standard position that passes through $(1, 1)$ and $(2, \sqrt{11/2})$ and has its transverse axis on the x axis.

9. The hyperbola with vertices $(-3, 2)$ and $(1, 2)$ whose asymptotes are perpendicular to one another.

10. The hyperbola with foci $(12, -3)$ and $(-8, -3)$ and asymptote $y + 3 = \frac{4}{3}(x - 2)$.

11. The conic section with eccentricity 2, directrix $x = -4$, and corresponding vertex $(-2, 0)$.

12. The conic section with eccentricity $\frac{1}{3}$ and foci $(0, 0)$ and $(0, -2)$.

In Exercises 13–18 sketch the graph of the equation.

13. $49x^2 - 9y^2 + 98x - 36y = 428$

14. $4x^2 - 16x + y^2 - 6y = 0$

15. $9x^2 - 36x + 5y + 21 = 0$

16. $x^2 - 2x - 4y^2 - 12y = 10$

17. $4x^2 - 24xy + 11y^2 + 40x + 30y - 45 = 0$

18. $73x^2 + 72xy + 52y^2 = 25$

19. Determine an equation for the collection of points (x, y) such that the distance from (x, y) to the point $(2, 4)$ is e times the distance from (x, y) to the x axis, where
   a. $e = 3$ b. $e = 1$ c. $e = \frac{1}{2}$

In Exercises 20–22 indicate the type of conic section represented by the given equation.

20. $r = \dfrac{3}{1 + 4\sin\theta}$

21. $r = \dfrac{3}{4 - \cos\theta}$

22. $r = \dfrac{3}{4 - 4\sin\theta}$

23. Find all points on the parabola $y^2 = -8x$ that are closest to the point $(-10, 0)$.

24. Let $x^2 = 4cy$, with $c > 0$, and let (a, b) be outside the parabola (that is, assume that $a^2 > |4cb|$). Find equations for the two lines that are tangent to the parabola and pass through the point (a, b).

25. Show that half the length of the minor axis of an ellipse is the geometric mean of the two lengths into which a focus divides the major axis (see Exercise 81 of Section 1.1).

Cumulative Review, Chapters 1–9

In Exercises 1–2 evaluate the limit.

1. $\displaystyle\lim_{x \to 0^+} \frac{\arcsin x^2}{\sqrt{x^3}}$

2. $\displaystyle\lim_{x \to \infty} \frac{\sinh x}{1 + \cosh x}$

3. Let $f(t) = \dfrac{\sqrt{3 + t} - \sqrt{3}}{t}$.
   a. Find the domain of f.
   b. Find $\lim_{t \to 0} f(t)$.

4. Let $f(x) = \ln(x + e^{-x})$. Show that $f''(0) = 1$.

5. Suppose that $y^2 + x \arcsin y = 1$. Find $\dfrac{dy}{dx}$.

6. The length of one side of a right triangle is 1 and the length of the hypotenuse increases at the rate of three inches per second. How fast is the length of the other leg increasing when the length of the hypotenuse is $\sqrt{5}$?

7. The strength S of a rectangular beam is proportional to the product of the width w and the square of the depth d of the beam. Find the dimensions of the strongest beam that can be cut from a cylindrical log 84 centimeters in diameter.

8. Let $f(x) = \sqrt{x}/\sqrt{1 - x}$. Sketch the graph of f, noting all relevant information.

9. Let $f(x) = \int_0^x e^{-t^2/2}\, dt$ for all real values of x. Find the values of x such that the line tangent to the graph of f at $(x, f(x))$ has slope $\frac{1}{4}$.

10. Let $f(x) = \int_0^x e^{-t^2/2}\, dt$ for all real values of x.
   a. Show that f is an odd function.
   b. Using the fact that there is a number B with $\lim_{x \to \infty} f(x) = B$, sketch the graph of f. Note all pertinent information.

11. Let $f(x) = \int_0^x e^{-t^2/2}\,dt$ for all real values of x.
    a. Show that f^{-1} exists.
    b. Let $\int_0^1 e^{-t^2/2}\,dt = A$. Find $(f^{-1})'(A)$.

12. Let $f(x) = \int_0^x e^{-t^2/2}\,dt$ for all real values of x.
    a. Use Simpson's Rule with $n = 4$ to determine an approximate value of $f(1)$.
    b. Find an upper bound for the error introduced by approximating the integral for $f(1)$ by Simpson's Rule with $n = 4$.

In Exercises 13–15 evaluate the integral.

13. $\displaystyle\int \frac{\sec x \tan x}{\sec^2 x + 1}\,dx$

14. $\displaystyle\int \frac{1}{x^4 + x^3 + x^2}\,dx$

15. $\displaystyle\int x^5 \sin x^3\,dx$

16. Let $f(x) = (1/x^2)\sin(1/x)$ on $[2/\pi, \infty)$. Find the area A of the region between the graph of f and the x axis on $[2/\pi, \infty)$.

17. Find the area A of the region between the graphs of $y = x^2$ and $y = 2x$ on $[-1, 2]$.

18. Let D be the solid whose base lies on the xy plane and is bounded by the graphs of $y = 2x^2$ and $y = 3 - x^2$, and whose cross sections perpendicular to the x axis are square. Find the volume V of D.

19. The vertical cross sections of a vat have the shape of the graph of $y = x^2$ for $-2 \le x \le 2$. The vat has horizontal length 10 feet, and is full of water. Find the work W required to pump the water to a level 4 feet above the top of the vat.

20. Consider the sequence $\left\{\dfrac{\ln n}{n + 1}\right\}_{n=1}^{\infty}$
    a. Show that the sequence is decreasing for $n > e^2$.
    b. Show that $\displaystyle\lim_{n \to \infty} \frac{\ln n}{n + 1} = 0$.

In Exercises 21–22 determine whether the series converges.

21. $\displaystyle\sum_{n=1}^{\infty} \frac{\sin(e^n)}{(n + 1)^{3/2}}$

22. $\displaystyle\sum_{n=0}^{\infty} \frac{n^{2n}}{(2n)!}$

In Exercises 23–24 find the interval of convergence of the given power series.

23. $\displaystyle\sum_{n=1}^{\infty} \frac{(-1)^n 2^n}{\sqrt[3]{n + 1}} x^n$

24. $\displaystyle\sum_{n=1}^{\infty} \frac{\ln n}{n + 1} x^n$

25. Let $f(x) = xe^x$. Find the Taylor series of f about 1.

11

VECTORS, LINES AND PLANES

Many physical and abstract quantities have only magnitude and thus can be described by numbers. Examples we have encountered are mass, cost, profit, speed, area, length, volume, and moments about an axis. Many other quantities, however, have both magnitude and direction. The most notable example is velocity, which involves not only the speed of an object but also the direction of its motion. Quantities that have both magnitude and direction are described mathematically by vectors. In this chapter we will study vectors and their applications, including the description of lines and planes in space.

Our first task will be to set up a coordinate system in space, much as we did for the plane. Using the new coordinate system, we will define the concept of vector and present various elementary properties of vectors.

11.1
CARTESIAN COORDINATES IN SPACE

We begin by considering three mutually perpendicular lines that pass through a point O, called the **origin** (Figure 11.1). The three lines are named the **x axis**, the **y axis**, and the **z axis**. For each of these axes we set up a correspondence between the points on the axis and the set of real numbers, letting the origin O correspond to the number 0. Normally we will only draw the positive portions of the axes, as in Figure 11.1, with the positive x axis pointing toward the viewer, the positive y axis pointing toward the right, and the positive z axis pointing upward.

If P is any point in space, then the three planes through P perpendicular to the three axes intersect the x axis, the y axis, and the z axis at points corresponding to numbers x, y, and z, respectively (Figure 11.2). Therefore we

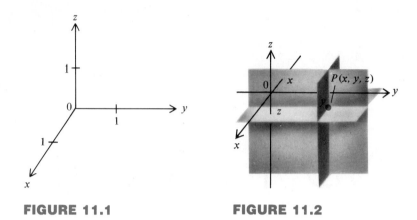

FIGURE 11.1   **FIGURE 11.2**

associate P with the ordered triple of numbers (x, y, z) and call (x, y, z) the **rectangular coordinates** (or **Cartesian coordinates**) of P. Figure 11.3 exhibits a few points in space, along with their Cartesian coordinates.

Under the association we have just described, each point in space is identified with an ordered triple of numbers. Conversely, each ordered triple (x, y, z) is identified with a single point whose coordinates are (x, y, z). This correspondence will enable us to describe geometric objects in space by means of equations and inequalities. Moreover, because each point in space has three coordinates, we sometimes refer to space as **three-dimensional space**.

In three-dimensional space there are three **coordinate planes**: the xy plane, which contains the x and y axes; the xz plane, which contains the x and z axes; and the yz plane, which contains the y and z axes. Since each plane divides space into two parts, the three coordinate planes together divide space into eight regions, called **octants** (Figure 11.4). The octant containing the positive x, y, and z axes is called the **first octant**.

Just as with points in the plane, the notion of distance between two points in space is fundamental. The formula for the **distance** $|PQ|$ between two points

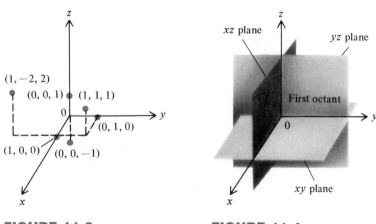

FIGURE 11.3   **FIGURE 11.4**

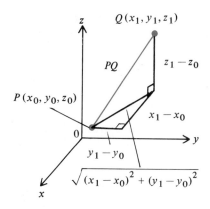

FIGURE 11.5

$P = (x_0, y_0, z_0)$ and $Q = (x_1, y_1, z_1)$ is

$$|PQ| = \sqrt{(x_1 - x_0)^2 + (y_1 - y_0)^2 + (z_1 - z_0)^2} \qquad (1)$$

as you can prove by two successive applications of the Pythagorean Theorem (Figure 11.5). If $Q = O$, then (1) reduces to

$$|PO| = |OP| = \sqrt{x_0^2 + y_0^2 + z_0^2}$$

Example 1   Let $P = (-1, 3, 6)$ and $Q = (4, 0, 5)$. Find $|PQ|$, $|PO|$, and $|OQ|$.

Solution   From (1) we calculate that

$$|PQ| = \sqrt{(4 - (-1))^2 + (0 - 3)^2 + (5 - 6)^2} = \sqrt{35}$$
$$|PO| = \sqrt{(-1)^2 + 3^2 + 6^2} = \sqrt{46}$$
$$|OQ| = \sqrt{4^2 + 0^2 + 5^2} = \sqrt{41} \quad \square$$

The three basic laws governing the distance between two points in space are

$$|PQ| = 0 \quad \text{if and only if} \quad P = Q$$

$$|PQ| = |QP|$$

$$|PQ| \le |PR| + |RQ| \quad \text{for any third point } R$$

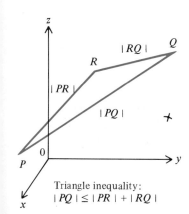

Triangle inequality:
$|PQ| \le |PR| + |RQ|$

FIGURE 11.6

The first two laws follow directly from (1). The third law, which is known as the **triangle inequality**, implies that the length of any side of a triangle does not exceed the sum of the lengths of the other two sides (Figure 11.6). (See Exercise 26 of Section 11.3 for a proof of the triangle inequality.)

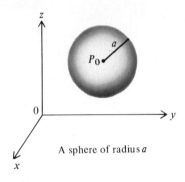

FIGURE 11.7

A sphere of radius a

The **sphere** with **center** $P_0 = (x_0, y_0, z_0)$ and **radius** a is defined to be the set of all points P such that

$$|P_0 P| = a$$

(Figure 11.7). Thus a point $P = (x, y, z)$ lies on that sphere if and only if

$$\sqrt{(x - x_0)^2 + (y - y_0)^2 + (z - z_0)^2} = a$$

or equivalently, if and only if

$$(x - x_0)^2 + (y - y_0)^2 + (z - z_0)^2 = a^2$$

This is an equation of a sphere in space.

Example 2 Show that

$$x^2 + y^2 + z^2 = 2x + 4y - 6z$$

is an equation of a sphere. Find the center and the radius of the sphere.

Solution We transpose terms from the right side to the left side of the equation, complete the squares, and obtain

$$(x^2 - 2x + 1) + (y^2 - 4y + 4) + (z^2 + 6z + 9) = 1 + 4 + 9$$

or $\qquad (x - 1)^2 + (y - 2)^2 + (z + 3)^2 = 14$

This is an equation of the sphere with center $(1, 2, -3)$ and radius $\sqrt{14}$. $\square$

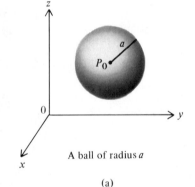

A ball of radius a

(a)

The **ball** with center $P_0 = (x_0, y_0, z_0)$ and radius a is the collection of points $P = (x, y, z)$ such that $|P_0 P| \leq a$, or such that

$$\sqrt{(x - x_0)^2 + (y - y_0)^2 + (z - z_0)^2} \leq a$$

(Figure 11.8(a)). This is equivalent to

$$(x - x_0)^2 + (y - y_0)^2 + (z - z_0)^2 \leq a^2$$

Thus if $P_0 = (1, 2, -3)$, then $P = (x, y, z)$ is in the ball with center P_0 and radius 3 provided that $|P_0 P| \leq 3$, that is,

$$\sqrt{(x - 1)^2 + (y - 2)^2 + (z + 3)^2} \leq 3$$

A ball is the solid region in space enclosed by a sphere. The corresponding region in the plane is enclosed by a circle, and we call such a region a **disk**. The disk with center $P_0 = (x_0, y_0)$ and radius a is the collection of points $P = (x, y)$ such that $|P_0 P| \leq a$, or such that

$$\sqrt{(x - x_0)^2 + (y - y_0)^2} \leq a$$

A disk of radius a

(b)

FIGURE 11.8

(Figure 11.8(b)). This is equivalent to

$$(x - x_0)^2 + (y - y_0)^2 \leq a^2$$

EXERCISES 11.1

In Exercises 1–8 find the distance D between the points P and Q.

1. $P = (\sqrt{2}, 0, 0), Q = (0, 1, 1)$

2. $P = (2, -1, -2), Q = (3, 1, 0)$

3. $P = (-3, 4, -5), Q = (0, 8, 7)$

4. $P = (4, -1, 3), Q = (4, 5, 11)$

5. $P = (-1, 3, 6), Q = (4, 2, 7)$

6. $P = (1, 0, -\frac{1}{2}), Q = (\frac{1}{2}, \frac{1}{2}\sqrt{2}, 0)$

7. $P = (2 \sin x, \cos x, \tan x), Q = (\sin x, 2 \cos x, 0)$

8. $P = (e^x, 0, 2\sqrt{2}), Q = (0, e^{-x}, \sqrt{2})$

9. Show that the point $(3, 0, 2)$ is equidistant from the points $(1, -1, 5)$ and $(5, 1, -1)$.

10. Find the perimeter of the triangle with vertices $(-1, 1, 2)$, $(2, 0, 3)$, and $(3, 4, 5)$.

11. Find an equation of the sphere with radius 5 and center $(2, 1, -7)$.

12. Find an equation of the sphere with radius $\sqrt{2}$ and center $(-1, 0, 3)$.

13. Show that
$$x^2 + y^2 + z^2 - 2x - 4y + 6z = -10$$
is an equation of a sphere. Find the radius and the center of the sphere.

14. Show that
$$x^2 + y^2 + z^2 + 6x + 8y - 4z + 4 = 0$$
is an equation of a sphere. Find the radius and the center of the sphere.

15. Find an inequality satisfied by all points in the ball with radius 6 and center $(0, -2, -3)$.

16. Find an inequality satisfied by all points in the disk that has radius $\sqrt{3}$ and center $(\frac{1}{2}, -1)$.

17. Let $P = (1, -1, 1)$, $Q = (2, 1, -1)$, and $R = (0, 0, 0)$. By computing the lengths of the sides, show that the triangle PQR is a right triangle.

18. Let $P = (3, 0, 3)$, $Q = (2, 0, -1)$, and $R = (c, 1, 2)$. Determine the two values of c that make triangle PQR a right triangle with hypotenuse PQ.

19. Find an equation of the set of points equidistant from the points $(2, 1, 0)$ and $(4, -1, -3)$.

20. Find an equation of the set of points twice as far from the origin as from the point $(-1, 1, 1)$. Show that the set of points is a sphere and find its center.

21. Show that the midpoint of the line segment joining the points (x_0, y_0, z_0) and (x_1, y_1, z_1) is $(\frac{1}{2}(x_0 + x_1), \frac{1}{2}(y_0 + y_1), \frac{1}{2}(z_0 + z_1))$.

22. Find the midpoint of the line segment joining the points $(3, 7, 11)$ and $(-9, 8, 31)$.

23. Suppose the points $(2, -1, 3)$ and $(4, 1, 7)$ are diametrically opposite each other on a sphere. Find an equation of the sphere.

11.2
VECTORS IN SPACE

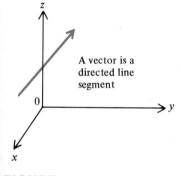

A vector is a directed line segment

FIGURE 11.9

Intuitively, a vector is a directed line segment in space, often described by an arrow (Figure 11.9). Vectors appear constantly in the study of motion in space. For instance, consider a particle moving along the path shown in Figure 11.10. If at a given time t the particle is at point P, we can assign to P a vector, called the **velocity vector**, which points in the direction of the particle's motion and has length equal to the speed of the particle. Vectors are also used to describe force; the vector describing a force points in the direction in which the force acts and has length equal to the magnitude of the force.

One obvious way to describe a vector is simply to give the coordinates of the initial and terminal points of the directed line segment associated with it. But as our comments on velocity and force might suggest, one is normally concerned more with the direction and length of a vector than with its initial and terminal

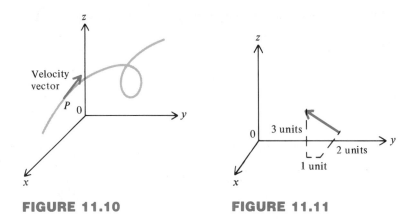

FIGURE 11.10   **FIGURE 11.11**

points. The vector appearing in Figure 11.11 represents a change of 2 units in the positive x direction, 1 unit in the negative y direction, and 3 units in the positive z direction. These three numbers identify the direction and length of the vector, and hence we can conveniently identify the vector with the ordered triple $(2, -1, 3)$. More generally, if a directed line segment has initial point (x_0, y_0, z_0) and terminal point (x_1, y_1, z_1), then the corresponding vector represents a change of $x_1 - x_0$ units in the positive x direction, $y_1 - y_0$ units in the positive y direction, and $z_1 - z_0$ units in the positive z direction. We group these three differences into the ordered triple $(x_1 - x_0, y_1 - y_0, z_1 - z_0)$ and identify the vector with this ordered triple.

DEFINITION 11.1

A **vector** is an ordered triple (a_1, a_2, a_3) of numbers. The numbers a_1, a_2, and a_3 are called the **components** of the vector. The vector associated with the directed line segment with initial point $P = (x_0, y_0, z_0)$ and terminal point $Q = (x_1, y_1, z_1)$ is $(x_1 - x_0, y_1 - y_0, z_1 - z_0)$ and is denoted $\overrightarrow{PQ}$.

Two vectors (a_1, a_2, a_3) and (b_1, b_2, b_3) are equal if and only if their components are equal, that is,

$$a_1 = b_1, \qquad a_2 = b_2, \quad \text{and} \quad a_3 = b_3$$

In the arrow notation, $\overrightarrow{PQ} = \overrightarrow{RS}$ if $\overrightarrow{PQ}$ and $\overrightarrow{RS}$ are represented by the same ordered triple.

Example 1 Let $P = (1, 3, 7)$, $Q = (-1, 0, 6)$, $R = (0, -1, -2)$, and $S = (-2, -4, -3)$. Show that $\overrightarrow{PQ}$ and $\overrightarrow{RS}$ are the same vector.

Solution Applying Definition 11.1 to the components of P, Q, R, and S, we find that

$$\overrightarrow{PQ} = (-2, -3, -1) = \overrightarrow{RS} \quad \square$$

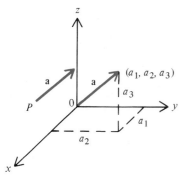

FIGURE 11.12

In print, vectors are almost always represented by boldface letters, although the choice of the particular letters used for vectors varies widely (depending on the context). To distinguish vectors from numbers, which are also called **scalars**, we will normally denote vectors by lowercase boldface letters near the beginning of the alphabet, such as **a**, **b**, and **c**. Other letters denote special vectors; for example, the zero vector $(0, 0, 0)$ is denoted **0**. Since it is difficult to write a boldface letter by hand, vectors are usually written by placing an arrow over a symbol or expression. Thus we would write **a** as $\vec{a}$.

Each vector $\mathbf{a} = (a_1, a_2, a_3)$ can be associated with a directed line segment having an arbitrary initial point P (Figure 11.12), and different initial points in space give us different representations of the same vector. If P is the origin, then the vector $\mathbf{a} = (a_1, a_2, a_3)$ is associated with the point (a_1, a_2, a_3) in space and with the directed line segment from the origin to (a_1, a_2, a_3) (Figure 11.12).

A natural way to assign a length to a vector $\mathbf{a} = (a_1, a_2, a_3)$ is to assign it the length of the directed line segment from the origin to the point (a_1, a_2, a_3).

DEFINITION 11.2

The **length** (or **norm**) of a vector $\mathbf{a} = (a_1, a_2, a_3)$ is denoted $\|\mathbf{a}\|$ and is defined by

$$\|\mathbf{a}\| = \sqrt{a_1^2 + a_2^2 + a_3^2}$$

A **unit vector** is a vector having length 1.

For example, if $\mathbf{a} = (-1, 3, 6)$, then

$$\|\mathbf{a}\| = \sqrt{(-1)^2 + 3^2 + 6^2} = \sqrt{46}$$

You can check that this is the distance between the point $(-1, 3, 6)$ and the origin. In general, the length $\|\overrightarrow{PQ}\|$ of the vector $\overrightarrow{PQ}$ is the same as the distance $|PQ|$ between the points P and Q.

There are three special unit vectors that will simplify describing and operating on vectors:

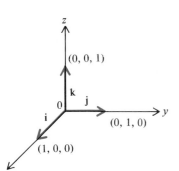

FIGURE 11.13

$$\mathbf{i} = (1, 0, 0), \qquad \mathbf{j} = (0, 1, 0), \qquad \mathbf{k} = (0, 0, 1)$$

(Figure 11.13). It follows from the definition of length that

$$\|\mathbf{i}\| = \|\mathbf{j}\| = \|\mathbf{k}\| = 1$$

so that **i**, **j**, and **k** are indeed unit vectors.

Combinations of Vectors

Numbers can be added, subtracted, and multiplied; vectors can be combined in similar ways.

DEFINITION 11.3

Let $\mathbf{a} = (a_1, a_2, a_3)$ and $\mathbf{b} = (b_1, b_2, b_3)$ be vectors, and let c be a number. Then we define the **sum $\mathbf{a} + \mathbf{b}$**, the **difference $\mathbf{a} - \mathbf{b}$**, and the **scalar multiple $c\mathbf{a}$** by

$$\mathbf{a} + \mathbf{b} = (a_1 + b_1, a_2 + b_2, a_3 + b_3)$$

$$\mathbf{a} - \mathbf{b} = (a_1 - b_1, a_2 - b_2, a_3 - b_3)$$

$$c\mathbf{a} = (ca_1, ca_2, ca_3)$$

Sometimes we write $\mathbf{a}/c$ for $(1/c)\mathbf{a}$. Thus we might express $\frac{1}{5}\mathbf{a}$ as $\mathbf{a}/5$.

Two types of products of vectors will be defined in Sections 11.3 and 11.4. However, we will not define any quotients of vectors.

There are many laws resulting from Definition 11.3. For example,

$$\mathbf{0} + \mathbf{a} = \mathbf{a} + \mathbf{0} = \mathbf{a} \qquad \mathbf{a} + \mathbf{b} = \mathbf{b} + \mathbf{a}$$

$$\mathbf{a} - \mathbf{b} = \mathbf{a} + (-1)\mathbf{b} \qquad c(\mathbf{a} + \mathbf{b}) = c\mathbf{a} + c\mathbf{b}$$

$$0\mathbf{a} = \mathbf{0} \qquad 1\mathbf{a} = \mathbf{a}$$

$$\mathbf{a} + (\mathbf{b} + \mathbf{c}) = (\mathbf{a} + \mathbf{b}) + \mathbf{c}$$

Example 2   Let $\mathbf{a} = (1, -3, 2)$ and $\mathbf{b} = (-4, -1, 0)$. Find $\mathbf{a} + \mathbf{b}$, $\mathbf{a} - \mathbf{b}$, and $-\frac{1}{2}\mathbf{a}$.

Solution   From Definition 11.3 we find that

$$\mathbf{a} + \mathbf{b} = (1 + (-4), -3 + (-1), 2 + 0) = (-3, -4, 2)$$

$$\mathbf{a} - \mathbf{b} = (1 - (-4), -3 - (-1), 2 - 0) = (5, -2, 2)$$

$$-\frac{1}{2}\mathbf{a} = \left(-\frac{1}{2}, \frac{3}{2}, -1\right) \quad \square$$

Using addition and scalar multiplication of vectors, we can express any vector $\mathbf{a} = (a_1, a_2, a_3)$ as a combination of the special unit vectors $\mathbf{i}$, $\mathbf{j}$, and $\mathbf{k}$. Indeed,

$$\mathbf{a} = (a_1, a_2, a_3) = (a_1, 0, 0) + (0, a_2, 0) + (0, 0, a_3)$$
$$= a_1(1, 0, 0) + a_2(0, 1, 0) + a_3(0, 0, 1)$$
$$= a_1\mathbf{i} + a_2\mathbf{j} + a_3\mathbf{k}$$

In other words,

$$\mathbf{a} = a_1\mathbf{i} + a_2\mathbf{j} + a_3\mathbf{k} \tag{1}$$

For example,

$$(1, -3, 2) = \mathbf{i} - 3\mathbf{j} + 2\mathbf{k}$$

If $P = (x_0, y_0, z_0)$ and $Q = (x_1, y_1, z_1)$, then we can express $\overrightarrow{PQ}$ in the form of (1):

$$\overrightarrow{PQ} = (x_1 - x_0)\mathbf{i} + (y_1 - y_0)\mathbf{j} + (z_1 - z_0)\mathbf{k}$$

In our later study of calculus we will often find it useful to write a vector in the form of (1). Since the components of a vector **a** are the coefficients of the unit vectors $\mathbf{i}, \mathbf{j}$, and $\mathbf{k}$ in (1), we sometimes call a_1 the **i** component, a_2 the **j** component, and a_3 the **k** component of **a**.

Using the notation in (1), we restate the formulas for the length, sum, difference, and scalar multiple of vectors:

$$\|a_1\mathbf{i} + a_2\mathbf{j} + a_3\mathbf{k}\| = \sqrt{a_1^2 + a_2^2 + a_3^2} \tag{2}$$

$$(a_1\mathbf{i} + a_2\mathbf{j} + a_3\mathbf{k}) + (b_1\mathbf{i} + b_2\mathbf{j} + b_3\mathbf{k}) = (a_1 + b_1)\mathbf{i} + (a_2 + b_2)\mathbf{j} + (a_3 + b_3)\mathbf{k}$$

$$(a_1\mathbf{i} + a_2\mathbf{j} + a_3\mathbf{k}) - (b_1\mathbf{i} + b_2\mathbf{j} + b_3\mathbf{k}) = (a_1 - b_1)\mathbf{i} + (a_2 - b_2)\mathbf{j} + (a_3 - b_3)\mathbf{k}$$

$$c(a_1\mathbf{i} + a_2\mathbf{j} + a_3\mathbf{k}) = ca_1\mathbf{i} + ca_2\mathbf{j} + ca_3\mathbf{k}$$

It follows from (2) that

$$|a_1| \le \|a_1\mathbf{i} + a_2\mathbf{j} + a_3\mathbf{k}\|, \qquad |a_2| \le \|a_1\mathbf{i} + a_2\mathbf{j} + a_3\mathbf{k}\|,$$

$$|a_3| \le \|a_1\mathbf{i} + a_2\mathbf{j} + a_3\mathbf{k}\|$$

In summary, we list four ways of describing a vector:

1. as (a_1, a_2, a_3), an ordered triple of numbers
2. as (a_1, a_2, a_3), a point in space
3. as a directed line segment with initial point (x_0, y_0, z_0) and terminal point $(x_0 + a_1, y_0 + a_2, z_0 + a_3)$
4. as $a_1\mathbf{i} + a_2\mathbf{j} + a_3\mathbf{k}$

Geometric Interpretations of Vector Operations

The many geometric and physical meanings that can be attached to combinations of vectors make vectors very powerful tools for scientists. To interpret the sum of two vectors geometrically, we begin by letting $\mathbf{a} = a_1\mathbf{i} + a_2\mathbf{j} + a_3\mathbf{k}$ and $\mathbf{b} = b_1\mathbf{i} + b_2\mathbf{j} + b_3\mathbf{k}$. Then we can think of $\mathbf{a}, \mathbf{b}$, and $\mathbf{a} + \mathbf{b}$ as directed line segments from the origin to the points P, Q, and R having coordinates (a_1, a_2, a_3), (b_1, b_2, b_3), and $(a_1 + b_1, a_2 + b_2, a_3 + b_3)$, respectively (Figure 11.14). Now observe that if the directed line segment representing $\mathbf{b}$ is placed so that its initial point is P, then its terminal point will be R. Thus the vector $\mathbf{a} + \mathbf{b}$ can be obtained by placing the initial point of a representative of $\mathbf{b}$ on the terminal point of $\mathbf{a}$.

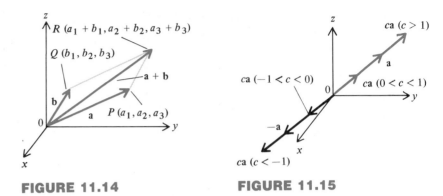

FIGURE 11.14   **FIGURE 11.15**

Notice that the two representations of **a** and **b** (in dark color and in light color) shown in Figure 11.14 determine a parallelogram whose diagonal is **a** + **b**.

Next we let $\mathbf{a} = a_1\mathbf{i} + a_2\mathbf{j} + a_3\mathbf{k}$ be a vector and c be any number. It follows from the definition of length that

$$\|c\mathbf{a}\| = |c|\,\|\mathbf{a}\| \tag{3}$$

When **a** is multiplied by a positive number c, its length is multiplied by c and its direction does not change (Figure 11.15). Moreover, the vector $-\mathbf{a}$ (which denotes $(-1)\mathbf{a}$) has the same length as **a** but has the opposite direction (Figure 11.15). Thus if **a** is multiplied by a negative number c, its length is multiplied by $|c|$ and its direction is reversed (Figure 11.15). Two nonzero vectors whose initial points are the origin are considered parallel only if they lie on the same line through the origin, and in that case they are multiples of one another. Hence we make the following definition.

DEFINITION 11.4

Two nonzero vectors **a** and **b** are **parallel** if and only if there is a number c such that $\mathbf{b} = c\mathbf{a}$.

Example 3   Let $\mathbf{a} = 6\mathbf{i} - 2\mathbf{j} + 4\mathbf{k}$ and $\mathbf{b} = -3\mathbf{i} + \mathbf{j} - 2\mathbf{k}$.   Determine whether **a** and **b** are parallel.

Solution   You can check that $\mathbf{b} = -\frac{1}{2}\mathbf{a}$; consequently **a** and **b** are parallel.   □

The **unit vector in the direction of a nonzero vector a** is $\mathbf{a}/\|\mathbf{a}\|$. Such vectors will be particularly important in Chapter 12.

Example 4   Find the unit vector in the direction of $4\mathbf{i} - \mathbf{j} - 3\mathbf{k}$.

Solution   Since

$$\|4\mathbf{i} - \mathbf{j} - 3\mathbf{k}\| = \sqrt{4^2 + (-1)^2 + (-3)^2} = \sqrt{26}$$

it follows that the required vector is

$$\frac{1}{\sqrt{26}}(4\mathbf{i} - \mathbf{j} - 3\mathbf{k}) \quad \square$$

The difference $\mathbf{a} - \mathbf{b}$ is the sum $\mathbf{a} + (-\mathbf{b})$. Therefore our interpretations of sum and constant multiples of vectors tell us that if the directed line segments representing $\mathbf{a}$ and $\mathbf{b}$ both have the same initial point P (such as the origin), then the directed line segment from the terminal point of $\mathbf{b}$ to the terminal point of $\mathbf{a}$ represents the vector $\mathbf{a} - \mathbf{b}$ (Figure 11.16(a)). From this and our interpretation of the sum of $\mathbf{a}$ and $\mathbf{b}$ we see that any nonzero vectors $\mathbf{a}$ and $\mathbf{b}$ with the same initial point determine a parallelogram whose diagonals are $\mathbf{a} + \mathbf{b}$ and $\mathbf{a} - \mathbf{b}$ (Figure 11.16(b)).

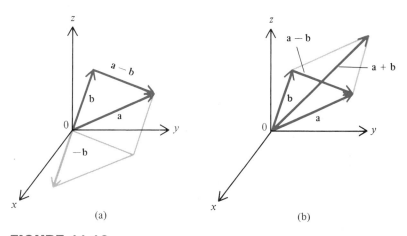

(a)   (b)

FIGURE 11.16

Applications of Vector Addition

Vector addition has been defined as in Definition 11.3 because so many physical quantities combine according to vector addition. For example, if several forces act at the same point P on an object, then the object reacts as though a single force equal to the vector sum of the several forces acts on the object.

Example 5 Two children pull a sled by ropes 4 feet long attached to the front center of the sled. The smaller child holds the rope 2 feet above the sled and 2 feet to the side of the point of attachment, and the larger child holds the rope 3 feet above the sled and 2 feet to the opposite side of the point of attachment (Figure 11.17(a)). The smaller child exerts a force $\mathbf{F}_1$ of magnitude 5 pounds, and the taller child exerts a force $\mathbf{F}_2$ of magnitude 7 pounds. Find the resultant force $\mathbf{F}_1 + \mathbf{F}_2$ on the sled.

Solution First we set up a coordinate system, as shown in Figure 11.17(b), with $\mathbf{a}$ representing the direction the smaller child pulls, and $\mathbf{b}$ representing

Road

(a)

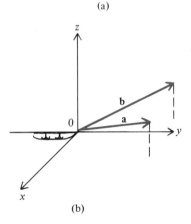

(b)

FIGURE 11.17

the direction the larger child pulls. By the hypotheses, we find that

$$\mathbf{a} = 2\mathbf{i} + a_2\mathbf{j} + 2\mathbf{k} \quad \text{and} \quad \mathbf{b} = -2\mathbf{i} + b_2\mathbf{j} + 3\mathbf{k}$$

where the constants a_2 and b_2 are to be determined. Since each rope is 4 feet long, we have

$$4^2 = \|\mathbf{a}\|^2 = 2^2 + a_2^2 + 2^2 \quad \text{and} \quad 4^2 = \|\mathbf{b}\|^2 = (-2)^2 + b_2^2 + 3^2$$

so that $a_2 = 2\sqrt{2}$ and $b_2 = \sqrt{3}$. By hypothesis $\|\mathbf{F}_1\| = 5$; then since $\mathbf{a}$ and $\mathbf{F}_1$ are parallel and $\|\mathbf{a}\| = 4$, we deduce that

$$\mathbf{F}_1 = \frac{5}{4}\mathbf{a} = \frac{5}{4}(2\mathbf{i} + 2\sqrt{2}\mathbf{j} + 2\mathbf{k})$$

Similarly, $\mathbf{F}_2$ and $\mathbf{b}$ are parallel, with $\|\mathbf{F}_2\| = 7$ and $\|\mathbf{b}\| = 4$. Thus

$$\mathbf{F}_2 = \frac{7}{4}\mathbf{b} = \frac{7}{4}(-2\mathbf{i} + \sqrt{3}\mathbf{j} + 3\mathbf{k})$$

Consequently

$$\mathbf{F}_1 + \mathbf{F}_2 = \frac{5}{4}(2\mathbf{i} + 2\sqrt{2}\mathbf{j} + 2\mathbf{k}) + \frac{7}{4}(-2\mathbf{i} + \sqrt{3}\mathbf{j} + 3\mathbf{k})$$

$$= \frac{1}{4}[-4\mathbf{i} + (10\sqrt{2} + 7\sqrt{3})\mathbf{j} + 31\mathbf{k}] \quad \square$$

Our next illustration of vector addition is taken from electrostatics. Experiments confirm that a particle may be charged positively (as a proton is), or negatively (as an electron is). In either case the charge produces an electric force on any other charged particle. If distance is measured in meters, force in newtons, and charge in coulombs, then the electric force $\mathbf{F}$ exerted by a charge q_1 at the point Q_1 on a charge q located at Q is given by **Coulomb's Law**:

$$\mathbf{F} = \frac{q_1 q}{4\pi\varepsilon_0 r^2}\mathbf{u} \tag{4}$$

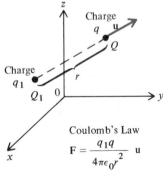

Coulomb's Law

$$\mathbf{F} = \frac{q_1 q}{4\pi\epsilon_0 r^2}\mathbf{u}$$

FIGURE 11.18

where $\mathbf{u}$ is the unit vector in the direction of $\overrightarrow{Q_1 Q}$ (Figure 11.18), r is the distance between the two charges, and ε_0 is the "permittivity of empty space," equal to 8.854×10^{-12}. If q and q_1 are like charges (both positive or both negative), then the coefficient of $\mathbf{u}$ in (4) is positive and $\mathbf{F}$ points in the direction of $\mathbf{u}$. This means that q is forced away from q_1. In contrast, if q and q_1 are opposite charges, then the coefficient is negative and $\mathbf{F}$ has the opposite direction; hence q is forced toward q_1. Of course, we can interchange q and q_1 and obtain similar results. Thus we have the familiar fact that "like charges repel each other and opposite charges attract each other."

If charges $q_1, q_2, \ldots, q_n$ are located at the points $Q_1, Q_2, \ldots, Q_n$, then a **principle of superposition** states that the total electric force $\mathbf{F}$ exerted by these charges on a charge q at Q is the sum of the individual forces exerted by $q_1, q_2, \ldots, q_n$, given by

$$\mathbf{F} = \sum_{j=1}^{n} \frac{q_j q}{4\pi\varepsilon_0 r_j^2} \mathbf{u}_j \tag{5}$$

where

$$r_j = \|\overrightarrow{Q_j Q}\| \quad \text{and} \quad \mathbf{u}_j = \frac{\overrightarrow{Q_j Q}}{\|\overrightarrow{Q_j Q}\|}$$

Example 6   Suppose the charge of a proton is 1.6×10^{-19} coulombs and the charge of an electron is -1.6×10^{-19} coulombs. Find the electric force $\mathbf{F}$ exerted on a positive unit charge at the origin if the proton is located at $(3 \times 10^{-11}, 4 \times 10^{-11}, 0)$ and the electron is located at $(0, 0, 10^{-11})$.

Solution   For the unit charge at the origin let $q = 1$ and $Q = (0, 0, 0)$, and for the proton and electron let

$$q_1 = 1.6 \times 10^{-19} \qquad Q_1 = (3 \times 10^{-11}, 4 \times 10^{-11}, 0)$$

$$q_2 = -1.6 \times 10^{-19} \qquad Q_2 = (0, 0, 10^{-11})$$

Then

$$r_1 = \|\overrightarrow{Q_1 Q}\| = \sqrt{(3 \times 10^{-11})^2 + (4 \times 10^{-11})^2} = 10^{-11}\sqrt{3^2 + 4^2} = 5 \times 10^{-11}$$

$$r_2 = \|\overrightarrow{Q_2 Q}\| = 10^{-11}$$

Therefore

$$\mathbf{u}_1 = \frac{\overrightarrow{Q_1 Q}}{\|\overrightarrow{Q_1 Q}\|} = \frac{-3 \times 10^{-11}\mathbf{i} - 4 \times 10^{-11}\mathbf{j}}{5 \times 10^{-11}} = -\frac{3}{5}\mathbf{i} - \frac{4}{5}\mathbf{j}$$

$$\mathbf{u}_2 = \frac{\overrightarrow{Q_2 Q}}{\|\overrightarrow{Q_2 Q}\|} = \frac{-10^{-11}\mathbf{k}}{10^{-11}} = -\mathbf{k}$$

Consequently by (5),

$$\mathbf{F} = \frac{q_1 q}{4\pi\varepsilon_0 r_1^2}\mathbf{u}_1 + \frac{q_2 q}{4\pi\varepsilon_0 r_2^2}\mathbf{u}_2$$

$$= \frac{(1.6 \times 10^{-19})(1)}{4\pi\varepsilon_0(5 \times 10^{-11})^2}\left(-\frac{3}{5}\mathbf{i} - \frac{4}{5}\mathbf{j}\right) + \frac{(-1.6 \times 10^{-19})(1)}{4\pi\varepsilon_0(10^{-11})^2}(-\mathbf{k})$$

$$= \frac{1.6 \times 10^{-19}}{4\pi\varepsilon_0 10^{-22}}\left(-\frac{3}{125}\mathbf{i} - \frac{4}{125}\mathbf{j} + \mathbf{k}\right)$$

$$\approx 1.150 \times 10^{11}(-3\mathbf{i} - 4\mathbf{j} + 125\mathbf{k}) \quad \square$$

Vectors in the Plane   Any vector $\mathbf{a} = a_1\mathbf{i} + a_2\mathbf{j} + a_3\mathbf{k}$ for which $a_3 = 0$ can be written more easily as

$$\mathbf{a} = a_1\mathbf{i} + a_2\mathbf{j} \tag{6}$$

The point (a_1, a_2) corresponding to the vector in (6) lies in the xy plane (Figure 11.19). If the vector is nonzero, then it can also be described by its length $\|\mathbf{a}\|$

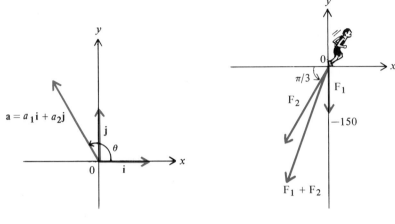

FIGURE 11.19   **FIGURE 11.20**

and the angle θ it makes with the positive x axis:

$$\mathbf{a} = \|\mathbf{a}\|(\cos\theta\,\mathbf{i} + \sin\theta\,\mathbf{j}) \qquad (7)$$

(see Exercise 19(a)). If $\mathbf{a}$ and $\mathbf{b}$ are vectors in the same plane and c is any number, then $\mathbf{a} + \mathbf{b}$, $\mathbf{a} - \mathbf{b}$, and $c\mathbf{a}$ are also in that plane. In solving problems that are two-dimensional in nature we may use vectors represented by either (6) or (7).

Example 7   A high jumper weighing 150 pounds exerts a force of magnitude 300 pounds on the ground at an angle of $\pi/3$ with the horizontal (Figure 11.20). Assuming that the force of the jumper's weight is exerted directly downward on the ground, find the total force $\mathbf{F}$ exerted on the ground.

Solution   We set up a coordinate system as shown in Figure 11.20. Since $\mathbf{F}_1$ is directed downward and the angle θ made by $\mathbf{F}_2$ with respect to the positive x axis is $\pi + \pi/3 = 4\pi/3$, we have

$$\mathbf{F}_1 = -150\mathbf{j}$$
$$\mathbf{F}_2 = \|\mathbf{F}_2\|\left(\cos\frac{4\pi}{3}\mathbf{i} + \sin\frac{4\pi}{3}\mathbf{j}\right) = 300\left(-\frac{1}{2}\mathbf{i} - \frac{\sqrt{3}}{2}\mathbf{j}\right) = -150(\mathbf{i} + \sqrt{3}\mathbf{j})$$

Therefore the total force $\mathbf{F}$ exerted on the ground is given by

$$\mathbf{F} = \mathbf{F}_1 + \mathbf{F}_2 = -150\mathbf{j} - 150(\mathbf{i} + \sqrt{3}\mathbf{j}) = -150\mathbf{i} - 150(1 + \sqrt{3})\mathbf{j} \quad \square$$

Example 8   Show that the diagonals of a parallelogram bisect each other.

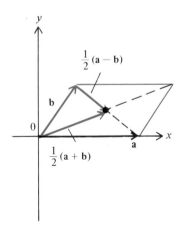

FIGURE 11.21

Solution   Assume that the parallelogram lies in the xy plane, as shown in Figure 11.21, and let $\mathbf{a}$ and $\mathbf{b}$ be vectors along the two sides that contain the origin. Then $\mathbf{a} + \mathbf{b}$ and $\mathbf{a} - \mathbf{b}$ lie along the diagonals. The vector $\frac{1}{2}(\mathbf{a} + \mathbf{b})$ joins the origin to the midpoint of the diagonal $\mathbf{a} + \mathbf{b}$, and the vector $\mathbf{b} + \frac{1}{2}(\mathbf{a} - \mathbf{b})$ joins the origin to the midpoint of the other diagonal. Since

$$\frac{1}{2}(\mathbf{a} + \mathbf{b}) = \mathbf{b} + \frac{1}{2}(\mathbf{a} - \mathbf{b})$$

it follows that the two midpoints coincide. Consequently the two diagonals bisect one another. ☐

EXERCISES 11.2

In Exercises 1–4 write $\overrightarrow{PQ}$ as a vector in the form $a\mathbf{i} + b\mathbf{j} + c\mathbf{k}$.

1. $P = (0, 0, 0), Q = (3, -4, 10)$
2. $P = (1, -3, 2), Q = (3, -1, 3)$
3. $P = (0, 1, 0), Q = (3, -1, \sqrt{7})$
4. $P = (2, 1, 1), Q = (2, 1 + \sqrt{2}, 1 + \sqrt{3})$

In Exercises 5–8 find $\mathbf{a} + \mathbf{b}$, $\mathbf{a} - \mathbf{b}$, and $c\mathbf{a}$.

5. $\mathbf{a} = 2\mathbf{i} - 5\mathbf{j} + 10\mathbf{k}, \mathbf{b} = -\mathbf{i} + 2\mathbf{j} - 9\mathbf{k}, c = 2$
6. $\mathbf{a} = \mathbf{i} + \mathbf{j} - 3\mathbf{k}, \mathbf{b} = \frac{1}{2}\mathbf{i} - \frac{1}{2}\mathbf{j} - 3\mathbf{k}, c = -1$
7. $\mathbf{a} = 2\mathbf{i}, \mathbf{b} = \mathbf{j} + \mathbf{k}, c = \frac{1}{3}$
8. $\mathbf{a} = \mathbf{i} + 2\mathbf{j}, \mathbf{b} = -2\mathbf{j} + \mathbf{k}, c = \pi$

In Exercises 9–13 find the length of the vector.

9. $\mathbf{a} = \mathbf{i} - \mathbf{j} + \mathbf{k}$
10. $\mathbf{a} = 2\mathbf{i} + \mathbf{j} - 2\mathbf{k}$
11. $\mathbf{b} = -3\mathbf{i} + 4\mathbf{j} - 12\mathbf{k}$
12. $\mathbf{b} = 4\mathbf{i} - 8\mathbf{j} + 8\mathbf{k}$
13. $\mathbf{c} = \sqrt{2}\mathbf{i} - \mathbf{j} + \mathbf{k}$

In Exercises 14–17 find a unit vector having the same direction as the given vector.

14. $\mathbf{a} = \mathbf{i} + \mathbf{j} - \mathbf{k}$
15. $\mathbf{a} = -3\mathbf{i} + 4\mathbf{j} - 12\mathbf{k}$
16. $\mathbf{b} = 7\mathbf{i} + 12\sqrt{2}\mathbf{j} - 12\sqrt{2}\mathbf{k}$
17. $\mathbf{b} = 2\mathbf{i} - 3\mathbf{j}$

18. Let $OABCDEFG$ be a cube. Show that $\overrightarrow{OB} + \overrightarrow{OD} + \overrightarrow{OF}$ is parallel to $\overrightarrow{OG}$ (Figure 11.22).

19. a. Let $\mathbf{a}$ be a nonzero vector in the xy plane, and let θ be the angle from the positive x axis to $\mathbf{a}$ in the counterclockwise direction. Show that

$$\mathbf{a} = \|\mathbf{a}\|(\cos\theta\mathbf{i} + \sin\theta\mathbf{j})$$

b. Suppose that $\mathbf{u}$ is a unit vector in the plane. Using part (a), show that

$$\mathbf{u} = \cos\theta\mathbf{i} + \sin\theta\mathbf{j}$$

for some number θ.

20. Let $\mathbf{a}$ and $\mathbf{b}$ be two vectors with initial points at the origin and terminal points P and Q, respectively.

a. Show that the vector $\mathbf{c}$ directed from the origin to the midpoint of PQ is $\frac{1}{2}(\mathbf{a} + \mathbf{b})$ (Figure 11.23(a)). (*Hint:* $\mathbf{c} = \mathbf{a} + \frac{1}{2}(\mathbf{b} - \mathbf{a})$.)

b. Show that the vector $\mathbf{c}$ directed from the origin to the point on PQ two thirds of the way from P to Q is $\frac{1}{3}\mathbf{a} + \frac{2}{3}\mathbf{b}$ (Figure 11.23(b)).

21. Using vectors, prove that a quadrilateral $PQRS$ is a

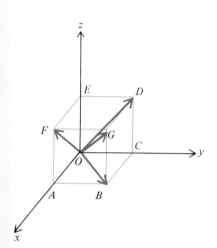

FIGURE 11.22

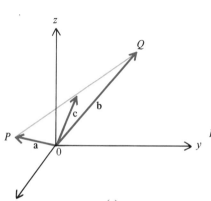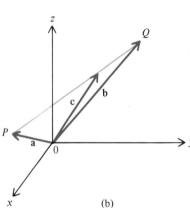

FIGURE 11.23

parallelogram if the diagonals PR and QS bisect each other.

*22. Using vectors, prove that the midpoints of the sides of any quadrilateral form a parallelogram. (*Hint:* Use Exercise 21. To prove that the diagonals of the quadrilateral formed by the midpoints bisect each other, pattern your proof after the solution to Example 8.)

23. Three children tug on a ball located at O (Figure 11.24). One child pulls with a force of 20 pounds in the direction of the negative y axis. Another child pulls with a force of 100 pounds at an angle of $\pi/3$ with the positive x axis. If the total force exerted on the ball is to be **0**, find the force **F** with which the third child should pull and the tangent of the angle θ at which the third child should pull.

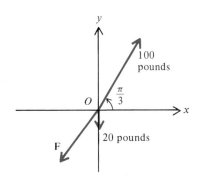

FIGURE 11.24

24. Two tugboats are pulling an ocean freighter as shown in Figure 11.25. If one tugboat exerts a force of 1000 pounds on the cable tied at A, what force must be exerted on the cable tied to the other tugboat at B if the freighter is to move along the line l?

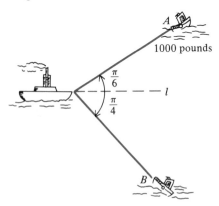

FIGURE 11.25

25. Suppose an airplane is flying in the xy plane with its body oriented at an angle of $\pi/6$ with respect to the positive x

axis. If the air is moving parallel to the positive y axis at 20 miles per hour and the speed of the airplane with respect to the air is 300 miles per hour, what is the speed of the airplane with respect to the ground? (*Hint:* The velocity of the plane with respect to the ground is equal to the sum of the velocity of the plane with respect to the air and the velocity of the air with respect to the ground.)

*26. When a boat travels due north at 8 miles per hour, a vane on the boat points due northwest. When the boat anchors, the vane points due west. What is the speed of the wind? (*Hint:* See the hint for Exercise 25 and use the fact that the vane points into the wind.)

27. Suppose the electron in Example 6 were located at $(10^{-12}, 10^{-12}, 0)$ and the proton at $(0, 10^{-11}, 10^{-11})$. Find the total electric force **F** exerted by the two particles on a positive unit charge located at the origin.

28. Find the total electric force **F** exerted by the two particles in Example 6 on a positive unit charge located at the point $(0, 10^{-11}, 0)$.

*29. Suppose that point masses $m_1, m_2, \ldots, m_n$ are located at points $P_1(x_1, y_1), P_2(x_2, y_2), \ldots, P_n(x_n, y_n)$, respectively, in the xy plane. Show that the centroid of these point masses is the point $P(\bar{x}, \bar{y})$ such that

$$m_1 \overrightarrow{PP_1} + m_2 \overrightarrow{PP_2} + \cdots + m_n \overrightarrow{PP_n} = \mathbf{0}$$

(*Hint:* Write out the two components of $m_1 \overrightarrow{PP_1} + m_2 \overrightarrow{PP_2} + \cdots + m_n \overrightarrow{PP_n}$, and then refer to the definitions of $\bar{x}$ and $\bar{y}$ in Section 6.7.)

30. a. Suppose that the xy plane is a mirror and that a ray of light is reflected by the mirror (Figure 11.26). Let **a** and **b** be unit vectors pointing along the paths of the incident and the reflected rays. Using the law of reflection, which states that the angle of incidence is equal to the angle of reflection, show that if $\mathbf{a} = a_1\mathbf{i} + a_2\mathbf{j} + a_3\mathbf{k}$, then $\mathbf{b} = a_1\mathbf{i} + a_2\mathbf{j} - a_3\mathbf{k}$.

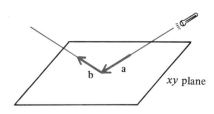

FIGURE 11.26

*b. Suppose a ray of light is reflected in turn from each of three mutually perpendicular mirrors, which you may think of as the portions of the coordinate planes in the first octant. Show that the reflected ray is parallel to the incident ray. (*Hint:* Use (a) and the analogous results for the yz plane and the xz plane.)

11.3
THE DOT PRODUCT

Now that we have defined the sum and difference of two vectors, it is natural to ask whether a useful product of two vectors can be defined. One obvious possibility is to multiply the vectors componentwise, just as we add and subtract vectors componentwise. But such a product has little physical significance and almost never arises in applications. However, two other products of vectors, known as the dot product and the cross product, have profound physical significance. First we consider the dot product, defined as follows.

DEFINITION 11.5

Let $\mathbf{a} = a_1\mathbf{i} + a_2\mathbf{j} + a_3\mathbf{k}$ and $\mathbf{b} = b_1\mathbf{i} + b_2\mathbf{j} + b_3\mathbf{k}$ be two vectors. The **dot product** (or **scalar product** or **inner product**) of $\mathbf{a}$ and $\mathbf{b}$ is the number $\mathbf{a} \cdot \mathbf{b}$ defined by

$$\mathbf{a} \cdot \mathbf{b} = a_1 b_1 + a_2 b_2 + a_3 b_3 \qquad (1)$$

From (1) we deduce that

$$\mathbf{a} \cdot \mathbf{i} = \mathbf{i} \cdot \mathbf{a} = a_1, \qquad \mathbf{a} \cdot \mathbf{j} = \mathbf{j} \cdot \mathbf{a} = a_2, \quad \text{and} \quad \mathbf{a} \cdot \mathbf{k} = \mathbf{k} \cdot \mathbf{a} = a_3$$

In particular

$$\mathbf{i} \cdot \mathbf{j} = \mathbf{j} \cdot \mathbf{k} = \mathbf{k} \cdot \mathbf{i} = 0 \quad \text{and} \quad \mathbf{i} \cdot \mathbf{i} = \mathbf{j} \cdot \mathbf{j} = \mathbf{k} \cdot \mathbf{k} = 1$$

Other dot products are almost as simple to calculate.

Example 1 Let $\mathbf{a} = 3\mathbf{i} - \mathbf{j} - 2\mathbf{k}$ and $\mathbf{b} = 2\mathbf{i} - 3\mathbf{j} + \frac{1}{2}\mathbf{k}$. Find $\mathbf{a} \cdot \mathbf{b}$.

Solution By (1),

$$\mathbf{a} \cdot \mathbf{b} = 3(2) + (-1)(-3) + (-2)\left(\frac{1}{2}\right) = 8 \qquad \square$$

The dot product satisfies many of the laws that hold for real numbers. For example,

$$\mathbf{a} \cdot \mathbf{b} = \mathbf{b} \cdot \mathbf{a} \qquad (c\mathbf{a}) \cdot \mathbf{b} = c(\mathbf{a} \cdot \mathbf{b}) = \mathbf{a} \cdot (c\mathbf{b})$$

$$\mathbf{a} \cdot (\mathbf{b} + \mathbf{c}) = \mathbf{a} \cdot \mathbf{b} + \mathbf{a} \cdot \mathbf{c} \qquad (\mathbf{a} + \mathbf{b}) \cdot \mathbf{c} = \mathbf{a} \cdot \mathbf{c} + \mathbf{b} \cdot \mathbf{c}$$

The **angle** between two nonzero vectors $\mathbf{a}$ and $\mathbf{b}$ is defined to be the angle θ, where $0 \le \theta \le \pi$, formed by the corresponding directed line segments whose initial points are the origin (Figure 11.27). The relationship between the dot product and the angle between two vectors is described in the following theorem.

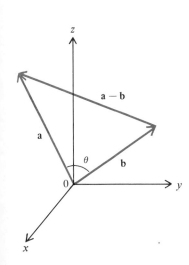

FIGURE 11.27

THEOREM 11.6

> Let $\mathbf{a} = a_1\mathbf{i} + a_2\mathbf{j} + a_3\mathbf{k}$ and $\mathbf{b} = b_1\mathbf{i} + b_2\mathbf{j} + b_3\mathbf{k}$ be nonzero vectors, and let θ be the angle between $\mathbf{a}$ and $\mathbf{b}$. Then
>
> $$\mathbf{a} \cdot \mathbf{b} = \|\mathbf{a}\|\,\|\mathbf{b}\|\cos\theta \tag{2}$$

Proof   Formula (2) is most easily proved by applying the Law of Cosines to the triangle formed by $\mathbf{a}$, $\mathbf{b}$, and $\mathbf{a} - \mathbf{b}$ (Figure 11.27). From this law we obtain

$$\|\mathbf{a} - \mathbf{b}\|^2 = \|\mathbf{a}\|^2 + \|\mathbf{b}\|^2 - 2\|\mathbf{a}\|\,\|\mathbf{b}\|\cos\theta \tag{3}$$

where θ is the angle between $\mathbf{a}$ and $\mathbf{b}$. Using the definition of the length of a vector and the fact that $\mathbf{a} - \mathbf{b} = (a_1 - b_1)\mathbf{i} + (a_2 - b_2)\mathbf{j} + (a_3 - b_3)\mathbf{k}$, we can rewrite (3) as

$$(a_1 - b_1)^2 + (a_2 - b_2)^2 + (a_3 - b_3)^2$$
$$= (a_1^2 + a_2^2 + a_3^2) + (b_1^2 + b_2^2 + b_3^2) - 2\|\mathbf{a}\|\,\|\mathbf{b}\|\cos\theta$$

By expanding the squared terms in this equation and then canceling like terms on both sides of the equation, we obtain

$$-2a_1b_1 - 2a_2b_2 - 2a_3b_3 = -2\|\mathbf{a}\|\,\|\mathbf{b}\|\cos\theta$$

from which we conclude that

$$\mathbf{a} \cdot \mathbf{b} = a_1b_1 + a_2b_2 + a_3b_3 = \|\mathbf{a}\|\,\|\mathbf{b}\|\cos\theta \quad\blacksquare$$

COROLLARY 11.7

> **a.** Two nonzero vectors $\mathbf{a}$ and $\mathbf{b}$ are perpendicular if and only if $\mathbf{a} \cdot \mathbf{b} = 0$.
> **b.** For any vector $\mathbf{a}$, $\mathbf{a} \cdot \mathbf{a} = \|\mathbf{a}\|^2$.

Example 2   Show that the vectors $\mathbf{a} = -4\mathbf{i} + 5\mathbf{j} + 7\mathbf{k}$ and $\mathbf{b} = \mathbf{i} - 2\mathbf{j} + 2\mathbf{k}$ are perpendicular.

Solution   We simply compute the dot product by using (1):

$$\mathbf{a} \cdot \mathbf{b} = (-4)(1) + (5)(-2) + 7(2) = 0$$

It follows from Corollary 11.7(a) that $\mathbf{a}$ and $\mathbf{b}$ are perpendicular. $\square$

Solving (2) for $\cos\theta$, we obtain

$$\cos\theta = \frac{\mathbf{a} \cdot \mathbf{b}}{\|\mathbf{a}\|\,\|\mathbf{b}\|} \tag{4}$$

Since $0 \le \theta \le \pi$, the angle between $\mathbf{a}$ and $\mathbf{b}$ is uniquely determined by (4).

Example 3   Find the angle between the vectors $\mathbf{a} = 2\mathbf{i} - \mathbf{j} + 2\mathbf{k}$ and $\mathbf{b} = \mathbf{i} - \mathbf{j}$.

Solution   Since

$$\|\mathbf{a}\| = \sqrt{2^2 + (-1)^2 + 2^2} = 3 \quad \text{and} \quad \|\mathbf{b}\| = \sqrt{1^2 + (-1)^2 + 0^2} = \sqrt{2}$$

and since

$$\mathbf{a} \cdot \mathbf{b} = 2(1) + (-1)(-1) + 2(0) = 3$$

we infer from (4) that

$$\cos \theta = \frac{3}{3\sqrt{2}} = \frac{1}{\sqrt{2}} = \frac{\sqrt{2}}{2}$$

which means that $\theta = \pi/4$.   □

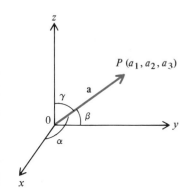

FIGURE 11.28

The angles α, β, and γ (between 0 and π inclusive) that a nonzero vector $\mathbf{a}$ makes with the positive x, y, and z axes are called the **direction angles** of $\mathbf{a}$ (Figure 11.28). If $\mathbf{a} = a_1\mathbf{i} + a_2\mathbf{j} + a_3\mathbf{k}$, then by (4) and (1),

$$\cos \alpha = \frac{\mathbf{a} \cdot \mathbf{i}}{\|\mathbf{a}\| \|\mathbf{i}\|} = \frac{a_1}{\|\mathbf{a}\|}, \quad \cos \beta = \frac{\mathbf{a} \cdot \mathbf{j}}{\|\mathbf{a}\| \|\mathbf{j}\|} = \frac{a_2}{\|\mathbf{a}\|}, \quad \cos \gamma = \frac{\mathbf{a} \cdot \mathbf{k}}{\|\mathbf{a}\| \|\mathbf{k}\|} = \frac{a_3}{\|\mathbf{a}\|}$$

and $\cos \alpha$, $\cos \beta$, and $\cos \gamma$ are the **direction cosines** of $\mathbf{a}$. It follows that

$$a_1 = \|\mathbf{a}\| \cos \alpha, \qquad a_2 = \|\mathbf{a}\| \cos \beta, \qquad a_3 = \|\mathbf{a}\| \cos \gamma$$

so that we may write $\mathbf{a}$ in the alternative form

$$\mathbf{a} = \|\mathbf{a}\|(\cos \alpha \mathbf{i} + \cos \beta \mathbf{j} + \cos \gamma \mathbf{k})$$

Example 4   Let $\mathbf{a} = -2\mathbf{i} + 3\mathbf{j} - 5\mathbf{k}$. Find the direction cosines of $\mathbf{a}$.

Solution   Since

$$\|\mathbf{a}\| = \sqrt{(-2)^2 + 3^2 + (-5)^2} = \sqrt{4 + 9 + 25} = \sqrt{38}$$

a direct calculation shows that

$$\cos \alpha = \frac{-2}{\sqrt{38}}, \qquad \cos \beta = \frac{3}{\sqrt{38}}, \qquad \cos \gamma = \frac{-5}{\sqrt{38}} \quad □$$

The Projection of One Vector onto Another

Suppose that two nonzero vectors $\mathbf{a}$ and $\mathbf{b}$ are positioned as in Figure 11.29(a), and that the sun casts a shadow on the line containing the vector $\mathbf{a}$. Informally we think of the shadow as determining a vector parallel to $\mathbf{a}$, which we call the

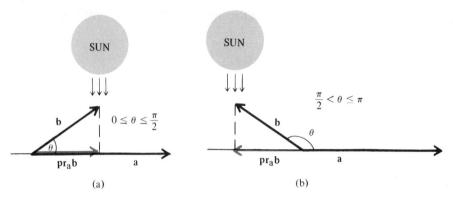

FIGURE 11.29

projection of **b** onto **a** and denote $\mathbf{pr_a b}$. Since $\mathbf{pr_a b}$ is parallel to **a** or is **0**, it must be a scalar multiple of **a**. The length of $\mathbf{pr_a b}$ is evidently $\|\mathbf{b}\| |\cos \theta|$, where θ is the angle between **a** and **b** (so that $0 \le \theta \le \pi$). If $0 \le \theta \le \pi/2$, it follows that

$$\mathbf{pr_a b} = \|\mathbf{b}\| \cos \theta \frac{1}{\|\mathbf{a}\|} \mathbf{a} = \|\mathbf{b}\| \left(\frac{\mathbf{a} \cdot \mathbf{b}}{\|\mathbf{a}\| \|\mathbf{b}\|} \right) \frac{1}{\|\mathbf{a}\|} \mathbf{a} = \frac{\mathbf{a} \cdot \mathbf{b}}{\|\mathbf{a}\|^2} \mathbf{a}$$

If $\pi/2 < \theta \le \pi$, then Figure 11.29(b) applies, and in a similar way we find that

$$\mathbf{pr_a b} = \|\mathbf{b}\| (-\cos \theta) \frac{1}{\|\mathbf{a}\|} (-\mathbf{a}) = \|\mathbf{b}\| \cos \theta \frac{1}{\|\mathbf{a}\|} \mathbf{a} = \frac{\mathbf{a} \cdot \mathbf{b}}{\|\mathbf{a}\|^2} \mathbf{a}$$

which is the same expression for $\mathbf{pr_a b}$. Thus we can define the projection by a single formula involving the dot product, whether $0 \le \theta \le \pi/2$ or $\pi/2 < \theta \le \pi$.

DEFINITION 11.8

> Let **a** be a nonzero vector. The **projection** of a vector **b** onto **a** is the vector $\mathbf{pr_a b}$ defined by
>
> $$\mathbf{pr_a b} = \frac{\mathbf{a} \cdot \mathbf{b}}{\|\mathbf{a}\|^2} \mathbf{a} \qquad (5)$$

Example 5 Let $\mathbf{a} = 3\mathbf{i} - \mathbf{j} - 2\mathbf{k}$ and $\mathbf{b} = 2\mathbf{i} - 3\mathbf{j} + \frac{1}{2}\mathbf{k}$, as in Example 1. Find $\mathbf{pr_a b}$.

Solution From Example 1 we know that $\mathbf{a} \cdot \mathbf{b} = 8$, and a simple calculation shows that $\|\mathbf{a}\|^2 = 14$. Consequently by (5) we have

$$\mathbf{pr_a b} = \frac{8}{14} (3\mathbf{i} - \mathbf{j} - 2\mathbf{k}) = \frac{4}{7} (3\mathbf{i} - \mathbf{j} - 2\mathbf{k}) \quad \square$$

$$= \frac{8}{14} (3i' - j' - 2k)$$
$$= \frac{12}{7} i' - \frac{4}{7} j - \frac{8}{7} k$$

For future reference we derive a formula for the length of $\mathbf{pr_a b}$ in terms of $\mathbf{a}$ and $\mathbf{b}$. Using (5), we find that

$$\|\mathbf{pr_a b}\| = \left\| \frac{\mathbf{a} \cdot \mathbf{b}}{\|\mathbf{a}\|^2} \mathbf{a} \right\| = \frac{|\mathbf{a} \cdot \mathbf{b}|}{\|\mathbf{a}\|^2} \|\mathbf{a}\| = \frac{|\mathbf{a} \cdot \mathbf{b}|}{\|\mathbf{a}\|}$$

Thus
$$\|\mathbf{pr_a b}\| = \frac{|\mathbf{a} \cdot \mathbf{b}|}{\|\mathbf{a}\|} \tag{6}$$

Resolution of a Vector

If $\mathbf{a}$ and $\mathbf{a}'$ are perpendicular, then any nonzero vector $\mathbf{b}$ lying in the same plane as $\mathbf{a}$ and $\mathbf{a}'$ can be expressed as the sum of the two vectors $\mathbf{pr_a b}$ and $\mathbf{pr_{a'} b}$, which are parallel to $\mathbf{a}$ and $\mathbf{a}'$, respectively:

$$\mathbf{b} = \mathbf{pr_a b} + \mathbf{pr_{a'} b}$$

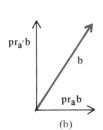

(a)

(Figure 11.30). We say that $\mathbf{b}$ has been **resolved** into vectors parallel to $\mathbf{a}$ and $\mathbf{a}'$. Geometrically, $\mathbf{pr_a b}$ and $\mathbf{pr_{a'} b}$ are the legs of the right triangle that has hypotenuse $\mathbf{b}$ and has legs parallel to $\mathbf{a}$ and $\mathbf{a}'$. It is not necessary to compute both projections directly, because if we know $\mathbf{b}$ and $\mathbf{pr_a b}$, then we can find $\mathbf{pr_{a'} b}$ by subtraction:

$$\mathbf{pr_{a'} b} = \mathbf{b} - \mathbf{pr_a b} \tag{7}$$

(b)

FIGURE 11.30

Example 6 Let $\mathbf{a} = \mathbf{i} + \mathbf{j}$, $\mathbf{a}' = \mathbf{i} - \mathbf{j}$, and $\mathbf{b} = 3\mathbf{i} - 4\mathbf{j}$. Resolve $\mathbf{b}$ into vectors parallel to $\mathbf{a}$ and $\mathbf{a}'$.

Solution Observe that all three vectors lie in the xy plane and that $\mathbf{a}$ and $\mathbf{a}'$ are perpendicular, since $\mathbf{a} \cdot \mathbf{a}' = 0$. To resolve $\mathbf{b}$, we notice that $\mathbf{a} \cdot \mathbf{b} = -1$ and $\|\mathbf{a}\|^2 = 2$, so that by (5),

$$\mathbf{pr_a b} = \frac{\mathbf{a} \cdot \mathbf{b}}{\|\mathbf{a}\|^2} \mathbf{a} = -\frac{1}{2}\mathbf{a} = -\frac{1}{2}(\mathbf{i} + \mathbf{j})$$

Then (7) yields

$$\mathbf{pr_{a'} b} = \mathbf{b} - \mathbf{pr_a b} = (3\mathbf{i} - 4\mathbf{j}) - \left(-\frac{1}{2}(\mathbf{i} + \mathbf{j}) \right) = \frac{7}{2}\mathbf{i} - \frac{7}{2}\mathbf{j} = \frac{7}{2}\mathbf{a}'$$

Therefore $\mathbf{b} = -\frac{1}{2}\mathbf{a} + \frac{7}{2}\mathbf{a}'$. ☐

Work Done by a Constant Force

One application of projections of vectors arises in the definition of the work done by a force on a moving object. Suppose that a constant force $\mathbf{F}$ acts on an object moving along the line from P to Q and that the direction of the force is parallel to that line. By our definition of work in Section 8.6, the work W done is either $\|\mathbf{F}\| \|\overrightarrow{PQ}\|$ or $-\|\mathbf{F}\| \|\overrightarrow{PQ}\|$, depending on whether $\mathbf{F}$ acts in the direction of $\overrightarrow{PQ}$ or in the opposite direction.

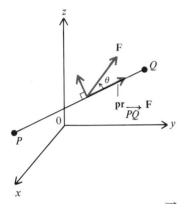

F is the sum of a vector parallel to $\overrightarrow{PQ}$ and a vector perpendicular to $\overrightarrow{PQ}$

FIGURE 11.31

If **F** is not parallel to $\overrightarrow{PQ}$, then we resolve **F** into a vector parallel to $\overrightarrow{PQ}$ and a vector perpendicular to $\overrightarrow{PQ}$ (Figure 11.31). The vector $\mathbf{pr}_{\overrightarrow{PQ}}\mathbf{F}$ parallel to $\overrightarrow{PQ}$ affects the speed of the object and hence does work on the object. In contrast, the vector perpendicular to $\overrightarrow{PQ}$ does not affect the speed, so we ignore it with regard to work. Accordingly, we define the work W done by the force **F** as the work done by the force $\mathbf{pr}_{\overrightarrow{PQ}}\mathbf{F}$ parallel to $\overrightarrow{PQ}$. Thus if θ is the angle between **F** and $\overrightarrow{PQ}$, then

$$W = \begin{cases} \|\mathbf{pr}_{\overrightarrow{PQ}}\mathbf{F}\|\,\|\overrightarrow{PQ}\| & \text{for } 0 \le \theta \le \dfrac{\pi}{2} \\[2mm] -\|\mathbf{pr}_{\overrightarrow{PQ}}\mathbf{F}\|\,\|\overrightarrow{PQ}\| & \text{for } \dfrac{\pi}{2} < \theta \le \pi \end{cases} \tag{8}$$

But by (6),

$$\|\mathbf{pr}_{\overrightarrow{PQ}}\mathbf{F}\| = \frac{|\mathbf{F}\cdot\overrightarrow{PQ}|}{\|\overrightarrow{PQ}\|} = \begin{cases} \dfrac{\mathbf{F}\cdot\overrightarrow{PQ}}{\|\overrightarrow{PQ}\|} & \text{for } 0 \le \theta \le \dfrac{\pi}{2} \\[2mm] \dfrac{-\mathbf{F}\cdot\overrightarrow{PQ}}{\|\overrightarrow{PQ}\|} & \text{for } \dfrac{\pi}{2} < \theta \le \pi \end{cases} \tag{9}$$

Formulas (8) and (9) together imply that

$$W = \mathbf{F}\cdot\overrightarrow{PQ} \tag{10}$$

Using (2), we can write (10) equivalently as

$$W = \|\mathbf{F}\|\,\|\overrightarrow{PQ}\|\cos\theta \tag{11}$$

Hence if a force with given magnitude is applied to an object in motion, then the nearer the direction of the force is to the direction of motion, the more work the force does on the object. This is why it is easier to shovel snow with the handle held nearly horizontal than with it held up.

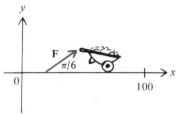

FIGURE 11.32

Example 7 Suppose that a force **F** of 50 pounds is exerted upward on a wheelbarrow at an angle of $\pi/6$ with the horizontal and that the wheelbarrow travels 100 feet in the horizontal direction (Figure 11.32). Find the work W done by the force on the wheelbarrow.

Solution Let us assume that the wheelbarrow travels along the x axis from 0 to 100 and that the force **F** acts in the xy plane (Figure 11.32), so that $P = (0, 0)$ and $Q = (100, 0)$. It follows that $\overrightarrow{PQ} = 100\mathbf{i}$, so that $\|\overrightarrow{PQ}\| = 100$. Since $\|\mathbf{F}\| = 50$ and the angle θ between **F** and $\overrightarrow{PQ}$ is $\pi/6$, we find from (11) that

$$W = \|\mathbf{F}\|\,\|\overrightarrow{PQ}\|\cos\frac{\pi}{6} = 50\cdot100\cdot\frac{\sqrt{3}}{2} = 2500\sqrt{3} \text{ (foot pounds)} \qquad \square$$

EXERCISES 11.3

In Exercises 1–4 find $\mathbf{a} \cdot \mathbf{b}$ and the cosine of the angle θ between $\mathbf{a}$ and $\mathbf{b}$.

1. $\mathbf{a} = \mathbf{i} + \mathbf{j} - \mathbf{k}, \mathbf{b} = 2\mathbf{i} - 3\mathbf{j} + 4\mathbf{k}$

2. $\mathbf{a} = \dfrac{1}{2}\mathbf{i} + \dfrac{1}{3}\mathbf{j} - 2\mathbf{k}, \mathbf{b} = 2\mathbf{i} - 2\mathbf{j} + \mathbf{k}$

3. $\mathbf{a} = \sqrt{2}\mathbf{i} + 4\mathbf{j} + \sqrt{3}\mathbf{k}, \mathbf{b} = -\sqrt{2}\mathbf{i} - \sqrt{3}\mathbf{j} + 2\mathbf{k}$

4. $\mathbf{a} = 4\mathbf{i} - 2\mathbf{j}, \mathbf{b} = -\frac{1}{2}\mathbf{i} - \mathbf{j} + \sqrt{3}\mathbf{k}$

5. Find the angle θ between the vector $\sqrt{6}\mathbf{i} + \mathbf{j} - \mathbf{k}$ and the positive x axis.

In Exercises 6–9 determine whether $\mathbf{a}$ and $\mathbf{b}$ are perpendicular.

6. $\mathbf{a} = \mathbf{i}, \mathbf{b} = \mathbf{j}$

7. $\mathbf{a} = \mathbf{i} - \mathbf{j}, \mathbf{b} = \mathbf{i} + \mathbf{j}$

8. $\mathbf{a} = -4\mathbf{i} + 2\mathbf{j} - 5\mathbf{k}, \mathbf{b} = \frac{1}{2}\mathbf{i} + 6\mathbf{j} + 2\mathbf{k}$

9. $\mathbf{a} = \sqrt{2}\mathbf{i} + 3\mathbf{j} + \mathbf{k}, \mathbf{b} = -\mathbf{i} + \sqrt{2}\mathbf{j} + 5\mathbf{k}$

10. Show that the vectors

$$2\mathbf{i} + \mathbf{j} - \mathbf{k}, \quad 3\mathbf{i} + 7\mathbf{j} + 13\mathbf{k}, \quad \text{and} \quad 20\mathbf{i} - 29\mathbf{j} + 11\mathbf{k}$$

are mutually perpendicular.

In Exercises 11–14 find $\mathbf{pr_a b}$.

11. $\mathbf{a} = 2\mathbf{i} - \mathbf{j} + 2\mathbf{k}, \mathbf{b} = \mathbf{i}$

12. $\mathbf{a} = \mathbf{i}, \mathbf{b} = -2\mathbf{i} + 3\mathbf{j} - 4\mathbf{k}$

13. $\mathbf{a} = \mathbf{i} + \mathbf{j}, \mathbf{b} = -\mathbf{i} + 2\mathbf{j} + 4\mathbf{k}$

14. $\mathbf{a} = \sqrt{3}\mathbf{i} + 2\mathbf{j} - 3\mathbf{k}, \mathbf{b} = 4\mathbf{i} - \mathbf{j} + 2\mathbf{k}$

In Exercises 15–17 $\mathbf{a}$, $\mathbf{a}'$, and $\mathbf{b}$ lie in the same plane. Show that $\mathbf{a}$ and $\mathbf{a}'$ are perpendicular, and resolve $\mathbf{b}$ into vectors parallel to $\mathbf{a}$ and $\mathbf{a}'$.

15. $\mathbf{a} = \mathbf{i} + 2\mathbf{j} - \mathbf{k}, \mathbf{a}' = \mathbf{j} + 2\mathbf{k}, \mathbf{b} = 3\mathbf{i} + \mathbf{j} - 13\mathbf{k}$

16. $\mathbf{a} = \mathbf{i} + \mathbf{j} - \mathbf{k}, \mathbf{a}' = 2\mathbf{i} - 3\mathbf{j} - \mathbf{k}, \mathbf{b} = -5\mathbf{j} + \mathbf{k}$

17. $\mathbf{a} = 2\mathbf{i} - 4\mathbf{j} + 5\mathbf{k}, \mathbf{a}' = 2\mathbf{i} + 6\mathbf{j} + 4\mathbf{k}, \mathbf{b} = \mathbf{i} + 13\mathbf{j} + \mathbf{k}$

18. Show by example that there exist nonzero vectors $\mathbf{a}$, $\mathbf{b}$, and $\mathbf{c}$ such that $\mathbf{a} \cdot \mathbf{b} = \mathbf{a} \cdot \mathbf{c}$, but $\mathbf{b} \neq \mathbf{c}$.

In Exercises 19–22 use the definition of the dot product to prove the statement.

19. $\mathbf{a} \cdot \mathbf{a} = \|\mathbf{a}\|^2$ for any vector $\mathbf{a}$.

20. $\mathbf{a} \cdot \mathbf{b} = \mathbf{b} \cdot \mathbf{a}$ for any vectors $\mathbf{a}$ and $\mathbf{b}$.

21. a. $\mathbf{a} \cdot (\mathbf{b} + \mathbf{c}) = \mathbf{a} \cdot \mathbf{b} + \mathbf{a} \cdot \mathbf{c}$ for any vectors $\mathbf{a}$, $\mathbf{b}$, and $\mathbf{c}$.
    b. If $\mathbf{a}$ is perpendicular to $\mathbf{b}$ and to $\mathbf{c}$, then $\mathbf{a}$ is perpendicular to $\mathbf{b} + \mathbf{c}$.

22. $(c\mathbf{a}) \cdot \mathbf{b} = c(\mathbf{a} \cdot \mathbf{b})$ for any vectors $\mathbf{a}$ and $\mathbf{b}$ and any number c.

23. Let $\mathbf{a}$ and $\mathbf{b}$ be unit vectors and let θ be the angle between $\mathbf{a}$ and $\mathbf{b}$.

a. For what value of θ in $[0, \pi]$ is $\mathbf{a} \cdot \mathbf{b}$ maximum?
b. For what value of θ in $[0, \pi]$ is $\mathbf{a} \cdot \mathbf{b}$ minimum?
c. For what value of θ in $[0, \pi]$ is $|\mathbf{a} \cdot \mathbf{b}|$ minimum?

24. a. Prove the **Cauchy-Schwarz Inequality**:

$$|\mathbf{a} \cdot \mathbf{b}| \leq \|\mathbf{a}\| \|\mathbf{b}\|$$

b. Conclude from (a) that

$$(a_1 b_1 + a_2 b_2 + a_3 b_3)^2 \leq (a_1^2 + a_2^2 + a_3^2)(b_1^2 + b_2^2 + b_3^2)$$

for any numbers $a_1, a_2, a_3, b_1, b_2,$ and b_3.

c. Let $b_1 = b_2 = b_3 = 1$ in (b) and conclude that

$$\left(\frac{a_1 + a_2 + a_3}{3}\right)^2 \leq \frac{a_1^2 + a_2^2 + a_3^2}{3}$$

for any numbers $a_1, a_2,$ and a_3. (This means that the square of the average of three numbers does not exceed the average of the squares of the three numbers.)

25. a. Using the equation $\|\mathbf{a} + \mathbf{b}\|^2 = (\mathbf{a} + \mathbf{b}) \cdot (\mathbf{a} + \mathbf{b})$, prove that $\|\mathbf{a} + \mathbf{b}\|^2 = \|\mathbf{a}\|^2 + 2\mathbf{a} \cdot \mathbf{b} + \|\mathbf{b}\|^2$.

b. Using part (a), prove the **Pythagorean Theorem**:

$$\|\mathbf{a} + \mathbf{b}\|^2 = \|\mathbf{a}\|^2 + \|\mathbf{b}\|^2 \quad \text{if and only if } \mathbf{a} \text{ and } \mathbf{b} \text{ are perpendicular}$$

*c. Using part (a) and Exercise 24(a), prove the **triangle inequality**:

$$\|\mathbf{a} + \mathbf{b}\| \leq \|\mathbf{a}\| + \|\mathbf{b}\| \tag{12}$$

(The inequality in (12) is illustrated in Figure 11.33.)

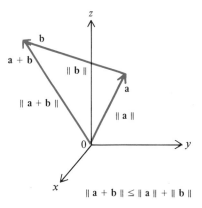

FIGURE 11.33

26. Let P, Q, and R be three points in space. Using Exercise 25 and the fact that

$$\overrightarrow{PQ} = \overrightarrow{PR} + \overrightarrow{RQ}$$

prove the alternate form of the triangle inequality:

$$\|\overrightarrow{PQ}\| \leq \|\overrightarrow{PR}\| + \|\overrightarrow{RQ}\|$$

27. a. Using the equations $\|\mathbf{a} + \mathbf{b}\|^2 = (\mathbf{a} + \mathbf{b}) \cdot (\mathbf{a} + \mathbf{b})$ and $\|\mathbf{a} - \mathbf{b}\|^2 = (\mathbf{a} - \mathbf{b}) \cdot (\mathbf{a} - \mathbf{b})$, prove the **parallelogram law**:

$$\|\mathbf{a} + \mathbf{b}\|^2 + \|\mathbf{a} - \mathbf{b}\|^2 = 2\|\mathbf{a}\|^2 + 2\|\mathbf{b}\|^2$$

(The geometric interpretation of this result is as follows: The sum of the squares of the lengths of the diagonals of the parallelogram determined by $\mathbf{a}$ and $\mathbf{b}$ is equal to the sum of the squares of the lengths of the four sides (Figure 11.34).)

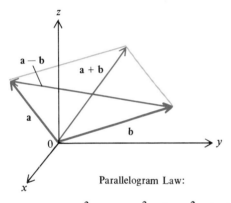

Parallelogram Law:

$$\| \mathbf{a} + \mathbf{b} \|^2 + \| \mathbf{a} - \mathbf{b} \|^2 = 2 \| \mathbf{a} \|^2 + 2 \| \mathbf{b} \|^2$$

FIGURE 11.34

b. Using the ideas of part (a), prove the **polarization identity**:

$$\|\mathbf{a} + \mathbf{b}\|^2 - \|\mathbf{a} - \mathbf{b}\|^2 = 4\mathbf{a} \cdot \mathbf{b}$$

c. Using part (b), prove that the diagonals of a parallelogram have equal length if and only if the parallelogram is a rectangle.

28. Prove that the diagonals of a rhombus (a parallelogram whose sides have equal length) are perpendicular.

29. Approximate the angle between the line segments that join the center of a cube to any two adjacent vertices of the cube.

30. Consider a rectangular parallelepiped whose base is a square of area 1 and whose height is c. Determine the value of c for which the angle formed by the lines from the parallelepiped's center to any pair of adjacent vertices on the base is equal to $\pi/3$.

31. Let $\mathbf{a}$ and $\mathbf{b}$ be nonzero vectors, and let $\mathbf{c} = \|\mathbf{b}\|\mathbf{a} + \|\mathbf{a}\|\mathbf{b}$. Prove that if $\mathbf{c} \neq \mathbf{0}$, then $\mathbf{c}$ bisects the angle formed by $\mathbf{a}$ and $\mathbf{b}$. (*Hint:* Compute the cosine of the angle between $\mathbf{a}$ and $\mathbf{c}$ and the cosine of the angle between $\mathbf{c}$ and $\mathbf{b}$.)

32. A person pulls a sled 100 feet with a rope that makes an angle of $\pi/4$ with the horizontal ground. Find the work W done on the sled if the tension in the rope is 5 pounds.

33. A person exerts a force $\mathbf{F}$ of 100 pounds on a wheelbarrow at an angle of $\pi/6$ with respect to the ground and pushes the wheelbarrow 500 feet. Compute the work W done on the wheelbarrow.

34. A person exerts a horizontal force $\mathbf{F}$ of 20 pounds on a box and pushes it up a ramp that is 15 feet long and inclined at an angle of $\pi/6$. How much work W is done on the box?

11.4
THE CROSS PRODUCT AND TRIPLE PRODUCTS

The second type of product of two vectors we will study is the cross product.

The Cross Product   Unlike the dot product, the cross product of two vectors is a vector.

DEFINITION 11.9

The **cross product** (or **vector product**) $\mathbf{a} \times \mathbf{b}$ of two vectors $\mathbf{a} = a_1\mathbf{i} + a_2\mathbf{j} + a_3\mathbf{k}$ and $\mathbf{b} = b_1\mathbf{i} + b_2\mathbf{j} + b_3\mathbf{k}$ is defined by

$$\mathbf{a} \times \mathbf{b} = (a_2b_3 - a_3b_2)\mathbf{i} + (a_3b_1 - a_1b_3)\mathbf{j} + (a_1b_2 - a_2b_1)\mathbf{k} \quad (1)$$

An easy way to remember (1) is to express the right-hand side as a determinant:

$$\begin{vmatrix} \mathbf{i} & \mathbf{j} & \mathbf{k} \\ a_1 & a_2 & a_3 \\ b_1 & b_2 & b_3 \end{vmatrix}$$

This determinant is evaluated by first repeating the first and second columns to obtain

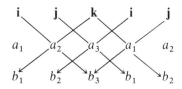

and then subtracting the sum of the products of the "southwest" diagonals from the sum of the products of the "southeast" diagonals.

Example 1 Show that

$$\mathbf{i} \times \mathbf{j} = \mathbf{k}, \quad \mathbf{j} \times \mathbf{k} = \mathbf{i}, \quad \text{and} \quad \mathbf{k} \times \mathbf{i} = \mathbf{j}$$

Solution By (1),

$$\mathbf{i} \times \mathbf{j} = \begin{vmatrix} \mathbf{i} & \mathbf{j} & \mathbf{k} \\ 1 & 0 & 0 \\ 0 & 1 & 0 \end{vmatrix} = [(0)(0) - (0)(1)]\mathbf{i} + [(0)(0) - (1)(0)]\mathbf{j} + [(1)(1) - (0)(0)]\mathbf{k} = \mathbf{k}$$

The remaining two formulas follow in a similar fashion. □

Example 2 Let $\mathbf{a} = 2\mathbf{i} - \mathbf{j} + 3\mathbf{k}$ and $\mathbf{b} = -\mathbf{i} - 2\mathbf{j} + 4\mathbf{k}$. Calculate $\mathbf{a} \times \mathbf{b}$ and $\mathbf{b} \times \mathbf{a}$.

Solution From the definition of the cross product, we obtain

$$\mathbf{a} \times \mathbf{b} = \begin{vmatrix} \mathbf{i} & \mathbf{j} & \mathbf{k} \\ 2 & -1 & 3 \\ -1 & -2 & 4 \end{vmatrix} = \begin{aligned} &[(-1)(4) - (3)(-2)]\mathbf{i} + [(3)(-1) - (2)(4)]\mathbf{j} \\ &+ [(2)(-2) - (-1)(-1)]\mathbf{k} \\ &= 2\mathbf{i} - 11\mathbf{j} - 5\mathbf{k} \end{aligned}$$

and

$$\mathbf{b} \times \mathbf{a} = \begin{vmatrix} \mathbf{i} & \mathbf{j} & \mathbf{k} \\ -1 & -2 & 4 \\ 2 & -1 & 3 \end{vmatrix} = \begin{aligned} &[(-2)(3) - (4)(-1)]\mathbf{i} + [(4)(2) - (-1)(3)]\mathbf{j} \\ &+ [(-1)(-1) - (-2)(2)]\mathbf{k} \\ &= -2\mathbf{i} + 11\mathbf{j} + 5\mathbf{k} \quad □ \end{aligned}$$

Notice that the vectors $\mathbf{a} \times \mathbf{b}$ and $\mathbf{b} \times \mathbf{a}$ in Example 2 are negatives of one another. This is no accident; in fact it follows directly from the definition of the cross product that if $\mathbf{a}$ and $\mathbf{b}$ are any two vectors, then

$$\mathbf{a} \times \mathbf{b} = -(\mathbf{b} \times \mathbf{a})$$

Other properties of the cross product that follow readily from the definition are

$$\mathbf{a} \times \mathbf{a} = \mathbf{0} \qquad\qquad \mathbf{a} \times (\mathbf{b} + \mathbf{c}) = (\mathbf{a} \times \mathbf{b}) + (\mathbf{a} \times \mathbf{c})$$

$$(c\mathbf{a}) \times \mathbf{b} = c(\mathbf{a} \times \mathbf{b}) = \mathbf{a} \times (c\mathbf{b}) \qquad (\mathbf{a} + \mathbf{b}) \times \mathbf{c} = (\mathbf{a} \times \mathbf{c}) + (\mathbf{b} \times \mathbf{c})$$

We also observe that

$$\mathbf{i} \times (\mathbf{i} \times \mathbf{k}) = \mathbf{i} \times (-\mathbf{j}) = -\mathbf{k} \quad \text{and} \quad (\mathbf{i} \times \mathbf{i}) \times \mathbf{k} = \mathbf{0} \times \mathbf{k} = \mathbf{0}$$

so that

$$\mathbf{i} \times (\mathbf{i} \times \mathbf{k}) \neq (\mathbf{i} \times \mathbf{i}) \times \mathbf{k}$$

Therefore $\mathbf{a} \times (\mathbf{b} \times \mathbf{c})$ and $(\mathbf{a} \times \mathbf{b}) \times \mathbf{c}$ are usually not equal.

THEOREM 11.10

Let $\mathbf{a}$ and $\mathbf{b}$ be two nonzero vectors.
a. Then

$$\mathbf{a} \cdot (\mathbf{a} \times \mathbf{b}) = 0 \quad \text{and} \quad \mathbf{b} \cdot (\mathbf{a} \times \mathbf{b}) = 0$$

Consequently if $\mathbf{a} \times \mathbf{b} \neq \mathbf{0}$, then $\mathbf{a} \times \mathbf{b}$ is perpendicular to both $\mathbf{a}$ and $\mathbf{b}$.
b. If θ is the angle between $\mathbf{a}$ and $\mathbf{b}$ (so that $0 \leq \theta \leq \pi$), then

$$\|\mathbf{a} \times \mathbf{b}\| = \|\mathbf{a}\|\,\|\mathbf{b}\| \sin\theta$$

Proof To prove (a), we simply apply the definitions of the dot product and the cross product:

$$
\begin{aligned}
\mathbf{a} \cdot (\mathbf{a} \times \mathbf{b}) &= a_1(a_2 b_3 - a_3 b_2) + a_2(a_3 b_1 - a_1 b_3) + a_3(a_1 b_2 - a_2 b_1) \\
&= a_1 a_2 b_3 - a_1 a_3 b_2 + a_2 a_3 b_1 - a_2 a_1 b_3 + a_3 a_1 b_2 - a_3 a_2 b_1 \\
&= 0
\end{aligned}
$$

and

$$
\begin{aligned}
\mathbf{b} \cdot (\mathbf{a} \times \mathbf{b}) &= b_1(a_2 b_3 - a_3 b_2) + b_2(a_3 b_1 - a_1 b_3) + b_3(a_1 b_2 - a_2 b_1) \\
&= b_1 a_2 b_3 - b_1 a_3 b_2 + b_2 a_3 b_1 - b_2 a_1 b_3 + b_3 a_1 b_2 - b_3 a_2 b_1 \\
&= 0
\end{aligned}
$$

For part (b) we use the definitions of the cross product and the length of a vector to deduce that

$$\|\mathbf{a} \times \mathbf{b}\|^2 = (a_2 b_3 - a_3 b_2)^2 + (a_3 b_1 - a_1 b_3)^2 + (a_1 b_2 - a_2 b_1)^2$$
$$= a_2^2 b_3^2 - 2a_2 a_3 b_2 b_3 + a_3^2 b_2^2 + a_3^2 b_1^2 - 2a_1 a_3 b_1 b_3 + a_1^2 b_3^2$$
$$+ a_1^2 b_2^2 - 2a_1 a_2 b_1 b_2 + a_2^2 b_1^2$$
$$= (a_1^2 + a_2^2 + a_3^2)(b_1^2 + b_2^2 + b_3^2) - (a_1 b_1 + a_2 b_2 + a_3 b_3)^2$$
$$= \|\mathbf{a}\|^2 \|\mathbf{b}\|^2 - (\mathbf{a} \cdot \mathbf{b})^2 = \|\mathbf{a}\|^2 \|\mathbf{b}\|^2 - (\|\mathbf{a}\| \|\mathbf{b}\| \cos \theta)^2$$
$$= \|\mathbf{a}\|^2 \|\mathbf{b}\|^2 (1 - \cos^2 \theta)$$

Therefore

$$\|\mathbf{a} \times \mathbf{b}\|^2 = \|\mathbf{a}\|^2 \|\mathbf{b}\|^2 \sin^2 \theta$$

Since $\sin \theta \geq 0$ for $0 \leq \theta \leq \pi$, we can take the square root of each side of the equation and obtain

$$\|\mathbf{a} \times \mathbf{b}\| = \|\mathbf{a}\| \|\mathbf{b}\| \sin \theta \quad \blacksquare$$

COROLLARY 11.11

Two nonzero vectors **a** and **b** are parallel if and only if $\mathbf{a} \times \mathbf{b} = \mathbf{0}$.

As we saw in Section 11.3, the dot product is especially effective for finding the angle between two vectors and for finding the projection of one vector onto another. By Theorem 11.10(a) the cross product yields a vector perpendicular to two nonzero vectors. This will be invaluable to us in finding equations of planes.

Example 3 Let $\mathbf{a} = 4\mathbf{i} - \mathbf{j} + 3\mathbf{k}$ and $\mathbf{b} = 2\mathbf{i} + 3\mathbf{j} - \mathbf{k}$. Find a vector perpendicular to **a** and **b**.

Solution By Theorem 11.10(a) the cross product $\mathbf{a} \times \mathbf{b}$ is one such vector:

$$\mathbf{a} \times \mathbf{b} = \begin{vmatrix} \mathbf{i} & \mathbf{j} & \mathbf{k} \\ 4 & -1 & 3 \\ 2 & 3 & -1 \end{vmatrix} = (1 - 9)\mathbf{i} + (6 + 4)\mathbf{j} + (12 + 2)\mathbf{k}$$
$$= -8\mathbf{i} + 10\mathbf{j} + 14\mathbf{k} \quad \square$$

From Theorem 11.10(b) we conclude that

$$\|\mathbf{a} \times \mathbf{b}\| = \text{the area of the parallelogram with adjacent sides } \mathbf{a} \text{ and } \mathbf{b} \quad (2)$$

(Figure 11.35). After all, the parallelogram has a base of length $\|\mathbf{a}\|$ and an altitude of length $\|\mathbf{b}\| \sin \theta$. Consequently the area of the parallelogram is

$$\|\mathbf{a}\| \|\mathbf{b}\| \sin \theta = \|\mathbf{a} \times \mathbf{b}\|$$

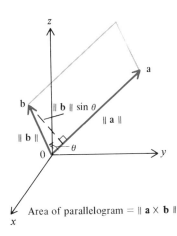

Area of parallelogram $= \|\mathbf{a} \times \mathbf{b}\|$

FIGURE 11.35

Because the triangle having **a** and **b** as two of its sides has an area one half the area of the parallelogram, determined by **a** and **b**, it follows from (2) that the area of the triangle equals $\frac{1}{2}\|\mathbf{a} \times \mathbf{b}\|$.

Example 4   Let $P = (3, -2, 1), Q = (7, -3, 4)$, and $R = (5, 1, 0)$. Find the area A of triangle PQR.

Solution   Let $\mathbf{a} = \overrightarrow{PQ} = 4\mathbf{i} - \mathbf{j} + 3\mathbf{k}$ and $\mathbf{b} = \overrightarrow{PR} = 2\mathbf{i} + 3\mathbf{j} - \mathbf{k}$. In Example 3 we found that

$$\mathbf{a} \times \mathbf{b} = -8\mathbf{i} + 10\mathbf{j} + 14\mathbf{k}$$

Consequently

$$A = \frac{1}{2}\|\mathbf{a} \times \mathbf{b}\| = \frac{1}{2}\sqrt{(-8)^2 + (10)^2 + (14)^2} = 3\sqrt{10} \quad \square$$

$$\frac{\sqrt{360}}{2}$$

Triple Products

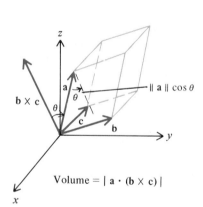

Volume = | **a** · (**b** × **c**) |

FIGURE 11.36

Let $\mathbf{a} = a_1\mathbf{i} + a_2\mathbf{j} + a_3\mathbf{k}$, $\mathbf{b} = b_1\mathbf{i} + b_2\mathbf{j} + b_3\mathbf{k}$, and $\mathbf{c} = c_1\mathbf{i} + c_2\mathbf{j} + c_3\mathbf{k}$. The products $\mathbf{a} \cdot (\mathbf{b} \times \mathbf{c})$, $(\mathbf{a} \times \mathbf{b}) \cdot \mathbf{c}$, $(\mathbf{a} \times \mathbf{b}) \times \mathbf{c}$, and $\mathbf{a} \times (\mathbf{b} \times \mathbf{c})$ occasionally arise in physical applications. The first two products are called **triple scalar products**, since they are scalars (that is, numbers). The last two are called **triple vector products**, since they are vectors.

In Theorem 11.10(a) we considered two triple scalar products, $\mathbf{a} \cdot (\mathbf{a} \times \mathbf{b})$ and $\mathbf{b} \cdot (\mathbf{a} \times \mathbf{b})$. We observed that they are always 0 and consequently are equal to each other. In fact, for any **a**, **b**, and **c** we find that

$$\mathbf{a} \cdot (\mathbf{b} \times \mathbf{c}) = a_1(b_2 c_3 - b_3 c_2) + a_2(b_3 c_1 - b_1 c_3) + a_3(b_1 c_2 - b_2 c_1)$$
$$(\mathbf{a} \times \mathbf{b}) \cdot \mathbf{c} = (a_2 b_3 - a_3 b_2)c_1 + (a_3 b_1 - a_1 b_3)c_2 + (a_1 b_2 - a_2 b_1)c_3$$

and we see by multiplying out that the right-hand sides are equal. It follows that

$$\mathbf{a} \cdot (\mathbf{b} \times \mathbf{c}) = (\mathbf{a} \times \mathbf{b}) \cdot \mathbf{c} \tag{3}$$

The quantity $\mathbf{a} \cdot (\mathbf{b} \times \mathbf{c})$ is useful because

$$|\mathbf{a} \cdot (\mathbf{b} \times \mathbf{c})| = \text{the volume of the parallelepiped with sides } \mathbf{a}, \mathbf{b}, \text{ and } \mathbf{c} \tag{4}$$

(Figure 11.36). Indeed, $\|\mathbf{b} \times \mathbf{c}\|$ is the area of the base of this parallelepiped, and if θ is the angle between **a** and $\mathbf{b} \times \mathbf{c}$, then $\|\mathbf{a}\|\,|\cos \theta|$ is the height, so that the volume is $\|\mathbf{a}\|\,|\cos \theta|\,\|\mathbf{b} \times \mathbf{c}\| = |\mathbf{a} \cdot (\mathbf{b} \times \mathbf{c})|$.

We can use the determinant to facilitate computation of $\mathbf{a} \cdot (\mathbf{b} \times \mathbf{c})$:

$$\mathbf{a} \cdot (\mathbf{b} \times \mathbf{c}) = \begin{vmatrix} a_1 & a_2 & a_3 \\ b_1 & b_2 & b_3 \\ c_1 & c_2 & c_3 \end{vmatrix} = \begin{matrix} (a_1 b_2 c_3 + a_2 b_3 c_1 + a_3 b_1 c_2) \\ - (a_3 b_2 c_1 + a_2 b_1 c_3 + a_1 b_3 c_2) \end{matrix} \tag{5}$$

Example 5   Let $\mathbf{a} = \mathbf{i} - \mathbf{j}$, $\mathbf{b} = 2\mathbf{i} + 3\mathbf{j} - \mathbf{k}$, and $\mathbf{c} = -\mathbf{i} + 2\mathbf{k}$. Find the volume V of the parallelepiped determined by **a**, **b**, and **c**.

Solution   By (4), $V = |\mathbf{a} \cdot (\mathbf{b} \times \mathbf{c})|$, and by (5),

$$\mathbf{a} \cdot (\mathbf{b} \times \mathbf{c}) = \begin{vmatrix} 1 & -1 & 0 \\ 2 & 3 & -1 \\ -1 & 0 & 2 \end{vmatrix} = (6 - 1 + 0) - (0 - 4 + 0) = 9$$

Thus $V = 9$.   □

For reference we list three additional formulas, which will occasionally be useful. The first is

$$\mathbf{a} \times (\mathbf{b} \times \mathbf{c}) = (\mathbf{a} \cdot \mathbf{c})\mathbf{b} - (\mathbf{a} \cdot \mathbf{b})\mathbf{c} \qquad (6)$$

which can be verified by expanding each side (see Exercise 18). This equation is sometimes written in the equivalent form

$$\mathbf{a} \times (\mathbf{b} \times \mathbf{c}) = \mathbf{b}(\mathbf{a} \cdot \mathbf{c}) - \mathbf{c}(\mathbf{a} \cdot \mathbf{b}) \qquad (7)$$

known as "the *bac − cab* rule." A special case of (7) occurs when $\mathbf{a} = \mathbf{b}$ and $\mathbf{c}$ is perpendicular to $\mathbf{a}$. In that case we have

$$\mathbf{a} \times (\mathbf{a} \times \mathbf{c}) = \mathbf{a}(0) - \mathbf{c}(\mathbf{a} \cdot \mathbf{a})$$

or more simply

$$\mathbf{a} \times (\mathbf{a} \times \mathbf{c}) = -\|\mathbf{a}\|^2 \mathbf{c} \qquad (8)$$

EXERCISES 11.4

In Exercises 1–5 find $\mathbf{a} \times \mathbf{b}$ and $\mathbf{c} \cdot (\mathbf{a} \times \mathbf{b})$.

1. $\mathbf{a} = \mathbf{i} + \mathbf{j}, \mathbf{b} = \mathbf{j} + \mathbf{k}, \mathbf{c} = -\mathbf{i} - 3\mathbf{j} + 4\mathbf{k}$

2. $\mathbf{a} = \mathbf{i} + \mathbf{j} + \mathbf{k}, \mathbf{b} = \mathbf{i} - \mathbf{k}, \mathbf{c} = \mathbf{i} + \mathbf{j} - \mathbf{k}$

3. $\mathbf{a} = 2\mathbf{i} + 3\mathbf{j} - \mathbf{k}, \mathbf{b} = -\mathbf{i} + 4\mathbf{j} + 5\mathbf{k}, \mathbf{c} = 2\mathbf{i} + 3\mathbf{j} + 4\mathbf{k}$

4. $\mathbf{a} = 3\mathbf{i} + 4\mathbf{j} + 12\mathbf{k}, \;\; \mathbf{b} = 3\mathbf{i} + 4\mathbf{j} - 12\mathbf{k}, \;\; \mathbf{c} = \frac{1}{8}\mathbf{i} - \frac{1}{12}\mathbf{j} + \frac{1}{16}\mathbf{k}$

5. $\mathbf{a} = 3\mathbf{i} + 4\mathbf{j} + 12\mathbf{k}, \mathbf{b} = 3\mathbf{i} + 4\mathbf{j} + 12\mathbf{k}, \mathbf{c} = \mathbf{i} + \mathbf{j}$

6. Using the cross product, find the sine of the angle between the vectors $\mathbf{a}$ and $\mathbf{b}$ in Exercise 4.

7. From the definition of the cross product prove that $\mathbf{a} \times \mathbf{b} = -(\mathbf{b} \times \mathbf{a})$.

8. From the definition of the cross product prove that $\mathbf{a} \times (\mathbf{b} + \mathbf{c}) = \mathbf{a} \times \mathbf{b} + \mathbf{a} \times \mathbf{c}$.

9. Suppose that $\mathbf{a} + \mathbf{b} + \mathbf{c} = \mathbf{0}$. Show that $\mathbf{a} \times \mathbf{b} = \mathbf{b} \times \mathbf{c} = \mathbf{c} \times \mathbf{a}$. (*Hint:* Expand $\mathbf{a} \times (\mathbf{a} + \mathbf{b} + \mathbf{c})$ and $\mathbf{b} \times (\mathbf{a} + \mathbf{b} + \mathbf{c})$.)

10. Find a vector perpendicular to both $\mathbf{i} - 3\mathbf{j} + 2\mathbf{k}$ and $-2\mathbf{i} + \mathbf{j} - 5\mathbf{k}$.

11. Find the volume V of the parallelepiped determined by the vectors $2\mathbf{i} - 3\mathbf{j} + 4\mathbf{k}, \mathbf{i} + \mathbf{j} - \mathbf{k}$, and $4\mathbf{i} - \mathbf{j} - \mathbf{k}$.

12. Find nonzero vectors $\mathbf{a}, \mathbf{b}$, and $\mathbf{c}$ in space such that $\mathbf{a} \times \mathbf{b} = \mathbf{a} \times \mathbf{c}$ but $\mathbf{b} \neq \mathbf{c}$.

13. Assume that $\mathbf{a} \neq \mathbf{0}, \mathbf{a} \cdot \mathbf{b} = \mathbf{a} \cdot \mathbf{c}$, and $\mathbf{a} \times \mathbf{b} = \mathbf{a} \times \mathbf{c}$. Does it follow that $\mathbf{b} = \mathbf{c}$? Support your answer.

14. Show that

$$\|\mathbf{a} \times \mathbf{b}\|^2 = \|\mathbf{a}\|^2 \|\mathbf{b}\|^2 - (\mathbf{a} \cdot \mathbf{b})^2$$

by using Theorems 11.6 and 11.10.

In Exercises 15–16 find $\mathbf{a} \times (\mathbf{b} \times \mathbf{c})$ directly, and then use the *bac − cab* rule (7) to compute $\mathbf{a} \times (\mathbf{b} \times \mathbf{c})$.

15. $\mathbf{a} = 2\mathbf{i} - 3\mathbf{j} + 4\mathbf{k}, \mathbf{b} = \frac{1}{2}\mathbf{i} + \mathbf{j} - \mathbf{k}, \mathbf{c} = 4\mathbf{i} - 5\mathbf{j} + 6\mathbf{k}$

16. $\mathbf{a} = \mathbf{i} - 4\mathbf{j} + 2\mathbf{k}, \mathbf{b} = 3\mathbf{i} - 7\mathbf{j} + 2\mathbf{k}, \mathbf{c} = 2\mathbf{i} - 5\mathbf{j}$

17. Use (3) and (8) to show that if $\mathbf{u}$ is perpendicular to $\mathbf{a}$ or to $\mathbf{b}$, then

$$(\mathbf{u} \times \mathbf{a}) \cdot (\mathbf{u} \times \mathbf{b}) = \|\mathbf{u}\|^2 (\mathbf{a} \cdot \mathbf{b})$$

*18. Prove (6), and hence the *bac − cab* rule. (*Hint:* Using the definition of the cross product, find an expression for $\mathbf{b} \times \mathbf{c}$ and then one for $\mathbf{a} \times (\mathbf{b} \times \mathbf{c})$.)

19. Using the $bac - cab$ rule, prove that

$$\mathbf{a} \times (\mathbf{b} \times \mathbf{c}) + \mathbf{b} \times (\mathbf{c} \times \mathbf{a}) + \mathbf{c} \times (\mathbf{a} \times \mathbf{b}) = \mathbf{0}$$

20. Express $(\mathbf{a} \times \mathbf{b}) \times \mathbf{c}$ as the difference of a vector parallel to $\mathbf{a}$ and a vector parallel to $\mathbf{b}$.

In Exercises 21–22 we will refer to the **moment M** of a force $\mathbf{F}$ at a point Q about another point P, which is defined by

$$\mathbf{M} = \overrightarrow{PQ} \times \mathbf{F}$$

The magnitude of $\mathbf{M}$ measures the tendency of the force $\mathbf{F}$ to make a body attached at P begin rotating about P. The body tends to rotate about the line that passes through P and is parallel to $\mathbf{M}$; this line is called the **axis of rotation**.

21. Suppose the arm of a stapler is resting at an angle of $\pi/6$ from the horizontal and is $\frac{3}{2}$ feet long (Figure 11.37). If a downward force of 32 pounds is applied to the tip of the arm, determine the moment $\mathbf{M}$ of the force about the axis of the stapler.

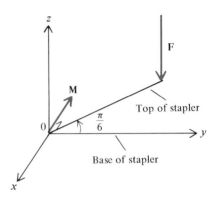

FIGURE 11.37

22. A 5-pound fish hangs from a fishing line on a 4-foot fishing pole held at an angle of $\pi/4$ with respect to the horizontal. Find the moment $\mathbf{M}$ due to the weight of the fish about the handle end of the pole.

11.5
LINES IN SPACE

Since we often think of vectors as directed line segments, it should not be surprising that vectors and lines are intimately related. In this section we will use vectors to describe lines.

Equations of Lines in Space

We say that a vector $\mathbf{L}$ and a line l are **parallel** if $\mathbf{L}$ is parallel to the vector $\overrightarrow{P_0 P}$ joining any two distinct points P_0 and P on l (Figure 11.38). It follows from Euclidean geometry that a line l in space is uniquely determined by a point $P_0 = (x_0, y_0, z_0)$ on l and a vector $\mathbf{L}$ parallel to the line. Thus a point $P = (x, y, z)$ is on l if and only if $\overrightarrow{P_0 P}$ is parallel to $\mathbf{L}$. By Definition 11.4 this means that

$$\overrightarrow{P_0 P} = t\mathbf{L} \tag{1}$$

for a suitable number t. If $\mathbf{r}_0 = x_0 \mathbf{i} + y_0 \mathbf{j} + z_0 \mathbf{k}$ and $\mathbf{r} = x\mathbf{i} + y\mathbf{j} + z\mathbf{k}$, then $\overrightarrow{P_0 P} = \mathbf{r} - \mathbf{r}_0$, so that (1) can be rewritten

$$\mathbf{r} = \mathbf{r}_0 + t\mathbf{L} \tag{2}$$

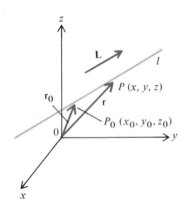

FIGURE 11.38

This is a **vector equation** of l. Since $\mathbf{r}_0$ can be any vector that joins the origin to a point on l, and since $\mathbf{L}$ can be any vector parallel to l, there are many different vector equations of a given line l.

Example 1 Find a vector equation of the line that contains $(-1, 3, 0)$ and is parallel to $2\mathbf{i} - 3\mathbf{j} - \mathbf{k}$.

Solution From (2) we obtain

$$\mathbf{r} = (-\mathbf{i} + 3\mathbf{j}) + t(2\mathbf{i} - 3\mathbf{j} - \mathbf{k})$$

or alternatively,

$$\mathbf{r} = (-1 + 2t)\mathbf{i} + (3 - 3t)\mathbf{j} - t\mathbf{k} \quad \square$$

Suppose we let $\mathbf{L} = a\mathbf{i} + b\mathbf{j} + c\mathbf{k}$. Then (2) can be rewritten

$$x\mathbf{i} + y\mathbf{j} + z\mathbf{k} = (x_0\mathbf{i} + y_0\mathbf{j} + z_0\mathbf{k}) + t(a\mathbf{i} + b\mathbf{j} + c\mathbf{k})$$

or equivalently,

$$x = x_0 + at, \qquad y = y_0 + bt, \qquad z = z_0 + ct \tag{3}$$

These equations are called **parametric equations** of l, and t is called a **parameter**. If $z_0 = 0$ and $c = 0$, then l lies in the xy plane and has parametric equations

$$x = x_0 + at, \qquad y = y_0 + bt, \qquad z = 0$$

which are compatible with the parametric equations for a line given in (3) of Section 6.4.

Example 2 Find parametric equations of the line that contains $(2, -4, 1)$ and is parallel to $3\mathbf{i} + \frac{1}{2}\mathbf{j} - \mathbf{k}$.

Solution Taking $(x_0, y_0, z_0) = (2, -4, 1)$ and $a\mathbf{i} + b\mathbf{j} + c\mathbf{k} = 3\mathbf{i} + \frac{1}{2}\mathbf{j} - \mathbf{k}$ in (3) gives us the parametric equations

$$x = 2 + 3t, \qquad y = -4 + \frac{1}{2}t, \qquad z = 1 - t \quad \square$$

Example 3 Find parametric equations of the line l that passes through the point $(1, 2, 3)$ for $t = 0$ and through the point $(3, 0, 4)$ for $t = 1$.

Solution If $x = x_0 + at$, then since $x = 1$ for $t = 0$ and $x = 3$ for $t = 1$, we find that

$$1 = x_0 + a(0) \quad \text{and} \quad 3 = x_0 + a(1)$$

From the first equation, $x_0 = 1$. Therefore from the second equation, we find that $a = 3 - x_0 = 3 - 1 = 2$. Therefore $x = 1 + 2t$. Next, if $y = y_0 + bt$, then

$$2 = y_0 + b(0) \quad \text{and} \quad 0 = y_0 + b(1)$$

From the first equation we have $y_0 = 2$. Thus from the second equation we find that $b = -2$. Therefore $y = 2 - 2t$. Finally if $z = z_0 + ct$, then

$$3 = z_0 + c(0) \quad \text{and} \quad 4 = z_0 + c(1)$$

In the same manner we find that $z_0 = 3$ and $c = 1$, so that $z = 3 + t$. Consequently parametric equations of l satisfying the given conditions are

$$x = 1 + 2t, \qquad y = 2 - 2t, \qquad z = 3 + t \quad \square$$

Another set of equations for l is obtained by eliminating t from the equations in (3). If a, b, and c are all nonzero, we can solve each of the equations in (3) for t and then equate the results, obtaining

$$\frac{x - x_0}{a} = \frac{y - y_0}{b} = \frac{z - z_0}{c} \tag{4}$$

These are called **symmetric equations** of l. Notice that the coordinates x_0, y_0, z_0 of the point P_0 on l appear in the numerators in (4) and that the components a, b, c of a vector $\mathbf{L}$ parallel to l appear in the denominators in (4).

Example 4 Find symmetric equations of the line l containing the points $P_1 = (4, -6, 5)$ and $P_2 = (2, -3, 0)$.

Solution In order to use (4) we find a vector $\mathbf{L}$ parallel to l. Since P_1 and P_2 are distinct points lying on l, the vector $\overrightarrow{P_1P_2}$ will serve. Therefore we let

$$\mathbf{L} = \overrightarrow{P_1P_2} = -2\mathbf{i} + 3\mathbf{j} - 5\mathbf{k}$$

For this choice of $\mathbf{L}$, if we use the hypothesis that $(4,\ 6, 5)$ is on l, then by (4) symmetric equations of l are

$$\frac{x - 4}{-2} = \frac{y + 6}{3} = \frac{z - 5}{-5}$$

Similarly, if we use the hypothesis that $(2, -3, 0)$ is on l, then symmetric equations of l are

$$\frac{x - 2}{-2} = \frac{y + 3}{3} = \frac{z}{-5} \quad \square$$

Example 5 Let l be the line with equations

$$\frac{x - 3}{-2} = \frac{y - 1}{-1} = z + 2$$

Find a vector parallel to l, and find two points on l.

Solution The numbers in the denominators of the given equations (where $z + 2 = (z + 2)/1$) are the components of a vector $\mathbf{L}$ parallel to the line l:

$$\mathbf{L} = -2\mathbf{i} - \mathbf{j} + \mathbf{k}$$

The point $P_0 = (3, 1, -2)$ lies on l, as we can see by letting $x = 3$, $y = 1$, and $z = -2$ in the given symmetric equations of l. To find a second point P_1 on l, we can let $x = 1$. Then the given set of equations becomes

$$\frac{1-3}{-2} = \frac{y-1}{-1} = z + 2$$

so that $y = 0$ and $z = -1$. Therefore $P_1 = (1, 0, -1)$ is also on l. □

We obtained symmetric equations of a line l from the parametric equations of l under the assumption that the numbers a, b, and c in (3) were all nonzero. If one or more of these numbers is 0, then the symmetric equations of the line in (4) must be altered to avoid dividing by 0. Suppose $a = 0$ but b and c are nonzero. Then (3) becomes

$$x = x_0, \qquad y = y_0 + bt, \qquad z = z_0 + ct$$

This means that the x coordinate of every point on the line is x_0. By solving for t in the last two parametric equations above we obtain

$$x = x_0, \qquad \frac{y - y_0}{b} = \frac{z - z_0}{c}$$

for symmetric equations of l. For example, the line passing through $(-2, -1, 1)$ parallel to $2\mathbf{j} - 3\mathbf{k}$ is described by

$$x = -2, \qquad \frac{y + 1}{2} = \frac{z - 1}{-3}$$

Going a step further, we observe that if $a = b = 0$ and $c \neq 0$, then (3) becomes

$$x = x_0, \qquad y = y_0, \qquad z = z_0 + ct$$

Consequently the symmetric equations of l are simply

$$x = x_0, \qquad y = y_0$$

This implies that each point on l has x coordinate x_0 and y coordinate y_0. Hence l is the vertical line passing through $(x_0, y_0, 0)$. Similar analyses can be made if $b = 0$ or $c = 0$.

Example 6 Find symmetric equations of the line l that contains the point $(10, -1, 1)$ and is parallel to the vector $2\mathbf{i} - 3\mathbf{k}$.

Solution In this case $b = 0$ and (3) becomes

$$x = 10 + 2t, \qquad y = -1 + 0t, \qquad z = 1 - 3t$$

Solving for t in the first and third equations yields the symmetric equations

$$y = -1, \qquad \frac{x-10}{2} = \frac{z-1}{-3} \quad \square$$

Distance from a Point to a Line   Many otherwise long and difficult computations can be simplified with the help of vector methods. A good example is the computation of the distance between a given line and a given point not on the line.

THEOREM 11.12

Let l be a line parallel to a vector $\mathbf{L}$, and let P_1 be a point *not* on l. Then the distance D between P_1 and l is given by

$$D = \frac{\|\mathbf{L} \times \overrightarrow{P_0 P_1}\|}{\|\mathbf{L}\|} \tag{5}$$

where P_0 is any point on l.

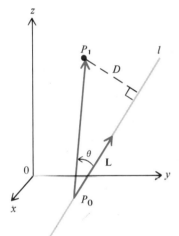

FIGURE 11.39

Proof   Let θ be the angle between $\mathbf{L}$ and $\overrightarrow{P_0 P_1}$, so that $0 \le \theta \le \pi$ (Figure 11.39). Then

$$D = \|\overrightarrow{P_0 P_1}\| \sin \theta$$

Since

$$\|\mathbf{L} \times \overrightarrow{P_0 P_1}\| = \|\mathbf{L}\| \, \|\overrightarrow{P_0 P_1}\| \sin \theta$$

it follows that

$$D = \frac{\|\mathbf{L} \times \overrightarrow{P_0 P_1}\|}{\|\mathbf{L}\|}$$

as we wished to prove.   ■

Example 7   Find the distance D from the point $(2, 1, -1)$ to the line with parametric equations $x = 3t$, $y = 1 + 2t$, $z = -5 - t$.

Solution   Notice that $(0, 1, -5)$ is on the line, $(2, 1, -1)$ is not on the line, and $3\mathbf{i} + 2\mathbf{j} - \mathbf{k}$ is parallel to the line. Therefore we let $P_0 = (0, 1, -5)$, $P_1 = (2, 1, -1)$, and $\mathbf{L} = 3\mathbf{i} + 2\mathbf{j} - \mathbf{k}$ in (5) and thereby obtain

$$D = \frac{\|\mathbf{L} \times \overrightarrow{P_0 P_1}\|}{\|\mathbf{L}\|} = \frac{\|(3\mathbf{i} + 2\mathbf{j} - \mathbf{k}) \times (2\mathbf{i} + 4\mathbf{k})\|}{\|3\mathbf{i} + 2\mathbf{j} - \mathbf{k}\|}$$

Now $\|3\mathbf{i} + 2\mathbf{j} - \mathbf{k}\| = \sqrt{3^2 + 2^2 + (-1)^2} = \sqrt{14}$, and

$$(3\mathbf{i} + 2\mathbf{j} - \mathbf{k}) \times (2\mathbf{i} + 4\mathbf{k}) = \begin{vmatrix} \mathbf{i} & \mathbf{j} & \mathbf{k} \\ 3 & 2 & -1 \\ 2 & 0 & 4 \end{vmatrix} = 8\mathbf{i} - 14\mathbf{j} - 4\mathbf{k}$$

so that

$$\|(3\mathbf{i} + 2\mathbf{j} - \mathbf{k}) \times (2\mathbf{i} + 4\mathbf{k})\| = \|8\mathbf{i} - 14\mathbf{j} - 4\mathbf{k}\|$$
$$= \sqrt{8^2 + (-14)^2 + (-4)^2} = \sqrt{276}$$

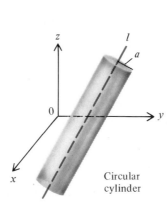

Circular
cylinder

FIGURE 11.40

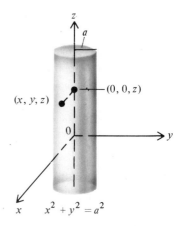

$x^2 + y^2 = a^2$

FIGURE 11.41

Consequently

$$D = \frac{\|(3\mathbf{i} + 2\mathbf{j} - \mathbf{k}) \times (2\mathbf{i} + 4\mathbf{k})\|}{\|3\mathbf{i} + 2\mathbf{j} - \mathbf{k}\|} = \frac{\sqrt{276}}{\sqrt{14}} = \frac{\sqrt{138}}{\sqrt{7}} \quad \square$$

The set of points whose distances from a given line l are all the same number $a > 0$ is called a **circular cylinder** with axis l and radius a (Figure 11.40). Since the distance from any point $P = (x, y, z)$ to the z axis is $\sqrt{x^2 + y^2}$, it follows that P lies on the cylinder whose axis is the z axis and whose radius is a (Figure 11.41) if and only if

$$\sqrt{x^2 + y^2} = a$$

or equivalently

$$x^2 + y^2 = a^2 \tag{6}$$

Caution: Be sure to distinguish between the graph of (6) in *three-dimensional space*, which is a cylinder, and the graph of (6) in *two-dimensional space*, which is a circle.

EXERCISES 11.5

In Exercises 1–7 find a vector equation, parametric equations, and symmetric equations for the line that contains the given point and is parallel to the vector **L**.

1. $(-2, 1, 0)$; $\mathbf{L} = 3\mathbf{i} - \mathbf{j} + 5\mathbf{k}$
2. $(0, 0, 0)$; $\mathbf{L} = 11\mathbf{i} - 13\mathbf{j} - 15\mathbf{k}$
3. $(3, 4, 5)$; $\mathbf{L} = \frac{1}{2}\mathbf{i} - \frac{1}{3}\mathbf{j} + \frac{1}{6}\mathbf{k}$

4. $(-3, 6, 2)$; $\mathbf{L} = \mathbf{i} - \mathbf{j}$ 5. $(2, 0, 5)$; $\mathbf{L} = 2\mathbf{j} + 3\mathbf{k}$
6. $(7, -1, 2)$; $\mathbf{L} = \mathbf{k}$ 7. $(4, 2, -1)$; $\mathbf{L} = \mathbf{j}$

8. Find parametric equations for the line that contains the point $(3, -1, 2)$ and is parallel to the line with equations

$$\frac{x - 1}{4} = \frac{y + 3}{2} = z$$

9. Find parametric equations for the line containing the points $(-1, 1, 0)$ and $(-2, 5, 7)$.

10. Find parametric equations for the line containing the points $(-1, 1, 0)$ and $(-1, 5, 7)$.

11. Find symmetric equations for the line containing the points $(-1, 1, 0)$ and $(-1, 1, 7)$.

12. Show that the line containing the points $(1, 7, 5)$ and $(3, 2, -1)$ is parallel to the line containing $(2, -2, 5)$ and $(-2, 8, 17)$.

13. Show that the line containing the points $(2, -1, 3)$ and $(0, 7, 9)$ is perpendicular to the line containing the points $(-1, 0, 4)$ and $(2, 3, 1)$.

14. Show that the line containing the points $(5, 7, 9)$ and $(4, 11, 9)$ is parallel to the line with equations

$$\frac{x - 1}{-3} = \frac{y - 2}{12}, \qquad z = 5$$

15. Show that the line containing the points $(0, 0, 5)$ and $(1, -1, 4)$ is perpendicular to the line with equations

$$\frac{x}{7} = \frac{y - 3}{4} = \frac{z + 9}{3}$$

In Exercises 16–18 let l be the line that passes through $P_1 = (-1, -2, -3)$ and $P_2 = (2, -1, 0)$. Find parametric equations for l for which the given conditions are satisfied.

16. P_1 corresponds to $t = 0$ and P_2 corresponds to $t = 1$.

17. P_1 corresponds to $t = 0$ and P_2 corresponds to $t = 2$.

18. P_1 corresponds to $t = -1$ and P_2 corresponds to $t = 4$.

19. Find the distance D from the point $(5, 0, -4)$ to the line with equations

$$x - 1 = \frac{y + 2}{-2} = \frac{z + 1}{3}$$

20. Find the distance D from the point $(2, 1, 0)$ to the line with equations

$$x = -2, \qquad y + 1 = z$$

21. Find the distance D from the origin to the line that contains the point $(-3, -3, 3)$ and is parallel to the vector $2\mathbf{i} - 3\mathbf{j} + 5\mathbf{k}$.

22. Find the distance D between the lines $x - 2y = 1$ and $2x - 4y = 3$ in the xy plane.

23. Find the distance D between the parallel lines

$$\frac{x - 1}{2} = \frac{y + 1}{-1} = \frac{z - 2}{-2} \quad \text{and} \quad \frac{x}{2} = \frac{y - 2}{-1} = \frac{z - 3}{-2}$$

In Exercises 24–26 find an equation of the cylinder.

24. The cylinder with radius 3 whose axis is the x axis.

25. The cylinder with radius $\sqrt{2}$ whose axis is the y axis.

*26. The cylinder with radius 5 whose axis is the line with equations $x = y = z$. (*Hint:* Use (5) to find a formula for the distance from any point (x, y, z) on the cylinder to the axis of the cylinder.)

27. Are $(1, 4, 2)$, $(4, -3, -5)$, and $(-5, -10, -8)$ points on the same line? Explain your answer.

28. Find numbers x and y such that the point $(x, y, 1)$ lies on the line passing through $(2, 5, 7)$ and $(0, 3, 2)$.

29. Let $\mathbf{L}$ be a vector parallel to a given line l. Show that if $\mathbf{a}$ and $\mathbf{b}$ have initial points at the origin and terminal points on l, then $\mathbf{L} \times \mathbf{a} = \mathbf{L} \times \mathbf{b}$ (Figure 11.42).

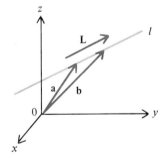

FIGURE 11.42

11.6
PLANES IN SPACE

There is only one plane that contains a given point and is perpendicular to a given line. Similarly, there is only one plane that contains a given point and is perpendicular to a given nonzero vector (Figure 11.43). In other words, a plane is determined by a point and a nonzero vector. For this reason vectors are indispensable to the analysis of planes.

Equations of Planes   Let $P_0 = (x_0, y_0, z_0)$ be a given point and $\mathbf{N} = a\mathbf{i} + b\mathbf{j} + c\mathbf{k}$ a nonzero vector. Then a point $P = (x, y, z)$ lies on the plane $\mathscr{P}$ that contains P_0 and is perpendicular to $\mathbf{N}$ if and only if the vector

$$\overrightarrow{P_0P} = (x - x_0)\mathbf{i} + (y - y_0)\mathbf{j} + (z - z_0)\mathbf{k}$$

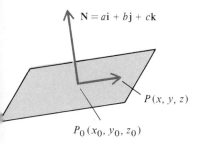

is perpendicular to $\mathbf{N}$ (Figure 11.43). This means simply that $\mathbf{N} \cdot \overrightarrow{P_0P} = 0$, which is equivalent to

$$a(x - x_0) + b(y - y_0) + c(z - z_0) = 0 \tag{1}$$

FIGURE 11.43

By expanding the left side of (1) and letting $d = ax_0 + by_0 + cz_0$, we obtain the equivalent equation

$$ax + by + cz = d \tag{2}$$

Both (1) and (2) are equations of the plane $\mathscr{P}$ that is perpendicular to $\mathbf{N}$ and contains P_0. The vector $\mathbf{N}$ is said to be **normal** to $\mathscr{P}$. Observe that the components of $\mathbf{N}$ appear as coefficients in (1) and (2), and the coordinates of the point P_0 in the plane appear inside the parentheses in (1).

Example 1   Find an equation of the plane that contains the point $(-2, 4, 5)$ and has normal vector $7\mathbf{i} - 6\mathbf{k}$.

Solution   We substitute

$$x_0 = -2, \quad y_0 = 4, \quad z_0 = 5 \quad \text{and} \quad a = 7, \quad b = 0, \quad c = -6$$

into (1) and obtain the equation

$$7(x + 2) + 0(y - 4) - 6(z - 5) = 0$$

Collecting terms, we have

$$7x - 6z = -44 \quad \square$$

Example 2   Find an equation of the plane $\mathscr{P}$ that contains the point $(\frac{1}{2}, 0, 3)$ and is perpendicular to the line l with equations

$$\frac{x + 1}{4} = \frac{y - 2}{-1} = \frac{z}{5}$$

Solution   Since l is perpendicular to $\mathscr{P}$, any vector parallel to l (such as $4\mathbf{i} - \mathbf{j} + 5\mathbf{k}$) is normal to $\mathscr{P}$. Therefore we may let $\mathbf{N} = 4\mathbf{i} - \mathbf{j} + 5\mathbf{k}$, and hence an equation of the plane is

$$4\left(x - \frac{1}{2}\right) + (-1)(y - 0) + 5(z - 3) = 0$$

or

$$4x - y + 5z = 17 \quad \square$$

Example 3   Find a vector normal to the plane $2x - 3y + 7z = -35$.

Solution   One such vector can be read directly from the equation: It is $2\mathbf{i} - 3\mathbf{j} + 7\mathbf{k}$. $\square$

Example 4   Sketch the plane $3x + 2y + z = 6$.

Solution   The plane is determined by any three distinct points on it. The points we will use are the x, y, and z intercepts. To find the x intercept we set y and z equal to 0 and solve for x in the given equation:

$$3x + 2(0) + 0 = 6$$
$$x = 2$$

For the y intercept we set x and z equal to 0 and solve for y:

$$3(0) + 2y + 0 = 6$$
$$y = 3$$

Similarly, to find the z intercept we set x and y equal to 0 and solve for z:

$$3(0) + 2(0) + z = 6$$
$$z = 6$$

Thus the points $(2, 0, 0)$, $(0, 3, 0)$ and $(0, 0, 6)$ are on the plane. With this information we are able to sketch the plane (Figure 11.44). $\square$

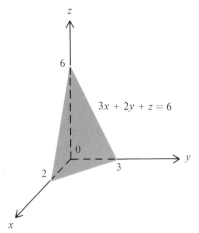

FIGURE 11.44

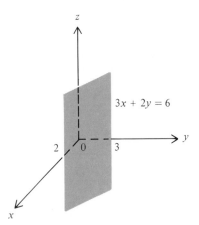

FIGURE 11.45

Example 5   Sketch the plane $3x + 2y = 6$.

Solution   Proceeding as in Example 4, we find that the x intercept is 2 and the y intercept is 3, so that the points $(2, 0, 0)$ and $(0, 3, 0)$ are on the plane. If we try to find a z intercept by setting x and y equal to 0, we obtain

$$3(0) + 2(0) \stackrel{?}{=} 6$$

Thus the plane contains no z intercept. It follows that the plane is perpendicular to the xy plane. Using this fact and the two intercepts already found, we can sketch the plane (Figure 11.45). $\square$

Three noncollinear points P_0, P_1, and P_2 determine a plane. To find a vector normal to the plane, we first form the two vectors $\overrightarrow{P_0 P_1}$ and $\overrightarrow{P_0 P_2}$ and compute their cross product. Since $\overrightarrow{P_0 P_1}$ and $\overrightarrow{P_0 P_2}$ are parallel to the plane, their cross product is perpendicular to both vectors and hence to the plane (Figure 11.46).

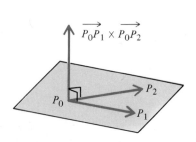

FIGURE 11.46

Example 6   Find an equation of the plane containing the points $P_0 = (1, 0, 2)$, $P_1 = (-1, 3, 4)$, and $P_2 = (3, 5, 7)$.

Solution   First we notice that the vectors $\overrightarrow{P_0P_1} = -2\mathbf{i} + 3\mathbf{j} + 2\mathbf{k}$ and $\overrightarrow{P_0P_2} = 2\mathbf{i} + 5\mathbf{j} + 5\mathbf{k}$ are not parallel (and thus P_0, P_1, and P_2 determine a plane). For a normal vector we can therefore take

$$\mathbf{N} = \overrightarrow{P_0P_1} \times \overrightarrow{P_0P_2} = \begin{vmatrix} \mathbf{i} & \mathbf{j} & \mathbf{k} \\ -2 & 3 & 2 \\ 2 & 5 & 5 \end{vmatrix} = 5\mathbf{i} + 14\mathbf{j} - 16\mathbf{k}$$

Since $\mathbf{N}$ is normal to the plane and $(1, 0, 2)$ lies on the plane, we use (1) to conclude that an equation of the plane is

$$5(x - 1) + 14(y - 0) - 16(z - 2) = 0$$

or more simply,

$$5x + 14y - 16z = -27 \quad \square$$

Distance from a Point to a Plane

Vector methods greatly simplify the calculation of distances between points and planes, just as between points and lines.

THEOREM 11.13

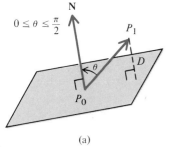

$0 \le \theta \le \dfrac{\pi}{2}$

(a)

Let $\mathscr{P}$ be a plane with normal $\mathbf{N}$, and let P_1 be any point not on $\mathscr{P}$. Then the distance D between P_1 and $\mathscr{P}$ is given by

$$D = \frac{|\mathbf{N} \cdot \overrightarrow{P_0P_1}|}{\|\mathbf{N}\|} \qquad (3)$$

where P_0 is any point on $\mathscr{P}$.

Proof   Let θ be the angle between $\mathbf{N}$ and $\overrightarrow{P_0P_1}$, so that $0 \le \theta \le \pi$. If $0 \le \theta \le \pi/2$, then

$$D = \|\overrightarrow{P_0P_1}\| \cos \theta$$

(Figure 11.47(a)), whereas if $\pi/2 < \theta \le \pi$, then

$$D = \|\overrightarrow{P_0P_1}\| \cos(\pi - \theta) = \|\overrightarrow{P_0P_1}\|(-\cos \theta)$$

(Figure 11.47(b)). It follows that whatever θ is, the distance D is given by

$$D = \|\overrightarrow{P_0P_1}\| |\cos \theta|$$

However

$$|\mathbf{N} \cdot \overrightarrow{P_0P_1}| = \|\mathbf{N}\| \, \|\overrightarrow{P_0P_1}\| \, |\cos \theta|$$

so that

$$D = \|\overrightarrow{P_0P_1}\| |\cos \theta| = \frac{|\mathbf{N} \cdot \overrightarrow{P_0P_1}|}{\|\mathbf{N}\|} \quad \blacksquare$$

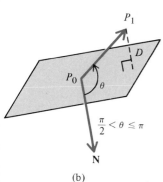

$\dfrac{\pi}{2} < \theta \le \pi$

(b)

FIGURE 11.47

Example 7 Calculate the distance D between the point $P_1 = (-1, 1, 2)$ and the plane

$$3x - 2y + z = 1$$

Solution The equation of the plane yields a vector $\mathbf{N}$ normal to the plane:

$$\mathbf{N} = 3\mathbf{i} - 2\mathbf{j} + \mathbf{k}$$

From the equation of the plane we can also find a point on the plane by letting $x = y = 0$. Then $z = 1$, so that $P_0 = (0, 0, 1)$ is on the plane. Straightforward computations yield

$$\overrightarrow{P_0P_1} = (-1 - 0)\mathbf{i} + (1 - 0)\mathbf{j} + (2 - 1)\mathbf{k} = -\mathbf{i} + \mathbf{j} + \mathbf{k}$$

and

$$\mathbf{N} \cdot \overrightarrow{P_0P_1} = (3)(-1) + (-2)(1) + (1)(1) = -3 - 2 + 1 = -4$$

Therefore (3) implies that

$$D = \frac{|\mathbf{N} \cdot \overrightarrow{P_0P_1}|}{\|\mathbf{N}\|} = \frac{|-4|}{\sqrt{(3)^2 + (-2)^2 + (1)^2}} = \frac{4}{\sqrt{14}} \quad \square$$

EXERCISES 11.6

In Exercises 1–5 find an equation of the plane that contains P_0 and has normal vector $\mathbf{N}$.

1. $P_0 = (-1, 2, 3)$, $\mathbf{N} = -4\mathbf{i} + 15\mathbf{j} - \frac{1}{2}\mathbf{k}$

2. $P_0 = (\pi, 0, -\pi)$, $\mathbf{N} = 2\mathbf{i} + 3\mathbf{j} - 4\mathbf{k}$

3. $P_0 = (9, 17, -7)$, $\mathbf{N} = 2\mathbf{i} - 3\mathbf{k}$

4. $P_0 = (-1, -1, -1)$, $\mathbf{N} = \frac{1}{\sqrt{2}}(\mathbf{i} + \mathbf{j} - \mathbf{k})$

5. $P_0 = (2, 3, -5)$, $\mathbf{N} = \mathbf{j}$

6. Find an equation of the plane that contains the points $(2, -1, 4)$, $(5, 3, 5)$, and $(2, 4, 3)$.

7. Find an equation of the plane that contains the point $(1, -1, 2)$ and the line with symmetric equations

$$x + 2 = y + 1 = \frac{z + 5}{2}$$

8. Find an equation of the plane that contains the two parallel lines

$$\frac{x - 1}{3} = \frac{y + 1}{2} = \frac{z - 5}{4}$$

and

$$\frac{x + 3}{3} = \frac{y - 4}{2} = \frac{z}{4}$$

9. Find an equation of the plane that contains the point $(2, \frac{1}{2}, \frac{1}{3})$ and is perpendicular to the line having parametric equations

$$x = \pi + 2t, \qquad y = 2\pi + 5t, \qquad z = 9t$$

10. Find parametric equations for the line that passes through $(2, -1, 0)$ and is perpendicular to the plane $2x - 3y + 4z = 5$.

11. Let l be the intersection of the two planes

$$2x - 3y + 4z = 2 \quad \text{and} \quad x - z = 1$$

a. Find a vector equation of l.
b. Find an equation of the plane that is perpendicular to l and contains the point $(-9, 12, 14)$.

12. Let l be the line

$$\frac{x + 1}{2} = \frac{y + 3}{3} = -z$$

and let $\mathscr{P}$ be the plane

$$3x - 2y + 4z = -1$$

a. Find the point of intersection P_0 of l and $\mathscr{P}$.
b. Find an equation of the plane perpendicular to l at P_0.
c. Find symmetric equations of the line perpendicular to $\mathscr{P}$ at P_0.

13. Find the distance D between the point $(3, -1, 4)$ and the plane $2x - y + z = 5$.

14. Find the distance D between the point $(2, 0, -4)$ and the plane $x + 2y + 4z - 3 = 0$.

15. Show that the distance D between the origin and the plane $ax + by + cz = d$ is $|d|/\sqrt{a^2 + b^2 + c^2}$.

16. Show that the planes

$$a_1 x + b_1 y + c_1 z = d_1 \quad \text{and} \quad a_2 x + b_2 y + c_2 z = d_2$$

are perpendicular if and only if $a_1 a_2 + b_1 b_2 + c_1 c_2 = 0$.

17. The set of all points equidistant from $(3, 1, 5)$ and $(5, -1, 3)$ is a plane. Find an equation of that plane.

In Exercises 18–21 sketch the plane.

18. $2x + y + z = 4$

19. $-\frac{1}{2}x + \frac{1}{3}y - z = 1$

20. $2x - z = 1$

21. $4y + 3z = 6$

22. Show that an equation of the plane having x intercept a, y intercept b, and z intercept c is

$$\frac{x}{a} + \frac{y}{b} + \frac{z}{c} = 1$$

provided that a, b, and c are all nonzero (Figure 11.48).

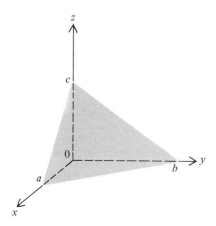

FIGURE 11.48

23. Show that the points $(2, 3, 2)$, $(1, -1, -3)$, $(1, 0, -1)$, and $(5, 9, 5)$ all lie on the same plane.

24. Let l be the line

$$\frac{x - 1}{2} = \frac{y + 1}{3} = \frac{z + 5}{7}$$

and $\mathcal{P}$ be the plane

$$2(x - 1) + 2(y + 3) - z = 0$$

Find the two points on l at a distance 3 from $\mathcal{P}$.

25. Suppose planes $\mathcal{P}_1$ and $\mathcal{P}_2$ intersect, and let $\mathbf{a}$ and $\mathbf{b}$ lie on $\mathcal{P}_1$ and $\mathbf{c}$ and $\mathbf{d}$ on $\mathcal{P}_2$. Show that $(\mathbf{a} \times \mathbf{b}) \times (\mathbf{c} \times \mathbf{d})$ is parallel to the intersection of $\mathcal{P}_1$ and $\mathcal{P}_2$.

26. Which of the following planes are identical, which are parallel to each other, and which are perpendicular?
   a. $x + 2y - 3z = 2$
   b. $15x - 9y + z = 2$
   c. $-2x - 4y + 6z + 4 = 0$
   d. $5x - 3y + \frac{1}{3}z - 1 = 0$

27. Which of the following planes are identical, which are parallel to each other, and which are perpendicular?
   a. $x + y - z + 3 = 0$
   b. $x - y - 2 = 0$
   c. $y - z = 2$
   d. $x + y = -5$

28. Find an equation of the plane that contains the point $(-1, 2, 3)$ and is
   a. parallel to the xy plane.
   b. perpendicular to the x axis.
   c. perpendicular to the y axis.

29. Find an equation of the plane that contains the point $(-4, -5, -3)$ and is
   a. parallel to the yz plane.
   b. parallel to the xz plane.
   c. perpendicular to the z axis.

30. Find an equation of the plane that contains the points $(-2, 1, 4)$ and $(0, 3, 1)$ and contains a line parallel to the vector $2\mathbf{i} - 4\mathbf{j} + 6\mathbf{k}$.

In Exercises 31–34 find all points of intersection of the three planes.

31. $x + y = 1$; $y + z = 2$; $x + z = 3$

32. $x + y - z = 2$; $-x + 2y - z = 3$; $x - y - z = 0$

33. $2x - 3y - z = 1$; $x + 3y + z = -2$; $y + z = 2$

34. $2x + 4y - 6z = 4$; $x - y - z - 2 = 0$;
    $x + 2y - 3z - 2 = 0$

In Exercises 35–37 find the distance D between the pair of parallel planes.

35. $x - y + z = 2$ and $3x - 3y + 3z = 1$

36. $y - 2z = 4$ and $-2y + 4z = 6$

37. $2x - 3y + 4z = 5$ and $4x - 6y + 8z = -1$

 C The angle between two planes $\mathcal{P}_1$ and $\mathcal{P}_2$ is the angle in $[0, \pi/2]$ between vectors normal to $\mathcal{P}_1$ and $\mathcal{P}_2$. In Exercises 38–40 approximate the angle θ.

38. $2x - y + z = 0$ and $x + y + 3z = 1$

39. $y - z = 2$ and $4x - y - 2z = -2$

40. $x + 2y - 3z = 4$ and $6 - 2x + 4y + 6z = 0$

Key Terms and Expressions

Three-dimensional space
Vector
  component
  length
Unit vector

Dot product; cross product
Resolution of a vector
Projection of one vector onto another
Vector, parametric, and symmetric equations of a line
Normal to a plane

Key Formulas

$$|PQ| = \sqrt{(x_1 - x_0)^2 + (y_1 - y_0)^2 + (z_1 - z_0)^2}$$

$$\vec{PQ} = (x_1 - x_0)\mathbf{i} + (y_1 - y_0)\mathbf{j} + (z_1 - z_0)\mathbf{k}$$

$$\|a_1\mathbf{i} + a_2\mathbf{j} + a_3\mathbf{k}\| = \sqrt{a_1^2 + a_2^2 + a_3^2}$$

$$\mathbf{a} \cdot \mathbf{b} = a_1b_1 + a_2b_2 + a_3b_3$$

$$\mathbf{a} \cdot \mathbf{b} = \|\mathbf{a}\|\,\|\mathbf{b}\| \cos \theta$$

$$\mathbf{pr_a b} = \frac{\mathbf{a} \cdot \mathbf{b}}{\|\mathbf{a}\|^2}\mathbf{a}$$

$$\mathbf{a} \times \mathbf{b} = (a_2b_3 - a_3b_2)\mathbf{i} + (a_3b_1 - a_1b_3)\mathbf{j} + (a_1b_2 - a_2b_1)\mathbf{k}$$

$$\|\mathbf{a} \times \mathbf{b}\| = \|\mathbf{a}\|\,\|\mathbf{b}\| \sin \theta$$

$$\mathbf{a} \times (\mathbf{b} \times \mathbf{c}) = \mathbf{b}(\mathbf{a} \cdot \mathbf{c}) - \mathbf{c}(\mathbf{a} \cdot \mathbf{b})$$

Vector equation of a line: $\mathbf{r} = \mathbf{r}_0 + t\mathbf{L}$

Parametric equations of a line:
$$x = x_0 + at, \, y = y_0 + bt, \, z = z_0 + ct$$

Symmetric equations of a line:
$$\frac{x - x_0}{a} = \frac{y - y_0}{b} = \frac{z - z_0}{c} \quad \text{if } a \neq 0, b \neq 0, c \neq 0$$

Distance between point P_1 and a line l parallel to $\mathbf{L}$:
$$\frac{\|\mathbf{L} \times \vec{P_0P_1}\|}{\|\mathbf{L}\|}, \quad \text{where } P_0 \text{ is on } l$$

Equation of a plane: $ax + by + cz = d$

Distance between a point P_1 and a plane $\mathscr{P}$:
$$\frac{|\mathbf{N} \cdot \vec{P_0P_1}|}{\|\mathbf{N}\|}, \quad \text{where } P_0 \text{ is on } \mathscr{P}$$

REVIEW EXERCISES

In Exercises 1–3 find $2\mathbf{a} + \mathbf{b} - 3\mathbf{c}$, $\mathbf{a} \times \mathbf{b}$, $\mathbf{c} \cdot (\mathbf{a} \times \mathbf{b})$, and $\mathbf{a} \times (\mathbf{b} \times \mathbf{c})$.

1. $\mathbf{a} = 2\mathbf{i} - 3\mathbf{j} + \mathbf{k}$, $\mathbf{b} = \mathbf{i} - \mathbf{j}$, $\mathbf{c} = \mathbf{j} - 3\mathbf{k}$

2. $\mathbf{a} = \frac{1}{2}\mathbf{i} - \mathbf{j} + 2\mathbf{k}$, $\mathbf{b} = 2\mathbf{i} - 4\mathbf{j} + 6\mathbf{k}$, $\mathbf{c} = \mathbf{i} - 5\mathbf{j} + 6\mathbf{k}$

3. $\mathbf{a} = 3\mathbf{i} - 2\mathbf{j} + \mathbf{k}$, $\mathbf{b} = 5\mathbf{i} - 2\mathbf{j} + \mathbf{k}$, $\mathbf{c} = \mathbf{j} - \mathbf{k}$

4. Find the cosine of the angle between the vectors $3\mathbf{i} - 4\mathbf{j} + 12\mathbf{k}$ and $\mathbf{i} - \mathbf{k}$.

5. Let $P = (1, -2, 3)$ and let $\mathbf{a} = 2\mathbf{i} - 2\mathbf{j} + \mathbf{k}$. Find a point Q such that $\vec{PQ}$ and $\mathbf{a}$ are the same vector.

6. Let $\mathbf{a} = 2\mathbf{i} - 3\mathbf{j} + 5\mathbf{k}$ and $\mathbf{b} = 5\mathbf{i} + 3\mathbf{j} - 7\mathbf{k}$.
   a. Find $\mathbf{a} \cdot \mathbf{b}$.
   b. Find $\mathbf{a} \times \mathbf{b}$.
   c. Find $\mathbf{pr_a b}$.

7. Resolve the vector $2\mathbf{i} - \mathbf{j} - \mathbf{k}$ into vectors parallel to $2\mathbf{j} + \mathbf{k}$ and $-20\mathbf{i} - 2\mathbf{j} + 4\mathbf{k}$. (Note that all three vectors lie in the same plane.)

8. Find the area A of the triangle with vertices $(1, 1, 1)$, $(2, 3, 5)$, and $(-1, 3, 1)$.

9. Show that $(\frac{1}{2}, \frac{1}{3}, 0)$, $(1, 1, -1)$, and $(-2, -3, 5)$ are collinear points, and find symmetric equations for the line containing them.

10. Let $P = (2, 5, -7)$ and $Q = (4, 3, 8)$.
    a. Find $\vec{PQ}$.
    b. Find $\|\vec{PQ}\|$.
    c. For the line l through P and Q, find
       i. a vector equation of l.
       ii. parametric equations of l.
       iii. symmetric equations of l.

11. Find a vector equation for the line that contains $(-3, -3, 1)$ and is perpendicular to the plane $2x - 3y + 4z = 7$.

12. Find an equation for the plane that contains the point $(-1, 3, 2)$ and has normal $\mathbf{N} = 2\mathbf{i} + \mathbf{j} - \mathbf{k}$.

13. Show that the points $(-1, 1, 1)$, $(0, 2, 1)$, $(0, 0, \frac{3}{2})$, and $(13, -1, 5)$ lie on the same plane.

14. Find an equation of the plane that contains the points $(1, 0, -1)$, $(-5, 3, 2)$, and $(2, -1, 4)$.

15. Find an equation of the plane that is parallel to the z axis and contains the points $(3, -1, 5)$ and $(7, 9, 4)$. (A plane is parallel to a line if any vector normal to the plane is perpendicular to the line.)

16. Show that the line

$$\frac{x+5}{7} = \frac{y-11}{9} = \frac{z}{45}$$

is parallel to the plane with equation $9x - 2y - z = 0$.

17. Find the distance D from the point $(1, -2, 5)$ to the plane $3(x-1) - 4(y+2) + 12z = 0$.

18. Find the distance D from the point $(1, -2, 5)$ to the line

$$x = 1 + 3t, \qquad y = -2 - 4t, \qquad z = 12t$$

19. In each of parts (a)–(c) determine whether or not the given planes have either a point or a line in common. If they do, then find the point or the line.
    a. $3x - y + z = 2$, $y = 4$, $z = 1$
    b. $2x + y - 2z = 1$, $3x + y - z = 2$, $x - y + z = 0$
    c. $2x - 11y + 6z = -2$, $2x - 3y + 2z = 2$, $2x - 9y + 5z = -1$

20. Suppose $a \neq 0$, $b \neq 0$, and $c \neq 0$. Find symmetric equations for the line that contains the point (a, b, c) and is parallel to the vector $a\mathbf{i} + b\mathbf{j} + c\mathbf{k}$. Show that the origin lies on the line.

21. Suppose the point (a, b, c) is not the origin. Find an equation of the plane that contains the point (a, b, c) and is perpendicular to $a\mathbf{i} + b\mathbf{j} + c\mathbf{k}$.

*22. Let $\mathscr{P}_0$ and $\mathscr{P}_1$ be the two parallel planes given respectively by $2x - 3y + 4z = 2$ and $2x - 3y + 4z = 6$. Let l be the line with symmetric equations

$$\frac{x-5}{3} = \frac{y+3}{2} = \frac{z}{6}$$

   a. Without finding the points of intersection, show that l intersects both $\mathscr{P}_0$ and $\mathscr{P}_1$.
   b. Assuming that the points of intersection are Q_0 and Q_1, determine $|Q_0 Q_1|$ without finding either Q_0 or Q_1. (*Hint:* First find the distance between the two planes.)

23. Use vectors to show that any angle inscribed in a semicircle is a right angle.

24. Let $\mathbf{a}$, $\mathbf{b}$, and $\mathbf{c}$ be the vectors pointing from the three vertices of a triangle to the midpoints of the opposite sides. Show that $\mathbf{a} + \mathbf{b} + \mathbf{c} = \mathbf{0}$.

25. Let $\mathbf{a}$ and $\mathbf{b}$ be nonzero vectors.
    a. Show that if $\mathbf{a} \cdot \mathbf{c} = \mathbf{b} \cdot \mathbf{c}$ for every vector $\mathbf{c}$, then $\mathbf{a} = \mathbf{b}$. (*Hint:* Take $\mathbf{c} = \mathbf{a} - \mathbf{b}$, and show that $(\mathbf{a} - \mathbf{b}) \cdot (\mathbf{a} - \mathbf{b}) = 0$.)
    b. Show that if $\mathbf{a} \times \mathbf{c} = \mathbf{b} \times \mathbf{c}$ for every vector $\mathbf{c}$, then $\mathbf{a} = \mathbf{b}$. (*Hint:* Take $\mathbf{c}$ to be a nonzero vector perpendicular to $\mathbf{a} - \mathbf{b}$.)

26. Prove that two nonzero vectors $\mathbf{a}$ and $\mathbf{b}$ are perpendicular if and only if $\|\mathbf{a}\| \leq \|\mathbf{a} + c\mathbf{b}\|$ for every number c. (*Hint:* $\|\mathbf{a}\|^2 = \mathbf{a} \cdot \mathbf{a}$ and $\|\mathbf{a} + c\mathbf{b}\|^2 = (\mathbf{a} + c\mathbf{b}) \cdot (\mathbf{a} + c\mathbf{b})$.)

27. Using the dot product, prove the trigonometric identity

$$\cos(x + y) = \cos x \cos y - \sin x \sin y$$

   (*Hint:* Use the unit vectors $\mathbf{a} = \cos x\mathbf{i} + \sin x\mathbf{j}$ and $\mathbf{b} = \cos y\mathbf{i} - \sin y\mathbf{j}$.)

28. a. Show that

$$\|\mathbf{a} + \mathbf{b}\|^2 \|\mathbf{a} - \mathbf{b}\|^2 = (\|\mathbf{a}\|^2 + \|\mathbf{b}\|^2)^2 - 4(\mathbf{a} \cdot \mathbf{b})^2$$

   (*Hint:* $\|\mathbf{a} + \mathbf{b}\|^2 = (\mathbf{a} + \mathbf{b}) \cdot (\mathbf{a} + \mathbf{b})$ and $\|\mathbf{a} - \mathbf{b}\|^2 = (\mathbf{a} - \mathbf{b}) \cdot (\mathbf{a} - \mathbf{b})$)
   b. Use part (a) to show that

$$\|\mathbf{a} + \mathbf{b}\| \|\mathbf{a} - \mathbf{b}\| \leq \|\mathbf{a}\|^2 + \|\mathbf{b}\|^2$$

29. Suppose that $\mathbf{a}$, $\mathbf{b}$, and $\mathbf{c}$ are mutually perpendicular unit vectors and that $\mathbf{d} = a\mathbf{a} + b\mathbf{b} + c\mathbf{c}$. Show that

$$\|\mathbf{d}\| = \sqrt{a^2 + b^2 + c^2}$$

   (*Hint:* $\|\mathbf{d}\|^2 = \mathbf{d} \cdot \mathbf{d}$.)

30. Let $\mathbf{u}$ be a unit vector, ϕ the angle between $\mathbf{u}$ and the positive z axis, and θ as in Figure 11.49. Show that

$$\mathbf{u} = \cos\theta \sin\phi\mathbf{i} + \sin\theta \sin\phi\mathbf{j} + \cos\phi\mathbf{k}$$

   (The angles ϕ and θ will play a major role in Section 14.6.)

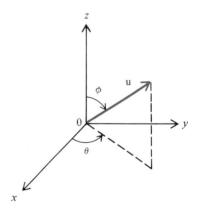

FIGURE 11.49

31. Suppose that forces of 500 pounds and 300 pounds are applied at the same point and that the angle between the

two forces is $\pi/3$. Find the magnitude of the resultant force and the cosine of the angle θ the resultant force makes with respect to the 500 pound force.

32. To provide traction for a broken leg, a horizontal force of 5 pounds is to be obtained by means of a 5-pound weight as shown in Figure 11.50. What angle θ should be chosen to accomplish this?

33. A river $\frac{1}{2}$ mile wide flows parallel to the shore at the rate of 5 miles per hour. If a motorboat can move at 10 miles per hour in still water, at what angle with respect to the shore should the boat be pointed in order to travel perpendicular to the shore? How long will it take to cross the river?

34. A jet pointed north travels with a motor speed of 500 miles per hour. The jet stream is clocked at 100 miles per hour and flows east. What is the ground speed of the jet?

35. Suppose particles whose charges are 3.2×10^{-19}, -6.4×10^{-19}, and 4.8×10^{-19} are located, respectively, at $(10^{-12}, 0, 0)$, $(0, 2 \times 10^{-12}, 0)$, and $(0, 0, 3 \times 10^{-12})$. Find the total electric force **F** exerted by the three charged particles on a positive unit charge located at the origin.

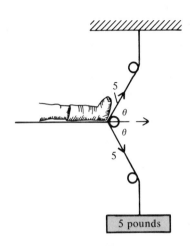

FIGURE 11.50

Cumulative Review, Chapters 1–10

1. Let $f(x) = \sqrt{\dfrac{1-2x}{1-3x}}$.
   a. Find the domain of f.
   b. Find $\lim_{x \to -\infty} f(x)$.

In Exercises 2–3 find the limit.

2. $\lim_{x \to 0^+} x(\ln x)^3$

3. $\lim_{x \to \infty} x^{\tan(1/x)}$

4. Let $f(x) = (1-2x)/(1+3x)$. Find all values of x for which $f'(x) = -20$.

5. Let $f(x) = \ln(\pi/2 + \arctan x)$.
   a. Show that f^{-1} exists, and determine the domain and range of f^{-1}.
   b. Find $(f^{-1})'(\ln(\pi/3))$.
   c. Find a formula for f^{-1}.

6. Recall that the line normal to the graph of a function f at $(a, f(a))$ is perpendicular to the line tangent at $(a, f(a))$. Find an equation for the line in the xy plane normal to the graph of $y = \sin x$ at $(\pi/4, \sqrt{2}/2)$.

7. Consider the portion of the cycloid parametrized by $x = t - \sin t$, $y = 1 - \cos t$ for $0 < t < \pi$.
   a. Use the fact that
   $$\frac{dy}{dx} = \frac{dy}{dt} \frac{1}{\frac{dx}{dt}}$$

to show that
   $$\frac{dy}{dx} = \frac{\sin t}{1 - \cos t} = \frac{\sqrt{2y - y^2}}{y}$$
   b. Using the parametric equations, show that $x = \arccos(1 - y) - \sqrt{2y - y^2}$.
   c. Use implicit differentiation on the equation of part (b) to show that $dy/dx = \sqrt{2y - y^2}/y$.

8. At the moment that the sun's elevation is $30°$, a ball is dropped from the top of a pole 96 feet tall. Assuming no air friction and an unimpaired shadow from the sun, determine how fast the shadow is moving
   a. after 1 second
   b. the moment the ball hits the ground

9. Let $f(x) = x(x^2 - 5)^2$. Sketch the graph of f, indicating all pertinent information.

10. A window with a perimeter of 16 feet has the shape of a rectangle with an isosceles right triangle attached to the top. Assume that the hypotenuse of the triangle is the top side of the rectangular part of the window. Show that the maximum amount of light enters the window if the length of each leg of the triangle equals the length of the vertical sides of the rectangle.

11. Let $f(x) = x + 1/x$. Find the area A of the region between the graph of f and the line $y = \frac{5}{2}$ on $[1, 3]$.

In Exercises 12–14 evaluate the integral.

12. $\displaystyle\int \frac{1}{t^2}\csc\frac{1}{t}\tan^2\frac{1}{t}\,dt$

13. $\displaystyle\int \frac{x^3}{x^2 - x + 1}\,dx$

14. $\displaystyle\int \frac{1}{x^6\sqrt{2x^2 - 1}}\,dx$

15. Determine whether $\int_2^\infty xe^{-3x^2}\,dx$ converges or diverges. If it converges, determine its value.

16. Let $f(x) = x^2/\sqrt{x^2 + 9}$ for $0 \le x \le 3$, and let R be the region between the graph of f and the x axis. Find the volume V of the solid obtained by revolving R about the y axis.

17. Let C be described parametrically by $x = e^{-t}\sin 2t$ and $y = e^{-t}\cos 2t$ for $-1 \le t \le 1$. Find the length $\mathscr{L}$ of C.

18. Find the area A of the region common to the cardioids $r = 1 + \cos\theta$ and $r = 1 - \cos\theta$.

19. Find all values of c for which $\lim_{n \to \infty}(\sqrt{n + c} - \sqrt{n})$ exists as a real number.

20. Find the numerical value of $\displaystyle\sum_{n=1}^\infty \frac{4}{n(n + 2)}$.

In Exercises 21–22 determine whether the series converges absolutely, converges conditionally, or diverges.

21. $\displaystyle\sum_{n=0}^\infty (-1)^n \frac{n^2}{\sqrt{n^4 + 2}}$

22. $\displaystyle\sum_{n=1}^\infty \frac{2\cdot 4\cdot 6\cdots(2n - 2)(2n)}{n^n}$

23. Let $f(x) = \sum_{n=1}^\infty x^n/(n + 2)$. Find the Taylor series about 0 for $\int_0^x f(t)\,dt$, and determine the interval of convergence for the Taylor series.

12

VECTOR-VALUED FUNCTIONS

Until now we have studied functions whose domains and ranges both consist of real numbers. In this chapter we will investigate a new type of function, whose domain consists of real numbers and whose range consists of vectors. Such functions, which are called vector-valued functions, have many applications. For example, a convenient way to describe the motion of a satellite is by means of a vector-valued function whose domain is an interval representing time and whose value at any time t is a vector that represents the position of the satellite at time t.

12.1
DEFINITIONS AND EXAMPLES

A correspondence between one set of numbers and another is a function. In the study of motion we frequently encounter a correspondence between a set of numbers and a set of vectors. Such an association determines a vector-valued function.

DEFINITION 12.1

A **vector-valued function** consists of two parts: a **domain**, which is a collection of numbers, and a **rule**, which assigns to each number in the domain one and only one vector.

The numbers in the domain of a vector-valued function will usually be denoted t; the reason is that in most applications the domain of such a function

662

will represent an interval of time. The vector-valued functions themselves will normally be denoted by the boldface letters **F**, **G**, and **H**. The total collection of vectors assigned by a vector-valued function to members of its domain is called the **range** of the function. Unless otherwise indicated, when a vector-valued function is defined by a formula, the domain consists of all numbers for which the formula is meaningful. For instance, vector-valued functions are defined by the formulas

$$\mathbf{F}(t) = (2 + 3t)\mathbf{i} + (-1 + t)\mathbf{j} + 2\mathbf{k}$$

$$\mathbf{G}(t) = \mathbf{i} + \sqrt{1 + t}\,\mathbf{j} + t\mathbf{k}$$

$$\mathbf{H}(t) = a(t - \sin t)\mathbf{i} + a(1 - \cos t)\mathbf{j} \quad \text{for } 0 \le t \le 4\pi$$

and the domains of **F**, **G**, and **H** are $(-\infty, \infty)$, $[-1, \infty)$, and $[0, 4\pi]$, respectively.

From now on we will refer to any function whose domain and range are sets of real numbers as a **real-valued function**, in order to distinguish it from a vector-valued function. Every vector-valued function **F** corresponds to three real-valued functions f_1, f_2, and f_3 in the following way: for each number t in the domain of **F**, let $f_1(t)$, $f_2(t)$, and $f_3(t)$ denote the **i**, **j**, and **k** components, respectively, of **F**(t). Then the domain of each of the functions f_1, f_2, and f_3 is the same as the domain of **F**, and

$$\mathbf{F}(t) = f_1(t)\mathbf{i} + f_2(t)\mathbf{j} + f_3(t)\mathbf{k} \quad \text{for } t \text{ in the domain of } \mathbf{F}$$

The functions f_1, f_2, and f_3 are the **component functions** of **F**.

Example 1   Let

$$\mathbf{F}(t) = \ln t\,\mathbf{i} + \sqrt{1 - t}\,\mathbf{j} + t^4\mathbf{k}$$

Determine the domain of **F** and its component functions.

Solution   The domain of **F** consists of those values of t for which $\ln t$, $\sqrt{1 - t}$, and t^4 are all defined, which is the interval $(0, 1]$. By observation we see that

$$f_1(t) = \ln t, \qquad f_2(t) = \sqrt{1 - t}, \quad \text{and} \quad f_3(t) = t^4 \quad \square$$

Usually it is impossible to draw the graph of a vector-valued function, because in theory we would need four dimensions to do so (one dimension for the domain and three dimensions for the range). However, we can depict a vector-valued function **F** by drawing only its range. If we think of **F**(t) as a point in space, then as t increases, **F**(t) traces out a curve C in space (Figure 12.1). Thus the range of **F** is associated with the curve C. If

$$\mathbf{F}(t) = f_1(t)\mathbf{i} + f_2(t)\mathbf{j} + f_3(t)\mathbf{k} \quad \text{for } t \text{ in the domain of } \mathbf{F}$$

then we can describe C by the parametric equations

$$x = f_1(t), \qquad y = f_2(t), \qquad z = f_3(t)$$

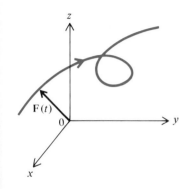

FIGURE 12.1

Thus a curve in space can be represented either by a vector-valued function or by a set of parametric equations.

The arrow on the curve C in Figure 12.1 indicates the direction in which the curve is traced out as t increases.

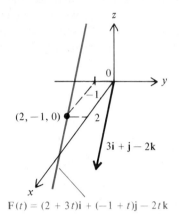

$F(t) = (2 + 3t)\mathbf{i} + (-1 + t)\mathbf{j} - 2t\mathbf{k}$

FIGURE 12.2

Example 2 Let

$$\mathbf{F}(t) = (2 + 3t)\mathbf{i} + (-1 + t)\mathbf{j} - 2t\mathbf{k}$$

Sketch the curve traced out by $\mathbf{F}$.

Solution If we let (x, y, z) be the point on the curve corresponding to $\mathbf{F}(t)$, then

$$x = 2 + 3t, \qquad y = -1 + t, \quad \text{and} \quad z = -2t$$

From (3) of Section 11.5 we recognize these as parametric equations of the line containing the point $(2, -1, 0)$ and parallel to the vector $3\mathbf{i} + \mathbf{j} - 2\mathbf{k}$ (Figure 12.2). □

More generally, let

$$\mathbf{F}(t) = (x_0 + at)\mathbf{i} + (y_0 + bt)\mathbf{j} + (z_0 + ct)\mathbf{k}$$

where $x_0, y_0, z_0, a, b,$ and c are constants and $a, b,$ and c are not all 0. An analysis similar to that in Example 2 shows that the curve traced out by $\mathbf{F}$ is the straight line passing through the point (x_0, y_0, z_0) and parallel to the vector $a\mathbf{i} + b\mathbf{j} + c\mathbf{k}$. For that reason $\mathbf{F}$ is called a **linear function**.

Example 3 Let

$$\mathbf{F}(t) = \cos t\mathbf{i} + \sin t\mathbf{j}$$

Show that $\mathbf{F}$ traces out, in the counterclockwise direction, the unit circle in the xy plane with center at the origin.

Solution The curve traced out by $\mathbf{F}$ lies in the xy plane because the $\mathbf{k}$ component of $\mathbf{F}$ is 0. Since

$$\|\mathbf{F}(t)\| = \sqrt{\cos^2 t + \sin^2 t} = 1 \quad \text{for all } t$$

it follows that the curve lies on the unit circle. Moreover, since $\mathbf{F}(t)$ makes an angle of t radians with the positive x axis and t assumes all real values (Figure 12.3), every point on the circle is $\mathbf{F}(t)$ for some value of t. Consequently the whole circle is traced out by $\mathbf{F}$. Finally, as t increases, the vector $\mathbf{F}(t)$ moves counterclockwise around the circle. □

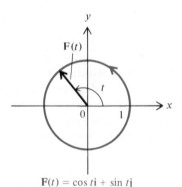

$F(t) = \cos t\mathbf{i} + \sin t\mathbf{j}$

FIGURE 12.3

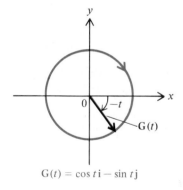

$G(t) = \cos t\,\mathbf{i} - \sin t\,\mathbf{j}$

FIGURE 12.4

From now on we will refer to the circle described in Example 3 as the **standard unit circle**. It will appear several times in this chapter.

The unit circle is traced out in the clockwise direction by the function **G**, where

$$\mathbf{G}(t) = \cos t\,\mathbf{i} - \sin t\,\mathbf{j} \tag{1}$$

(Figure 12.4).

Example 4 Let

$$\mathbf{F}(t) = \cos t\,\mathbf{i} + \cos t\,\mathbf{j} + \sqrt{2}\sin t\,\mathbf{k}$$

Sketch the curve traced out by **F**.

Solution This curve is also a circle. To see this, let (x, y, z) be the point on the curve corresponding to $\mathbf{F}(t)$. Then

$$x = \cos t, \qquad y = \cos t, \quad \text{and} \quad z = \sqrt{2}\sin t$$

From these equations we find that

$$x^2 + y^2 + z^2 = \cos^2 t + \cos^2 t + 2\sin^2 t = 2$$

and also $\qquad\qquad\qquad\qquad x = y$

Therefore the point (x, y, z) lies on a sphere of radius $\sqrt{2}$ and on the plane $x = y$. It follows that the curve lies on a circle, and again it can be shown that every point on the circle is actually traced out by **F** (Figure 12.5). □

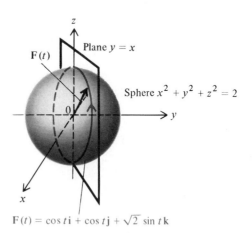

$\mathbf{F}(t) = \cos t\,\mathbf{i} + \cos t\,\mathbf{j} + \sqrt{2}\,\sin t\,\mathbf{k}$

FIGURE 12.5

Combinations of Vector-Valued Functions

Before sketching additional curves traced out by vector-valued functions, we define several combinations of vector-valued functions.

DEFINITION 12.2

Let **F** and **G** be vector-valued functions and f and g real-valued functions. Then the functions $\mathbf{F} + \mathbf{G}, \mathbf{F} - \mathbf{G}, f\mathbf{F}, \mathbf{F} \cdot \mathbf{G}, \mathbf{F} \times \mathbf{G}$, and $\mathbf{F} \circ g$ are defined by

$$(\mathbf{F} + \mathbf{G})(t) = \mathbf{F}(t) + \mathbf{G}(t) \qquad (\mathbf{F} \cdot \mathbf{G})(t) = \mathbf{F}(t) \cdot \mathbf{G}(t)$$

$$(\mathbf{F} - \mathbf{G})(t) = \mathbf{F}(t) - \mathbf{G}(t) \qquad (\mathbf{F} \times \mathbf{G})(t) = \mathbf{F}(t) \times \mathbf{G}(t)$$

$$(f\mathbf{F})(t) = f(t)\mathbf{F}(t) \qquad (\mathbf{F} \circ g)(t) = \mathbf{F}(g(t))$$

Example 5   Let

$$\mathbf{F}(t) = \cos t\,\mathbf{i} + \sin t\,\mathbf{j} + t\mathbf{k} \quad \text{and} \quad \mathbf{G}(t) = -\sin t\,\mathbf{i} + \cos t\,\mathbf{j} + t\mathbf{k}$$

If $g(t) = \sqrt{t}$, find $\mathbf{F} + \mathbf{G}, \mathbf{F} - \mathbf{G}, g\mathbf{F}, \mathbf{F} \cdot \mathbf{G}, \mathbf{F} \times \mathbf{G}$, and $\mathbf{F} \circ g$.

Solution   From the definitions we find that

$$(\mathbf{F} + \mathbf{G})(t) = (\cos t - \sin t)\mathbf{i} + (\sin t + \cos t)\mathbf{j} + 2t\mathbf{k}$$

$$(\mathbf{F} - \mathbf{G})(t) = (\cos t + \sin t)\mathbf{i} + (\sin t - \cos t)\mathbf{j}$$

$$(g\mathbf{F})(t) = \sqrt{t} \cos t\,\mathbf{i} + \sqrt{t} \sin t\,\mathbf{j} + t^{3/2}\mathbf{k}$$

$$(\mathbf{F} \cdot \mathbf{G})(t) = -\cos t \sin t + \sin t \cos t + t^2 = t^2$$

$$(\mathbf{F} \times \mathbf{G})(t) = t(\sin t - \cos t)\mathbf{i} - t(\sin t + \cos t)\mathbf{j} + \mathbf{k}$$

$$(\mathbf{F} \circ g)(t) = \mathbf{F}(g(t)) = \cos \sqrt{t}\,\mathbf{i} + \sin \sqrt{t}\,\mathbf{j} + \sqrt{t}\,\mathbf{k}$$

The domain of each function except $g\mathbf{F}$ and $\mathbf{F} \circ g$ consists of all real numbers; the domains of $g\mathbf{F}$ and $\mathbf{F} \circ g$ consist of $[0, \infty)$, since this is the domain of g.   ☐

In the next example we decompose a vector-valued function into the sum of two other vector-valued functions in order to draw the curve traced out.

Example 6   Let

$$\mathbf{H}(t) = \cos t\,\mathbf{i} + \sin t\,\mathbf{j} + t\mathbf{k}$$

Sketch the curve traced out by **H**.

Solution   If we let

$$\mathbf{F}(t) = \cos t\,\mathbf{i} + \sin t\,\mathbf{j} \quad \text{and} \quad \mathbf{G}(t) = t\mathbf{k}$$

then we have

$$\mathbf{H}(t) = \mathbf{F}(t) + \mathbf{G}(t) = \mathbf{F}(t) + t\mathbf{k}$$

so that the point corresponding to $\mathbf{H}(t)$ lies $|t|$ units directly above or below the point corresponding to $\mathbf{F}(t)$. Since we already know that the curve traced out by **F** is the standard unit circle (see Figure 12.3), the curve traced out by **H** is as shown in Figure 12.6.   ☐

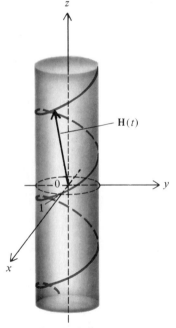

Circular helix
$\mathbf{H}(t) = \cos t\,\mathbf{i} + \sin t\,\mathbf{j} + t\mathbf{k}$

FIGURE 12.6

The curve described in Example 6 is called a **circular helix**. Notice that it lies on the circular cylinder $x^2 + y^2 = 1$.

For our final example we turn to the **cycloid**, the curve traced out by a point on a circle as the circle rolls along a line. We initially discussed this curve in Section 6.4.

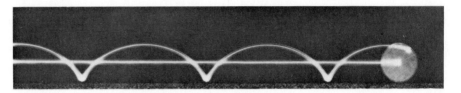

The path traced out by a point on the rim of a rolling wheel is a cycloid. For this photograph, taken with the camera shutter held open, one light was attached to the rim of the wheel and another light to its center.

Example 7   A circle of radius r rolls along the x axis in the positive direction, rotating at the rate of 1 radian per unit of time. Let P be the point on the circle that is at the origin at time 0. Find a vector-valued function $\mathbf{F}$ that traces out the curve described by P.

Solution   As can be seen from Figure 12.7, the vector $\mathbf{F}(t)$ from the origin to P at time t can be written as $\mathbf{a} + \mathbf{b}$, where $\mathbf{a}$ points from the origin to the center C of the circle, and $\mathbf{b}$ points from C to P. Since the circle rotates t radians in t units of time and has radius r, the circle travels a distance rt along the x axis in t units of time (see (1) of Section 1.7). Thus the $\mathbf{i}$ component of $\mathbf{a}$ is rt. Since C is r units from the x axis, the $\mathbf{j}$ component of $\mathbf{a}$ is r. Consequently $\mathbf{a} = rt\mathbf{i} + r\mathbf{j}$. In order to express $\mathbf{b}$ in terms of t, we first use the hypothesis that the circle rotates at the rate of 1 radian per unit of time, so that t is the number of radians through which $\overrightarrow{CP}$ has rotated since time 0. Therefore at time t the vector $\overrightarrow{CP}$ makes an angle of $(3\pi/2 - t)$ radians with the positive x axis (Figure 12.8). Since the length

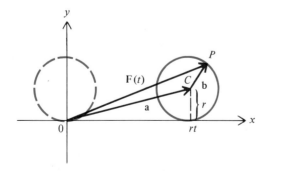

FIGURE 12.7

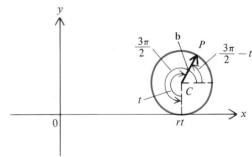

FIGURE 12.8

of **b** is the radius r of the circle, we conclude that $\mathbf{b} = \overrightarrow{CP} = r\cos(3\pi/2 - t)\mathbf{i} + r\sin(3\pi/2 - t)\mathbf{j}$. Consequently

$$\mathbf{F}(t) = \mathbf{a} + \mathbf{b} = (rt\mathbf{i} + r\mathbf{j}) + r\cos\left(\frac{3}{2}\pi - t\right)\mathbf{i} + r\sin\left(\frac{3}{2}\pi - t\right)\mathbf{j}$$

$$= r(t - \sin t)\mathbf{i} + r(1 - \cos t)\mathbf{j} \quad \square$$

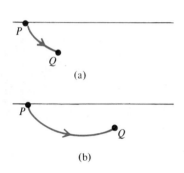

(a)

(b)

FIGURE 12.9

Although the usefulness of the cycloid was noticed by Galileo, it was Johann Bernoulli (1667–1748) who made the cycloid famous by his solution to the **brachistochrone*** (shortest time) **problem**. This is the problem of finding the curve along which a particle subject only to the force of gravity should slide, in order to go from a given point P to a lower point Q in the shortest possible time. Bernoulli showed that the curve requiring minimal time is an arc of a cycloid (Figure 12.9). This is a surprising result, because for certain locations of Q the minimal time is achieved by sliding to a point on the cycloid lower than Q and moving upward to Q (Figure 12.9(b)).

* **Brachistochrone:** Pronounced "bra-*kist*-o-krone."

EXERCISES 12.1

In Exercises 1–12 determine the domain and the component functions of the given function.

1. $\mathbf{F}(t) = t\mathbf{i} + t^2\mathbf{j} + t^3\mathbf{k}$

2. $\mathbf{F}(t) = \sqrt{t + 1}\mathbf{i} + \sqrt{1 - t}\mathbf{j} + \mathbf{k}$

3. $\mathbf{F}(t) = \tanh t\mathbf{i} - \dfrac{1}{t^2 - 4}\mathbf{k}$

4. $\mathbf{F}(t) = [(t^2 - 1)\mathbf{i} + \ln t\mathbf{j} + \cot t\mathbf{k}]$
   $$\times \left[(4 - t^2)\mathbf{i} + e^{-5t}\mathbf{j} + \frac{1}{t}\mathbf{k}\right]$$

5. $\mathbf{F}(t) = (t\mathbf{i} + \mathbf{j}) \times (\mathbf{i} - t^2\mathbf{j} + 2\sqrt{t}\mathbf{k})$

6. $\mathbf{F} - \mathbf{G}$, where $\mathbf{F}(t) = 2t\mathbf{i} + t^2\mathbf{j} - \ln t\mathbf{k}$ and $\mathbf{G}(t) = e^t\mathbf{i} + e^{-t}\mathbf{j} + 2t\mathbf{k}$

7. $2\mathbf{F} - 3\mathbf{G}$, where $\mathbf{F}(t) = t\mathbf{i} + t^2\mathbf{j} + t^3\mathbf{k}$ and $\mathbf{G}(t) = \cos t\mathbf{i} + \sin t\mathbf{j} + \mathbf{k}$

8. $\mathbf{F} \times \mathbf{G}$, where $\mathbf{F}(t) = t\mathbf{i} + t^2\mathbf{j} + t^3\mathbf{k}$ and $\mathbf{G}(t) = \cos t\mathbf{i} + \sin t\mathbf{j} + \mathbf{k}$

9. $\mathbf{F} \times \mathbf{G}$, where $\mathbf{F}(t) = t\mathbf{j} - (1/\sqrt{t})\mathbf{k}$ and $\mathbf{G}(t) = (t - \sin t)\mathbf{i} + (1 - \cos t)\mathbf{j}$

10. $f\mathbf{F}$, where $\mathbf{F}(t) = \ln t\mathbf{i} - 4e^{2t}\mathbf{j} + (\sqrt{t - 1}/t)\mathbf{k}$ and $f(t) = \sqrt{t}$

11. $\mathbf{F} \circ g$, where $\mathbf{F}(t) = \cos t\mathbf{i} + \sin t\mathbf{j} + \sqrt{t + 2}\mathbf{k}$ and $g(t) = t^{1/3}$

12. $\mathbf{F} \circ g$, where $\mathbf{F}(t) = e^{-2t}\mathbf{i} + e^{(t^2)}\mathbf{j} + t^3\mathbf{k}$ and $g(t) = \ln t$

In Exercises 13–26 sketch the curve traced out by the vector-valued function. Indicate the direction in which the curve is traced out.

13. $\mathbf{F}(t) = t\mathbf{i}$ for $-1 \le t \le \frac{1}{2}$

14. $\mathbf{F}(t) = \cos \pi t\mathbf{k}$ for $-1 \le t \le \frac{1}{3}$

15. $\mathbf{F}(t) = t\mathbf{i} + t\mathbf{j} + t\mathbf{k}$

16. $\mathbf{F}(t) = 2t\mathbf{i} - 3t\mathbf{j} + \mathbf{k}$

17. $\mathbf{F}(t) = (2t - 1)\mathbf{i} + (t + 1)\mathbf{j} + 3t\mathbf{k}$

18. $\mathbf{F}(t) = (1 - t)\mathbf{i} + (3t - \frac{1}{2})\mathbf{j} - (-4 + t)\mathbf{k}$

19. $\mathbf{F}(t) = -16t^2\mathbf{k}$ for $t \ge 0$

20. $\mathbf{F}(t) = t\mathbf{j} + t^2\mathbf{k}$

21. $\mathbf{F}(t) = \cos t\mathbf{i} + \sin t\mathbf{j}$ for $0 \le t \le \pi/2$

22. $\mathbf{F}(t) = \cos 3t\mathbf{i} + \sin 3t\mathbf{j}$ for $0 \le t \le \pi/2$

23. $\mathbf{F}(t) = 2\cos t\mathbf{i} - \sin t\mathbf{j} - 3\mathbf{k}$ for $-\pi \le t \le 0$

24. $\mathbf{F}(t) = \cos t\mathbf{i} + \sin t\mathbf{j} + t^2\mathbf{k}$

25. $\mathbf{F}(t) = 3\sin t\mathbf{i} + 3\sin t\mathbf{j} - 3\sqrt{2}\cos t\mathbf{k}$

26. $\mathbf{F}(t) = \sqrt{2}\cos t\mathbf{i} - 2\sin t\mathbf{j} + \sqrt{2}\cos t\mathbf{k}$

27. Let $\mathbf{F}(t) = 2t\mathbf{i} + (t + 1)\mathbf{j} - 3t\mathbf{k}$,
    $\mathbf{G}(t) = (t + 1)\mathbf{i} + (3t - 2)\mathbf{j} + t\mathbf{k}$,

    and $g(t) = \cos t$. Show that the curves traced out by the following functions are lines or line segments by finding parametric equations for each.

    a. $\mathbf{F} - \mathbf{G}$ b. $\mathbf{F} + 3\mathbf{G}$ c. $\mathbf{F} \circ g$

28. Find the points of intersection of the cylinder $x^2 + y^2 = 4$ and the curve traced out by

$$\mathbf{F}(t) = t \cos \pi t \mathbf{i} + t \sin \pi t \mathbf{j} + t \mathbf{k}$$

29. Find the points of intersection of the sphere $x^2 + y^2 + z^2 = 10$ and the curve traced out by

$$\mathbf{F}(t) = \cos \pi t \mathbf{i} + \sin \pi t \mathbf{j} + t \mathbf{k}$$

*30. A disk of radius r rolls at the rate of 1 radian per unit of time along a line. The curve traced out by a point P on the disk located at a distance b from the center of the disk is called a **trochoid**. (If $b = r$, the trochoid is a cycloid.) Using the method employed in Example 7, show that the trochoid is traced out by the vector-valued function $\mathbf{F}$ defined by

$$\mathbf{F}(t) = (rt - b \sin t)\mathbf{i} + (r - b \cos t)\mathbf{j}$$

(*Hint*: Assume that at time $t = 0$, both the point P and the center of the disk lie on the positive y axis, as shown in Figure 12.10.)

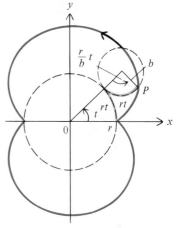

Epicycloid with $b = \frac{1}{2}r$

FIGURE 12.11

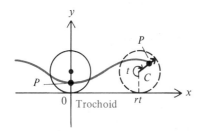

FIGURE 12.10

*31. Suppose that a circle of radius b rolls at the rate of 1 radian per unit of time along the outside of a circle of radius r. The curve traced out by a point on the circumference of the rolling circle is called an **epicycloid** (Figure 12.11). Show that the epicycloid is traced out by the vector-valued function $\mathbf{F}$, where

$$\mathbf{F}(t) = \left[(r + b) \cos t - b \cos \left(\frac{r + b}{b} t \right) \right] \mathbf{i}$$

$$+ \left[(r + b) \sin t - b \sin \left(\frac{r + b}{b} t \right) \right] \mathbf{j}$$

If $b = r$, the epicycloid is called a **cardioid** (Figure 12.12).

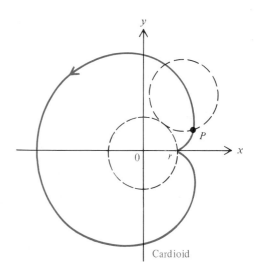

Cardioid

FIGURE 12.12

32. A helical staircase circles once around a cylindrical water tower 200 feet tall and 100 feet in diameter. Find a formula for a vector-valued function that represents the staircase. (*Hint*: Place the coordinate system so that the axis of the cylinder is the z axis.)

12.2
LIMITS AND CONTINUITY OF VECTOR-VALUED FUNCTIONS

The definitions of limits and continuity for vector-valued functions are virtually identical with the corresponding definitions for real-valued functions.

DEFINITION 12.3

Let $\mathbf{F}$ be a vector-valued function defined at each point in some open interval containing t_0, except possibly at t_0 itself. A vector $\mathbf{L}$ is the **limit of $\mathbf{F}(t)$ as t approaches t_0** (or $\mathbf{L}$ is the **limit of $\mathbf{F}$ at t_0**) if for every $\varepsilon > 0$ there is a number $\delta > 0$ such that

$$\text{if } 0 < |t - t_0| < \delta, \quad \text{then} \quad \|\mathbf{F}(t) - \mathbf{L}\| < \varepsilon$$

In this case we write

$$\lim_{t \to t_0} \mathbf{F}(t) = \mathbf{L}$$

and say that $\lim_{t \to t_0} \mathbf{F}(t)$ **exists**.

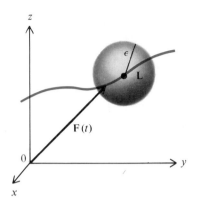

FIGURE 12.13

To say that $\lim_{t \to t_0} \mathbf{F}(t) = \mathbf{L}$ means intuitively that we can make $\mathbf{F}(t)$ as close to $\mathbf{L}$ as we wish by taking t sufficiently close to t_0. If we think of $\mathbf{F}(t)$ and $\mathbf{L}$ as points in space, then we have $\|\mathbf{F}(t) - \mathbf{L}\| < \varepsilon$ if and only if $\mathbf{F}(t)$ lies in the interior of the ball of radius ε and center $\mathbf{L}$ (Figure 12.13). Therefore Definition 12.3 can be interpreted geometrically as follows: $\lim_{t \to t_0} \mathbf{F}(t) = \mathbf{L}$ if for any ball with center $\mathbf{L}$ there is an open interval I about t_0 such that $\mathbf{F}$ assigns to each number in I (except possibly t_0) a point interior to the ball.

In the next theorem we establish the useful fact that the limit of a vector-valued function can be determined from the limits of its component functions.

THEOREM 12.4

Let $\mathbf{F}(t) = f_1(t)\mathbf{i} + f_2(t)\mathbf{j} + f_3(t)\mathbf{k}$. Then $\mathbf{F}$ has a limit at t_0 if and only if f_1, f_2, and f_3 have limits at t_0. In that case

$$\lim_{t \to t_0} \mathbf{F}(t) = \left[\lim_{t \to t_0} f_1(t) \right] \mathbf{i} + \left[\lim_{t \to t_0} f_2(t) \right] \mathbf{j} + \left[\lim_{t \to t_0} f_3(t) \right] \mathbf{k}$$

Proof   First assume that $\lim_{t \to t_0} \mathbf{F}(t) = \mathbf{L} = a\mathbf{i} + b\mathbf{j} + c\mathbf{k}$. This means that for any $\varepsilon > 0$ there is a number $\delta > 0$ such that

$$\text{if } 0 < |t - t_0| < \delta, \quad \text{then} \quad \|\mathbf{F}(t) - \mathbf{L}\| < \varepsilon$$

For such values of t we find that

$$|f_1(t) - a| = \sqrt{(f_1(t) - a)^2} \leq \sqrt{(f_1(t) - a)^2 + (f_2(t) - b)^2 + (f_3(t) - c)^2}$$
$$= \|\mathbf{F}(t) - \mathbf{L}\| < \varepsilon$$

Therefore if $0 < |t - t_0| < \delta$, then $|f_1(t) - a| < \varepsilon$. This verifies that $\lim_{t \to t_0} f_1(t) = a$. Similar arguments can be used to show that $\lim_{t \to t_0} f_2(t) = b$ and $\lim_{t \to t_0} f_3(t) = c$. Conversely, assume that

$$\lim_{t \to t_0} f_1(t) = a, \qquad \lim_{t \to t_0} f_2(t) = b, \qquad \lim_{t \to t_0} f_3(t) = c$$

and let

$$\mathbf{L} = a\mathbf{i} + b\mathbf{j} + c\mathbf{k}$$

This means that for any $\varepsilon > 0$ there is a number $\delta > 0$ such that if $0 < |t - t_0| < \delta$, then

$$|f_1(t) - a| < \frac{\varepsilon}{\sqrt{3}}, \qquad |f_2(t) - b| < \frac{\varepsilon}{\sqrt{3}}, \quad \text{and} \quad |f_3(t) - c| < \frac{\varepsilon}{\sqrt{3}}$$

Hence if $0 < |t - t_0| < \delta$, then

$$\|\mathbf{F}(t) - \mathbf{L}\| = \sqrt{(f_1(t) - a)^2 + (f_2(t) - b)^2 + (f_3(t) - c)^2}$$
$$< \sqrt{\left(\frac{\varepsilon}{\sqrt{3}}\right)^2 + \left(\frac{\varepsilon}{\sqrt{3}}\right)^2 + \left(\frac{\varepsilon}{\sqrt{3}}\right)^2} = \varepsilon$$

It follows from Definition 12.3 that $\lim_{t \to t_0} \mathbf{F}(t)$ exists. Moreover,

$$\lim_{t \to t_0} \mathbf{F}(t) = \mathbf{L} = a\mathbf{i} + b\mathbf{j} + c\mathbf{k} = \lim_{t \to t_0} f_1(t)\mathbf{i} + \lim_{t \to t_0} f_2(t)\mathbf{j} + \lim_{t \to t_0} f_3(t)\mathbf{k} \quad \blacksquare$$

Example 1   Find

$$\lim_{t \to 0} \left(2 \cos t\mathbf{i} + \frac{\sin t}{t}\mathbf{j} + t^2\mathbf{k} \right)$$

Solution   By Theorem 12.4 we may evaluate the limit componentwise:

$$\lim_{t \to 0} \left(2 \cos t\mathbf{i} + \frac{\sin t}{t}\mathbf{j} + t^2\mathbf{k} \right) = \left(\lim_{t \to 0} 2 \cos t \right)\mathbf{i} + \left(\lim_{t \to 0} \frac{\sin t}{t} \right)\mathbf{j} + \left(\lim_{t \to 0} t^2 \right)\mathbf{k}$$
$$= 2\mathbf{i} + \mathbf{j} \quad \square$$

Theorem 12.4 also provides an easy way of finding formulas for limits of combinations of vector-valued functions.

THEOREM 12.5

Let $\mathbf{F}$ and $\mathbf{G}$ be vector-valued functions, and let f and g be real-valued functions. Assume that $\lim_{t \to t_0} \mathbf{F}(t)$ and $\lim_{t \to t_0} \mathbf{G}(t)$ exist and that $\lim_{t \to t_0} f(t)$ exists and $\lim_{s \to s_0} g(s) = t_0$. Then

a. $\displaystyle\lim_{t \to t_0} (\mathbf{F} + \mathbf{G})(t) = \lim_{t \to t_0} \mathbf{F}(t) + \lim_{t \to t_0} \mathbf{G}(t)$

b. $\displaystyle\lim_{t \to t_0} (\mathbf{F} - \mathbf{G})(t) = \lim_{t \to t_0} \mathbf{F}(t) - \lim_{t \to t_0} \mathbf{G}(t)$

c. $\displaystyle\lim_{t \to t_0} f\mathbf{F}(t) = \lim_{t \to t_0} f(t) \lim_{t \to t_0} \mathbf{F}(t)$

d. $\displaystyle\lim_{t \to t_0} (\mathbf{F} \cdot \mathbf{G})(t) = \lim_{t \to t_0} \mathbf{F}(t) \cdot \lim_{t \to t_0} \mathbf{G}(t)$

e. $\displaystyle\lim_{t \to t_0} (\mathbf{F} \times \mathbf{G})(t) = \lim_{t \to t_0} \mathbf{F}(t) \times \lim_{t \to t_0} \mathbf{G}(t)$

f. $\displaystyle\lim_{s \to s_0} (\mathbf{F} \circ g)(s) = \lim_{t \to t_0} \mathbf{F}(t)$

Proof   We prove only (d); the proofs of (a), (b), (c), (e), and (f) are similar. Let

$$\mathbf{F}(t) = f_1(t)\mathbf{i} + f_2(t)\mathbf{j} + f_3(t)\mathbf{k} \quad \text{and} \quad \mathbf{G}(t) = g_1(t)\mathbf{i} + g_2(t)\mathbf{j} + g_3(t)\mathbf{k}$$

To evaluate $\lim_{t \to t_0} (\mathbf{F} \cdot \mathbf{G})(t)$, we note that by Theorem 12.4, f_1, f_2, f_3, g_1, g_2, and g_3 have limits at t_0, so we can employ the limit rules of Chapter 2:

$$\lim_{t \to t_0} (\mathbf{F} \cdot \mathbf{G})(t) = \lim_{t \to t_0} [f_1(t)g_1(t) + f_2(t)g_2(t) + f_3(t)g_3(t)]$$

$$= \lim_{t \to t_0} f_1(t) \lim_{t \to t_0} g_1(t) + \lim_{t \to t_0} f_2(t) \lim_{t \to t_0} g_2(t) + \lim_{t \to t_0} f_3(t) \lim_{t \to t_0} g_3(t)$$

To evaluate $\lim_{t \to t_0} \mathbf{F}(t) \cdot \lim_{t \to t_0} \mathbf{G}(t)$, we first apply Theorem 12.4:

$$\lim_{t \to t_0} \mathbf{F}(t) \cdot \lim_{t \to t_0} \mathbf{G}(t) = \left(\lim_{t \to t_0} f_1(t)\mathbf{i} + \lim_{t \to t_0} f_2(t)\mathbf{j} + \lim_{t \to t_0} f_3(t)\mathbf{k} \right)$$

$$\cdot \left(\lim_{t \to t_0} g_1(t)\mathbf{i} + \lim_{t \to t_0} g_2(t)\mathbf{j} + \lim_{t \to t_0} g_3(t)\mathbf{k} \right)$$

Multiplying out the dot product on the right verifies that

$$\lim_{t \to t_0} (\mathbf{F} \cdot \mathbf{G})(t) = \lim_{t \to t_0} \mathbf{F}(t) \cdot \lim_{t \to t_0} \mathbf{G}(t) \quad \blacksquare$$

Example 2   Let

$$\mathbf{F}(t) = \cos \pi t \mathbf{i} + 2 \sin \pi t \mathbf{j} + 4t^2 \mathbf{k} \quad \text{and} \quad \mathbf{G}(t) = t\mathbf{i} + t^3 \mathbf{k}$$

Find $\lim_{t \to 1} (\mathbf{F} \cdot \mathbf{G})(t)$ and $\lim_{t \to 1} (\mathbf{F} \times \mathbf{G})(t)$.

Solution   For each of the required limits we have a choice of methods. We can (1) find the product of $\mathbf{F}$ and $\mathbf{G}$ and take the limit of the product, or (2) find $\lim_{t \to 1} \mathbf{F}(t)$ and $\lim_{t \to 1} \mathbf{G}(t)$, take the product of these limits, and apply Theorem 12.5. For $\lim_{t \to 1} (\mathbf{F} \cdot \mathbf{G})(t)$ we use method (1). Since

$$(\mathbf{F} \cdot \mathbf{G})(t) = (\cos \pi t \mathbf{i} + 2 \sin \pi t \mathbf{j} + 4t^2 \mathbf{k}) \cdot (t\mathbf{i} + t^3 \mathbf{k}) = t \cos \pi t + 4t^5$$

it follows that

$$\lim_{t \to 1} (\mathbf{F} \cdot \mathbf{G})(t) = \lim_{t \to 1} (t \cos \pi t + 4t^5) = \cos \pi + 4 = 3$$

For $\lim_{t \to 1} (\mathbf{F} \times \mathbf{G})(t)$ we use method (2). By Theorem 12.4,

$$\lim_{t \to 1} \mathbf{F}(t) = \cos \pi \mathbf{i} + 2 \sin \pi \mathbf{j} + 4\mathbf{k} = -\mathbf{i} + 4\mathbf{k}$$

and

$$\lim_{t \to 1} \mathbf{G}(t) = \mathbf{i} + \mathbf{k}$$

Therefore by Theorem 12.5(e) we deduce that

$$\lim_{t \to 1} (\mathbf{F} \times \mathbf{G})(t) = (-\mathbf{i} + 4\mathbf{k}) \times (\mathbf{i} + \mathbf{k}) = 5\mathbf{j} \quad \square$$

Now that we have the concept of limit for vector-valued functions, we can define continuity for such functions. Again, our definition is essentially the same as for real-valued functions.

DEFINITION 12.6

A vector-valued function $\mathbf{F}$ is **continuous** at a point t_0 in its domain if

$$\lim_{t \to t_0} \mathbf{F}(t) = \mathbf{F}(t_0)$$

Since continuity is defined in terms of limits and since limits can be computed componentwise (by Theorem 12.4), you can easily prove the following theorem.

THEOREM 12.7

A vector-valued function $\mathbf{F}$ is continuous at t_0 if and only if each of its component functions is continuous at t_0.

Theorem 12.7 makes it easy to define continuity on an open or closed interval I: We say that a vector-valued function $\mathbf{F}$ is **continuous on I** if the component functions of $\mathbf{F}$ are continuous on I (see Definition 2.12).

EXERCISES 12.2

In Exercises 1–10 compute the limit or explain why it does not exist.

1. $\lim_{t \to 4} (\mathbf{i} - \mathbf{j} + \mathbf{k})$

2. $\lim_{t \to -1} (3\mathbf{i} + t\mathbf{j} + t^5\mathbf{k})$

3. $\lim_{t \to \pi} (\tan t\mathbf{i} + 3t\mathbf{j} - 4\mathbf{k})$

4. $\lim_{t \to 0} \left(\dfrac{\sin t}{t}\mathbf{i} + e^t\mathbf{j} + (t + \sqrt{2})\mathbf{k} \right)$

5. $\lim_{t \to 2} \mathbf{F}(t)$, where

$$\mathbf{F}(t) = \begin{cases} 5\mathbf{i} - \sqrt{2t^2 + 2t + 4}\mathbf{j} + e^{-(t-2)}\mathbf{k} & \text{for } t < 2 \\ (t^2 + 1)\mathbf{i} + (4 - t^3)\mathbf{j} + \mathbf{k} & \text{for } t > 2 \end{cases}$$

6. $\lim_{t \to 0} \mathbf{F}(t)$, where $\mathbf{F}(t) = \begin{cases} t\mathbf{i} + e^{-1/t^2}\mathbf{j} + t^2\mathbf{k} & \text{for } t \neq 0 \\ \mathbf{j} & \text{for } t = 0 \end{cases}$

7. $\lim_{t \to 0} (\mathbf{F} - \mathbf{G})(t)$, where $\mathbf{F}(t) = e^{-1/t^2}\mathbf{i} + \cos t\mathbf{j} + t^3\mathbf{k}$ and

$$\mathbf{G}(t) = -\pi\mathbf{i} + \dfrac{1 + \cos t}{t}\mathbf{j}$$

8. $\lim_{t \to 1} (\mathbf{F} \cdot \mathbf{G})(t)$, where

$$\mathbf{F}(t) = \dfrac{\sin(t-1)}{t-1}\mathbf{i} + \dfrac{t+3}{t-2}\mathbf{j} + \cos \pi t\mathbf{k} \quad \text{and}$$

$$\mathbf{G}(t) = (t^2 + 1)\mathbf{i} - \dfrac{t-2}{t+3}\mathbf{j} - \sqrt{t^2 + 1}\mathbf{k}$$

9. $\lim_{t \to 3} \left(\dfrac{t^2 - 5t + 6}{t - 3}\mathbf{i} + \dfrac{t^2 - 2t - 3}{t - 3}\mathbf{j} + \dfrac{t^2 + 4t - 21}{t - 3}\mathbf{k} \right)$

10. $\lim_{t \to 1} \left(\dfrac{t^2 + 1}{t - 1}\mathbf{i} + \dfrac{t^2 - 1}{t + 1}\mathbf{j} + \dfrac{t^2 + 7t - 8}{t - 1}\mathbf{k} \right)$

In Exercises 11–15 determine the intervals on which the function is continuous.

11. $\mathbf{F}(t) = 4\mathbf{j} - \sqrt{5}\mathbf{k}$

12. $\mathbf{F}(t) = 3\mathbf{i} + (t - 5)\mathbf{j} + \cos t\mathbf{k}$

13. $\mathbf{F}(t) = e^t\mathbf{i} + e^{-t}\mathbf{j} + \sqrt{2t}\mathbf{k}$

14. $\mathbf{F}(t) = \sqrt{2 - t}\mathbf{i} + \dfrac{1}{1 - t^2}\mathbf{j} - \dfrac{1}{3t - 1}\mathbf{k}$

15. $\mathbf{F}(t) = \begin{cases} \dfrac{\sin t}{t}\mathbf{i} + \mathbf{j} + \mathbf{k} & \text{for } t < 0 \\ (t^2 + 1)\mathbf{i} + \ln(t + 3)\mathbf{j} + \mathbf{k}. & \text{for } t \geq 0 \end{cases}$

16. a. Formulate a definition of the right-hand limit $\lim_{t \to t_0^+} \mathbf{F}(t)$ analogous to Definition 12.3, and show that the limit can be computed componentwise.
    b. Formulate a definition of the left-hand limit $\lim_{t \to t_0^-} \mathbf{F}(t)$ analogous to Definition 12.3, and show that the limit can be computed componentwise.

17. Use Exercise 16 to compute the limits that exist.
    a. $\lim_{t \to 0^+} \left(t\mathbf{i} + 2t^{1/4}\mathbf{j} - \dfrac{\ln t}{t}\mathbf{k} \right)$
    b. $\lim_{t \to 1^+} (e^{1/(1-t)}\mathbf{i} + \sqrt{t - 1}\mathbf{j} + \ln t\mathbf{k})$
    c. $\lim_{t \to 1^-} [\sqrt{1 - t}\mathbf{i} - (1 - t)\ln(1 - t)\mathbf{j}]$

18. a. Formulate a definition of $\lim_{t \to \infty} \mathbf{F}(t)$, and show that the limit can be computed componentwise.

b. Evaluate

$$\lim_{t \to \infty} \left(\frac{1}{t} \mathbf{i} + \frac{t-1}{t+1} \mathbf{j} + \frac{\sin t^3}{t^2} \mathbf{k} \right)$$

19. Let $\mathbf{F}$ and $\mathbf{G}$ be continuous at t_0, and let c be a number. Prove that the following functions are continuous at t_0.

a. $\mathbf{F} + \mathbf{G}$ b. $c\mathbf{F}$

c. $\|\mathbf{F}\|$ (where $\|\mathbf{F}\|(t) = \|\mathbf{F}(t)\|$ for all t in the domain of $\mathbf{F}$)

d. $\mathbf{F} \cdot \mathbf{G}$ e. $\mathbf{F} \times \mathbf{G}$

12.3
DERIVATIVES AND INTEGRALS OF VECTOR-VALUED FUNCTIONS

Our definition of the derivative of a vector-valued function, like the definitions of limits and continuity of vector-valued functions, is closely related to the corresponding definition for real-valued functions.

DEFINITION 12.8

Let t_0 be a number in the domain of a vector-valued function $\mathbf{F}$. If

$$\lim_{t \to t_0} \frac{\mathbf{F}(t) - \mathbf{F}(t_0)}{t - t_0}$$

exists, we call this limit the **derivative** of $\mathbf{F}$ at t_0 and write

$$\mathbf{F}'(t_0) = \lim_{t \to t_0} \frac{\mathbf{F}(t) - \mathbf{F}(t_0)}{t - t_0} .$$

In this case we say that $\mathbf{F}$ **has a derivative at** t_0, that $\mathbf{F}$ **is differentiable at** t_0, or that $\mathbf{F}'(t_0)$ **exists.**

We define the **derivative** of $\mathbf{F}$ to be the vector-valued function $\mathbf{F}'$ whose domain consists of all numbers t such that $\mathbf{F}$ is differentiable at t, and whose value at any number t in its domain is $\mathbf{F}'(t)$. The Leibniz notation for the derivative of $\mathbf{F}$ is $d\mathbf{F}/dt$.

We can interpret $\mathbf{F}'(t_0)$ geometrically as follows. Denote by C the curve traced out by $\mathbf{F}$. Let P and P_0 be points on C corresponding to $\mathbf{F}(t)$ and $\mathbf{F}(t_0)$, respectively. Then $\mathbf{F}(t) - \mathbf{F}(t_0)$ has the same direction as the "secant vector" $\overrightarrow{P_0 P}$. Thus if $t > t_0$, then

$$\frac{\mathbf{F}(t) - \mathbf{F}(t_0)}{t - t_0}$$

also has the same direction as $\overrightarrow{P_0 P}$ (Figure 12.14(a)), whereas if $t < t_0$, then it has the opposite direction, that is, the same direction as $\overrightarrow{PP_0}$ (Figure 12.14(b)). Therefore if $\mathbf{F}'(t_0)$ exists, it is the limit of vectors parallel to secant vectors passing through P_0 and pointing in the same general direction in which C is traced out by $\mathbf{F}$. Accordingly, it is reasonable to say that $\mathbf{F}'(t_0)$ is tangent to C at P_0 (Figure 12.14(c)).

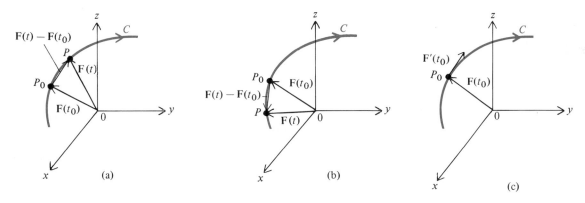

FIGURE 12.14

If $\mathbf{F}(t) = f_1(t)\mathbf{i} + f_2(t)\mathbf{j} + f_3(t)\mathbf{k}$, then

$$\frac{\mathbf{F}(t) - \mathbf{F}(t_0)}{t - t_0} = \frac{f_1(t) - f_1(t_0)}{t - t_0}\mathbf{i} + \frac{f_2(t) - f_2(t_0)}{t - t_0}\mathbf{j} + \frac{f_3(t) - f_3(t_0)}{t - t_0}\mathbf{k}$$

Using Theorem 12.4, we find that $\mathbf{F}'(t_0)$ can be obtained directly from the derivatives of the component functions.

THEOREM 12.9

Let $\mathbf{F}(t) = f_1(t)\mathbf{i} + f_2(t)\mathbf{j} + f_3(t)\mathbf{k}$. Then $\mathbf{F}$ is differentiable at t_0 if and only if $f_1, f_2,$ and f_3 are differentiable at t_0. In that case,

$$\mathbf{F}'(t_0) = f'_1(t_0)\mathbf{i} + f'_2(t_0)\mathbf{j} + f'_3(t_0)\mathbf{k}$$

As a result, finding derivatives of vector-valued functions is as easy (or as difficult) as finding derivatives of real-valued functions. First we discuss the derivative of a constant vector-valued function.

Example 1   Let $\mathbf{F}(t) = a\mathbf{i} + b\mathbf{j} + c\mathbf{k}$ for all t. Show that $\mathbf{F}' = \mathbf{0}$.

Solution   Since $f_1(t) = a$, $f_2(t) = b$, and $f_3(t) = c$, the component functions are all constant functions. Consequently

$$\mathbf{F}'(t) = 0\mathbf{i} + 0\mathbf{j} + 0\mathbf{k} = \mathbf{0} \quad \square$$

We can express the result of Example 1 by saying that the derivative of a constant vector-valued function is $\mathbf{0}$. Next we turn to the derivative of a linear vector-valued function.

Example 2   Let $\mathbf{F}(t) = (x_0 + at)\mathbf{i} + (y_0 + bt)\mathbf{j} + (z_0 + ct)\mathbf{k}$.   Show   that $d\mathbf{F}/dt = a\mathbf{i} + b\mathbf{j} + c\mathbf{k}$.

Solution   In this case $f_1(t) = x_0 + at$, $f_2(t) = y_0 + bt$, and $f_3(t) = z_0 + ct$. Therefore Theorem 12.9 tells us immediately that

$$\frac{d\mathbf{F}}{dt} = a\mathbf{i} + b\mathbf{j} + c\mathbf{k} \quad \square$$

Thus the derivative of a linear vector-valued function is a constant vector-valued function.

Example 3   Let $\mathbf{F}(t) = t\cos t\,\mathbf{i} + t\sin t\,\mathbf{j} + t\mathbf{k}$. Find $\mathbf{F}'(\pi)$.

Solution   First we differentiate componentwise to obtain

$$\mathbf{F}'(t) = (\cos t - t\sin t)\mathbf{i} + (\sin t + t\cos t)\mathbf{j} + \mathbf{k}$$

Then we substitute the value π for t:

$$\mathbf{F}'(\pi) = [-1 - \pi(0)]\mathbf{i} + [0 + \pi(-1)]\mathbf{j} + \mathbf{k} = -\mathbf{i} - \pi\mathbf{j} + \mathbf{k} \quad \square$$

In the remainder of this chapter we will most often encounter vector-valued functions defined on intervals. If I is an interval, then we say that $\mathbf{F}$ is **differentiable on I** if the components of $\mathbf{F}$ are differentiable on I. If I is a closed interval $[a, b]$, this implies in particular that the component functions have appropriate one-sided derivatives at a and b (see (5) of Section 3.2). For example, the function $\mathbf{F}$ defined by

$$\mathbf{F}(t) = |t|\mathbf{i} + |1 - t|\mathbf{j}$$

is differentiable on $[0, 1]$ but is not differentiable on $[-1, 1]$, since $\mathbf{F}'(0)$ does not exist, or on $[0, 2]$, since $\mathbf{F}'(1)$ does not exist.

Almost all the differentiation rules proved in Chapter 3 have counterparts for vector-valued functions. Since these rules can be proved with simple modifications of the proofs used in Section 3.3, we state them here without proof.

THEOREM 12.10

Let $\mathbf{F}$, $\mathbf{G}$, and f be differentiable at t_0, and let g be differentiable at s_0 with $g(s_0) = t_0$. Then the following hold.
a. $(\mathbf{F} + \mathbf{G})'(t_0) = \mathbf{F}'(t_0) + \mathbf{G}'(t_0)$
b. $(\mathbf{F} - \mathbf{G})'(t_0) = \mathbf{F}'(t_0) - \mathbf{G}'(t_0)$
c. $(f\mathbf{F})'(t_0) = f'(t_0)\mathbf{F}(t_0) + f(t_0)\mathbf{F}'(t_0)$
d. $(\mathbf{F} \cdot \mathbf{G})'(t_0) = \mathbf{F}'(t_0) \cdot \mathbf{G}(t_0) + \mathbf{F}(t_0) \cdot \mathbf{G}'(t_0)$
e. $(\mathbf{F} \times \mathbf{G})'(t_0) = \mathbf{F}'(t_0) \times \mathbf{G}(t_0) + \mathbf{F}(t_0) \times \mathbf{G}'(t_0)$
f. $(\mathbf{F} \circ g)'(s_0) = \mathbf{F}'(g(s_0))g'(s_0) = \mathbf{F}'(t_0)g'(s_0)$

Example 4   Let $\mathbf{F}(t) = \arctan t\,\mathbf{i} + 5\mathbf{k}$ and $\mathbf{G}(t) = \mathbf{i} + \ln t\,\mathbf{j} - 2t\mathbf{k}$.   Find $(\mathbf{F} \cdot \mathbf{G})'(t)$ and $(\mathbf{F} \times \mathbf{G})'(t)$.

Solution Two methods are available for finding each required derivative. We can (1) calculate $(\mathbf{F} \cdot \mathbf{G})(t)$ or $(\mathbf{F} \times \mathbf{G})(t)$ and then differentiate it, or (2) find $\mathbf{F}'(t)$ and $\mathbf{G}'(t)$ and then use Theorem 12.10 to find $(\mathbf{F} \cdot \mathbf{G})'$ or $(\mathbf{F} \times \mathbf{G})'$. We use method (1) to find $(\mathbf{F} \cdot \mathbf{G})'(t)$. After calculating that $(\mathbf{F} \cdot \mathbf{G})(t) = \arctan t - 10t$, we conclude that

$$(\mathbf{F} \cdot \mathbf{G})'(t) = \frac{1}{t^2 + 1} - 10$$

To find $(\mathbf{F} \times \mathbf{G})'(t)$, we use method (2). To that end we first find that

$$\mathbf{F}'(t) = \frac{1}{t^2 + 1}\mathbf{i} \quad \text{and} \quad \mathbf{G}'(t) = \frac{1}{t}\mathbf{j} - 2\mathbf{k}$$

Then

$$\mathbf{F}'(t) \times \mathbf{G}(t) = \begin{vmatrix} \mathbf{i} & \mathbf{j} & \mathbf{k} \\ \dfrac{1}{t^2 + 1} & 0 & 0 \\ 1 & \ln t & -2t \end{vmatrix} = \frac{2t}{t^2 + 1}\mathbf{j} + \frac{\ln t}{t^2 + 1}\mathbf{k}$$

and

$$\mathbf{F}(t) \times \mathbf{G}'(t) = \begin{vmatrix} \mathbf{i} & \mathbf{j} & \mathbf{k} \\ \arctan t & 0 & 5 \\ 0 & \dfrac{1}{t} & -2 \end{vmatrix} = -\frac{5}{t}\mathbf{i} + 2\arctan t\,\mathbf{j} + \frac{\arctan t}{t}\mathbf{k}$$

Now Theorem 12.10(e) yields

$$(\mathbf{F} \times \mathbf{G})'(t) = \mathbf{F}'(t) \times \mathbf{G}(t) + \mathbf{F}(t) \times \mathbf{G}'(t)$$

$$= -\frac{5}{t}\mathbf{i} + \left(\frac{2t}{t^2 + 1} + 2\arctan t\right)\mathbf{j} + \left(\frac{\ln t}{t^2 + 1} + \frac{\arctan t}{t}\right)\mathbf{k} \quad \square$$

An elementary but significant consequence of Theorem 12.10 is the following.

COROLLARY 12.11

> Let $\mathbf{F}$ be differentiable on an interval I, and assume that there is a number c such that
>
> $$\|\mathbf{F}(t)\| = c \quad \text{for } t \text{ in } I$$
>
> Then $\qquad\qquad \mathbf{F}(t) \cdot \mathbf{F}'(t) = 0 \quad \text{for } t \text{ in } I$

Proof Since $\|\mathbf{F}(t)\| = c$ by hypothesis, it follows that

$$(\mathbf{F} \cdot \mathbf{F})(t) = \mathbf{F}(t) \cdot \mathbf{F}(t) = \|\mathbf{F}(t)\|^2 = c^2 \quad \text{for } t \text{ in } I$$

Therefore $\mathbf{F} \cdot \mathbf{F}$ is a constant real-valued function, which by Example 1 has derivative 0. As a result, Theorem 12.10(d) implies that

$$\mathbf{F}'(t) \cdot \mathbf{F}(t) + \mathbf{F}(t) \cdot \mathbf{F}'(t) = (\mathbf{F} \cdot \mathbf{F})'(t) = 0$$

so that

$$\mathbf{F}(t) \cdot \mathbf{F}'(t) = 0 \quad \text{for } t \text{ in } I \quad \blacksquare$$

Corollary 12.11 tells us that if $\|\mathbf{F}\|$ is a constant real-valued function, then for each t in the domain of $\mathbf{F}$, one of the following is true:

1. $\mathbf{F}(t) = \mathbf{0}$
2. $\mathbf{F}'(t) = \mathbf{0}$
3. $\mathbf{F}(t)$ and $\mathbf{F}'(t)$ are perpendicular

In particular, if S is a sphere centered at the origin and if $\mathbf{F}$ represents the position of an object moving on S, then $\|\mathbf{F}\|$ is a constant function, so that $\mathbf{F}(t)$ is perpendicular to $\mathbf{F}'(t)$ for all t for which $\mathbf{F}'(t) \neq \mathbf{0}$.

The **second derivative** of $\mathbf{F}$ is defined to be the derivative of $\mathbf{F}'$, denoted $\mathbf{F}''$. If

$$\mathbf{F}(t) = f_1(t)\mathbf{i} + f_2(t)\mathbf{j} + f_3(t)\mathbf{k}$$

then two applications of Theorem 12.9 yield

$$\mathbf{F}''(t) = f_1''(t)\mathbf{i} + f_2''(t)\mathbf{j} + f_3''(t)\mathbf{k}$$

For example, if $\mathbf{F}(t) = (t^{1/2} + t)\mathbf{i} + (t^3 + 1)\mathbf{j} + e^{2t}\mathbf{k}$, then

$$\mathbf{F}''(t) = -\frac{1}{4}t^{-3/2}\mathbf{i} + 6t\mathbf{j} + 4e^{2t}\mathbf{k}$$

Velocity and Acceleration

The most important physical applications of derivatives of vector-valued functions arise in the study of motion. As an object moves through space, the coordinates x, y, and z of its location are functions of time. Let us assume that these functions are twice differentiable and define the position, velocity, speed, and acceleration as follows:

$$\text{position:} \quad \mathbf{r}(t) = x(t)\mathbf{i} + y(t)\mathbf{j} + z(t)\mathbf{k}$$

$$\text{velocity:} \quad \mathbf{v}(t) = \frac{d\mathbf{r}}{dt} = \frac{dx}{dt}\mathbf{i} + \frac{dy}{dt}\mathbf{j} + \frac{dz}{dt}\mathbf{k}$$

$$\text{speed:} \quad \|\mathbf{v}(t)\| = \sqrt{\left(\frac{dx}{dt}\right)^2 + \left(\frac{dy}{dt}\right)^2 + \left(\frac{dz}{dt}\right)^2}$$

$$\text{acceleration:} \quad \mathbf{a}(t) = \frac{d\mathbf{v}}{dt} = \frac{d^2\mathbf{r}}{dt^2} = \frac{d^2x}{dt^2}\mathbf{i} + \frac{d^2y}{dt^2}\mathbf{j} + \frac{d^2z}{dt^2}\mathbf{k}$$

The position vector $\mathbf{r}(t)$ is often called the **radial vector** or **radius vector** (this is why the symbol $\mathbf{r}$ is used). Usually $\mathbf{r}(t)$ is represented by a directed line segment from the origin to the position of the object at time t. If t_0 is fixed and $t \neq t_0$, then the vector $\mathbf{r}(t) - \mathbf{r}(t_0)$, sometimes called the **displacement vector**, is directed from the position of the object at time t_0 to the position at time t (Figure 12.15). The **average velocity** is then defined as

$$\frac{\mathbf{r}(t) - \mathbf{r}(t_0)}{t - t_0}$$

Since
$$\mathbf{v}(t_0) = \mathbf{r}'(t_0) = \lim_{t \to t_0} \frac{\mathbf{r}(t) - \mathbf{r}(t_0)}{t - t_0}$$

velocity is the limit of average velocity.

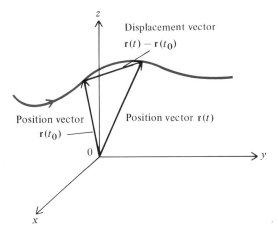

FIGURE 12.15

Example 5 Suppose an object remains at rest at the point (a, b, c). Show that $\mathbf{v} = \mathbf{a} = \mathbf{0}$.

Solution The position is constant and therefore is given by

$$\mathbf{r}(t) = a\mathbf{i} + b\mathbf{j} + c\mathbf{k}$$

Consequently

$$\mathbf{v}(t) = \frac{d\mathbf{r}}{dt} = \mathbf{0} \quad \text{and} \quad \mathbf{a}(t) = \frac{d\mathbf{v}}{dt} = \mathbf{0} \quad \square$$

Example 6 Suppose the position vector of an object is given by

$$\mathbf{r}(t) = (x_0 + at)\mathbf{i} + (y_0 + bt)\mathbf{j} + (z_0 + ct)\mathbf{k}$$

Determine the velocity, speed, and acceleration of the object.

Solution Differentiating, we find that

$$\mathbf{v}(t) = \frac{d\mathbf{r}}{dt} = a\mathbf{i} + b\mathbf{j} + c\mathbf{k}$$

and hence that

$$\mathbf{a}(t) = \frac{d\mathbf{v}}{dt} = \mathbf{0}$$

The speed is the length of the velocity vector, which in this case means that

$$\|\mathbf{v}(t)\| = \sqrt{a^2 + b^2 + c^2} \quad \square$$

In Example 6 the object moves along the line whose parametric equations are

$$x = x_0 + at, \qquad y = y_0 + bt, \qquad z = z_0 + ct$$

with constant velocity and zero acceleration. If we had assumed that

$$\mathbf{r}(t) = (x_0 + at^2)\mathbf{i} + (y_0 + bt^2)\mathbf{j} + (z_0 + ct^2)\mathbf{k}$$

with a, b, or c different from 0, then the object would still have moved along the same line. However, the velocity and acceleration would be given by

$$\mathbf{v}(t) = 2at\mathbf{i} + 2bt\mathbf{j} + 2ct\mathbf{k}$$

and

$$\mathbf{a}(t) = 2a\mathbf{i} + 2b\mathbf{j} + 2c\mathbf{k}$$

In this case the object would move faster as t increases with $t > 0$, and the acceleration, while constant, would not be $\mathbf{0}$.

In our next example we study the motion of an object that traces out a cycloidal path (see Example 7 of Section 12.1). You might visualize the object as a point on the rim of an automobile tire that is moving along a straight line with constant speed.

Example 7   Let

$$\mathbf{r}(t) = r(t - \sin t)\mathbf{i} + r(1 - \cos t)\mathbf{j}$$

Find those values of t for which $\mathbf{v}(t) = \mathbf{0}$.

Solution   By differentiating $\mathbf{r}$ we find that

$$\mathbf{v}(t) = \frac{d\mathbf{r}}{dt} = r(1 - \cos t)\mathbf{i} + r \sin t\mathbf{j}$$

Therefore

$$\mathbf{v}(t) = \mathbf{0} \quad \text{if and only if} \quad r(1 - \cos t) = 0 = r \sin t$$

The last condition means that t is an integral multiple of 2π. $\square$

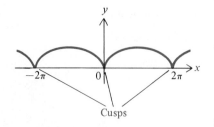

FIGURE 12.16

If, in Example 7, t is an integral multiple of 2π, then $\mathbf{r}(t)$ corresponds to one of the isolated points on the cycloid called **cusps** that lie on the x axis (Figure 12.16). It is at precisely such points on the cycloid that the velocity is $\mathbf{0}$.

Example 8 An object moves counterclockwise along a circle of radius $r_0 > 0$ with a constant speed $v_0 > 0$. Show that the position vector of the object is given by the equation

$$\mathbf{r}(t) = r_0\left(\cos\frac{v_0 t}{r_0}\mathbf{i} + \sin\frac{v_0 t}{r_0}\mathbf{j}\right)$$

Find formulas for the velocity and acceleration of the object.

Solution We set up the coordinate system so that the circle lies in the xy plane with the origin as its center and so that the object is on the positive x axis at time 0 and moves counterclockwise around the circle. For any time t let $\theta(t)$ be the angle from the positive x axis to the vector $\mathbf{r}(t)$ (Figure 12.17(a)). Then

$$\mathbf{r}(t) = r_0[\cos\theta(t)\mathbf{i} + \sin\theta(t)\mathbf{j}] \qquad (1)$$

To obtain an expression for $\theta(t)$ in terms of t, r_0 and v_0, we differentiate both sides of (1) and obtain

$$\mathbf{v}(t) = r_0\theta'(t)[-\sin\theta(t)\mathbf{i} + \cos\theta(t)\mathbf{j}] \qquad (2)$$

Because the object moves counterclockwise, θ is increasing; consequently $\theta'(t) > 0$. Therefore since $r_0 > 0$, (2) implies that

$$\|\mathbf{v}(t)\| = |r_0\theta'(t)|\sqrt{(-\sin\theta(t))^2 + (\cos\theta(t))^2} = |r_0\theta'(t)| = r_0\theta'(t) \qquad (3)$$

But $\|\mathbf{v}(t)\| = v_0$ by hypothesis, so that (3) becomes $\theta'(t) = v_0/r_0$. Since $\theta(0) = 0$ by assumption, we can integrate to obtain $\theta(t) = v_0 t/r_0$. Substitution into (1) now yields

$$\mathbf{r}(t) = r_0\left(\cos\frac{v_0 t}{r_0}\mathbf{i} + \sin\frac{v_0 t}{r_0}\mathbf{j}\right)$$

By differentiating twice in succession, we find that

$$\mathbf{v}(t) = v_0\left(-\sin\frac{v_0 t}{r_0}\mathbf{i} + \cos\frac{v_0 t}{r_0}\mathbf{j}\right)$$

and

$$\mathbf{a}(t) = -\frac{v_0^2}{r_0}\left(\cos\frac{v_0 t}{r_0}\mathbf{i} + \sin\frac{v_0 t}{r_0}\mathbf{j}\right) = -\frac{v_0^2}{r_0^2}\mathbf{r}(t)$$

(Figure 12.17(b)). □

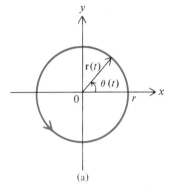

(a)

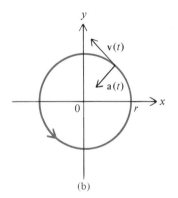

(b)

FIGURE 12.17

Notice that although the speed v of the object in Example 8 is constant, the velocity $\mathbf{v}$ is not constant. Notice also that the acceleration is directed toward the center of the circle. Such an acceleration is called a **centripetal acceleration**, and any force producing a centripetal acceleration is called a **centripetal force**. Since $\|\mathbf{r}(t)\| = r_0$ by (1), the magnitude of the acceleration is given by

$$\|\mathbf{a}(t)\| = \frac{v_0^2}{r_0^2}\|\mathbf{r}(t)\| = \frac{v_0^2}{r_0} \qquad (4)$$

Now suppose a satellite travels in a circular orbit around the earth and is influenced only by the earth's gravity. We will derive a formula relating the speed

v_0 (in feet per second) of the satellite to the radius r_0 (in feet) of the orbit. To that end, we recall that by Newton's Law of Gravitation, the magnitude $\|\mathbf{a}\|$ of the acceleration of an object r_0 feet from the center of the earth and influenced only by the earth's gravity is given by

$$\|\mathbf{a}(t)\| = \frac{C}{r_0^2} \tag{5}$$

where C is a constant. Combining this equation with (4), we find that

$$\frac{v_0^2}{r_0} = \frac{C}{r_0^2}, \quad \text{so that} \quad v_0^2 = \frac{C}{r_0} \tag{6}$$

Since the Law of Gravitation holds even if the object is at the earth's surface, that is, if r_0 is equal to the radius of the earth (which is approximately 3960 miles), and since the acceleration due to gravity on the surface of the earth is approximately 32 feet per second per second, it follows from (5) that

$$32 \approx \frac{C}{(3960)^2 (5280)^2}$$

Thus

$$C \approx 32(3960)^2 (5280)^2$$

Now suppose a weather satellite moves with constant speed v_0 in a circular orbit 4100 miles from the center of the earth (that is, some 140 miles above ground). Then by (6),

$$v_0 = \sqrt{\frac{C}{r_0}} = \sqrt{\frac{(32)(3960)^2 (5280)^2}{(4100)(5280)}} \approx 25{,}421.2 \text{ (feet per second)}$$

which is approximately 17,332.6 miles per hour.

Integrals of Vector-Valued Functions

Since a vector-valued function $\mathbf{F}$ is determined by its component functions, we define the integral of $\mathbf{F}$ in terms of its component functions.

DEFINITION 12.12

Let

$$\mathbf{F}(t) = f_1(t)\mathbf{i} + f_2(t)\mathbf{j} + f_3(t)\mathbf{k}$$

where f_1, f_2, and f_3 are continuous on $[a, b]$. Then the **definite integral** $\int_a^b \mathbf{F}(t)\, dt$ and the **indefinite integral** $\int \mathbf{F}(t)\, dt$ are defined by

$$\int_a^b \mathbf{F}(t)\, dt = \left(\int_a^b f_1(t)\, dt \right)\mathbf{i} + \left(\int_a^b f_2(t)\, dt \right)\mathbf{j} + \left(\int_a^b f_3(t)\, dt \right)\mathbf{k}$$

and

$$\int \mathbf{F}(t)\, dt = \left(\int f_1(t)\, dt \right)\mathbf{i} + \left(\int f_2(t)\, dt \right)\mathbf{j} + \left(\int f_3(t)\, dt \right)\mathbf{k}$$

Example 9   Let

$$\mathbf{F}(t) = t\mathbf{i} + t^2\mathbf{j} + \sin t\mathbf{k}$$

Find $\int \mathbf{F}(t)\,dt$ and $\int_0^\pi \mathbf{F}(t)\,dt$.

Solution   Integrating componentwise, we find that

$$\int \mathbf{F}(t)\,dt = \left(\int t\,dt\right)\mathbf{i} + \left(\int t^2\,dt\right)\mathbf{j} + \left(\int \sin t\,dt\right)\mathbf{k}$$

$$= \left(\frac{1}{2}t^2 + C_1\right)\mathbf{i} + \left(\frac{1}{3}t^3 + C_2\right)\mathbf{j} + (-\cos t + C_3)\mathbf{k}$$

Therefore

$$\int_0^\pi \mathbf{F}(t)\,dt = \left(\frac{1}{2}t^2\Big|_0^\pi\right)\mathbf{i} + \left(\frac{1}{3}t^3\Big|_0^\pi\right)\mathbf{j} + \left(-\cos t\Big|_0^\pi\right)\mathbf{k}$$

$$= \frac{1}{2}\pi^2\mathbf{i} + \frac{1}{3}\pi^3\mathbf{j} + 2\mathbf{k} \quad \square$$

If

$$\mathbf{F}(t) = f_1(t)\mathbf{i} + f_2(t)\mathbf{j} + f_3(t)\mathbf{k}$$

where $f_1, f_2,$ and f_3 are continuously differentiable, then by Theorem 12.9 and Definition 12.12,

$$\int \mathbf{F}'(t)\,dt = \int [f_1'(t)\mathbf{i} + f_2'(t)\mathbf{j} + f_3'(t)\mathbf{k}]\,dt$$

$$= \left(\int f_1'(t)\,dt\right)\mathbf{i} + \left(\int f_2'(t)\,dt\right)\mathbf{j} + \left(\int f_3'(t)\,dt\right)\mathbf{k}$$

$$= [f_1(t) + C_1]\mathbf{i} + [f_2(t) + C_2]\mathbf{j} + [f_3(t) + C_3]\mathbf{k}$$

$$= (f_1(t)\mathbf{i} + f_2(t)\mathbf{j} + f_3(t)\mathbf{k}) + (C_1\mathbf{i} + C_2\mathbf{j} + C_3\mathbf{k})$$

$$= \mathbf{F}(t) + (C_1\mathbf{i} + C_2\mathbf{j} + C_3\mathbf{k})$$

This proves that

$$\int \mathbf{F}'(t)\,dt = \mathbf{F}(t) + \mathbf{C} \tag{7}$$

where $\mathbf{C}$ is a constant vector. Formula (7) is valuable in the study of motion. Indeed, since the velocity $\mathbf{v}$ is the derivative of the position $\mathbf{r}$ and the acceleration $\mathbf{a}$ is the derivative of the velocity $\mathbf{v}$, it follows from (7) that

$$\int \mathbf{v}(t)\,dt = \int \mathbf{r}'(t)\,dt = \mathbf{r}(t) + \mathbf{C}$$

and

$$\int \mathbf{a}(t)\,dt = \int \mathbf{v}'(t)\,dt = \mathbf{v}(t) + \mathbf{C}$$

We will use these formulas in discussing Newton's Second Law of Motion.

The most basic physical law pertaining to motion is Newton's Second Law of Motion. Suppose that at any time t an object of mass m experiences a force $\mathbf{F}(t)$ and undergoes an acceleration $\mathbf{a}(t)$. Then Newton's Second Law of Motion states that

$$\mathbf{F}(t) = m\mathbf{a}(t) \tag{8}$$

From this law we can determine the acceleration of an object once its mass and the force acting on it are known. Since the velocity is the integral of the acceleration and the position is the integral of the velocity, it is possible (at least in principle) to determine the position of an object of known mass, initial position, and initial velocity once the force acting on it is known. Normally the initial position and velocity are taken at time $t = 0$.

For an elementary application of these ideas, we consider the motion of an object that remains near some point P on the earth's surface and moves only under the influence of the earth's gravity. By Newton's Law of Gravitation the gravitational force on the object is nearly constant and is exerted downward (toward the center of the earth). We choose a coordinate system so that the positive z axis emanates from the center of the earth and points upward, passing through P. Consequently the acceleration is $-g\mathbf{k}$, where g is a constant approximately equal to 32 feet (or 9.8 meters) per second per second.

Example 10   An object has initial position $\mathbf{r}_0$ and initial velocity $\mathbf{v}_0$ and undergoes a constant acceleration $-g\mathbf{k}$. Show that the position of the object at any time t is given by

$$\mathbf{r}(t) = -\frac{1}{2}gt^2\mathbf{k} + t\mathbf{v}_0 + \mathbf{r}_0$$

Solution   By hypothesis $\mathbf{a}(t) = -g\mathbf{k}$. Successive integrations yield

$$\mathbf{v}(t) = \int \mathbf{a}(t)\,dt = \int -g\mathbf{k}\,dt = -gt\mathbf{k} + \mathbf{C} \tag{9}$$

and

$$\mathbf{r}(t) = \int \mathbf{v}(t)\,dt = -\frac{1}{2}gt^2\mathbf{k} + t\mathbf{C} + \mathbf{C}_1 \tag{10}$$

where $\mathbf{C}$ and $\mathbf{C}_1$ are constant vectors. By assumption $\mathbf{v}(0) = \mathbf{v}_0$, so that by letting $t = 0$ in (9) we find that $\mathbf{C} = \mathbf{v}_0$. Similarly, $\mathbf{r}(0) = \mathbf{r}_0$, so that $\mathbf{C}_1 = \mathbf{r}_0$ by (10). Therefore (10) can be rewritten

$$\mathbf{r}(t) = -\frac{1}{2}gt^2\mathbf{k} + t\mathbf{v}_0 + \mathbf{r}_0 \quad \square$$

Example 11   A baseball is hit 4 feet above the ground at 100 feet per second and at an angle of $\pi/6$ with respect to the ground. When does the baseball hit the ground?

Solution We choose a coordinate system so that the xy plane represents the ground, the ball travels in the yz plane with the $\mathbf{j}$ component of its position vector increasing, and the initial position of the ball is $\mathbf{r}_0 = 4\mathbf{k}$. Since the initial angle is $\pi/6$ and the initial speed of the ball is 100 feet per second, so that $\|\mathbf{v}_0\| = 100$, the initial velocity is given by

$$\mathbf{v}_0 = 100\left(\cos\frac{\pi}{6}\mathbf{j} + \sin\frac{\pi}{6}\mathbf{k}\right) = 100\left(\frac{\sqrt{3}}{2}\mathbf{j} + \frac{1}{2}\mathbf{k}\right) = 50(\sqrt{3}\mathbf{j} + \mathbf{k})$$

From Example 10 we know that the position of the baseball is given by

$$\mathbf{r}(t) = -\frac{1}{2}gt^2\mathbf{k} + t\mathbf{v}_0 + \mathbf{r}_0$$

Hank Aaron hitting a home run in 1973

Substituting for $\mathbf{r}_0$ and $\mathbf{v}_0$, we find that

$$\mathbf{r}(t) = -\frac{1}{2}gt^2\mathbf{k} + t[50(\sqrt{3}\mathbf{j} + \mathbf{k})] + 4\mathbf{k}$$

$$= 50\sqrt{3}\,t\mathbf{j} + \left(4 + 50t - \frac{1}{2}gt^2\right)\mathbf{k}$$

When the ball hits the ground, the k component of $\mathbf{r}(t)$ must be 0. This happens at the time $t > 0$ such that

$$4 + 50t - \frac{gt^2}{2} = 0$$

which means that

$$t = \frac{50 + \sqrt{2500 + 8g}}{g}$$

Taking $g = 32$, we find that $t \approx 3.2$, so the ball hits the ground after approximately 3.2 seconds. $\square$

EXERCISES 12.3

In Exercises 1–16 find the derivative of the function.

1. $\mathbf{F}(t) = \mathbf{i} + t\mathbf{j} + t^5\mathbf{k}$

2. $\mathbf{F}(t) = 3\mathbf{i} + (t^2 + t)\mathbf{j} - t\mathbf{k}$

3. $\mathbf{F}(t) = (1 + t)^{3/2}\mathbf{i} - (1 - t)^{3/2}\mathbf{j} + \frac{3}{2}t\mathbf{k}$

4. $\mathbf{F}(t) = t^2\cos t\mathbf{i} + t^3\sin t\mathbf{j} + t^4\mathbf{k}$

5. $\mathbf{F}(t) = \tan t\mathbf{i} + \mathbf{j} + \sec t\mathbf{k}$

6. $\mathbf{F}(t) = e^t\cos t\mathbf{i} - e^t\sin t\mathbf{k}$

7. $\mathbf{F}(t) = \cosh t\mathbf{i} + \sinh t\mathbf{j} - \sqrt{t}\mathbf{k}$

8. $\mathbf{F}(t) = \arcsin 4t\mathbf{i} - 3\arctan(2t - 1)\mathbf{j} + 7\ln 3t^2\mathbf{k}$

9. $4\mathbf{F} - 2\mathbf{G}$, where $\mathbf{F}(t) = 2\sec t\mathbf{i} - 3\mathbf{j} + \csc t\mathbf{k}$ and $\mathbf{G}(t) = 3t\mathbf{i} - t^2\mathbf{j} - 4\csc t\mathbf{k}$

10. $\mathbf{F}\cdot\mathbf{G}$, where $\mathbf{F}(t) = 2\sec t\mathbf{i} - 3\mathbf{j} + \csc t\mathbf{k}$ and $\mathbf{G}(t) = 3t\mathbf{i} - t^2\mathbf{j} - 4\csc t\mathbf{k}$

11. $\mathbf{F}\times\mathbf{G}$, where $\mathbf{F}(t) = 2\sec t\mathbf{i} - 3\mathbf{j} + \csc t\mathbf{k}$ and $\mathbf{G}(t) = \ln t\mathbf{k}$

12. $\mathbf{F}\cdot\mathbf{G}$, where $\mathbf{F}(t) = (3/t)\mathbf{i} - \mathbf{j}$ and $\mathbf{G}(t) = t\mathbf{i} - e^{-t}\mathbf{j}$

13. $\mathbf{F}\times\mathbf{G}$, where $\mathbf{F}(t) = (3/t)\mathbf{i} - \mathbf{j}$ and $\mathbf{G}(t) = t\mathbf{i} - e^{-t}\mathbf{j}$

14. $\mathbf{F}\cdot\mathbf{G}$, where

$$\mathbf{F}(t) = \frac{-2t}{1 + t^2}\mathbf{i} + \frac{1 + 2t^2}{1 + t^2}\mathbf{j}$$

and

$$\mathbf{G}(t) = \frac{-2 - 4t^2}{(1 + t^2)^2}\mathbf{i} - \frac{4t}{(1 + t^2)^2}\mathbf{j}$$

15. $\mathbf{F} \circ g$, where $\mathbf{F}(t) = \ln t\mathbf{i} - 4e^{2t}\mathbf{j} + \dfrac{t-1}{t}\mathbf{k}$ and $g(t) = \sqrt{t}$

16. $\mathbf{F} \circ g$, where $\mathbf{F}(t) = t^3\mathbf{i} - \sqrt{3}t\mathbf{j} + \dfrac{1}{t^2}\mathbf{k}$ and $g(t) = \cos t$

In Exercises 17–21 evaluate the integral.

17. $\displaystyle\int \left(t^2\mathbf{i} - (3t-1)\mathbf{j} - \dfrac{1}{t^3}\mathbf{k} \right) dt$

18. $\displaystyle\int (t\cos t\mathbf{i} + t\sin t\mathbf{j} + 3t^4\mathbf{k})\, dt$

19. $\displaystyle\int_0^1 (e^t\mathbf{i} + e^{-t}\mathbf{j} + 2t\mathbf{k})\, dt$

20. $\displaystyle\int_0^1 (\cosh t\mathbf{i} + \sinh t\mathbf{j} + \mathbf{k})\, dt$

21. $\displaystyle\int_{-1}^1 [(1+t)^{3/2}\mathbf{i} + (1-t)^{3/2}\mathbf{j}]\, dt$

In Exercises 22–27 find the velocity, speed, and acceleration of an object having the given position function.

22. $\mathbf{r}(t) = 3t\mathbf{i} + 2t\mathbf{j} - 16t^2\mathbf{k}$

23. $\mathbf{r}(t) = \cos t\mathbf{i} + \sin t\mathbf{j} - 16t^2\mathbf{k}$

24. $\mathbf{r}(t) = e^{-t}\mathbf{i} + e^{-t}\mathbf{j}$

25. $\mathbf{r}(t) = 2t\mathbf{i} + t^2\mathbf{j} + \ln t\mathbf{k}$

26. $\mathbf{r}(t) = \cosh t\mathbf{i} + \sinh t\mathbf{j} + t\mathbf{k}$

27. $\mathbf{r}(t) = e^t\sin t\mathbf{i} + e^t\cos t\mathbf{j} + e^t\mathbf{k}$

In Exercises 28–32 find the position, velocity, and speed of an object having the given acceleration, initial velocity, and initial position.

28. $\mathbf{a}(t) = -32\mathbf{k}$; $\mathbf{v}_0 = \mathbf{0}$; $\mathbf{r}_0 = \mathbf{0}$

29. $\mathbf{a}(t) = -32\mathbf{k}$; $\mathbf{v}_0 = \mathbf{i} + \mathbf{j}$; $\mathbf{r}_0 = \mathbf{0}$

30. $\mathbf{a}(t) = -32\mathbf{k}$; $\mathbf{v}_0 = 3\mathbf{i} - 2\mathbf{j} + \mathbf{k}$; $\mathbf{r}_0 = 5\mathbf{j} + 2\mathbf{k}$

31. $\mathbf{a}(t) = -\cos t\mathbf{i} - \sin t\mathbf{j}$; $\mathbf{v}_0 = \mathbf{k}$; $\mathbf{r}_0 = \mathbf{i}$

32. $\mathbf{a}(t) = e^t\mathbf{i} + e^{-t}\mathbf{j}$; $\mathbf{v}_0 = \mathbf{i} - \mathbf{j} + \sqrt{2}\mathbf{k}$; $\mathbf{r}_0 = \mathbf{i} + \mathbf{j}$

33. Let $\mathbf{F}(t) = \displaystyle\int_0^t (u\tan u^3\mathbf{i} + \cos e^u\mathbf{j} + e^{(u^2)}\mathbf{k})\, du$.
Find $\mathbf{F}'(t)$.

34. Let $\mathbf{G}(t) = \displaystyle\int_0^{t^2} (\cos u\mathbf{i} + e^{-(u^2)}\mathbf{j} + \tan u\mathbf{k})\, du$.
Find $\mathbf{G}'(t)$.

35. Let

$$\mathbf{F}(t) = \dfrac{4t}{1+4t^2}\mathbf{i} + \dfrac{1-4t^2}{1+4t^2}\mathbf{j}$$

Using Corollary 12.11, show that $\mathbf{F}(t) \cdot \mathbf{F}'(t) = 0$ for all t.

36. Let $\mathbf{F}(t) = \cos t\mathbf{i} + \sin t\mathbf{j}$. Show that there is a value of t in $(0, \pi)$ such that $[\mathbf{F}(\pi) - \mathbf{F}(0)]/(\pi - 0)$ is parallel to $\mathbf{F}'(t)$,

but there is no value of t in $(0, \pi)$ such that

$$\mathbf{F}'(t) = \dfrac{\mathbf{F}(\pi) - \mathbf{F}(0)}{\pi - 0}$$

37. Let $\mathbf{F}(t) = \sin t\mathbf{i} - \cos t\mathbf{j}$. Show that for all t the vectors $\mathbf{F}(t)$ and $\mathbf{F}''(t)$ are parallel. Determine whether there is a value of t for which $\mathbf{F}(t)$ and $\mathbf{F}''(t)$ have the same direction.

38. Let $\mathbf{F}(t) = e^{-2t}\mathbf{i} + e^{2t}\mathbf{k}$. Show that for all t the vectors $\mathbf{F}(t)$ and $\mathbf{F}''(t)$ are parallel. Determine whether there is a value of t for which $\mathbf{F}(t)$ and $\mathbf{F}''(t)$ have the same direction.

39. Show that if $\mathbf{F}''(t)$ exists for all t, then

$$\dfrac{d}{dt}(\mathbf{F} \times \mathbf{F}') = \mathbf{F}(t) \times \mathbf{F}''(t)$$

40. Suppose $\mathbf{F}(t)$ is parallel to $\mathbf{F}''(t)$ for all t. Show that $\mathbf{F} \times \mathbf{F}'$ is constant. (*Hint:* Use Exercise 39.)

41. Let $\mathbf{F}$, $\mathbf{G}$, and $\mathbf{H}$ be differentiable. Using the differentiation rules for the dot and cross products of vector-valued functions, verify that

$$\dfrac{d}{dt}[\mathbf{F} \cdot (\mathbf{G} \times \mathbf{H})] = \dfrac{d\mathbf{F}}{dt} \cdot (\mathbf{G} \times \mathbf{H}) + \mathbf{F} \cdot \left(\dfrac{d\mathbf{G}}{dt} \times \mathbf{H} \right)$$
$$+ \mathbf{F} \cdot \left(\mathbf{G} \times \dfrac{d\mathbf{H}}{dt} \right)$$

42. If an object has mass m and position $\mathbf{r}(t)$ at time t, then its **kinetic energy** $K(t)$ at time t is defined by

$$K(t) = \dfrac{1}{2}m\|\mathbf{v}(t)\|^2$$

Suppose a ball is dropped from a height of 96 feet and later is thrown straight down from the same height with an initial speed of 80 feet per second. How much larger is the kinetic energy when the ball hits the ground the second time than the first time?

43. Suppose the position of a golf ball is given by

$$\mathbf{r}(t) = 90\sqrt{2}t\mathbf{i} + 90\sqrt{2}t\mathbf{j} + (64t - 16t^2)\mathbf{k} \quad \text{for } t \geq 0$$

a. Find the initial position and the initial velocity of the golf ball.

b. Show that the golf ball strikes the ground at time $t = 4$, and determine the distance from its initial position.

44. An object leaves the point $(0, 0, 1)$ with initial velocity $\mathbf{v}_0 = 2\mathbf{i} + 3\mathbf{k}$. Thereafter it is subject only to the force of gravity. Find a formula for the position of the object at any time $t > 0$.

45. A cylindrical container with water H feet deep stands on the floor and has a spout h feet below the water line (Figure 12.18). Torricelli's Theorem (presented in Chapter 4 Review Exercise 57) states that the speed of the water as it flows through the spout is $\sqrt{2gh}$.

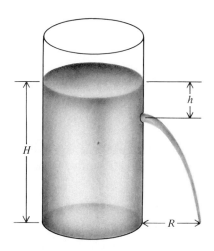

FIGURE 12.18

a. Find a formula for the position $\mathbf{r}(t)$ of a water drop t seconds after it leaves the spout.
b. Determine the value of h that maximizes the distance R as shown in Figure 12.18.
c. Find the maximum possible value of R.

46. A ping pong ball rolls off a table 2.6 feet high with an initial speed of 2 feet per second.

a. How long after it leaves the table does it hit the floor? (Note that the velocity vector is horizontal when the ball leaves the table.)
b. At what speed does it hit the floor?

47. The *Nimbus* weather satellite orbits the earth in a circular orbit approximately 500 miles above the surface of the earth. Find the speed of the satellite.

48. A satellite can remain stationary over a fixed point on earth if it orbits the earth in a circular orbit with a period of 24 hours. This is the basic principle of a communications satellite. Find the speed and the radius of the orbit of such a satellite. (*Hint:* Use (6) and the fact that the circumference of the orbit equals both $2\pi r_0$ and $24(3600)v_0$.)

49. A bobsled moving at a constant speed of 60 miles per hour rounds a circular turn with a radius of 100 feet. Find the magnitude of the bobsled's centripetal acceleration.

50. The accelerator at the Fermi National Accelerator Laboratory in Batavia, Illinois, is circular with a radius of 1 kilometer. Find the magnitude of the centripetal acceleration of a proton moving around the accelerator with a constant speed of
a. 2.5×10^5 kilometers per second.
b. 2.9×10^5 kilometers per second.

12.4
SPACE CURVES AND THEIR LENGTHS

In Section 12.1 we discussed curves traced out by vector-valued functions. From now on we will use the word "curve" in a more restricted sense.

DEFINITION 12.13

A **space curve** (or simply **curve**) is the range of a continuous vector-valued function on an interval of real numbers.

All the vector-valued functions we have encountered are continuous, and consequently their ranges are curves. In particular, any point, line, line segment, circle, parabola, cycloid, or circular helix is a curve. We will generally use C to denote a curve and $\mathbf{r}$ to denote a vector-valued function whose range is a curve C. In that case we say that C is **parametrized** by $\mathbf{r}$, or that $\mathbf{r}$ is a **parametrization** of C. However, we will occasionally refer to $\mathbf{r}$ as a curve and say, for example, "the curve $\mathbf{r}(t) = e^t\mathbf{i} + \cos t\mathbf{j} + 3\mathbf{k}$." The component functions of a vector-valued function $\mathbf{r}$ are usually denoted x, y, and z. Thus

$$\mathbf{r}(t) = x(t)\mathbf{i} + y(t)\mathbf{j} + z(t)\mathbf{k}$$

Corresponding parametric equations for the curve are

$$x = x(t), \qquad y = y(t), \qquad z = z(t)$$

Since we are accustomed to thinking of the graph of any continuous real-valued function as a curve, we now verify that Definition 12.13 is broad enough to include such curves.

Example 1   Let f be a continuous real-valued function on an interval I. Show that the graph of f is a curve, and find a parametrization of that curve.

Solution   The vector-valued function $\mathbf{r}$ defined by

$$\mathbf{r}(t) = t\mathbf{i} + f(t)\mathbf{j} \quad \text{for } t \text{ in } I$$

is continuous and traces out the graph of f (Figure 12.19). Thus the graph of f is the range of $\mathbf{r}$, so by Definition 12.13 the graph of f is a curve.   □

FIGURE 12.19

As we have just seen, it is possible to parametrize the graph of any continuous real-valued function. For instance, if $f(t) = \ln t$, then

$$\mathbf{r}(t) = t\mathbf{i} + \ln t\mathbf{j} \quad \text{for } t > 0$$

parametrizes the graph of f.

Properties of Curves   We will discuss three of the many properties a curve can possess: the properties of being closed, smooth, and piecewise smooth.

DEFINITION 12.14

> A curve C is **closed** if it has a parametrization whose domain is a closed interval $[a, b]$ such that $\mathbf{r}(a) = \mathbf{r}(b)$.

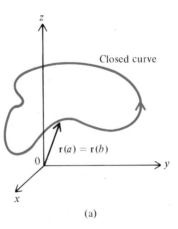

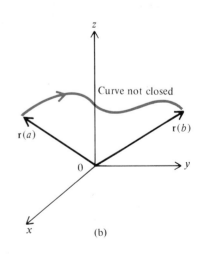

FIGURE 12.20

To put it informally, a curve is closed if its initial and terminal points coincide (Figure 12.20(a)). For example, since

$$\mathbf{r}(t) = \cos t\mathbf{i} + \sin t\mathbf{j} \quad \text{for } 0 \leq t \leq 2\pi$$

parametrizes the standard unit circle and since $\mathbf{r}(0) = \mathbf{r}(2\pi)$, the standard unit circle is closed. In fact, any circle or ellipse is closed. In contrast, the curve sketched in Figure 12.20(b) is not closed. Line segments, circular helixes, and cycloids also are not closed.

Next we consider smoothness of curves.

DEFINITION 12.15

> **a.** A vector-valued function $\mathbf{r}$ defined on an interval I is **smooth** if $\mathbf{r}$ has a continuous derivative on I and $\mathbf{r}'(t) \neq \mathbf{0}$ for each interior point t. A curve C is **smooth** if it has a smooth parametrization.
>
> **b.** A continuous vector-valued function $\mathbf{r}$ defined on an interval I is **piecewise smooth** if I is composed of a finite number of subintervals on each of which $\mathbf{r}$ is smooth and if $\mathbf{r}$ has one-sided derivatives at each interior point of I. A curve C is **piecewise smooth** if it has a piecewise smooth parametrization.

Intuitively, a curve is smooth if it is composed of one piece with no sharp corners (Figure 12.21(a)). A curve is piecewise smooth if it is composed of one piece and has at most a finite number of sharp corners (Figure 12.21(b)). Of course, smooth curves are always piecewise smooth, but the converse is not true. We will show in Example 4 that the curve composed of the two arches of the cycloid is piecewise smooth (although it can be shown not to be smooth).

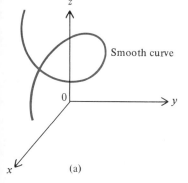

Smooth curve

(a)

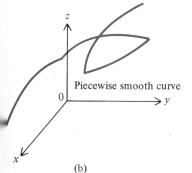

Piecewise smooth curve

(b)

FIGURE 12.21

Example 2 Show that the standard unit circle is smooth.

Solution The circle can be parametrized by

$$\mathbf{r}(t) = \cos t\mathbf{i} + \sin t\mathbf{j} \quad \text{for } 0 \leq t \leq 2\pi$$

The function $\mathbf{r}$ is differentiable on $[0, 2\pi]$, and

$$\mathbf{r}'(t) = -\sin t\mathbf{i} + \cos t\mathbf{j} \quad \text{for } 0 \leq t \leq 2\pi$$

Therefore $\mathbf{r}'$ is continuous on $[0, 2\pi]$, and

$$\|\mathbf{r}'(t)\| = \sqrt{(-\sin t)^2 + (\cos t)^2} = 1$$

From this we conclude that $\mathbf{r}'(t) \neq \mathbf{0}$ for each t in $[0, 2\pi]$. It follows that the circle is smooth. $\square$

Example 3 Show that the helix

$$\mathbf{r}(t) = \cos t\mathbf{i} + \sin t\mathbf{j} + t\mathbf{k}$$

is smooth.

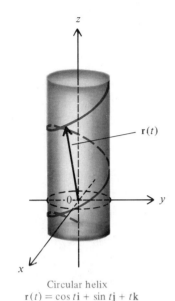

Circular helix
$\mathbf{r}(t) = \cos t\mathbf{i} + \sin t\mathbf{j} + t\mathbf{k}$

FIGURE 12.22

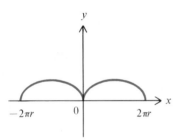

Two arches of a cycloid

FIGURE 12.23

Solution   Since $\mathbf{r}'(t) = -\sin t\mathbf{i} + \cos t\mathbf{j} + \mathbf{k}$, it follows that $\mathbf{r}'$ is continuous and $\mathbf{r}'(t) \neq \mathbf{0}$ for every t. Therefore the helix is smooth (see Figure 12.22).   □

Example 4   Show that the curve

$$\mathbf{r}(t) = r(t - \sin t)\mathbf{i} + r(1 - \cos t)\mathbf{j} \quad \text{for } -2\pi \leq t \leq 2\pi$$

which parametrizes two arches of a cycloid, is piecewise smooth.

Solution   We find that

$$\mathbf{r}'(t) = r(1 - \cos t)\mathbf{i} + r\sin t\mathbf{j}$$

Even though $\mathbf{r}'$ is continuous, $\mathbf{r}$ is not smooth because $\mathbf{r}'(0) = \mathbf{0}$. However, $\mathbf{r}'(t) \neq \mathbf{0}$ if t is not $-2\pi, 0,$ or 2π. Therefore $\mathbf{r}$ is smooth on $[-2\pi, 0]$ and on $[0, 2\pi]$, and hence $\mathbf{r}$ is piecewise smooth (Figure 12.23).   □

Suppose (x_0, y_0, z_0) and (x_1, y_1, z_1) are distinct points in space, and consider the parametric equations

$$x = x_0 + (x_1 - x_0)t, \qquad y = y_0 + (y_1 - y_0)t, \qquad z = z_0 + (z_1 - z_0)t \quad (1)$$

Since $(x, y, z) = (x_0, y_0, z_0)$ for $t = 0$ and $(x, y, z) = (x_1, y_1, z_1)$ for $t = 1$, it follows that (1) gives parametric equations for the line through (x_0, y_0, z_0) and (x_1, y_1, z_1). Moreover,

$$\mathbf{r}(t) = [x_0 + (x_1 - x_0)t]\mathbf{i} + [y_0 + (y_1 - y_0)t]\mathbf{j} + [z_0 + (z_1 - z_0)t]\mathbf{k} \quad \text{for } 0 \leq t \leq 1$$

is a smooth parametrization of the line segment from (x_0, y_0, z_0) to (x_1, y_1, z_1).

Example 5   Find a smooth parametrization of the line segment from $(4, 3, 5)$ to $(2, 8, 5)$.

Solution   We apply the general method just described, with $x_0 = 4$, $y_0 = 3$, $z_0 = 5$, $x_1 = 2$, $y_1 = 8$, and $z_1 = 5$. We find that

$$\mathbf{r}(t) = (4 - 2t)\mathbf{i} + (3 + 5t)\mathbf{j} + 5\mathbf{k} \quad \text{for } 0 \leq t \leq 1$$

is a smooth parametrization of the line segment.   □

Length of a Curve   By means of a method similar to the one we employed in Sections 6.3 and 6.4 to define length for curves in the plane, we will now assign length to curves in space.

In studying the length of curves we must be careful to avoid any parametrization that traces out a curve more than once. For example, the standard unit circle is traced out twice by the parametrization

$$\mathbf{r}(t) = \cos t\mathbf{i} + \sin t\mathbf{j} \quad \text{for } 0 \leq t \leq 4\pi$$

If we employed $\mathbf{r}$ to compute the length (circumference) of the circle, our answer would be twice the length we normally attribute to the circle. An admissible

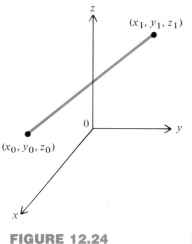

FIGURE 12.24

parametrization will trace out a curve only once, except that finitely many points may be traced out more than once, as happens with the parametrization

$$\mathbf{r}(t) = \cos t\mathbf{i} + \sin t\mathbf{j} \quad \text{for } 0 \le t \le 2\pi$$

of the standard unit circle. In this case the single point $(1,0)$ is traced out twice, for $t = 0$ and $t = 2\pi$.

We begin by finding the length of the simplest type of curve. Let C be the line segment joining the points (x_0, y_0, z_0) and (x_1, y_1, z_1) in space (Figure 12.24). Then we define the length $\mathscr{L}$ of C as the distance between the two points by the formula for distance given in (1) of Section 11.1:

$$\mathscr{L} = \sqrt{(x_1 - x_0)^2 + (y_1 - y_0)^2 + (z_1 - z_0)^2} \tag{2}$$

Next, let C be a smooth curve and

$$\mathbf{r}(t) = x(t)\mathbf{i} + y(t)\mathbf{j} + z(t)\mathbf{k} \quad \text{for } a \le t \le b$$

be a smooth parametrization of C. Let $\mathscr{P} = \{t_0, t_1, t_2, \ldots, t_n\}$ be any partition of $[a, b]$, and let L be the (black) polygonal curve shown in Figure 12.25. We denote the length of the kth portion of C by $\Delta\mathscr{L}_k$. If Δt_k is small, then $\Delta\mathscr{L}_k$ is approximately equal to the length of the corresponding line segment. In other words,

$$\Delta\mathscr{L}_k \approx \sqrt{[x(t_k) - x(t_{k-1})]^2 + [y(t_k) - y(t_{k-1})]^2 + [z(t_k) - z(t_{k-1})]^2}$$

By the Mean Value Theorem there are numbers u_k, v_k, and w_k in $[t_{k-1}, t_k]$ such that

$$x(t_k) - x(t_{k-1}) = x'(u_k)\Delta t_k, \qquad y(t_k) - y(t_{k-1}) = y'(v_k)\Delta t_k,$$

$$z(t_k) - z(t_{k-1}) = z'(w_k)\Delta t_k$$

where $\Delta t_k = t_k - t_{k-1}$. Consequently

$$\Delta\mathscr{L}_k \approx \sqrt{(x'(u_k))^2(\Delta t_k)^2 + (y'(v_k))^2(\Delta t_k)^2 + (z'(w_k))^2(\Delta t_k)^2}$$

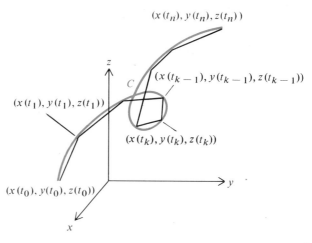

FIGURE 12.25

or equivalently,

$$\Delta \mathscr{L}_k \approx \sqrt{(x'(u_k))^2 + (y'(v_k))^2 + (z'(w_k))^2} \, \Delta t_k \tag{3}$$

Therefore the total length $\mathscr{L}$ of C, which is the sum of $\Delta \mathscr{L}_1, \Delta \mathscr{L}_2, \ldots, \Delta \mathscr{L}_n$, should be approximately

$$\sum_{k=1}^{n} \sqrt{(x'(u_k))^2 + (y'(v_k))^2 + (z'(w_k))^2} \, \Delta t_k \tag{4}$$

which itself is approximately

$$\int_a^b \sqrt{(x'(t))^2 + (y'(t))^2 + (z'(t))^2} \, dt$$

Since

$$\sqrt{(x'(t))^2 + (y'(t))^2 + (z'(t))^2} = \|\mathbf{r}'(t)\|$$

it follows that $\mathscr{L}$ should be approximately $\int_a^b \|\mathbf{r}'(t)\| \, dt$. The same approximation occurs when C is only piecewise smooth. This leads us to the definition of the length of C.

DEFINITION 12.16

> Let C be a curve with a piecewise smooth parametrization $\mathbf{r}$ defined on $[a, b]$. Then the **length** $\mathscr{L}$ of C is defined by
>
> $$\mathscr{L} = \int_a^b \|\mathbf{r}'(t)\| \, dt = \int_a^b \left\| \frac{d\mathbf{r}}{dt} \right\| dt \tag{5}$$

If

$$\mathbf{r}(t) = x(t)\mathbf{i} + y(t)\mathbf{j} + z(t)\mathbf{k} \quad \text{for } a \leq t \leq b$$

then (5) can be rewritten as

$$\mathscr{L} = \int_a^b \sqrt{(x'(t))^2 + (y'(t))^2 + (z'(t))^2} \, dt$$

$$= \int_a^b \sqrt{\left(\frac{dx}{dt}\right)^2 + \left(\frac{dy}{dt}\right)^2 + \left(\frac{dz}{dt}\right)^2} \, dt \tag{6}$$

Example 6   Find the length $\mathscr{L}$ of the segment of the circular helix

$$\mathbf{r}(t) = \cos t\mathbf{i} + \sin t\mathbf{j} + t\mathbf{k} \quad \text{for } 0 \leq t \leq 2\pi$$

Solution   Applying (6), we have

$$\mathscr{L} = \int_0^{2\pi} \sqrt{(-\sin t)^2 + (\cos t)^2 + 1^2} \, dt = \int_0^{2\pi} \sqrt{2} \, dt = 2\sqrt{2}\pi \quad \square$$

Example 7   Find the length $\mathscr{L}$ of the curve

$$\mathbf{r}(t) = t\mathbf{i} + \frac{\sqrt{6}}{2}t^2\mathbf{j} + t^3\mathbf{k} \quad \text{for } -1 \le t \le 1$$

Solution   By (6),

$$\mathscr{L} = \int_{-1}^{1} \sqrt{1 + 6t^2 + 9t^4}\, dt = \int_{-1}^{1} (1 + 3t^2)\, dt = (t + t^3)\Big|_{-1}^{1} = 4 \quad \square$$

If the function in Example 7 were slightly different, say,

$$\mathbf{r}_1(t) = t\mathbf{i} + t^2\mathbf{j} + t^3\mathbf{k} \quad \text{for } -1 \le t \le 1$$

then the length of the corresponding curve would be

$$\mathscr{L} = \int_{-1}^{1} \sqrt{1 + 4t^2 + 9t^4}\, dt$$

Unfortunately, it is impossible to evaluate this integral by elementary means; its value can only be approximated, for instance by the Trapezoidal Rule or Simpson's Rule. In fact, the lengths of most curves can only be approximated.

We close the discussion by noting that if C is composed of two curves C_1 and C_2, each of which is smooth, then the length of C is the sum of the lengths of C_1 and C_2.

Independence of Parametrization

Every curve has many parametrizations. For example, the line segment from $(\frac{1}{4}, \frac{1}{4}, \frac{1}{4})$ to $(1, 1, 1)$ is parametrized by each of the following functions:

$$\mathbf{r}_1(t) = t\mathbf{i} + t\mathbf{j} + t\mathbf{k} \quad \text{for } \tfrac{1}{4} \le t \le 1$$

$$\mathbf{r}_2(t) = (t - 2)\mathbf{i} + (t - 2)\mathbf{j} + (t - 2)\mathbf{k} \quad \text{for } \tfrac{9}{4} \le t \le 3$$

$$\mathbf{r}_3(t) = t^2\mathbf{i} + t^2\mathbf{j} + t^2\mathbf{k} \quad \text{for } \tfrac{1}{2} \le t \le 1$$

If we compute the length of the line segment by applying Definition 12.16 to $\mathbf{r}_1$, $\mathbf{r}_2$, and $\mathbf{r}_3$, we obtain the same length for all three parametrizations. It can be shown that we would obtain the same length for any piecewise smooth curve by using any of its piecewise smooth parametrizations. We express this by saying that the length of the curve is **independent of parametrization**.

Length of a Polar Graph

With the help of Definition 12.16 we can derive a formula for the length of the polar graph of an equation $r = f(\theta)$ in the plane, where f' is continuous on a closed interval $[\alpha, \beta]$. The polar graph of $r = f(\theta)$ is just the curve parametrized by

$$\mathbf{r}(\theta) = \overbrace{r\cos\theta}^{x}\mathbf{i} + \overbrace{r\sin\theta}^{y}\mathbf{j} = f(\theta)\cos\theta\,\mathbf{i} + f(\theta)\sin\theta\,\mathbf{j}$$

Then

$$\mathbf{r}'(\theta) = [f'(\theta)\cos\theta - f(\theta)\sin\theta]\mathbf{i} + [f'(\theta)\sin\theta + f(\theta)\cos\theta]\mathbf{j}$$

so it follows that

$$\|\mathbf{r}'(\theta)\|^2 = [f'(\theta)\cos\theta - f(\theta)\sin\theta]^2 + [f'(\theta)\sin\theta + f(\theta)\cos\theta]^2$$
$$= [f(\theta)]^2 + [f'(\theta)]^2$$

Therefore by Definition 12.16 the length $\mathscr{L}$ of the polar graph is given by

$$\mathscr{L} = \int_\alpha^\beta \sqrt{(f(\theta))^2 + (f'(\theta))^2}\, d\theta \qquad (7)$$

Example 8 Find the length $\mathscr{L}$ of the circle $r = 3$.

Solution A polar equation of the circle is $r = 3$ for $0 \le \theta \le 2\pi$. Therefore

$$f(\theta) = 3 \quad \text{and} \quad f'(\theta) = 0$$

Consequently (7) implies that

$$\mathscr{L} = \int_0^{2\pi} \sqrt{3^2 + 0^2}\, d\theta = 6\pi \quad \square$$

Example 9 Find the length $\mathscr{L}$ of the cardioid

$$r = 1 - \cos\theta \quad \text{for } 0 \le \theta \le 2\pi$$

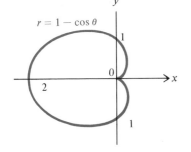

$r = 1 - \cos\theta$

FIGURE 12.26

(Figure 12.26).

Solution Since $f(\theta) = 1 - \cos\theta$ and $f'(\theta) = \sin\theta$, it follows from (7) that

$$\mathscr{L} = \int_0^{2\pi} \sqrt{(1 - \cos\theta)^2 + \sin^2\theta}\, d\theta = \int_0^{2\pi} \sqrt{2(1 - \cos\theta)}\, d\theta$$

Using the trigonometric identity $\frac{1}{2}(1 - \cos\theta) = \sin^2\theta/2$ and the fact that $\sin\theta/2 \ge 0$ for $0 \le \theta \le 2\pi$, we find that

$$\mathscr{L} = \int_0^{2\pi} \sqrt{4\sin^2\frac{\theta}{2}}\, d\theta = 2\int_0^{2\pi} \sin\frac{\theta}{2}\, d\theta = 4\left(-\cos\frac{\theta}{2}\right)\Big|_0^{2\pi} = 8 \quad \square$$

The Arc Length Function

Let C be a smooth curve parametrized on an interval I by

$$\mathbf{r}(t) = x(t)\mathbf{i} + y(t)\mathbf{j} + z(t)\mathbf{k} \quad \text{for } t \text{ in } I$$

and let a be a fixed number in I. We define the **arc length function** s by

$$s(t) = \int_a^t \|\mathbf{r}'(u)\|\, du = \int_a^t \sqrt{(x'(u))^2 + (y'(u))^2 + (z'(u))^2}\, du \quad \text{for } t \text{ in } I \qquad (8)$$

Notice that if $t \ge a$, then $s(t)$ is the length of that portion of the curve between $\mathbf{r}(a)$ and $\mathbf{r}(t)$ (Figure 12.27), and if $\mathbf{r}(t)$ denotes the position of an object at time $t \ge a$, then $s(t)$ is the distance traveled by the object between time a and time t.

If we differentiate the expressions in (8) with respect to t, we obtain

$$s'(t) = \|\mathbf{r}'(t)\| = \sqrt{(x'(t))^2 + (y'(t))^2 + (z'(t))^2}$$

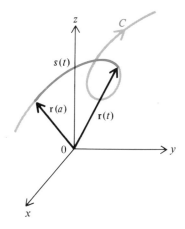

FIGURE 12.27

or equivalently,

$$\frac{ds}{dt} = \left\| \frac{d\mathbf{r}}{dt} \right\| = \sqrt{\left(\frac{dx}{dt}\right)^2 + \left(\frac{dy}{dt}\right)^2 + \left(\frac{dz}{dt}\right)^2} \tag{9}$$

For later reference we mention that if $\mathbf{r}$ denotes the position of an object in motion and $\mathbf{v}$ its velocity, then from (9) and the definition of speed in Section 12.3 we have

$$\frac{ds}{dt} = \|\mathbf{r}'(t)\| = \|\mathbf{v}(t)\| \geq 0 \tag{10}$$

Moreover, since $\mathbf{r}$ is smooth, it follows that $ds/dt > 0$ at each interior point of I, so that t can be regarded as a function of s (see Theorem 7.4). Therefore any quantity depending on t also depends on s. In particular, if C is parametrized by $\mathbf{r}(t)$ for t in $[a, b]$ and if C has length $\mathcal{L}$, then C can be parametrized by $\mathbf{r}(t(s))$ for s in $[0, \mathcal{L}]$.

We mentioned earlier that it is usually impossible to compute the length of a curve. Consequently it is usually impossible as well to find a formula for the arc length function. However, ds/dt can often be computed by means of equation (9), without computing any lengths of curves, and in Section 12.5 we will need to do so.

Example 10 Suppose that

$$\mathbf{r}(t) = t\mathbf{i} + t^2\mathbf{j} + t^3\mathbf{k}$$

Find ds/dt.

Solution Using (9), we find that

$$\frac{ds}{dt} = \sqrt{\left(\frac{dx}{dt}\right)^2 + \left(\frac{dy}{dt}\right)^2 + \left(\frac{dz}{dt}\right)^2}$$
$$= \sqrt{1 + (2t)^2 + (3t^2)^2} = \sqrt{1 + 4t^2 + 9t^4} \quad \square$$

EXERCISES 12.4

In Exercises 1–10 determine which of the parametrizations are smooth, which are piecewise smooth, and which are neither.

1. $\mathbf{r}(t) = t\mathbf{i} + t^2\mathbf{j} + t^3\mathbf{k}$

2. $\mathbf{r}(t) = (t-1)\mathbf{i} + (t-1)\mathbf{j} + (t-1)\mathbf{k}$

3. $\mathbf{r}(t) = |t|\mathbf{i} + t\mathbf{j} + t\mathbf{k}$

4. $\mathbf{r}(t) = t^{2/3}\mathbf{i} + t\mathbf{j} + t^2\mathbf{k}$

5. $\mathbf{r}(t) = (1+t)^{3/2}\mathbf{i} + (1-t)^{3/2}\mathbf{j} + \dfrac{3t}{2}\mathbf{k}$

6. $\mathbf{r}(t) = \sin t\mathbf{i} + \cos t\mathbf{j} + t^2\mathbf{k}$

7. $\mathbf{r}(t) = \cos^2 t\mathbf{i} + \sin^2 t\mathbf{j} + t^2\mathbf{k}$

8. $\mathbf{r}(t) = e^t\mathbf{i} + e^{-t}\mathbf{j} + 2t\mathbf{k}$

9. $\mathbf{r}(t) = (e^t - t)\mathbf{i} + t^2\mathbf{j} + t^3\mathbf{k}$

10. $\mathbf{r}(t) = 2t\mathbf{i} + t^2\mathbf{j} + \ln t\mathbf{k}$

In Exercises 11–20 find a smooth parametrization of the curve described.

11. The straight line from $(-3, 2, 1)$ to $(4, 0, 5)$

12. The straight line from $(0, 3, -2)$ to $(6, \frac{1}{2}, -2)$

13. The circle in the xy plane centered at the origin with radius 6

14. The circle in the plane $z = -1$ centered at $(2, 4, -1)$ with radius $\frac{5}{2}$

15. The semicircle in the xy plane that passes through $(1, 0)$, $(0, 1)$, and $(-1, 0)$

16. The quarter circle in the xy plane whose endpoints are $(1, 0)$ and $(0, -1)$ and whose center is the origin

17. The quarter circle in the plane $z = 4$ whose endpoints are $(\sqrt{2}/2, \sqrt{2}/2, 4)$ and $(\sqrt{2}/2, -\sqrt{2}/2, 4)$ and whose center is $(0, 0, 4)$

18. The graph of f, where $f(x) = x^2 + 1$

19. The graph of $y = \tan x$ for $0 \le x \le \pi/4$

20. The graph of $y = x^5 - x^2 + 5$ for $-1 \le x \le 0$

In Exercises 21–28 find the length $\mathscr{L}$ of the curve.

21. $\mathbf{r}(t) = \cos^3 t\mathbf{i} + \sin^3 t\mathbf{j}$ for $0 \le t \le 2\pi$

22. $\mathbf{r}(t) = 2t\mathbf{i} + t^2\mathbf{j} + \ln t\mathbf{k}$ for $1 \le t \le 2$

23. $\mathbf{r}(t) = \frac{1}{3}(1 + t)^{3/2}\mathbf{i} + \frac{1}{3}(1 - t)^{3/2}\mathbf{j} + \frac{1}{2}t\mathbf{k}$ for $-1 \le t \le 1$

24. $\mathbf{r}(t) = \cosh t\mathbf{i} + \sinh t\mathbf{j} + t\mathbf{k}$ for $0 \le t \le 1$

25. $\mathbf{r}(t) = e^t\mathbf{i} + e^{-t}\mathbf{j} + \sqrt{2}t\mathbf{k}$ for $0 \le t \le 1$

26. $\mathbf{r}(t) = \cos t\mathbf{i} + \sin t\mathbf{j} + t^{3/2}\mathbf{k}$ for $0 \le t \le \frac{20}{3}$

27. $\mathbf{r}(t) = 2(t^2 - 1)^{3/2}\mathbf{i} + 3t^2\mathbf{j} + 3t^2\mathbf{k}$ for $1 \le t \le \sqrt{8}$

28. $\mathbf{r}(t) = (3t - t^3)\mathbf{i} + 3t^2\mathbf{j} + (3t + t^3)\mathbf{k}$ for $0 \le t \le 1$

In Exercises 29–33 find ds/dt.

29. $\mathbf{r}(t) = \sin 2t\mathbf{i} + \cos 2t\mathbf{j} + \frac{2}{3}t^{3/2}\mathbf{k}$

30. $\mathbf{r}(t) = \frac{1}{3}t^3\mathbf{i} + \frac{\sqrt{2}}{2}t^2\mathbf{j} + t\mathbf{k}$

31. $\mathbf{r}(t) = t \cos t\mathbf{i} + t \sin t\mathbf{j} + t\mathbf{k}$

32. $\mathbf{r}(t) = 2t\mathbf{i} + t^2\mathbf{j} + \frac{1}{3}t^3\mathbf{k}$

33. $\mathbf{r}(t) = (t - \sin t)\mathbf{i} + (1 - \cos t)\mathbf{j} + t\mathbf{k}$

In Exercises 34–36 find the length $\mathscr{L}$ of the polar graph.

34. $r = a \cos \theta$ for $-\pi/2 \le \theta \le \pi/2$, where $a > 0$

35. $r = a \sin \theta$ for $0 \le \theta \le \pi$, where $a > 0$

36. $r = \theta$ for $0 \le \theta \le 1$

In Exercises 37–39 use Simpson's Rule with $n = 4$ to approximate the length $\mathscr{L}$ of the curve.

37. $\mathbf{r}(t) = \sin t\mathbf{i} + \cos t\mathbf{j} + \frac{1}{3}t^3\mathbf{k}$ for $-1 \le t \le 1$

38. $\mathbf{r}(t) = \cos 2t\mathbf{i} + \sin 2t\mathbf{j} + \frac{4}{5}t^{5/2}\mathbf{k}$ for $0 \le t \le 2$

39. $\mathbf{r}(t) = t\mathbf{i} + t^2\mathbf{j} + t^3\mathbf{k}$ for $-1 \le t \le 1$

40. Consider one arch of the cycloid described by

$$\mathbf{r}(t) = r(t - \sin t)\mathbf{i} + r(1 - \cos t)\mathbf{j} \quad \text{for } 0 \le t \le 2\pi$$

Find a formula for the arc length function of the cycloid.

*41. Show that the complete circle $x^2 + y^2 = 1$ except the point $(-1, 0)$ is parametrized in the counterclockwise direction by

$$\mathbf{r}(t) = \frac{1 - t^2}{1 + t^2}\mathbf{i} + \frac{2t}{1 + t^2}\mathbf{j}$$

*42. Show that both branches of the hyperbola $x^2 - y^2 = 1$, except the point $(1, 0)$, are parametrized by

$$\mathbf{r}(t) = \frac{t^2 + 1}{t^2 - 1}\mathbf{i} + \frac{2t}{t^2 - 1}\mathbf{j} \quad \text{for } |t| \ne 1$$

43. The **Cauchy–Crofton formula** gives an estimate of the length of a plane curve. To present the formula, let $\mathscr{F}$ be a family of parallel lines spaced a distance $d > 0$ apart, and consider the family $\mathscr{G}$ of lines obtained by rotating each line in $\mathscr{F}$ through angles of 0, $\pi/4$, $\pi/2$, and $3\pi/4$ about a fixed point in the plane. If C is the curve whose length $\mathscr{L}$ we wish to estimate, then let n be the total number of intersections of the lines in $\mathscr{G}$ with C, counting each time each line crosses C. Then

$$\mathscr{L} \approx \frac{\pi n d}{8} \tag{11}$$

a. Show that (11) yields the exact circumference of a circle if two lines tangent to the circle are rotated about the center of the circle.

*b. Using (11), approximate the length $\int_{-1}^{1} \sqrt{1 + x^4}\, dx$ of the curve $y = x^3/3$ for $-1 \le x \le 1$. Rotate the lines $x = -1$, $x = -\frac{2}{3}$, $x = -\frac{1}{3}$, $x = 0$, $x = \frac{1}{3}$, $x = \frac{2}{3}$, and $x = 1$ about the origin.

12.5
TANGENTS AND NORMALS TO CURVES

Tangent lines and normal lines to the graph of a real-valued function f are defined by means of the derivative of f. Now we will use the derivatives of vector-valued functions to make analogous definitions for curves in space. These tangents and normals will be vectors rather than lines.

Tangents to Curves Because any smooth curve necessarily has a smooth parametrization, we will assume in this section that any parametrization we encounter for a smooth curve is smooth.

DEFINITION 12.17

Let C be a smooth curve and $\mathbf{r}$ a (smooth) parametrization of C defined on an interval I. Then for any interior point t of I, the **tangent vector** $\mathbf{T}(t)$ at the point $\mathbf{r}(t)$ is defined by

$$\mathbf{T}(t) = \frac{\mathbf{r}'(t)}{\|\mathbf{r}'(t)\|} = \frac{d\mathbf{r}/dt}{\|d\mathbf{r}/dt\|} \tag{1}$$

By definition $\mathbf{T}(t)$ is a unit vector; therefore it is uniquely determined by its direction. In Section 12.3 we saw that the vector $\mathbf{r}'(t)$ can be regarded as tangent to the curve traced out by $\mathbf{r}$. Since $\mathbf{r}'(t)/\|\mathbf{r}'(t)\|$ has the same direction as $\mathbf{r}'(t)$, our name for $\mathbf{T}(t)$ is reasonable (Figure 12.28).

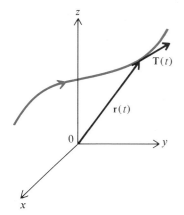

FIGURE 12.28

Example 1 Consider the circle of radius r parametrized by

$$\mathbf{r}(t) = r\cos t\,\mathbf{i} + r\sin t\,\mathbf{j} \quad \text{for } 0 \le t \le 2\pi$$

Find a formula for the tangent vector $\mathbf{T}(t)$, and calculate $\mathbf{T}(\pi/3)$.

Solution Since

$$\mathbf{r}'(t) = -r\sin t\,\mathbf{i} + r\cos t\,\mathbf{j}$$

we have

$$\|\mathbf{r}'(t)\| = \sqrt{(-r\sin t)^2 + (r\cos t)^2} = r$$

Therefore by (1),

$$\mathbf{T}(t) = \frac{\mathbf{r}'(t)}{\|\mathbf{r}'(t)\|} = -\sin t\,\mathbf{i} + \cos t\,\mathbf{j}$$

Consequently

$$\mathbf{T}\left(\frac{\pi}{3}\right) = -\frac{\sqrt{3}}{2}\mathbf{i} + \frac{1}{2}\mathbf{j} \quad \square$$

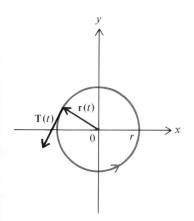

FIGURE 12.29

Notice that for the position function $\mathbf{r}(t)$ in Example 1 we have

$$\mathbf{T}(t) \cdot \mathbf{r}(t) = (-\sin t)(r\cos t) + (\cos t)(r\sin t) = 0 \quad \text{for all } t$$

This means that the tangent vector is perpendicular to the position vector for all t (Figure 12.29).

Example 2 Find a formula for the tangent $\mathbf{T}(t)$ to the circular helix

$$\mathbf{r}(t) = 2\cos t\,\mathbf{i} + 2\sin t\,\mathbf{j} + 3t\mathbf{k}$$

Solution First we calculate the derivative and its length:

$$\frac{d\mathbf{r}}{dt} = -2\sin t\mathbf{i} + 2\cos t\mathbf{j} + 3\mathbf{k}$$

$$\left\|\frac{d\mathbf{r}}{dt}\right\| = \sqrt{(-2\sin t)^2 + (2\cos t)^2 + 3^2} = \sqrt{13}$$

Hence we conclude from (1) that

$$\mathbf{T}(t) = \frac{d\mathbf{r}/dt}{\|d\mathbf{r}/dt\|} = -\frac{2}{\sqrt{13}}\sin t\mathbf{i} + \frac{2}{\sqrt{13}}\cos t\mathbf{j} + \frac{3}{\sqrt{13}}\mathbf{k}$$

(Figure 12.30).   □

Normals to Curves The line normal to the graph of a real-valued function at a given point is by definition perpendicular to the tangent at that point. Similarly, in defining a vector normal to a curve in space at a given point on the curve, we wish the normal to be perpendicular to the tangent vector. However, since there are many vectors in space that are perpendicular to any given vector (Figure 12.31), we need a way of choosing only one of them.

Now suppose a smooth curve C has a parametrization $\mathbf{r}$ that not only is smooth but also has a smooth derivative $\mathbf{r}'$. Then the tangent $\mathbf{T}$, which is defined in terms of $\mathbf{r}'$, is differentiable. Moreover, $\|\mathbf{T}(t)\| = 1$ for all t in the domain of $\mathbf{T}$. It follows from Corollary 12.11 that whenever $\mathbf{T}'(t)$ exists, it satisfies

$$\mathbf{T}'(t) \cdot \mathbf{T}(t) = 0$$

Therefore if $\mathbf{T}'(t) \neq 0$ then $\mathbf{T}'(t)$ is perpendicular to $\mathbf{T}(t)$, and the unit vector in the direction of $\mathbf{T}'(t)$ is our choice for the normal to C at $\mathbf{r}(t)$.

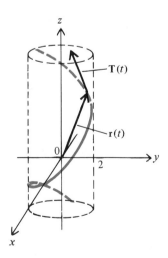

$\mathbf{r}(t) = 2\cos t\mathbf{i} + 2\sin t\mathbf{j} + 3t\mathbf{k}$

FIGURE 12.30

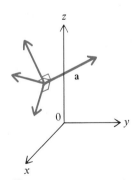

There are many vectors in space perpendicular to a given vector $\mathbf{a}$

FIGURE 12.31

DEFINITION 12.18

Let C be a smooth curve, and let $\mathbf{r}$ be a (smooth) parametrization of C defined on an interval I such that $\mathbf{r}'$ is smooth. Then for any interior point t of I for which $\mathbf{T}'(t) \neq \mathbf{0}$, the **normal vector** $\mathbf{N}(t)$ at the point $\mathbf{r}(t)$ is defined by

$$\mathbf{N}(t) = \frac{\mathbf{T}'(t)}{\|\mathbf{T}'(t)\|} = \frac{d\mathbf{T}/dt}{\|d\mathbf{T}/dt\|} \tag{2}$$

Example 3   Find a formula for the normal $\mathbf{N}(t)$ to the circle

$$\mathbf{r}(t) = r\cos t\,\mathbf{i} + r\sin t\,\mathbf{j}$$

Solution   From Example 1 we know that

$$\mathbf{T}(t) = -\sin t\,\mathbf{i} + \cos t\,\mathbf{j}$$

Therefore

$$\mathbf{T}'(t) = -\cos t\,\mathbf{i} - \sin t\,\mathbf{j}$$

Since $\|\mathbf{T}'(t)\| = 1$ for all t, it follows from (2) that

$$\mathbf{N}(t) = \frac{T'(t)}{\|\mathbf{T}'(t)\|} = -\cos t\,\mathbf{i} - \sin t\,\mathbf{j} = -\frac{1}{r}\mathbf{r}(t) \quad \square$$

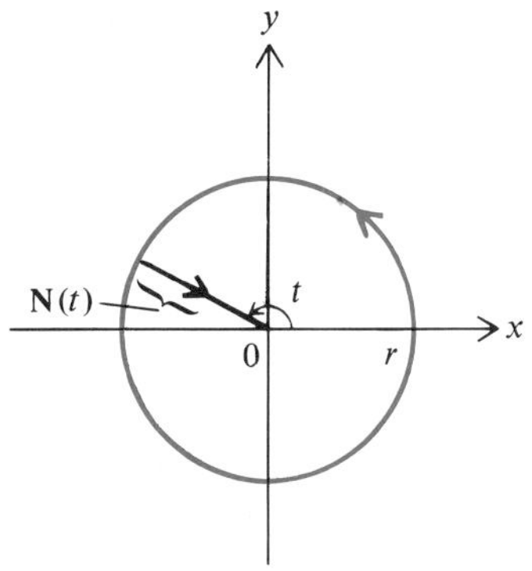

The normal to a circle points toward the center

FIGURE 12.32

We can conclude from Example 3 that the normal to the path of an object moving counterclockwise about a circle points toward the center of the circle, opposite to the radial vector (Figure 12.32). If the circle is traversed clockwise, then the resulting normal also points toward the center of the circle. In either case, at any point on the circle the normal vector is perpendicular to the tangent vector.

Example 4   Find a formula for the normal $\mathbf{N}(t)$ for the parabola

$$\mathbf{r}(t) = t\,\mathbf{i} + t^2\,\mathbf{j}$$

Solution   First we find that

$$\frac{d\mathbf{r}}{dt} = \mathbf{i} + 2t\,\mathbf{j} \quad \text{and} \quad \left\|\frac{d\mathbf{r}}{dt}\right\| = \sqrt{1 + 4t^2}$$

Therefore

$$\mathbf{T}(t) = \frac{1}{\sqrt{1 + 4t^2}}\,\mathbf{i} + \frac{2t}{\sqrt{1 + 4t^2}}\,\mathbf{j}$$

Then

$$\mathbf{T}'(t) = \frac{-4t}{(1 + 4t^2)^{3/2}}\,\mathbf{i} + \frac{2}{(1 + 4t^2)^{3/2}}\,\mathbf{j} \quad \text{and} \quad \|\mathbf{T}'(t)\| = \frac{2}{1 + 4t^2}$$

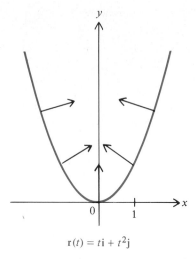

$$\mathbf{r}(t) = t\mathbf{i} + t^2\mathbf{j}$$

FIGURE 12.33

so we obtain the formula

$$\mathbf{N}(t) = \frac{-2t}{\sqrt{1 + 4t^2}}\mathbf{i} + \frac{1}{\sqrt{1 + 4t^2}}\mathbf{j}$$

In Figure 12.33 we have drawn a few normal vectors for the parabola. □

The curves we investigated in Examples 3 and 4 lie in the xy plane. At any point at which either of the curves is concave upward, the normal vector has a positive $\mathbf{j}$ component (that is, the normal vector points upward), and at any point at which either is concave downward, the normal vector has a negative $\mathbf{j}$ component (that is, the normal vector points downward). It is possible to show that these properties are shared by any smooth curve in the xy plane. Thus the normal vectors for the graph of the sine function are as shown in Figure 12.34.

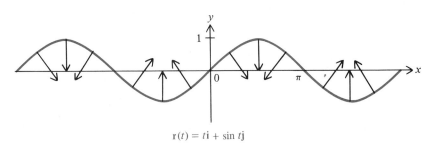

$$\mathbf{r}(t) = t\mathbf{i} + \sin t\mathbf{j}$$

FIGURE 12.34

Example 5 Find a formula for the normal $\mathbf{N}(t)$ to the circular helix

$$\mathbf{r}(t) = 2\cos t\mathbf{i} + 2\sin t\mathbf{j} + 3t\mathbf{k}$$

Solution From Example 2 we have

$$\mathbf{T}(t) = -\frac{2}{\sqrt{13}}\sin t\mathbf{i} + \frac{2}{\sqrt{13}}\cos t\mathbf{j} + \frac{3}{\sqrt{13}}\mathbf{k}$$

Therefore

$$\frac{d\mathbf{T}}{dt} = -\frac{2}{\sqrt{13}}\cos t\mathbf{i} - \frac{2}{\sqrt{13}}\sin t\mathbf{j} \quad \text{and} \quad \left\|\frac{d\mathbf{T}}{dt}\right\| = \frac{2}{\sqrt{13}}$$

Consequently from (2) we deduce that

$$\mathbf{N}(t) = \frac{d\mathbf{T}/dt}{\|d\mathbf{T}/dt\|} = \frac{-(2/\sqrt{13})\cos t\mathbf{i} - (2/\sqrt{13})\sin t\mathbf{j}}{2/\sqrt{13}}$$

$$= -\cos t\mathbf{i} - \sin t\mathbf{j} \quad \square$$

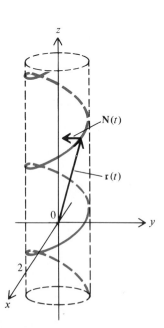

FIGURE 12.35

We see from Example 5 that the vector normal to the helix is always perpendicular to and directed toward the z axis (Figure 12.35).

Tangential and Normal Components of Acceleration

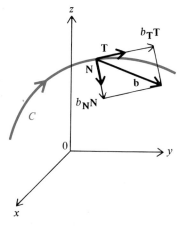

FIGURE 12.36

Since the tangent vector $\mathbf{T}$ and the normal vector $\mathbf{N}$ at any point on a smooth curve C are perpendicular, it follows from our discussion in Section 11.3 of the resolution of a vector that any vector $\mathbf{b}$ in the plane determined by $\mathbf{T}$ and $\mathbf{N}$ can be expressed in the form

$$\mathbf{b} = b_{\mathbf{T}}\mathbf{T} + b_{\mathbf{N}}\mathbf{N}$$

(Figure 12.36). We call $b_{\mathbf{T}}$ and $b_{\mathbf{N}}$ the **tangential** and **normal components** of $\mathbf{b}$, respectively. Our next tasks will be to show that the velocity and acceleration vectors of an object moving along C lie in the plane determined by $\mathbf{T}$ and $\mathbf{N}$ and to find their tangential and normal components.

To consider the velocity $\mathbf{v}$, we let $\mathbf{r}$ denote the position of the object and assume that $\mathbf{T}$ and $\mathbf{N}$ exist. Then by (1) we have

$$\mathbf{v} = \frac{d\mathbf{r}}{dt} = \left\|\frac{d\mathbf{r}}{dt}\right\|\mathbf{T} = \|\mathbf{v}\|\,\mathbf{T} \tag{3}$$

Thus the tangential component of $\mathbf{v}$ is the speed $\|\mathbf{v}\|$ of the object, and the normal component of the velocity is 0.

To obtain the acceleration $\mathbf{a}$, we differentiate the left-hand and right-hand expressions in (3), obtaining

$$\mathbf{a} = \frac{d\mathbf{v}}{dt} = \frac{d\|\mathbf{v}\|}{dt}\mathbf{T} + \|\mathbf{v}\|\frac{d\mathbf{T}}{dt}$$

Since $d\mathbf{T}/dt = \|d\mathbf{T}/dt\|\,\mathbf{N}$ from (2), the equation for $\mathbf{a}$ becomes

$$\mathbf{a} = \frac{d\|\mathbf{v}\|}{dt}\mathbf{T} + \|\mathbf{v}\|\left\|\frac{d\mathbf{T}}{dt}\right\|\mathbf{N}$$

Hence

$$\mathbf{a} = a_{\mathbf{T}}\mathbf{T} + a_{\mathbf{N}}\mathbf{N} \tag{4}$$

where

$$a_{\mathbf{T}} = \frac{d\|\mathbf{v}\|}{dt} \quad\text{and}\quad a_{\mathbf{N}} = \|\mathbf{v}\|\left\|\frac{d\mathbf{T}}{dt}\right\| \tag{5}$$

The numbers $a_{\mathbf{T}}$ and $a_{\mathbf{N}}$ are the **tangential** and **normal components of acceleration**. Since $\mathbf{T}$ and $\mathbf{N}$ are mutually perpendicular unit vectors, we conclude from (4) that

$$\|\mathbf{a}\|^2 = \mathbf{a}\cdot\mathbf{a} = (a_{\mathbf{T}}\mathbf{T} + a_{\mathbf{N}}\mathbf{N})\cdot(a_{\mathbf{T}}\mathbf{T} + a_{\mathbf{N}}\mathbf{N}) = a_{\mathbf{T}}^2\|\mathbf{T}\|^2 + a_{\mathbf{N}}^2\|\mathbf{N}\|^2$$
$$= a_{\mathbf{T}}^2 + a_{\mathbf{N}}^2$$

It follows that $a_{\mathbf{N}}$, which is often quite difficult to calculate by (5), can be calculated from $\mathbf{a}$ and $a_{\mathbf{T}}$:

$$a_{\mathbf{N}} = \sqrt{\|\mathbf{a}\|^2 - a_{\mathbf{T}}^2} \tag{6}$$

Example 6   Let

$$\mathbf{r}(t) = t^2\mathbf{i} + t\mathbf{j} + t^2\mathbf{k}$$

Find the tangential and normal components of acceleration.

Solution   By differentiating $\mathbf{r}$ we find that

$$\mathbf{v} = 2t\mathbf{i} + \mathbf{j} + 2t\mathbf{k}$$

and thus
$$\|\mathbf{v}\| = \sqrt{(2t)^2 + 1 + (2t)^2} = \sqrt{8t^2 + 1}$$

From this and (5) it follows that

$$a_{\mathbf{T}} = \frac{d\|\mathbf{v}\|}{dt} = \frac{8t}{\sqrt{8t^2 + 1}}$$

In order to calculate $a_{\mathbf{N}}$, we first notice that

$$\mathbf{a} = \frac{d\mathbf{v}}{dt} = 2\mathbf{i} + 2\mathbf{k}$$

so that
$$\|\mathbf{a}\| = \sqrt{2^2 + 2^2} = 2\sqrt{2}$$

Therefore we conclude from (6) that

$$a_{\mathbf{N}} = \sqrt{\|\mathbf{a}\|^2 - a_{\mathbf{T}}^2} = \sqrt{8 - \frac{64t^2}{8t^2 + 1}} = \frac{2\sqrt{2}}{\sqrt{8t^2 + 1}} \quad \square$$

Orientations of Curves   Any piecewise smooth parametrization $\mathbf{r} = x\mathbf{i} + y\mathbf{j} + z\mathbf{k}$ of a curve C determines a tangent vector at all but finitely many points of C. Since each tangent vector points in the direction in which the curve is traced out by $\mathbf{r}$, we say that $\mathbf{r}$ determines an **orientation** (or **direction**) of C (Figure 12.37(a)).

Usually there are two orientations for a given piecewise smooth curve, much as there are two directions in which one can drive on a highway. The tangent vectors for one orientation have directions opposite to the tangent vectors for the other orientation. Once a piecewise smooth curve C has a given orientation, the tangent vectors to C are uniquely defined, independent of any parametrization $\mathbf{r}$ of C.

Suppose $\mathbf{r}$ is a piecewise smooth parametrization of C on $[a, b]$, and let

$$\mathbf{r}_1(t) = \mathbf{r}(a + b - t) \quad \text{for } a \le t \le b$$

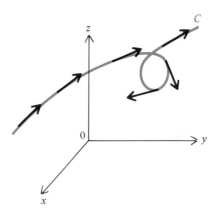

An orientation of a piecewise
smooth curve C

(a)

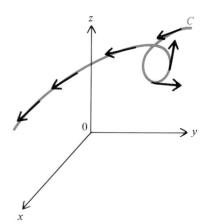

The opposite orientation of C

(b)

FIGURE 12.37

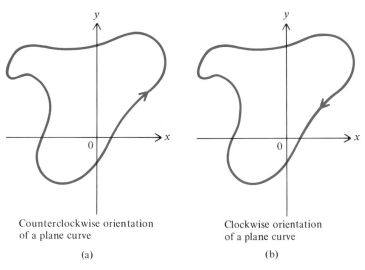

Counterclockwise orientation of a plane curve
(a)

Clockwise orientation of a plane curve
(b)

FIGURE 12.38

Then $\mathbf{r}_1$ is also a piecewise smooth parametrization of C and determines an orientation opposite to the orientation determined by $\mathbf{r}$ (Figure 12.37(b)). When we consider C with a given orientation, we designate the curve with the opposite orientation by $-C$. In the case of a plane curve that is closed and does not "cross itself," the two orientations are called the **counterclockwise orientation** and the **clockwise orientation** (Figure 12.38).

An **oriented curve** is a piecewise smooth curve with a particular orientation associated with it. In Chapter 15 we will make use of oriented curves.

EXERCISES 12.5

In Exercises 1–10 first show that $\|\mathbf{r}'(t)\|$ is as given. Then find the tangent and the normal of the curve parametrized by $\mathbf{r}$.

1. $\mathbf{r}(t) = (t^2 + 4)\mathbf{i} + 2t\mathbf{j}; \; \|\mathbf{r}'(t)\| = 2\sqrt{t^2 + 1}$

2. $\mathbf{r}(t) = \cos^3 t\mathbf{i} + \sin^3 t\mathbf{j}$ for $\pi/6 \le t \le \pi/3$;
   $\|\mathbf{r}'(t)\| = 3 \sin t \cos t$

3. $\mathbf{r}(t) = \cos t\mathbf{i} + \cos t\mathbf{j} + \sqrt{2} \sin t\mathbf{k}; \; \|\mathbf{r}'(t)\| = \sqrt{2}$

4. $\mathbf{r}(t) = \dfrac{1}{3}(1 + t)^{3/2}\mathbf{i} + \dfrac{1}{3}(1 - t)^{3/2}\mathbf{j} + \dfrac{\sqrt{2}}{2}t\mathbf{k}; \; \|\mathbf{r}'(t)\| = 1$

5. $\mathbf{r}(t) = 2t\mathbf{i} + t^2\mathbf{j} + \frac{1}{3}t^3\mathbf{k}; \; \|\mathbf{r}'(t)\| = 2 + t^2$

6. $\mathbf{r}(t) = \frac{4}{5}\cos t\mathbf{i} + (1 - \sin t)\mathbf{j} - \frac{3}{5}\cos t\mathbf{k}; \; \|\mathbf{r}'(t)\| = 1$

7. $\mathbf{r}(t) = e^t\mathbf{i} + e^{-t}\mathbf{j} + \sqrt{2}t\mathbf{k}; \; \|\mathbf{r}'(t)\| = e^t + e^{-t}$

8. $\mathbf{r}(t) = \cosh t\mathbf{i} + \sinh t\mathbf{j} + t\mathbf{k}; \; \|\mathbf{r}'(t)\| = \sqrt{2}\cosh t$

9. $\mathbf{r}(t) = 2t\mathbf{i} + t^2\mathbf{j} + \ln t\mathbf{k}; \; \|\mathbf{r}'(t)\| = \dfrac{2t^2 + 1}{t}$

10. $\mathbf{r}(t) = 2t^{9/2}\mathbf{i} + \frac{3}{2}\sqrt{2}t^3\mathbf{j} + \frac{3}{2}\sqrt{2}t^3\mathbf{k}$;
    $\|\mathbf{r}'(t)\| = 9t^2\sqrt{t^3 + 1}$

In Exercises 11–15 find the tangential and normal components $a_{\mathbf{T}}$ and $a_{\mathbf{N}}$ of acceleration.

11. $\mathbf{r}(t) = r(t - \sin t)\mathbf{i} + r(1 - \cos t)\mathbf{j}$

12. $\mathbf{r}(t) = 2 \cos t\mathbf{i} + 3 \sin t\mathbf{j}$

13. $\mathbf{r}(t) = 2t\mathbf{i} + t^2\mathbf{j} + \frac{1}{3}t^3\mathbf{k}$

14. $\mathbf{r}(t) = \frac{4}{5}\cos t\mathbf{i} + (1 - \sin t)\mathbf{j} - \frac{3}{5}\cos t\mathbf{k}$

15. $\mathbf{r}(t) = e^t\mathbf{i} + e^{-t}\mathbf{j} + \sqrt{2}t\mathbf{k}$

In Exercises 16–21 find a piecewise smooth (smooth if possible) parametrization with the given orientation for the curve.

16. The circle in the plane $x = -2$ centered at the point $(-2, -2, -1)$, with radius 3 and with a clockwise orientation as viewed from the yz plane.

17. The semicircle in the yz plane that begins at $(0, 0, 4)$, passes through $(0, -4, 0)$, and terminates at $(0, 0, -4)$.

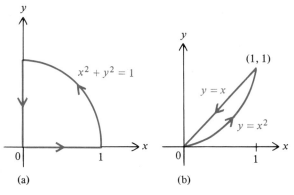

FIGURE 12.39

18. The curve shown in Figure 12.39(a)

19. The curve shown in Figure 12.39(b)

20. The curve shown in Figure 12.39(c)

21. The curve shown in Figure 12.39(d)

22. Show that the curves parametrized by

$$\mathbf{r}_1(t) = \left(\frac{1}{2}t^2 + t\right)\mathbf{i} + (t + 1)\mathbf{j} - t\mathbf{k}$$

and $\quad \mathbf{r}_2(t) = \sin t\,\mathbf{i} + e^t\mathbf{j} - \tan t\,\mathbf{k}$

intersect at the point $(0, 1, 0)$ and that the vectors tangent to the two curves at $(0, 1, 0)$ are parallel.

23. Show that the curves parametrized by

$$\mathbf{r}_1(t) = t\mathbf{i} + 2t\mathbf{j} + t^2\mathbf{k}$$

and $\quad \mathbf{r}_2(t) = t^2\mathbf{i} + (1 - t)\mathbf{j} + (2 - t^2)\mathbf{k}$

intersect at the point $(1, 2, 1)$ and that the vectors tangent to the two curves at $(1, 2, 1)$ are perpendicular. (*Hint:* For what values of t do $\mathbf{r}_1$ and $\mathbf{r}_2$ trace out the point $(1, 2, 1)$?)

24. Suppose an object moves in such a way that its acceleration vector is always perpendicular to its velocity vector. Show that the speed of the object is constant.

25. Prove that the tangential component of acceleration of an object is 0 if and only if the speed is constant.

26. In Example 1 we showed that a vector tangent to a circle with center at the origin is perpendicular to the position vector. Of course, such a circle lies on the surface of a sphere centered at the origin. Using Corollary 12.11, prove more generally that the vector tangent to any smooth curve lying on the surface of a sphere centered at the origin is perpendicular to the position vector.

27. Show that the graph of the sine function has no normal vector at $(n\pi, 0)$ for any integer n (see Figure 12.34).

28. Let $\mathbf{r}(t) = x(t)\mathbf{i} + y(t)\mathbf{j}$ for $a \le t \le b$ be a smooth parametrization of a curve C. Find a formula for $\mathbf{T}(t)$ in terms of x and y and their derivatives.

29. Let the curve C traced out by a vector-valued function $\mathbf{r}$ have tangent $\mathbf{T}$ and normal $\mathbf{N}$. The vector-valued function $\mathbf{B}$ defined by

$$\mathbf{B} = \mathbf{T} \times \mathbf{N}$$

is called the **binormal** of C. Notice that $\mathbf{B}$, $\mathbf{T}$, and $\mathbf{N}$ are mutually perpendicular. Find the binormal of the helix in Example 5.

*30. The **rated speed** v_R of a banked curve on a road is the maximum speed a car can attain on the curve without skidding outward, under the assumption that there is no friction between the road and the tires (under icy road conditions, for example). Suppose that the curve is circular with radius ρ and that it is banked at an angle θ

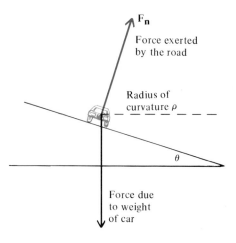

FIGURE 12.40

with respect to the horizontal (Figure 12.40). When a car of mass m traverses the curve with a constant speed v, there are two forces acting on the car: the vertical force (directed downward) due to the weight of the car, and a force $\mathbf{F_n}$ exerted by and normal to the road. At the rated speed the vertical component of $\mathbf{F_n}$ balances the weight of the car, so that

$$\|\mathbf{F_n}\| \cos \theta = mg \qquad (7)$$

The horizontal component of $\mathbf{F_n}$ produces a centripetal force on the car, so that by Newton's Second Law of Motion and (4) in Section 12.3,

$$\|\mathbf{F_n}\| \sin \theta = \frac{mv_{\mathbf{R}}^2}{\rho} \qquad (8)$$

a. Using (7) and (8), show that

$$v_{\mathbf{R}}^2 = \rho g \tan \theta$$

b. Find the rated speed of a circular curve with radius 500 feet that is banked at an angle of $\pi/12$.

c. What radius of the curve will result in doubling the rated speed found in (b) for the same angle θ of banking?

12.6
CURVATURE

Let C be a smooth curve. The length of the tangent vector is always 1; the direction can vary from point to point, according to the nature of the curve. For example, if the curve is a straight line, as in Figure 12.41(a), then the direction of the tangent vector is constant. If the curve undulates gently, as in Figure 12.41(b), then the tangent vector changes direction slowly along the curve. Finally, if the curve twists as in Figure 12.41(c), then the tangent vector changes direction rapidly. Figure 12.41(a)–(c) suggests that the rate of change $d\mathbf{T}/ds$ of the tangent

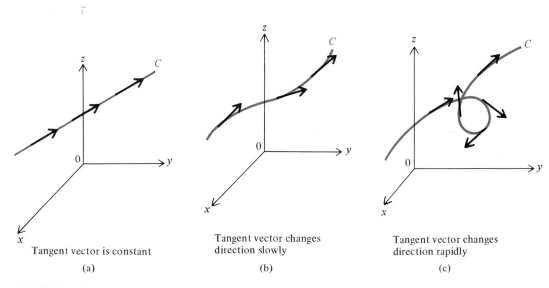

Tangent vector is constant

(a)

Tangent vector changes direction slowly

(b)

Tangent vector changes direction rapidly

(c)

FIGURE 12.41

vector with respect to the arc length function s is closely related to the rate at which the curve twists and turns. Since

$$\frac{d\mathbf{T}}{dt} = \frac{d\mathbf{T}}{ds}\frac{ds}{dt}$$

it follows that

$$\frac{d\mathbf{T}}{ds} = \frac{d\mathbf{T}/dt}{ds/dt} = \frac{d\mathbf{T}/dt}{\|d\mathbf{r}/dt\|}$$

This suggests the following definition of the curvature κ of a curve (κ is the Greek letter kappa).

DEFINITION 12.19

Let C have a smooth parametrization $\mathbf{r}$ such that $\mathbf{r}'$ is differentiable. Then the **curvature** κ of C is defined by the formula

$$\kappa(t) = \frac{\|\mathbf{T}'(t)\|}{\|\mathbf{r}'(t)\|} = \frac{\|d\mathbf{T}/dt\|}{\|d\mathbf{r}/dt\|} \tag{1}$$

By definition $\kappa(t)$ is a nonnegative number. It is possible for $\kappa(t)$ to be 0; in fact, the curvature of a straight line is 0 at every point, since the tangent vector is constant. We can also readily determine the curvature of a circle.

Example 1   Show that the curvature κ of a circle of radius r is $1/r$.

Solution   If the center of the circle is (x_0, y_0), then we parametrize the circle by

$$\mathbf{r}(t) = (x_0 + r\cos t)\mathbf{i} + (y_0 + r\sin t)\mathbf{j}$$

Then

$$\mathbf{r}'(t) = -r\sin t\,\mathbf{i} + r\cos t\,\mathbf{j}$$

Moreover,

$$\|\mathbf{r}'(t)\| = r \quad\text{and}\quad \mathbf{T}(t) = -\sin t\,\mathbf{i} + \cos t\,\mathbf{j}$$

Therefore $\mathbf{T}'(t) = -\cos t\,\mathbf{i} - \sin t\,\mathbf{j}$, so that by (1),

$$\kappa(t) = \frac{\|\mathbf{T}'(t)\|}{\|\mathbf{r}'(t)\|} = \frac{1}{r}\|-\cos t\,\mathbf{i} - \sin t\,\mathbf{j}\| = \frac{1}{r}$$

A circle of radius r has constant curvature $1/r$

FIGURE 12.42

(see Figure 12.42).   □

It follows from Example 1 that the larger the radius of a circle is, the smaller its curvature is. This explains why the earth was once thought to be flat; its very large radius gives it a curvature imperceptible to anyone on it.

Because the curvature of a circle with radius r is $1/r$, we define the **radius of curvature** $\rho(t)$ of a curve C at a point P corresponding to t to be

$$\rho(t) = \frac{1}{\kappa(t)}$$

Thus the radius of curvature of a circle is the radius. The radius of curvature is important in engineering problems.

In our next example the curve has a variable curvature.

Example 2   Find the curvature κ of the parabola parametrized by

$$\mathbf{r}(t) = t\mathbf{j} + t^2\mathbf{j}$$

Solution   First we make the necessary computations:

$$\frac{d\mathbf{r}}{dt} = \mathbf{i} + 2t\mathbf{j}$$

$$\left\|\frac{d\mathbf{r}}{dt}\right\| = \sqrt{1 + 4t^2}$$

$$\mathbf{T}(t) = \frac{1}{\sqrt{1 + 4t^2}}\mathbf{i} + \frac{2t}{\sqrt{1 + 4t^2}}\mathbf{j}$$

$$\frac{d\mathbf{T}}{dt} = \frac{-4t}{(1 + 4t^2)^{3/2}}\mathbf{i} + \frac{2}{(1 + 4t^2)^{3/2}}\mathbf{j}$$

$$\left\|\frac{d\mathbf{T}}{dt}\right\| = \frac{2}{1 + 4t^2}$$

Finally, we use (1) to obtain

$$\kappa(t) = \frac{\|d\mathbf{T}/dt\|}{\|d\mathbf{r}/dt\|} = \frac{2/(1 + 4t^2)}{(1 + 4t^2)^{1/2}} = \frac{2}{(1 + 4t^2)^{3/2}} \quad \square$$

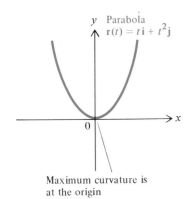

y   Parabola
$\mathbf{r}(t) = t\mathbf{i} + t^2\mathbf{j}$

Maximum curvature is at the origin

FIGURE 12.43

Notice that the curvature of the parabola is largest when $t = 0$ and gradually diminishes as $|t|$ increases (Figure 12.43). Since the curve is symmetric with respect to the y axis, naturally the curvature is the same for t and for $-t$.

Alternative Formulas for Curvature

By using the decomposition of the velocity and acceleration vectors into their tangential and normal components, we can obtain a formula for curvature that gives rise to simpler computations than (1) does.

Let $\mathbf{r}$ be a smooth parametrization of a smooth curve C with tangent $\mathbf{T}$ and normal $\mathbf{N}$. Recall from (3) and (4) in Section 12.5 that the velocity and acceleration of an object moving along C with position $\mathbf{r}$ are given by

$$\mathbf{v} = \|\mathbf{v}\|\mathbf{T} \quad \text{and} \quad \mathbf{a} = a_{\mathbf{T}}\mathbf{T} + a_{\mathbf{N}}\mathbf{N}$$

Then since $\mathbf{T} \times \mathbf{T} = \mathbf{0}$ we find that

$$\begin{aligned}
\mathbf{v} \times \mathbf{a} &= (\|\mathbf{v}\|\mathbf{T}) \times (a_{\mathbf{T}}\mathbf{T} + a_{\mathbf{N}}\mathbf{N}) \\
&= [(\|\mathbf{v}\|\mathbf{T}) \times (a_{\mathbf{T}}\mathbf{T})] + [(\|\mathbf{v}\|\mathbf{T}) \times (a_{\mathbf{N}}\mathbf{N})] \\
&= (\|\mathbf{v}\| a_{\mathbf{N}})(\mathbf{T} \times \mathbf{N})
\end{aligned} \tag{2}$$

Since $a_{\mathbf{N}} = \|\mathbf{v}\| \|d\mathbf{T}/dt\|$ by (5) in Section 12.5, and since $\mathbf{T}$ and $\mathbf{N}$ are perpendicular unit vectors, it follows from (2) that

$$\|\mathbf{v} \times \mathbf{a}\| = \|\mathbf{v}\| a_{\mathbf{N}} = \|\mathbf{v}\|^2 \left\|\frac{d\mathbf{T}}{dt}\right\|$$

However,

$$\kappa = \frac{\|d\mathbf{T}/dt\|}{\|d\mathbf{r}/dt\|} = \frac{\|d\mathbf{T}/dt\|}{\|\mathbf{v}\|}$$

so that

$$\kappa = \frac{\|\mathbf{v} \times \mathbf{a}\|}{\|\mathbf{v}\|^3} \tag{3}$$

Example 3   Using (3), find the curvature κ of the curve

$$\mathbf{r}(t) = \frac{1}{3}t^3\mathbf{i} + \frac{\sqrt{2}}{2}t^2\mathbf{j} + t\mathbf{k}$$

Solution   We have

$$\mathbf{v} = \frac{d\mathbf{r}}{dt} = t^2\mathbf{i} + \sqrt{2}t\mathbf{j} + \mathbf{k}$$

$$\|\mathbf{v}\| = \sqrt{t^4 + 2t^2 + 1} = t^2 + 1$$

$$\mathbf{a} = \frac{d\mathbf{v}}{dt} = 2t\mathbf{i} + \sqrt{2}\mathbf{j}$$

Therefore

$$\mathbf{v} \times \mathbf{a} = \begin{vmatrix} \mathbf{i} & \mathbf{j} & \mathbf{k} \\ t^2 & \sqrt{2}t & 1 \\ 2t & \sqrt{2} & 0 \end{vmatrix} = -\sqrt{2}\mathbf{i} + 2t\mathbf{j} - \sqrt{2}t^2\mathbf{k}$$

and thus

$$\|\mathbf{v} \times \mathbf{a}\| = \sqrt{(-\sqrt{2})^2 + (2t)^2 + (-\sqrt{2}t^2)^2}$$
$$= \sqrt{2 + 4t^2 + 2t^4} = \sqrt{2}(t^2 + 1)$$

Consequently by (3),

$$\kappa = \frac{\|\mathbf{v} \times \mathbf{a}\|}{\|\mathbf{v}\|^3} = \frac{\sqrt{2}(t^2 + 1)}{(t^2 + 1)^3} = \frac{\sqrt{2}}{(t^2 + 1)^2} \quad \square$$

In the event that $\mathbf{r}$ represents an object moving on a curve in the xy plane, we have

$$\mathbf{r} = x\mathbf{i} + y\mathbf{j}$$

$$\mathbf{v} = \frac{dx}{dt}\mathbf{i} + \frac{dy}{dt}\mathbf{j}$$

$$\mathbf{a} = \frac{d^2x}{dt^2}\mathbf{i} + \frac{d^2y}{dt^2}\mathbf{j}$$

By calculating the cross product in (3) we obtain the formula

$$\kappa = \frac{\left| \dfrac{dx}{dt}\dfrac{d^2y}{dt^2} - \dfrac{d^2x}{dt^2}\dfrac{dy}{dt} \right|}{\left[\left(\dfrac{dx}{dt}\right)^2 + \left(\dfrac{dy}{dt}\right)^2 \right]^{3/2}} \tag{4}$$

If $x = t$ (so that y is also a function of x), then

$$\kappa = \frac{|d^2y/dx^2|}{[1 + (dy/dx)^2]^{3/2}} \tag{5}$$

Example 4 Let one arch of a cycloid be described by

$$\mathbf{r}(t) = r(t - \sin t)\mathbf{i} + r(1 - \cos t)\mathbf{j} \quad \text{for } 0 < t < 2\pi$$

Find the curvature κ by means of (4).

Solution Here we have

$$x(t) = r(t - \sin t) \qquad\qquad y(t) = r(1 - \cos t)$$

$$\frac{dx}{dt} = r(1 - \cos t) \qquad\qquad \frac{dy}{dt} = r \sin t$$

$$\frac{d^2x}{dt^2} = r \sin t \qquad\qquad \frac{d^2y}{dt^2} = r \cos t$$

Then by (4),

$$\kappa = \frac{|r(1 - \cos t)(r \cos t) - (r \sin t)(r \sin t)|}{[r^2(1 - \cos t)^2 + r^2 \sin^2 t]^{3/2}}$$

$$= \frac{r^2(1 - \cos t)}{[r^2(2 - 2\cos t)]^{3/2}} = \frac{1}{2^{3/2} r \sqrt{1 - \cos t}} \quad \square$$

EXERCISES 12.6

In Exercises 1–10 use (1) to find the curvature of the curve traced out by $\mathbf{r}$.

1. $\mathbf{r}(t) = (t^2 + 4)\mathbf{i} + 2t\mathbf{j}$

2. $\mathbf{r}(t) = \cos^3 t\,\mathbf{i} + \sin^3 t\,\mathbf{j}$ for $\pi/6 \le t \le \pi/3$

3. $\mathbf{r}(t) = \cos t\,\mathbf{i} + \cos t\,\mathbf{j} + \sqrt{2} \sin t\,\mathbf{k}$

4. $\mathbf{r}(t) = \dfrac{1}{3}(1 + t)^{3/2}\mathbf{i} + \dfrac{1}{3}(1 - t)^{3/2}\mathbf{j} + \dfrac{\sqrt{2}}{2} t\mathbf{k}$

5. $\mathbf{r}(t) = 2t\mathbf{i} + t^2\mathbf{j} + \frac{1}{3}t^3\mathbf{k}$

6. $\mathbf{r}(t) = \frac{4}{5}\cos t\,\mathbf{i} + (1 - \sin t)\mathbf{j} - \frac{3}{5}\cos t\mathbf{k}$

7. $\mathbf{r}(t) = e^t\mathbf{i} + e^{-t}\mathbf{j} + \sqrt{2}\,t\mathbf{k}$

8. $\mathbf{r}(t) = \cosh t\,\mathbf{i} + \sinh t\,\mathbf{j} + t\mathbf{k}$

9. $\mathbf{r}(t) = 2t\mathbf{i} + t^2\mathbf{j} + \ln t\mathbf{k}$

10. $\mathbf{r}(t) = 2t^{9/2}\mathbf{i} + \frac{3}{2}\sqrt{2}t^3\mathbf{j} + \frac{3}{2}\sqrt{2}t^3\mathbf{k}$

In Exercises 11–15 use (3) to find the curvature κ of the curve parametrized by **r**.

11. $\mathbf{r}(t) = (2t - 1)\mathbf{i} + (t^2 + 1)\mathbf{j}$

12. $\mathbf{r}(t) = t\cos t\,\mathbf{i} + t\sin t\,\mathbf{j}$

13. $\mathbf{r}(t) = e^t\sin t\,\mathbf{i} + e^t\cos t\,\mathbf{j} + t\mathbf{k}$

14. $\mathbf{r}(t) = \dfrac{1}{3}(t^2 - 1)^{3/2}\mathbf{i} + \dfrac{\sqrt{2}}{4}t^2\mathbf{j} + \dfrac{\sqrt{2}}{4}t^2\mathbf{k}$

15. $\mathbf{r}(t) = \sin t\,\mathbf{i} + \cos t\,\mathbf{j} + \frac{2}{3}t^{3/2}\mathbf{k}$

In Exercises 16–19 use (4) to find the curvature κ of the plane curve parametrized by **r**. Then find the radius of curvature at the point on the curve corresponding to the given value of t_0.

16. $\mathbf{r}(t) = 2\cos t\,\mathbf{i} + 3\sin t\,\mathbf{j}; \; t_0 = 0$

17. $\mathbf{r}(t) = 2\cos t\,\mathbf{i} + 3\sin t\,\mathbf{j}; \; t_0 = \pi/2$

18. $\mathbf{r}(t) = 2\cosh t\,\mathbf{i} + 3\sinh t\,\mathbf{j}; \; t_0 = 0$

19. $\mathbf{r}(t) = t\mathbf{i} + \frac{1}{3}t^3\mathbf{j}; \; t_0 = 1$

In Exercises 20–23 use (5) to find the curvature κ of the graph of the given equation.

20. $y = \sin x$ 21. $y = \ln x$

22. $y = x^{1/3}$ for $x > 0$ 23. $y = \dfrac{1}{x}$ for $x < 0$

24. Find all points on the ellipse $4x^2 + 9y^2 = 36$ at which the curvature is maximum and all points at which it is minimum. (*Hint:* The ellipse is parametrized by $\mathbf{r}(t) = 3\cos t\,\mathbf{i} + 2\sin t\,\mathbf{j}$ for $0 \le t \le 2\pi$.)

25. Find the point on the graph of $y = e^x$ at which the curvature is maximum.

26. Let g be a polynomial function, and define

$$f(x) = \begin{cases} -x & \text{for } x < -1 \\ g(x) & \text{for } -1 \le x \le 1 \\ x & \text{for } x > 1 \end{cases}$$

    a. Show that if $g(x) = -\frac{1}{8}x^4 + \frac{3}{4}x^2 + \frac{3}{8}$, then the curvature of the graph of f is continuous.

    b. Show that if g is *any* 3rd-degree polynomial, then the curvature of the graph of f is not continuous simultaneously at $x = -1$ and $x = 1$.

27. Show that the helix $\mathbf{r}(t) = \cos t\,\mathbf{i} + \sin t\,\mathbf{j} + t\mathbf{k}$ has constant curvature.

28. Suppose the graph of f has an inflection point at $(a, f(a))$, and assume that $f''(a)$ exists. Show that the curvature of the graph at $(a, f(a))$ is 0.

29. a. Using (5) in Section 12.5 and Definition 12.19, show that $a_{\mathbf{N}} = \kappa\|\mathbf{v}\|^2$.

    b. Use the result of part (a) and Exercise 28 to show that if a particle traverses the graph of the sine function in the xy plane, then the normal component of the particle's acceleration at the point $(\pi, 0)$ is 0.

30. Assume that f has a second derivative. Show that the curvature of the polar graph of the equation $r = f(\theta)$ is given by

$$\kappa(\theta) = \frac{|2[f'(\theta)]^2 - f(\theta)f''(\theta) + [f(\theta)]^2|}{[(f'(\theta))^2 + (f(\theta))^2]^{3/2}}$$

(*Hint:* Remember that $x = r\cos\theta$ and $y = r\sin\theta$. Then apply (4) to the parametrization $r(\theta) = f(\theta)\cos\theta\,\mathbf{i} + f(\theta)\sin\theta\,\mathbf{j}$.)

31. Using the result of Exercise 30, find the curvature of the single leaf of the three-leaved rose $r = \sin 3\theta$ for $0 < \theta < \pi/3$.

32. Using the result of Exercise 30, find the curvature of the cardioid $r = 1 - \cos\theta$ for $0 < \theta < 2\pi$.

*33. a. Let κ be any nonnegative function that is differentiable on an interval I, and let a be any interior point of I. Let

$$\theta(t) = \int_a^t \kappa(u)\,du$$

and

$$\mathbf{r}(t) = \left[\int_a^t \cos\theta(u)\,du\right]\mathbf{i} + \left[\int_a^t \sin\theta(u)\,du\right]\mathbf{j}$$

Show that the curvature of the curve traced out by **r** is κ. (Thus we have a way of describing a curve with any prescribed curvature.)

    b. Using (a), find a parametrization on $(-1, 1)$ of a curve whose curvature is

$$\kappa(t) = \frac{1}{\sqrt{1 - t^2}} \quad \text{for } -1 < t < 1$$

    c. Using (a), find a parametrization of a curve whose curvature is

$$\kappa(t) = \frac{1}{1 + t^2}$$

34. Railroad tracks are curved in such a way that the curvature exists and is continuous at each point. Which of the following functions could trace out a railroad track?
    a. $\mathbf{r}(t) = t\mathbf{i}$ for $t < 0$, and $\mathbf{r}(t) = t\mathbf{i} + t^2\mathbf{j}$ for $t \ge 0$
    b. $\mathbf{r}(t) = t\mathbf{i}$ for $t < 0$, and $\mathbf{r}(t) = t\mathbf{i} + t^{7/3}\mathbf{j}$ for $t \ge 0$
    c. $\mathbf{r}(t) = t\mathbf{i}$ for $t < 0$, and $\mathbf{r}(t) = t\mathbf{i} + t^3\mathbf{j}$ for $t \ge 0$

35. A car travels around a circular track whose radius is 729 feet. At an instant when the speed of the car is 81 feet per second, the brakes are applied, reducing the speed of the car to 0 in 9 seconds at a constant rate (that is, $d\|\mathbf{v}\|/dt$ is constant). Find the tangential and normal components of acceleration of the car, and find the magnitude of the acceleration. (*Hint:* Take $t = 0$ at the instant when the brakes are applied, express $\|\mathbf{v}\|$ as a function of t, and use Exercise 29(a).)

36. A ball weighing 2 pounds swings back and forth at the end of a string 3 feet long. Assume that when the string makes an angle of $\pi/6$ with respect to the vertical, the magnitude of the force exerted on the ball by the string is 4 pounds (Figure 12.44). Determine the acceleration and speed of the ball. (*Hint:* To determine the acceleration **a**, use Newton's Second Law of Motion $\mathbf{F} = m\mathbf{a}$, where $m = \text{weight}/g = \frac{2}{32} = \frac{1}{16}$ is the mass of the ball and **F** is the force acting on the ball. Then $\mathbf{F} = \mathbf{F}_1 + \mathbf{F}_2$, where

$$\mathbf{F}_1 = -mg\mathbf{k} = 2\mathbf{k}$$

is the force due to gravity and

$$\mathbf{F}_2 = 4\left(\cos\frac{2\pi}{3}\mathbf{j} + \sin\frac{2\pi}{3}\mathbf{k}\right) = -2\mathbf{j} + 2\sqrt{3}\mathbf{k}$$

is the force exerted on the ball by the string. To determine the speed, find $a_{\mathbf{N}}$ by using the formula $a_{\mathbf{N}} = \mathbf{a} \cdot \mathbf{N}$ and then use Exercise 29(a).)

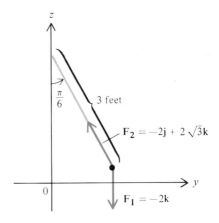

FIGURE 12.44

12.7
KEPLER'S LAWS OF MOTION

The motion of celestial bodies has intrigued people for centuries. However, the first person to state precise and accurate laws regarding the orbits of the planets in our solar system was the German mathematician and astronomer Johannes Kepler (1571–1630). After many years of calculations based primarily on the extremely detailed observations recorded by the Danish astronomer Tycho Brahe (1546–1601), Kepler formulated three laws that describe planetary motion. We will derive these three laws from two fundamental laws of Newton: the Second Law of Motion and the Law of Gravitation.

Consider one object moving around another, much larger, object. You might think of a planet or comet traveling around a star, or an artificial satellite or the moon traveling around the earth. For simplicity we will call the larger object a sun and the smaller one a planet. We assume that the sun's gravitational force is so much larger than all other forces acting on the planet that such other forces produce a negligible effect on the planet's path. Thus we ignore all objects in the universe except a sun and one planet moving around it.

Let us select a coordinate system with the sun at the origin, and let **r** be the position vector from the sun to the planet. Newton's laws governing **r** are

$$\text{Newton's Second Law of Motion:} \qquad \mathbf{F} = m\frac{d^2\mathbf{r}}{dt^2} \qquad (1)$$

$$\text{Newton's Law of Gravitation:} \qquad \mathbf{F} = \frac{-GMm}{r^3}\mathbf{r} = \frac{-GMm}{r^2}\frac{\mathbf{r}}{r} \qquad (2)$$

where **F** is the gravitational force on the planet with mass m, M is the mass of the sun, G is the universal gravitational constant, and $r = \|\mathbf{r}\|$. Notice that **r** is a (not necessarily constant) function of t. Before deriving Kepler's laws we will show that the planet moves in a single plane.

Notice from (2) that $\mathbf{F}$ and $\mathbf{r}$ are parallel, so that $\mathbf{r} \times \mathbf{F} = \mathbf{0}$. Combining this fact with (1), we find that

$$\mathbf{r} \times \frac{d^2\mathbf{r}}{dt^2} = \mathbf{r} \times \left(\frac{1}{m}\mathbf{F}\right) = \frac{1}{m}(\mathbf{r} \times \mathbf{F}) = \mathbf{0}$$

Consequently

$$\frac{d}{dt}\left(\mathbf{r} \times \frac{d\mathbf{r}}{dt}\right) = \left(\frac{d\mathbf{r}}{dt} \times \frac{d\mathbf{r}}{dt}\right) + \left(\mathbf{r} \times \frac{d^2\mathbf{r}}{dt^2}\right) = \mathbf{0} + \mathbf{0} = \mathbf{0} \tag{3}$$

This means that the vector-valued function $\mathbf{r} \times (d\mathbf{r}/dt)$ is a constant vector, a result that holds provided only that the sun is located at the origin of the coordinate system. If we now restrict our attention to a coordinate system whose positive z axis points in the same direction as $\mathbf{r} \times (d\mathbf{r}/dt)$, it follows that

$$\mathbf{r} \times \frac{d\mathbf{r}}{dt} = p\mathbf{k} \tag{4}$$

where ρ is a suitable positive constant. From (4) we infer that at any time t, $\mathbf{r}$ is perpendicular to $\mathbf{k}$, and consequently the object is constrained to move in a single plane perpendicular to $\mathbf{k}$. In particular:

Planetary motion is planar.

We will henceforth assume that the planet moves in the xy plane, so that $\mathbf{r}$ lies in the xy plane.

Kepler's First Law Next we show that the orbit in which the planet moves is a conic section. Let $\mathbf{u}$ be the unit vector pointing in the direction of $\mathbf{r}$, so that $\mathbf{r} = r\mathbf{u}$ and consequently $\mathbf{u}$ is perpendicular to $\mathbf{k}$ and $r > 0$ for all t. Differentiation of $\mathbf{r}$ with respect to t yields

$$\frac{d\mathbf{r}}{dt} = \frac{d}{dt}(r\mathbf{u}) = \frac{dr}{dt}\mathbf{u} + r\frac{d\mathbf{u}}{dt} \tag{5}$$

From (4) and (5) we deduce that

$$p\mathbf{k} = \mathbf{r} \times \frac{d\mathbf{r}}{dt} = (r\mathbf{u}) \times \left(\frac{dr}{dt}\mathbf{u} + r\frac{d\mathbf{u}}{dt}\right)$$

$$= r\frac{dr}{dt}\overbrace{(\mathbf{u} \times \mathbf{u})}^{=0} + r^2\left(\mathbf{u} \times \frac{d\mathbf{u}}{dt}\right)$$

Therefore

$$p\mathbf{k} = r^2\left(\mathbf{u} \times \frac{d\mathbf{u}}{dt}\right) \tag{6}$$

Next we combine (1) and (2), which gives

$$\frac{d^2\mathbf{r}}{dt^2} = \frac{\mathbf{F}}{m} = -\frac{GM}{r^2}\frac{\mathbf{r}}{r}$$

and then we substitute $\mathbf{u}$ for $\mathbf{r}/r$ to obtain

$$\frac{d^2\mathbf{r}}{dt^2} = \frac{-GM}{r^2}\mathbf{u} \tag{7}$$

Formulas (6) and (7) together yield

$$\frac{d^2\mathbf{r}}{dt^2} \times p\mathbf{k} = \frac{-GM}{r^2}\mathbf{u} \times r^2\left(\mathbf{u} \times \frac{d\mathbf{u}}{dt}\right) = -GM\mathbf{u} \times \left(\mathbf{u} \times \frac{d\mathbf{u}}{dt}\right)$$

But $\|\mathbf{u}\| = 1$, and consequently $d\mathbf{u}/dt$ is perpendicular to $\mathbf{u}$ (see Corollary 12.11). Therefore the formula $\mathbf{a} \times (\mathbf{a} \times \mathbf{c}) = -\|\mathbf{a}\|^2\mathbf{c}$ (which is (8) of Section 11.4) yields

$$\frac{d^2\mathbf{r}}{dt^2} \times p\mathbf{k} = -GM\left(-\|\mathbf{u}\|^2\frac{d\mathbf{u}}{dt}\right) = GM\frac{d\mathbf{u}}{dt} \tag{8}$$

Since $p\mathbf{k}$ is a constant function of t, the left-hand side of (8) is the derivative of $(d\mathbf{r}/dt) \times p\mathbf{k}$. Thus

$$\frac{d}{dt}\left(\frac{d\mathbf{r}}{dt} \times p\mathbf{k}\right) = \frac{d^2\mathbf{r}}{dt^2} \times p\mathbf{k} = \frac{d}{dt}(GM\mathbf{u})$$

and it follows that

$$\frac{d\mathbf{r}}{dt} \times p\mathbf{k} = GM\mathbf{u} + \mathbf{w}_1 \tag{9}$$

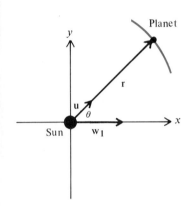

where $\mathbf{w}_1$ is a suitable constant vector. Since $(d\mathbf{r}/dt) \times p\mathbf{k}$ is perpendicular to $\mathbf{k}$, as is $\mathbf{u}$, it follows that $\mathbf{w}_1$ is perpendicular to $\mathbf{k}$ and hence lies in the xy plane. Therefore we may make the further assumption that the coordinate system is situated with the positive x axis pointing in the direction of $\mathbf{w}_1$. If we let θ be the angle from $\mathbf{w}_1$ to $\mathbf{u}$ at any given time, then (r, θ) is a set of polar coordinates of the position of the planet at that time (Figure 12.45). Now we let $w = \|\mathbf{w}_1\|$, so that

$$\mathbf{u} \cdot \mathbf{w}_1 = \|\mathbf{u}\|\,\|\mathbf{w}_1\|\cos\theta = w\cos\theta \tag{10}$$

FIGURE 12.45

From (4), (9), and (10), as well as the formula $(\mathbf{a} \times \mathbf{b}) \cdot \mathbf{c} = \mathbf{a} \cdot (\mathbf{b} \times \mathbf{c})$ (which is (3) of Section 11.4), we obtain

$$p^2 = (p\mathbf{k}) \cdot (p\mathbf{k}) \overset{(4)}{=} \left(\mathbf{r} \times \frac{d\mathbf{r}}{dt}\right) \cdot p\mathbf{k} = \mathbf{r} \cdot \left(\frac{d\mathbf{r}}{dt} \times p\mathbf{k}\right)$$

$$\overset{(9)}{=} (r\mathbf{u}) \cdot (GM\mathbf{u} + \mathbf{w}_1) = rGM(\mathbf{u} \cdot \mathbf{u}) + r(\mathbf{u} \cdot \mathbf{w}_1) \overset{(10)}{=} GMr + wr\cos\theta$$

Thus we have

$$p^2 = GMr + wr\cos\theta \tag{11}$$

In order to interpret (11), we change from polar coordinates to rectangular coordinates and rewrite (11) as

$$p^2 = GM\sqrt{x^2 + y^2} + wx$$

so that

$$GM\sqrt{x^2 + y^2} = p^2 - wx$$

Squaring both sides yields

$$G^2M^2(x^2 + y^2) = p^4 - 2wp^2x + w^2x^2 \tag{12}$$

A composite photograph of Saturn taken by Voyager 1 from a distance of 18,000,000 kilometers. The rings are composed of many small particles in elliptical orbit around the planet.

Equation (12) has the form of a conic section, regardless of the constant w, and it only remains to determine which conic sections arise for different values of w.

If $w = 0$, then (12) yields

$$x^2 + y^2 = \frac{p^4}{G^2 M^2} \tag{13}$$

which is an equation of a circle with radius p^2/GM.

If $w = GM$, then we can rewrite (12) as

$$G^2 M^2 y^2 = p^4 - 2GM\, p^2 x$$

which is an equation of a parabola.

Finally, if $w \neq 0$ and $w \neq GM$, then we may rewrite (12) as

$$(G^2 M^2 - w^2)x^2 + 2wp^2 x + G^2 M^2 y^2 = p^4 \tag{14}$$

This is an equation of an ellipse if $G^2 M^2 > w^2$ and an equation of a hyperbola if $G^2 M^2 < w^2$.

Therefore, regardless of the value of w, the orbit of the object around its sun is a conic section. If the orbiting object returns to its initial position periodically, as planets actually do, then the orbit is necessarily elliptical. In this case it can be shown that one of the foci is at the origin (see Exercise 7), where the sun is located. Thus we have Kepler's First Law of planetary motion, announced in 1609:

A planet revolves around the sun in an elliptical orbit.

As we noted above, the sun is located at a focus of the elliptical orbit, not at the center of the orbit as one might conjecture. It turns out that if the orbit is described by (14), then the points on the ellipse closest to and farthest from the sun occur on the x axis (Figure 12.46), to the right and left of the origin, respectively (see Exercise 2). The nearest point is called the **perihelion** of the orbit, and the farthest point is called the **aphelion**.* For an object moving in an elliptical orbit about the earth, the terms are **perigee** and **apogee**, respectively.[†]

Theoretically, a comet can have an elliptical, parabolic, or hyperbolic orbit around the sun. If the orbit is parabolic or hyperbolic, then we can view the comet at most twice. However, if the orbit is elliptical, then the comet can return to view periodically. The most illustrious example of a comet with elliptical orbit is Halley's comet, which has a period of about 76 years. Halley's comet reappeared during 1985–1986, its perihelion occurring on February 9, 1986. Artificial satellites can have any of the three types of orbits. Certain earth satellites, such as those in the Syncom series, have been placed in circular orbits around the earth.

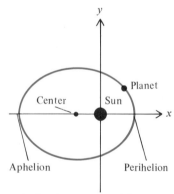

A planet moves around the sun in an elliptical orbit

FIGURE 12.46

Kepler's Second Law

As a planet orbits the sun, the position vector $\mathbf{r}$ from the sun to the planet describes an elliptical region. Let A denote the area swept out by $\mathbf{r}$ from an initial time t_0 to time t (Figure 12.47). Let α and θ be the angles made by the vectors $\mathbf{r}(t_0)$

---

* **Perihelion**: Pronounced "per-i-*heel*-yon."
  Aphelion: Pronounced "a-*feel*-yon."
[†] **Perigee**: Pronounced "*per*-i-jee."
  Apogee: Pronounced "*ap*-o-jee."

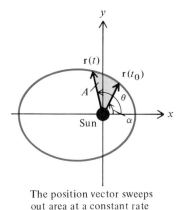

r(t)

r(t_0)

A

θ

Sun

α

The position vector sweeps
out area at a constant rate

FIGURE 12.47

and $\mathbf{r}(t)$, respectively, with the positive x axis. Finally, let $r = f(\phi)$ for $\alpha \leq \phi \leq \theta$. Then from (2) of Section 8.10,

$$A = \int_{\alpha}^{\theta} \frac{1}{2}[f(\phi)]^2 \, d\phi$$

Differentiating both sides of this equation with respect to θ, we find that

$$\frac{dA}{d\theta} = \frac{1}{2}(f(\theta))^2 = \frac{1}{2}r^2$$

and consequently

$$\frac{dA}{dt} = \frac{dA}{d\theta}\frac{d\theta}{dt} = \frac{1}{2}r^2\frac{d\theta}{dt} \tag{15}$$

Since $\mathbf{u}$ is a unit vector and θ is the angle between $\mathbf{u}$ and the positive x axis, we deduce that

$$\mathbf{u} = \cos\theta\mathbf{i} + \sin\theta\mathbf{j}$$

Therefore

$$\frac{d\mathbf{u}}{dt} = \frac{d\mathbf{u}}{d\theta}\frac{d\theta}{dt} = \frac{d\theta}{dt}(-\sin\theta\mathbf{i} + \cos\theta\mathbf{j})$$

Since $(\cos\theta\mathbf{i} + \sin\theta\mathbf{j}) \times (-\sin\theta\mathbf{i} + \cos\theta\mathbf{j}) = \mathbf{k}$

it follows that

$$\mathbf{u} \times \frac{d\mathbf{u}}{dt} = (\cos\theta\mathbf{i} + \sin\theta\mathbf{j}) \times \frac{d\theta}{dt}(-\sin\theta\mathbf{i} + \cos\theta\mathbf{j}) = \frac{d\theta}{dt}\mathbf{k} \tag{16}$$

Substituting (16) into (6), we obtain

$$p\mathbf{k} = r^2\left(\mathbf{u} \times \frac{d\mathbf{u}}{dt}\right) = r^2\frac{d\theta}{dt}\mathbf{k}$$

Therefore $$p = r^2\frac{d\theta}{dt} \tag{17}$$

Combining (15) and (17), we conclude that

$$\frac{dA}{dt} = \frac{1}{2}r^2\frac{d\theta}{dt} = \frac{p}{2} \tag{18}$$

Equation (18) brings us to Kepler's Second Law, also announced in 1609:

> *The position vector from the sun to a planet sweeps out area at a constant rate.*

An immediate consequence of this law is the fact that when a planet is near its sun, it moves more swiftly than when it is farther away.

Kepler's Third Law   Since the area swept out by the position vector changes at a constant rate, we find by integrating the expressions in (18) that

$$A(t_1) - A(t_0) = \int_{t_0}^{t_1} \frac{dA}{dt} \, dt = \int_{t_0}^{t_1} \frac{p}{2} \, dt = \frac{p(t_1 - t_0)}{2}$$

If T denotes the period of the planet about the sun, then

$$A(t_0 + T) - A(t_0) = \frac{pT}{2} \qquad (19)$$

But during an interval of duration T, the position vector sweeps out the complete ellipse, whose area is given by (19). If the major and minor axes of the ellipse have lengths $2a$ and $2b$, respectively, then the area of the ellipse is πab (see the comment after Example 2 of Section 8.3). Consequently

$$\pi ab = A(t_0 + T) - A(t_0) = \frac{pT}{2}$$

so that

$$T = \frac{2\pi ab}{p} \qquad (20)$$

A suitable alteration of (20) will yield Kepler's Third Law.

First we put (14) into standard form by completing the square and dividing by the right-hand side:

$$\frac{\left(x + \dfrac{wp^2}{G^2 M^2 - w^2}\right)^2}{\dfrac{p^4 G^2 M^2}{(G^2 M^2 - w^2)^2}} + \frac{y^2}{\dfrac{p^4}{G^2 M^2 - w^2}} = 1 \qquad (21)$$

This implies that

$$a^2 = \frac{p^4 G^2 M^2}{(G^2 M^2 - w^2)^2} \quad \text{and} \quad b^2 = \frac{p^4}{G^2 M^2 - w^2} \qquad (22)$$

Therefore by (20),

$$T^2 = \frac{4\pi^2 a^2 b^2}{p^2} = \frac{4\pi^2 p^6 G^2 M^2}{(G^2 M^2 - w^2)^3} = \frac{4\pi^2 a^3}{GM} \qquad (23)$$

Equation (23) yields Kepler's Third Law, announced in 1619:

The square of the period of a planet is $4\pi^2/GM$ times the cube of half the length of the major axis of its orbit.

Observe that the number $4\pi^2/GM$ is independent of the planet orbiting the sun. Thus the period of a planet about its sun depends only on the length of the major axis of the orbit.

In our derivation of Kepler's Laws we ignored all objects in the universe except one given planet and the sun. However, the planets Neptune and Uranus are sufficiently close together to affect each other's orbits measurably. In fact,

Neptune was discovered only after astronomers observed certain irregularities in the nearly elliptical orbit of Uranus and tried to explain them by the existence of a nearby planet. Working independently, both John Adams in Britain and Urbain Le Verrier in France calculated what the position of the unknown planet in the sky would have to be in order to produce the observed irregularities in the orbit of Uranus. In 1846, using Le Verrier's predictions, the German astronomer Johann Galle was able to locate Neptune with his telescope in a few hours.

EXERCISES 12.7

The exercises for this section will employ the following data.

r_0 = minimum distance from an object to the sun

v_0 = speed of an object when its distance from the sun is r_0

$a^2 = b^2 + c^2$

eccentricity = $\dfrac{c}{a}$

$GM_e = 1.237 \times 10^{12}$ (miles cubed per hour squared), where M_e is the mass of the earth.

1. a. From (6) show that
$$p = r^2 \left\| \frac{d\mathbf{u}}{dt} \right\|$$

(*Hint:* Because $\mathbf{u}$ is a unit vector, $\mathbf{u}$ and $d\mathbf{u}/dt$ are perpendicular to one another.)

b. Whatever kind of conic section constitutes the orbit of an object about the sun, the distance r from the object to the sun is a differentiable function of time t, and has a minimum value for some special value of t. From (5) show that
$$\frac{d\mathbf{r}}{dt} = r \frac{d\mathbf{u}}{dt}$$
when r is minimum.

c. Using (a), (b), and the definitions of r_0 and v_0, show that
$$p = r_0 v_0$$

2. a. Solve (11) for r and use Exercise 1(c) to show that
$$r = \frac{r_0^2 v_0^2}{GM + w \cos \theta}$$

b. Find a value of θ that produces the minimum value r_0 of r, and calculate r_0 in terms of v_0, G, M, and w.

c. Assuming that the orbit is elliptical, find a value of θ which produces the maximum value of r.

3. Suppose an object moves in a circular orbit. Use (13) and Exercise 1(c) to show that
$$v_0 = \sqrt{\frac{GM}{r_0}}$$

4. Find the speed of a satellite orbiting the earth in a circular orbit with radius 5000 miles. (*Hint:* Take M equal to the mass M_e of the earth.)

5. Using Kepler's Third Law with $M = M_e$, find the period of a satellite orbiting the earth in a circular orbit with radius 5000 miles.

6. The earth makes one complete revolution around its axis in approximately 23.9344 hours. If we wish a satellite to have a circular orbit with period equal to 23.9344 hours, what would be its velocity and its distance from the surface of the earth (3960 miles from the center of the earth)? (*Hint:* Use Kepler's Third Law, with a denoting the distance between the satellite and the center of the earth. Also use Exercise 3 with $M = M_e$.)

7. Show that the sun is located at one of the foci of the elliptical orbit of any of its planets. (*Hint:* Using (22), compute $c = \sqrt{a^2 - b^2}$ and compare your answer with the first term in (21). Remember that the sun is located at the origin and that by (21) the center of the ellipse is $(-wp^2/(G^2M^2 - w^2), 0)$.)

8. Using Exercise 2, show that $w = r_0 v_0^2 - GM$.

9. a. Use (9) and the fact that $d\mathbf{r}/dt$ is perpendicular to $\mathbf{k}$ to show that the speed $\|d\mathbf{r}/dt\|$ of an object in orbit is given by
$$\left\| \frac{d\mathbf{r}}{dt} \right\| = \frac{1}{p} \| GM\mathbf{u} + \mathbf{w}_1 \|$$

b. The vector $\mathbf{w}_1$ has length w and is directed along the positive x axis. Use this information, along with Exercises 1(c) and 8 and the formula derived in part (a), to show that the maximum speed (the maximum value of $\|d\mathbf{r}/dt\|$) is v_0.

10. The earth takes 365.256 days to orbit the sun. The orbit has an eccentricity of 0.016732, and the value of a is approximately 92,955,821 miles.
    a. Using this information, first find b, and then use (20) to calculate the value of p for the earth.
    b. Using the fact that $r_0 = a - c$, calculate r_0, the minimum distance from the earth to the sun.
    c. Calculate v_0, the maximum speed of the earth as it orbits the sun. (*Hint:* Use (a), (b), and Exercise 1(c).)
    d. Use (23) to calculate the mass of the sun. (Note: Take $G = 3.024 \times 10^{-12}$ miles cubed per hour squared per slug. Your answer will be in slugs.)

11. Suppose the maximum and minimum altitudes above the surface of the earth of a satellite moving in an elliptical orbit about the earth are 3100 and 100 miles, respectively.
    a. Find the maximum speed of the satellite. (*Hint:* Recall that $r_0 = a - c$, and note that $2a = 3100 + 100 + 2(3960) = 11,120$. Calculate a, c, b, p, and then r_0, by means of (23), (20), and Exercise 1(c).)
    b. Find the minimum speed of the satellite. (*Hint:* Use v_0, p, and r_0 from part (a); then find w from Exercise 8. Finally, use the equation in Exercise 9(a). Assume that the minimum speed occurs at aphelion on the negative x axis.)

12. A satellite is launched from 100 miles above the earth at 20,000 miles per hour in a direction parallel to the surface of the earth that takes the satellite into an elliptical orbit whose minimum distance from the surface of the earth is 100 miles.
    a. Determine an equation for the orbit of the satellite. (*Hint:* From Exercises 1(c) and 8, find p and w, and then obtain the answer in the form of (14).)
    b. Compute the distance from the apogee to the surface of the earth.
    c. Compute the period of the satellite.

13. Suppose an object is in a parabolic orbit. Using Exercise 8 and the fact that $GM = w$, show that

$$v_0 = \sqrt{\frac{2GM}{r_0}}$$

14. If a satellite is close to the surface of the earth and its distance from the center of the earth is r_0, then the gravitational acceleration GM/r_0^2 of the satellite is approximately g. In that case, the value of v_0 found in Exercise 13 is approximately given by

$$v_0 = \sqrt{2gr_0}$$

If the satellite has this velocity, it enters a parabolic orbit and disappears. Accordingly, this value of v_0 is called the **escape velocity** of the satellite. Taking the radius of the earth to be 3960 miles and assuming that a satellite is to

achieve an escape velocity at an altitude of 100 miles, what will that velocity need to be? (Note: Take $g = 7.855 \times 10^4$ miles per hour squared.)

15. The moon has a mass of (approximately) 0.0123 times the mass of the earth and is (approximately) 240,000 miles from the earth. If a spacecraft is on a line between the moon and the earth, 4080 miles from the center of the moon, what is the ratio of the gravitational force of the earth on the spacecraft to the gravitational force of the moon on the spacecraft? (*Hint:* Use Newton's Law of Gravitation for the earth and for the moon.)

16. a. If a satellite is to orbit the moon in a circular orbit with radius 1200 miles, what must be the speed of the satellite? (*Hint:* See Exercise 15 for the mass of the moon, and apply Exercise 3.)
    b. To what value should the speed be increased if the satellite is to achieve escape velocity at a point the same distance from the center of the moon? (*Hint:* Apply Exercise 13.)

17. Let $\mathbf{r}$ and $\mathbf{v}$ be the position and velocity vectors, respectively, of an object moving about the sun, and let

$$\mathbf{L}(t) = \mathbf{r}(t) \times m\mathbf{v}(t) \quad \text{for all } t$$

where m is the mass of the object. Then $\mathbf{L}$ is called the **angular momentum** of the object about its sun. Use (3) to show that $\mathbf{L}'(t) = \mathbf{0}$ for all t, and conclude that $\mathbf{L}$ is a constant function. This fact is often referred to as the **conservation of angular momentum** of the object.

18. Suppose a planet with mass m and speed v moves in a circular orbit with radius r. Then the gravitational force $\mathbf{F}$ is a centripetal force. From (4) of Section 12.3, the magnitude $\|\mathbf{F}\|$ of this force is given by

$$\|\mathbf{F}\| = m \|\mathbf{a}\| = \frac{mv^2}{r} \tag{24}$$

a. Without using any formulas from the present section, show that the period T of the planet is given by

$$T = \frac{2\pi r}{v} \tag{25}$$

b. Observe that Kepler's Third Law has the form

$$T^2 = cr^3 \tag{26}$$

where c is a suitable positive constant. Using (24)–(26), show that

$$\|\mathbf{F}\| = \frac{4\pi^2}{c} \frac{m}{r^2} \tag{27}$$

(Before Newton announced his Law of Gravitation, the validity of (27) was noticed by Robert Hooke and was mentioned about 1679 in a recently discovered letter from Hooke to Newton.)

Key Terms and Expressions

Vector-valued function; component function
Limit of a vector-valued function
Vector-valued function continuous at t_0; continuity on an interval
Derivative of a vector-valued function; differentiability on an interval
Definite integral of a vector-valued function
Space curve; parametrization of a curve

Smooth curve; piecewise smooth curve
Length of a curve
Tangent vector
Normal vector
Orientation of a curve; oriented curve
Curvature
Tangential and normal components of acceleration

Key Formulas

$$\mathbf{v} = \frac{d\mathbf{r}}{dt}, \qquad \mathbf{a} = \frac{d\mathbf{v}}{dt}, \qquad \|\mathbf{v}\| = \frac{ds}{dt}$$

$$\mathscr{L} = \int_a^b \sqrt{\left(\frac{dx}{dt}\right)^2 + \left(\frac{dy}{dt}\right)^2 + \left(\frac{dz}{dt}\right)^2}\, dt$$

$$\left\|\frac{d\mathbf{r}}{dt}\right\| = \frac{ds}{dt} = \sqrt{\left(\frac{dx}{dt}\right)^2 + \left(\frac{dy}{dt}\right)^2 + \left(\frac{dz}{dt}\right)^2}$$

$$\mathbf{T}(t) = \frac{\mathbf{r}'(t)}{\|\mathbf{r}'(t)\|} = \frac{d\mathbf{r}/dt}{\|d\mathbf{r}/dt\|}$$

$$\mathbf{N}(t) = \frac{\mathbf{T}'(t)}{\|\mathbf{T}'(t)\|} = \frac{d\mathbf{T}/dt}{\|d\mathbf{T}/dt\|}$$

$$\kappa(t) = \frac{\|\mathbf{T}'(t)\|}{\|\mathbf{r}'(t)\|} = \frac{\|\mathbf{v} \times \mathbf{a}\|}{\|\mathbf{v}\|^3}$$

$$a_{\mathbf{T}} = \frac{d\|\mathbf{v}\|}{dt} \quad \text{and} \quad a_{\mathbf{N}} = \sqrt{\|\mathbf{a}\|^2 - a_{\mathbf{T}}^2}$$

REVIEW EXERCISES

In Exercises 1–3 sketch the curve traced out by the function.

1. $\mathbf{F}(t) = \sin 2t\,\mathbf{i} - \cos 2t\,\mathbf{j}$ for $0 \le t \le 4\pi$

2. $\mathbf{F}(t) = 4\sqrt{2}\cos t\,\mathbf{i} - 4\sin t\,\mathbf{j} - 4\sin t\,\mathbf{k}$

3. $[(\mathbf{F} \times \mathbf{G}) \times \mathbf{H}](t)$, where $\mathbf{F}(t) = t\mathbf{i} + \mathbf{j}$, $\mathbf{G}(t) = \mathbf{j} + t\mathbf{k}$, and $\mathbf{H}(t) = t\mathbf{j}$

4. Let $\mathbf{F}(t) = e^t\mathbf{i} + t^2\mathbf{j} + e^{-t}\mathbf{k}$ and $\mathbf{G}(t) = t\mathbf{i} + e^{-t}\mathbf{j} + e^t\mathbf{k}$.
   a. Find $(\mathbf{F} \cdot \mathbf{G})'(t)$.   b. Find $(\mathbf{F} \times \mathbf{G})'(t)$.

5. Let $\mathbf{F}(t) = \frac{1}{t}\mathbf{i} + t\mathbf{j}$ and $\mathbf{G}(t) = t^2\mathbf{j} - \frac{1}{t^2}\mathbf{k}$.
   a. Find $(\mathbf{F} \cdot \mathbf{G})'(t)$.   b. Find $(\mathbf{F} \times \mathbf{G})'(t)$.

6. Let $\mathbf{F}(t) = \ln t\,\mathbf{i} + t\ln t\,\mathbf{j} - \ln 6t\,\mathbf{k}$ and $g(t) = e^{2t}$. Find $(\mathbf{F} \circ g)'(t)$.

7. Find

$$\int\left(\tan 2\pi t\,\mathbf{i} + \sec^2 2\pi t\,\mathbf{j} + \frac{4}{1 + t^2}\mathbf{k}\right) dt$$

8. Let

$$\mathbf{F}(t) = \frac{2t}{1 + t^2}\mathbf{i} + \frac{1 - t^2}{1 + t^2}\mathbf{j}$$

Use Corollary 12.11 to show that $\mathbf{F}$ is perpendicular to $\mathbf{F}'$.

9. Let $\mathbf{r}(t) = e^t\cos t\,\mathbf{i} + e^t\sin t\,\mathbf{j}$. Find the length of the curve that $\mathbf{r}$ traces out from
   a. $t = 0$ to $t = 3\pi$ b. $t = -2\pi$ to $t = 1$

10. An object moves along the curve $y = \ln(\sec x)$ for $-\pi/2 < x < \pi/2$, in such a manner that its x coordinate at any time t is given by $x(t) = e^t$. Find the speed of the object at time 0.

11. Suppose a car traverses the curve $y = \frac{2}{3}x^{3/2}$ in such a manner that its x coordinate at any time $t > \frac{3}{2}$ is given by

$$x(t) = \left(\frac{3}{2}t\right)^{2/3} - 1$$

Show that the speed of the car is constant.

12. Show that the **Folium of Descartes**

$$\mathbf{r}(t) = \frac{3t}{1 + t^3}\mathbf{i} + \frac{3t^2}{1 + t^3}\mathbf{j}$$

passes through the origin and that the tangent to the curve at the origin is parallel to the x axis.

13. Let

$$\mathbf{r}(t) = (3t - t^3)\mathbf{i} + 3t^2\mathbf{j} + (3t + t^3)\mathbf{k}$$

Find the curvature κ of the curve that $\mathbf{r}$ traces out.

14. Show that the curvature κ of the catenary $\mathbf{r}(t) = t\mathbf{i} + \cosh t\mathbf{j}$ is given by the formula

$$\kappa(t) = \frac{1}{\cosh^2 t}$$

15. Let C be the graph of the equation $xy = 1$ for $x > 0$.
    a. Find a formula for the curvature of C.
    b. Find the maximum value of the curvature of C.
    c. Determine the radius of curvature of C at $(1, 1)$.

16. Let $\mathbf{r}(t) = t\mathbf{i} + (2\sqrt{2}/3)t^{3/2}\mathbf{j} + (1/2)t^2\mathbf{k}$ be a parametrization of a curve.
    a. Determine whether the curve is smooth, piecewise smooth, or neither.
    b. Find the length $\mathcal{L}$ of the portion of the curve that lies between $\mathbf{r}(0)$ and $\mathbf{r}(1)$.
    c. Find $\mathbf{v}$, $\|\mathbf{v}\|$, and $\mathbf{a}$.
    d. Find a_T and a_N.
    e. Find $\kappa(t)$.

17. Let the position of an object in motion be traced out by

$$\mathbf{r}(t) = e^t \cos t\mathbf{i} + e^t \sin t\mathbf{j} + e^t\mathbf{k}$$

Find the velocity and acceleration of the object, and then compute the curvature and the radius of curvature of the curve traced out.

In Exercises 18–19 find the tangent, normal, and curvature of the curve traced out by $\mathbf{r}$.

18. $\mathbf{r}(t) = e^{2t}\mathbf{i} + 2\sqrt{2}e^t\mathbf{j} + 2t\mathbf{k}$

19. $\mathbf{r}(t) = (t - \sin t)\mathbf{i} + (1 - \cos t)\mathbf{j} + 4\sin\dfrac{t}{2}\mathbf{k}$

*20. The curve

$$\mathbf{r}(t) = \sin t\mathbf{i} + \left[\cos t + \ln\left(\tan\frac{t}{2}\right)\right]\mathbf{j} \quad \text{for } 0 < t < \pi$$

is called a **tractrix**. Show that the length of the segment of the line tangent to the tractrix from any point of tangency to the point of intersection with the y axis is always 1. (*Hint:* Use the fact that $2(\sin t/2)(\cos t/2) = \sin t$.)

21. Assume that an object moves so that its velocity and acceleration vectors are always unit vectors. Prove that the curvature of the curve traversed by the object is always 1.

Cumulative Review, Chapters 1–11

In Exercises 1–2 find the limit.

1. $\displaystyle\lim_{x\to-3^+}\left(\frac{1}{x+3} - \frac{1}{|x-3|}\right)$ 2. $\displaystyle\lim_{x\to0^+}(\cos x)^{1/x^2}$

3. Find $\displaystyle\lim_{y\to0}\frac{x^2 - y^3}{x^2 + y^2}$. (*Hint:* There are two cases: $x = 0$ and $x \neq 0$.)

4. Let $f(x) = \begin{cases} \dfrac{1}{x}\sin x^2 & \text{for } x \neq 0 \\ 0 & \text{for } x = 0 \end{cases}$
    a. Show that f is continuous at 0.
    b. Show that f is differentiable at 0 and find $f'(0)$.

5. Let $f(x) = \arcsin\left(\dfrac{\sin x}{5}\right)$. Find $f'(x)$.

6. Determine the points on the ellipse $x^2/4 + y^2/16 = 1$ at which the tangents have slope 1.

7. Consider the equation $(x^3 - 2y^3)/(x^2 + y^2) = 6$. Use implicit differentiation to determine all points on the graph of the equation at which the tangent is horizontal. (*Hint:* -4 is a solution of $z^3 + 3z^2 + 16 = 0$.)

8. Use the tangent line approximation to estimate the value of $\sqrt{63}$.

9. Two sets of railroad tracks meet at an angle of $\pi/3$. At noon a train on one set of tracks is 100 miles from the junction and is traveling toward the junction at 60 miles per hour. On the other tracks a second train is 120 miles from the junction and is traveling toward it at 80 miles per hour. Determine the rate at which the trains are approaching each other at 1 p.m. (*Hint:* Use the Law of Cosines.)

10. Let $f(x) = 2x/\sqrt{x^2 + 4}$. Sketch the graph of f, indicating all pertinent information.

11. Suppose that there are 77 feet of fencing, to be placed along three sides of a rectangular garden with a house comprising the fourth side. Assuming that a 3-foot wide gate is to be made out of other material, find the dimensions that will maximize the area of the garden. (*Hint:* There are two cases, depending on the side on which the gate is placed.)

12. Among all circular sectors with a perimeter of 10 centimeters, determine the one with maximum area.

13. Suppose that after 5 years only 10% of a given radioactive substance has decayed. Find the half-life of the substance.

In Exercises 14–16 evaluate the integral.

14. $\displaystyle\int \sqrt{x}\,\cos\sqrt{x}\,dx$

15. $\displaystyle\int \frac{x}{(x+2)(x^2+6)}\,dx$

16. $\displaystyle\int 27t^2\sqrt{3t-2}\,dt$

17. Determine whether $\displaystyle\int_0^{\pi^2/4} \frac{\cos\sqrt{x}}{\sqrt{x}}\,dx$ converges. If it converges, determine its value.

18. The base of a solid is the semicircle $x^2 + y^2 = 1$ with $y \geq 0$. Suppose the cross sections of the solid perpendicular to the x axis are isosceles right triangles with hypotenuse on the base. Determine the volume V of the solid.

19. A conical tank 10 feet high and 5 feet in radius with vertex at the top has water to a depth of 4 feet. Determine the work W done in raising the water up to a height 3 feet above the top of the cone.

20. Let $f(x) = \frac{1}{4}x^2 - \frac{1}{2}\ln x$ for $1 \leq x \leq 3$. Find the length $\mathscr{L}$ of the graph of f.

21. Determine whether $\left\{\dfrac{(-1)^n n}{n+1}\right\}_{n=1}^{\infty}$ converges or diverges.

In Exercises 22–23 determine whether the series converges absolutely, converges conditionally, or diverges.

22. $\displaystyle\sum_{n=1}^{\infty} \frac{\sinh n}{n^2}$

23. $\displaystyle\sum_{n=0}^{\infty} \frac{n^n}{3^n n!}$

24. Find the interval of convergence of $\displaystyle\sum_{n=1}^{\infty} \frac{3^n}{4^{2n}} x^{2n}$.

25. Let $f(x) = -2x/(1+x^2)^2$. Find a power series representation for f, and determine the interval of convergence for the power series.

26. A unit vector $\mathbf{v}$ is perpendicular to both the line
$$\frac{x-1}{2} = \frac{y+1}{-4} = z$$
and the vector $2\mathbf{i} - 3\mathbf{j} - \mathbf{k}$, and has negative $\mathbf{k}$-component. Find $\mathbf{v}$.

27. Determine an equation of the plane that passes through the points $(1, -1, 2)$, $(2, 3, -1)$, and $(0, 2, 0)$.

13
PARTIAL DERIVATIVES

Until now we have studied functions of a single variable (both real-valued and vector-valued). Many phenomena in the physical world can be described by such functions, but most quantities actually depend on more than one variable. For example, the volume of a rectangular box depends on its length, width, and height; the temperature at a point on a metal plate depends on the coordinates of the point (and possibly on time as well); the cost of insulation for an attic floor depends on the area of the attic floor and the thickness of the insulation. Any quantity that depends on several other quantities can be thought of as determining a function of several variables. The primary concern of this chapter is the extension of the concept of differentiation to functions of several variables. In Chapter 14 we will discuss integration of such functions.

13.1
FUNCTIONS OF SEVERAL VARIABLES

A vector-valued function assigns vectors (that is, points in the plane or in space) to real numbers. A function of several variables does the opposite; it assigns real numbers to points in the plane or in space.

DEFINITION 13.1

A **function of several variables** consists of two parts: a **domain**, which is a collection of points in the plane or in space, and a **rule**, which assigns to each member of the domain one and only one number.

A function of several variables is called a **function of two variables** if its domain is a set of points in the plane and a **function of three variables** if its domain is a set of points in space. Although we will not discuss functions of more than three variables in this book, it is possible to extend the theory in this chapter to such functions. To distinguish functions of several variables from those whose domains are sets of real numbers, we will refer to the latter type as **functions of a single variable**.

A function of several variables is usually denoted f or g. The value of a function f of two variables at a point (x, y) is denoted $f(x, y)$, and the value of a function f of three variables at a point (x, y, z) is denoted $f(x, y, z)$. As with functions of a single variable, we frequently specify a function of several variables by one or more formulas for its values at the various points in the domain. Unless otherwise indicated, the domain of a function so specified consists of all points in space (or in the plane for a function of two variables) for which the formula is meaningful. For example, if

$$f(x, y) = \sqrt{9 - 4x^2 - y^2}$$

then the domain of f is the region in the xy plane bounded by the ellipse $4x^2 + y^2 = 9$, because the square root is defined only for nonnegative numbers. If

$$g(x, y, z) = \sqrt{x^2 + y^2 + z^2}$$

then the domain of g consists of all points (x, y, z) in space, because $x^2 + y^2 + z^2 \geq 0$ for all (x, y, z). In contrast, if

$$f(x, y, z) = xyz \quad \text{for } x \geq 0, \, y \geq 0, \, z \geq 0$$

then the domain of f is the first octant.

The following are some functions of several variables that we will encounter later in this chapter, along with geometric or physical interpretations.

1. $f(x, y) = xy$ for $x \geq 0$ and $y \geq 0$ — area of a rectangle
2. $f(x, y, z) = xyz$ for $x \geq 0$, $y \geq 0$, and $z \geq 0$ — volume of a rectangular parallelepiped
3. $f(x, y, z) = 2xy + 2yz + 2xz$ for $x \geq 0$, $y \geq 0$, and $z \geq 0$ — surface area of a rectangular parallelepiped
4. $f(x, y, z) = \dfrac{c}{x^2 + y^2 + z^2}$, where c is a positive constant — magnitude of the gravitational force exerted by the sun, located at the origin, on a unit mass at (x, y, z)
5. $f(x, y, z) = \dfrac{c}{\sqrt{x^2 + y^2}}$, where c is a constant — strength of the electric field at (x, y, z) due to an infinitely long wire lying along the z axis

Combinations of Functions of Several Variables

The sum, product, and quotient of two functions f and g of several variables are defined exactly as you would expect. For functions of two variables the formulas are

$$(f + g)(x, y) = f(x, y) + g(x, y)$$
$$(f - g)(x, y) = f(x, y) - g(x, y)$$
$$(fg)(x, y) = f(x, y)g(x, y)$$
$$\left(\frac{f}{g}\right)(x, y) = \frac{f(x, y)}{g(x, y)}$$

The formulas for functions of three variables are analogous. The domains of $f + g$, $f - g$, and fg consist of all points simultaneously in the domain of f and the domain of g, whereas the domain of f/g consists of all points simultaneously in the domain of f and the domain of g at which g does not assume the value 0. For example, if

$$F(x, y) = axy + \frac{by}{x}$$

then F may be thought of as the sum of the functions f and g, where

$$f(x, y) = axy \quad \text{and} \quad g(x, y) = \frac{by}{x}$$

A function f of two variables x and y is a **polynomial function** if it is a sum of functions of the form $cx^m y^n$, where c is a number and m and n are nonnegative integers. A **rational function** is, as with functions of one variable, the quotient of two polynomial functions. Similar terminology is used for polynomial and rational functions of three variables. Thus if

$$f(x, y) = 3x^2y^2 + 4x^2 - 7y^2 + 3x + \sqrt{2}$$

and

$$g(x, y, z) = 4xyz^5 - 6$$

then f and g are polynomial functions, and if

$$f(x, y) = \frac{x^3 + y^3}{x^2 + y^2}$$

and

$$g(x, y, z) = \frac{xy}{1 + x^2 + y^2 + z^2}$$

then f and g are rational functions.

It is also possible to form composites using functions of several variables. If f is a function of two variables and g is a function of a single variable, then the function $g \circ f$ is defined by

$$(g \circ f)(x, y) = g(f(x, y))$$

for all (x, y) in the domain of f such that $f(x, y)$ is in the domain of g. The definition of the composite of a function of three variables and a function of a single variable is similar.

Example 1 Let $F(x, y) = \sqrt{\ln(4 - x^2 - y^2)}$. Find a function f of two variables and a function g of one variable such that $F = g \circ f$. Find the domain of F.

Solution There are many ways to express F as a composite. For example, if

$$f(x, y) = \ln(4 - x^2 - y^2) \quad \text{and} \quad g(t) = \sqrt{t}$$

then $F = g \circ f$. The domain of f consists of all (x, y) for which $4 - x^2 - y^2 > 0$, and the domain of g consists of all nonnegative numbers. Therefore the domain of $g \circ f$ consists of all (x, y) such that $4 - x^2 - y^2 > 0$ and $\ln(4 - x^2 - y^2) \geq 0$. Since $\ln(4 - x^2 - y^2) \geq 0$ if and only if $4 - x^2 - y^2 \geq 1$, we conclude that the domain of $F = g \circ f$ consists of all (x, y) for which $4 - x^2 - y^2 \geq 1$, or equivalently, $x^2 + y^2 \leq 3$. $\square$

Graphs of Functions of Two Variables

The **graph** of a function f of two variables is the collection of points $(x, y, f(x, y))$ for which (x, y) is in the domain of f (Figure 13.1). It is customary to let $z = f(x, y)$; then the graph of f consists of all points (x, y, z) such that $z = f(x, y)$.

In sketching the graph of a function f of two variables, it is often helpful to determine the intersections of the graph of f with planes of the form $z = c$ (Figure 13.2). We call each such intersection the **trace** of the graph of f in the plane $z = c$. Thus the trace of the graph of f in the plane $z = c$ is the collection of points (x, y, c) such that $f(x, y) = c$. If we think of the graph of f as the surface of a mountain, then the trace in a plane $z = c$ is a curve of constant altitude. A mountain climber walking along such a trace would neither ascend nor descend.

Closely related to the trace is the notion of level curve. The set of points (x, y) in the xy plane such that $f(x, y) = c$ is a **level curve** of f (Figure 13.2); we identify the level curve with the equation $f(x, y) = c$ and call the equation a level curve of

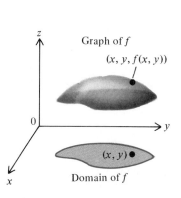

FIGURE 13.1

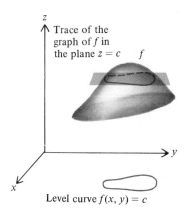

FIGURE 13.2

f. Notice that the trace of *f* in the plane $z = c$ either lies directly above or directly below the level curve $f(x, y) = c$ or coincides with it. The idea behind level curves is to provide three-dimensional information in a two-dimensional setting. For example, level curves are employed in contour maps to indicate elevations and depths of points on the surface of the earth (Figure 13.3(a)); a single level curve represents points of identical altitude. On a weather map a level curve represents points with identical temperature or barometric pressure (Figure 13.3(b)). A level curve can even be used to show places having the same number of days per year with thunderstorms (Figure 13.3(c)).

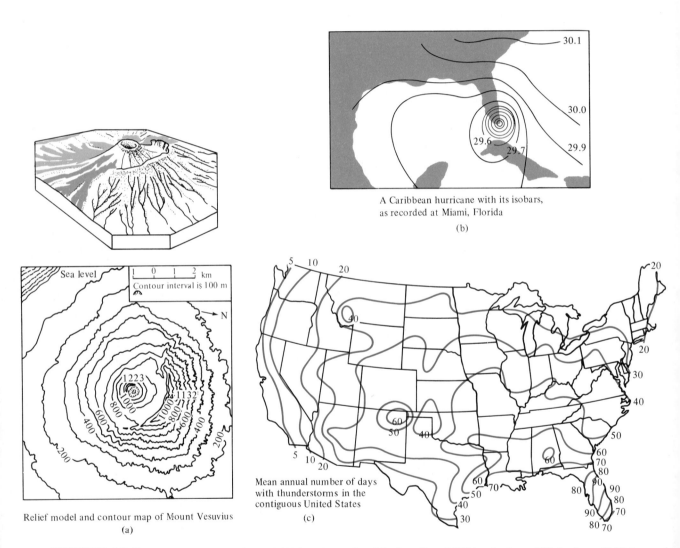

A Caribbean hurricane with its isobars, as recorded at Miami, Florida

(b)

Relief model and contour map of Mount Vesuvius

(a)

Mean annual number of days with thunderstorms in the contiguous United States

(c)

FIGURE 13.3

Figure 13.3(a) from *Principles of Geology*, Fourth Edition, by James Gilluly, Aaron C. Waters, and A. O. Woodford. W. H. Freeman and Company. Copyright © 1975, p. 591. (b) from *An Introduction to Climate*, Fourth Edition, by Glenn T. Trewartha. Copyright © 1968, p. 223. McGraw-Hill Book Company. Used with permission of the publisher. (c) from Howard J. Critchfield, *General Climatology*, Fourth Edition, © 1983, p. 187. Adapted by permission of Prentice-Hall, Inc., Englewood Cliffs, New Jersey.

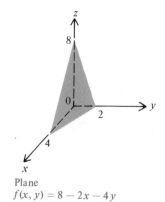

Plane
$f(x, y) = 8 - 2x - 4y$

FIGURE 13.4

Example 2   Let $f(x, y) = 8 - 2x - 4y$. Sketch the graph of f, and determine the level curves.

Solution   If we let $z = f(x, y)$, then the equation becomes

$$z = 8 - 2x - 4y$$

This is an equation of a plane with x intercept 4, y intercept 2, and z intercept 8. The plane is sketched in Figure 13.4. For any value of c, the level curve $f(x, y) = c$ is the line in the xy plane with equation $2x + 4y = 8 - c$. $\square$

It is possible to show that any plane not perpendicular to the xy plane is the graph of a function f of the form $f(x, y) = ax + by + c$, and conversely, that the graph of any function of this form is a plane. The level curves of such a function are lines unless $a = b = 0$, in which case the level curve is the whole plane.

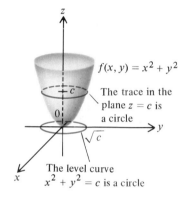

$f(x, y) = x^2 + y^2$

The trace in the plane $z = c$ is a circle

The level curve $x^2 + y^2 = c$ is a circle

FIGURE 13.5

Example 3   Let $f(x, y) = x^2 + y^2$. Sketch the graph of f and determine the level curves.

Solution   If $c > 0$, then the level curve $f(x, y) = c$ is given by

$$x^2 + y^2 = c$$

which is a circle with radius $\sqrt{c}$. Therefore the trace of the graph in the plane $z = c$ is also a circle with radius $\sqrt{c}$. The level curve $f(x, y) = 0$ is the point $(0, 0)$. For $c < 0$ the level curve $f(x, y) = c$ contains no points. The intersections of the graph with the planes $x = 0$ and $y = 0$ are both parabolas. The graph is sketched in Figure 13.5. $\square$

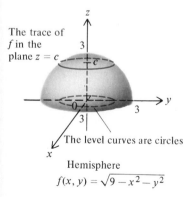

The trace of f in the plane $z = c$

The level curves are circles

Hemisphere
$f(x, y) = \sqrt{9 - x^2 - y^2}$

FIGURE 13.6

Example 4   Let $f(x, y) = \sqrt{9 - x^2 - y^2}$. Sketch the graph of f and indicate the level curves.

Solution   If we let $z = f(x, y)$, the equation of f becomes

$$z = \sqrt{9 - x^2 - y^2} \tag{1}$$

Squaring both sides of this equation and transposing x^2 and y^2, we obtain

$$x^2 + y^2 + z^2 = 9$$

which we recognize as the equation of a sphere centered at the origin with radius 3. Since (1) holds only for $z \geq 0$, we conclude that the graph of f is the hemisphere sketched in Figure 13.6. In this case the level curve $f(x, y) = c$ is a circle if $0 \leq c < 3$ and is the point $(0, 0)$ if $c = 3$. $\square$

Example 5   Let $f(x, y) = \sin y$. Sketch the graph of f.

Solution   Since the equation of f is independent of x, it follows that if $(0, y, \sin y)$ is a point on the graph, so is the entire line parallel to the x axis through

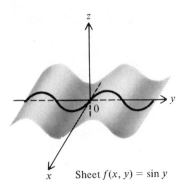

Sheet $f(x, y) = \sin y$

FIGURE 13.7

$(0, y, \sin y)$. Thus the whole surface is determined by a sine curve in the yz plane (Figure 13.7).   ☐

We call the graph of a function like that in Example 5 a **sheet**, or more precisely, a sheet parallel to the x axis. The graph of the function whose equation is $f(x, y) = -x^3$ is a sheet parallel to the y axis (Figure 13.8).

It is important not to be misled by the simplicity of the graphs in Figures 13.4–13.8. The problems inherent in graphing functions of two variables are many times as great as in graphing functions of a single variable. To see the difficulties that can arise, you might try sketching the graph of the innocuous-looking function defined by

$$f(x, y) = \sin x + \sin y$$

A computer-drawn sketch of the graph of f appears on the facing page, along with several other computer-drawn graphs.

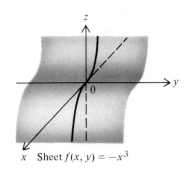

Sheet $f(x, y) = -x^3$

FIGURE 13.8

Level Surfaces

Although we can sketch the graphs of many functions of two variables, it is impossible to sketch the graph of any function of three variables, for that would entail four dimensions. However, we can gain information about a function f of three variables from what we call its level surfaces. For any number c the set of points (x, y, z) for which $f(x, y, z) = c$ is called a **level surface** of f, and we identify a level surface with the corresponding equation $f(x, y, z) = c$. Level surfaces of functions of three variables are analogous to level curves of functions of two variables. If $f(x, y, z)$ denotes the temperature at any point (x, y, z) in a region in space, then the level surface $f(x, y, z) = c$ is the surface on which the temperature is constantly c, and is called an **isothermal surface**. Similarly, if $V(x, y, z)$ is the voltage (or potential) at (x, y, z), then the level surface $V(x, y, z) = c$ is called an **equipotential surface**.

In Chapter 11 we encountered three kinds of level surfaces. First, spheres centered at the origin are level surfaces of the function

$$f(x, y, z) = x^2 + y^2 + z^2$$

since $x^2 + y^2 + z^2 = c^2$ is an equation of such a sphere for $c \neq 0$. Second, any cylinder whose axis is the z axis is a level surface of the function

$$f(x, y, z) = x^2 + y^2$$

because $x^2 + y^2 = c^2$ is an equation of such a cylinder if $c \neq 0$. Finally, planes are level surfaces of functions of the form

$$f(x, y, z) = ax + by + cz$$

provided that a, b, and c are not all 0.

We also observe that the graph of any function f of two variables is a level surface. We need only let

$$g(x, y, z) = z - f(x, y)$$

and notice that

$$g(x, y, z) = 0 \quad \text{if and only if} \quad z = f(x, y)$$

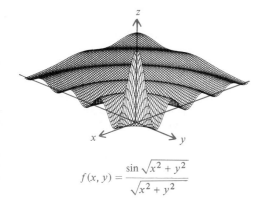

$$f(x, y) = \frac{\sin \sqrt{x^2 + y^2}}{\sqrt{x^2 + y^2}}$$

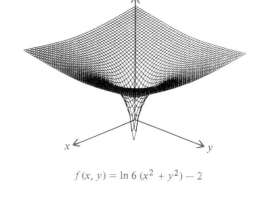

$$f(x, y) = \ln 6 \, (x^2 + y^2) - 2$$

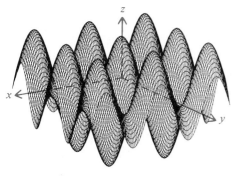

$$f(x, y) = \sin x + \sin y$$

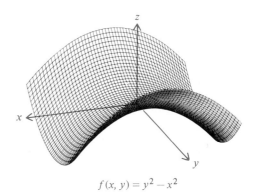

$$f(x, y) = y^2 - x^2$$

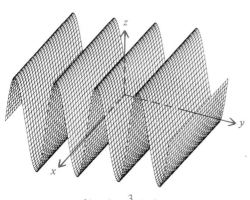

$$f(x, y) = \frac{3}{4} \sin 2y$$

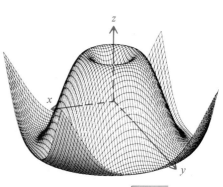

$$f(x, y) = \sin \sqrt{x^2 + y^2}$$

Thus the level surface $g(x, y, z) = 0$ is the graph of f, or equivalently, the graph of the equation $z = f(x, y)$. This is why we often call the graph of a function of, or an equation in, two variables a **surface**.

In sketching a level surface we will use the intersections of the level surface with planes of the form $x = c$ or $y = c$, as well as those of the form $z = c$. In each case the intersection of the level surface with the plane is called the **trace** of the level surface in that plane. The most important level surfaces are those called quadric surfaces, which we discuss and graph next.

Quadric Surfaces   A **quadric surface** is a level surface of a polynomial function f given by

$$f(x, y, z) = Ax^2 + By^2 + Cz^2 + Dxy + Exz + Fyz + Gx + Hy + Iz + J$$

Quadric surfaces fall into nine major classes. We will list these classes and sketch one surface in each class, assuming in each case that the constants a, b, and c are positive.

$$\textbf{Ellipsoid:} \quad \frac{x^2}{a^2} + \frac{y^2}{b^2} + \frac{z^2}{c^2} = 1$$

The trace of the ellipsoid in any plane parallel to a coordinate plane is either an ellipse, a point, or empty (Figure 13.9). If $a = b = c$, the equation becomes $x^2 + y^2 + z^2 = a^2$, and the surface is a sphere.

$$\textbf{Elliptic cylinder:} \quad \frac{x^2}{a^2} + \frac{y^2}{b^2} = 1$$

The trace of the elliptic cylinder in any plane parallel to the plane $z = 0$ is the ellipse $x^2/a^2 + y^2/b^2 = 1$ (Figure 13.10). If $a = b$, the surface is a **circular cylinder**.

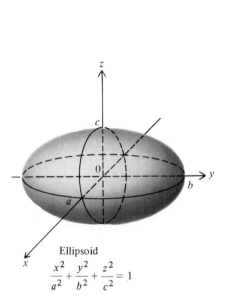

Ellipsoid

$$\frac{x^2}{a^2} + \frac{y^2}{b^2} + \frac{z^2}{c^2} = 1$$

FIGURE 13.9

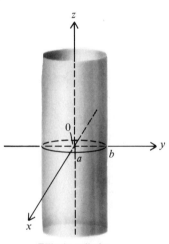

Elliptic cylinder

$$\frac{x^2}{a^2} + \frac{y^2}{b^2} = 1$$

FIGURE 13.10

Elliptic double cone: $\dfrac{x^2}{a^2} + \dfrac{y^2}{b^2} = \dfrac{z^2}{c^2}$

The trace of the cone in any plane parallel to the xy plane is either an ellipse (a circle if $a = b$) or a point (Figure 13.11). The trace in either of the planes $x = 0$ and $y = 0$ consists of two lines through the origin. If $a = b$, the surface is called a **circular double cone**.

Elliptic paraboloid: $\dfrac{x^2}{a^2} + \dfrac{y^2}{b^2} = \dfrac{z}{c}$

The trace of the paraboloid in any plane parallel to the xy plane is either an ellipse (a circle if $a = b$), a point, or empty. The traces in the planes $x = 0$ and $y = 0$ are parabolas (Figure 13.12). If $a = b$, the surface is called a **circular paraboloid**. (The surface in Figure 13.5 is a circular paraboloid.)

Parabolic sheet (or **parabolic cylinder**): $z = ax^2$

The trace of the sheet in the plane $y = 0$ is the parabola $z = ax^2$ (Figure 13.13).

Hyperbolic paraboloid: $\dfrac{y^2}{b^2} - \dfrac{x^2}{a^2} = \dfrac{z}{c}$

The traces in the planes $x = 0$ and $y = 0$ are parabolas, the former opening upward and the latter opening downward (Figure 13.14). The trace in the plane $z = 0$ consists of two intersecting lines. The trace in any other plane parallel to the xy plane is a hyperbola. The surface has the appearance of a saddle.

Two hyperbolic sheets (or **hyperbolic cylinder**): $\dfrac{y^2}{b^2} - \dfrac{x^2}{a^2} = 1$

The trace in any plane parallel to the xy plane is the hyperbola $y^2/b^2 - x^2/a^2 = 1$ (Figure 13.15).

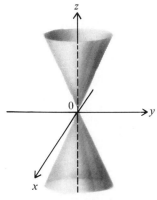

Elliptic double cone

$$\frac{x^2}{a^2} + \frac{y^2}{b^2} = \frac{z^2}{c^2}$$

FIGURE 13.11

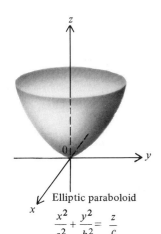

Elliptic paraboloid

$$\frac{x^2}{a^2} + \frac{y^2}{b^2} = \frac{z}{c}$$

FIGURE 13.12

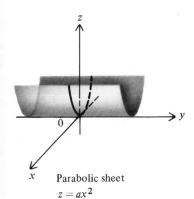

Parabolic sheet

$z = ax^2$

FIGURE 13.13

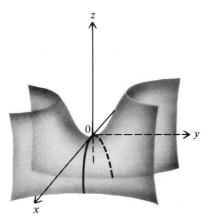

Hyperbolic paraboloid

$$\frac{y^2}{b^2} - \frac{x^2}{a^2} = \frac{z}{c}$$

FIGURE 13.14

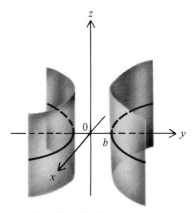

Two hyperbolic sheets

$$\frac{y^2}{b^2} - \frac{x^2}{a^2} = 1$$

FIGURE 13.15

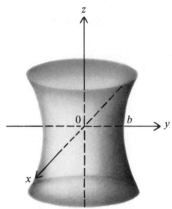

Hyperboloid of one sheet

$$\frac{x^2}{a^2} + \frac{y^2}{b^2} - \frac{z^2}{c^2} = 1$$

FIGURE 13.16

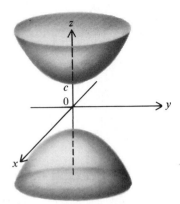

Hyperboloid of two sheets

$$\frac{z^2}{c^2} - \frac{x^2}{a^2} - \frac{y^2}{b^2} = 1$$

FIGURE 13.17

Hyperboloid of one sheet: $\dfrac{x^2}{a^2} + \dfrac{y^2}{b^2} - \dfrac{z^2}{c^2} = 1$

The traces in the planes $x = 0$ and $y = 0$ are hyperbolas. The trace in any plane parallel to the xy plane is an ellipse (a circle if $a = b$) (Figure 13.16).

Hyperboloid of two sheets: $\dfrac{z^2}{c^2} - \dfrac{x^2}{a^2} - \dfrac{y^2}{b^2} = 1$

The trace in any plane parallel to the xz or the yz plane is a hyperbola. The trace in any plane parallel to the xy plane is an ellipse (a circle if $a = b$), a point, or empty (Figure 13.17).

EXERCISES 13.1

In Exercises 1–12 find the domain of the function.

1. $f(x, y) = \sqrt{x} + \sqrt{y}$

2. $f(x, y) = \sqrt{x + y}$

3. $f(x, y) = \dfrac{y}{x} - \dfrac{x}{y}$

4. $f(x, y) = \sin \dfrac{1}{xy}$

5. $g(x, y) = \sqrt{x^2 + y^2 - 25}$

6. $g(x, y) = \sqrt{25 - x^2 - y^2}$

7. $f(x, y) = \dfrac{1}{x + y}$

8. $f(u, v) = \ln \dfrac{u^2 + v^2}{(u^2 - v^2)^2}$

9. $f(x, y, z) = \sqrt{1 - x^2 - y^2 - z^2}$

10. $f(x, y, z) = \dfrac{1}{xyz}$

11. $g(x, y, z) = \dfrac{x}{y} - \dfrac{y}{z} + \dfrac{z}{x}$

12. $f(x, y, z) = \dfrac{xyz}{(x + y)^3 - (x + z)^3}$

In Exercises 13–18 sketch the level curve $f(x, y) = c$.

13. $f(x, y) = 3x - y; c = 2, 3$

14. $f(x, y) = 6x^2; c = 6, 24$

15. $f(x, y) = x^2 + 4y^2; c = 1, 4$

16. $f(x, y) = x^2 - y; c = -2, 2$

17. $f(x, y) = x^2 - y^2; c = -1, 0, 1$

18. $f(x, y) = 2y - \cos x; c = 0, 1, 2$

In Exercises 19–23 sketch the graph of f.

19. $f(x, y) = x + 2y$ 20. $f(x, y) = 2x - 3y + 4$
21. $f(x, y) = \sqrt{4 - x^2 - y^2}$ 22. $f(x, y) = \sqrt{4x^2 + 9y^2}$
23. $f(x, y) = x^{1/3}$

In Exercises 24–32 sketch the graph of the equation.

24. $z = 2$ 25. $x = -3$
26. $z = y^2$ 27. $z = x^3 + 1$
28. $z = y^3 - 1$ 29. $x = \sqrt{1 - y^2}$
30. $x = \sqrt{4 - y^2 - z^2}$ 31. $x = \sqrt{y^2 + 4z^2}$
32. $y = \sqrt{1 - x^2 - z^2}$

In Exercises 33–37 sketch the level surface $f(x, y, z) = c$.

33. $f(x, y, z) = 2x - 4y + z; c = -1$
34. $f(x, y, z) = x^2 + y^2 + z^2; c = 2$
35. $f(x, y, z) = 4x^2 + 4y^2 + z^2; c = 1$
36. $f(x, y, z) = x^2 + y^2 - z^2; c = 0$
37. $f(x, y, z) = z - 1 - x^2 - y^2; c = 2$

In Exercises 38–55 sketch the quadric surface.

38. $\dfrac{x^2}{4} + y^2 + \dfrac{z^2}{9} = 1$ 39. $x^2 + 2y^2 + 3z^2 = 6$

40. $x^2 + z^2 = 4$ 41. $y^2 + z^2 = 9$

42. $z = x^2 + \dfrac{y^2}{9}$ 43. $x = y^2 + \dfrac{z^2}{4}$

44. $z^2 = x^2 + 4y^2$ 45. $x^2 = 9y^2 + 4z^2$
46. $y = 1 - x^2$ 47. $x = z^2 + 3$
48. $z = y^2 - 4x^2$ 49. $x = 4z^2 - y^2$
50. $y^2 - x^2 = 4$ 51. $z^2 - y^2 = 9$
52. $z^2 + 4y^2 - 2x^2 = 1$ 53. $4x^2 + y^2 - z^2 = 16$
54. $z^2 - 4y^2 - x^2 = 1$ 55. $x^2 - 9y^2 - 4z^2 = 36$

In Exercises 56–60 determine a function f of two variables (different from F) and a function g of one variable such that $F = g \circ f$.

56. $F(x, y) = \sqrt{4 - x^2 - y^2}$
57. $F(x, y) = \ln(x^2 + 6y)$
58. $F(x, y) = e^{x\sqrt{y}}$
59. $F(x, y) = x^2 y^2 + 5xy + 10$
60. $F(x, y) = \left(\sin \dfrac{x^2}{x^2 + y^2}\right)^{2/3} - 3$

61. Express the height h of a right circular cylinder as a function of the volume V and radius r.

62. Express the radius r of the base of a right circular cone as a function of the volume V and height h.

63. Express the surface area S of a rectangular box with no top as a function of the dimensions x, y, and z.

64. Express the amount A of metal required to make a storage box in the shape of a rectangular parallelepiped as a function of the length x, width y, and height z if the box is to have 12 compartments in 2 rows of 6 each and no top (Figure 13.18).

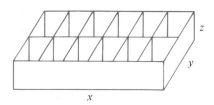

FIGURE 13.18

65. Express the cost C of painting a rectangular wall as a function of the dimensions x and y (in meters) if the cost per square meter is \$0.30.

66. Express the cost C of painting a rectangular wall as a function of the dimensions x and y (in meters) if the cost per square meter is \$0.30 and the wall contains a window 1 square meter in area.

67. The strength of the electric field at (x, y, z) due to an infinitely long charged wire lying along the z axis is given by

$$E(x, y, z) = \frac{c}{\sqrt{x^2 + y^2}}$$

where c is a positive constant. Describe the level surfaces of E.

68. The magnitude of the gravitational force exerted on a unit mass at (x, y, z) by a point mass located at the origin is given by

$$F(x, y, z) = \frac{c}{x^2 + y^2 + z^2}$$

where c is a positive constant. Describe the level surfaces of F.

69. Suppose a thin metal plate occupies the first quadrant of the xy plane and the temperature at (x, y) is given by

$$T(x, y) = xy$$

Describe the isothermal curves, that is, the level curves of T.

70. Let $f(x, y) = (x + 1)(y + 2)$ for $x \geq 0$ and $y \geq 0$. Sketch the level curves $f(x, y) = 3$ and $f(x, y) = 4$. (If f represents a utility function for two competing goods such as beer and wine, then the level curves are called **indifference curves**.)

13.2
LIMITS AND CONTINUITY

The definitions of the limit of a real-valued function in Section 2.2 and the limit of a vector-valued function in Section 12.2 provide the models for our definition of the limit of a function of several variables. Intuitively, L is the limit of $f(x, y)$ as (x, y) approaches (x_0, y_0) if $f(x, y)$ is as close to L as we wish whenever (x, y) is close enough to (x_0, y_0). Similarly, L is the limit of $f(x, y, z)$ as (x, y, z) approaches (x_0, y_0, z_0) if $f(x, y, z)$ is as close to L as we wish whenever (x, y, z) is close enough to (x_0, y_0, z_0). To formalize these ideas, we recall that the distance between the points (x, y) and (x_0, y_0) in the plane is less than δ if

$$\sqrt{(x - x_0)^2 + (y - y_0)^2} < \delta$$

and the distance between the points (x, y, z) and (x_0, y_0, z_0) in space is less than δ if

$$\sqrt{(x - x_0)^2 + (y - y_0)^2 + (z - z_0)^2} < \delta$$

DEFINITION 13.2

Let f be defined throughout a set containing a disk centered at (x_0, y_0), except possibly at (x_0, y_0) itself, and let L be a number. Then L is the **limit of f at (x_0, y_0)** if for every $\varepsilon > 0$ there is a $\delta > 0$ such that

$$\text{if } 0 < \sqrt{(x - x_0)^2 + (y - y_0)^2} < \delta, \quad \text{then} \quad |f(x, y) - L| < \varepsilon$$

In this case we write

$$\lim_{(x, y) \to (x_0, y_0)} f(x, y) = L$$

and say that $\lim_{(x, y) \to (x_0, y_0)} f(x, y)$ exists. Similarly, let f be defined throughout a set containing a ball centered at (x_0, y_0, z_0), except possibly at (x_0, y_0, z_0) itself. Then L is the **limit of f at (x_0, y_0, z_0)** if for every $\varepsilon > 0$ there is a $\delta > 0$ such that

$$\text{if } 0 < \sqrt{(x - x_0)^2 + (y - y_0)^2 + (z - z_0)^2} < \delta,$$
$$\text{then} \quad |f(x, y, z) - L| < \varepsilon$$

In this case we write

$$\lim_{(x, y, z) \to (x_0, y_0, z_0)} f(x, y, z) = L$$

and say that $\lim_{(x, y, z) \to (x_0, y_0, z_0)} f(x, y, z)$ exists.

The geometric interpretation of $\lim_{(x, y) \to (x_0, y_0)} f(x, y) = L$ is presented in Figure 13.19; you should compare it with the analogous geometric interpretation in Section 2.2. There is no analogous pictorial interpretation of $\lim_{(x, y, z) \to (x_0, y_0, z_0)} f(x, y, z) = L$, because that would require four dimensions.

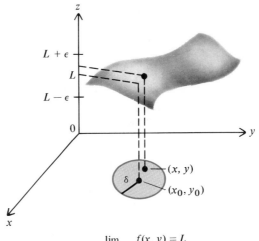

$$\lim_{(x,\,y)\,\to\,(x_0,\,y_0)} f(x,y) = L$$

FIGURE 13.19

Example 1 Show that

$$\lim_{(x,\,y)\to(x_0,\,y_0)} x = x_0 \quad \text{and} \quad \lim_{(x,\,y)\to(x_0,\,y_0)} y = y_0$$

Solution   Let $\varepsilon > 0$. Observe that

$$\sqrt{(x - x_0)^2} \leq \sqrt{(x - x_0)^2 + (y - y_0)^2}$$

Therefore if we let $\delta = \varepsilon$, it follows that

if $0 < \sqrt{(x - x_0)^2 + (y - y_0)^2} < \delta$,   then $|x - x_0| = \sqrt{(x - x_0)^2} < \delta = \varepsilon$

This proves that $\lim_{(x,\,y)\to(x_0,\,y_0)} x = x_0$. The second limit is established analogously. $\square$

Example 2 Show that

$$\lim_{(x,\,y,\,z)\to(x_0,\,y_0,\,z_0)} x = x_0, \qquad \lim_{(x,\,y,\,z)\to(x_0,\,y_0,\,z_0)} y = y_0, \quad \text{and} \quad \lim_{(x,\,y,\,z)\to(x_0,\,y_0,\,z_0)} z = z_0$$

Solution   Let $\varepsilon > 0$. Since

$$\sqrt{(x - x_0)^2} \leq \sqrt{(x - x_0)^2 + (y - y_0)^2 + (z - z_0)^2}$$

we can let $\delta = \varepsilon$ and deduce that

$$\text{if } 0 < \sqrt{(x - x_0)^2 + (y - y_0)^2 + (z - z_0)^2} < \delta$$
$$\text{then } |x - x_0| = \sqrt{(x - x_0)^2} < \delta = \varepsilon$$

Thus the first limit is verified. The proofs for the other two limits are similar. $\square$

From Theorem 2.5 we know that if f is a function of one variable, then in order that $\lim_{x \to a} f(x) = L$, both one-sided limits $\lim_{x \to a^-} f(x)$ and $\lim_{x \to a^+} f(x)$ must exist and equal L. Now if f is a function of two variables and $\lim_{(x,y) \to (x_0, y_0)} f(x, y) = L$, then Definition 13.2 implies that $f(x, y)$ must approach L as (x, y) approaches (x_0, y_0) along *each line* (or curve) through (x_0, y_0). Thus to show that $\lim_{(x,y) \to (x_0, y_0)} f(x, y)$ does *not* exist, it suffices to show that $f(x, y)$ approaches two different numbers as (x, y) approaches (x_0, y_0) along two distinct lines through (x_0, y_0).

Example 3 Let

$$f(x, y) = \frac{y^2 - x^2}{y^2 + x^2}$$

Show that $\lim_{(x,y) \to (0,0)} f(x, y)$ does not exist.

Solution Notice that if $y = 0$ and $x \neq 0$ then $f(x, y) = -1$, whereas if $x = 0$ and $y \neq 0$ then $f(x, y) = 1$. Thus $f(x, y)$ approaches -1 as (x, y) approaches $(0, 0)$ along the line $y = 0$, and $f(x, y)$ approaches 1 as (x, y) approaches $(0, 0)$ along the line $x = 0$. Therefore f does not have a limit at $(0, 0)$. The computer-drawn graph of f appearing in Figure 13.20 supports the claim that f does not have a limit at $(0, 0)$. □

FROM:
R. ELLIS +
D. Gulick
Calculus

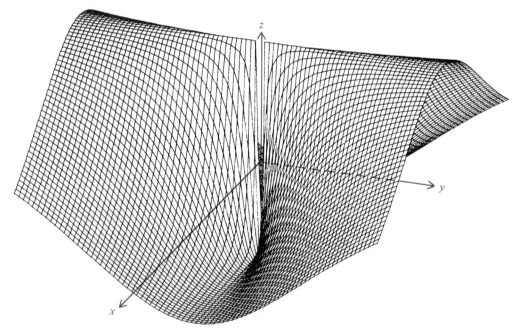

FIGURE 13.20

The limit formulas for sums, products, and quotients have counterparts for functions of several variables. For functions of two variables, they are as follows.

If $\lim_{(x,y)\to(x_0,y_0)} f(x,y)$ and $\lim_{(x,y)\to(x_0,y_0)} g(x,y)$ exist, then

$$\lim_{(x,y)\to(x_0,y_0)} (f+g)(x,y) = \lim_{(x,y)\to(x_0,y_0)} f(x,y) + \lim_{(x,y)\to(x_0,y_0)} g(x,y)$$

$$\lim_{(x,y)\to(x_0,y_0)} (f-g)(x,y) = \lim_{(x,y)\to(x_0,y_0)} f(x,y) - \lim_{(x,y)\to(x_0,y_0)} g(x,y)$$

$$\lim_{(x,y)\to(x_0,y_0)} (fg)(x,y) = \lim_{(x,y)\to(x_0,y_0)} f(x,y) \ \lim_{(x,y)\to(x_0,y_0)} g(x,y)$$

$$\lim_{(x,y)\to(x_0,y_0)} \left(\frac{f}{g}\right)(x,y) = \frac{\lim_{(x,y)\to(x_0,y_0)} f(x,y)}{\lim_{(x,y)\to(x_0,y_0)} g(x,y)} \quad \text{provided} \quad \lim_{(x,y)\to(x_0,y_0)} g(x,y) \neq 0$$

The formulas for limits of functions of three variables are similar.

Example 4   Show that

$$\lim_{(x,y)\to(-1,2)} \frac{x^3+y^3}{x^2+y^2} = \frac{7}{5} \quad \text{and} \quad \lim_{(x,y)\to(0,0)} \frac{x^3+y^3}{x^2+y^2} = 0$$

Solution   Since

$$\lim_{(x,y)\to(-1,2)} x = -1 \quad \text{and} \quad \lim_{(x,y)\to(-1,2)} y = 2$$

the product formula yields

$$\lim_{(x,y)\to(-1,2)} x^3 = -1, \qquad \lim_{(x,y)\to(-1,2)} y^3 = 8,$$

$$\lim_{(x,y)\to(-1,2)} x^2 = 1, \qquad \lim_{(x,y)\to(-1,2)} y^2 = 4$$

Then the sum and quotient formulas combine to yield

$$\lim_{(x,y)\to(-1,2)} \frac{x^3+y^3}{x^2+y^2} = \frac{\lim_{(x,y)\to(-1,2)} (x^3+y^3)}{\lim_{(x,y)\to(-1,2)} (x^2+y^2)}$$

$$= \frac{\lim_{(x,y)\to(-1,2)} x^3 + \lim_{(x,y)\to(-1,2)} y^3}{\lim_{(x,y)\to(-1,2)} x^2 + \lim_{(x,y)\to(-1,2)} y^2} = \frac{-1+8}{1+4} = \frac{7}{5}$$

We cannot use quite the same procedure to verify the second limit, since $\lim_{(x,y)\to(0,0)}(x^2+y^2) = 0$, so the quotient formula does not apply. To verify the second limit, we will show first that

$$\lim_{(x,y)\to(0,0)} \frac{x^3}{x^2+y^2} = 0 \tag{1}$$

For this limit we observe that

$$0 \leq \left|\frac{x^3}{x^2+y^2}\right| \leq \left|\frac{x^3}{x^2}\right| = |x|$$

Since $\lim_{(x,y)\to(0,0)} x = 0$ by Example 2, a version of the Squeezing Theorem for functions of two variables yields (1). A similar argument shows that

$$\lim_{(x,y)\to(0,0)} \frac{y^3}{x^2 + y^2} = 0 \tag{2}$$

We can use the sum formula to combine the limits in (1) and (2):

$$\lim_{(x,y)\to(0,0)} \frac{x^3 + y^3}{x^2 + y^2} = \lim_{(x,y)\to(0,0)} \frac{x^3}{x^2 + y^2} + \lim_{(x,y)\to(0,0)} \frac{y^3}{x^2 + y^2} = 0 + 0 = 0 \quad \Box$$

An alternative way to show that

$$\lim_{(x,y)\to(0,0)} \frac{x^3 + y^3}{x^2 + y^2} = 0 \tag{3}$$

uses polar coordinates. Notice that if $x = r\cos\theta$ and $y = r\sin\theta$, then (x, y) approaches $(0,0)$ if and only if r approaches 0. Since

$$\frac{x^3 + y^3}{x^2 + y^2} = \frac{r^3\cos^3\theta + r^3\sin^3\theta}{r^2} = r(\cos^3\theta + \sin^3\theta)$$

and since $\lim_{r\to 0} r(\cos^3\theta + \sin^3\theta) = 0$, it follows that (3) is valid.

There are also versions of the Substitution Rule for functions of several variables. For the two-variable case we suppose that

$$\lim_{(x,y)\to(x_0, y_0)} f(x, y) = L$$

and that g is a function of a single variable which is continuous at L. Then

$$\lim_{(x,y)\to(x_0, y_0)} g(f(x, y)) = g(L)$$

Example 5 Find $\lim_{(x,y)\to(e, 1)} \ln\frac{x}{y}$.

Solution First we let

$$f(x, y) = \frac{x}{y} \quad \text{and} \quad g(t) = \ln t$$

Then by the quotient formula for limits,

$$\lim_{(x,y)\to(e, 1)} f(x, y) = \frac{\lim_{(x,y)\to(e, 1)} x}{\lim_{(x,y)\to(e, 1)} y} = \frac{e}{1} = e$$

Since g is continuous at e and $g(e) = 1$, it follows from the substitution formula above that

$$\lim_{(x,y)\to(e, 1)} \ln\frac{x}{y} = \lim_{(x,y)\to(e, 1)} g(f(x, y)) = g(e) = 1 \quad \Box$$

Continuity   Now that limits have been defined, continuity is defined in the standard way.

DEFINITION 13.3

> **a.** A function f of two variables is **continuous at** (x_0, y_0) if
> $$\lim_{(x, y) \to (x_0, y_0)} f(x, y) = f(x_0, y_0)$$
> **b.** A function f of three variables is **continuous at** (x_0, y_0, z_0) if
> $$\lim_{(x, y, z) \to (x_0, y_0, z_0)} f(x, y, z) = f(x_0, y_0, z_0)$$
> **c.** A function of several variables is **continuous** if it is continuous at each point in its domain.

Example 1 tells us that if $f(x, y) = x$ and $g(x, y) = y$, then f and g are continuous functions. Similarly, if $f(x, y, z) = x$, $g(x, y, z) = y$, and $h(x, y, z) = z$, then by Example 2, f, g, and h are continuous functions. Sums, products, and quotients of continuous functions are continuous. Thus polynomial functions and rational functions are continuous.

Finally, we observe that if f is a function of two variables and g is a function of a single variable, then the continuity of f at (x_0, y_0) and the continuity of g at $f(x_0, y_0)$ together imply that $g \circ f$ is continuous at (x_0, y_0). A similar result holds when f is a function of three variables.

Example 6   Let

$$F(x, y) = \sin \frac{xy}{1 + x^2 + y^2}$$

Show that F is a continuous function.

Solution   First we let

$$f(x, y) = \frac{xy}{1 + x^2 + y^2} \quad \text{and} \quad g(t) = \sin t$$

Then $F = g \circ f$. Furthermore, f and g are continuous functions, and consequently so is their composite F. $\square$

Continuity on a Set   Let R be a set in the plane. Then for any point P in R, one of the following conditions holds:

1. There is a disk centered at P and contained in R.
2. Every disk centered at P contains points outside R.

Points satisfying condition 1 are **interior points** of R, and points satisfying condition 2 are **boundary points** of R (Figure 13.21). More generally, a point P (in R or not) is called a **boundary point** of a set R if every disk centered at P contains at least one point inside R and at least one point outside R. The **boundary** of R is the collection of its boundary points.

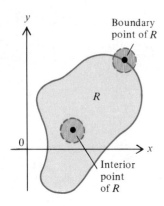

FIGURE 13.21

The boundaries of most regions we will encounter are easily described. For example, the boundary of the disk $x^2 + y^2 \le r^2$ is the circle $x^2 + y^2 = r^2$ (Figure 13.22(a)), and the boundary of a rectangular region is the rectangle (Figure 13.22(b)). Finally, a line segment or any other piecewise smooth curve defined on a closed interval $[a, b]$ is its own boundary (Figure 13.23).

Suppose a plane region R contains its boundary; then we would like to define continuity of a function f on the set R. Observe that continuity of f at an interior point (x_0, y_0) of R has already been defined (Definition 13.3). For continuity of f at a boundary point (x_0, y_0) of R, we first modify Definition 13.2 slightly:

A number L is the **limit of f at a boundary point (x_0, y_0)** of R if for every $\varepsilon > 0$ there is a number $\delta > 0$ such that

$$\text{if } (x, y) \text{ is in } R \text{ and } 0 < \sqrt{(x - x_0)^2 + (y - y_0)^2} < \delta,$$

$$\text{then } |f(x, y) - L| < \varepsilon$$

In this case we say that the limit of f at (x_0, y_0) exists and write

$$\lim_{(x, y) \to (x_0, y_0)} f(x, y) = L$$

We say that f is **continuous at a boundary point (x_0, y_0)** of R if

$$\lim_{(x, y) \to (x_0, y_0)} f(x, y) = f(x_0, y_0)$$

Finally, f is **continuous on a set R** containing its boundary if it is continuous at each point of R.

Example 7   Let R be the rectangular region consisting of all points (x, y) such that $0 \le x \le 1$ and $0 \le y \le 2$. Let

$$f(x, y) = \begin{cases} 4 - x - y & \text{for } (x, y) \text{ in } R \\ 0 & \text{for } (x, y) \text{ not in } R \end{cases}$$

Show that f is continuous on R but f is not a continuous function.

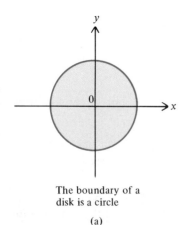

The boundary of a
disk is a circle

(a)

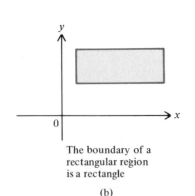

The boundary of a
rectangular region
is a rectangle

(b)

FIGURE 13.22

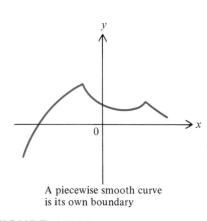

A piecewise smooth curve
is its own boundary

FIGURE 13.23

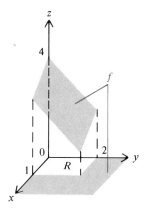

FIGURE 13.24

Solution Since the polynomial $4 - x - y$ is continuous, both the polynomial and f are continuous on R. Next, since $|f(x, y)| \geq 1$ for all (x, y) in R and $f(x, y) = 0$ for (x, y) not in R, it follows from Definition 13.2 that f has no limit at any boundary point of R and thus is not continuous (Figure 13.24). $\square$

Our notion of continuity on a set containing its boundary is a two-dimensional analogue of continuity of a function of a single variable on a closed interval (see Section 2.6). The boundary of a solid region in space is defined in a way similar to the method for plane regions (but with balls replacing disks). For example, the boundary of a ball is a sphere, and the boundary of a rectangular parallelepiped consists of the six faces enclosing it.

EXERCISES 13.2

In Exercises 1–15 evaluate the limit.

1. $\displaystyle\lim_{(x, y) \to (2, 4)} (x + \tfrac{1}{2})$

2. $\displaystyle\lim_{(x, y) \to (1, -2)} (2x^3 - 4xy + 5y^2)$

3. $\displaystyle\lim_{(x, y) \to (1, 0)} \frac{x^2 - xy + 1}{x^2 + y^2}$

4. $\displaystyle\lim_{(x, y, z) \to (-1, 2, 0)} (x^2 + 3y - 4z^2 + 2)$

5. $\displaystyle\lim_{(x, y, z) \to (2, 1, -1)} \frac{2x^2 y - xz^2}{y^2 - xz}$

6. $\displaystyle\lim_{(x, y) \to (-1, 1)} \frac{x^2 + 2xy^2 + y^4}{1 + y^2}$

7. $\displaystyle\lim_{(x, y) \to (2, 1)} \frac{x^3 + 2x^2 y - xy - 2y^2}{x + 2y}$

8. $\displaystyle\lim_{(x, y) \to (\ln 2, 0)} e^{2x + y^2}$

9. $\displaystyle\lim_{(x, y, z) \to (\pi/2, -\pi/2, 0)} \cos(x + y + z)$

10. $\displaystyle\lim_{(x, y) \to (0, 1)} \frac{\sin xy}{y}$

11. $\displaystyle\lim_{(x, y) \to (0, 0)} \frac{\sin(x^2 + y^2)}{x^2 + y^2}$

12. $\displaystyle\lim_{(x, y) \to (1, 0)} xe^{-1/|y|}$

13. $\displaystyle\lim_{(x, y) \to (0, 0)} xy \frac{x^2 - y^2}{x^2 + y^2}$

14. $\displaystyle\lim_{(x, y) \to (0, 0)} \frac{xy}{\sqrt{x^2 + y^2}}$

15. $\displaystyle\lim_{(x, y, z) \to (0, 0, 0)} \frac{x^3 + y^3 + z^3}{x^2 + y^2 + z^2}$

16. Show that $\displaystyle\lim_{(x, y) \to (0, 0)} \frac{y}{x}$ does not exist.

17. Show that $\displaystyle\lim_{(x, y) \to (0, 0)} \frac{xy}{x^2 + y^2}$ does not exist.

18. Show that $\displaystyle\lim_{(x, y) \to (0, 0)} |y|^x$ does not exist.

19. Suppose $f(x, y) = \arcsin(x - y)$ and $g(t) = t^2 - 1$. Find $\displaystyle\lim_{(x, y) \to (1, 1)} g(f(x, y))$.

20. Suppose $f(x, y) = e^{x^2 y}$ and $g(t) = \ln t$. Find $\displaystyle\lim_{(x, y) \to (-1, 1)} g(f(x, y))$.

In Exercises 21–26 explain why f is continuous.

21. $f(x, y) = xy^2$

22. $f(x, y, z) = 3x^2 z - \pi \dfrac{xy}{z}$

23. $f(x, y) = \dfrac{x^2 + y}{x^2 + y^2 - 1}$

24. $f(x, y, z) = \sin(xyz - 1)$

25. $f(x, y, z) = \ln(e^x + e^{yz})$

26. $f(x, y) = \begin{cases} \dfrac{x^2 y^2}{x^2 + y^2} & \text{for } (x, y) \neq (0, 0) \\ 0 & \text{for } (x, y) = (0, 0) \end{cases}$

27. Let

$$f(x, y) = \begin{cases} \dfrac{xy^2}{x^3 + y^3} & \text{for } (x, y) \neq (0, 0) \\ 0 & \text{for } (x, y) = (0, 0) \end{cases}$$

a. Show that f is continuous in each variable separately at $(0, 0)$, that is, $f(x, 0)$ is a continuous function of x at 0, and $f(0, y)$ is a continuous function of y at 0.
b. Show that f is not continuous at $(0, 0)$.

*28. Let

$$f(x, y) = \begin{cases} \dfrac{\sin xy}{x^2 + y^2} & \text{for } (x, y) \neq (0, 0) \\ 0 & \text{for } (x, y) = (0, 0) \end{cases}$$

Show that f is not continuous at $(0, 0)$.

*29. Let

$$f(x, y) = \begin{cases} \dfrac{x^3 y^3}{x^{12} + y^4} & \text{for } (x, y) \neq (0, 0) \\ 0 & \text{for } (x, y) = (0, 0) \end{cases}$$

Show that f is not continuous at $(0, 0)$.

30. Find formulas for the boundaries of the following regions.
a. The disk with center $(-3, 2)$ and radius 6
b. The rectangular region with vertices $(0, 0), (2, 0), (0, -3)$, and $(2 - 3)$
c. The triangular region with vertices $(-1, 1), (1, 1)$, and $(0, -5)$
d. The upper half of the xy plane, consisting of all (x, y) such that $y \geq 0$
e. The graph of the parabola $y = 4x^2$
f. The entire plane except the origin

31. Let R be the entire plane except the origin, and let

$$f(x, y) = \frac{x^2}{x^2 + y^2} \quad \text{for } (x, y) \text{ in } R$$

Determine whether $\lim_{(x, y) \to (0, 0)} f(x, y)$ exists.

32. Let R be the entire plane except the origin, and let

$$f(x, y) = \frac{x^2 \sin x}{x^2 + y^2} \quad \text{for } (x, y) \text{ in } R$$

Determine whether $\lim_{(x, y) \to (0, 0)} f(x, y)$ exists.

33. Let R be the set of all (x, y) such that $|y| < |x^3|$, and let

$$f(x, y) = \frac{y}{x} \quad \text{for } (x, y) \text{ in } R$$

Determine whether $\lim_{(x, y) \to (0, 0)} f(x, y)$ exists. Compare your answer with the one to Exercise 16.

In Exercises 34–38 determine whether f is continuous on the given region R.

34. $f(x, y) = \dfrac{6x^2 - 5xy}{(x - 3)^2}$; R is the disk $x^2 + y^2 \leq 4$

35. $f(x, y) = \begin{cases} x + y + 2 & \text{for } x \leq 0 \\ 0 & \text{for } x > 0 \end{cases}$
R is the set of all (x, y) for which $x \geq -1$

36. $f(x, y) = \begin{cases} 1 & \text{for } x^2 + y^2 \leq 9 \\ 0 & \text{for } x^2 + y^2 > 9 \end{cases}$
R is the disk $x^2 + y^2 \leq 9$

37. f is as in Exercise 36; R is the disk $x^2 + y^2 \leq 4$

38. f is as in Exercise 36; R is the disk $x^2 + y^2 \leq 16$

39. a. Give a definition of the boundary of a set in space.
b. Give a definition of the limit of a function at a boundary point of a given set in space.
c. Give a definition of continuity of a function on a set in space.

40. When x moles of sulfuric acid are mixed with y moles of water, the heat $Q(x, y)$ produced is given by

$$Q(x, y) = \frac{17,860xy}{(1.798)x + y} \quad \text{for } x > 0 \text{ and } y > 0$$

Determine whether Q has a limit at $(0, 0)$, and if so, compute its value.

41. The function f defined by

$$f(x, y) = \frac{100ax}{ax + by} \quad \text{for } x > 0 \text{ and } y > 0$$

where a and b are positive constants, appears in the study of the relationship of blood flow through the right lung to the total blood flow in the system. Determine whether f has a limit at $(0, 0)$.

13.3
PARTIAL DERIVATIVES

Let f be a function of two variables. If we fix one of the two variables, say $y = y_0$, the function whose values are $f(x, y_0)$ is a function of x alone; if that function has a derivative at x_0, we call the derivative a partial derivative at (x_0, y_0).

DEFINITION 13.4

> Let f be a function of two variables, and let (x_0, y_0) be in the domain of f. The **partial derivative of f with respect to x at (x_0, y_0)** is defined by
>
> $$f_x(x_0, y_0) = \lim_{h \to 0} \frac{f(x_0 + h, y_0) - f(x_0, y_0)}{h}$$
>
> provided that this limit exists. The **partial derivative of f with respect to y at (x_0, y_0)** is defined by
>
> $$f_y(x_0, y_0) = \lim_{h \to 0} \frac{f(x_0, y_0 + h) - f(x_0, y_0)}{h}$$
>
> provided that this limit exists.

The functions f_x and f_y that arise through partial differentiation and are defined by

$$f_x(x, y) = \lim_{h \to 0} \frac{f(x + h, y) - f(x, y)}{h}$$

and

$$f_y(x, y) = \lim_{h \to 0} \frac{f(x, y + h) - f(x, y)}{h}$$

are called the **partial derivatives** of f and are frequently denoted $\partial f / \partial x$ and $\partial f / \partial y$*. If we specify a function by an equation of the form $z = f(x, y)$, then we write $\partial z / \partial x$ and $\partial z / \partial y$. Moreover, if we use other variables, such as u and v, rather than x and y, to denote points in the plane, then the partial derivatives of f would be denoted f_u and f_v, or $\partial f / \partial u$ and $\partial f / \partial v$.

Finding partial derivatives is no more difficult than finding derivatives of functions of a single variable.

Example 1   Let $f(x, y) = 24xy - 6x^2 y$. Find f_x and f_y, and evaluate f_x and f_y at $(1, 2)$.

Solution   By holding y constant and differentiating f with respect to x, we find that

$$f_x(x, y) = 24y - 12xy$$

so that $f_x(1, 2) = 48 - 24 = 24$. By holding x constant and differentiating f with respect to y, we find that

$$f_y(x, y) = 24x - 6x^2$$

so that $f_y(1, 2) = 24 - 6 = 18$.   □

---

* The symbol ∂, which is *not* a Greek letter, was invented in the mid-nineteenth century to distinguish partial derivatives from other types of derivatives. We read $\partial f / \partial x$ as "the partial of f with respect to x."

Example 2   Let $z = x^2 \cos y$. Find $\dfrac{\partial z}{\partial x}$ and $\dfrac{\partial z}{\partial y}$.

Solution   We find immediately that

$$\frac{\partial z}{\partial x} = 2x \cos y \quad \text{and} \quad \frac{\partial z}{\partial y} = -x^2 \sin y \quad \square$$

The sum, product, and quotient rules for derivatives have counterparts for partial derivatives. Thus if f and g have partial derivatives, then

$$(f + g)_x = f_x + g_x \qquad \text{and} \quad (f + g)_y = f_y + g_y$$

$$(f - g)_x = f_x - g_x \qquad \text{and} \quad (f - g)_y = f_y - g_y$$

$$(fg)_x = f_x g + f g_x \quad \text{and} \qquad (fg)_y = f_y g + f g_y$$

$$\left(\frac{f}{g}\right)_x = \frac{f_x g - f g_x}{g^2} \quad \text{and} \qquad \left(\frac{f}{g}\right)_y = \frac{f_y g - f g_y}{g^2}$$

It follows that a polynomial or rational function has partial derivatives at each point in its domain.

Example 3   Let

$$f(x, y) = \frac{x^3 y - xy^3}{x^2 + y^2}$$

Find f_x and f_y.

Solution   Since f is a rational function, $f_x(x, y)$ and $f_y(x, y)$ are defined for all x and y such that the denominator $x^2 + y^2$ is not 0, that is, for all points except the origin. To find $f_x(x, y)$ we hold y constant and use the quotient rule for partial derivatives to differentiate with respect to x:

$$f_x(x, y) = \frac{(3x^2 y - y^3)(x^2 + y^2) - (x^3 y - xy^3)2x}{(x^2 + y^2)^2}$$

$$= \frac{x^4 y + 4x^2 y^3 - y^5}{(x^2 + y^2)^2}$$

For $f_y(x, y)$ we hold x constant and use the quotient rule for partial derivatives to differentiate with respect to y:

$$f_y(x, y) = \frac{x^5 - 4x^3 y^2 - xy^4}{(x^2 + y^2)^2} \quad \square$$

In the next example we must refer directly to Definition 13.4 in order to find a partial derivative.

Example 4 Let

$$f(x, y) = \begin{cases} \dfrac{x^3 y - xy^3}{x^2 + y^2} & \text{for } (x, y) \neq (0, 0) \\ 0 & \text{for } (x, y) = (0, 0) \end{cases}$$

Show that $f_x(0, 0) = f_y(0, 0) = 0$

Solution   Since $f(h, 0) = 0$ for all h, we have

$$f_x(0, 0) = \lim_{h \to 0} \frac{f(h, 0) - f(0, 0)}{h - 0} = \lim_{h \to 0} \frac{0 - 0}{h - 0} = 0$$

Similarly, $f(0, h) = 0$ for all h, so that

$$f_y(0, 0) = \lim_{h \to 0} \frac{f(0, h) - f(0, 0)}{h - 0} = \lim_{h \to 0} \frac{0 - 0}{h - 0} = 0$$

The computer-drawn graph of f appearing in Figure 13.25 lends credence to the fact that the partial derivatives of f at $(0, 0)$ are zero.   □

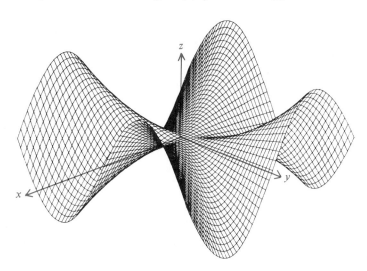

FIGURE 13.25

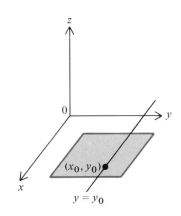

FIGURE 13.26

We can think of the partial derivative $f_x(x_0, y_0)$ as the rate of change of $f(x, y)$ at (x_0, y_0) with respect to x when y is held constant. For example, suppose $f(x, y)$ is the temperature at any point (x, y) on a flat metal plate lying on the xy plane. Then $f_x(x_0, y_0)$ is the rate at which the temperature changes at (x_0, y_0) along the line through (x_0, y_0) parallel to the x axis (Figure 13.26). If the temperature increases as x increases, then $f_x(x_0, y_0) > 0$, whereas if the temperature decreases as x increases, then $f_x(x_0, y_0) < 0$. Similarly, the partial derivative $f_y(x_0, y_0)$ is the rate at which the temperature changes at (x_0, y_0) along the line through (x_0, y_0) parallel to the y axis.

Geometrically, the partial derivatives $f_x(x_0, y_0)$ and $f_y(x_0, y_0)$ describe how the graph of f is slanted near the point $(x_0, y_0, f(x_0, y_0))$. Since $f_x(x_0, y_0)$ is obtained by holding y fixed at y_0, let us consider the curve C determined by the

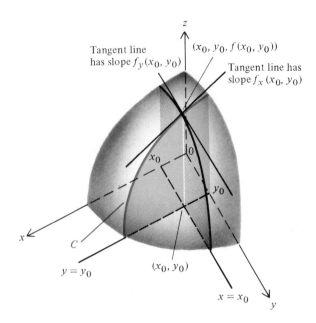

FIGURE 13.27

intersection of the graph of f and the plane $y = y_0$ (Figure 13.27). If $h \neq 0$, then

$$\frac{f(x_0 + h, y_0) - f(x_0, y_0)}{h}$$

is the slope of the secant line through the points $(x_0, y_0, f(x_0, y_0))$ and $(x_0 + h, y_0, f(x_0 + h, y_0))$ on C in the plane $y = y_0$. Since

$$f_x(x_0, y_0) = \lim_{h \to 0} \frac{f(x_0 + h, y_0) - f(x_0, y_0)}{h}$$

it follows that $f_x(x_0, y_0)$ is the slope of the line tangent to C at $(x_0, y_0, f(x_0, y_0))$. This implies that the vector $\mathbf{i} + f_x(x_0, y_0)\mathbf{k}$ is tangent to C at $(x_0, y_0, f(x_0, y_0))$, a fact that we will use later. An analogous argument shows that $f_y(x_0, y_0)$ is the slope of the line tangent to the curve obtained by intersecting the graph of f and the plane $x = x_0$ (Figure 13.27). The corresponding tangent vector is $\mathbf{j} + f_y(x_0, y_0)\mathbf{k}$.

For functions of three variables, partial derivatives at (x_0, y_0, z_0) are defined as follows:

$$f_x(x_0, y_0, z_0) = \lim_{h \to 0} \frac{f(x_0 + h, y_0, z_0) - f(x_0, y_0, z_0)}{h}$$

$$f_y(x_0, y_0, z_0) = \lim_{h \to 0} \frac{f(x_0, y_0 + h, z_0) - f(x_0, y_0, z_0)}{h} \tag{1}$$

$$f_z(x_0, y_0, z_0) = \lim_{h \to 0} \frac{f(x_0, y_0, z_0 + h) - f(x_0, y_0, z_0)}{h}$$

provided that these limits exist. The formulas in (1) give rise to the partial derivative functions f_x, f_y, and f_z. Alternative notations for these partial derivatives are $\partial f/\partial x$, $\partial f/\partial y$, and $\partial f/\partial z$, or $\partial w/\partial x$, $\partial w/\partial y$, and $\partial w/\partial z$ if $w = f(x, y, z)$.

Example 5   Let $f(x, y, z) = e^{2x}\cos z + e^{3y}\sin z$. Find the partial derivatives of f.

Solution   First, we hold y and z constant and differentiate with respect to x:

$$\frac{\partial f}{\partial x} = 2e^{2x}\cos z$$

Next, we hold x and z constant and differentiate with respect to y:

$$\frac{\partial f}{\partial y} = 3e^{3y}\sin z$$

Finally, we hold x and y constant and differentiate with respect to z:

$$\frac{\partial f}{\partial z} = -e^{2x}\sin z + e^{3y}\cos z \quad \square$$

Higher-Order Partial Derivatives   A function of one variable may have second, third, and higher derivatives. There is an analogue for functions of several variables. If f is a function of the variables x and y, then the functions f_x and f_y may each have two partial derivatives. In this case the partial derivatives of f_x and f_y would generate four new partial derivatives:

$$(f_x)_x, \quad \text{usually denoted} \quad f_{xx} \quad \text{or} \quad \frac{\partial^2 f}{\partial x^2}$$

$$(f_x)_y, \quad \text{usually denoted} \quad f_{xy} \quad \text{or} \quad \frac{\partial^2 f}{\partial y\,\partial x}$$

$$(f_y)_x, \quad \text{usually denoted} \quad f_{yx} \quad \text{or} \quad \frac{\partial^2 f}{\partial x\,\partial y}$$

$$(f_y)_y, \quad \text{usually denoted} \quad f_{yy} \quad \text{or} \quad \frac{\partial^2 f}{\partial y^2}$$

The partial derivatives just described are called **second partial derivatives of** f; f_{xy} and f_{yx} are usually called **mixed partial derivatives of** f, or more briefly, **mixed partials**. To avoid confusion we will often refer to f_x, f_y, and f_z as **first partial derivatives**, or **first partials**, or **partials**.

Caution: Notice that the order in which x and y appear in the expression f_{xy} is opposite to the order in which they appear in $\partial^2 f/\partial y\,\partial x$.

Example 6 Let $f(x, y) = \sin xy^2$. Find all second partial derivatives of f.

Solution The first partials are given by

$$f_x(x, y) = y^2 \cos xy^2 \quad \text{and} \quad f_y(x, y) = 2xy \cos xy^2$$

We obtain the second partials by computing the partial derivatives of the first partials:

$$f_{xx}(x, y) = -y^4 \sin xy^2$$
$$f_{xy}(x, y) = 2y \cos xy^2 - 2xy^3 \sin xy^2$$
$$f_{yx}(x, y) = 2y \cos xy^2 - 2xy^3 \sin xy^2$$
$$f_{yy}(x, y) = 2x \cos xy^2 - 4x^2 y^2 \sin xy^2 \quad \square$$

As with first partial derivatives, we sometimes must refer directly to Definition 13.4 to determine a second partial derivative.

Example 7 Let

$$f(x, y) = \begin{cases} \dfrac{x^3 y - xy^3}{x^2 + y^2} & \text{for } (x, y) \neq (0,0) \\ 0 & \text{for } (x, y) = (0,0) \end{cases}$$

Show that $f_{xy}(0,0) \neq f_{yx}(0,0)$.

Solution From Example 3 we deduce that

$$f_x(0, y) = \frac{0^4 y + 4(0^2)y^3 - y^5}{(0^2 + y^2)^2} = -y \quad \text{for } y \neq 0$$

$$f_y(x, 0) = \frac{x^5 - 4x^3(0^2) - x(0^4)}{(x^2 + 0^2)^2} = x \quad \text{for } x \neq 0$$

Furthermore, $f_x(0,0) = 0 = f_y(0,0)$ by Example 4. These formulas imply that

$$f_{xy}(0,0) = \lim_{h \to 0} \frac{f_x(0,h) - f_x(0,0)}{h - 0} = \lim_{h \to 0} \frac{-h - 0}{h - 0} = -1$$

$$f_{yx}(0,0) = \lim_{h \to 0} \frac{f_y(h,0) - f_y(0,0)}{h - 0} = \lim_{h \to 0} \frac{h - 0}{h - 0} = 1$$

Consequently $f_{xy}(0,0) \neq f_{yx}(0,0)$. $\quad \square$

The mixed partials computed in Example 6 are equal, whereas those in Example 7 are not equal. However, the mixed partials of the functions we usually encounter are equal, thanks to the following theorem. We omit the proof of this theorem.

THEOREM 13.5

Let f be a function of two variables, and assume that f_{xy} and f_{yx} are continuous at (x_0, y_0). Then

$$f_{xy}(x_0, y_0) = f_{yx}(x_0, y_0)$$

Second partials of a function of three variables are defined in the same way as for a function of two variables. Moreover, if f is such a function and if f_{xy} and f_{yx} are continuous at a point (x_0, y_0, z_0), then

$$f_{xy}(x_0, y_0, z_0) = f_{yx}(x_0, y_0, z_0)$$

Analogous statements hold for f_{xz} and f_{zx}, and for f_{yz} and f_{zy}.

Analysis of a Rainbow

Each ray of sunlight contains all the colors in the visible spectrum, from red to violet. We will refer to a single such color in a given ray as a **monochromatic ray**. When a monochromatic ray from the sun hits a droplet of water, the ray is refracted in accordance with Snell's Law:

$$\sin i = \mu \sin r \tag{2}$$

where i is the angle of incidence, r is the angle of refraction, and μ is the **index of refraction** of water for the monochromatic ray (Figure 13.28(a)). The index μ depends only on the color, and is defined by the formula

$$\mu = \frac{\text{speed of the monochromatic ray in a vacuum}}{\text{speed of the monochromatic ray in pure water}}$$

For the visible spectrum μ varies from approximately 1.330 for red to 1.342 for violet. The main rainbow one sees in the sky is created by various monochromatic rays that have been refracted, then reflected, and finally refracted again by droplets of water, as in Figure 13.28(b). The **scattering angle** θ formed by the path

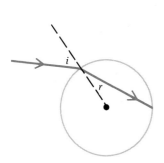

Refraction of a monochromatic ray at the surface of a water droplet

(a)

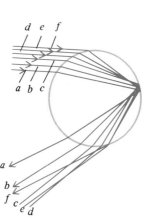

(b)

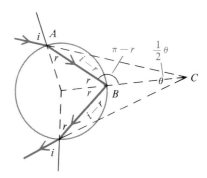

In triangle ABC,

$$(i - r) + (\pi - r) + \frac{1}{2}\theta = \pi$$

so $\theta = 4r - 2i$

(c)

FIGURE 13.28

of a monochromatic ray before the ray enters the droplet and the path after the ray leaves the droplet depends on both μ and i. The angles θ, r, and i are related by

$$\theta = 4r - 2i$$

(Figure 13.28(c)). Solving for r in (2), we find that

$$r = \arcsin \frac{\sin i}{\mu}$$

so that we obtain the formula

$$\theta(\mu, i) = 4 \arcsin \frac{\sin i}{\mu} - 2i \tag{3}$$

for θ as a function of μ and i.

Monochromatic rays emanating from the sun travel in almost, but not quite, parallel paths, and a human eye can receive rays from a single droplet through a tiny but positive angle (Figure 13.29). These two facts imply that if an observer is located so that a ray with angle of incidence i enters the eye, then other rays with nearly the same angle of incidence can still enter the eye. In 1637 Descartes observed that for rays of a single color, the less θ changes for small changes in i, the more intense is the impression of that color on the eye. This condition corresponds to $|\partial\theta/\partial i|$ being small. It follows that the strongest optical impression occurs when

$$\frac{\partial\theta}{\partial i}(\mu, i) = 0 \tag{4}$$

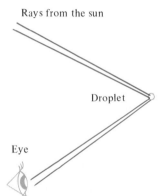

Rays from the sun

Droplet

Eye

FIGURE 13.29

and this effect is so pronounced that the color with index of refraction μ is perceived only when i satisfies equation (4). Let us refer to such an angle i as i_μ. In Example 8 we show that i_μ is unique (for a given value of μ) and find a formula for i_μ in terms of μ. Then from (3) it is easy to calculate the corresponding value $\theta(\mu, i_\mu)$ of the scattering angle through which a ray passes when it is observed by the eye.

Example 8   Let

$$\theta(\mu, i) = 4 \arcsin \frac{\sin i}{\mu} - 2i$$

For any given μ, find the angle i_μ for which

$$\frac{\partial\theta}{\partial i}(\mu, i_\mu) = 0$$

Solution   We find that

$$\frac{\partial\theta}{\partial i}(\mu, i) = 4\frac{(\cos i)/\mu}{\sqrt{1 - (\sin^2 i)/\mu^2}} - 2 = 4\frac{\cos i}{\sqrt{\mu^2 - \sin^2 i}} - 2$$

Consequently

$$\frac{\partial\theta}{\partial i}(\mu, i) = 0$$

provided that

$$4\frac{\cos i}{\sqrt{\mu^2 - \sin^2 i}} = 2$$

or

$$\sqrt{\mu^2 - \sin^2 i} = 2\cos i \tag{5}$$

Squaring both sides of (5), substituting $1 - \sin^2 i$ for $\cos^2 i$, and solving for $\sin i$, we obtain

$$\sin i = \sqrt{\frac{4 - \mu^2}{3}}$$

Consequently

$$i_\mu = \arcsin\sqrt{\frac{4 - \mu^2}{3}} \qquad \square$$

Using the values of μ given earlier for red and for violet along with the formula just obtained for i_μ, we calculate that, in degrees,

$$i_{red} \approx 59.6° \quad \text{and} \quad i_{violet} \approx 58.9°$$

and by (3) this means that

$$\theta(red, i_{red}) \approx 42.5° \quad \text{and} \quad \theta(violet, i_{violet}) \approx 40.8°$$

Thus the largest scattering angle θ is approximately $42.5°$. This tells us that for a person standing on the ground to see a rainbow, the sun can be no higher than $42.5°$ above the horizon, because otherwise the terminal rays would point upward, rather than downward toward the observer's eye.

Next we show that $\theta(\mu, i_\mu)$ is a decreasing function of μ and then determine the order of the colors in the rainbow. Let $f(\mu) = \theta(\mu, i_\mu)$. From (3) we have

$$f(\mu) = 4\arcsin\frac{\sin i_\mu}{\mu} - 2i_\mu$$

Therefore

$$\frac{df}{d\mu} = \frac{4}{\sqrt{1 - (\sin^2 i_\mu)/\mu^2}}\left(\frac{\mu(\cos i_\mu)(di_\mu/d\mu) - \sin i_\mu}{\mu^2}\right) - 2\frac{di_\mu}{d\mu}$$

$$= \frac{4\cos i_\mu(di_\mu/d\mu)}{\sqrt{\mu^2 - \sin^2 i_\mu}} - \frac{4\sin i_\mu}{\mu\sqrt{\mu^2 - \sin^2 i_\mu}} - 2\frac{di_\mu}{d\mu}$$

$$\overset{(5)}{=} \frac{4\cos i_\mu(di_\mu/d\mu)}{2\cos i_\mu} - \frac{4\sin i_\mu}{2\mu\cos i_\mu} - 2\frac{di_\mu}{d\mu}$$

$$= -\frac{2}{\mu}\tan i_\mu$$

Since i_μ is an acute angle, $df/d\mu$ is negative, so that θ is a decreasing function of μ. In other words, $\theta(\mu, i_\mu)$ is smallest for the largest value of μ, which corresponds to violet, and is largest for the smallest value of μ, which corresponds to red. It follows that the red is at the top and violet is at the bottom of the rainbow (Figure 13.30).

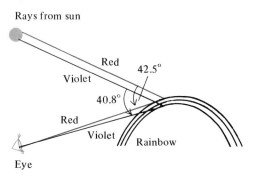

FIGURE 13.30

Sometimes one can see a second, fainter rainbow, higher in the sky than the first, created by monochromatic rays that have been reflected *twice* instead of once within the droplet of water. It turns out that the colors of the second rainbow are reversed, that is, violet is at the top and red at the bottom (see Exercise 66).

EXERCISES 13.3

In Exercises 1–19 find the first partial derivatives of the function.

1. $f(x, y) = \frac{2}{3}x^{3/2}$

2. $f(x, y) = 9 - x^2 - 4y^2$

3. $f(x, y) = 2x + 3x^2 y^4$

4. $g(x, y) = x^3 e^{2y}$

5. $g(u, v) = \dfrac{u^3 + v^3}{u^2 + v^2}$

6. $f(x, y) = \sqrt{x^2 + y^2}$

7. $f(x, y) = \sqrt{4 - x^2 - 9y^2}$

8. $z = \sqrt{\frac{1}{4}x^2 - y^2}$

9. $z = \sqrt{(1 - x^{2/3})^3 - y^2}$

10. $w = \cos\dfrac{u}{v}$

11. $z = (\sin x^2 y)^3$

12. $z = x^y$

13. $f(x, y, z) = x^2 y^5 + xz^2$

14. $f(x, y, z) = x(\cos y)e^z$

15. $f(x, y, z) = \dfrac{x + y + z}{xy + yz + zx}$

16. $f(u, v, w) = \dfrac{1}{\sqrt{u^2 + v^2 + w^2}}$

17. $w = e^x(\cos y + \sin z)$

18. $w = \left(\dfrac{x}{y}\right)^z$

19. $w = \arcsin\dfrac{1}{1 + xyz^2}$

In Exercises 20–25 find the first partial derivatives of f at the given point.

20. $f(x, y) = x^4 - 6x^2 - 3xy^2 + 17; (-1, 2)$

21. $f(x, y) = \sqrt{4x^2 + y^2}; (2, -3)$

22. $f(x, y, z) = xy^2 \sin z; (-1, 2, 0)$

23. $f(x, y, z) = e^{2x - 4y - z}; (0, -1, 1)$

24. $f(x, y) = \begin{cases} \dfrac{x^3 + y^3}{x^2 + y^2} & \text{for } (x, y) \neq (0, 0) \\ 0 & \text{for } (x, y) = (0, 0) \end{cases}$ $(0, 0)$

25. $f(x, y) = \begin{cases} \dfrac{x^2 y^3}{x^2 + 4y^3} & \text{for } (x, y) \neq (0, 0) \\ 0 & \text{for } (x, y) = (0, 0) \end{cases}$ $(0, 0)$

26. Let $f(x, y) = \displaystyle\int_1^x P(t)\, dt + \int_1^y Q(t)\, dt$. Find f_x and f_y.

27. Let $f(x, y) = \displaystyle\int_\pi^{x^2 + y^2} \sin t^2\, dt$. Find f_x and f_y.

28. Let

$f(x, y) = (\tan x)(y^{xy - \sin y})e^{\cos y} + \ln (1 + x^2) \cos (x + 1)^y$

Find $f_y(0, 1)$. (*Hint:* Use Definition 13.4.)

In Exercises 29–33 find f_{xy} and f_{yx}.

29. $f(x, y) = 3x^2 - \sqrt{2}xy^2 + y^5 - 2$

30. $f(x, y) = \dfrac{x^2 - y^2}{x^2 + y^2}$

31. $f(x, y) = \sqrt{x^2 + y^2}$

32. $f(x, y, z) = x^4 - 2x^2 y\sqrt{z} + 3yz^4 + 2$

33. $f(x, y, z) = z \cos xy$

In Exercises 34–38 find f_{xx}, f_{yy}, and f_{zz} (where applicable).

34. $f(x, y) = \sqrt{16 - 9x^2 - 4y^2}$

35. $f(x, y) = e^{x - 2y}$

36. $f(x, y) = \int_0^x \sin t^2 \, dt \int_0^y \cos t^2 \, dt$

37. $f(x, y, z) = \sqrt{x^2 + y^2 + z^2}$

38. $f(x, y, z) = e^{x^2} \sin yz + \ln(x^2 + y^2 + z^2)$

39. Find symmetric equations for the line that lies in the plane $y = 1$ and is tangent to the intersection of the plane and the paraboloid $z = x^2 + 16y^2$ at $(-3, 1, 25)$.

40. Find symmetric equations for the line that lies in the plane $x = 2$ and is tangent to the intersection of the plane and the cone $z = \sqrt{x^2 + y^2}$ at $(2, 2\sqrt{3}, 4)$.

In Exercises 41–44 find $\sqrt{f_x^2 + f_y^2 + 1}$.

41. $f(x, y) = 1 - x$ \qquad 42. $f(x, y) = 4 - y^2$

43. $f(x, y) = \sqrt{x^2 + y^2}$ \qquad 44. $f(x, y) = \sqrt{1 - x^2 - y^2}$

45. Show that the function in Exercise 27 of Section 13.2 has partial derivatives at $(0, 0)$ even though it is not continuous at $(0, 0)$.

46. Rectangular and polar coordinates in the plane are related by the equations $x = r \cos \theta$, $y = r \sin \theta$, $r = \sqrt{x^2 + y^2}$, $\theta = \arctan y/x$. Find the following partial derivatives.

   a. $\dfrac{\partial x}{\partial r}$ \quad b. $\dfrac{\partial x}{\partial \theta}$ \quad c. $\dfrac{\partial y}{\partial r}$ \quad d. $\dfrac{\partial y}{\partial \theta}$

   e. $\dfrac{\partial r}{\partial x}$ \quad f. $\dfrac{\partial r}{\partial y}$ \quad g. $\dfrac{\partial \theta}{\partial x}$ \quad h. $\dfrac{\partial \theta}{\partial y}$

47. If $z = e^{-ay} \cos ax$, show that

$$\frac{\partial^2 z}{\partial x^2} = a \frac{\partial z}{\partial y}$$

48. Let $z = x^c e^{-y/x}$, where c is a constant. Find the value of c such that

$$\frac{\partial z}{\partial x} = y \frac{\partial^2 z}{\partial y^2} + \frac{\partial z}{\partial y}$$

In Exercises 49–51 show that the functions u and v satisfy the **Cauchy–Riemann equations** $u_x = v_y$ and $u_y = -v_x$.

49. $u = x^2 - y^2$, $v = 2xy$

50. $u = x^3 - 3xy^2$, $v = 3x^2 y - y^3$

51. $u = e^x \cos y$, $v = e^x \sin y$

A function z satisfies **Laplace's equation** if

$$\frac{\partial^2 z}{\partial x^2} + \frac{\partial^2 z}{\partial y^2} = 0$$

In Exercises 52–53 show that the function satisfies Laplace's equation.

52. $z = x^4 - 6x^2 y^2 + y^4$

53. $z = \ln(x^2 + y^2)$

54. Show that if u and v satisfy the Cauchy–Riemann equations (as in Exercises 49–51) and the mixed partials of u and v are continuous, then u and v satisfy Laplace's equation.

55. Show that Example 7 does not contradict Theorem 13.5 by computing $f_{xy}(x, 0)$ for $x \neq 0$, and then verifying that f_{xy} is not continuous at $(0, 0)$.

56. Let f be a function of two variables with partials of all orders. Then f has four second partials, f_{xx}, f_{xy}, f_{yx}, and f_{yy}. If the second partials are continuous, only three of them can be distinct.

   a. How many third-order partials does f have? If these are continuous, what is the maximum number that can be distinct?

   b. Generalize (a) to nth-order partials. (*Hint:* Observe that if all nth-order partials are continuous, then any nth-order partial is equal to another nth-order partial in which all the differentiations with respect to x are performed first and the differentiations with respect to y are done last.)

57. Let M have continuous partials on a rectangle bounded by $x = a$, $x = b$, $y = c$, and $y = d$. Show that

$$\int_a^b \frac{\partial M}{\partial x}(x, y) \, dx = M(b, y) - M(a, y) \quad \text{for } c \leq y \leq d$$

and

$$\int_c^d \frac{\partial M}{\partial y}(x, y) \, dy = M(x, d) - M(x, c) \quad \text{for } a \leq x \leq b$$

58. If c is a constant, then an equation of the form

$$\frac{\partial^2 u}{\partial x^2} = c^2 \frac{\partial^2 u}{\partial t^2}$$

is called a **wave equation**. Show that if $u = \sin ax \sin bt$, where a and b are constants, then u satisfies the wave equation with $c = a/b$.

59. If c is a constant, then an equation of the form

$$\frac{\partial u}{\partial t} = c \frac{\partial^2 u}{\partial x^2}$$

is called a **diffusion equation**.

   a. Show that if $u = e^{ax + bt}$, where a and b are constant, then u satisfies the diffusion equation with $c = b/a^2$.

   *b. Show that if

$$u = u(x, t) = \frac{1}{\sqrt{t}} e^{-x^2/at}$$

where a is a constant, then u satisfies a diffusion equation. [The number $u(x, t)$ might represent the concentration of a drug at a point x in a muscle at time t. For each value of t the graph of u (considered as a function of x) is a bell-shaped curve. As t increases, the curve becomes flatter (Figure 13.31).]

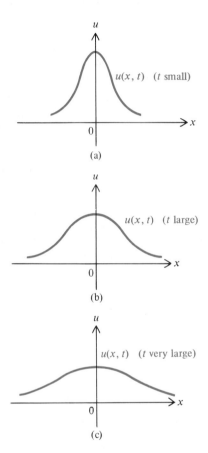

(a)

(b)

(c)

FIGURE 13.31

60. Let R be the rectangle consisting of all (x, y) for which $a \leq x \leq b$ and $c \leq y \leq d$. Suppose f is a function of two variables defined on R such that f and f_x are continuous on R. Then $\int_c^d f(x, y)\, dy$ is a differentiable function of x and

$$\frac{d}{dx} \int_c^d f(x, y)\, dy = \int_c^d \frac{\partial f}{\partial x}(x, y)\, dy \qquad (6)$$

Using (6), find the derivatives with respect to x of the following functions.

a. $\displaystyle\int_1^{\sqrt{3}} \ln(x^2 + y^2)\, dy$ b. $\displaystyle\int_1^2 \frac{1}{y} \arctan \frac{x}{y}\, dy$

61. Let m_1 and m_2 be masses, with $m_1 \geq m_2$, and assume that they are connected to an apparatus called an Atwood machine (see Figure 13.32). The acceleration a of the mass m_1 downward is given by

$$a = \frac{m_1 - m_2}{m_1 + m_2} g$$

where g is the acceleration due to gravity. Show that

$$m_1 \frac{\partial a}{\partial m_1} + m_2 \frac{\partial a}{\partial m_2} = 0$$

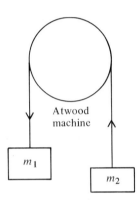

Atwood machine

m_1

m_2

FIGURE 13.32

62. When two resistors having resistances R_1 and R_2 are connected in parallel, the combined resistance R is given by $R = R_1 R_2 / (R_1 + R_2)$. Show that

$$\frac{\partial^2 R}{\partial R_1^2} \frac{\partial^2 R}{\partial R_2^2} = \frac{4R^2}{(R_1 + R_2)^4}$$

63. The kinetic energy K of a body with mass m and velocity v is given by $K = \frac{1}{2} mv^2$. Show that

$$\frac{\partial K}{\partial m} \frac{\partial^2 K}{\partial v^2} = K$$

64. The ideal gas law states that if n moles of a gas has volume V and temperature T and is under pressure p, then $pV = nkT$, where k is the universal gas constant. Show that

$$\frac{\partial V}{\partial T} \frac{\partial T}{\partial p} \frac{\partial p}{\partial V} = -1$$

65. The index of refraction of water at 20°C for yellow light from a sodium flame is 1.333. Determine the angle $\theta(\mu, i_\mu)$ for $\mu = 1.333$.

66. For the secondary rainbow the angle θ is given by

$$\theta(\mu, i) = 2\pi + 2i - 6 \arcsin \frac{\sin i}{\mu}$$

a. Show that for any given μ, the angle i_μ for which

$$\frac{\partial \theta}{\partial i}(\mu, i_\mu) = 0$$

is given by

$$i_\mu = \arcsin \sqrt{\frac{9 - \mu^2}{8}}$$

b. Using the values of μ for red and violet given after (2), deduce that

$$\theta(\text{red}, i_{\text{red}}) \approx 230.1° \quad \text{and} \quad \theta(\text{violet}, i_{\text{violet}}) \approx 233.2°$$

and thus that red lies below violet in the secondary rainbow.

67. Let D be a solid region whose boundary is a cylinder of radius r and height h, capped on each end by a hemisphere. The volume V and the surface area S of D are given by

$$V = \pi r^2 h + 2\left(\frac{2}{3}\pi r^3\right) = \pi r^2\left(h + \frac{4}{3}r\right)$$

$$S = 2\pi r h + 2(2\pi r^2) = 2\pi r(h + 2r)$$

If D represents a bacterium, then the rate R at which a chemical substance can be absorbed into D is given by $R = c(S/V)$, where c is a positive constant. Show that $\partial R / \partial r < 0$ and $\partial R / \partial h < 0$. (The result implies that an increase in either the radius or height of the bacterium decreases the rate of chemical absorption.)

68. When an X-ray beam passes through an object, the intensity of the beam at a distance x from the point of entry is given by

$$I(x, \sigma) = I_0 e^{-\sigma x}$$

where I_0 is the intensity at entry and σ depends on the frequency of the X-rays (Figure 13.33). Suppose we wish to check the uniformity of thickness of an aluminum bar

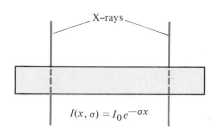

FIGURE 13.33

1 centimeter thick by X-raying it. Then

$$\frac{\partial I}{\partial x}(1, \sigma)$$

is a measure of the variation of intensity with small variations of thickness. Thus

$$\left|\frac{\partial I}{\partial x}(1, \sigma)\right|$$

is a measure of the "contrast" of the X-ray.

a. Find the value of σ that maximizes

$$\left|\frac{\partial I}{\partial x}(1, \sigma)\right|$$

(From that value one could determine the X-ray frequency that would provide the best contrast.)

b. Suppose a doctor X-rays a bone x_0 centimeters thick to search for a fracture. Find the value of σ that maximizes

$$\left|\frac{\partial I}{\partial x}(x_0, \sigma)\right|$$

69. If $f(x, y)$ is the amount of a commodity produced from x units of capital and y units of labor, then f is called a **production function**. If

$$f(x, y) = x^\alpha y^\beta \quad \text{for } x > 0 \text{ and } y > 0$$

where α and β are positive constants less than 1, then f is called a **Cobb–Douglas production function**.

a. Show that $f(tx, ty) = t^{\alpha + \beta} f(x, y)$.

b. If $z = f(x, y)$, show that

$$\frac{1}{z}\frac{\partial z}{\partial x} = \frac{\alpha}{x} \quad \text{and} \quad \frac{1}{z}\frac{\partial z}{\partial y} = \frac{\beta}{y}$$

and that

$$x\frac{\partial z}{\partial x} + y\frac{\partial z}{\partial y} = (\alpha + \beta)z$$

70. If f is a production function (see Exercise 69), then $\partial f / \partial x$ and $\partial f / \partial y$ are called the **marginal productivity of capital** and the **marginal productivity of labor**, respectively. If

$$f(x, y) = 60x^{1/2}y^{2/3} \quad \text{for } x \geq 0 \text{ and } y \geq 0$$

find the pairs (x, y) at which the marginal productivity of capital is equal to the marginal productivity of labor.

71. It seems reasonable that an increase in taxation on a commodity would decrease the production of that commodity. The following argument supports that claim. Assume that all required derivatives exist. Let $P_0(x)$ be the profit before taxes on x units produced, for $x \geq 0$. Let

$P(x, t)$ denote the profit after taxes on x units produced with tax t on each unit. Assume that at any tax rate t the company will maximize its profits by producing $f(t)$ units so that

$$\frac{\partial P}{\partial x}(f(t), t) = 0 \tag{7}$$

$$\frac{\partial^2 P}{\partial x^2}(f(t), t) < 0 \tag{8}$$

(The conditions in (7) and (8) are just those required for the Second Derivative Test.)

a. Show that $P(x, t) = P_0(x) - tx$.

b. Using (a), show that

$$\frac{\partial P}{\partial x}(x, t) = P_0'(x) - t \quad \text{and} \quad \frac{\partial^2 P}{\partial x^2}(x, t) = P_0''(x)$$

c. From (7) and (b), show that $P_0'(f(t)) - t = 0$.

d. By differentiating both sides of the equation in (c) and by using (b) and (8), show that

$$f'(t) = \frac{1}{P_0''(f(t))} = \frac{1}{\dfrac{\partial^2 P}{\partial x^2}(f(t), t)} < 0$$

(Thus the production tends to decrease as the tax rate increases.)

13.4
THE CHAIN RULE

The Chain Rule for functions of one variable provides a formula for differentiating the composite of two functions f and g. If $u = f(x)$ and $y = g(u) = g(f(x))$, the Chain Rule can be expressed in the Leibniz notation as

$$\frac{dy}{dx} = \frac{dy}{du}\frac{du}{dx}$$

Composites involving functions of several variables have their own versions of the Chain Rule, which involve derivatives and partial derivatives. In this section we will introduce various versions of the Chain Rule, apply them in several examples, and then present the theoretical basis for the Chain Rule, which includes the notion of differentiability for functions of several variables.

Versions of the Chain Rule   First we present two versions of the Chain Rule for functions of two variables. In each statement we assume that all functions have the required derivatives.

a. Let $z = f(x, y)$, $x = g_1(t)$, and $y = g_2(t)$. Then $z = f(g_1(t), g_2(t))$, and

$$\frac{dz}{dt} = \frac{\partial z}{\partial x}\frac{dx}{dt} + \frac{\partial z}{\partial y}\frac{dy}{dt} \tag{1}$$

b. Let $z = f(x, y)$, $x = g_1(u, v)$, and $y = g_2(u, v)$. Then $z = f(g_1(u, v), g_2(u, v))$, and

$$\frac{\partial z}{\partial u} = \frac{\partial z}{\partial x}\frac{\partial x}{\partial u} + \frac{\partial z}{\partial y}\frac{\partial y}{\partial u}$$

$$\frac{\partial z}{\partial v} = \frac{\partial z}{\partial x}\frac{\partial x}{\partial v} + \frac{\partial z}{\partial y}\frac{\partial y}{\partial v} \tag{2}$$

Example 1   Let $z = x^2 e^y$, $x = \sin t$, and $y = t^3$. Find dz/dt.

Solution   Using (1), we find that

$$\frac{dz}{dt} = \frac{\partial z}{\partial x}\frac{dx}{dt} + \frac{\partial z}{\partial y}\frac{dy}{dt} = 2xe^y \cos t + x^2 e^y (3t^2)$$

$$= 2(\sin t)e^{(t^3)}\cos t + 3(\sin^2 t)e^{(t^3)}t^2 \quad \square$$

Notice that we completed the solution by writing the answer in terms of t, since we desired dz/dt.

Example 2   Let $z = x \ln y$, $x = u^2 + v^2$, and $y = u^2 - v^2$. Find $\partial z/\partial u$ and $\partial z/\partial v$.

Solution   Using (2), we find that

$$\frac{\partial z}{\partial u} = \frac{\partial z}{\partial x}\frac{\partial x}{\partial u} + \frac{\partial z}{\partial y}\frac{\partial y}{\partial u} = (\ln y)(2u) + \left(\frac{x}{y}\right)(2u)$$

$$= 2u \ln(u^2 - v^2) + 2u\frac{u^2 + v^2}{u^2 - v^2}$$

and

$$\frac{\partial z}{\partial v} = \frac{\partial z}{\partial x}\frac{\partial x}{\partial v} + \frac{\partial z}{\partial y}\frac{\partial y}{\partial v} = (\ln y)(2v) + \left(\frac{x}{y}\right)(-2v)$$

$$= 2v \ln(u^2 - v^2) - 2v\frac{u^2 + v^2}{u^2 - v^2} \quad \square$$

Rather than state additional versions of the Chain Rule, we will give a general procedure for finding derivatives of composite functions of several variables. First we draw a diagram that indicates how the variables are related. For example, the diagram corresponding to (2) is

$$z \overset{\partial z/\partial x}{\underset{\partial z/\partial y}{\diagup \diagdown}} \begin{matrix} x \overset{\partial x/\partial u}{\underset{\partial x/\partial v}{\diagup \diagdown}} \begin{matrix} u \\ v \end{matrix} \\ y \overset{\partial y/\partial u}{\underset{\partial y/\partial v}{\diagup \diagdown}} \begin{matrix} u \\ v \end{matrix} \end{matrix}$$

The diagram is read from left to right and shows that z depends on x and y, which in turn depend on u and v. To obtain $\partial z/\partial u$ we find the product of the partial derivatives along each individual path from z to u (that is, $(\partial z/\partial x)(\partial x/\partial u)$ and $(\partial z/\partial y)(\partial y/\partial u)$), and then add these products:

$$\frac{\partial z}{\partial u} = \frac{\partial z}{\partial x}\frac{\partial x}{\partial u} + \frac{\partial z}{\partial y}\frac{\partial y}{\partial u}$$

This is the first equation in (2). The second equation in (2), which gives $\partial z/\partial v$, is obtained by using all paths from z to v. Analogous diagrams apply to other versions of the Chain Rule, and are illustrated in the following examples.

Example 3 Let $w = x \cos yz^2$, $x = \sin t$, $y = t^2$ and $z = e^t$. Find dw/dt.

Solution The appropriate diagram is

$$w \begin{array}{l} \xrightarrow{\partial w/\partial x} x \xrightarrow{dx/dt} t \\ \xrightarrow{\partial w/\partial y} y \xrightarrow{dy/dt} t \\ \xrightarrow{\partial w/\partial z} z \xrightarrow{dz/dt} t \end{array}$$

Since there are three paths from w to t, we find that

$$\frac{dw}{dt} = \frac{\partial w}{\partial x}\frac{dx}{dt} + \frac{\partial w}{\partial y}\frac{dy}{dt} + \frac{\partial w}{\partial z}\frac{dz}{dt}$$

$$= \cos yz^2 \cos t - xz^2(\sin yz^2)(2t) - 2xyz(\sin yz^2)(e^t)$$

$$= \cos(t^2 e^{2t})\cos t - 2te^{2t}\sin t \sin(t^2 e^{2t}) - 2t^2 e^{2t}\sin t \sin(t^2 e^{2t}) \quad \square$$

Example 4 Let $w = \sqrt{x} + y^2 z^3$, $x = 1 + u^2 + v^2$, $y = uv$, and $z = 3u$. Find $\partial w/\partial u$ and $\partial w/\partial v$.

Solution The diagram that applies here is

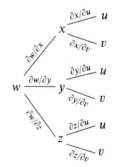

It follows that

$$\frac{\partial w}{\partial u} = \frac{\partial w}{\partial x}\frac{\partial x}{\partial u} + \frac{\partial w}{\partial y}\frac{\partial y}{\partial u} + \frac{\partial w}{\partial z}\frac{\partial z}{\partial u}$$

$$= \frac{1}{2\sqrt{x}}(2u) + 2yz^3(v) + 3y^2 z^2(3) = \frac{u}{\sqrt{1 + u^2 + v^2}} + 54u^4 v^2 + 81u^4 v^2$$

$$= \frac{u}{\sqrt{1 + u^2 + v^2}} + 135u^4 v^2$$

and

$$\frac{\partial w}{\partial v} = \frac{\partial w}{\partial x}\frac{\partial x}{\partial v} + \frac{\partial w}{\partial y}\frac{\partial y}{\partial v} + \frac{\partial w}{\partial z}\frac{\partial z}{\partial v}$$

$$= \frac{1}{2\sqrt{x}}(2v) + 2yz^3(u) + 3y^2z^2(0) = \frac{v}{\sqrt{1 + u^2 + v^2}} + 54u^5v \quad \square$$

Example 5 Suppose that $w = f(x, y)$, $x = g(u, v)$, $y = h(u, v)$, $u = j(t)$, and $v = k(t)$. Find a formula for dw/dt.

Solution We draw the diagram

$$w \overset{\partial w/\partial x}{\underset{\partial w/\partial y}{\diagdown}} \begin{matrix} x \overset{\partial x/\partial u}{\underset{\partial x/\partial v}{\diagdown}} & u \xrightarrow{du/dt} t \\ & v \xrightarrow{dv/dt} t \\ y \overset{\partial y/\partial u}{\underset{\partial y/\partial v}{\diagdown}} & u \xrightarrow{du/dt} t \\ & v \xrightarrow{dv/dt} t \end{matrix}$$

Using the four paths leading from w to t, we find that

$$\frac{dw}{dt} = \frac{\partial w}{\partial x}\frac{\partial x}{\partial u}\frac{du}{dt} + \frac{\partial w}{\partial x}\frac{\partial x}{\partial v}\frac{dv}{dt} + \frac{\partial w}{\partial y}\frac{\partial y}{\partial u}\frac{du}{dt} + \frac{\partial w}{\partial y}\frac{\partial y}{\partial v}\frac{dv}{dt} \quad \square$$

Example 6 Let $z = f(u - v, v - u)$. Show that

$$\frac{\partial z}{\partial u} + \frac{\partial z}{\partial v} = 0$$

Solution If we let $x = u - v$ and $y = v - u$, then $z = f(x, y)$, and (2) applies:

$$\frac{\partial z}{\partial u} = \frac{\partial z}{\partial x}\frac{\partial x}{\partial u} + \frac{\partial z}{\partial y}\frac{\partial y}{\partial u} = \frac{\partial z}{\partial x}(1) + \frac{\partial z}{\partial y}(-1)$$

$$\frac{\partial z}{\partial v} = \frac{\partial z}{\partial x}\frac{\partial x}{\partial v} + \frac{\partial z}{\partial y}\frac{\partial y}{\partial v} = \frac{\partial z}{\partial x}(-1) + \frac{\partial z}{\partial y}(1)$$

Therefore $$\frac{\partial z}{\partial u} + \frac{\partial z}{\partial v} = 0 \quad \square$$

The Chain Rule for functions of several variables is particularly well adapted to related rates problems.

Example 7 Suppose the volume V of a conical sand pile grows at a rate of 4 cubic inches per second and the radius r of the circular base grows at a rate of e^{-r}

inches per second. Find the rate at which the height h of the cone is growing at the instant when the volume is 60 cubic inches and the radius is 6 inches.

Solution We wish to find dh/dt at the instant when $V = 60$ and $r = 6$. The formula for the volume of the cone is

$$V = \frac{1}{3}\pi r^2 h$$

so the formula for the height is

$$h = \frac{3V}{\pi r^2}$$

The diagram for dh/dt is

$$h \begin{array}{c} \overset{\partial h/\partial V}{\diagup} V \xrightarrow{dV/dt} t \\ \underset{\partial h/\partial r}{\diagdown} r \xrightarrow{dr/dt} t \end{array}$$

Since $dV/dt = 4$ and $dr/dt = e^{-r}$ by hypothesis, we find that

$$\frac{dh}{dt} = \frac{\partial h}{\partial V}\frac{dV}{dt} + \frac{\partial h}{\partial r}\frac{dr}{dt} = \frac{3}{\pi r^2}(4) + \frac{-6V}{\pi r^3}(e^{-r}) = \frac{6}{\pi r^2}\left(2 - \frac{V}{r}e^{-r}\right)$$

If t_0 denotes the time it takes for the volume to become 60 cubic inches and the radius to become 6 inches, then

$$\left.\frac{dh}{dt}\right|_{t=t_0} = \frac{6}{\pi 6^2}\left(2 - \frac{60}{6}e^{-6}\right) = \frac{1}{3\pi}(1 - 5e^{-6}) \quad \text{(inches per second)} \quad \square$$

Implicit Differentiation Through the Chain Rule we can more completely describe the process of implicit differentiation, which we introduced in Section 3.6. The next example, which we will use in the discussion of implicit differentiation, is useful in its own right.

Example 8 Suppose that $w = f(x, y)$ and $y = g(x)$. Find a formula for dw/dx.

Solution The appropriate diagram is as follows:

$$w \begin{array}{c} \overset{\partial w/\partial x}{\diagup} x \\ \underset{\partial w/\partial y}{\diagdown} y \xrightarrow{dy/dx} x \end{array}$$

It follows that

$$\frac{dw}{dx} = \frac{\partial w}{\partial x} + \frac{\partial w}{\partial y}\frac{dy}{dx} \tag{3}$$

(In this formula, $\partial w/\partial x$ refers to the partial derivative of $z = w(x, y)$ with respect to x, whereas dw/dx refers to the derivative of $z = f(x, g(x))$ with respect to x.) $\square$

Now suppose that f is a function of two variables that has partial derivatives, and assume that the equation $f(x, y) = 0$ defines a differentiable function $y = g(x)$ of x, so that $f(x, g(x)) = 0$. If $w = f(x, y)$, then by assumption,

$$\frac{dw}{dx} = \frac{d}{dx}(f(x, g(x))) = \frac{d}{dx}(0) = 0$$

Therefore by (3),

$$0 = \frac{dw}{dx} = \frac{\partial w}{\partial x} + \frac{\partial w}{\partial y}\frac{dy}{dx}$$

Finally, if $\partial w/\partial y \neq 0$, then by solving for dy/dx, we obtain

$$\frac{dy}{dx} = \frac{-\partial w/\partial x}{\partial w/\partial y} \tag{4}$$

Now we apply (4) to rework Example 1 of Section 3.6.

Example 9   Let $x^3 + y^3 = 2xy$. Find dy/dx.

Solution   Let $w = x^3 + y^3 - 2xy$. Then

$$\frac{\partial w}{\partial x} = 3x^2 - 2y \quad \text{and} \quad \frac{\partial w}{\partial y} = 3y^2 - 2x$$

and (4) implies that

$$\frac{dy}{dx} = \frac{-\partial w/\partial x}{\partial w/\partial y} = \frac{-(3x^2 - 2y)}{3y^2 - 2x} = \frac{2y - 3x^2}{3y^2 - 2x} \quad \square$$

Differentiability and the Theory Behind the Chain Rule

In proving one version of the Chain Rule we will use the formula that appears in the following theorem and gives rise to the concept of differentiability for functions of several variables.

THEOREM 13.6

Let f be a function having partial derivatives throughout a set containing a disk D centered at (x_0, y_0). If f_x and f_y are continuous at (x_0, y_0) then there are functions ε_1 and ε_2 of two variables such that

$$f(x, y) - f(x_0, y_0) = f_x(x_0, y_0)(x - x_0) + f_y(x_0, y_0)(y - y_0)$$
$$+ \varepsilon_1(x, y)(x - x_0) + \varepsilon_2(x, y)(y - y_0)$$
$$\text{for } (x, y) \text{ in } D$$

where $\lim_{(x, y) \to (x_0, y_0)} \varepsilon_1(x, y) = 0$ and $\lim_{(x, y) \to (x_0, y_0)} \varepsilon_2(x, y) = 0$.

Proof   Let (x, y) be any point in D. First we write
$$f(x, y) - f(x_0, y_0) = [f(x, y) - f(x_0, y)] + [f(x_0, y) - f(x_0, y_0)]$$
and let
$$g(t) = f(t, y) \quad \text{and} \quad G(t) = f(x_0, t)$$

Applying the Mean Value Theorem to g on the interval with endpoints x and x_0 and to G on the interval with endpoints y and y_0, we obtain a number $u(x)$ between x and x_0 and a number $v(y)$ between y and y_0 such that

$$\begin{aligned} f(x, y) - f(x_0, y) &= g(x) - g(x_0) = g'(u(x))(x - x_0) \\ &= f_x(u(x), y)(x - x_0) \end{aligned} \tag{5}$$

and

$$\begin{aligned} f(x_0, y) - f(x_0, y_0) &= G(y) - G(y_0) = G'(v(y))(y - y_0) \\ &= f_y(x_0, v(y))(y - y_0) \end{aligned} \tag{6}$$

Equations (5) and (6) yield

$$\begin{aligned} f(x, y) - f(x_0, y_0) &= [f(x, y) - f(x_0, y)] + [f(x_0, y) - f(x_0, y_0)] \\ &= f_x(u(x), y)(x - x_0) + f_y(x_0, v(y))(y - y_0) \\ &= f_x(x_0, y_0)(x - x_0) + \overbrace{[f_x(u(x), y) - f_x(x_0, y_0)]}^{\varepsilon_1(x, y)}(x - x_0) \\ &\quad + f_y(x_0, y_0)(y - y_0) + \overbrace{[f_y(x_0, v(y)) - f_y(x_0, y_0)]}^{\varepsilon_2(x, y)}(y - y_0) \\ &= f_x(x_0, y_0)(x - x_0) + f_y(x_0, y_0)(y - y_0) + \varepsilon_1(x, y)(x - x_0) \\ &\quad + \varepsilon_2(x, y)(y - y_0) \end{aligned}$$

Since f_x and f_y are continuous at (x_0, y_0), and since $\lim_{(x, y) \to (x_0, y_0)} u(x) = x_0$ and $\lim_{(x, y) \to (x_0, y_0)} v(y) = y_0$, both $\varepsilon_1(x, y)$ and $\varepsilon_2(x, y)$ approach 0 as (x, y) approaches (x_0, y_0). ∎

We use the formula in Theorem 13.6 in the following definition.

DEFINITION 13.7

A function f of two variables is **differentiable at (x_0, y_0)** if there exist a disk D centered at (x_0, y_0) and functions ε_1 and ε_2 of two variables such that

$$\begin{aligned} f(x, y) - f(x_0, y_0) &= f_x(x_0, y_0)(x - x_0) + f_y(x_0, y_0)(y - y_0) \\ &\quad + \varepsilon_1(x, y)(x - x_0) + \varepsilon_2(x, y)(y - y_0) \end{aligned} \tag{7}$$

for (x, y) in D

where $\lim_{(x, y) \to (x_0, y_0)} \varepsilon_1(x, y) = 0$ and $\lim_{(x, y) \to (x_0, y_0)} \varepsilon_2(x, y) = 0$.

In terms of differentiability, Theorem 13.6 states that if f has partial derivatives on a disk D centered at (x_0, y_0) and if f_x and f_y are continuous at (x_0, y_0), then f is differentiable at (x_0, y_0). It is usually easier to verify that a function f has continuous partial derivatives throughout a disk centered at (x_0, y_0) than to apply Definition 13.7 to prove that f is differentiable at (x_0, y_0). Because most of the functions we will encounter have continuous partial derivatives at all or almost all points in their respective domains, they are differentiable at all (or almost all) points in their domains. In particular, every rational function is differentiable at every point in its domain.

Now we will use differentiability to prove the version of the Chain Rule associated with (1). Since we will use the Chain Rule in the proofs of many theorems, differentiability will often occur as a hypothesis in the theorems.

THEOREM 13.8
CHAIN RULE
(FIRST VERSION)

Let f be a function of two variables that is differentiable at (x_0, y_0), and let g_1 and g_2 be functions of one variable that are differentiable at t_0. Suppose $x_0 = g_1(t_0)$ and $y_0 = g_2(t_0)$, and let

$$F(t) = f(g_1(t), g_2(t))$$

Then F is differentiable at t_0, and

$$F'(t_0) = f_x(x_0, y_0)g_1'(t_0) + f_y(x_0, y_0)g_2'(t_0)$$

Proof   Since g_1 and g_2 are differentiable at t_0, they are continuous at t_0, so that

$$\lim_{t \to t_0} g_1(t) = x_0 \quad \text{and} \quad \lim_{t \to t_0} g_2(t) = y_0$$

Because f is differentiable at (x_0, y_0) by hypothesis, the functions ε_1 and ε_2 appearing in (7) have the limit 0 at (x_0, y_0); hence a two-variable substitution yields

$$\lim_{t \to t_0} \varepsilon_1(g_1(t), g_2(t)) = 0 \quad \text{and} \quad \lim_{t \to t_0} \varepsilon_2(g_1(t), g_2(t)) = 0$$

Using (7) with x replaced by $g_1(t)$ and y replaced by $g_2(t)$, we find that

$$F'(t_0) = \lim_{t \to t_0} \frac{F(t) - F(t_0)}{t - t_0} = \lim_{t \to t_0} \left(\frac{f(g_1(t), g_2(t)) - f(g_1(t_0), g_2(t_0))}{t - t_0} \right)$$

$$= \lim_{t \to t_0} \left(f_x(x_0, y_0) \frac{g_1(t) - g_1(t_0)}{t - t_0} + f_y(x_0, y_0) \frac{g_2(t) - g_2(t_0)}{t - t_0} \right.$$

$$\left. + \varepsilon_1(g_1(t), g_2(t)) \frac{g_1(t) - g_1(t_0)}{t - t_0} + \varepsilon_2(g_1(t), g_2(t)) \frac{g_2(t) - g_2(t_0)}{t - t_0} \right)$$

$$= f_x(x_0, y_0)g_1'(t_0) + f_y(x_0, y_0)g_2'(t_0) + 0 \cdot g_1'(t_0) + 0 \cdot g_2'(t_0)$$

$$= f_x(x_0, y_0)g_1'(t_0) + f_y(x_0, y_0)g_2'(t_0) \quad \blacksquare$$

We state the version of the Chain Rule associated with (2), but omit its proof, which follows along the same lines as that for the first version.

THEOREM 13.9
CHAIN RULE
(SECOND VERSION)

Let f be a function of two variables that is differentiable at (x_0, y_0), and let g_1 and g_2 be functions of two variables having partial derivatives at (u_0, v_0). Suppose $x_0 = g_1(u_0, v_0)$ and $y_0 = g_2(u_0, v_0)$, and let

$$F(u, v) = f(g_1(u, v), g_2(u, v))$$

Then F has partial derivatives at (u_0, v_0), and

$$F_u(u_0, v_0) = f_x(x_0, y_0)(g_1)_u(u_0, v_0) + f_y(x_0, y_0)(g_2)_u(u_0, v_0)$$

$$F_v(u_0, v_0) = f_x(x_0, y_0)(g_1)_v(u_0, v_0) + f_y(x_0, y_0)(g_2)_v(u_0, v_0)$$

The concept of differentiability extends to functions of three variables.

DEFINITION 13.10

A function f of three variables is **differentiable at** (x_0, y_0, z_0) if there exist a ball B centered at (x_0, y_0, z_0) and functions ε_1, ε_2, and ε_3 of three variables such that

$$f(x, y, z) - f(x_0, y_0, z_0) = f_x(x_0, y_0, z_0)(x - x_0) + f_y(x_0, y_0, z_0)(y - y_0)$$
$$+ f_z(x_0, y_0, z_0)(z - z_0) + \varepsilon_1(x, y, z)(x - x_0)$$
$$+ \varepsilon_2(x, y, z)(y - y_0) + \varepsilon_3(x, y, z)(z - z_0)$$
$$\text{for all } (x, y, z) \text{ in } B$$

where $\lim_{(x, y, z) \to (x_0, y_0, z_0)} \varepsilon_1(x, y, z) = 0$, $\lim_{(x, y, z) \to (x_0, y_0, z_0)} \varepsilon_2(x, y, z) = 0$, and $\lim_{(x, y, z) \to (x_0, y_0, z_0)} \varepsilon_3(x, y, z) = 0$.

A theorem analogous to Theorem 13.6 implies that if f has partial derivatives on a ball centered at (x_0, y_0, z_0) and if f_x, f_y, and f_z are continuous at (x_0, y_0, z_0), then f is differentiable at (x_0, y_0, z_0). Again, most of the functions of three variables we will encounter will be differentiable at all or at almost all points in their domains. Moreover, if f is differentiable at (x_0, y_0, z_0), then theorems analogous to Theorems 13.8 and 13.9 hold.

EXERCISES 13.4

In Exercises 1–5 compute dz/dt.

1. $z = 2x^2 - 3y^3$; $x = \sqrt{t}$, $y = e^{2t}$
2. $z = \ln(3x^2 + y^3)$; $x = e^{2t}$, $y = t^{1/3}$
3. $z = \sin x + \cos xy$; $x = t^2$, $y = 1$
4. $z = \arctan(y^2 - x^2)$; $x = \sin t$, $y = \cos t$
5. $z = \sqrt{2x - 4y}$; $x = \ln t$, $y = 1 - 3t^3$

In Exercises 6-10 compute $\partial z/\partial u$ and $\partial z/\partial v$.

6. $z = \dfrac{x}{y^2}$; $x = u + v - 1$, $y = u - v - 1$

7. $z = \dfrac{4}{xy} - \dfrac{x}{y}$; $x = u^2$, $y = uv$

8. $z = 16 - 4x^2 - y^2$; $x = u \sin v$, $y = v \cos u$
9. $z = \ln(x^2 - y^2)$; $x = u - v$, $y = u^2 + v^2$
10. $z = 2e^{x^2y}$; $x = \sqrt{uv}$, $y = 1/u$

In Exercises 11–14 compute $\partial z/\partial r$ and $\partial z/\partial s$.

11. $z = \sin 2u \cos 3v$; $u = (r + s)^2$, $v = (r - s)^2$
12. $z = \ln u + \ln v$; $u = 4^{rs}$, $v = 4^{r/s}$
13. $z = ue^v + ve^{-u}$; $u = \ln r$, $v = s \ln r$
14. $z = 2^{u-v}$; $u = r \cos s$, $v = r \sin s$

In Exercises 15–20 compute dw/dt.

15. $w = \dfrac{x}{y} - \dfrac{z}{x}$; $x = \sin t$, $y = \cos t$, $z = \tan t$

16. $w = \dfrac{z}{xy^2} - 3$; $x = \dfrac{1}{t^2}$, $y = -5t$, $z = \sqrt{t}$

17. $w = \sqrt{x^2 + y^2 + z^2}$; $x = e^t$, $y = e^{-t}$, $z = 2t$
18. $w = \ln(x^2 + y^2 + z^2)$; $x = \sin t$, $y = \cos t$, $z = e^{-t^2}$
19. $w = \sin xy^2z^3$; $x = 3t$, $y = t^{1/2}$, $z = t^{1/3}$
20. $w = \sqrt{x^2 + y^2} - \sqrt{y^3 - z^3}$; $x = t^2$, $y = t^3$, $z = -t^3$

In Exercises 21–24 find $\partial w/\partial u$ and $\partial w/\partial v$.

21. $w = \dfrac{yz}{x^2 + xy}$; $x = u^2$, $y = v^2$, $z = u^2 - v^2$

22. $w = x^2 - 2y - 7z$; $x = v \cos(\pi - u)$, $y = u \sin(\pi - v)$, $z = uv$

23. $w = y \ln xz$; $x = ve^u$, $y = u^2v^4$, $z = ue^v$

24. $w = e^{x/y} + e^{z/x}$; $x = \dfrac{\ln u}{v}$, $y = \ln u$, $z = \dfrac{\ln u}{uv}$

In Exercises 25–29 find dy/dx by implicit differentiation.

25. $x^3 + 4x^2y - 3xy^2 + 2y^3 + 5 = 0$

26. $x^{2/3} + y^{2/3} = 2$

27. $x^2 + y^2 + \sin xy^2 = 0$

28. $e^{x/y} + \ln y/x + 15 = 0$

29. $x^2 = \dfrac{y^2}{y^2 - 1}$

In Exercises 30–32 assume that the equation defines z implicitly as a function of x and y, and use "implicit partial differentiation" to find $\partial z/\partial x$ and $\partial z/\partial y$.

30. $x^2 z^2 - 2xyz + z^3 y^2 = 3$

31. $x - yz + \cos xyz = 2$

32. $\dfrac{1}{z} + \dfrac{1}{y + z} + \dfrac{1}{x + y + z} = \dfrac{1}{2}$

33. Let $z = f(x - y)$. Show that $\dfrac{\partial z}{\partial x} = -\dfrac{\partial z}{\partial y}$.

34. Let $w = f(x - y, y - z, z - x)$. Show that

$$\frac{\partial w}{\partial x} + \frac{\partial w}{\partial y} + \frac{\partial w}{\partial z} = 0$$

35. Let $z = f(y + ax) + g(y - ax)$, with $a \neq 0$. Show that z satisfies the wave equation

$$\frac{\partial^2 z}{\partial x^2} = a^2 \frac{\partial^2 z}{\partial y^2}$$

*36. Let $z = f(x, y)$, $x = r \cos \theta$, and $y = r \sin \theta$.

a. Show that $\dfrac{\partial z}{\partial x} = \dfrac{\partial z}{\partial r} \cos \theta - \dfrac{\partial z}{\partial \theta} \dfrac{\sin \theta}{r}$

and $\dfrac{\partial z}{\partial y} = \dfrac{\partial z}{\partial r} \sin \theta + \dfrac{\partial z}{\partial \theta} \dfrac{\cos \theta}{r}$

b. Show that $\left(\dfrac{\partial z}{\partial x}\right)^2 + \left(\dfrac{\partial z}{\partial y}\right)^2 = \left(\dfrac{\partial z}{\partial r}\right)^2 + \dfrac{1}{r^2}\left(\dfrac{\partial z}{\partial \theta}\right)^2$

*37. Let $w = f(x, y)$, $x = e^s \cos t$, and $y = e^s \sin t$. Assuming that the second partials of f exist, show that

$$\frac{\partial^2 w}{\partial x^2} + \frac{\partial^2 w}{\partial y^2} = e^{-2s}\left(\frac{\partial^2 w}{\partial s^2} + \frac{\partial^2 w}{\partial t^2}\right)$$

38. A function f of two variables is **homogeneous of degree n** if for any real number t we have

$$f(tx, ty) = t^n f(x, y) \tag{8}$$

Show that in this case

$$xf_x(x, y) + yf_y(x, y) = nf(x, y)$$

(*Hint:* Differentiate both sides of (8) with respect to t, and then set $t = 1$.)

39. Let

$$f(x, y) = \tan\frac{x^2 + y^2}{xy}$$

Use Exercise 38 to show that

$$xf_x(x, y) + yf_y(x, y) = 0$$

40. Show that if f is differentiable at (x_0, y_0), then f is continuous at (x_0, y_0). (*Hint:* Using (7), show that $\lim_{(x, y) \to (x_0, y_0)} f(x, y) = f(x_0, y_0)$.)

41. Let

$$f(x, y) = \begin{cases} \dfrac{xy}{x^2 + y^2} & \text{for } (x, y) \neq (0, 0) \\ 0 & \text{for } (x, y) = (0, 0) \end{cases}$$

Use the result of Exercise 40 to show that f is not differentiable at $(0, 0)$. (Notice that f has partial derivatives at $(0, 0)$ despite its nondifferentiability there.)

42. A tree trunk may be considered a circular cylinder. Suppose the diameter of the trunk increases 1 inch per year and the height of the trunk increases 6 inches per year. How fast is the volume of wood in the trunk increasing when it is 100 inches high and 5 inches in diameter?

43. The time rate Q of flow of fluid through a cylindrical tube (such as a windpipe) with radius r and height l is given by

$$Q = \frac{\pi p r^4}{8 l \eta}$$

where η is the viscosity of the fluid and p is the difference in pressure at the two ends of the tube. Suppose the length of the tube remains constant, while the radius increases at the rate of $\frac{1}{10}$ and the pressure decreases at the rate of $\frac{1}{5}$. Find the rate of change of Q with respect to time.

44. A car moving at 20 miles per hour approaches an intersection with a train track along a road perpendicular to the track. If a train approaches the intersection at 100 miles per hour, at what rate is the distance between the car and the train changing when the car is 0.5 miles from the intersection and the train is 1.2 miles from the intersection?

45. The mass of a rocket lifting off from earth is decreasing (due to fuel consumption) at the rate of 40 kilograms per second. How fast is the magnitude F of the force of gravity decreasing when the rocket is 6400 kilometers from the center of the earth and is rising with a velocity of 100 kilometers per second? (*Hint:* By Newton's Law of Gravitation, $F = GMm/r^2$, where G is the universal gravitational constant, M is the mass of the earth, m is the mass of the rocket, and r is the distance between the rocket and the center of the earth.)

13.5
DIRECTIONAL DERIVATIVES

We may think of the partial derivatives of a function as describing the rate of change of the function along lines parallel to the coordinate axes. Now we discuss rates of change along lines that are not necessarily parallel to any coordinate axis.

DEFINITION 13.11

Let f be a function defined on a set containing a disk D centered at (x_0, y_0), and let $\mathbf{u} = a_1\mathbf{i} + a_2\mathbf{j}$ be a unit vector. Then the **directional derivative** of f at (x_0, y_0) in the direction of $\mathbf{u}$, denoted $D_\mathbf{u}f(x_0, y_0)$, is defined by

$$D_\mathbf{u}f(x_0, y_0) = \lim_{h \to 0} \frac{f(x_0 + ha_1, y_0 + ha_2) - f(x_0, y_0)}{h}$$

provided that this limit exists.

Observe that if $\mathbf{u} = \mathbf{i}$, then

$$D_\mathbf{u}f(x_0, y_0) = \lim_{h \to 0} \frac{f(x_0 + h, y_0) - f(x_0, y_0)}{h} = f_x(x_0, y_0)$$

and if $\mathbf{u} = \mathbf{j}$, then

$$D_\mathbf{u}f(x_0, y_0) = \lim_{h \to 0} \frac{f(x_0, y_0 + h) - f(x_0, y_0)}{h} = f_y(x_0, y_0)$$

Thus the first partial derivatives of f are special cases of directional derivatives—in the directions of the positive coordinate axes. It turns out that if f is differentiable at (x_0, y_0), then any directional derivative can be evaluated by means of the two partial derivatives.

THEOREM 13.12

Let f be differentiable at (x_0, y_0). Then f has a directional derivative at (x_0, y_0) in every direction. Moreover, if $\mathbf{u} = a_1\mathbf{i} + a_2\mathbf{j}$ is a unit vector, then

$$D_\mathbf{u}f(x_0, y_0) = f_x(x_0, y_0)a_1 + f_y(x_0, y_0)a_2 \tag{1}$$

Proof   Let

$$F(h) = f(x_0 + ha_1, y_0 + ha_2)$$

Then

$$\frac{F(h) - F(0)}{h - 0} = \frac{f(x_0 + ha_1, y_0 + ha_2) - f(x_0, y_0)}{h}$$

so that $D_\mathbf{u}f(x_0, y_0)$ exists if and only if $F'(0)$ exists. If we let

$$g_1(h) = x_0 + ha_1 \quad \text{and} \quad g_2(h) = y_0 + ha_2$$

then $F(h) = f(g_1(h), g_2(h))$, and also $g_1(0) = x_0$ and $g_2(0) = y_0$. With h replacing t and 0 replacing t_0, the hypotheses of the first version of the Chain Rule (Theorem 13.8) are satisfied. Consequently $F'(0)$ exists, and

$$D_{\mathbf{u}}f(x_0, y_0) = F'(0) = f_x(x_0, y_0)g_1'(0) + f_y(x_0, y_0)g_2'(0)$$
$$= f_x(x_0, y_0)a_1 + f_y(x_0, y_0)a_2 \quad \blacksquare$$

Example 1 Let $f(x, y) = 6 - 3x^2 - y^2$, and let $\mathbf{u} = (1/\sqrt{2})\mathbf{i} - (1/\sqrt{2})\mathbf{j}$. Find $D_{\mathbf{u}}f(1, 2)$.

Solution Notice that $\mathbf{u}$ is a unit vector. We will compute $D_{\mathbf{u}}f(1, 2)$ by means of (1). First we calculate the partial derivatives of f:

$$f_x(x, y) = -6x \quad \text{and} \quad f_y(x, y) = -2y$$

Therefore $f_x(1, 2) = -6$ and $f_y(1, 2) = -4$, so by (1),

$$D_{\mathbf{u}}f(1, 2) = f_x(1, 2)\left(\frac{1}{\sqrt{2}}\right) + f_y(1, 2)\left(-\frac{1}{\sqrt{2}}\right)$$

$$= (-6)\left(\frac{1}{\sqrt{2}}\right) + (-4)\left(-\frac{1}{\sqrt{2}}\right) = -\sqrt{2} \quad \square$$

In the expression $D_{\mathbf{u}}f(x_0, y_0)$, $\mathbf{u}$ represents a unit vector. The directional derivative $D_{\mathbf{a}}f(x_0, y_0)$ in the direction of an arbitrary nonzero vector $\mathbf{a}$ is defined to be $D_{\mathbf{u}}f(x_0, y_0)$, where $\mathbf{u}$ is the unit vector in the direction of $\mathbf{a}$, that is, $\mathbf{u} = \mathbf{a}/\|\mathbf{a}\|$.

Example 2 Let $f(x, y) = xy^2$ and let $\mathbf{a} = \mathbf{i} - 2\mathbf{j}$. Find the directional derivative of f at $(-3, 1)$ in the direction of $\mathbf{a}$.

Solution In this case $\|\mathbf{a}\| = \sqrt{1^2 + (-2)^2} = \sqrt{5}$, so we will find $D_{\mathbf{u}}f(-3, 1)$, where

$$\mathbf{u} = \frac{1}{\|\mathbf{a}\|}\mathbf{a} = \frac{1}{\sqrt{5}}\mathbf{i} - \frac{2}{\sqrt{5}}\mathbf{j}$$

Since

$$f_x(x, y) = y^2 \quad \text{and} \quad f_y(x, y) = 2xy$$

it follows that $f_x(-3, 1) = 1$ and $f_y(-3, 1) = -6$, so by (1),

$$D_{\mathbf{a}}f(-3, 1) = D_{\mathbf{u}}f(-3, 1) = f_x(-3, 1)\left(\frac{1}{\sqrt{5}}\right) + f_y(-3, 1)\left(\frac{-2}{\sqrt{5}}\right)$$

$$= 1\left(\frac{1}{\sqrt{5}}\right) + (-6)\left(\frac{-2}{\sqrt{5}}\right) = \frac{13}{5}\sqrt{5} \quad \square$$

In defining the directional derivative for functions of three variables we make only the necessary changes in notation. Let $\mathbf{u} = a_1\mathbf{i} + a_2\mathbf{j} + a_3\mathbf{k}$ be a unit vector in space. The **directional derivative** $D_{\mathbf{u}}f(x_0, y_0, z_0)$ is defined by

$$D_{\mathbf{u}}f(x_0, y_0, z_0) = \lim_{h \to 0} \frac{f(x_0 + ha_1, y_0 + ha_2, z_0 + ha_3) - f(x_0, y_0, z_0)}{h}$$

provided that this limit exists. Moreover, if f is differentiable at (x_0, y_0, z_0), then the analogue of Theorem 13.12 holds:

$$D_{\mathbf{u}}f(x_0, y_0, z_0) = f_x(x_0, y_0, z_0)a_1 + f_y(x_0, y_0, z_0)a_2 + f_z(x_0, y_0, z_0)a_3 \qquad (2)$$

As before, if $\mathbf{a}$ is any nonzero vector in space, then we define the directional derivative of f at (x_0, y_0, z_0) in the direction of $\mathbf{a}$ to be $D_{\mathbf{u}}f(x_0, y_0, z_0)$, where $\mathbf{u} = \mathbf{a}/\|\mathbf{a}\|$.

Example 3 Let $f(x, y, z) = xe^{y^2z}$, and let $\mathbf{a} = \mathbf{i} - \mathbf{j} + \sqrt{2}\mathbf{k}$. Find the directional derivative of f at $(2, 1, 0)$ in the direction of $\mathbf{a}$.

Solution First we find the partial derivatives of f:

$$f_x(x, y, z) = e^{y^2z}, \qquad f_y(x, y, z) = 2xyze^{y^2z}, \qquad f_z(x, y, z) = xy^2e^{y^2z}$$

Since $\|\mathbf{a}\| = \sqrt{1^2 + (-1)^2 + (\sqrt{2})^2} = 2$, we apply (2) with

$$\mathbf{u} = \frac{1}{2}\mathbf{a} = \frac{1}{2}\mathbf{i} - \frac{1}{2}\mathbf{j} + \frac{\sqrt{2}}{2}\mathbf{k}$$

and obtain

$$D_{\mathbf{a}}f(2, 1, 0) = D_{\mathbf{u}}f(2, 1, 0) = f_x(2, 1, 0)\left(\frac{1}{2}\right) + f_y(2, 1, 0)\left(-\frac{1}{2}\right) + f_z(2, 1, 0)\left(\frac{\sqrt{2}}{2}\right)$$

$$= 1\left(\frac{1}{2}\right) + 0\left(-\frac{1}{2}\right) + 2\left(\frac{\sqrt{2}}{2}\right) = \frac{1}{2} + \sqrt{2} \quad \square$$

EXERCISES 13.5

In Exercises 1–14 find the directional derivative of f at the point P in the direction of $\mathbf{a}$.

1. $f(x, y) = 2x^2 - 3xy + y^2 + 15$; $P = (1, 1)$;
   $$\mathbf{a} = \frac{1}{\sqrt{2}}\mathbf{i} + \frac{1}{\sqrt{2}}\mathbf{j}$$

2. $f(x, y) = x^2 + y^2$; $P = (1, 2)$; $\mathbf{a} = \frac{1}{\sqrt{3}}\mathbf{i} - \frac{\sqrt{2}}{\sqrt{3}}\mathbf{j}$

3. $f(x, y) = \dfrac{x^2 - y^2}{x^2 + y^2}$; $P = (3, 4)$; $\mathbf{a} = \dfrac{1}{2}\mathbf{i} - \dfrac{\sqrt{3}}{2}\mathbf{j}$

4. $f(x, y) = x - y^2$; $P = (2, -3)$; $\mathbf{a} = \mathbf{i} + 2\mathbf{j}$

5. $f(x, y) = e^{4y}$; $P = (\frac{1}{2}, \frac{1}{4})$; $\mathbf{a} = 4\mathbf{i}$

6. $f(x, y) = \sin xy^2$; $P = (1/\pi, \pi)$; $\mathbf{a} = \mathbf{i} - 3\mathbf{j}$

7. $f(x, y) = \tan(x + 2y)$; $P = (0, \pi/6)$; $\mathbf{a} = -4\mathbf{i} + 5\mathbf{j}$

8. $f(x, y, z) = 3x - 2y + 4z$; $P = (1, -1, 2)$;
   $\mathbf{a} = \mathbf{i} + \mathbf{j} + \mathbf{k}$

9. $f(x, y, z) = x^3 y^2 z$; $P = (2, -1, 2)$; $\mathbf{a} = 2\mathbf{i} - \mathbf{j} - 2\mathbf{k}$

10. $f(x, y, z) = xy - yz + 3xz$; $P = (1, -1, 3)$;
    $\mathbf{a} = -\mathbf{i} + 3\mathbf{j} + 2\mathbf{k}$

11. $f(x, y, z) = \dfrac{x - y - z}{x + y + z}$; $P = (2, 1, -1)$;
    $\mathbf{a} = -2\mathbf{i} - \mathbf{j} - \mathbf{k}$

12. $f(x, y, z) = e^{x^2 + y^2 + z^2}$; $P = (0, 0, 0)$; $\mathbf{a} = -\mathbf{i} + \mathbf{j} - \mathbf{k}$

13. $f(x, y, z) = yz2^x$; $P = (1, -1, 1)$; $\mathbf{a} = 2\mathbf{j} - \mathbf{k}$

14. $f(x, y, z) = y \arcsin xz$; $P = (1/\sqrt{2}, 0, 1/\sqrt{2})$;
    $\mathbf{a} = -2\mathbf{i} - 2\mathbf{j} - 2\mathbf{k}$

15. Let $\mathbf{a} = \mathbf{i} - \mathbf{j}$ and $\mathbf{b} = 3\mathbf{i} + 3\mathbf{j}$. If $D_\mathbf{a} f(1, 2) = 6\sqrt{2}$ and $D_\mathbf{b} f(1, 2) = -2\sqrt{2}$, find $f_x(1, 2)$ and $f_y(1, 2)$.

16. Let $\mathbf{a} = \mathbf{i} - \mathbf{j} + \mathbf{k}$, $\mathbf{b} = \mathbf{j} - 3\mathbf{k}$, and $\mathbf{c} = 2\mathbf{i} - \mathbf{j}$. Suppose $D_\mathbf{a} f(x_0, y_0, z_0) = 2\sqrt{3}$, $D_\mathbf{b} f(x_0, y_0, z_0) = -\sqrt{10}$, and $D_\mathbf{c} f(x_0, y_0, z_0) = \sqrt{5}$. Find $f_x(x_0, y_0, z_0)$, $f_y(x_0, y_0, z_0)$, and $f_z(x_0, y_0, z_0)$.

17. Let $f(x, y) = \sinh(y \ln x)$. Find the directional derivative of f at $(2, -1)$ in the direction away from the origin.

18. Let $f(x, y) = 2x + x^2 y + y \sin y$ and let $\mathbf{u} = a\mathbf{i} + b\mathbf{j}$ be a unit vector.
    a. Express $D_\mathbf{u} f(1, 0)$ in terms of a and b.
    b. Using the result of part (a), find the values of a and b for which $D_\mathbf{u} f(1, 0)$ is maximum.

13.6
THE GRADIENT

By means of the first partials of a function we can define a vector called the gradient. This vector is important in the analysis of directional derivatives, plays a crucial role in the definition of the plane tangent to the graph of a function of two variables, and has special significance in physical applications.

DEFINITION 13.13

a. Let f be a function of two variables that has partial derivatives at (x_0, y_0). Then the **gradient** of f at (x_0, y_0), which is denoted grad $f(x_0, y_0)$ or $\nabla f(x_0, y_0)$, is defined by

$$\text{grad } f(x_0, y_0) = \nabla f(x_0, y_0) = f_x(x_0, y_0)\mathbf{i} + f_y(x_0, y_0)\mathbf{j}$$

b. Let f be a function of three variables that has partial derivatives at (x_0, y_0, z_0). Then the **gradient** of f at (x_0, y_0, z_0), which is denoted grad $f(x_0, y_0, z_0)$ or $\nabla f(x_0, y_0, z_0)$, is defined by

$$\text{grad } f(x_0, y_0, z_0) = \nabla f(x_0, y_0, z_0)$$
$$= f_x(x_0, y_0, z_0)\mathbf{i} + f_y(x_0, y_0, z_0)\mathbf{j} + f_z(x_0, y_0, z_0)\mathbf{k}$$

The symbol ∇, which is read "del," is an inverted Greek capital delta.

Example 1   Let $f(x, y) = \sin xy$. Find a formula for the gradient of f, and in particular find grad $f(\pi/3, 1)$.

Solution   Definition 13.13 tells us that

$$\text{grad } f(x, y) = f_x(x, y)\mathbf{i} + f_y(x, y)\mathbf{j} = y \cos xy\,\mathbf{i} + x \cos xy\,\mathbf{j}$$

Consequently

$$\text{grad } f\left(\frac{\pi}{3}, 1\right) = \cos\frac{\pi}{3}\mathbf{i} + \frac{\pi}{3}\cos\frac{\pi}{3}\mathbf{j} = \frac{1}{2}\mathbf{i} + \frac{\pi}{6}\mathbf{j} \quad \square$$

Example 2 Let $f(x, y, z) = 1/\sqrt{x^2 + y^2 + z^2}$. Find a formula for the gradient of f, and in particular find $\nabla f(2\sqrt{2}, 2\sqrt{2}, -3)$.

Solution By definition,

$$\nabla f(x, y, z) = f_x(x, y, z)\mathbf{i} + f_y(x, y, z)\mathbf{j} + f_z(x, y, z)\mathbf{k}$$

$$= -\frac{x}{(x^2 + y^2 + z^2)^{3/2}}\mathbf{i} - \frac{y}{(x^2 + y^2 + z^2)^{3/2}}\mathbf{j} - \frac{z}{(x^2 + y^2 + z^2)^{3/2}}\mathbf{k}$$

$$= \frac{-1}{(x^2 + y^2 + z^2)^{3/2}}(x\mathbf{i} + y\mathbf{j} + z\mathbf{k})$$

Since $(2\sqrt{2})^2 + (2\sqrt{2})^2 + (-3)^2 = 25$, we find that

$$\nabla f(2\sqrt{2}, 2\sqrt{2}, -3) = \frac{1}{125}(-2\sqrt{2}\mathbf{i} - 2\sqrt{2}\mathbf{j} + 3\mathbf{k}) \quad \square$$

If f is a function of two variables that is differentiable at (x_0, y_0), and if $\mathbf{u} = a_1\mathbf{i} + a_2\mathbf{j}$ is a unit vector in the xy plane, then

$$\text{grad } f(x_0, y_0) \cdot \mathbf{u} = \text{grad } f(x_0, y_0) \cdot (a_1\mathbf{i} + a_2\mathbf{j}) = f_x(x_0, y_0)a_1 + f_y(x_0, y_0)a_2$$

Thus by (1) of Section 13.5,

$$D_\mathbf{u}f(x_0, y_0) = [\text{grad } f(x_0, y_0)] \cdot \mathbf{u} \tag{1}$$

This formula shows the relationship of the directional derivative to the gradient and the direction $\mathbf{u}$. If $\text{grad } f(x_0, y_0) = \mathbf{0}$, then (1) implies that $D_\mathbf{u}f(x_0, y_0) = 0$ for every choice of $\mathbf{u}$. If $\text{grad } f(x_0, y_0) \neq \mathbf{0}$, then we can determine the largest value of $D_\mathbf{u}f(x_0, y_0)$ as a function of $\mathbf{u}$; moreover, we can determine the direction $\mathbf{u}$ that produces this value of $D_\mathbf{u}f(x_0, y_0)$. Let $\mathbf{u}$ be any unit vector in the plane, and let ϕ be the angle between $\mathbf{u}$ and $\text{grad } f(x_0, y_0)$. From (1) we deduce that

$$D_\mathbf{u}f(x_0, y_0) = [\text{grad } f(x_0, y_0)] \cdot \mathbf{u} = \|\mathbf{u}\| \|\text{grad } f(x_0, y_0)\| \cos \phi$$
$$= \|\text{grad } f(x_0, y_0)\| \cos \phi$$

From this we see that the largest value of $D_\mathbf{u}f(x_0, y_0)$ is $\|\text{grad } f(x_0, y_0)\|$, and this value is assumed when $\phi = 0$, that is, when $\mathbf{u}$ points in the same direction as $\text{grad } f(x_0, y_0)$.

We summarize the results of the preceding paragraph as a theorem:

THEOREM 13.14

Let f be a function of two variables that is differentiable at (x_0, y_0).
a. For any unit vector $\mathbf{u}$,

$$D_\mathbf{u}f(x_0, y_0) = [\text{grad } f(x_0, y_0)] \cdot \mathbf{u}$$

b. The maximum value of $D_\mathbf{u}f(x_0, y_0)$ is $\|\text{grad } f(x_0, y_0)\|$.
c. If $\text{grad } f(x_0, y_0) \neq \mathbf{0}$, then $D_\mathbf{u}f(x_0, y_0)$, regarded as a function of $\mathbf{u}$, attains its maximum value when $\mathbf{u}$ points in the same direction as $\text{grad } f(x_0, y_0)$.

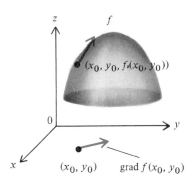

At (x_0, y_0), f increases most rapidly
in the direction of grad $f(x_0, y_0)$

FIGURE 13.34

Theorem 13.14(c) could also be interpreted as saying that at (x_0, y_0), f **increases most rapidly** in the direction of grad $f(x_0, y_0)$. Pictorially, Theorem 13.14(c) says that at $(x_0, y_0, f(x_0, y_0))$ the graph of f has the steepest incline in the direction of the gradient (Figure 13.34); thus if the graph represents the surface of a mountain, then at the point $(x_0, y_0, f(x_0, y_0))$ on the side of the mountain, the incline would be steepest in the direction of the gradient.

Example 3 Let $f(x, y) = 6 - 3x^2 - y^2$. In what direction is f increasing most rapidly at $(1, 2)$?

Solution Recall from Example 1 of Section 13.5 that

$$f_x(x, y) = -6x \quad \text{and} \quad f_y(x, y) = -2y$$

Therefore

$$\text{grad } f(1, 2) = f_x(1, 2)\mathbf{i} + f_y(1, 2)\mathbf{j} = -6\mathbf{i} - 4\mathbf{j}$$

so f increases most rapidly at $(1, 2)$ in the direction of $-6\mathbf{i} - 4\mathbf{j}$. □

At a point on the side of a mountain, the incline is steepest in the direction of the gradient. We may also say that the decline is steepest in the direction opposite the gradient (see Exercises 21–23).

There is a theorem for functions of three variables analogous to Theorem 13.14. Since the proofs of both theorems are so similar, we state the three-variable version without proof.

THEOREM 13.15

Let f be a function of three variables that a differentiable at (x_0, y_0, z_0).
a. For any unit vector $\mathbf{u}$ in space,

$$D_{\mathbf{u}}f(x_0, y_0, z_0) = [\text{grad } f(x_0, y_0, z_0)] \cdot \mathbf{u}$$

b. The largest value of $D_{\mathbf{u}}f(x_0, y_0, z_0)$ is $\|\text{grad } f(x_0, y_0, z_0)\|$.
c. If grad $f(x_0, y_0, z_0) \neq \mathbf{0}$, then $D_{\mathbf{u}}f(x_0, y_0, z_0)$, regarded as a function of $\mathbf{u}$, attains its maximum value when $\mathbf{u}$ points in the same direction as grad $f(x_0, y_0, z_0)$.

The Gradient as a Normal Vector

Suppose a smooth curve C is a level curve of a function f that is differentiable at a point (x_0, y_0) on C. As a point (x, y) moves along C, the value $f(x, y)$ is constant by hypothesis and hence does not change. This suggests that at (x_0, y_0) the rate of change of f in the direction of a unit vector $\mathbf{u}$ tangent to C is 0, that is, $D_{\mathbf{u}}f(x_0, y_0) = 0$ (Figure 13.35). However, (1) implies that $D_{\mathbf{u}}f(x_0, y_0) = 0$ for any unit vector $\mathbf{u}$ perpendicular to grad $f(x_0, y_0)$. Combining these results, we can reasonably conjecture that grad $f(x_0, y_0)$ is perpendicular, or equivalently, normal to C at (x_0, y_0) (Figure 13.35).

THEOREM 13.16

Let f be differentiable at (x_0, y_0), and let $f(x_0, y_0) = c$. Also let C be the level curve $f(x, y) = c$ that passes through (x_0, y_0). If C is smooth and grad $f(x_0, y_0) \neq \mathbf{0}$, then grad $f(x_0, y_0)$ is normal to C at (x_0, y_0).

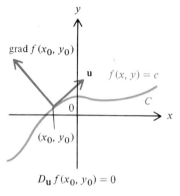

FIGURE 13.35

Proof Let I be an interval, and

$$\mathbf{r}(t) = x(t)\mathbf{i} + y(t)\mathbf{j} \quad \text{for } t \text{ in } I$$

a smooth parametrization of C. Since $f(x, y) = f(x(t), y(t))$ for t in I, the Chain Rule (Theorem 13.8) implies that

$$\frac{d}{dt} f(x(t), y(t)) = \frac{\partial f}{\partial x}\frac{dx}{dt} + \frac{\partial f}{\partial y}\frac{dy}{dt} = [\text{grad } f(x(t), y(t))] \cdot \frac{d\mathbf{r}}{dt}$$

But $f(x(t), y(t)) = c$ for t in I, so that

$$\frac{d}{dt} f(x(t), y(t)) = 0$$

and consequently

$$0 = [\text{grad } f(x(t), y(t))] \cdot \frac{d\mathbf{r}}{dt}$$

Since $\mathbf{r}$ is smooth by assumption, it follows that $d\mathbf{r}/dt$, which is tangent to C, is nonzero. Therefore grad $f(x_0, y_0)$, which is also nonzero by assumption, is perpendicular to the vector tangent to C at (x_0, y_0), and hence is normal to C at (x_0, y_0). ∎

Example 4 Assuming that the curve $x^2 - xy + 3y^2 = 5$ is smooth, find a unit vector that is perpendicular to the curve at $(1, -1)$.

Solution Let $f(x, y) = x^2 - xy + 3y^2$, so that the given curve is the level curve $f(x, y) = 5$. Since $f(1, -1) = 5$, the point $(1, -1)$ lies on the given level curve. By Theorem 13.16, grad $f(1, -1)$ is perpendicular to the given curve at $(1, -1)$. We find that

$$\text{grad } f(x, y) = (2x - y)\mathbf{i} + (-x + 6y)\mathbf{j}$$

and thus

$$\text{grad } f(1, -1) = 3\mathbf{i} - 7\mathbf{j}$$

grad $f(x_0, y_0, z_0)$

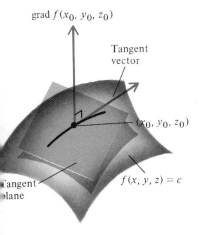

Tangent
vector

(x_0, y_0, z_0)

Tangent
plane

$f(x, y, z) = c$

FIGURE 13.36

Therefore the unit vector

$$\frac{1}{\|3\mathbf{i} - 7\mathbf{j}\|}(3\mathbf{i} - 7\mathbf{j}) = \frac{1}{\sqrt{58}}(3\mathbf{i} - 7\mathbf{j})$$

is the desired vector perpendicular to the given curve. $\square$

Now let f be a function of three variables that is differentiable at a point (x_0, y_0, z_0), and let $f(x_0, y_0, z_0) = c$. If C is any smooth curve on the level surface $f(x, y, z) = c$ and if C passes through (x_0, y_0, z_0), then an argument similar to the one used in proving Theorem 13.16 shows that if grad $f(x_0, y_0, z_0) \neq \mathbf{0}$, then grad $f(x_0, y_0, z_0)$ is perpendicular to the tangent vector of C at (x_0, y_0, z_0) (Figure 13.36). Because grad $f(x_0, y_0, z_0)$ is perpendicular to the tangent vector to any such curve C at (x_0, y_0, z_0), all such tangents lie in a single plane called a tangent plane.

DEFINITION 13.17

Let f be differentiable at a point (x_0, y_0, z_0) on a level surface S of f. If grad $f(x_0, y_0, z_0) \neq \mathbf{0}$, then the plane through (x_0, y_0, z_0) whose normal is grad $f(x_0, y_0, z_0)$ is the plane **tangent** to S at (x_0, y_0, z_0), and grad $f(x_0, y_0, z_0)$ is **normal** to S.

Since

$$\text{grad } f(x_0, y_0, z_0) = f_x(x_0, y_0, z_0)\mathbf{i} + f_y(x_0, y_0, z_0)\mathbf{j} + f_z(x_0, y_0, z_0)\mathbf{k}$$

and since grad $f(x_0, y_0, z_0)$ is normal to the tangent plane at (x_0, y_0, z_0), an equation of the tangent plane is

$$f_x(x_0, y_0, z_0)(x - x_0) + f_y(x_0, y_0, z_0)(y - y_0) + f_z(x_0, y_0, z_0)(z - z_0) = 0$$

Example 5   Find an equation of the plane tangent to the sphere $x^2 + y^2 + z^2 = 4$ at the point $(-1, 1, \sqrt{2})$.

Solution   The sphere is the level surface $f(x, y, z) = 4$, where $f(x, y, z) = x^2 + y^2 + z^2$. The partials of f at $(-1, 1, \sqrt{2})$ are given by

$$f_x(-1, 1, \sqrt{2}) = -2, \qquad f_y(-1, 1, \sqrt{2}) = 2 \qquad f_z(-1, 1, \sqrt{2}) = 2\sqrt{2}$$

Therefore an equation of the plane tangent at $(-1, 1, \sqrt{2})$ is

$$-2(x - (-1)) + 2(y - 1) + 2\sqrt{2}(z - \sqrt{2}) = 0$$

or equivalently,

$$-x + y + \sqrt{2}z = 4 \quad \square$$

Suppose f is a function of two variables that is differentiable at (x_0, y_0), and let

$$g(x, y, z) = f(x, y) - z$$

Then the graph of f is the level surface $g(x, y, z) = 0$. Accordingly, we define the plane tangent to the graph of f at $(x_0, y_0, f(x_0, y_0))$ to be the plane tangent to the level surface $g(x, y, z) = 0$ at $(x_0, y_0, f(x_0, y_0))$. Since

$$\text{grad } g(x_0, y_0, z_0) = f_x(x_0, y_0)\mathbf{i} + f_y(x_0, y_0)\mathbf{j} - \mathbf{k} \qquad (2)$$

an equation of the tangent plane of f at $(x_0, y_0, f(x_0, y_0))$ is

$$f_x(x_0, y_0)(x - x_0) + f_y(x_0, y_0)(y - y_0) - [z - f(x_0, y_0)] = 0 \qquad (3)$$

or equivalently,

$$z = f(x_0, y_0) + f_x(x_0, y_0)(x - x_0) + f_y(x_0, y_0)(y - y_0) \qquad (4)$$

and the vector described in (2) is said to be **normal** to the graph of f at $(x_0, y_0, f(x_0, y_0))$.

Example 6   Let $f(x, y) = 6 - 3x^2 - y^2$. Find a vector normal to the graph of f at $(1, 2, -1)$, and find an equation of the plane tangent to the graph of f at $(1, 2, -1)$.

Solution   First we calculate that

$$f_x(x, y) = -6x \quad \text{and} \quad f_y(x, y) = -2y$$

so that

$$f_x(1, 2) = -6 \quad \text{and} \quad f_y(1, 2) = -4$$

Therefore by (2) a vector normal to the graph of f at $(1, 2, -1)$ is given by

$$f_x(1, 2)\mathbf{i} + f_y(1, 2)\mathbf{j} - \mathbf{k} = -6\mathbf{i} - 4\mathbf{j} - \mathbf{k}$$

Then (3) implies that an equation of the tangent plane of f at $(1, 2, -1)$ is

$$-6(x - 1) - 4(y - 2) - (z + 1) = 0$$

or

$$6x + 4y + z = 13 \quad \square$$

The gradient arises in physical situations. Let us recall from Section 13.1 that if $T(x, y, z)$ is the temperature at any point (x, y, z), then the level surfaces of T are isothermal surfaces. On an isothermal surface the temperature is constant, and no heat flows along such a surface. Instead, heat flows in a direction perpendicular to an isothermal surface; more precisely, it flows in the direction of the gradient. Similarly, if $V(x, y, z)$ represents the voltage at the point (x, y, z), then $-\text{grad } V(x, y, z)$ turns out to be the electric force that would be exerted on a positive unit charge at (x, y, z). This force is perpendicular to the equipotential surface at (x, y, z).

EXERCISES 13.6

In Exercises 1–8 find the gradient of the function.

1. $f(x, y) = 3x - 5y$

2. $f(x, y) = y^2 + x \sin x^2 y$

3. $g(x, y) = e^{-2x} \ln(y - 4)$

4. $f(x, y) = \dfrac{xy - 1}{x^2 + y^2}$

5. $f(x, y, z) = 2x^2 - y^2 - 4z^2$

6. $f(x, y, z) = (2x + y^2 + z^3)^{5/2}$

7. $g(x, y, z) = \dfrac{-x + y}{-x + z}$

8. $g(x, y, z) = -x^2 y^3 e^{(z^2)}$

In Exercises 9–14 find the gradient of the function at the given point.

9. $f(x, y) = \dfrac{x + 3y}{5x + 2y}; (-1, \frac{3}{2})$

10. $f(x, y) = x \cos xy; (1, -\pi)$

11. $g(x, y) = x \ln(x + y); (-2, 3)$

12. $f(x, y, z) = z - \sqrt{x^2 + y^2}; (3, -4, 7)$

13. $f(x, y, z) = z e^{-x} \tan y; (0, \pi, -2)$

14. $g(x, y, z) = e^x (\sin y + \sin z); (1, \pi/2, \pi/2)$

In Exercises 15–20 find the direction in which f increases most rapidly at the given point.

15. $f(x, y) = e^x (\cos y + \sin y); (0, 0)$

16. $f(x, y) = e^{2x} (\cos y - \sin y); (\frac{1}{6}, -\pi/2)$

17. $f(x, y) = 3x^2 + 4y^2; (-1, 1)$

18. $f(x, y, z) = \ln(x^2 + y^2 + z^2); (2, 0, 1)$

19. $f(x, y, z) = e^x + e^y + e^{2z}; (1, 1, -1)$

20. $f(x, y, z) = \cos xyz; (\frac{1}{3}, \frac{1}{2}, \pi)$

From (1) it follows that the directional derivative of a function f at a point is smallest in the direction opposite to the gradient of f at that point. Thus we say that a function **decreases most rapidly** in the direction opposite the gradient. In Exercises 21–23 find the direction in which the function decreases most rapidly at the given point.

21. $f(x, y) = \sin \pi xy; (\frac{1}{2}, \frac{2}{3})$

22. $f(x, y) = \arctan(x - y); (2, -2)$

23. $f(x, y, z) = \dfrac{x - z}{y + z}; (-1, 1, 3)$

In Exercises 24–26 find a vector that is normal to the graph of the equation at the given point. Assume that each curve is smooth.

24. $x^3 - 3x^2 y + y^2 = 5; (1, -1)$

25. $\sin \pi xy = \sqrt{3}/2; (\frac{1}{6}, 2)$

26. $e^{x^2 y} = 2; (1, \ln 2)$

In Exercises 27–30 find a vector that is normal to the graph of f at the given point.

27. $f(x, y) = 3x^2 + 4y^2; (-2, 1, 16)$

28. $f(x, y) = \sqrt{4 - x^2 - y^2}; (-1, 1, \sqrt{2})$

29. $f(x, y) = 1 - x^2; (0, 2, 1)$

30. $f(x, y) = y^2 e^x; (0, -3, 9)$

In Exercises 31–38 find an equation of the plane tangent to the graph of the given function at the indicated point(s).

31. $f(x, y) = xy - x + y + 5; (0, 2, 7)$

32. $f(x, y) = \dfrac{x + 2}{y + 1}; (2, 3, 1)$

33. $g(x, y) = \sin \pi xy; (-\sqrt{2}, \sqrt{2}, 0)$ and $(-\frac{1}{2}, \frac{1}{3}, -\frac{1}{2})$

34. $f(x, y) = e^{x + y^2}; (-1, 0, e^{-1})$ and $(0, 1, e)$

35. $f(x, y) = (2 + x - y)^2; (3, -1, 36)$

36. $g(x, y) = 4x^2 + y^2 - 1; (2, 1, 16)$

37. $f(x, y) = \ln(x^2 + y^2); (-1, 0, 0)$ and $(-1, 1, \ln 2)$

38. $f(x, y) = \begin{cases} \dfrac{x^3 - y^3}{x^2 + y^2} & \text{for } (x, y) \neq (0, 0) \\ 0 & \text{for } (x, y) = (0, 0) \end{cases}$ $(1, 0, 1)$

In Exercises 39–45 find an equation of the plane tangent to the given level surface at the given point.

39. $x^2 + y^2 + z^2 = 1; (\frac{1}{2}, -\frac{1}{2}, -1/\sqrt{2})$

40. $\dfrac{x^2}{4} + \dfrac{y^2}{9} + \dfrac{z^2}{16} = 1; (0, 0, -4)$

41. $xyz = 1; (\frac{1}{2}, -2, -1)$

42. $z^2 + \sin xy = 2; (\pi, \frac{1}{2}, -1)$

43. $y e^{xy} + z^2 = 0; (0, -1, 1)$

44. $\dfrac{x^2 - y^2}{x^2 + y^2} = 0; (-1, -1, -1)$

45. $\ln \sqrt{x^2 + y^2 + z^2} = 0; (0, -1, 0)$

46. Find the point on the hyperbolic paraboloid $z = x^2 - 3y^2$ at which the tangent plane is parallel to the plane $8x + 3y - z = 4$.

47. Find the point on the paraboloid $z = 9 - 4x^2 - y^2$ at which the tangent plane is parallel to the plane $z = 4y$.

48. Show that the surfaces $z = \sqrt{x^2 + y^2}$ and $10z = 25 + x^2 + y^2$ have the same tangent plane at $(3, 4, 5)$.

49. Show that the surfaces $z = xy - 2$ and $x^2 + y^2 + z^2 = 3$ have the same tangent plane at $(1, 1, -1)$.

We say that two surfaces are **normal** at a given point if their tangent planes at that point are perpendicular to one another. In Exercises 50–51 show that the pair of surfaces are normal at the given point.

50. $x^2 + y^2 + z^2 = 16$ and $z^2 = x^2 + y^2$; $(2, 2, 2\sqrt{2})$

51. $z = x^2 + 4y^2 - 12$ and $8z = 4x + y^2 + 19$; $(-3, -1, 1)$

*52. Show that the line determined by the intersection of the plane $z = 0$ and the plane tangent to the surface $z^2(x^2 + y^2) = 4$ at a point of the form $(2 \cos \theta, 2 \sin \theta, 1)$ is tangent to the circle $x^2 + y^2 = 16$ at the point $(4 \cos \theta, 4 \sin \theta)$.

53. Show that an equation of the plane tangent to the ellipsoid

$$\frac{x^2}{a^2} + \frac{y^2}{b^2} + \frac{z^2}{c^2} = 1$$

at the point (x_0, y_0, z_0) is

$$\frac{xx_0}{a^2} + \frac{yy_0}{b^2} + \frac{zz_0}{c^2} = 1$$

54. Let $c \neq 0$. Show that an equation of the plane tangent to the paraboloid

$$cz = \frac{x^2}{a^2} + \frac{y^2}{b^2}$$

at the point (x_0, y_0, z_0) is

$$c(z + z_0) = \frac{2xx_0}{a^2} + \frac{2yy_0}{b^2}$$

55. Let g be a differentiable function of one variable and let $f(x, y) = xg(y/x)$. Show that every plane tangent to the graph of f passes through the origin.

56. Show that every line normal to the sphere $x^2 + y^2 + z^2 = 1$ passes through the origin.

57. Show that every line normal to the double cone $z^2 = x^2 + y^2$ intersects the z axis.

58. Show that there is exactly one plane tangent to the paraboloid $z = x^2 + y^2$ and parallel to any given non-vertical plane.

In Exercises 59–61 use implicit differentiation to find $\partial z/\partial x$ and $\partial z/\partial y$ at the given point. Then find an equation of the plane tangent to the level surface at that point.

59. $x^2 - y^2 - z^2 = 1$; $(\sqrt{2}, 0, 1)$

60. $xyz = 1$; $(2, -3, -\frac{1}{6})$

61. $\ln x + \ln y + \ln z = 1$; $(1, 1, e)$

62. A mountain climber's oxygen mask is leaking. If the surface of the mountain is represented by $z = 5 - x^2 - 2y^2$ and the climber is at $(\frac{1}{2}, -\frac{1}{2}, \frac{17}{4})$, in what direction should the climber turn to descend most rapidly?

63. A metal plate with vertices $(1, 1)$, $(5, 1)$, $(1, 3)$ and $(5, 3)$ is heated by a flame at the origin, and the temperature at a point on the plate is inversely proportional to the distance from the origin. If an ant is located at the point $(3, 2)$, in what direction should the ant crawl to cool the fastest?

64. Suppose the quadric surface $x^2 - y^2 = z$ is an equipotential surface. Show that the electric force on a positive unit charge at the origin is perpendicular to the xy plane.

65. Suppose $T(x, y, z) = x^3y + 3x^2y^2z$. Find the directional derivative of T at $(1, 1, -1)$ in the direction of the gradient.

66. Let f be a function of three variables with a nonzero gradient at the point (a, b, c). Find $D_\mathbf{a} f(a, b, c)$, where $\mathbf{a} = \operatorname{grad} f(a, b, c)$.

13.7
TANGENT PLANE APPROXIMATIONS AND DIFFERENTIALS

Suppose f is differentiable at (x_0, y_0). Then by Definition 13.7,

$$f(x, y) - f(x_0, y_0) = f_x(x_0, y_0)(x - x_0) + f_y(x_0, y_0)(y - y_0) + \varepsilon_1(x, y)(x - x_0) + \varepsilon_2(x, y)(y - y_0) \tag{1}$$

where

$$\lim_{(x, y) \to (x_0, y_0)} \varepsilon_1(x, y) = \lim_{(x, y) \to (x_0, y_0)} \varepsilon_2(x, y) = 0 \tag{2}$$

Since the limits in (2) are 0, the two numbers $f(x, y) - f(x_0, y_0)$ and $f_x(x_0, y_0)(x - x_0) + f_y(x_0, y_0)(y - y_0)$ are approximately equal when (x, y) is close to (x_0, y_0):

$$f(x, y) - f(x_0, y_0) \approx f_x(x_0, y_0)(x - x_0) + f_y(x_0, y_0)(y - y_0) \tag{3}$$

or equivalently,

$$f(x, y) \approx f(x_0, y_0) + f_x(x_0, y_0)(x - x_0) + f_y(x_0, y_0)(y - y_0) \qquad (4)$$

Now recall from (4) of Section 13.6 that any point (x, y, z) on the plane tangent to the graph of f at $(x_0, y_0, f(x_0, y_0))$ satisfies

$$z = f(x_0, y_0) + f_x(x_0, y_0)(x - x_0) + f_y(x_0, y_0)(y - y_0) \qquad (5)$$

Since the right sides of (4) and (5) are identical, we can use z from (5) to approximate $f(x, y)$ if (x, y) is close to (x_0, y_0). Pictorially, this amounts to using (x, y, z) on the tangent plane to approximate $(x, y, f(x, y))$ on the graph of f (Figure 13.37). For this reason the approximation of $f(x, y)$ by (4) is called a **tangent plane approximation** and is reminiscent of the tangent line approximations introduced in Section 3.8.

In order to emphasize that we will consider only points (x, y) that are close to (x_0, y_0), we replace x by $x_0 + h$ and y by $y_0 + k$ in (4), which becomes

$$f(x_0 + h, y_0 + k) \approx f(x_0, y_0) + f_x(x_0, y_0)h + f_y(x_0, y_0)k \qquad (6)$$

Although we will not discuss estimates of the error introduced by using the right side of (6) to approximate $f(x_0 + h, y_0 + k)$, the following example shows how the approximation is carried out.

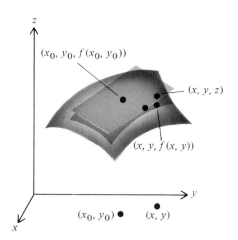

FIGURE 13.37

Example 1 Approximate $\sqrt{(3.012)^2 + (3.997)^2}$.

Solution Since 3.012 is close to 3 and 3.997 is close to 4, and since $\sqrt{3^2 + 4^2} = 5$, we let

$$f(x, y) = \sqrt{x^2 + y^2}$$

and take $x_0 = 3$, $y_0 = 4$, $h = 0.012$, and $k = -0.003$. Next we find that

$$f_x(x, y) = \frac{x}{\sqrt{x^2 + y^2}} \quad \text{and} \quad f_y(x, y) = \frac{y}{\sqrt{x^2 + y^2}}$$

so that $f_x(3, 4) = \frac{3}{5}$ and $f_y(3, 4) = \frac{4}{5}$. Then by (6) we have

$$\sqrt{(3.012)^2 + (3.997)^2} = f(3.012, 3.997)$$

$$\approx f(3, 4) + f_x(3, 4)(0.012) + f_y(3, 4)(-0.003)$$

$$= 5 + \frac{3}{5}(0.012) + \frac{4}{5}(-0.003) = 5.0048 \quad \square$$

The actual value of the given square root is 5.00481, accurate to 6 places. The accuracy of the estimate is excellent, and finding the estimate involved little computation.

There is a similar approximation for a function f of three variables that is differentiable at a point (x_0, y_0, z_0). By Definition 13.10,

$$f(x, y, z) - f(x_0, y_0, z_0) = f_x(x_0, y_0, z_0)(x - x_0) + f_y(x_0, y_0, z_0)(y - y_0)$$
$$+ f_z(x_0, y_0, z_0)(z - z_0) + \varepsilon_1(x, y, z)(x - x_0)$$
$$+ \varepsilon_2(x, y, z)(y - y_0) + \varepsilon_3(x, y, z)(z - z_0)$$

where

$$\lim_{(x, y, z) \to (x_0, y_0, z_0)} \varepsilon_1(x, y, z) = \lim_{(x, y, z) \to (x_0, y_0, z_0)} \varepsilon_2(x, y, z)$$

$$= \lim_{(x, y, z) \to (x_0, y_0, z_0)} \varepsilon_3(x, y, z) = 0$$

This leads to the approximation

$$f(x_0 + h, y_0 + k, z_0 + l) \approx f(x_0, y_0, z_0) + f_x(x_0, y_0, z_0)h$$
$$+ f_y(x_0, y_0, z_0)k + f_z(x_0, y_0, z_0)l \tag{7}$$

Example 2 Suppose a cardboard box in the shape of a rectangular parallelepiped has outer dimensions 14, 14, and 28 inches. If the cardboard is $\frac{1}{8}$ inch thick, approximate the volume of cardboard.

Solution Let $V(x, y, z) = xyz$. Then $V(14, 14, 28)$ is the volume of a box with outer dimensions 14, 14, and 28. If we let $x_0 = y_0 = 14$, $z_0 = 28$, and $h = k = l = -\frac{1}{4}$, then we seek

$$V(14, 14, 28) - V\left(14 - \frac{1}{4}, 14 - \frac{1}{4}, 28 - \frac{1}{4}\right)$$

But by (7) we find that

$$V(14, 14, 28) - V\left(14 - \frac{1}{4}, 14 - \frac{1}{4}, 28 - \frac{1}{4}\right)$$

$$\approx -V_x(14, 14, 28)h - V_y(14, 14, 28)k - V_z(14, 14, 28)l$$

$$= -14 \cdot 28\left(-\frac{1}{4}\right) - 14 \cdot 28\left(-\frac{1}{4}\right) - 14 \cdot 14\left(-\frac{1}{4}\right) = 245$$

Thus the volume of the cardboard is approximately 245 cubic inches. $\square$

The actual volume of cardboard is 241.516 cubic inches, accurate to 6 places, so the estimate is in error by less than 4 cubic inches.

Differentials If f is a function of two variables, we can replace (x_0, y_0) by any point (x, y) in the domain of f at which f is differentiable and transform (6) into

$$f(x + h, y + k) - f(x, y) \approx f_x(x, y)h + f_y(x, y)k \tag{8}$$

The number $f_x(x, y)h + f_y(x, y)k$ on the right side of (8) is usually called the **differential** (or **total differential**) of f (at (x, y) with increments h and k) and is denoted df. Thus

$$df = f_x(x, y)h + f_y(x, y)k \tag{9}$$

Of course, df depends on x, y, h, and k, even though they are not indicated in the notation df.

If $g_1(x, y) = x$ and $g_2(x, y) = y$, then the differential dg_1 is denoted dx, and the differential dg_2 is denoted dy. Since

$$(g_1)_x(x, y) = 1, \qquad (g_1)_y(x, y) = 0, \qquad (g_2)_x(x, y) = 0, \qquad (g_2)_y(x, y) = 1$$

we have

$$dx = dg_1 = 1 \cdot h + 0 \cdot k = h \quad \text{and} \quad dy = dg_2 = 0 \cdot h + 1 \cdot k = k$$

Therefore we can rewrite (9) as

$$df = f_x(x, y)\, dx + f_y(x, y)\, dy \quad \text{or} \quad df = \frac{\partial f}{\partial x}\, dx + \frac{\partial f}{\partial y}\, dy \tag{10}$$

Example 3 Let $f(x, y) = xy^2 + y \sin x$. Find df.

Solution Since

$$\frac{\partial f}{\partial x} = y^2 + y \cos x \quad \text{and} \quad \frac{\partial f}{\partial y} = 2xy + \sin x$$

we deduce from (10) that

$$df = (y^2 + y\cos x)\,dx + (2xy + \sin x)\,dy \quad \square$$

If f is a function of three variables that is differentiable at (x_0, y_0, z_0), then the **differential** df is defined by

$$df = f_x(x, y, z)h + f_y(x, y, z)k + f_z(x, y, z)l \tag{11}$$

The more usual form for df is

or

$$df = f_x(x, y, z)\,dx + f_y(x, y, z)\,dy + f_z(x, y, z)\,dz$$

$$df = \frac{\partial f}{\partial x}\,dx + \frac{\partial f}{\partial y}\,dy + \frac{\partial f}{\partial z}\,dz \tag{12}$$

Example 4   Let $f(x, y, z) = x^2 \ln(y - z)$. Find df.

Solution   Since

$$\frac{\partial f}{\partial x} = 2x \ln(y - z), \qquad \frac{\partial f}{\partial y} = \frac{x^2}{y - z}, \qquad \frac{\partial f}{\partial z} = \frac{-x^2}{y - z} = \frac{x^2}{z - y}$$

(12) tells us that

$$df = 2x \ln(y - z)\,dx + \frac{x^2}{y - z}\,dy + \frac{x^2}{z - y}\,dz \quad \square$$

EXERCISES 13.7

In Exercises 1–8 approximate the value of f at the given point.

1. $f(x, y) = \sqrt{x^2 + y^2}$; (3.01, 4.03)
2. $f(x, y) = \sqrt{x^2 + y}$; (3.02, −4.98)
3. $f(x, y) = \ln(x^2 + y^2)$; (−0.03, 0.98)
4. $f(x, y) = \sin \pi xy$; (−1.97, 2.005)
5. $f(x, y) = \tan xy$; (0.99π, 0.24)
6. $f(x, y) = \sqrt{6 - x^2 - y^2}$; (0.987, 1.013)
7. $f(x, y, z) = \sqrt{x^2 + y^2 + z^2}$; (3.01, 4.02, 11.98)
8. $f(x, y, z) = xyz^2$; (−2.1, 1.01, 0.989)

In Exercises 9–12 approximate the number.

9. $\sqrt[4]{(1.9)^3 + (2.1)^3}$
10. $(16.05)^{1/4}(7.95)^{2/3}$
11. $e^{0.1} \ln 0.9$
12. $\sin \dfrac{9\pi}{20} \cos \dfrac{9\pi}{30}$

In Exercises 13–22 determine df.

13. $f(x, y) = 3x^3 - x^2 y + y + 17$
14. $f(x, y) = y \ln \dfrac{1 + x}{1 - x}$
15. $f(x, y) = x^2 + y^2$
16. $f(x, y) = \cos(x + y) + \cos(x - y)$
17. $f(x, y) = x \tan y + y \cot x$
18. $f(x, y, z) = x^2 + y^2 + z^2$
19. $f(x, y, z) = z^2 \sqrt{1 + x^2 + y^2}$
20. $f(x, y, z) = \ln \sqrt{x^2 + y^2 + z^2}$
21. $f(x, y, z) = xe^{y^2 - z^2}$
22. $f(x, y, z) = \dfrac{x}{x^2 + y^2 + z^2}$

23. When two resistors having resistances R_1 and R_2 are connected in parallel, the resistance of the combination is given by

$$R = \frac{R_1 R_2}{R_1 + R_2}$$

Approximate the resistance when resistors having resistances of 2.013 ohms and 5.972 ohms are connected in parallel.

24. Approximate the length of the hypotenuse and the area of a right triangle with legs having lengths 5.011 and 11.877.

25. Approximate the surface area of a rectangular box with dimensions 3.019, 3.979, and 11.973, assuming that the box has a top.

26. If fiberglass sheeting costs $3 per square foot, approximate the cost of a sheet of fiberglass 3.012 feet wide and 5.982 feet long.

13.8
EXTREME VALUES

Like functions of one variable, functions of several variables can have maximum and minimum values on a given set. As in the one-variable case, the notion of relative extreme values will facilitate the study of extreme values of functions of several variables. The problem of recognizing and classifying relative extreme values is much more difficult for functions of several variables than for functions of one variable, so in this section we will limit our discussion to the two-variable case.

DEFINITION 13.18

Let f be a function of two variables, and let R be a set contained in the domain of f. Then f **has a maximum value** (respectively, a **minimum value**) **on R at (x_0, y_0)** if $f(x, y) \leq f(x_0, y_0)$ (respectively, $f(x, y) \geq f(x_0, y_0)$) for all (x, y) in R. If R is the domain of f, we say that f has a maximum value (respectively, a minimum value) at (x_0, y_0). Furthermore, f **has a relative maximum value** (respectively, a **relative minimum value**) **at (x_0, y_0)** if there is a disk D centered at (x_0, y_0) and contained in the domain of f such that $f(x, y) \leq f(x_0, y_0)$ (respectively, $f(x, y) \geq f(x_0, y_0)$) for all (x, y) in D.

We combine maximum and minimum values under the heading **extreme values** and relative maximum and relative minimum values under the heading **relative extreme values**. If there is a disk centered at (x_0, y_0) that is contained in the domain of f, then an extreme value of f at (x_0, y_0) is also a relative extreme value.

Relative extreme values correspond to the hilltops and valley bottoms on the graph of a function (Figure 13.38). Recall that a relative extreme value of a function f of a single variable occurs at a critical point—a number x_0 such that either $f'(x_0) = 0$ or $f'(x_0)$ does not exist. As a first step in identifying relative extreme values of functions of two variables, we assume that f has a relative extreme value at (x_0, y_0) and let

$$g(x) = f(x, y_0) \quad \text{and} \quad G(y) = f(x_0, y)$$

Then g has a relative extreme value at x_0, and G has a relative extreme value at y_0

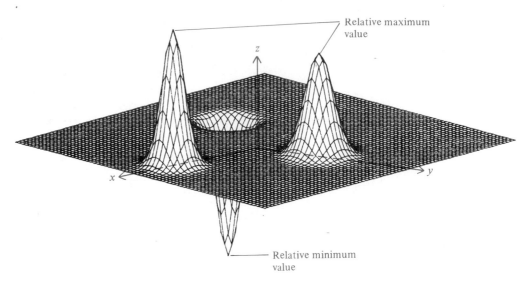

FIGURE 13.38

(Figure 13.39). Therefore if $f_x(x_0, y_0)$ and $f_y(x_0, y_0)$ exist, then

$$f_x(x_0, y_0) = g'(x_0) = 0 \quad \text{and} \quad f_y(x_0, y_0) = G'(y_0) = 0$$

This proves the following theorem.

THEOREM 13.19

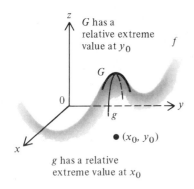

g has a relative
extreme value at x_0

FIGURE 13.39

> Let f have a relative extreme value at (x_0, y_0). If f has partial derivatives at (x_0, y_0), then
> $$f_x(x_0, y_0) = f_y(x_0, y_0) = 0$$

From Theorem 13.19 we see that relative extreme values of f occur only at those points at which the partial derivatives of f exist and are 0, or at which one or both of the partial derivatives does not exist. We say that f has a **critical point** at a point (x_0, y_0) in the domain of f if $f_x(x_0, y_0) = f_y(x_0, y_0) = 0$, or if one of the partial derivatives does not exist. Then we can interpret Theorem 13.19 as saying that f has relative extreme values only at critical points in its domain.

Example 1   Let $f(x, y) = 3 - x^2 + 2x - y^2 - 4y$. Find all critical points of f.

Solution   The partial derivatives of f exist at every point in the domain of f, so relative extreme values can occur only at points at which both partial derivatives are 0. The partial derivatives are

$$f_x(x, y) = -2x + 2 \quad \text{and} \quad f_y(x, y) = -2y - 4$$

Therefore $f_x(x, y) = 0$ only if $x = 1$, and $f_y(x, y) = 0$ only if $y = -2$. This means that $f_x(x, y) = f_y(x, y) = 0$ only if $(x, y) = (1, -2)$, and thus $(1, -2)$ is the only critical point of f. $\square$

We can determine whether the function in Example 1 has a relative extreme value at $(1, -2)$ by completing squares in the formula for $f(x, y)$. Thus

$$3 - x^2 + 2x - y^2 - 4y = 3 - (x^2 - 2x + 1) - (y^2 + 4y + 4) + 1 + 4$$
$$= 8 - (x - 1)^2 - (y + 2)^2$$

Since $(x - 1)^2 \geq 0$ and $(y + 2)^2 \geq 0$ for all x and y, it is apparent that $f(1, -2) = 8$ is a relative maximum value of f; in fact, we can even conclude that 8 is the maximum value of f.

Example 2 Let $f(x, y) = \sqrt{x^2 + y^2}$. Determine all critical points and all relative extreme values of f.

Solution We find that

$$f_x(x, y) = \frac{x}{\sqrt{x^2 + y^2}} \quad \text{and} \quad f_y(x, y) = \frac{y}{\sqrt{x^2 + y^2}}$$

Since the partial derivatives exist at all points except the origin, and since the origin is in the domain of f, the origin is a critical point of f. Because $f_x(x, y) = 0$ only if $x = 0$ and $f_y(x, y) = 0$ only if $y = 0$, there is no point (x, y) such that $f_x(x, y) = f_y(x, y) = 0$. Consequently $(0, 0)$ is the only critical point of f. Since $f(0, 0) = 0$ and $f(x, y) \geq 0$ for all (x, y), it follows that 0 is the only (relative) minimum value of f, and there is no relative maximum value. □

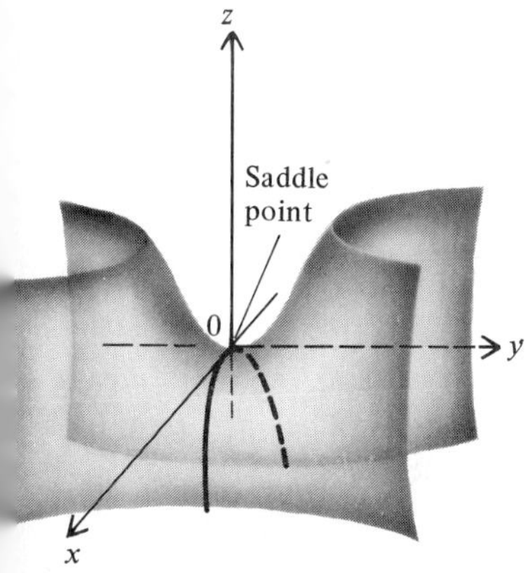

The next example illustrates the fact that a function need not have a relative extreme value at a critical point.

Example 3 Let $f(x, y) = y^2 - x^2$. Show that the origin is the only critical point but that $f(0, 0)$ is not a relative extreme value of f.

Solution In this case

$$f_x(x, y) = -2x \quad \text{and} \quad f_y(x, y) = 2y$$

so that $(0, 0)$ is the unique critical point of f. However, $f(0, 0) = 0$ is not a relative extreme value of f, because $f(x, 0) = -x^2 < 0$ for $x \neq 0$, and $f(0, y) = y^2 > 0$ for $y \neq 0$. Consequently f has no relative extreme values (Figure 13.40). □

$f(x, y) = y^2 - x^2$

FIGURE 13.40

If we consider the values of the function f in Example 3 at points along the x axis, we obtain a function of x that has its maximum value at the origin. However, if we consider the values of f at points along the y axis, we obtain a function of y that has its minimum value at the origin. Thus the graph of f resembles a saddle (see Figure 13.40), and that is why f does not have a relative extreme value at $(0, 0)$, despite the fact that $f_x(0, 0) = f_y(0, 0) = 0$. More generally, if f is a function for which $f_x(x_0, y_0) = f_y(x_0, y_0) = 0$, we say that f has a **saddle point** at (x_0, y_0) if there is a disk centered at (x_0, y_0) such that the following condition holds: f

assumes its maximum value on one diameter of the disk only at (x_0, y_0), and assumes its minimum value on another diameter of the disk only at (x_0, y_0). From this definition and our comments above we conclude that if $f(x, y) = y^2 - x^2$, then f has a saddle point at the origin.

The Second Partials Test

Suppose (x_0, y_0) is a critical point of a function f. How can we determine whether f has a relative extreme value or a saddle point at (x_0, y_0)? The next theorem, which is closely related to the Second Derivative Test and which we present without proof, helps answer this question.

THEOREM 13.20
SECOND PARTIALS TEST

Assume that f has a critical point at (x_0, y_0) and that f has continuous second partial derivatives in a disk centered at (x_0, y_0). Let

$$D(x_0, y_0) = f_{xx}(x_0, y_0)f_{yy}(x_0, y_0) - [f_{xy}(x_0, y_0)]^2$$

a. If $D(x_0, y_0) > 0$ and $f_{xx}(x_0, y_0) < 0$ (or $f_{yy}(x_0, y_0) < 0$), then f has a relative maximum value at (x_0, y_0).
b. If $D(x_0, y_0) > 0$ and $f_{xx}(x_0, y_0) > 0$ (or $f_{yy}(x_0, y_0) > 0$), then f has a relative minimum value at (x_0, y_0).
c. If $D(x_0, y_0) < 0$, then f has a saddle point at (x_0, y_0).

Finally, if $D(x_0, y_0) = 0$, then f may or may not have a relative extreme value at (x_0, y_0).

The expression $D(x_0, y_0)$ in the second Partials Test is called the **discriminant of f at (x_0, y_0)**. It can also be given in determinant form:

$$D(x_0, y_0) = \begin{vmatrix} f_{xx}(x_0, y_0) & f_{xy}(x_0, y_0) \\ f_{xy}(x_0, y_0) & f_{yy}(x_0, y_0) \end{vmatrix}$$

Notice that if $D(x_0, y_0) > 0$, then $f_{xx}(x_0, y_0)$ and $f_{yy}(x_0, y_0)$ have the same sign, that is, $f_{xx}(x_0, y_0) > 0$ and $f_{yy}(x_0, y_0) > 0$, or $f_{xx}(x_0, y_0) < 0$ and $f_{yy}(x_0, y_0) < 0$. Thus if $D(x_0, y_0) > 0$, then the sign of either $f_{xx}(x_0, y_0)$ or $f_{yy}(x_0, y_0)$ determines the nature of the relative extreme value.

A critical point (x_0, y_0) is said to be **degenerate** if $D(x_0, y_0) = 0$ and **nondegenerate** otherwise. If a critical point is nondegenerate, the Second Partials Test determines the nature of the critical point. In contrast, the test implies nothing about a degenerate critical point. Indeed, f may have a relative extreme value or a saddle point or neither at a degenerate critical point. For example, let

$$f(x, y) = y^4 - x^4, \qquad g(x, y) = x^4 + y^4, \quad \text{and} \quad h(x, y) = x^5 + x^3 + y^3$$

Then the point $(0, 0)$ is a critical point of f, g, and h, and in each case,

$$D(0,0) = 0$$

However, a simple calculation shows that f has a saddle point at $(0, 0)$, whereas $g(x, y) \geq 0$ for all (x, y) and $g(0, 0) = 0$, so that g has a minimum value at $(0, 0)$. Finally, it turns out that on no line through the origin in the xy plane does h have a relative extreme value; thus h has neither a relative extreme value nor a saddle point at $(0, 0)$.

Example 4 Let

$$f(x, y) = x^2 - 2xy + \frac{1}{3}y^3 - 3y$$

Using the Second Partials Test, determine at which points f has relative extreme values and at which points f has saddle points.

Solution We find that

$$f_x(x, y) = 2x - 2y \quad \text{and} \quad f_y(x, y) = -2x + y^2 - 3$$

Observe that $f_x(x, y) = 0$ if $x = y$ and $f_y(x, y) = 0$ if $-2x + y^2 - 3 = 0$. Thus (x, y) is a critical point if

$$x = y \quad \text{and} \quad -2x + y^2 - 3 = 0$$

By substituting y for x we can transform the second equation into $y^2 - 2y - 3 = 0$. The two solutions of this equation are $y = 3$ and $y = -1$. Thus the critical points of f are $(3, 3)$ and $(-1, -1)$. For the second partials of f we find that

$$f_{xx}(x, y) = 2, \qquad f_{yy}(x, y) = 2y, \quad \text{and} \quad f_{xy}(x, y) = -2$$

Therefore

$$D(3, 3) = f_{xx}(3, 3)f_{yy}(3, 3) - [f_{xy}(3, 3)]^2 = (2)(6) - (-2)^2 = 8 > 0$$

and

$$D(-1, -1) = f_{xx}(-1, -1)f_{yy}(-1, -1) - [f_{xy}(-1, -1)]^2$$
$$= (2)(-2) - (-2)^2 = -8 < 0$$

Since $D(3, 3) > 0$ and $f_{xx}(3, 3) = 2 > 0$, f has a relative minimum value at $(3, 3)$. Since $D(-1, -1) < 0$, f has a saddle point at $(-1, -1)$ (Figure 13.41). $\square$

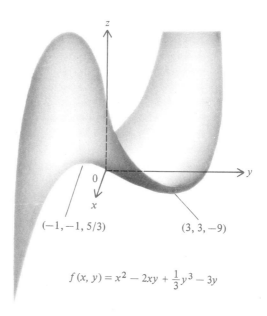

$(-1, -1, 5/3)$ $(3, 3, -9)$

$$f(x, y) = x^2 - 2xy + \frac{1}{3}y^3 - 3y$$

FIGURE 13.41

Extreme Values on a Set

Let R be a set in the plane. Assume that R is **bounded** (that is, R is contained in some rectangle) and contains its boundary (see Section 13.2). Under these conditions there is a theorem (which we state without proof) for functions of two variables analogous to the Maximum–Minimum Theorem (Theorem 4.2).

THEOREM 13.21
MAXIMUM–MINIMUM THEOREM FOR TWO VARIABLES

Let R be a bounded set in the plane that contains its boundary, and let f be continuous on R. Then f has both a maximum value and a minimum value on R.

By Theorem 13.19 and the comments that follow it, if R is bounded and contains its boundary and if f has an extreme value on R at (x_0, y_0), then (x_0, y_0) is either a critical point of f or a boundary point of R. This observation, along with the Maximum–Minimum Theorem for Two Variables, provides us with a method of finding extreme values:

1. Find the critical points of f in R, and compute the values of f at these points.
2. Find the extreme values of f on the boundary of R.
3. The maximum value of f on R will be the largest of the values computed in steps 1 and 2, and the minimum value of f on R will be the smallest of those values.

Example 5 Let $f(x, y) = xy - x^2$, and let R be the square region shown in Figure 13.42. Find the extreme values of f on R.

Solution By the Maximum–Minimum Theorem f has extreme values on R. Since

$$f_x(x, y) = y - 2x \quad \text{and} \quad f_y(x, y) = x$$

it follows that $f_x(x, y) = 0$ if $y = 2x$ and $f_y(x, y) = 0$ if $x = 0$. Thus the only critical point of f is $(0, 0)$, which happens to be a boundary point of R. Therefore the extreme values of f on R must occur on the boundary of R, which is composed of the four line segments l_1, l_2, l_3, and l_4 (Figure 13.42). On $l_1, x = 0$ and $0 \leq y \leq 1$, and since $f(0, y) = 0$, the maximum and minimum values of f on l_1 are both 0. On $l_2, y = 0$ and $0 \leq x \leq 1$, and since

$$f(x, 0) = -x^2$$

the maximum value of f on l_2 is 0 and the minimum value is -1. On $l_3, x = 1$ and $0 \leq y \leq 1$, and since

$$f(1, y) = y - 1$$

the maximum value of f on l_3 is 0 and the minimum value is -1. Finally, on l_4, $y = 1$ and $0 \leq x \leq 1$, and since

$$f(x, 1) = x - x^2$$

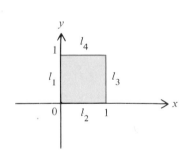

FIGURE 13.42

the maximum value of f on l_4 is $\frac{1}{4}$ and the minimum value is 0, as you can verify by the methods of Chapter 4. By comparing the extreme values of f on l_1, l_2, l_3, and l_4, we conclude that the maximum value of f on R is $\frac{1}{4}$ and the minimum value is -1. $\square$

Example 6 Under present Post Office regulations a package in the shape of a rectangular parallelepiped can be mailed parcel post only if the sum of the length and girth (Figure 13.43) of the package is not more than 108 inches. Find the largest volume V of such a package.

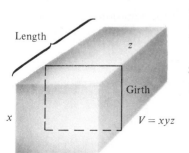

Solution If x, y, and z represent the dimensions of the package, with z denoting the length of the largest side, then $2x + 2y$ is the girth (Figure 13.43), and

$$V = xyz \quad \text{and} \quad \overbrace{2x + 2y}^{\text{girth}} + \overbrace{z}^{\text{length}} \le 108$$

where we assume that $x \ge 0$, $y \ge 0$, and $z \ge 0$. Since we wish the largest possible volume, we may assume that

$$2x + 2y + z = 108 \tag{1}$$

Solving for z and substituting in the equation for V, we obtain

$$V = xy(108 - 2x - 2y) = 108xy - 2x^2y - 2xy^2$$

Now we wish to find the maximum value of V as a function of x and y. Since x, y, and z must be nonnegative, so that $2x + 2y \le 108$ by (1), our goal is to find the maximum value of V on the triangular region R (which includes its boundary) shown in Figure 13.44. Such a maximum value exists by the Maximum–Minimum Theorem. If $x = 0$ or $y = 0$ or $2x + 2y = 108$ (in which case $z = 0$ by (1)), then the volume is 0. Therefore the maximum value of V on R does not occur on the boundary of R and hence must occur at a critical point in the interior of R. To find the critical points in the interior of R, we take partial derivatives of V:

$$\frac{\partial V}{\partial x} = 108y - 4xy - 2y^2 \quad \text{and} \quad \frac{\partial V}{\partial y} = 108x - 2x^2 - 4xy$$

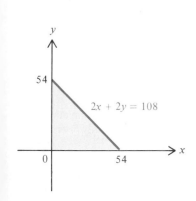

FIGURE 13.44

Then $\partial V/\partial x = 0$ if $108y - 4xy - 2y^2 = 0$, that is, if $y = 0$ or if $108 - 4x - 2y = 0$. Similarly, $\partial V/\partial y = 0$ if $108x - 2x^2 - 4xy = 0$, that is, if $x = 0$ or if $108 - 2x - 4y = 0$. Since (x, y) can be a critical point in the interior of R only if $x \ne 0$ and $y \ne 0$, it follows that (x, y) is such a critical point only if

$$108 - 4x - 2y = 0 \quad \text{and} \quad 108 - 2x - 4y = 0 \tag{2}$$

Solving for y in the first equation, we obtain $y = 54 - 2x$. Substituting for y in the second equation gives us

$$0 = 108 - 2x - 4(54 - 2x) = -108 + 6x$$

so that $x = 18$. Then

$$y = 54 - 2(18) = 18$$

Thus $(18, 18)$ is the only critical point in the interior of R. We conclude that

FIGURE 13.43

$V(18, 18) = 11,664$ (cubic inches) is the maximum volume of a mailable rectangular parallelepiped. □

When we solved the equations $\partial V/\partial x = 0$ and $\partial V/\partial y = 0$ in Example 6, we obtained two equations involving x and y (see (2)). Then we solved for y in one equation and substituted for it in the other equation to find the possible values for x. This is the general method of attack: When you obtain two equations in two unknowns, try to solve for one variable in one of the equations; then substitute for that variable in the other equation and find the required value of the remaining variable.

EXERCISES 13.8

In Exercises 1–23 find all critical points. Determine whether each critical point yields a relative maximum value, a relative minimum value, or a saddle point.

1. $f(x, y) = x^2 + 2y^2 - 6x + 8y - 1$
2. $f(x, y) = x^2 - 2y^2 - 6x + 8y + 3$
3. $f(x, y) = x^2 + 6xy + 2y^2 - 6x + 10y - 2$
4. $g(x, y) = x^2 - xy - 2y^2 + 7x - 8y + 3$
5. $k(x, y) = -x^2 - 2xy - 2y^2 + 6x - 10y + 5$
6. $f(x, y) = -x^2 + 4xy + y^2 - 2x + 9$
7. $f(x, y) = x^2y - 2xy + 2y^2 - 15y$
8. $f(x, y) = x^3 - 6x^2 - 3y^2$
9. $f(x, y) = 3x^2 - 3xy^2 + y^3 + 3y^2$
10. $f(u, v) = u^3 + v^3 - 6uv$
11. $f(x, y) = 4xy + 2x^2y - xy^2$
12. $f(x, y) = \dfrac{1}{x} + \dfrac{1}{y} + xy$
13. $f(x, y) = x^2 - e^{y^2}$
14. $f(x, y) = (y - 2)\ln xy$
15. $k(x, y) = e^x \sin y$
16. $f(x, y) = e^x(\sin y - 1)$
17. $f(u, v) = |u| + |v|$
18. $g(u, v) = 3 - |u - 2| + |v + 1|$
19. $f(x, y) = e^{xy}$
20. $f(x, y) = \sin x + \sin y$ for $0 < x < \pi/2, 0 < y < \pi/2$
*21. $f(x, y) = \sin x + \sin y$
22. $k(u, v) = (u + v)^2$
23. $f(x, y) = (y + ax + b)^2$, where a and b are constants
24. Let a and b be nonzero and $f(x, y) = (ax + by)e^{-x^2-y^2}$. Show that all the critical points of f lie on a line.

In Exercises 25–28 find the extreme values of f on R.

25. $f(x, y) = x^2 - y^2$; R is the disk $x^2 + y^2 \le 1$.
26. $f(x, y) = ye^{-x}$; R is the rectangular region with vertices $(0, 0)$, $(\ln 2, 0)$, $(\ln 2, 3)$, $(0, 3)$.
27. $f(x, y) = 2\sin x + 3\cos y$; R is the square region with vertices $(0, -\pi/2)$, $(\pi, -\pi/2)$, $(\pi, \pi/2)$, $(0, \pi/2)$.
28. $f(x, y) = e^{x^2-y^2}$; R is the ring bounded by the circles $x^2 + y^2 = \frac{1}{2}$ and $x^2 + y^2 = 2$.

In the remaining exercises in this section, assume that the required extreme values exist.

29. Find the three positive numbers whose sum is 48 and whose product is as large as possible. Calculate that product.

30. Find the three positive numbers whose product is 48 and whose sum is as small as possible. Calculate that sum.

31. Show that the box in the shape of a rectangular parallelepiped whose volume is the largest of any inscribed in a given sphere is a cube.

32. An open-topped rectangular box is to have a volume of 32 cubic meters. Find the dimensions of such a box having the smallest possible surface area.

33. Find the point in space the sum of whose coordinates is 48 and whose distance from the origin is minimum.

34. Find a vector in space whose length is 16 and whose components have the largest possible sum.

35. Show that if f has a critical point at (x_0, y_0) and if $f_{xx}(x_0, y_0) > 0$ and $f_{yy}(x_0, y_0) < 0$, then f has a saddle point at (x_0, y_0).

One measure of the closeness of a line l with equation $y = mx + b$ to the point (x_1, y_1) is the square $[y_1 - (mx_1 + b)]^2$ of the distance between (x_1, y_1) and l measured along the vertical line $x = x_1$ (Figure 13.45). For n fixed points (x_1, y_1),

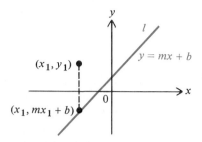

FIGURE 13.45

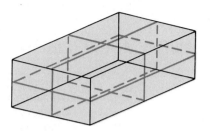

FIGURE 13.47

$(x_2, y_2),\ldots,(x_n, y_n)$, the corresponding measure of the closeness of the line l to these points is

$$f(m, b) = \sum_{k=1}^{n} (y_k - (mx_k + b))^2$$

It can be shown that f has a minimum value $f(m_0, b_0)$, which can be determined by solving the equations $f_m(m, b) = 0$ and $f_b(m, b) = 0$ for m and b. The line $y = m_0 x + b_0$ is called the **line of best fit** for the n given points (Figure 13.46). (Statisticians also call this line the **line of regression**. This terminology originated from a statistical study of the tendency of the height of the offspring of tall parents to regress toward the average height.) The method of determining the line of best fit to n given points is called the **method of least squares**. In Exercises 36–37 use the method of least squares to determine the line of best fit for the given collection of points.

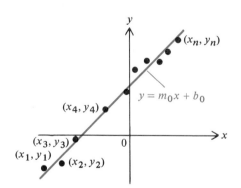

FIGURE 13.46

36. (1, 1), (2, 3), (3, 4)

37. (0, 0), (1, −1), (−2, 1)

38. A rectangular box is tied once each way around with a string of length l, and without knots (Figure 13.47). Find the maximum possible volume of the package.

39. A gymnasium in the shape of a rectangular parallelepiped with volume 960,000 cubic feet is to be erected. Assume that because of decorations, the front wall will cost twice as much per square foot as the side and back walls and the floor, and the roof will cost $\frac{3}{2}$ as much as the side walls. Find the dimensions of the gymnasium that will minimize the cost.

*40. A park ranger must walk from a certain spot in a thicket back to the ranger station, first through the thicket and then through marshland, and finally along a road, as in Figure 13.48. Suppose the ranger can proceed through the thicket at 3 kilometers per hour, through the marshland at 4 kilometers per hour, and on the road at 5 kilometers per hour. What is the most expeditious route?

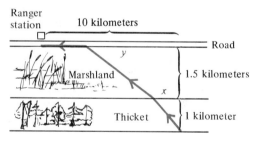

FIGURE 13.48

*41. A rectangular piece of tin with width l is to be bent as shown in Figure 13.49. Show that the maximal cross-sectional area is obtained if $x = \frac{1}{3}l$ and $\theta = \pi/3$. (*Hint:* After taking partial derivatives, eliminate l from the equations you must solve.)

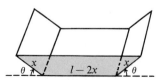

FIGURE 13.49

13.9
LAGRANGE MULTIPLIERS

In solving Example 6 of Section 13.8 we found the maximum value of the function xyz subject to the condition $2x + 2y + z = 108$. Such a condition on the values of x, y, and z is known as a **constraint** (or **side condition**). In this section we describe a general method of finding the extreme values of a function subject to a constraint. The method was introduced by the French mathematician Joseph Louis Lagrange (1736–1813).

The Lagrange Method for Functions of Two Variables

Let us consider the problem of finding an extreme value of a function f of two variables subject to a constraint of the form $g(x, y) = c$; that is, we seek an extreme value of f on the level curve $g(x, y) = c$ (rather than on the entire domain of f). If f has an extreme value on the level curve at the point (x_0, y_0), then under certain conditions there exists a number λ such that

$$\text{grad } f(x_0, y_0) = \lambda \, \text{grad } g(x_0, y_0) \tag{1}$$

To suggest why this equation holds, let us consider Figure 13.50, which shows the level curve C of g along with several level curves of f, and assume that $\text{grad } f(x_0, y_0) \neq \mathbf{0}$. Let (x_0, y_0) be the point in Figure 13.50 at which f assumes its maximum value 6 on C. Observe that C cannot cross the level curve $f(x, y) = 6$ because if it did, C would intersect level curves of f corresponding to values of f larger than 6, and that would contradict the assumption that 6 is the maximum value of f on C. Therefore C and the level curve $f(x, y) = 6$ have the same tangent at (x_0, y_0). Consequently grad $g(x_0, y_0)$, which by Theorem 13.16 is normal to C at (x_0, y_0), is parallel to grad $f(x_0, y_0)$, which is normal to the level curve

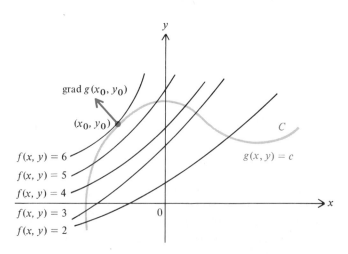

FIGURE 13.50

$f(x, y) = 6$ at (x_0, y_0). It follows that if neither gradient is zero, they are parallel to each other. This is the content of (1).

Now we state and prove the result just discussed.

THEOREM 13.22

Let f and g be differentiable at (x_0, y_0). Let C be the level curve $g(x, y) = c$ that contains (x_0, y_0). Assume that C is smooth, and that (x_0, y_0) is not an endpoint of the curve. If grad $g(x_0, y_0) \neq \mathbf{0}$ and if f has an extreme value on C at (x_0, y_0), then there is a number λ such that

$$\text{grad } f(x_0, y_0) = \lambda \text{ grad } g(x_0, y_0) \qquad (2)$$

Proof If grad $f(x_0, y_0) = \mathbf{0}$, then (2) is satisfied with $\lambda = 0$. Thus for the rest of the proof we will assume that grad $f(x_0, y_0) \neq \mathbf{0}$. Let I be an interval and

$$\mathbf{r}(t) = x(t)\mathbf{i} + y(t)\mathbf{j} \quad \text{for } t \text{ in } I$$

a smooth parametrization of C. Let t_0 be such that $\mathbf{r}(t_0)$ corresponds to the point (x_0, y_0). Then t_0 is not an endpoint of I since (x_0, y_0) is not an endpoint of C. Finally, let F be defined by

$$F(t) = f(x(t), y(t)) \quad \text{for } t \text{ in } I$$

As in the proof of Theorem 13.16, the Chain Rule yields

$$F'(t) = \frac{dF}{dt} = \frac{\partial f}{\partial x}\frac{dx}{dt} + \frac{\partial f}{\partial y}\frac{dy}{dt} = \text{grad } f(x(t), y(t)) \cdot \mathbf{r}'(t)$$

Since f has an extreme value on C at $(x_0, y_0) = (x(t_0), y(t_0))$, it follows that F has an extreme value on I at t_0. Since F is differentiable on I and t_0 is not an endpoint of I it follows that $F'(t_0) = 0$. Therefore

$$0 = F'(t_0) = \text{grad } f(x_0, y_0) \cdot \mathbf{r}'(t_0)$$

But $\mathbf{r}'(t_0) \neq 0$ since $\mathbf{r}$ is a smooth parametrization of I, and grad $f(x_0, y_0) \neq \mathbf{0}$ by assumption. Thus grad $f(x_0, y_0)$ is perpendicular to $\mathbf{r}'(t_0)$, which itself is tangent to C. Therefore grad $f(x_0, y_0)$ is normal to C. But grad $g(x_0, y_0)$ is normal to C by Theorem 13.16. Consequently grad $f(x_0, y_0)$ and grad $g(x_0, y_0)$ are parallel; this yields (2). ■

The number λ in (2) is called a **Lagrange multiplier** for f and g. Observe that the equation in (2) is equivalent to the pair of equations

$$f_x(x_0, y_0) = \lambda g_x(x_0, y_0) \quad \text{and} \quad f_y(x_0, y_0) = \lambda g_y(x_0, y_0)$$

The method of determining extreme values by means of Lagrange multipliers proceeds as follows:

---

1. Assume that f has an extreme value on the level curve $g(x, y) = c$.
2. Solve the equations

constraint $g(x, y) = c$

$$\operatorname{grad} f(x, y) = \lambda \operatorname{grad} g(x, y) \qquad \begin{cases} f_x(x, y) = \lambda g_x(x, y) \\ f_y(x, y) = \lambda g_y(x, y) \end{cases}$$

3. Calculate the value of f at each point (x, y) that arises in step 2, and at each endpoint (if any) of the curve. If f has a maximum value on the level curve $g(x, y) = c$, it will be the largest of the values computed; if f has a minimum value on the level curve, it will be the smallest of the values computed.

Example 1   Let $f(x, y) = x^2 + 4y^3$. Find the extreme values of f on the ellipse $x^2 + 2y^2 = 1$.

Solution   Let

$$g(x, y) = x^2 + 2y^2$$

so that the constraint is $g(x, y) = x^2 + 2y^2 = 1$. Since

$$\operatorname{grad} f(x, y) = 2x\mathbf{i} + 12y^2\mathbf{j} \quad \text{and} \quad \operatorname{grad} g(x, y) = 2x\mathbf{i} + 4y\mathbf{j}$$

the equations we will use to find x and y are

constraint $x^2 + 2y^2 = 1$ (3)

$$\operatorname{grad} f(x, y) = \lambda \operatorname{grad} g(x, y) \qquad \begin{cases} 2x = 2x\lambda & (4) \\ 12y^2 = 4y\lambda & (5) \end{cases}$$

By (4), either $x = 0$ or $\lambda = 1$. If $x = 0$, then (3) implies that $2y^2 = 1$, so that either $y = 1/\sqrt{2}$ or $y = -1/\sqrt{2}$. If $\lambda = 1$, then (5) becomes $12y^2 = 4y$, which means that $y = 0$ or $y = \frac{1}{3}$. By (3),

$$\text{if } y = 0, \text{ then } x^2 + 2(0)^2 = 1, \quad \text{so} \quad x = 1 \text{ or } x = -1$$

$$\text{if } y = \frac{1}{3}, \text{ then } x^2 + 2\left(\frac{1}{3}\right)^2 = 1, \quad \text{so} \quad x = \frac{\sqrt{7}}{3} \text{ or } x = -\frac{\sqrt{7}}{3}$$

Thus the only possible extreme values of f occur at $(0, 1/\sqrt{2})$, $(0, -1/\sqrt{2})$, $(1, 0)$, $(-1, 0)$, $(\sqrt{7}/3, \frac{1}{3})$, and $(-\sqrt{7}/3, \frac{1}{3})$. Since

$$f\left(0, \frac{1}{\sqrt{2}}\right) = \sqrt{2} \qquad\qquad f(1, 0) = 1 = f(-1, 0)$$

$$f\left(0, -\frac{1}{\sqrt{2}}\right) = -\sqrt{2} \qquad f\left(\frac{\sqrt{7}}{3}, \frac{1}{3}\right) = \frac{25}{27} = f\left(-\frac{\sqrt{7}}{3}, \frac{1}{3}\right)$$

we conclude that the maximum value $\sqrt{2}$ of f occurs at $(0, 1/\sqrt{2})$ and the minimum value $-\sqrt{2}$ of f occurs at $(0, -1/\sqrt{2})$. ☐

Example 2 Let $f(x, y) = 3x^2 + 2y^2 - 4y + 1$. Find the extreme values of f on the disk $x^2 + y^2 \leq 16$.

Solution By the Maximum–Minimum Theorem f has extreme values on the disk, and it can have them on the boundary $x^2 + y^2 = 16$ and in the interior $x^2 + y^2 < 16$.

First, we use Lagrange multipliers to find the possible extreme values of f on the circle $x^2 + y^2 = 16$. Let $g(x, y) = x^2 + y^2$, so that the constraint is $g(x, y) = x^2 + y^2 = 16$. Since

$$\text{grad } f(x, y) = 6x\mathbf{i} + (4y - 4)\mathbf{j} \quad \text{and} \quad \text{grad } g(x, y) = 2x\mathbf{i} + 2y\mathbf{j}$$

the equations we will use to find x and y are

$$\text{constraint} \qquad\qquad\qquad x^2 + y^2 = 16 \qquad\qquad (6)$$

$$\text{grad } f(x, y) = \lambda \text{ grad } g(x, y) \qquad \begin{cases} 6x = 2x\lambda & (7) \\ 4y - 4 = 2y\lambda & (8) \end{cases}$$

By (7), either $x = 0$ or $\lambda = 3$. If $x = 0$, then it follows from (6) that $y = 4$ or $y = -4$. If $\lambda = 3$, then (8) becomes

$$4y - 4 = 6y, \quad \text{so that} \quad y = -2$$

Then (6) implies that

$$x^2 + (-2)^2 = 16, \quad \text{so that} \quad x = \sqrt{12} \quad \text{or} \quad x = -\sqrt{12}$$

Thus f can have its extreme values on the circle $x^2 + y^2 = 16$ only at $(0, 4)$, $(0, -4)$, $(\sqrt{12}, -2)$, or $(-\sqrt{12}, -2)$.

Turning to the interior of the disk, we find that

$$f_x(x, y) = 6x \quad \text{and} \quad f_y(x, y) = 4y - 4$$

so that $f_x(x, y) = 0 = f_y(x, y)$ only if $x = 0$ and $y = 1$. Thus f can also have an extreme value on the disk at $(0, 1)$. Finally, we calculate that

$$f(0, 4) = 17 \qquad f(\sqrt{12}, -2) = 53 = f(-\sqrt{12}, -2)$$

$$f(0, -4) = 49 \qquad\qquad f(0, 1) = -1$$

Our conclusion is that the maximum value of f on the disk $x^2 + y^2 \leq 16$ is 53 and the minimum value is -1. ☐

The Lagrange Method for Functions of Three Variables

Next we consider the problem of finding extreme values of a function of three variables subject to a constraint of the form $g(x, y, z) = c$. By an argument similar to that used for functions of two variables, it is possible to show that if f has such an extreme value at (x_0, y_0, z_0), then grad $f(x_0, y_0, z_0)$ and grad $g(x_0, y_0, z_0)$, if not $\mathbf{0}$, are both normal to the level surface $g(x, y, z) = c$ at (x_0, y_0, z_0), and hence are parallel to each other. Thus there is a number λ, again called a Lagrange multiplier, such that

$$\operatorname{grad} f(x_0, y_0, z_0) = \lambda \operatorname{grad} g(x_0, y_0, z_0)$$

To find the extreme values of f subject to the constraint $g(x, y, z) = c$, we follow the same approach as in steps 1–3 for functions of two variables:

1. Assume that f has an extreme value on the level surface $g(x, y, z) = c$.
2. Solve the equations

   constraint $\qquad\qquad\qquad\qquad\qquad g(x, y, z) = c$

   $$\operatorname{grad} f(x, y, z) = \lambda \operatorname{grad} g(x, y, z) \quad \begin{cases} f_x(x, y, z) = \lambda g_x(x, y, z) \\ f_y(x, y, z) = \lambda g_y(x, y, z) \\ f_z(x, y, z) = \lambda g_z(x, y, z) \end{cases}$$

3. Calculate $f(x, y, z)$ for each point (x, y, z) that arises from step 2. If f has a maximum (minimum) value on the level surface, it will be the largest (smallest) of the values computed.

In the next example we rework Example 6 of Section 13.8 using Lagrange multipliers.

Example 3   Let $V(x, y, z) = xyz$ for $x \geq 0$, $y \geq 0$, and $z \geq 0$. Find the maximum value of V subject to the constraint $2x + 2y + z = 108$.

Solution   Let $g(x, y, z) = 2x + 2y + z$, so the constraint is $g(x, y, z) = 2x + 2y + z = 108$. Because

$$\operatorname{grad} V(x, y, z) = yz\mathbf{i} + xz\mathbf{j} + xy\mathbf{k} \quad \text{and} \quad \operatorname{grad} g(x, y, z) = 2\mathbf{i} + 2\mathbf{j} + \mathbf{k}$$

the equations we will use to find x, y, and z are

constraint $\qquad\qquad\qquad\qquad\qquad 2x + 2y + z = 108 \qquad (9)$

$$\operatorname{grad} f(x, y, z) = \lambda \operatorname{grad} g(x, y, z) \quad \begin{cases} yz = 2\lambda & (10) \\ xz = 2\lambda & (11) \\ xy = \lambda & (12) \end{cases}$$

First we solve for λ in terms of x, y, and z in (10)–(12), obtaining

$$\lambda = \frac{yz}{2} = \frac{xz}{2} = xy \qquad (13)$$

Since $V(x, y, z) = 0$ if x, y, or z is 0, and since 0 is obviously not the maximum value of V subject to (9), we can assume that x, y, and z are different from 0. Then

(13) tells us that $x = y$ and $z = 2y$. Substituting for x and z in (9) yields

$$2y + 2y + 2y = 108, \quad \text{so that} \quad y = 18$$

Thus $x = 18$ and $z = 2(18) = 36$, and therefore $V(18, 18, 36)$ is the only possible extreme value of V subject to the constraint. Since we are assuming that V has a maximum value subject to the constraint, we conclude that $V(18, 18, 36) = 11{,}664$ is that value. $\square$

Example 4 Find the minimum distance from a point on the surface $xy + 2xz = 5\sqrt{5}$ to the origin.

Solution We could let

$$f_1(x, y, z) = \sqrt{x^2 + y^2 + z^2}$$

which represents the distance from (x, y, z) to the origin, and seek the minimum value of f_1 on the surface. However, the computations involved in using Lagrange multipliers will be simplified if we let

$$f(x, y, z) = x^2 + y^2 + z^2$$

and minimize f subject to the constraint

$$g(x, y, z) = xy + 2xz = 5\sqrt{5}$$

Notice that f and f_1 have extreme values at identical points, so using f instead of f_1 will not alter the point we find whose distance from the origin is minimum. Because

$$\operatorname{grad} f(x, y, z) = 2x\mathbf{i} + 2y\mathbf{j} + 2z\mathbf{k}$$

and

$$\operatorname{grad} g(x, y, z) = (y + 2z)\mathbf{i} + x\mathbf{j} + 2x\mathbf{k}$$

the equations we will use to find x, y, and z are

$$\text{constraint} \qquad\qquad\qquad xy + 2xz = 5\sqrt{5} \qquad (14)$$

$$\operatorname{grad} f(x, y, z) = \lambda \operatorname{grad} g(x, y, z) \qquad \begin{cases} 2x = (y + 2z)\lambda & (15) \\ 2y = x\lambda & (16) \\ 2z = 2x\lambda & (17) \end{cases}$$

Now λ cannot be 0, for then $x = y = z = 0$ by (15)–(17), so (14) would not hold. Since $\lambda \neq 0$, (16) can be rewritten as

$$x = \frac{2y}{\lambda} \qquad (18)$$

Substituting for x in (17) yields

$$2z = 2\left(\frac{2y}{\lambda}\right)\lambda$$

so that

$$z = 2y \qquad (19)$$

If y were 0, then by (18) and (19) we would have $x = 0$ and $z = 0$, so (14) would not hold. Thus $y \neq 0$. Using (18) and (19) to substitute for x and z in (15), we find that

$$2\left(\frac{2y}{\lambda}\right) = [y + 2(2y)]\lambda$$

so that since $y \neq 0$,

$$\frac{4}{\lambda} = 5\lambda, \quad \text{or} \quad \lambda^2 = \frac{4}{5} \tag{20}$$

Using (18) and (19) to substitute for x and z in (14), we obtain

$$5\sqrt{5} = \left(\frac{2y}{\lambda}\right)y + 2\left(\frac{2y}{\lambda}\right)(2y) = \frac{10}{\lambda}y^2 \tag{21}$$

From (21) we see that λ is positive. Therefore (20) implies that $\lambda = 2/\sqrt{5}$, and then (21) implies that $y = 1$ or $y = -1$. By (18) and (19),

$$\text{if } y = 1, \quad \text{then} \quad x = \frac{2y}{\lambda} = \sqrt{5} \text{ and } z = 2y = 2$$

$$\text{if } y = -1, \quad \text{then} \quad x = \frac{2y}{\lambda} = -\sqrt{5} \text{ and } z = 2y = -2$$

Consequently the only points on the surface $xy + 2xy = 5\sqrt{5}$ that can have a minimum distance from the origin are $(\sqrt{5}, 1, 2)$ and $(-\sqrt{5}, -1, -2)$. Since by assumption such a minimum distance exists and is the minimum value of f_1 on the surface, and since

$$f_1(\sqrt{5}, 1, 2) = \sqrt{10} = f_1(-\sqrt{5}, -1, -2)$$

the minimum distance from a point on the surface to the origin is $\sqrt{10}$.   □

EXERCISES 13.9

In Exercises 1–6 find the extreme values of f subject to the given constraint. In each case assume that the extreme values exist.

1. $f(x, y) = x + y^2; x^2 + y^2 = 4$
2. $f(x, y) = xy; (x + 1)^2 + y^2 = 1$
3. $f(x, y) = x^3 + 2y^3; x^2 + y^2 = 1$
4. $f(x, y, z) = y^3 + xz^2; x^2 + y^2 + z^2 = 1$
5. $f(x, y, z) = xyz; x^2 + y^2 + 4z^2 = 6$
6. $f(x, y, z) = xy + yz; x^2 + y^2 + z^2 = 8$

In Exercises 7–10 find the minimum value of f subject to the given constraint. In each case assume that the minimum value exists.

7. $f(x, y) = 4x^2 + y^3 + 3y + 7; 2x^2 + \frac{3}{2}y^2 = \frac{3}{2}$
8. $f(x, y, z) = x^2 + 2y^2 + z^2; x + y + z = 4$
9. $f(x, y, z) = x^4 + 8y^4 + 27z^4; x + y + z = \frac{11}{12}$
10. $f(x, y, z) = 3z - x - 2y; z = x^2 + 4y^2$

In Exercises 11–14 find the extreme values of f in the region described by the given inequalities. In each case assume that the extreme values exist.

11. $f(x, y) = 2x^2 + y^2 + 2y - 3; x^2 + y^2 \leq 4$
12. $f(x, y) = x^3 + x^2 + \frac{y^2}{3}; x^2 + y^2 \leq 36$
13. $f(x, y) = xy; 2x^2 + y^2 \leq 4$
14. $f(x, y) = 16 - x^2 - 4y^2; x^4 + 2y^4 \leq 1$

In the remaining exercises in this section, assume that the required extreme values exist.

15. Find the points on the surface $x^2 - yz = 1$ that are closest to the origin.
16. Find the points on the sphere $x^2 + y^2 + z^2 = 1$ that are closest to or farthest from the point $(4, 2, 1)$.
17. Let x and y denote the acute angles of a right triangle. Find the maximum value of $\sin x \sin y$.

18. Let x, y, and z denote the angles of an arbitrary triangle. Find the maximum value of $\sin x \sin y \sin z$.

19. Find the minimum volume of a tetrahedron bounded by the planes $x = 0$, $y = 0$, $z = 0$, and a plane tangent to the sphere $x^2 + y^2 + z^2 = 1$. (*Hint:* If the plane is tangent to the sphere at the point (x_0, y_0, z_0), then the volume of the tetrahedron is $1/(6x_0y_0z_0)$.)

20. A rectangular parallelepiped lies in the first octant, with three sides on the coordinate planes and one vertex on the plane $2x + y + 4z = 12$. Find the maximum possible volume of the parallelepiped.

21. A rectangular storage box with volume 12 cubic inches is to be made in the form shown in Figure 13.51. Find the dimensions that will minimize the total surface area of the box. (*Hint:* Solve for λ in each equation you obtain.)

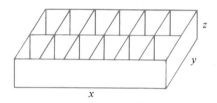

FIGURE 13.51

22. A rectangular box, open at the top, is to have a volume of 1728 cubic inches. Find the dimensions that will minimize the cost of the box if
    a. the material for the bottom costs 16 times as much per unit area as the material for the sides
    b. the material for the bottom costs twice as much per unit area as the material for the sides

23. A cylindrical pipe of radius r and length l must be slid on the floor from one corridor to a perpendicular corridor, each 3 meters wide (Figure 13.52). Find the dimensions of

the pipe that maximize its volume V. (*Hint:* As in Figure 13.52, assume that the opposite ends of the pipe touch the walls when the angle between the pipe and the wall is $\pi/4$.)

24. Suppose that on your vacation you plan to spend x days in San Francisco, y days in your home town, and z days in New York. You calculate that your total enjoyment $f(x, y, z)$ will be given by

$$f(x, y, z) = 2x + y + 2z$$

If plans and financial limitations dictate that

$$x^2 + y^2 + z^2 = 225$$

how long should each stay be to maximize your enjoyment?

25. The ground state energy $E(x, y, z)$ of a particle of mass m in a rectangular box with dimensions x, y, and z is given by

$$E(x, y, z) = \frac{h^2}{8m}\left(\frac{1}{x^2} + \frac{1}{y^2} + \frac{1}{z^2}\right)$$

where h is a constant. Assuming that the volume V of the box is fixed, find the values of x, y, and z that minimize the value of E.

26. The object distance p, image distance q, and local length f of a simple lens satisfy the equation

$$\frac{1}{p} + \frac{1}{q} = \frac{1}{f}$$

Determine the minimum distance $p + q$ between the object and the image for a given focal length.

27. Fermat's Principle states that light always travels the path between points requiring the least time. Suppose light travels from a point A in one medium in which it has velocity v to a point B in a second medium in which it has velocity u. Using the fact that in a single medium, light travels in a straight line, and using Figure 13.53, show that

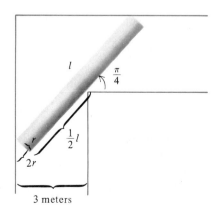

FIGURE 13.52

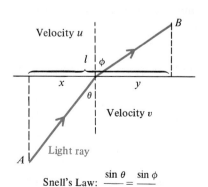

FIGURE 13.53

the light is bent according to Snell's Law:

$$\frac{\sin\theta}{v} = \frac{\sin\phi}{u}$$

(*Hint:* The constraint is $x + y = l$, where x and y are the distances indicated in Figure 13.53.)

28. A pharmaceutical company plans to make capsules containing a given volume V of medicine. One executive would like to have the capsules in the form of a right circular cylinder having length h and base radius r with a hemisphere at each end (see Figure 13.54). A second executive objects to the wastefulness of materials, and contends that the same volume could be contained in a spherical capsule having a smaller surface area. Which executive should get the next promotion?

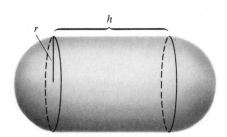

FIGURE 13.54

29. Let x represent capital and y labor in the manufacture of $f(x, y)$ units of a given product. Assume that capital costs a dollars per unit and labor costs b dollars per unit and that there are c dollars available, so that $ax + by = c$.
   a. Using Lagrange multipliers, show that production is maximum at the point (x_0, y_0) such that

   $$\frac{f_x(x_0, y_0)}{a} = \frac{f_y(x_0, y_0)}{b} = \lambda$$

   where λ is the Lagrange multiplier, called the **equimarginal productivity** of the production function f.

b. Let f be the Cobb–Douglas production function given by

$$f(x, y) = x^\alpha y^\beta \quad \text{for } x > 0 \text{ and } y > 0$$

where α and β are positive constants less than 1. Using (a), show that at maximum production $f(x_0, y_0)$ we have

$$\frac{y_0}{x_0} = \frac{\beta a}{\alpha b}$$

which is independent of the money available.

*30. Let x, y, and z be positive numbers. Show that the geometric mean $(xyz)^{1/3}$ of x, y, and z is less than or equal to their arithmetic mean $(x + y + z)/3$. (*Hint:* Maximize $(xyz)^{1/3}$ subject to the constraint $x + y + z = c$, where c is a fixed number.)

Lagrange multipliers can be used to find the extreme values of functions subject to more than one constraint. Suppose we wish to find the extreme values of a function f of three variables satisfying the two constraints

$$g_1(x, y, z) = c_1 \quad \text{and} \quad g_2(x, y, z) = c_2$$

where c_1 and c_2 are constants. The method is to solve the equations

$$\text{grad } f(x, y, z) = \lambda \text{ grad } g_1(x, y, z) + \mu \text{ grad } g_2(x, y, z)$$

for (x, y, z) and for λ and μ if necessary and then to determine the largest and smallest values of f. Both λ and μ are called Lagrange multipliers. Use this method to solve Exercises 31–32.

31. Find the minimum distance between the origin and a point on the intersection of the paraboloid $z = \frac{3}{2} - x^2 - y^2$ and the plane $x + 2y = 1$.

32. Find the distance from the point $(2, -2, 3)$ to the intersection of the planes

$$2x - y + 3z = 1 \quad \text{and} \quad -x + 3y + z = -3$$

Key Terms and Expressions

Function of several variables; graph of a function of several variables
Trace in a plane
Level curve; level surface
Quadric surface
Limits and continuity of a function of several variables; continuity on a set

Partial derivative; second partial derivative; mixed partial derivative
Differentiability at a point
Directional derivative
Gradient
Direction of most rapid increase
Tangent plane

Tangent plane approximation
Differential
Extreme value of a function of several variables; extreme value on a set; relative extreme value

Critical point of a function of several variables
Saddle point
Lagrange multiplier

Key Formulas

$$f_x(x_0, y_0) = \lim_{h \to 0} \frac{f(x_0 + h, y_0) - f(x_0, y_0)}{h}$$

$$f_y(x_0, y_0) = \lim_{h \to 0} \frac{f(x_0, y_0 + h) - f(x_0, y_0)}{h}$$

$$f_x(x_0, y_0, z_0) = \lim_{h \to 0} \frac{f(x_0 + h, y_0, z_0) - f(x_0, y_0, z_0)}{h}$$

$$f_y(x_0, y_0, z_0) = \lim_{h \to 0} \frac{f(x_0, y_0 + h, z_0) - f(x_0, y_0, z_0)}{h}$$

$$f_z(x_0, y_0, z_0) = \lim_{h \to 0} \frac{f(x_0, y_0, z_0 + h) - f(x_0, y_0, z_0)}{h}$$

$$\text{grad } f(x_0, y_0) = \nabla f(x_0, y_0) = f_x(x_0, y_0)\mathbf{i} + f_y(x_0, y_0)\mathbf{j}$$

$$\text{grad } f(x_0, y_0, z_0) = \nabla f(x_0, y_0, z_0) = f_x(x_0, y_0, z_0)\mathbf{i} + f_y(x_0, y_0, z_0)\mathbf{j} + f_z(x_0, y_0, z_0)\mathbf{k}$$

$$D_{\mathbf{u}}f(x_0, y_0) = [\text{grad } f(x_0, y_0)] \cdot \mathbf{u}$$

$$D_{\mathbf{u}}f(x_0, y_0, z_0) = [\text{grad } f(x_0, y_0, z_0)] \cdot \mathbf{u}$$

$$z = f(x_0, y_0) + f_x(x_0, y_0)(x - x_0) + f_y(x_0, y_0)(y - y_0)$$

Key Theorems

Chain Rule
Second Partials Test
Maximum–Minimum Theorem for Two Variables

REVIEW EXERCISES

In Exercises 1–3 find the domain of the function.

1. $f(x, y) = \sqrt{1 - \frac{1}{4}x^2 - \frac{1}{25}y^2}$

2. $g(u, v) = \arcsin(u - v)$

3. $k(x, y, z) = \ln(x - y + z)$

In Exercises 4–6 find the level curve $f(x, y) = c$.

4. $f(x, y) = y - x + 4$; $c = -1, 6$

5. $f(x, y) = xy$; $c = 1, -1$

6. $f(x, y) = \dfrac{x^2 - y^2}{x^2 + y^2}$; $c = \frac{1}{2}$

In Exercises 7–8 sketch the graph of the level surface $f(x, y, z) = c$.

7. $f(x, y, z) = \sqrt{4 - x^2 - y^2} - z$; $c = 1, -1$

8. $f(x, y, z) = 4z - x^2 - y^2$; $c = 2, 0$

In Exercises 9–13 sketch the quadric surface.

9. $x^2 = 4y^2 + z^2$

10. $z = 2x^2 - y^2$

11. $z = 2x^2 + y^2$

12. $x = y^2$

13. $y^2 - x^2 - z^2 = 8$

In Exercises 14–16 evaluate the limit.

14. $\displaystyle \lim_{(x, y) \to (0, 0)} \frac{y^2}{\sqrt{x^2 + y^2}}$

15. $\displaystyle \lim_{(x, y) \to (-2, \sqrt{2})} \frac{x^4 + x^2y^2 - 6y^4}{x^2 - 2y^2}$

16. $\displaystyle \lim_{(x, y, z) \to (-1, 1, 2)} \frac{2x^2 + 4xy - 6x^3z^2}{xyz^2}$

In Exercises 17–21 find the first partial derivatives.

17. $f(x, y) = 4x^3 - 3y^2$

18. $g(x, y) = \dfrac{x - y}{x^2 + 2y^2}$

19. $f(x, y, z) = e^{x^2} \ln(y^2 - 3z)$

20. $f(x, y, z) = \dfrac{\cos z^4}{xy^2}$

21. $k(x, y, z) = [\sqrt{z} \tan(x^2 + y)]^{5/2}$

In Exercises 22–23 show that $f_y = g_x$.

22. $f(x, y) = x - \cos y$, $g(x, y) = x \sin y$

23. $f(x, y) = y + 2xe^y$, $g(x, y) = x + x^2 e^y$

In Exercises 24–26 find all second partials.

24. $f(x, y) = \arcsin(x^2 - y^2)$

25. $g(u, v) = \ln \dfrac{u^2}{e^v}$

26. $f(x, y, z) = x \sin yz^2$

27. Let $z = e^{-ay} \cos bx$. Show that

$$a^2 \frac{\partial^2 z}{\partial x^2} + b^2 \frac{\partial^2 z}{\partial y^2} = 0$$

28. Let $z = \sqrt{x^2 + y^2}$. Show that

$$\frac{\partial^2 z}{\partial x^2} \frac{\partial^2 z}{\partial y^2} = \left(\frac{\partial^2 z}{\partial x \, \partial y} \right)^2$$

29. Compute dz/dt if $z = e^{x^2} - e^{y/2}$, $x = t^2$, $y = t^3 - t$.

30. Compute $\partial z/\partial u$ and $\partial z/\partial v$ if $z = \sin xy - y^2 \cos x$, $x = u^2 v$, $y = 1/v$.

31. Compute dw/dt if $w = \sqrt{x^2 + y^2 z^4}$, $x = 2t$, $y = t^3$, $z = 1/t$.

32. Compute $\partial w/\partial u$ and $\partial w/\partial v$ if $w = x^2 + y \sin yz$, $x = u^2 + v^2$, $y = uv$, $z = u^2 - v^2$.

33. Let $z = f(x^2 + y^2)$. Show that

$$y \frac{\partial z}{\partial x} - x \frac{\partial z}{\partial y} = 0$$

34. Let $\arctan xy + \arcsin xy = \pi/2$. Find dy/dx by implicit differentiation.

35. Let $xyz + 1/xyz = z^3$. Use implicit partial differentiation to find $\partial z/\partial x$ and $\partial z/\partial y$.

In Exercises 36–38 find the directional derivative of f at the given point P in the direction of $\mathbf{a}$.

36. $f(x, y) = 4 - x^2 + 3y^2 + y$; $P = (-1, 0)$;
$\mathbf{a} = (-1/\sqrt{2})\mathbf{i} + (1/\sqrt{2})\mathbf{j}$

37. $f(x, y, z) = 1/(x^2 + y^2 + z^2)$; $P = (-1, 0, 2)$;
$\mathbf{a} = \mathbf{i} - \mathbf{j} - \mathbf{k}$

38. $f(x, y, z) = \csc(yz + x)$; $P = (\pi, -\pi/4, 1)$;
$\mathbf{a} = -3\mathbf{i} - \mathbf{j} - 2\mathbf{k}$

In Exercises 39–40 find the gradient of the function at the given point.

39. $f(x, y) = e^{2x} \ln y$; $(0, 1)$

40. $f(x, y, z) = x^2 \cos y \sin\left(\dfrac{\pi}{2} \sin z \right)$; $(1, -\pi/6, \pi/3)$

41. Let $f(x, y) = \cos xy$. Find the direction in which f is increasing most rapidly at $(\frac{1}{2}, \pi)$.

42. Let $f(x, y, z) = xye^z$. Find the direction in which f is increasing most rapidly at $(2, 3, 0)$.

In Exercises 43–44 find a vector normal to the graph of the equation at the given point. Assume in each case that the graph is a smooth curve.

43. $\arctan(x^2 + y) = \pi/4$; $(\frac{1}{2}, \frac{3}{4})$

44. $x^4 - 3x^2 y + 2y^3 = 11$; $(-1, 2)$

In Exercises 45–47 find a vector normal to the graph of f at the given point and an equation of the plane tangent to the graph of f at the given point.

45. $f(x, y) = 1/x - 1/y$; $(-\frac{1}{2}, \frac{1}{3}, -5)$

46. $f(x, y) = ye^x$; $(0, 1, 1)$

47. $f(x, y) = \sqrt{3x^2 + 2y^2 + 2}$; $(4, -5, 10)$

In Exercises 48–50 find an equation of the plane tangent to the level surface at the given point.

48. $x^2 - y^2 - z^2 = 1$; $(3, -2, 2)$

49. $xe^{yz} - 2y = -1$; $(1, 1, 0)$

50. $\sin xy + \cos yz = 0$; $(0, \pi/2, 1)$

51. Find the points on the surface $2x^3 + y - z^2 = 5$ at which the tangent plane is parallel to the plane $24x + y - 6z = 3$.

52. Show that there are exactly two planes tangent to the sphere $x^2 + y^2 + z^2 = 1$ and parallel to any given plane.

53. Show that the surfaces $z = x^2 + 4y^2$ and $z = 4x + y^2 - 4$ have the same tangent plane at $(2, 0, 4)$.

54. Show that the plane tangent to the elliptic hyperboloid

$$\frac{x^2}{a^2} - \frac{y^2}{b^2} + \frac{z^2}{c^2} = 1$$

at (x_0, y_0, z_0) has equation

$$\frac{xx_0}{a^2} - \frac{yy_0}{b^2} + \frac{zz_0}{c^2} = 1$$

55. Let $f(x, y) = \arctan \dfrac{x}{1 + y}$. Approximate $f(0.97, 0.05)$.

56. Let $f(x, y, z) = \sqrt{3x^2 + y^2 + 5z^2}$. Find an approximation for $f(2.9, 2.1, 0.9)$.

57. Let $f(x, y) = \ln x/y$. Find df.

58. Let $f(x, y, z) = x^2 y - e^{(z^2)} \cos yz$. Find df.

In Exercises 59–62 find all critical points. Specify which yield relative maximum values, which yield relative minimum values, and which yield saddle points.

59. $f(x, y) = x^2 + y^2 - 2x - 4y + 5$

60. $f(x, y) = x^2 + y - \dfrac{1}{2y^2}$

61. $f(x, y) = xy + \dfrac{8}{x^2} + \dfrac{8}{y^2}$

62. $f(x, y) = -2x^2 + xy + y^2 - 4x + 3y - 1$

In Exercises 63–66 find all extreme values of f subject to the given constraint. Assume that the extreme values exist.

63. $f(x, y) = 3x^2 - xy + y^2$; $3x^2 + y^2 = 3$

64. $f(x, y) = x^2 + y^2 - 2x - 4y - 6$; $x^2 + y^2 \le 16$

65. $f(x, y, z) = \dfrac{x^2 + y^2}{z^2 + 5}$; $x^2 + y^2 - z = 2$

66. $f(x, y, z) = \dfrac{x^2 + y^2}{z^2 + 5}$; $x^2 + y^2 - 2 \le z \le 6$

67. Find the dimensions of the rectangular parallelepiped with faces parallel to the coordinate planes whose volume is the largest of any inscribed in the ellipsoid $x^2 + 4y^2 + 9z^2 = 36$. Assume that there exists a largest volume.

68. The image distance q from a lens is related to the object distance p and the focal length f of the lens by

$$q = \frac{pf}{p - f}$$

Show that $q_f(p, f) > q_p(p, f)$ for all (p, f) in the domain of q.

69. Consider the pendulum shown in Figure 13.55. The angle θ of deflection depends on time; for convenience we denote $d\theta/dt$ by $\dot\theta$ and $d^2\theta/dt^2$ by $\ddot\theta$ in this exercise. If the pendulum has small vibrations, then the kinetic energy T and the potential energy V of the pendulum are given

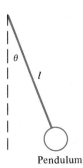

Pendulum

FIGURE 13.55

(approximately) by

$$T = \frac{ml^2\dot\theta^2}{2} \quad \text{and} \quad V = \frac{mgl\theta^2}{2} \tag{1}$$

where g is the acceleration due to gravity, l is the length of the pendulum cord, and m is the mass of the object at the bottom of the pendulum. It can be shown that T and V satisfy **Lagrange's equation**

$$\frac{d}{dt}\frac{\partial T}{\partial \dot\theta} + \frac{\partial V}{\partial \theta} = 0$$

Using (1), show that Lagrange's equation reduces to

$$\ddot\theta + \frac{g}{l}\theta = 0$$

*70. A farmer wishes to employ tomato pickers to harvest tomatoes. Each picker can harvest 625 tomatoes per hour and is paid \$6 per hour. In addition, the farmer must pay a supervisor \$10 per hour and the union \$10 for each picker employed. Finally, if V tomatoes are picked, then a service charge of $\$50{,}000/\sqrt{V}$ is levied against the farmer. Show that the total cost to the farmer is minimum if five pickers are employed, and determine the volume V that would be picked at this minimum cost.

Cumulative Review, Chapters 1–12

In Exercises 1–2 find the limit.

1. $\displaystyle\lim_{h \to 0} \frac{\sqrt{x - h} - \sqrt{x}}{h}$

2. $\displaystyle\lim_{x \to 0^+} (\arctan x)^x$

3. Consider the parabola $y = 4x^2$. Find an equation for each tangent to the parabola that passes through the point $(0, -2)$.

4. Let $f(x) = \ln(x + \sqrt{1 + x^2})$. Show that $\displaystyle\lim_{x \to \infty} xf'(x) = 1$.

5. Let n be a positive integer, and $f(x) = (\sin nx)(\sin^n x)$. Show that $f'(x) = n(\sin^{n-1} x)(\sin(n + 1)x)$.

6. Let $f(x) = \int_{x^2}^{\pi/2} \sin \sqrt{t}\, dt$. Find $f'(x)$.

7. Suppose that $\sqrt{x} + \sqrt{y} = 9$. Use implicit differentiation to find d^2y/dx^2.

8. The sides of an equilateral triangle grow at the rate of 2 centimeters per minute. An inscribed circle is attached to the triangle and grows with the triangle. Determine

the rate at which the area of the region between the circle and the triangle grows when the sides of the triangle are 12 centimeters long. (*Hint:* The radius of the circle is $\sqrt{3}/6$ times as long as a side of the triangle.)

9. Let

$$f(x) = \frac{2 - x^2}{1 - x^2}$$

Sketch the graph of f, indicating all pertinent information.

10. A manufacturer plans to create a rectangular box with square bottom and top out of 960 square inches of cardboard. If the bottom will have 3 layers of cardboard and the top 2 layers, determine the dimensions that will maximize the volume V. (Assume that the cardboard has negligible thickness.)

11. Find the area A of the region between the curves $y = 2x^3 + 2x^2 + 10x$ and $y = x^3 - 3x^2 + 4x$.

In Exercises 12–14 evaluate the integral.

12. $\displaystyle\int x^2 \arctan x \, dx$ 13. $\displaystyle\int (\cos^2 \theta - \cos^3 \theta) \, d\theta$

14. $\displaystyle\int \left[x(8 - 2x^2)\sqrt{4 - x^2} - \frac{2}{3}(4 - x^2)^{3/2} \right] dx$

15. Determine whether

$$\int_1^\infty \frac{x^3}{(3 + x^4)^{3/2}} \, dx$$

converges or diverges. If it converges, determine its value.

16. A football-like solid is obtained by revolving around the x axis the graph of $y = x \cos x$ for $0 \le x \le \pi/2$. Determine the volume V of the solid.

17. Let R be the region in the xy plane between the graphs of $y = 2 - x^2$ and $y = 1$. Determine the center of gravity of R.

18. Sketch the polar graph of $r = 1 - 2 \sin \theta$.

19. Find the area A of the region inside the circle $r = \sin \theta$ and outside the circle $r = \sqrt{3} \cos \theta$.

20. Evaluate $\displaystyle\lim_{n \to \infty} \frac{(\sin n!) \ln n}{n}$.

21. Find the numerical value of $\displaystyle\sum_{n=1}^\infty \left(\frac{1}{\sqrt{n}} - \frac{1}{\sqrt{n+1}} \right)$.

22. Determine the interval of convergence of $\displaystyle\sum_{n=1}^\infty \frac{x^n}{n 4^n}$.

23. Let $f(x) = 2x^3 e^{-(x^3)}$. Find the Taylor series of f and f', and determine their radii of convergence.

24. Determine which, if either, is farther from the plane $x - 2y + 3z = 6$: the point $(-1, 3, 4)$ or the point $(1, 2, 0)$.

25. A curve C is parametrized by $\mathbf{r}(t) = t\mathbf{i} + t^2\mathbf{j} + \frac{2}{3}t^3\mathbf{k}$.
   a. Find the value of t for which the tangent vector $\mathbf{T}(t)$ is parallel to the vector $\mathbf{i} - \mathbf{j} + \frac{1}{2}\mathbf{k}$.
   b. Find the curvature $\kappa(t)$ for an arbitrary value of t, and show that it is maximum for $t = 0$.

14

MULTIPLE INTEGRALS

Our definition in Chapter 5 of the integral $\int_a^b f(x)\,dx$ of a function that is continuous on an interval $[a, b]$ was motivated by means of area. Integrals of functions of two or three variables, which we will define in this chapter, will enable us to compute areas of more complicated regions, volumes of many types of solid regions, and masses and centers of gravity of two- and three-dimensional objects. The proofs of the theorems stated in this chapter are complicated and normally are given only in advanced calculus books; thus we will omit them.

14.1
DOUBLE INTEGRALS

From now on we will refer to the definite integral $\int_a^b f(x)\,dx$ of a function f that is continuous on an interval $[a, b]$ as a **single integral**. In this section we will define the double integral of a function that is continuous on a certain type of plane region. The definition of the double integral will be motivated by means of certain elementary properties we expect volume to possess.

Volume and the Double Integral

Consider a region R in the xy plane, a function f that is nonnegative and continuous on R, and the solid region D shown in Figure 14.1, bounded below by R, above by the graph of f, and on the sides by the vertical surface passing through the boundary of R. We call D the **solid region between the graph of f and R**. Our goal is to define the volume of D.

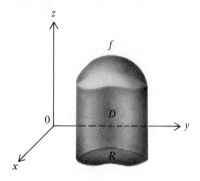

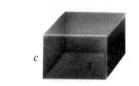

FIGURE 14.1

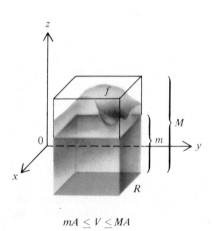

FIGURE 14.2

The first and basic assumption is that if a rectangular parallelepiped has height c and base area A, then its volume V is given by

$$V = cA$$

(Figure 14.2). This is a three-dimensional version of the Rectangle Property of single integrals.

Next we consider the case in which f is nonnegative and continuous on a rectangle R. Then f has a minimum value m and a maximum value M on R. Consequently the solid region D between the graph of f and R contains a rectangular parallelepiped having height m and base R and is contained in a rectangular parallelepiped having height M and base R (Figure 14.3). Accordingly, the volume V of D should lie between the volumes of these two parallelepipeds. Thus if A is the area of R, then

$$mA \le V \le MA$$

This feature of volume is analogous to the Comparison Property for single integrals.

Next we partition R into subrectangles $R_1, R_2, \ldots, R_n$. For each integer k between 1 and n, let m_k be the minimum value and M_k the maximum value of f on R_k, and let ΔA_k be the area of R_k. Then the rectangular parallelepiped with height m_k and base R_k is entirely contained in D (Figure 14.4(a)). Consequently the volume V of D should be at least as large as the sum of the volumes of the n parallelepipeds:

$$m_1 \Delta A_1 + m_2 \Delta A_2 + \cdots + m_n \Delta A_n \le V \tag{1}$$

On the other hand, the rectangular parallelepipeds with bases $R_1, R_2, \ldots, R_n$ and heights $M_1, M_2, \ldots, M_n$ together contain D (Figure 14.4(b)), so the sum of their volumes should be at least as large as the volume V of D:

$$V \le M_1 \Delta A_1 + M_2 \Delta A_2 + \cdots + M_n \Delta A_n \tag{2}$$

Combining (1) and (2), we find that

$$m_1 \Delta A_1 + m_2 \Delta A_2 + \cdots + m_n \Delta A_n \le V \le M_1 \Delta A_1 + M_2 \Delta A_2 + \cdots + M_n \Delta A_n$$

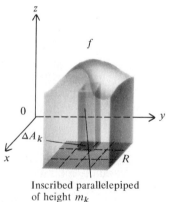

$$mA \le V \le MA$$

FIGURE 14.3

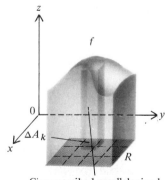

Inscribed parallelepiped of height m_k

(a)

Circumscribed parallelepiped of height M_k

(b)

FIGURE 14.4

Here we use the fact that we can add the volumes of separate solid regions to obtain the volume of the combined region. It seems reasonable to assume that we can make the two sums in the preceding inequalities as close to V as we wish by taking small enough subrectangles of R.

Finally, we relax the assumption that R is a rectangle and assume only that R is bounded, that is, that R is contained in a rectangle R'. We partition R' into a collection $\mathscr{P}$ of rectangles. In general, some of the rectangles in $\mathscr{P}$ will be entirely contained in R, some only partially contained in R, and some will contain no points of R (Figure 14.5(a)). We number the rectangles in $\mathscr{P}$ so that $R_1, R_2, \ldots, R_n$ are those entirely contained in R, while $R_{n+1}, R_{n+2}, \ldots, R_p$ are those only partially contained in R and the remaining rectangles contain no points of R. If R_k is entirely contained in R, let m_k be the minimum value of f on R_k, and if R_k is partially or totally contained in R, let M_k be the maximum value of f on R_k. Finally, let ΔA_k be the area of any rectangle R_k. Then the volume V of D should be at least as large as the sum of the volumes of the n inscribed parallelepipeds (Figure 14.5(a)) and no larger than the sum of the volumes of the p parallelepipeds that together circumscribe R (Figure 14.5(b)):

$$m_1 \Delta A_1 + m_2 \Delta A_2 + \cdots + m_n \Delta A_n \leq V \leq M_1 \Delta A_1 + M_2 \Delta A_2 + \cdots + M_p \Delta A_p$$

The sum on the left side is called the **lower sum** of f for $\mathscr{P}$ and is denoted $L_f(\mathscr{P})$. The sum on the right side is called the **upper sum** of f for $\mathscr{P}$ and is denoted $U_f(\mathscr{P})$. In this notation, the inequalities can be rewritten

$$L_f(\mathscr{P}) \leq V \leq U_f(\mathscr{P}) \tag{3}$$

If there is a unique number that lies between every lower sum and every upper sum, then it is natural to define the volume V of D to be that number.

It is possible to extend the definitions of $L_f(\mathscr{P})$ and $U_f(\mathscr{P})$ to cover the case in which f is not necessarily nonnegative, and in such a way that $L_f(\mathscr{P}) \leq U_f(\mathscr{P})$. We can therefore make the following definition.

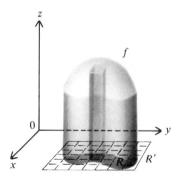

Inscribed parallelepiped

(a)

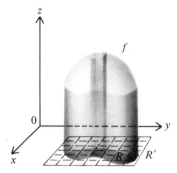

Circumscribed parallelepiped

(b)

FIGURE 14.5

DEFINITION 14.1

Let R be a bounded region in the xy plane and f a function continuous on R.

a. If there is a unique number I satisfying

$$L_f(\mathscr{P}) \leq I \leq U_f(\mathscr{P})$$

for every partition $\mathscr{P}$ of any rectangle R' containing R, then f is **integrable** on R. We denote the unique number by

$$\iint\limits_R f(x, y)\, dA$$

and call it the **double integral** of f over R.

b. If f is nonnegative and integrable on R, then the **volume** V of the solid region between the graph of f and R is given by

$$V = \iint\limits_R f(x, y)\, dA$$

The sums $L_f(\mathscr{P})$ and $U_f(\mathscr{P})$ approximate $\iint_R f(x, y)\,dA$. Other sums also approximate $\iint_R f(x, y)\,dA$. More specifically, let $\mathscr{P}$ be a partition of a rectangle containing R into subrectangles, numbered so that $R_1, R_2, \ldots, R_n$ are those entirely contained in R. For each integer k between 1 and n, let (x_k, y_k) be a point in R_k. Then the sum

$$\sum_{k=1}^{n} f(x_k, y_k)\,\Delta A_k$$

is called a **Riemann sum** for f on R. Notice that lower sums are Riemann sums, but upper sums need not be, since the definition of Riemann sum does not provide for rectangles that are only partially contained in R.

The following theorem, which resembles Theorem 5.6, states that Riemann sums can approximate the double integral as accurately as we like if we choose the subrectangles small enough.

THEOREM 14.2

Let f be integrable on a bounded region R, and let R' be a rectangle containing R. For any $\varepsilon > 0$ there is a number $\delta > 0$ such that the following statement holds.

If $\mathscr{P}$ is a partition of R' into subrectangles whose dimensions are all less than δ, and if $R_1, R_2, \ldots, R_n$ are those subrectangles contained in R, then

$$\left| \iint_R f(x, y)\,dA - \sum_{k=1}^{n} f(x_k, y_k)\,\Delta A_k \right| < \varepsilon$$

where the point (x_k, y_k) is arbitrarily chosen in R_k for $1 \leq k \leq n$.

The conclusion of Theorem 14.2 is usually expressed as

$$\iint_R f(x, y)\,dA = \lim_{\|\mathscr{P}\| \to 0} \sum_{k=1}^{n} f(x_k, y_k)\,\Delta A_k$$

where $\|\mathscr{P}\|$ denotes the largest of the dimensions of the rectangles in $\mathscr{P}$ and is called the **norm of the partition** $\mathscr{P}$.

It is practically impossible, except in very special cases, to evaluate a double integral by computing lower and upper sums, or even Riemann sums. The next example will give an idea of the complexity of such computations.

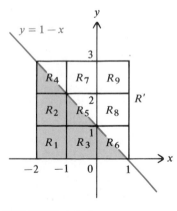

FIGURE 14.6

Example 1   Let R be the triangular region between the graph of $y = 1 - x$ and the x axis on $[-2, 1]$ (shaded in Figure 14.6), and suppose that

$$f(x, y) = 4 - y \quad \text{for } (x, y) \text{ in } R$$

Let R' and the partition $\mathscr{P} = \{R_1, R_2, \ldots, R_9\}$ of R' be as shown in Figure 14.6. Find $L_f(\mathscr{P})$ and $U_f(\mathscr{P})$. Then find the Riemann sum for f that results from choosing (x_k, y_k) to be the midpoint of R_k for $1 \leq k \leq 3$.

Solution   Notice that each of the rectangles has area 1, so that $\Delta A_k = 1$ for all k. But of the nine rectangles, only R_1, R_2, and R_3 are entirely contained in R, and only $R_1, R_2, \ldots, R_8$ contain points of R. Thus $n = 3$ and $p = 8$. Since

$$m_1 = 3, \qquad m_2 = 2, \qquad m_3 = 3$$

it follows that

$$L_f(\mathscr{P}) = \sum_{k=1}^{3} m_k \, \Delta A_k = 3 \cdot 1 + 2 \cdot 1 + 3 \cdot 1 = 8$$

Since
$$M_1 = 4, \qquad M_2 = 3, \qquad M_3 = 4, \qquad M_4 = 2,$$
$$M_5 = 3, \qquad M_6 = 4, \qquad M_7 = 2, \qquad M_8 = 3$$

we deduce that

$$U_f(\mathscr{P}) = \sum_{k=1}^{8} M_k \, \Delta A_k$$
$$= 4 \cdot 1 + 3 \cdot 1 + 4 \cdot 1 + 2 \cdot 1 + 3 \cdot 1 + 4 \cdot 1 + 2 \cdot 1 + 3 \cdot 1 = 25$$

Since only the three rectangles R_1, R_2, and R_3 in the partition are contained in R, the required Riemann sum is

$$\sum_{k=1}^{3} f(x_k, y_k) \, \Delta A_k = f\left(-\frac{3}{2}, \frac{1}{2}\right) \Delta A_1 + f\left(-\frac{3}{2}, \frac{3}{2}\right) \Delta A_2 + f\left(-\frac{1}{2}, \frac{1}{2}\right) \Delta A_3$$
$$= \frac{7}{2} \cdot 1 + \frac{5}{2} \cdot 1 + \frac{7}{2} \cdot 1 = \frac{19}{2} \quad \square$$

Vertically and Horizontally Simple Regions

As we have mentioned, it is difficult to compute double integrals by lower sums, upper sums, and Riemann sums; it also happens that there are regions on which not every continuous function is integrable. Hence we restrict our attention to two particular types of regions, on which every continuous function is integrable and for which we have a straightforward method of computing double integrals.

DEFINITION 14.3

> **a.** A plane region R is **vertically simple** if there are two continuous functions g_1 and g_2 on an interval $[a, b]$ such that $g_1(x) \leq g_2(x)$ for $a \leq x \leq b$ and such that R is the region between the graphs of g_1 and g_2 on $[a, b]$ (Figure 14.7). In this case we say that R is **the vertically simple region between the graphs of g_1 and g_2 on $[a, b]$**.
>
> **b.** A plane region R is **horizontally simple** if there are two continuous functions h_1 and h_2 on an interval $[c, d]$ such that $h_1(y) \leq h_2(y)$ for $c \leq y \leq d$ and such that R is the region between the graphs of h_1 and h_2 on $[c, d]$ (Figure 14.8). In this case we say that R is **the horizontally simple region between the graphs of h_1 and h_2 on $[c, d]$**.
>
> **c.** A plane region R is **simple** if it is both vertically simple and horizontally simple (Figure 14.9).

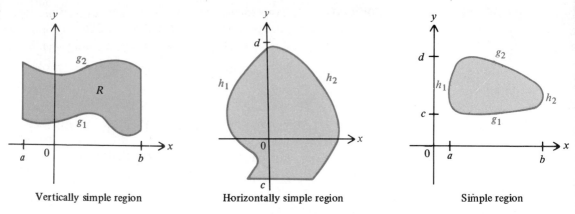

Vertically simple region

FIGURE 14.7

Horizontally simple region

FIGURE 14.8

Simple region

FIGURE 14.9

Vertical lines intersect the boundary of a vertically simple region R at most twice, except for those vertical lines composing part of the boundary of R. Rectangles, triangles, and circles are vertically simple, as are the regions depicted in Figure 14.10(a) and (b); however, the regions in Figure 14.10(c) and (d) are not vertically simple. Similarly, horizontal lines intersect the boundary of a horizontally simple region R at most twice, except for those horizontal lines comprising part of the boundary of R. Again rectangles, triangles, and circles are horizontally simple, as are the regions drawn in Figure 14.10(a) and (c). However, the regions in Figure 14.10(b) and (d) are not horizontally simple. Finally, the region in Figure 14.10(a) is simple, and those in Figure 14.10(b)–(d) are not simple.

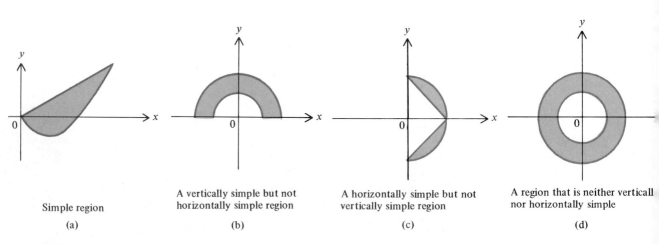

Simple region

(a)

A vertically simple but not horizontally simple region

(b)

A horizontally simple but not vertically simple region

(c)

A region that is neither verticall nor horizontally simple

(d)

FIGURE 14.10

Example 2   Let R be the region·between the graphs of $y = x^2$ and $y = x + 6$. Show that R is simple.

Solution   First we determine where the two graphs intersect. Observe that

$$y = x^2 \quad \text{and} \quad y = x + 6$$

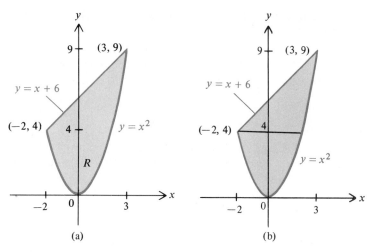

FIGURE 14.11

only if
$$y = x^2 = x + 6$$

which means that $x = -2$ or $x = 3$. Thus the graphs intersect at the points $(-2, 4)$ and $(3, 9)$ (Figure 14.11(a)). Therefore if $g_1(x) = x^2$ and $g_2(x) = x + 6$, then $g_1 \le g_2$ on $[-2, 3]$, so R is the vertically simple region between the graphs of g_1 and g_2 on $[-2, 3]$. To prove that R is horizontally simple, we notice that R is composed of two portions, one for which $0 \le y \le 4$ and the other for which $4 \le y \le 9$ (Figure 14.11(b)). Thus (x, y) is in R provided that either

$$0 \le y \le 4 \quad \text{and} \quad -\sqrt{y} \le x \le \sqrt{y}$$

or
$$4 \le y \le 9 \quad \text{and} \quad y - 6 \le x \le \sqrt{y}$$

Consequently if we let

$$h_1(y) = \begin{cases} -\sqrt{y} & \text{for } 0 \le y \le 4 \\ y - 6 & \text{for } 4 \le y \le 9 \end{cases} \quad \text{and} \quad h_2(y) = \sqrt{y}$$

then $h_1 \le h_2$ on $[0, 9]$, so R is the horizontally simple region between the graphs of h_1 and h_2 on $[0, 9]$. Since R is both vertically and horizontally simple, R is simple by definition. $\square$

Although the region in Example 2 is both vertically and horizontally simple, it is more easily described as a vertically simple region than as a horizontally simple region.

Later, when we actually evaluate a double integral on R, we will often describe R as the region between the graphs of

$$y = g_1(x) \quad \text{and} \quad y = g_2(x) \quad \text{for } a \le x \le b$$

if R is vertically simple, or as the region between the graphs of

$$x = h_1(y) \quad \text{and} \quad x = h_2(y) \quad \text{for } c \le y \le d$$

if R is horizontally simple.

Evaluation of Double Integrals

To explain how we can evaluate double integrals over vertically simple regions, let f be nonnegative and continuous on a vertically simple region R between the graphs of g_1 and g_2 on $[a, b]$. Let D be the solid region between the graph of f and R. Since f is continuous in each variable separately, the cross-sectional area $A(x)$ of D at any x in $[a, b]$ is given by

$$A(x) = \int_{g_1(x)}^{g_2(x)} f(x, y)\,dy$$

(Figure 14.12). It can be shown that A is continuous on $[a, b]$; hence by the cross-sectional definition of volume in Section 8.1, the volume V of D is given by

$$V = \int_a^b A(x)\,dx = \int_a^b \left[\int_{g_1(x)}^{g_2(x)} f(x, y)\,dy \right] dx$$

But by Definition 14.1, V is also given by

$$V = \iint_R f(x, y)\,dA$$

It turns out that the two formulas for V yield the same result. Therefore

$$\iint_R f(x, y)\,dA = \int_a^b \left[\int_{g_1(x)}^{g_2(x)} f(x, y)\,dy \right] dx$$

Similarly, if R is the horizontally simple region between the graphs of h_1 and h_2 on $[c, d]$, then

$$\iint_R f(x, y)\,dA = \int_c^d \left[\int_{h_1(y)}^{h_2(y)} f(x, y)\,dx \right] dy$$

The integrals

$$\int_a^b \left[\int_{g_1(x)}^{g_2(x)} f(x, y)\,dy \right] dx \quad \text{and} \quad \int_c^d \left[\int_{h_1(y)}^{h_2(y)} f(x, y)\,dx \right] dy \tag{4}$$

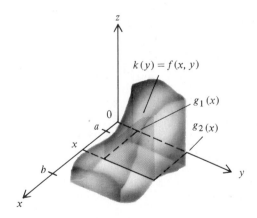

FIGURE 14.12

are called **iterated integrals**, because they are performed iteratively. That is, to evaluate $\int_a^b \left[\int_{g_1(x)}^{g_2(x)} f(x, y) \, dy \right] dx$, we first evaluate $\int_{g_1(x)}^{g_2(x)} f(x, y) \, dy$ with x fixed and then integrate the resulting function with respect to x. Normally we omit the brackets appearing in (4) and write

$$\int_a^b \int_{g_1(x)}^{g_2(x)} f(x, y) \, dy \, dx \quad \text{for} \quad \int_a^b \left[\int_{g_1(x)}^{g_2(x)} f(x, y) \, dy \right] dx$$

and

$$\int_c^d \int_{h_1(y)}^{h_2(y)} f(x, y) \, dx \, dy \quad \text{for} \quad \int_c^d \left[\int_{h_1(y)}^{h_2(y)} f(x, y) \, dx \right] dy$$

Our results are summarized in the following theorem.

THEOREM 14.4

Let f be continuous on a region R in the xy plane.

a. If R is the vertically simple region between the graphs of g_1 and g_2 on $[a, b]$, then f is integrable on R, and

$$\iint_R f(x, y) \, dA = \int_a^b \int_{g_1(x)}^{g_2(x)} f(x, y) \, dy \, dx$$

b. If R is the horizontally simple region between the graphs of h_1 and h_2 on $[c, d]$, then f is integrable on R, and

$$\iint_R f(x, y) \, dA = \int_c^d \int_{h_1(y)}^{h_2(y)} f(x, y) \, dx \, dy$$

Observe that if R is simple, then by Theorem 14.4(a) and (b),

$$\iint_R f(x, y) \, dA = \int_a^b \int_{g_1(x)}^{g_2(x)} f(x, y) \, dy \, dx = \int_c^d \int_{h_1(y)}^{h_2(y)} f(x, y) \, dx \, dy$$

Example 3 Let R be the rectangular region bounded by the lines $x = -1$, $x = 2$, $y = 0$, and $y = 2$. Find $\iint_R x^2 y \, dA$.

Solution The region R is the vertically simple region between the graphs of

$$y = 0 \quad \text{and} \quad y = 2 \quad \text{for} \quad -1 \leq x \leq 2$$

Therefore

$$\iint_R f(x, y) \, dA = \int_{-1}^2 \int_0^2 x^2 y \, dy \, dx$$

To evaluate the iterated integral, we first compute $\int_0^2 x^2 y \, dy$ for each x in $[-1, 2]$.

We obtain

$$\int_0^2 x^2 y \, dy = x^2 \int_0^2 y \, dy = x^2 \left(\frac{1}{2} y^2 \right) \Big|_0^2 = 2x^2$$

because x is held constant when we integrate with respect to y. We conclude that

$$\iint_R f(x, y) \, dA = \int_{-1}^2 \int_0^2 x^2 y \, dy \, dx = \int_{-1}^2 2x^2 \, dx$$

$$= \frac{2}{3} x^3 \Big|_{-1}^2 = 6 \quad \square$$

We can also evaluate the double integral in Example 3 by regarding R as the horizontally simple region between the graphs of

$$x = -1 \quad \text{and} \quad x = 2 \quad \text{for } 0 \le y \le 2$$

Using this approach, we obtain

$$\iint_R f(x, y) \, dA = \int_0^2 \int_{-1}^2 x^2 y \, dx \, dy = \int_0^2 y \left(\frac{1}{3} x^3 \Big|_{-1}^2 \right) dy$$

$$= \int_0^2 y \left(\frac{8}{3} - \left(-\frac{1}{3} \right) \right) dy = \int_0^2 3y \, dy$$

$$= \frac{3}{2} y^2 \Big|_0^2 = 6$$

Example 4   Let R and f be as in Example 1. Evaluate $\iint_R f(x, y) \, dA$.

Solution   Again the region is simple (Figure 14.13), so that we can calculate the double integral by means of two different iterated integrals. Let us regard R as the vertically simple region between the graphs of

$$y = 0 \quad \text{and} \quad y = 1 - x \quad \text{for } -2 \le x \le 1$$

It follows that

$$\iint_R f(x, y) \, dA = \int_{-2}^1 \int_0^{1-x} (4 - y) \, dy \, dx = \int_{-2}^1 \left(4y - \frac{y^2}{2} \right) \Big|_0^{1-x} dx$$

$$= \int_{-2}^1 \left(4(1 - x) - \frac{(1 - x)^2}{2} \right) dx = \int_{-2}^1 \left(\frac{7}{2} - 3x - \frac{x^2}{2} \right) dx$$

$$= \left(\frac{7}{2} x - \frac{3}{2} x^2 - \frac{x^3}{6} \right) \Big|_{-2}^1 = \frac{27}{2} \quad \square$$

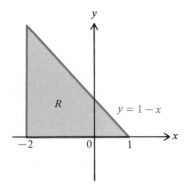

FIGURE 14.13

When the region R in Example 4 is regarded as the horizontally simple region between the graphs of

$$x = -2 \quad \text{and} \quad x = 1 - y \quad \text{for } 0 \le y \le 3$$

we find that

$$\iint_R f(x, y)\, dA = \int_0^3 \int_{-2}^{1-y} (4 - y)\, dx\, dy = \int_0^3 (4 - y) x \Big|_{-2}^{1-y} dy$$

$$= \int_0^3 (4 - y)[(1 - y) - (-2)]\, dy = \int_0^3 (12 - 7y + y^2)\, dy$$

$$= \left(12y - \frac{7}{2} y^2 + \frac{y^3}{3}\right)\Big|_0^3 = \frac{27}{2}$$

Thus we obtain the same value for $\iint_R f(x, y)\, dA$ whether we consider R as a vertically simple region or a horizontally simple region.

Example 5 Let R be the region between the graphs of $y = x^2$ and $y = x + 6$. Evaluate $\iint_R x\, dA$.

Solution As we saw in Example 2, the graphs of the two equations intersect at $(-2, 4)$ and $(3, 9)$, and $x^2 \le x + 6$ for $-2 \le x \le 3$. Consequently

$$\iint_R x\, dA = \int_{-2}^3 \int_{x^2}^{x+6} x\, dy\, dx = \int_{-2}^3 xy \Big|_{x^2}^{x+6} dx$$

$$= \int_{-2}^3 (x^2 + 6x - x^3)\, dx$$

$$= \left(\frac{x^3}{3} + 3x^2 - \frac{x^4}{4}\right)\Big|_{-2}^3 = \frac{125}{12} \quad \square$$

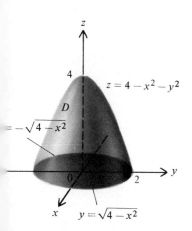

FIGURE 14.14

Example 6 Find the volume V of the solid region D bounded above by the paraboloid $z = 4 - x^2 - y^2$ and below by the xy plane.

Solution The intersection of D and the xy plane is the region R bounded by the circle $x^2 + y^2 = 4$ (Figure 14.14). By Definition 14.1(b),

$$V = \iint_R (4 - x^2 - y^2)\, dA$$

Since R is the vertically simple region between the graphs of

$$y = -\sqrt{4 - x^2} \quad \text{and} \quad y = \sqrt{4 - x^2} \quad \text{for } -2 \le x \le 2$$

we conclude that

$$V = \int_{-2}^{2} \int_{-\sqrt{4-x^2}}^{\sqrt{4-x^2}} (4 - x^2 - y^2) \, dy \, dx$$

$$= \int_{-2}^{2} \left((4 - x^2)y - \frac{y^3}{3} \right) \Bigg|_{-\sqrt{4-x^2}}^{\sqrt{4-x^2}} dx$$

$$= 2 \int_{-2}^{2} \left((4 - x^2)\sqrt{4 - x^2} - \frac{(4 - x^2)^{3/2}}{3} \right) dx$$

$$= \frac{4}{3} \int_{-2}^{2} (4 - x^2)^{3/2} \, dx \overset{x = 2\sin\theta}{=} \frac{64}{3} \int_{-\pi/2}^{\pi/2} \cos^4\theta \, d\theta$$

$$\overset{\substack{(14) \text{ of} \\ \text{Section 8.1}}}{=} \frac{64}{3} \left(\frac{1}{4}\cos^3\theta \sin\theta + \frac{3}{8}\cos\theta \sin\theta + \frac{3}{8}\theta \right) \Bigg|_{-\pi/2}^{\pi/2} = 8\pi \quad \square$$

In Section 14.2 we will discuss the use of polar coordinates to evaluate double integrals. With polar coordinates the calculations needed in order to find the volume of the region in Example 6 will be greatly reduced.

We define the area A of a plane region R by

$$A = \iint_R 1 \, dA \tag{5}$$

When R is the region between the graphs of two continuous functions g_1 and g_2 on $[a, b]$ such that $g_1 \leq g_2$, we have two definitions of the area A of R: (5) above and (1) of Section 5.8. However, since Theorem 14.4 implies that

$$A = \iint_R 1 \, dA = \int_a^b \int_{g_1(x)}^{g_2(x)} 1 \, dy \, dx = \int_a^b [g_2(x) - g_1(x)] \, dx$$

both definitions yield the same value for A. For instance, if R is the circle of radius 3 centered at the origin, then

$$A = \iint_R 1 \, dA = \int_{-3}^{3} \int_{-\sqrt{9-x^2}}^{\sqrt{9-x^2}} 1 \, dy \, dx$$

$$= 2 \int_{-3}^{3} \sqrt{9 - x^2} \, dx = 18 \int_{-\pi/2}^{\pi/2} \cos^2\theta \, d\theta$$

$$= 18 \int_{-\pi/2}^{\pi/2} \left(\frac{1}{2} + \frac{1}{2}\cos 2\theta \right) d\theta$$

$$= 9 \left(\theta + \frac{1}{2}\sin 2\theta \right) \Bigg|_{-\pi/2}^{\pi/2} = 9\pi$$

Reversing the Order of Integration

In case R is a simple region, $\iint_R f(x, y)\,dA$ can be evaluated as either $\int_a^b \int_{g_1(x)}^{g_2(x)} f(x, y)\,dy\,dx$ or $\int_c^d \int_{h_1(y)}^{h_2(y)} f(x, y)\,dx\,dy$. Which iterated integral we use depends on the situation—the integrand, the limits of integration, and our convenience. It sometimes happens that one iterated integral is either difficult or impossible to evaluate, whereas the other iterated integral can be evaluated easily. The change from one iterated integral to the other is called **reversing the order of integration**, since it involves changing from $dy\,dx$ to $dx\,dy$, or vice versa.

Example 7   By reversing the order of integration, evaluate

$$\int_0^9 \int_{\sqrt{y}}^3 \sin \pi x^3\, dx\, dy$$

Solution   To reverse the order of integration we must determine the region R over which the integration is performed. From the limits on the given iterated integral we infer that R is the horizontally simple region between the graphs of

$$x = \sqrt{y} \quad \text{and} \quad x = 3 \quad \text{for } 0 \le y \le 9$$

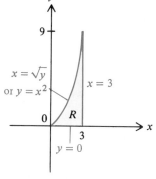

(Figure 14.15). Since R is also the vertically simple region between the graphs of

$$y = 0 \quad \text{and} \quad y = x^2 \quad \text{for } 0 \le x \le 3$$

we find that

$$\int_0^9 \int_{\sqrt{y}}^3 \sin \pi x^3\, dx\, dy = \iint_R \sin \pi x^3\, dA = \int_0^3 \int_0^{x^2} \sin \pi x^3\, dy\, dx$$

$$= \int_0^3 y \sin \pi x^3 \Big|_0^{x^2}\, dx = \int_0^3 x^2 \sin \pi x^3\, dx$$

$$= \frac{-1}{3\pi} \cos \pi x^3 \Big|_0^3 = \frac{2}{3\pi} \quad \square$$

FIGURE 14.15

Reversing the order of integration applies only to simple regions. If such a region R is given as a vertically simple (respectively, horizontally simple) region, then reversing the order of integration amounts to reformulating R as a horizontally simple (respectively, vertically simple) region.

Double Integration Over More General Regions

If R is composed of two or more vertically or horizontally simple subregions R_1, $R_2, \ldots, R_n$, with the property that any two subregions have only boundaries in common, then any function f that is continuous on R is integrable on R, and

$$\iint_R f(x, y)\,dA = \iint_{R_1} f(x, y)\,dA + \iint_{R_2} f(x, y)\,dA + \cdots + \iint_{R_n} f(x, y)\,dA \qquad (6)$$

Example 8 Let R be the region between the graphs of $y = x$ and $y = x^3$. Evaluate $\iint_R (x - 1)\, dA$.

Solution The graphs of the two equations intersect at (x, y) if

$$x = y = x^3$$

which happens if $x = -1$, $x = 0$, or $x = 1$. Therefore R is composed of the two vertically simple regions R_1 and R_2, as depicted in Figure 14.16. Consequently

$$\iint_R (x - 1)\, dA = \iint_{R_1} (x - 1)\, dA + \iint_{R_2} (x - 1)\, dA$$

$$= \int_{-1}^{0} \int_{x}^{x^3} (x - 1)\, dy\, dx + \int_{0}^{1} \int_{x^3}^{x} (x - 1)\, dy\, dx$$

$$= \int_{-1}^{0} (x - 1)y \Big|_{x}^{x^3} dx + \int_{0}^{1} (x - 1)y \Big|_{x^3}^{x} dx$$

$$= \int_{-1}^{0} (x - 1)(x^3 - x)\, dx + \int_{0}^{1} (x - 1)(x - x^3)\, dx$$

$$= \int_{-1}^{0} (x^4 - x^3 - x^2 + x)\, dx - \int_{0}^{1} (x^4 - x^3 - x^2 + x)\, dx$$

$$= \left(\frac{x^5}{5} - \frac{x^4}{4} - \frac{x^3}{3} + \frac{x^2}{2} \right) \Big|_{-1}^{0} - \left(\frac{x^5}{5} - \frac{x^4}{4} - \frac{x^3}{3} + \frac{x^2}{2} \right) \Big|_{0}^{1}$$

$$= -\frac{1}{2} \quad \square$$

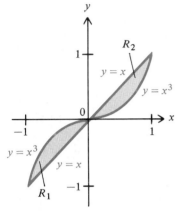

FIGURE 14.16

FIGURE 14.17

Occasionally we encounter regions that are neither vertically nor horizontally simple but that can be subdivided into regions that are vertically (or horizontally) simple (Figure 14.17). In such cases the formula in (6) is still applicable.

EXERCISES 14.1

1. Let R, f, R', and $\mathscr{P}$ be as in Example 1. Find the Riemann sum that employs
   a. the lower left vertices of R_1, R_2, and R_3.
   b. the upper right vertices of R_1, R_2, and R_3.

2. Let R be the triangular region bounded by the lines $y = 2x$, $x = 0$, and $y = 4$, and let R' be the rectangle whose sides are the lines $x = 0$, $x = 2$, $y = 0$, and $y = 4$. Suppose the partition $\mathscr{P}$ of R' consists of the squares whose sides are 1 unit long. If $f(x, y) = x + y$ for (x, y) in R, find
   a. $L_f(\mathscr{P})$.   b. $U_f(\mathscr{P})$.

c. the Riemann sum of f that uses the midpoints of the rectangles of $\mathscr{P}$ that are contained in R.

In Exercises 3–19 evaluate the iterated integral.

3. $\displaystyle\int_{0}^{1} \int_{-1}^{1} x\, dy\, dx$ 4. $\displaystyle\int_{-5}^{-3} \int_{-2}^{3} y\, dy\, dx$

5. $\displaystyle\int_{0}^{1} \int_{0}^{1} e^{x+y}\, dy\, dx$ 6. $\displaystyle\int_{0}^{1} \int_{1}^{5} \frac{1}{r}\, dr\, ds$

7. $\displaystyle\int_{0}^{1} \int_{x}^{x^2} 1\, dy\, dx$ 8. $\displaystyle\int_{0}^{1} \int_{0}^{3} x\sqrt{x^2 + y}\, dy\, dx$

9. $\displaystyle\int_0^1\int_0^y x\sqrt{y^2-x^2}\,dx\,dy$

10. $\displaystyle\int_0^2\int_0^{\sqrt{4-y^2}} y\,dx\,dy$

11. $\displaystyle\int_0^2\int_0^{\sqrt{4-y^2}} x\,dx\,dy$

12. $\displaystyle\int_0^{2\pi}\int_0^1 r\sin\theta\,dr\,d\theta$

13. $\displaystyle\int_0^{2\pi}\int_0^1 r\sqrt{1-r^2}\,dr\,d\theta$

14. $\displaystyle\int_0^{2\pi}\int_0^{1+\cos\theta} r\,dr\,d\theta$

15. $\displaystyle\int_1^3\int_0^x \frac{2}{x^2+y^2}\,dy\,dx$

16. $\displaystyle\int_1^e\int_1^{\ln y} e^x\,dx\,dy$

17. $\displaystyle\int_0^1\int_0^x e^{(x^2)}\,dy\,dx$

18. $\displaystyle\int_{\ln(\pi/6)}^{\ln(\pi/2)}\int_0^{e^y}\cos e^y\,dx\,dy$

19. $\displaystyle\int_0^{\pi/4}\int_0^{\cos y} e^x\sin y\,dx\,dy$

In Exercises 20–33 express the double integral as an iterated integral and evaluate it.

20. $\iint_R(x+y)\,dA$; R is the rectangular region bounded by the lines $x=2$, $x=3$, $y=4$, and $y=6$.

21. $\iint_R(x+y)\,dA$; R is the triangular region bounded by the lines $y=2x$, $x=0$, and $y=4$.

22. $\iint_R xy^2\,dA$; R is the region between the parabola $y=4-x^2$ and the x axis on $[0,1]$.

23. $\iint_R x\,dA$; R is the trapezoidal region bounded by the lines $x=3$, $x=5$, $y=1$, and $y=x$.

24. $\iint_R x(x-1)e^{xy}\,dA$; R is the triangular region bounded by the lines $x=0$, $y=0$, and $x+y=2$.

25. $\iint_R(3x-5)\,dA$; R is the triangular region bounded by the lines $y=5+x$, $y=-x+7$, and $x=10$.

26. $\iint_R xy\,dA$; R is the region bounded by the graphs of $y=x$ and $y=x^2$.

27. $\iint_R 1\,dA$; R is the region between the graphs of $y=1+x$ and $y=\sin x$ on $[\pi,2\pi]$.

28. $\iint_R(4+x^2)\,dA$; R is the region bounded by the graphs of $y=1+x^2$ and $y=9-x^2$.

29. $\iint_R(1-y)\,dA$; R is the region bounded by the graphs of $x=y^2$ and $x=2-y$.

30. $\iint_R x\,dA$; R is the portion of the disk $x^2+y^2\le16$ in the second quadrant.

31. $\iint_R xy^2\,dA$; R is the region above the line $y=1-x$ and inside the circle $x^2+y^2=1$.

32. $\iint_R 2y\,dA$; R is the region between the graph of $y=\sin x$ and the x axis on $[0,3\pi/2]$.

33. $\iint_R x^2\,dA$; R is the region between the graphs of $y=x^3+x^2+1$ and $y=x^3+x+1$ on $[-1,1]$.

In Exercises 34–44 find the volume V of the region, using the methods of this section.

34. The solid region bounded by the planes $z=1+x+y$, $x=2$, $y=1$ and the coordinate planes

35. The solid region bounded by the plane $x+2y+3z=6$ and the coordinate planes

36. The solid region in the first octant bounded by the coordinate planes and the planes $x+y=1$ and $x+4y+2z=6$

37. The solid region in the first octant bounded by the paraboloid $z=x^2+y^2$, the plane $x+y=1$, and the coordinate planes

38. The solid region in the first octant bounded by the parabolic sheet $z=x^2$ and the planes $x=2y$, $y=0$, $z=0$, and $x=2$

39. The solid region in the first octant bounded by the cylinder $x^2+y^2=4$, the plane $z=y$, the xy plane, and the yz plane

40. The solid region bounded above by the parabolic sheet $z=y^2$, on the sides by the sheet $x=1-(y-1)^2$ and the plane $y=x$, and on the bottom by the xy plane

41. The solid region in the first octant that is common to the cylinders $x^2+y^2=1$ and $x^2+z^2=1$ (*Hint:* Over what region in the xy plane does the region lie? Integrate first with respect to y.)

42. The region bounded above by the sphere $x^2+y^2+z^2=1$ and below by the xy plane

43. The solid region bounded above by the surface $z=xy$, below by the xy plane, and on the sides by the plane $y=x$ and the surface $y=x^3$ (*Hint:* There are two parts to the region in the xy plane.)

44. The region bounded above by the plane $z=10+2x+3y$, below by the xy plane, and on the sides by the surfaces $y=x^2$ and $y=x$

In Exercises 45–50 the iterated integral represents the volume of a solid region D. Sketch the region D.

45. $\displaystyle\int_{-2}^1\int_1^4 3\,dy\,dx$

46. $\displaystyle\int_{-3}^3\int_{-\sqrt{9-x^2}}^{\sqrt{9-x^2}} 5\,dy\,dx$

47. $\displaystyle\int_{-5}^5\int_{-\sqrt{25-x^2}}^{\sqrt{25-x^2}}\sqrt{25-x^2-y^2}\,dy\,dx$

48. $\displaystyle\int_0^5\int_0^{\sqrt{25-x^2}}\sqrt{25-x^2-y^2}\,dy\,dx$

49. $\displaystyle\int_{-2}^2\int_{-\sqrt{4-y^2}}^{\sqrt{4-y^2}}(16-4x^2-4y^2)\,dx\,dy$

50. $\displaystyle\int_{-8}^8\int_{-\sqrt{64-x^2}}^{\sqrt{64-x^2}}(16-\sqrt{x^2+y^2})\,dy\,dx$

In Exercises 51–56 find the area A of the region in the xy plane by the methods of this section.

51. The region bounded by the parabolas $y=x^2$ and $y=\sqrt{x}$ on $[1,4]$

52. The region bounded by the graphs of the equations $y = \cosh x$ and $y = \sinh x$ on $[-1, 1]$ (Leave your answer in terms of cosh or sinh.)

53. The region between the parabolas $x = y^2$ and $x = 32 - y^2$

54. The region bounded by the parabolas $x = y^2$ and $x = 4y^2 - 3$

55. The region in the first quadrant that is bounded by the graph of $x = 2 - y^2$ and by the lines $x = 0$ and $x = y$

56. The region in the first quadrant that is bounded by the graphs of $x + y = 4$ and $y = 3/x$

In Exercises 57–64 reverse the order of integration and evaluate the resulting integral.

57. $\displaystyle\int_0^1 \int_y^1 e^{(x^2)} \, dx \, dy$

58. $\displaystyle\int_1^4 \int_{\sqrt{y}}^2 \sin\left(\frac{x^3}{3} - x\right) dx \, dy$

59. $\displaystyle\int_0^2 \int_{1+y^2}^5 y e^{(x-1)^2} \, dx \, dy$

60. $\displaystyle\int_0^1 \int_{\arcsin y}^{\pi/2} \sec^2(\cos x) \, dx \, dy$

61. $\displaystyle\int_1^e \int_0^{\ln x} y \, dy \, dx$

62. $\displaystyle\int_1^e \int_{1/e}^{1/y} \cos(x - \ln x) \, dx \, dy$

63. $\displaystyle\int_0^{\pi^{1/3}} \int_{y^2}^{\pi^{2/3}} \sin x^{3/2} \, dx \, dy$

64. $\displaystyle\int_0^{\sqrt{\pi/2}} \int_y^{\sqrt{\pi/2}} y^2 \sin x^2 \, dx \, dy$

(*Hint:* After changing the order of integration, make the substitution $u = x^2$.)

65. a. Suppose R is a rectangle and f is continuous on R. If $L_f(\mathscr{P}) = U_f(\mathscr{P})$ for some partition $\mathscr{P}$ of R, what can you say about f and about $\iint_R f(x, y) \, dA$?

    b. Suppose R is a vertically or horizontally simple region and f is continuous on R. Let R' be a rectangle containing R. If there is a partition $\mathscr{P}$ of R' for which $L_f(\mathscr{P}) = U_f(\mathscr{P})$, what can you say about f and about $\iint_R f(x, y) \, dA$?

66. Suppose f is continuous on a disk R with center (x_0, y_0). If $\iint_{R_0} f(x, y) \, dA = 0$ for every rectangle R_0 contained in R, show that $f(x_0, y_0) = 0$. (*Hint:* If $f(x_0, y_0) > 0$, then $f(x, y) \geq \frac{1}{2} f(x_0, y_0)$ for all (x, y) in some rectangle R about (x_0, y_0).)

14.2
DOUBLE INTEGRALS IN POLAR COORDINATES

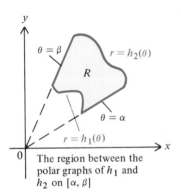

The region between the polar graphs of h_1 and h_2 on $[\alpha, \beta]$

FIGURE 14.18

Certain curves in the xy plane, such as circles and cardioids, can be described more easily in polar coordinates than in rectangular coordinates. Thus it is reasonable to expect double integrals over regions enclosed by such curves to be more easily evaluated by means of polar coordinates.

The type of region over which we will integrate in polar coordinates can be described as follows. Suppose that h_1 and h_2 are continuous on an interval $[\alpha, \beta]$, where $0 \leq \beta - \alpha \leq 2\pi$, and that

$$0 \leq h_1(\theta) \leq h_2(\theta) \quad \text{for } \alpha \leq \theta \leq \beta$$

Let R be the region in the xy plane bounded by the lines $\theta = \alpha$ and $\theta = \beta$ and by the polar graphs of $r = h_1(\theta)$ and $r = h_2(\theta)$ (Figure 14.18). We say that R is **the region between the polar graphs of h_1 and h_2 on $[\alpha, \beta]$**. Such a region R need not be vertically or horizontally simple. Nevertheless, it turns out that every function that is continuous on R is integrable on R, so $\iint_R f(x, y) \, dA$ exists.

To explain intuitively how to evaluate $\iint_R f(x, y) \, dA$, we assume that f is nonnegative on R, so that $\iint_R f(x, y) \, dA$ is the volume of the solid region D between the graph of f and R. Although the definition of $\iint_R f(x, y) \, dA$ was based on partitions of R by rectangles, it is also possible to partition R by other types of regions. When working in polar coordinates, we first circumscribe R by a circular arch R', which is the region between two circular arcs (Fig-

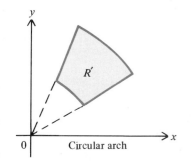

FIGURE 14.19

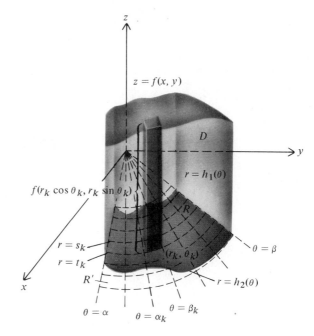

FIGURE 14.20

ure 14.19), and then partition R' into circular subarches, numbered so that $R_1, R_2, \ldots, R_n$ are the ones entirely contained in R. For each k between 1 and n, let R_k be bounded by the lines $\theta = \alpha_k$ and $\theta = \beta_k$ and by the circular arcs $r = s_k$ and $r = t_k$ (Figure 14.20). Furthermore, let

$$r_k = \frac{1}{2}(s_k + t_k) \quad \text{and} \quad \theta_k = \frac{1}{2}(\alpha_k + \beta_k)$$

so that (r_k, θ_k) are polar coordinates of the "center" of R_k. From (1) of Section 8.10, we find that the area of R_k is given by

$$\frac{1}{2}t_k^2(\beta_k - \alpha_k) - \frac{1}{2}s_k^2(\beta_k - \alpha_k) = \frac{1}{2}(t_k + s_k)(t_k - s_k)(\beta_k - \alpha_k)$$

$$= r_k(t_k - s_k)(\beta_k - \alpha_k)$$

The volume of the solid region with height $f(r_k \cos \theta_k, r_k \sin \theta_k)$ and base R_k is approximately the volume of the portion of D that lies over R_k (Figure 14.20). Therefore

$$\iint\limits_{R} f(x, y)\, dA \approx \sum_{k=1}^{n} f(r_k \cos \theta_k, r_k \sin \theta_k) r_k (t_k - s_k)(\beta_k - \alpha_k) \qquad (1)$$

However, R corresponds to the horizontally simple region S in the $r\theta$ plane between the graphs of $r = h_1(\theta)$ and $r = h_2(\theta)$ on $[\alpha, \beta]$ (Figure 14.21), and each circular arch R_k into which we partitioned R' corresponds to the rectangle S_k in the $r\theta$ plane bounded by the lines $\theta = \alpha_k, \theta = \beta_k, r = s_k,$ and $r = t_k$. The area ΔA_k of S_k is $(t_k - s_k)(\beta_k - \alpha_k)$. Consequently

$$\sum_{k=1}^{n} f(r_k \cos \theta_k, r_k \sin \theta_k) r_k (t_k - s_k)(\beta_k - \alpha_k) = \sum_{k=1}^{n} f(r_k \cos \theta_k, r_k \sin \theta_k) r_k\, \Delta A_k \qquad (2)$$

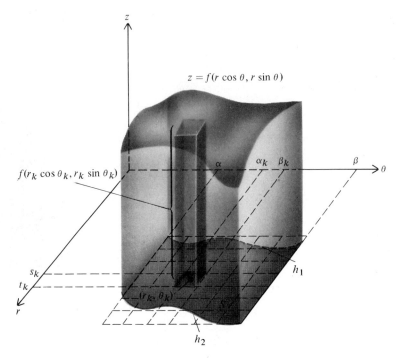

FIGURE 14.21

and the right side of (2) is a Riemann sum associated with the double integral $\iint_S f(r\cos\theta, r\sin\theta)r\,dA$, so that

$$\sum_{k=1}^{n} f(r_k\cos\theta, r_k\sin\theta)r_k\,\Delta A_k \approx \iint_S f(r\cos\theta, r\sin\theta)r\,dA \qquad (3)$$

Since S is horizontally simple, we know from Theorem 14.4(b) that

$$\iint_S f(r\cos\theta, r\sin\theta)r\,dA = \int_\alpha^\beta \int_{h_1(\theta)}^{h_2(\theta)} f(r\cos\theta, r\sin\theta)r\,dr\,d\theta \qquad (4)$$

Equations (1)–(4) suggest the following theorem.

THEOREM 14.5

Suppose that h_1 and h_2 are continuous on $[\alpha, \beta]$, where $0 \leq \beta - \alpha \leq 2\pi$, and that $0 \leq h_1(\theta) \leq h_2(\theta)$ for $\alpha \leq \theta \leq \beta$. Let R be the region between the polar graphs of

$$r = h_1(\theta) \quad \text{and} \quad r = h_2(\theta) \quad \text{for } \alpha \leq \theta \leq \beta$$

If f is continuous on R, then

$$\iint_R f(x, y)\,dA = \int_\alpha^\beta \int_{h_1(\theta)}^{h_2(\theta)} f(r\cos\theta, r\sin\theta)r\,dr\,d\theta \qquad (5)$$

In the event that f is nonnegative on R, the volume V of the region between the graph of f and R is given by

$$V = \int_{\alpha}^{\beta} \int_{h_1(\theta)}^{h_2(\theta)} f(r\cos\theta, r\sin\theta) r\, dr\, d\theta$$

and the area A of R is given by

$$A = \int_{\alpha}^{\beta} \int_{h_1(\theta)}^{h_2(\theta)} r\, dr\, d\theta$$

These two formulas follow from (5) and the formulas for volume and area presented in Section 14.1.

Example 1 Suppose R is the region bounded by the circles $r = 1$ and $r = 2$ and the lines $\theta = \alpha$ and $\theta = \beta$, where $0 \le \beta - \alpha \le 2\pi$. Express

$$\iint_R (3x + 8y^2)\, dA$$

as an iterated integral in polar coordinates. Then evaluate the iterated integral for

    a. $\alpha = 0$ and $\beta = \pi/2$ (R is a quarter-ring)
    b. $\alpha = 0$ and $\beta = \pi$ (R is a half-ring)
    c. $\alpha = 0$ and $\beta = 2\pi$ (R is a ring)

Solution By Theorem 14.5 we have

$$\iint_R (3x + 8y^2)\, dA = \int_{\alpha}^{\beta} \int_{1}^{2} (3r\cos\theta + 8r^2\sin^2\theta) r\, dr\, d\theta$$

If R is the quarter-ring (Figure 14.22(a)), we find that

$$\iint_R (3x + 8y^2)\, dA = \int_{0}^{\pi/2} \int_{1}^{2} (3r\cos\theta + 8r^2\sin^2\theta) r\, dr\, d\theta$$

$$= \int_{0}^{\pi/2} \int_{1}^{2} (3r^2\cos\theta + 8r^3\sin^2\theta)\, dr\, d\theta$$

$$= \int_{0}^{\pi/2} (r^3\cos\theta + 2r^4\sin^2\theta)\Big|_{1}^{2}\, d\theta$$

$$= \int_{0}^{\pi/2} (7\cos\theta + 30\sin^2\theta)\, d\theta$$

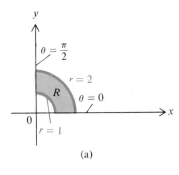

(a)

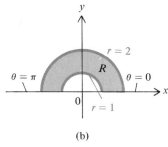

(b)

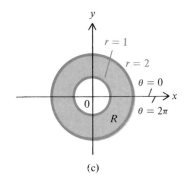

(c)

FIGURE 14.22

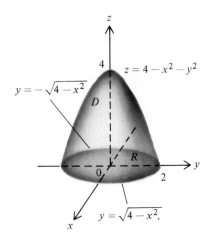

FIGURE 14.23

$$= \int_0^{\pi/2} (7\cos\theta + 15 - 15\cos 2\theta)\,d\theta$$

$$= \left(7\sin\theta + 15\theta - \frac{15}{2}\sin 2\theta\right)\Big|_0^{\pi/2} = 7 + \frac{15}{2}\pi$$

For the half-ring (Figure 14.22(b)) only the outer limit $\pi/2$ of integration is replaced by π, so we obtain

$$\iint_R (3x + 8y^2)\,dA = \int_0^\pi \int_1^2 (3r\cos\theta + 8r^2\sin^2\theta)r\,dr\,d\theta$$

$$= \left(7\sin\theta + 15\theta - \frac{15}{2}\sin 2\theta\right)\Big|_0^\pi = 15\pi$$

For the ring (Figure 14.22(c)), we substitute 2π for π in the outer limits of integration, obtaining

$$\iint_R (3x + 8y^2)\,dA = \int_0^{2\pi} \int_1^2 (3r\cos\theta + 8r^2\sin^2\theta)r\,dr\,d\theta$$

$$= \left(7\sin\theta + 15\theta - \frac{15}{2}\sin 2\theta\right)\Big|_0^{2\pi} = 30\pi \quad \square$$

Example 2   Let D be the solid region bounded above by the paraboloid $z = 4 - x^2 - y^2$ and below by the xy plane. Find the volume V of D.

Solution   From Example 6 of Section 14.1, the region R over which the integral is to be taken is bounded by the circle $x^2 + y^2 = 4$, whose equation in polar coordinates is $r = 2$; therefore D can be described as the region between the paraboloid and the disk $x^2 + y^2 \le 4$ (Figure 14.23). As a result

$$V = \iint_R (4 - x^2 - y^2)\,dA = \int_0^{2\pi} \int_0^2 (4 - r^2)r\,dr\,d\theta$$

$$= \int_0^{2\pi} \int_0^2 (4r - r^3)\,dr\,d\theta = \int_0^{2\pi} \left(2r^2 - \frac{r^4}{4}\right)\Big|_0^2 d\theta$$

$$= \int_0^{2\pi} 4\,d\theta = 8\pi \quad \square$$

Although the volume we just calculated is the same as the one calculated in Example 6 of Section 14.1, the use of polar coordinates simplified the solution considerably.

Example 3   Suppose D is the solid region bounded on the sides by the cylinder $r = \cos\theta$, above by the cone $z = 16 - \sqrt{x^2 + y^2}$, and below by the xy plane (Figure 14.24). Find the volume V of D.

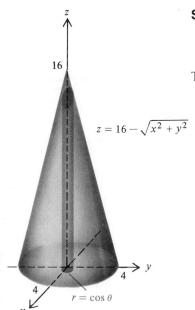

FIGURE 14.24

Solution Observe that R is the region between the polar graphs of

$$r = 0 \quad \text{and} \quad r = \cos\theta \quad \text{for} \quad -\frac{\pi}{2} \le \theta \le \frac{\pi}{2}$$

Therefore

$$V = \iint_R (16 - \sqrt{x^2 + y^2})\, dA = \int_{-\pi/2}^{\pi/2} \int_0^{\cos\theta} (16 - r) r\, dr\, d\theta$$

$$= \int_{-\pi/2}^{\pi/2} \int_0^{\cos\theta} (16r - r^2)\, dr\, d\theta = \int_{-\pi/2}^{\pi/2} \left(8r^2 - \frac{r^3}{3} \right)\Big|_0^{\cos\theta} d\theta$$

$$= \int_{-\pi/2}^{\pi/2} \left(8\cos^2\theta - \frac{\cos^3\theta}{3} \right) d\theta$$

$$= \int_{-\pi/2}^{\pi/2} \left(4 + 4\cos 2\theta - \frac{\cos\theta}{3} + \frac{\cos\theta \sin^2\theta}{3} \right) d\theta$$

$$= \left(4\theta + 2\sin 2\theta - \frac{\sin\theta}{3} + \frac{\sin^3\theta}{9} \right)\Big|_{-\pi/2}^{\pi/2}$$

$$= 4\pi - \frac{4}{9} \quad \square$$

Caution: When a double integral is evaluated by means of formula (5), h_1 and h_2 must be nonnegative on $[\alpha, \beta]$. Notice that the circle that bounds the disk in Example 3 can also be described as the graph of $r = \cos\theta$ for $0 \le \theta \le \pi$. However, $\cos\theta$ is negative for $\pi/2 \le \theta \le \pi$, and you can show that

$$\int_0^\pi \int_0^{\cos\theta} (16 - r) r\, dr\, d\theta \ne \int_{-\pi/2}^{\pi/2} \int_0^{\cos\theta} (16 - r) r\, dr\, d\theta$$

We also remark that the reason for our stipulation that $0 \le \beta - \alpha \le 2\pi$ at the beginning of this section was to ensure that each point in R corresponds to exactly one point in the $r\theta$ plane. However, certain other regions, such as the region between two appropriately chosen spirals, have this property, even if we take $\alpha = 0$ and $\beta = 3\pi$ (in which case $\beta - \alpha > 2\pi$).

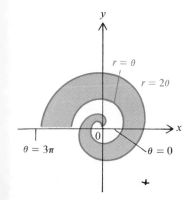

FIGURE 14.25

Example 4 Let R be the region between the polar graphs of $r = \theta$ and $r = 2\theta$ for $0 \le \theta \le 3\pi$ (Figure 14.25). Evaluate $\iint_R (x^2 + y^2)\, dA$.

Solution We find that

$$\iint_R (x^2 + y^2)\, dA = \int_0^{3\pi} \int_\theta^{2\theta} (r^2) r\, dr\, d\theta = \int_0^{3\pi} \int_\theta^{2\theta} r^3\, dr\, d\theta$$

$$= \int_0^{3\pi} \frac{r^4}{4}\Big|_\theta^{2\theta} d\theta = \frac{15}{4} \int_0^{3\pi} \theta^4\, d\theta$$

$$= \frac{15}{4} \frac{\theta^5}{5}\Big|_0^{3\pi} = \frac{729}{4} \pi^5 \quad \square$$

Evaluating the double integral in Example 4 with the iterated integrals of Section 14.1 would be very difficult.

EXERCISES 14.2

In Exercises 1–6 express the integral as an iterated integral in polar coordinates, and then evaluate it.

1. $\iint_R xy\,dA$, where R is the region bounded by the circle $r = 5$

2. $\iint_R y\,dA$, where R is the region bounded by the circle $r = \cos\theta$

3. $\iint_R (x + y)\,dA$, where R is the region in the first quadrant bounded by the lines $y = 0$ and $y = \sqrt{3}x$ and the circle $r = 2$

4. $\iint_R (x^2 + y^2)^{1/2}\,dA$, where R is the region bounded by the limaçon $r = 2 + \cos\theta$

5. $\iint_R (x^2 + y^2)\,dA$, where R is the region bounded by the cardioid $r = 2(1 + \sin\theta)$

6. $\iint_R x^2\,dA$, where R is the region bounded by the circle $r = 4\sin\theta$

In Exercises 7–13 find the volume V of the region by using iterated integrals in polar coordinates.

7. The solid region bounded by a hemisphere with radius 3

8. The solid region bounded by the planes $z = 0$ and $z = 4$ and the cylinder $r = 2\cos\theta$

9. The solid region bounded above by the plane $z = 4 + x + 2y$, on the sides by the cylinder $x^2 + y^2 = 1$, and below by the xy plane

10. The solid region bounded above by the plane $z = x$, on the sides by the cylinder $x^2 + y^2 = x$, and below by the xy plane

11. The solid region inside the sphere $x^2 + y^2 + z^2 = 4$, outside the cylinder $x^2 + y^2 = 1$, and above the xy plane

12. The solid region bounded by the paraboloid $z = 4 - x^2 - y^2$, the plane $z = 3$, and the xy plane

13. The solid region above the xy plane bounded on the sides by the cylinder $x^2 + y^2 - 4x = 0$ and above by the cone $z^2 = x^2 + y^2$

In Exercises 14–21 find the area A of the region in the xy plane by means of iterated integrals in polar coordinates.

14. The circular sector bounded by the graph of $r = 1$ on $[0, \alpha]$, where $0 \le \alpha \le 2\pi$

15. The region bounded by the limaçon $r = 2 + \sin\theta$

16. One leaf of the three-leaved rose bounded by the graph of $r = 2\sin 3\theta$

17. The region bounded by the lemniscate $r^2 = 4\cos 2\theta$ (*Hint:* First calculate the area of the portion for which $-\pi/4 \le \theta \le \pi/4$.)

18. The region inside the cardioid $r = 1 + \cos\theta$ and outside the circle $r = \frac{1}{2}$

19. The region inside the large loop and outside the small loop of $r = 1 + 2\sin\theta$

20. The region bounded by the graph of $r = \sin\theta - \cos\theta$ (*Hint:* Be careful with the limits of integration.)

21. The region between the spirals $r = e^\theta$ and $r = e^{2\theta}$ on $[0, 3\pi]$

In Exercises 22–29 change the integral to an iterated integral in polar coordinates, and then evaluate it.

22. $\displaystyle\int_0^1 \int_0^{\sqrt{1-x^2}} 1\,dy\,dx$ 23. $\displaystyle\int_0^1 \int_y^{\sqrt{2-y^2}} 1\,dx\,dy$

24. $\displaystyle\int_{3/\sqrt{2}}^3 \int_0^{\sqrt{9-x^2}} \frac{1}{\sqrt{x^2 + y^2}}\,dy\,dx$

25. $\displaystyle\int_0^1 \int_0^{\sqrt{1-y^2}} \sin(x^2 + y^2)\,dx\,dy$

26. $\displaystyle\int_0^1 \int_0^{\sqrt{1-x^2}} e^{\sqrt{x^2+y^2}}\,dy\,dx$

27. $\displaystyle\int_0^1 \int_0^{\sqrt{1-x^2}} e^{-(x^2+y^2)}\,dy\,dx$

28. $\displaystyle\int_0^1 \int_{\sqrt{x-x^2}}^{\sqrt{1-x^2}} 1\,dy\,dx$

29. $\displaystyle\int_0^1 \int_{-\sqrt{x-x^2}}^{\sqrt{x-x^2}} (x^2 + y^2)\,dy\,dx$

30. a. By changing to polar coordinates, evaluate the improper integral

$$\int_0^\infty \int_0^\infty e^{-(x^2+y^2)}\,dx\,dy$$

as $\displaystyle\lim_{b\to\infty} \iint_{R_b} e^{-(x^2+y^2)}\,dA$

where R_b is the region in the first quadrant bounded by the x axis, the y axis, and the circle $x^2 + y^2 = b^2$.

b. Assume that

$$\int_0^\infty e^{-x^2}\,dx \int_0^\infty e^{-y^2}\,dy = \int_0^\infty \int_0^\infty e^{-x^2}e^{-y^2}\,dx\,dy$$

$$= \int_0^\infty \int_0^\infty e^{-(x^2+y^2)}\,dx\,dy$$

Using (a), show that

$$\int_0^\infty e^{-x^2}\,dx = \frac{1}{2}\sqrt{\pi}$$

c. Using (b), show that

$$\int_{-\infty}^\infty e^{-x^2}\,dx = \sqrt{\pi}$$

31. Using the ideas of Exercise 30, show that

$$\int_{-\infty}^\infty e^{-x^2/2}\,dx = \sqrt{2\pi}$$

(Thus the region between the graph of $y = e^{-x^2/2}$ and the x axis has area $\sqrt{2\pi}$, a result fundamental to statistics.)

32. Let S be the region in the $r\theta$ plane bounded by the straight lines $r = s$ and $r = t$ with $s < t$, and $\theta = \alpha$ and $\theta = \beta$ with $0 \le \beta - \alpha \le 2\pi$. Let R be the region in the xy plane bounded by the polar graphs $r = s$, $r = t$, $\theta = \alpha$, and $\theta = \beta$. By what factor must the area of S be multiplied to yield the area of R?

14.3
SURFACE AREA

In Section 6.5 we presented formulas for the surface area of the surface generated by revolving a curve in the xy plane about the x axis. The double integral will play a prominent role in this section as we define the surface area S of more general kinds of three-dimensional surfaces.

Let Σ be the graph in space of a function f that is continuous on a vertically or horizontally simple region R in the xy plane, and assume that f has continuous partial derivatives on R. If R' is a rectangle containing R, then let us consider a partition $\mathscr{P}$ of R' into subrectangles, numbered so that $R_1, R_2, \ldots, R_n$ are the ones entirely contained in R. For any integer k between 1 and n let (x_k, y_k) be the corner of R_k with the smallest values of x_k and y_k, as in Figure 14.26. Let Δx_k and Δy_k be the lengths of two adjacent sides of R_k, and let ΔA_k be the area of R_k. The

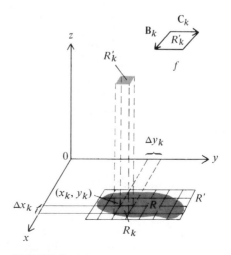

FIGURE 14.26

projection of R_k onto the plane tangent to Σ at the point $(x_k, y_k, f(x_k, y_k))$ yields a parallelogram R'_k (Figure 14.26). We think of the area $\Delta A'_k$ of R'_k as an approximation to the area of the portion of the surface Σ that corresponds to R_k. The vectors $\mathbf{i} + f_x(x_k, y_k)\mathbf{k}$ and $\mathbf{j} + f_y(x_k, y_k)\mathbf{k}$ are parallel to two adjacent sides of R'_k (see Section 13.3). This implies that

$$\mathbf{B}_k = \Delta x_k \mathbf{i} + f_x(x_k, y_k)\Delta x_k \mathbf{k} \quad \text{and} \quad \mathbf{C}_k = \Delta y_k \mathbf{j} + f_y(x_k, y_k)\Delta y_k \mathbf{k}$$

form two adjacent sides of R'_k. By the results of Section 11.4, the area $\Delta A'_k$ of R'_k is the length of the cross product of the sides $\mathbf{B}_k$ and $\mathbf{C}_k$. As a result,

$$\begin{aligned}
\Delta A'_k &= \|\mathbf{B}_k \times \mathbf{C}_k\| \\
&= \|-f_x(x_k, y_k)\Delta x_k \Delta y_k \mathbf{i} - f_y(x_k, y_k)\Delta x_k \Delta y_k \mathbf{j} + \Delta x_k \Delta y_k \mathbf{k}\| \\
&= \sqrt{[f_x(x_k, y_k)]^2 + [f_y(x_k, y_k)]^2 + 1}\; \Delta x_k \Delta y_k \\
&= \sqrt{[f_x(x_k, y_k)]^2 + [f_y(x_k, y_k)]^2 + 1}\; \Delta A_k
\end{aligned}$$

Therefore

$$\sum_{k=1}^{n} \Delta A'_k = \sum_{k=1}^{n} \sqrt{[f_x(x_k, y_k)]^2 + [f_y(x_k, y_k)]^2 + 1}\; \Delta A_k$$

The sum $\sum_{k=1}^{n} \Delta A'_k$ should approximate the surface area S of Σ, while the sum on the right-hand side of the equation is a Riemann sum approximating

$$\iint\limits_{R} \sqrt{[f_x(x, y)]^2 + [f_y(x, y)]^2 + 1}\; dA$$

This suggests the following definition of surface area.

DEFINITION 14.6

> Let R be a vertically or horizontally simple region, and let f have continuous partial derivatives on R. If Σ is the graph of f on R, then the **surface area** S of Σ is defined by
>
> $$S = \iint\limits_{R} \sqrt{[f_x(x, y)]^2 + [f_y(x, y)]^2 + 1}\; dA \tag{1}$$

It is usually difficult or impossible to compute surface area by this formula. However, in some cases it is possible.

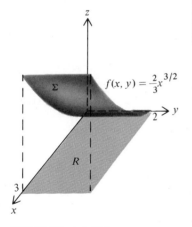

FIGURE 14.27

Example 1 Let R be the rectangular region bounded by the lines

$$x = 0, \qquad x = 3, \qquad y = 0, \qquad y = 2$$

and let $f(x, y) = \frac{2}{3}x^{3/2}$. Find the surface area S of the portion of the graph of f that lies over R (Figure 14.27).

Solution   Notice that

$$f_x(x, y) = x^{1/2} \quad \text{and} \quad f_y(x, y) = 0 \quad \text{for } (x, y) \text{ in } R$$

Consequently (1) implies that

$$S = \iint_R \sqrt{(x^{1/2})^2 + 0 + 1} \, dA = \int_0^3 \int_0^2 \sqrt{x + 1} \, dy \, dx$$

$$= 2 \int_0^3 \sqrt{x + 1} \, dx = \frac{4}{3}(x + 1)^{3/2} \Big|_0^3 = \frac{28}{3} \quad \square$$

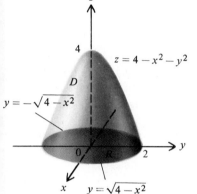

FIGURE 14.28

Example 2   Find the surface area S of the portion of the paraboloid

$$z = 4 - x^2 - y^2$$

that lies above the xy plane (see Figure 14.28).

Solution   The given surface lies over the region R in the xy plane bounded by the circle $x^2 + y^2 = 4$. If $f(x, y) = 4 - x^2 - y^2$, then

$$f_x(x, y) = -2x \quad \text{and} \quad f_y(x, y) = -2y$$

By (1),

$$S = \iint_R \sqrt{4x^2 + 4y^2 + 1} \, dA$$

This double integral is tailor-made for evaluation by polar coordinates:

$$S = \int_0^{2\pi} \int_0^2 \sqrt{4r^2 + 1} \, r \, dr \, d\theta = \frac{1}{12} \int_0^{2\pi} (4r^2 + 1)^{3/2} \Big|_0^2 \, d\theta$$

$$= \frac{1}{12} \int_0^{2\pi} (17^{3/2} - 1) \, d\theta = \frac{1}{6}\pi(17^{3/2} - 1) \quad \square$$

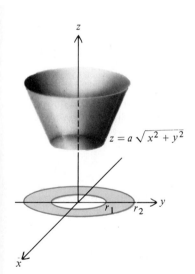

FIGURE 14.29

Example 3   Let $a > 0$. Find the surface area S of the frustum of the cone

$$z = a\sqrt{x^2 + y^2}$$

with minimum and maximum radii r_1 and r_2, respectively (Figure 14.29).

Solution   The given surface lies over the region R in the xy plane bounded by the annulus

$$r_1^2 \le x^2 + y^2 \le r_2^2$$

If $f(x, y) = a\sqrt{x^2 + y^2}$, then

$$f_x(x, y) = \frac{ax}{\sqrt{x^2 + y^2}} \quad \text{and} \quad f_y(x, y) = \frac{ay}{\sqrt{x^2 + y^2}}$$

so that by (1),

$$S = \iint_R \sqrt{\left(\frac{ax}{\sqrt{x^2 + y^2}}\right)^2 + \left(\frac{ay}{\sqrt{x^2 + y^2}}\right)^2 + 1} \, dA$$

$$= \iint_R \sqrt{\frac{a^2 x^2}{x^2 + y^2} + \frac{a^2 y^2}{x^2 + y^2} + 1} \, dA = \iint_R \sqrt{a^2 + 1} \, dA$$

$$= \sqrt{a^2 + 1} \iint_R 1 \, dA = \sqrt{a^2 + 1} \quad \text{(area of } R)$$

Since the area of the annulus R is $\pi r_2^2 - \pi r_1^2 = \pi(r_2^2 - r_1^2)$, we find that

$$S = \pi \sqrt{a^2 + 1} \, (r_2^2 - r_1^2) \quad \square$$

The formula for the surface area of the frustum obtained in Example 3 looks rather different from the one given in (1) of Section 6.5, which involved the slant height l of the frustum. We now determine the slant height of the frustum in Example 3.

The intersection of the frustum with the first quadrant of the yz plane is the line segment in the yz plane having equation $z = ay$ for $r_1 \leq y \leq r_2$. The slant height l, which is the length of this line segment, is given by

$$l = \sqrt{(r_2 - r_1)^2 + (ar_2 - ar_1)^2} = \sqrt{(1 + a^2)(r_2 - r_1)^2}$$
$$= \sqrt{1 + a^2} \, (r_2 - r_1) = \sqrt{a^2 + 1} \, (r_2 - r_1)$$

Thus the surface area S can be written as follows:

$$S = \pi \overbrace{\sqrt{a^2 + 1} \, (r_2 - r_1)}^{l}(r_2 + r_1) = \pi(r_1 + r_2)l$$

This is consistent with the formula appearing in (1) of Section 6.5.

The formula for surface area in this section can be shown to be compatible with the formula in (4) of Section 6.5 when the surface is obtained by revolving the graph of a continuous, nonnegative function about the x axis. The details are tedious, so we omit them.

EXERCISES 14.3

In Exercises 1–10 find the surface area of the given surface.

1. The portion of the plane $x + 2y + 3z = 6$ in the first octant

2. The portion of the paraboloid $z = 9 - x^2 - y^2$ above the xy plane

3. The portion of the paraboloid $z = 9 - x^2 - y^2$ above the plane $z = 5$

4. The portion of the sphere $x^2 + y^2 + z^2 = 4$ that is inside the cylinder $x^2 + y^2 = 1$

5. The portion of the sphere $x^2 + y^2 + z^2 = 16$ that is inside the cylinder $x^2 - 4x + y^2 = 0$

6. The portion of the cylinder $x^2 + z^2 = 9$ that is directly above the triangle with vertices $(0, 0, 0)$, $(1, 0, 0)$, and $(1, 1, 0)$

7. The portion of the parabolic sheet $z = x^2$ directly above the triangle with vertices $(0, 0, 0)$, $(1, 0, 0)$, and $(1, 1, 0)$

8. The portion of the sphere $x^2 + y^2 + z^2 = 9$ that is inside the paraboloid $x^2 + y^2 = 8z$

9. The portion of the sphere $x^2 + y^2 + z^2 = 14z$ that is inside the paraboloid $x^2 + y^2 = 5z$

10. The portion of the graph of $z = \frac{2}{3}\sqrt{2}x^{3/2} + \frac{2}{3}y^{3/2}$ directly over the region in the xy plane between the graph of $y = x^2$ and the x axis on $[0, 1]$

14.4
TRIPLE INTEGRALS

An analogue of single and double integrals, called the triple integral, can be defined for functions of three variables.

Definition of the Triple Integral

Let R be a vertically or horizontally simple region in the xy plane, and let F_1 and F_2 be continuous on R and satisfy

$$F_1(x, y) \le F_2(x, y) \quad \text{for } (x, y) \text{ in } R$$

Let D denote the solid region consisting of all points (x, y, z) such that

$$(x, y) \text{ is in } R \quad \text{and} \quad F_1(x, y) \le z \le F_2(x, y)$$

(Figure 14.30). We refer to D as the **solid region between the graphs of F_1 and F_2 on R**. In examples it will be convenient to refer to D as the solid region between the graphs of $z = F_1(x, y)$ and $z = F_2(x, y)$ for (x, y) in R.

In order to define integrals of continuous functions on D, it is necessary to make further assumptions on the boundary of D that normally appear only in advanced calculus texts. We will assume that every solid region that appears in the remainder of this book satisfies these assumptions.

In preparation for the definition of the triple integral, let D' be a rectangular parallelepiped containing D, and let $\mathscr{P}$ partition D' into smaller rectangular parallelepipeds (Figure 14.31), numbered so that $D_1, D_2, \ldots, D_n$ are the ones entirely contained in D, and $D_{n+1}, D_{n+2}, \ldots, D_p$ are the ones only partially contained in D. In addition, we assume that f is nonnegative and continuous on D, and we let m_k be the minimum value of f on D_k for $1 \le k \le n$ and M_k the maximum value of f on D_k for $1 \le k \le p$. If ΔV_k denotes the volume of D_k and

$$L_f(\mathscr{P}) = \sum_{k=1}^{n} m_k \Delta V_k \quad \text{and} \quad U_f(\mathscr{P}) = \sum_{k=1}^{p} M_k \Delta V_k$$

then it can be proved that there is exactly one number I such that

$$L_f(\mathscr{P}) \le I \le U_f(\mathscr{P})$$

for every partition $\mathscr{P}$ of any parallelepiped D' containing D.

It is possible to extend the definitions of $L_f(\mathscr{P})$ and $U_f(\mathscr{P})$ to cover the case in which f is not necessarily nonnegative, and in such a way that $L_f(\mathscr{P}) \le U_f(\mathscr{P})$. We can therefore make the following definition.

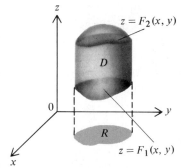

The solid region between the graphs of F_1 and F_2 on R

FIGURE 14.30

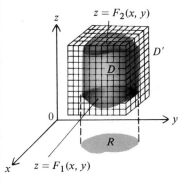

FIGURE 14.31

DEFINITION 14.7

Let D be the solid region between the graphs of two continuous functions F_1 and F_2 on a vertically or horizontally simple region R in the xy plane. If f is continuous on D, we write

$$\iiint_D f(x, y, z)\, dV$$

for the unique number that lies between $L_f(\mathscr{P})$ and $U_f(\mathscr{P})$ for every partition $\mathscr{P}$ of any parallelepiped containing D. The number $\iiint_D f(x, y, z)\, dV$ is called the **triple integral** of f on D.

Now let (x_k, y_k, z_k) be an arbitrary point in D_k for $1 \leq k \leq n$. Then the sum

$$\sum_{k=1}^{n} f(x_k, y_k, z_k)\, \Delta V_k$$

is a **Riemann sum** for f on D, and it approximates $\iiint_D f(x, y, z)\, dV$ in the following sense.

THEOREM 14.8

Let f be continuous on the solid region D between the graphs of two continuous functions, and let D' be a parallelepiped containing D. For any $\varepsilon > 0$ there is a number $\delta > 0$ such that the following statement holds: If $\mathscr{P}$ is a partition of D' into subparallelepipeds whose dimensions are all less then δ, and if $D_1, D_2, \ldots, D_n$ are the subparallelepipeds in D' entirely contained in D, then

$$\left| \iiint_D f(x, y, z)\, dV - \sum_{k=1}^{n} f(x_k, y_k, z_k) \Delta V_k \right| < \varepsilon$$

where (x_k, y_k, z_k) is an arbitrary point in D_k for $1 \leq k \leq n$.

The conclusion of Theorem 14.8 is usually expressed as

$$\iiint_D f(x, y, z)\, dV = \lim_{\|\mathscr{P}\| \to 0} \sum_{k=1}^{n} f(x_k, y_k, z_k)\, \Delta V_k$$

where $\|\mathscr{P}\|$ is the largest of the dimensions of the subparallelepipeds in $\mathscr{P}$ and is called the **norm of the partition** $\mathscr{P}$.

Evaluation of Triple Integrals

In general, lower sums, upper sums, and Riemann sums are not very effective in evaluating a triple integral $\iiint_D f(x, y, z)\, dV$. Once again, iterated integrals (this time three integrals in succession) provide a method of evaluating triple integrals.

THEOREM 14.9

Let D be the solid region between the graphs of two continuous functions F_1 and F_2 on a vertically or horizontally simple region R in the xy plane, and let f be continuous on D. Then

$$\iiint_D f(x, y, z)\, dV = \iint_R \left(\int_{F_1(x, y)}^{F_2(x, y)} f(x, y, z)\, dz \right) dA$$

We evaluate $\int_{F_1(x, y)}^{F_2(x, y)} f(x, y, z)\, dz$ by integrating with respect to z while both x and y are held fixed, thus obtaining a number depending on x and y. If R is the vertically simple region between the graphs of g_1 and g_2 on $[a, b]$, we evaluate the double integral over R by using Theorem 14.4(a), obtaining

$$\iiint_D f(x, y, z)\, dV = \int_a^b \left[\int_{g_1(x)}^{g_2(x)} \left(\int_{F_1(x, y)}^{F_2(x, y)} f(x, y, z)\, dz \right) dy \right] dx \qquad (1)$$

In contrast, if R is the horizontally simple region between the graphs of h_1 and h_2 on $[c, d]$, then from Theorem 14.4(b) we obtain

$$\iiint_D f(x, y, z)\, dV = \int_c^d \left[\int_{h_1(y)}^{h_2(y)} \left(\int_{F_1(x, y)}^{F_2(x, y)} f(x, y, z)\, dz \right) dx \right] dy \qquad (2)$$

The integrals on the right sides of (1) and (2) are called (triple) **iterated integrals**. Normally we omit the parentheses and brackets and write

$$\int_a^b \int_{g_1(x)}^{g_2(x)} \int_{F_1(x, y)}^{F_2(x, y)} f(x, y, z)\, dz\, dy\, dx$$

for

$$\int_a^b \left[\int_{g_1(x)}^{g_2(x)} \left(\int_{F_1(x, y)}^{F_2(x, y)} f(x, y, z)\, dz \right) dy \right] dx$$

and

$$\int_c^d \int_{h_1(y)}^{h_2(y)} \int_{F_1(x, y)}^{F_2(x, y)} f(x, y, z)\, dz\, dx\, dy$$

for

$$\int_c^d \left[\int_{h_1(y)}^{h_2(y)} \left(\int_{F_1(x, y)}^{F_2(x, y)} f(x, y, z)\, dz \right) dx \right] dy$$

Example 1 Let R be the rectangular region in the xy plane bounded by the lines $x = 2$, $x = \frac{5}{2}$, $y = 0$, and $y = \pi$, and let D be the parallelepiped between the

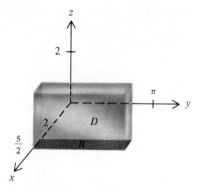

FIGURE 14.32

graphs of

$$z = 0 \quad \text{and,} \quad z = 2 \quad \text{on } R$$

(Figure 14.32). Evaluate $\iiint_D zx \sin xy \, dV$.

Solution   By (1) we have

$$\iiint_D zx \sin xy \, dV = \int_2^{5/2} \int_0^\pi \int_0^2 zx \sin xy \, dz \, dy \, dx$$

$$= \int_2^{5/2} \int_0^\pi \left(\frac{z^2 x}{2} \sin xy \right) \Big|_0^2 dy \, dx$$

$$= \int_2^{5/2} \int_0^\pi 2x \sin xy \, dy \, dx = \int_2^{5/2} -2 \cos xy \Big|_0^\pi dx$$

$$= \int_2^{5/2} (2 - 2 \cos \pi x) \, dx = \left(2x - \frac{2}{\pi} \sin \pi x \right) \Big|_2^{5/2}$$

$$= \left(5 - \frac{2}{\pi} \sin \frac{5\pi}{2} \right) - \left(4 - \frac{2}{\pi} \sin 2\pi \right) = 1 - \frac{2}{\pi} \quad \square$$

Example 2   Let R be the triangular region in the xy plane between the graphs of $y = 0$ and $y = x$ for $0 \le x \le 1$, and let D be the solid region between the graphs of the surfaces

$$z = -y^2 \quad \text{and} \quad z = x^2 \quad \text{for } (x, y) \text{ in } R$$

(Figure 14.33). Evaluate $\iiint_D (x + 1) \, dV$.

Solution   By (1) we have

$$\iiint_D (x + 1) \, dV = \int_0^1 \int_0^x \int_{-y^2}^{x^2} (x + 1) \, dz \, dy \, dx$$

$$= \int_0^1 \int_0^x (x + 1)z \Big|_{-y^2}^{x^2} dy \, dx$$

$$= \int_0^1 \int_0^x (x + 1)(x^2 + y^2) \, dy \, dx$$

$$= \int_0^1 \int_0^x (x^3 + x^2 + xy^2 + y^2) \, dy \, dx$$

$$= \int_0^1 \left(x^3 y + x^2 y + \frac{xy^3}{3} + \frac{y^3}{3} \right) \Big|_0^x dx$$

$$= \frac{4}{3} \int_0^1 (x^4 + x^3) \, dx$$

$$= \frac{4}{3} \left(\frac{x^5}{5} + \frac{x^4}{4} \right) \Big|_0^1 = \frac{3}{5} \quad \square$$

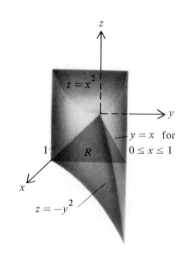

FIGURE 14.33

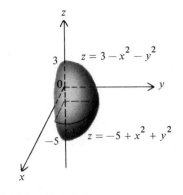

FIGURE 14.34

Example 3 Let D be the solid region bounded by the portions of the two circular paraboloids

$$z = 3 - x^2 - y^2 \quad \text{and} \quad z = -5 + x^2 + y^2$$

for which $x \geq 0$ and $y \geq 0$ (Figure 14.34). Evaluate $\iiint_D y \, dV$.

Solution To be able to use (2), we must determine the region R in the xy plane such that D is the solid region between the two paraboloids on R. For this purpose we first determine where the two paraboloids intersect. At any point (x, y, z) of intersection, x and y must satisfy

$$3 - x^2 - y^2 = -5 + x^2 + y^2$$

which is equivalent to $x^2 + y^2 = 4$. But if $x^2 + y^2 = 4$, then

$$z = 3 - x^2 - y^2 = 3 - 4 = -1$$

so the intersection lies in the plane $z = -1$. We find that the corresponding region R in the xy plane is the horizontally simple region in the first quadrant that lies inside the circle $x^2 + y^2 = 4$, and hence between the graphs of $x = 0$ and $x = \sqrt{4 - y^2}$ for $0 \leq y \leq 2$. Since

$$3 - x^2 - y^2 \geq -5 + x^2 + y^2 \quad \text{for } (x, y) \text{ in } R$$

we have

$$\iiint_D y \, dV = \int_0^2 \int_0^{\sqrt{4-y^2}} \int_{-5+x^2+y^2}^{3-x^2-y^2} y \, dz \, dx \, dy$$

$$= \int_0^2 \int_0^{\sqrt{4-y^2}} yz \Big|_{-5+x^2+y^2}^{3-x^2-y^2} \, dx \, dy$$

$$= \int_0^2 \int_0^{\sqrt{4-y^2}} y(8 - 2x^2 - 2y^2) \, dx \, dy$$

$$= \int_0^2 \left[y(8 - 2y^2)x - \frac{2}{3}yx^3 \right]\Big|_0^{\sqrt{4-y^2}} \, dy$$

$$= \int_0^2 \left[y(8 - 2y^2)\sqrt{4 - y^2} - \frac{2}{3}y(4 - y^2)^{3/2} \right] dy$$

$$= \frac{4}{3} \int_0^2 y(4 - y^2)^{3/2} \, dy = \frac{-4}{15}(4 - y^2)^{5/2}\Big|_0^2 = \frac{128}{15} \quad \square$$

Volume by Triple Integration

Let D be the solid region between the graphs of two continuous functions F_1 and F_2 on a region R in the xy plane. The volume V of D is defined by

$$V = \iiint_D 1 \, dV \tag{3}$$

By Theorem 14.9 we can rewrite the integral for the volume as a double integral over R:

$$V = \iiint\limits_D 1 \, dV = \iint\limits_R \left[\int_{F_1(x,\,y)}^{F_2(x,\,y)} 1 \, dz \right] dA = \iint\limits_R [F_2(x, y) - F_1(x, y)] \, dA$$

In case F_1 is the constant function 0, this double integral is just the double integral appearing in Definition 14.1(b) with F_2 replacing f. Consequently (3) defines volume for a larger class of solid regions than Definition 14.1(b) does (because F_1 need not be 0).

Example 4   Let D be the solid region in the first octant between the graphs of

$$z = x^2 + 2y + 1 \quad \text{and} \quad z = y + 2$$

Find the volume V of D.

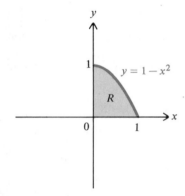

FIGURE 14.35

Solution   First we must determine the region R. The surfaces $z = x^2 + 2y + 1$ and $z = y + 2$ intersect in the curve

$$x^2 + 2y + 1 = y + 2$$

or equivalently

$$y = 1 - x^2$$

Since D is in the first octant, R must be in the first quadrant. Therefore R is the vertically simple region between the graphs of $y = 0$ and $y = 1 - x^2$ on $[0, 1]$ (Figure 14.35). Since $x^2 + 2y + 1 \le y + 2$ for (x, y) in R, it follows from (3) that

$$V = \iiint\limits_D 1 \, dV = \int_0^1 \int_0^{1-x^2} \int_{x^2+2y+1}^{y+2} 1 \, dz \, dy \, dx$$

$$= \int_0^1 \int_0^{1-x^2} z \bigg|_{x^2+2y+1}^{y+2} \, dy \, dx$$

$$= \int_0^1 \int_0^{1-x^2} (1 - x^2 - y) \, dy \, dx$$

$$= \int_0^1 \left[(1 - x^2)y - \frac{1}{2} y^2 \right] \bigg|_0^{1-x^2} \, dx$$

$$= \int_0^1 \left[(1 - x^2)^2 - \frac{1}{2}(1 - x^2)^2 \right] dx$$

$$= \int_0^1 \frac{1}{2}(1 - x^2)^2 \, dx = \frac{1}{2}\left(x - \frac{2x^3}{3} + \frac{x^5}{5} \right) \bigg|_0^1 = \frac{4}{15} \quad \square$$

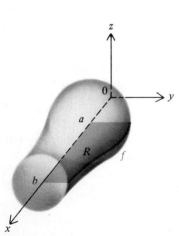

FIGURE 14.36

Example 5   Suppose that f is a nonnegative continuous function on $[a, b]$ and R is the region in the xy plane between the graph of f and the x axis on $[a, b]$. Let D be the solid region generated by revolving R about the x axis (Figure 14.36).

Show that the formula for the volume V of D given in (3) yields formula (3) of Section 6.1.

Solution Let R_1 be the vertically simple region in the xy plane between the graphs of $-f$ and f on $[a, b]$. Then D is the solid region between the surfaces

$$z = -\sqrt{[f(x)]^2 - y^2} \quad \text{and} \quad z = \sqrt{[f(x)]^2 - y^2} \quad \text{on } R_1$$

Thus (3) implies that

$$V = \iiint_D 1 \, dV = \int_a^b \int_{-f(x)}^{f(x)} \int_{-\sqrt{[f(x)]^2 - y^2}}^{\sqrt{[f(x)]^2 - y^2}} 1 \, dz \, dy \, dx$$

$$= \int_a^b \int_{-f(x)}^{f(x)} z \Big|_{-\sqrt{[f(x)]^2 - y^2}}^{\sqrt{[f(x)]^2 - y^2}} dy \, dx$$

$$= \int_a^b \int_{-f(x)}^{f(x)} 2\sqrt{[f(x)]^2 - y^2} \, dy \, dx$$

Making the substitution

$$y = f(x) \sin \theta, \quad \text{so that} \quad dy = f(x) \cos \theta \, d\theta$$

we obtain

$$\int_a^b \int_{-f(x)}^{f(x)} 2\sqrt{[f(x)]^2 - y^2} \, dy \, dx = \int_a^b \int_{-\pi/2}^{\pi/2} [2f(x) \cos \theta] f(x) \cos \theta \, d\theta \, dx$$

$$= \int_a^b \int_{-\pi/2}^{\pi/2} 2[f(x)]^2 \cos^2 \theta \, d\theta \, dx$$

$$= \int_a^b 2[f(x)]^2 \left(\frac{1}{2}\theta + \frac{1}{4}\sin 2\theta \right) \Big|_{-\pi/2}^{\pi/2} dx$$

$$= \int_a^b \pi[f(x)]^2 \, dx$$

Thus
$$V = \int_a^b \pi[f(x)]^2 \, dx$$

The final integral is the one that appeared in (3) of Section 6.1. $\square$

Example 5 shows that our formulas for volume here and in Section 6.1 are compatible. Of course, the one presented here is much more general, because the solid region need not be generated by revolving a region about the x axis. It is also possible to show that the formula in Section 6.2 for the volume of a solid of revolution about the y axis is compatible with our new definition.

It is sometimes advantageous to interchange the roles of x, y, and z; instead of regarding a given region D as the solid region between the graphs of two functions of x and y, we occasionally wish to regard D as the solid region between two functions of x and z or two functions of y and z.

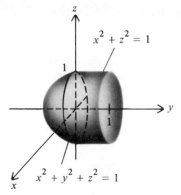

FIGURE 14.37

Example 6   Evaluate $\iiint_D y\sqrt{1 - x^2}\,dV$, where D is the region depicted in Figure 14.37.

Solution   It is convenient to think of D as the region between the graphs of $y = -\sqrt{1 - x^2 - z^2}$ and $y = 1$ on the region in the xz plane bounded by the circle $x^2 + z^2 = 1$. Using this approach, we find that

$$\iiint_D y\sqrt{1 - x^2}\,dV = \int_{-1}^{1}\int_{-\sqrt{1-x^2}}^{\sqrt{1-x^2}}\int_{-\sqrt{1-x^2-z^2}}^{1} y\sqrt{1 - x^2}\,dy\,dz\,dx$$

$$= \int_{-1}^{1}\int_{-\sqrt{1-x^2}}^{\sqrt{1-x^2}}\left(\frac{1}{2}\sqrt{1 - x^2}\,y^2\Big|_{-\sqrt{1-x^2-z^2}}^{1}\right)dz\,dx$$

$$= \int_{-1}^{1}\int_{-\sqrt{1-x^2}}^{\sqrt{1-x^2}}\frac{1}{2}\sqrt{1 - x^2}(x^2 + z^2)\,dz\,dx$$

$$= \int_{-1}^{1}\frac{1}{2}\sqrt{1 - x^2}\left(x^2 z + \frac{1}{3}z^3\right)\Big|_{-\sqrt{1-x^2}}^{\sqrt{1-x^2}}dx$$

$$= \int_{-1}^{1}\left[x^2(1 - x^2) + \frac{1}{3}(1 - x^2)^2\right]dx$$

$$= \int_{-1}^{1}\left(-\frac{2}{3}x^4 + \frac{1}{3}x^2 + \frac{1}{3}\right)dx$$

$$= \left(-\frac{2}{15}x^5 + \frac{1}{9}x^3 + \frac{1}{3}x\right)\Big|_{-1}^{1} = \frac{28}{45} \quad \square$$

Suppose D is composed of two or more subregions $D_1, D_2, \ldots, D_n$ of the type appearing in Definition 14.7 and having at most boundaries in common. Then we define $\iiint_D f(x, y, z)\,dV$ by the formula

$$\iiint_D f(x, y, z)\,dV = \iiint_{D_1} f(x, y, z)\,dV + \iiint_{D_2} f(x, y, z)\,dV + \cdots$$

$$+ \iiint_{D_n} f(x, y, z)\,dV$$

(4)

Mass and Charge   According to the molecular theory of matter, any piece of matter is just a collection of molecules and consequently has a mass equal to the sum of the masses of its molecules. However, the molecules are so numerous that finding such a sum would defy even modern computers. In order to calculate the mass of an object like a cube of ice or a steel beam, one idealizes the mass and thinks of it as being "spread everywhere" throughout the object (not only at the locations of the molecules). If the mass m of the object is also distributed uniformly, or homogeneously, throughout the object and the volume V of the object is not 0, then we can define the **mass density** δ of the object by the equation

$$\delta = \frac{m}{V} \tag{5}$$

This equation applies to any such object, which might be a quart of oil, a cube of ice, or a steel beam.

In contrast, consider a cube of frozen orange juice. Due to settling, the bottom portion of the cube would probably be heavier than the top, in which case the mass would not be distributed uniformly throughout the cube. In order to describe the mass in the different parts of the cube, we define the **mass density** $\delta(x, y, z)$ at the point (x, y, z) in the cube by

$$\delta(x, y, z) = \lim_{\|\Delta V\| \to 0} \frac{\Delta m}{\Delta V} \tag{6}$$

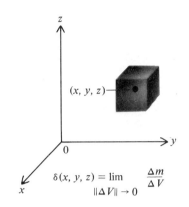

$\delta(x, y, z) = \lim_{\|\Delta V\| \to 0} \dfrac{\Delta m}{\Delta V}$

FIGURE 14.38

provided that this limit exists, where Δm is the mass and ΔV the volume of a parallelepiped centered at (x, y, z) and where $\|\Delta V\|$ is the largest dimension of the parallelepiped (Figure 14.38). Formula (6) applies to any object for which the limit exists. Our assumption that the mass is "spread everywhere" throughout the object is formalized as the assumption that δ is defined and continuous on D. Under this hypothesis, scientists are able to describe many physical phenomena with remarkable accuracy. Our immediate goal is to show that under this hypothesis the total mass m of the object is given by

$$m = \iiint_D \delta(x, y, z)\, dV$$

Let D' be any parallelepiped containing D, and let $\mathscr{P}$ be a partition of D' into subparallelepipeds, numbered so that $D_1, D_2, \ldots, D_n$ are the ones entirely contained in D. For each k between 1 and n, choose an arbitrary point (x_k, y_k, z_k) in D_k, and let Δm_k be the mass of D_k and ΔV_k the volume of D_k. Formula (6) suggests that $\delta(x_k, y_k, z_k) \Delta V_k$ approximates Δm_k if the dimensions of D_k are small. Since m is approximately $\Delta m_1 + \Delta m_2 + \cdots + \Delta m_n$, m is approximated by

$$\sum_{k=1}^{n} \delta(x_k, y_k, z_k)\, \Delta V_k$$

which is a Riemann sum for δ on D. From Theorem 14.8 we conclude that the **mass** of D is given by

$$m = \iiint_D \delta(x, y, z)\, dV \tag{7}$$

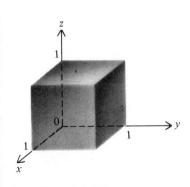

FIGURE 14.39

Example 7   A cubical object occupying the solid region D shown in Figure 14.39 has a density given by $\delta(x, y, z) = 1 + xyz$ for (x, y, z) in D. Find the mass m of the object.

Solution   By (7) we find that

$$m = \iiint\limits_{D} \delta(x, y, z)\, dV = \iiint\limits_{D} (1 + xyz)\, dV$$

$$= \int_0^1 \int_0^1 \int_0^1 (1 + xyz)\, dz\, dy\, dx$$

$$= \int_0^1 \int_0^1 \left(z + \frac{1}{2} xyz^2 \right)\Big|_0^1 dy\, dx$$

$$= \int_0^1 \int_0^1 \left(1 + \frac{1}{2} xy \right) dy\, dx = \int_0^1 \left(y + \frac{1}{4} xy^2 \right)\Big|_0^1 dx$$

$$= \int_0^1 \left(1 + \frac{1}{4} x \right) dx = \left(x + \frac{1}{8} x^2 \right)\Big|_0^1 = \frac{9}{8} \quad \square$$

If an object such as a copper ball has a homogeneous charge distribution (that is, if any two portions of the object having the same volume have the same charge), then we define the **charge density** ρ of the object by

$$\rho = \frac{q}{V}$$

where q is the charge in the object and V is its volume. If an object does not have a homogeneous charge distribution, the **charge density** $\rho(x, y, z)$ at any point (x, y, z) in the object is defined by

$$\rho(x, y, z) = \lim_{\|\Delta V\| \to 0} \frac{\Delta q}{\Delta V}$$

provided that this limit exists, where Δq is the charge in a parallelepiped centered at (x, y, z), ΔV is the volume of the parallelepiped, and $\|\Delta V\|$ is the largest dimension of the parallelepiped. The same argument we used to derive (7) shows that the **total charge** q in a charged object is given by

$$q = \iiint\limits_{D} \rho(x, y, z)\, dV \tag{8}$$

where D is the region occupied by the object.

Example 8   Suppose the charge density $\rho(x, y, z)$ at any point (x, y, z) occupying the ball $x^2 + y^2 + z^2 \le 1$ is equal to the distance from (x, y, z) to the xy plane. Find the total charge q in the ball.

Solution   Let D be the ball. We know that $\rho(x, y, z) = |z|$, so an application of (8) yields

$$
q = \iiint_D \rho(x, y, z)\, dV
$$

$$
= \int_{-1}^{1} \int_{-\sqrt{1-x^2}}^{\sqrt{1-x^2}} \int_{-\sqrt{1-x^2-y^2}}^{\sqrt{1-x^2-y^2}} |z|\, dz\, dy\, dx
$$

$$
= \int_{-1}^{1} \int_{-\sqrt{1-x^2}}^{\sqrt{1-x^2}} \frac{1}{2} z|z| \Big|_{-\sqrt{1-x^2-y^2}}^{\sqrt{1-x^2-y^2}}\, dy\, dx
$$

$$
= \int_{-1}^{1} \int_{-\sqrt{1-x^2}}^{\sqrt{1-x^2}} (1 - x^2 - y^2)\, dy\, dx
$$

$$
= \int_{-1}^{1} \left((1 - x^2)y - \frac{y^3}{3} \right) \Big|_{-\sqrt{1-x^2}}^{\sqrt{1-x^2}}\, dx
$$

$$
= \frac{4}{3} \int_{-1}^{1} (1 - x^2)^{3/2}\, dx
$$

$$
\overset{x=\sin\theta}{=} \frac{4}{3} \int_{-\pi/2}^{\pi/2} \cos^4 \theta\, d\theta
$$

$$
\overset{\underset{\text{Section 8.1}}{(14)\text{ of}}}{=} \frac{4}{3} \left(\frac{1}{4} \cos^3 \theta \sin \theta + \frac{3}{8} \cos \theta \sin \theta + \frac{3}{8} \theta \right) \Big|_{-\pi/2}^{\pi/2}
$$

$$
= \frac{\pi}{2} \quad \square
$$

EXERCISES 14.4

In Exercises 1–10 evaluate the iterated integral.

1. $\displaystyle\int_0^3 \int_{-1}^{1} \int_2^4 (y - xz)\, dz\, dy\, dx$

2. $\displaystyle\int_0^3 \int_{-1}^{1} \int_2^4 (y - xz)\, dy\, dx\, dz$

3. $\displaystyle\int_{-1}^{1} \int_0^x \int_{x-y}^{x+y} (z - 2x - y)\, dz\, dy\, dx$

4. $\displaystyle\int_0^{\pi/2} \int_0^1 \int_0^{\sqrt{1-x^2}} x \cos z\, dy\, dx\, dz$

5. $\displaystyle\int_0^{\ln 3} \int_0^1 \int_0^y (z^2 + 1)e^{(y^2)}\, dx\, dz\, dy$

6. $\displaystyle\int_0^{\sqrt{\pi/6}} \int_0^y \int_0^y (1 + y^2 z \cos xz)\, dx\, dz\, dy$

7. $\displaystyle\int_{-15}^{13} \int_1^e \int_0^{1/\sqrt{x}} z(\ln x)^2\, dz\, dx\, dy$

8. $\displaystyle\int_{-1}^{2} \int_1^{y+2} \int_e^{e^2} \frac{x+y}{z}\, dz\, dx\, dy$

9. $\displaystyle\int_0^{\pi/2} \int_0^{\pi/2} \int_0^{\sin z} x^2 \sin y\, dx\, dy\, dz$

10. $\displaystyle\int_{-\pi/2}^{\pi/2} \int_{-\cos z}^{\cos z} \int_{-\cos zy}^{\cos zy} x \cos zy\, dx\, dy\, dz$

In Exercises 11–19 evaluate the integral.

11. $\iiint_D e^y\, dV$, where D is the solid region bounded by the planes $y = 1$, $z = 0$, $y = x$, $y = -x$, and $z = y$

12. $\iiint_D 1/x\, dV$, where D is the prism bounded by the planes $x + y + z = 4$, $y = x$, $x = 1$, $x = 2$, $z = 0$, and $y = 0$

13. $\iiint_D ye^{xy}\, dV$, where D is the cube bounded by the planes $x = 1$, $x = 3$, $y = 0$, $y = 2$, $z = -2$, and $z = 0$

14. $\iiint_D xy\, dV$, where D is the solid region in the first octant bounded above by the hemisphere $z = \sqrt{4 - x^2 - y^2}$ and on the sides and bottom by the coordinate planes

15. $\iiint_D zy \, dV$, where D is the solid region in the first octant bounded above by the plane $z = 1$ and below by the cone $z = \sqrt{x^2 + y^2}$

16. $\iiint_D (x + z) \, dV$, where D is the solid region bounded by the cylinder $x^2 + z^2 = 1$ and by the planes $y = -4$, $y = 5$, $x = 0$, and $x = 1$

17. $\iiint_D z \, dV$, where D is the solid region bounded above by the sphere $x^2 + y^2 + z^2 = 9$, below by the plane $z = 0$, and on the sides by the planes $x = -1$, $x = 1$, $y = -1$, and $y = 1$

18. $\iiint_D xz \, dV$, where D is the solid region in the first octant bounded above by the sphere $x^2 + y^2 + z^2 = 4$, below by the plane $z = 0$, and on the sides by the planes $x = 0$ and $y = 0$ and the cylinder $x^2 + y^2 = 1$

19. $\iiint_D 3xy \, dV$, where D is the solid region bounded below by the cone $z = \sqrt{x^2 + y^2}$ and above by the cylinder $x^2 + z^2 = 1$

In Exercises 20–26 find the volume V of the region.

20. The solid region bounded by the plane $x + 3y + 6z = 1$ and the three coordinate planes

21. The solid region in the first octant bounded by the planes $z = 10 + x + y$, $y = 2 - x$, $y = x$, $z = 0$, and $x = 0$

22. The solid region in the first octant bounded above by the plane $z = 2x$, below by the xy plane, and on the sides by the elliptic cylinder $2x^2 + y^2 = 1$ and the plane $y = 0$

23. The solid region bounded above by the circular paraboloid $z = 4(x^2 + y^2)$, below by the plane $z = -2$, and on the sides by the parabolic sheet $y = x^2$ and the plane $y = x$

24. The solid region bounded above by the elliptic paraboloid $z = 2x^2 + 3y^2$, on the sides by the parabolic sheet $y^2 = 1 - x$ and the plane $x = 0$, and below by the plane $z = 0$ (*Hint:* Integrate with respect to x before y.)

*25. The solid region bounded above by the plane $z = h$ ($h > 0$) and below by the cone $z = h\sqrt{x^2 + y^2}$ (The answer is the volume of a cone whose height is h and whose base has radius 1.)

*26. A right pyramid with height h and a square base having sides of length s (*Hint:* Place the base on the xy plane with center at the origin and vertices on the axes. Compute the volume of the portion in the first octant and multiply by 4.)

27. An object occupies the tetrahedron in the first octant bounded by the coordinate planes and the plane $2x + y + z = 1$. Find the total mass m of the object if the mass density at a given point in the object is
   a. equal to the distance from the xy plane.
   b. equal to twice the distance from the xz plane.
   c. equal to the distance from the yz plane.

28. An object occupies a cube of volume 8 with three of its faces on the coordinate planes. If the mass density at any point in the cube is equal to the square of its distance from the origin, find the total mass m of the object.

29. An object occupies the solid region bounded above by the sphere $x^2 + y^2 + z^2 = 9$ and below by the xy plane. If the charge density of the object at any point is equal to its distance from the xy plane, find the total charge q on the object.

30. An object occupies the solid region in the first octant bounded by the coordinate planes and the two cylinders $x^2 + y^2 = 4$ and $y^2 + z^2 = 4$. If the charge density at any point (x, y, z) is x, find the total charge q on the object.

31. Suppose D is the solid region between the graphs of two continuous functions and has nonzero volume V. If f is continuous on D, then the **average value of f on D** is defined to be $(1/V) \iiint_D f(x, y, z) \, dV$. Under the following conditions, find the average value of f on D.
   a. $f(x, y, z) = x + y + z$; D is the solid region bounded by the sheet $z = x^2$ and by the planes $z = x$, $y = 0$, and $y = 1$.
   b. $f(x, y, z) = xy$; D is the solid region in the first octant bounded by the circular paraboloids $z = 2 - x^2 - y^2$ and $z = x^2 + y^2$ and the planes $x = 0$ and $y = 0$.

14.5
TRIPLE INTEGRALS IN CYLINDRICAL COORDINATES

Just as certain double integrals are easier to evaluate by means of polar coordinates than by rectangular coordinates, certain triple integrals are easier to evaluate by coordinates other than rectangular coordinates. In Sections 14.5 and 14.6 we introduce two new types of coordinates—cylindrical and spherical—and explain how to evaluate triple integrals by means of these coordinates.

Cylindrical Coordinates

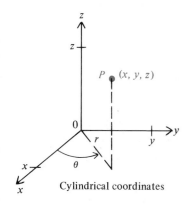

Cylindrical coordinates

FIGURE 14.40

Let (x, y, z) be the rectangular coordinates of a point P in space. If (r, θ) is a set of polar coordinates for the point (x, y), then we call (r, θ, z) a set of **cylindrical coordinates** for P (Figure 14.40). Given the rectangular coordinates (x, y, z) of a point P, we can determine a set of cylindrical coordinates for P with the aid of the formulas

$$x^2 + y^2 = r^2 \quad \text{and} \quad \tan \theta = \frac{y}{x} \quad (\text{if } x \neq 0)$$

Conversely, from any set (r, θ, z) of cylindrical coordinates of a point P we can determine the rectangular coordinates (x, y, z) of P by the formulas

$$x = r \cos \theta \quad \text{and} \quad y = r \sin \theta$$

Let us compare equations in rectangular coordinates and in cylindrical coordinates for several surfaces (Table 14.1). These surfaces, which are displayed in Figure 14.41(a)–(d), will recur frequently in this and the next chapter. The simplicity of the formula $r = a$ for the cylinder suggests a reason for the name "cylindrical coordinates." We usually refer to the double cone simply as a cone, and to its upper and lower portions as its **upper nappe** and **lower nappe**.

TABLE 14.1

Surface	Rectangular	Cylindrical
Cylinder	$x^2 + y^2 = a^2$	$r = a$
Sphere	$x^2 + y^2 + z^2 = a^2$	$r^2 + z^2 = a^2$
Double circular cone	$x^2 + y^2 = a^2 z^2$	$r = az$ or $z = r \cot \phi$
Circular paraboloid	$x^2 + y^2 = az$	$r^2 = az$

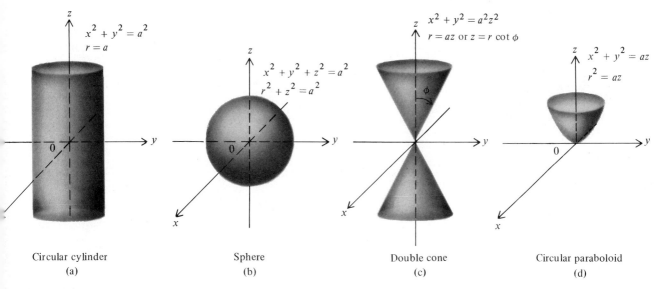

Circular cylinder
(a)

Sphere
(b)

Double cone
(c)

Circular paraboloid
(d)

FIGURE 14.41

Triple Integrals in Cylindrical Coordinates

Theorem 14.9 asserts that under suitable conditions

$$\iiint\limits_{D} f(x, y, z)\, dV = \iint\limits_{R} \left(\int_{F_1(x, y)}^{F_2(x, y)} f(x, y, z)\, dz \right) dA$$

We also know how to evaluate double integrals by means of polar coordinates, thanks to Theorem 14.5. Together these two results give us a method of evaluating triple integrals by means of cylindrical coordinates.

THEOREM 14.10

Let D be the solid region between the graphs of F_1 and F_2 on R, where R is the plane region between the polar graphs of h_1 and h_2 on $[\alpha, \beta]$, with $0 \le \beta - \alpha \le 2\pi$ and $0 \le h_1(\theta) \le h_2(\theta)$ for $\alpha \le \theta \le \beta$. If f is continuous on D, then

$$\iiint\limits_{D} f(x, y, z)\, dV = \int_{\alpha}^{\beta} \int_{h_1(\theta)}^{h_2(\theta)} \int_{F_1(r\cos\theta, r\sin\theta)}^{F_2(r\cos\theta, r\sin\theta)} f(r\cos\theta, r\sin\theta, z)\, r\, dz\, dr\, d\theta$$

Integration by means of cylindrical coordinates is especially effective when expressions containing $x^2 + y^2$ appear in the integrand or in the limits of integration and the region over which the integration is taken is easily described by polar coordinates.

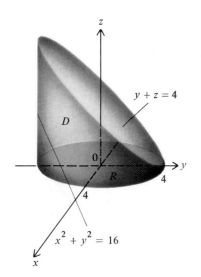

FIGURE 14.42

Example 1 Let D be the solid region bounded above by the plane $y + z = 4$, below by the xy plane, and on the sides by the cylinder $x^2 + y^2 = 16$ (Figure 14.42). Evaluate $\iiint_{D} \sqrt{x^2 + y^2}\, dV$.

Solution Observe that D is the solid region between the graphs of $z = 0$ and $z = 4 - y$ on R, where R is the disk $x^2 + y^2 \le 16$. In polar coordinates R is the region between the polar graphs of

$$r = 0 \quad \text{and} \quad r = 4 \quad \text{for } 0 \le \theta \le 2\pi$$

Consequently in cylindrical coordinates D is the solid region between the graphs of

$$z = 0 \quad \text{and} \quad z = 4 - r\sin\theta \quad \text{for } (r, \theta) \text{ in } R$$

Then Theorem 14.10 implies that

$$\iiint\limits_{D} \sqrt{x^2 + y^2}\, dV = \int_{0}^{2\pi} \int_{0}^{4} \int_{0}^{4 - r\sin\theta} r \cdot r\, dz\, dr\, d\theta$$

$$= \int_{0}^{2\pi} \int_{0}^{4} r^2 z \Big|_{0}^{4 - r\sin\theta}\, dr\, d\theta$$

$$= \int_{0}^{2\pi} \int_{0}^{4} (4r^2 - r^3 \sin\theta)\, dr\, d\theta$$

$$= \int_0^{2\pi} \left(\frac{4}{3}r^3 - \frac{r^4}{4}\sin\theta \right)\Big|_0^4 \, d\theta$$

$$= \int_0^{2\pi} \left(\frac{256}{3} - 64\sin\theta \right) d\theta$$

$$= \left(\frac{256}{3}\theta + 64\cos\theta \right)\Big|_0^{2\pi} = \frac{512}{3}\pi \quad \square$$

Example 2 Find the volume V of the solid region that the cylinder $r = a\cos\theta$ cuts out of the sphere of radius a centered at the origin (Figure 14.43).

Solution In this case R is bounded by the polar graphs of

$$r = 0 \quad \text{and} \quad r = a\cos\theta \quad \text{for} \ -\frac{\pi}{2} \le \theta \le \frac{\pi}{2}$$

and D is the solid region determined by the sphere between the graphs of

$$z = -\sqrt{a^2 - r^2} \quad \text{and} \quad z = \sqrt{a^2 - r^2} \quad \text{for} \ (r, \theta) \text{ in } R$$

Using symmetry, we find that

$$V = \iiint_D 1 \, dV = \int_{-\pi/2}^{\pi/2} \int_0^{a\cos\theta} \int_{-\sqrt{a^2-r^2}}^{\sqrt{a^2-r^2}} 1 \cdot r \, dz \, dr \, d\theta$$

$$= 4 \int_0^{\pi/2} \int_0^{a\cos\theta} \int_0^{\sqrt{a^2-r^2}} r \, dz \, dr \, d\theta = 4 \int_0^{\pi/2} \int_0^{a\cos\theta} r\sqrt{a^2-r^2} \, dr \, d\theta$$

$$= 4 \int_0^{\pi/2} -\frac{1}{3}(a^2-r^2)^{3/2}\Big|_0^{a\cos\theta} \, d\theta = \frac{4}{3} \int_0^{\pi/2} [a^3 - (a^2 - a^2\cos^2\theta)^{3/2}] \, d\theta$$

$$= \frac{4}{3}a^3 \int_0^{\pi/2} (1 - \sin^3\theta) \, d\theta = \frac{4}{3}a^3 \int_0^{\pi/2} [1 - (1 - \cos^2\theta)\sin\theta] \, d\theta$$

$$= \frac{4}{3}a^3 \left(\theta + \cos\theta - \frac{1}{3}\cos^3\theta \right)\Big|_0^{\pi/2} = \frac{2}{9}a^3(3\pi - 4) \quad \square$$

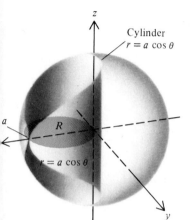

Cylinder
$r = a\cos\theta$

a

R

$r = a\cos\theta$

y

Sphere $x^2 + y^2 + z^2 = a^2$

FIGURE 14.43

Example 3 Let D be the solid region between the sphere $r^2 + z^2 = a^2$ and the nappe of the cone $z = r\cot\phi$ that makes an angle of ϕ with the positive z axis. Show that the volume V of D is given by

$$V = \frac{2\pi a^3}{3}(1 - \cos\phi)$$

Solution First we consider the case $0 \le \phi \le \pi/2$. Then the nappe in question is the upper nappe (Figure 14.44). To determine where the sphere and the cone intersect, we substitute $r\cot\phi$ for z in the equation $r^2 + z^2 = a^2$ of the sphere and obtain

$$r^2(1 + \cot^2\phi) = a^2$$

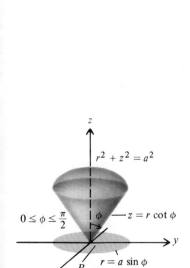

$r^2 + z^2 = a^2$

$0 \le \phi \le \frac{\pi}{2}$

ϕ —— $z = r\cot\phi$

y

R

$r = a\sin\phi$

x

FIGURE 14.44

This implies that

$$r \csc \phi = a, \quad \text{or} \quad r = a \sin \phi$$

It follows that D is the solid region between the cone and the sphere on the plane region R bounded by the circle $r = a \sin \phi$ (see Figure 14.44). As a result,

$$V = \iiint\limits_{D} 1 \, dV = \int_0^{2\pi} \int_0^{a \sin \phi} \int_{r \cot \phi}^{\sqrt{a^2 - r^2}} 1 \cdot r \, dz \, dr \, d\theta$$

$$= \int_0^{2\pi} \int_0^{a \sin \phi} rz \Big|_{r \cot \phi}^{\sqrt{a^2 - r^2}} dr \, d\theta$$

$$= \int_0^{2\pi} \int_0^{a \sin \phi} \left(r\sqrt{a^2 - r^2} - r^2 \cot \phi \right) dr \, d\theta$$

$$= \int_0^{2\pi} \left(-\frac{1}{3}(a^2 - r^2)^{3/2} - \frac{r^3}{3} \cot \phi \right) \Big|_0^{a \sin \phi} d\theta$$

$$= \int_0^{2\pi} \left(-\frac{1}{3}(a^2 - a^2 \sin^2 \phi)^{3/2} - \frac{a^3 \sin^3 \phi}{3} \cot \phi + \frac{a^3}{3} \right) d\theta$$

$$= \frac{a^3}{3}(-\cos^3 \phi - \sin^2 \phi \cos \phi + 1) \int_0^{2\pi} 1 \, d\theta$$

$$= \frac{2\pi a^3}{3}[-(\cos^2 \phi + \sin^2 \phi)\cos \phi + 1] = \frac{2\pi a^3}{3}(1 - \cos \phi)$$

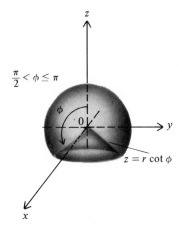

FIGURE 14.45

If $\pi/2 < \phi \le \pi$, then the nappe under consideration is the lower nappe (Figure 14.45). The volume V we seek is the difference between the volume $4\pi a^3/3$ of the complete sphere and the volume of the solid region bounded below by the sphere and above by the lower nappe of the cone. It follows from the first part of the solution that the latter volume is

$$\frac{2\pi a^3}{3}[1 - \cos(\pi - \phi)]$$

Thus

$$V = \frac{4\pi a^3}{3} - \frac{2\pi a^3}{3}[1 - \cos(\pi - \phi)]$$

$$= \frac{4\pi a^3}{3} - \frac{2\pi a^3}{3}(1 + \cos \phi)$$

$$= \frac{2\pi a^3}{3}(1 - \cos \phi) \quad \square$$

It is not uncommon to encounter a triple iterated integral in rectangular coordinates that could be evaluated more easily by changing to cylindrical coordinates. The procedure is first to describe in cylindrical coordinates the region over which the integration is to be performed and then to evaluate the corresponding iterated integral in cylindrical coordinates.

Example 4 Evaluate $\int_{-2}^{2}\int_{-\sqrt{4-x^2}}^{\sqrt{4-x^2}}\int_{(x^2+y^2)^2}^{1} x^2\,dz\,dy\,dx$.

Solution The limits -2 and 2 on the first integral and $-\sqrt{4-x^2}$ and $\sqrt{4-x^2}$ on the second integral tell us that those two integrals are taken over the region bounded by the circle $x^2+y^2=4$, or $r=2$, in the xy plane. It follows that

$$\int_{-2}^{2}\int_{-\sqrt{4-x^2}}^{\sqrt{4-x^2}}\int_{(x^2+y^2)^2}^{1} x^2\,dz\,dy\,dx = \int_{0}^{2\pi}\int_{0}^{2}\int_{r^4}^{1}(r\cos\theta)^2 r\,dz\,dr\,d\theta$$

$$= \int_{0}^{2\pi}\int_{0}^{2}\left[(r^3\cos^2\theta)z\right]\Big|_{r^4}^{1}\,dr\,d\theta$$

$$= \int_{0}^{2\pi}\int_{0}^{2}(r^3-r^7)\cos^2\theta\,dr\,d\theta$$

$$= \int_{0}^{2\pi}\left(\frac{r^4}{4}-\frac{r^8}{8}\right)\cos^2\theta\Big|_{0}^{2}\,d\theta$$

$$= -28\int_{0}^{2\pi}\cos^2\theta\,d\theta$$

$$= -28\int_{0}^{2\pi}\left(\frac{1}{2}+\frac{1}{2}\cos 2\theta\right)d\theta$$

$$= -14\left(\theta+\frac{1}{2}\sin 2\theta\right)\Big|_{0}^{2\pi} = -28\pi \quad \square$$

EXERCISES 14.5

In Exercises 1–8 write the equation in cylindrical coordinates, and sketch its graph.

1. $y=-4$
2. $x=5z$
3. $x+y+z=3$
4. $x^2+y^2+z^2=16$
5. $x^2+y^2+z=1$
6. $x^2+y^2+z^2=0$
7. $4x^2+4y^2-z^2=0$
8. $x^2+y^2+3z^2=9$

In Exercises 9–13 evaluate the iterated integral.

9. $\int_{0}^{2\pi}\int_{1}^{2}\int_{0}^{5} e^z r\,dz\,dr\,d\theta$

10. $\int_{0}^{\pi/2}\int_{0}^{1}\int_{0}^{\sqrt{1-r^2}} r\sin\theta\,dz\,dr\,d\theta$

11. $\int_{-\pi/2}^{0}\int_{0}^{2\sin\theta}\int_{0}^{r^2} r^2\cos\theta\,dz\,dr\,d\theta$

12. $\int_{0}^{\pi/4}\int_{0}^{1+\cos\theta}\int_{0}^{r} 1\,dz\,dr\,d\theta$

13. $\int_{-\pi/4}^{\pi/4}\int_{0}^{1-2\cos^2\theta}\int_{0}^{1} r\sin\theta\,dz\,dr\,d\theta$

In Exercises 14–18 express the triple integral as an iterated integral in cylindrical coordinates. Then evaluate it.

14. $\iiint_D (x^2+y^2)\,dV$, where D is the solid region bounded by the cylinder $x^2+y^2=1$ and the planes $z=0$ and $z=4$

15. $\iiint_D z\,dV$, where D is the portion of the ball $x^2+y^2+z^2\le 1$ that lies in the first octant

16. $\iiint_D y^2\,dV$, where D is the solid region common to the cylinder $x^2+y^2=1$ and the sphere $x^2+y^2+z^2=4$

17. $\iiint_D xz\,dV$, where D is the portion of the ball $x^2+y^2+z^2\le 4$ in the first octant

18. $\iiint_D yz\,dV$, where D is the solid region in the first octant bounded by the sphere $x^2+y^2+z^2=1$, the circular cylinder $r=\cos\theta$, and the planes $y=0$ and $z=0$

In Exercises 19–31 find the volume V of the region.

19. The solid region bounded below by the surface $z=\sqrt{r}$ and above by the plane $z=1$

20. The solid region bounded above by the sphere $x^2 + y^2 + z^2 = 2$ and below by the circular paraboloid $z = x^2 + y^2$

21. The solid region bounded above by the surface $z = e^{-x^2 - y^2}$, below by the xy plane, and on the sides by the cylinder $x^2 + y^2 = 1$

22. The solid region above the plane $z = 1$ and inside the sphere $x^2 + y^2 + z^2 = 2$

23. The solid region inside the sphere $x^2 + y^2 + z^2 = 4$ and above the upper nappe of the cone $z^2 = 3x^2 + 3y^2$

24. The solid region bounded above by the paraboloid $z = 1 - x^2 - y^2$ and below by the plane $z = -3$

25. The solid region inside the cone $z = r$ and between the planes $z = 1$ and $z = 2$

26. The solid region bounded above by the cone $z = 8 - \sqrt{x^2 + y^2}$, on the sides by the cylinder $x^2 + y^2 = 2x$, and below by the xy plane

27. The solid region bounded above by the plane $z = y$ and below by the paraboloid $z = x^2 + y^2$

28. The solid region inside the sphere $x^2 + y^2 + z^2 = 4a^2$ and the cylinder $x^2 + y^2 = a^2$

29. The solid region inside the sphere $r^2 + z^2 = a^2$ and the cylinder $r = a \sin \theta$

30. The solid region bounded on the sides by the surface $r = 1 + \cos \theta$, above by the upper nappe of the cone $z = r$, and below by the xy plane

31. The solid region in the first octant bounded by the coordinate planes, the circular paraboloid $z = r^2$, and the surface $r^2 = 4 \cos \theta$

32. An object occupies the solid region bounded by the cylinder $x^2 + y^2 = 9$ and the planes $z = 0$ and $z = 5$. If the mass density at any point is equal to the distance from the point to the axis of the cylinder, find the total mass m of the object.

33. An object occupies the solid region bounded by the upper nappe of the cone $z^2 = 9x^2 + 9y^2$ and the plane $z = 9$. Find the total mass m of the object if the mass density at (x, y, z) is equal to the distance from (x, y, z) to the top.

34. A cylindrical hole of radius b is bored out of the center of a spherical ball of radius a, where $0 < b < a$. How much volume is retained? (This exercise is related to Review Exercise 8 of Chapter 6.)

35. Find the volume V of the solid region bounded above by the plane $z = h$ ($h > 0$) and below by the upper nappe of the cone $z^2 = h^2(x^2 + y^2)$.

14.6
TRIPLE INTEGRALS IN SPHERICAL COORDINATES

The final coordinate system we discuss is the spherical coordinate system. This system simplifies evaluation of triple integrals over solid regions bounded by surfaces such as spheres and cones.

Spherical Coordinates Let (x, y, z) and (r, θ, z) be, respectively, sets of rectangular and cylindrical coordinates (see Figure 14.46) for a point P in space, with $r \geq 0$. We let

$$\rho = \text{the length of the line segment } OP = x^2 + y^2 + z^2$$

If $\rho \neq 0$, then we let

$$\phi = \text{the angle } OP \text{ makes with the positive } z \text{ axis, with } 0 \leq \phi \leq \pi$$

If $\rho = 0$, then ϕ may be chosen arbitrarily. Finally, we continue to let

$$\theta = \text{the angle } OQ \text{ makes with the positive } x \text{ axis}$$

(Figure 14.46). The point P is specified by the three quantities ρ, ϕ, and θ, and we call the triple (ρ, ϕ, θ) a set of **spherical coordinates** for P. Figure 14.47 depicts the points A, B, and C with spherical coordinates $(1, \pi/6, \pi/4)$, $(1, \pi/3, \pi/4)$, and $(1, 2\pi/3, \pi/4)$, respectively.

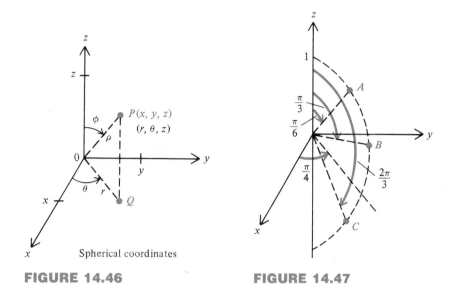

FIGURE 14.46

Spherical coordinates

FIGURE 14.47

From trigonometry we find that

$$r = \rho \sin \phi \quad \text{and} \quad z = \rho \cos \phi$$

These equations, along with the polar coordinate formulas

$$x = r \cos \theta \quad \text{and} \quad y = r \sin \theta$$

yield the following formulas for converting from spherical coordinates to rectangular coordinates:

$$x = r \cos \theta = \rho \sin \phi \cos \theta$$
$$y = r \sin \theta = \rho \sin \phi \sin \theta$$
$$z = \rho \cos \phi$$

Example 1 A point P has spherical coordinates $(8, 2\pi/3, -\pi/6)$. Find the rectangular coordinates for P.

Solution The rectangular coordinates for P are given by

$$x = 8 \sin\left(\frac{2\pi}{3}\right) \cos\left(-\frac{\pi}{6}\right) = 8\left(\frac{\sqrt{3}}{2}\right)\left(\frac{\sqrt{3}}{2}\right) = 6$$

$$y = 8 \sin\left(\frac{2\pi}{3}\right) \sin\left(-\frac{\pi}{6}\right) = 8\left(\frac{\sqrt{3}}{2}\right)\left(-\frac{1}{2}\right) = -2\sqrt{3}$$

$$z = 8 \cos\left(\frac{2\pi}{3}\right) = 8\left(-\frac{1}{2}\right) = -4 \quad \square$$

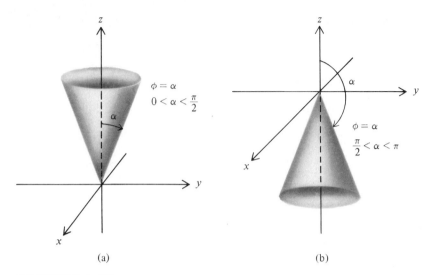

FIGURE 14.48

Spheres centered at the origin have particularly simple equations in spherical coordinates. Indeed, from the definition of ρ we find that an equation in spherical coordinates of the sphere centered at the origin with radius a is $\rho = a$.

Analogously, if $0 < \alpha < \pi$ and $\alpha \neq \pi/2$, then the graph of the equation $\phi = \alpha$ is a cone whose angle with respect to the positive z axis is α. Notice that if $0 < \alpha < \pi/2$, then the sides of the cone open upward, whereas if $\pi/2 < \alpha < \pi$, then the sides open downward (Figure 14.48(a) and (b)). The equations $\phi = 0$, $\phi = \pi$, and $\phi = \pi/2$ yield the positive z axis, negative z axis, and xy plane, respectively.

In Table 14.2 we list the surfaces on which the spherical coordinates ρ, ϕ, and θ are constant.

Formulas in spherical coordinates for horizontal planes are a little less simple than those for vertical half-planes.

TABLE 14.2

Surface	Equation
Sphere	$\rho = a$
Cone	$\phi = a$
Vertical half-plane	$\theta = a$

Example 2   Find an equation in spherical coordinates for the plane $z = 2$.

Solution   Since $z = \rho \cos \phi$, the equation $z = 2$ becomes

$$\rho \cos \phi = 2$$

Dividing by $\cos \phi$, which cannot be 0 if $\rho \cos \phi = 2$, we find that

$$\rho = \frac{2}{\cos \phi} = 2 \sec \phi$$

Thus an equation in spherical coordinates for the plane $z = 2$ is $\rho = 2 \sec \phi$.   □

Triple Integrals in Spherical Coordinates   Just as parallelepipeds form the basic three-dimensional solids for integration in rectangular coordinates, spherical wedges form the basic solids for integration in spherical coordinates. A **spherical wedge** D is a solid region bounded by the surfaces

$$\rho = \rho_0 \quad \text{and} \quad \rho = \rho_1, \qquad \phi = \phi_0 \quad \text{and} \quad \phi = \phi_1, \qquad \theta = \theta_0 \quad \text{and} \quad \theta = \theta_1$$

where

$$0 \le \rho_0 \le \rho_1, \qquad 0 \le \phi_0 \le \phi_1 \le \pi, \quad \text{and} \quad 0 \le \theta_1 - \theta_0 \le 2\pi$$

(Figure 14.49).

Our next goal is to estimate the volume V of D, under the conditions that $\rho_1 - \rho_0$, $\phi_1 - \phi_0$, and $\theta_1 - \theta_0$ are each small. In that case adjacent sides of D are nearly perpendicular to each other, so that if the sides are labeled as in Figure 14.49, then

$$V \approx abc$$

The formula for a is easy:

$$a = \rho_1 - \rho_0$$

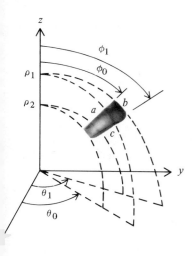

FIGURE 14.49

Using the fact that b is the length of an arc of angle $\phi_1 - \phi_0$ in a circle of radius ρ_1, we deduce that

$$b = \rho_1(\phi_1 - \phi_0)$$

Finally, c is approximately the length of an arc of angle $\theta_1 - \theta_0$ in a circle of radius $r_1 = \rho_1 \sin \phi_1$, so that

$$c \approx r_1(\theta_1 - \theta_0) = (\rho_1 \sin \phi_1)(\theta_1 - \theta_0)$$

Consequently

$$V \approx abc \approx \overbrace{(\rho_1 - \rho_0)}^{a}[\overbrace{\rho_1(\phi_1 - \phi_0)}^{b}][\overbrace{(\rho_1 \sin \phi_1)(\theta_1 - \theta_0)}^{\approx c}]$$

$$= (\rho_1^2 \sin \phi_1)(\rho_1 - \rho_0)(\phi_1 - \phi_0)(\theta_1 - \theta_0)$$

It is possible to show that if ρ_1^* and ϕ_1^* are chosen appropriately, with $\rho_0 \le \rho_1^* \le \rho_1$ and $\phi_0 \le \phi_1^* \le \phi_1$, then

$$V = (\rho_1^*)^2(\sin \phi_1^*)(\rho_1 - \rho_0)(\phi_1 - \phi_0)(\theta_1 - \theta_0) \tag{1}$$

This formula will be of fundamental use in deriving a formula for evaluating triple integrals by means of spherical coordinates.

Let h_1, h_2, F_1, and F_2 be continuous, and let D be the collection of all points in space with spherical coordinates (ρ, ϕ, θ) such that

$$\alpha \le \theta \le \beta, \qquad 0 \le h_1(\theta) \le \phi \le h_2(\theta) \le \pi, \quad \text{and} \quad F_1(\phi, \theta) \le \rho \le F_2(\phi, \theta)$$

with $0 \le \beta - \alpha \le 2\pi$. Our objective is to find a way of evaluating $\iiint_D f(x, y, z)\, dV$, where f is continuous on D.

First circumscribe D with a spherical wedge D', and then partition D' into smaller spherical wedges, of which $D_1, D_2, \ldots, D_n$ are entirely contained in D. For each k between 1 and n, let D_k have dimensions $\Delta\rho_k$, $\Delta\phi_k$, and $\Delta\theta_k$ in spherical coordinates. By (1) the volume V_k of the wedge D_k is given by

$$V_k = (\rho_k^*)^2 \sin \phi_k^* \, \Delta\rho_k \, \Delta\phi_k \, \Delta\theta_k$$

where $(\rho_k^*, \phi_k^*, \theta_k^*)$ are spherical coordinates of a point in D_k. This suggests that the triple integral $\iiint_D f(x, y, z)\, dV$ is approximately

$$\sum_{k=1}^{n} f(\rho_k^* \sin \phi_k^* \cos \theta_k^*, \rho_k^* \sin \phi_k^* \sin \theta_k^*, \rho_k^* \cos \phi_k^*)(\rho_k^*)^2 \sin \phi_k^* \, \Delta\rho_k \, \Delta\phi_k \, \Delta\theta_k$$

By an argument similar to the one used in Section 14.2, it can be shown that this sum is a Riemann sum for the triple integral

$$\int_\alpha^\beta \int_{h_1(\theta)}^{h_2(\theta)} \int_{F_1(\phi,\theta)}^{F_2(\phi,\theta)} f(\rho \sin \phi \cos \theta, \rho \sin \phi \sin \theta, \rho \cos \phi) \rho^2 \sin \phi \, d\rho \, d\phi \, d\theta$$

This leads us to the following theorem.

THEOREM 14.11

Let α and β be real numbers with $\alpha \le \beta \le \alpha + 2\pi$. Let h_1, h_2, F_1, and F_2 be continuous functions with $0 \le h_1 \le h_2 \le \pi$ and $0 \le F_1 \le F_2$. Let D be the solid region consisting of all points in space whose spherical coordinates (ρ, ϕ, θ) satisfy

$$\alpha \le \theta \le \beta$$

$$h_1(\theta) \le \phi \le h_2(\theta)$$

$$F_1(\phi, \theta) \le \rho \le F_2(\phi, \theta)$$

If f is continuous on D, then

$$\iiint\limits_D f(x, y, z) \, dV$$

$$= \int_\alpha^\beta \int_{h_1(\theta)}^{h_2(\theta)} \int_{F_1(\phi,\theta)}^{F_2(\phi,\theta)} f(\rho \sin \phi \cos \theta, \rho \sin \phi \sin \theta, \rho \cos \phi) \rho^2 \sin \phi \, d\rho \, d\phi \, d\theta$$

When you find the limits of integration for a triple integral in spherical coordinates, keep in mind that

ρ measures distance from the origin, so $\rho \ge 0$
ϕ measures the angle from the positive z axis, and $0 \le \phi \le \pi$
θ measures the angle from the positive x axis

Example 3   Let D be the solid region between the spheres $\rho = 1$ and $\rho = 2$. Evaluate $\iiint_D z^2 \, dV$.

Solution   Observe that D is the collection of points with spherical coordinates (ρ, ϕ, θ) such that

$$0 \le \theta \le 2\pi, \qquad 0 \le \phi \le \pi, \quad \text{and} \quad 1 \le \rho \le 2$$

Since $z = \rho \cos \phi$, Theorem 14.11 tells us that

$$\iiint\limits_D z^2 \, dV = \int_0^{2\pi} \int_0^\pi \int_1^2 (\rho^2 \cos^2 \phi) \rho^2 \sin \phi \, d\rho \, d\phi \, d\theta$$

$$= \int_0^{2\pi} \int_0^\pi \int_1^2 \rho^4 \cos^2 \phi \sin \phi \, d\rho \, d\phi \, d\theta$$

$$= \int_0^{2\pi} \int_0^\pi \frac{1}{5} \rho^5 \cos^2 \phi \sin \phi \Big|_1^2 d\phi \, d\theta$$

$$= \int_0^{2\pi} \int_0^\pi \frac{31}{5} \cos^2 \phi \sin \phi \, d\phi \, d\theta$$

$$= \int_0^{2\pi} \frac{31}{5} \left[\frac{1}{3} (-\cos^3 \phi) \Big|_0^\pi \right] d\theta$$

$$= \frac{62}{15} \int_0^{2\pi} 1 \, d\theta = \frac{124}{15} \pi \quad \square$$

Example 4   Find the volume V of the solid region D between the spheres $x^2 + y^2 + z^2 = 1$ and $x^2 + y^2 + z^2 = 9$, and above the upper nappe of the cone $z^2 = 3(x^2 + y^2)$ (Figure 14.50).

Solution   In spherical coordinates the equations of the given spheres are $\rho = 1$ and $\rho = 3$. Next, recall that $z^2 = \rho^2 \cos^2 \phi$ and $x^2 + y^2 = r^2 = \rho^2 \sin^2 \phi$. Therefore the equation $z^2 = 3(x^2 + y^2)$ becomes

$$\rho^2 \cos^2 \phi = 3\rho^2 \sin^2 \phi$$

It follows that $\rho = 0$ or $\cos^2 \phi = 3 \sin^2 \phi$, so that if $\rho \neq 0$, then

$$\tan^2 \phi = \frac{1}{3}$$

Since D lies above the upper nappe of the cone $z^2 = 3(x^2 + y^2)$, we have $0 \leq \phi \leq \pi/2$, so $\tan \phi = 1/\sqrt{3}$. Thus $\phi = \pi/6$. Consequently D is the collection of

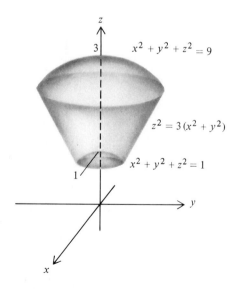

FIGURE 14.50

points with spherical coordinates (ρ, ϕ, θ) such that

$$1 \le \rho \le 3, \qquad 0 \le \phi \le \frac{\pi}{6} \quad \text{and} \quad 0 \le \theta \le 2\pi$$

Therefore

$$V = \iiint\limits_{D} 1 \, dV = \int_0^{2\pi} \int_0^{\pi/6} \int_1^3 \rho^2 \sin\phi \, d\rho \, d\phi \, d\theta$$

$$= \int_0^{2\pi} \int_0^{\pi/6} \left[\frac{\rho^3}{3} \Big|_1^3 \right] \sin\phi \, d\phi \, d\theta = \int_0^{2\pi} \int_0^{\pi/6} \frac{26}{3} \sin\phi \, d\phi \, d\theta$$

$$= \frac{26}{3} \int_0^{2\pi} (-\cos\phi) \Big|_0^{\pi/6} d\theta = \frac{26}{3} \left(1 - \frac{\sqrt{3}}{2} \right) \int_0^{2\pi} 1 \, d\theta$$

$$= \frac{26}{3} \pi (2 - \sqrt{3}) \quad \square$$

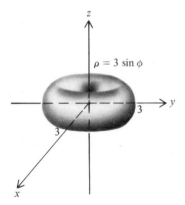

$\rho = 3\sin\phi$

FIGURE 14.51

Example 5 Find the volume V of the solid region enclosed by the torus $\rho = 3\sin\phi$ (Figure 14.51).

Solution In this case D is the collection of points with spherical coordinates (ρ, ϕ, θ) such that

$$0 \le \theta \le 2\pi, \qquad 0 \le \phi \le \pi, \quad \text{and} \quad 0 \le \rho \le 3\sin\phi$$

Consequently

$$V = \iiint\limits_{D} 1 \, dV = \int_0^{2\pi} \int_0^{\pi} \int_0^{3\sin\phi} \rho^2 \sin\phi \, d\rho \, d\phi \, d\theta$$

$$= \int_0^{2\pi} \int_0^{\pi} \frac{\rho^3}{3} \sin\phi \Big|_0^{3\sin\phi} d\phi \, d\theta = \int_0^{2\pi} \int_0^{\pi} 9 \sin^4\phi \, d\phi \, d\theta$$

$$\overset{\underset{\text{(13) of}}{\text{Section 8.1}}}{=} \int_0^{2\pi} 9 \left(-\frac{1}{4} \sin^3\phi \cos\phi - \frac{3}{8} \sin\phi \cos\phi + \frac{3}{8}\phi \right) \Big|_0^{\pi} d\theta$$

$$= \int_0^{2\pi} \frac{27\pi}{8} \, d\theta = \frac{27\pi^2}{4} \quad \square$$

Example 6 A star occupies a spherical region D centered at the origin with radius a. Its density is given by

$$\delta(x, y, z) = ce^{-[(x^2 + y^2 + z^2)/a^2]^{3/2}}$$

where c is a constant. Determine the total mass m of the star.

Solution   Using (7) of Section 14.4 and using spherical coordinates, we find that

$$m = \iiint_D \delta(x, y, z)\, dV = \iiint_D ce^{-[(x^2 + y^2 + z^2)/a^2]^{3/2}}\, dV$$

$$= \int_0^{2\pi} \int_0^{\pi} \int_0^a ce^{-\rho^3/a^3}\, \rho^2 \sin\phi\, d\rho\, d\phi\, d\theta$$

$$= \int_0^{2\pi} \int_0^{\pi} -\frac{a^3 c}{3} e^{-\rho^3/a^3} \sin\phi \Big|_0^a\, d\phi\, d\theta$$

$$= \int_0^{2\pi} \int_0^{\pi} \left[\frac{a^3 c}{3}\left(1 - \frac{1}{e}\right) \right] \sin\phi\, d\phi\, d\theta$$

$$= \frac{a^3 c}{3}\left(1 - \frac{1}{e}\right) \int_0^{2\pi} (-\cos\phi)\Big|_0^{\pi}\, d\theta$$

$$= \frac{a^3 c}{3}\left(1 - \frac{1}{e}\right) \int_0^{2\pi} 2\, d\theta = \frac{4}{3}\pi a^3 c \left(1 - \frac{1}{e}\right) \quad \square$$

EXERCISES 14.6

1. Plot the points having the following spherical coordinates. Then give their rectangular coordinates.

a. $\left(1, \frac{\pi}{2}, \frac{\pi}{6}\right)$

b. $\left(2, \pi, \frac{\pi}{2}\right)$

c. $\left(3, \frac{\pi}{4}, \frac{4\pi}{3}\right)$

d. $\left(\frac{1}{2}, \frac{\pi}{3}, \frac{5\pi}{4}\right)$

e. $\left(1, 0, \frac{7\pi}{6}\right)$

f. $\left(5, \frac{\pi}{2}, 0\right)$

2. Give a set of spherical coordinates of the points having the following rectangular coordinates.
   a. $(1, 0, 1)$
   b. $(3, 0, 0)$
   c. $(2, 2, 2\sqrt{2}/\sqrt{3})$
   d. $(2\sqrt{2}, -2\sqrt{2}, -4\sqrt{3})$

In Exercises 3–7 evaluate the iterated integral.

3. $\int_0^{2\pi} \int_0^{\pi/4} \int_0^1 \rho^2 \sin\phi\, d\rho\, d\phi\, d\theta$

4. $\int_0^{2\pi} \int_0^{\pi/2} \int_1^3 \rho^3 \cos\phi \sin\phi\, d\rho\, d\phi\, d\theta$

5. $\int_0^{\pi} \int_{\pi/2}^{\pi} \int_1^2 \rho^4 \sin^2\phi \cos^2\theta\, d\rho\, d\phi\, d\theta$

6. $\int_0^{\pi} \int_0^{\pi/2} \int_0^{\sin\phi} \rho^2 \sin\phi\, d\rho\, d\phi\, d\theta$

7. $\int_{\pi/4}^{\pi/3} \int_0^{\theta} \int_0^{9\sec\phi} \rho \cos^2\phi \cos\theta\, d\rho\, d\phi\, d\theta$

In Exercises 8–13 express the integral as an iterated integral in spherical coordinates. Then evaluate it.

8. $\iiint_D (x^2 + y^2)\, dV$, where D is the ball $x^2 + y^2 + z^2 \le 1$

9. $\iiint_D x^2\, dV$, where D is the solid region between the spheres $x^2 + y^2 + z^2 = 4$ and $x^2 + y^2 + z^2 = 9$

10. $\iiint_D (x^2 + y^2 + z^2)^2\, dV$, where D is the solid region bounded above by the sphere $x^2 + y^2 + z^2 = 1$ and below by the upper nappe of the cone $z^2 = x^2 + y^2$

11. $\iiint_D \frac{1}{x^2 + y^2 + z^2}\, dV$, where D is the solid region above the xy plane bounded by the cone $z = \sqrt{3x^2 + 3y^2}$ and the spheres $x^2 + y^2 + z^2 = 9$ and $x^2 + y^2 + z^2 = 81$

12. $\iiint_D (z^2 + 1)\, dV$, where D is the solid region in the first octant bounded by the spheres $x^2 + y^2 + z^2 = 1$ and $x^2 + y^2 + z^2 = 2$ and by the coordinate planes

13. $\iiint_D \sqrt{z}\, dV$, where D is the solid region in the first octant bounded by the sphere $x^2 + y^2 + z^2 = 16$ and the planes $z = 0$, $x = \sqrt{3}y$, and $x = y$

In Exercises 14–21 find the volume V of the region.

14. The solid region in the first octant bounded by the spheres $\rho = 1$ and $\rho = 3$ and by the coordinate planes

15. The solid region bounded above by the sphere $x^2 + y^2 + z^2 = 4$ and below by the upper nappe of the cone $z^2 = x^2 + y^2$

16. The solid region bounded above by the sphere $x^2 + y^2 + z^2 = 4$ and below by the upper nappe of the cone $z^2 = 3(x^2 + y^2)$

17. The solid region between the spheres $x^2 + y^2 + z^2 = 1$ and $x^2 + y^2 + z^2 = 4$ and below the upper nappe of the cone $3z^2 = x^2 + y^2$

18. The solid region bounded above by the sphere $x^2 + y^2 + z^2 = 4z$ and below by the upper nappe of the cone $z^2 = x^2 + y^2$

19. The solid region bounded above by the upper nappe of the cone $x^2 + y^2 = z^2$, on the sides by the cylinder $x^2 + y^2 = 4$, and below by the xy plane

20. The smaller of the two solid regions bounded by the sphere $\rho = 5$ and the half-planes $\theta = \pi/6$ and $\theta = \pi/3$

*21. The solid region bounded above by the sphere $x^2 + y^2 + z^2 = 64$ and below by the plane $z = -4\sqrt{3}$

22. Find the volume V of the solid region bounded below by the xy plane and above by the surface $\rho = 1 + \cos\phi$.

23. Find the volume V of the solid region bounded by the surface $\rho = \cos\phi$.

24. Find the total mass m of an object occupying the solid region bounded above by the sphere $x^2 + y^2 + z^2 = 4$ and below by the upper nappe of the cone $x^2 + y^2 = z^2$. Assume that the mass density at the point (x, y, z) is equal to the distance from (x, y, z) to the origin.

25. Find the total mass m of an object occupying the solid region bounded by the spheres $x^2 + y^2 + z^2 = 4$ and $x^2 + y^2 + z^2 = 16$, with mass density at (x, y, z) equal to the reciprocal of the distance from (x, y, z) to the origin.

14.7
MOMENTS AND CENTERS OF GRAVITY

In Section 6.7 we defined the moments and center of gravity of a plane region. In this section we will use double integrals to define moments and centers of gravity of more general plane regions. Then we will turn to solid regions and define their moments (about the three coordinate planes) and centers of gravity.

Moments and Centers of Gravity of Plane Regions

Let R be a rectangle with center (x, y) and area A. Recall from Section 6.7 that the moments $\mathcal{M}_x$ and $\mathcal{M}_y$ of R about the x axis and y axis are given by

$$\mathcal{M}_x = yA \quad \text{and} \quad \mathcal{M}_y = xA$$

More generally, if R is a vertically or horizontally simple region, let R' be a rectangle that circumscribes R. Partition R' into subrectangles, numbered so that $R_1, R_2, \ldots, R_n$ are the ones entirely contained in R. For any k between 1 and n, let (x_k, y_k) be the center of R_k and ΔA_k the area of R_k. It follows that the moment $\mathcal{M}_k$ of R_k about the x axis is $y_k \Delta A_k$. Moreover, the moment $\mathcal{M}_x$ of R about the x axis should be approximately the moment of the region comprised of the rectangles $R_1, R_2, \ldots, R_n$, which means the sum of the moments $\mathcal{M}_1, \mathcal{M}_2, \ldots, \mathcal{M}_n$. Consequently $\mathcal{M}_x$ should be approximately $\sum_{k=1}^{n} y_k \Delta A_k$, which is a Riemann sum for the function y on R. Accordingly we should define $\mathcal{M}_x$ by the formula

$$\mathcal{M}_x = \iint\limits_{R} y\, dA$$

Similar reasoning applies to $\mathcal{M}_y$. Thus we make the following definition.

DEFINITION 14.12

Let R be a vertically or horizontally simple region. The **moment** $\mathcal{M}_x$ of R **about the x axis** and the **moment** $\mathcal{M}_y$ of R **about the y axis** are defined by

$$\mathcal{M}_x = \iint\limits_R y \, dA \quad \text{and} \quad \mathcal{M}_y = \iint\limits_R x \, dA$$

If R has positive area A, then the **center of gravity** (or **center of mass**, or **centroid**) of R is the point $(\bar{x}, \bar{y})$ defined by

$$\bar{x} = \frac{\mathcal{M}_y}{A} = \frac{\iint_R x \, dA}{\iint_R 1 \, dA} \quad \text{and} \quad \bar{y} = \frac{\mathcal{M}_x}{A} = \frac{\iint_R y \, dA}{\iint_R 1 \, dA}$$

Definition 14.12 actually applies to any bounded region R on which every continuous function is integrable, although we will not use this fact. In the event that R is vertically simple, we can obtain our original formulas for $\mathcal{M}_x$ and $\mathcal{M}_y$ (see Definition 6.4) from those in Definition 14.12. Indeed, if R is the region between the graphs of f and g on $[a, b]$ and if $g(x) \leq f(x)$ for $a \leq x \leq b$, then

$$\mathcal{M}_x = \iint\limits_R y \, dA = \int_a^b \int_{g(x)}^{f(x)} y \, dy \, dx = \int_a^b \frac{1}{2} y^2 \Big|_{g(x)}^{f(x)} dx$$

$$= \frac{1}{2} \int_a^b \left\{ [f(x)]^2 - [g(x)]^2 \right\} dx$$

and

$$\mathcal{M}_y = \iint\limits_R x \, dA = \int_a^b \int_{g(x)}^{f(x)} x \, dy \, dx = \int_a^b xy \Big|_{g(x)}^{f(x)} dx = \int_a^b x[f(x) - g(x)] \, dx$$

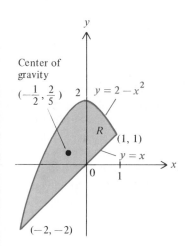

Center of gravity

$\left(-\dfrac{1}{2}, \dfrac{2}{5}\right)$

$y = 2 - x^2$

R $(1, 1)$

$y = x$

$(-2, -2)$

FIGURE 14.52

Example 1 Let R be the plane region bounded by the line $y = x$ and the parabola $y = 2 - x^2$ (Figure 14.52). Find the moments of R about the x and y axes and determine the center of gravity of R.

Solution The line and the parabola intersect at the points (x, y) satisfying

$$x = y = 2 - x^2, \quad \text{that is,} \quad x^2 + x - 2 = 0$$

This means that $x = -2$ or $x = 1$, and thus the points of intersection are $(-2, -2)$ and $(1, 1)$. Consequently by Definition 14.12,

$$\mathcal{M}_x = \iint\limits_R y \, dA = \int_{-2}^1 \int_x^{2-x^2} y \, dy \, dx = \int_{-2}^1 \frac{y^2}{2} \Big|_x^{2-x^2} dx$$

$$= \frac{1}{2} \int_{-2}^1 (x^4 - 5x^2 + 4) \, dx = \frac{1}{2} \left(\frac{x^5}{5} - \frac{5}{3} x^3 + 4x \right) \Big|_{-2}^1 = \frac{9}{5}$$

and

$$\mathcal{M}_y = \iint\limits_R x\,dA = \int_{-2}^{1} \int_x^{2-x^2} x\,dy\,dx = \int_{-2}^{1} xy\,\Big|_x^{2-x^2}\,dx$$

$$= \int_{-2}^{1} x(2 - x^2 - x)\,dx = \left(x^2 - \frac{x^4}{4} - \frac{x^3}{3}\right)\Big|_{-2}^{1} = -\frac{9}{4}$$

Since the area A of R is given by

$$A = \iint\limits_R 1\,dA = \int_{-2}^{1} \int_x^{2-x^2} 1\,dy\,dx = \int_{-2}^{1} (2 - x^2 - x)\,dx$$

$$= \left(2x - \frac{x^3}{3} - \frac{x^2}{2}\right)\Big|_{-2}^{1} = \frac{9}{2}$$

it follows that

$$\bar{x} = \frac{\mathcal{M}_y}{A} = \frac{-9/4}{9/2} = -\frac{1}{2} \quad \text{and} \quad \bar{y} = \frac{\mathcal{M}_x}{A} = \frac{9/5}{9/2} = \frac{2}{5}$$

Consequently the center of gravity of R is $(-\frac{1}{2}, \frac{2}{5})$ (Figure 14.52).   □

Moments and Centers of Gravity of Solid Regions

If a point mass having mass m is located at (x, y, z), we define its moments $\mathcal{M}_{xy}$, $\mathcal{M}_{xz}$, and $\mathcal{M}_{yz}$ about the coordinate planes as follows:

$$\text{moment about the } xy \text{ plane} = \mathcal{M}_{xy} = zm$$

$$\text{moment about the } xz \text{ plane} = \mathcal{M}_{xz} = ym \qquad (1)$$

$$\text{moment about the } yz \text{ plane} = \mathcal{M}_{yz} = xm$$

Now consider an object occupying a solid region D and having a continuous mass density δ. To define the moments and center of gravity of the object, we first circumscribe D with a rectangular parallelepiped D' and then partition D' into rectangular parallelepipeds, of which $D_1, D_2, \ldots, D_n$ are entirely contained in D. For each k between 1 and n, let ΔV_k be the volume of D_k, and choose any point (x_k, y_k, z_k) in D_k. If all the dimensions of D_k are small, then since δ is by assumption continuous on D_k, (5) of Section 14.4 implies that the mass in D_k is approximately $\delta(x_k, y_k, z_k)\Delta V_k$, and the distance from any point in D_k to the xy plane is approximately z_k. Accordingly, the moment of D_k about the xy plane should be approximately equal to the moment $z_k \delta(x_k, y_k, z_k)\Delta V_k$ of a point mass located at (x_k, y_k, z_k) and having mass $\delta(x_k, y_k, z_k)\Delta V_k$. It follows that the moment of the whole solid region D about the xy plane should be approximately equal to

$$\sum_{k=1}^{n} z_k \delta(x_k, y_k, z_k)\Delta V_k$$

which is a Riemann sum for the function $z\delta$ on D. Consequently we are led to define the moment $\mathcal{M}_{xy}$ of D about the xy plane by the formula

$$\mathcal{M}_{xy} = \iiint\limits_D z\delta(x, y, z)\, dV$$

Similar considerations apply to the moments about the xz and yz planes.

DEFINITION 14.13

Suppose an object with continuous mass density δ occupies a solid region D. Then the object's **moment $\mathcal{M}_{xy}$ about the xy plane**, its **moment $\mathcal{M}_{xz}$ about the xz plane**, and its **moment $\mathcal{M}_{yz}$ about the yz plane** are defined by

$$\mathcal{M}_{xy} = \iiint\limits_D z\delta(x, y, z)\, dV$$

$$\mathcal{M}_{xz} = \iiint\limits_D y\delta(x, y, z)\, dV$$

$$\mathcal{M}_{yz} = \iiint\limits_D x\delta(x, y, z)\, dV$$

If the mass m of the object is positive, then the **center of gravity** (or **center of mass**) $(\bar{x}, \bar{y}, \bar{z})$ of the object is defined by

$$\bar{x} = \frac{\mathcal{M}_{yz}}{m} = \frac{\iiint_D x\delta(x, y, z)\, dV}{\iiint_D \delta(x, y, z)\, dV}$$

$$\bar{y} = \frac{\mathcal{M}_{xz}}{m} = \frac{\iiint_D y\delta(x, y, z)\, dV}{\iiint_D \delta(x, y, z)\, dV}$$

$$\bar{z} = \frac{\mathcal{M}_{xy}}{m} = \frac{\iiint_D z\delta(x, y, z)\, dV}{\iiint_D \delta(x, y, z)\, dV}$$

Notice that a point mass with mass m located at the center of gravity of a solid region D has the same moments about the coordinate planes as an object with constant mass density that occupies D. When the mass density is constant, the object's center of gravity is often referred to as its **centroid**; in that case the center of gravity (or centroid) is independent of the mass density. As a result one often speaks of the centroid of a solid region.

If the mass density δ and the region D an object occupies are both symmetric with respect to a plane, then the center of gravity lies on that plane. In particular, if the mass density is constant and D is symmetric with respect to all the coordinate planes (as is a sphere centered at the origin), then the center of gravity of D lies on each of the coordinate planes and consequently is the origin.

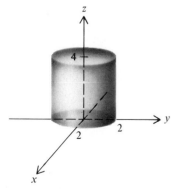

FIGURE 14.53

Example 2 Suppose an object occupying the solid region D bounded by the circular cylinder shown in Figure 14.53 has a mass density δ given by

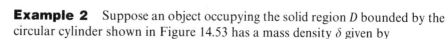

$$\delta(x, y, z) = 20 - z^2$$

Compute the moments $\mathcal{M}_{xy}$, $\mathcal{M}_{xz}$, and $\mathcal{M}_{yz}$, and then determine the center of gravity of the object.

Solution   By definition,

$$\mathcal{M}_{xz} = \iiint\limits_{D} y\delta(x,y,z)\,dV = \iiint\limits_{D} y(20-z^2)\,dV$$

and you can verify that the triple integral is 0 either by carrying out the integration or by noticing that D is symmetric with respect to the xz plane and that the integrand is an odd function of y. By a similar argument,

$$\mathcal{M}_{yz} = \iiint\limits_{D} x\delta(x,y,z)\,dV = \iiint\limits_{D} x(20-z^2)\,dV = 0$$

For the computation of $\mathcal{M}_{xy}$ it is convenient to use cylindrical coordinates. By doing so we obtain

$$\mathcal{M}_{xy} = \iiint\limits_{D} z\delta(x,y,z)\,dV = \iiint\limits_{D} z(20-z^2)\,dV$$

$$= \int_0^{2\pi}\int_0^2\int_0^4 (20z-z^3)r\,dz\,dr\,d\theta$$

$$= \int_0^{2\pi}\int_0^2 \left(10z^2 r - \frac{z^4 r}{4}\right)\Big|_0^4\,dr\,d\theta$$

$$= \int_0^{2\pi}\int_0^2 96r\,dr\,d\theta = \int_0^{2\pi} 48r^2\Big|_0^2\,d\theta$$

$$= \int_0^{2\pi} 192\,d\theta = 384\pi$$

To determine the center of gravity, we must still compute the mass m of the object and then gather our information:

$$m = \iiint\limits_{D} \delta(x,y,z)\,dV = \iiint\limits_{D} (20-z^2)\,dV$$

$$= \int_0^{2\pi}\int_0^2\int_0^4 (20-z^2)r\,dz\,dr\,d\theta$$

$$= \int_0^{2\pi}\int_0^2 \left(20z - \frac{z^3}{3}\right)\Big|_0^4 r\,dr\,d\theta$$

$$= \int_0^{2\pi}\int_0^2 \frac{176}{3}r\,dr\,d\theta = \frac{176}{3}\int_0^{2\pi}\frac{r^2}{2}\Big|_0^2\,d\theta$$

$$= \frac{352}{3}\int_0^{2\pi} 1\,d\theta = \frac{704}{3}\pi$$

Therefore the center of gravity is $(\bar{x}, \bar{y}, \bar{z})$, where

$$\bar{x} = \frac{\mathcal{M}_{yz}}{m} = 0, \qquad \bar{y} = \frac{\mathcal{M}_{xz}}{m} = 0, \qquad \bar{z} = \frac{\mathcal{M}_{xy}}{m} = \frac{384\pi}{704\pi/3} = \frac{18}{11} \quad \square$$

Notice that the center of gravity of the object occupying the cylinder is $(0, 0, \frac{18}{11})$, whereas the centroid of the solid region enclosed by the cylinder is the center $(0, 0, 2)$ of the region.

EXERCISES 14.7

In Exercises 1–6 determine the center of gravity of the plane region. Use symmetry where applicable.

1. The region between the graphs of $y = 5$ and $y = 1 + x^2$

2. The region between the graphs of $y = x^2$ and $y = x^4$

3. The region inside the circles $(x - 1)^2 + y^2 = 1$ and $x^2 + (y - 1)^2 = 1$

4. The leaf of the four-leaved rose $r = \sin 2\theta$ that lies in the first quadrant

5. The region bounded by the cardioid $r = 1 + \cos \theta$

6. The region inside the circle $r = 2 \sin \theta$ and outside the circle $r = 1$

In Exercises 7–13 find the centroid of the region. Use symmetry wherever possible to reduce calculations.

7. The solid region bounded above by the sphere $x^2 + y^2 + z^2 = a^2$ and below by the xy plane

8. The solid region bounded below by the paraboloid $z = 4x^2 + 4y^2$ and above by the plane $z = 2$

9. The solid region bounded above by the plane $z = 1$ and below by the upper nappe of the cone $z^2 = 9x^2 + 9y^2$

10. The solid region bounded above by the sphere $x^2 + y^2 + z^2 = 1$ and below by the cone $z = \sqrt{x^2 + y^2}$

11. The solid region in the first octant bounded by the planes $x = 0$ and $y = 0$ and the paraboloids $z = 1 - x^2 - y^2$ and $z = x^2 + y^2$.

12. The solid region in the first octant bounded by the coordinate planes and the planes $z = 1 + x + 2y$, $x = 1$, and $y = 1$

13. The pyramid with vertices $(1, 0, 0)$, $(0, 1, 0)$, $(-1, 0, 0)$, $(0, -1, 0)$, and $(0, 0, 2)$

In Exercises 14–20 find the center of gravity of an object that occupies the given region and has the given mass density.

14. The ball $x^2 + y^2 + z^2 \le 4$; $\delta(x, y, z)$ is equal to
    a. the distance from (x, y, z) to the origin.
    b. $1 + z^2$.

15. The solid region inside the sphere $x^2 + y^2 + z^2 = 4$ and the cylinder $x^2 + y^2 = 2$; $\delta(x, y, z) = z^2 + 1$.

16. The solid region bounded by the paraboloids $z = 1 - x^2 - y^2$ and $z = x^2 + y^2$; $\delta(x, y, z) = 2 - z$.

17. The cube in the first octant with sides of length 2 and one vertex at the origin; $\delta(x, y, z) = 1 + x$.

18. The solid region bounded by the sheet $z = 1 - x^2$ and the planes $z = 0$, $y = -1$, and $y = 1$; $\delta(x, y, z) = z(y + 2)$.

19. The solid region bounded above by the sphere $x^2 + y^2 + z^2 = 9$ and below by the xy plane; $\delta(x, y, z)$ is equal to the distance from (x, y, z) to the z axis.

20. The solid region bounded above by the sphere $x^2 + y^2 + z^2 = 4$ and below by the upper nappe of the cone $z^2 = x^2 + y^2$; $\delta(x, y, z) = z^2(x^2 + y^2 + z^2)$.

21. A cylindrical can containing pineapple juice is 20 centimeters tall and 8 centimeters in diameter. Set up a coordinate system with the xy plane at the bottom of the can, and assume that due to settling, the density of the pineapple juice is given by

$$\delta(x, y, z) = a(40 - z)$$

where a is a positive constant. Determine the center of gravity of the pineapple juice.

22. A cube with side length 1 and constant density 1 is located on top of another cube with side length 1 and constant density 10. Determine the center of gravity of the combination of cubes.

The moments of inertia about the three coordinate axes of an object that occupies a solid region D and has mass density δ are defined as follows:

moment of inertia about the x axis

$$= I_x = \iiint_D (y^2 + z^2) \delta(x, y, z) \, dV$$

moment of inertia about the y axis

$$= I_y = \iiint_D (x^2 + z^2)\delta(x, y, z)\, dV$$

moment of inertia about the z axis

$$= I_z = \iiint_D (x^2 + y^2)\delta(x, y, z)\, dV$$

In Exercises 23–26 find the moments of inertia about the coordinates axes for the given region and mass density.

23. The ball $x^2 + y^2 + z^2 \le 25$; $\delta(x, y, z) = 5$

24. The ball $x^2 + y^2 + z^2 \le 4$; $\delta(x, y, z) = x^2 + y^2 + z^2$

25. The solid region bounded by the cylinder $x^2 + y^2 = 4$ and the planes $z = 0$ and $z = 6$; $\delta(x, y, z) = 2$

26. The solid region bounded above by the plane $z = 4$ and below by the upper nappe of the cone $x^2 + y^2 = z^2$; $\delta(x, y, z) = x^2 + y^2$

14.8
CHANGE OF VARIABLES IN MULTIPLE INTEGRALS

In Section 14.2 we used the formula

$$\iint_R f(x, y)\, dA = \int_\alpha^\beta \int_{h_1(\theta)}^{h_2(\theta)} f(r\cos\theta, r\sin\theta)\, r\, dr\, d\theta$$

to evaluate integrals by means of polar coordinates rather than rectangular coordinates. Other formulas arose in Sections 14.5 and 14.6 for evaluating integrals by means of cylindrical or spherical coordinates instead of rectangular coordinates. Such formulas are called **change of variables formulas**, because they allow us to evaluate a multiple integral in a coordinate system other than the given coordinate system.

The present section is devoted to more general change of variables formulas—formulas that will allow us to integrate with respect to a variety of other coordinates rather than with respect to rectangular coordinates. The usefulness of such a formula becomes apparent when we seek to evaluate the integral $\iint_R 7xy\, dA$, where R is the region in Figure 14.54(a). Evaluating the

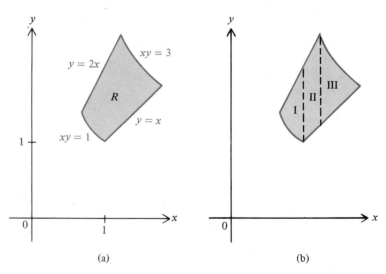

(a)   (b)

FIGURE 14.54

double integral by means of rectangular coordinates could be achieved, but it would necessitate evaluating three separate iterated integrals, one over each of the portions labeled I, II, and III in Figure 14.54(b). However, it is possible (as you will see in Example 3), to make a change of variables that converts $\iint_R 7xy\, dA$ to an integral over the region S shown in Figure 14.55; this new double integral is easily evaluated.

Change of Variables for Double Integrals

Before we derive the change of variables formula for double integrals, let us return to the formula

$$\iint_R f(x, y)\, dA = \int_\alpha^\beta \int_{h_1(\theta)}^{h_2(\theta)} f(r\cos\theta, r\sin\theta)\, r\, dr\, d\theta$$

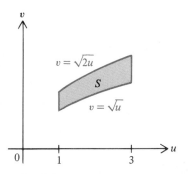

FIGURE 14.55

In order to explain the origin of the extra r on the right side, recall that x, y, r, and θ are related by the formulas

$$x = r\cos\theta \quad \text{and} \quad y = r\sin\theta \tag{1}$$

Now the area of a rectangle in the $r\theta$ plane of dimensions Δr and $\Delta\theta$ is $\Delta r\, \Delta\theta$ (Figure 14.56(a)), whereas the area of the associated arched region in the xy plane (Figure 14.56(b)) derived by using the equations in (1) is $r\, \Delta r\, \Delta\theta$, where r is the average radius of the arched region. One can think of r as the "magnification factor" by which the area of a sufficiently small region in the $r\theta$ plane must be multiplied to obtain the area of an associated region in the xy plane.

More generally, let R be a given region in the xy plane, and consider a new set of variables u and v. Suppose that for (u, v) in a set S in the uv plane, u and v are related to x and y by the formulas

$$x = g_1(u, v) \quad \text{and} \quad y = g_2(u, v) \tag{2}$$

where g_1 and g_2 are continuously differentiable (and (x, y) is in R). Our goal is to find an appropriate function $J(u, v)$ such that

$$\iint_R f(x, y)\, dA = \int_\alpha^\beta \int_{h_1(\theta)}^{h_2(\theta)} f(g_1(u, v), g_2(u, v))\, J(u, v)\, dv\, du \tag{3}$$

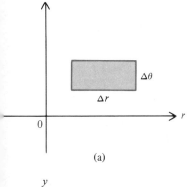

(a)

irrespective of the continuous function f. The function defined by (2) is called a **transformation**; we will denote it by T. Thus T is defined by

$$T(u, v) = (x, y) \quad \text{for } (u, v) \text{ in } S$$

We will also assume that each point (x, y) in R is assigned to exactly one point (u, v) in S.

For future reference we will need a special combination of partial derivatives called the **Jacobian** of T and defined by

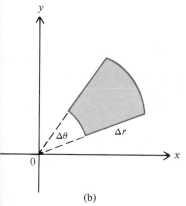

(b)

FIGURE 14.56

$$\frac{\partial(x, y)}{\partial(u, v)} = \begin{vmatrix} \dfrac{\partial x}{\partial u} & \dfrac{\partial x}{\partial v} \\ \dfrac{\partial y}{\partial u} & \dfrac{\partial y}{\partial v} \end{vmatrix} = \frac{\partial x}{\partial u}\frac{\partial y}{\partial v} - \frac{\partial x}{\partial v}\frac{\partial y}{\partial u}$$

Example 1 Suppose the transformation T is defined by

$$x = \frac{u}{v} \quad \text{and} \quad y = v$$

Find the Jacobian of T.

Solution By definition,

$$\frac{\partial(x, y)}{\partial(u, v)} = \begin{vmatrix} \dfrac{\partial x}{\partial u} & \dfrac{\partial x}{\partial v} \\[2ex] \dfrac{\partial y}{\partial u} & \dfrac{\partial y}{\partial v} \end{vmatrix} = \begin{vmatrix} \dfrac{1}{v} & -\dfrac{u}{v^2} \\[2ex] 0 & 1 \end{vmatrix} = \frac{1}{v} \quad \square$$

Now let T be a transformation that maps points in the uv plane into points in the xy plane and is given by (2). Assume that S is a rectangle with vertices (u, v), $(u + h, v), (u + h, v + k)$, and $(u, v + k)$ (Figure 14.57(a)), so that the area A_S of S is hk. If h and k are small, the continuity of g_1 and g_2 implies that the image R of S is approximately a parallelogram in the xy plane three of whose vertices are $(x, y) = (g_1(u, v), g_2(u, v)), (g_1(u + h, v), g_2(u + h, v))$, and $(g_1(u, v + k), g_2(u, v + k))$. But since v is constant on the line joining (u, v) and $(u + h, v)$, it follows from the definition of the partial derivatives $\partial g_1/\partial u$ and $\partial g_2/\partial u$ that if h is small, then

$$g_1(u + h, v) \approx g_1(u, v) + \frac{\partial g_1}{\partial u} h = x + \frac{\partial g_1}{\partial u} h \tag{4}$$

$$g_2(u + h, v) \approx g_2(u, v) + \frac{\partial g_2}{\partial u} h = y + \frac{\partial g_2}{\partial u} h \tag{5}$$

Similarly, u is constant on the line joining (u, v) and $(u, v + k)$, so that if k is small, we have

$$g_1(u, v + k) \approx g_1(u, v) + \frac{\partial g_1}{\partial v} k = x + \frac{\partial g_1}{\partial v} k \tag{6}$$

$$g_2(u, v + k) \approx g_2(u, v) + \frac{\partial g_2}{\partial v} k = y + \frac{\partial g_2}{\partial v} k \tag{7}$$

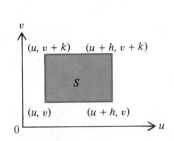

A rectangular region S in the uv plane

(a)

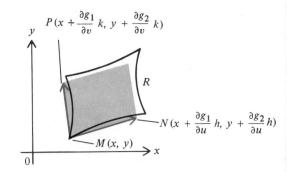

The image R of S in the xy plane

(b)

FIGURE 14.57

Using (4)–(7) we find that if h and k are small, the parallelogram is very close to the parallelogram determined by the points

$$M = (x, y), \qquad N = \left(x + \frac{\partial g_1}{\partial u} h, y + \frac{\partial g_2}{\partial u} h \right), \qquad P = \left(x + \frac{\partial g_1}{\partial v} k, y + \frac{\partial g_2}{\partial v} k \right)$$

(Figure 14.57(b)). The area A of the latter parallelogram is the length of the cross product $\overrightarrow{MN} \times \overrightarrow{MP}$. (See Section 11.4.) Since

$$\overrightarrow{MN} \times \overrightarrow{MP} = \begin{vmatrix} \mathbf{i} & \mathbf{j} & \mathbf{k} \\ \frac{\partial g_1}{\partial u} h & \frac{\partial g_2}{\partial u} h & 0 \\ \frac{\partial g_1}{\partial v} k & \frac{\partial g_2}{\partial v} k & 0 \end{vmatrix} = \left[\left(\frac{\partial g_1}{\partial u} h \right)\left(\frac{\partial g_2}{\partial v} k \right) - \left(\frac{\partial g_2}{\partial u} h \right)\left(\frac{\partial g_1}{\partial v} k \right) \right] \mathbf{k}$$

it follows that

$$\| \overrightarrow{MN} \times \overrightarrow{MP} \| = \left| \frac{\partial g_1}{\partial u} \frac{\partial g_2}{\partial v} - \frac{\partial g_2}{\partial u} \frac{\partial g_1}{\partial v} \right| hk = \left| \frac{\partial(x, y)}{\partial(u, v)} \right| hk = \left| \frac{\partial(x, y)}{\partial(u, v)} \right| A_S$$

We conclude that

$$A \approx \left| \frac{\partial(x, y)}{\partial(u, v)} \right| A_S \tag{8}$$

Therefore the area of the image R of the rectangle S is approximately the product of the area of S and the absolute value of the Jacobian of T.

Now we no longer assume that R is rectangular but do assume that R has a piecewise smooth boundary and that f is continuous on R. Since we can approximate the double integral $\iint_R f(x, y) \, dA$ as accurately as we wish by Riemann sums of the form $\sum_{k=1}^{n} f(x_k, y_k) \Delta A_k$, where ΔA_k is the area of a suitable small rectangle R_k contained in R for $k = 1, \dots, n$, one can use (8) to prove the following theorem.

THEOREM 14.14

Suppose S and R are sets in the uv and xy planes, respectively, each with a piecewise smooth boundary. Let T be a transformation from S to R, defined by

$$x = g_1(u, v) \quad \text{and} \quad y = g_2(u, v)$$

where g_1 and g_2 have continuous partial derivatives on S. Suppose also that each point (x, y) in R is the image of a unique point (u, v) in S and that $\partial(x, y)/\partial(u, v) \neq 0$ throughout S except possibly at finitely many points. Finally, assume that f is continuous on R. Then

$$\iint_R f(x, y) \, dA = \iint_S f(g_1(u, v), g_2(u, v)) \left| \frac{\partial(x, y)}{\partial(u, v)} \right| dA \tag{9}$$

Thus the function $J(u, v)$ we sought in (3) is given by

$$J(u, v) = \left| \frac{\partial(x, y)}{\partial(u, v)} \right|$$

To change from rectangular to polar coordinates, we have

$$x = r \cos \theta \quad \text{and} \quad y = r \sin \theta \qquad (10)$$

For the transformation defined by (10) we find that

$$\frac{\partial(x, y)}{\partial(r, \theta)} = \begin{vmatrix} \dfrac{\partial x}{\partial r} & \dfrac{\partial x}{\partial \theta} \\ \dfrac{\partial y}{\partial r} & \dfrac{\partial y}{\partial \theta} \end{vmatrix} = \begin{vmatrix} \cos \theta & -r \sin \theta \\ \sin \theta & r \cos \theta \end{vmatrix} = r \cos^2 \theta + r \sin^2 \theta = r$$

Therefore if a region S in the $r\theta$ plane corresponds to a region R under the transformation given in (10), then (9) yields

$$\iint\limits_{R} f(x, y) \, dA = \iint\limits_{S} f(r \cos \theta, r \sin \theta) r \, dA$$

which is the polar coordinate change of variable formula in (1).

To implement (9) when the transformation T is specified, we need first to calculate $\partial(x, y)/\partial(u, v)$, which is generally easy, and to find S, which may not be so easy. It is usually simplest to find S by determining the curves (or lines) in the uv plane corresponding to the boundary of R; then S will be the region bounded by these curves and lines.

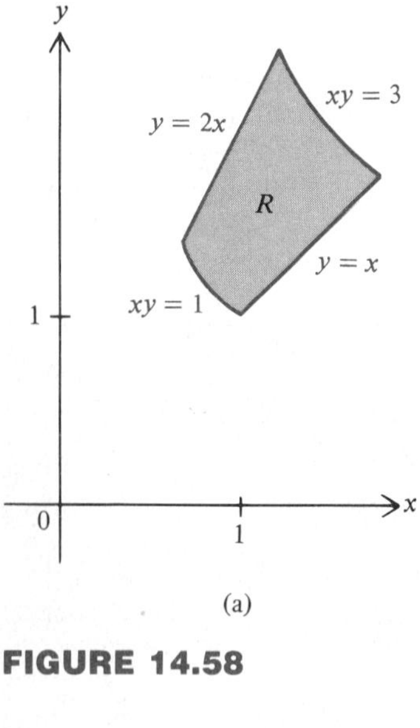

Example 2   Evaluate $\iint_R x^3 y \, dA$, where R is the region in the first quadrant bounded by the lines $y = x$ and $y = 2x$, and by the hyperbolas $xy = 1$ and $xy = 3$ (Figure 14.58). Let T be defined by

$$x = \frac{u}{v} \quad \text{and} \quad y = v \qquad (11)$$

Solution   As we computed in Example 1,

$$\frac{\partial(x, y)}{\partial(u, v)} = \frac{1}{v}$$

Next we determine the region S in the uv plane that is mapped onto R by T. For points on the hyperbola $xy = 1$ we substitute from (11) to obtain

$$1 = xy = \left(\frac{u}{v}\right) v = u$$

so that T maps the line $u = 1$ onto the hyperbola $xy = 1$. Similarly, for the hyperbola $xy = 3$ we obtain

$$3 = xy = \left(\frac{u}{v}\right) v = u$$

so that T maps the line $u = 3$ onto the hyperbola $xy = 3$. Next, for points on the line $y = x$, we substitute from (11) to obtain

$$v = y = x = \frac{u}{v}$$

FIGURE 14.58

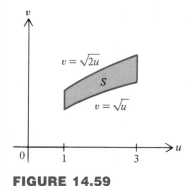

FIGURE 14.59

so that $v^2 = u$, or equivalently, $v = \sqrt{u}$. Similarly, for points on the line $y = 2x$ we obtain

$$v = y = 2x = 2\left(\frac{u}{v}\right)$$

so that $v^2 = 2u$, or equivalently, $v = \sqrt{2u}$. Consequently S is the region in the uv plane bounded by the lines $u = 1$ and $u = 3$, and by the parabolas $v = \sqrt{u}$ and $v = \sqrt{2u}$ (Figure 14.59). Now we can apply (9) to conclude that

$$\iint_R x^3 y \, dA \overset{(9)}{=} \iint_S \left(\frac{u}{v}\right)^3 v \left|\frac{\partial(x, y)}{\partial(u, v)}\right| dA = \int_1^3 \int_{\sqrt{u}}^{\sqrt{2u}} \frac{u^3}{v^3} v \frac{1}{v} \, dv \, du$$

$$= \int_1^3 \int_{\sqrt{u}}^{\sqrt{2u}} \frac{u^3}{v^3} \, dv \, du = \int_1^3 -\frac{1}{2} \frac{u^3}{v^2} \bigg|_{\sqrt{u}}^{\sqrt{2u}} du$$

$$= \int_1^3 \frac{1}{4} u^2 \, du = \frac{1}{12} u^3 \bigg|_1^3 = \frac{13}{6} \quad \square$$

Suppose we have a double integral $\iint_R f(x, y) \, dA$ in which either f or R is complicated, so that the integral is hard to evaluate, and suppose no transformation is prescribed. Then we proceed in the following way:

1. Use the information in the integral to define a transformation T_R from R into the uv plane.
2. Determine the image S of R under the transformation.
3. Solve for x and y from the equations defining T_R in order to obtain a transformation T from S to R.
4. Compute $\dfrac{\partial(x, y)}{\partial(u, v)}$.
5. Evaluate $\iint_R f(x, y) \, dA$ by means of (9).

The examples below use the procedure just outlined.

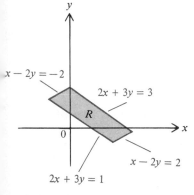

FIGURE 14.60

Example 3   Evaluate $\iint_R 7xy \, dA$, where the boundary of R is the parallelogram determined by the lines $2x + 3y = 1$, $2x + 3y = 3$, $x - 2y = 2$, and $x - 2y = -2$ (Figure 14.60).

Solution   As you can see from the figure, the region R is not so easy to integrate over. Since $2x + 3y$ and $x - 2y$ appear in the description of R, we let T_R be defined by

$$u = 2x + 3y \quad \text{and} \quad v = x - 2y \tag{12}$$

To find the image S of R under T_R, we notice that by our definition of u the images of the lines $2x + 3y = 1$ and $2x + 3y = 3$ are the lines $u = 1$ and $u = 3$ in the uv plane. Similarly, the images of the lines $x - 2y = 2$ and $x - 2y = -2$ are the lines $v = 2$ and $v = -2$, respectively. Therefore S is the rectangular region bounded by

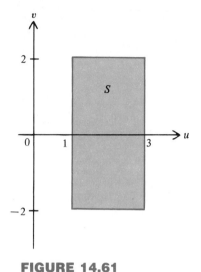

FIGURE 14.61

the lines $u = 1, u = 3, v = 2$, and $v = -2$ (Figure 14.61). Next we solve for x and y in (12). We find that

$$u - 2v = (2x + 3y) - 2(x - 2y) = 7y$$

and

$$2u + 3v = 2(2x + 3y) + 3(x - 2y) = 7x$$

Therefore

$$x = \frac{1}{7}(2u + 3v) \quad \text{and} \quad y = \frac{1}{7}(u - 2v)$$

which defines a transformation T from S onto R. Since

$$\frac{\partial(x, y)}{\partial(u, v)} = \begin{vmatrix} \dfrac{\partial x}{\partial u} & \dfrac{\partial x}{\partial v} \\ \dfrac{\partial y}{\partial u} & \dfrac{\partial y}{\partial v} \end{vmatrix} = \begin{vmatrix} \dfrac{2}{7} & \dfrac{3}{7} \\ \dfrac{1}{7} & -\dfrac{2}{7} \end{vmatrix} = \left(\frac{2}{7}\right)\left(-\frac{2}{7}\right) - \left(\frac{3}{7}\right)\left(\frac{1}{7}\right) = -\frac{1}{7}$$

it follows from (9) that

$$\iint\limits_{R} 7xy \, dA = \iint\limits_{S} [(2u + 3v)]\left[\frac{1}{7}(u - 2v)\right]\left|-\frac{1}{7}\right| dA$$

$$= \iint\limits_{S} \frac{1}{49}(2u^2 - uv - 6v^2) \, dA$$

$$= \int_{1}^{3} \int_{-2}^{2} \frac{1}{49}(2u^2 - uv - 6v^2) \, dv \, du$$

$$= \frac{1}{49}\int_{1}^{3} \left(2u^2 v - \frac{1}{2}uv^2 - 2v^3\right)\Bigg|_{-2}^{2} du = \frac{1}{49}\int_{1}^{3}(8u^2 - 32) \, du$$

$$= \frac{1}{49}\left(\frac{8}{3}u^3 - 32u\right)\Bigg|_{1}^{3} = \frac{1}{49}\left(72 - 96 - \frac{8}{3} + 32\right) = \frac{16}{147} \quad \square$$

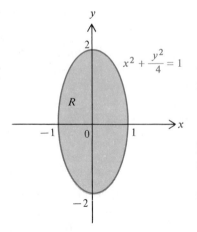

FIGURE 14.62

Integration over an elliptical region is most often facilitated by transforming the region into a circular one and then proceeding by polar coordinates (when feasible). The following example illustrates this method.

Example 4   Evaluate $\iint_{R} (x^2 + y^2) \, dA$, where R is the region bounded by the ellipse $x^2 + y^2/4 = 1$ (Figure 14.62).

Solution   Let T_R be the transformation defined by

$$u = x \quad \text{and} \quad v = \frac{1}{2}y \tag{13}$$

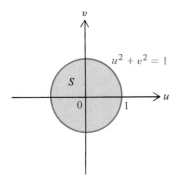

FIGURE 14.63

This transformation maps the ellipse into the unit circle, because

$$1 = x^2 + \frac{y^2}{4} = u^2 + v^2$$

Thus S is the unit disk $u^2 + v^2 \leq 1$ (Figure 14.63). Solving for x and y in (13) is trivial, and yields

$$x = u \quad \text{and} \quad y = 2v$$

We find that

$$\frac{\partial(x, y)}{\partial(u, v)} = \begin{vmatrix} \dfrac{\partial x}{\partial u} & \dfrac{\partial x}{\partial v} \\[2mm] \dfrac{\partial y}{\partial u} & \dfrac{\partial y}{\partial v} \end{vmatrix} = \begin{vmatrix} 1 & 0 \\ 0 & 2 \end{vmatrix} = 2$$

Therefore by (9),

$$\iint_R (x^2 + y^2)\, dA = \iint_S (u^2 + (2v)^2) \left| \frac{\partial(x, y)}{\partial(u, v)} \right| dA = \iint_S (u^2 + 4v^2) 2\, dA$$

Converting to polar coordinates with $u = r \cos \theta$ and $v = r \sin \theta$, we obtain

$$\iint_S (u^2 + 4v^2) 2\, dA = 2 \iint_S (u^2 + v^2 + 3v^2)\, dA$$

$$= 2 \int_0^{2\pi} \int_0^1 (r^2 + 3r^2 \sin^2 \theta) r\, dr\, d\theta$$

$$= 2 \int_0^{2\pi} \int_0^1 (1 + 3 \sin^2 \theta) r^3\, dr\, d\theta$$

$$= 2 \int_0^{2\pi} (1 + 3 \sin^2 \theta) \frac{r^4}{4} \bigg|_0^1 d\theta$$

$$= \frac{1}{2} \int_0^{2\pi} (1 + 3 \sin^2 \theta)\, d\theta$$

$$= \frac{1}{2} \left[\theta + 3 \left(\frac{1}{2} \theta - \frac{1}{4} \sin 2\theta \right) \right] \bigg|_0^{2\pi} = \frac{5}{2} \pi \quad \square$$

As we will see in the next example, sometimes it is not necessary to solve for x and y directly in the integrand. Since it is possible to prove that

$$\frac{\partial(x, y)}{\partial(u, v)} = \frac{1}{\dfrac{\partial(u, v)}{\partial(x, y)}} \tag{14}$$

we can dispense with actually computing T from T_R in the evaluation of the integral.

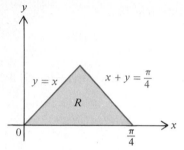

FIGURE 14.64

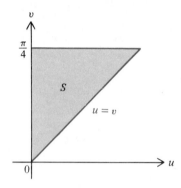

FIGURE 14.65

Example 5   Evaluate $\displaystyle\iint\limits_{R} \frac{\sin(x-y)}{\cos(x+y)}\,dA$, where R is the triangular region bounded by the lines $y = 0$, $y = x$, and $x + y = \pi/4$ (Figure 14.64).

Solution   This time the integrand is complicated, so that direct integration appears to be impossible. In an effort to simplify the integrand we let T_R be defined by

$$u = x - y \quad \text{and} \quad v = x + y$$

Now we compute the region S. We notice that if $y = 0$, then $u = x = v$, so the image of the line $y = 0$ is the line $u = v$. Next, if $y = x$, then $u = x - y = 0$, so the image of the line $y = x$ is the line $u = 0$. Finally, if $x + y = \pi/4$, then $v = x + y = \pi/4$, so the image of the line $x + y = \pi/4$ is the line $v = \pi/4$. Therefore the image S of R is the triangular region bounded by the lines $u = v$, $u = 0$, and $v = \pi/4$ (Figure 14.65). By definition,

$$\frac{\partial(u,v)}{\partial(x,y)} = \begin{vmatrix} \dfrac{\partial u}{\partial x} & \dfrac{\partial u}{\partial y} \\[2mm] \dfrac{\partial v}{\partial x} & \dfrac{\partial v}{\partial y} \end{vmatrix} = \begin{vmatrix} 1 & -1 \\ 1 & 1 \end{vmatrix} = (1)(1) - (-1)(1) = 2$$

Therefore it follows from (9) and (14) that

$$\iint\limits_{R} \frac{\sin(x-y)}{\cos(x+y)}\,dA \overset{(9)}{=} \iint\limits_{S} \frac{\sin u}{\cos v}\left|\frac{\partial(x,y)}{\partial(u,v)}\right|dA \overset{(14)}{=} \int_{0}^{\pi/4}\int_{0}^{v} \frac{\sin u}{\cos v}\frac{1}{2}\,du\,dv$$

$$= \frac{1}{2}\int_{0}^{\pi/4} -\frac{\cos u}{\cos v}\Big|_{0}^{v}\,dv = \frac{1}{2}\int_{0}^{\pi/4}\left(-1 + \frac{1}{\cos v}\right)dv$$

$$= \frac{1}{2}\int_{0}^{\pi/4}(\sec v - 1)\,dv = \frac{1}{2}(\ln|\sec v + \tan v| - v)\Big|_{0}^{\pi/4}$$

$$= \frac{1}{2}\left[\ln(\sqrt{2}+1) - \frac{\pi}{4}\right] \quad \square$$

Change of Variables for Triple Integrals

The change of variables formula for triple integrals is similar to the formula for double integrals. If a transformation T is defined by

$$x = g_1(u, v, w), \qquad y = g_2(u, v, w), \quad \text{and} \quad z = g_3(u, v, w)$$

and if T maps a region E in uvw space onto a region D in xyz space, then the formula for triple integrals corresponding to (9) is

$$\iiint\limits_{D} f(x, y, z)\,dV = \iiint\limits_{E} f(g_1(u, v, w), g_2(u, v, w), g_3(u, v, w))\left|\frac{\partial(x, y, z)}{\partial(u, v, w)}\right|dV \quad (15)$$

where

$$\frac{\partial(x, y, z)}{\partial(u, v, w)} = \frac{\partial x}{\partial u}\left(\frac{\partial y}{\partial v}\frac{\partial z}{\partial w} - \frac{\partial y}{\partial w}\frac{\partial z}{\partial v}\right) + \frac{\partial x}{\partial v}\left(\frac{\partial y}{\partial w}\frac{\partial z}{\partial u} - \frac{\partial y}{\partial u}\frac{\partial z}{\partial w}\right) + \frac{\partial x}{\partial w}\left(\frac{\partial y}{\partial u}\frac{\partial z}{\partial v} - \frac{\partial y}{\partial v}\frac{\partial z}{\partial u}\right)$$

We abbreviate the preceding equation by

$$\frac{\partial(x, y, z)}{\partial(u, v, w)} = \begin{vmatrix} \dfrac{\partial x}{\partial u} & \dfrac{\partial x}{\partial v} & \dfrac{\partial x}{\partial w} \\[2mm] \dfrac{\partial y}{\partial u} & \dfrac{\partial y}{\partial v} & \dfrac{\partial y}{\partial w} \\[2mm] \dfrac{\partial z}{\partial u} & \dfrac{\partial z}{\partial v} & \dfrac{\partial z}{\partial w} \end{vmatrix} \tag{16}$$

The expression in (16) is called the **Jacobian** of the transformation T. By change of variables we will derive the formula for integrals in spherical coordinates.

Example 6 Using (15), derive the formula for integrals in spherical coordinates that appeared in Theorem 14.11.

Solution In this case we have

$$x = \rho \sin \phi \cos \theta, \qquad y = \rho \sin \phi \sin \theta, \quad \text{and} \quad z = \rho \cos \phi$$

Therefore by (16),

$$\frac{\partial(x, y, z)}{\partial(\rho, \phi, \theta)} = \begin{vmatrix} \sin \phi \cos \theta & \rho \cos \phi \cos \theta & -\rho \sin \phi \sin \theta \\ \sin \phi \sin \theta & \rho \cos \phi \sin \theta & \rho \sin \phi \cos \theta \\ \cos \phi & -\rho \sin \phi & 0 \end{vmatrix}$$

$$= \sin \phi \cos \theta (0 + \rho^2 \sin^2 \phi \cos \theta)$$
$$\quad + \rho \cos \phi \cos \theta (\rho \sin \phi \cos \phi \cos \theta - 0)$$
$$\quad - \rho \sin \phi \sin \theta (-\rho \sin^2 \phi \sin \theta - \rho \cos^2 \phi \sin \theta)$$

$$= \rho^2 \sin \phi (\sin^2 \phi \cos^2 \theta + \cos^2 \phi \cos^2 \theta + \sin^2 \phi \sin^2 \theta + \cos^2 \phi \sin^2 \theta)$$

$$= \rho^2 \sin \phi$$

Since $0 \leq \phi \leq \pi$ and thus $\rho^2 \sin \phi \geq 0$, it follows from (15) that

$$\iiint_D f(x, y, z)\, dV = \iiint_E f(\rho \sin \phi \cos \theta, \rho \sin \phi \sin \theta, \rho \cos \phi) \rho^2 \sin \phi\, dV$$

If the region E in $\rho \phi \theta$ space is defined by

$$\alpha \leq \theta \leq \beta, \qquad h_1(\theta) \leq \phi \leq h_2(\theta), \quad \text{and} \quad F_1(\phi, \theta) \leq \rho \leq F_2(\phi, \theta)$$

where h_1, h_2, F_1, and F_2 are continuous and where $0 \leq \beta - \alpha \leq 2\pi$, $0 \leq h_1 \leq h_2 \leq \pi$, and $0 \leq F_1 \leq F_2$, then (15) becomes

$$\iiint_D f(x, y, z)\, dV$$

$$= \int_\alpha^\beta \int_{h_1(\theta)}^{h_2(\theta)} \int_{F_1(\phi, \theta)}^{F_2(\phi, \theta)} f(\rho \sin \phi \cos \theta, \rho \sin \phi \sin \theta, \rho \cos \phi) \rho^2 \sin \phi\, d\rho\, d\phi\, d\theta$$

which is the formula appearing in Theorem 14.11. □

EXERCISES 14.8

In Exercises 1–8 find the Jacobian of the transformation.

1. $x = 3u - 4v, y = \dfrac{1}{2}u + \dfrac{1}{6}v$

2. $x = 3u - 6v, y = -2u + 4v$

3. $x = uv, y = u^2 + v^2$

4. $x = \cos u + \sin v, y = -\sin u + \cos v$

5. $x = e^v, y = ue^v$

6. $x = u - \ln v, y = v + \ln u$

7. $x = au, y = bv, z = w$

8. $x = \dfrac{u}{v^2}, y = \dfrac{v}{w^2}, z = \dfrac{w}{u^2}$

In Exercises 9–17 evaluate the integral by using the given transformation.

9. $\displaystyle\iint_R \dfrac{y}{x - 3y}\, dA$, where R is the region bounded by the lines $y = 1, y = \frac{1}{4}x$, and $x - 3y = e$; let $x = 3u + v, y = u$.

10. $\iint_R y^2\, dA$, where R is the region bounded by the ellipse $4x^2 + 9y^2 = 1$; let $x = \frac{1}{2}u, y = \frac{1}{3}v$.

11. $\iint_R xy^2\, dA$, where R is the region bounded by the lines $x - y = 2,\ x - y = -1,\ 2x + 3y = 1,$ and $2x + 3y = 0$; let $x = \frac{1}{5}(3u + v), y = \frac{1}{5}(v - 2u)$.

12. $\iint_R e^{(y - x)^2}\, dA$, where R is the region bounded by the lines $y = \frac{3}{2}x, y = 2x$, and $y = x + 1$; let $x = u + v, y = u + 2v$.

13. $\iint_R \dfrac{y}{x}e^{x^2 - y^2}\, dA$, where R is the region in the first quadrant bounded by the hyperbolas $x^2 - y^2 = 1$ and $x^2 - y^2 = 4$ and by the lines $x = 2y$ and $x = \sqrt{2}y$; let $x = u \sec v$ and $y = u \tan v$ for $u > 0$ and $0 < v < \pi/2$.

14. $\iint_R y \cos xy\, dA$, where R is bounded by the curves $xy = \pi/2,\ xy = \pi,\ y(2 - x) = 2,$ and $y(2 - x) = 4$; let $x = \dfrac{2v}{u + v}, y = u + v$.

15. $\iint_R e^{x^2 - y^2}\, dA$, where R is the region in the first quadrant bounded by the curves $x^2 - y^2 = 1, x^2 - y^2 = 4, y = 0,$ and $y = \frac{3}{5}x$; let $x = u \cosh v, y = u \sinh v$ for $u > 0$.

16. $\iiint_D x\, dV$, where D is the solid region in the first octant bounded by the ellipsoid

$$\dfrac{x^2}{4} + \dfrac{y^2}{9} + z^2 = 1$$

Let $x = 2u, y = 3v, z = w$.

17. $\iiint_D (x^2 + y^2)\, dV$, where D is the solid region in the first octant bounded by the coordinate planes, the paraboloids $z = x^2 + y^2$ and $z = 4(x^2 + y^2)$, and the planes $z = 1$ and $z = 4$; let $x = (v/u)\cos w, y = (v/u)\sin w, z = v^2$ for $u > 0$, $v \geq 0$, and $0 \leq w \leq \pi/2$.

In Exercises 18–25 evaluate the integral by a suitable change of variables.

18. $\iint_R 49x^2y\, dA$, where R is the region bounded by the lines $2x - y = 1,\ 2x - y = -2,\ x + 3y = 0,$ and $x + 3y = 1$

19. $\displaystyle\iint_R \left(\dfrac{x - 2y}{x + 2y}\right)^3 dA$, where R is the region bounded by the lines $x - 2y = 1,\ x - 2y = 2,\ x + 2y = 1,$ and $x + 2y = 3$

20. $\displaystyle\iint_R \left(1 + \dfrac{x^2}{16} + \dfrac{y^2}{25}\right)^{3/2} dA$, where R is the region bounded by the ellipse $\dfrac{x^2}{16} + \dfrac{y^2}{25} = 1$

21. $\iint_R x^2\, dA$, where R is the region bounded by the ellipse $\dfrac{x^2}{a^2} + \dfrac{y^2}{b^2} = 1$

22. $\iint_R \cos(x^2 + 4y^2 + \pi - 1)\, dA$, where R is the region above the x axis bounded by the ellipse $x^2 + 4y^2 = 1$

23. $\displaystyle\iint_R \sin\left[\pi\left(\dfrac{y - x}{y + x}\right)\right] dA$, where R is the region bounded by the lines $x + y = 1, x + y = 2, x = 0,$ and $y = 0$

24. $\iint_R (x - y)^2 \sin^2(x + y)\, dA$, where R is the region bounded by the parallelogram with vertices $(\pi, 0), (2\pi, \pi),$ $(\pi, 2\pi),$ and $(0, \pi)$

25. $\iint_R e^{(2x - y)/(x + y)}\, dA$, where R is the region bounded by the lines $x = 2y, y = 2x, x + y = 1,$ and $x + y = 2$

In Exercises 26–27 find the area A of R by using the given transformation.

26. R is the region in the first quadrant bounded by the curves $xy = 1, xy = 2, y = x,$ and $y = 4x$. Let $x = u/v$ and $y = v$.

27. R is the region in the first quadrant bounded by $x^2 - y^2 = a^2, x^2 - y^2 = b^2,\ y = 0,$ and $y = \frac{1}{2}x$, where $b \geq a > 0$. Let $x = u \cosh v$ and $y = u \sinh v$.

28. Using a change of variables, determine the volume V of the solid region D bounded by the ellipsoid $x^2 + 2y^2 + 4z^2 = 1$.

Key Terms and Expressions

Double integral
Volume; area
Vertically simple region; horizontally simple region; simple region
Iterated integral
Surface area
Solid region between the graphs of F_1 and F_2 on R
Triple integral
Mass; mass density

Charge; charge density
Cylindrical coordinates
Spherical coordinates
Moment about an axis; center of gravity of a plane region
Moment about a plane; center of gravity of a solid region
Jacobian

Key Formulas

$$A = \iint_R 1 \, dA$$

$$V = \iiint_D 1 \, dV$$

$$S = \iint_R \sqrt{[f_x(x, y)]^2 + [f_y(x, y)]^2 + 1} \, dA$$

$$\iint_R f(x, y) \, dA = \int_a^b \int_{g_1(x)}^{g_2(x)} f(x, y) \, dy \, dx$$

$$\iint_R f(x, y) \, dA = \int_c^d \int_{h_1(y)}^{h_2(y)} f(x, y) \, dx \, dy$$

$$\iint_R f(x, y) \, dA = \int_\alpha^\beta \int_{h_1(\theta)}^{h_2(\theta)} f(r \cos \theta, r \sin \theta) r \, dr \, d\theta$$

$$\iiint_D f(x, y, z) \, dV = \int_a^b \int_{g_1(x)}^{g_2(x)} \int_{F_1(x, y)}^{F_2(x, y)} f(x, y, z) \, dz \, dy \, dx$$

$$\iiint_D f(x, y, z) \, dV = \int_c^d \int_{h_1(y)}^{h_2(y)} \int_{F_1(x, y)}^{F_2(x, y)} f(x, y, z) \, dz \, dx \, dy$$

$$\iiint_D f(x, y, z) \, dV = \int_\alpha^\beta \int_{h_1(\theta)}^{h_2(\theta)} \int_{F_1(r \cos \theta, r \sin \theta)}^{F_2(r \cos \theta, r \sin \theta)} f(r \cos \theta, r \sin \theta, z) r \, dz \, dr \, d\theta$$

$$\iiint_D f(x, y, z) \, dV = \int_\alpha^\beta \int_{h_1(\theta)}^{h_2(\theta)} \int_{F_1(\phi, \theta)}^{F_2(\phi, \theta)} f(\rho \sin \phi \cos \theta, \rho \sin \phi \sin \theta, \rho \cos \phi) \rho^2 \sin \phi \, d\rho \, d\phi \, d\theta$$

$$\iint_R f(x, y) \, dA = \iint_S f(g_1(u, v), g_2(u, v)) \left| \frac{\partial(x, y)}{\partial(u, v)} \right| dA$$

REVIEW EXERCISES

In Exercises 1–4 evaluate the iterated integrals.

1. $\displaystyle \int_0^1 \int_x^{3x} y e^{(x^3)} \, dy \, dx$

2. $\displaystyle \int_0^{\sqrt{\pi}} \int_0^x \sin x^2 \, dy \, dx$

3. $\displaystyle \int_{-1}^1 \int_0^2 \int_{2x}^{5x} e^{xy} \, dz \, dy \, dx$

4. $\displaystyle \int_1^e \int_0^x \int_0^{1/(x+y)} \ln(x + y) \, dz \, dy \, dx$

In Exercises 5–7 reverse the order of integration, and then evaluate.

5. $\displaystyle \int_0^1 \int_{\sqrt{x}}^1 e^{(y^3)} \, dy \, dx$

6. $\displaystyle \int_1^9 \int_{\sqrt{y}}^3 \frac{e^{(x^2 - 2x)}}{x + 1} \, dx \, dy$

7. $\int_1^{\sqrt{3}} \int_x^{\sqrt{3}} \frac{x}{(x^2 + y^2)^{3/2}} \, dy \, dx$

In Exercises 8–11 find the area A of the region in the xy plane by means of double integrals.

8. The region bounded by the graphs of $y^2 = x$ and $y = x^3$

9. The region bounded by the graphs of $\sqrt{x} + \sqrt{y} = \sqrt{a}$ and $x + y = a$

10. The region inside the cardioid $r = 1 + \sin\theta$ and outside the cardioid $r = 1 + \cos\theta$

11. The region outside the limaçon $r = 3 - \sin\theta$ and inside the circle $r = 5\sin\theta$

In Exercises 12–17 evaluate the multiple integral.

12. $\iint_R \sin(x + y) \, dA$, where R is the region in the xy plane bounded by the lines $y = x$, $y = 0$, and $x = \pi/2$

13. $\iint_R (3x - 5) \, dA$, where R is the region bounded by the lines $y = 5 + x$, $y = -x + 7$, $x = 0$, and $x = 1$

14. $\iint_R (4 + x^2) \, dA$, where R is the region between the parabolas $y = 1 + x^2$ and $y = 3 - x^2$

15. $\iiint_D (z^2 + 1) \, dV$, where D is the solid region bounded below by the upper nappe of the cone $z^2 = 3x^2 + 3y^2$ and above by the sphere $x^2 + y^2 + z^2 = 4$

16. $\iiint_D xy \, dV$, where D is the solid region in the first octant bounded by the coordinate planes and the cylinders $x^2 + y^2 = 1$ and $x^2 + z^2 = 1$

17. $\iiint_D xyz \, dV$, where D is the solid region bounded below by the hemisphere $z = -\sqrt{9 - x^2 - y^2}$ and above by the xy plane

In Exercises 18–20 find the surface area S of the surface.

18. The portion of the parabolic sheet $z = \frac{1}{2}y^2$ cut out by the planes $y = x$, $y = 2\sqrt{2}$, and $x = 0$

19. The portion of the surface $z = xy$ that is inside the cylinder $x^2 + y^2 = 1$

20. The portion of the surface $z = \frac{2}{3}(x^{3/2} + y^{3/2})$ cut out by the planes $x = 0$, $y = 1$, and $x = 7y$

In Exercises 21–30 find the volume V of the region.

21. The solid region bounded by the paraboloid $z = x^2 + y^2$ and the upper nappe of the cone $x^2 + y^2 = z^2$

22. The solid region bounded by the surface $z = e^x$ and the planes $x = y$, $y = 0$, $x = 1$, and $z = 0$

23. The solid region bounded on the sides by the cylinder $r = 4\sin\theta$, above by the cone $r = z$, and below by the xy plane

24. The solid region in the first octant bounded by the coordinate planes, the cylinder $x^2 + y^2 = 5y$, and the sphere $x^2 + y^2 + z^2 = 25$

25. The solid region bounded below by the paraboloid $z = 4x^2 + y^2$ and above by the parabolic sheet $z = 16 - 3y^2$

26. The solid region bounded by the paraboloids $\frac{1}{2}z = -9 + x^2 + y^2$ and $x^2 + y^2 + z = 9$

27. The solid region in the first octant bounded by the planes $z = x + 2y$ and $6 = x + 3y$, and by the coordinate planes

28. The solid region in the first octant bounded above by the plane $z = 3$, below by the upper nappe of the cone $x^2 + y^2 = 3z^2$, and on the sides by the planes $y = 0$ and $x = \sqrt{3}\,y$

29. The solid region bounded above by the sphere $x^2 + y^2 + z^2 = 49$ and below by the paraboloid $x^2 + y^2 = 3z + 21$

30. The solid region bounded on the sides by the cylinder $x^2 + y^2 = 6x$, above by the paraboloid $x^2 + y^2 = 4z$, and below by the plane $z = -2$

31. Find the center of mass of an object occupying the region bounded by the cone $z = \sqrt{x^2 + y^2}$ and the plane $z = 3$, if the mass density at any point in the object is equal to the distance from the point to the xy plane.

32. Find the total mass and the center of mass of a body that is bounded on the sides by the cylinder $x^2 + y^2 = 4$, above by the cone $z = \sqrt{x^2 + y^2}$, and below by the xy plane, if the mass density at (x, y, z) is given by $\delta(x, y, z) = z + 3$.

33. Find the centroid of the solid region bounded below by the paraboloid $z = x^2 + y^2$ and above by the upper nappe of the cone $z^2 = x^2 + y^2$.

34. Find the mass of an object occupying a ball of radius a if the mass density at any point is equal to its distance from the outer boundary of the ball.

In Exercises 35–36 evaluate the integral by using the given transformation.

35. $\iint_R x \, dA$, where R is the region bounded by $xy = 1$, $xy = 2$, $x(1 - y) = 1$, and $x(1 - y) = 2$; let $x = u + v$, $y = v/(u + v)$.

36. $\iint_R (x - y^2) \, dA$, where R is the region bounded by $x = y^2 - y$, $x = 2y + y^2$, and $y = 2$; let $x = 2u - v + (u + v)^2$ and $y = u + v$.

In Exercises 37–38 evaluate the integral by using a suitable change of variables.

37. $\iint_R (2x - y^2) \, dA$, where R is the region bounded by the lines $x - y = 1$, $x - y = 3$, $x + y = 2$, and $x + y = 4$

38. $\iint_R \cos\left(\frac{y - x}{y + x}\right) dA$, where R is the region bounded by the lines $x + y = 2$, $x + y = 4$, $x = 0$, and $y = 0$

Cumulative Review, Chapters 1–13

In Exercises 1–2 find the limit.

1. $\displaystyle\lim_{h\to 0}\frac{2x^2h - 5xh^2 - 6h^3}{3xh + 2h^2}$ 2. $\displaystyle\lim_{x\to\infty}\left(1 - \frac{2}{x^2}\right)^{4x^2}$

3. Let $f(x) = \sqrt{x + \sqrt{x^2 - 4}}$.
   a. Find the domain of f.
   b. Find $f'(x)$.

4. Let $f(x) = x^{(\sin x)/x}$. Find $f'(\pi)$.

5. Find the two points on the graph of the equation $x^2 - xy + y^2 = 4$ at which the slope of the tangent line is 1.

6. A pyramid whose base is a square 4 inches on a side and whose height is 6 inches is being filled with water at a constant rate. Determine the rate that will make the water level rise $\frac{1}{2}$ inch per minute when the water level is 2 inches above the base. (*Hint:* The volume V of a pyramid of base side a and height h is given by $V = \frac{1}{3}a^2h$.)

7. A rectangular toolshed 7 feet tall is to be made from cedar, and is to have a volume of 1512 cubic feet. If the front and back are one and a half times as expensive per square foot as the two sides, and if the top is twice as expensive per square foot as the two sides, determine the dimensions that will minimize the cost of the toolshed.

8. Let l denote the portion of a line tangent to the curve $y = 1/(2x)$ that lies in the first quadrant. Determine the shortest possible length for l.

9. Let

$$f(x) = \frac{x}{\ln x}$$

Sketch the graph of f, indicating all pertinent information.

In Exercises 10–12 evaluate the integral.

10. $\displaystyle\int (2x + 9x^3)\sqrt{1 + 4x^2 + 9x^4}\,dx$

11. $\displaystyle\int \tan x \sin^2 x \cos^5 x\,dx$

12. $\displaystyle\int \frac{2x^3 + 3}{x^3 - 2x^2 + x}\,dx$

13. Consider

$$\int_{1/2}^{1} \frac{1}{\sqrt{1 - x^2}}\,dx$$

Determine whether the region between the graph of the integrand and the x axis on $[\frac{1}{2}, 1]$ has finite area. If it does, calculate the area.

14. The base of a solid is a square of side length 1, and the cross sections perpendicular to any fixed diagonal of the square are semicircular. Find the volume V of the solid.

15. The curve C is parametrized by $x = \frac{2}{3}\sin^{3/2} t$, $y = \sin t$ for $0 \leq t \leq \pi/2$. Determine the length $\mathscr{L}$ of C.

16. Sketch the graph of $r = \cos 5\theta$.

17. Show that $\left\{\dfrac{(k!)^3 27^k}{(3k)!}\right\}_{k=1}^{\infty}$ is an increasing sequence.

18. Determine whether $\left\{1 + 4 + \dfrac{4^2}{2!} + \dfrac{4^3}{3!} + \cdots + \dfrac{4^n}{n!}\right\}_{n=1}^{\infty}$ converges or diverges.

19. Find the interval of convergence of $\displaystyle\sum_{n=1}^{\infty} \frac{(-1)^n}{n + 1} x^{2n}$.

20. Use Taylor series to approximate $\int_0^1 \sin(x^4)\,dx$ to within 10^{-5} of the exact value.

21. A person exerts a force of 100 pounds at a $45°$ angle from the vertical in raising a sack of concrete mix 6 feet vertically. Determine the work W done on the sack.

22. A projectile is fired from the back of a truck 4 feet above the ground. If the projectile has an initial speed of 1500 feet per second and the angle of elevation is $30°$, find the position and velocity of the projectile at any time t until it hits the ground.

23. Let $\mathbf{r}(t) = t^3\mathbf{i} + 6t\mathbf{j} + 3t^2\mathbf{k}$.
   a. Find the tangent vector $\mathbf{T}(t)$, and determine any value(s) of t for which $\mathbf{T}(t)$ is parallel to the y axis.
   b. Find the normal vector $\mathbf{N}(t)$ for any value of t.
   c. Find the unit vector perpendicular to both $\mathbf{T}(1)$ and $\mathbf{N}(1)$ and having a negative $\mathbf{j}$-component.

24. Let $g(x, y) = \begin{cases} \dfrac{\sin(x^2(y^2 + 1))}{x} & \text{for } x \neq 0 \\ 0 & \text{for } x = 0 \end{cases}$
   a. Find $g_x(0, 0)$ and $g_y(0, 0)$.
   b. Determine whether or not g is continuous at $(0, 0)$.

25. Consider the paraboloid $z = 1 - x^2 - y^2$.
   a. Sketch the paraboloid.
   b. Find symmetric equations for the line l that is normal to the paraboloid and passes through the point $(-1, 1, -1)$.

26. Determine the dimensions of the parallelepiped with maximum volume that lies inside the first octant and inside the ellipsoid $x^2 + 4y^2 + 8z^2 = 24$, and three of whose sides are in coordinate planes. Assume the desired parallelepiped exists.

15

CALCULUS OF VECTOR FIELDS

In this chapter we study calculus of a type of function called a vector field, which assigns vectors to points in space. The gravitational field of the earth is an example of a vector field.

The chapter opens with an introduction to vector fields. In Section 15.2 we define the line integral and employ it to give a formula for the work done by a force on an object as it travels through space. The following section is devoted to the Fundamental Theorem of Line Integrals, which in spirit is similar to the Fundamental Theorem of Calculus.

The surface integral is defined in Section 15.5 and is used to calculate the amount of fluid flowing through a surface, such as a membrane. The chapter culminates in three important higher-dimensional analogues of the Fundamental Theorem of Calculus: Green's Theorem, Stokes's Theorem, and the Divergence Theorem.

15.1
VECTOR FIELDS

The gravitational field of the earth associates with the point (x, y, z) in space the force that the earth would exert on a unit mass located at (x, y, z). Analogously, the electric field due to a given charge associates with the point (x, y, z) in space the electric force that the given charge would exert on a positive unit charge located at (x, y, z). In both cases we associate a vector with each point of a given region in space. Such associations are called vector fields.

874

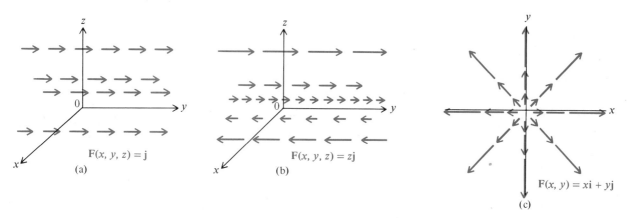

FIGURE 15.1

DEFINITION 15.1

A **vector field** $\mathbf{F}$ consists of two parts: a collection D of points in space, called the **domain**, and a **rule**, which assigns to each point (x, y, z) in D one and only one vector $\mathbf{F}(x, y, z)$.

Of course, it is impossible to graph a vector field, because this would require six dimensions (three for the domain and three for the range). However, it is possible to represent a vector field $\mathbf{F}$ graphically by drawing the vector $\mathbf{F}(x, y, z)$ as an arrow emanating from (x, y, z). Although we cannot show $\mathbf{F}(x, y, z)$ for every point (x, y, z), we can usually obtain a good impression of the vector field by graphing $\mathbf{F}(x, y, z)$ for several choices of (x, y, z). For example, if $\mathbf{F}(x, y, z) = \mathbf{j}$, then $\mathbf{F}$ could be represented as in Figure 15.1(a), and if $\mathbf{F}(x, y, z) = z\mathbf{j}$, then the graph in Figure 15.1(b) would represent $\mathbf{F}$. In the special case in which the domain and range of $\mathbf{F}$ are contained in the xy plane, we let (x, y) and $\mathbf{F}(x, y)$ denote the points in the domain and range, respectively, and we represent $\mathbf{F}$ by vectors in the xy plane. For example, if $\mathbf{F}(x, y) = x\mathbf{i} + y\mathbf{j}$, then $\mathbf{F}$ could be represented as in Figure 15.1(c).

Next we obtain formulas for the gravitational and electric fields. According to Newton's Law of Gravitation, the gravitational force $\mathbf{F}(x, y, z)$ exerted by a given point mass m at the origin on a unit point mass located at a point (x, y, z) other than the origin is given by

$$\mathbf{F}(x, y, z) = \frac{Gm}{x^2 + y^2 + z^2} \mathbf{u}(x, y, z)$$

where G is the universal gravitational constant and $\mathbf{u}(x, y, z)$ is the unit vector emanating from (x, y, z) and directed toward the origin. The vector field $\mathbf{F}$ is called the **gravitational field** of the point mass. Because $\mathbf{u}(x, y, z)$ points from (x, y, z) to the origin, it has the same direction as $-x\mathbf{i} - y\mathbf{j} - z\mathbf{k}$. Since $\mathbf{u}(x, y, z)$ is a unit vector, it can be written in the form

$$\mathbf{u}(x, y, z) = \frac{-1}{\sqrt{x^2 + y^2 + z^2}} (x\mathbf{i} + y\mathbf{j} + z\mathbf{k})$$

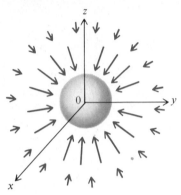

The gravitational field of the earth

FIGURE 15.2

and consequently

$$\mathbf{F}(x, y, z) = \frac{-Gm}{(x^2 + y^2 + z^2)^{3/2}}(x\mathbf{i} + y\mathbf{i} + z\mathbf{k}) \tag{1}$$

The gravitational field $\mathbf{F}$ has the properties that $\mathbf{F}(x, y, z)$ always points toward the origin and, moreover, that the magnitude of $\mathbf{F}(x, y, z)$ is the same for all points (x, y, z) located the same distance from the origin (Figure 15.2). A vector field that represents force and has these properties is called a **central force field**.

In physics a point (x, y, z) in space is often represented by the vector

$$\mathbf{r} = x\mathbf{i} + y\mathbf{j} + z\mathbf{k}$$

Thus in terms of the vector $\mathbf{r}$, the gravitational field can be written

$$\mathbf{F}(\mathbf{r}) = \frac{-Gm}{\|\mathbf{r}\|^3}\mathbf{r}$$

If a charge q is located at the origin, then according to Coulomb's Law the electric force $\mathbf{E}(x, y, z)$ exerted by the charge on a positive unit charge located at a point (x, y, z) other than the origin is given by the equation

$$\mathbf{E}(x, y, z) = \frac{q}{4\pi\varepsilon_0(x^2 + y^2 + z^2)}\mathbf{u}(x, y, z)$$

or

$$\mathbf{E}(x, y, z) = \frac{q}{4\pi\varepsilon_0(x^2 + y^2 + z^2)^{3/2}}(x\mathbf{i} + y\mathbf{j} + z\mathbf{k})$$

where $\mathbf{u}$ is the unit vector emanating from the origin and directed toward (x, y, z) and ε_0 is a constant called the permittivity of empty space (see (4) of Section 11.2). We also assume that distance is measured in meters, charge in coulombs, and force in newtons. The vector field $\mathbf{E}$ is called the **electric field** of the point charge. Like the gravitational field, the electric field of a point charge is a central force field.

The **electric field** at any point (x, y, z) due to a finite collection of charges is defined to be the total force the charges would exert on a positive unit charge at (x, y, z). Since the total force exerted is the vector sum of the forces exerted by the individual charges, it follows that the electric field of the collection of charges is equal to the sum of the electric fields of the individual charges. This result is known as a **superposition principle**.

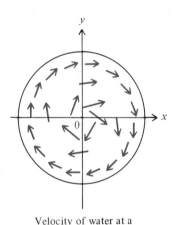

Velocity of water at a
few points in a sink

FIGURE 15.3

Vector fields can also describe the motion of a fluid. If at each point (x, y, z) in a given region we let $\mathbf{v}(x, y, z)$ denote the velocity of the fluid at (x, y, z), then $\mathbf{v}$ is a vector field, called the **velocity field** of the fluid. For example, Figure 15.3 might represent the velocity of water in a sink.

Just as a vector can be expressed in terms of its three components, a vector field $\mathbf{F}$ can be expressed in terms of three **component functions** M, N, and P:

$$\mathbf{F}(x, y, z) = M(x, y, z)\mathbf{i} + N(x, y, z)\mathbf{j} + P(x, y, z)\mathbf{k}$$

or in condensed form,

$$\mathbf{F} = M\mathbf{i} + N\mathbf{j} + P\mathbf{k}$$

Using components, we can rewrite formula (1) for the gravitational field as

$$\mathbf{F}(x, y, z) = \overbrace{\frac{-Gmx}{(x^2 + y^2 + z^2)^{3/2}}}^{M(x, y, z)}\mathbf{i} + \overbrace{\frac{-Gmy}{(x^2 + y^2 + z^2)^{3/2}}}^{N(x, y, z)}\mathbf{j} + \overbrace{\frac{-Gmz}{(x^2 + y^2 + z^2)^{3/2}}}^{P(x, y, z)}\mathbf{k} \quad (2)$$

Now let $\mathbf{F} = M\mathbf{i} + N\mathbf{j} + P\mathbf{k}$ be a vector field. We say that $\mathbf{F}$ is **continuous** at (x, y, z) if and only if M, N, and P are continuous at (x, y, z). The gravitational field $\mathbf{F}$ and the electric field $\mathbf{E}$ defined above are continuous at every point in their domains. In fact, every vector field we will consider is continuous at every point in its domain.

The Gradient as a Vector Field

Suppose f is a differentiable function of three variables. Then the gradient of f, defined in Section 13.6, is actually a vector field, denoted grad f or ∇f and given by

$$\text{grad } f(x, y, z) = \nabla f(x, y, z) = \frac{\partial f}{\partial x}(x, y, z)\mathbf{i} + \frac{\partial f}{\partial y}(x, y, z)\mathbf{j} + \frac{\partial f}{\partial z}(x, y, z)\mathbf{k}$$

If a vector field $\mathbf{F}$ is equal to grad f for some differentiable function f of several variables, then $\mathbf{F}$ is called a **conservative vector field**, and f is a **potential function*** for $\mathbf{F}$. Many vector fields that arise in physics are conservative. For example, we can show that the gravitational field $\mathbf{F}$ of a point mass is conservative. Indeed, we know from Example 2 of Section 13.6 that if

$$f_1(x, y, z) = \frac{1}{(x^2 + y^2 + z^2)^{1/2}}$$

then

$$\text{grad } f_1(x, y, z) = \frac{-1}{(x^2 + y^2 + z^2)^{3/2}}(x\mathbf{i} + y\mathbf{j} + z\mathbf{k})$$

Consequently if

$$f(x, y, z) = \frac{Gm}{(x^2 + y^2 + z^2)^{1/2}}$$

then the formula in (1) for $\mathbf{F}$ implies that

$$\text{grad } f(x, y, z) = \frac{-Gm}{(x^2 + y^2 + z^2)^{3/2}}(x\mathbf{i} + y\mathbf{j} + z\mathbf{k}) = \mathbf{F}(x, y, z) \quad (3)$$

Therefore f is a potential function for $\mathbf{F}$, and thus $\mathbf{F}$ is conservative. Because the electric field of a point charge differs from the gravitational field of a point mass by only a constant factor, the electric field is also conservative.

The Divergence of a Vector Field

There are two types of derivatives of a vector field, one that is a real-valued function and one that is a vector field. We begin with the real-valued derivative.

---

* In physics the potential function for $\mathbf{F}$ is taken to be the function f such that $\mathbf{F} = -\text{grad } f$.

DEFINITION 15.2

Let $\mathbf{F} = M\mathbf{i} + N\mathbf{j} + P\mathbf{k}$ be a vector field such that $\partial M/\partial x$, $\partial N/\partial y$, and $\partial P/\partial z$ exist. Then the **divergence** of $\mathbf{F}$, denoted div $\mathbf{F}$ or $\nabla \cdot \mathbf{F}$, is the function defined by

$$\text{div } \mathbf{F}(x, y, z) = \nabla \cdot \mathbf{F}(x, y, z)$$

$$= \frac{\partial M}{\partial x}(x, y, z) + \frac{\partial N}{\partial y}(x, y, z) + \frac{\partial P}{\partial z}(x, y, z)$$

Example 1   Let $\mathbf{F}$ be the gravitational field given by (2). Show that div $\mathbf{F} = 0$.

Solution   We find that

$$\frac{\partial M}{\partial x}(x, y, z) = \frac{-Gm(x^2 + y^2 + z^2)^{3/2} - (-Gmx)(\frac{3}{2})(2x)(x^2 + y^2 + z^2)^{1/2}}{(x^2 + y^2 + z^2)^3}$$

$$= \frac{Gm(2x^2 - y^2 - z^2)}{(x^2 + y^2 + z^2)^{5/2}}$$

In a similar fashion we find that

$$\frac{\partial N}{\partial y}(x, y, z) = \frac{Gm(2y^2 - z^2 - x^2)}{(x^2 + y^2 + z^2)^{5/2}}$$

and

$$\frac{\partial P}{\partial z}(x, y, z) = \frac{Gm(2z^2 - x^2 - y^2)}{(x^2 + y^2 + z^2)^{5/2}}$$

Therefore

$$\text{div } \mathbf{F} = \frac{\partial M}{\partial x} + \frac{\partial N}{\partial y} + \frac{\partial P}{\partial z} = 0 \quad \square$$

If div $\mathbf{F} = 0$, then $\mathbf{F}$ is said to be **divergence free** or **solenoidal**. Example 1 shows that the gravitational field is divergence free.

A revealing interpretation of div $\mathbf{F}$ arises from the study of fluid flowing through a region. Suppose $\mathbf{v}$ represents the velocity field of a fluid, such as air, flowing through a surface, such as a screen. Then in physical terms, div $\mathbf{v}(x, y, z)$ represents the rate (with respect to time) of mass flow per unit volume of the fluid from the point (x, y, z). (We will examine this interpretation of div $\mathbf{v}$ further in Section 15.8.) A point (x, y, z) is a **source** if div $\mathbf{v}(x, y, z) > 0$; this means that there is a positive mass flow *from* the point (x, y, z). By contrast, (x, y, z) is a **sink** if div $\mathbf{v}(x, y, z) < 0$; this means that there is a positive flow *to* the point (x, y, z). Finally, if div $\mathbf{v}(x, y, z) = 0$ for all (x, y, z) in a region, then there are neither sources nor sinks in the region. A fluid whose velocity field is divergence free is called **incompressible**.

Example 2   Let $\mathbf{v}(x, y, z) = x^3yz^2\mathbf{i} + x^2y^2z^2\mathbf{j} + x^2yz^3\mathbf{k}$. Determine which points in space are sources and which are sinks.

Solution   A straightforward calculation shows that

$$\text{div } \mathbf{v}(x, y, z) = 3x^2yz^2 + 2x^2yz^2 + 3x^2yz^2 = 8x^2yz^2$$

Thus div $\mathbf{v}(x, y, z) = 0$ if (x, y, z) lies on any of the coordinate planes. Moreover, div $\mathbf{v}(x, y, z) > 0$ if $y > 0$ and x and z are not 0, whereas div $\mathbf{v}(x, y, z) < 0$ if $y < 0$ and x and z are not 0. Consequently the sources lie to the right of the xz plane and the sinks to the left.   ☐

The Curl of a Vector Field

The second type of derivative of a vector field is a vector field.

DEFINITION 15.3

Let $\mathbf{F} = M\mathbf{i} + N\mathbf{j} + P\mathbf{k}$ be a vector field such that the first partial derivatives of M, N, and P all exist. Then the **curl** of $\mathbf{F}$, which is denoted curl $\mathbf{F}$ or $\nabla \times \mathbf{F}$, is defined by

$$\text{curl } \mathbf{F}(x, y, z) = \nabla \times \mathbf{F}(x, y, z)$$

$$= \left(\frac{\partial P}{\partial y} - \frac{\partial N}{\partial z}\right)\mathbf{i} + \left(\frac{\partial M}{\partial z} - \frac{\partial P}{\partial x}\right)\mathbf{j} + \left(\frac{\partial N}{\partial x} - \frac{\partial M}{\partial y}\right)\mathbf{k}$$

We often express curl $\mathbf{F}$ symbolically as

$$\begin{vmatrix} \mathbf{i} & \mathbf{j} & \mathbf{k} \\ \dfrac{\partial}{\partial x} & \dfrac{\partial}{\partial y} & \dfrac{\partial}{\partial z} \\ M & N & P \end{vmatrix}$$

Example 3   Let $\mathbf{F}(x, y, z) = xz\mathbf{i} + xy^2z\mathbf{j} - e^{2y}\mathbf{k}$. Find curl $\mathbf{F}$.

Solution   By definition

$$\text{curl } \mathbf{F}(x, y, z) = \begin{vmatrix} \mathbf{i} & \mathbf{j} & \mathbf{k} \\ \dfrac{\partial}{\partial x} & \dfrac{\partial}{\partial y} & \dfrac{\partial}{\partial z} \\ xz & xy^2z & -e^{2y} \end{vmatrix}$$

$$= (-2e^{2y} - xy^2)\mathbf{i} + (x - 0)\mathbf{j} + (y^2z - 0)\mathbf{k}$$

$$= (-2e^{2y} - xy^2)\mathbf{i} + x\mathbf{j} + y^2z\mathbf{k}   \quad \square$$

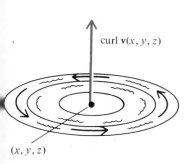

curl $\mathbf{v}(x, y, z)$

(x, y, z)

FIGURE 15.4

Suppose that $\mathbf{v}$ represents the velocity field of a fluid flowing through a solid region. Then it turns out that curl $\mathbf{v}$ measures the tendency of the fluid to curl, or rotate, about an axis. More specifically, particles in the fluid tend to rotate about the axis that points in the direction of curl $\mathbf{v}(x, y, z)$, and the length of curl $\mathbf{v}(x, y, z)$ measures the swiftness of the motion of the particles around the axis (Figure 15.4). (We will return to this interpretation of curl $\mathbf{v}$ in Section 15.7.) Formerly the term for curl **F** was "rot **F**," where "rot" is an abbreviation for "rotation." If curl **F** = **0**, then **F** is said to be **irrotational**, whether or not **F** represents a velocity field.

Among the several useful relations between the gradient, divergence, and curl, the two that occur most often are

$$\text{div}\,(\text{curl}\,\mathbf{F}) = 0 \tag{4}$$

and

$$\text{curl}\,(\text{grad}\,f) = \mathbf{0} \tag{5}$$

The proofs of (4) and (5) involve performing the required differentiations, and they depend on the equality of the mixed partials that arise (see Exercises 27 and 28).

Another important formula is

$$\text{div}\,(\text{grad}\,f) = \frac{\partial^2 f}{\partial x^2} + \frac{\partial^2 f}{\partial y^2} + \frac{\partial^2 f}{\partial z^2}$$

The right side of this formula is the **Laplacian** of f, usually denoted $\nabla^2 f$. A function that satisfies the equation

$$\nabla^2 f = 0$$

which is known as **Laplace's equation,*** is said to be **harmonic**. Harmonic functions are important in physics.

Let f, M, and N be functions of two variables, and let $\mathbf{F} = M\mathbf{i} + N\mathbf{j}$. Then the two-dimensional versions of the gradient, divergence, curl, and Laplacian, which we will employ later, are

$$\text{grad}\,f(x, y) = \frac{\partial f}{\partial x}\mathbf{i} + \frac{\partial f}{\partial y}\mathbf{j} \qquad \text{curl}\,\mathbf{F}(x, y) = \left(\frac{\partial N}{\partial x} - \frac{\partial M}{\partial y}\right)\mathbf{k}$$

$$\text{div}\,\mathbf{F}(x, y) = \frac{\partial M}{\partial x} + \frac{\partial N}{\partial y} \qquad \nabla^2 f(x, y) = \frac{\partial^2 f}{\partial x^2} + \frac{\partial^2 f}{\partial y^2}$$

Recovering a Function from Its Gradient

Just as a function of a single variable can be recovered from its derivative by integration, a function of several variables can sometimes be recovered from its gradient by successive integrations. We illustrate the procedure in the next two examples.

Example 4   Find a function f of two variables such that

$$\text{grad}\,f(x, y) = y^3\mathbf{i} + 3xy^2\mathbf{j}$$

Solution   Since

$$\frac{\partial f}{\partial x}\mathbf{i} + \frac{\partial f}{\partial y}\mathbf{j} = \text{grad}\,f(x, y) = y^3\mathbf{i} + 3xy^2\mathbf{j}$$

---

* After the French mathematician Pierre Simone Laplace (1749–1827).

we have
$$\frac{\partial f}{\partial x} = y^3 \quad \text{and} \quad \frac{\partial f}{\partial y} = 3xy^2 \tag{6}$$

By integrating both sides of the first equation in (6) with respect to x we obtain
$$f(x, y) = xy^3 + g(y)$$

where $g(y)$ is constant with respect to x. Taking partial derivatives of both sides with respect to y, we find that
$$\frac{\partial f}{\partial y} = 3xy^2 + \frac{dg}{dy} \tag{7}$$

Comparison of (7) and the second equation in (6) reveals that
$$\frac{dg}{dy} = 0 \quad \text{so that} \quad g(y) = C$$

where C is a constant. Therefore
$$f(x, y) = xy^3 + C \quad \square$$

The constant C in the solution of Example 4 corresponds to the constant of integration in the indefinite integral of a function of a single variable.

Example 5 Find a function f of three variables such that
$$\text{grad } f(x, y, z) = (2xy + z^2)\mathbf{i} + x^2\mathbf{j} + (2xz + \pi \cos \pi z)\mathbf{k}$$

Solution Since
$$\frac{\partial f}{\partial x}\mathbf{i} + \frac{\partial f}{\partial y}\mathbf{j} + \frac{\partial f}{\partial z}\mathbf{k} = \text{grad } f(x, y, z) = (2xy + z^2)\mathbf{i} + x^2\mathbf{j} + (2xz + \pi \cos \pi z)\mathbf{k}$$

we have
$$\frac{\partial f}{\partial x} = 2xy + z^2, \quad \frac{\partial f}{\partial y} = x^2, \quad \text{and} \quad \frac{\partial f}{\partial z} = 2xz + \pi \cos \pi z \tag{8}$$

Integrating both sides of the first equation in (8) with respect to x, we obtain
$$f(x, y, z) = x^2y + xz^2 + g(y, z) \tag{9}$$

where g is constant with respect to x. Differentiation of both sides of (9) with respect to y yields
$$\frac{\partial f}{\partial y} = x^2 + \frac{\partial g}{\partial y} \tag{10}$$

Comparing (10) with the second equation in (8), we find that
$$\frac{\partial g}{\partial y} = 0$$

so that g is constant with respect to y. Thus (9) can be rewritten as

$$f(x, y, z) = x^2 y + xz^2 + h(z) \tag{11}$$

for an appropriate function h of z. Next we differentiate both sides of (11) with respect to z and obtain

$$\frac{\partial f}{\partial z} = 2xz + \frac{dh}{dz} \tag{12}$$

Comparing (12) with the third equation in (8), we find that

$$\frac{dh}{dz} = \pi \cos \pi z$$

Thus $h(z) = \sin \pi z + C$ for some constant C. Consequently (11) becomes

$$f(x, y, z) = x^2 y + xz^2 + \sin \pi z + C \quad \square$$

If $\mathbf{F} = M\mathbf{i} + N\mathbf{j} + P\mathbf{k}$ is a vector field such that M, N, and P have continuous partial derivatives, and if there is a function f such that $\mathbf{F} = \operatorname{grad} f$, then (5) implies that

$$\operatorname{curl} \mathbf{F} = \operatorname{curl} (\operatorname{grad} f) = \mathbf{0}$$

But $\operatorname{curl} \mathbf{F} = \mathbf{0}$ is equivalent to

$$\frac{\partial P}{\partial y} = \frac{\partial N}{\partial z}, \quad \frac{\partial M}{\partial z} = \frac{\partial P}{\partial x}, \quad \text{and} \quad \frac{\partial N}{\partial x} = \frac{\partial M}{\partial y} \tag{13}$$

This argument is not reversible. If the equations in (13) hold for a vector field $\mathbf{F} = M\mathbf{i} + N\mathbf{j} + P\mathbf{k}$, then $\mathbf{F}$ is not necessarily the gradient of a function. (See Exercise 10 of Section 15.3.) However, if the domain D of $\mathbf{F}$ is all of three-dimensional space or a ball or a parallelepiped (or more generally, if D contains no "holes"), then the argument is reversible. We summarize these results in the next theorem.

THEOREM 15.4

Let $\mathbf{F} = M\mathbf{i} + N\mathbf{j} + P\mathbf{k}$ be a vector field. If there is a function f having continuous mixed partials whose gradient is $\mathbf{F}$, then

$$\frac{\partial P}{\partial y} = \frac{\partial N}{\partial z}, \quad \frac{\partial M}{\partial z} = \frac{\partial P}{\partial x}, \quad \text{and} \quad \frac{\partial N}{\partial x} = \frac{\partial M}{\partial y} \tag{14}$$

If the domain of $\mathbf{F}$ is all of three-dimensional space and if the equations in (14) are satisfied, then there is a function f such that $\mathbf{F} = \operatorname{grad} f$.

Example 6   Let

$$\mathbf{F}(x, y, z) = 2xyz\mathbf{i} + x^2 z\mathbf{j} + (x^2 y + 1)\mathbf{k}$$

and

$$\mathbf{G}(x, y, z) = yz \cos xy\mathbf{i} + xz \cos xy\mathbf{j} + \cos xy\mathbf{k}$$

Show that **F** is the gradient of some function but that **G** is not the gradient of any function.

Solution It is routine to verify that for **F** we have

$$\frac{\partial P}{\partial y} = x^2 = \frac{\partial N}{\partial z}, \quad \frac{\partial M}{\partial z} = 2xy = \frac{\partial P}{\partial x}, \quad \text{and} \quad \frac{\partial N}{\partial x} = 2xz = \frac{\partial M}{\partial y}$$

Since the domain of **F** is all of three-dimensional space, Theorem 15.4 implies that **F** is the gradient of some function. However, for **G** we have

$$\frac{\partial P}{\partial y} = -x \sin xy \quad \text{and} \quad \frac{\partial N}{\partial z} = x \cos xy$$

so that the first equation in (14) is not satisfied. By Theorem 15.4, **G** cannot be the gradient of any function. □

In case a vector field **F** is given by

$$\mathbf{F}(x, y) = M(x, y)\mathbf{i} + N(x, y)\mathbf{j}$$

the conditions in (14) reduce to

$$\frac{\partial N}{\partial x} = \frac{\partial M}{\partial y}$$

and the corresponding statement in Theorem 15.4 holds for such vector fields.

Example 7 Let

$$\mathbf{F}(x, y) = y^2 e^{xy}\mathbf{i} + (1 + xy)e^{xy}\mathbf{j}$$

and

$$\mathbf{G}(x, y) = \frac{x}{y}\mathbf{i} + \frac{y}{x}\mathbf{j}$$

Show that **F** is the gradient of some function but that **G** is not the gradient of any function.

Solution For **F** we have

$$\frac{\partial N}{\partial x} = ye^{xy} + (1 + xy)ye^{xy} = (2y + xy^2)e^{xy}$$

and

$$\frac{\partial M}{\partial y} = 2ye^{xy} + y^2 xe^{xy} = (2y + xy^2)e^{xy}$$

Since $\partial N/\partial x = \partial M/\partial y$ and the domain of **F** is the xy plane, we conclude from the two-dimensional version of Theorem 15.4 that **F** is the gradient of some function.

For **G** we find that

$$\frac{\partial N}{\partial x} = \frac{-y}{x^2} \quad \text{and} \quad \frac{\partial M}{\partial y} = \frac{-x}{y^2}$$

so that **G** is not the gradient of any function. ☐

EXERCISES 15.1

In Exercises 1–10 find the curl and the divergence of the given vector field.

1. $\mathbf{F}(x, y) = x\mathbf{i} + y\mathbf{j}$

2. $\mathbf{F}(x, y) = \dfrac{x}{x^2 + y^2}\mathbf{i} + \dfrac{y}{x^2 + y^2}\mathbf{j}$

3. $\mathbf{F}(x, y, z) = y\mathbf{i} + z\mathbf{j} + x\mathbf{k}$

4. $\mathbf{F}(x, y, z) = yz\mathbf{i} + zx\mathbf{j} + xy\mathbf{k}$

5. $\mathbf{F}(x, y, z) = x^2\mathbf{i} + y^2\mathbf{j} + z^2\mathbf{k}$

6. $\mathbf{F}(x, y, z) = \cos x\mathbf{i} + \sin y\mathbf{j} + e^{xy}\mathbf{k}$

7. $\mathbf{F}(x, y, z) = \dfrac{-x}{z}\mathbf{i} - \dfrac{y}{z}\mathbf{j} + \dfrac{1}{z}\mathbf{k}$

8. $\mathbf{F}(x, y, z) = (y + z)\mathbf{i} + (z + x)\mathbf{j} + (x + y)\mathbf{k}$

9. $\mathbf{F}(x, y, z) = e^x \cos y\mathbf{i} + e^x \sin y\mathbf{j} + z\mathbf{k}$

10. $\mathbf{F}(x, y, z) = \dfrac{x}{(x^2 + y^2)^{3/2}}\mathbf{i} + \dfrac{y}{(x^2 + y^2)^{3/2}}\mathbf{j} + \mathbf{k}$

In Exercises 11–14 show that f satisfies Laplace's equation.

11. $f(x, y) = x^2 - y^2$

12. $f(x, y) = \arctan \dfrac{y}{x}$

13. $f(x, y, z) = x^2 + y^2 - 2z^2$

14. $f(x, y, z) = \dfrac{1}{\sqrt{x^2 + y^2 + z^2}}$

In Exercises 15–24 determine whether **F** is the gradient of some function f. If it is, find such a function f.

15. $\mathbf{F}(x, y) = e^y\mathbf{i} + (xe^y + y)\mathbf{j}$

16. $\mathbf{F}(x, y) = y^2e^{xy}\mathbf{i} + (1 + xy)e^{xy}\mathbf{j}$

17. $\mathbf{F}(x, y) = (\sin xy)\mathbf{i} + (\cos xy)\mathbf{j}$

18. $\mathbf{F}(x, y) = (3x^2y^2 + 3y)\mathbf{i} + (2x^3y + 3x)\mathbf{j}$

19. $\mathbf{F}(x, y, z) = 2xyz\mathbf{i} + x^2z\mathbf{j} + (x^2y + 1)\mathbf{k}$

20. $\mathbf{F}(x, y, z) = yz\mathbf{i} + xz\mathbf{j} + xy\mathbf{k}$

21. $\mathbf{F}(x, y, z) = xz\mathbf{i} + yz\mathbf{j} + xz\mathbf{k}$

22. $\mathbf{F}(x, y, z) = (2xz + 1)\mathbf{i} + 2y(z + 1)\mathbf{j} + (x^2 + y^2 + 3z^2)\mathbf{k}$

23. $\mathbf{F}(x, y, z) = (y^2 + x^2)\mathbf{i} + (z^2 + y^2)\mathbf{j} + (x^2 + z^2)\mathbf{k}$

24. $\mathbf{F}(x, y, z) = yze^{xy}\mathbf{i} + xze^{xy}\mathbf{j} + (e^{xy} + \cos z)\mathbf{k}$

25. Let f and g be functions of several variables, and let **F** and **G** be vector fields. Decide which of the following ex-

pressions represent vector fields, which represent functions of several variables, and which are meaningless.

a. grad (fg)
b. grad **F**
c. curl (grad f)
d. grad (div **F**)
e. curl (curl **F**)
f. div (grad f)
g. (grad f) × (curl **F**)
h. div (curl (grad f))
i. curl (div (grad f))

26. Let **F** and **G** be vector fields, and let f be a function of three variables. Then $f\mathbf{F}$, $\mathbf{F} \cdot \mathbf{G}$, and $\mathbf{F} \times \mathbf{G}$ are defined by the following formulas.

$$(f\mathbf{F})(x, y, z) = f(x, y, z)\mathbf{F}(x, y, z)$$

$$(\mathbf{F} \cdot \mathbf{G})(x, y, z) = \mathbf{F}(x, y, z) \cdot \mathbf{G}(x, y, z)$$

$$(\mathbf{F} \times \mathbf{G})(x, y, z) = \mathbf{F}(x, y, z) \times \mathbf{G}(x, y, z)$$

a. Use components of the vector field **F** to prove that $f\mathbf{F}$ is a continuous vector field if f and **F** are continuous.
b. Use components of the vector fields **F** and **G** to prove that $\mathbf{F} \cdot \mathbf{G}$ is a continuous function of several variables if **F** and **G** are continuous.
c. Use components of the vector fields **F** and **G** to prove that $\mathbf{F} \times \mathbf{G}$ is a continuous vector field if **F** and **G** are continuous.

In Exercises 27–31 verify the identity. Use the formulas in Exercise 26 for $f\mathbf{F}$, $\mathbf{F} \cdot \mathbf{G}$, and $\mathbf{F} \times \mathbf{G}$, and assume that the required partial derivatives exist and are continuous.

27. div (curl **F**) = 0 (Thus the curl of the vector field is solenoidal.)

28. curl (grad f) = **0** (Thus the gradient of a function is irrotational.)

29. div $(f\mathbf{F}) = f$ div **F** + (grad f) · **F**

30. div $(\mathbf{F} \times \mathbf{G}) = (\text{curl } \mathbf{F}) \cdot \mathbf{G} - \mathbf{F} \cdot (\text{curl } \mathbf{G})$ (Thus the cross product of two irrotational vector fields is solenoidal.)

31. curl $(f\mathbf{F}) = f(\text{curl } \mathbf{F}) + (\text{grad } f) \times \mathbf{F}$

32. a. Suppose f is continuous on a simple region R in the xy plane. If grad $f(x, y) = \mathbf{0}$ for all (x, y) in R, show that f is constant.
b. Suppose f is continuous on a simple solid region D. If grad $f(x, y, z) = \mathbf{0}$ for all (x, y, z) in D, show that f is constant.

In Exercises 33–36 assume that all functions and all components of vector fields have the required continuous partial derivatives.

33. Let **F** be a constant vector, and let $G(x, y, z) = x\mathbf{i} + y\mathbf{j} + z\mathbf{k}$. Show that curl $(\mathbf{F} \times \mathbf{G}) = 2\mathbf{F}$.

34. Let **F** be a vector field, f a function of several variables, and $\mathbf{G} = \mathbf{F} + \operatorname{grad} f$. Prove that curl $\mathbf{G} = \operatorname{curl} \mathbf{F}$. (*Hint:* First show that if $\mathbf{F}_1$ and $\mathbf{F}_2$ are vector fields, then curl $(\mathbf{F}_1 + \mathbf{F}_2) = \operatorname{curl} \mathbf{F}_1 + \operatorname{curl} \mathbf{F}_2$.)

35. Show that if **F** and **G** are conservative, then $\mathbf{F} + \mathbf{G}$ is also conservative.

36. Let $\mathbf{F}(x, y, z) = M(y, z)\mathbf{i} + N(x, z)\mathbf{j} + P(x, y)\mathbf{k}$. Show that **F** is solenoidal.

37. An object having mass m is spun around in a circular orbit with angular velocity ω and is subject to a centrifugal force **F** given by

$$\mathbf{F}(x, y, z) = m\omega^2(x\mathbf{i} + y\mathbf{j} + z\mathbf{k})$$

Show that the function f defined by

$$f(x, y, z) = \frac{m\omega^2}{2}(x^2 + y^2 + z^2)$$

is a potential function for **F**.

38. Suppose that the potential function f for an electric field **E** produced by an electric dipole at the origin is given by

$$f(x, y, z) = \frac{ax + by + cz}{(x^2 + y^2 + z^2)^{3/2}}$$

where a, b, and c are constants. Find the electric field **E**.

39. A vector field **G** is a **vector potential** of a vector field **F** if $\mathbf{F} = \operatorname{curl} \mathbf{G}$. Use Exercise 27 to show that if **F** has a vector potential, then **F** is solenoidal.

40. Suppose the magnetic induction field **B** associated with a current in a wire is given by

$$\mathbf{B}(x, y, z) = I\left(\frac{-y}{x^2 + y^2}\mathbf{i} + \frac{x}{x^2 + y^2}\mathbf{j}\right)$$

where I is a constant. Let

$$\mathbf{G}(x, y, z) = \frac{-I}{2}\ln(x^2 + y^2)\mathbf{k}$$

Show that **G** is a vector potential for **B**.

41. Let $\mathbf{F}(x, y, z) = 8\mathbf{i}$. Find a vector potential for **F**.

42. If **v** represents the velocity field of a homogeneous fluid that rotates at a constant angular velocity ω about the z axis, then

$$\mathbf{v}(x, y, z) = -\omega y\mathbf{i} + \omega x\mathbf{j}$$

Show that curl $\mathbf{v}(x, y, z)$ depends only on ω and not on (x, y, z).

43. Suppose the temperature in a region is given by

$$T(x, y, z) = 30 - 2x^2 - y^2 - 4z^2$$

a. Show that grad T (called the **temperature gradient**) is continuous.

b. Determine whether grad T is a central force field.

15.2
LINE INTEGRALS

In this section we define a single integral that is more general than the single integral introduced in Chapter 5. Instead of integrating a function over an interval $[a, b]$, we will integrate over a curve C in space. Such an integral is called a line integral, although the term "curve integral" would be more appropriate. Line integrals have many physical applications, and we will motivate the definition of the line integral by one such application: the mass of a thin wire with a known mass density.

Suppose that a wire lies along a piecewise smooth curve C of finite length and that the mass density (mass per unit length) of the wire at any point (x, y, z) on C is $f(x, y, z)$. Consider any partition $\mathscr{P}$ of C obtained by choosing points P_0, P_1, $P_2, \ldots, P_n$ of subdivision along C (Figure 15.5). For each integer k between 1 and n let (x_k, y_k, z_k) be any point on C between P_{k-1} and P_k, and let Δs_k be the length of the portion of C between P_{k-1} and P_k. If Δs_k is small, then the mass of the portion of the wire between P_{k-1} and P_k should be approximately $f(x_k, y_k, z_k)\Delta s_k$, which is the product of mass density at (x_k, y_k, z_k) and the length Δs_k. Therefore the mass m of the whole wire should be approximately $\sum_{k=1}^{n} f(x_k, y_k, z_k)\Delta s_k$. It seems

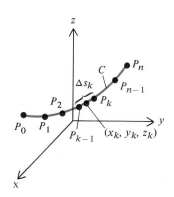

FIGURE 15.5

reasonable to expect the sum

$$\sum_{k=1}^{n} f(x_k, y_k, z_k) \Delta s_k \qquad (1)$$

to approach m as the largest of the lengths $\Delta s_1, \Delta s_2, \ldots, \Delta s_n$, which we denote $\|\mathscr{P}\|$, approaches 0. It can be proved that if f is continuous at every point on C, then the sum in (1) does in fact approach a limit, and we call that limit the line integral of f over C.

DEFINITION 15.5

> Let f be continuous on a piecewise smooth curve C with finite length. Then the **line integral** $\int_C f(x, y, z)\, ds$ of f over C is defined by
>
> $$\int_C f(x, y, z)\, ds = \lim_{\|\mathscr{P}\| \to 0} \sum_{k=1}^{n} f(x_k, y_k, z_k) \Delta s_k \qquad (2)$$

In case f is the mass density of a wire, we compute the mass m of the wire by the formula

$$m = \int_C f(x, y, z)\, ds \qquad (3)$$

Evaluation of the line integral $\int_C f(x, y, z)\, ds$ by means of the sum in (2) is an unwieldy process. It is usually simpler to employ a parametrization of the curve C. For the present let us assume that C is parametrized by a smooth vector-valued function $\mathbf{r} = x\mathbf{i} + y\mathbf{j} + z\mathbf{k}$ on an interval $[a, b]$. By using the fact that $\|d\mathbf{r}/dt\| = ds/dt$, one can show that

$$\int_C f(x, y, z)\, ds = \int_a^b f(x(t), y(t), z(t)) \left\| \frac{d\mathbf{r}}{dt} \right\| dt \qquad (4)$$

and that any smooth parametrization $\mathbf{r}$ of the curve C yields the same value of $\int_C f(x, y, z)\, ds$.

Example 1 Let C be the line segment from $(0, 0, 0)$ to $(1, -3, 2)$. Find $\int_C (x + y^2 - 2z)\, ds$.

Solution We parametrize C by

$$\mathbf{r}(t) = t\mathbf{i} - 3t\mathbf{j} + 2t\mathbf{k} \quad \text{for } 0 \le t \le 1$$

Then since $x(t) = t$, $y(t) = -3t$, $z(t) = 2t$, and

$$\left\| \frac{d\mathbf{r}}{dt} \right\| = \sqrt{1^2 + (-3)^2 + 2^2} = \sqrt{14}$$

we deduce from (4) that

$$\int_C (x + y^2 - 2z)\,ds = \int_0^1 [t + (-3t)^2 - 2(2t)]\sqrt{14}\,dt$$

$$= \sqrt{14}\int_0^1 (-3t + 9t^2)\,dt$$

$$= \sqrt{14}\left(\frac{-3}{2}t^2 + 3t^3\right)\Big|_0^1 = \frac{3}{2}\sqrt{14} \quad \square$$

Example 2 Let C be the twisted cubic curve parametrized by

$$\mathbf{r}(t) = t\mathbf{i} + t^2\mathbf{j} + t^3\mathbf{k} \quad \text{for } 0 \le t \le \frac{1}{2}$$

Find

$$\int_C (8x + 36z)\,ds$$

Solution First we notice that $x(t) = t$, $y(t) = t^2$, $z(t) = t^3$, and

$$\frac{d\mathbf{r}}{dt} = \mathbf{i} + 2t\mathbf{j} + 3t^2\mathbf{k}$$

Then

$$\left\|\frac{d\mathbf{r}}{dt}\right\| = \sqrt{1^2 + (2t)^2 + (3t^2)^2} = \sqrt{1 + 4t^2 + 9t^4}$$

It follows from (4) that

$$\int_C (8x + 36z)\,ds = \int_0^{1/2} (8t + 36t^3)\sqrt{1 + 4t^2 + 9t^4}\,dt$$

$$= \frac{2}{3}(1 + 4t^2 + 9t^4)^{3/2}\Big|_0^{1/2}$$

$$= \frac{2}{3}\left(\frac{41\sqrt{41}}{64} - 1\right) \quad \square$$

You might imagine in Example 2 that we are computing the mass of a wire bent in the shape of a twisted cubic curve.

The line segment C from $(a, 0, 0)$ to $(b, 0, 0)$ on the x axis is parametrized by

$$\mathbf{r}(t) = t\mathbf{i} \quad \text{for } a \le t \le b$$

Consequently if f is continuous on C, formula (4) becomes

$$\int_C f(x, y, z)\,ds = \int_a^b f(x(t), y(t), z(t))\left\|\frac{d\mathbf{r}}{dt}\right\|\,dt$$

$$= \int_a^b f(t, 0, 0)(1)\,dt = \int_a^b f(t, 0, 0)\,dt$$

By identifying f with f_0, where

$$f_0(t) = f(t, 0, 0) \quad \text{for } a \le t \le b$$

we find that

$$\int_C f(x, y, z)\, ds = \int_a^b f(t, 0, 0)\, dt = \int_a^b f_0(t)\, dt$$

Thus the line integral of a function f that is continuous on a closed interval on the x axis is the same as the single integral of the associated function f_0 of a single variable.

We observe also that if $\mathbf{r}$ parametrizes C on $[a, b]$, then

$$\int_C 1\, ds = \int_a^b \left\| \frac{d\mathbf{r}}{dt} \right\|\, dt$$

and the expression on the right is the length of the curve C, by Definition 12.16 of Section 12.4. Consequently the length of a smooth curve in space may be given in terms of a line integral.

It follows from the definition of line integral that if f is continuous on a piecewise smooth curve C composed of smooth curves $C_1, C_2, \ldots, C_n$, then

$$\int_C f(x, y, z)\, ds = \int_{C_1} f(x, y, z)\, ds + \int_{C_2} f(x, y, z)\, ds + \cdots + \int_{C_n} f(x, y, z)\, ds$$

Example 3   Let C be composed of the two curves C_1 and C_2 shown in Figure 15.6. Evaluate $\int_C (1 + xy)\, ds$.

Solution   The curves C_1 and C_2 can be parametrized as follows.

$$C_1: \quad \mathbf{r}_1(t) = 2\cos t\, \mathbf{i} + 2\sin t\, \mathbf{j} \quad \text{for } -\frac{\pi}{2} \le t \le \frac{\pi}{2}$$

$$C_2: \quad \mathbf{r}_2(t) = -2t\mathbf{j} \quad \text{for } -1 \le t \le 1$$

Consequently

$$\left\| \frac{d\mathbf{r}_1}{dt} \right\| = \sqrt{(-2\sin t)^2 + (2\cos t)^2} = 2 \quad \text{and} \quad \left\| \frac{d\mathbf{r}_2}{dt} \right\| = 2$$

so that

$$\int_{C_1} (1 + xy)\, ds = \int_{-\pi/2}^{\pi/2} [1 + (2\cos t)(2\sin t)]2\, dt = 2(t + 2\sin^2 t)\Big|_{-\pi/2}^{\pi/2} = 2\pi$$

and

$$\int_{C_2} (1 + xy)\, ds = \int_{-1}^{1} [1 + 0(-2t)]2\, dt = 2t\Big|_{-1}^{1} = 4$$

Therefore

$$\int_C (1 + xy)\, ds = \int_{C_1} (1 + xy)\, ds + \int_{C_2} (1 + xy)\, ds = 2\pi + 4 \quad \square$$

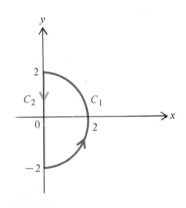

FIGURE 15.6

**Line Integrals
of Vector Fields**

Recall from Section 11.3 that if a constant force $\mathbf{F}$ is applied to an object moving along a straight line from a point P to a point Q in space, then the work W done on the object by the force is given by

$$W = \mathbf{F} \cdot \overrightarrow{PQ}$$

Now assume that the object moves along a smooth curve C of finite length in space and that the vector field $\mathbf{F}$ represents a continuous force on the object. Our task is to define the work done by the force on the object as it traverses C.

We orient C in the direction in which the object moves and consider any partition $\mathscr{P}$ of C obtained by choosing points $P_0, P_1, P_2, \ldots, P_n$ of subdivision along C (Figure 15.7). For each integer k between 1 and n, let Δs_k be the length of the portion of the curve between P_{k-1} and P_k, and let (x_k, y_k, z_k) be a point on that portion of the curve. If Δs_k is small, then as the object moves along C from P_{k-1} to P_k, it proceeds very nearly in the direction of the tangent vector $\mathbf{T}(x_k, y_k, z_k)$ at (x_k, y_k, z_k). Since by hypothesis $\mathbf{F}$ is continuous on C, the values of $\mathbf{F}$ along the curve from P_{k-1} to P_k are close to $\mathbf{F}(x_k, y_k, z_k)$. Consequently the amount of work done on the object on this portion of the curve C should be approximately

$$\mathbf{F}(x_k, y_k, z_k) \cdot [\Delta s_k \mathbf{T}(x_k, y_k, z_k)]$$

which is the same as $[\mathbf{F}(x_k, y_k, z_k) \cdot \mathbf{T}(x_k, y_k, z_k)] \Delta s_k$. It follows that the total amount of work done on the object as it moves along the entire curve C should be approximately

$$\sum_{k=1}^{n} [\mathbf{F}(x_k, y_k, z_k) \cdot \mathbf{T}(x_k, y_k, z_k)] \Delta s_k$$

Since better and better approximations should result from choosing partitions with smaller and smaller norms, we replace f in Definition 15.5 by $\mathbf{F} \cdot \mathbf{T}$ and define the work W done by the force to be

$$W = \int_C \mathbf{F}(x, y, z) \cdot \mathbf{T}(x, y, z)\, ds \qquad (5)$$

Integrals in the form of (5) will appear in other contexts. For that reason we introduce an abbreviated notation for the integral.

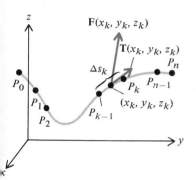

FIGURE 15.7

DEFINITION 15.6

Let $\mathbf{F}$ be a continuous vector field defined on a smooth oriented curve C. Then the **line integral** of $\mathbf{F}$ over C, denoted $\int_C \mathbf{F} \cdot d\mathbf{r}$, is defined by

$$\int_C \mathbf{F} \cdot d\mathbf{r} = \int_C \mathbf{F}(x, y, z) \cdot \mathbf{T}(x, y, z)\, ds$$

where $\mathbf{T}(x, y, z)$ is the tangent vector at (x, y, z) for the given orientation of C.

Caution: Notice that the line integral $\int_C f(x, y, z)\, ds$ in Definition 15.5 does not require C to be oriented. However, for the line integral $\int_C \mathbf{F} \cdot d\mathbf{r}$ of the vector field $\mathbf{F}$, the curve C must be oriented.

If $\mathbf{F}$ represents the force on an object moving along a curve C that is oriented in the direction of motion, then the **work** W done by the force on the object as it traverses C is given by

$$W = \int_C \mathbf{F} \cdot d\mathbf{r} \qquad (6)$$

The integral $\int_C \mathbf{F} \cdot d\mathbf{r}$ depends on the orientation of C, but not on any particular smooth parametrization $\mathbf{r}$ that induces the orientation of C. To prepare $\int_C \mathbf{F} \cdot d\mathbf{r}$ for evaluation, let $\mathbf{r} = x\mathbf{i} + y\mathbf{j} + z\mathbf{k}$ be a parametrization of C with domain $[a, b]$, and assume that the parametrization induces the given orientation on C. Then we know from (1) of Section 12.5 that

$$\mathbf{T}(x(t), y(t), z(t)) = \frac{d\mathbf{r}/dt}{\|d\mathbf{r}/dt\|}$$

so it follows from (4) and Definition 15.6 that

$$\int_C \mathbf{F} \cdot d\mathbf{r} = \int_C \left[\mathbf{F}(x(t), y(t), z(t)) \cdot \frac{d\mathbf{r}/dt}{\|d\mathbf{r}/dt\|} \right] \left\| \frac{d\mathbf{r}}{dt} \right\| dt$$

$$= \int_C \mathbf{F}(x(t), y(t), z(t)) \cdot \frac{d\mathbf{r}}{dt} dt$$

Thus
$$\int_C \mathbf{F} \cdot d\mathbf{r} = \int_a^b \mathbf{F}(x(t), y(t), z(t)) \cdot \frac{d\mathbf{r}}{dt} dt \qquad (7)$$

The right-hand integral in (7) is the integral we use to evaluate $\int_C \mathbf{F} \cdot d\mathbf{r}$.

Example 4 A particle moves upward along the circular helix C_1 parametrized by

$$\mathbf{r}(t) = \cos t\mathbf{i} + \sin t\mathbf{j} + t\mathbf{k} \quad \text{for } 0 \le t \le 2\pi$$

(Figure 15.8), under a force given by

$$\mathbf{F}(x, y, z) = -zy\mathbf{i} + zx\mathbf{j} + xy\mathbf{k}$$

Find the work W done on the particle by the force.

Solution We have $x(t) = \cos t$, $y(t) = \sin t$, $z(t) = t$, and

$$\frac{d\mathbf{r}}{dt} = -\sin t\mathbf{i} + \cos t\mathbf{j} + \mathbf{k}$$

Then we use (6) and (7) to conclude that

$$W = \int_{C_1} \mathbf{F} \cdot d\mathbf{r} = \int_0^{2\pi} \mathbf{F}(x(t), y(t), z(t)) \cdot \frac{d\mathbf{r}}{dt} dt$$

$$= \int_0^{2\pi} (-t\sin t\mathbf{i} + t\cos t\mathbf{j} + \cos t\sin t\mathbf{k}) \cdot (-\sin t\mathbf{i} + \cos t\mathbf{j} + \mathbf{k}) dt$$

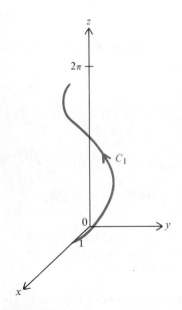

FIGURE 15.8

$$= \int_0^{2\pi} (t\sin^2 t + t\cos^2 t + \cos t \sin t)\, dt$$

$$= \int_0^{2\pi} (t + \cos t \sin t)\, dt = \left(\frac{1}{2}t^2 + \frac{1}{2}\sin^2 t\right)\Big|_0^{2\pi} = 2\pi^2 \quad \square$$

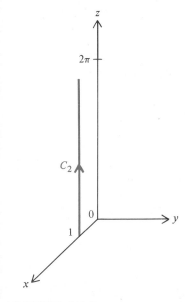

FIGURE 15.9

Example 5 Assume that the particle in Example 4 moves under the same force and with the same initial and terminal points, but along the line segment C_2 parametrized by

$$\mathbf{r}(t) = \mathbf{i} + t\mathbf{k} \quad \text{for } 0 \le t \le 2\pi$$

(Figure 15.9). Find the work W done on the particle by the force.

Solution Here $x(t) = 1$, $y(t) = 0$, $z(t) = t$, and $d\mathbf{r}/dt = \mathbf{k}$. Consequently

$$W = \int_{C_2} \mathbf{F} \cdot d\mathbf{r} = \int_0^{2\pi} \mathbf{F}(x(t), y(t), z(t)) \cdot \frac{d\mathbf{r}}{dt}\, dt$$

$$= \int_0^{2\pi} (0\mathbf{i} + t\mathbf{j} + 0\mathbf{k}) \cdot \mathbf{k}\, dt = \int_0^{2\pi} 0\, dt = 0 \quad \square$$

One implication of the last two examples is that a given force may perform different amounts of work if the paths along which it acts are different.

The orientation of the curve $-C$ is by definition opposite to the orientation of C. Since the tangent vector at any point on $-C$ is the negative of the tangent vector at the same point on C, Definition 15.6 tells us that

$$\int_{-C} \mathbf{F} \cdot d\mathbf{r} = -\int_C \mathbf{F} \cdot d\mathbf{r}$$

For example, if $\mathbf{F}$ and C_1 are as in Example 4, then

$$\int_{-C_1} \mathbf{F} \cdot d\mathbf{r} = -\int_{C_1} \mathbf{F} \cdot d\mathbf{r} = -2\pi^2$$

Alternative Form of the Line Integral Let $\mathbf{F} = M\mathbf{i} + N\mathbf{j} + P\mathbf{k}$ be a continuous vector field defined on a smooth oriented curve C parametrized by

$$\mathbf{r}(t) = x(t)\mathbf{i} + y(t)\mathbf{j} + z(t)\mathbf{k} \quad \text{for } a \le t \le b$$

Then by (7),

$$\int_C \mathbf{F} \cdot d\mathbf{r} = \int_a^b \mathbf{F}(x(t), y(t), z(t)) \cdot \frac{d\mathbf{r}}{dt}\, dt$$

$$= \int_a^b [M(x(t), y(t), z(t))\mathbf{i} + N(x(t), y(t), z(t))\mathbf{j}$$

$$+ P(x(t), y(t), z(t))\mathbf{k}] \cdot \left[\frac{dx}{dt}\mathbf{i} + \frac{dy}{dt}\mathbf{j} + \frac{dz}{dt}\mathbf{k} \right] dt$$

$$= \int_a^b \left[M(x(t), y(t), z(t))\frac{dx}{dt} + N(x(t), y(t), z(t))\frac{dy}{dt} \right.$$

$$\left. + P(x(t), y(t), z(t))\frac{dz}{dt} \right] dt$$

It is common to write the final integral above in the abbreviated form

$$\int_C M(x, y, z)\,dx + N(x, y, z)\,dy + P(x, y, z)\,dz \tag{8}$$

or even more briefly as

$$\int_C M\,dx + N\,dy + P\,dz$$

Thus the integral in (8) is just another notation for $\int_C \mathbf{F} \cdot d\mathbf{r}$ and is usually evaluated by means of the formula

$$\int_C M(x, y, z)\,dx + N(x, y, z)\,dy + P(x, y, z)\,dz$$

$$= \int_a^b \left[M(x(t), y(t), z(t))\frac{dx}{dt} + N(x(t), y(t), z(t))\frac{dy}{dt} \right. \tag{9}$$

$$\left. + P(x(t), y(t), z(t))\frac{dz}{dt} \right] dt$$

Example 6 Let C be the twisted cubic curve parametrized by

$$\mathbf{r}(t) = t\mathbf{i} + t^2\mathbf{j} + t^3\mathbf{k} \quad \text{for } 0 \le t \le 1$$

Evaluate $\int_C xy\,dx + 3zx\,dy - 5x^2yz\,dz$.

Solution Notice that

$$x(t) = t, \qquad \frac{dx}{dt} = 1$$

$$y(t) = t^2, \qquad \frac{dy}{dt} = 2t$$

$$z(t) = t^3, \qquad \frac{dz}{dt} = 3t^2$$

Consequently (9) implies that

$$\int_C xy\,dx + 3zx\,dy - 5x^2yz\,dz = \int_0^1 [(t)(t^2)(1) + 3(t^3)(t)(2t) - 5(t)^2(t^2)(t^3)(3t^2)]\,dt$$

$$= \int_0^1 (t^3 + 6t^5 - 15t^9)\,dt$$

$$= \left(\frac{t^4}{4} + t^6 - \frac{3t^{10}}{2}\right)\Big|_0^1 = -\frac{1}{4} \quad \square$$

In the event that $\mathbf{F} = M\mathbf{i}$, so that $N = P = 0$, we write the expression in (8) as

$$\int_C M(x, y, z)\,dx, \quad \text{or} \quad \int_C M\,dx$$

The integrals $\int_C N(x, y, z)\,dy$ and $\int_C P(x, y, z)\,dz$ have analogous interpretations. When we evaluate these reduced line integrals, we do so by the formulas

$$\int_C M(x, y, z)\,dx = \int_a^b \left[M(x(t), y(t), z(t))\frac{dx}{dt} \right] dt$$

$$\int_C N(x, y, z)\,dy = \int_a^b \left[N(x(t), y(t), z(t))\frac{dy}{dt} \right] dt \qquad (10)$$

$$\int_C P(x, y, z)\,dz = \int_a^b \left[P(x(t), y(t), z(t))\frac{dz}{dt} \right] dt$$

Example 7 Let C be the unit circle in the xz plane, oriented by the parametrization

$$\mathbf{r}(t) = \cos t\,\mathbf{i} + \sin t\,\mathbf{k} \quad \text{for } 0 \le t \le 2\pi$$

Find $\int_C (y + z)\,dx$.

Solution Here $x(t) = \cos t$, $y(t) = 0$, $z(t) = \sin t$, and $dx/dt = -\sin t$. The first formula in (10) tells us that

$$\int_C (y + z)\,dx = \int_0^{2\pi} (0 + \sin t)(-\sin t)\,dt$$

$$= -\int_0^{2\pi} \sin^2 t\,dt = -\int_0^{2\pi} \left(\frac{1}{2} - \frac{1}{2}\cos 2t\right) dt$$

$$= -\left(\frac{1}{2}t - \frac{1}{4}\sin 2t\right)\Big|_0^{2\pi} = -\pi \quad \square$$

If an oriented curve C is not smooth but is piecewise smooth, composed of smooth curves $C_1, C_2, \ldots, C_n$, then we define the line integral $\int_C M(x, y, z)\, dx + N(x, y, z)\, dy + P(x, y, z)\, dz$ by

$$\int_C M(x, y, z)\, dx + N(x, y, z)\, dy + P(x, y, z)\, dz$$

$$= \sum_{k=1}^{n} \int_{C_k} M(x, y, z)\, dx + N(x, y, z)\, dy + P(x, y, z)\, dz \tag{11}$$

Each of the line integrals on the right side of (11) is evaluated by means of (9). In our original notation for line integrals of vector fields, (11) becomes

$$\int_C \mathbf{F} \cdot d\mathbf{r} = \sum_{k=1}^{n} \int_{C_k} \mathbf{F} \cdot d\mathbf{r} \tag{12}$$

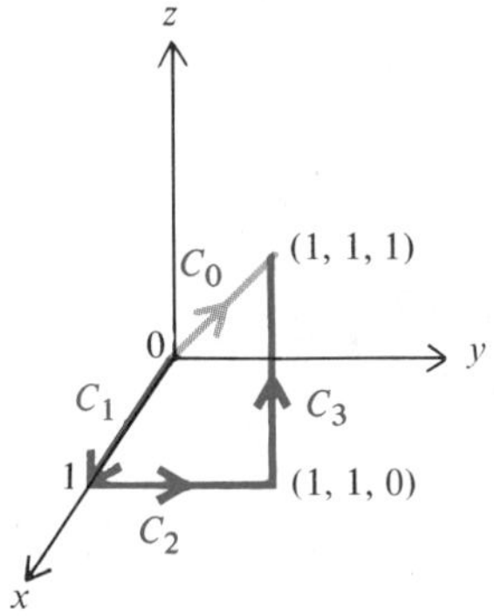

Example 8 Suppose C_0, C_1, C_2, and C_3 are as shown in Figure 15.10, and let C be composed of C_1, C_2, and C_3. Show that

$$\int_C yz\, dx + xz\, dy + xy\, dz = \int_{C_0} yz\, dx + xz\, dy + xy\, dz$$

FIGURE 15.10

Solution The curves are parametrized as follows:

C_0: $\mathbf{r}_0(t) = t\mathbf{i} + t\mathbf{j} + t\mathbf{k}$ for $0 \le t \le 1$ $\left(\dfrac{dx}{dt} = \dfrac{dy}{dt} = \dfrac{dz}{dt} = 1\right)$

C_1: $\mathbf{r}_1(t) = t\mathbf{i}$ for $0 \le t \le 1$ $\left(\dfrac{dx}{dt} = 1, \dfrac{dy}{dt} = \dfrac{dz}{dt} = 0\right)$

C_2: $\mathbf{r}_2(t) = \mathbf{i} + t\mathbf{j}$ for $0 \le t \le 1$ $\left(\dfrac{dx}{dt} = \dfrac{dz}{dt} = 0, \dfrac{dy}{dt} = 1\right)$

C_3: $\mathbf{r}_3(t) = \mathbf{i} + \mathbf{j} + t\mathbf{k}$ for $0 \le t \le 1$ $\left(\dfrac{dx}{dt} = \dfrac{dy}{dt} = 0, \dfrac{dz}{dt} = 1\right)$

Consequently we infer from (9) that

$$\int_{C_0} yz\, dx + xz\, dy + xy\, dz = \int_0^1 [(t \cdot t)1 + (t \cdot t)1 + (t \cdot t)1]\, dt = \int_0^1 3t^2\, dt = 1$$

$$\int_{C_1} yz\, dx + xz\, dy + xy\, dz = \int_0^1 [(0 \cdot 0)1 + (t \cdot 0)0 + (t \cdot 0)0]\, dt = \int_0^1 0\, dt = 0$$

$$\int_{C_2} yz\, dx + xz\, dy + xy\, dz = \int_0^1 [(t \cdot 0)0 + (1 \cdot 0)1 + (1 \cdot t)0]\, dt = \int_0^1 0\, dt = 0$$

$$\int_{C_3} yz\, dx + xz\, dy + xy\, dz = \int_0^1 [(1 \cdot t)0 + (1 \cdot t)0 + (1 \cdot 1)1]\, dt = \int_0^1 1\, dt = 1$$

Combining these equations according to (11), we find that

$$\int_C yz\, dx + xz\, dy + xy\, dz = 1 = \int_{C_0} yz\, dx + xz\, dy + xy\, dz \quad \square$$

In many physical applications one wishes to integrate a function of two variables x and y over a curve that lies in the xy plane. Of course, such a plane curve is a space curve, and consequently the procedures we have discussed for evaluating line integrals still apply. In particular, the two-dimensional versions of (7) and (9) are

$$\int_C \mathbf{F} \cdot d\mathbf{r} = \int_a^b \mathbf{F}(x(t), y(t)) \cdot \frac{d\mathbf{r}}{dt}\, dt$$

and

$$\int_C M(x, y)\, dx + N(x, y)\, dy = \int_a^b \left[M(x(t), y(t))\frac{dx}{dt} + N(x(t), y(t))\frac{dy}{dt} \right] dt$$

Example 9 Evaluate $\displaystyle\int_C y\, dx + xy\, dy$, where C is the circle $x^2 + y^2 = 4$, oriented counterclockwise.

Solution The circle is parametrized by

$$\mathbf{r}(t) = 2\cos t\,\mathbf{i} + 2\sin t\,\mathbf{j} \quad \text{for } 0 \le t \le 2\pi$$

Consequently

$$\int_C y\, dx + xy\, dy = \int_0^{2\pi} [(2\sin t)(-2\sin t) + (2\cos t)(2\sin t)(2\cos t)]\, dt$$

$$= 4\int_0^{2\pi} (-\sin^2 t + 2\cos^2 t \sin t)\, dt$$

$$= 4\int_0^{2\pi} \left(-\frac{1}{2} + \frac{1}{2}\cos 2t + 2\cos^2 t \sin t \right) dt$$

$$= 4\left(-\frac{1}{2}t + \frac{1}{4}\sin 2t - \frac{2}{3}\cos^3 t \right)\Big|_0^{2\pi} = -4\pi \quad \square$$

EXERCISES 15.2

In Exercises 1–10 evaluate the line integral.

1. $\int_C (9 + 8y^{1/2})\, ds$, where C is parametrized by $\mathbf{r}(t) = 2t^{3/2}\mathbf{i} + t^2\mathbf{j}$ for $0 \le t \le 1$

2. $\int_C xy\, ds$, where C is the circle $x^2 + y^2 = 4$

3. $\int_C y\, ds$, where C is parametrized by $\mathbf{r}(t) = t\mathbf{i} + t^3\mathbf{j}$ for $-1 \le t \le 0$

4. $\int_C (x^3 + y^3)\, ds$, where C is parametrized by $\mathbf{r}(t) = \cos^3 t\,\mathbf{i} + \sin^3 t\,\mathbf{j}$ for $0 \le t \le \pi/2$

5. $\int_C 2xyz\,ds$, where C is parametrized by $\mathbf{r}(t) = e^t\mathbf{i} + e^{-t}\mathbf{j} + \sqrt{2}\,t\mathbf{k}$ for $0 \le t \le 1$

6. $\int_C (x - z^2)\,ds$, where C is parametrized by $\mathbf{r}(t) = \ln t\,\mathbf{i} - t^2\mathbf{j} + 2t\mathbf{k}$ for $1 \le t \le 2$

7. $\int_C (1 + \frac{9}{4}z^{2/3})^{1/4}\,ds$, where C is parametrized by $\mathbf{r}(t) = \cos t\,\mathbf{i} + \sin t\,\mathbf{j} + t^{3/2}\mathbf{k}$ for $0 \le t \le \frac{20}{3}$

8. $\int_C (2xy - 5yz)\,ds$, where C is the line segment from $(1, 0, 1)$ to $(0, 3, 2)$

9. $\int_C (y + 2z)\,ds$, where C is the triangular path from $(-1, 0, 0)$ to $(0, 1, 0)$, from $(0, 1, 0)$ to $(0, 0, 1)$, and from $(0, 0, 1)$ to $(-1, 0, 0)$

10. $\int_C (3x - 2y + z)\,ds$, where C is as shown in Figure 15.11

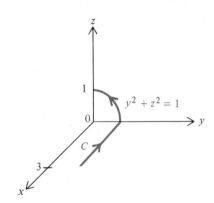

FIGURE 15.11

In Exercises 11–17 evaluate $\int_C \mathbf{F} \cdot d\mathbf{r}$, where C is parametrized by $\mathbf{r}(t)$.

11. $\mathbf{F}(x, y, z) = z\mathbf{i} - y\mathbf{j} - x\mathbf{k}$;
    $\mathbf{r}(t) = 5\mathbf{i} - \sin t\,\mathbf{j} - \cos t\,\mathbf{k}$ for $0 \le t \le \pi/4$

12. $\mathbf{F}(x, y, z) = y\mathbf{i} + x\mathbf{j} + z^3\mathbf{k}$;
    $\mathbf{r}(t) = (1 - t)\mathbf{i} + t\mathbf{j} + \pi t\mathbf{k}$ for $0 \le t \le 1$

13. $\mathbf{F}(x, y, z) = y\mathbf{i} + xy\mathbf{j} + z^3\mathbf{k}$;
    $\mathbf{r}(t) = \cos t\,\mathbf{i} + \sin t\,\mathbf{j} + 2t\mathbf{k}$ for $0 \le t \le \pi/2$

14. $\mathbf{F}(x, y, z) = -z\mathbf{i} + x\mathbf{k}$;
    $\mathbf{r}(t) = \cos t\,\mathbf{i} + \sin t\,\mathbf{k}$ for $0 \le t \le \pi$

15. $\mathbf{F}(x, y, z) = -z\mathbf{i} + x\mathbf{k}$; $\mathbf{r}$ parametrizes the curve in Exercise 14 in the opposite direction

16. $\mathbf{F}(x, y, z) = ye^{(x^2)}\mathbf{i} + xe^{(y^2)}\mathbf{j} + \cosh xy^3\mathbf{k}$;
    $\mathbf{r}(t) = e^t\mathbf{i} + e^t\mathbf{j} + 3\mathbf{k}$ for $0 \le t \le 1$

17. $\mathbf{F}(x, y, z) = 5e^{\sin \pi x}\mathbf{i} - 4e^{\cos \pi x}\mathbf{j}$;
    $\mathbf{r}(t) = \frac{1}{2}\mathbf{i} + 2\mathbf{j} - \ln(\cosh t)\mathbf{k}$ for $0 \le t \le \pi/6$

18. Let $\mathbf{F}(x, y, z) = xy\mathbf{i} + 2z\mathbf{j} + (y + z)\mathbf{k}$. Find $\int_C \mathbf{F} \cdot d\mathbf{r}$, where
    a. C is composed of the line segments from $(0, 0, 0)$ to $(0, -1, 0)$ and from $(0, -1, 0)$ to $(1, 1, 2)$.
    b. C is composed of the line segments from $(0, 0, 0)$ to $(0, 1, 1)$ and from $(0, 1, 1)$ to $(1, 1, 2)$.

c. C is the parabolic curve parametrized by $\mathbf{r}(t) = t\mathbf{i} + t\mathbf{j} + 2t^2\mathbf{k}$ for $0 \le t \le 1$.

In Exercises 19–28 evaluate the line integral.

19. $\int_C y\,dx - x\,dy + xyz^2\,dz$, where C is parametrized by $\mathbf{r}(t) = e^{-t}\mathbf{i} + e^t\mathbf{j} + t\mathbf{k}$ for $0 \le t \le 1$

20. $\int_C e^x\,dx + xy\,dy + xyz\,dz$, where C is parametrized by $\mathbf{r}(t) = t\mathbf{i} + t\mathbf{j} + 2t\mathbf{k}$ for $-1 \le t \le 1$

21. $\int_C e^x\,dx + xy\,dy + xyz\,dz$, where C is the curve opposite to the one in Exercise 20

22. $\int_C y(x^2 + y^2)\,dx - x(x^2 + y^2)\,dy + xy\,dz$, where C is parametrized by $\mathbf{r}(t) = \cos t\,\mathbf{i} + \sin t\,\mathbf{j} + t\mathbf{k}$ for $-\pi \le t \le \pi$

23. $\int_C xy\,dx + (x + z)\,dy + z^2\,dz$, where C is parametrized by $\mathbf{r}(t) = (t + 1)\mathbf{i} + (t - 1)\mathbf{j} + t^2\mathbf{k}$ for $-1 \le t \le 2$

24. $\int_C x\,dx + y\,dy + xy\,dz$, where C is parametrized by $\mathbf{r}(t) = \cos t\,\mathbf{i} + \sin t\,\mathbf{j} + \cos t\,\mathbf{k}$ for $-\pi/2 \le t \le 0$

25. $\int_C \dfrac{1}{1 + x^2}\,dx + \dfrac{2}{1 + y^2}\,dy$, where C is the quarter unit circle from $(1, 0)$ to $(0, 1)$

26. $\int_C z\,dx + xy\,dy$, where C is the line segment from $(-1, 1, 0)$ to $(2, 1, 0)$

27. $\int_C x\ln(xz/y)\,dx + \cos(\pi xy/z)\,dy$, where C is parametrized by $\mathbf{r}(t) = t\mathbf{i} + t^2\mathbf{j} + t^3\mathbf{k}$ for $1 \le t \le 2$

28. $\int_C x\ln(xz/y)\,dx + \cos(\pi xy/z)\,dy$, where C is composed of the line segments from $(1, 1, 1)$ to $(2, 2, 2)$ and from $(2, 2, 2)$ to $(4, 2, 2)$

29. Evaluate $\int_C y\,dx + z\,dy + x\,dz$, where C is composed of
    a. the line segments from $(0, 0, 0)$ to $(0, -5, 0)$ and from $(0, -5, 0)$ to $(0, 1, 1)$.
    b. the line segments from $(0, 0, 0)$ to $(1, 0, 0)$ and from $(1, 0, 0)$ to $(0, 1, 1)$.

30. Let C_1 be parametrized by
$$\mathbf{r}_1(t) = t\mathbf{i} + t\mathbf{j} + t\mathbf{k} \quad \text{for } 0 \le t \le \frac{1}{2}$$
and C_2 by
$$\mathbf{r}_2(t) = \sin t\,\mathbf{i} + \sin t\,\mathbf{j} + \sin t\,\mathbf{k} \quad \text{for } 0 \le t \le \frac{\pi}{6}$$
Evaluate $\int_{C_1} (xy + z)\,ds$ and $\int_{C_2} (xy + z)\,ds$. Are the answers the same? Explain why or why not.

31. Suppose the first component of a parametrization of a piecewise smooth curve C is constant, and let M be continuous on C. Show that $\int_C M(x, y, z)\,dx = 0$.

32. Suppose the mass density at a given point on a thin wire is equal to the square of the distance from that point to the x axis. If the wire is helical and is parametrized by
$$\mathbf{r}(t) = \sin t\,\mathbf{i} - \cos t\,\mathbf{j} + 4t\mathbf{k} \quad \text{for } \pi \le t \le 2\pi$$
find the mass m of the wire.

33. Suppose the mass density at a given point on the wire in Exercise 32 is equal to the square of the distance from that point to the origin. Find the mass m of the wire.

34. Let $\mathbf{F}(x, y, z) = xy\mathbf{i} + yz\mathbf{j} + xz\mathbf{k}$. Find the work done by the force $\mathbf{F}$ on an object as it traverses the twisted quartic curve C parametrized by
$$\mathbf{r}(t) = t\mathbf{i} + t^2\mathbf{j} + t^4\mathbf{k} \quad \text{for } 0 \le t \le 1$$

35. Let $\mathbf{F}(x, y, z) = (2x - y)\mathbf{i} + 2z\mathbf{j} + (y - z)\mathbf{k}$.
    a. Find the work W done by the force $\mathbf{F}$ on an object moving from $(0, 0, 0)$ to $(1, 1, 1)$ along a straight line.
    b. Find the work W done as the object moves along the curve parametrized by
$$\mathbf{r}(t) = \sin\frac{\pi t}{2}\mathbf{i} + \sin\frac{\pi t}{2}\mathbf{j} + t\mathbf{k} \quad \text{for } 0 \le t \le 1$$

36. A painter weighing 120 pounds carries a pail of paint weighing 30 pounds up a helical staircase surrounding a circular cylindrical water tower. If the tower is 200 feet tall and 100 feet in diameter and the painter makes exactly four revolutions during the ascent to the top, how much work is done by gravity on the painter and the pail during the ascent?

37. If the pail in Exercise 36 steadily leaks 1 pound of paint for every 10 feet the painter rises, how much less work would be done by gravity on the painter and the pail during the ascent than would be done if there were no leak?

38. A satellite weighing 6000 pounds orbits the earth in a circular orbit 4400 miles from the center of the earth. Find the work W done on the satellite by gravity during
    a. half a revolution.
    b. one complete revolution.

39. a. Suppose a continuous force acts on an object in a direction normal to the path of the object. Show that the work done by the force on the object is 0.
    b. Redo Exercise 38 using part (a) of this exercise.

15.3
THE FUNDAMENTAL THEOREM OF LINE INTEGRALS

This section is devoted to a generalization of the Fundamental Theorem of Calculus that is applicable to line integrals of the form $\int_C \mathbf{F} \cdot d\mathbf{r}$ for certain vector fields $\mathbf{F}$ and for certain curves C. Recall from Section 12.5 that an oriented curve is by definition piecewise smooth.

THEOREM 15.7
FUNDAMENTAL THEOREM OF LINE INTEGRALS

Let C be an oriented curve with initial point (x_0, y_0, z_0) and terminal point (x_1, y_1, z_1). Let f be a function of three variables that is differentiable at every point on C, and assume that grad f is continuous on C. Then
$$\int_C \text{grad } f \cdot d\mathbf{r} = f(x_1, y_1, z_1) - f(x_0, y_0, z_0)$$

Proof First we consider the case in which C is smooth. Let $\mathbf{r}$ be a smooth parametrization of C defined on an interval $[a, b]$. Then by (7) of Section 15.2, the Chain Rule, and the Fundamental Theorem of Calculus,

$$\int_C \text{grad } f \cdot d\mathbf{r} = \int_a^b [\text{grad } f(x(t), y(t), z(t))] \cdot \frac{d\mathbf{r}}{dt}\, dt$$
$$= \int_a^b \left[\frac{\partial f}{\partial x}(x(t), y(t), z(t))\frac{dx}{dt} + \frac{\partial f}{\partial y}(x(t), y(t), z(t))\frac{dy}{dt} \right.$$
$$\left. + \frac{\partial f}{\partial z}(x(t), y(t), z(t))\frac{dz}{dt} \right] dt$$

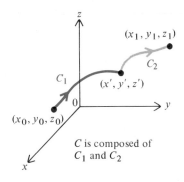

C is composed of C_1 and C_2

FIGURE 15.12

$$= \int_a^b \frac{d}{dt}[f(x(t), y(t), z(t))]\, dt$$

$$= f(x(b), y(b), z(b)) - f(x(a), y(a), z(a))$$

$$= f(x_1, y_1, z_1) - f(x_0, y_0, z_0)$$

If C is only piecewise smooth, we can prove the result by combining line integrals over the smooth portions of the curve. We will carry out the proof for the case in which C is composed of two smooth curves C_1 and C_2, where C_1 joins (x_0, y_0, z_0) to (x', y', z'), and C_2 joins (x', y', z') to (x_1, y_1, z_1) (Figure 15.12). Using the result we just proved for smooth curves, we have

$$\int_C \text{grad } f \cdot d\mathbf{r} = \int_{C_1} \text{grad } f \cdot d\mathbf{r} + \int_{C_2} \text{grad } f \cdot d\mathbf{r}$$

$$= [f(x', y', z') - f(x_0, y_0, z_0)] + [f(x_1, y_1, z_1) - f(x', y', z')]$$

$$= f(x_1, y_1, z_1) - f(x_0, y_0, z_0) \quad \blacksquare$$

If a continuous vector field $\mathbf{F}$ is the gradient of a function f, then

$$\int_C \mathbf{F} \cdot d\mathbf{r} = \int_C \text{grad } f \cdot d\mathbf{r}$$

so that the formula in the Fundamental Theorem of Line Integrals becomes

$$\int_C \mathbf{F} \cdot d\mathbf{r} = f(x_1, y_1, z_1) - f(x_0, y_0, z_0) \tag{1}$$

This is reminiscent of the formula

$$\int_a^b f'(x)\, dx = f(b) - f(a)$$

for functions of one variable, which follows from the Fundamental Theorem of Calculus. If C is piecewise smooth but we do not have a parametrization of C, nevertheless equation (1) enables us to evaluate $\int_C \mathbf{F} \cdot d\mathbf{r}$ provided that we know the end points of C and can find a potential function for $\mathbf{F}$. Finding a potential function f for $\mathbf{F}$, as described in Section 15.1, thus becomes the central problem in the evaluation of $\int_C \mathbf{F} \cdot d\mathbf{r}$ by means of (1), as the following examples indicate.

Example 1 Let C be the curve from $(1, -1, -\frac{1}{2})$ to $(1, 1, \frac{1}{2})$ parametrized by

$$\mathbf{r}(t) = -\cos \pi t^4 \mathbf{i} + t^{5/3}\mathbf{j} + \frac{t}{t^2 + 1}\mathbf{k} \quad \text{for } -1 \le t \le 1$$

and let $\mathbf{F}(x, y, z) = (2xy + z^2)\mathbf{i} + x^2\mathbf{j} + (2xz + \pi \cos \pi z)\mathbf{k}$. Find $\int_C \mathbf{F} \cdot d\mathbf{r}$.

Solution By Example 5 of Section 15.1, $\mathbf{F} = \text{grad } f$, where

$$f(x, y, z) = x^2 y + xz^2 + \sin \pi z$$

As a result, the Fundamental Theorem of Line Integrals implies that

$$\int_C \mathbf{F} \cdot d\mathbf{r} = f(1, 1, \tfrac{1}{2}) - f(1, -1, -\tfrac{1}{2})$$
$$= (1 + \tfrac{1}{4} + 1) - (-1 + \tfrac{1}{4} - 1) = 4 \quad \square$$

Example 2   Let C be a piecewise smooth curve from $(1, -2, 3)$ to $(1, 0, 0)$ and $\mathbf{F}$ the gravitational field of a point mass m at the origin. Find the work W done by $\mathbf{F}$ on a unit point mass that traverses C.

Solution   The force on the unit mass is equal to the gravitational field $\mathbf{F}$, and by (3) of Section 15.1 we know that the gravitational field $\mathbf{F}$ is the gradient of f, where

$$f(x, y, z) = \frac{Gm}{(x^2 + y^2 + z^2)^{1/2}}$$

Consequently

$$W = \int_C \mathbf{F} \cdot d\mathbf{r} = f(1, 0, 0) - f(1, -2, 3)$$
$$= Gm - \frac{Gm}{(1 + 4 + 9)^{1/2}} = Gm\left(1 - \frac{1}{\sqrt{14}}\right) \quad \square$$

Naturally, the Fundamental Theorem of Line Integrals can be formulated in two dimensions: Let f be a function of two variables that is differentiable at every point on an oriented curve C with initial point (x_0, y_0) and terminal point (x_1, y_1), and let grad f be continuous on C. Then

$$\int_C \text{grad } f \cdot d\mathbf{r} = f(x_1, y_1) - f(x_0, y_0)$$

Moreover, if $\mathbf{F}$ is a continuous vector field such that $\mathbf{F} = \text{grad } f$, then

$$\int_C \mathbf{F} \cdot d\mathbf{r} = f(x_1, y_1) - f(x_0, y_0)$$

Example 3   Let C be the portion of the parabola $y = x^2$ with initial point $(0, 0)$ and terminal point $(2, 4)$, and let

$$\mathbf{F}(x, y) = y^3\mathbf{i} + 3xy^2\mathbf{j}$$

Find $\int_C \mathbf{F} \cdot d\mathbf{r}$.

Solution   By Example 4 of Section 15.1, $\mathbf{F} = \text{grad } f$ if $f(x, y) = xy^3$. Consequently

$$\int_C \mathbf{F} \cdot d\mathbf{r} = f(2, 4) - f(0, 0) = 2 \cdot 4^3 - 0 = 128 \quad \square$$

**Independence
of Path**

The Fundamental Theorem of Line Integrals says that if $\mathbf{F}$ is the gradient of a differentiable function f and has domain D, then the value of a line integral $\int_C \mathbf{F} \cdot d\mathbf{r}$ depends only on the initial and terminal points of the oriented curve C in D. Thus if C_1 is another oriented curve lying in D and having the same initial and terminal points as C, then

$$\int_{C_1} \mathbf{F} \cdot d\mathbf{r} = \int_C \mathbf{F} \cdot d\mathbf{r}$$

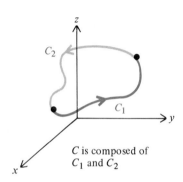

C is composed of
C_1 and C_2

FIGURE 15.13

More generally, if $\mathbf{F}$ is a continuous vector field with domain D and $\int_{C_1} \mathbf{F} \cdot d\mathbf{r} = \int_{C_2} \mathbf{F} \cdot d\mathbf{r}$ for any two oriented curves C_1 and C_2 lying in D and having the same initial and terminal points, we say that $\int_C \mathbf{F} \cdot d\mathbf{r}$ is **independent of path**. Rephrased in this terminology, the Fundamental Theorem of Line Integrals says that if $\mathbf{F}$ is continuous and is the gradient of some differentiable function, then $\int_C \mathbf{F} \cdot d\mathbf{r}$ is independent of path.

The condition that $\int_C \mathbf{F} \cdot d\mathbf{r}$ is independent of path for oriented curves in a region D implies that if C is any closed, oriented curve in D, then $\int_C \mathbf{F} \cdot d\mathbf{r} = 0$. The reason is that C can be regarded as composed of two oriented curves C_1 and C_2, where the terminal point of C_1 is the initial point of C_2 and the inital point of C_1 is the terminal point of C_2 (Figure 15.13). Since C_1 and $-C_2$ have the same initial and terminal points, the independence of path implies that

$$\int_C \mathbf{F} \cdot d\mathbf{r} = \int_{C_1} \mathbf{F} \cdot d\mathbf{r} + \int_{C_2} \mathbf{F} \cdot d\mathbf{r} = \int_{C_1} \mathbf{F} \cdot d\mathbf{r} - \int_{-C_2} \mathbf{F} \cdot d\mathbf{r} = 0$$

A somewhat more technical argument shows that if $\int_C \mathbf{F} \cdot d\mathbf{r} = 0$ for every closed, oriented curve in D, then $\mathbf{F} = \text{grad } f$ for some function f. Hence we have the following three equivalent conditions.

1. $\mathbf{F} = \text{grad } f$ for some function f; that is, $\mathbf{F}$ is conservative.
2. $\int_C \mathbf{F} \cdot d\mathbf{r}$ is independent of path.
3. $\int_C \mathbf{F} \cdot d\mathbf{r} = 0$ for every closed, oriented curve C lying in the domain of $\mathbf{F}$.

By (5) of Section 15.1, condition 1 implies a fourth condition:

4. curl $\mathbf{F} = \mathbf{0}$.

If the domain of $\mathbf{F}$ is all of three-dimensional space, or any region with no "holes," then condition 4 implies condition 1, so that conditions 1–4 are equivalent.

Because condition 1 implies condition 3, the work done by a conservative, continuous force on an object that traverses a closed, piecewise smooth curve is 0. In particular, this is true of the gravitational and electric fields described in Section 15.1, since both are conservative.

Conservation of Energy

Now we use Newton's Second Law of Motion to discuss further the work done by a force on an object and to derive the Law of Conservation of Energy.

Suppose an object moves under the influence of a continuous force $\mathbf{F}$ along a smooth curve C parametrized by

$$\mathbf{r}(t) = x(t)\mathbf{i} + y(t)\mathbf{j} + z(t)\mathbf{k} \quad \text{for } a \leq t \leq b$$

For the sake of brevity we write

$$\mathbf{F}(\mathbf{r}(t)) \quad \text{for} \quad \mathbf{F}(x(t), y(t), z(t))$$

as is often done. Newton's Second Law of Motion states that the force $\mathbf{F}$ and the acceleration $\mathbf{a}$ of the object are related by

$$\mathbf{F}(\mathbf{r}(t)) = m\mathbf{a}(t) = m\mathbf{r}''(t) \quad \text{for } a \leq t \leq b \tag{2}$$

The work W done by the force $\mathbf{F}$ on the object is given by

$$W = \int_C \mathbf{F} \cdot d\mathbf{r} = \int_a^b \mathbf{F}(\mathbf{r}(t)) \cdot \mathbf{r}'(t)\,dt = \int_a^b m\mathbf{r}''(t) \cdot \mathbf{r}'(t)\,dt$$

$$= \frac{m}{2} \int_a^b \frac{d}{dt}[\mathbf{r}'(t) \cdot \mathbf{r}'(t)]\,dt = \frac{m}{2} \int_a^b \frac{d}{dt}\|\mathbf{r}'(t)\|^2\,dt$$

By the Fundamental Theorem of Calculus,

$$\frac{m}{2} \int_a^b \frac{d}{dt}\|\mathbf{r}'(t)\|^2\,dt = \frac{m}{2}\|\mathbf{r}'(t)\|^2\Big|_a^b = \frac{m}{2}\|\mathbf{r}'(b)\|^2 - \frac{m}{2}\|\mathbf{r}'(a)\|^2$$

Thus
$$W = \frac{m}{2}\|\mathbf{r}'(b)\|^2 - \frac{m}{2}\|\mathbf{r}'(a)\|^2 \tag{3}$$

For any number t in $[a, b]$, the quantity

$$\frac{m}{2}\|\mathbf{r}'(t)\|^2$$

is called the **kinetic energy** of the object. (Notice that $\|\mathbf{r}'(t)\|$ is the speed of the object, so that the kinetic energy is one half the product of the mass and the square of the speed.) Therefore (3) expresses the fact that the work done by the force on the object from the beginning to the end of the curve is equal to the change in the kinetic energy of the object at the two endpoints of the curve. It also follows from (3) that the work depends only on the speeds at the two end points of the curve and on the mass.

Now let us assume that **F** is a conservative force. Then there is a function f such that $\mathbf{F} = -\operatorname{grad} f$ (here the minus sign is included, as is done in physics). The function f is called the **potential energy function** of the object corresponding to **F**. It follows from the Chain Rule that

$$\frac{d}{dt} f(\mathbf{r}(t)) = \frac{d}{dt} f(x(t), y(t), z(t)) = \frac{\partial f}{\partial x}\frac{dx}{dt} + \frac{\partial f}{\partial y}\frac{dy}{dt} + \frac{\partial f}{\partial z}\frac{dz}{dt}$$

$$= \left(\frac{\partial f}{\partial x}\mathbf{i} + \frac{\partial f}{\partial y}\mathbf{j} + \frac{\partial f}{\partial z}\mathbf{k}\right) \cdot \left(\frac{dx}{dt}\mathbf{i} + \frac{dy}{dt}\mathbf{j} + \frac{dz}{dt}\mathbf{k}\right)$$

$$= [\operatorname{grad} f(\mathbf{r}(t))] \cdot \mathbf{r}'(t)$$

Therefore
$$\frac{d}{dt} f(\mathbf{r}(t)) = [\operatorname{grad} f(\mathbf{r}(t))] \cdot \mathbf{r}'(t) \tag{4}$$

Using (2) and (4), we find that

$$\frac{d}{dt}\left(\frac{m}{2}\|\mathbf{r}'(t)\|^2 + f(\mathbf{r}(t))\right) = \frac{d}{dt}\left(\frac{m}{2}\mathbf{r}'(t)\cdot\mathbf{r}'(t) + f(\mathbf{r}(t))\right)$$

$$\stackrel{(4)}{=} m\mathbf{r}''(t)\cdot\mathbf{r}'(t) + [\operatorname{grad} f(\mathbf{r}(t))]\cdot\mathbf{r}'(t)$$

$$= [m\mathbf{r}''(t) + \operatorname{grad} f(\mathbf{r}(t))]\cdot\mathbf{r}'(t)$$

$$= [m\mathbf{r}''(t) - \mathbf{F}(\mathbf{r}(t))]\cdot\mathbf{r}'(t)$$

$$\stackrel{(2)}{=} \mathbf{0}\cdot\mathbf{r}'(t) = 0$$

Thus
$$\frac{d}{dt}\left(\frac{m}{2}\|\mathbf{r}'(t)\|^2 + f(\mathbf{r}(t))\right) = 0$$

When we integrate both sides, we obtain

$$\frac{m}{2}\|\mathbf{r}'(t)\|^2 + f(\mathbf{r}(t)) = C \tag{5}$$

for some constant C. Correctly interpreted, (5) expresses the **Law of Conservation of Energy** of Newtonian mechanics:

The sum of the kinetic energy and the potential energy of an object due to a conservative force is constant.

EXERCISES 15.3

In Exercises 1–7 show that the line integral is independent of path, and evaluate the integral.

1. $\int_C (e^x + y)\,dx + (x + 2y)\,dy$; C is any piecewise smooth curve in the xy plane from $(0, 1)$ to $(2, 3)$.

2. $\int_C (2xy^2 + 1)\,dx + 2x^2y\,dy$; C is any piecewise smooth curve in the xy plane from $(-1, 2)$ to $(2, 3)$.

3. $\int_C y\,dx + (x + z)\,dy + y\,dz$; C is parametrized by

$$\mathbf{r}(t) = \frac{t^2 + 1}{t^2 - 1}\mathbf{i} + \cos\pi t\,\mathbf{j} + 2t\sin\pi t\,\mathbf{k} \quad \text{for } 0 \le t \le \tfrac{1}{2}$$

4. $\int_C (\cos x + 2yz)\,dx + (\sin y + 2xz)\,dy + (z + 2xy)\,dz$; C is any piecewise smooth curve from $(0, 0, 0)$ to $(\pi, \pi, 1/\pi)$.

5. $\displaystyle\int_c \frac{x}{1 + x^2 + y^2 + z^2} dx + \frac{y}{1 + x^2 + y^2 + z^2} dy$
$$+ \frac{z}{1 + x^2 + y^2 + z^2} dz;$$

   C is parametrized by $\mathbf{r}(t) = t\mathbf{i} + t^2\mathbf{j} + t^4\mathbf{k}$ for $0 \le t \le 1$.

6. $\int_C (y + 2xe^y) dx + (x + x^2 e^y) dy$; C is parametrized by $\mathbf{r}(t) = t^{1/2}\mathbf{i} + \ln t\mathbf{j} + t\mathbf{k}$ for $1 \le t \le 4$.

7. $\displaystyle\int_c e^{-x} \ln y\, dx - \frac{e^{-x}}{y} dy + z\, dz$; C is parametrized by $\mathbf{r}(t) = (t - 1)\mathbf{i} + e^{t^4}\mathbf{j} + (t^2 + 1)\mathbf{k}$ for $0 \le t \le 1$.

8. Evaluate $\int_C (x + \cos \pi y) dx - \pi x \sin \pi y\, dy$, where C is the curve parametrized by $\mathbf{r}(t) = e^{t^2}\mathbf{i} + t\mathbf{j} + (\tan \pi t/4)\mathbf{k}$ for $0 \le t \le 1$.

9. Suppose that f, g, and h are continuous functions. Prove that $\int_C f(x) dx + g(y) dy + h(z) dz$ is independent of path.

10. Let
$$\mathbf{F}(x, y) = \frac{y}{x^2 + y^2}\mathbf{i} - \frac{x}{x^2 + y^2}\mathbf{j}$$

   a. Show that curl $\mathbf{F} = \mathbf{0}$.
   b. Let R be any region in the plane that contains the circle $x^2 + y^2 = 1$ but does not contain the origin. Show that $\int_C \mathbf{F} \cdot d\mathbf{r}$ is not independent of path in R, and hence $\mathbf{F}$ is not conservative. (*Hint:* Calculate first $\int_{C_1} \mathbf{F} \cdot d\mathbf{r}$ and then $\int_{C_2} \mathbf{F} \cdot d\mathbf{r}$, where C_1 is the semicircle parametrized by $\mathbf{r}_1(t) = \cos t\mathbf{i} + \sin t\mathbf{j}$ for $0 \le t \le \pi$ and C_2 is the semicircle parametrized by $\mathbf{r}_2(t) = \cos t\mathbf{i} - \sin t\mathbf{j}$ for $0 \le t \le \pi$.)

11. Let g be a continuous function of one variable, and let
$$\mathbf{F}(x, y, z) = [g(x^2 + y^2 + z^2)](x\mathbf{i} + y\mathbf{j} + z\mathbf{k}) \quad (6)$$

   a. Show that $\mathbf{F}$ is conservative. (*Hint:* Show that $\mathbf{F} = $ grad f, where $f(x, y, z) = \frac{1}{2}h(x^2 + y^2 + z^2)$ and $h(u) = \int g(u) du$.)
   b. Show that $\mathbf{F}$ is irrotational.
   (Because central force fields are in the form of (6), parts (a) and (b) show that they are both conservative and irrotational.)

12. Find the work W done on a rocket weighing 5000 pounds by the earth's gravitational field when the rocket descends to earth from a distance of 7460 miles from the center of the earth. (*Hint:* Take the radius of the earth to be 3960 miles, and recall that the gravitational field is conservative. The weight of an object is the magnitude of the gravitational force on it at the earth's surface.)

13. Assume that an electron is located at the origin and has a charge of $(-1.6) \times 10^{-19}$ coulombs. Find the work W done by the electric field on a positive unit charge that moves from a distance of 10^{-11} meters to a distance of 10^{-12} meters from the electron. (*Hint:* The electric field is conservative. Take $4\pi\varepsilon_0 = 1.113 \times 10^{-10}$. Your answer will be in joules.)

14. The force $\mathbf{F}(x, y)$ exerted at the point (x, y) by a two-dimensional linear oscillator at the origin is given by
$$\mathbf{F}(x, y) = -ax\mathbf{i} - ay\mathbf{j}$$

   where a is a positive constant. Find the work W done by the force on an object that moves from $(3, -6, 0)$ to $(1, -2, 0)$.

15. Suppose the speed of an object having a mass of 5 kilograms and moving in a conservative force field decreases from 50 meters per second to 10 meters per second. Find the increase in potential energy of the object. (Your answer will be in joules.)

15.4
GREEN'S THEOREM

If f is continuous on a closed interval $[a, b]$ and if F is an antiderivative of f on $[a, b]$, then the Fundamental Theorem of Calculus tells us that

$$\int_a^b f(x) dx = F(b) - F(a)$$

An alternative way of interpreting the Fundamental Theorem is that we can evaluate $\int_a^b f(x) dx$ by evaluating an antiderivative F at the boundary points a and b of the interval $[a, b]$. In the preceding section we discovered a similar result for certain line integrals. Now we will present an analogue of the Fundamental Theorem that applies to regions in the plane.

Although the result we will derive applies to almost any plane region you are likely to encounter, the proof is complicated when very general regions are involved. Hence we will state the theorem, which is usually called Green's Theorem,* only for simple regions in the plane.

THEOREM 15.8
GREEN'S THEOREM

Let R be a simple region in the xy plane with a piecewise smooth boundary C oriented counterclockwise. Let M and N be functions of two variables having continuous partial derivatives on R. Then

$$\int_C M(x, y)\, dx + N(x, y)\, dy = \iint_R \left(\frac{\partial N}{\partial x} - \frac{\partial M}{\partial y} \right) dA$$

Proof It is sufficient to prove that

$$\int_C N(x, y)\, dy = \iint_R \frac{\partial N}{\partial x}\, dA \qquad (1)$$

and

$$\int_C M(x, y)\, dx = - \iint_R \frac{\partial M}{\partial y}\, dA \qquad (2)$$

since the result follows from adding these two equations. We will prove only (2), because (1) follows by a similar argument. To prove (2), we assume that R is the region between the graphs of g_1 and g_2 on $[a, b]$. On the one hand, this means that

$$-\iint_R \frac{\partial M}{\partial y}\, dA = -\int_a^b \int_{g_1(x)}^{g_2(x)} \frac{\partial M}{\partial y}\, dy\, dx = -\int_a^b M(x, y) \Big|_{g_1(x)}^{g_2(x)} dx$$

$$= -\int_a^b [M(x, g_2(x)) - M(x, g_1(x))]\, dx$$

$$= \int_a^b [M(x, g_1(x)) - M(x, g_2(x))]\, dx$$

On the other hand, the boundary C of R is composed of the curves C_1, C_2, C_3, and C_4 (Figure 15.14), parametrized as follows:

C_1: $\mathbf{r}_1(t) = t\mathbf{i} + g_1(t)\mathbf{j}$ for $a \le t \le b$

C_2: $\mathbf{r}_2(t) = b\mathbf{i} + t\mathbf{j}$ for $g_1(b) \le t \le g_2(b)$

C_3: $\mathbf{r}_3(t) = (a + b - t)\mathbf{i} + g_2(a + b - t)\mathbf{j}$ for $a \le t \le b$

C_4: $\mathbf{r}_4(t) = a\mathbf{i} + [g_1(a) + g_2(a) - t]\mathbf{j}$ for $g_1(a) \le t \le g_2(a)$

(C_2 or C_4 or both may contain only one point.) We will use (10) of Section 15.2 to evaluate $\int_{C_i} M(x, y)\, dx$ for $i = 1, 2, 3, 4$. First we notice that x is constant on C_2 and on C_4, so that $dx/dt = 0$ and thus

$$\int_{C_2} M(x, y)\, dx = 0 \quad \text{and} \quad \int_{C_4} M(x, y)\, dx = 0$$

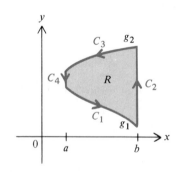

FIGURE 15.14

* George Green (1793–1841) was a self-taught mathematician who published privately the book *An Essay on the Application of Mathematical Analysis to the Theories of Electricity and Magnetism*. This treatise contained results related to the theorem that now bears his name. Some years later he entered Cambridge University as a student.

Therefore

$$\int_C M(x, y)\, dx = \int_{C_1} M(x, y)\, dx + \int_{C_2} M(x, y)\, dx + \int_{C_3} M(x, y)\, dx + \int_{C_4} M(x, y)\, dx$$

$$= \int_{C_1} M(x, y)\, dx + \int_{C_3} M(x, y)\, dx$$

$$= \int_a^b M(t, g_1(t))(1)\, dt + \int_a^b M(a + b - t, g_2(a + b - t))(-1)\, dt$$

$$\overset{u = a + b - t}{=} \int_a^b M(t, g_1(t))\, dt + \int_b^a M(u, g_2(u))\, du$$

$$= \int_a^b [M(x, g_1(x)) - M(x, g_2(x))]\, dx$$

Notice that the last integral appears at the end of our calculation of the double integral above. Thus

$$\int_C M(x, y)\, dx = -\iint\limits_R \frac{\partial M}{\partial y}\, dA \quad \blacksquare$$

The assumption in Green's Theorem that C is the boundary of R and is oriented counterclockwise implies that if we traverse C in the direction defined by the orientation of C, the region R will always lie to our left (as in Figure 15.14).

By using Green's Theorem we can sometimes evaluate a line integral $\int_C M(x, y)\, dx + N(x, y)\, dy$ without using a parametrization of C. Example 1 illustrates this consequence.

Example 1 Let $M(x, y) = -x^2 y$ and $N(x, y) = x^3$, and let C be the circle $x^2 + y^2 = 4$, oriented counterclockwise. Find $\int_C M(x, y)\, dx + N(x, y)\, dy$.

Solution We can find the line integral directly, but it is simpler to use Green's Theorem, where R is the disk $x^2 + y^2 \leq 4$. Since

$$\frac{\partial N}{\partial x} = 3x^2 \quad \text{and} \quad \frac{\partial M}{\partial y} = -x^2$$

we have

$$\iint\limits_R \left(\frac{\partial N}{\partial x} - \frac{\partial M}{\partial y}\right) dA = \iint\limits_R (3x^2 + x^2)\, dA = \int_0^{2\pi} \int_0^2 4(r\cos\theta)^2 r\, dr\, d\theta$$

$$= \int_0^{2\pi} \int_0^2 4r^3 \cos^2\theta\, dr\, d\theta = \int_0^{2\pi} (\cos^2\theta) r^4 \Big|_0^2 d\theta$$

$$= 16 \int_0^{2\pi} \cos^2\theta\, d\theta$$

$$= 16\left(\frac{\theta}{2} + \frac{1}{4}\sin 2\theta\right)\Big|_0^{2\pi} = 16\pi$$

Thus by Green's Theorem,

$$\int_C M(x, y)\,dx + N(x, y)\,dy = 16\pi \quad \square$$

Example 2   Let C be the closed curve described in Figure 15.15, oriented counterclockwise. Evaluate

$$\int_C 2y^3\,dx + (x^4 + 6y^2x)\,dy$$

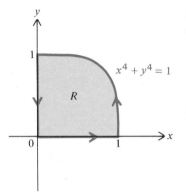

$x^4 + y^4 = 1$

R

FIGURE 15.15

Solution   Again, we can either evaluate the line integral or use Green's Theorem and evaluate a double integral. As before, the latter is more efficient:

$$\int_C 2y^3\,dx + (x^4 + 6y^2x)\,dy = \iint_R \left[\frac{\partial}{\partial x}(x^4 + 6y^2x) - \frac{\partial}{\partial y}(2y^3) \right] dA$$

$$= \int_0^1 \int_0^{(1-x^4)^{1/4}} 4x^3\,dy\,dx$$

$$= \int_0^1 4x^3(1 - x^4)^{1/4}\,dx$$

$$= -\frac{4}{5}(1 - x^4)^{5/4} \Big|_0^1 = \frac{4}{5} \quad \square$$

Green's Theorem can be used in the reverse way, to evaluate a double integral by evaluating a line integral. In particular, suppose we wish to evaluate $\iint_R 1\,dA$, which represents the area of R. If R is a simple region with a piecewise smooth boundary C and if M and N are chosen so that

$$\frac{\partial N}{\partial x} - \frac{\partial M}{\partial y} = 1 \qquad (3)$$

then by Green's Theorem,

$$\iint_R 1\,dA = \iint_R \left(\frac{\partial N}{\partial x} - \frac{\partial M}{\partial y} \right) dA = \int_C M(x, y)\,dx + N(x, y)\,dy$$

where C is the boundary of R, oriented counterclockwise. Of the many possible choices of M and N satisfying equation (3), the three most commonly used are

$$M(x, y) = 0 \qquad \text{and} \quad N(x, y) = x$$

$$M(x, y) = -y \qquad \text{and} \quad N(x, y) = 0$$

$$M(x, y) = -\frac{1}{2}y \quad \text{and} \quad N(x, y) = \frac{1}{2}x$$

These three sets of equations for M and N yield the following formulas for the area A of R:

$$A = \int_C x\,dy = -\int_C y\,dx = \frac{1}{2}\int_C x\,dy - y\,dx \qquad (4)$$

Of course, the first two integrals appear to be simpler than the third. But there are occasions when the third is the easiest to evaluate.

Example 3 Use (4) to compute the area A of the region R enclosed by the ellipse

$$\frac{x^2}{a^2} + \frac{y^2}{b^2} = 1$$

Solution The ellipse is oriented counterclockwise by the parametrization

$$\mathbf{r}(t) = a\cos t\,\mathbf{i} + b\sin t\,\mathbf{j} \quad \text{for } 0 \le t \le 2\pi$$

Therefore by (4),

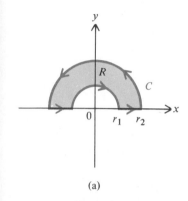

(a)

$$A = \frac{1}{2}\int_C x\,dy - y\,dx = \frac{1}{2}\int_0^{2\pi}\left[(a\cos t)(b\cos t) - (b\sin t)(-a\sin t)\right]dt$$

$$= \frac{1}{2}\int_0^{2\pi} ab\,dt = \pi ab \quad \square$$

Green's Theorem applies to many regions that are not simple but can be broken up into collections of simple regions.

Example 4 Show that Green's Theorem holds for the semiannular region R shown in Figure 15.16(a).

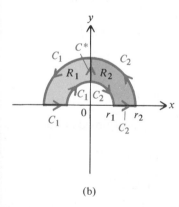

(b)

Solution The region R is vertically simple but not horizontally simple, so Green's Theorem as stated does not apply. However, let us insert the line C^* indicated in Figure 15.16(b). Then R is divided into two simple subregions R_1 and R_2, with C_1 and C^* composing the boundary of R_1 and C_2 and $-C^*$ composing the boundary of R_2 (Figure 15.16(c)). Since each boundary is oriented counterclockwise, we can use Green's Theorem on R_1 and R_2:

$$\iint_R \left(\frac{\partial N}{\partial x} - \frac{\partial M}{\partial y}\right)dA = \iint_{R_1}\left(\frac{\partial N}{\partial x} - \frac{\partial M}{\partial y}\right)dA + \iint_{R_2}\left(\frac{\partial N}{\partial x} - \frac{\partial M}{\partial y}\right)dA$$

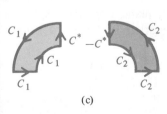

(c)

FIGURE 15.16

$$= \int_{C_1} M(x,y)\,dx + N(x,y)\,dy + \int_{C^*} M(x,y)\,dx + N(x,y)\,dy$$

$$+ \int_{C_2} M(x,y)\,dx + N(x,y)\,dy$$

$$+ \int_{-C^*} M(x,y)\,dx + N(x,y)\,dy$$

$$\iint\limits_R \left(\frac{\partial N}{\partial x} - \frac{\partial M}{\partial y}\right) dA = \int_{C_1} M(x, y)\, dx + N(x, y)\, dy + \int_{C_2} M(x, y)\, dx + N(x, y)\, dy$$

$$= \int_C M(x, y)\, dx + N(x, y)\, dy \quad \square$$

Example 5 Let R be the semiannular region shown in Figure 15.16(a), where $r_1 = 1$ and $r_2 = 2$. Find

$$\int_C y^3\, dx - x^3\, dy$$

Solution By Example 4 we may apply Green's Theorem to obtain

$$\int_C y^3\, dx - x^3\, dy = \iint\limits_R \left[\frac{\partial}{\partial x}(-x^3) - \frac{\partial}{\partial y}(y^3)\right] dA = -\iint\limits_R 3(x^2 + y^2)\, dA$$

$$= -\int_0^\pi \int_1^2 (3r^2) r\, dr\, d\theta = -\int_0^\pi \frac{3r^4}{4}\Big|_1^2\, d\theta$$

$$= -\int_0^\pi \frac{45}{4}\, d\theta = -\frac{45}{4}\pi \quad \square$$

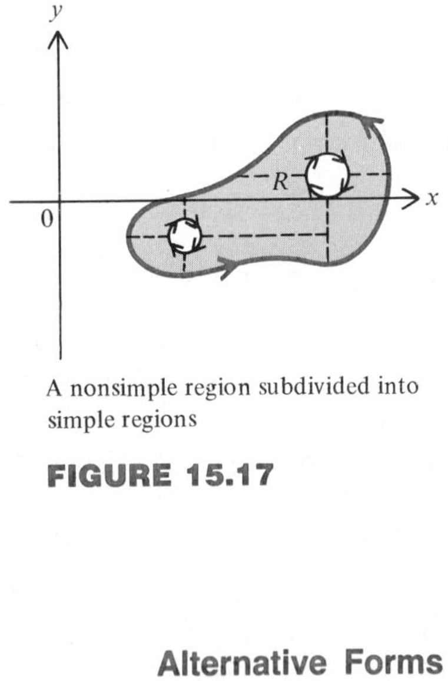

To evaluate the integral in Example 5 by parametrizing the boundary of R and then evaluating the line integral would be tedious, because there would be four integrals to evaluate.

The procedure followed in Example 4—dividing a region into simple subregions—can be applied to almost any plane region that one ordinarily encounters (Figure 15.17). Thus Green's Theorem applies to very general regions. We now have three ways of evaluating line integrals of the form $\int_C \mathbf{F} \cdot d\mathbf{r}$:

1. Parametrize C and then use the formula

$$\int_C \mathbf{F} \cdot d\mathbf{r} = \int_C M\, dx + N\, dy + P\, dz$$

2. Use the Fundamental Theorem of Line Integrals (provided that $\mathbf{F}$ is a gradient).
3. Use Green's Theorem (provided that C is closed).

A nonsimple region subdivided into simple regions

FIGURE 15.17

Alternative Forms of Green's Theorem

To conclude, we discuss two other forms of Green's Theorem. Let $\mathbf{F} = M\mathbf{i} + N\mathbf{j}$ be a continuous vector field defined on a simple region R in the xy plane, and assume that the range of $\mathbf{F}$ is contained in the xy plane. Furthermore, let the boundary C of R have its counterclockwise orientation. By Green's Theorem,

$$\int_C \mathbf{F} \cdot d\mathbf{r} = \int_C M(x, y)\, dx + N(x, y)\, dy = \iint\limits_R \left(\frac{\partial N}{\partial x} - \frac{\partial M}{\partial y}\right) dA$$

Since

$$\text{curl } \mathbf{F} = \left(\frac{\partial N}{\partial x} - \frac{\partial M}{\partial y}\right) \mathbf{k}$$

it follows that

$$(\text{curl } \mathbf{F}) \cdot \mathbf{k} = \left[\left(\frac{\partial N}{\partial x} - \frac{\partial M}{\partial y} \right) \mathbf{k} \right] \cdot \mathbf{k} = \frac{\partial N}{\partial x} - \frac{\partial M}{\partial y}$$

so that

$$\int_C \mathbf{F} \cdot d\mathbf{r} = \iint_R (\text{curl } \mathbf{F}) \cdot \mathbf{k} \, dA \tag{5}$$

To obtain the second alternative form of Green's Theorem, we assume that the boundary C of a simple region R is oriented counterclockwise by a smooth parametrization

$$\mathbf{r}(t) = x(t)\mathbf{i} + y(t)\mathbf{j} \quad \text{for } a \le t \le b$$

Then the tangent $\mathbf{T}$ of C is given by

$$\mathbf{T}(t) = \frac{x'(t)}{\|\mathbf{r}'(t)\|} \mathbf{i} + \frac{y'(t)}{\|\mathbf{r}'(t)\|} \mathbf{j}$$

Let

$$\mathbf{n}(t) = \frac{y'(t)}{\|\mathbf{r}'(t)\|} \mathbf{i} - \frac{x'(t)}{\|\mathbf{r}'(t)\|} \mathbf{j}$$

Then $\mathbf{n}$ lies in the xy plane and is perpendicular to $\mathbf{T}$, so $\mathbf{n}$ is parallel to the normal vector $\mathbf{N}$ of C. (It can be shown that $\mathbf{n}$ points "out of" the region R.) If $\mathbf{F} = M\mathbf{i} + N\mathbf{j}$ is a continuous vector field, then by (4) of Section 15.2,

$$\int_C \mathbf{F} \cdot \mathbf{n} \, ds = \int_a^b (\mathbf{F} \cdot \mathbf{n})(t) \|\mathbf{r}'(t)\| \, dt$$

$$= \int_a^b [M(x(t), y(t))\mathbf{i} + N(x(t), y(t))\mathbf{j}] \cdot \left(\frac{y'(t)}{\|\mathbf{r}'(t)\|} \mathbf{i} - \frac{x'(t)}{\|\mathbf{r}'(t)\|} \mathbf{j} \right) \|\mathbf{r}'(t)\| \, dt$$

$$= \int_a^b [M(x(t), y(t))y'(t) - N(x(t), y(t))x'(t)] \, dt$$

$$= \int_C M(x, y) \, dy - N(x, y) \, dx$$

Then using Green's Theorem and the fact that div $\mathbf{F}(x, y) = \partial M / \partial x + \partial N / \partial y$, we find that

$$\int_C \mathbf{F} \cdot \mathbf{n} \, ds = \int_C M(x, y) \, dy - N(x, y) \, dx = \iint_R \left(\frac{\partial M}{\partial x} + \frac{\partial N}{\partial y} \right) dA$$

$$= \iint_R \text{div } \mathbf{F}(x, y) \, dA$$

Thus

$$\int_C \mathbf{F} \cdot \mathbf{n} \, ds = \iint_R \text{div } \mathbf{F}(x, y) \, dA \tag{6}$$

Formulas (5) and (6) are related to formulas appearing in two other generalizations of the Fundamental Theorem, which we will present in Sections 15.7 and 15.8.

EXERCISES 15.4

In Exercises 1–5 find $\int_C M(x, y)\,dx + N(x, y)\,dy$, where C is oriented counterclockwise.

1. $M(x, y) = y$, $N(x, y) = 0$; C is composed of the portion of the quarter circle $x^2 + y^2 = 4$ in the first quadrant and of the intervals $[0, 2]$ on the x and y axes.

2. $M(x, y) = 0$, $N(x, y) = x$; C is the circle $x^2 + y^2 = 9$.

3. $M(x, y) = xy$, $N(x, y) = x^{3/2} + y^{3/2}$; C is the square with vertices $(0, 0)$, $(1, 0)$, $(1, 1)$, and $(0, 1)$.

4. $M(x, y) = y \cos x$, $N(x, y) = x \sin y$; C is the triangle with vertices $(0, 0)$, $(\pi/2, 0)$, and $(0, \pi/2)$.

5. $M(x, y) = (x^2 + y^2)^{3/2} = N(x, y)$; C is the circle $x^2 + y^2 = 1$.

In Exercises 6–14 use Green's Theorem to evaluate the line integral. Assume that each curve is oriented counterclockwise.

6. $\int_C (y^3 + y)\,dx + 3y^2 x\,dy$; C is the circle $x^2 + y^2 = 100$.

7. $\int_C y\,dx - x\,dy$; C is the cardioid $r = 1 - \cos\theta$.

8. $\int_C (2xy^3 + \cos x)\,dx + (3x^2 y^2 + 5x)\,dy$; C is the circle $x^2 + y^2 = 64$.

9. $\int_C e^x \sin y\,dx + e^x \cos y\,dy$; C is composed of the graph of $\sqrt{x} + \sqrt{y} = 5$ and the intervals $[0, 25]$ on the x and y axes.

10. $\int_C (x \cos^2 x - y)\,dx + (x - e^y)\,dy$; C is the square with vertices $(-1, 1)$, $(-1, 0)$, $(0, 0)$, and $(0, 1)$.

11. $\int_C xy\,dx + (\frac{1}{2}x^2 + xy)\,dy$; C is composed of the interval $[-1, 1]$ on the x axis and the top half of the ellipse $x^2 + 4y^2 = 1$.

12. $\int_C xy^2\,dx + (x^4 + x^2 y)\,dy$; C is the graph of $x^4 + y^4 = 1$.

13. $\int_C (\cos^3 x + e^x)\,dx + e^y\,dy$; C is the graph of $x^6 + y^8 = 1$.

14. $\int_C \ln(x^2 + y^2)\,dx + \ln(x^2 + y^2)\,dy$; C is the boundary of the semiannular region in Figure 15.16(a), with $r_1 = 1$ and $r_2 = 2$.

In Exercises 15–18 use Green's Theorem to evaluate $\int_C \mathbf{F} \cdot d\mathbf{r}$, where C is oriented counterclockwise.

15. $\mathbf{F}(x, y) = y\mathbf{i} + 3x\mathbf{j}$; C is the circle $x^2 + y^2 = 4$.

16. $\mathbf{F}(x, y) = y^4\mathbf{i} + x^3\mathbf{j}$; C is the square with vertices $(-2, -2)$, $(-2, 2)$, $(2, -2)$, and $(2, 2)$.

17. $\mathbf{F}(x, y) = y \sin x\mathbf{i} - \cos x\mathbf{j}$; C is composed of the semicircle $x^2 + y^2 = 9$ for $y \geq 0$, and the line $y = 0$ for $-3 \leq x \leq 3$.

18. $\mathbf{F}(x, y) = y(x^2 + y^2)\mathbf{i} - x(x^2 + y^2)\mathbf{j}$; C is the unit circle $x^2 + y^2 = 1$.

In Exercises 19–21 use Green's Theorem to find the area A of the given region.

19. The region bounded below by the x axis and above by one arch of the cycloid parametrized by

$$\mathbf{r}(t) = (t - \sin t)\mathbf{i} + (1 - \cos t)\mathbf{j} \quad \text{for } 0 \leq t \leq 2\pi$$

(*Hint*: To obtain the area of the region you must use the correct orientation of the boundary.)

20. The region bounded by the hypocycloid parametrized by

$$\mathbf{r}(t) = \cos^3 t\mathbf{i} + \sin^3 t\mathbf{j} \quad \text{for } 0 \leq t \leq 2\pi$$

21. The region bounded by the y axis, the line $y = \frac{1}{4}$, and the curve parametrized by

$$\mathbf{r}(t) = \sin \pi t\mathbf{i} + t(1 - t)\mathbf{j} \quad \text{for } 0 \leq t \leq \frac{1}{2}$$

*22. Let

$$M(x, y) = \frac{-y}{x^2 + y^2} \quad \text{and} \quad N(x, y) = \frac{x}{x^2 + y^2}$$

a. Verify that

$$\int_C M(x, y)\,dx + N(x, y)\,dy = \iint_R \left(\frac{\partial N}{\partial x} - \frac{\partial M}{\partial y}\right)dA$$

if R is the annulus whose boundary C is composed of the circle $x^2 + y^2 = 4$ oriented counterclockwise and the circle $x^2 + y^2 = 1$ oriented clockwise.

b. Show that

$$\int_C M(x, y)\,dx + N(x, y)\,dy \neq \iint_R \left(\frac{\partial N}{\partial x} - \frac{\partial M}{\partial y}\right)dA$$

if R is the disk whose boundary C is the circle $x^2 + y^2 = 1$.

c. Why does the result of part (b) not contradict Green's Theorem?

23. Assume that R is a simple region and that C_1 and C_2 are piecewise smooth closed curves in R, both oriented counterclockwise. Suppose $\partial N/\partial x = \partial M/\partial y$ on R. Use Green's Theorem to prove that

$$\int_{C_1} M(x, y)\,dx + N(x, y)\,dy = \int_{C_2} M(x, y)\,dx + N(x, y)\,dy$$

*24. Prove the formula in (1), which yields the second half of the proof of Green's Theorem.

25. Let g be a function and $\mathbf{F}$ a vector field, and assume that g and $\mathbf{F}$ are defined on a simple region R in the xy plane with a piecewise smooth boundary C oriented counterclockwise. Assume that the partial derivatives of g and of the component functions of $\mathbf{F}$ exist and are continuous throughout R.

a. Using the identity

$$\text{div}(g\mathbf{F}) = g\,\text{div}\,\mathbf{F} + (\text{grad}\,g)\cdot\mathbf{F}$$

and the alternative form of Green's Theorem in (6), show that

$$\int_C g\mathbf{F}\cdot\mathbf{n}\,ds = \iint_R [g\,\text{div}\,\mathbf{F} + (\text{grad}\,g)\cdot\mathbf{F}]\,dA \quad (7)$$

b. Suppose f is a function of two variables having continuous partial derivatives throughout R. Show that

$$\int_C g(\text{grad}\,f)\cdot\mathbf{n}\,ds$$

$$= \iint_R [g\nabla^2 f + (\text{grad}\,g)\cdot(\text{grad}\,f)]\,dA \quad (8)$$

Equation (8) is known as **Green's Theorem in the first form**.

c. Interchange the roles of f and g in (8), subtract the two sides of the resulting equation from (8), and deduce that

$$\int_C (g\,\text{grad}\,f - f\,\text{grad}\,g)\cdot\mathbf{n}\,ds$$

$$= \iint_R (g\nabla^2 f - f\nabla^2 g)\,dA \quad (9)$$

Equation (9) is known as **Green's Theorem in the second form**. (Formulas (8) and (9), which appear in a three-dimensional form in Green's book, are two of the formulas usually known as "Green's identities." But they are the closest of the results in his book to the theorem called Green's Theorem.)

*26. Let f be defined on a simple region R with piecewise smooth boundary C, and assume that $f(x, y) = 0$ for (x, y) on C. Assume also that $\nabla^2 f = 0$ on R.

a. Taking $f = g$ in (8), prove that

$$\iint_{R_0} \|\text{grad}\,f\|^2\,dA = 0$$

for every rectangle R_0 contained in R.

b. Use (a) and Exercise 66 of Section 14.1 to deduce that $\text{grad}\,f(x, y) = \mathbf{0}$ for (x, y) in R.

c. Use (b) and Exercise 32(a) of Section 15.1 to conclude that f is constant on R.

27. a. Let C be the line segment from a point (x_1, y_1) to a point (x_2, y_2) in the plane. Show that

$$\frac{1}{2}\int_C x\,dy - y\,dx = \frac{1}{2}(x_1 y_2 - x_2 y_1)$$

b. Suppose the vertices of a polygon, labeled counterclockwise, are $(x_1, y_1), (x_2, y_2), \ldots, (x_n, y_n)$. Using part (a), show that the area A of the polygon is given by

$$A = \frac{1}{2}(x_1 y_2 - x_2 y_1) + \frac{1}{2}(x_2 y_3 - x_3 y_2) + \cdots$$

$$+ \frac{1}{2}(x_{n-1} y_n - x_n y_{n-1}) + \frac{1}{2}(x_n y_1 - x_1 y_n)$$

c. Find the area of the quadrilateral with vertices $(0, 0)$, $(1, 0)$, $(2, 3)$, and $(-1, 1)$.

28. Suppose f is a function of two variables that is harmonic throughout a simple region R. Use Green's Theorem to show that $\displaystyle\int_C - f_y\,dx + f_x\,dy$ is independent of path in R.

*29. In this exercise you will use techniques of this section to rederive Kepler's Second Law, which describes the rate at which the position vector from the sun to a planet sweeps out area.

a. Let $\mathbf{r} = x\mathbf{i} + y\mathbf{j}$. Using (4) of Section 12.7, with τ replacing t, show that

$$x\frac{dy}{d\tau} - y\frac{dx}{d\tau} = p$$

b. Let (x_0, y_0) be the position of a planet at a fixed time t_0, and (x, y) the position at an arbitrary later time t. Let $A(t)$ be the area of the region bounded by the curve C that is composed of the curves C_1, C_2, and C_3 in Figure 15.18. Show that

$$A(t) = \frac{1}{2}\int_C x\,dy - y\,dx = \frac{1}{2}\int_{t_0}^{t} \left(x\frac{dy}{d\tau} - y\frac{dx}{d\tau}\right)d\tau$$

(*Hint:* First use Exercise 27(a) to show that $\frac{1}{2}\int_{C_1} x\,dy - y\,dx = \frac{1}{2}\int_{C_2} x\,dy - y\,dx = 0$.)

c. Using (a) and (b), show that $dA/dt = p/2$. (This yields the same result as (18) of Section 12.7, which led us to Kepler's Second Law.)

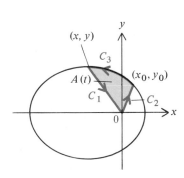

FIGURE 15.18

15.5
SURFACE INTEGRALS

Now we turn to a higher-dimensional analogue of line integrals, called surface integrals. These too help to describe certain physical phenomena, and they appear in the statements of two more generalizations of the Fundamental Theorem of Calculus: Stokes's Theorem and the Divergence Theorem, which we will present in the last two sections of this chapter.

Let Σ be the graph of a function having continuous partial derivatives and defined on a region R in the xy plane that is composed of a finite number of vertically or horizontally simple regions. Let g be continuous on Σ. You might think of Σ as representing a thin metal plate and $g(x, y, z)$ as the mass density (mass per unit area) of Σ at the point (x, y, z). To determine the mass m of the plate we select a rectangle R' that circumscribes R and let $\mathscr{P}$ partition R' into subrectangles, numbered so that $R_1, R_2, \ldots, R_n$ are those entirely contained in R. Let $\Sigma_1, \Sigma_2, \ldots, \Sigma_n$ be the projections of $R_1, R_2, \ldots, R_n$, respectively, onto Σ (Figure 15.19), and let $\Delta S_1, \Delta S_2, \ldots, \Delta S_n$ be their respective surface areas (as defined in Section 14.3). For each integer k between 1 and n, let (x_k, y_k, z_k) be an arbitrary point on Σ_k. If the dimensions of all the subrectangles in $\mathscr{P}$ are small enough, then m should be approximately

$$\sum_{k=1}^{n} g(x_k, y_k, z_k) \Delta S_k \tag{1}$$

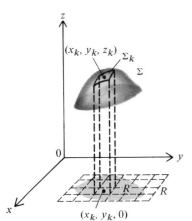

FIGURE 15.19

Whether or not g represents mass density, it can be proved that as the maximum $\|\mathscr{P}\|$ of the dimensions of the subrectangles tends to 0, the sum in (1) tends to a limit. This leads us to the following definition.

DEFINITION 15.9

> Let Σ be the graph of a function having continuous partial derivatives and defined on a region R in the xy plane that is composed of a finite number of vertically or horizontally simple regions. Let g be continuous on Σ. The **surface integral** of g over Σ, denoted $\iint_\Sigma g(x, y, z)\, dS$, is defined by
>
> $$\iint_\Sigma g(x, y, z)\, dS = \lim_{\|\mathscr{P}\| \to 0} \sum_{k=1}^{n} g(x_k, y_k, z_k)\, \Delta S_k$$

If Σ represents a metal plate of negligible thickness, and if $g(x, y, z)$ represents the mass density of the plate at the point (x, y, z) on Σ, then we compute the mass m of the plate by the formula

$$m = \iint_\Sigma g(x, y, z)\, dS \tag{2}$$

Using the ideas we employed while deriving the formula for surface area in Section 14.3, one can show that if Σ is the graph of f on R, then

$$\iint_\Sigma g(x, y, z)\, dS = \iint_R g(x, y, f(x, y))\sqrt{[f_x(x, y)]^2 + [f_y(x, y)]^2 + 1}\, dA \quad (3)$$

Observe that in the double integral on the right, the variable z in $g(x, y, z)$ is replaced by $f(x, y)$, where $z = f(x, y)$ is an equation of Σ. We also observe that if $g(x, y, z) = 1$ for all (x, y, z) on Σ, then the double integral in (3) represents the surface area of Σ given in Definition 14.6.

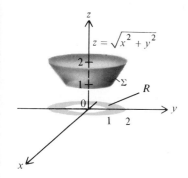

FIGURE 15.20

Example 1 Evaluate $\iint_\Sigma z^2\, dS$, where Σ is the portion of the cone $z = \sqrt{x^2 + y^2}$ for which $1 \le x^2 + y^2 \le 4$ (Figure 15.20).

Solution If R is the ring $1 \le x^2 + y^2 \le 4$ and if

$$f(x, y) = \sqrt{x^2 + y^2} \quad \text{for } (x, y) \text{ in } R$$

then Σ is the graph of f on R. Because

$$\sqrt{f_x^2(x, y) + f_y^2(x, y) + 1} = \sqrt{\frac{x^2}{x^2 + y^2} + \frac{y^2}{x^2 + y^2} + 1} = \sqrt{2}$$

(3) implies that

$$\iint_\Sigma z^2\, dS = \iint_R (x^2 + y^2)\sqrt{2}\, dA = \sqrt{2}\int_0^{2\pi}\int_1^2 r^2 r\, dr\, d\theta$$

$$= \sqrt{2}\int_0^{2\pi}\int_1^2 r^3\, dr\, d\theta = \sqrt{2}\int_0^{2\pi}\left.\frac{r^4}{4}\right|_1^2 d\theta$$

$$= \frac{15\sqrt{2}}{4}\int_0^{2\pi} 1\, d\theta = \frac{15\pi\sqrt{2}}{2} \quad \square$$

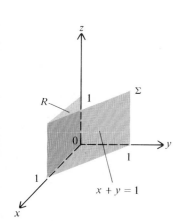

FIGURE 15.21

If the surface Σ in Example 1 represents a thin metal funnel with mass density given by $g(x, y, z) = z^2$, then according to (2) and the calculation in Example 1, the mass of the funnel is $15\pi\sqrt{2}/2$.

Some surfaces cannot be expressed as graphs of functions of x and y but can be expressed as graphs of functions of x and z or of y and z. In such cases we can still apply (3), with the roles of x, y, and z interchanged.

Example 2 Evaluate $\iint_\Sigma (x + y + z)\, dS$, where Σ is the portion of the plane $x + y = 1$ in the first octant for which $0 \le z \le 1$ (Figure 15.21).

Solution   Notice that there is no function f of x and y whose graph is Σ. However, if R is the square in the xz plane consisting of all (x, z) for which $0 \le x \le 1$ and $0 \le z \le 1$, and if

$$f(x, z) = 1 - x$$

then Σ is the graph of $y = f(x, z)$ on R. Interchanging the roles of y and z in (3), we obtain

$$\sqrt{f_x^2(x, z) + f_z^2(x, z) + 1} = \sqrt{(-1)^2 + 0 + 1} = \sqrt{2}$$

and finally

$$\iint_\Sigma (x + y + z)\, dS = \iint_R (x + (1 - x) + z)\sqrt{2}\, dA$$

$$= \sqrt{2} \int_0^1 \int_0^1 (1 + z)\, dx\, dz = \sqrt{2} \int_0^1 (1 + z)\, dz$$

$$= \sqrt{2} \left(z + \frac{z^2}{2} \right) \Big|_0^1 = \frac{3\sqrt{2}}{2} \quad \square$$

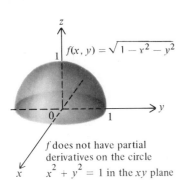

$f(x, y) = \sqrt{1 - x^2 - y^2}$

f does not have partial derivatives on the circle $x^2 + y^2 = 1$ in the xy plane

FIGURE 15.22

Occasionally (3) is not strictly applicable because Σ is the graph of a function f that does not have partial derivatives at every point in R. This occurs, for example, if Σ is the hemisphere $z = \sqrt{1 - x^2 - y^2}$. Here $f(x, y) = \sqrt{1 - x^2 - y^2}$; however, f does not have partial derivatives at any point on the circle $x^2 + y^2 = 1$ (Figure 15.22). Nevertheless, formula (3) frequently remains valid if the double integral over R in (3) as interpreted as an improper integral.

Example 3   Evaluate $\iint_\Sigma (1 + z)\, dS$, where Σ is the hemisphere $z = \sqrt{1 - x^2 - y^2}$.

Solution   If

$$f(x, y) = \sqrt{1 - x^2 - y^2}$$

then Σ is the graph of f, and the domain R of f is the disk bounded by the circle $x^2 + y^2 = 1$. We find that

$$\sqrt{f_x^2(x, y) + f_y^2(x, y) + 1} = \sqrt{\frac{x^2}{1 - x^2 - y^2} + \frac{y^2}{1 - x^2 - y^2} + 1}$$

$$= \frac{1}{\sqrt{1 - x^2 - y^2}} \quad \text{for } x^2 + y^2 < 1$$

These formulas do not hold on the circle $x^2 + y^2 = 1$, but for any number b strictly between 0 and 1 they hold on the disk R_b bounded by the circle $x^2 + y^2 = b^2$. Therefore we can evaluate $\iint_\Sigma (1 + z)\, dS$ by (3) as follows:

$$\iint_\Sigma (1 + z)\, dS = \lim_{b \to 1^-} \iint_{R_b} (1 + \sqrt{1 - x^2 - y^2}) \left(\frac{1}{\sqrt{1 - x^2 - y^2}} \right) dA$$

$$= \lim_{b \to 1^-} \iint_{R_b} \left(\frac{1}{\sqrt{1 - x^2 - y^2}} + 1 \right) dA$$

$$= \lim_{b \to 1^-} \int_0^{2\pi} \int_0^b \left(\frac{1}{\sqrt{1 - r^2}} + 1 \right) r \, dr \, d\theta$$

$$= \lim_{b \to 1^-} \int_0^{2\pi} \left(-\sqrt{1 - r^2} + \frac{r^2}{2} \right) \Big|_0^b \, d\theta$$

$$= \lim_{b \to 1^-} \int_0^{2\pi} \left(-\sqrt{1 - b^2} + \frac{b^2}{2} + 1 \right) d\theta$$

$$= \lim_{b \to 1^-} 2\pi \left(-\sqrt{1 - b^2} + \frac{b^2}{2} + 1 \right) = 3\pi \quad \square$$

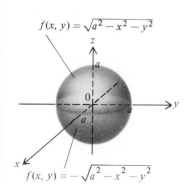

$f(x, y) = \sqrt{a^2 - x^2 - y^2}$

$f(x, y) = -\sqrt{a^2 - x^2 - y^2}$

FIGURE 15.23

It is sometimes necessary to integrate over a surface composed of several subsurfaces, each of which is the graph of a function. A sphere is an excellent example of such a surface. Although it is not the graph of a single function, a sphere is composed of two hemispheres, each of which is the graph of a function (Figure 15.23). In general, if a surface Σ is composed of several subsurfaces $\Sigma_1, \Sigma_2, \ldots, \Sigma_n$, we write $\Sigma = \Sigma_1 + \Sigma_2 + \cdots + \Sigma_n$. If g is continuous on Σ, then as you might expect,

$$\iint_{\Sigma} g(x, y, z) \, dS = \iint_{\Sigma_1} g(x, y, z) \, dS + \iint_{\Sigma_2} g(x, y, z) \, dS + \cdots + \iint_{\Sigma_n} g(x, y, z) \, dS \quad (4)$$

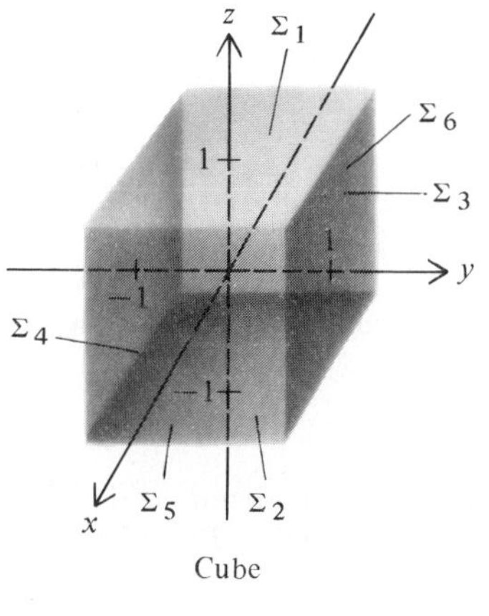

FIGURE 15.24

Σ_1 Σ_6 Σ_3 Σ_4 Σ_5 Σ_2

Cube

Example 4 Evaluate $\iint_{\Sigma}(x + y + z) \, dS$, where Σ is the cube centered at the origin (Figure 15.24).

Solution Since $\Sigma = \Sigma_1 + \Sigma_2 + \cdots + \Sigma_6$, it follows from (4) that

$$\iint_{\Sigma}(x + y + z) \, dS = \sum_{k=1}^{6} \iint_{\Sigma_k}(x + y + z) \, dS$$

and it is apparent that the solution involves six integrals. As a first step we observe that if

$$f_1(x, y) = 1 \quad \text{for } -1 \le x \le 1 \text{ and } -1 \le y \le 1$$

then the top Σ_1 is the graph of f_1, and thus by (3),

$$\iint_{\Sigma_1}(x + y + z) \, dS = \int_{-1}^{1} \int_{-1}^{1} (x + y + 1)\sqrt{0 + 0 + 1} \, dy \, dx$$

$$= \int_{-1}^{1} \left(xy + \frac{y^2}{2} + y \right) \Big|_{-1}^{1} \, dx$$

$$= \int_{-1}^{1} (2x + 2) \, dx = (x^2 + 2x) \Big|_{-1}^{1} = 4$$

Second, if

$$f_2(x, y) = -1 \quad \text{for } -1 \le x \le 1 \text{ and } -1 \le y \le 1$$

then the bottom Σ_2 is the graph of f_2, and thus by (3),

$$\iint_{\Sigma_2} (x + y + z)\, dS = \int_{-1}^{1} \int_{-1}^{1} (x + y - 1)\sqrt{0 + 0 + 1}\, dy\, dx$$

$$= \int_{-1}^{1} \left(xy + \frac{1}{2}y^2 - y \right)\Big|_{-1}^{1} dx$$

$$= \int_{-1}^{1} (2x - 2)\, dx = (x^2 - 2x)\Big|_{-1}^{1} = -4$$

Instead of computing the remaining four integrals, we observe that the sides of the cube Σ are symmetric with respect to the coordinate planes and that the integrand $x + y + z$ remains unchanged when x, y, and z are interchanged. Consequently

$$\iint_{\Sigma_1} (x + y + z)\, dS = \iint_{\Sigma_3} (x + y + z)\, dS = \iint_{\Sigma_5} (x + y + z)\, dS = 4$$

and

$$\iint_{\Sigma_2} (x + y + z)\, dS = \iint_{\Sigma_4} (x + y + z)\, dS = \iint_{\Sigma_6} (x + y + z)\, dS = -4$$

Therefore we conclude that

$$\iint_{\Sigma} (x + y + z)\, dS = \sum_{k=1}^{6} \iint_{\Sigma_k} (x + y + z)\, dS$$

$$= 4 - 4 + 4 - 4 + 4 - 4 = 0 \quad \square$$

EXERCISES 15.5

In Exercises 1–12 evaluate $\iint_{\Sigma} g(x, y, z)\, dS$.

1. $g(x, y, z) = x$; Σ is the part of the plane $2x + 3y + z = 6$ in the first octant.

2. $g(x, y, z) = z^2$; Σ is the part of the cone $z = \sqrt{x^2 + y^2}$ between the planes $z = 1$ and $z = 3$.

3. $g(x, y, z) = 2x^2 + 1$; Σ is the part of the plane $z = 3x - 2$ inside the cylinder $x^2 + y^2 = 4$.

4. $g(x, y, z) = z^2$; Σ is the part of the sphere $x^2 + y^2 + z^2 = 9$ in the first octant.

5. $g(x, y, z) = \sqrt{4x^2 + 4y^2 + 1}$; Σ is the part of the paraboloid $z = x^2 + y^2$ below the plane $y = z$.

6. $g(x, y, z) = xy$; Σ is the part of the paraboloid $z = 4 - x^2 - y^2$ that lies above the xy plane. (*Hint:* Use rectangular coordinates.)

7. $g(x, y, z) = y$; Σ is the part of the parabolic sheet $z = 4 - y^2$ for which $0 \le x \le 3$ and $0 \le y \le 2$.

8. $g(x, y, z) = x^2 z$; Σ is the part of the cylinder $x^2 + z^2 = 1$ between the planes $y = -1$ and $y = 2$ and above the xy plane.

9. $g(x, y, z) = z(x^2 + y^2)$; Σ is the hemisphere $\sqrt{x^2 + y^2 + z^2} = 2$.

10. $g(x, y, z) = x^2 + y^2$; Σ is composed of the part of the paraboloid $z = 1 - x^2 - y^2$ above the xy plane, and the part of the xy plane that lies inside the circle $x^2 + y^2 = 1$.

11. $g(x, y, z) = x + y$; Σ is the cube with vertices $(0, 0, 0)$, $(1, 0, 0)$, $(1, 1, 0)$, $(0, 1, 0)$, $(0, 0, 1)$, $(1, 0, 1)$, $(1, 1, 1)$, and $(0, 1, 1)$.

12. $g(x, y, z) = x + 1$; Σ is the tetrahedron with vertices $(0, 0, 0)$, $(1, 0, 0)$, $(0, 1, 0)$, $(0, 0, 1)$.

13. Suppose that a thin metal funnel has the shape of the conical surface $z = 2\sqrt{x^2 + y^2}$ for $\frac{1}{2} \le z \le 4$ and that the mass density of the funnel is given by $\delta(x, y, z) = 6 - z$. Compute the mass of the funnel.

14. Suppose the mass density at any point (x, y, z) of a thin spherical metal shell with radius 5 is given by $\delta(x, y, z) = 1 + z^2$. Compute the mass of the shell.

Let D be a solid region filled with water and Σ a surface in contact with the water (Figure 15.25). The hydrostatic pressure $p(x, y, z)$ at any point (x, y, z) on Σ is defined to be the force per unit surface area of Σ, that is,

$$p(x, y, z) = \lim_{\varepsilon \to 0} \frac{\Delta F_\varepsilon}{\Delta S_\varepsilon}$$

where ΔF_ε is the hydrostatic force exerted by the water on the portion Σ_ε of Σ that is at a distance less than ε from (x, y, z), and where ΔS_ε is the surface area of Σ_ε. It can be shown that

$$p(x, y, z) = 62.5(z_0 - z)$$

where z_0 is the z coordinate of the highest point(s) in D, and that the hydrostatic force F exerted by the water on the entire

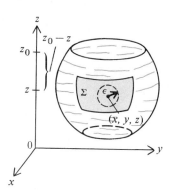

FIGURE 15.25

surface Σ is given by

$$F = \iint_\Sigma p(x, y, z) \, dS = \iint_\Sigma 62.5(z_0 - z) \, dS$$

(We assume that length is measured in feet and force in pounds. Water weighs approximately 62.5 pounds per cubic foot.)

In Exercises 15–17 assume that the given water tank is full of water, and compute the hydrostatic force on the tank.

15. A hemispherical tank having a radius of 10 feet and a flat bottom

16. A semicylindrical tank having a radius of 4 feet, a length of 12 feet, a flat top, and a round bottom

17. A spherical tank having a radius of 10 feet (*Hint:* Integrate separately over the top and the bottom hemispheres.)

15.6
INTEGRALS OVER ORIENTED SURFACES

In Section 15.2 we first defined the line integral $\int_C f(x, y, z) \, ds$ of a function over a curve C and then used it to define the line integral $\int_C \mathbf{F} \cdot d\mathbf{r}$ of a vector field over an oriented curve C. Now that we have defined the surface integral $\iint_\Sigma g(x, y, z) \, dS$ of a function over a surface, we will introduce the notion of orientation of a surface and then define the surface integral of a vector field over an oriented surface.

For surface integrals of vector fields we must further limit the surfaces over which we integrate to those having two sides—which for many surfaces (such as a sphere) you might think of as the outside and the inside. In order to identify two sides of a surface Σ, we assume that Σ has a tangent plane at each of its nonboundary points. At such a point on the surface two (unit) normal vectors exist, and they have opposite directions (Figure 15.26(a)). If it is possible to select one normal at each nonboundary point in such a way that the chosen normal varies continuously on Σ, then Σ is said to be an **orientable surface**, and the selection of the normal gives an **orientation** to Σ and thus makes Σ an **oriented**

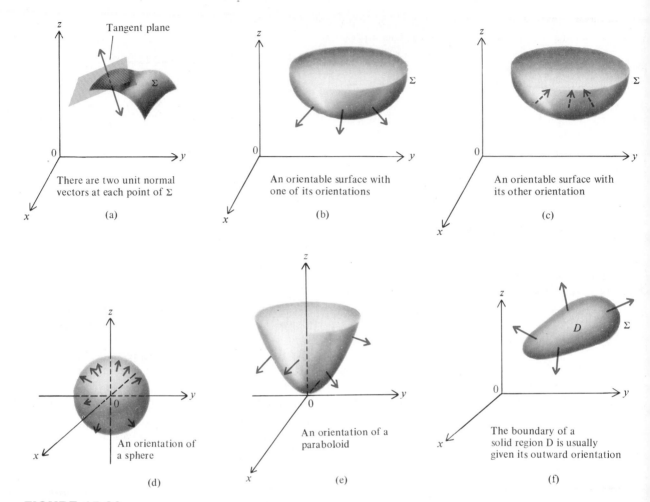

FIGURE 15.26

surface. In such a case there are two possible orientations (Figure 15.26(b) and (c)). Spheres, paraboloids, and all the other surfaces we have encountered thus far are orientable (Figure 15.26(d) and (e)). When Σ is the boundary of a solid region D in space, we customarily choose the normal to Σ that is directed outward from D (Figure 15.26(f)).

Some surfaces are not orientable; the most celebrated one is the Möbius band* (Figure 15.27). If you think of the normal at a given point of the Möbius band as a toy soldier, and if you lead the soldier around the band along a curve whose initial and terminal points are the same, you will find that the soldier's head points in one direction at the outset and in the opposite direction at the

---

* **Möbius:** This name is German and is pronounced, approximately, "*Mer*-bius," but without the "r" sound.
The Möbius band is named after Augustus Ferdinand Möbius (1790–1868), a prominent nineteenth-century geometer.

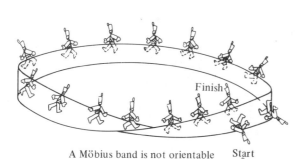

A Möbius band is not orientable   Start

FIGURE 15.27

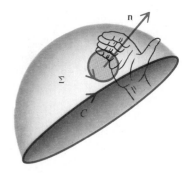

FIGURE 15.28

end (Figure 15.27). This suggests (but does not prove) that the band has no orientation.

In the remainder of this chapter we will assume that all surfaces under consideration are orientable. If Σ is an oriented surface bounded by a curve C, then the orientation of Σ induces an orientation on C. To see how this is done, imagine placing the side of your right hand at any point on Σ, with the thumb pointed in the direction of the normal that gives Σ its orientation. If you move your right hand toward C, always keeping your thumb pointed in the direction of the normal, then the remaining fingers naturally curl in a way that defines a direction, or orientation, on C (Figure 15.28). That orientation is called the **induced orientation** on C. Notice that if the orientation of Σ is reversed, then the orientation induced on C is also reversed.

Flux Integrals

Suppose Σ is an oriented surface, which means that we can choose a unit normal vector $\mathbf{n}$ at each nonboundary point of Σ so that $\mathbf{n}$ varies continuously over Σ. For the present, think of Σ as a membrane through which a fluid of constant density δ is flowing with a constant velocity $\mathbf{v}$. We wish to determine the rate with respect to time at which mass is flowing in the direction of $\mathbf{n}$ through the surface Σ (Figure 15.29). In the event that Σ is a plane surface having surface area S, with $\mathbf{n}$ pointing in the same direction as $\mathbf{v}$, the volume of fluid that passes through Σ between a time t and a later time $t + h$ is $h\|\mathbf{v}\|S$, the volume of the cylinder in Figure 15.30(a). Therefore the mass passing through Σ in the direction of $\mathbf{n}$ during that time interval is $\delta h\|\mathbf{v}\|S$, and hence the rate at which mass is flowing through Σ in the direction of $\mathbf{n}$ is

$$\frac{\delta h\|\mathbf{v}\|S}{h} = \delta\|\mathbf{v}\|S$$

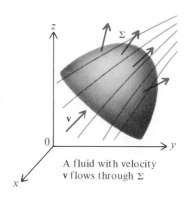

A fluid with velocity $\mathbf{v}$ flows through Σ

FIGURE 15.29

If Σ is a plane surface but $\mathbf{n}$ does not necessarily have the same direction as $\mathbf{v}$, let θ be the angle between $\mathbf{n}$ and $\mathbf{v}$, so that by convention $0 \le \theta \le \pi$. Then the volume of fluid that passes through Σ is $h\|\mathbf{v}\|(\cos\theta)S$ if $0 \le \theta \le \pi/2$ and is $-h\|\mathbf{v}\|(\cos\theta)S$ if $\pi/2 < \theta \le \pi$ (Figure 15.30(b) and (c)), for each of these numbers is the volume of the solid in the respective figure. If $0 \le \theta < \pi/2$, then the mass flows through the surface in the general direction of $\mathbf{n}$; if $\pi/2 < \theta \le \pi$, then the

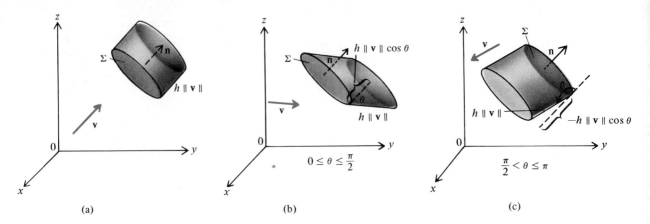

(a)   (b)   (c)

FIGURE 15.30

mass flows in the opposite direction. Consequently in either case the amount of mass flowing though Σ *in the general direction of* $\mathbf{n}$ is $\delta\|\mathbf{v}\|\,(\cos\theta)S = \delta h(\mathbf{v}\cdot\mathbf{n})S$. (If this number is negative, mass is actually flowing in the general direction opposite to $\mathbf{n}$.) It follows that the rate of mass flow through Σ in the general direction of $\mathbf{n}$ is

$$\frac{\delta h(\mathbf{v}\cdot\mathbf{n})S}{h} = \delta(\mathbf{v}\cdot\mathbf{n})S$$

It is not necessary for the surface Σ to be a plane region or for the velocity of the fluid to be constant. Let us dispense with these hypotheses and assume only that Σ is a surface with a given orientation and surface area S. Assume also that the velocity $\mathbf{v}$ is a continuous vector field on Σ. We partition Σ into subsurfaces $\Sigma_1, \Sigma_2, \ldots, \Sigma_n$, each of which is so small that it is nearly a plane surface. If ΔS_k is the surface area of Σ_k and (x_k, y_k, z_k) is an arbitrary point on Σ_k, then the discussion in the preceding paragraph shows that the amount of mass flowing through Σ_k in the general direction of $\mathbf{n}$ between a time t and a later time $t + h$ should be approximately

$$\delta h[\mathbf{v}(x_k, y_k, z_k)\cdot\mathbf{n}(x_k, y_k, z_k)]\,\Delta S_k$$

This means that the amount of mass flowing through the entire surface Σ in the general direction of $\mathbf{n}$ during that time should be approximately

$$\sum_{k=1}^{n} \delta h[\mathbf{v}(x_k, y_k, z_k)\cdot\mathbf{n}(x_k, y_k, z_k)]\,\Delta S_k$$

Consequently the rate of mass flow should be approximately

$$\sum_{k=1}^{n} \delta[\mathbf{v}(x_k, y_k, z_k)\cdot\mathbf{n}(x_k, y_k, z_k)]\,\Delta S_k$$

We conclude that the rate of mass flow of the liquid through the whole surface Σ

should be

$$\iint_{\Sigma} \delta[\mathbf{v}(x, y, z) \cdot \mathbf{n}(x, y, z)] \, dS$$

or in abbreviated form,

$$\iint_{\Sigma} \delta \mathbf{v} \cdot \mathbf{n} \, dS \tag{1}$$

Because (1) is derived in terms of fluid motion, the integral in (1) is called a **flux integral**. In fact, any integral of the form

$$\iint_{\Sigma} \mathbf{F} \cdot \mathbf{n} \, dS \tag{2}$$

is called a **flux integral**, whether or not $\mathbf{F} = \delta\mathbf{v}$.

Caution: Notice that the surface integral $\iint_{\Sigma} g(x, y, z) \, dS$ does not require Σ to be oriented. However, for the flux integral $\iint_{\Sigma} \mathbf{F} \cdot \mathbf{n} \, dS$, Σ *must* be oriented.

Although flux integrals might look very difficult to evaluate, this is not the case when Σ is the graph of a function f with continuous partials on a region $\mathbf{R}$ in the xy plane that is composed of vertically and horizontally simple regions. First, recall from (2) of Section 13.6 that $f_x \mathbf{i} + f_y \mathbf{j} - \mathbf{k}$ is normal to Σ. The normal $\mathbf{n}$ to Σ that we choose has unit length and is given by

$$\mathbf{n} = \frac{-f_x \mathbf{i} - f_y \mathbf{j} + \mathbf{k}}{\sqrt{(f_x)^2 + (f_y)^2 + 1}} \tag{3}$$

or

$$\mathbf{n} = \frac{f_x \mathbf{i} + f_y \mathbf{j} - \mathbf{k}}{\sqrt{(f_x)^2 + (f_y)^2 + 1}} \tag{4}$$

depending on the orientation of Σ. When $\mathbf{n}$ is directed upward (that is, its $\mathbf{k}$ component is positive), $\mathbf{n}$ has the form in (3), whereas when $\mathbf{n}$ is directed downward, $\mathbf{n}$ has the form in (4). Let $\mathbf{F} = M\mathbf{i} + N\mathbf{j} + P\mathbf{k}$. If $\mathbf{n}$ is given by (3), we find from (3) of Section 15.5 that

$$\iint_{\Sigma} \mathbf{F} \cdot \mathbf{n} \, dS = \iint_{R} (M\mathbf{i} + N\mathbf{j} + P\mathbf{k}) \cdot \left(\frac{-f_x \mathbf{i} - f_y \mathbf{j} + \mathbf{k}}{\sqrt{(f_x)^2 + (f_y)^2 + 1}} \right) \sqrt{(f_x)^2 + (f_y)^2 + 1} \, dA$$

and consequently

$$\iint_{\Sigma} \mathbf{F} \cdot \mathbf{n}\, dS = \iint_{R} [-M(x,y,f(x,y))f_x(x,y)$$
$$- N(x,y,f(x,y))f_y(x,y) + P(x,y,f(x,y))]\, dA \qquad (5)$$

if the normal $\mathbf{n}$ is given by (3)

In contrast, if $\mathbf{n}$ is given by (4), then again by (3) of Section 15.5,

$$\iint_{\Sigma} \mathbf{F} \cdot \mathbf{n}\, dS = \iint_{R} [M(x,y,f(x,y))f_x(x,y)$$
$$+ N(x,y,f(x,y))f_y(x,y) - P(x,y,f(x,y))]\, dA \qquad (6)$$

if the normal $\mathbf{n}$ is given by (4)

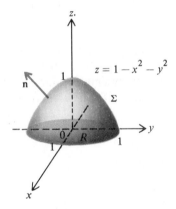

FIGURE 15.31

Example 1   Suppose Σ is the part of the paraboloid $z = 1 - x^2 - y^2$ that lies above the xy plane and is oriented by the normal directed upward (Figure 15.31). Assume that the velocity of a fluid with constant density δ is given by $\mathbf{v}(x,y,z) = x\mathbf{i} + y\mathbf{j} + 2z\mathbf{k}$. Determine the rate of mass flow through Σ in the direction of $\mathbf{n}$.

Solution   According to (1), the rate of mass flow is equal to the surface integral

$$\iint_{\Sigma} \delta \mathbf{v} \cdot \mathbf{n}\, dS$$

Since the normal is directed upward, we will evaluate this integral by using (5), with

$$\mathbf{F}(x,y,z) = \delta \mathbf{v}(x,y,z) = \delta x\mathbf{i} + \delta y\mathbf{j} + 2\delta z\mathbf{k}$$

so that

$$M(x,y,z) = \delta x \qquad N(x,y,z) = \delta y, \quad \text{and} \quad P(x,y,z) = 2\delta z$$

Let R denote the disk with boundary $x^2 + y^2 = 1$. If

$$f(x,y) = 1 - x^2 - y^2$$

then Σ is the graph of $z = f(x,y)$ on R. Since

$$f_x(x,y) = -2x \quad \text{and} \quad f_y(x,y) = -2y$$

(5) implies that

$$\iint_{\Sigma} \delta \mathbf{v} \cdot \mathbf{n}\, dS = \iint_{R} [-\delta x(-2x) - \delta y(-2y) + 2\delta(1 - x^2 - y^2)]\, dA$$

$$= \delta \iint_R 2 \, dA = \delta \int_0^{2\pi} \int_0^1 2r \, dr \, d\theta = \delta \int_0^{2\pi} r^2 \Big|_0^1 \, d\theta$$

$$= \delta \int_0^{2\pi} 1 \, d\theta = 2\pi\delta \quad \square$$

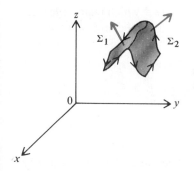

FIGURE 15.32

Flux integrals can be defined for a surface Σ composed of several oriented surfaces $\Sigma_1, \Sigma_2, \dots, \Sigma_n$, provided that the surfaces induce opposite orientations on the common curves that bind them together (Figure 15.32). We simply let

$$\iint_\Sigma \mathbf{F} \cdot \mathbf{n} \, dS = \iint_{\Sigma_1} \mathbf{F} \cdot \mathbf{n} \, dS + \iint_{\Sigma_2} \mathbf{F} \cdot \mathbf{n} \, dS + \cdots + \iint_{\Sigma_n} \mathbf{F} \cdot \mathbf{n} \, dS \qquad (7)$$

Example 2 Let Σ be the unit sphere $x^2 + y^2 + z^2 = 1$, oriented with the normal that is directed outward, and let $\mathbf{F}(x, y, z) = z\mathbf{k}$. Find $\iint_\Sigma \mathbf{F} \cdot \mathbf{n} \, dS$.

Solution Since $\mathbf{F}(x, y, z) = z\mathbf{k}$, we have

$$M = N = 0 \quad \text{and} \quad P(x, y, z) = z$$

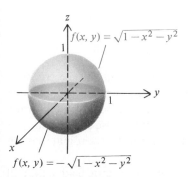

FIGURE 15.33

We divide Σ into the upper hemisphere Σ_1 and the lower hemisphere Σ_2 (Figure 15.33). Now Σ_1 is the graph of $z = \sqrt{1 - x^2 - y^2}$ on the region R bounded by the circle $x^2 + y^2 = 1$, and the normal to Σ_1 points upward. Therefore if $f(x, y) = \sqrt{1 - x^2 - y^2}$, then by (5),

$$\iint_{\Sigma_1} \mathbf{F} \cdot \mathbf{n} \, dS = \iint_R P(x, y, f(x, y)) \, dA = \iint_R \sqrt{1 - x^2 - y^2} \, dA$$

$$= \int_0^{2\pi} \int_0^1 \sqrt{1 - r^2} \, r \, dr \, d\theta = \int_0^{2\pi} \frac{-1}{3}(1 - r^2)^{3/2} \Big|_0^1 \, d\theta$$

$$= \int_0^{2\pi} \frac{1}{3} \, d\theta = \frac{2}{3}\pi$$

In the same way, Σ_2 is the graph of $z = -\sqrt{1 - x^2 - y^2}$ on R, and the normal to Σ_2 points downward. Hence if we let $f(x, y) = -\sqrt{1 - x^2 - y^2}$, we find by (6) and the calculations above that

$$\iint_{\Sigma_2} \mathbf{F} \cdot \mathbf{n} \, dS = \iint_R -P(x, y, f(x, y)) \, dA = \iint_R -(-\sqrt{1 - x^2 - y^2}) \, dA$$

$$= \iint_R \sqrt{1 - x^2 - y^2} \, dA = \frac{2}{3}\pi$$

Therefore (7) implies that

$$\iint_\Sigma \mathbf{F} \cdot \mathbf{n} \, dS = \iint_{\Sigma_1} \mathbf{F} \cdot \mathbf{n} \, dS + \iint_{\Sigma_2} \mathbf{F} \cdot \mathbf{n} \, dS = \frac{2}{3}\pi + \frac{2}{3}\pi = \frac{4}{3}\pi \quad \square$$

Gauss's Law   One of the basic laws in electrostatics is **Gauss's Law**,* which relates the electric field $\mathbf{E}$ on the boundary Σ of a solid region D to the total charge q in D:

$$\iint_\Sigma \mathbf{E} \cdot \mathbf{n} \, dS = \frac{q}{\varepsilon_0} \tag{8}$$

where ε_0 is the permittivity of empty space (see Section 11.2 for the value of ε_0). Physicists call this flux integral the **flux of the electric field through Σ**. From Gauss's Law, if one knows the electric field $\mathbf{E}$ on Σ, one can compute the total charge q in D. For example, if the vector field $\mathbf{F}$ appearing in Example 2 represents the electric field over the unit sphere, we would be able to deduce directly from the solution to Example 2 and Gauss's Law that

$$\frac{4}{3}\pi = \frac{q}{\varepsilon_0}$$

and hence the total charge q in the unit ball is

$$q = \frac{4\pi\varepsilon_0}{3}$$

Conversely, it is possible to use Gauss's Law to find the electric field produced by a charged object whose charge distribution possesses sufficient symmetry. We illustrate the method by finding the electric field of a uniformly charged, infinitely long wire (an idealized telephone wire) whose charge density λ (charge per unit length) is constant. For simplicity we select a coordinate system so that the wire lies on the z axis.

Let (x, y, z) be any point not on the wire. There is an electric field at (x, y, z) due to the portion of the wire below (x, y, z), and there is an electric field due to the portion of the wire above (x, y, z) (Figure 15.34). Since the charge on the wire is uniformly distributed, these two electric fields have the same magnitude and make the same acute angle θ with the wire. It follows from the superposition principle that the electric field at (x, y, z) due to the entire wire is the sum of these two electric fields and, by vector addition, is perpendicular to the wire (Figure 15.34). We deduce from the symmetry of the charge distribution that the

---

* **Gauss:** Pronounced "Gowss."
The German mathematician Karl Friedrich Gauss (1777–1855) ranks with the handful of towering mathematical geniuses of all time. A child prodigy, he is said to have noticed an error in his father's bookkeeping while only three years old. Gauss made fundamental contributions to electricity and magnetism, astronomy, and many branches of mathematics.

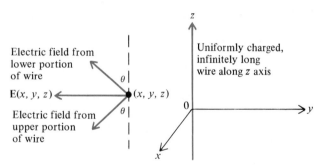

FIGURE 15.34

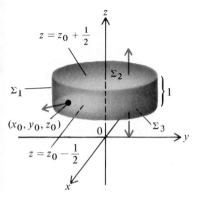

FIGURE 15.35

magnitude of the electric field at any two points having the same distance from the wire must be equal. Consequently the magnitude of the electric field at any point depends only on the distance from the point to the wire. As a result there exists a real-valued function f of a single variable (the distance from the z axis) such that for any point (x, y, z) not on the z axis, the electric field at (x, y, z) is given by

$$\mathbf{E}(x, y, z) = f(\sqrt{x^2 + y^2}) \frac{x\mathbf{i} + y\mathbf{j}}{\sqrt{x^2 + y^2}}$$

Next we determine the value of $f(\sqrt{x_0^2 + y_0^2})$ for any specific point (x_0, y_0, z_0) not on the z axis. Let Σ be the surface shown in Figure 15.35, composed of a portion Σ_1 of a circular cylinder with height 1, a top Σ_2, and a bottom Σ_3. Then $\mathbf{E} \cdot \mathbf{n} = 0$ on Σ_2 and Σ_3, since $\mathbf{E}$ is perpendicular to the z axis and since on Σ_2 and Σ_3 the normal $\mathbf{n}$ is parallel to the z axis. Consequently

$$\iint_\Sigma \mathbf{E} \cdot \mathbf{n}\, dS = \iint_{\Sigma_1} \mathbf{E} \cdot \mathbf{n}\, dS + \iint_{\Sigma_2} \mathbf{E} \cdot \mathbf{n}\, dS + \iint_{\Sigma_3} \mathbf{E} \cdot \mathbf{n}\, dS = \iint_{\Sigma_1} \mathbf{E} \cdot \mathbf{n}\, dS$$

On Σ_1 we have

$$\mathbf{E}(x, y, z) = f(\sqrt{x^2 + y^2}) \frac{x\mathbf{i} + y\mathbf{j}}{\sqrt{x^2 + y^2}} = f(\sqrt{x^2 + y^2})\mathbf{n}$$

Using the fact that $f(\sqrt{x^2 + y^2})$ is constant on Σ_1, we deduce that

$$\iint_\Sigma \mathbf{E} \cdot \mathbf{n}\, dS = \iint_{\Sigma_1} \mathbf{E} \cdot \mathbf{n}\, dS = \iint_{\Sigma_1} [f(\sqrt{x^2 + y^2})\mathbf{n}] \cdot \mathbf{n}\, dS$$

$$= \iint_{\Sigma_1} f(\sqrt{x^2 + y^2})\, dS = \iint_{\Sigma_1} f(\sqrt{x_0^2 + y_0^2})\, dS$$

$$= f(\sqrt{x_0^2 + y_0^2}) \cdot \text{surface area of } \Sigma_1$$

$$= f(\sqrt{x_0^2 + y_0^2})(2\pi\sqrt{x_0^2 + y_0^2})$$

Thus if q is the total charge inside Σ, then by Gauss's Law,

$$f(\sqrt{x_0^2 + y_0^2})2\pi\sqrt{x_0^2 + y_0^2} = \iint_{\Sigma} \mathbf{E} \cdot \mathbf{n}\, dS = \frac{q}{\varepsilon_0}$$

However, the total charge q inside Σ is simply the charge on the portion of the wire inside Σ. By hypothesis the length of the wire inside Σ is 1 and the charge per unit length of wire is λ; it follows that $q = \lambda \cdot 1 = \lambda$. Consequently

$$f(\sqrt{x_0^2 + y_0^2}) = \frac{\lambda}{2\pi\varepsilon_0}\frac{1}{\sqrt{x_0^2 + y_0^2}}$$

We conclude that the electric field at (x_0, y_0, z_0) is given by

$$\mathbf{E}(x_0, y_0, z_0) = \frac{\lambda}{2\pi\varepsilon_0}\frac{1}{\sqrt{x_0^2 + y_0^2}}\frac{x_0\mathbf{i} + y_0\mathbf{j}}{\sqrt{x_0^2 + y_0^2}} = \frac{\lambda}{2\pi\varepsilon_0}\frac{x_0\mathbf{i} + y_0\mathbf{j}}{x_0^2 + y_0^2}$$

Thus the formula for the electric field is completely determined once we know the charge density λ along the wire.

EXERCISES 15.6

In Exercises 1–4 sketch the boundary of the surface and indicate the induced orientation on the boundary.

1. The surface shown in Figure 15.36(a)

2. The surface shown in Figure 15.36(b)

3. The surface shown in Figure 15.36(c)

4. The surface shown in Figure 15.36(d)

In Exercises 5–12 evaluate $\iint_{\Sigma} \mathbf{F} \cdot \mathbf{n}\, dS$.

5. $\mathbf{F}(x, y, z) = y\mathbf{i} - x\mathbf{j} + 8\mathbf{k}$; Σ is the part of the paraboloid $z = 9 - x^2 - y^2$ above the xy plane; $\mathbf{n}$ is directed upward.

6. $\mathbf{F}(x, y, z) = y\mathbf{i} - x\mathbf{j} + z^2\mathbf{k}$; Σ is the part of the cone $z = \sqrt{x^2 + y^2}$ above the square with vertices $(0, 0, 0)$, $(1, 0, 0)$, $(1, 1, 0)$, and $(0, 1, 0)$; $\mathbf{n}$ is directed upward.

7. $\mathbf{F}(x, y, z) = \mathbf{i} + \mathbf{j} + 2\mathbf{k}$; Σ is the hemisphere $z = -\sqrt{1 - x^2 - y^2}$; $\mathbf{n}$ is directed upward.

8. $\mathbf{F}(x, y, z) = x\mathbf{i} - \mathbf{j} + 2x^2\mathbf{k}$; Σ is the part of the paraboloid $z = x^2 + y^2$ above the region in the xy plane bounded by the parabolas $x = 1 - y^2$ and $x = y^2 - 1$; $\mathbf{n}$ is directed downward.

9. $\mathbf{F}(x, y, z) = x\mathbf{i} + y\mathbf{j} + z\mathbf{k}$; Σ is the cube with vertices $(0, 0, 0)$, $(1, 0, 0)$, $(1, 1, 0)$, $(0, 1, 0)$, $(0, 0, 1)$, $(1, 0, 1)$, $(1, 1, 1)$, and $(0, 1, 1)$; $\mathbf{n}$ is directed outward from the cube.

10. $\mathbf{F}(x, y, z) = x\mathbf{i} + y\mathbf{j} + z\mathbf{k}$; Σ is the part of the cylinder $x^2 + z^2 = 1$ between the planes $y = -2$ and $y = 1$; $\mathbf{n}$ is directed away from the y axis.

11. $\mathbf{F}(x, y, z) = -y\mathbf{i} + x\mathbf{j} + z^4\mathbf{k}$; Σ is the sphere $x^2 + y^2 + z^2 = 4$; $\mathbf{n}$ is directed outward from the sphere.

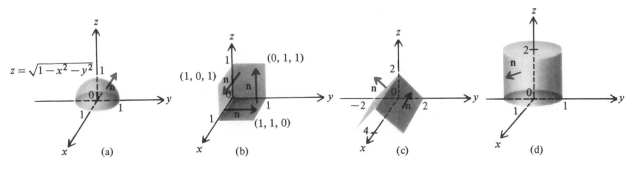

FIGURE 15.36

12. $F(x, y, z) = \sin(x^2 + y^2)(\mathbf{i} + y\mathbf{j})$; Σ is composed of the part of the cylinder $x^2 + y^2 = \pi/2$ between the planes $z = 0$ and $z = 1$, including both the bottom ($z = 0$) and the top ($z = 1$); $\mathbf{n}$ is directed outward from the cylinder.

13. Suppose a fluid having constant density 50 flows with velocity $\mathbf{v} = x\mathbf{i} + y\mathbf{j} + z\mathbf{k}$. Determine the rate of mass flow through the sphere $x^2 + y^2 + z^2 = 10$ in the direction of the outward normal.

14. The cross sections of a straight canal are rectangles 40 feet wide and 10 feet high. Set up a coordinate system with the z axis vertical and the y axis running down the center of the canal bottom. Suppose the canal is full of water flowing with velocity

$$\mathbf{v}(x, y, z) = z(400 - x^2)\mathbf{j}$$

where time is measured in minutes. Calculate the rate of flow of mass through a cross section perpendicular to the y axis. The mass density of water is approximately 1.95 slugs per cubic foot.

15. Suppose an electric field is given by $\mathbf{E}(x, y, z) = x\mathbf{i} + y\mathbf{j}$. Let Σ be the part of the cone $x^2 + z^2 = y^2$ that lies above the xy plane and between the planes $y = 0$ and $y = 1$. Find the flux of $\mathbf{E}$ through Σ in the direction of the normal that points upward.

16. Suppose an electric field is given by

$$\mathbf{E}(x, y, z) = 2x\mathbf{i} + 2y\mathbf{j} + 4z\mathbf{k}$$

Use Gauss's Law to find the total charge q inside the cube whose center is the origin, whose volume is 8, and whose sides are parallel to the coordinate planes.

17. Suppose the xy plane is uniformly charged with charge density σ (charge per unit area). Use Gauss's Law and symmetry to show that the electric field is given by

$$\mathbf{E}(x, y, z) = \begin{cases} \dfrac{\sigma}{2\varepsilon_0}\mathbf{k} & \text{for } z > 0 \\[2mm] \dfrac{-\sigma}{2\varepsilon_0}\mathbf{k} & \text{for } z < 0 \end{cases}$$

(*Hint*: Integrate over the surface of a rectangular parallelepiped, as shown in Figure 15.37).

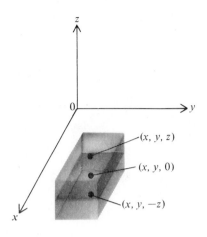

Rectangular parallelepiped with cross-sectional area 1

FIGURE 15.37

*18. Suppose a sphere with radius a and center at the origin is uniformly charged with charge density σ (charge per unit area). Use Gauss's Law and symmetry to show that the electric field is given by

$$\mathbf{E}(x, y, z) = \begin{cases} \mathbf{0} & \text{for } 0 \le x^2 + y^2 + z^2 < a^2 \\[2mm] \dfrac{a^2\sigma}{\varepsilon_0(x^2 + y^2 + z^2)^{3/2}}(x\mathbf{i} + y\mathbf{j} + z\mathbf{k}) \\[2mm] & \text{for } x^2 + y^2 + z^2 \ge a^2 \end{cases}$$

*19. Suppose charge is distributed uniformly throughout the interior of a sphere of radius a, with charge density ρ_0 (charge per unit volume). Use Gauss's Law and symmetry to show that the electric field is given by

$$\mathbf{E}(x, y, z) = \begin{cases} \dfrac{\rho_0}{3\varepsilon_0}(x\mathbf{i} + y\mathbf{j} + z\mathbf{k}) \\[2mm] & \text{for } 0 \le x^2 + y^2 + z^2 \le a^2 \\[2mm] \dfrac{\rho_0}{3\varepsilon_0}\dfrac{a^3}{(x^2 + y^2 + z^2)^{3/2}}(x\mathbf{i} + y\mathbf{j} + z\mathbf{k}) \\[2mm] & \text{for } x^2 + y^2 + z^2 > a^2 \end{cases}$$

15.7
STOKES'S THEOREM

Stokes's Theorem is a three-dimensional version of Green's Theorem, involving three-dimensional surfaces and their boundaries rather than plane regions and their boundaries.

THEOREM 15.10
STOKES'S THEOREM

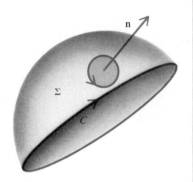

FIGURE 15.38

Let Σ be an oriented surface with normal $\mathbf{n}$ and finite surface area. Assume that Σ is bounded by a closed, piecewise smooth curve C whose orientation is induced by Σ (Figure 15.38). Let $\mathbf{F}$ be a continuous vector field defined on Σ, and assume that the component functions of $\mathbf{F}$ have continuous partial derivatives at each nonboundary point of Σ. Then

$$\int_C \mathbf{F} \cdot d\mathbf{r} = \iint_\Sigma (\operatorname{curl} \mathbf{F}) \cdot \mathbf{n} \, dS \tag{1}$$

If $\mathbf{F} = M\mathbf{i} + N\mathbf{j} + P\mathbf{k}$, then

$$\int_C M(x,y,z)\,dx + N(x,y,z)\,dy + P(x,y,z)\,dz = \iint_\Sigma (\operatorname{curl} \mathbf{F}) \cdot \mathbf{n} \, dS \tag{2}$$

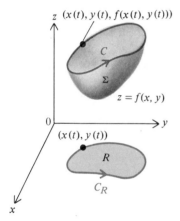

FIGURE 15.39

Theorem 15.10 is named after George Gabriel Stokes (1819–1903), professor at Cambridge University, who introduced it publicly by asking students to prove it in the Smith Prize Examination of 1854. However, Stokes indicated that it was William Thomson (1824–1907), usually known as Lord Kelvin, who first stated the theorem in a letter to Stokes in 1850.

That Stokes's Theorem is a genuine extension of Green's Theorem follows directly from the alternative form of Green's Theorem given in (5) of Section 15.4, because the normal vector $\mathbf{k}$ induces the counterclockwise orientation on the boundary of any surface in the xy plane.

The proof of Stokes's Theorem is too complicated for this book, but we will now give an idea of how to prove (2) in the special case that Σ is the graph of a function f with continuous partials on a simple region R in the xy plane whose boundary C_R is piecewise smooth, and for which the image of C_R is C (Figure 15.39). We will also assume that on Σ the normal $\mathbf{n}$ points upward.

Since

$$\operatorname{curl} \mathbf{F} = \left(\frac{\partial P}{\partial y} - \frac{\partial N}{\partial z} \right)\mathbf{i} + \left(\frac{\partial M}{\partial z} - \frac{\partial P}{\partial x} \right)\mathbf{j} + \left(\frac{\partial N}{\partial x} - \frac{\partial M}{\partial y} \right)\mathbf{k}$$

it follows from (5) of Section 15.6 that

$$\iint_\Sigma (\operatorname{curl} \mathbf{F}) \cdot \mathbf{n} \, dS = \iint_R \left[-\left(\frac{\partial P}{\partial y} - \frac{\partial N}{\partial z} \right)f_x - \left(\frac{\partial M}{\partial z} - \frac{\partial P}{\partial x} \right)f_y + \left(\frac{\partial N}{\partial x} - \frac{\partial M}{\partial y} \right) \right] dA$$

where all partial derivatives are to be evaluated at $(x, y, f(x, y))$. Thus (2) can be rewritten

$$\int_C M\,dx + N\,dy + P\,dz$$

$$= \iint_R \left[-\left(\frac{\partial P}{\partial y} - \frac{\partial N}{\partial z} \right)f_x - \left(\frac{\partial M}{\partial z} - \frac{\partial P}{\partial x} \right)f_y + \left(\frac{\partial N}{\partial x} - \frac{\partial M}{\partial y} \right) \right] dA \tag{3}$$

Now we will show that

$$\int_C M \, dx = -\iint_R \left(\frac{\partial M}{\partial y} + \frac{\partial M}{\partial z} f_y \right) dA \qquad (4)$$

If we let $M_R(x, y) = M(x, y, f(x, y))$ for (x, y) in R, then by the Chain Rule,

$$\frac{\partial M_R}{\partial y} = \frac{\partial M}{\partial y} + \frac{\partial M}{\partial z} f_y \qquad (5)$$

Next we let $\mathbf{r}(t) = x(t)\mathbf{i} + y(t)\mathbf{j}$ for $a \le t \le b$ be a parametrization of C_R. Then a parametrization of C is given by

$$\mathbf{r}_1(t) = x(t)\mathbf{i} + y(t)\mathbf{j} + f(x(t), y(t))\mathbf{k} \quad \text{for } a \le t \le b$$

Therefore

$$\int_C M(x, y, z) \, dx = \int_a^b M(x(t), y(t), f(x(t), y(t)))x'(t) \, dt$$

$$= \int_a^b M_R(x(t), y(t))x'(t) \, dt$$

$$= \int_{C_R} M_R(x, y) \, dx$$

so that more succinctly,

$$\int_C M \, dx = \int_{C_R} M_R \, dx \qquad (6)$$

Using (2) in the proof of Green's Theorem to relate integrals involving R and C_R, we find that

$$\int_C M \, dx \overset{(6)}{=} \int_{C_R} M_R \, dx \overset{\text{Green's Theorem}}{=} -\iint_R \frac{\partial M_R}{\partial y} \, dA \overset{(5)}{=} -\iint_R \left(\frac{\partial M}{\partial y} + \frac{\partial M}{\partial z} f_y \right) dA$$

which yields (4). Using the same ideas, one can show that

$$\int_C N \, dy = \iint_R \left(\frac{\partial N}{\partial x} + \frac{\partial N}{\partial z} f_x \right) dA \qquad (7)$$

and with a little more work that

$$\int_C P \, dz = \iint_R \left(\frac{\partial P}{\partial x} f_y - \frac{\partial P}{\partial y} f_x \right) dA \qquad (8)$$

The sum of the left sides of (4), (7), and (8) is the left side of (2), and similarly for the right sides. This completes the proof of Stokes's Theorem for the special case.

Example 1 Let C be the oriented triangle described in Figure 15.40, which lies in the plane $z = y/2$. If

$$\mathbf{F}(x, y, z) = -3y^2\mathbf{i} + 4z\mathbf{j} + 6x\mathbf{k}$$

calculate $\int_C \mathbf{F} \cdot d\mathbf{r}$.

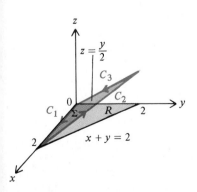

FIGURE 15.40

Solution Direct calculation of the line integral would require evaluation of three separate line integrals, one over each of the line segments $C_1, C_2,$ and C_3 composing C. However, if we apply (1) then we need only evaluate $\iint_\Sigma (\text{curl } \mathbf{F}) \cdot \mathbf{n}\, dS$, where Σ is the triangular surface having boundary C and is oriented by the normal directed upward. First we find that

$$\text{curl } \mathbf{F}(x, y, z) = \begin{vmatrix} \mathbf{i} & \mathbf{j} & \mathbf{k} \\ \dfrac{\partial}{\partial x} & \dfrac{\partial}{\partial y} & \dfrac{\partial}{\partial z} \\ -3y^2 & 4z & 6x \end{vmatrix} = -4\mathbf{i} - 6\mathbf{j} + 6y\mathbf{k}$$

If $f(x, y) = y/2$, then Σ is the graph of f on the triangular region R in the first quadrant bounded by the coordinate axes and the line $x + y = 2$. Now we use (5) of Section 15.6, with curl $\mathbf{F}$ replacing $\mathbf{F}$, to obtain

$$\iint_\Sigma (\text{curl } \mathbf{F}) \cdot \mathbf{n}\, dS = \int_0^2 \int_0^{2-x} \left[-(-4)0 - (-6)\frac{1}{2} + 6y \right] dy\, dx$$

$$= \int_0^2 \int_0^{2-x} (3 + 6y)\, dy\, dx$$

$$= 3\int_0^2 (y + y^2)\Big|_0^{2-x} dx$$

$$= 3\int_0^2 (6 - 5x + x^2)\, dx$$

$$= 3\left(6x - \frac{5}{2}x^2 + \frac{x^3}{3} \right)\Big|_0^2 = 14$$

Thus $\int_C \mathbf{F} \cdot d\mathbf{r} = 14$. ☐

Example 2 Let C be the intersection of the paraboloid $z = x^2 + y^2$ and the plane $z = y$, and give C its counterclockwise orientation as viewed from the positive z axis (Figure 15.41). Evaluate $\int_C xy\, dx + x^2\, dy + z^2\, dz$.

Solution Let

$$\mathbf{F}(x, y, z) = xy\mathbf{i} + x^2\mathbf{j} + z^2\mathbf{k}$$

and let Σ be the portion of the plane $z = y$ that lies inside the paraboloid. By (2) we can evaluate the given line integral by evaluating

$$\iint_\Sigma (\text{curl } \mathbf{F}) \cdot \mathbf{n}\, dS$$

Notice that if (x, y, z) is on C, then $x^2 + y^2 = y$, and this is an equation of the circular cylinder having equation $r = \sin\theta$ in cylindrical coordinates. Therefore if R is the region in the xy plane bounded by the circle $r = \sin\theta$, then Σ is the graph of $z = y$ on R. When we orient Σ by the normal directed upward, the induced

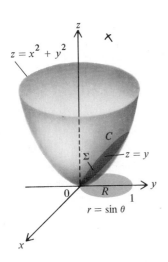

FIGURE 15.41

orientation on C is counterclockwise, as prescribed. Since

$$\operatorname{curl} \mathbf{F}(x, y, z) = \begin{vmatrix} \mathbf{i} & \mathbf{j} & \mathbf{k} \\ \dfrac{\partial}{\partial x} & \dfrac{\partial}{\partial y} & \dfrac{\partial}{\partial z} \\ xy & x^2 & z^2 \end{vmatrix} = x\mathbf{k}$$

and $f(x, y) = y$, we conclude from (2) above and (5) of Section 15.6 that

$$\int_C xy\, dx + x^2\, dy + z^2\, dz = \iint_\Sigma (\operatorname{curl} \mathbf{F}) \cdot \mathbf{n}\, dS = \iint_R [-0(0) - 0(1) + x]\, dA$$

$$= \iint_R x\, dA = \int_0^\pi \int_0^{\sin\theta} (r\cos\theta)r\, dr\, d\theta$$

$$= \int_0^\pi \frac{r^3}{3}\cos\theta \Big|_0^{\sin\theta} d\theta = \frac{1}{3}\int_0^\pi \sin^3\theta \cos\theta\, d\theta$$

$$= \frac{1}{12}\sin^4\theta \Big|_0^\pi = 0 \quad \square$$

Suppose two oriented surfaces Σ_1 and Σ_2 are bounded by the same curve C and induce the same orientation on C. If $\mathbf{n}_1$ and $\mathbf{n}_2$ denote the normals of Σ_1 and Σ_2, respectively, then by Stokes's Theorem we infer that

$$\iint_{\Sigma_1} (\operatorname{curl} \mathbf{F}) \cdot \mathbf{n}_1\, dS = \int_C \mathbf{F} \cdot d\mathbf{r} = \iint_{\Sigma_2} (\operatorname{curl} \mathbf{F}) \cdot \mathbf{n}_2\, dS \tag{9}$$

for any vector field $\mathbf{F}$ whose components have continuous partial derivatives on both Σ_1 and Σ_2. In cases where it is difficult to integrate over Σ_1, it may be easier to integrate over Σ_2.

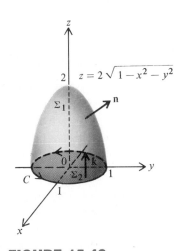

FIGURE 15.42

Example 3   Let Σ_1 be the semiellipsoid $z = 2\sqrt{1 - x^2 - y^2}$, oriented so that the normal $\mathbf{n}$ is directed upward (Figure 15.42), and let

$$\mathbf{F}(x, y, z) = x^2\mathbf{i} + y^2\mathbf{j} + z^2 \tan xy\mathbf{k}$$

Evaluate $\iint_{\Sigma_1} (\operatorname{curl} \mathbf{F}) \cdot \mathbf{n}\, dS$.

Solution   First we compute curl $\mathbf{F}$:

$$\operatorname{curl} \mathbf{F}(x, y, z) = \begin{vmatrix} \mathbf{i} & \mathbf{j} & \mathbf{k} \\ \dfrac{\partial}{\partial x} & \dfrac{\partial}{\partial y} & \dfrac{\partial}{\partial z} \\ x^2 & y^2 & z^2 \tan xy \end{vmatrix} = xz^2 \sec^2 xy\mathbf{i} - yz^2 \sec^2 xy\mathbf{j}$$

We could determine the normal $\mathbf{n}$ of Σ_1 and then evaluate $\iint_{\Sigma_1} (\operatorname{curl} \mathbf{F}) \cdot \mathbf{n} \, dS$. But notice from Figure 15.42 that the unit disk Σ_2 in the xy plane has the same boundary as Σ_1 has, and when Σ_2 is oriented with normal $\mathbf{n}$ directed upward, the induced orientations on the common boundary of Σ_1 and Σ_2 are identical. Thus by (9),

$$\iint_{\Sigma_1} (\operatorname{curl} \mathbf{F}) \cdot \mathbf{n} \, dS = \iint_{\Sigma_2} (\operatorname{curl} \mathbf{F}) \cdot \mathbf{n} \, dS$$

The integral over Σ_2 is easily evaluated, since the normal to Σ_2 is $\mathbf{k}$, which is perpendicular to curl $\mathbf{F}$. We obtain

$$\iint_{\Sigma_1} (\operatorname{curl} \mathbf{F}) \cdot \mathbf{n} \, dS = \iint_{\Sigma_2} (\operatorname{curl} \mathbf{F}) \cdot \mathbf{n} \, dS = \iint_{\Sigma_2} 0 \, dS = 0 \quad \square$$

Imagine that a fluid flows through a cylinder capped by the surface $z = 2\sqrt{1 - x^2 - y^2}$ and that the vector field curl $\mathbf{F}$ represents the velocity of the fluid. Then $\iint_{\Sigma_1} \delta(\operatorname{curl} \mathbf{F}) \cdot \mathbf{n} \, dS$ would represent the rate of mass flow through Σ_1. Stokes's Theorem tells us that the rate of mass flow through Σ_1 is the same as the rate of mass flow through the disk Σ_2 with the same boundary.

Now suppose two oriented surfaces Σ_1 and Σ_2 are bounded by the same curve C but induce opposite orientations on C. Then Stokes's Theorem implies that

$$\iint_{\Sigma_1} (\operatorname{curl} \mathbf{F}) \cdot \mathbf{n} \, dS = - \iint_{\Sigma_2} (\operatorname{curl} \mathbf{F}) \cdot \mathbf{n} \, dS \tag{10}$$

Example 4   Let Σ be the unit sphere $x^2 + y^2 + z^2 = 1$, oriented by the normal $\mathbf{n}$ that is directed outward. Let $\mathbf{F}$ be any vector field whose component functions have continuous partial derivatives on Σ. Show that $\iint_\Sigma (\operatorname{curl} \mathbf{F}) \cdot \mathbf{n} \, dS = 0$.

Solution   Let Σ_1 and Σ_2 be the upper and lower hemispheres of Σ (Figure 15.43). Then the normals that are directed outward from Σ_1 and Σ_2 induce opposite orientations on the circle $x^2 + y^2 = 1$ in the xy plane, which is the boundary of both Σ_1 and Σ_2. By (10) we conclude that

$$\iint_{\Sigma} (\operatorname{curl} \mathbf{F}) \cdot \mathbf{n} \, dS = \iint_{\Sigma_1} (\operatorname{curl} \mathbf{F}) \cdot \mathbf{n} \, dS + \iint_{\Sigma_2} (\operatorname{curl} \mathbf{F}) \cdot \mathbf{n} \, dS = 0 \quad \square$$

Stokes's Theorem can be extended to many surfaces that are not covered by our original version, in much the same way that we extended Green's Theorem. Let us show that Stokes's Theorem holds for the circular cylinder Σ in Figure 15.44(a), with normal directed outward. Stokes's Theorem does not apply directly, because the boundary of Σ is composed of two disjoint circles C_1 and C_2.

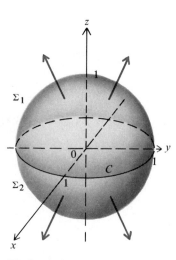

The hemispheres Σ_1 and Σ_2 induce opposite orientations on C

FIGURE 15.43

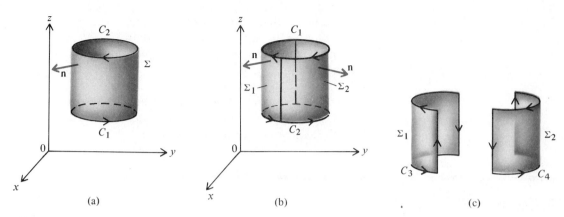

FIGURE 15.44

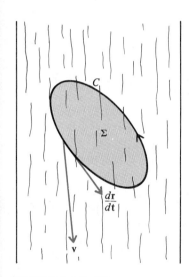

FIGURE 15.45

But suppose we "cut" the cylinder along two lines, as shown in Figure 15.44(b). Then we can apply Stokes's Theorem to each of the resulting half-cylinders Σ_1 and Σ_2 and the corresponding curves C_3 and C_4 which bound them (Figure 15.44(c)). Since the lines forming the cut each receive opposite orientations from the two half-cylinders, the line integrals over them cancel each other. Therefore we obtain

$$\iint_\Sigma (\text{curl}\,\mathbf{F}) \cdot \mathbf{n}\, dS = \iint_{\Sigma_1} (\text{curl}\,\mathbf{F}) \cdot \mathbf{n}_1\, dS + \iint_{\Sigma_2} (\text{curl}\,\mathbf{F}) \cdot \mathbf{n}_2\, dS$$

$$= \int_{C_3} \mathbf{F} \cdot d\mathbf{r} + \int_{C_4} \mathbf{F} \cdot d\mathbf{r} = \int_{C_1} \mathbf{F} \cdot d\mathbf{r} + \int_{C_2} \mathbf{F} \cdot d\mathbf{r}$$

Thus Stokes's Theorem applies if we add the line integrals over both parts C_1 and C_2 of the boundary, each with its induced orientation.

Stokes's Theorem enables us to interpret curl $\mathbf{v}$ when $\mathbf{v}$ represents the velocity of a fluid in motion. Let Σ be an oriented surface whose boundary C has an induced counterclockwise orientation, and let $\mathbf{v}$ denote the velocity of a fluid. If $\mathbf{r}$ parametrizes C on $[a, b]$, then the closer the directions of $\mathbf{v}$ and $d\mathbf{r}/dt$ are to one another, the larger $\mathbf{v} \cdot d\mathbf{r}/dt$ is (Figure 15.45). Since $d\mathbf{r}/dt$ is always parallel to the tangent of C and since

$$\int_C \mathbf{v} \cdot d\mathbf{r} = \int_a^b \mathbf{v} \cdot \frac{d\mathbf{r}}{dt}\, dt$$

it follows that the closer the directions of $\mathbf{v}$ and the tangent are, the larger $\int_C \mathbf{v} \cdot d\mathbf{r}$ is. Consequently $\int_C \mathbf{v} \cdot d\mathbf{r}$ measures the tendency of the fluid to move counterclockwise around C. As a result, $\int_C \mathbf{v} \cdot d\mathbf{r}$ is often called the **circulation of the fluid around C**.

If (x_0, y_0, z_0) is on Σ and if Σ is very small and has area S, then $\iint_\Sigma (\text{curl}\,\mathbf{v}) \cdot \mathbf{n}\, dS$ is very nearly $[\text{curl}\,\mathbf{v}(x_0, y_0, z_0)] \cdot \mathbf{n}(x_0, y_0, z_0)S$. Because Stokes's Theorem tells us that

$$\int_C \mathbf{v} \cdot d\mathbf{r} = \iint_\Sigma (\text{curl}\,\mathbf{v}) \cdot \mathbf{n}\, dS$$

we conclude that for small Σ, the circulation is approximately $[\text{curl } \mathbf{v}(x_0, y_0, z_0)] \cdot \mathbf{n}(x_0, y_0, z_0)S$. Thus $[\text{curl } \mathbf{v}(x_0, y_0, z_0)] \cdot \mathbf{n}(x_0, y_0, z_0)$ is approximately the rate of circulation per unit area in the counterclockwise direction at (x_0, y_0, z_0). The larger the circulation is per unit area, the larger the rotation, or curling, around (x_0, y_0, z_0). Moreover, the closer to one another the directions of $\mathbf{n}(x_0, y_0, z_0)$ and curl $\mathbf{v}(x_0, y_0, z_0)$ are, the larger $[\text{curl } \mathbf{v}(x_0, y_0, z_0)] \cdot \mathbf{n}(x_0, y_0, z_0)$ is. Consequently the curling effect is greatest about the axis parallel to curl $\mathbf{v}(x_0, y_0, z_0)$.

EXERCISES 15.7

In Exercises 1–5 use Stokes's Theorem to compute $\int_C \mathbf{F} \cdot d\mathbf{r}$, where C is the curve that bounds Σ and that has the induced orientation from Σ.

1. $\mathbf{F}(x, y, z) = z\mathbf{i} + x\mathbf{j} + y\mathbf{k}$; Σ is the part of the paraboloid $z = 1 - x^2 - y^2$ in the first octant; $\mathbf{n}$ is directed downward.

2. $\mathbf{F}(x, y, z) = z^2\mathbf{i} - y^2\mathbf{j}$; Σ is composed of the three squares shown with their normals in Figure 15.46.

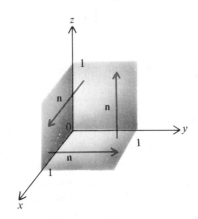

FIGURE 15.46

3. $\mathbf{F}(x, y, z) = 2y\mathbf{i} + 3z\mathbf{j} - 2x\mathbf{k}$; Σ is the part of the sphere $x^2 + y^2 + z^2 = 1$ in the first octant; $\mathbf{n}$ is directed upward.

4. $\mathbf{F}(x, y, z) = y^2\mathbf{i} + xy\mathbf{j} - 2xz\mathbf{k}$; Σ is the hemisphere $z = \sqrt{4 - x^2 - y^2}$; $\mathbf{n}$ is directed upward.

5. $\mathbf{F}(x, y, z) = y\mathbf{i} - x\mathbf{j} + z\mathbf{k}$; Σ is composed of the part of the cylinder $x^2 + y^2 = 1$ between the planes $z = 0$ and $z = 1$ and the part of the plane $z = 1$ inside the cylinder $x^2 + y^2 = 1$; $\mathbf{n}$ is directed away from the z axis on the cylinder and upward on the plane.

In Exercises 6–11 use Stokes's Theorem to evaluate $\int_C \mathbf{F} \cdot d\mathbf{r}$. In each case assume that C has its counterclockwise orientation as viewed from above.

6. $\mathbf{F}(x, y, z) = \dfrac{1}{\sqrt{x^2 + y^2 + z^2 + 1}}(x\mathbf{i} + y\mathbf{j} + z\mathbf{k})$
C is the intersection of the paraboloid $2z = x^2 + y^2$ and the cylinder $x^2 + y^2 = 2x$.

7. $\mathbf{F}(x, y, z) = xz\mathbf{i} + y^2\mathbf{j} + x^2\mathbf{k}$; C is the intersection of the plane $x + y + z = 5$ and the elliptic cylinder $x^2 + y^2/4 = 1$.

8. $\mathbf{F}(x, y, z) = 3y\mathbf{i} + 2z\mathbf{j} - x\mathbf{k}$; C is the triangle with vertices $(1, 0, 0)$, $(0, 1, 0)$, and $(0, 0, 1)$.

9. $\mathbf{F}(x, y, z) = y(x^2 + y^2)\mathbf{i} - x(x^2 + y^2)\mathbf{j}$; C is the rectangle with vertices $(0, 0, 0)$, $(1, 0, 0)$, $(1, 1, 1)$, and $(0, 1, 1)$.

10. $\mathbf{F}(x, y, z) = \ln(y^2 + 1)\mathbf{i} + x\mathbf{j} + (x + y)\mathbf{k}$; C is the rectangle in the xy plane with vertices $(0, 0)$, $(3, 0)$, $(3, 1)$, and $(0, 1)$.

11. $\mathbf{F}(x, y, z) = (z - y)\mathbf{i} + y\mathbf{j} + x\mathbf{k}$; C is the intersection of the circular cylinder $r = \cos\theta$ and the part of the sphere $x^2 + y^2 + z^2 = 1$ above the xy plane.

12. Let $\mathbf{F}(x, y, z) = yz\mathbf{k}$, and let C be the boundary of the part of the cone $z = \sqrt{x^2 + y^2}$ in the first octant between the planes $z = 2$ and $z = 3$. Give the counterclockwise orientation to the part of C in the plane $z = 2$ and the clockwise orientation to the part of C in the plane $z = 3$, as seen from above. Use Stokes's Theorem to evaluate $\int_C \mathbf{F} \cdot d\mathbf{r}$.

13. Let $\mathbf{F}(x, y, z) = (x^2 + z)\mathbf{i} + (y^2 + x)\mathbf{j} + (z^2 + y)\mathbf{k}$, and let C be the intersection of the sphere $x^2 + y^2 + z^2 = 1$ and the cone $z = \sqrt{x^2 + y^2}$. Use Stokes's Theorem and (9) to evaluate $\int_C \mathbf{F} \cdot d\mathbf{r}$, where C has its counterclockwise orientation as seen from above.

In Exercises 14–20 use Stokes's Theorem, and (9) where appropriate, to evaluate $\iint_\Sigma (\text{curl } \mathbf{F}) \cdot \mathbf{n}\, dS$.

14. $\mathbf{F}(x, y, z) = xz\mathbf{i} + (y^2 + 2x)\mathbf{j} + x\mathbf{k}$; Σ is the part of the paraboloid $z = 9 - x^2 - y^2$ above the xy plane; $\mathbf{n}$ is directed upward.

15. $\mathbf{F}(x, y, z) = x\mathbf{i} + (x^2 + y^2 + z^2)\mathbf{j} + z(y^4 - 1)\mathbf{k}$; Σ is composed of the four upper sides of the pyramid whose apex is $(0, 0, 6)$ and whose base is the square in the xy

plane with vertices $(-1, -1)$, $(1, -1)$, $(1, 1)$, and $(-1, 1)$; **n** is directed upward.

16. $\mathbf{F}(x, y, z) = z^2 y \mathbf{i} - x \mathbf{j} + z \sin x^2 y^2 z \mathbf{k}$; Σ is the part of the cylinder $x^2 + y^2 = 15$ between the planes $z = -1$ and $z = 2$; **n** is directed away from the z axis.

17. $\mathbf{F}(x, y, z) = x \sin z \mathbf{i} + xy \mathbf{j} + yz \mathbf{k}$; Σ is composed of all faces of the cube with vertices $(0, 0, 0)$, $(1, 0, 0)$, $(1, 1, 0)$, $(0, 1, 0)$, $(0, 0, 1)$, $(1, 0, 1)$, $(1, 1, 1)$, and $(0, 1, 1)$ except the face in the plane $z = 1$; **n** is directed outward from the cube.

18. $\mathbf{F}(x, y, z) = (3y^2 z + y)\mathbf{i} + y \mathbf{j} + 3xy^2 \mathbf{k}$; Σ is the upper half of the torus obtained by rotating the circle $r = \sin \theta$ about the x axis; **n** is directed outward from the torus.

19. $\mathbf{F}(x, y, z) = xz^2 \mathbf{i} + x^3 \mathbf{j} + \cos xz \mathbf{k}$; Σ is the part of the ellipsoid $x^2 + y^2 + 3z^2 = 1$ below the xy plane; **n** is directed outward from the ellipsoid.

20. $\mathbf{F}(x, y, z) = (z + 1)y^3 \mathbf{i} - (z + 1)xy \mathbf{j} + e^{x^2 y^2} \mathbf{k}$; Σ is the part of the surface $x^2 + y^2 + z^6 = 1$ above the xy plane; **n** is directed upward.

21. Suppose Σ is an oriented surface with oriented boundary C and $\mathbf{F}$ is a constant vector field defined on Σ. Show that $\int_C \mathbf{F} \cdot d\mathbf{r} = 0$.

22. Suppose Σ is an ellipsoid with normal directed outward and $\mathbf{F}$ a vector field whose component functions have continuous partial derivatives on Σ. Show that $\iint_\Sigma (\text{curl } \mathbf{F}) \cdot \mathbf{n} \, dS = 0$.

23. Let Σ be the portion of the sphere $x^2 + y^2 + z^2 = 1$ above the plane $z = y$, and orient Σ by the normal directed outward from the sphere. Assume that the velocity **v** of a fluid passing through Σ is given by

$$\mathbf{v}(x, y, z) = x^3 \mathbf{i} - zy \mathbf{j} + x \mathbf{k}$$

Find the circulation of the fluid around the boundary of Σ.

24. Let Σ consist of the portion of the paraboloid

$$2z = x^2 + y^2$$

below the plane $z = x$, and orient Σ by the normal directed downward. Suppose a fluid is passing through Σ with a velocity **v** given by

$$\mathbf{v}(x, y, z) = (\sin xz + 2yz)\mathbf{i}$$
$$+ (\cosh y + 2xz)\mathbf{j} + (e^{x^2 z^2} + 5y)\mathbf{k}$$

Find the circulation of the fluid around the boundary of Σ.

15.8
THE DIVERGENCE THEOREM

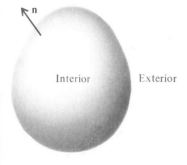

Interior   Exterior

Simple solid region

FIGURE 15.47

The Divergence Theorem is our second and final higher-dimensional analogue of Green's Theorem. To simplify the statement of the Divergence Theorem, we will call a solid region D a **simple solid region** if D is the solid region between the graphs of two functions F_1 and F_2 on a simple region R in the xy plane and if D has the corresponding properties with respect to the xz plane and the yz plane. Regions bounded by spheres, hemispheres, ellipsoids, cubes, and tetrahedrons are simple solid regions. A simple solid region has an interior and an exterior, separated by a boundary surface (Figure 15.47). We assume from now on that any simple solid region D we consider has an orientable boundary surface, and we will orient it by the normal directed outward from D (Figure 15.47). This brings us to the Divergence Theorem, also called Gauss's Theorem.

THEOREM 15.11
THE DIVERGENCE THEOREM

Let D be a simple solid region whose boundary surface Σ is oriented by the normal **n** directed outward from D, and let $\mathbf{F}$ be a vector field whose component functions have continuous partial derivatives on D. Then

$$\iint_\Sigma \mathbf{F} \cdot \mathbf{n} \, dS = \iiint_D \text{div } \mathbf{F}(x, y, z) \, dV$$

Proof Let $\mathbf{F} = M\mathbf{i} + N\mathbf{j} + P\mathbf{k}$. Then the formula $\iint_\Sigma \mathbf{F} \cdot \mathbf{n}\, dS = \iiint_D \operatorname{div} \mathbf{F}(x, y, z)\, dV$ is equivalent to

$$\iint_\Sigma M\mathbf{i} \cdot \mathbf{n}\, dS + \iint_\Sigma N\mathbf{j} \cdot \mathbf{n}\, dS + \iint_\Sigma P\mathbf{k} \cdot \mathbf{n}\, dS$$

$$= \iiint_D \frac{\partial M}{\partial x}\, dV + \iiint_D \frac{\partial N}{\partial y}\, dV + \iiint_D \frac{\partial P}{\partial z}\, dV$$

Therefore it suffices to show that

$$\iint_\Sigma M\mathbf{i} \cdot \mathbf{n}\, dS = \iiint_D \frac{\partial M}{\partial x}\, dV, \qquad \iint_\Sigma N\mathbf{j} \cdot \mathbf{n}\, dS = \iiint_D \frac{\partial N}{\partial y}\, dV$$

$$\iint_\Sigma P\mathbf{k} \cdot \mathbf{n}\, dS = \iiint_D \frac{\partial P}{\partial z}\, dV$$

Since all three formulas are proved in analogous ways, we prove only the third:

$$\iint_\Sigma P\mathbf{k} \cdot \mathbf{n}\, dS = \iiint_D \frac{\partial P}{\partial z}\, dV \tag{1}$$

By hypothesis D is a simple solid region; hence there exist a simple region R in the xy plane and continuous functions F_1 and F_2 on R such that D consists of all points (x, y, z) satisfying

$$(x, y) \text{ is in } R \quad \text{and} \quad F_1(x, y) \le z \le F_2(x, y)$$

For the triple integral in (1) we obtain

$$\iiint_D \frac{\partial P}{\partial z}\, dV = \iint_R \left[\int_{F_1(x, y)}^{F_2(x, y)} \frac{\partial P}{\partial z}(x, y, z)\, dz \right] dA$$

$$= \iint_R \left[P(x, y, F_2(x, y)) - P(x, y, F_1(x, y)) \right] dA \tag{2}$$

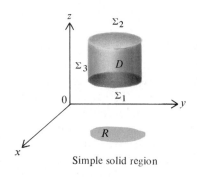

Simple solid region

FIGURE 15.48

For the surface integral in (1) we first observe that Σ is composed of three subsurfaces shown in Figure 15.48:

the bottom Σ_1, consisting of all points $(x, y, F_1(x, y))$ for (x, y) in R
the top Σ_2, consisting of all points $(x, y, F_2(x, y))$ for (x, y) in R
the sides Σ_3, consisting of all points (x, y, z) for which (x, y) is on the boundary of R and

$$F_1(x, y) \le z \le F_2(x, y)$$

(In some cases, such as when Σ is a sphere, Σ_3 will not appear.) Now on Σ_1 the normal $\mathbf{n}$ is directed downward. Hence by (6) of Section 15.6,

$$\iint_{\Sigma_1} P\mathbf{k} \cdot \mathbf{n}\, dS = -\iint_R P(x, y, F_1(x, y))\, dA$$

Next, on Σ_2 the normal $\mathbf{n}$ is directed upward. Hence by (5) of Section 15.6,

$$\iint\limits_{\Sigma_2} P\mathbf{k} \cdot \mathbf{n}\, dS = \iint\limits_{R} P(x, y, F_2(x, y))\, dA$$

Finally, on Σ_3 the normal $\mathbf{n}$ is horizontal and hence is perpendicular to $\mathbf{k}$. Consequently

$$\iint\limits_{\Sigma_3} P\mathbf{k} \cdot \mathbf{n}\, dS = \iint\limits_{\Sigma_3} 0\, dS = 0$$

Putting our information about the three surface integrals together, we conclude that

$$\iint\limits_{\Sigma} P\mathbf{k} \cdot \mathbf{n}\, dS = -\iint\limits_{R} P(x, y, F_1(x, y))\, dA + \iint\limits_{R} P(x, y, F_2(x, y))\, dA + 0$$

$$= \iint\limits_{R} [P(x, y, F_2(x, y)) - P(x, y, F_1(x, y))]\, dA$$

This is the same expression we found in (2) for

$$\iiint\limits_{D} \frac{\partial P}{\partial z}(x, y, z)\, dV \quad \blacksquare$$

Again we have a theorem that relates an integral over a region to another integral over its boundary; in that sense it is quite similar to the Fundamental Theorem of Calculus. Moreover, observe that the Divergence Theorem is a higher-dimensional analogue of the second alternative form of Green's Theorem, given in (6) of Section 15.4.

Example 1 Suppose Σ is the sphere $x^2 + y^2 + z^2 = 4$, and let $\mathbf{F}(x, y, z) = 3x\mathbf{i} + 4y\mathbf{j} + 5z\mathbf{k}$. Evaluate $\iint_{\Sigma} \mathbf{F} \cdot \mathbf{n}\, dS$.

Solution Evaluation of $\iint_{\Sigma} \mathbf{F} \cdot \mathbf{n}\, dS$ itself would involve calculating two surface integrals, one each over the upper and the lower hemispheres of Σ. However, if we use the Divergence Theorem, we need only evaluate $\iiint_{D} \operatorname{div} \mathbf{F}(x, y, z)\, dV$, where D is the ball $x^2 + y^2 + z^2 \leq 4$. Since

$$\operatorname{div} \mathbf{F}(x, y, z) = 3 + 4 + 5 = 12$$

it follows that

$$\iint\limits_{\Sigma} \mathbf{F} \cdot \mathbf{n}\, dS = \iiint\limits_{D} \operatorname{div} \mathbf{F}(x, y, z)\, dV = \iiint\limits_{D} 12\, dV = 12\left(\frac{4}{3}\pi 2^3\right) = 128\pi \quad \square$$

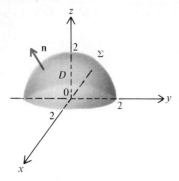

FIGURE 15.49

Example 2   Let D be the region bounded by the xy plane and the hemisphere shown in Figure 15.49, and let $\mathbf{F}(x, y, z) = x^3\mathbf{i} + y^3\mathbf{j} + z^3\mathbf{k}$. Evaluate $\iint_\Sigma \mathbf{F} \cdot \mathbf{n}\, dS$, where Σ is the boundary of D.

Solution   Since a direct calculation of the surface integral would be complicated, we turn to the Divergence Theorem. In order to use it, we calculate that

$$\operatorname{div} \mathbf{F}(x, y, z) = 3x^2 + 3y^2 + 3z^2 = 3(x^2 + y^2 + z^2)$$

and conclude that

$$\iint_\Sigma \mathbf{F} \cdot \mathbf{n}\, dS = \iiint_D \operatorname{div} \mathbf{F}(x, y, z)\, dV = \iiint_D 3(x^2 + y^2 + z^2)\, dV$$

$$= \int_0^{2\pi} \int_0^{\pi/2} \int_0^2 (3\rho^2)\rho^2 \sin\phi\, d\rho\, d\phi\, d\theta$$

$$= 3 \int_0^{2\pi} \int_0^{\pi/2} \int_0^2 \rho^4 \sin\phi\, d\rho\, d\phi\, d\theta$$

$$= 3 \int_0^{2\pi} \int_0^{\pi/2} \left.\frac{\rho^5}{5}\right|_0^2 \sin\phi\, d\phi\, d\theta = \frac{96}{5} \int_0^{2\pi} \int_0^{\pi/2} \sin\phi\, d\phi\, d\theta$$

$$= \frac{96}{5} \int_0^{2\pi} \left.-\cos\phi\right|_0^{\pi/2} d\theta = \frac{96}{5} \int_0^{2\pi} 1\, d\theta = \frac{192}{5}\pi \quad \square$$

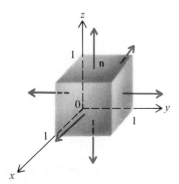

FIGURE 15.50

Example 3   Let Σ be the cube shown in Figure 15.50, and let $\mathbf{F}(x, y, z) = x\mathbf{i} + y\mathbf{j} + z^2\mathbf{k}$. Find $\iint_\Sigma \mathbf{F} \cdot \mathbf{n}\, dS$.

Solution   Direct evaluation of the surface integral would involve six integrals, one over each of the six faces of the cube. However, if we apply the Divergence Theorem, we need evaluate only one triple integral. To calculate the triple integral, we first find that

$$\operatorname{div} \mathbf{F}(x, y, z) = 1 + 1 + 2z = 2(1 + z)$$

Letting D denote the solid region whose boundary is Σ and applying the Divergence Theorem, we compute that

$$\iint_\Sigma \mathbf{F} \cdot \mathbf{n}\, dS = \iiint_D 2(1 + z)\, dV = \int_0^1 \int_0^1 \int_0^1 2(1 + z)\, dz\, dy\, dx$$

$$= \int_0^1 \int_0^1 \left.(1 + z)^2\right|_0^1 dy\, dx = \int_0^1 \int_0^1 3\, dy\, dx = 3 \quad \square$$

By means of the Divergence Theorem it is possible to reinterpret the divergence of a vector field in terms of fluid motion. Suppose (x_0, y_0, z_0) is a fixed point in space and D is a ball centered at (x_0, y_0, z_0) with boundary Σ and volume V. Assume that the vector field $\mathbf{v}$ represents the velocity of a fluid with density 1. If

D is very small, then $\iiint_D \text{div } \mathbf{v}(x, y, z) \, dV$ is very nearly $[\text{div } \mathbf{v}(x_0, y_0, z_0)]V$. Therefore the Divergence Theorem implies that

$$\iint_\Sigma \mathbf{v} \cdot \mathbf{n} \, dS = \iiint_D \text{div } \mathbf{v}(x, y, z) \, dV \approx [\text{div } \mathbf{v}(x_0, y_0, z_0)] V$$

so that

$$\text{div } \mathbf{v}(x_0, y_0, z_0) \approx \frac{1}{V} \iint_\Sigma \mathbf{v} \cdot \mathbf{n} \, dS \tag{3}$$

Now recall that $\iint_\Sigma \mathbf{v} \cdot \mathbf{n} \, dS$ equals the rate (with respect to time) of mass flow from D outward through Σ. Thus (3) suggests that div $\mathbf{v}(x_0, y_0, z_0)$ should be the rate of mass flow per unit volume at (x_0, y_0, z_0) in the direction of the outward normal. Hence the name "divergence" of $\mathbf{v}$ at a point.

Derivation of Gauss's Law The British physicist James Clerk Maxwell (1831–1879) formulated four equations, now known as Maxwell's equations, which supposedly explain all classical electromagnetic phenomena. One of these equations is

$$\text{div } \mathbf{E} = \frac{1}{\varepsilon_0} \rho \tag{4}$$

which relates the electric field $\mathbf{F}$ produced by electric charges in a solid region to the charge density ρ (charge per unit volume) in that region. Assuming that ρ is continuous in a given solid region, and using the Divergence Theorem, we will show how to use (4) to derive Gauss's Law, which was given in (8) of Section 15.6.

Let D be a simple solid region, Σ its boundary oriented by the normal $\mathbf{n}$ that is directed outward, and q the total charge in D. Then by the Divergence Theorem, (4) above, and (8) of Section 14.4,

$$\iint_\Sigma \mathbf{E} \cdot \mathbf{n} \, dS = \iiint_D \text{div } \mathbf{E}(x, y, z) \, dV = \iiint_D \frac{1}{\varepsilon_0} \rho(x, y, z) \, dV$$

$$= \frac{1}{\varepsilon_0} \iiint_D \rho(x, y, z) \, dV = \frac{1}{\varepsilon_0} q$$

It is also possible to reverse the procedure and derive (4) from Gauss's Law; hence Gauss's Law and (4) are equivalent.

EXERCISES 15.8

In Exercises 1–8 determine whether the given region is a simple solid region.

1. The solid region inside the cylinder $x^2 + y^2 = 1$ and between the planes $z = 0$ and $z = 1$

2. The solid region inside the ellipsoid $x^2 + y^2 + 2z^2 = 1$

3. The solid region bounded by the planes $x = 0$, $y = 0$, $z = 0$, and $x + y + z = 1$

4. The solid region bounded by the paraboloids $z = 1 - x^2 - y^2$ and $z = x^2 + y^2 - 1$

5. The solid region inside the double cone $z^2 = x^2 + y^2$ and between the planes $z = -1$ and $z = 1$

6. The solid region bounded by the parabolic sheet $y = z^2$ and the planes $x = 0$, $x = 1$, and $y = 1$

7. The solid region bounded by the surface $y = \sin x$ and the planes $x = -2\pi$, $x = 2\pi$, $y = 1$, $z = 0$, and $z = 3$

8. The solid region bounded by the surface $z = y^3 - y$ and the planes $z = 2$, $x = 0$, $x = 1$, $y = -2$, and $y = 1$

In Exercises 9–23 use the Divergence Theorem to compute $\iint_\Sigma \mathbf{F} \cdot \mathbf{n} \, dS$, where $\mathbf{n}$ is the normal to Σ that is directed outward.

9. $\mathbf{F}(x, y, z) = x^2\mathbf{i} + xy\mathbf{j} - 2xz\mathbf{k}$; Σ is the tetrahedron with vertices $(0, 0, 0)$, $(1, 0, 0)$, $(0, 1, 0)$, and $(0, 0, 1)$.

10. $\mathbf{F}(x, y, z) = x^2\mathbf{i} + y^2\mathbf{j} + z^2\mathbf{k}$; Σ is the parallelepiped with vertices $(0, 0, 0)$, $(1, 0, 0)$, $(1, 2, 0)$, $(0, 2, 0)$, $(0, 0, 3)$, $(1, 0, 3)$, $(1, 2, 3)$, and $(0, 2, 3)$.

11. $\mathbf{F}(x, y, z) = x\mathbf{i} + y\mathbf{j} + z\mathbf{k}$; Σ is the boundary of the solid region in the first octant that is inside the cylinder $x^2 + y^2 = 1$ and between the planes $z = 0$ and $z = 1$.

12. $\mathbf{F}(x, y, z) = 2x\mathbf{i} + xy\mathbf{j} + xz\mathbf{k}$; Σ is the sphere $x^2 + y^2 + z^2 = 1$.

13. $\mathbf{F}(x, y, z) = x\mathbf{i} + y\mathbf{j} + z\mathbf{k}$; Σ is composed of the hemisphere $z = \sqrt{1 - x^2 - y^2}$ and the disk in the xy plane bounded by the circle $x^2 + y^2 = 1$.

14. $\mathbf{F}(x, y, z) = (x - \cos x)\mathbf{i} + (y - y\sin x)\mathbf{j} + 2z\mathbf{k}$; Σ is the tetrahedron described in Exercise 9.

15. $\mathbf{F}(x, y, z) = x^2\mathbf{i} + y^2\mathbf{j} + z^2\mathbf{k}$; Σ is the boundary of the solid region inside the cylinder $x^2 + y^2 = 4$ and between the planes $z = 0$ and $z = 2$.

16. $\mathbf{F}(x, y, z) = 3x\mathbf{i} - 2y\mathbf{j} + z\mathbf{k}$; Σ is the sphere $x^2 + y^2 + z^2 = 4$.

17. $\mathbf{F}(x, y, z) = y(x^2 + y^2)^{3/2}\mathbf{i} - x(x^2 + y^2)^{3/2}\mathbf{j} + (z + 1)\mathbf{k}$; Σ is the boundary of the solid region bounded above by the plane $z = 2x$ and below by the paraboloid $z = x^2 + y^2$.

18. $\mathbf{F}(x, y, z) = yz\mathbf{i} + xy\mathbf{j} + xz\mathbf{k}$; Σ is the boundary of the solid region inside the cylinder $x^2 + z^2 = 1$ and between the planes $y = -1$ and $y = 1$.

19. $\mathbf{F}(x, y, z) = -2x\mathbf{i} + 4y\mathbf{j} - 7z\mathbf{k}$; Σ is the boundary of the solid region inside the sphere $x^2 + y^2 + z^2 = 4$ and outside the cylinder $x^2 + y^2 = 1$.

20. $\mathbf{F}(x, y, z) = y\mathbf{i} + x\mathbf{j} + 8\mathbf{k}$; Σ is the boundary of the solid region bounded above by the paraboloid $z = 1 - x^2 - y^2$ and below by the xy plane.

21. $\mathbf{F}(x, y, z) = x^2\mathbf{i} + y\mathbf{j} - 2z^2\mathbf{k}$; Σ is the boundary of the solid region bounded below by the xy plane, above by the plane $z = x$, and on the sides by the parabolic sheet $y^2 = 2 - x$.

22. $\mathbf{F}(x, y, z) = x^2y\mathbf{i} + yz\mathbf{j} + z^2\mathbf{k}$; Σ is the boundary of the solid region bounded by the planes $z = 1$ and $x + y = 1$ and the coordinate planes.

23. $\mathbf{F}(x, y, z) = (x^2 + y^2 + z^2)(x\mathbf{i} + y\mathbf{j})$; Σ is the sphere $x^2 + y^2 + z^2 = 9$.

24. Let

$$\mathbf{F}(x, y, z) = x^3y^2z^2\mathbf{i} - x^4yz^2\mathbf{j}$$

Evaluate $\iiint_D \operatorname{div} \mathbf{F}(x, y, z) \, dV$ by using the Divergence Theorem, where D is the ball $x^2 + y^2 + z^2 \le 1$.

25. Let $\mathbf{F}$ and $\mathbf{G}$ be vector fields defined on a simple solid region D with boundary Σ. Assume that $\mathbf{F} = \operatorname{curl} \mathbf{G}$ on D and that the components of $\mathbf{F}$ have continuous partial derivatives on D. Show that $\iint_\Sigma \mathbf{F} \cdot \mathbf{n} \, dS = 0$.

26. Let $\mathbf{F}(x, y, z) = x\mathbf{i} + y\mathbf{j} + z\mathbf{k}$, and let D be a simple solid region with boundary Σ and normal $\mathbf{n}$ directed outward. Show that the volume V of D is given by the formula

$$V = \frac{1}{3} \iint_\Sigma \mathbf{F} \cdot \mathbf{n} \, dS$$

27. Use Exercise 26 to find the volume of a circular cone with height h and radius a.

28. Let f and g be functions of three variables whose partial derivatives are continuous on a simple solid region D having boundary Σ.
    a. Prove that

$$\iint_\Sigma (g \operatorname{grad} f) \cdot \mathbf{n} \, dS$$

$$= \iiint_D [g\nabla^2 f + (\operatorname{grad} g) \cdot (\operatorname{grad} f)] \, dV$$

(*Hint:* Use the identity

$$\operatorname{div}(g\mathbf{F}) = g \operatorname{div} \mathbf{F} + (\operatorname{grad} g) \cdot \mathbf{F}.)$$

b. Using part (a), prove that

$$\iint_\Sigma (g \operatorname{grad} f - f \operatorname{grad} g) \cdot \mathbf{n} \, dS$$

$$= \iiint_D (g\nabla^2 f - f\nabla^2 g) \, dV$$

(The formulas in (a) and (b) are those Green actually presented in his book on electricity and magnetism. See also Exercise 25 of Section 15.4.)

29. Suppose **F** is a constant vector field defined on a simple solid region D having boundary Σ. Prove that $\iint_\Sigma \mathbf{F} \cdot \mathbf{n}\, dS = 0$.

30. The magnetic field of an object like the earth is a vector field and is often denoted **B**. One of Maxwell's equations implies that $\mathbf{B} = \operatorname{curl} \mathbf{F}$ for a suitable vector field **F**. Let D be a simple solid region having boundary Σ. Find the fallacy in the following steps, which "prove" that $\mathbf{B} = \mathbf{0}$ on D (that is, that magnetic fields do not exist).
   1. $\operatorname{div} \mathbf{B} = 0$
   2. $\iint_\Sigma \mathbf{B} \cdot \mathbf{n}\, dS = \iiint_D \operatorname{div} \mathbf{B}\, dV = 0$ by the Divergence Theorem

3. Let C be any closed, oriented curve on Σ. Then by Stokes's Theorem,

$$\int_C \mathbf{F} \cdot d\mathbf{r} = \iint_\Sigma (\operatorname{curl} \mathbf{F}) \cdot \mathbf{n}\, dS = \iint_\Sigma \mathbf{B} \cdot \mathbf{n}\, dS = 0$$

4. By step 3 of this exercise and by the equivalence of statements (1) and (3) of Section 15.3, $\mathbf{F} = \operatorname{grad} f$ for a suitable function f of several variables.

5. $\mathbf{B} = \operatorname{curl} \mathbf{F} = \operatorname{curl}(\operatorname{grad} f) = \mathbf{0}$.

(Adapted from George Arfken, "Magnetic Fields Are Not Real," *American Journal of Physics* 27 (1959), p. 526.)

Key Terms and Expressions

Vector field; conservative vector field
Gradient
Divergence
Curl
Potential function
Line integral

Independence of path
Surface integral
Flux integral
Orientation of a surface; oriented surface; induced orientation
Simple solid region

Key Formulas

$$\int_C f(x,y,z)\, ds = \int_a^b f(x(t),y(t),z(t))\left\|\frac{d\mathbf{r}}{dt}\right\| dt$$

$$\int_C \mathbf{F} \cdot d\mathbf{r} = \int_a^b \mathbf{F}(x(t),y(t),z(t)) \cdot \frac{d\mathbf{r}}{dt}\, dt$$

$$\int_C M\, dx + N\, dy + P\, dz = \int_a^b \left[M(x(t),y(t),z(t))\frac{dx}{dt} \right.$$
$$\left. + N(x(t),y(t),z(t))\frac{dy}{dt} + P(x(t),y(t),z(t))\frac{dz}{dt} \right] dt$$

$$\int_C \operatorname{grad} f \cdot d\mathbf{r} = f(x_1,y_1,z_1) - f(x_0,y_0,z_0)$$

$$\int_C M(x,y)\, dx + N(x,y)\, dy = \iint_R \left(\frac{\partial N}{\partial x} - \frac{\partial M}{\partial y} \right) dA$$

$$\iint_\Sigma g(x,y,z)\, dS =$$

$$\iint_R g(x,y,f(x,y))\sqrt{[f_x(x,y)]^2 + [f_y(x,y)]^2 + 1}\, dA$$

$$\iint_\Sigma \mathbf{F} \cdot \mathbf{n}\, dS = \pm \iint_R [M(x,y,f(x,y))f_x(x,y)$$
$$+ N(x,y,f(x,y))f_y(x,y) - P(x,y,f(x,y))]\, dA$$

$$\int_C \mathbf{F} \cdot d\mathbf{r} = \iint_\Sigma (\operatorname{curl} \mathbf{F}) \cdot \mathbf{n}\, dS$$

$$\iint_\Sigma \mathbf{F} \cdot \mathbf{n}\, dS = \iiint_D \operatorname{div} \mathbf{F}(x,y,z)\, dV$$

Key Theorems

Fundamental Theorem of Line Integrals
Green's Theorem

Stokes's Theorem
Divergence Theorem

REVIEW EXERCISES

1. Let **a** and **b** be constant vectors, and let $\mathbf{r}(x, y, z) = x\mathbf{i} + y\mathbf{j} + z\mathbf{k}$. Show that
$$\mathbf{a} \times \mathbf{b} = \text{grad}[\mathbf{a} \cdot (\mathbf{b} \times \mathbf{r})]$$

2. Let f be a function of several variables whose component functions have continuous second derivatives. Show that
$$\text{curl}(f \text{ grad } f) = \mathbf{0}$$
(*Hint:* Use Exercises 28 and 31 of Section 15.1.)

In Exercises 3–4 find a function f having the given gradient.

3. $\text{grad } f(x, y) = (y^2 - y \sin xy)\mathbf{i} + (2xy - x \sin xy)\mathbf{j}$

4. $\text{grad } f(x, y, z) = ye^z\mathbf{i} + (xe^z + e^y)\mathbf{j} + (xy + 1)e^z\mathbf{k}$

In Exercises 5–26 evaluate the integral.

5. $\int_C (xy + z^2) \, ds$, where C is parametrized by $\mathbf{r}(t) = \cos t\mathbf{i} + \sin t\mathbf{j} + t\mathbf{k}$ for $\pi/4 \le t \le 3\pi/4$

6. $\int_C (2xy - 3yz) \, ds$, where C is composed of the line segments from $(1, 0, 4)$ to $(0, 3, 2)$ and from $(0, 3, 2)$ to $(0, 0, 0)$

7. $\int_C xy \, dx + z \cos x \, dy + z \, dz$, where C is parametrized by $\mathbf{r}(t) = t\mathbf{i} + \cos t\mathbf{j} + \sin t\mathbf{k}$ for $0 \le t \le \pi/2$

8. $\int_C y \sin 2x \, dx + \sin^2 x \, dy$, where C is parametrized by $\mathbf{r}(t) = t\mathbf{i} + \sin t\mathbf{j}$ for $0 \le t \le \pi/3$

9. $\int_C x \, ds$, where C is the graph of $y = x^2$ for $0 \le x \le 1$

10. $\int_C (1 + y \sin z) \, dx + (1 + x \sin z) \, dy + xy \cos z \, dz$, where C is the curve parametrized by $\mathbf{r}(t) = \tan^5 t\mathbf{i} + \cos^4 t\mathbf{j} + t\mathbf{k}$ for $0 \le t \le \pi/4$

11. $\int_C e^x \cos z \, dx + y \, dy - e^x \sin z \, dz$, where C is parametrized by $\mathbf{r}(t) = e^{t^2(t^2-1)}\mathbf{i} + t\mathbf{j} + te^{t^{17}-1}\mathbf{k}$ for $0 \le t \le 1$

12. $\iint_\Sigma \mathbf{F} \cdot \mathbf{n} \, dS$, where
$$\mathbf{F}(x, y, z) = x\sqrt{x^2 + y^2 + z^2}\mathbf{i} + y\sqrt{x^2 + y^2 + z^2}\mathbf{j} + z\sqrt{x^2 + y^2 + z^2}\mathbf{k}$$
and where Σ is the torus $\rho = \sin \phi$ oriented by the normal directed outward

13. $\iint_\Sigma \mathbf{F} \cdot \mathbf{n} \, dS$, where $\mathbf{F}(x, y, z) = x\mathbf{i} + xy\mathbf{j} + (1 + z)\mathbf{k}$ and where Σ is the boundary of the solid region in the first octant bounded by the coordinate planes, the plane $z = 1 + x$, and the parabolic sheet $x = 1 - y^2$ (Σ oriented by the normal directed outward)

14. $\int_C (e^x - 4y \sin^2 x) \, dx + (2x + \sin 2x) \, dy$, where C is the square with vertices $(2, 2)$, $(2, -2)$, $(-2, 2)$, and $(-2, -2)$, oriented counterclockwise

15. $\iint_\Sigma \mathbf{F} \cdot \mathbf{n} \, dS$, where $\mathbf{F}(x, y, z) = xyz\mathbf{i} + 2yz\mathbf{j} + x^3y\mathbf{k}$, and where Σ is the boundary of the part of the cylindrical region $x^2 + y^2 \le 4$ between the planes $z = -2$ and $z = 3$ (Σ oriented by the normal directed outward)

16. $\int_C x(x^2 + y^2)^{1/2} \, dx + y(x^2 + y^2)^{1/2} \, dy$, where C is the graph of the equation $(x - 2)^8 + (y - 2)^8 = 1$, oriented counterclockwise

17. $\iint_\Sigma \mathbf{F} \cdot \mathbf{n} \, dS$, where $\mathbf{F}(x, y, z) = y\mathbf{i} - x\mathbf{j} + z\mathbf{k}$ and where Σ is the portion of the paraboloid $z = -1 + x^2 + y^2$ below the plane $z = 1$ (Σ oriented by the normal directed downward)

18. $\int_C \mathbf{F} \cdot d\mathbf{r}$, where $\mathbf{F}(x, y, z) = xyz\mathbf{i} + 2x^2z\mathbf{j} + y^6\mathbf{k}$ and where C is the boundary of the rectangle in the plane $z = y$ for which $-1 \le x \le 1$ and $0 \le y \le 2$, oriented counterclockwise as viewed from above

19. $\iint_\Sigma (\text{curl } \mathbf{F}) \cdot \mathbf{n} \, dS$, where $\mathbf{F}(x, y, z) = x^3y\mathbf{i} - y^3\mathbf{j} + z \sec xyz\mathbf{k}$ and where Σ is the portion of the sphere $x^2 + y^2 + z^2 = 2$ above the plane $z = 1$ (Σ oriented by the normal directed downward)

20. $\int_C \sin x \sin y \, dx - \cos x \cos y \, dy$, where C is the circle $x^2 + y^2 = 4$, oriented clockwise

21. $\iint_\Sigma \mathbf{F} \cdot \mathbf{n} \, dS$, where
$$\mathbf{F}(x, y, z) = \tan x\mathbf{i} - y(1 + \tan^2 x)\mathbf{j} - 6z\mathbf{k}$$
and where Σ is composed of the hemisphere $z = \sqrt{4 - x^2 - y^2}$ and the disk $x^2 + y^2 \le 4$ in the xy plane (Σ oriented by the normal directed outward)

22. $\int_C (y^4 + x^3y^2) \, dx + x^2y^3 \, dy$, where C is the boundary of the region between the graphs of $y = x^2$ and $y = 1$, oriented counterclockwise

23. $\int_C y \, dx + y \, dy + x^2 \, dz$, where C is the intersection of the surfaces $z = x^2 + y^2$ and $z = 1 - y^2$, oriented counterclockwise as viewed from above

24. $\iint_\Sigma \mathbf{F} \cdot \mathbf{n} \, dS$, where $\mathbf{F}(x, y, z) = x^2\mathbf{i} + 2yz\mathbf{j} - z^2\mathbf{k}$ and where Σ is the boundary of the solid region in the first octant bounded by the coordinate planes, by the surface $z = y^2$, and by the cylinder $r = \sin \theta$ (Σ oriented by the normal directed outward)

25. $\int_C \mathbf{F} \cdot d\mathbf{r}$, where $\mathbf{F}(x, y, z) = x^4\mathbf{i} - y^2\mathbf{j} + z\mathbf{k}$ and where C is parametrized by $\mathbf{r}(t) = t\mathbf{i} - t^3\mathbf{k}$ for $-1 \le t \le 1$

26. $\iint_\Sigma \mathbf{F} \cdot \mathbf{n} \, dS$, where $\mathbf{F}(x, y, z) = yz\mathbf{k}$ and where Σ is the boundary of the solid region in the first octant bounded by the cone $z^2 = x^2 + y^2$ and the planes $x = 0$, $y = 0$, $z = 2$, and $z = 3$ (Σ oriented by the normal directed outward)

27. Let a force **F** be given by
$$\mathbf{F}(x, y, z) = y\mathbf{i} + x\mathbf{j} + z^3\mathbf{k}$$
Find the work W done by **F** on an object that moves from $(1, 0, 0)$ to $(0, 1, \pi)$
a. along a straight line.
b. along a helix parametrized by $\mathbf{r}(t) = \cos t\mathbf{i} + \sin t\mathbf{j} + 2t\mathbf{k}$ for $0 \le t \le \pi/2$.

28. Find the area A of the region enclosed by the curve parametrized by $\mathbf{r}(t) = \cos t\mathbf{i} + \sin 2t\mathbf{j}$ for $-\pi/2 \le t \le \pi/2$.

Cumulative Review, Chapters 1–14

In Exercises 1–2 find the limit.

1. $\lim\limits_{x\to-\infty} \dfrac{1}{x^2+2} \sqrt{\dfrac{x^5-1}{2x+1}}$ 2. $\lim\limits_{x\to\pi/2} \dfrac{\ln(\sin x)}{(\pi-2x)^2}$

3. Let $f(x) = \sin(x^3)$. Find the smallest positive integer n such that $f^{(n)}(0) \neq 0$.

4. Determine whether there is a nonzero value of c for which the parabola $y = cx^2$ and the circle $x^2 + y^2 = 4$ intersect at right angles.

5. A carpenter drills downward through the center of the top of a 2×4 inch plank at the rate of $\frac{1}{12}$ inch per second. If the drill bit is conical with a radius of $\frac{1}{2}$ inch and a height of 1 inch, determine the rate at which the volume of the drilled hole increases when the point of the drill is $\frac{2}{3}$ inch into the wood.

6. A right circular cone is inscribed in a sphere of radius r. Find the height h of the cone with the largest possible volume.

7. Let

$$f(x) = \dfrac{x}{x^2 + 2x + 2}$$

Sketch the graph of f, indicating all pertinent information.

8. Determine the area A of the region bounded by the curves $y = 2x^2$ and $16x = y^4$.

In Exercises 9–10 evaluate the integral.

9. $\displaystyle\int x^2 \sec^2 x \tan x \, dx$ 10. $\displaystyle\int \dfrac{3x+4}{x^3+4x} \, dx$

11. Determine whether

$$\int_0^2 \dfrac{x^2}{\sqrt{4-x^2}} \, dx$$

converges or diverges. If it converges, find its value.

12. Let R be the region bounded by the curve $y = x^2 + 1$ and the line $y = 2x + 1$. Find the volume V of the solid generated by revolving R about the
    a. x axis   b. y axis

13. Let $(x^2 + y^2)^3 = 3x^2 y^2$.
    a. Convert the equation to polar coordinates, and then sketch its graph.
    b. Determine the maximum distance between a point on the graph of the equation and the origin.

14. Find the sum of the series $\displaystyle\sum_{n=2}^{\infty} 3(.3)^{2n}$

15. Determine whether

$$\sum_{n=1}^{\infty} (-1)^n \left(1 + \dfrac{1}{n}\right)^{-n^2}$$

converges absolutely, converges conditionally, or diverges.

16. Find the interval of convergence of $\displaystyle\sum_{n=0}^{\infty} (\cosh n)x^n$.

17. Show that

$$\dfrac{5}{6} \le \int_0^1 e^{-t^2/2} \, dt \le \dfrac{103}{120}$$

(*Hint:* Use the Taylor series expansion for $e^{-t^2/2}$.)

18. Consider the point $P = (0, 1, 2)$ and the plane $4x - y + z = 12$.
    a. Find symmetric equations for the line l that passes through P and is perpendicular to the plane.
    b. Find the distance D between the plane and P.

19. Let $\mathbf{r}(t) = t\mathbf{i} + \frac{4}{3}t^{3/2}\mathbf{j} + t^2\mathbf{k}$ represent the motion of a particle traversing a curve C.
    a. Find the positive value of b that makes the length $\mathcal{L}$ of the curve C from $t = 0$ to $t = b$ exactly 30 units.
    b. Show that the tangential component of acceleration is constant, and find that constant.
    c. Find the normal component of acceleration, and show that it is a decreasing function on $(0, \infty)$.

20. Let $f(x, y) = \sqrt{2x^2 + y^2}$.
    a. Find $f_{yy}(x, y)$.
    b. Find the unit vector $\mathbf{u}$ such that at the point $(1, \sqrt{2})$, f decreases most rapidly in the direction of $\mathbf{u}$.

21. Consider the two surfaces $x^2 - 2y^2 + z^2 = 0$ and $xyz = 1$. Show that at each point of their intersection their tangent planes are perpendicular to one another. (*Hint:* Do not find any points of intersection.)

22. Assuming that the minimum distance exists, find the minimum distance from the origin $(0, 0, 0)$ to the surface $xy^2z^4 = 32$.

23. Evaluate $\displaystyle\int_1^8 \int_{y^{1/3}}^2 \cos\left(\dfrac{x^4}{4} - x\right) dx \, dy$.

24. Determine the surface area S of that portion of the surface $z = x^2 - y^2$ inside the cylinder $x^2 + y^2 = 4$.

25. Compute $\displaystyle\iiint_D z \, dV$, where D is the solid region bounded below by the paraboloid $2z = x^2 + y^2$ and above by the sphere $x^2 + y^2 + z^2 = 8$.

26. Evaluate $\displaystyle\iint_R (y - x)\sin(y + x)^3 \, dA$, where R is the triangular region with vertices $(0, 0)$, $(2, 2)$ and $(0, 4)$.

16

DIFFERENTIAL EQUATIONS

In this optional chapter we discuss differential equations, which are equations involving an (unknown) function and some of its derivatives. The primary objective is to find all functions that satisfy such an equation. Actually, we have already encountered differential equations. For example, the equation

$$f'(x) = kf(x)$$

appearing in Section 7.5 on exponential growth and decay is a differential equation.

Through most of the chapter, standard techniques are developed for finding functions satisfying various types of differential equations. The applications include electric circuits and the motion of springs. In the final section we analyze graphs of functions satisfying certain differential equations without actually finding formulas for the functions.

16.1
BASIC NOTIONS IN DIFFERENTIAL EQUATIONS

Equations that involve functions of x and y and various derivatives of y are called **differential equations**. Some examples of differential equations are

$$\frac{dy}{dx} = -32x \tag{1}$$

$$\frac{dy}{dx} + 5y = 12e^{7x} \tag{2}$$

$$\frac{d^2y}{dx^2} + 8\frac{dy}{dx} + 16y = 0 \tag{3}$$

$$\frac{d^2y}{dx^2} + \sin y = 0 \tag{4}$$

$$\left(\frac{d^3y}{dx^3}\right)^3 + \frac{d^2y}{dx^2}\left(\frac{dy}{dx}\right)^4 - \sqrt{2}y = \ln(x + e^{xy}) \tag{5}$$

Differential equations can describe many phenomena, such as the motion of a falling body, as in (1), change in the size of a population, as in (2), flow of current in an electric circuit, as in (3), or the motion of a pendulum, as in (4).

The **order** of a differential equation is the order of the highest-order derivative of y appearing in the equation. Equations (1) and (2) are first-order differential equations; (3) and (4) are of second order; and (5) is of third order. In Chapter 16 we will study special kinds of first- and second-order differential equations, with emphasis on those that are related to the topics and concepts encountered in Chapters 1–15.

If y is a function that satisfies a given differential equation, we say that y is a **solution** of the differential equation.

Example 1   Show that if $y = -16x^2$, then y is a solution of

$$\frac{dy}{dx} = -32x$$

Solution   Differentiation of y yields immediately that $dy/dx = -32x$, so y is a solution of the given differential equation. $\square$

Example 2   Let $y = e^{7x}$. Show that y is a solution of

$$\frac{dy}{dx} + 5y = 12e^{7x}$$

Solution   Notice that $dy/dx = 7e^{7x}$, so that

$$\frac{dy}{dx} + 5y = 7e^{7x} + 5e^{7x} = 12e^{7x}$$

Thus y satisfies the given equation. $\square$

Example 3   Let $y = 3e^{-4x} - xe^{-4x}$. Show that y is a solution of

$$\frac{d^2y}{dx^2} + 8\frac{dy}{dx} + 16y = 0$$

Solution First we observe that

$$\frac{dy}{dx} = -12e^{-4x} - e^{-4x} + 4xe^{-4x} = -13e^{-4x} + 4xe^{-4x}$$

$$\frac{d^2y}{dx^2} = 52e^{-4x} + 4e^{-4x} - 16xe^{-4x} = 56e^{-4x} - 16xe^{-4x}$$

Therefore

$$\frac{d^2y}{dx^2} + 8\frac{dy}{dx} + 16y = (56e^{-4x} - 16xe^{-4x}) + 8(-13e^{-4x} + 4xe^{-4x})$$

$$+ 16(3e^{-4x} - xe^{-4x})$$

$$= 0$$

Consequently y satisfies the given equation. ☐

Not all solutions of differential equations can be expressed so simply. For example, the equation

$$x^2 + xy^3 + \frac{1}{2}e^{-2y} = 0 \tag{6}$$

gives a solution of the differential equation

$$2x + y^3 + (3xy^2 - e^{-2y})\frac{dy}{dx} = 0 \tag{7}$$

in the sense that there is a function y defined implicitly by (6), and we can obtain the differential equation (7) by differentiating both sides of equation (6) implicitly. The solution is given by $F(x, y) = 0$, where $F(x, y) = x^2 + xy^3 + \frac{1}{2}e^{-2y}$, and we say that $F(x, y) = 0$ is an implicit solution of (7). More generally, if an equation of the form $F(x, y) = C$ (where C is a constant) defines at least one function y implicitly, and if y satisfies a given differential equation by means of implicit differentiation, then $F(x, y) = C$ is called an **implicit solution** of the differential equation. In contrast, a solution given in the form $y = f(x)$ is called an **explicit solution**; the solutions derived in Examples 1–3 are explicit solutions. Finally, there may be no simple formula for the solutions of a given differential equation. This is the case with equation (4).

It follows from Example 1 that if C is an arbitrary constant and

$$y = -16x^2 + C \tag{8}$$

then y is a solution of (1). By Theorem 4.7, every solution of (1) has the form of equation (8). For that reason, we say that equation (8) is the general solution of (1). Similarly, if C, C_1, and C_2 are arbitrary constants, then it can be shown that the functions defined by

$$y = e^{7x} + Ce^{-5x} \tag{9}$$

and
$$y = C_1e^{-4x} + C_2xe^{-4x} \tag{10}$$

are general solutions of (2) and (3), in the sense that any solution of those differential equations can be derived from (9) and (10), respectively, by choosing the constants C, C_1, and C_2 appropriately. We will call a given solution, whether explicit or implicit, a **general solution** if each solution can be obtained from the given solution by choosing the constants suitably.

A solution in which all the constants (like C, C_1, and C_2 in (9) and (10)) are replaced by specific numbers is a **particular solution**. Thus the solutions we found in Examples 1–3 are particular solutions. When we wish to obtain a particular solution from the general solution, we often require that the solution satisfy **initial** (or **boundary**) **conditions**, which specify the value of the solution or certain of its derivatives at specific numbers in the domain. Thus if we wish the solution of (2) such that $y(1) = 3$, then the condition $y(1) = 3$ is called an initial condition. Similarly, the solution of (3) for which $y(0) = 2$ and $y'(0) = -4$ has the initial conditions $y(0) = 2$ and $y'(0) = -4$. The problem of finding a particular solution of a differential equation that satisfies one or more initial conditions is called an **initial value problem**. All of the initial value problems in this book have unique solutions. Thus when we ask for the particular solution satisfying a given initial condition, there will be only one such solution.

Example 4   Let $y = e^{\sin x}$. Show that y is the particular solution of

$$\frac{dy}{dx} = y \cos x$$

such that $y(\pi/2) = e$.

Solution   Differentiating y, we find that

$$\frac{dy}{dx} = e^{\sin x}(\cos x) = y \cos x$$

Therefore y satisfies the differential equation. Since

$$y\left(\frac{\pi}{2}\right) = e^{\sin \pi/2} = e^1 = e$$

it follows that y also satisfies the initial condition $y(\pi/2) = e$.   □

EXERCISES 16.1

In Exercises 1–12 verify that the function y satisfies the given differential equation.

1. $\dfrac{dy}{dx} = e^{3x};\ y = \dfrac{1}{3}e^{3x}$

2. $\dfrac{dy}{dx} = 2x + 3y;\ y = 5e^{3x} - \dfrac{2}{3}x - \dfrac{2}{9}$

3. $2\dfrac{dy}{dx} - y^2 = 1;\ y = \tan x + \sec x$ for $0 < x < \pi/2$

4. $x\left(\dfrac{dy}{dx}\right)^2 - y\dfrac{dy}{dx} + 1 = 0;\ y = 3 + \dfrac{x}{3}$

5. $\dfrac{d^2y}{dx^2} = y;\ y = \sinh x$

6. $\dfrac{d^2y}{dx^2} + 4y = 0;\ y = \sin 2x - \cos 2x$

7. $\dfrac{d^2y}{dx^2} + y = 2e^{-x};\ y = e^{-x} + \sin x$

8. $\dfrac{d^2y}{dx^2} + 4\dfrac{dy}{dx} + 4y = 0;\ y = xe^{-2x}$

9. $\dfrac{d^2y}{dx^2} - 2a\dfrac{dy}{dx} + (a^2 + b^2)y = 0;\ y = e^{ax}\sin bx$

10. $x^2\dfrac{d^2y}{dx^2} + x\dfrac{dy}{dx} + y = 0;\ y = \sin(\ln x) + \cos(\ln x)$

11. $\dfrac{d^3y}{dt^3} + 64y = 0;\ y = e^{-4t}$

12. $\dfrac{d^4y}{dt^4} - 81y = 0;\ y = \cos 3t$

In Exercises 13–17 verify that y is the particular solution that satisfies the initial conditions.

13. $\dfrac{dy}{dx} + 5y = -4e^{-3x};\ y(0) = -2;\ y = -2e^{-3x}$

14. $\dfrac{d^2y}{dx^2}\ 2\dfrac{dy}{dx} + 2y = 0;\ y(0) = 0,\ y'(0) = 1;\ y = e^x\sin x$

15. $x^3\dfrac{d^2y}{dx^2} + x^2\dfrac{dy}{dx} - xy = x;\ y(1) = 0,\ y'(1) = -1;$

$y = \dfrac{1}{x} - 1$

16. $\left(\dfrac{dy}{dx}\right)^2 = 4(y + 1);\ y(0) = y(2) = 0;\ y = x^2 - 2x$

17. $x\dfrac{dy}{dx} - y = x^2\sqrt{1 + x^4};\ y(0) = y'(0) = 0;$

$y = x\displaystyle\int_0^x \sqrt{1 + t^4}\,dt$

18. Show that if $y = \sqrt{r^2 - x^2}$, the graph of which is a semicircle, then y satisfies the differential equation

$$y\dfrac{dy}{dx} + x = 0$$

19. Show that if $y^2 = 8x + 16$, then y satisfies the differential equation

$$y = 2x\dfrac{dy}{dx} + y\left(\dfrac{dy}{dx}\right)^2 \quad \text{for } y \neq 0$$

20. Show that if $y = c\cosh x/c$, the graph of which is a catenary, then y satisfies the differential equation

$$\dfrac{d^2y}{dx^2} = \dfrac{1}{c}\sqrt{1 + \left(\dfrac{dy}{dx}\right)^2}$$

21. Suppose that the rate at which the population of a country increases because of births is proportional to the population and that there is a constant rate c of immigration. Then the population N satisfies the differential equation

$$\dfrac{dN}{dt} = kN + c$$

where c and k are positive constants. Show that

$$N = N_0 e^{kt} + \dfrac{c}{k}(e^{kt} - 1)$$

is a solution of the differential equation, where N_0 is a positive constant representing the initial population.

16.2
SEPARABLE DIFFERENTIAL EQUATIONS

The easiest kind of differential equation to solve is a **separable differential equation**, which is one that can be written

$$P(x) + Q(y)\dfrac{dy}{dx} = 0 \tag{1}$$

or in the more symmetric differential notation,

$$P(x)\,dx + Q(y)\,dy = 0, \quad \text{or} \quad Q(y)\,dy = -P(x)\,dx \tag{2}$$

(Notice that the x's and y's in (2) are separated from one another, whence the name "separable differential equation.") Integrating both sides of (1) with respect to x, we see that

$$\int P(x)\,dx + \int Q(y(x))\frac{dy}{dx}\,dx = C$$

or

$$\int P(x)\,dx + \int Q(y)\,dy = C \qquad (3)$$

This is also equivalent to

$$\int Q(y)\,dy = -\int P(x)\,dx \qquad (4)$$

Equation (3) or (4) gives the general (implicit) solution of (1).

Example 1   Find the general solution of

$$\frac{x^3}{y^2} + \frac{dy}{dx} = 0$$

Solution   First we rewrite the equation as

$$x^3 + y^2\frac{dy}{dx} = 0$$

which is a separable differential equation. By (3), the general solution is

$$\int x^3\,dx + \int y^2\,dy = C$$

Performing the integrations, we obtain the implicit solution

$$\frac{x^4}{4} + \frac{y^3}{3} = C$$

which can be made explicit by solving for y:

$$y = \left(3C - \frac{3}{4}x^4\right)^{1/3} \qquad \square$$

Example 2   Find the particular solution of

$$\frac{dy}{dx} = \frac{2x}{3y^2 + 1}$$

for which $y(0) = 3$.

Solution   The differential equation is equivalent to

$$(3y^2 + 1)\,dy = 2x\,dx$$

which is separable. By (4) the general (implicit) solution is

$$\int (3y^2 + 1)\, dy = \int 2x\, dx$$

which by integration becomes

$$y^3 + y = x^2 + C \qquad (5)$$

For the particular solution we substitute $x = 0$ and $y = 3$ into (5) to obtain

$$3^3 + 3 = 0^2 + C$$

Thus $C = 30$, so that the required particular solution is

$$y^3 + y = x^2 + 30 \quad \square$$

A special type of separable differential equation occurs if $Q = -1$. In that case (1) becomes

$$P(x) - \frac{dy}{dx} = 0$$

or

$$\frac{dy}{dx} = P(x)$$

This is an **integrable differential equation**, and as we know from the definition of indefinite integrals, the general solution is

$$y = \int P(x)\, dx$$

For example,

$$\sec^2 x - \frac{dy}{dx} = 0$$

has the general solution

$$y = \int \sec^2 x\, dx = \tan x + C$$

Now we will solve a differential equation of a different variety.

Example 3   Let a, b, and r be constants, with $b < a$. Show that the general solution of

$$\frac{dy}{dx} = r(y - a)(y - b)$$

has the form

$$y = \frac{a + bCe^{(a-b)rx}}{1 + Ce^{(a-b)rx}}$$

Solution   The differential equation can be written

$$\frac{1}{(y-a)(y-b)} dy = r\,dx$$

and by (4) the general solution is

$$\int \frac{1}{(y-a)(y-b)} dy = \int r\,dx = rx + C_1 \tag{6}$$

where C_1 is a constant. Since the partial fraction decomposition of $1/[(y-a)(y-b)]$ is given by

$$\frac{1}{(y-a)(y-b)} = \frac{1}{a-b}\left(\frac{1}{y-a} - \frac{1}{y-b}\right)$$

it follows that

$$\int \frac{1}{(y-a)(y-b)} dy = \frac{1}{a-b}\int\left(\frac{1}{y-a} - \frac{1}{y-b}\right) dy$$

$$= \frac{1}{a-b}(\ln|y-a| - \ln|y-b|) + C_2$$

$$= \frac{1}{a-b}\ln\left|\frac{y-a}{y-b}\right| + C_2$$

where C_2 is a constant. Therefore the general solution given in (6) can be transformed into

$$\frac{1}{a-b}\ln\left|\frac{y-a}{y-b}\right| = rx + C_3$$

or

$$\ln\left|\frac{y-a}{y-b}\right| = (a-b)rx + (a-b)C_3$$

or

$$\left|\frac{y-a}{y-b}\right| = e^{(a-b)rx} e^{(a-b)C_3}$$

where C_3 is a constant. Observe that $(y-a)/(y-b) > 0$ for $y < b$ or $y > a$, whereas $(y-a)/(y-b) < 0$ for $b < y < a$. Thus if we seek a solution all of whose values lie in $(-\infty, b)$ or in (a, ∞), we let $C = -e^{(a-b)C_3}$, and if we seek a solution all of whose values lie in (b, a), we let $C = e^{(a-b)C_3}$. Then we solve for y and conclude that

$$y = \frac{a + bCe^{(a-b)rx}}{1 + Ce^{(a-b)rx}} \quad \square$$

One special case of the separable differential equation in Example 3 is particularly noteworthy: If $r = -k$ and $b = 0$, then the equation becomes

$$\frac{dy}{dx} = -ky(y-a)$$

The solution can be written

$$y = \frac{a}{1 + Ce^{-kax}}$$

which represents the function that appeared in Exercise 66 of Section 6.3, whose graph is the logistic curve.

EXERCISES 16.2

In Exercises 1–10 find the general solution of the differential equation.

1. $\dfrac{dy}{dx} = \dfrac{x}{y}$

2. $\dfrac{dy}{dx} = \dfrac{y}{x}$ for $x > 0$

3. $\dfrac{2y}{y^2 + 1}\dfrac{dy}{dx} = \dfrac{1}{x^2}$;

4. $\dfrac{dy}{dx} = \dfrac{\sin x - \cos x}{y^4 + y}$

5. $(y^2 - 3)\dfrac{dy}{dt} = 1$

6. $t^3\dfrac{dy}{dt} = \sqrt{t^2 - y^2 t^2}$ for $t > 0$

7. $(1 + x^2)\,dy = (1 + y^2)\,dx$

8. $dy = (y + y^2)\,dx$

9. $\dfrac{1 + e^x}{1 - e^{-y}}\,dy + e^{x+y}\,dx = 0$

10. $e^{x+y^2}\,dy = \dfrac{x}{y}\,dx$

In Exercises 11–16 find the particular solution satisfying the initial condition.

11. $y^2 x\dfrac{dy}{dx} - x + 1 = 0$; $y(1) = 3$

12. $(\ln y)^2\dfrac{dy}{dx} = x^2 y$; $y(2) = 1$

13. $\sqrt{x^2 + 1}\dfrac{dy}{dx} = \dfrac{x}{y}$; $y(\sqrt{3}) = 2$

14. $\dfrac{dy}{dx} = x \sin x$; $y(\pi/2) = 0$

15. $e^{-2y}\,dy = (x - 2)\,dx$; $y(0) = 0$

16. $\dfrac{1}{9}\,dy = \dfrac{e^x}{ey^2 + y^2 e^2}\,dx$; $y(1) = 3$

17. A differential equation $dy/dx = f(x, y)$ is **homogeneous** if there is a function g of one variable such that $f(x, y) = g(y/x)$.
    a. Let $v = y/x$. Show that
    $$\frac{dy}{dx} = x\frac{dv}{dx} + v \tag{7}$$
    b. Use (7) to show that a homogeneous differential equation $dy/dx = f(x, y)$ can be rewritten in the form
    $$\frac{1}{g(v) - v}\,dv = \frac{1}{x}\,dx \tag{8}$$
    which is a separable differential equation.

In Exercises 18–19 show that the differential equation is homogeneous. Then find the general solution by using (8).

18. $\dfrac{dy}{dx} = \dfrac{y^2 + 2xy}{x^2}$

19. $\dfrac{dy}{dx} = \dfrac{x + y}{x}$

16.3
EXACT DIFFERENTIAL EQUATIONS

Now we turn to differential equations of the form

$$M(x, y) + N(x, y)\frac{dy}{dx} = 0 \tag{1}$$

Such an equation is an **exact differential equation** if there is a function f of two variables with continuous partial derivatives such that

$$\frac{\partial f}{\partial x} = M \quad \text{and} \quad \frac{\partial f}{\partial y} = N \tag{2}$$

Frequently equation (1) is given in the equivalent, symmetric form

$$M(x, y)\, dx + N(x, y)\, dy = 0$$

Every separable differential equation

$$P(x) + Q(y)\frac{dy}{dx} = 0$$

is exact, because if

$$f(x, y) = \int P(x)\, dx + \int Q(y)\, dy$$

then $\partial f/\partial x = P$ and $\partial f/\partial y = Q$.

Suppose f is a function satisfying (2), and assume that y is a differentiable function of x satisfying (1). Then

$$\frac{df}{dx} = \frac{\partial f}{\partial x} + \frac{\partial f}{\partial y}\frac{dy}{dx} = M(x, y) + N(x, y)\frac{dy}{dx} = 0$$

so that integration yields $f(x, y) = C$ as an implicit solution of (1) for any constant C. It can be proved that the general (implicit) solution of (1) is given by $f(x, y) = C$.

If M and N and their first partial derivatives are continuous in a rectangular region in the plane (or in any plane region without "holes"), then it turns out that equation (1) is exact if and only if $\partial N/\partial x = \partial M/\partial y$.*

Example 1   Show that the differential equation

$$2x^3y^2 + x^4y\frac{dy}{dx} = 0 \tag{3}$$

is exact, and find the general solution.

Solution   Since

$$\frac{\partial}{\partial x}(x^4y) = 4x^3y = \frac{\partial}{\partial y}(2x^3y^2)$$

the differential equation in (3) is exact, which implies that there is a function f such that

$$\frac{\partial f}{\partial x} = 2x^3y^2 \quad \text{and} \quad \frac{\partial f}{\partial y} = x^4y \tag{4}$$

---

* This constitutes part of the two-dimensional version of Theorem 15.4. Moreover, finding a function f whose partial derivatives are M and N is equivalent to finding a (potential) function whose gradient is $M\mathbf{i} + N\mathbf{j}$, and a method for determining f is described in Section 15.1.

Integrating both sides of the first equation in (4) with respect to x, we obtain

$$f(x, y) = \frac{1}{2}x^4 y^2 + g(y) \tag{5}$$

where g is constant with respect to x. Therefore when we differentiate both sides of (5) with respect to y, we find that

$$\frac{\partial f}{\partial y} = x^4 y + \frac{dg}{dy} \tag{6}$$

Comparing (6) with the second equation in (4), we observe that

$$x^4 y = x^4 y + \frac{dg}{dy}$$

so that

$$\frac{dg}{dy} = 0$$

Therefore

$$g(y) = C_1$$

for some constant C_1. Thus by (5),

$$f(x, y) = \frac{1}{2}x^4 y^2 + C_1$$

so the general solution of the differential equation in (3) is given by

$$\frac{1}{2}x^4 y^2 = C \quad \square$$

Example 2 Show that the differential equation

$$(2x + y^3)\,dx + (3xy^2 - e^{-2y})\,dy = 0 \tag{7}$$

is exact, and find the particular solution such that $y(-1) = 0$.

Solution Since

$$\frac{\partial}{\partial x}(3xy^2 - e^{-2y}) = 3y^2 = \frac{\partial}{\partial y}(2x + y^3)$$

the given differential equation is exact, so there exists a function f such that

$$\frac{\partial f}{\partial x} = 2x + y^3 \quad \text{and} \quad \frac{\partial f}{\partial y} = 3xy^2 - e^{-2y} \tag{8}$$

Integrating both sides of the first equation of (8) with respect to x yields

$$f(x, y) = x^2 + xy^3 + g(y)$$

where g is constant with respect to x. Therefore

$$\frac{\partial f}{\partial y} = 3xy^2 + \frac{dg}{dy} \tag{9}$$

Comparing (9) with the second equation of (8), we find that

$$\frac{dg}{dy} = -e^{-2y}$$

By integration we obtain

$$g(y) = \frac{1}{2}e^{-2y} + C_1$$

Thus $$f(x, y) = x^2 + xy^3 + \frac{1}{2}e^{-2y} + C_1$$

and consequently the general solution of (7) is

$$x^2 + xy^3 + \frac{1}{2}e^{-2y} = C$$

For the solution y such that $y(-1) = 0$ we have

$$(-1)^2 + (-1)(0)^3 + \frac{1}{2}e^{-2(0)} = C$$

or more simply,

$$1 + 0 + \frac{1}{2} = C$$

Therefore $C = \frac{3}{2}$, so the particular solution is given by

$$x^2 + xy^3 + \frac{1}{2}e^{-2y} = \frac{3}{2} \quad \square$$

Notice that it is not feasible to solve

$$x^2 + xy^3 + \frac{1}{2}e^{-2y} = C$$

for y; we have found implicit, but not explicit, solutions of (7).

EXERCISES 16.3

In Exercises 1–14 determine whether the differential equation is exact. If it is, find the general solution.

1. $2x - y^3 - 3xy^2\dfrac{dy}{dx} = 0$

2. $\sin y - (y - x\cos y)\dfrac{dy}{dx} = 0$

3. $(2xy^3 - ye^{-x})\,dx + (3x^2y^2 + e^{-x} - 4)\,dy = 0$

4. $(\cos x^2 y)\left(2xy + x^2\dfrac{dy}{dx}\right) = 0$

5. $\left(\dfrac{1}{x} + \cos 2y\right)\dfrac{dy}{dx} + 2x - \dfrac{y}{x^2} = 0$

6. $(2x - 5y)\dfrac{dy}{dx} = 6x - 2y$

7. $\pi y + (\pi x + \arcsin y)\dfrac{dy}{dx} = \sin x$

8. $(3x^2 - 3y)\,dx - (3y^2 + 3x)\,dy = 0$

9. $\dfrac{\ln y}{x}\,dx + \left(\dfrac{\ln x}{y} + \sin y\right)dy = 0$

10. $(y^2 e^{xy} - \pi)\,dx + (1 + xy)e^{xy}\,dy = 0$

11. $\sec^2 x \sec y\,dx + (1 + \tan x \sec y \tan y)\,dy = 0$ for $-\pi/2 < x < \pi/2$

12. $y^2 \cos xy\,dx + (\sin xy + xy \cos xy)\,dy = 0$

13. $(x^3 y - e^x \cos y)\,dx + (e^x \sin y + \frac{1}{4}x^4 + \sec^2 y)\,dy = 0$

14. $(2xye^{x^2 y} + \sin y)\,dx + (x^2 e^{x^2 y} + x \cos y - y)\,dy = 0$

In Exercises 15–18 find the particular solution satisfying the initial condition.

15. $4y^2 + 8xy\dfrac{dy}{dx} = 0;\ y(3) = \dfrac{\sqrt{2}}{2}$

16. $(e^y + ye^x)\,dx + (e^x + xe^y)\,dy = 0;\ y(1) = 0$

17. $\ln(1 + y^2)\,dx + \dfrac{2xy}{1 + y^2}\,dy = 0;\ y(2) = \sqrt{e - 1}$

18. $y \cosh xy\,dx + (x \cosh xy - y)\,dy = 0;\ y(0) = \sqrt{5}$

In Exercises 19–20 find the positive value of a that makes the differential equation exact. Using that value of a, find the general solution of the differential equation.

19. $axe^{x^2} \sin 3y + \dfrac{1}{4}e^{x^2} \cos 3y\dfrac{dy}{dx} = 0$

20. $ax^5 y^{a+1}\,dx + 2x^6 y^a\,dy = 0$

21. Find the general solution of the differential equation

$$\frac{y^2}{x^2 + y^2} + \frac{2xy}{x^2 + y^2}\frac{dy}{dx} = 0$$

which is not exact but can be made exact by multiplying each side by a suitable expression in x and y.

16.4
LINEAR FIRST-ORDER DIFFERENTIAL EQUATIONS

A **linear first-order differential equation** has the form

$$\frac{dy}{dx} + P(x)y = Q(x) \tag{1}$$

where P and Q are continuous. The equation

$$f'(x) = kf(x)$$

which in the notation of differential equations is

$$\frac{dy}{dx} = ky$$

is the special case of (1) for which $P(x) = -k$ and $Q(x) = 0$. We studied this equation in Section 7.5 in connection with exponential growth and decay. There we found solutions defined in $[0, \infty)$.

The solution of (1) is obtained by letting S be an antiderivative of P, so that $S'(x) = P(x)$, and then multiplying both sides of (1) by $e^{S(x)}$. This yields

$$e^{S(x)}\frac{dy}{dx} + e^{S(x)}P(x)y = e^{S(x)}Q(x) \tag{2}$$

Now observe that

$$\frac{d}{dx}(ye^{S(x)}) = \frac{dy}{dx}e^{S(x)} + ye^{S(x)}S'(x)$$

$$= e^{S(x)}\frac{dy}{dx} + e^{S(x)}P(x)y$$

Therefore by (2),

$$\frac{d}{dx}(ye^{S(x)}) = e^{S(x)}Q(x)$$

Integrating both sides of this equation, we obtain

$$ye^{S(x)} = \int e^{S(x)}Q(x)\,dx$$

so that

$$y = e^{-S(x)} \int e^{S(x)}Q(x)\,dx, \quad \text{where } S(x) = \int P(x)\,dx \tag{3}$$

The expression $e^{S(x)}$, by which we multiplied (1) in order to obtain (2), is called an **integrating factor** for the differential equation, because the left side of (2) is easily integrated.

Example 1 Let a and b be constants, with $a \neq 0$. Find the general solution of $dy/dx + ay = b$.

Solution This equation has the form of (1) with $P(x) = a$ and $Q(x) = b$. Since ax is an antiderivative of P, $S(x) = ax$, so (3) implies that

$$y = e^{-ax} \int e^{ax}b\,dx = e^{-ax}\left(\frac{b}{a}e^{ax} + C\right) = \frac{b}{a} + Ce^{-ax} \quad \square$$

Under certain ideal laboratory conditions, the surface temperature y of an object changes in time at a rate proportional to the difference between the temperature of the object and that of the surrounding medium, which we will assume is constantly y_0. This gives rise to the differential equation

$$\frac{dy}{dt} = k(y - y_0)$$

where k is a negative constant. This equation, known as **Newton's Law of Cooling**, is equivalent to the equation in Example 1 if $a = -k$ and $b = -ky_0$. Consequently the general solution is

$$y = \frac{-ky_0}{-k} + Ce^{kt} = y_0 + Ce^{kt}$$

Example 2 Find the particular solution of

$$x^2\frac{dy}{dx} - 2xy = 3x^6 \quad \text{for } x > 0$$

for which $y(2) = 20$.

Solution First we transform the given equation into

$$\frac{dy}{dx} - \frac{2}{x}y = 3x^4$$

which has the form of (1) with $P(x) = -2/x$ and $Q(x) = 3x^4$. Since $-2\ln x$ is an antiderivative of P, $S(x) = -2\ln x$, so (3) implies that

$$y = e^{2\ln x} \int e^{-2\ln x} 3x^4 \, dx = x^2 \left(\int 3x^2 \, dx \right) = x^2(x^3 + C) = x^5 + Cx^2$$

Finally, to determine the particular solution, we notice that

$$20 = y(2) = 2^5 + C(2^2) = 32 + 4C$$

so that $C = -3$. Therefore the solution satisfying $y(2) = 20$ is given by

$$y = x^5 - 3x^2 \quad \square$$

Electric Circuits Assume that a closed electric circuit contains electric current I (in amperes), a resistance R (in ohms), an inductance L (in henrys), and an applied electromotive force E (in volts), usually called a voltage (Figure 16.1). You may think of the resistance as tending to impede the flow of electricity, the inductance as tending to keep the electricity flowing at a constant rate, and the voltage as tending to increase or decrease the flow of electricity, depending on the direction attached to the voltage. Kirchhoff's Second Law* says that the applied voltage is equal to the sum of the voltage drops in the rest of the circuit. This law gives rise to the differential equation

$$L\frac{dI}{dt} + RI = E(t)$$

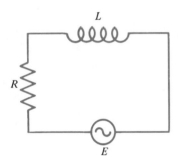

FIGURE 16.1

In the following example we will assume that the inductance and resistance are constant and that the voltage varies sinusoidally with time.

Example 3 Let L and R be nonzero constants. Find the general solution of $L(dI/dt) + RI = \sin t$.

Solution The equation is equivalent to

$$\frac{dI}{dt} + \frac{R}{L}I = \frac{1}{L}\sin t$$

which conforms to (1) if $P(t) = R/L$ and $Q(t) = (1/L)\sin t$. Since $(R/L)t$ is an antiderivative of P, $S(t) = (R/L)t$, so we conclude from (3) that

$$I = e^{-(R/L)t} \int e^{(R/L)t} \frac{1}{L} \sin t \, dt$$

$$= \frac{1}{L} e^{-(R/L)t} \int e^{(R/L)t} \sin t \, dt$$

---

* Gustav Kirchhoff (1824–1887) was a German physicist famous for his work in electricity and spectroscopy.

We evaluate the integral by parts as follows:

$$\int e^{(R/L)t} \sin t \, dt = -e^{(R/L)t} \cos t + \frac{R}{L} \int e^{(R/L)t} \cos t \, dt$$

$$= -e^{(R/L)t} \cos t + \frac{R}{L} e^{(R/L)t} \sin t - \frac{R^2}{L^2} \int e^{(R/L)t} \sin t \, dt$$

Combining the integrals involving $\sin t$, we obtain

$$\int e^{(R/L)t} \sin t \, dt = \frac{L^2}{L^2 + R^2} \left(-e^{(R/L)t} \cos t + \frac{R}{L} e^{(R/L)t} \sin t \right) + C$$

$$= \frac{L}{L^2 + R^2} e^{(R/L)t} (R \sin t - L \cos t) + C$$

Thus

$$I = \frac{1}{L} e^{-(R/L)t} \left(\frac{L}{L^2 + R^2} e^{(R/L)t} (R \sin t - L \cos t) + C \right)$$

$$= \frac{1}{L^2 + R^2} (R \sin t - L \cos t) + \frac{C}{L} e^{-(R/L)t} \quad \square$$

EXERCISES 16.4

In Exercises 1–14 find the general solution of the differential equation.

1. $\dfrac{dy}{dx} + \dfrac{1}{x^2} y = 0$

2. $\dfrac{dy}{dx} - (\sinh x) y = 0$

3. $\dfrac{dy}{dx} + 2y = 4$

4. $\dfrac{dy}{dx} - 2y = x$

5. $\dfrac{dy}{dx} - ay = f(x),$
   where f is continuous

6. $y + 2xy = 4x$

7. $y' + 6x^5 y = x^5$

8. $y' = \dfrac{1}{x^2}(xy + 1)$
   for $x > 0$

9. $y' - y = \dfrac{1}{1 - e^{-x}}$

10. $\dfrac{dy}{dx} + y \cos x = \cos x$

11. $\dfrac{dy}{dx} + y \tan x = \tan x$
    for $-\pi/2 < x < \pi/2$

12. $\dfrac{dy}{dx} - y \tan x = e^{\sin x}$
    for $-\pi/2 < x < \pi/2$

13. $\dfrac{dy}{dt} + \dfrac{1}{t} y - \sin t^2 = 0$
    for $t > 0$

14. $t \dfrac{dy}{dt} + y = t \sin t^2 + 5t$
    for $t > 0$

In Exercises 15–18 find the particular solution satisfying the initial condition.

15. $\dfrac{dy}{dx} + 5y = -4e^{-3x}; \; y(0) = -4$

16. $x \dfrac{dy}{dx} - 4y = x; \; y(1) = 1$

17. $(\cos x) \dfrac{dy}{dx} + y = 1$ for $0 < x < \pi/2; \; y(\pi/4) = 2$

18. $\dfrac{dy}{dx} - 2xy = x; \; y(0) = 0$

In Exercises 19–20 find the general solution of the differential equation

$$L \dfrac{dI}{dt} + RI = E(t)$$

given in the text, where the voltage $E(t)$ is as given.

19. $E(t) = e^t$

20. $E(t) = \cos t$

21. Find the particular solution of the differential equation

$$L \dfrac{dI}{dt} + RI = \cos t$$

for which $I(0) = R/(L^2 + R^2)$.

22. The velocity v of an object of mass m falling in the earth's atmosphere with air resistance, which is assumed to be

proportional to the velocity, satisfies the differential equation

$$\frac{dv}{dt} - \frac{p}{m}v = -g \tag{4}$$

where g is the acceleration due to gravity (assumed to be constant), and the constant of proportionality p for the air resistance is negative.

a. Let the initial velocity v_0 be given. Find the particular solution of (4) for which $v(0) = v_0$.

b. Find $\lim_{t \to \infty} |v(t)|$, where $v(t)$ is the solution found in part (a). (This limit is called the **terminal speed** of the object; under the stated hypotheses, the speed of the object cannot exceed this value.)

Even before a skydiver's parachute opens to slow his descent, there is a limiting value that the speed of his fall cannot exceed.

16.5
HOMOGENEOUS SECOND-ORDER LINEAR DIFFERENTIAL EQUATIONS

So far we have studied first-order differential equations. In this section we turn to second-order differential equations of the form

$$\frac{d^2y}{dx^2} + b\frac{dy}{dx} + cy = g(x) \tag{1}$$

where b and c are constants and g is continuous. Our analysis of this type of differential equation will be in two parts. In this section we discuss equations of the form (1) with $g = 0$. In this case (1) is said to be **homogeneous**. In Section 16.6 we discuss equations of the form (1) with $g \neq 0$. In that case (1) is said to be **nonhomogeneous**.

We now set out to find solutions of the second-order differential equation

$$\frac{d^2y}{dx^2} + b\frac{dy}{dx} + cy = 0 \tag{2}$$

If y_1 and y_2 are solutions of (1) and if C_1 and C_2 are any numbers, then

$$y = C_1 y_1 + C_2 y_2 \tag{3}$$

is also a solution of (2) (see Exercise 21). Moreover, it can be shown that if y_1 is not a constant multiple of y_2 (and *vice versa*), then any solution of (2) has the form of (3). As a result, if we find two solutions of (2) that are not multiples of one another, we can immediately find the general solution by (3). Thus our objective is to find two such solutions.

Let $y = e^{sx}$, where s is a fixed but arbitrary constant, so that

$$\frac{dy}{dx} = se^{sx} \quad \text{and} \quad \frac{d^2y}{dx^2} = s^2 e^{sx}$$

Then y is a solution of (2) if and only if

$$(s^2 e^{sx}) + b(se^{sx}) + ce^{sx} = 0 \tag{4}$$

Since $e^{sx} \neq 0$ for all x, (4) is equivalent to

$$s^2 + bs + c = 0$$

which is called the **characteristic** (or **auxiliary**) **equation** of (2). It follows that e^{sx} is a solution of (2) if and only if s is a root of the characteristic equation. By the quadratic formula, the roots are given by

$$s_1 = \frac{-b + \sqrt{b^2 - 4c}}{2} \quad \text{and} \quad s_2 = \frac{-b - \sqrt{b^2 - 4c}}{2} \tag{5}$$

if $b^2 - 4c \geq 0$. There are no real roots if $b^2 - 4c < 0$. Thus we consider three cases, depending on the value of $b^2 - 4c$.

Case 1: $b^2 - 4c > 0$. In this case, the characteristic equation has two distinct roots s_1 and s_2 given by (5), and we deduce that $e^{s_1 x}$ and $e^{s_2 x}$ are two distinct solutions that are not multiples of each other. By (3), the general solution of (2) is

$$y = C_1 e^{s_1 x} + C_2 e^{s_2 x} \tag{6}$$

Example 1   Find the general solution of

$$\frac{d^2y}{dx^2} + 5\frac{dy}{dx} + 6y = 0$$

Solution   The characteristic equation is $s^2 + 5s + 6 = 0$, whose roots are $s = -2$ and $s = -3$. Therefore by (6) the general solution is

$$y = C_1 e^{-2x} + C_2 e^{-3x} \quad \square$$

Example 2   Find the particular solution of

$$\frac{d^2y}{dx^2} - 3\frac{dy}{dx} - 4y = 0$$

for which $y(0) = 2$ and $y'(0) = -3$.

Solution   The characteristic equation is $s^2 - 3s - 4 = 0$, whose roots are $s = 4$ and $s = -1$. Consequently the general solution is

$$y = C_1 e^{4x} + C_2 e^{-x}$$

If $y(0) = 2$ and $y'(0) = -3$, then

$$2 = y(0) = C_1 e^{4 \cdot 0} + C_2 e^{-0} = C_1 + C_2$$

$$-3 = y'(0) = 4C_1 e^{4 \cdot 0} - C_2 e^{-0} = 4C_1 - C_2$$

To solve the equations

$$2 = C_1 + C_2 \quad \text{and} \quad -3 = 4C_1 - C_2$$

for C_1 and C_2, we notice from the first equation that $C_1 = 2 - C_2$, so that by substituting for C_1 in the second equation we obtain

$$-3 = 4(2 - C_2) - C_2 = 8 - 5C_2$$

Therefore $C_2 = \frac{11}{5}$, and consequently $C_1 = 2 - C_2 = -\frac{1}{5}$. Thus the required particular solution is

$$y = -\frac{1}{5} e^{4x} + \frac{11}{5} e^{-x} \quad \square$$

Case 2: $b^2 - 4c = 0$. In this case the characteristic equation has one (double) root $s = -b/2$. Thus

$$s^2 + bs + c = 0 \quad \text{and} \quad 2s + b = 0 \tag{7}$$

which we will use below. We know that e^{sx} is a solution of (2) from our remarks preceding Case 1. Let us show that xe^{sx} is also a solution of (2). Substituting xe^{sx} for y and carrying out the differentiation, we obtain

$$\frac{d^2 y}{dx^2} + b\frac{dy}{dx} + cy = (2se^{sx} + s^2 xe^{sx}) + b(e^{sx} + sxe^{sx}) + cxe^{sx}$$

$$= (s^2 + bs + c)xe^{sx} + (2s + b)e^{sx} = 0$$

where the last equality follows from the two equations in (7). It follows that xe^{sx} is a solution of (2). Since e^{sx} and xe^{sx} are not multiples of one another, the general solution of (2) for Case 2 is

$$y = C_1 e^{sx} + C_2 xe^{sx} \tag{8}$$

Example 3   Find the particular solution of

$$\frac{d^2 y}{dx^2} + 8\frac{dy}{dx} + 16y = 0$$

for which $y(2) = 3e^{-8}$ and $y'(2) = -10e^{-8}$.

Solution   The characteristic equation is $s^2 + 8s + 16 = 0$, which has the double root $s = -4$. Thus by (8) the general solution is

$$y = C_1 e^{-4x} + C_2 xe^{-4x}$$

If $y(2) = 3e^{-8}$ and $y'(2) = -10e^{-8}$, then

$$3e^{-8} = y(2) = C_1 e^{-4 \cdot 2} + 2C_2 e^{-4 \cdot 2} = (C_1 + 2C_2)e^{-8}$$

and

$$-10e^{-8} = y'(2) = -4C_1e^{-4\cdot2} + C_2e^{-4\cdot2} - 8C_2e^{-4\cdot2} = (-4C_1 - 7C_2)e^{-8}$$

Therefore

$$3 = C_1 + 2C_2 \quad \text{and} \quad 10 = 4C_1 + 7C_2$$

Solving for C_1 in the first equation, we obtain $C_1 = 3 - 2C_2$. Then the second equation becomes

$$10 = 4(3 - 2C_2) + 7C_2 = 12 - C_2$$

Consequently $C_2 = 2$ and thus $C_1 = -1$. We conclude that

$$y = -e^{-4x} + 2xe^{-4x}$$

is the particular solution we sought. ☐

Case 3: $b^2 - 4c < 0$. In this case the characteristic equation has no real roots, so solutions of the given differential equation must be of a different type. Let

$$u = -\frac{b}{2} \quad \text{and} \quad v = \frac{1}{2}\sqrt{4c - b^2} \tag{9}$$

For later use we observe that

$$u^2 - v^2 + bu + c = 0 \quad \text{and} \quad 2uv + bv = 0 \tag{10}$$

We will prove that if

$$y_1 = e^{ux}\sin vx$$

then y_1 is a solution of (2). The first and second derivatives of y_1 are given by

$$\frac{dy_1}{dx} = ue^{ux}\sin vx + e^{ux}v\cos vx = e^{ux}(u\sin vx + v\cos vx)$$

and

$$\frac{d^2y_1}{dx^2} = ue^{ux}(u\sin vx + v\cos vx) + e^{ux}(uv\cos vx - v^2\sin vx)$$
$$= e^{ux}[(u^2 - v^2)\sin vx + 2uv\cos vx]$$

Therefore

$$\frac{d^2y_1}{dx^2} + b\frac{dy_1}{dx} + cy_1 = e^{ux}[(u^2 - v^2)\sin vx + 2uv\cos vx]$$
$$+ be^{ux}(u\sin vx + v\cos vx) + ce^{ux}\sin vx$$
$$= e^{ux}[(u^2 - v^2 + bu + c)\sin vx + (2uv + bv)\cos vx] = 0$$

where the last equality follows from (10). Thus y_1 is a solution of (2).

Similarly, if

$$y_2 = e^{ux}\cos vx$$

then y_2 is a solution of (2). Since y_1 and y_2 are not multiples of one another, a general solution of (2) is given by

$$y = C_1e^{ux}\sin vx + C_2e^{ux}\cos vx \tag{11}$$

Example 4   Find the general solution of

$$\frac{d^2y}{dt^2} + y = 0$$

Solution   The characteristic equation is $s^2 + 1 = 0$, and since $b = 0$ and $c = 1$, it follows that $b^2 - 4c = -4 < 0$, so that Case 3 applies. Using (9), we let

$$u = -\frac{b}{2} = 0 \quad \text{and} \quad v = \frac{1}{2}\sqrt{4c - b^2} = \frac{1}{2}\sqrt{4} = 1$$

Then by (11) the general solution is given by

$$y = C_1 e^{0 \cdot t} \sin(1 \cdot t) + C_2 e^{0 \cdot t} \cos(1 \cdot t)$$

that is, $y = C_1 \sin t + C_2 \cos t$ ☐

Example 5   Find the particular solution of

$$\frac{d^2y}{dx^2} + \frac{dy}{dx} + y = 0$$

for which $y(0) = 0$ and $y'(0) = 2$.

Solution   Since the characteristic equation is $s^2 + s + 1 = 0$, it follows that $b = 1 = c$ and thus $b^2 - 4c = -3 < 0$, so that Case 3 applies. Let

$$u = -\frac{b}{2} = -\frac{1}{2} \quad \text{and} \quad v = \frac{1}{2}\sqrt{4c - b^2} = \frac{1}{2}\sqrt{4 - 1} = \frac{1}{2}\sqrt{3}$$

Then by (11) the general solution is given by

$$y = C_1 e^{-x/2} \sin\frac{1}{2}\sqrt{3}x + C_2 e^{-x/2} \cos\frac{1}{2}\sqrt{3}x$$

To find the particular solution required, we first compute that

$$0 = y(0) = C_1 \cdot 0 + C_2 \cdot 1 = C_2$$

Thus $y = C_1 e^{-x/2} \sin\frac{1}{2}\sqrt{3}x$

Next we find that

$$\frac{dy}{dx} = -\frac{1}{2}C_1 e^{-x/2} \sin\frac{1}{2}\sqrt{3}x + \frac{1}{2}\sqrt{3}C_1 e^{-x/2} \cos\frac{1}{2}\sqrt{3}x$$

so that $2 = y'(0) = \frac{1}{2}\sqrt{3}C_1$

and therefore $C_1 = \frac{4}{3}\sqrt{3}$. Consequently the particular solution is given by

$$y = \frac{4}{3}\sqrt{3}e^{-x/2} \sin\frac{1}{2}\sqrt{3}x$$ ☐

Electric Circuits   Suppose a closed electric circuit contains a resistor, an inductor, and a capacitor in series, with no external voltage applied to the circuit. Let R be the resistance (in ohms), L the inductance (in henrys), and C the capacitance (in farads). It follows from Kirchhoff's Second Law that the charge q on the capacitor satisfies the differential equation

$$L\frac{d^2q}{dt^2} + R\frac{dq}{dt} + \frac{1}{C}q = 0 \qquad (12)$$

If we divide each side of the equation by L, then the resulting equation has the form in (2) and can be solved by our analysis.

Example 6   Suppose a closed electric circuit has resistance 4 ohms, inductance 0.04 henrys, and capacitance 0.002 farads. Assume that initially there is no current (so that $q'(0) = 0$) but that the initial charge is 20 coulombs. Find a formula for the charge at any time $t \geq 0$.

Solution   From (12) with $R = 4$, $L = 0.04$, and $C = 0.002$, we have

$$0.04\frac{d^2q}{dt^2} + 4\frac{dq}{dt} + \frac{1}{0.002}q = 0 \qquad (13)$$

or equivalently,

$$\frac{d^2q}{dt^2} + 100\frac{dq}{dt} + 12,500q = 0$$

The characteristic equation is $s^2 + 100s + 12,500 = 0$, which falls under Case 3 with $b = 100$ and $c = 12,500$. Using (9), we find that

$$u = -\frac{b}{2} = -\frac{100}{2} = -50$$

and

$$v = \frac{1}{2}\sqrt{4c - b^2} = \frac{1}{2}\sqrt{4(12,500) - (100)^2} = \frac{1}{2}\sqrt{40,000} = 100$$

Thus by (11) the general solution of (13) is given by

$$q = C_1 e^{-50t}\sin 100t + C_2 e^{-50t}\cos 100t$$

By hypothesis $q(0) = 20$, so that $C_2 = 20$ and thus

$$q = C_1 e^{-50t}\sin 100t + 20e^{-50t}\cos 100t$$

To determine C_1, we first differentiate q to obtain

$$q'(t) = C_1(-50e^{-50t}\sin 100t + 100e^{-50t}\cos 100t)$$
$$+ 20(-50e^{-50t}\cos 100t - 100e^{-50t}\sin 100t)$$

Since $q'(0) = 0$ by hypothesis,

$$0 = q'(0) = C_1(100) + 20(-50)$$

which means that $C_1 = 10$. Consequently

$$q = 10e^{-50t}\sin 100t + 20e^{-50t}\cos 100t \quad \square$$

EXERCISES 16.5

In Exercises 1–14 find the general solution of the differential equation.

1. $\dfrac{d^2y}{dx^2} - 5\dfrac{dy}{dx} - 14y = 0$ 2. $\dfrac{d^2y}{dx^2} - 25y = 0$

3. $\dfrac{d^2y}{dx^2} + 2\dfrac{dy}{dx} - 24y = 0$ 4. $\dfrac{d^2y}{dx^2} - 6\dfrac{dy}{dx} + 9y = 0$

5. $\dfrac{d^2y}{dx^2} + 10\dfrac{dy}{dx} + 25y = 0$ 6. $\dfrac{d^2y}{dx^2} + 2\sqrt{2}\dfrac{dy}{dx} + 2y = 0$

7. $\dfrac{d^2y}{dx^2} + 9y = 0$ 8. $\dfrac{d^2y}{dx^2} + 2y = 0$

9. $\dfrac{d^2y}{dx^2} + 3\dfrac{dy}{dx} + 3y = 0$ 10. $\dfrac{d^2y}{dt^2} + 2\dfrac{dy}{dt} + 5y = 0$

11. $6\dfrac{d^2y}{dt^2} - 4y = 0$ 12. $2\dfrac{d^2y}{dt^2} + 5\dfrac{dy}{dt} + 2y = 0$

13. $4\dfrac{d^2y}{dx^2} + 12\dfrac{dy}{dx} + 9y = 0$ 14. $2\dfrac{d^2y}{dx^2} + 4\dfrac{dy}{dx} + 5y = 0$

In Exercises 15–20 find the particular solution satisfying the given initial conditions.

15. $\dfrac{d^2y}{dx^2} - 2\dfrac{dy}{dx} - 15y = 0$; $y(0) = 1$, $y'(0) = -1$

16. $\dfrac{d^2y}{dx^2} + 16y = 0$; $y(0) = 3$, $y'(0) = 12$

17. $\dfrac{d^2y}{dx^2} - 10\dfrac{dy}{dx} + 25y = 0$; $y(1) = 0$, $y'(1) = e^5$

18. $4\dfrac{d^2y}{dx^2} - 4\dfrac{dy}{dx} + y = 0$; $y(0) = 0$, $y'(0) = 1$

19. $3\dfrac{d^2y}{dx^2} + 8\dfrac{dy}{dx} - 3y = 0$; $y(0) = 2$, $y'(0) = -2$

20. $\dfrac{d^2y}{dx^2} + 4\dfrac{dy}{dx} + 5y = 0$; $y(0) = 1$, $y'(0) = 0$

21. Show that if y_1 and y_2 are solutions of the differential equation
$$\frac{d^2y}{dx^2} + b\frac{dy}{dx} + cy = 0$$
then for any constants C_1 and C_2, the function $y = C_1y_1 + C_2y_2$ is also a solution.

22. A closed electric circuit has resistance 15 ohms, inductance 0.05 henry, and capacitance 0.001 farad. Suppose the initial charge on the capacitor is 5 coulombs and the initial current is -200 amps. Find a formula for the charge at any time $t \geq 0$.

23. A closed electric circuit has a resistance 2 ohms, inductance 0.01 henry, and capacitance 0.005 farad. Suppose the initial current is 100 amps and the initial charge on the capacitor is 1 coulomb. Find a formula for the charge at any time $t \geq 0$.

16.6
NONHOMOGENEOUS SECOND-ORDER LINEAR DIFFERENTIAL EQUATIONS

Next we consider differential equations of the form
$$\frac{d^2y}{dx^2} + b\frac{dy}{dx} + cy = g(x)$$
where b and c are constants and g is a nonzero continuous function. For simplicity in the discussion below, we will write the differential equation more succinctly in the form
$$y'' + by' + cy = g \tag{1}$$
Our goal is to find the general solution of (1).

The first step is to look at the associated homogeneous equation
$$y'' + by' + cy = 0 \tag{2}$$

Formulas (6), (8), and (11) of Section 16.5 together imply that the general solution of (2) has the form

$$y = C_1 y_1 + C_2 y_2 \tag{3}$$

where y_1 and y_2 are solutions of (2), neither being a constant multiple of the other.

Now suppose that y_p is a particular solution of the nonhomogeneous equation (1) and that y is a second such solution. Then

$$
\begin{aligned}
(y - y_p)'' + b(y - y_p)' + c(y - y_p) &= y'' - y_p'' + by' - by_p' + cy - cy_p \\
&= (y'' + by' + cy) - (y_p'' + by_p' + cy_p) \\
&= g(x) - g(x) = 0
\end{aligned}
$$

so that $y - y_p$ is a solution of the associated homogeneous equation (2). Consequently by (3) we know that

$$y - y_p = C_1 y_1 + C_2 y_2$$

for appropriate constants C_1 and C_2. Therefore

$$y = y_p + C_1 y_1 + C_2 y_2 \tag{4}$$

is the general solution of (2). This gives us a general procedure for solving nonhomogeneous linear equations in the form of (1):

1. Find the general solution

$$y = C_1 y_1 + C_2 y_2$$

   of the associated homogeneous equation in (2).
2. Find a single solution y_p of (2).
3. Express the general solution of (2) in the form

$$y = y_p + C_1 y_1 + C_2 y_2$$

The general solution $y = C_1 y_1 + C_2 y_2$ of the associated homogeneous equation is called the **complementary solution**, and the solution y_p is a **particular solution**. By (4) we see that the general solution of a nonhomogeneous linear differential equation is the sum of a particular solution and the complementary solution. Since we already know from the preceding section how to find the complementary solution, the problem of solving a nonhomogeneous second-order linear differential equation reduces to finding a particular solution.

Variation of Parameters

We will now present a method for determining a particular solution y_p of (1). The solution y_p will appear in the form

$$y_p = u_1 y_1 + u_2 y_2 \tag{5}$$

where y_1 and y_2 are solutions of the homogeneous equation that appeared in (3) and u_1 and u_2 are functions that must be determined. It can be shown that such y_1 and y_2 satisfy

$$y_1(x)y_2'(x) - y_1'(x)y_2(x) \neq 0 \tag{6}$$

for all x in the domains of y_1 and y_2. We will use this information to derive formulas for u_1 and u_2 so that $y_p = u_1 y_1 + u_2 y_2$ is indeed a solution of (1). Using the product rule to differentiate y_p, we find that

$$y_p' = u_1' y_1 + u_1 y_1' + u_2' y_2 + u_2 y_2' = (u_1 y_1' + u_2 y_2') + (u_1' y_1 + u_2' y_2) \quad (7)$$

Differentiating y_p', we find that

$$y_p'' = (u_1' y_1' + u_1 y_1'' + u_2' y_2' + u_2 y_2'') + (u_1' y_1 + u_2' y_2)' \quad (8)$$

Substituting in turn (5), (7), and (8) for y, y', and y'' in (1), we obtain

$$[(u_1' y_1' + u_1 y_1'' + u_2' y_2' + u_2 y_2'') + (u_1' y_1 + u_2' y_2)']$$
$$+ b[(u_1 y_1' + u_2 y_2') + (u_1' y_1 + u_2' y_2)] + c(u_1 y_1 + u_2 y_2) = g$$

Rearranging, we obtain

$$u_1(y_1'' + by_1' + cy_1) + u_2(y_2'' + by_2' + cy_2) + u_1' y_1' + u_2' y_2'$$
$$+ (u_1' y_1 + u_2' y_2)' + b(u_1' y_1 + u_2' y_2) = g$$

Notice that since y_1 and y_2 are solutions to the homogeneous equation (2), the expressions in the first two parentheses are 0. It follows that y_p satisfies (1) provided that

$$u_1' y_1' + u_2' y_2' + (u_1' y_1 + u_2' y_2)' + b(u_1' y_1 + u_2' y_2) = g$$

which automatically happens if

$$\begin{cases} u_1' y_1 + u_2' y_2 = 0 \\ u_1' y_1' + u_2' y_2' = g \end{cases}$$

This is a system of two equations in two unknowns, u_1' and u_2', and because of (6) we can find unique formulas for u_1' and u_2':

$$u_1' = \frac{-y_2 g}{y_1 y_2' - y_1' y_2} \quad (9)$$

and

$$u_2' = \frac{y_1 g}{y_1 y_2' - y_1' y_2} \quad (10)$$

To obtain formulas for u_1 and u_2, we need only integrate both sides of (9) and (10), which, in theory at least, we can do. (In practice one might well be unable to find a reasonable formula for u_1 from (9) or u_2 from (10), without it entailing an integral sign.) At any rate, we now have formally devised a method for finding a particular solution y_p of the original differential equation. The method is called **variation of parameters** because u_1 and u_2 are used as variables or parameters in (5).

Example 1   Find a particular solution y_p of

$$y'' + y = \tan x$$

Solution   By Example 4 of Section 16.5,

$$y_1 = \sin x \quad \text{and} \quad y_2 = \cos x$$

are solutions of the homogeneous equation $y'' + y = 0$. Thus by (5) a particular solution y_p is given by

$$y_p = u_1(x) \sin x + u_2(x) \cos x \tag{11}$$

where u_1 and u_2 are to be determined from (9) and (10), with $g(x) = \tan x$. We obtain

$$u_1'(x) = \frac{-\cos x \tan x}{(\sin x)(-\sin x) - (\cos x)(\cos x)} = \sin x$$

and

$$u_2'(x) = \frac{\sin x \tan x}{(\sin x)(-\sin x) - (\cos x)(\cos x)} = \frac{-\sin^2 x}{\cos x}$$

$$= \frac{-(1 - \cos^2 x)}{\cos x} = -\sec x + \cos x$$

By integration we deduce that

$$u_1(x) = -\cos x \quad \text{and} \quad u_2(x) = -\ln|\sec x + \tan x| + \sin x$$

Thus it follows from (11) that

$$y_p = -\cos x \sin x + (-\ln|\sec x + \tan x| + \sin x)(\cos x)$$
$$= -\cos x \ln|\sec x + \tan x| \quad \square$$

Example 2 Find the general solution of $\dfrac{d^2y}{dx^2} + 5\dfrac{dy}{dx} + 6y = e^{-x}$.

Solution From Example 1 of Section 16.5,

$$y_1 = e^{-2x} \quad \text{and} \quad y_2 = e^{-3x}$$

are solutions of the homogeneous equation

$$\frac{d^2y}{dx^2} + 5\frac{dy}{dx} + 6y = 0$$

Thus by (5) a particular solution y_p is given by

$$y_p = u_1(x)e^{-2x} + u_2(x)e^{-3x} \tag{12}$$

where u_1 and u_2 are to be determined from (9) and (10). We obtain

$$u_1'(x) = \frac{-e^{-3x}e^{-x}}{(e^{-2x})(-3e^{-3x}) - (-2e^{-2x})(e^{-3x})} = e^x$$

and

$$u_2'(x) = \frac{e^{-2x}e^{-x}}{(e^{-2x})(-3e^{-3x}) - (-2e^{-2x})(e^{-3x})} = -e^{2x}$$

By integration we deduce that

$$u_1(x) = e^x \quad \text{and} \quad u_2(x) = -\frac{1}{2}e^{2x}$$

Thus it follows from (12) that

$$y_p = e^x e^{-2x} - \frac{1}{2} e^{2x} e^{-3x} = \frac{1}{2} e^{-x}$$

We conclude from (4) that the general solution of the given differential equation is given by

$$y = \frac{1}{2} e^{-x} + C_1 e^{-2x} + C_2 e^{-3x} \quad \square$$

In the theory of circuits that contain resistance R, inductance L, and capacitance C, as well as voltage E, Kirchhoff's Second Law still holds, and the charge q on the capacitor satisfies the differential equation

$$L\frac{d^2q}{dt^2} + R\frac{dq}{dt} + \frac{1}{C}q = E(t)$$

This is one of the fundamental equations in the theory of electricity. Provided that E is exponential or sinusoidal, one can solve for q explicitly by means of variation of parameters.

EXERCISES 16.6

In Exercises 1–16 find the general solution of the differential equation.

1. $\dfrac{d^2y}{dx^2} + 5\dfrac{dy}{dx} + 4y = 3$

2. $\dfrac{d^2y}{dx^2} - 4y = e^{3x}$

3. $\dfrac{d^2y}{dx^2} - 2\dfrac{dy}{dx} - 3y = e^x$

4. $\dfrac{d^2y}{dx^2} + 8\dfrac{dy}{dx} + 15y = e^{-x}$

5. $\dfrac{d^2y}{dx^2} - \dfrac{dy}{dx} = 2x - 3$

6. $\dfrac{d^2y}{dx^2} + 4\dfrac{dy}{dx} + 4y = e^{-x}$

7. $\dfrac{d^2y}{dx^2} + y = \csc x \cot x$
   for $0 < x < \pi/2$

8. $\dfrac{d^2y}{dx^2} + y = \cos x$
   for $0 < x < \pi$

9. $\dfrac{d^2y}{dt^2} + 9y = 3t$

10. $\dfrac{d^2y}{dt^2} - 9y = \sin t$

11. $\dfrac{d^2y}{dt^2} + 9y = \sin 3t$

12. $\dfrac{d^2y}{dx^2} - y = xe^x$

13. $\dfrac{d^2y}{dx^2} - 2\dfrac{dy}{dx} + y = \dfrac{1}{x}e^x$

14. $\dfrac{d^2y}{dx^2} + 6\dfrac{dy}{dx} + 9y = e^{-4x}$

15. $\dfrac{d^2y}{dx^2} - 5\dfrac{dy}{dx} + 6y = x^3 e^{2x}$

16. $\dfrac{d^2y}{dx^2} + 5\dfrac{dy}{dx} - 6y = 7e^x$

17. Suppose a closed electric circuit has a resistance 5 ohms, inductance 0.04 henry, and capacitance 0.01 farad. Find a formula for the charge at any time $t \geq 0$ if the voltage $E(t)$ is given by

a. $E(t) = e^{-50t}$ b. $E(t) = \sin 50t$

18. By Newton's Second Law of Motion, the position of an object of mass m falling in the earth's atmosphere satisfies the differential equation

$$m\frac{d^2x}{dt^2} = F \tag{13}$$

where F is the sum of the force due to gravity and the force F_1 due to air resistance. Assume that the object is near the earth's surface, so that the force due to gravity is approximately $-mg$. Assume also that F_1 is proportional to the speed dx/dt, with constant of proportionality p. Then (13) becomes

$$m\frac{d^2x}{dt^2} = -mg + p\frac{dx}{dt}$$

or equivalently,

$$\frac{d^2x}{dt^2} - \frac{p}{m}\frac{dx}{dt} = -g \tag{14}$$

Find the general solution of (14).

16.7
THE MOTION OF A SPRING

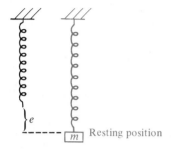

FIGURE 16.2

Hooke's Law states that if a spring is stretched (or compressed) x units from its natural length l, then the spring exerts a force $h(x)$ that tends to restore the spring to its natural length and is proportional to x. Thus there is a positive constant k, called the **spring constant**, such that

$$h(x) = -kx \qquad (1)$$

The spring constant does not depend on x, but it does vary from spring to spring.

Suppose that a rigid object (such as a ball or a weight) with mass m is attached to a spring, and let e be the resulting elongation of the spring when the object and spring are at rest (Figure 16.2). If the weight of the spring is assumed to be negligible in comparison with the weight mg of the object, then the weight of the object is (essentially) balanced by the force of the spring, which by (1) is ke. Thus

$$ke = mg \qquad (2)$$

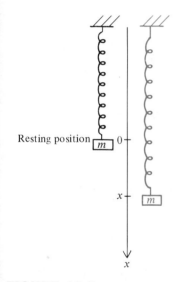

FIGURE 16.3

Let us set up a vertical axis with origin at the resting position of the top of the object and with the positive direction pointed downward. If the object is displaced from its resting position and subsequently moves under the influence of gravity, the spring force, and possibly other forces, then the position x of the object at time t depends on t (Figure 16.3). The velocity of the object is dx/dt, and its acceleration is d^2x/dt^2. Newton's Second Law of Motion states that

$$m\frac{d^2x}{dt^2} = f(x) \qquad (3)$$

where $f(x)$ is the total force exerted on the object when its position on the vertical axis is x. Occasionally the motion of the object and spring suffers resistance due to the medium (such as air or a liquid) in which they are located. Such a force is called a **damping force**, and it produces a **damped motion**. Otherwise the motion is **undamped**. If the only forces acting on the spring and object are gravity, the force of the spring, and a damping force that is proportional to the velocity (as is often assumed), then the vibration of the object on the spring is said to be a **free vibration**.

Free, Undamped Vibration

Suppose now that the only forces acting on an object attached to a spring are the force of the spring and the weight mg of the object. When the top of the object is located at x on the (vertical) x axis, the spring is stretched (or compressed) $x + e$ units from its natural length (Figure 16.4), so by Hooke's Law the force $h(x)$ exerted by the spring is $-k(x + e)$, and thus the total force $f(x)$ acting on the object is given by

$$f(x) = -k(x + e) + mg$$

Since $ke = mg$ by (2), we have

$$f(x) = -k(x + e) + mg = -kx - ke + mg = -kx$$

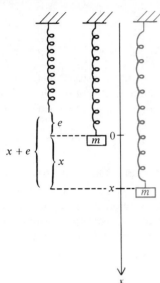

FIGURE 16.4

Thus it follows from (3) that

$$m\frac{d^2x}{dt^2} = -kx$$

or equivalently,

$$\frac{d^2x}{dt^2} + \frac{k}{m}x = 0 \qquad (4)$$

This is an equation for free, undamped vibration.

Example 1 Suppose a 2-pound weight attached to a spring stretches the spring $\frac{1}{2}$ foot. If the spring is stretched an additional $\frac{1}{4}$ foot and is released with initial velocity 0, find a formula for the position of the weight at any subsequent time.

Solution By (2),

$$\frac{k}{m} = \frac{g}{e} = \frac{32}{\frac{1}{2}} = 64$$

so that by (4) the position of the weight satisfies the equation

$$\frac{d^2x}{dt^2} + 64x = 0 \qquad (5)$$

The equation in (5) falls under Case 3 of Section 16.5 with $b = 0$ and $c = 64$. As in (9) of that section, we let

$$u = -\frac{b}{2} = 0 \quad \text{and} \quad v = \frac{1}{2}\sqrt{4c - b^2} = \frac{1}{2}\sqrt{256} = 8$$

Thus by (11) of Section 16.5 the general solution of (5) is given by

$$x = C_1 e^{0 \cdot t} \sin 8t + C_2 e^{0 \cdot t} \cos 8t$$

or more simply,

$$x = C_1 \sin 8t + C_2 \cos 8t \qquad (6)$$

Using the initial condition $y(0) = \frac{1}{4}$, we obtain

$$\frac{1}{4} = x(0) = C_2$$

so that

$$x = C_1 \sin 8t + \frac{1}{4}\cos 8t$$

Next,

$$\frac{dx}{dt} = 8C_1 \cos 8t - 2\sin 8t$$

By hypothesis the initial velocity is 0, which means that $x'(0) = 0$. Therefore

$$0 = x'(0) = 8C_1, \quad \text{so that} \quad C_1 = 0$$

Consequently

$$x = \frac{1}{4}\cos 8t$$

is the particular solution we sought.   □

Free, Damped Vibration   In addition to the force of gravity and the force of the spring, let us henceforth assume that the object experiences a damping force that is proportional to the velocity dx/dt and is directed opposite to the velocity. This means that we assume the existence of a positive constant p, called the **damping constant**, such that the damping force equals $-p\,dx/dt$. Then (3) becomes

$$m\frac{d^2x}{dt^2} = -k(x + e) + mg - p\frac{dx}{dt} \tag{7}$$

and since $ke = mg$ by (2), we obtain

$$m\frac{d^2x}{dt^2} = -kx - p\frac{dx}{dt}$$

or equivalently,

$$\frac{d^2x}{dt^2} + \frac{p}{m}\frac{dx}{dt} + \frac{k}{m}x = 0 \tag{8}$$

Example 2   Suppose the spring constant for a given spring is 6, and a certain weight attached to the spring extends it 4 inches. If the damping constant is $\frac{11}{8}$, determine a formula for the position of the weight at any subsequent time.

Solution   We have $k = 6$, $e = \frac{1}{3}$ (foot), and $p = \frac{11}{8}$. In order to find m, we apply (2), which yields

$$m = \frac{ke}{g} = \frac{6(\frac{1}{3})}{32} = \frac{1}{16}$$

Therefore (8) becomes

$$\frac{d^2x}{dt^2} + \frac{\frac{11}{8}\,dx}{\frac{1}{16}\,dt} + \frac{6}{\frac{1}{16}}x = 0$$

or equivalently,

$$\frac{d^2x}{dt^2} + 22\frac{dx}{dt} + 96x = 0 \tag{9}$$

The characteristic equation is

$$s^2 + 22s + 96 = 0$$

which can be factored to give

$$(s + 16)(s + 6) = 0$$

Thus the characteristic equation falls under Case 1 of Section 16.5, and therefore by (6) of Section 16.5 the general solution of (9) is given by

$$x(t) = C_1 e^{-16t} + C_2 e^{-6t} \quad \square$$

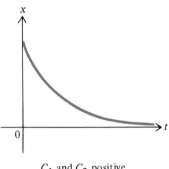

C_1 and C_2 positive

FIGURE 16.5

Observe that

$$\lim_{t \to \infty} (C_1 e^{-16t} + C_2 e^{-6t}) = 0$$

Thus the object on the spring approaches its resting position as time progresses. However, the motion at the outset depends on the signs of C_1 and C_2. On the one hand, if C_1 and C_2 are both positive or both negative, then $x(t) \neq 0$ for all t, so the object never quite returns to its resting position (Figure 16.5). On the other hand, if C_1 and C_2 have opposite signs, then $x(t) = 0$ for at most one value of t (see Exercise 9). In this case the object can return to its resting position once, overshoot it, and then slowly tend back towards its resting position (Figure 16.6). The motion is similar to that of a fender of an automobile when one pushes down hard on it and lets go, provided that the shock absorbers are in good shape. In either case this type of vibration is said to be **overdamped.**

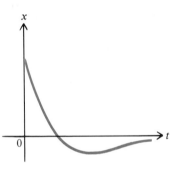

C_1 and C_2 of opposite sign

FIGURE 16.6

Example 3 Suppose that a weight of 2 pounds is attached to a spring and the spring constant is 4. Find a formula for the motion of the spring if the damping constant is

a. 1   b. $\frac{1}{4}$

Solution For part (a) we have $k = 4$, $m = \frac{2}{32} = \frac{1}{16}$, and $p = 1$, so (8) becomes

$$\frac{d^2 x}{dt^2} + 16 \frac{dx}{dt} + 64x = 0 \tag{10}$$

Since the characteristic equation is $s^2 + 16s + 64 = 0$, or equivalently, $(s + 8)^2 = 0$, it falls under Case 2 of Section 16.5 with $b = 16$. Thus by (8) of Section 16.5 the general solution of (10) is given by

$$y = C_1 e^{-8t} + C_2 t e^{-8t} = e^{-8t}(C_1 + C_2 t) \tag{11}$$

For part (b) we have $k = 4$, $m = \frac{1}{16}$, and $p = \frac{1}{4}$, so (8) becomes

$$\frac{d^2 x}{dt^2} + 4 \frac{dx}{dt} + 64x = 0 \tag{12}$$

Since the characteristic equation is $s^2 + 4s + 64 = 0$ and since $4^2 - 4 \cdot 64 < 0$, the characteristic equation falls under Case 3 of Section 16.5 with $b = 4$ and $c = 64$. Letting

$$u = -\frac{b}{2} = -2$$

and

$$v = \frac{1}{2}\sqrt{4c - b^2} = \frac{1}{2}\sqrt{4 \cdot 64 - 4^2} = \frac{1}{2}\sqrt{256 - 16} = 2\sqrt{15}$$

we find from (11) of Section 16.5 that the general solution of (12) is given by

$$x = C_1 e^{-2t} \sin 2\sqrt{15}t + C_2 e^{-2t} \cos 2\sqrt{15}t \quad \square \tag{13}$$

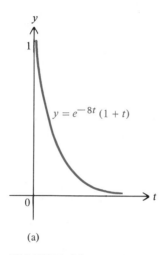

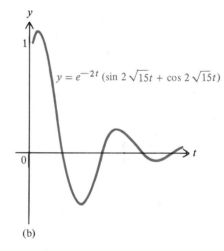

(a)   (b)

FIGURE 16.7

The graphs of the functions in (11) and (13) obtained by letting $C_1 = C_2 = 1$ are shown in Figure 16.7(a) and (b). Notice that in each case, $\lim_{t \to \infty} x(t) = 0$. As we see in Figure 16.7(a), the motion corresponding to (11) is very much like overdamped vibration and is said to be **critically damped**. In contrast, from Figure 16.7(b) we see that for the motion corresponding to (13) the object passes through its resting position at regular intervals, as the amplitude shrinks toward 0. This kind of vibration is **underdamped**.

Examples 1–3 describe the four basic kinds of vibration, and the graphs that arise from the general solutions of the differential equations can well be different from one another. Table 16.1 lists the four categories related to the differential equations, along with the general types of solutions.

TABLE 16.1

Type of vibration	Relationship of k, m, p	General type of solution
undamped, free vibration	$p = 0$	$C_1 \sin at + C_2 \cos at$
overdamped vibration	$p^2 - 4km > 0$	$C_1 e^{-at} + C_2 e^{-bt}$
critically damped vibration	$p^2 = 4km$	$e^{-at}(C_1 + C_2 t)$
underdamped vibration	$p^2 - 4km < 0$	$C_1 e^{-at} \sin bt + C_2 e^{-at} \cos bt$

EXERCISES 16.7

1. A 2-pound weight stretches a spring 6 inches. If the weight is pulled down an additional 2 inches and then released with initial velocity 0, find a formula for the position of the weight at any subsequent time, assuming that the motion is undamped.

2. Suppose that when the weight in Exercise 1 is released, it is traveling downward with a speed of 1 inch per second. Find a formula for the position of the weight at any subsequent time, again assuming that the motion is undamped.

3. Suppose that in Exercise 2 the initial speed of the object were 2 inches per second in the upward direction. Find the resulting formula for the position of the weight at any subsequent time.

4. A weight weighing 1.6 pounds is attached to a spring with spring constant $\frac{1}{4}$. If the weight is raised 1 inch above its resting position and released with initial velocity 0, find a formula for the position of the weight at any subsequent time, assuming that the damping constant for air resistance is 0.2.

5. Suppose that when the weight in Exercise 4 is released, it is traveling upward with speed 2 inches per second. Find a formula for the position of the weight at any subsequent time, again assuming that the damping constant is 0.2.

6. Suppose that the weight in Exercise 4 is reduced to 0.96 pound but the spring and damping constants remain the same. Find a formula for the position of the weight at any subsequent time.

7. Determine the weight that would be required to produce critical damping for the spring in Exercise 4, again assuming that the damping constant is 0.2.

8. A weight of 8 pounds is to be suspended from a spring with spring constant 0.25. If critically damped motion is to be obtained, what must the damping constant be?

9. Let $f(t) = C_1 e^{r_1 t} + C_2 e^{r_2 t}$, where $r_1 \neq r_2$.
   a. Show that if C_1 and C_2 have the same sign, then $f(t) \neq 0$ for all t.
   b. Show that if C_1 and C_2 have opposite signs, then $f(t) = 0$ for exactly one value of t, and determine that value of t.

10. Let $f(t) = (C_1 + C_2 t)e^{rt}$, with $C_2 \neq 0$.
    a. Show that $f(t) = 0$ for exactly one value of t, and determine that value of t.
    b. Under what conditions is the value found in part (a) positive?

16.8
SERIES SOLUTIONS OF DIFFERENTIAL EQUATIONS

Most differential equations are not as simple to solve as the ones we have studied so far. In fact, many differential equations do not have a solution for which we can give a simple formula. Two well-known such equations are

$$x^2 \frac{d^2 y}{dx^2} + x \frac{dy}{dx} + x^2 y = 0 \quad \text{and} \quad \frac{d^2 y}{dx^2} - 2x \frac{dy}{dx} + y = 0$$

However, it is possible to find solutions to these and many other differential equations in terms of power series. Such a solution is called a **series solution** of the differential equation.

In our study of series solutions of differential equations, we will need to use results from Chapter 9 on power series. In particular, we will add power series, differentiate power series, and use the fact stated in Corollary 9.27 that if

$$\sum_{n=0}^{\infty} a_n x^n = \sum_{n=0}^{\infty} b_n x^n$$

for all x in an open interval about 0, then $a_n = b_n$ for all n.

To see how one can obtain a series solution for a differential equation, we begin by analyzing the differential equation

$$\frac{dy}{dx} = 3y$$

whose general solution is Ce^{3x} (either by the results of Section 7.5 or by those of Section 16.4).

Example 1 Find a series solution of the differential equation $dy/dx = 3y$.

Solution Let us assume that $y = \sum_{n=0}^{\infty} c_n x^n$ is a solution, and then determine values of c_n for all n. To that end we differentiate y to obtain

$$\frac{dy}{dx} = \sum_{n=1}^{\infty} n c_n x^{n-1} = \sum_{n=0}^{\infty} (n+1)c_{n+1} x^n$$

Since $dy/dx = 3y$, it follows that

$$\sum_{n=0}^{\infty} (n+1)c_{n+1} x^n = 3 \sum_{n=0}^{\infty} c_n x^n = \sum_{n=0}^{\infty} 3c_n x^n$$

Corollary 9.27 then implies that corresponding coefficients are equal, that is,

$$(n+1)c_{n+1} = 3c_n \quad \text{for all } n$$

so that

$$c_{n+1} = \frac{3}{n+1} c_n \quad \text{for all } n$$

The coefficients are thus defined recursively, once c_0 is determined. In particular,

$$c_1 = \frac{3}{1}c_0 \qquad\qquad c_3 = \frac{3}{3}c_2 = \frac{3^3}{3!}c_0$$

$$c_2 = \frac{3}{2}c_1 = \frac{3^2}{2}c_0 \qquad c_4 = \frac{3}{4}c_3 = \frac{3^4}{4!}c_0$$

and in general,

$$c_n = \frac{3^n}{n!} c_0$$

Therefore the series solution is given by

$$y = \sum_{n=0}^{\infty} c_n x^n = \sum_{n=0}^{\infty} \frac{3^n}{n!} c_0 x^n = c_0 \sum_{n=0}^{\infty} \frac{3^n}{n!} x^n \quad \square$$

Since

$$\sum_{n=0}^{\infty} \frac{3^n}{n!} x^n = \sum_{n=0}^{\infty} \frac{(3x)^n}{n!} = e^{3x}$$

we see that the series we obtained is indeed a solution of the differential equation in Example 1. Because we assumed at the outset of the solution of Example 1 that a series solution existed, it was necessary actually to check that the series we obtained was a genuine solution of the differential equation.

In the next example we will find series solutions of a famous differential equation.

Example 2 Find a series solution of the Bessel equation*

$$x^2 \frac{d^2y}{dx^2} + x \frac{dy}{dx} + x^2 y = 0$$

Solution We assume that $y = \sum_{n=0}^{\infty} c_n x^n$ is a solution, and we set out to determine values for the constants $c_0, c_1, \ldots$, by substituting series for y, dy/dx, and d^2y/dx^2 in the equation. To that end we notice that

$$\frac{dy}{dx} = \sum_{n=1}^{\infty} n c_n x^{n-1}$$

and

$$\frac{d^2y}{dx^2} = \sum_{n-2}^{\infty} n(n-1) c_n x^{n-2}$$

Therefore the differential equation can be rewritten as

$$x^2 \sum_{n=2}^{\infty} n(n-1) c_n x^{n-2} + x \sum_{n=1}^{\infty} n c_n x^{n-1} + x^2 \sum_{n=0}^{\infty} c_n x^n = 0$$

This equation is equivalent to the following equations:

$$\sum_{n=2}^{\infty} n(n-1) c_n x^n + \sum_{n=1}^{\infty} n c_n x^n + \sum_{n=0}^{\infty} c_n x^{n+2} = 0$$

$$\sum_{n=2}^{\infty} n(n-1) c_n x^n + \left(c_1 x + \sum_{n=2}^{\infty} n c_n x^n \right) + \sum_{n=2}^{\infty} c_{n-2} x^n = 0$$

$$c_1 x + \sum_{n=2}^{\infty} [n(n-1) c_n + n c_n + c_{n-2}] x^n = 0$$

This equation is satisfied if $c_1 = 0$ and

$$n(n-1) c_n + n c_n + c_{n-2} = 0$$

that is,

$$(n^2 - n + n) c_n + c_{n-2} = 0$$

or more simply,

$$c_n = -\frac{c_{n-2}}{n^2} \tag{1}$$

We can also write (1) as

$$c_{n+2} = -\frac{c_n}{(n+2)^2} \tag{2}$$

---

* The equation is named after the German mathematician F. B. Bessel (1784–1846), who was especially interested in celestial mechanics. The equation occurs often in mathematical physics.

The fact that $c_1 = 0$ implies by (2) that in turn $c_3 = 0$, $c_5 = 0$, and in general, $c_n = 0$ for all odd positive integers n. For even n we deduce from (2) that

$$c_2 = -\frac{1}{2^2}c_0 \qquad\qquad c_6 = -\frac{1}{6^2}c_4 = \frac{(-1)^3}{2^2 4^2 6^2}c_0$$

$$c_4 = -\frac{1}{4^2}c_2 = \frac{(-1)^2}{2^2 4^2}c_0 \qquad c_8 = -\frac{1}{8^2}c_6 = \frac{(-1)^4}{2^2 4^2 6^2 8^2}c_0$$

and in general,

$$c_{2n} = \frac{(-1)^n}{2^2 \cdot 4^2 \cdot 6^2 \cdots (2n)^2}c_0 = \frac{(-1)^n}{2^{2n}(n!)^2}c_0$$

Consequently a series solution is given by

$$y = \sum_{n=0}^{\infty}\frac{(-1)^n}{2^{2n}(n!)^2}c_0 x^{2n} = c_0 \sum_{n=0}^{\infty}\frac{(-1)^n}{2^{2n}(n!)^2}x^{2n} \tag{3}$$

By making the necessary differentiations of y you can check that the series in (3) is actually a solution to the given Bessel equation. $\square$

Generally it is difficult to determine explicit formulas for the coefficients in a series solution, as Example 2 might indicate.

EXERCISES 16.8

In Exercises 1–6 find a nonzero series solution of the differential equation.

1. $\dfrac{dy}{dx} - 5y = 0$

2. $\dfrac{dy}{dx} + 4xy = 0$

3. $\dfrac{d^2y}{dx^2} + y = 0$

4. $\dfrac{d^2y}{dx^2} - 16y = 0$

5. $\dfrac{d^2y}{dx^2} + x\dfrac{dy}{dx} + y = 0$

6. $x^2\dfrac{d^2y}{dx^2} + x\dfrac{dy}{dx} - 49y = 0$

In Exercises 7–12 find the particular series solution that satisfies the given initial condition(s).

7. $\dfrac{dy}{dx} = xy$; $y(0) = 1$

8. $\dfrac{dy}{dx} = x^2 y$; $y(0) = 2$

9. $x^2\dfrac{d^2y}{dx^2} - 6y = 0$; $y(1) = 5$

10. $\dfrac{d^2y}{dx^2} - 2x\dfrac{dy}{dx} - 2y = 0$; $y(0) = 1$, $y'(0) = 0$

11. $\dfrac{d^2y}{dx^2} - x\dfrac{dy}{dx} - y = 0$; $y(0) = 1$, $y'(0) = 0$

12. $\dfrac{d^2y}{dx^2} - x\dfrac{dy}{dx} - y = 0$; $y(0) = 2$, $y'(0) = 1$

16.9
AUTONOMOUS DIFFERENTIAL EQUATIONS

In Sections 16.1–16.8 we derived formulas for solutions of various differential equations. In this section we will analyze the general nature of solutions of certain differential equations without actually solving the differential equations. We will find that we can determine the general appearance of the graph of the solutions even without solving the equation.

The differential equations we will study are those of the form

$$y' = f(y) \tag{1}$$

where f is a continuously differentiable function. Such an equation is called an **autonomous differential equation**, because the independent variable (customarily t or x) does not appear explicitly in the equation. Normally in physical applications the independent variable is t, and for that reason the equation in (1) is usually written in the form

$$\frac{dy}{dt} = f(y) \tag{2}$$

Examples of autonomous differential equations are

$$\frac{dy}{dt} = ky, \qquad \frac{dy}{dt} = y^2 - 2y - 3, \quad \text{and} \quad \frac{dy}{dt} = \sin y \tag{3}$$

the first equation of which we studied in Section 6.5 on exponential growth and decay. From our discussion below we will be able to sketch the graphs of solutions of the equations in (3). For convenience we will restrict our attention to graphs on $[0, \infty)$.

Now we begin our analysis of solutions of autonomous differential equations. The first result is the following.

THEOREM 16.1

> The constant solutions of $dy/dt = f(y)$ are exactly those functions $y = c$ for which $f(c) = 0$.

Proof On the one hand, if $y = c$ is a constant solution of the equation, then its derivative is 0, so that

$$f(c) = f(y) = \frac{dy}{dt} = 0$$

On the other hand, if $f(c) = 0$ and we let $y = c$, then

$$\frac{dy}{dt} = 0 = f(c) = f(y)$$

so that the function $y = c$ is a solution. Thus the constant solutions of the differential equation are exactly the functions $y = c$ for which $f(c) = 0$. ■

Example 1 Find the constant solutions of

a. $\dfrac{dy}{dt} = y^2 - 2y - 3$

b. $\dfrac{dy}{dt} = e^{-y}$

Solution For (a) we notice that

$$y^2 - 2y - 3 = (y + 1)(y - 3) = 0$$

if $y = -1$ or $y = 3$. By Theorem 16.1, the constant solutions of the differential equation are the functions $y = -1$ and $y = 3$. For (b) we observe that $e^{-y} > 0$ for all y, and thus by Theorem 16.1 there is no constant solution of the differential equation. $\square$

The next theorem gives us much information about the nonconstant solutions of autonomous differential equations.

THEOREM 16.2

> Let $dy/dt = f(y)$ be an autonomous differential equation.
> **a.** The graphs of two distinct solutions cannot intersect.
> **b.** Let y be a solution. If $f(y(0)) > 0$, then y is a strictly increasing function, whereas if $f(y(0)) < 0$, then y is a strictly decreasing function.
> **c.** Let y be a solution. If $f(y(0)) > 0$, then the graph of y is concave upward (downward) on any open interval on which $f'(y(t)) > 0$ ($f'(y(t)) < 0$). If $f(y(0)) < 0$, then the graph of y is concave downward (upward) on any open interval on which $f'(y(t)) > 0$ ($f'(y(t)) < 0$).

Proof To prove (a), suppose that the graphs of two solutions y_1 and y_2 intersect, which means that $y_1(t_0) = y_2(t_0)$ for some value of t_0. Recalling from Section 16.1 that there is at most one solution with the value $y_1(t_0)$ at t_0, we conclude that $y_1 = y_2$. Thus two distinct solutions have graphs that do not intersect. Turning to (b), we suppose that y is a solution and that $f(y(0)) > 0$, so by (1), $y'(0) = f(y(0)) > 0$. We will prove that y is a strictly increasing function. If that were not true, then there would be a value of t_0 for which $y'(t_0) < 0$. The continuity of f and the differential equation $dy/dt = f(y)$ imply that y' is also continuous, so by the Intermediate Value Theorem there would need to be a number t_1 in $[0, t_0]$ for which $y'(t_1) = 0$. Therefore by (1),

$$f(y(t_1)) = y'(t_1) = 0$$

so $y(t_1)$ is a zero of f. It follows from Theorem 16.1 that the constant function $y_1 = y(t_1)$ is a solution. But since $y(t_1) = y_1(t_1)$, and since there is at most one solution with value $y_1(t_1)$ at t_1, we deduce that $y = y_1$. Therefore y is constant, which contradicts the hypothesis that $y'(0) > 0$. As a result, if $y'(0) > 0$, then y is a strictly increasing function. The proof for $y'(0) < 0$ is analogous, so we omit it. To prove (c), we differentiate both sides of (2), obtaining

$$\frac{d^2y}{dt^2} = f'(y(t))\frac{dy}{dt} \tag{4}$$

Since the graph of y is concave upward on an open interval if $d^2y/dt^2 > 0$ on that interval, and similarly the graph is concave downward if $d^2y/dt^2 < 0$, and since dy/dt has the same sign as $f(y(0))$ by part (b), the proof is completed by noting the signs of d^2y/dt^2, $f'(y(t))$, and dy/dt in (4). ∎

The implications of Theorems 16.1 and 16.2 for solutions of the equation $dy/dt = f(y)$ are profound. They tell us that the graphs of the constant solutions split the plane into horizontal strips, and the graph of each nonconstant solution lies in exactly one of the strips. Furthermore, any nonconstant solution is either strictly increasing or strictly decreasing, and the concavity of its graph depends on the sign of f'.

Continuing with our discussion on the concavity of the graph of a solution, we use the fact that dy/dt does *not* change sign to deduce from (4) that if $f'(y(t))$ changes sign at t_0, then so does d^2y/dt^2. This means that the concavity of the graph of y changes at $(t_0, f(t_0))$, that is, there is an inflection point at $(t_0, f(t_0))$. It follows that an inflection point (if any) in the graph of a solution corresponds to a relative extreme value of f.

Finally, it can be proved that a point $(t, y(t))$ on the graph of any nonconstant solution either approaches the graph of a constant solution as t increases without bound, or else gets arbitrarily far away from the t axis.

The preceding discussion yields the following procedure for sketching the graph of a solution of an autonomous equation

$$\frac{dy}{dt} = f(y)$$

for which the initial value $y(0)$ is given.

1. Determine the zeros of f. Each zero c yields a constant solution $y = c$ of the differential equation. Draw the horizontal lines corresponding to the constant solutions on a single ty coordinate system.
2. Compute $f(y(0))$. If $f(y(0)) > 0$ then y is strictly increasing; if $f(y(0)) < 0$, then y is strictly decreasing. Notice that $f(y(0))$ is the slope of the graph of y at 0.
3. From the equation

$$\frac{d^2y}{dt^2} = f'(y(t))\frac{dy}{dt} \tag{5}$$

determine the intervals on which the graph of the solution is concave upward and those on which it is concave downward. If f' changes sign at $y(t_0)$, then the graph of the solution has an inflection point at the point $(t_0, y(t_0))$.
4. Beginning with the point $(0, y(0))$ on the y axis, sketch the graph of y on $[0, \infty)$, using the results of steps 2 and 3. The graph cannot cross any of the horizontal lines drawn in step 1. As t approaches ∞, the point $(t, y(t))$ on the graph will either approach one of the horizontal lines drawn in step 1 or will get arbitrarily far away from the t axis.

Example 2   Sketch the graph of the solution of the differential equation $dy/dt = y^2 - 2y - 3$ for which

a. $y(0) = 2$ b. $y(0) = 4$

Solution   First we let $f(y) = y^2 - 2y - 3$, so that

$$f'(y) = 2y - 2 \tag{6}$$

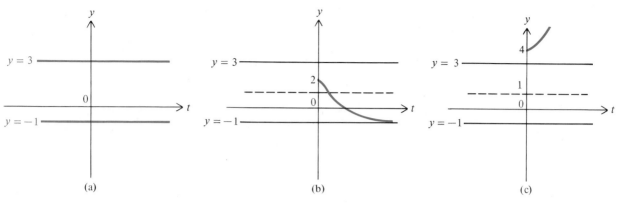

FIGURE 16.8

Since $y^2 - 2y - 3 = (y + 1)(y - 3)$, the zeros of f are -1 and 3. As we observed in Example 1, this means that the constant solutions are $y = -1$ and $y = 3$. Thus we draw the horizontal lines $y = -1$ and $y = 3$ as in Figure 16.8(a). Now in part (a) we have $y(0) = 2$, so that

$$f(y(0)) = f(2) = 4 - 4 - 3 < 0$$

and therefore y is a strictly decreasing function. Consequently $dy/dt < 0$ and $y(t) < y(0) = 2$ for $t > 0$. From (5) and (6) we have

$$\frac{d^2y}{dt^2} = f'(y(t))\frac{dy}{dt} = (2y(t) - 2)\frac{dy}{dt} \tag{7}$$

Since $dy/dt < 0$, it follows that $d^2y/dt^2 < 0$ if $y(t) > 1$, and $d^2y/dt^2 > 0$ if $y(t) < 1$. Thus the part of the graph for which $y(t) > 1$ is concave downward, and the part for which $y(t) < 1$ is concave upward. Consequently as the graph crosses the line $y = 1$, it has an inflection point. Since y is decreasing and the graph of y must lie in the strip bounded by $y = -1$ and $y = 3$, we know from step 4 that the line $y = -1$ is a horizontal asymptote of the graph of y. The graph of·y is sketched in Figure 16.8(b).

For part (b) we draw the same horizontal lines $y = -1$ and $y = 3$ as in part (a). This time, however, since $y(0) = 4$ by hypothesis, we have

$$f(y(0)) = f(4) = 16 - 8 - 3 = 5 > 0$$

so that y is a strictly increasing function. Therefore $dy/dt > 0$ and $y(t) > y(0) = 4$ for $t > 0$. As a result, $2y(t) - 2 > 0$ for all $t > 0$, and thus from (7) we see that $d^2y/dt^2 > 0$ for $t > 0$. Consequently the graph of y is concave upward on $(0, \infty)$. Since y is increasing and the graph of y lies above the horizontal line $y = 3$ and is concave upward, by step 4 we find that the graph is as in Figure 16.8(c).   ☐

A similar analysis would show that if $y(0) = 0$ or $y(0) = -2$ in Example 2, then the corresponding graph of y would be as in Figure 16.9(a). Since all

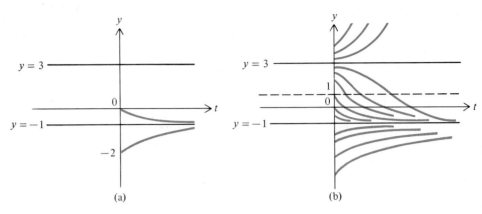

FIGURE 16.9

solutions whose graphs lie in a suitably defined strip have like characteristics, we can represent the whole collection of solutions of an autonomous differential equation such as $dy/dt = y^2 - 2y - 3$ on one graph (Figure 16.9(b)).

EXERCISES 16.9

In Exercises 1–6 find the constant solutions of the differential equation.

1. $\dfrac{dy}{dt} = 2y + 5$

2. $\dfrac{dy}{dt} = 2y(4 - 3y)$

3. $\dfrac{dy}{dt} = y^2 - 2y - 15$

4. $\dfrac{dy}{dt} = 4y^2 - y$

5. $\dfrac{dy}{dt} = y^2 e^{-y}$

6. $\dfrac{dy}{dt} = \sin y$

For each initial condition in Exercises 7–20 sketch the graph of the particular solution that satisfies the initial condition and is defined on $[0, \infty)$.

7. $\dfrac{dy}{dt} = 1 - y;\ y(0) = 2;\ y(0) = -2$

8. $\dfrac{dy}{dt} = y + 2;\ y(0) = -\frac{3}{2};\ y(0) = -3;\ y(0) = -2$

9. $\dfrac{dy}{dt} = y^2 - 2y;\ y(0) = \frac{5}{2};\ y(0) = \frac{3}{2};\ y(0) = \frac{1}{2};\ y(0) = -\frac{1}{2}$

10. $\dfrac{dy}{dt} = -y^2 + 2y;\ y(0) = \frac{5}{2};\ y(0) = \frac{3}{2};\ y(0) = \frac{1}{2};\ y(0) = -\frac{1}{2}$

11. $\dfrac{dy}{dt} = y^2 + 2y + 4;\ y(0) = -\frac{2}{3};\ y(0) = -3$

12. $\dfrac{dy}{dt} = 2y^2 - y - 1;\ y(0) = 2;\ y(0) = \frac{1}{2};\ y(0) = \frac{1}{8};\ y(0) = -2$

13. $\dfrac{dy}{dt} = y^3;\ y(0) = 1;\ y(0) = -1$

14. $\dfrac{dy}{dt} = \dfrac{1}{y};\ y(0) = 1;\ y(0) = -2$

15. $\dfrac{dy}{dt} = y^2(y - 1);\ y(0) = \frac{3}{2};\ y(0) = \frac{3}{4};\ y(0) = \frac{1}{6};\ y(0) = -\frac{1}{2}$

16. $\dfrac{dy}{dt} = y^2(y - 2)^2;\ y(0) = 3;\ y(0) = \frac{3}{2};\ y(0) = \frac{1}{2};\ y(0) = -\frac{1}{2}$

17. $\dfrac{dy}{dt} = y^3(y - 2)^3;\ y(0) = \frac{5}{2};\ y(0) = \frac{3}{2};\ y(0) = \frac{1}{2};\ y(0) = -\frac{1}{2}$

18. $\dfrac{dy}{dt} = \sin y;\ y(0) = \pi/2;\ y(0) = 11\pi/6;\ y(0) = -\pi;$ $y(0) = -5\pi/2$

19. $\dfrac{dy}{dt} = \sin^2 y;\ y(0) = \pi/4;\ y(0) = 3\pi/4;\ y(0) = 5\pi/4;$ $y(0) = 7\pi/4$

20. $\dfrac{dy}{dt} = e^{-y};\ y(0) = 0$

21. Suppose f is continuous, and consider the differential equation

$$\frac{dy}{dt} = f(y)$$

Sketch the graphs of the particular solutions of the

differential equation for which $y(0) = -1$, $y(0) = 1$, and $y(0) = 3$, where the graph of f is as in

a. Figure 16.10(a)   b. Figure 16.10(b)

c. Figure 16.10(c)

22. Newton's Law of Cooling states that if the surrounding temperature is held at a constant temperature $T_1 > 0$, then under ideal conditions the temperature T of an object satisfies the equation

$$\frac{dT}{dt} = -k(T - T_1)$$

where k is a positive constant. Assuming that $T_1 = 10$, sketch the solution of the differential equation with initial condition

a. $T(0) = 15$ b. $T(0) = 5$

23. Let y be the population of a given species. If we assume that the rate of growth dy/dt of the population is proportional both to y and to $a - y$, where a is a positive constant, we obtain the differential equation

$$\frac{dy}{dt} = ky(a - y) \qquad (8)$$

where k is also a positive constant. The number a is called the carrying capacity.

a. Sketch the graph of a solution y whose initial condition satisfies $0 < y(0) < a/2$. (The graph is called a logistic curve.)

b. Sketch the graph of a solution y whose initial condition satisfies $a/2 < y(0) < a$.

*24. In an autocatalytic reaction one substance is converted into another substance that controls its own formation. Assume that the amount y of the new substance satisfies the differential equation

$$\frac{dy}{dt} = k(a - y)(b + y)$$

where k, a, and b are positive constants. Assuming that $a > b$, sketch the graph of the solution of the differential equation with initial condition $y(0) = b$. (*Hint:* There are two cases, depending on whether $a \le 3b$ or $a > 3b$.)

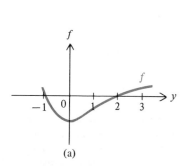

(a)

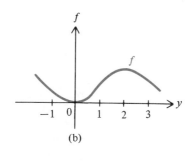

(b)

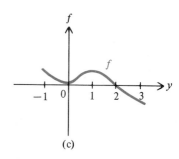

(c)

FIGURE 16.10

Key Terms and Expressions

Differential equation

Solution; implicit solution; explicit solution; general solution; particular solution; series solution

Initial condition

Separable differential equation

Exact differential equation

Linear first-order differential equation

Second-order linear differential equation; homogeneous equation; nonhomogeneous equation

Characteristic equation

Complementary solution; particular solution

Variation of parameters

Autonomous differential equation

REVIEW EXERCISES

In Exercises 1–14 find the general solution of the differential equation.

1. $\dfrac{dy}{dx} + \dfrac{2}{x}y = x^2 + 6$ for $x > 0$

2. $\dfrac{d^2y}{dx^2} + 6\dfrac{dy}{dx} + 2y = 0$

3. $\dfrac{x}{\sqrt{x^2 + y^2}}\,dx + \dfrac{y}{\sqrt{x^2 + y^2}}\,dy = 0$

4. $x\sqrt{1 - y^2} + y\sqrt{1 - x^2}\dfrac{dy}{dx} = 0$

5. $\dfrac{d^2y}{dx^2} - 4\dfrac{dy}{dx} + 8y = 0$

6. $\dfrac{dy}{dx} - y\cot x = \csc x$ for $0 < x < \pi$

7. $\dfrac{dy}{dx} + y\cot x = \csc x$ for $0 < x < \pi$

8. $(x + \sin y - \cos y)\,dx + x(\cos y + \sin y)\,dy = 0$

9. $\cosh 2x \cosh 2y\,dx + \sinh 2x \sinh 2y\,dy = 0$

10. $\dfrac{d^2y}{dx^2} + 8\dfrac{dy}{dx} + 16y = 0$

11. $e^{y^2}\,dx + x^2 y\,dy = 0$

12. $\dfrac{d^2y}{dx^2} + y = \sec^2 x$ for $-\pi/2 < x < \pi/2$

13. $\dfrac{d^2y}{dx^2} + 4y = \cos 2x$ 14. $\dfrac{d^2y}{dx^2} - \dfrac{dy}{dx} - 2y = 2e^{-x}$

In Exercises 15–24 find the particular solution that satisfies the given initial condition(s).

15. $2x(y + 1)\,dx = y\,dy;\ y(0) = -2$

16. $\dfrac{dy}{dx} = xy + x;\ y(1) = 2$

17. $\dfrac{d^2y}{dx^2} - 2\dfrac{dy}{dx} + 3y = 0;\ y(0) = 1,\ y'(0) = 3$

18. $\dfrac{d^2y}{dx^2} + 2\dfrac{dy}{dx} - 3y = 0;\ y(0) = 1,\ y'(0) = 3$

19. $y(1 + x^2)\,dy + (y^2 + 1)\,dx = 0;\ y(0) = \sqrt{3}$

20. $(e^y + y\cos x)\,dx + (xe^y + \sin x + 1)\,dy = 0;$ $y(3\pi/2) = -1$

21. $\dfrac{dy}{dx} + y\tan x = \sec x$ for $-\pi/2 < x < \pi/2;\ y(0) = \pi/4$

22. $\dfrac{d^2y}{dx^2} - 2y = 0;\ y(0) = \tfrac{1}{2}\sqrt{2},\ y'(0) = -\tfrac{1}{2}$

23. $\left(\dfrac{3}{2}x^2 - 3y\right)dy = x(x^4 - 3y)\,dx;\ y(-1) = 2$

24. $\dfrac{d^2y}{dx^2} - 5\dfrac{dy}{dx} + 6y = 8e^x;\ y(0) = 0,\ y'(0) = 0$

25. Find the general series solution of
$$\dfrac{d^2y}{dx^2} - 3x\dfrac{dy}{dx} - 3y = 0.$$

26. Find the particular series solution of
$$x^2\dfrac{d^2y}{dx^2} - x\dfrac{dy}{dx} + y = 0$$
that satisfies the initial conditions $y(0) = 0$ and $y'(0) = -7$.

For each initial condition in Exercises 27–30 sketch the graph of the particular solution that satisfies the initial condition and is defined on $[0, \infty)$.

27. $\dfrac{dy}{dt} = 2y^2 - 7y + 5;\ y(0) = \tfrac{1}{2};\ y(0) = \tfrac{3}{2};\ y(0) = 2;\ y(0) = 3$

28. $\dfrac{dy}{dt} = y^3(y - 2);\ y(0) = -1;\ y(0) = 1;\ y(0) = \tfrac{7}{4};\ y(0) = \tfrac{5}{2}$

29. $\dfrac{dy}{dt} = ye^y;\ y(0) = -2;\ y(0) = -\tfrac{1}{2};\ y(0) = \tfrac{1}{2}$

30. $\dfrac{dy}{dt} = \cos y;\ y(0) = 1;\ y(0) = -1$

31. A 3-pound weight stretches a spring 6 inches. If the weight is pulled down an additional inch and then released with an initial velocity of 0.5 inch per second, find a formula for the position of the weight at any subsequent time, assuming that the motion is undamped.

32. An 8-pound weight stretches a spring 2 inches. Suppose the weight is set in motion and experiences a damping force with damping constant 4.1. Will the motion be underdamped, critically damped, or overdamped? Explain your answer.

33. Let y be the population of a given species. Assume that the death rate is proportional to y, and the birth rate is proportional to the product of the number of males and the number of females (assumed to be equal). Then
$$\dfrac{dy}{dt} = ay^2 - by$$
where a and b are positive constants. Show that if $0 < y(0) < b/a$, then the species will become extinct.

APPENDIX

PROOFS OF SELECTED THEOREMS

In the Appendix we present the proofs of all theorems that were not proved in Chapters 1–9, with the exception of Theorem 7.15. The first collection of theorems will be proved from the definition of limit and other definitions and results already appearing in the text.

First we need an elementary observation about the limits of a pair of functions. In finding the limits of two functions at a point by use of ε and δ, we may choose the same δ for both functions. We will prove this in the following lemma.

LEMMA A.1

Assume that $\lim_{x \to a} f(x) = L$ and $\lim_{x \to a} g(x) = M$. For any $\varepsilon > 0$, there is a common $\delta > 0$ such that if $0 < |x - a| < \delta$, then both

$$|f(x) - L| < \varepsilon \quad \text{and} \quad |g(x) - M| < \varepsilon$$

Proof   For any $\varepsilon > 0$, there is a $\delta_1 > 0$ such that

$$\text{if } 0 < |x - a| < \delta_1, \quad \text{then} \quad |f(x) - L| < \varepsilon$$

and a $\delta_2 > 0$ such that

$$\text{if } 0 < |x - a| < \delta_2, \quad \text{then} \quad |g(x) - M| < \varepsilon$$

Let δ be the minimum of δ_1 and δ_2; then $\delta > 0$. Moreover, if $0 < |x - a| < \delta$, then we certainly have both $0 < |x - a| < \delta_1$ and $0 < |x - a| < \delta_2$. Therefore

$$|f(x) - L| < \varepsilon \quad \text{and} \quad |g(x) - M| < \varepsilon \quad \blacksquare$$

We now turn to the limit theorems. First we prove that if f has a limit at a, then f has *only one* limit at a.

THEOREM A.2
UNIQUENESS OF LIMITS

If the limit of a function f at a exists, then this limit is unique. Equivalently, if L and M are both limits of f at a, then $L = M$.

Proof   Let $\varepsilon > 0$, and let $f = g$ in Lemma A.1. The lemma says that there is a common $\delta > 0$ such that if $0 < |x - a| < \delta$, then

$$|f(x) - L| < \varepsilon \quad \text{and} \quad |f(x) - M| < \varepsilon$$

If we choose an x such that $0 < |x - a| < \delta$, we can conclude that

$$|L - M| = |L - f(x) + f(x) - M|$$
$$\leq |L - f(x)| + |f(x) - M| < \varepsilon + \varepsilon = 2\varepsilon$$

Consequently for *any* positive ε, the distance from L to M is less than 2ε. This implies that $L = M$.   ∎

THEOREM A.3
SQUEEZING THEOREM
(THEOREM 2.3)

Assume that $f(x) \leq g(x) \leq h(x)$ for all x in some open interval I about a, except possibly a itself. If $\lim_{x \to a} f(x) = \lim_{x \to a} h(x) = L$, then $\lim_{x \to a} g(x)$ exists and $\lim_{x \to a} g(x) = L$.

Proof   Let $\varepsilon > 0$. By Lemma A.1 there is a common $\delta > 0$ such that if $0 < |x - a| < \delta$, then x is in I and

$$|f(x) - L| < \varepsilon \quad \text{and} \quad |h(x) - L| < \varepsilon$$

Since $f(x) \leq g(x)$, the first inequality above implies that

$$L - \varepsilon < f(x) \leq g(x) \tag{1}$$

Similarly, since $|h(x) - L| < \varepsilon$ and $g(x) \leq h(x)$,

$$g(x) \leq h(x) < L + \varepsilon \tag{2}$$

Together (1) and (2) imply that

$$L - \varepsilon < g(x) < L + \varepsilon$$

and thus

$$|g(x) - L| < \varepsilon \quad ∎$$

THEOREM A.4
SUM RULE
FOR LIMITS
(THEOREM 2.2)

Assume that $\lim_{x \to a} f(x)$ and $\lim_{x \to a} g(x)$ both exist. Then $\lim_{x \to a} [f(x) + g(x)]$ exists, and

$$\lim_{x \to a} [f(x) + g(x)] = \lim_{x \to a} f(x) + \lim_{x \to a} g(x)$$

Proof   Let $\lim_{x \to a} f(x) = L$, $\lim_{x \to a} g(x) = M$, and $\varepsilon > 0$. We will show that there is a $\delta > 0$ such that if $0 < |x - a| < \delta$, then $|[f(x) + g(x)] - (L + M)| < \varepsilon$. Suppose we replace ε by $\varepsilon/2$ in Lemma A.1. The lemma implies that there is a common $\delta > 0$ such that if $0 < |x - a| < \delta$, then

$$|f(x) - L| < \frac{\varepsilon}{2} \quad \text{and} \quad |g(x) - M| < \frac{\varepsilon}{2}$$

Therefore

$$|[f(x) + g(x)] - (L + M)| = |[f(x) - L] + [g(x) - M]|$$
$$\leq |f(x) - L| + |g(x) - M|$$
$$< \frac{\varepsilon}{2} + \frac{\varepsilon}{2} = \varepsilon \quad \blacksquare$$

THEOREM A.5
CONSTANT MULTIPLE RULE FOR LIMITS
(THEOREM 2.2)

Assume that $\lim_{x \to a} f(x)$ exists and that c is any real number. Then $\lim_{x \to a} cf(x)$ exists, and

$$\lim_{x \to a} cf(x) = c \lim_{x \to a} f(x)$$

Proof If $c = 0$, the proof is trivial, because both sides of the equation are 0. If $c \neq 0$, we let $\lim_{x \to a} f(x) = L$ and choose an arbitrary $\varepsilon > 0$. We will show that for some $\delta > 0$, if $0 < |x - a| < \delta$, then $|cf(x) - cL| < \varepsilon$. Note that $\varepsilon/|c| > 0$, so that corresponding to this positive quantity there is a $\delta > 0$ such that if $0 < |x - a| < \delta$, we have

$$|f(x) - L| < \varepsilon/|c|$$

Consequently

$$|cf(x) - cL| = |c||f(x) - L| < |c|\frac{\varepsilon}{|c|} = \varepsilon$$

and thus

$$\lim_{x \to a} cf(x) = cL = c \lim_{x \to a} f(x) \quad \blacksquare$$

To prove that the limit of a product is the product of the limits, we first consider the special case in which the limits are 0.

LEMMA A.6

Assume that $\lim_{x \to a} f(x) = 0$ and $\lim_{x \to a} g(x) = 0$. Then $\lim_{x \to a} f(x)g(x)$ exists, and

$$\lim_{x \to a} f(x)g(x) = 0$$

Proof Let $\varepsilon > 0$. We may assume that $0 < \varepsilon < 1$. Let 0 replace both L and M in Lemma A.1. The lemma says that we can find a common $\delta > 0$ such that if $0 < |x - a| < \delta$, then

$$|f(x) - 0| < \varepsilon \quad \text{and} \quad |g(x) - 0| < \varepsilon$$

Thus

$$|f(x)g(x) - 0| = |f(x) - 0||g(x) - 0| < \varepsilon \cdot \varepsilon = \varepsilon^2 < \varepsilon \quad \blacksquare$$

THEOREM A.7
PRODUCT RULE
FOR LIMITS
(THEOREM 2.2)

Assume that $\lim_{x \to a} f(x)$ and $\lim_{x \to a} g(x)$ both exist. Then $\lim_{x \to a} f(x)g(x)$ exists, and

$$\lim_{x \to a} f(x)g(x) = \lim_{x \to a} f(x) \lim_{x \to a} g(x)$$

Proof   Let $\lim_{x \to a} f(x) = L$ and $\lim_{x \to a} g(x) = M$. Then from the Sum Rule we deduce that

$$\lim_{x \to a} [f(x) - L] = \lim_{x \to a} [f(x) + (-L)] = \lim_{x \to a} f(x) + \lim_{x \to a} (-L)$$

$$= L + (-L) = 0$$

Similarly,

$$\lim_{x \to a} [g(x) - M] = 0$$

Next, the product $f(x)g(x)$ can be written in the form

$$f(x)g(x) = [f(x) - L][g(x) - M] + [Lg(x) + Mf(x)] - LM$$

as can be verified by multiplying out $[f(x) - L][g(x) - M]$. Now by Lemma A.6,

$$\lim_{x \to a} [f(x) - L][g(x) - M] = 0$$

Moreover, the Constant Multiple Rule yields

$$\lim_{x \to a} Lg(x) = LM, \qquad \lim_{x \to a} f(x)M = LM, \quad \text{and} \quad \lim_{x \to a} (-LM) = -LM$$

By the Sum Rule we obtain

$$\lim_{x \to a} f(x)g(x) = \lim_{x \to a} [f(x) - L][g(x) - M] + \lim_{x \to a} Lg(x)$$

$$+ \lim_{x \to a} f(x)M + \lim_{x \to a} (-LM)$$

$$= 0 + LM + LM - LM$$

$$= LM \quad \blacksquare$$

Before proving the Quotient Rule for Limits, we prove a preliminary result.

LEMMA A.8

If $\lim_{x \to a} g(x) = M > 0$, then there is a $\delta > 0$ such that

$$\text{if } 0 < |x - a| < \delta, \quad \text{then} \quad g(x) > \frac{1}{2}M$$

If $\lim_{x \to a} g(x) = M < 0$, then there is a $\delta > 0$ such that

$$\text{if } 0 < |x - a| < \delta, \quad \text{then} \quad g(x) < \frac{1}{2}M$$

Proof   Suppose that $\lim_{x \to a} g(x) = M > 0$. Then by Definition 2.1 with $\varepsilon = \frac{1}{2}M$, there is a $\delta > 0$ such that

$$\text{if } 0 < |x - a| < \delta, \quad \text{then} \quad |g(x) - M| < \frac{1}{2}M$$

Consequently if $0 < |x - a| < \delta$, then

$$g(x) = M - (M - g(x)) \geq M - |M - g(x)| > M - \frac{1}{2}M = \frac{1}{2}M$$

Similarly, if $\lim_{x \to a} g(x) = M < 0$, then there is a $\delta > 0$ such that

$$\text{if } 0 < |x - a| < \delta, \quad \text{then} \quad |g(x) - M| < -\frac{1}{2}M$$

Consequently if $0 < |x - a| < \delta$, then

$$g(x) = (g(x) - M) + M \leq |g(x) - M| + M < -\frac{1}{2}M + M = \frac{1}{2}M \quad \blacksquare$$

It follows directly from Lemma A.8 that

$$\text{if } \lim_{x \to a} g(x) > 0, \quad \text{then} \quad g(x) > 0 \text{ for all } x \text{ sufficiently close to } a$$

and

$$\text{if } \lim_{x \to a} g(x) < 0, \quad \text{then} \quad g(x) < 0 \text{ for all } x \text{ sufficiently close to } a$$

These are the statements in (9) and (10) of Section 2.3, which were to be proved here in the Appendix.

Now we are prepared to prove the Quotient Rule for Limits.

THEOREM A.9
QUOTIENT RULE
FOR LIMITS
(THEOREM 2.2)

Suppose that $\lim_{x \to a} f(x)$ and $\lim_{x \to a} g(x)$ exist and that $\lim_{x \to a} g(x) \neq 0$. Then $\lim_{x \to a} f(x)/g(x)$ exists, and

$$\lim_{x \to a} \frac{f(x)}{g(x)} = \frac{\lim_{x \to a} f(x)}{\lim_{x \to a} g(x)}$$

Proof   Let $\lim_{x \to a} f(x) = L$ and $\lim_{x \to a} g(x) = M \neq 0$. From Lemma A.8 it follows that there is a $\delta > 0$ such that if $0 < |x - a| < \delta$, then $|g(x)| > |M|/2$. For such x we have

$$\left| \frac{1}{Mg(x)} \right| < \frac{2}{M^2}$$

Consequently

$$\left| \frac{f(x)}{g(x)} - \frac{L}{M} \right| = \left| \frac{Mf(x) - Lg(x)}{Mg(x)} \right| = |Mf(x) - Lg(x)| \left| \frac{1}{Mg(x)} \right|$$

$$\leq |Mf(x) - Lg(x)| \left(\frac{2}{M^2} \right)$$

As a result, for $0 < |x - a| < \delta$, the function

$$\left| \frac{f(x)}{g(x)} - \frac{L}{M} \right|$$

is squeezed between the function 0 and the function

$$|Mf(x) - Lg(x)| \left(\frac{2}{M^2} \right)$$

But by the Sum and Constant Multiple Rules,

$$\lim_{x \to a} [Mf(x) - Lg(x)] = M \lim_{x \to a} f(x) - L \lim_{x \to a} g(x)$$

$$= ML - LM = 0$$

Thus

$$\lim_{x \to a} |Mf(x) - Lg(x)| \left(\frac{2}{M^2} \right) = 0.$$

Therefore by the Squeezing Theorem we have

$$\lim_{x \to a} \left| \frac{f(x)}{g(x)} - \frac{L}{M} \right| = 0$$

which is equivalent to

$$\lim_{x \to a} \frac{f(x)}{g(x)} = \frac{L}{M} \quad \blacksquare$$

THEOREM A.10
SUBSTITUTION RULE
FOR LIMITS

Suppose $\lim_{x \to a} f(x) = c$ and $f(x) \neq c$ for all x in some open interval about a, with the possible exception of a itself. Suppose also that $\lim_{y \to c} g(y)$ exists. Then

$$\lim_{x \to a} g(f(x)) = \lim_{y \to c} g(y)$$

Proof   Let $\varepsilon > 0$. Suppose $\lim_{y \to c} g(y) = L$. Then there is a $\delta_1 > 0$ such that

$$\text{if } 0 < |y - c| < \delta_1, \quad \text{then} \quad |g(y) - L| < \varepsilon \tag{3}$$

Since $\lim_{x \to a} f(x) = c$ and $\delta_1 > 0$, there is a $\delta > 0$ such that

$$\text{if } 0 < |x - a| < \delta, \quad \text{then} \quad |f(x) - c| < \delta_1$$

By hypothesis, δ may be chosen so small that if $0 < |x - a| < \delta$, then $f(x) \neq c$. Thus

$$\text{if } 0 < |x - a| < \delta, \quad \text{then} \quad 0 < |f(x) - c| < \delta_1$$

and hence by (3),

$$|g(f(x)) - L| < \varepsilon$$

Consequently

$$\lim_{x \to a} g(f(x)) = L = \lim_{y \to c} g(y) \quad \blacksquare$$

Our final limit theorem shows how one-sided limits and two-sided limits are related.

THEOREM A.11
(THEOREM 2.5)

Let f be defined in an open interval about a, except possibly at a itself. Then $\lim_{x \to a} f(x)$ exists if and only if both one-sided limits, $\lim_{x \to a^+} f(x)$ and $\lim_{x \to a^-} f(x)$, exist and

$$\lim_{x \to a^+} f(x) = \lim_{x \to a^-} f(x)$$

In that case

$$\lim_{x \to a} f(x) = \lim_{x \to a^+} f(x) = \lim_{x \to a^-} f(x)$$

Proof From the definitions, if $\lim_{x \to a} f(x) = L$, then $\lim_{x \to a^+} f(x) = L$ and $\lim_{x \to a^-} f(x) = L$. Conversely, assume that $\lim_{x \to a^+} f(x) = L$ and $\lim_{x \to a^-} f(x) = L$, and let $\varepsilon > 0$. It follows from Definition 2.4 that there exist $\delta_1 > 0$ and $\delta_2 > 0$ such that

$$\text{if } 0 < x - a < \delta_1, \quad \text{then} \quad |f(x) - L| < \varepsilon$$

and

$$\text{if } -\delta_2 < x - a < 0, \quad \text{then} \quad |f(x) - L| < \varepsilon$$

Let $\delta > 0$ be the minimum of δ_1 and δ_2. It follows directly that

$$\text{if } 0 < |x - a| < \delta, \quad \text{then} \quad |f(x) - L| < \varepsilon$$

This implies that $\lim_{x \to a} f(x) = L$. $\quad \blacksquare$

Next we prove the Chain Rule for derivatives.

THEOREM A.12
CHAIN RULE
(THEOREM 3.8)

Let f be differentiable at a, and let g be differentiable at $f(a)$. Then $g \circ f$ is differentiable at a, and

$$(g \circ f)'(a) = g'(f(a))f'(a)$$

Proof Let

$$G(x) = \begin{cases} \dfrac{g(x) - g(f(a))}{x - f(a)} & \text{for } x \neq f(a) \\[2ex] g'(f(a)) & \text{for } x = f(a) \end{cases}$$

By the definition of the derivative of g at $f(a)$ and by the Substitution Rule,

$$\lim_{x \to a} G(f(x)) = \lim_{y \to f(a)} G(y) = \lim_{y \to f(a)} \frac{g(y) - g(f(a))}{y - f(a)} = g'(f(a))$$

By considering the cases $f(x) = f(a)$ and $f(x) \neq f(a)$ separately, you can show that

$$\frac{g(f(x)) - g(f(a))}{x - a} = G(f(x)) \left(\frac{f(x) - f(a)}{x - a} \right)$$

By the Product Rule we conclude that

$$\lim_{x \to a} \frac{g(f(x)) - g(f(a))}{x - a} = \lim_{x \to a} G(f(x)) \lim_{x \to a} \left(\frac{f(x) - f(a)}{x - a} \right)$$

$$= g'(f(a)) f'(a)$$

Thus $(g \circ f)'(a) = g'(f(a)) f'(a)$. ■

The Least Upper Bound Axiom and Its Consequences

The proofs of the theorems in this subsection rely on a property of the real numbers that has not yet been discussed in this book. Before introducing this property, we extend the notion of boundedness that was introduced for intervals in Section 1.1. We make the blanket assumption that any set discussed in this Appendix contains at least one number.

DEFINITION A.13

a. A set S of numbers is **bounded above** if there is a number M such that $x \leq M$ for every x in S. Any such number M is called an **upper bound** of S.
b. A set S of numbers is **bounded below** if there is a number m such that $m \leq x$ for every x in S. Any such number m is called a **lower bound** of S.
c. A set of numbers is **bounded** if it is bounded above and below.

Any interval of the form (a, b) $[a, b]$, $(a, b]$, or $[a, b)$ is bounded according to Definition A.13. In each case b is an upper bound and a is a lower bound of the interval. Any interval of the form $(-\infty, a)$ or $(-\infty, a]$ is bounded above but not below. Analogously, any interval of the form (a, ∞) or $[a, \infty)$ is bounded below but not above. Hence Definition A.13 implies that intervals of the form (a, ∞), $[a, \infty)$, $(-\infty, a)$, or $(-\infty, a]$ are not bounded.

If M_1 is an upper bound of a set S and if $M_2 > M_1$, then M_2 is also an upper bound of S. Thus if a set has an upper bound, then it has infinitely many upper bounds, and an upper bound of such a set may be chosen arbitrarily large. Our interest will center on the possibility that there exists a *smallest* upper bound of a set.

DEFINITION A.14

a. A number M is a **least upper bound** of a set S if M is an upper bound of S and if no upper bound of S is less than M.
b. A number m is a **greatest lower bound** of a set S if m is a lower bound of S and if no lower bound of S is greater than m.

If M_1 and M_2 are both least upper bounds of S, then by definition neither can be less than the other, and thus $M_1 = M_2$. We conclude that a set can have at most one least upper bound. Similarly, a set can have at most one greatest lower bound. The least upper bound of an interval of the form (a, b), $[a, b]$, $(a, b]$, or $[a, b)$ is b, and the greatest lower bound is a. From these examples we see that the least upper bound, and also the greatest lower bound, may or may not be in a given set. For an example not involving an interval, let S be the set of all rational numbers in the open interval $(0, 1)$ that have decimal expansions containing only finitely many zeros and nines and no other integers. Some examples of numbers in S are 0.9999, 0.900009, amd 0.90900909. The greatest lower bound of S is 0, and the least upper bound of S is 1. However, neither 0 nor 1 belongs to S. In contrast, an unbounded interval like (a, ∞) and $[a, \infty)$ has no least upper bound, and neither $(-\infty, a)$ nor $(-\infty, a]$ has a greatest lower bound.

Now we state the property of real numbers mentioned earlier.

LEAST UPPER BOUND AXIOM

Every set of real numbers that is bounded above has a least upper bound.

It is possible to prove that the Least Upper Bound Axiom is equivalent to the property that every set of real numbers that is bounded below has a greatest lower bound (see Exercise 11). It follows that any set that is contained in a bounded interval has a least upper bound and a greatest lower bound, each of which may or may not be contained in the set itself.

We are now in a position to prove the theorems that rely on the Least Upper Bound Axiom.

THEOREM A.15
INTERMEDIATE VALUE THEOREM
(THEOREM 2.13)

Suppose f is continuous on a closed interval $[a, b]$. Let p be any number between $f(a)$ and $f(b)$, so that $f(a) \le p \le f(b)$ or $f(b) \le p \le f(a)$. Then there exists a number c in $[a, b]$ such that $f(c) = p$.

Proof If $f(a) = p$ or $f(b) = p$, then let $c = a$ or $c = b$. Otherwise, either $f(a) < p < f(b)$ or $f(b) < p < f(a)$. The proof of the theorem is similar for the two cases, so we will only prove the theorem under the assumption that $f(a) < p < f(b)$. Let S be the collection of all x in $[a, b]$ such that $f(x) < p$. Then a is in S and b is an upper bound of S. By the Least Upper Bound Axiom, S has a least upper bound, which we will call c; notice that $a \le c \le b$. Either $f(c) < p$, $f(c) > p$, or $f(c) = p$. If $f(c) < p$, it follows that $c < b$, since $p < f(b)$ by prior assumption, and that $f(c) + \varepsilon < p$ for sufficiently small ε. Since f is continuous at c, there is a $\delta > 0$ such that if $|x - c| < \delta$, then x is in $[a, b]$ and $|f(c) - f(x)| < \varepsilon$. In particular, if we take $x = c + \delta/2$, then we have $|f(c) - f(c + \delta/2)| < \varepsilon$, which implies that

$$f\left(c + \frac{\delta}{2}\right) < f(c) + \varepsilon < p$$

But then $c + \delta/2$ is in S, so c is not the least upper bound of S. Consequently the assumption that $f(c) < p$ is false. Analogously, the assumption that $f(c) > p$ is false. Therefore $f(c) = p$. ∎

One implication of the Least Upper Bound Axiom that is particularly useful in proofs involves covers of sets. A collection $\mathscr{C}$ of open intervals is a **cover** of a set S, or **covers** S, if every point in S is in at least one of the intervals comprising $\mathscr{C}$. The following theorem, due to Eduard Heine (1821–1881) and Emile Borel (1871–1956), is derived from the Least Upper Bound Axiom.

THEOREM A.16

HEINE–BOREL THEOREM

Let $\mathscr{C}$ be a collection of open intervals that cover a closed interval $[a, b]$. Then there is a finite subcollection $\mathscr{C}_0$ of $\mathscr{C}$ that covers $[a, b]$.

Proof   Let S consist of all x in $[a, b]$ such that some finite subcollection of $\mathscr{C}$ covers $[a, x]$. Then a is in S and b is an upper bound of S. By the Least Upper Bound Axiom, S has a least upper bound c, and $a \leq c \leq b$. Since $\mathscr{C}$ covers $[a, b]$, one member of $\mathscr{C}$, say (d, e), contains c. Since $d < c$ and c is the least upper bound of S, d is not an upper bound of S. Consequently there is a number x in S such that $d < x \leq c$. Since x is in S, a finite subcollection $\mathscr{C}_1$ of $\mathscr{C}$ covers $[a, x]$. Let $\mathscr{C}_0$ consist of $\mathscr{C}_1$ and (d, e). Since (d, e) contains $[x, c]$, $\mathscr{C}_0$ covers $[a, c]$. It cannot be the case that $c < b$, because then there would be a number y in $[a, b]$ such that $c < y < e$, and thus $\mathscr{C}_0$ would cover $[a, y]$. That would mean that y is in S, so that c would not be an upper bound of S, a contradiction to our assumption. Thus $c = b$; since $c < e$, $\mathscr{C}_0$ covers $[a, b]$. ∎

We now use the Heine–Borel Theorem to prove a preliminary result on boundedness of functions. We say that a function f is **bounded on a set** S if there are numbers m and M such that

$$m \leq f(x) \leq M \quad \text{for all } x \text{ in } S$$

that is, if the values assumed by f on S form a bounded set.

THEOREM A.17

If f is continuous on a closed interval $[a, b]$, then f is bounded on $[a, b]$.

Proof   Since f is continuous on $[a, b]$, it follows that for each t in $[a, b]$ there is a number $\delta(t) > 0$ such that if x is in $[a, b]$ and $t - \delta(t) < x < t + \delta(t)$, then $f(t) - 1 < f(x) < f(t) + 1$. The collection of open intervals $(t - \delta(t), t + \delta(t))$, for t in $[a, b]$, forms a cover $\mathscr{C}$ of $[a, b]$. By the Heine–Borel Theorem there is a finite subcollection $\mathscr{C}_0$ of $\mathscr{C}$ that covers $[a, b]$. Let the intervals comprising $\mathscr{C}_0$ be $(t_1 - \delta(t_1), t_1 + \delta(t_1))$, $(t_2 - \delta(t_2), t_2 + \delta(t_2)), \ldots, (t_n - \delta(t_n), t_n + \delta(t_n))$, and let m be the smallest of the numbers $f(t_1) - 1, f(t_2) - 1, \ldots, f(t_n) - 1$, and M the largest of the numbers $f(t_1) + 1, f(t_2) + 1, \ldots, f(t_n) + 1$. Each x in $[a, b]$ lies in one of the subintervals $(t_k - \delta(t_k), t_k + \delta(t_k))$ comprising $\mathscr{C}_0$, so $m \leq f(t_k) - 1 < f(x) < f(t_k) + 1 \leq M$. Therefore f is bounded on $[a, b]$. ∎

THEOREM A.18
MAXIMUM–MINIMUM
THEOREM
(THEOREM 4.2)

Let f be continuous on a closed interval $[a, b]$. Then f has a maximum and a minimum value on $[a, b]$.

Proof Let S be the set of values assumed by f on $[a, b]$. By Theorem A.17, f is bounded, and thus S is bounded. Applying the Least Upper Bound Axiom and the remark following it, we conclude that S has both a least upper bound M and a greatest lower bound m. If $f(x) \neq M$ for all x in $[a, b]$, then let g be defined by

$$g(x) = \frac{1}{M - f(x)} \quad \text{for } a \leq x \leq b$$

By construction g is positive and continuous on $[a, b]$, so Theorem A.17 implies that g is bounded. Thus there is a positive number L such that

$$\frac{1}{M - f(x)} \leq L \quad \text{for } a \leq x \leq b$$

Solving for $f(x)$, we obtain

$$f(x) \leq M - \frac{1}{L} \quad \text{for } a \leq x \leq b$$

Therefore $M - 1/L$ is an upper bound of S, and this contradicts the assumption that M is the *least* upper bound of S. Consequently the assumption that $f(x) \neq M$ for all x in $[a, b]$ is false, so $f(c) = M$ for some number c in $[a, b]$. In other words, f has a maximum value on $[a, b]$. A similar argument proves that for an appropriate number d in $[a, b]$, we have $f(d) = m$, so that f has a minimum value on $[a, b]$. ■

Recall that a function f is continuous on an interval I if for every x in I and every $\varepsilon > 0$ there is a $\delta > 0$, depending on ε and x, such that

if y is in I and $|x - y| < \delta$, then $|f(x) - f(y)| < \varepsilon$

We emphasize that δ depends on both ε and x. There are functions f and intervals I for which δ need only depend on ε, and not on x.

DEFINITION A.19

A function f is **uniformly continuous on an interval** I if for every $\varepsilon > 0$ there is a $\delta > 0$ such that

if x and y are in I and $|x - y| < \delta$, then $|f(x) - f(y)| < \varepsilon$

Notice that any uniformly continuous function is continuous. The converse false, as Exercise 13 shows. However, it turns out that any function that is ˙ntinuous on a *closed, bounded* interval is also uniformly continuous on the terval. That is the content of the following theorem, which provides the key to ˙ır proof of the existence of the definite integral.

THEOREM A.20

> If f is continuous on a closed interval $[a, b]$, then f is uniformly continuous on $[a, b]$.

Proof Let $\varepsilon > 0$. Since f is continuous on $[a, b]$, for each t in $[a, b]$ there is a number $\delta(t) > 0$ such that

$$\text{if } x \text{ is in } [a, b] \text{ and } |t - x| < \delta(t), \quad \text{then} \quad |f(x) - f(t)| < \frac{\varepsilon}{2}$$

The collection $\mathscr{C}$ of open intervals $(t - \frac{1}{2}\delta(t), t + \frac{1}{2}\delta(t))$, for t in $[a, b]$, covers $[a, b]$. By the Heine–Borel Theorem a finite subcollection $\mathscr{C}_0$ covers $[a, b]$. Let the intervals comprising $\mathscr{C}_0$ be $(t_1 - \frac{1}{2}\delta(t_1), t_1 + \frac{1}{2}\delta(t_1)), (t_2 - \frac{1}{2}\delta(t_2), t_2 + \frac{1}{2}\delta(t_2)), \ldots,$ $(t_n - \frac{1}{2}\delta(t_n), t_n + \frac{1}{2}\delta(t_n))$, and let δ be the smallest of the numbers $\frac{1}{2}\delta(t_1), \frac{1}{2}\delta(t_2), \ldots,$ $\frac{1}{2}\delta(t_n)$. If x is in $[a, b]$, then x must lie in one such interval, say $(t_k - \frac{1}{2}\delta(t_k),$ $t_k + \frac{1}{2}\delta(t_k))$. Then $|x - t_k| < \frac{1}{2}\delta(t_k)$. Therefore if y is in $[a, b]$ and $|y - x| < \delta$, then

$$|y - t_k| = |y - x + x - t_k| \le |y - x| + |x - t_k| < \delta + \tfrac{1}{2}\delta(t_k) \le \delta(t_k)$$

By definition of $\delta(t_k)$ it follows that

$$|f(x) - f(t_k)| < \frac{\varepsilon}{2} \quad \text{and} \quad |f(y) - f(t_k)| < \frac{\varepsilon}{2}$$

Therefore

$$|f(x) - f(y)| = |f(x) - f(t_k) + f(t_k) - f(y)|$$
$$\le |f(x) - f(t_k)| + |f(t_k) - f(y)|$$
$$< \frac{\varepsilon}{2} + \frac{\varepsilon}{2} = \varepsilon$$

Thus f is uniformly continuous on $[a, b]$ by Definition A.19. ∎

In Section 5.2 we asserted that if $a < b$ and if f is continuous on $[a, b]$, then there is exactly one number I greater than or equal to every lower sum and smaller than or equal to every upper sum of f. We then defined $\int_a^b f(x)\, dx$ to be that number. Now we will substantiate the claim that I exists.

THEOREM A.21

> Let f be continuous on $[a, b]$. Then there is a unique number I satisfying
>
> $$L_f(\mathscr{P}) \le I \le U_f(\mathscr{P})$$
>
> for every partition $\mathscr{P}$ of $[a, b]$.

Proof We observed in Sections 5.1 and 5.2 that every lower sum of f on $[a, b]$ is less than or equal to every upper sum. Thus the collection $\mathscr{C}_1$ of all lower sums is bounded above (by any upper sum) and the collection $\mathscr{C}_2$ of all upper sums is bounded below (by any lower sum). By the Least Upper Bound Axiom, $\mathscr{C}_1$ has a least upper bound L and $\mathcal{C}_2$ has a greatest lower bound G. From our preceding remarks it follows that

$$L_f(\mathcal{P}) \leq L \leq G \leq U_f(\mathcal{P})$$

for each partition $\mathcal{P}$ of $[a, b]$. Moreover, any number I satisfying

$$L_f(\mathcal{P}) \leq I \leq U_f(\mathcal{P})$$

for each partition $\mathcal{P}$ of $[a, b]$ must satisfy

$$L \leq I \leq G$$

since L is the least upper bound of the lower sums and G is the greatest lower bound of the upper sums. Hence to complete the proof of the theorem it is enough to prove that $L = G$. Let $\varepsilon > 0$. Since f is continuous on $[a, b]$, it follows from Theorem A.20 that f is uniformly continuous on $[a, b]$. Thus there is a $\delta > 0$ such that

$$\text{if } x \text{ and } y \text{ are in } [a, b] \text{ and } |x - y| < \delta, \quad \text{then} \quad |f(x) - f(y)| < \frac{\varepsilon}{b - a}$$

Let $\mathcal{P} = \{x_0, x_1, x_2, \ldots, x_n\}$ be a partition of $[a, b]$ such that $\Delta x_k < \delta$ for $1 \leq k \leq n$, and let M_k and m_k be, respectively, the largest and smallest values of f on $[x_{k-1}, x_k]$. Then

$$
\begin{aligned}
U_f(\mathcal{P}) - L_f(\mathcal{P}) &= (M_1 \Delta x_1 + M_2 \Delta x_2 + \cdots + M_n \Delta x_n) \\
&\quad - (m_1 \Delta x_1 + m_2 \Delta x_2 + \cdots + m_n \Delta x_n) \\
&= (M_1 - m_1) \Delta x_1 + (M_2 - m_2) \Delta x_2 + \cdots + (M_n - m_n) \Delta x_n \\
&< \frac{\varepsilon}{b - a} (\Delta x_1 + \Delta x_2 + \cdots + \Delta x_n) \\
&= \frac{\varepsilon}{b - a} (b - a) = \varepsilon
\end{aligned}
$$

Since $L_f(\mathcal{P}) \leq L \leq G \leq U_f(\mathcal{P})$, it follows that

$$0 \leq G - L \leq U_f(\mathcal{P}) - L_f(\mathcal{P}) < \varepsilon$$

Because ε was arbitrary, we conclude that $L = G$. ∎

We now incorporate the proof of Theorem A.21 into a proof of a theorem concerning Riemann sums.

THEOREM A.22
(THEOREM 5.6)

Let f be continuous on $[a, b]$. For any $\varepsilon > 0$ there is a number $\delta > 0$ such that the following statement holds: If $\mathcal{P} = \{x_0, x_1, \ldots, x_n\}$ is any partition of $[a, b]$ each of whose subintervals has length less than δ, and if $x_{k-1} \leq t_k \leq x_k$ for each k between 1 and n, then the associated Riemann sum $\sum_{k=1}^{n} f(t_k) \Delta x_k$ satisfies

$$\left| \int_a^b f(x)\, dx - \sum_{k=1}^{n} f(t_k) \Delta x_k \right| < \varepsilon$$

Proof   For any $\varepsilon > 0$ choose $\delta > 0$ such that

if x and y are in $[a, b]$ and $|x - y| < \delta$,   then $|f(x) - f(y)| < \dfrac{\varepsilon}{b - a}$

If $\mathscr{P}$ is chosen so that $\Delta x_k < \delta$ for each k, then by the proof of Theorem A.21,

$$U_f(\mathscr{P}) - L_f(\mathscr{P}) < \varepsilon$$

Moreover, if $x_{k-1} \le t_k \le x_k$ for $1 \le k \le n$, then

$$m_k \le f(t_k) \le M_k \quad \text{for } 1 \le k \le n$$

It follows that

$$L_f(\mathscr{P}) = \sum_{k=1}^{n} m_k \, \Delta x_k \le \sum_{k=1}^{n} f(t_k) \, \Delta x_k \le \sum_{k=1}^{n} M_k \, \Delta x_k = U_f(\mathscr{P})$$

Since

$$L_f(\mathscr{P}) \le \int_a^b f(x) \, dx \le U_f(\mathscr{P})$$

we conclude that

$$\left| \int_a^b f(x) \, dx - \sum_{k=1}^{n} f(t_k) \, \Delta x_k \right| \le U_f(\mathscr{P}) - L_f(\mathscr{P}) < \varepsilon \quad \blacksquare$$

We observed in Section 7.1 that if f is strictly increasing or strictly decreasing, then f^{-1} exists. There is a converse to that result which applies to functions that are continuous on intervals, and whose proof uses the Intermediate Value Theorem.

LEMMA A.23

Let f be continuous on an interval I. If f has an inverse, then f is either strictly increasing or strictly decreasing on I.

Proof   Since f^{-1} exists, if a, b, and c are distinct numbers in I, then $f(a)$, $f(b)$, and $f(c)$ are also distinct numbers. Suppose f were neither strictly increasing nor strictly decreasing. Then there would necessarily be numbers a, b, and c in I such that $a < b < c$ and such that $f(b)$ does not lie between $f(a)$ and $f(c)$. This means that either

$$f(c) \quad \text{lies between} \quad f(a) \text{ and } f(b)$$

or
$$f(a) \quad \text{lies between} \quad f(b) \text{ and } f(c)$$

(Figure A.1(a) and (b)). If the first condition holds, then by the Intermediate Value Theorem there is a number d between a and b such that $f(d) = f(c)$ (Figure A.1(a)). If the second condition holds, then by the Intermediate Value Theorem there is a number d between b and c such that $f(d) = f(a)$ (Figure A.1(b)). In either case, f assigns the same value to two distinct points (d and c, or d and a). Consequently f cannot have an inverse, which contradicts the hypothesis. Thus f must be either strictly increasing or strictly decreasing. $\quad \blacksquare$

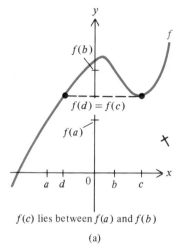

$f(c)$ lies between $f(a)$ and $f(b)$

(a)

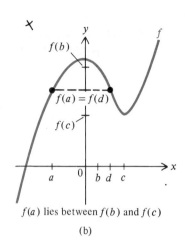

$f(a)$ lies between $f(b)$ and $f(c)$

(b)

FIGURE A.1

If f is continuous on an interval I, then by the Intermediate Value Theorem, the values assigned by f to the points in I form an interval J. We now prove that if in addition f has an inverse, then f^{-1} is continuous.

THEOREM A.24

(THEOREM 7.5)

> Let f be continuous on an interval I, and let the values assigned by f to the points in I form the interval J. If f has an inverse, then f^{-1} is continuous on J.

Proof   Let c be any interior point of J, and let $\varepsilon > 0$. We must find an $\delta > 0$ such that

$$\text{if } |y - c| < \delta, \quad \text{then} \quad |f^{-1}(y) - f^{-1}(c)| < \varepsilon$$

Because c is an interior point of J, and because f and hence f^{-1} are either strictly increasing or strictly decreasing by Lemma A.23, it follows that $f^{-1}(c)$ is an interior point of I. Therefore there exists an $\varepsilon_1 > 0$ such that $\varepsilon_1 < \varepsilon$ and such that the interval K described by

$$K = (f^{-1}(c) - \varepsilon_1, f^{-1}(c) + \varepsilon_1)$$

is contained in I. Since f is continuous and either strictly increasing or strictly decreasing on the interval K, the values f assigns to the points of K form an interval J_0 in J by the Intermediate Value Theorem. In particular J_0 contains an interval of the form $(c - \delta, c + \delta)$ about c. But then by the definition of K we conclude that

$$\text{if } |y - c| < \delta, \quad \text{then} \quad |f^{-1}(y) - f^{-1}(c)| < \varepsilon_1 < \varepsilon$$

Consequently f^{-1} is continuous at c, when c is an interior point of J. An analogous argument proves that f is continuous at any endpoint of J (if there are any). ∎

The final theorem we prove in this subsection concerns sequences.

THEOREM A.25
(THEOREM 9.7)

If a bounded sequence is either increasing or decreasing, then it converges.

Proof   Let $\{a_n\}_{n=1}^{\infty}$ be a bounded, increasing sequence. By the Least Upper Bound Axiom there is a least upper bound L of the set S consisting of the numbers $a_1, a_2, a_3, \ldots$. We will show that $\lim_{n \to \infty} a_n = L$. Let $\varepsilon > 0$. Since L is the least upper bound of S, $L - \varepsilon$ is not an upper bound of S. Thus there is a number a_N in S such that $L - \varepsilon < a_N \leq L$. Since the given sequence is increasing, we have $a_N \leq a_n$ and hence $L - \varepsilon < a_N \leq a_n \leq L$ for any $n \geq N$. Thus

$$|a_n - L| < \varepsilon \quad \text{for } n \geq N$$

Therefore

$$\lim_{n \to \infty} a_n = L$$

A similar proof shows that every bounded, decreasing sequence converges.   ∎

Power Series Theorems

We now prepare to prove the Differentiation Theorem for power series.

LEMMA A.26

The power series $\sum_{n=0}^{\infty} c_n x^n$ and $\sum_{n=1}^{\infty} n c_n x^{n-1}$ have the same radius of convergence.

Proof   Let s be a fixed nonzero number, and let $0 < |x| < |s|$. Assuming that $\sum_{n=0}^{\infty} c_n s^n$ converges, we will prove that $\sum_{n=1}^{\infty} n c_n x^{n-1}$ converges. First we notice that since $\sum_{n=0}^{\infty} c_n s^n$ converges, it follows that

$$\lim_{n \to \infty} \left| \frac{c_n s^n}{x} \right| = \frac{1}{x} \lim_{n \to \infty} |c_n s^n| = 0$$

Consequently there is a positive integer N such that

$$\left| \frac{c_n s^n}{x} \right| \leq 1 \quad \text{for } n \geq N$$

This means that

$$|n c_n x^{n-1}| = \left| n c_n \frac{x^n}{x} \frac{s^n}{s^n} \right| = \left| \frac{c_n s^n}{x} \right| n \left| \frac{x}{s} \right|^n \leq n \left| \frac{x}{s} \right|^n \quad \text{for } n \geq N$$

Since $|x| < |s|$ and thus $|x/s| < 1$, and since $\sum_{n=1}^{\infty} n x^n$ converges for $|x| < 1$ by Example 4 of Section 9.8, we know that $\sum_{n=N}^{\infty} n |x/s|^n$ converges. From the Comparison Test we conclude that $\sum_{n=N}^{\infty} n c_n x^{n-1}$ converges, and hence $\sum_{n=1}^{\infty} n c_n x^{n-1}$ also converges. Since x was arbitrary with $0 < |x| < |s|$, it follows that the radius of convergence of the series $\sum_{n=1}^{\infty} n c_n x^{n-1}$ is at least as large as the radius of convergence of the series $\sum_{n=0}^{\infty} c_n x^n$. Conversely, if $\sum_{n=1}^{\infty} n c_n s^{n-1}$ converges and $|x| < |s|$, then

$$\sum_{n=1}^{\infty} x n c_n x^{n-1} = \sum_{n=1}^{\infty} n c_n x^n$$

and hence $\sum_{n=1}^{\infty} nc_n x^n$ converges absolutely. Since

$$|c_n| \leq |nc_n| \quad \text{for al } n \geq 1$$

it follows from the Comparison Test that $\sum_{n=1}^{\infty} c_n x^n$ converges. Thus the radius of convergence of $\sum_{n=0}^{\infty} c_n x^n$ is at least as large as the radius of convergence of $\sum_{n=1}^{\infty} nc_n x^{n-1}$. This completes the proof of the converse. ∎

COROLLARY A.27

> The power series $\sum_{n=0}^{\infty} c_n x^n$, $\sum_{n=1}^{\infty} nc_n x^{n-1}$, and $\sum_{n=2}^{\infty} n(n-1)c_n x^{n-2}$ have the same radius of convergence.

Proof The proof consists merely of replacing $\sum_{n=0}^{\infty} c_n x^n$ and $\sum_{n=1}^{\infty} nc_n x^{n-1}$ in Lemma A.26 by $\sum_{n=1}^{\infty} nc_n x^{n-1}$ and $\sum_{n=2}^{\infty} n(n-1)c_n x^{n-2}$, respectively. The conclusion is that all three series have the same radius of convergence. ∎

As a final preparation for proving the Differentiation Theorem, we present two applications of the Mean Value Theorem, both of which will appear in the proof. First, if $f(z) = z^n$ and if t and x are two distinct numbers, then there is a number s_n between t and x such that

$$\frac{f(t) - f(x)}{t - x} = f'(s_n)$$

This implies that

$$\frac{t^n - x^n}{t - x} = ns_n^{n-1} \tag{4}$$

Second, if $g(z) = z^{n-1}$ and if x and s_n are two distinct numbers, then there is a number r_n between x and s_n such that

$$\frac{g(x) - g(s_n)}{x - s_n} = g'(r_n)$$

Substitution for values of g and g' yields

$$x^{n-1} - s_n^{n-1} = (x - s_n)(n - 1)r_n^{n-2} \tag{5}$$

Now we are ready to prove the Differentiation Theorem.

THEOREM A.28
DIFFERENTIATION
THEOREM FOR
POWER SERIES
(THEOREM 9.25)

> Let $\sum_{n=0}^{\infty} c_n x^n$ be a power series with radius of convergence $R > 0$. Then $\sum_{n=1}^{\infty} nc_n x^{n-1}$ has the same radius of convergence, and
>
> $$\frac{d}{dx}\left(\sum_{n=0}^{\infty} c_n x^n\right) = \sum_{n=1}^{\infty} nc_n x^{n-1} = \sum_{n=1}^{\infty} \frac{d}{dx}(c_n x^n) \quad \text{for } |x| < R \tag{6}$$

Proof Let x be any number in $(-R, R)$, and let $f(t) = \sum_{n=0}^{\infty} c_n t^n$, for $|t| < R$. We will show that

$$\lim_{t \to x}\left(\sum_{n=1}^{\infty} nc_n x^{n-1} - \frac{f(t) - f(x)}{t - x}\right) = 0$$

that is,

$$\lim_{t \to x} \left(\sum_{n=1}^{\infty} nc_n x^{n-1} - \frac{\sum_{n=0}^{\infty} c_n t^n - \sum_{n=0}^{\infty} c_n x^n}{t - x} \right) = 0$$

This will verify the first equality in (6). To that end, we first choose a number b such that $|x| < b < R$. Then $\sum_{n=0}^{\infty} |c_n| b^n$ converges, and thus by Corollary A.27 the series $\sum_{n=2}^{\infty} n(n-1)|c_n| b^{n-2}$ converges. For later use we let

$$c = \sum_{n=2}^{\infty} n(n-1)|c_n| b^{n-2}$$

Next, observe that for any t in $(-b, b)$ distinct from x we have

$$\sum_{n=1}^{\infty} nc_n x^{n-1} - \frac{\sum_{n=0}^{\infty} c_n t^n - \sum_{n=0}^{\infty} c_n x^n}{t - x} = \sum_{n=1}^{\infty} nc_n x^{n-1} - \sum_{n=0}^{\infty} c_n \frac{t^n - x^n}{t - x}$$

$$\overset{(4)}{=} \sum_{n=1}^{\infty} nc_n x^{n-1} - \sum_{n=1}^{\infty} c_n n s_n^{n-1}$$

$$= \sum_{n=1}^{\infty} nc_n (x^{n-1} - s_n^{n-1})$$

$$\overset{(5)}{=} \sum_{n=2}^{\infty} (x - s_n) n(n-1) c_n r_n^{n-2}$$

By the definition of s_n we know that

$$|x - s_n| \leq |x - t| \tag{7}$$

and by the definition of r_n we know that $|r_n| < b < R$. Consequently from our calculations above and from (7) we obtain

$$\lim_{t \to x} \left| \sum_{n=1}^{\infty} nc_n x^{n-1} - \frac{\sum_{n=0}^{\infty} c_n t^n - \sum_{n=0}^{\infty} c_n x^n}{t - x} \right| = \lim_{t \to x} \left| \sum_{n=2}^{\infty} (x - s_n) n(n-1) c_n r_n^{n-2} \right|$$

$$\leq \lim_{t \to x} |x - t| \sum_{n=2}^{\infty} n(n-1)|c_n| b^{n-2}$$

$$= \lim_{t \to x} |x - t| c = 0$$

This completes the proof that the first equality in (6) holds. Since

$$nc_n x^{n-1} = \frac{d}{dx} (c_n x^n)$$

the second quality in (6) also holds. ∎

We now use the Differentiation Theorem to prove the Integration Theorem for Power Series.

THEOREM A.29
INTEGRATION THEOREM
FOR POWER SERIES
(THEOREM 9.28)

Let $\sum_{n=0}^{\infty} c_n x^n$ be a power series with radius of convergence $R > 0$. Then for any x in $(-R, R)$ we have

$$\int_0^x \left(\sum_{n=0}^{\infty} c_n t^n \right) dt = \sum_{n=0}^{\infty} \frac{c_n}{n+1} x^{n+1} = \sum_{n=0}^{\infty} \left(\int_0^x c_n t^n \, dt \right) \tag{8}$$

Proof   Since $\sum_{n=0}^{\infty} c_n x^n$ converges absolutely for $-R < x < R$ and since

$$\left| \frac{c_n x^n}{n+1} \right| \leq |c_n x^n|$$

we know by the Comparison Test that $\sum_{n=0}^{\infty} c_n x^n/(n+1)$ also converges for $-R < x < R$, and hence

$$\sum_{n=0}^{\infty} \frac{c_n}{n+1} x^{n+1}$$

converges for $-R < x < R$. Since $\sum_{n=0}^{\infty} c_n t^n$ is continuous on $[0, x]$, it follows from the Fundamental Theorem of Calculus that

$$\frac{d}{dx} \left(\int_0^x \sum_{n=0}^{\infty} c_n t^n \, dt \right) = \sum_{n=0}^{\infty} c_n x^n$$

Moreover, from the Differentiation Theorem,

$$\frac{d}{dx} \left(\sum_{n=0}^{\infty} \frac{c_n}{n+1} x^{n+1} \right) = \sum_{n=0}^{\infty} \frac{(n+1)c_n}{n+1} x^n = \sum_{n=0}^{\infty} c_n x^n$$

As a result,

$$\int_0^x \left(\sum_{n=0}^{\infty} c_n t^n \right) dt = \sum_{n=0}^{\infty} \frac{c_n}{n+1} x^{n+1} + C$$

for an appropriate constant C. However, taking $x = 0$, we find that

$$0 = \int_0^0 \left(\sum_{n=0}^{\infty} c_n t^n \right) dt = \sum_{n=0}^{\infty} \frac{c_n}{n+1} 0^{n+1} + C = C$$

so that $C = 0$. This completes the proof of the first part of (8). The second part of (8) follows from the fact that

$$\int_0^x t^n \, dt = \frac{x^{n+1}}{n+1} \quad \blacksquare$$

EXERCISES

In Exercises 1–8 find the least upper bound and the greatest lower bound of the given set.

1. $(-1, 1)$ 2. $[\frac{1}{2}, 100]$ 3. $[0, \pi)$ 4. $(-9.9, \sqrt{2})$

5. The set consisting of $(0, 1)$ and $(2, 5]$

6. The set consisting of 4 and the interval $(-1, 3)$

7. The set consisting of 0 and all numbers of the form $1/n$, where n is a positive integer

8. The set S consisting of all numbers in $(\frac{3}{10}, \frac{1}{3})$ that have decimal expansions containing only finitely many zeros and threes after the decimal point and no other integers

9. Show that the set of positive integers has no upper bound. (*Hint:* If there were an upper bound, then there would be a least upper bound M. Using the fact that $n + 1$ is a positive integer if n is a positive integer, show that it would follow that $n \leq M - 1$ for every positive integer n, and obtain a contradiction.)

10. a. Show that if $a > 0$, then the set S consisting of the numbers of the form na, where n is a positive integer, has no upper bound. (*Hint:* Using the idea of the hint for Exercise 9, show that if M were a least upper bound of S, then $M - a$ would be an upper bound of S, which is a contradiction.)

b. Using (a), prove **Archimedes' Principle:** If b is any number and $a > 0$, there is a positive integer n such that $na > b$.

11. Using the Least Upper Bound Axiom, prove that every set that is bounded below has a greatest lower bound.

12. Using Exercise 9, prove that $\lim_{n \to \infty} 1/n = 0$. (*Hint:* Show that if the given result did not hold, then the set of positive integers would be bounded above.)

13. Show that the continuous function $1/x$ is not uniformly continuous on (0, 1). (*Hint:* Let $\varepsilon = 1$, and show that for any positive $\delta < 1$, if $x = \delta$ and $y = \frac{1}{2}\delta$, then $|x - y| < \delta$ but $|1/x - 1/y| \geq \varepsilon$.)

TABLES

TRIGONOMETRIC FUNCTIONS

Degrees	Radians	Sine	Tangent	Cotangent	Cosine		
0	0	0	0	—	1.0000	1.5708	90
1	0.0175	0.0175	0.0175	57.290	0.9998	1.5533	89
2	0.0349	0.0349	0.0349	28.636	0.9994	1.5359	88
3	0.0524	0.0523	0.0524	19.081	0.9986	1.5184	87
4	0.0698	0.0698	0.0699	14.301	0.9976	1.5010	86
5	0.0873	0.0872	0.0875	11.430	0.9962	1.4835	85
6	0.1047	0.1045	0.1051	9.5144	0.9945	1.4661	84
7	0.1222	0.1219	0.1228	8.1443	0.9925	1.4486	83
8	0.1396	0.1392	0.1405	7.1154	0.9903	1.4312	82
9	0.1571	0.1564	0.1584	6.3138	0.9877	1.4137	81
10	0.1745	0.1736	0.1763	5.6713	0.9848	1.3963	80
11	0.1920	0.1908	0.1944	5.1446	0.9816	1.3788	79
12	0.2094	0.2079	0.2126	4.7046	0.9781	1.3614	78
13	0.2269	0.2250	0.2309	4.3315	0.9744	1.3439	77
14	0.2443	0.2419	0.2493	4.0108	0.9703	1.3265	76
15	0.2618	0.2588	0.2679	3.7321	0.9659	1.3090	75
16	0.2793	0.2756	0.2867	3.4874	0.9613	1.2915	74
17	0.2967	0.2924	0.3057	3.2709	0.9563	1.2741	73
18	0.3142	0.3090	0.3249	3.0777	0.9511	1.2566	72
19	0.3316	0.3256	0.3443	2.9042	0.9455	1.2392	71
20	0.3491	0.3420	0.3640	2.7475	0.9397	1.2217	70
21	0.3665	0.3584	0.3839	2.6051	0.9336	1.2043	69
22	0.3840	0.3746	0.4040	2.4751	0.9272	1.1868	68
23	0.4014	0.3907	0.4245	2.3559	0.9205	1.1694	67
24	0.4189	0.4067	0.4452	2.2460	0.9135	1.1519	66
25	0.4363	0.4226	0.4663	2.1445	0.9063	1.1345	65
26	0.4538	0.4384	0.4877	2.0503	0.8988	1.1170	64
27	0.4712	0.4540	0.5095	1.9626	0.8910	1.0996	63
28	0.4887	0.4695	0.5317	1.8807	0.8829	1.0821	62
29	0.5061	0.4848	0.5543	1.8040	0.8746	1.0647	61
30	0.5236	0.5000	0.5774	1.7321	0.8660	1.0472	60
31	0.5411	0.5150	0.6009	1.6643	0.8572	1.0297	59
32	0.5585	0.5299	0.6249	1.6003	0.8480	1.0123	58
33	0.5760	0.5446	0.6494	1.5399	0.8387	0.9948	57
34	0.5934	0.5592	0.6745	1.4826	0.8290	0.9774	56
35	0.6109	0.5736	0.7002	1.4281	0.8192	0.9599	55
36	0.6283	0.5878	0.7265	1.3764	0.8090	0.9425	54
37	0.6458	0.6018	0.7536	1.3270	0.7986	0.9250	53
38	0.6632	0.6157	0.7813	1.2799	0.7880	0.9076	52
39	0.6807	0.6293	0.8098	1.2349	0.7771	0.8901	51
40	0.6981	0.6428	0.8391	1.1918	0.7660	0.8727	50
41	0.7156	0.6561	0.8693	1.1504	0.7547	0.8552	49
42	0.7330	0.6691	0.9004	1.1106	0.7431	0.8378	48
43	0.7505	0.6820	0.9325	1.0724	0.7314	0.8203	47
44	0.7679	0.6947	0.9657	1.0355	0.7193	0.8029	46
45	0.7854	0.7071	1.0000	1.0000	0.7071	0.7854	45
		Cosine	Cotangent	Tangent	Sine	Radians	Degrees

TRIGONOMETRIC FUNCTIONS (x in radians)

x (radians)	sin x	cos x	tan x	x (radians)	sin x	cos x	tan x	x (radians)	sin x	cos x	tan x
0.00	0.0000	1.0000	0.0000	0.56	0.5312	0.8473	0.6269	1.11	0.8957	0.4447	2.0143
0.01	0.0100	1.0000	0.0100	0.57	0.5396	0.8419	0.6410	1.12	0.9001	0.4357	2.0660
0.02	0.0200	0.9998	0.0200	0.58	0.5480	0.8365	0.6552	1.13	0.9044	0.4267	2.1198
0.03	0.0300	0.9996	0.0300	0.59	0.5564	0.8309	0.6696	1.14	0.9086	0.4176	2.1759
0.04	0.0400	0.9992	0.0400	0.60	0.5646	0.8253	0.6841	1.15	0.9128	0.4085	2.2345
0.05	0.0500	0.9988	0.0500	0.61	0.5729	0.8196	0.6989	1.16	0.9168	0.3993	2.2958
0.06	0.0600	0.9982	0.0601	0.62	0.5810	0.8139	0.7139	1.17	0.9208	0.3902	2.3600
0.07	0.0699	0.9976	0.0701	0.63	0.5891	0.8080	0.7291	1.18	0.9246	0.3809	2.4273
0.08	0.0799	0.9968	0.0802	0.64	0.5972	0.8021	0.7445	1.19	0.9284	0.3717	2.4979
0.09	0.0899	0.9960	0.0902	0.65	0.6052	0.7961	0.7602	1.20	0.9320	0.3624	2.5722
0.10	0.0998	0.9950	0.1003	0.66	0.6131	0.7900	0.7761	1.21	0.9356	0.3530	2.6503
0.11	0.1098	0.9940	0.1104	0.67	0.6210	0.7838	0.7923	1.22	0.9391	0.3436	2.7328
0.12	0.1197	0.9928	0.1206	0.68	.6288	0.7776	0.8087	1.23	0.9425	0.3342	2.8198
0.13	0.1296	0.9916	0.1307	0.69	0.6365	0.7712	0.8253	1.24	0.9458	0.3248	2.9119
0.14	0.1395	0.9902	0.1409	0.70	0.6442	0.7648	0.8423	1.25	0.9490	0.3153	3.0096
0.15	0.1494	0.9888	0.1511	0.71	0.6518	0.7584	0.8595	1.26	0.9521	0.3058	3.1133
0.16	0.1593	0.9872	0.1614	0.72	0.6594	0.7518	0.8771	1.27	0.9551	0.2963	3.2236
0.17	0.1692	0.9856	0.1717	0.73	0.6669	0.7452	0.8949	1.28	0.9580	0.2867	3.3413
0.18	0.1790	0.9838	0.1820	0.74	0.6743	0.7385	0.9131	1.29	0.9608	0.2771	3.4672
0.19	0.1889	0.9820	0.1923	0.75	0.6816	0.7317	0.9316	1.30	0.9636	0.2675	3.6021
0.20	0.1987	0.9801	0.2027	0.76	0.6889	0.7248	0.9505	1.31	0.9662	0.2579	3.7471
0.21	0.2085	0.9780	0.2131	0.77	0.6961	0.7179	0.9697	1.32	0.9687	0.2482	3.9033
0.22	0.2182	0.9759	0.2236	0.78	0.7033	0.7109	0.9893	1.33	0.9711	0.2385	4.0723
0.23	0.2280	0.9737	0.2341	0.79	0.7104	0.7038	1.0092	1.34	0.9735	0.2288	4.2556
0.24	0.2377	0.9713	0.2447	0.80	0.7174	0.6967	1.0296	1.35	0.9757	0.2190	4.4552
0.25	0.2474	0.9689	0.2553	0.81	0.7243	0.6895	1.0505	1.36	0.9779	0.2092	4.6734
0.26	0.2571	0.9664	0.2660	0.82	0.7311	0.6822	1.0717	1.37	0.9799	0.1994	4.9131
0.27	0.2667	0.9638	0.2768	0.83	0.7379	0.6749	1.0934	1.38	0.9819	0.1896	5.1774
0.28	0.2764	0.9611	0.2876	0.84	0.7446	0.6675	1.1156	1.39	0.9837	0.1798	5.4707
0.29	0.2860	0.9582	0.2984	0.85	0.7513	0.6600	1.1383	1.40	0.9854	0.1700	5.7979
0.30	0.2955	0.9553	0.3093	0.86	0.7578	0.6524	1.1616	1.41	0.9871	0.1601	6.1654
0.31	0.3051	0.9523	0.3203	0.87	0.7643	0.6448	1.1853	1.42	0.9887	0.1502	6.5811
0.32	0.3146	0.9492	0.3314	0.88	0.7707	0.6372	1.2097	1.43	0.9901	0.1403	7.0555
0.33	0.3240	0.9460	0.3425	0.89	0.7771	0.6294	1.2346	1.44	0.9915	0.1304	7.6018
0.34	0.3335	0.9428	0.3537	0.90	0.7833	0.6216	1.2602	1.45	0.9927	0.1205	8.2381
0.35	0.3429	0.9394	0.3650	0.91	0.7895	0.6137	1.2864	1.46	0.9939	0.1106	8.9886
0.36	0.3523	0.9359	0.3764	0.92	0.7956	0.6058	1.3133	1.47	0.9949	0.1006	9.8874
0.37	0.3616	0.9323	0.3879	0.93	0.8016	0.5978	1.3409	1.48	0.9959	0.0907	10.9834
0.38	0.3709	0.9287	0.3994	0.94	0.8076	0.5898	1.3692	1.49	0.9967	0.0807	12.3499
0.39	0.3802	0.9249	0.4111	0.95	0.8134	0.5817	1.3984	1.50	0.9975	0.0707	14.1014
0.40	0.3894	0.9211	0.4228	0.96	0.8192	0.5735	1.4284	1.51	0.9982	0.0608	16.4281
0.41	0.3986	0.9171	0.4346	0.97	0.8249	0.5653	1.4592	1.52	0.9987	0.0508	19.6695
0.42	0.4078	0.9131	0.4466	0.98	0.8305	0.5570	1.4910	1.53	0.9992	0.0408	24.4984
0.43	0.4169	0.9090	0.4586	0.99	0.8360	0.5487	1.5237	1.54	0.9995	0.0308	32.4611
0.44	0.4259	0.9048	0.4708	1.00	0.8415	0.5403	1.5574	1.55	0.9998	0.0208	48.0785
0.45	0.4350	0.9004	0.4831	1.01	0.8468	0.5319	1.5922	1.56	0.9999	0.0108	92.6205
0.46	0.4439	0.8961	0.4954	1.02	0.8521	0.5234	1.6281	1.57	1.0000	0.0008	1255.77
0.47	0.4529	0.8916	0.5080	1.03	0.8573	0.5148	1.6652	$\pi/2$	1.0000	0.0000	—
0.48	0.4618	0.8870	0.5206	1.04	0.8624	0.5062	1.7036				
0.49	0.4706	0.8823	0.5334	1.05	0.8674	0.4976	1.7433				
0.50	0.4794	0.8776	0.5463	1.06	0.8724	0.4889	1.7844				
0.51	0.4882	0.8727	0.5594	1.07	0.8772	0.4801	1.8270				
0.52	0.4969	0.8678	0.5726	1.08	0.8820	0.4713	1.8712				
0.53	0.5055	0.8628	0.5859	1.09	0.8866	0.4625	1.9171				
0.54	0.5141	0.8577	0.5994	1.10	0.8912	0.4536	1.9648				
0.55	0.5227	0.8525	0.6131								

EXPONENTIAL FUNCTIONS e^x AND e^{-x}

x	e^x	e^{-x}	x	e^x	e^{-x}
0.00	1.0000	1.0000	1.5	4.4817	0.2231
0.01	1.0101	0.9901	1.6	4.9530	0.2019
0.02	1.0202	0.9802	1.7	5.4739	0.1827
0.03	1.0305	0.9702	1.8	6.0496	0.1653
0.04	1.0408	0.9608	1.9	6.6859	0.1496
0.05	1.0513	0.9512	2.0	7.3891	0.1353
0.06	1.0618	0.9418	2.1	8.1662	0.1225
0.07	1.0725	0.9324	2.2	9.0250	0.1108
0.08	1.0833	0.9331	2.3	9.9742	0.1003
0.09	1.0942	0.9139	2.4	11.023	0.0907
0.10	1.1052	0.9048	2.5	12.182	0.0821
0.11	1.1163	0.8958	2.6	13.464	0.0743
0.12	1.1275	0.8869	2.7	14.880	0.0672
0.13	1.1388	0.8781	2.8	16.445	0.0608
0.14	1.1503	0.8694	2.9	18.174	0.0550
0.15	1.1618	0.8607	3.0	20.086	0.0498
0.16	1.1735	0.8521	3.1	22.198	0.0450
0.17	1.1853	0.8437	3.2	24.533	0.0408
0.18	1.1972	0.8353	3.3	27.113	0.0369
0.19	1.2092	0.8270	3.4	29.964	0.0334
0.20	1.2214	0.8187	3.5	33.115	0.0302
0.21	1.2337	0.8106	3.6	36.598	0.0273
0.22	1.2461	0.8025	3.7	40.447	0.0247
0.23	1.2586	0.7945	3.8	44.701	0.0224
0.24	1.2712	0.7866	3.9	49.402	0.0202
0.25	1.2840	0.7788	4.0	54.598	0.0183
0.30	1.3499	0.7408	4.1	60.340	0.0166
0.35	1.4191	0.7047	4.2	66.686	0.0150
0.40	1.4918	0.6703	4.3	73.700	0.0136
0.45	1.5683	0.6376	4.4	81.451	0.0123
0.50	1.6487	0.6065	4.5	90.017	0.0111
0.55	1.7333	0.5769	4.6	99.484	0.0101
0.60	1.8221	0.5488	4.7	109.95	0.0091
0.65	1.9155	0.5220	4.8	121.51	0.0082
0.70	2.0138	0.4966	4.9	134.29	0.0074
0.75	2.1170	0.4724	5.0	148.41	0.0067
0.80	2.2255	0.4493	5.5	244.69	0.0041
0.85	2.3396	0.4274	6.0	403.43	0.0025
0.90	2.4596	0.4066	6.5	665.14	0.0015
0.95	2.5857	0.3867	7.0	1096.6	0.0009
1.0	2.7183	0.3679	7.5	1808.0	0.0006
1.1	3.0042	0.3329	8.0	2981.0	0.0003
1.2	3.3201	0.3012	8.5	4914.8	0.0002
1.3	3.6693	0.2725	9.0	8103.1	0.0001
1.4	4.0552	0.2466	10.0	22026	0.00005

NATURAL LOGARITHMS OF NUMBERS (base e)

n	$\ln n$	n	$\ln n$	n	$\ln n$
		4.5	1.5041	9.0	2.1972
0.1	−2.3026	4.6	1.5261	9.1	2.2083
0.2	−1.6094	4.7	1.5476	9.2	2.2192
0.3	−1.2040	4.8	1.5486	9.3	2.2300
0.4	−0.9163	4.9	1.5892	9.4	2.2407
0.5	−0.6931	5.0	1.6094	9.5	2.2513
0.6	−0.5108	5.1	1.6292	9.6	2.2618
0.7	−0.3567	5.2	1.6487	9.7	2.2721
0.8	−0.2231	5.3	1.6677	9.8	2.2824
0.9	−0.1054	5.4	1.6864	9.9	2.2925
1.0	0.0000	5.5	1.7047	10	2.3026
1.1	0.0953	5.6	1.7228	11	2.3979
1.2	0.1823	5.7	1.7405	12	2.4849
1.3	0.2624	5.8	1.7579	13	2.5649
1.4	0.3365	5.9	1.7750	14	2.6391
1.5	0.4055	6.0	1.7918	15	2.7081
1.6	0.4700	6.1	1.8083	16	2.7726
1.7	0.5306	6.2	1.8245	17	2.8332
1.8	0.5878	6.3	1.8405	18	2.8904
1.9	0.6419	6.4	1.8563	19	2.9444
2.0	0.6931	6.5	1.8718	20	2.9957
2.1	0.7419	6.6	1.8871	25	3.2189
2.2	0.7885	6.7	1.9021	30	3.4012
2.3	0.8329	6.8	1.9169	35	3.5553
2.4	0.8755	6.9	1.9315	40	3.6889
2.5	0.9163	7.0	1.9459	45	3.8067
2.6	0.9555	7.1	1.9601	50	3.9120
2.7	0.9933	7.2	1.9741	55	4.0073
2.8	1.0296	7.3	1.9879	60	4.0943
2.9	1.0647	7.4	2.0015	65	4.1744
3.0	1.0986	7.5	2.0149	70	4.2485
3.1	1.1314	7.6	2.0281	75	4.3175
3.2	1.1632	7.7	2.0412	80	4.3820
3.3	1.1939	7.8	2.0541	85	4.4427
3.4	1.2238	7.9	2.0669	90	4.4998
3.5	1.2528	8.0	2.0794	100	4.6052
3.6	1.2809	8.1	2.0919	110	4.7005
3.7	1.3083	8.2	2.1041	120	4.7875
3.8	1.3350	8.3	2.1163	130	4.8676
3.9	1.3610	8.4	2.1282	140	4.9416
4.0	1.3863	8.5	2.1401	150	5.0106
4.1	1.4110	8.6	2.1518	160	5.0752
4.2	1.4351	8.7	2.1633	170	5.1358
4.3	1.4586	8.8	2.1748	180	5.1930
4.4	1.4816	8.9	2.1861	190	5.2470

TABLE OF INTEGRALS

Trigonometric Forms

1. $\int \sin x \, dx = -\cos x + C$

2. $\int \cos x \, dx = \sin x + C$

3. $\int \tan x \, dx = -\ln|\cos x| + C$

4. $\int \cot x \, dx = \ln|\sin x| + C$

5. $\int \sec x \, dx = \ln|\sec x + \tan x| + C$

6. $\int \csc x \, dx = -\ln|\csc x + \cot x| + C$

7. $\int \sec^2 x \, dx = \tan x + C$

8. $\int \csc^2 x \, dx = -\cot x + C$

9. $\int \sec x \tan x \, dx = \sec x + C$

10. $\int \csc x \cot x \, dx = -\csc x + C$

11. $\int \sin^2 x \, dx = \frac{1}{2}x - \frac{1}{4}\sin 2x + C$

12. $\int \cos^2 x \, dx = \frac{1}{2}x + \frac{1}{4}\sin 2x + C$

13. $\int \tan^2 x \, dx = \tan x - x + C$

14. $\int \sec^3 x \, dx = \frac{1}{2}\sec x \tan x + \frac{1}{2}\ln|\sec x + \tan x| + C$

15. $\int \sin^4 x \, dx = -\frac{1}{4}\sin^3 x \cos x - \frac{3}{8}\sin x \cos x + \frac{3}{8}x + C$

16. $\int \cos^4 x \, dx = \frac{1}{4}\cos^3 x \sin x + \frac{3}{8}\cos x \sin x + \frac{3}{8}x + C$

17. $\int \sin^n x \, dx = -\frac{1}{n}\sin^{n-1} x \cos x + \frac{n-1}{n}\int \sin^{n-2} x \, dx$

18. $\int \cos^n x \, dx = \frac{1}{n}\cos^{n-1} x \sin x + \frac{n-1}{n}\int \cos^{n-2} x \, dx$

19. $\int \tan^n x \, dx = \frac{1}{n-1}\tan^{n-1} x - \int \tan^{n-2} x \, dx$

20. $\int \cot^n x \, dx = -\frac{1}{n-1}\cot^{n-1} x - \int \cot^{n-2} x \, dx$

21. $\int \sec^n x \, dx = \frac{1}{n-1}\sec^{n-2} x \tan x + \frac{n-2}{n-1}\int \sec^{n-2} x \, dx$

22. $\int \csc^n x \, dx = -\frac{1}{n-1}\csc^{n-2} x \cot x + \frac{n-2}{n-1}\int \csc^{n-2} x \, dx$

23. $\int \sin ax \sin bx \, dx = -\frac{1}{2(a+b)}\sin(a+b)x + \frac{1}{2(a-b)}\sin(a-b)x + C$

24. $\int \cos ax \cos bx \, dx = \frac{1}{2(a+b)}\sin(a+b)x + \frac{1}{2(a-b)}\sin(a-b)x + C$

25. $\int \sin ax \cos bx \, dx = -\frac{1}{2(a+b)}\cos(a+b)x - \frac{1}{2(a-b)}\cos(a-b)x + C$

26. $\int x \sin x \, dx = -x \cos x + \sin x + C$

27. $\int x \cos x \, dx = x \sin x + \cos x + C$

28. $\int x^2 \sin x \, dx = -x^2 \cos x + 2x \sin x + 2 \cos x + C$

29. $\int x^2 \cos x \, dx = x^2 \sin x + 2x \cos x - 2 \sin x + C$

30. $\int x^n \sin x \, dx = -x^n \cos x + n \int x^{n-1} \cos x \, dx$

31. $\int x^n \cos x \, dx = x^n \sin x - n \int x^{n-1} \sin x \, dx$

32. $\int \sin^m x \cos^n x \, dx = -\dfrac{\sin^{m-1} x \cos^{n+1} x}{m+n} + \dfrac{m-1}{m+n} \int \sin^{m-2} x \cos^n x \, dx$

$= \dfrac{\sin^{m+1} x \cos^{n-1} x}{m+n} + \dfrac{n-1}{m+n} \int \sin^m x \cos^{n-2} x \, dx$

Inverse Trigonometric Forms

33. $\int \arcsin x \, dx = x \arcsin x + \sqrt{1-x^2} + C$

34. $\int \arccos x \, dx = x \arccos x - \sqrt{1-x^2} + C$

35. $\int \arctan x \, dx = x \arctan x - \frac{1}{2} \ln(x^2+1) + C$

36. $\int \text{arccot}\, x \, dx = x\, \text{arccot}\, x + \frac{1}{2} \ln(x^2+1) + C$

37. $\int \text{arcsec}\, x \, dx = x\, \text{arcsec}\, x - \ln|x + \sqrt{x^2-1}| + C$

38. $\int \text{arccsc}\, x \, dx = x\, \text{arccsc}\, x + \ln|x + \sqrt{x^2-1}| + C$

Exponential and Logarithmic Forms

39. $\int e^x \, dx = e^x + C$

40. $\int a^x \, dx = \dfrac{a^x}{\ln a} + C$

41. $\int xe^x \, dx = xe^x - e^x + C$

42. $\int x^n e^x \, dx = x^n e^x - n \int x^{n-1} e^x \, dx$

43. $\int x^n a^x \, dx = \dfrac{x^n a^x}{\ln a} - \dfrac{n}{\ln a} \int x^{n-1} a^x \, dx + C$

44. $\int \dfrac{1}{x} dx = \ln|x| + C$

45. $\int \ln x \, dx = x \ln x - x + C$

46. $\int (\ln x)^n \, dx = x(\ln x)^n - n \int (\ln x)^{n-1} dx$

47. $\int \dfrac{1}{x \ln x} dx = \ln|\ln x| + C$

48. $\int x^n \ln x \, dx = \dfrac{1}{n+1} x^{n+1} \ln x - \dfrac{1}{(n+1)^2} x^{n+1} + C$

49. $\int e^{ax} \sin bx \, dx = \dfrac{e^{ax}}{a^2+b^2}(a \sin bx - b \cos bx) + C$

50. $\int e^{ax} \cos bx \, dx = \dfrac{e^{ax}}{a^2+b^2}(a \cos bx + b \sin bx) + C$

Hyperbolic Forms

51. $\int \sinh x \, dx = \cosh x + C$

52. $\int \cosh x \, dx = \sinh x + C$

53. $\int \tanh x \, dx = \ln \cosh x + C$

54. $\int \coth x \, dx = \ln|\sinh x| + C$

55. $\int \text{sech}\, x \, dx = \arctan(\sinh x) + C$

56. $\int \text{csch}\, x \, dx - \ln|\tanh \frac{1}{2}x| + C$

57. $\int \operatorname{sech}^2 x\, dx = \tanh x + C$

59. $\int \operatorname{sech} x \tanh x\, dx = -\operatorname{sech} x + C$

58. $\int \operatorname{csch}^2 x\, dx = -\coth x + C$

60. $\int \operatorname{csch} x \coth x\, dx = -\operatorname{csch} x + C$

61. $\int \sinh^2 x\, dx = \frac{1}{4}\sinh 2x - \frac{1}{2}x + C$

62. $\int \cosh^2 x\, dx = \frac{1}{4}\sinh 2x + \frac{1}{2}x + C$

63. $\int e^{ax} \sinh bx\, dx = \dfrac{e^{ax}}{a^2 - b^2}(a \sinh bx - b \cosh bx) + C$

64. $\int e^{ax} \cosh bx\, dx = \dfrac{e^{ax}}{a^2 - b^2}(a \cosh bx - b \sinh bx) + C$

Forms Involving $a^2 - x^2$

65. $\int \dfrac{1}{a^2 - x^2}\, dx = \dfrac{1}{2a}\ln\left|\dfrac{x + a}{x - a}\right| + C$

66. $\int \dfrac{1}{\sqrt{a^2 - x^2}}\, dx = \arcsin\dfrac{x}{a} + C$

67. $\int \sqrt{a^2 - x^2}\, dx = \dfrac{x}{2}\sqrt{a^2 - x^2} + \dfrac{a^2}{2}\arcsin\dfrac{x}{a} + C$

68. $\int x^2 \sqrt{a^2 - x^2}\, dx = \dfrac{x}{8}(2x^2 - a^2)\sqrt{a^2 - x^2} + \dfrac{a^4}{8}\arcsin\dfrac{x}{a} + C$

69. $\int \dfrac{\sqrt{a^2 - x^2}}{x}\, dx = \sqrt{a^2 - x^2} - a\ln\left|\dfrac{a + \sqrt{a^2 - x^2}}{x}\right| + C$

70. $\int \dfrac{\sqrt{a^2 - x^2}}{x^2}\, dx = -\dfrac{1}{x}\sqrt{a^2 - x^2} - \arcsin\dfrac{x}{a} + C$

71. $\int \dfrac{x^2}{\sqrt{a^2 - x^2}}\, dx = -\dfrac{x}{2}\sqrt{a^2 - x^2} + \dfrac{a^2}{2}\arcsin\dfrac{x}{a} + C$

72. $\int \dfrac{1}{x\sqrt{a^2 - x^2}}\, dx = -\dfrac{1}{a}\ln\left|\dfrac{a + \sqrt{a^2 - x^2}}{x}\right| + C$

73. $\int \dfrac{1}{x^2\sqrt{a^2 - x^2}}\, dx = -\dfrac{1}{a^2 x}\sqrt{a^2 - x^2} + C$

74. $\int (a^2 - x^2)^{3/2}\, dx = -\dfrac{x}{8}(2x^2 - 5a^2)\sqrt{a^2 - x^2} + \dfrac{3a^4}{8}\arcsin\dfrac{x}{a} + C$

75. $\int \dfrac{1}{(a^2 - x^2)^{3/2}}\, dx = \dfrac{x}{a^2\sqrt{a^2 - x^2}} + C$

Forms Involving $x^2 + a^2$

76. $\int \dfrac{1}{x^2 + a^2}\, dx = \dfrac{1}{a}\arctan\dfrac{x}{a} + C$

77. $\int \sqrt{x^2 + a^2}\, dx = \dfrac{x}{2}\sqrt{x^2 + a^2} + \dfrac{a^2}{2}\ln|x + \sqrt{x^2 + a^2}| + C$

78. $\int x^2 \sqrt{x^2 + a^2}\, dx = \dfrac{x}{8}(2x^2 + a^2)\sqrt{x^2 + a^2} - \dfrac{a^4}{8}\ln|x + \sqrt{x^2 + a^2}| + C$

79. $\int \dfrac{\sqrt{x^2 + a^2}}{x}\, dx = \sqrt{x^2 + a^2} - a\ln\left|\dfrac{a + \sqrt{x^2 + a^2}}{x}\right| + C$

80. $\int \dfrac{\sqrt{x^2 + a^2}}{x^2}\, dx = -\dfrac{\sqrt{x^2 + a^2}}{x} + \ln|x + \sqrt{x^2 + a^2}| + C$

81. $\int (x^2 + a^2)^{3/2}\, dx = \frac{x}{8}(2x^2 + 5a^2)\sqrt{x^2 + a^2} + \frac{3a^4}{8}\ln|x + \sqrt{x^2 + a^2}| + C$

82. $\int \frac{1}{\sqrt{x^2 + a^2}}\, dx = \ln|x + \sqrt{x^2 + a^2}| + C = \sinh^{-1}\frac{x}{a} + C$

83. $\int \frac{x^2}{\sqrt{x^2 + a^2}}\, dx = \frac{x}{2}\sqrt{x^2 + a^2} - \frac{a^2}{2}\ln|x + \sqrt{x^2 + a^2}| + C$

84. $\int \frac{1}{x\sqrt{x^2 + a^2}}\, dx = -\frac{1}{a}\ln\left|\frac{a + \sqrt{x^2 + a^2}}{x}\right| + C = -\frac{1}{a}\sinh^{-1}\frac{a}{x} + C$

85. $\int \frac{1}{x^2\sqrt{x^2 + a^2}}\, dx = -\frac{\sqrt{x^2 + a^2}}{a^2 x} + C$

86. $\int \frac{1}{(x^2 + a^2)^{3/2}}\, dx = \frac{x}{a^2\sqrt{x^2 + a^2}} + C$

Forms Involving $x^2 - a^2$

87. $\int \frac{1}{x^2 - a^2}\, dx = \frac{1}{2a}\ln\left|\frac{x - a}{x + a}\right| + C$

88. $\int \frac{1}{x\sqrt{x^2 - a^2}}\, dx = \frac{1}{a}\operatorname{arcsec}\left|\frac{x}{a}\right| + C$

89. $\int \sqrt{x^2 - a^2}\, dx = \frac{x}{2}\sqrt{x^2 - a^2} - \frac{a^2}{2}\ln|x + \sqrt{x^2 - a^2}| + C$

90. $\int x^2\sqrt{x^2 - a^2}\, dx = \frac{x}{8}(2x^2 - a^2)\sqrt{x^2 - a^2} - \frac{a^4}{8}\ln|x + \sqrt{x^2 - a^2}| + C$

91. $\int \frac{\sqrt{x^2 - a^2}}{x}\, dx = \sqrt{x^2 - a^2} - a\operatorname{arcsec}\left|\frac{x}{a}\right| + C$

92. $\int \frac{\sqrt{x^2 - a^2}}{x^2}\, dx = -\frac{\sqrt{x^2 - a^2}}{x} + \ln|x + \sqrt{x^2 - a^2}| + C$

93. $\int (x^2 - a^2)^{3/2}\, dx = \frac{x}{8}(2x^2 - 5a^2)\sqrt{x^2 - a^2} + \frac{3a^4}{8}\ln|x + \sqrt{x^2 - a^2}| + C$

94. $\int \frac{1}{\sqrt{x^2 - a^2}}\, dx = \ln|x + \sqrt{x^2 - a^2}| + C = \cosh^{-1}\frac{x}{a} + C$

95. $\int \frac{x^2}{\sqrt{x^2 - a^2}}\, dx = \frac{x}{2}\sqrt{x^2 - a^2} + \frac{a^2}{2}\ln|x + \sqrt{x^2 - a^2}| + C$

96. $\int \frac{1}{x^2\sqrt{x^2 - a^2}}\, dx = \frac{\sqrt{x^2 - a^2}}{a^2 x} + C$

97. $\int \frac{1}{(x^2 - a^2)^{3/2}}\, dx = -\frac{x}{a^2\sqrt{x^2 - a^2}} + C$

Forms Involving $a + bx$

98. $\int \frac{x}{a + bx}\, dx = \frac{1}{b^2}(a + bx - a\ln|a + bx|) + C$

99. $\int \frac{x^2}{a + bx}\, dx = \frac{1}{b^3}\left[\frac{1}{2}(a + bx)^2 - 2a(a + bx) + a^2\ln|a + bx|\right] + C$

100. $\int \frac{x}{(a + bx)^2}\, dx = \frac{1}{b^2}\left(\frac{a}{a + bx} + \ln|a + bx|\right) + C$

101. $\int \frac{x^2}{(a + bx)^2}\, dx = \frac{1}{b^3}\left(a + bx - \frac{a^2}{a + bx} - 2a\ln|a + bx|\right) + C$

102. $\int \frac{1}{x(a + bx)} \, dx = \frac{1}{a} \ln \left| \frac{x}{a + bx} \right| + C$

103. $\int \frac{1}{x^2(a + bx)} \, dx = -\frac{1}{ax} + \frac{b}{a^2} \ln \left| \frac{a + bx}{x} \right| + C$

104. $\int \frac{1}{x(a + bx)^2} \, dx = \frac{1}{a(a + bx)} + \frac{1}{a^2} \ln \left| \frac{x}{a + bx} \right| + C$

Forms Involving $\sqrt{a + bx}$

105. $\int x \sqrt{a + bx} \, dx = \frac{2}{15b^2}(3bx - 2a)(a + bx)^{3/2} + C$

106. $\int x^n \sqrt{a + bx} \, dx = \frac{2x^n(a + bx)^{3/2}}{b(2n + 3)} - \frac{2an}{b(2n + 3)} \int x^{n-1} \sqrt{a + bx} \, dx$

107. $\int \frac{x}{\sqrt{a + bx}} \, dx = \frac{2}{3b^2}(bx - 2a) \sqrt{a + bx} + C$

108. $\int \frac{x^n}{\sqrt{a + bx}} \, dx = \frac{2x^n \sqrt{a + bx}}{b(2n + 1)} - \frac{2an}{b(2n + 1)} \int \frac{x^{n-1}}{\sqrt{a + bx}} \, dx$

109. $\int \frac{1}{x \sqrt{a + bx}} \, dx = \begin{cases} \dfrac{1}{\sqrt{a}} \ln \left| \dfrac{\sqrt{a + bx} - \sqrt{a}}{\sqrt{a + bx} + \sqrt{a}} \right| + C & \text{if } a > 0 \\[4mm] \dfrac{2}{\sqrt{-a}} \arctan \sqrt{\dfrac{a + bx}{-a}} + C & \text{if } a < 0 \end{cases}$

110. $\int \frac{1}{x^n \sqrt{a + bx}} \, dx = -\frac{\sqrt{a + bx}}{a(n - 1)x^{n-1}} - \frac{b(2n - 3)}{2a(n - 1)} \int \frac{1}{x^{n-1} \sqrt{a + bx}} \, dx$

111. $\int \frac{\sqrt{a + bx}}{x} \, dx = 2\sqrt{a + bx} + a \int \frac{1}{x \sqrt{a + bx}} \, dx$

112. $\int \frac{\sqrt{a + bx}}{x^n} \, dx = -\frac{(a + bx)^{3/2}}{a(n - 1)x^{n-1}} - \frac{b(2n - 5)}{2a(n - 1)} \int \frac{\sqrt{a + bx}}{x^{n-1}} \, dx$

Forms Involving $\sqrt{2ax - x^2}$

113. $\int \sqrt{2ax - x^2} \, dx = \frac{x - a}{2} \sqrt{2ax - x^2} + \frac{a^2}{2} \arccos \left(1 - \frac{x}{a} \right) + C$

114. $\int x \sqrt{2ax - x^2} \, dx = \frac{2x^2 - ax - 3a^2}{6} \sqrt{2ax - x^2} + \frac{a^3}{2} \arccos \left(1 - \frac{x}{a} \right) + C$

115. $\int \frac{\sqrt{2ax - x^2}}{x} \, dx = \sqrt{2ax - x^2} + a \arccos \left(1 - \frac{x}{a} \right) + C$

116. $\int \frac{\sqrt{2ax - x^2}}{x^2} \, dx = -\frac{2\sqrt{2ax - x^2}}{x} - \arccos \left(1 - \frac{x}{a} \right) + C$

117. $\int \frac{1}{\sqrt{2ax - x^2}} \, dx = \arccos \left(1 - \frac{x}{a} \right) + C$

118. $\int \frac{x}{\sqrt{2ax - x^2}} \, dx = -\sqrt{2ax - x^2} + a \arccos \left(1 - \frac{x}{a} \right) + C$

119. $\int \frac{x^2}{\sqrt{2ax - x^2}} \, dx = -\frac{x + 3a}{2} \sqrt{2ax - x^2} + \frac{3a^2}{2} \arccos \left(1 - \frac{x}{a} \right) + C$

120. $\int \frac{1}{x \sqrt{2ax - x^2}} \, dx = -\frac{\sqrt{2ax - x^2}}{ax} + C$

121. $\int \frac{1}{(2ax - x^2)^{3/2}} \, dx = \frac{x - a}{a^2 \sqrt{2ax - x^2}} + C$

Miscellaneous Forms

122. $\int x \sqrt{\dfrac{a+x}{a-x}}\, dx = -\dfrac{2a+x}{2}\sqrt{a^2 - x^2} + \dfrac{a^2}{2}\arcsin\dfrac{x}{a} + C$

123. $\int \sqrt{\dfrac{x}{x^3 - a}}\, dx = \dfrac{2}{3}\ln(x^{3/2} + \sqrt{x^3 - a}) + C$

124. $\int \sqrt{\dfrac{x}{x+a}}\, dx = \sqrt{x^2 + ax} - a\ln(\sqrt{x+a} + \sqrt{x}) + C$

125. $\int \dfrac{1+x^2}{(1-x^2)\sqrt{1+x^4}}\, dx = \dfrac{1}{\sqrt{2}}\ln\dfrac{\sqrt{2}\,x + \sqrt{1+x^4}}{1-x^2} + C$

126. $\int \dfrac{1}{\sqrt{ax+b}\,\sqrt{cx+d}}\, dx = \dfrac{2}{\sqrt{ac}}\operatorname{arctanh}\sqrt{\dfrac{c(ax+b)}{a(cx+d)}} + C$

127. $\int \dfrac{1}{1+\sin ax}\, dx = \dfrac{1}{a}\tan\left(\dfrac{\pi}{4} - \dfrac{ax}{2}\right) + C$

128. $\int \dfrac{1}{1+\cos ax}\, dx = \dfrac{1}{a}\tan\dfrac{ax}{2} + C$

129. $\int \dfrac{1}{1+\tan ax}\, dx = \dfrac{1}{2}\left[x + \dfrac{1}{a}\ln(\cos ax + \sin ax)\right] + C$

130. $\int \dfrac{1}{a\sin cx + b\cos cx}\, dx = \dfrac{1}{c\sqrt{a^2 + b^2}}\ln\tan\dfrac{1}{2}\left(cx + \arctan\dfrac{b}{a}\right) + C$

131. $\int \dfrac{a+e^x}{b+e^x}\, dx = \dfrac{ax}{b} - \dfrac{a-b}{b}\ln(b+e^x) + C$

132. $\int \dfrac{a + be^x + ce^{2x}}{d+e^x}\, dx = \dfrac{ax}{d} + ce^x - \dfrac{a - bd + cd^2}{d}\ln(d+e^x) + C$

133. $\int \dfrac{x}{(e^x - 1)^2}\, dx = -\dfrac{1}{2}\dfrac{e^x + 1}{e^x - 1} + C$

134. $\int \sqrt{1+e^{ax}}\, dx = \dfrac{2}{a}\sqrt{1+e^{ax}} + \dfrac{1}{a}\ln\dfrac{\sqrt{1+e^{ax}} + 1}{\sqrt{1+e^{ax}} - 1} + C$

135. $\int \dfrac{1}{\sqrt{e^x + a^2}}\, dx = \dfrac{2}{a}\ln(\sqrt{e^x + a^2} - a) - \dfrac{x}{a} + C$

136. $\int \dfrac{x^n}{\sqrt{x^3 + 1}}\, dx = \dfrac{2}{2n-1}x^{n-2}\sqrt{x^3 + 1} - \dfrac{2n-4}{2n-1}\int \dfrac{x^{n-3}}{\sqrt{x^3 + 1}}\, dx$

137. $\int \dfrac{x^n}{\sqrt{x^4 + 1}}\, dx = \dfrac{x^{n-3}}{n-1}\sqrt{x^4 + 1} - \dfrac{n-3}{n-1}\int \dfrac{x^{n-4}}{\sqrt{x^4 + 1}}\, dx$

ANSWERS TO ODD-NUMBERED EXERCISES

CHAPTER 1

Section 1.1

1. $a > b$ 3. $a > b$ 5. $\sqrt{2} > 1.41$

7. closed, bounded   9. open, unbounded

11. closed, unbounded   13. closed, unbounded

15. $(-3, 4)$ 17. $(1, \infty)$

19. $(-\infty, -\frac{7}{6})$ 21. $[1, \frac{7}{2})$

23. union of $(-\infty, -\frac{1}{2}]$ and $[1, \infty)$

25. union of $(-\infty, -\frac{1}{3})$ and $(0, \frac{2}{3})$

27. union of $(-\infty, -3)$ and $(-1, \infty)$

29. $(-\infty, \frac{3}{2}]$ 31. $(\frac{1}{2}, \infty)$

33. union of $(-\infty, -\sqrt{6}), (-2, 0)$, and $(2, \sqrt{6})$

35. union of $(-\infty, -2], (-1, 1)$, and $(1, \infty)$

37. $(-\infty, \frac{3}{2})$ 39. $(-1, -\frac{1}{3})$ 41. $[-3, 1)$

43. -3 45. 10 47. $-1, 1$

49. $-1, 3$ 51. $-\frac{5}{6}$ 53. $-1, 0, 1$

55. $0, -2$ 57. 0 59. $(1, 3)$

61. $(-1.01, -0.99)$

63. union of $(-\infty, -6]$ and $[0, \infty)$

65. union of $(-\infty, -1]$ and $[0, \infty)$

67. union of $(-\infty, -\frac{1}{6})$ and $(\frac{1}{2}, \infty)$

69. $(\frac{3}{2}, \frac{5}{2})$ 73. yes   75. no

Section 1.2

1.

3. 5

5. $4\sqrt{2}$

7. $9\sqrt{2}$

9. $\sqrt{6 - 2\sqrt{6}}$

11. $\sqrt{2}|b - a|$

13. $y = -\frac{2}{3}x$ 15. $y = -x + 2$

17. $y = 3x - 7$ 19. $y = -x + 1$

21. $y = -x$ 23. $y = 3x - 3$

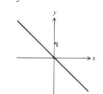

A-31

25. $m = -1, b = 0$

27. $m = -\frac{1}{3}, b = -4$

29. $m = 2, b = 7$

31. $m = -2, b = 4$

33. perpendicular; $\left(-\frac{8}{5}, -\frac{1}{5}\right)$ 35. parallel

37. perpendicular; $(1, -1)$ 39. parallel

41. parallel   43. $y = 3x - 7$

45. $y = -x$ 47. $y = \frac{2}{3}x - \frac{1}{3}$

49. $y = -\frac{1}{2}x - \frac{7}{2}$ 51. $y = \frac{3}{2}x$

53. $y = -\frac{1}{2}x - 3$

55. a. -4 b. $\frac{1}{2}\sqrt{2}$ c. -3 d. 5 e. -2 f. none

57. $\dfrac{x}{\frac{1}{2}} + \dfrac{y}{-2} = 1$ 59. $\dfrac{x}{-\frac{1}{2}} + \dfrac{y}{\frac{1}{3}} = 1$

61.

63.

65.

67.

69.

71. a. the x axis
    b. the y axis

Section 1.3

1. $\sqrt{3}, \sqrt{3}$ 3. $1, 1$ 5. $\frac{1}{4}$

7. $3, -\frac{1}{2}$ 9. $\frac{1}{8}$

11. all real numbers   13. $[-2, 8]$

15. $[-2, \infty)$ 17. all real numbers

19. union of $\left(-\infty, -\sqrt{3}/3\right)$ and $\left[\sqrt{3}/3, \infty\right)$

21. all real numbers

23. all real numbers except 1

25. all real numbers except -4 and 4

27. all real numbers   29. $[-4, -1]$ and $(0, 6)$

31. $\left[-3, -2\sqrt{2}\right]$ and $\left[2\sqrt{2}, 3\right]$

33. the set consisting of -1

35. $(-\infty, 10)$

37. all real numbers except 0

39. a, e, f, g, and i define functions.

41. $f_1 = f_5$ and $f_2 = f_3$

43. a. $(-\infty, -1]$ and $[1, \infty)$

45. $f(x) = -\sqrt{x}$ for $x \geq 0$

47. $-6, 0$ 49. $-1, 5$

51. none   53. $-3, -\frac{1}{2}$

57. $f(x) = \frac{1}{2}x^2 \sqrt[3]{x/5}$ for $x \geq 0$

59. $A(x) = \frac{1}{4}\sqrt{3}x^2$ for $x \geq 0$

61. $L(r) = (300 - 4\pi r^3)/(3\pi r^2)$ for $r > 0$

63. The ball has traveled 144 feet after 3 seconds. It hits the ground after 7 seconds.

65. $D(t) = \begin{cases} 400t & \text{for } 0 \leq t < 2 \\ |1600 - 400t| & \text{for } 2 \leq t \leq 5 \end{cases}$

67. a. $0; \dfrac{2}{c + 2d}$; no response   b. $x = \dfrac{cR(x)}{1 - dR(x)}$

Section 1.4

1.

3.

5.

7.

9.

11.

13.

15.

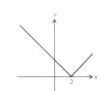

17.

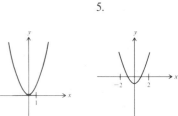

19.

21.

23.

25.

27.

29. not the graph of a function

31. the graph of a function

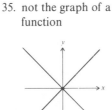

33. not the graph of a function

35. not the graph of a function

37. the x and y axes; not the graph of a function

39. not the graph of a function

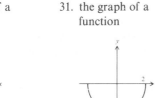

41. The graphs in a, d, e, and h are graphs of functions.
43. $P(x) = 22 - 17[-(x - 1)] = 22 - 17[1 - x]$ for $x > 0$

Section 1.5

1. y intercepts: $-\sqrt{\frac{2}{3}}, \sqrt{\frac{2}{3}}$; x intercept: -2; symmetry with respect to the x axis

3. no y intercepts; x intercepts: $-1, 1$; symmetry with respect to the x axis, y axis, origin

5. y intercept: 0; x intercept: 0; symmetry with respect to the y axis

7. no intercepts; symmetry with respect to the x axis, y axis, origin

9. no y intercepts; x intercepts: $-1, 1$; symmetry with respect to the origin

11. y intercept: 0; x intercepts: [0, 1); no symmetry

13. y intercept: 3; x intercepts: $-3, 3$; symmetry with respect to the y axis

15. y intercept: 1; x intercept: 1; no symmetry

17. y intercept: 0; x intercept: 0; symmetry with respect to the origin

19. y intercept: -3; x intercepts: $-\sqrt{3}, \sqrt{3}$; symmetry with respect to the y axis

21. y intercepts: $-1, 1$; no x intercepts; symmetry with respect to the x axis, y axis, origin

23. y intercepts: $-2, 2$; x intercept: 2; symmetry with respect to the x axis

25. y intercept: 0; x intercept: 0; symmetry with respect to the x axis, y axis, origin

27.

29.

31.

33.

35.

37.

39. a. odd   b. even   c. neither   d. neither   e. even
    f. odd   g. odd   h. even   i. odd

41.

43. The graph of g is d units above the graph of f if $d \geq 0$ and
    is $-d$ units below the graph if $d < 0$.
45. no
47. symmetry with respect to the line $x = c$

Section 1.6

1. -1 3. $-\frac{33}{4}$ 5. 6
7. $2x + 5$ 9. $\frac{529}{257}$ 11. $\frac{4}{41}\sqrt{3}$
13. 0
15. $(f + g)(x) = x + 4$ for all x
    $(fg)(x) = -2x^2 + 5x + 3$ for all x
    $\left(\dfrac{f}{g}\right)(x) = \dfrac{2x + 1}{3 - x}$ for $x \neq 3$
17. $(f + g)(x) = \dfrac{x^2 - 2x + 3}{x - 1}$ for $x \neq 1$
    $(fg)(x) = 2$ for $x \neq 1$
    $(f/g)(x) = 2/(x - 1)^2$ for $x \neq 1$
19. $(f + g)(t) = t^{3/4} + t^2 + 3$ for $t \geq 0$
    $(fg)(t) = t^{11/4} + 3t^{3/4}$ for $t \geq 0$
    $(f/g)(t) = t^{3/4}/(t^2 + 3)$ for $t \geq 0$
21. $(f + g)(t) = t^{2/3} + t^{3/5} - 1$ for all t
    $(fg)(t) = t^{19/15} - t^{2/3}$ for all t
    $(f/g)(t) = t^{2/3}/(t^{3/5} - 1)$ for $t \neq 1$

23. $(g \circ f)(x) = -2x + 7$ for all x
    $(f \circ g)(x) = -2x - 4$ for all x
25. $(g \circ f)(x) = \sqrt{x^2} = |x|$ for all x
    $(f \circ g)(x) = x$ for $x \geq 0$
27. $(g \circ f)(x) = x - 5\sqrt{x} + 6$ for $x \geq 0$
    $(f \circ g)(x) = \sqrt{x^2 - 5x + 6}$ for $x \leq 2$ or $x \geq 3$
29. $(g \circ f)(x) = (x - 1)/x$ for $x \neq 0, 1$
    $(f \circ g)(x) = -(x + 1)/x$ for $x \neq -1, 0$
31. $(g \circ f)(x) = [\sqrt{9 - x^2}\,]$ for $-3 \leq x \leq 3$
    $(f \circ g)(x) = \sqrt{9 - [x]^2}$ for $-3 \leq x < 4$
33. $f(x) = x - 3$; $g(x) = \sqrt{x}$
35. $f(x) = 3x^2 - 5\sqrt{x}$; $g(x) = x^{1/3}$
37. $f(x) = x + 3$; $g(x) = 1/(x^2 + 1)$ (or
    $f(x) = (x + 3)^2 + 1$; $g(x) = 1/x$)
39. $f(x) = \sqrt{x} - 1$; $g(x) = \sqrt{x}$
    (or $f(x) = \sqrt{x}$; $g(x) = \sqrt{x - 1}$)
41. $g(x) = -|x - 2|$ 43. $[-3, 1]$
45. all functions with range contained in $[0, \infty)$
47. all functions whose ranges do not contain 0
55. $P(x) = -\frac{1}{10}x^4 + x^2 + 24x - 38$; $P(1) = -13.1$;
    $P(2) = 12.4$
57. a. $V(r(s)) = \dfrac{1}{6\sqrt{\pi}}\, s^{3/2}$ for $s \geq 0$ b. $\sqrt{6/\pi}$

Section 1.7

1. a. $7\pi/6$ b. $7\pi/4$ c. $-9\pi/4$ d. 6π e. $\pi/180$
3. a. $-\frac{1}{2}$ b. $-\frac{1}{2}\sqrt{3}$ c. $-\frac{1}{2}\sqrt{2}$ d. $-\frac{1}{2}\sqrt{3}$ e. $\sqrt{3}$
   f. -1 g. $\sqrt{3}$ h. $\frac{1}{3}\sqrt{3}$ i. -1 j. 2   k. 1   l. $\frac{2}{3}\sqrt{3}$
5. $\tan x = -\frac{4}{3}$, $\cot x = -\frac{3}{4}$, $\sec x = -\frac{5}{3}$, $\csc x = \frac{5}{4}$
7. $\sin x = -\frac{2}{5}\sqrt{5}$, $\cos x = \frac{1}{5}\sqrt{5}$, $\cot x = -\frac{1}{2}$,
   $\csc x = -\frac{1}{2}\sqrt{5}$
9. $7\pi/6$, $11\pi/6$
11. $5\pi/6$, $7\pi/6$
13. 0, $2\pi/3$, $4\pi/3$
15. $[0, 7\pi/6)$ and $(11\pi/6, 2\pi)$
17. $[\pi/4, \pi/2)$ and $[5\pi/4, 3\pi/2)$
19. $(\pi/2, 3\pi/2)$
21. $(\pi/6, 5\pi/6)$ and $(7\pi/6, 11\pi/6)$
23. $(0, \pi/4]$, $(\pi/2, 3\pi/4]$, $(\pi, 5\pi/4]$, and $(3\pi/2, 7\pi/4]$
25. y intercept: -1; x intercepts: $\pi/2 + n\pi$ for any integer n;
    symmetric with respect to the y axis; even function

27. same as Exercise 25

29. no y intercept; x intercepts: $\pi/2 + n\pi$ for any integer n; symmetric with respect to the origin; odd function

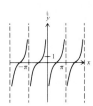

31. y intercept: 1; no x intercepts; symmetric with respect to the y axis; even function. The graph is the one in Figure 1.50(c).

33. y intercept: 0; x intercepts: $n\pi/2$ for any integer n; symmetric with respect to the origin; odd function

35. $\frac{1}{4}\sqrt{2}\,(\sqrt{3}+1)$ 37. $\frac{1}{4}\sqrt{2}\,(\sqrt{3}-1)$

39. $\dfrac{\sqrt{3}+1}{\sqrt{3}-1} = 2+\sqrt{3}$

41. $\pi/6 + 2n\pi$, $5\pi/6 + 2n\pi$, and $3\pi/2 + 2n\pi$ for any integer n

47. $\frac{14}{5}$

49. a. π b. π c. 2π d. 2π e. $2\pi/3$ f. π g. π h. π

51. $2\sec\theta$ 55. $60 + 4\sqrt{3} \approx 129.3$ feet

57. $V = 40\cos\theta\,(1 + \sin\theta)$ 59. b. $\dfrac{921}{2\sin\pi/5} \approx 783.4$ feet

Chapter 1 Review Exercises

1. union of $(-\infty, -\frac{3}{2})$ and $(4, \infty)$

3. $[\frac{5}{2}, 3)$ 5. $(\frac{7}{12}, \frac{3}{4})$ 7. a. a b. 0

9. $y = -4x + 1$ 11. $y = -\frac{1}{3}x + 6$

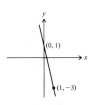

13. perpendicular

15. a. $y = -\frac{1}{2}x - \frac{3}{2}$ b. $y = 2x - 4$

19. all real numbers except -3, -1, and 0

21.   23.

25. y intercept: 1; x intercept: 1; no symmetry

27. y intercepts: $-\sqrt{3}, \sqrt{3}$; x intercepts: $-\sqrt{3}, \sqrt{3}$; symmetry with respect to the x axis, y axis, and origin

29.   31.

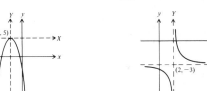

33. $(f-g)(x) = \dfrac{3}{(x-3)(x-1)}$ for $x \neq -1, 1$, and 3

$\left(\dfrac{f}{g}\right)(x) = \dfrac{x+2}{x-1}$ for $x \neq -1, 1$, and 3

35. a. Domain of f: $[-1, \infty)$; domain of g: $[-2, \infty)$; domain of h: $(-\infty, -2]$ and $[-1, \infty)$

b. Domain of fg: $[-1, \infty)$. The domain of fg contains fewer numbers than the domain of h.

41. $\cos x = \sqrt{5}/3$; $\cot x = -\sqrt{5}/2$; $\sec x = 3\sqrt{5}/5$; $\csc x = -\frac{3}{2}$

43. $(2\pi/3, 4\pi/3)$

45. y intercept: 1; x intercepts: $\pi/2 + 2n\pi$ for any integer n; not symmetric with respect to either axis or origin

47. y intercept: -1; x intercepts: $\pi/4 + n\pi$ for any integer n; not symmetric with respect to either axis or origin

49. even

51. a. $a = 2, b = \pi/3 + 2n\pi; a = -2, b = 4\pi/3 + 2n\pi$
    b. $a = 2, b = -\pi/6 + 2n\pi; a = -2, b = 5\pi/6 + 2n\pi$

53. no   55. $(1, 3), (5, 1),$ or $(-1, -1)$

CHAPTER 2

Section 2.1

1. 3   3. 0   5. $-\frac{1}{11}$ 7. $\frac{2}{5}$
9. -4 11. -6 13. 3   15. 3
17. 1
19. 1.0033467, 1.0000333, 1.0000003, 1.0033467, 1.0000333, 1.0000003; $\lim_{x \to 0} f(x) = 1$
21. 0.04995835, 0.00499996, 0.0005, -0.04995835, -0.00499996, -0.0005; $\lim_{x \to 0} f(x) = 0$
23. $x + 3$; 6   25. $2(x + 4)$; 16   27. b or d
29. a. -6 b. $-\frac{2}{3}$
31. $2(t + 2)$; 8 miles per minute

Section 2.2

1. -5 3. $\frac{1}{2}$ 5. 3
7. $\frac{3}{2}$ 9. 0   11. $y = \pi$
13. $y = 2$ 15. $y = 4x - \frac{1}{2}$ 17. $y = 5x - 1$
19. $y = -\frac{1}{5}$ 21. 2   23. -128
25. -56 27. b and c
35. -32 feet per second
37. a. -80 feet per second   b. 80 feet per second
39. -112 feet per second

Section 2.3

1. -2 3. $-\frac{15}{2}$ 5. 14   7. 21
9. $-\frac{9}{2}$ 11. 144   13. $-\frac{1}{12}$ 15. $\frac{3}{2}$
17. $-\frac{9}{16}$ 19. -2 21. 3   23. -8
25. 6   27. $\frac{2}{3}$ 29. -6 31. -1
33. 1   35. $y + 2 = 2(x + 1)$

37. $y - \frac{1}{2} = -\frac{1}{4}(x - 2)$ 39. $y - \frac{1}{2} = -\frac{1}{4}(x + 1)$
41. $y - 4 = \frac{1}{8}(x - 16)$
49. $f(x) = 1/(x - 1), g(x) = -1/(x - 1)$
51. $f(x) = \begin{cases} 0 & \text{for } x < 2 \\ 1 & \text{for } x \geq 2 \end{cases}$ and $g(x) = \begin{cases} 1 & \text{for } x < 2 \\ 0 & \text{for } x \geq 2 \end{cases}$
53. $-16t_0^2 + v_0 t_0 + h_0$

Section 2.4

1. $\frac{3}{2} - \frac{2}{3}\pi$ 3. $\frac{1}{6}\pi^2$ 5. $-\frac{3}{8}\sqrt{3}$ 7. 16
9. $\frac{1}{32}$ 11. $\frac{1}{2}$ 13. 0   15. $\frac{2}{9}$
17. $\cos^2 1$ 19. $\frac{3}{5}$ 21. 1   23. 0
25. 1   27. 0   29. $\frac{1}{2}$ 31. -2
33. -1 37. $y = 1$
39. $y - \sqrt{3} = -\frac{1}{3}\sqrt{3}(x - 1)$
43. c. $f(x) = \begin{cases} -1 & \text{for } x \leq 0 \\ 1 & \text{for } x > 0 \end{cases}$
45. a. 0   b. 0   49. c. $\dfrac{\sqrt{1 - x^2} - \sqrt{1 - a^2}}{x - a}$

Section 2.5

1. -19 3. 4   5. $\frac{5}{2}$ 7. -1 9. 2
11. $-\infty$ 13. $-\infty$ 15. $-\infty$ 17. $-\infty$ 19. ∞
21. The limit does not exist.
23. ∞ 25. 0   27. ∞ 29. ∞ 31. $-\infty$
33. ∞ 35. $-\infty$ 37. $-\infty$
39. The limit does not exist.
41. $-\infty$ 43. $\frac{1}{2}$ 45. $\frac{1}{6}$ 47. ∞ 49. -9
51. ∞ 53. $x = -4$
55. $x = -\sqrt{2}$ 57. $x = -2$ and $x = 2$
59. $x = -4$ and $x = 4$ 61. $x = 3$
63. $x = -2$ and $x = 0$ 65. $x = 0$
67. none
69. $x = \pi/2 + n\pi$ for every integer n 71. $x = 0$ 73. $x = 0$
75. two sided limit: (a); right-hand limit: (a), (e), (f); left-hand limit: (a), (b), (f); none: (c), (d)
79. $1/(a + b)$ 81. -80 feet per second

Section 2.6

1. continuous at 2   3. continuous at 0
5. discontinuous at -1 7. discontinuous at 0
9. continuous at $\frac{1}{2}$; continuous from the left at 1
11. continuous at 0; continuous from the left at 1
13. continuous from the right at $\pi/2$
15. continuous at $\pi/2$
31. a, b, and e 33. b and f

Section 2.7

9. union of $(-\infty, 2)$ and $(2, 5)$
11. union of $(-\infty, -3]$ and $[-1, 0]$
13. union of $(-\infty, -\frac{1}{2})$, $(\frac{1}{2}, 1]$ and $[3, \infty)$
15. $(1, 3)$ 17. $\frac{23}{16}$ 19. $\frac{23}{16}$
21. $\frac{11}{16}$ 23. $\frac{35}{16}$ 25. $\frac{13}{16}$

Chapter 2 Review Exercises

1. $-\frac{5}{3}$ 3. 2   5. -1
7. $\frac{15}{2}$ 9. $-\infty$ 11. 3
13. $\frac{3}{2}\sqrt{2}$ 15. The limit does not exist.
17. $x = 4$ 19. $x = -2$ and $x = 7$
21. $x = -2$
23. continuous from the left at $-\sqrt{13}$
25. continuous from the right at 2
31. 10
33. a. number b. The limit does not exist c. The limit does not exist d. The limit does not exist e. number f. ∞
35. union of $[-4, -2)$ and $(2, \infty)$
37. $\frac{23}{16}$ 39. $\frac{19}{64}\pi$
41. It does not contain the phrase "provided the limits of the two functions both exist."
47. $(4, 2)$
49. a. 576 feet b. 192 feet per second

CHAPTER 3

Section 3.1

1. 0   3. 2   5. -2
7. $-\frac{1}{4}$ 9. 1   11. 4
13. $f'(x) = 0$ 15. $f'(x) = -10x$
17. $g'(x) = 3x^2$ 19. $k'(x) = -2/x^3$
21. $dy/dx = 0$ 23. $dy/dx = 6x$
25. 0   27. 4
29. no derivative   31. no derivative
33. 1   35. no derivative
37. $y = -4x - 4$ 39. $y = \frac{1}{4}x + 1$
41. a. $f(x) = x^4$; $f'(2) = 32$ b. $f(x) = x^4$; $f'(2) = 32$
43. $-f'(a)$ 45. a. -2 b. 2   47. no
51. 3   53. $-\frac{3}{2500}$ 55. $\frac{225}{4}$
57. $\frac{17}{2}$ thousand dollars per thousand gallons; an increase of $\frac{1}{2}$ thousand dollars per thousand gallons
59. 25 knots
61. a. $\sqrt{8 - 4\sqrt{2}}$ meters per minute
    b. $3/\sqrt{2 - \sqrt{2}}$ meters per minute

Section 3.2

1. 0   3. 3; 0   5. 8
7. 10   9. $0; \frac{1}{2}\sqrt{3}$ 11. -2
13. $5x^4$ 15. $2x + 1$ 17. $4x/(x^2 + 1)^2$
19. $3\sin x$ 21. $\frac{2}{3}x^{-1/3}$ 23. $1/(2\sqrt{x - 1})$
31. a. -3 b. 5 c. 3 or -3
    d. $-\pi/6 + 2n\pi$ or $\pi/6 + 2n\pi$ for any integer n
33. $v(t) = -3\cos t$; $v(\pi/6) = -\frac{3}{2}\sqrt{3}$
37. $2\pi r$ 39. a. $\frac{1}{2}\sqrt{3}x$ b. 2
41. $F'(r) = -2k/r^3$
43. a. $y - 1 = \frac{1}{2}(x + 1)$ b. $y + 5 = -\frac{1}{2}(x + 1)$ c. $y = -x$

Section 3.3

1. $-12x^2$ 3. $6x - 4$
5. $16x^3 + 9x^2 + 4x + 1$ 7. $36/t^{10}$
9. $4x + 7$ 11. $-1/x^2 + 2/x^3$
13. $12x^{-4} - 2\sin x$ 15. $-6z^2 + 4\sec z \tan z$
17. $2z\sin z + z^2\cos z$ 19. $2\sin x\cos x = \sin 2x$
21. $-4\tan x\sec x - 4x\sec^3 x - 4x\tan^2 x\sec x$
23. $-14/(4x - 1)^2$ 25. $-1/(t^2 + 4t + 4)$
27. $-4(t^2 + 12t + 26)/(t^2 + t - 20)^2$
29. $-\csc^2 x$
31. $\frac{1}{2}(1/\sqrt{y})\sec y + \sqrt{y}\sec y\tan y$
33. $18x^2 - 26x - 5$ 35. $2x - 2/x^3$
37. $7/(5 - 3x)^2$
39. $(-x^6 + 4x^3 + 3x^2)/(x^4 + 1)^2$
41. $-\csc^2 x + \sec^2 x$
43. $\dfrac{(x^3 + x)\cos x + (1 - x^2)\sin x}{(x^2 + 1)^2}$
45. 63   47. $-16/\pi$ 49. $-\sqrt{2}\pi^{-3/2}$
51. $y = x - 8$ 53. $y = x + 1 - \pi/2$
55. $(f + g)(x) = 0$ for all x
61. a. $v(1) = -8$ feet per second; $v(2) = -40$ feet per second
    b. $v(1) = 16$ feet per second
63. a. 256 feet b. -128 feet per second
65. $v(t) = -32t - 16$
67. a. $\sqrt{1.5} \approx 1.225$ seconds
    b. $32\sqrt{1.5} \approx 39.19$ feet per second
    c. $40 - 32\sqrt{1.5} \approx 0.81$ feet per second
69. $F'(x) = \dfrac{-50\mu(\mu\cos x - \sin x)}{(\mu\sin x + \cos x)^2}$
71. $f'(x) = \dfrac{100nkx^{n-1}}{(1 + kx^n)^2}$ for $x > 0$
73. 9

Section 3.4

1. $\frac{9}{4}x^{5/4}$ 3. $\frac{2}{3}x^{-1/3} + \frac{7}{3}x^{-4/3}$
5. $-2400x(4 - 3x^2)^{399}$ 7. $6(x - 1)^2/(x + 1)^4$

9. $(2 - 14x^2)/\sqrt{2 - 7x^2}$ 11. $5\cos 5t$

13. $4\sin^3 t \cos t - 4\cos^3 t \sin t$

15. $6\tan^5 x \sec^2 x$

17. $(-\frac{1}{3}\cos x)(1 - \sin x)^{-2/3}$

19. $-[\sin(\sin x)]\cos x$

21. $\left(x^2 + \dfrac{1}{x^2}\right)^{-1/2}\left(x - \dfrac{1}{x^3}\right)$

23. $\dfrac{3x - 5}{x^2(5 - 2x)^{3/2}}$ 25. $\cos\dfrac{1}{x} + \dfrac{1}{x}\sin\dfrac{1}{x}$

27. $[2z - (2z)^{1/3}]^{-1/2}[1 - \frac{1}{3}(2z)^{-2/3}]$

29. $-36z^5 \cos(3z^6)\sin(3z^6)$

31. $\frac{1}{4}\sqrt[4]{\sec(\tan z)}\tan(\tan x)\sec^2 x$

33. $-6[\cot(2\sqrt{3x + 1})][\csc^2(2\sqrt{3x + 1})][1/\sqrt{3x + 1}]$

35. $-2x^{-5/3}$ 37. $-(1 + 6x^2)/\sqrt{1 + 3x^2}$

39. $-\dfrac{2}{3}\dfrac{\sin x + x\cos x}{(x\sin x)^{5/3}}$ 41. $\frac{3}{2}\tan^2(\frac{1}{2}x)\sec^2(\frac{1}{2}x)$

43. $5y^4\, dy/dx$ 45. $-\dfrac{2}{y^2}\dfrac{dy}{dx}$

47. $\left(\dfrac{1}{2\sqrt{y}}\cos y\right)\dfrac{dy}{dx}$ 49. $3x^2y^2 + 2x^3y\dfrac{dy}{dx}$

51. $\left(x + y\dfrac{dy}{dx}\right)/\sqrt{x^2 + y^2}$ 53. $y = -2x + 1$

55. $y = 2$

57. $(g \circ f)'(-3) = 4\sqrt{2}; (f \circ g)'(0) = -28;$
$(g \circ f)'(0) = 26; (f \circ g)'(2) = 2\sqrt{2};$
$(g \circ f)'(2) = 22$

59. $\dfrac{d}{dx}x^{1/n} = \dfrac{1}{n}x^{(1/n) - 1}$

61. $F'(t) = 14{,}000\sin\dfrac{\pi t}{24}$

63. a. $-\dfrac{40}{V}\left(1 - \dfrac{v}{V}\right)^{-0.6}$

    b. $\dfrac{40v}{V^2}\left(1 - \dfrac{v}{V}\right)^{-0.6}$

65. a. $v'(r) = -\dfrac{96{,}000}{r^2}\left(\dfrac{192{,}000}{r} + v_0^2 - 48\right)^{-1/2}$

    b. $-\sqrt{6}/72{,}000$ miles per second per mile

67. $dV/dt = 40\pi r^2$

69. $\dfrac{dA}{dh} = \frac{2}{3}\sqrt{3}h; \left.\dfrac{dA}{dh}\right|_{h=\sqrt{3}} = 2$

Section 3.5

1. 0

3. $-240x^3 + 6x^2 + \frac{1}{4}(1 - x)^{-3/2}$

5. $192(1 - 4x)^{-4}$ 7. $an(n + 1)x^{-n-2}$

9. $6x(2x^3 + 1)/(x^3 - 1)^3$

11. $\dfrac{15\pi}{4}x^{1/2} - \dfrac{\cos x}{x} + \dfrac{2\sin x}{x^2} + \dfrac{2\cos x}{x^3}$

13. $\sec x\tan^2 x + \sec^3 x$

15. $8\csc^2(-4x) + 32x\csc^2(-4x)\cot(-4x)$

17. $12\tan^2 2x\sec^2 2x + 24x(\tan 2x\sec^4 2x + \tan^3 2x\sec^2 2x)$

19. $\frac{3}{4}x^{-1/2}$

21. $6(x^4 - \tan x)(4x^3 - \sec^2 x)^2 +$
$3(x^4 - \tan x)^2(12x^2 - 2\sec^2 x\tan x)$

23. $2a$ 25. $2/(3 - x)^3$

27. $\csc x\cot^2 x + \csc^3 x$ 29. $-\sin x - \cos x$

31. $0.$

33. $-12x\sin x^2 - 8x^3\cos x^2$

35. $-6/x^4$ 37. $1440/(4x + 5)^4$

39. 0 41. $-\frac{3}{8}x^{-9/2}$

43. $\dfrac{1}{x^4}\cos\dfrac{1}{x}$ 45. $6a$

47. $5040x^4 + 270x^2 + \frac{135}{64}x^{-13/4} + 48x^{-5}$

49. $\pi^4\sin\pi x$

51. $v(t) = -32t + 3; a(t) = -32$

53. $v(t) = 2\cos t + 3\sin t; a(t) = -2\sin t + 3\cos t$

55. 0

57. a. $f^{(3)}(x) = f'(x)$ b. $f^{(35)}(x) = f^{(31)}(x)$

59. $f^{(n)}(x) = (-1)^n n!\left[\dfrac{1}{x^{n+1}} - \dfrac{1}{(x + 1)^{n+1}}\right]$

61. $h''(x) = g''(f(x))(f'(x))^2 + g'(f(x))f''(x)$

63. -29.2432 feet per second per second

Section 3.6

1. $-\dfrac{4x^3}{3y}$ 3. $\dfrac{2}{(2y + 1)(1 - x)^2}$

5. $\dfrac{\sec^2 x}{\sec y\tan y}$

7. $\dfrac{3(y^2 + 1)^2}{(y^2 + 1)\cos y - 2y\sin y}$

9. $\dfrac{-2x - 2xy^2}{2x^2y + 3y^2}$ 11. $\dfrac{x^4 + y^2}{xy - x^3y}$

13. $\dfrac{-y(xy)^{-1/2} - (x + 2y)^{-1/2}}{x(xy)^{-1/2} + (x + 2y)^{-1/2}}$

15. $\dfrac{2y^2 - x^3 - xy^2}{y^3 + x^2y + 2xy}$

17. 0 19. $-\frac{1}{2}$ 21. $-\frac{7}{2}$

23. $-\frac{1}{9}$ 25. -4

29. $y = \frac{3}{4}x - \frac{9}{2}$ 31. $y = -3x + \pi$

33. $\dfrac{2y^4 - 3x^2}{4y^7}$ 35. $\dfrac{(\tan 2y)(2\sec^2 2y + 1)}{x^2}$

37. $\dfrac{x}{y}\dfrac{dx}{dt}$ 39. $-\dfrac{\tan y}{x}\dfrac{dx}{dt}$

41. $-\dfrac{y^2 \sin(xy^2)\,\dfrac{dx}{dt}}{1 + 2xy \sin(xy^2)}$

43. a. $-\dfrac{1}{5y^4 + 1}$ b. $\dfrac{-3x^2}{5y^4 + 3y^2 + 1}$

c. $\dfrac{5}{3y^2 + \cos y + 1}$

45. tangent line: $y = -x + 3$

47. at $(\frac{1}{2}, \frac{1}{2}\sqrt{3})$, $\theta = \pi/3$; at $(\frac{1}{2}, -\frac{1}{2}\sqrt{3})$, $\theta = 2\pi/3$

49. $(\frac{1}{4}\sqrt{6}, \frac{1}{4}\sqrt{2}), (\frac{1}{4}\sqrt{6}, -\frac{1}{4}\sqrt{2}), (-\frac{1}{4}\sqrt{6}, \frac{1}{4}\sqrt{2}),$
$(-\frac{1}{4}\sqrt{6}, -\frac{1}{4}\sqrt{2})$

Section 3.7

1. decreasing at 32π cubic inches per minute
3. decreasing at $12/\pi^2$ inches per hour
5. increasing at $1/(64\pi^{1/3}3^{2/3})$ inches per second
7. increasing at $3/(20\pi)$ inches per minute
9. $\frac{4}{3}\sqrt{10}$ feet per second
11. a. sliding at $\frac{3}{2}$ feet per second
    b. decreasing at $\frac{7}{4}$ square feet per second
13. poured in at $2\pi/9$ cubic inches per second
15. 40π miles per minute
17. increasing at $\frac{3}{4}$ foot per second
19. pulled in at $\frac{24}{13}$ feet per second
21. moving at 1 foot per second
23. $\frac{6}{25}$ radians per second
25. $\frac{5}{156}\sqrt{39} \approx \frac{1}{5}$ (miles)
27. $5\sqrt{3}$ feet per second
29. increasing at $\frac{1500}{29}$ radians per hour
31. $\frac{1920\sqrt{11}}{121}$ feet per second
33. $\frac{3}{250}$
35. 160π feet per minute
37. a. decreasing at $\frac{9}{7}$ feet per second
    b. decreasing at $\frac{3}{4}$ feet per second

Section 3.8

1. 10.05
3. $3 + \frac{2}{27} \approx 3.07407$
5. 6.03
7. 85
9. $\frac{1}{2}\sqrt{3} + \pi/156 \approx 0.886164$
11. $\sqrt{2}(1 - \pi/68) \approx 1.34888$
13. 0.05
15. 0.02
17. $15x^2\,dx$
19. $-\sin x \cos(\cos x)\,dx$
21. $2x^3(1 + x^4)^{-1/2}\,dx$
25. 1.879385
27. 2.094552
29. -0.822876
31. 4.493410
33. 3.872983
35. 2.080084
37. $-0.18, 0.60$
39. $c_n = c_{n+1} = c_{n+2} = \cdots$
41. get approximate zero 0.123078 outside $[-2, 0]$
43. c_n increases without bound
47. $4\pi 5^2(.137) \approx 43.0398$ cubic inches

Chapter 3 Review Exercises

1. $-12x^2 - 4/x^3$
3. $59/(5x + 7)^2$
5. $-(2x + 1)/(2x - 1)^3$
7. $-\sin t \sin 2t + 2 \cos t \cos 2t$
9. $5 \tan t + 5t \sec^2 t + 9 \sec 3t \tan 3t$
11. $12x^2 - \sqrt{3} - 2/(5x^2)$
13. $-3x^2 \csc^2 x^3$
15. $\dfrac{\cos x - 1 + \tan^2 x}{(1 - \sec x)^2}$
17. $3x^2\sqrt{x^2 - 4} + x^4/\sqrt{x^2 - 4}$
19. $y = 5x$
21. $y = x$
23. $y = \frac{13}{4}x - \frac{25}{4}$
25. $y = 2x$
27. $33x^{10} - 180x^4$
29. $3(t^2 + 9)^{1/2} + 3t^2(t^2 + 9)^{-1/2}$
31. $2 \sec^4 x + 4 \tan^2 x \sec^2 x$
33. $\dfrac{8xy - y}{9y^2 - 4x^2 + x}$
35. $\dfrac{2\sqrt{x} - y}{2\sqrt{x}(\sqrt{x} + 1)}$
37. $-\dfrac{y \cos xy^2}{3y + 2x \cos xy^2}$
39. $\frac{6}{5}$
41. $-\dfrac{y}{x}\dfrac{dx}{dt}$
43. $(2x \cos x - x^2 \sin x)\,dx$
45. $7(x - \sin x)^6(1 - \cos x)\,dx$
47. $\frac{25}{6}$
49. $\sqrt{2}(1 + \pi/100)$
51. 1.895
55. $2a^{1/2}f'(a)$
59. a. $dV/dr = 2\pi rh/3$
    b. $dV/dh = \pi r^2/3$
    c. $dh/dV = 3/(\pi r^2)$
63. $\frac{1}{2}$

Cumulative Review Exercises (Chapters 1 and 2)

1. union of $(-\sqrt{3}, 0)$ and $(1, \sqrt{3})$
3. union of $(-1, -\frac{1}{5})$ and $(\frac{1}{5}, 1)$
5. union of $(-\infty, -\frac{1}{2})$ and $(\frac{2}{3}, \infty)$
7. $(-\infty, -1]$
9. -1
11. -1
13. 2
15.

CHAPTER 4

Section 4.1

1. -2
3. $-2, 0, 1$
5. $-1, 1$
7. 0
9. 0

11. $\pi/2 + n\pi$ for any integer n

13. $(2n + 1)\pi$ for any integer n

15. 2

17. minimum value: $f(\frac{1}{2}) = -\frac{1}{4}$; maximum value: $f(2) = 2$

19. minimum value: $g(1) = -3$; maximum value: $g(-4) = 272$

21. no extreme values

23. minimum value: $k(0) = 1$; maximum value: $k(3) = \sqrt{10}$

25. minimum value: $f(0) = 0$; no maximum value

27. minimum value: $f(\pi^3/8) = -1$; maximum value: $f(-\pi^3/27) = \frac{1}{2}\sqrt{3}$

29. no extreme values

31. relative minimum value: $f(2) = -4$; relative maximum value: $f(0) = 0$

33. relative minimum value: $f(1) = -7$; relative maximum value: $f(-3) = 25$

35. relative minimum value: $g(\sqrt{2}) = -16\sqrt{2}$; relative maximum value: $g(-\sqrt{2}) = 16\sqrt{2}$

37. relative minimum value: $f(\pi/2 + n\pi) = -4$ for any even integer n; relative maximum value: $f(\pi/2 + n\pi) = -2$ for any odd integer n

39. relative minimum value: $f(n) = 1$ for any even integer n; relative maximum value: $f(n) = -1$ for any odd integer n

41. $\frac{3}{16}$ 43. -3.195823

45. $-2.879385, -0.652704, 0.532089$

53. largest: $\frac{1}{4}$; smallest: 0

Section 4.2

1. 2   3. $-\frac{2}{3}\sqrt{3}$ 5. -1

7. $1 - \frac{1}{3}\sqrt{21}$ 9. $(\frac{7}{3})^{3/2}$

15. $9 < 28^{2/3} \le \frac{83}{9}$ 17. $|\sqrt{3} - 1.7| \le 0.03245$

19. yes

Section 4.3

1. C 3. $\frac{1}{2}x^2 + C$

5. $-\frac{1}{3}x^3 + C$ 7. $-3x^2 + 5x + C$

9. $-\cos x + C$ 11. $f(x) = -2x$

13. $f(x) = -\frac{1}{2}x^2 + \frac{1}{2} + \sqrt{2}$ 15. $f(x) = \frac{1}{3}x^3 - 5$

17. $f(x) = \sin x + 1 - \frac{1}{2}\sqrt{3}$

19. $f(x) = C_1 x + C_2$, where C_1 and C_2 are constants

21. $f(x) = -x + 2$

23. $f(x) = -\sin x - 3x + 4$

25. $f(x) = C_1 x^3 + C_2 x^2 + C_3 x + C_4$, where C_1, C_2, C_3, and C_4 are constants

27. strictly decreasing on $(-\infty, -\frac{1}{2}]$; strictly increasing on $[-\frac{1}{2}, \infty)$

29. strictly increasing on $(-\infty, \infty)$

31. strictly decreasing on $(-\infty, \frac{3}{2}]$; strictly increasing on $[\frac{3}{2}, \infty)$

33. strictly decreasing on $[-\sqrt{\frac{5}{2}}, \sqrt{\frac{5}{2}}]$; strictly increasing on $(-\infty, -\sqrt{\frac{5}{2}}]$ and $[\sqrt{\frac{5}{2}}, \infty)$

35. strictly decreasing on $[0, 4]$; strictly increasing on $[-4, 0]$

37. strictly decreasing on $(-\infty, -3)$ and $(-3, \infty)$

39. strictly decreasing on $[0, \infty)$; strictly increasing on $(-\infty, 0]$

41. strictly increasing on $(\pi/2 + n\pi, \pi/2 + (n + 1)\pi)$ for any integer n

43. strictly decreasing on $[-\pi/6 + 2n\pi, 7\pi/6 + 2n\pi]$ for any integer n; strictly increasing on $[7\pi/6 + 2n\pi, 11\pi/6 + 2n\pi]$ for any integer n

45. d. no   e. no   57. b. 0.450184   59. b.

Section 4.4

1. negative to positive at -3

3. negative to positive at -1 and 1; positive to negative at 0

5. negative to positive at 1; positive to negative at -1

7. negative to positive at $4\pi/3 + 2n\pi$ for any integer n; positive to negative at $2\pi/3 + 2n\pi$ for any integer n

9. relative maximum value: $f(\frac{1}{2}) = \frac{31}{4}$

11. relative minimum value: $f(0) = 4$; relative maximum value: $f(-2) = 8$

13. relative minimum value: $g(-\frac{1}{2}) = 3$

15. relative maximum value: $f(2) = \frac{1}{12}$

17. relative minimum value: $f(-\frac{1}{2}\sqrt{2}) = -\frac{1}{2}$; relative maximum value: $f(\frac{1}{2}\sqrt{2}) = \frac{1}{2}$

19. relative minimum values: $k(5\pi/6 + 2n\pi) = -\frac{1}{2}\sqrt{3} + 5\pi/12 + n\pi$ for any integer n; relative maximum values: $k(\pi/6 + 2n\pi) = \frac{1}{2}\sqrt{3} + \pi/12 + n\pi$ for any integer n

21. relative minimum value: $k(0) = 0$

23. relative maximum value: $f(\frac{3}{8}) = -\frac{7}{16}$

25. relative minimum value: $f(4) = -79$; relative maximum value: $f(-2) = 29$

27. relative minimum values: $f(-\frac{1}{2}) = \frac{1}{16}$ and $f(\frac{3}{2}) = -\frac{127}{16}$; relative maximum value: $f(0) = \frac{1}{2}$

29. relative minimum value: $f\left(\dfrac{1}{\sqrt[3]{2}}\right) = \dfrac{1}{2^{2/3}} + \sqrt[3]{2} + 1$

31. relative minimum value: $f(\pi/4 + n\pi) = -\sqrt{2}$ for any odd integer n; relative maximum value: $f(\pi/4 + n\pi) = \sqrt{2}$ for any even integer n

33. relative minimum value:
$f(-4) = -4$

35. no relative extreme values

35. $\sqrt[3]{V/2\pi}$
39. $2L$
43. $5\sqrt{5} \times 10^4$ units per day

37. $\frac{4}{27}\pi R^2 H$
41. 100 meters
45. a. 10   b. $800

Section 4.6

1. concave downward on $(-\infty, \infty)$

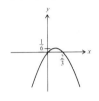

3. concave upward on $(0, \infty)$; concave downward on $(-\infty, 0)$

37. relative minimum value:
$f(-1) = -3$

5. concave upward on $(2, \infty)$; concave downward on $(-\infty, 2)$

7. concave upward on $(-\infty, \infty)$

39. relative minimum value:
$f(-1) = f(1) = 0$;
relative maximum value:
$f(0) = 1$

41. relative minimum value:
$f(-1) = f(2) = 0$
relative maximum value:
$f(\frac{1}{2}) = \frac{81}{16}$

43. 0   45. 0   47. 2

9. concave upward on $(0, \infty)$; concave downward on $(-\infty, 0)$

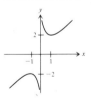

11. concave upward on $(\frac{4}{3}, \infty)$; concave downward on $(1, \frac{4}{3})$

Section 4.5

1. 9 and 9
3. base side length: 2 meters; height: 1 meter
7. triangle side length: $12/(6 - \sqrt{3}) \approx 2.8$ feet; height of rectangle: $(18 - 6\sqrt{3})/(6 - \sqrt{3}) \approx 1.8$ feet
9. approximately 2:16 P.M.
11. $a/2$
17. The company should not switch to bonds.
19. $a/\sqrt{2}$
21. $(3, 2)$
23. 2, 2, and $2\sqrt{2}$
25. The sides should each be 1 mile long.
27. minimum value of A: $\frac{1}{4}L^2/(4 + \pi)$; no maximum value of A if the wire is actually cut
29. a. $6\sqrt{7}/7$ miles down the road
   b. directly toward the car
   c. if $c > 6\sqrt{7}/7$, then to the point $6\sqrt{7}/7$ miles down the road; otherwise directly toward the car
31. $43 + 4\sqrt{70}$ square inches   33. 18

13. concave upward on $(n\pi + \pi/2, (n + 1)\pi)$ for any integer n; concave downward on $(n\pi, n\pi + \pi/2)$ for any integer n

15. concave upward on $(-\pi/2 + 2n\pi, \pi/2 + 2n\pi)$ for any integer n; concave downward on $(\pi/2 + 2n\pi, 3\pi/2 + 2n\pi)$ for any integer n

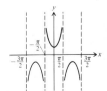

17. $(-2, 0)$

19. $(-1, 9)$

21. $(0,0)$ and $(-\frac{2}{3}, -\frac{16}{27})$

23. $(-1/\sqrt[3]{2}, \frac{11}{8})$, $(0, 0)$, and $(1/\sqrt[3]{2}, -\frac{11}{8})$

25. $(-\frac{2}{9}, \frac{4}{5}(\frac{2}{9})^{2/3})$

27. $(n\pi, 0)$ for any integer n

29. a.

b.

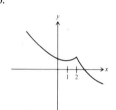

c.

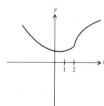

d.

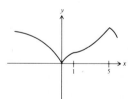

31. a. upward   b. downward
33. $f(x) = g(x) = x^2 - 1$, $c = 0$
39. $n - 2$

Section 4.7

1. 0	3. $\frac{1}{3}$	5. 2
7. ∞	9. 0	11. 0
13. $-\infty$	15. 0	

17. horizontal asymptote: $y = 0$

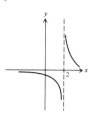

19. horizontal asymptote: $y = \frac{3}{2}$

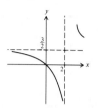

21. horizontal asymptote: $y = -2$

23. horizontal asymptote: $y = 1$

25. horizontal asymptote: $y = 1$

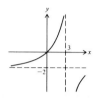

27. horizontal asymptote: $y = 0$

29. $y = 1$ and $y = -1$
35. $GMm/3960$

33. b

Section 4.8

1. inflection point: $(0, 2)$

3. relative minimum value: $f(0) = -3$

5. relative maximum value: $g(-2) = -4$; relative minimum value: $g(2) = 4$

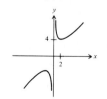

7. relative maximum value: $g(2) = \frac{1}{4}$; inflection point: $(3, \frac{2}{9})$

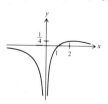

9.

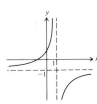

11. relative maximum
value: $k(\frac{1}{2}) = 1$;
relative minimum
value: $k(-\frac{1}{2}) = -1$;
inflection points:
$(-\frac{1}{2}\sqrt{3}, -\frac{1}{2}\sqrt{3})$,
$(0,0)$, and $(\frac{1}{2}\sqrt{3}, \frac{1}{2}\sqrt{3})$

29.

31. relative maximum
value: $f(1) = 1$

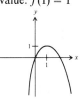

13. relative maximum
value: $f(0) = -1$

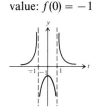

15. The graph is that of
Exercise 13 shifted
up one unit.

33. relative maximum value:
$g(\pi/2 + n\pi) = 1$ for
any integer n;
relative minimum value:
$g(n\pi) = 0$ for
any integer n

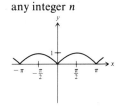

35. relative maximum value:
$g(\pi/3 + 2n\pi) = 2$ for
any integer n;
relative minimum value:
$g(-2\pi/3 + 2n\pi) = -2$
for any integer n;
inflection points:
$(-\pi/6 + 2n\pi, 0)$ and
$(5\pi/6 + 2n\pi, 0)$ for
any integer n

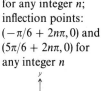

17. relative minimum
value: $f(0) = 0$;
inflection point: $(\frac{1}{2}, \frac{2}{9})$

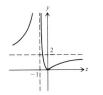

19. relative maximum value:
$f(\sqrt{3}) = -\frac{3}{2}\sqrt{3}$;
relative minimum value:
$f(-\sqrt{3}) = \frac{3}{2}\sqrt{3}$;
inflection point: $(0,0)$

37. relative maximum value:
$g(\pi/2 + n\pi) = 1$ for
any integer n;
relative minimum value:
$g(n\pi) = 0$ for
any integer n;
inflection points:
$(\pi/4 + n\pi/2, \frac{1}{2})$ for
any integer n

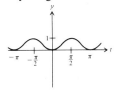

39. x intercepts:
approx. $-2.55, 0.55$;
relative maximum value:
$f(-1) = -1$;
relative minimum value:
$f(-2) = -2 = f(0)$;
inflection points:
$(-1 - \frac{1}{3}\sqrt{3}, -\frac{14}{9})$,
$(-1 + \frac{1}{3}\sqrt{3}, -\frac{14}{9})$

21. inflection points:
$(-\sqrt[3]{\frac{1}{2}}, -\frac{2}{3})$ and
$(0, -1)$

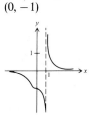

23.

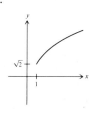

41. relative maximum value:
$f(1 + \sqrt{2}) = \dfrac{\sqrt{2}}{4 + 3\sqrt{2}}$
relative minimum value:
$f(1 - \sqrt{2}) = \dfrac{\sqrt{2}}{3\sqrt{2} - 4}$
inflection point:
approx. $(3.85, 0.15)$

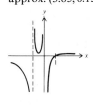

43.

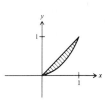

25. relative minimum
value: $f(0) = 0$

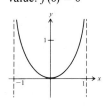

27. inflection point: $(0,0)$

45.

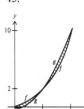

47.

 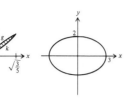

49.

51.

53.

55.

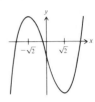

57.

59.

61.

63.

65.

Chapter 4 Review Exercises

1. $0, \frac{8}{5}, 2$
3. minimum value: $f(-\frac{1}{2}) = \frac{3}{4}$; maximum value: $f(2) = 7$
5. minimum value: $f(-\frac{1}{2}\sqrt{2}) = -\sqrt{2}$;
   maximum value: $f(1) = 1$
7. b. because f is not continuous at 0
9. $f(x) = \frac{1}{3}x^3 + \cos x + C$
11. $f(x) = \frac{1}{12}x^4 - 2x^2 + C_1 x + C_2$
13. $(-\infty, \infty)$
15. increasing on $[2n\pi - \pi/3,\ 2n\pi + \pi/3]$; decreasing on $[2n\pi + \pi/3,\ 2n\pi + \pi/2),\ (2n\pi + \pi/2,\ 2n\pi + 3\pi/2)$, and $(2n\pi + 3\pi/2, 2n\pi + 5\pi/3]$ for any integer n
19. relative minimum values: $f(0) = 3$ and $f(2) = -5$;
    relative maximum value: $f(\frac{1}{2}) = \frac{55}{16}$
21. relative minimum value: $f(-1) = f(2) = 0$;
    relative maximum value: $f(0) = 16$

23. concave upward on $(-\infty, -2)$ and $(1, \infty)$;
    concave downward on $(-2, 1)$
25. concave upward on $(-\infty, -\sqrt[4]{\frac{3}{5}})$ and $(\sqrt[4]{\frac{3}{5}}, \infty)$;
    concave downward on $(-\sqrt[4]{\frac{3}{5}}, \sqrt[4]{\frac{3}{5}})$
27. relative maximum value: $f(-\sqrt{2}) = 4\sqrt{2} - 1$;
    relative minimum value: $f(\sqrt{2}) = -4\sqrt{2} - 1$;
    inflection point: $(0, -1)$

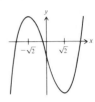

29. relative maximum value: $f(-\sqrt{3}) = \frac{2}{9}\sqrt{3}$;
    relative minimum value: $f(\sqrt{3}) = -\frac{2}{9}\sqrt{3}$;
    inflection points: $(-\sqrt{6}, \frac{5}{36}\sqrt{6})$ and $(\sqrt{6}, -\frac{5}{36}\sqrt{6})$

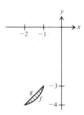

31. relative maximum
    value: $k(0) = -\frac{1}{4}$

33.

35. circle of radius 2 centered at the origin
37.   39.

41. 48 miles per hour   43. $(\frac{7}{2}, \sqrt{\frac{7}{2}})$
45. base 10 feet on a side, height 8 feet
47. $125,000   49.

51. 160 feet 53. $\sqrt{b/a}$ 55. 12%
57. $h = H/2$,
$R = H$

Cumulative Review Exercises (Chapters 1–3)

1. union of $(-1/\sqrt[3]{2}, 0)$ and $(1/\sqrt[3]{2}, \infty)$
3. union of $(-\sqrt[4]{15}, -1]$ and $[1, \sqrt[4]{15})$
5. a. union of $(-\infty, \frac{1}{3})$ and $(\frac{1}{2}, \infty)$

b. $(f \circ g)(x) = \sqrt{\dfrac{2x-1}{6x-2}}$

7. $-\infty$ 9. 2
11. a. It is. b. It is not.
13. $\dfrac{2x+5}{(x^2+2x+2)^{3/2}}$ 17. $\dfrac{4-y^2}{2xy+3}$
19. $\frac{1}{16}\pi$ square feet

CHAPTER 5

Section 5.1

1. $L_f(\mathscr{P}) = 21$; $U_f(\mathscr{P}) = 21$
3. $L_f(\mathscr{P}) = \frac{27}{4}$; $U_f(\mathscr{P}) = \frac{33}{4}$
5. $L_f(\mathscr{P}) = \frac{99}{32}$; $U_f(\mathscr{P}) = \frac{371}{32}$

7. $L_f(\mathscr{P}) = \dfrac{\pi\sqrt{2}}{8}$; $U_f(\mathscr{P}) = \dfrac{\pi}{4}\left(\dfrac{\sqrt{2}}{2} + 1\right)$

9. $L_f(\mathscr{P}) = \dfrac{\pi}{6}(1 + \sqrt{3})$; $U_f(\mathscr{P}) = \dfrac{\pi}{6}(\sqrt{3} + 2)$

11. $L_f(\mathscr{P}) = 10$; $U_f(\mathscr{P}) = 24$
13. $L_f(\mathscr{P}) \approx 0.603926$; $U_f(\mathscr{P}) \approx 0.715037$
15. $L_f(\mathscr{P}) \approx 2.07362$; $U_f(\mathscr{P}) \approx 2.33115$
17. $L_f(\mathscr{P}) = 2$; $L_f(\mathscr{P}') = 3$; $U_f(\mathscr{P}) = 7$; $U_f(\mathscr{P}') = 6$

19. $L_f(\mathscr{P}) = 0$; $L_f(\mathscr{P}') = \dfrac{\sqrt{2}\pi}{4}$; $U_f(\mathscr{P}) = \pi$;

$U_f(\mathscr{P}') = \dfrac{\pi}{4}(\sqrt{2} + 2)$

21. $L_f(\mathscr{P}) = \left(\dfrac{\pi}{2} + 1\right)\dfrac{\pi}{2}$; $L_f(\mathscr{P}') = \dfrac{\pi}{4}\left(\dfrac{3\pi}{2} + \sqrt{2} + 1\right)$;

$U_f(\mathscr{P}) = \left(\dfrac{3\pi}{2} + 1\right)\dfrac{\pi}{2}$; $U_f(\mathscr{P}') = \dfrac{\pi}{4}\left(\dfrac{5\pi}{2} + \sqrt{2} + 1\right)$

23. $\frac{869}{1800}$
27. There is no maximum value on any subinterval containing 0.

Section 5.2

1. $L_f(\mathscr{P}) = 4$; $U_f(\mathscr{P}) = 12$
3. $L_f(\mathscr{P}) = 4$; $U_f(\mathscr{P}) = 6$
5. $L_f(\mathscr{P}) = -\frac{3}{8}\sqrt{2}\pi$; $U_f(\mathscr{P}) = \frac{3}{8}\sqrt{2}\pi$
7. 20 9. $-\frac{2}{3}$ 11. -2π

13. 0 15. 0 17. -6
19. 0 21. $-\frac{35}{3}$ 23. $\frac{25}{2}$
25. $\frac{15}{2}$ 27. $\frac{26}{3}$
29. left sum: $0(1) + 2(1) = 2$; right sum: $2(1) + 6(1) = 8$;
midpoint sum: $\frac{3}{4}(1) + \frac{15}{4}(1) = \frac{9}{2}$
31. left sum: $0(\frac{1}{2}) + 1(\frac{1}{2}) + 0(1) = \frac{1}{2}$
right sum: $1(\frac{1}{2}) + 0(\frac{1}{2}) + 0(1) = \frac{1}{2}$

midpoint sum: $\dfrac{\sqrt{2}}{2}\left(\dfrac{1}{2}\right) + \dfrac{\sqrt{2}}{2}\left(\dfrac{1}{2}\right) + (-1)(1) = \dfrac{\sqrt{2}}{2} - 1 \approx$
$-.292893$
33. left sum: $1(1) + \frac{1}{2}(1) + \frac{1}{3}(1) + \frac{1}{4}(1) = \frac{25}{12} \approx 2.08333$;
right sum: $\frac{1}{2}(1) + \frac{1}{3}(1) + \frac{1}{4}(1) + \frac{1}{5}(1) = \frac{77}{60} \approx 1.28333$;
midpoint sum: $\frac{2}{3}(1) + \frac{2}{5}(1) + \frac{2}{7}(1) + \frac{2}{9}(1) = \frac{496}{315}$
≈ 1.57460

35. left sum: $\dfrac{2\sqrt{2}}{\pi}\left(\dfrac{\pi}{4}\right) + \dfrac{2}{\pi}\left(\dfrac{\pi}{4}\right) \approx 1.20711$; right sum: $\dfrac{2}{\pi}\left(\dfrac{\pi}{4}\right)$

$+ \dfrac{2\sqrt{2}}{3\pi}\left(\dfrac{\pi}{4}\right) \approx 0.735702$

37. left sum: $0\left(\dfrac{1}{2}\right) + \dfrac{\sqrt{2}}{2}\left(\dfrac{\sqrt{2}}{2} - \dfrac{1}{2}\right) + 1\left(\dfrac{\sqrt{3}}{2} - \dfrac{\sqrt{2}}{2}\right)$

$+ \dfrac{\sqrt{2}}{2}\left(1 - \dfrac{\sqrt{3}}{2}\right) \approx .400100$

right sum: $\dfrac{\sqrt{2}}{2}\left(\dfrac{1}{2}\right) + 1\left(\dfrac{\sqrt{2}}{2} - \dfrac{1}{2}\right) + \dfrac{\sqrt{2}}{2}\left(\dfrac{\sqrt{3}}{2} - \dfrac{\sqrt{2}}{2}\right)$

$+ 0\left(1 - \dfrac{\sqrt{3}}{2}\right) \approx .673033$

39. $0(\frac{1}{4}) + \frac{7}{8}(\frac{1}{4}) + 2(\frac{1}{4}) + \frac{27}{8}(\frac{1}{4}) = \frac{25}{16}$
41. $0(\frac{1}{2}) + \frac{1}{3}(\frac{1}{2}) + \frac{1}{2}(1) = \frac{2}{3}$
43. $1(\frac{1}{50}) + (\frac{51}{50})^2(\frac{1}{50}) + (\frac{52}{50})^2(\frac{1}{50}) + \cdots + (\frac{149}{50})^2(\frac{1}{50}) \approx 8.5868$

55. $\dfrac{1}{n+1}(b^{n+1} - a^{n+1})$

Section 5.3

1. 14 3. 83 5. 30
7. $\frac{1}{2} + \frac{3}{2} = 2$ 9. $-\frac{1}{3} + \frac{8}{3} = \frac{7}{3}$
11. $a = 3, b = 2$ 13. $a = 5, b = 1$
15. $m = M = 13$; $39 \leq \displaystyle\int_{-3}^{0} 13\,dx \leq 39$

17. $m = 0, M = 9$; $0 \leq \displaystyle\int_{-1}^{3} x^2\,dx \leq 36$

19. $m = \dfrac{1}{3}, M = \dfrac{1}{2}$; $\dfrac{1}{3} \leq \displaystyle\int_{2}^{3} \dfrac{1}{x}\,dx \leq \dfrac{1}{2}$

21. $m = \dfrac{1}{2}, M = \dfrac{\sqrt{2}}{2}$; $\dfrac{\pi}{24} \leq \displaystyle\int_{\pi/4}^{\pi/3} \cos x\,dx \leq \dfrac{\sqrt{2}\pi}{24}$

23. $m = 0, M = \sqrt{3}$; $0 \leq \displaystyle\int_{0}^{\pi/3} \tan x\,dx \leq \dfrac{\sqrt{3}\pi}{3}$

25. $\frac{5}{6}$

Section 5.4

1. $x(1 + x^3)^{29}$
3. $-1/y^3$
5. $2x^3 \sin x^2$
7. $-(1 + y^2)^{1/2} + 2y(1 + y^4)^{1/2}$
9. $\frac{512}{5}x(1 + 16x^2)^{-1/5}$
11. 4
13. -4
15. 12
17. $\frac{1}{101}$
19. 0
21. $\frac{9}{2}(4^{2/9} - 1)$
23. $10\pi - 2\pi^2 + 8.625$
25. $\frac{9}{2}$
27. $\frac{1}{2}\sqrt{3}$
29. $\frac{1}{2} - \frac{1}{2}\sqrt{2}$
31. $\frac{7}{24}$
33. $\sqrt{3}$
35. 1
37. $\frac{2}{5}$
39. $\frac{3}{2}$
41. $\frac{14}{3}$
43. 1
45. $\pi/4$
47. a. $\frac{1}{2}x^2$ b. $-\frac{2}{3}x^3$ c. $\cos x - 1$ d. $2x^5$
49. $\frac{25}{3}$
51. $\frac{770}{3600} \approx 0.213889$ miles
53. \$50.18
55. a. $5t^2 - \frac{1}{3}t^3$ b. $\frac{250}{3}$
57. $\dfrac{672{,}000}{\pi}$ tons
59. 1 mile
63. b. $\frac{2}{3}$ thousand dollars per thousand umbrellas
65. $781{,}250{,}000\pi$ foot-pounds
67. $\dfrac{1}{\pi}$
69. $\dfrac{4}{5\pi}$

Section 5.5

1. $x^2 - 7x + C$
3. $\frac{3}{2}x^{4/3} - \frac{12}{7}x^{7/4} + \frac{5}{7}x^{7/5} + C$
5. $\dfrac{t^6}{6} + \dfrac{1}{3t^3} + C$
7. $2 \sin x - \frac{5}{2}x^2 + C$
9. $-3 \cot x - \frac{1}{2}x^2 + C$
11. $\frac{4}{3}t^3 + 2t^2 + t + C$
13. $-\frac{15}{2}$
15. $3 - 5\sqrt{2}$
17. $\dfrac{9\pi^2}{32} + \dfrac{2}{\pi} + \dfrac{\sqrt{2}}{2}$
19. $-1 - \frac{5}{3}\sqrt{3}$
21. $-\frac{136}{3}$
23. $\pi + \dfrac{5}{\pi} + \dfrac{\pi^2}{4}$
25. $-\frac{142}{3}$
27. $\frac{5}{2}$
29. $\frac{613}{10}$
31. 2
33. $\int 20x(1 + x^2)^9 \, dx = (1 + x^2)^{10} + C$
35. $\int (x \cos x + 2 \sin x) \, dx = x \sin x - \cos x + C$
37. $\int 21 \sin^6 x \cos x \, dx = 3 \sin^7 x + C$
39. 10
41. 12
43. $3 - \sqrt{2}/2$
49. $\frac{1}{12}\sqrt{3}\pi$
57. a. $2\sqrt{2} - 1$ b. 1

Section 5.6

1. $\frac{1}{6}(4x - 5)^{3/2} + C$
3. $(\sin \pi x)/\pi + C$
5. $\frac{1}{2}\sin x^2 + C$
7. $\frac{1}{3}\cos^{-3}t + C$
9. $-\dfrac{2}{5} \cdot \dfrac{1}{(t^2 - 3t + 1)^{5/2}} + C$
11. $\frac{2}{5}(x + 1)^{5/2} - \frac{4}{3}(x + 1)^{3/2} + C$
13. $\frac{2}{3}(3 + \sec x)^{3/2} + C$
15. $\frac{1}{13}(x^3 + 1)^{13} + C$
17. $\frac{1}{6}(x^2 + 3x + 4)^6 + C$
19. $\frac{2}{9}(3x + 7)^{3/2} + C$
21. $\frac{1}{3}(1 + 2x + 4x^2)^{3/2} + C$
23. 0
25. $\frac{1}{7}\sin^7 t + C$
27. $\frac{1}{3}(\sin 2z)^{3/2} + C$
29. $\sqrt{2} - 1$
31. $2 \tan \sqrt{z} + C$
33. $\frac{1}{3}(w^2 + 1)^{3/2} + (w^2 + 1)^{1/2} + C$
35. $\frac{1}{2}(27 - 5\sqrt{5})$
37. $\frac{2}{5}(x + 2)^{5/2} - \frac{4}{3}(x + 2)^{3/2} + C$
39. $\frac{96}{5}$
41. $-\frac{1}{256}[\frac{1}{3}(1 - 8t)^{3/2} - \frac{2}{5}(1 - 8t)^{5/2} + \frac{1}{7}(1 - 8t)^{7/2}] + C$
43. $\frac{26}{15}$
45. $\frac{14}{3}$
47. $\frac{3}{20}$
49. $\frac{38}{3}$
53. b. $-(1/a)[\cos (ax + b)] + C$
   c. $(ax + b)^{n+1}/[a(n + 1)] + C$
57. a. 0 b. $\sqrt{2}$

Section 5.7

1. $\ln 3$
3. $\frac{1}{3}\ln \frac{4}{9}$
5. $2 \ln 3$
7. domain: $(-1, \infty)$; $f'(x) = \dfrac{1}{x + 1}$
9. domain: union of $(-\infty, -1)$ and $(1, \infty)$;
   $g'(x) = \ln(x^2 - 1) + \dfrac{2x^2}{x^2 - 1}$
11. domain: union of $(-\infty, 2)$ and $(3, \infty)$;
   $f'(x) = \dfrac{1}{2(x - 3)(x - 2)}$
13. domain: union of $(0, 1)$ and $(1, \infty)$;
   $f'(x) = \dfrac{x - 1 - x \ln x}{x(x - 1)^2}$
15. domain: $(0, \infty)$; $f'(t) = [\cos (\ln t)]\dfrac{1}{t}$
17. domain: $(1, \infty)$; $f'(x) = \dfrac{1}{x \ln x}$
19. $\dfrac{dy}{dx} = \dfrac{(y^2 + x)\ln (y^2 + x) + x}{5(y^2 + x) - 2xy}$
21.

23. relative maximum value: $g\left(\dfrac{\pi}{2} + 2n\pi\right) = \ln 3$; relative minimum value: $g\left(\dfrac{3\pi}{2} + 2n\pi\right) = 0$; inflection points: $\left(-\dfrac{\pi}{6} + 2n\pi, \ln \dfrac{3}{2}\right)$ and $\left(\dfrac{7\pi}{6} + 2n\pi, \ln \dfrac{3}{2}\right)$

25. relative minimum value: 27. 1.531584
$f(1) = 0$; inflection
point: $(e, 1)$

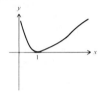

29. $\ln|x - 1| + C$
33. $\frac{1}{4}\ln|x^4 - 4| + C$
37. $2\ln\frac{3}{2}$
41. $\frac{1}{2}(\ln(\ln t))^2 + C$
45. $\ln|x\sin x + \cos x| + C$
49. $\frac{1}{2}\ln 2$
53. 16.3
65. b. $1/e$ c.

31. $\frac{1}{2}\ln(x^2 + 4) + C$
35. $-\frac{2}{3}\ln 2$
39. $\frac{1}{2}(\ln z)^2 + C$
43. $\ln|\sin t| + C$
47. 1
51. $\frac{8}{3} - \frac{2}{3}\sqrt{2}$

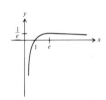

67. $0 \le \int_1^2 \ln x \, dx \le \frac{1}{2}$
73. a. approximately 89.7834 kilograms
   b. approximately 10.2204 kilograms

Section 5.8

1. $\frac{56}{3}$
5. $\frac{2}{3}$
11. $\frac{11}{6}$
17. $\frac{6}{5}\sqrt{3} + \frac{26}{15}$
23. $\frac{37}{12}$
29. 4

3. $2\sqrt{2} - \frac{3}{2} - \frac{1}{2}\sqrt{3}$
7. $\frac{1}{2}\ln\frac{3}{2}$
13. $\frac{2}{3}\sqrt{3} + 2$
25. 12
31. $\frac{1}{3}$

9. $\frac{27}{4}$
15. $\ln 2$
21. 36
27. $\frac{16}{3}$
33. $\frac{343}{24}$

Chapter 5 Review Exercises

1. $L_f(\mathscr{P}) = \frac{19}{20}$; $U_f(\mathscr{P}) = \frac{77}{60}$ 3. $\frac{5}{8}x^{8/5} - 3x^{8/3} + C$
5. $\frac{1}{4}x^4 - \frac{3}{2}x^2 + 2x - 2\ln|x| + C$
7. $(1 + \sqrt{x + 1})^2 + C$
9. $-\frac{1}{12}\cos^4 3t + C$
11. $\frac{4}{5}(1 + \sqrt{x})^{5/2} - \frac{4}{3}(1 + \sqrt{x})^{3/2} + C$

13. $-\frac{51}{40} - \frac{6}{5}(2^{2/3})$
17. $\frac{2}{5}\ln 2$
21. $2\ln 2 - \ln(2 - \sqrt{2})$
25. $68 - 6\sqrt{2}$
29. $\frac{128}{3}$
33.

15. $\frac{5}{2}(\sqrt{2} - \sqrt{3})$
19. 20
23. $\frac{195}{4}$
27. 4
31. $\frac{1}{2}$
35. $x\sqrt{1 + x^5}$

37. $\dfrac{1}{x\ln x}$ 39. $2\sin(\ln x)$

41. $\dfrac{\tan x}{\ln\cos x + 2\ln\sec x}$

43. a. ii. $\frac{1}{3}(x^2 + 6)^{3/2} + C$
   b. iii. $-2\cos\sqrt{x} + C$
   c. iii. $\frac{1}{2}[\ln(x + 1)]^2 + C$

47. c. $0 \le \displaystyle\int_1^2 x\ln x \, dx \le \frac{5}{6}$ 49. 242 feet

51. approximately 1.95 years   53. $6050\,R$
55. b. $e^{-(10^4)}$

Cumulative Review Exercises (Chapters 1–4)

1. union of $(-\infty, 2)$ and $(4, \infty)$
3. $-\infty$ 5. 2

7. $\dfrac{2x}{(x^2 + 1)^2}\csc^2\dfrac{1}{x^2 + 1}$

11. $c = -\dfrac{3}{16}$; root: -2 13.

15. relative minimum value: $f(2) = \frac{3}{4}$; inflection point: $(3, \frac{7}{9})$

17. a. -64 feet per second b. 2 seconds
19. \$3.50

CHAPTER 6

Section 6.1

1. $\pi/4$ 3. $\pi/6$ 5. $\pi/2$
7. π 9. $\frac{2}{9}\pi(27 - 2\sqrt{2})$ 11. $\pi(\frac{4}{3}\sqrt{3} - \frac{\pi}{2})$
13. $\frac{1}{8}\pi(\pi + 2)$ 15. $\frac{19}{4}\pi$ 17. 4π
19. π 21. 540π 23. $\frac{37}{15}\pi$
25. $\frac{8}{3}\pi$ 27. $\frac{16}{3}$ 29. $\frac{500}{3}\sqrt{3}$
31. $V = \displaystyle\int_a^b \pi[f(x) - c]^2\,dx$ 33. $\frac{3}{2}\pi^2 - 4\pi$
35. $\frac{7}{30}\pi$ 39. $\frac{1}{12}\sqrt{3}\,a^2 h$
41. $\frac{32,000}{3}$ cubic feet   43. $\frac{2}{3}\pi$ cubic centimeters
45. $-2/\pi$ inches per second

Section 6.2

1. $\frac{16}{3}\pi$ 3. $\frac{4}{3}\pi$ 5. $\pi(1 + \frac{\sqrt{2}}{2})$
7. $\frac{28}{5}\sqrt{3}\pi$ 9. $\pi(\ln 2)^2$ 11. $\frac{2}{3}\pi[1 - (\frac{7}{8})^{1/2}]$
13. $\frac{20}{3}\pi$ 15. $\pi(\sqrt{2} - 1)$
17. $\frac{1}{18}\pi(35 - 16\sqrt{2})$ 19. $\frac{8}{3}\pi$
21. $\frac{4}{3}\pi$ 23. $\frac{11}{15}\pi$
25. $V = \displaystyle\int_a^b 2\pi(x - c)[f(x) - g(x)]\,dx$
27. $\frac{81}{2}\pi$ 29. $\pi a^2 h/3$ 31. $2\pi^2 br^2$

Section 6.3

1. $4\sqrt{5}$ 3. $5 + \frac{1}{8}\ln\frac{3}{2}$ 5. 12
7. $\frac{256}{15}$ 9. $\frac{1923}{128}$ 11. $3 + \ln 2$
13. $\ln\frac{9}{2} + \frac{1}{6}$ 15. $\frac{17}{16} + \frac{1}{32}\pi$ 17. $\frac{5}{2}$
19. $\frac{4}{5}(10^{5/2} - 5^{5/2})$ 21. $\ln(\sqrt{2} + 1)$
23. $\sqrt{\dfrac{301}{220}} + \sqrt{\dfrac{261}{252}} \simeq 2.18739$
25. $(c, c^{2/3})$, where $c = \frac{4}{9}[(\frac{1}{2} + \frac{13}{16}\sqrt{13})^{2/3} - 1] \approx 0.566294$

Section 6.4

1.

3.

5.

7.

9.

11.

13.

15.

17. 14 19. $\frac{1}{6}(2\sqrt{2} - 1)$
21. $\frac{1}{8}\pi^2$ 23. $6r$

Section 6.5

1. 8π 3. $\frac{1}{6}\pi(17\sqrt{17} - 27)$
5. $\frac{1}{9}\pi(5\sqrt{5} - 1)$ 7. $\frac{67}{36}\pi$
9. $\frac{38}{3}\pi$ 11. $\frac{8}{9}\pi(216 - 10\sqrt{10})$
13. $\sqrt{2}\pi$ 15. $\frac{12}{5}\pi r^2$
17. a. $2\pi rh$ b. $2\pi rh$

Section 6.6

1. 3384 foot-pounds   3. 80 foot-pounds
5. 2×10^4 mile-pounds   7. $\frac{5}{72}$ foot-pounds
9. 405 foot-pounds   11. 24 foot-pounds
13. $265,625\pi$ foot-pounds   15. 6750π foot-pounds
17. $17,920$ foot-pounds
19. a. 1.4 foot-pounds   b. 0.2 foot-pound
    c. 0 foot-pounds
23. $\frac{375}{2}$ foot-pounds   25. $215,000$ foot-pounds
27. 16 foot-pounds

Section 6.7

1. rise
3. the 15-kilogram child with the 10-kilogram toddler, and the 20-kilogram child on the opposite side, each 2 meters from the axis of revolution
5. $(\frac{10}{9}, -\frac{4}{9})$ 7. $(\frac{2}{3}, 0)$ 9. $(\frac{14}{9}, \frac{7}{2})$
11. $(\frac{173}{85}, \frac{81}{34})$ 13. $(\frac{21}{5}, 1)$ 15. $(0, 7)$
17. $(\frac{1}{2}, \frac{3}{2})$ 19. $(3, -1)$ 21. $(0, 1)$
23. $(0, \frac{1}{3}h)$ 25. $(\frac{3}{4}a, \frac{3}{10}h)$ 29. $8\pi\sqrt{2}$
31. $\int_{b-r}^{b+r} 4\pi x \sqrt{r^2 - (x - b)^2}\, dx$
35. a. $\left(\frac{1}{6}, \frac{1}{3}\right)$ b. $\left(\frac{-2}{12 + 3\pi}, \frac{3\pi + 8}{6\pi + 24}\right)$ c. $\left(\frac{14}{15\pi}, \frac{9}{10}\right)$
   d. $\left(-\frac{1}{6}, \frac{14}{9\pi}\right)$ e. $\left(\frac{79}{38}, \frac{101}{38}\right)$
37. $(\frac{1}{2}r, 0)$

Section 6.8

1. 140.625 pounds
3. $\dfrac{62.5}{9}\sqrt{3} \approx 12.0281$ pounds
5. $\frac{125}{3}\sqrt{3}(\frac{9}{2} - \sqrt{3}) \approx 199.760$ pounds for each triangle pointing downward, and $\frac{125}{3}\sqrt{3}(\frac{9}{2} - \frac{1}{2}\sqrt{3}) \approx 262.260$ pounds for those pointing upward

Section 6.9

1. a. $(\frac{3}{2}\sqrt{2}, \frac{3}{2}\sqrt{2})$ b. $(-\sqrt{3}, 1)$
   c. $(\frac{3}{2}, \frac{3}{2}\sqrt{3})$ d. $(5, 0)$
   e. $(0, -2)$ f. $(0, 2)$
   g. $(-2\sqrt{2}, 2\sqrt{2})$ h. $(0, 0)$
   i. $(-\frac{1}{2}, \frac{1}{2}\sqrt{3})$ j. $(-\frac{1}{2}, -\frac{1}{2}\sqrt{3})$
   k. $(0, -1)$ l. $(-\frac{3}{2}\sqrt{3}, -\frac{3}{2})$
3. $r = \dfrac{4}{2\cos\theta + 3\sin\theta}$
5. $r = \dfrac{1}{\sqrt{8\sin^2\theta + 1}}$
7. $r^2 = \cos 2\theta$
9. $r^2 = \dfrac{r}{\cos\theta(\cos^2\theta - 3\sin^2\theta)}$
11. $r = \dfrac{3\cos^2\theta - \sin^2\theta}{\cos\theta}$ 13. $x^2 + y^2 = 3x$
15. $x = 3y$ 17. $(x^2 + y^2)^3 = 4x^2 y^2$

21.

23.

25.
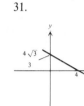

27.

29.

31.
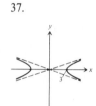

33.

35.

37.
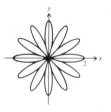

39.
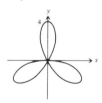

41. symmetry with respect to both axes and origin
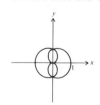

43. symmetry with respect to y axis

45. symmetry with respect to both axes and origin

47. symmetry with respect to both axes and origin

49. symmetry with respect to both axes and origin

51. symmetry with respect to both axes and origin

53. symmetry with respect to x axis

35.

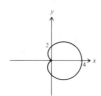

37.

39. $2\sqrt{2}$

41. $\sqrt{3} - \frac{1}{3}\pi$

55. symmetry with respect to both axes and origin

57. symmetry with respect to y axis

61. $r^2 = 2\cos 2\theta$

Cumulative Review Exercises (Chapters 1–5)

1. ∞ 3. 0   5. $\frac{9}{4}$
7. $f^{(24)}(x) = 24\sin x + (x-1)\cos x$
9. $\frac{9}{2}$ inches per minute
11. inflection points: $(-1, -\frac{8}{3}), (0, 0), (1, \frac{8}{3})$

13. $\pi/6$
19. $\ln 3$

15. $(-1, 1)$ and $(1, 1)$
21. $\frac{3}{8}(\ln t)^{8/3} + C$

Section 6.10

1. 2π 3. 16   5. 4
7. 16π 9. $9\pi/4$ 11. π
13. $81\pi/16$ 15. $\pi/16$ 17. 6π
19. $41\pi/2$ 21. 50   23. 24π
25. $3\pi/4$ 27. $\pi - 1$ 29. $17\pi/4$
31. $7\pi/12 - \sqrt{3}$ 33. $2\sqrt{2}$ 35. b. $2 - \pi/2$
37. $\frac{32}{5}\pi$

CHAPTER 7

Section 7.1

1. inverse exists;
   domain: $(-\infty, \infty)$;
   range: $(-\infty, \infty)$

3. inverse exists;
   domain: $(-\infty, \infty)$;
   range: $(-\infty, \infty)$

5. inverse does not exist

7. inverse exists;
   domain: $[0, \infty)$;
   range: $[0, \infty)$

Chapter 6 Review Exercises

1. $\frac{5}{2}\pi$ 3. 4π
5. a. $\frac{64}{5}\pi$ b. $(\frac{8}{5}, \frac{16}{7})$ c. $\frac{64}{5}\pi$
7. a. $4\pi c$ b. $4\pi c + \frac{2}{3}\pi c^2$
9. $20\sqrt{3}$ cubic feet   11. $\ln(7 + 4\sqrt{3})$
13.   15. 18

17. $\frac{\pi}{6}\left(\frac{10}{3} + \sqrt{2} + \sqrt{17}\right)$ 19. $\frac{1377}{200}\pi$
21. $\frac{2}{3}\pi(2\sqrt{2} - 1)$ 23. 20 foot-pounds
25. $243{,}000\pi$ foot-pounds   27. $(1, 1)$
29. $(\frac{57}{40}, \frac{33}{40})$

9. inverse exists;
   domain: $[0, \infty)$;
   range: $(-\infty, 4]$

11. inverse does not exist

13. inverse exists;
    domain: $(-\infty, \infty)$;
    range: $(-\infty, \infty)$

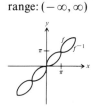

15. inverse does not exist

17. inverse exists;
    domain: $(1, \infty)$;
    range: $(0, \pi/2)$

25. $f^{-1}(x) = -\sqrt[3]{(x+1)/4}$
27. $g^{-1}(x) = x^2 - 1$ for $x \geq 0$
29. $f^{-1}(x) = \dfrac{1}{5x+2}$ 31. $k^{-1}(t) = \dfrac{t+1}{1-t}$
33. $(-\infty, 0]$ (or $[0, \infty)$)   35. $(-\infty, 0]$ (or $[0, \infty)$)
37. $(-\infty, -\sqrt{\frac{5}{3}}]$ (or $[-\sqrt{\frac{5}{3}}, \sqrt{\frac{5}{3}}]$ or $[\sqrt{\frac{5}{3}}, \infty)$)
39. $(-\infty, 0]$ (or $[0, \infty)$)
41. $(-\pi/2 + n\pi, \pi/2 + n\pi)$ for any integer n
43. $[-\pi/2 + n\pi, n\pi)$ (or $(n\pi, \pi/2 + n\pi])$ for any integer n
49. a. first quadrant   b. fourth quadrant
55. a. no   b. yes

Section 7.2

1. $\frac{1}{3}$ 3. $\frac{1}{2}$
5. $\frac{1}{4}$ 7. $\frac{1}{6}$
9. $1/(9x^8 + 7)$ 11. $(x^3 + 1)/3x^2$
13. $1/\cos x$ for $-\pi/2 < x < \pi/2$
15. $\frac{1}{2}\sqrt{2}$ 21. $\frac{16}{35}$

Section 7.3

1. $\frac{1}{2}$ 3. e
5. $3x$ 7. $4e^{4x}$
9. $5x^4 e^{(x^5)}$ 11. $-2e^x/(e^x - 1)^2$
13. $3e^{3z}\sec^2(e^{3z})$ 15. $e^{-z}(-\sin az + a\cos az)$
17. $(y - 2x^2 ye^y)/(x^3 ye^y - x)$ 19. $f^{(n)}(x) = 4^n e^{4x}$
23.

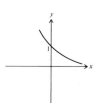

25. relative maximum value:
    $f(\pi(1 + 12n)/6\sqrt{3}) = (\sqrt{3}/2)e^{\pi(1 + 12n)/6\sqrt{3}}$;
    relative minimum value:
    $f(\pi(7 + 12n)/6\sqrt{3}) = -(\sqrt{3}/2)e^{\pi(7 + 12n)/6\sqrt{3}}$;
    inflection points: $(\pi(-1 + 6n)/6\sqrt{3}, (-1)^n(\sqrt{3}/2)e^{\pi(-1 + 6n)/6\sqrt{3}})$

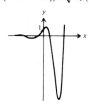

27. relative minimum value: $f(0) = \ln 2$

29. $-\frac{1}{4}e^{-4x} + C$ 31. $\frac{1}{2}(e^2 - 1)$
33. $e - 1$ 35. $\ln(e^t + 1) + C$
37. $\sqrt{e^{2t} - 4} + C$ 39. $-\frac{1}{2}[\ln(1 + e^{-t})]^2 + C$
41. $2e^{x/2} + C$ 43. $\ln\dfrac{e^x}{1 + e^x} + C$
45. $-e^{-e^x} + C$ 47. $y = -6x + 2$
49. $(\frac{1}{2}, e^{1/2})$
51. $(\frac{1}{4}\ln 2, 2^{3/4})$
55. $f^{-1}(x) = \ln[(1 + x)/(1 - x)]$
57. 0.567143 59. 1
61. $1 - e^{-1}$ 63. $\frac{1}{2}(e + e^{-1}) - 1$
67. $\pi(e + e^{-1} - 2)$ 69. $\sqrt{2}(e^\pi - 1)$
71. $e^{-14} \approx 8.31529 \times 10^{-7}$
73. a. 449   b. after approximately 5.28520 days
75. a. $\dfrac{c}{a-b}[e^{-[b/(a-b)]\ln(a/b)} - e^{-[a/(a-b)]\ln(a/b)}]$

    b. The values of y approach 0 as t becomes very large.

Section 7.4

1. $\frac{1}{2}$ 3. x 5. $2x$

7. $(\ln 5)5^x$ 9. $(5 \ln 3)3^{5x-7}$

11. $(\ln a)a^x \sin bx + ba^x \cos bx$

13. $\log_2 x + \dfrac{1}{\ln 2}$ 15. $t^t(\ln t + 1)$

17. $t^{2/t}\left(\dfrac{2 - 2\ln t}{t^2}\right)$

19. $-\sin x (\cos x)^{\cos x}(\ln \cos x + 1)$

21. $2\sqrt{2}(2x)^{\sqrt{2}-1}$ 23. Inflection point: $(0,0)$

25. $(1,0)$ 27. $\dfrac{2^x}{\ln 2} + C$

29. $8/\ln 3$ 31. $-\dfrac{5^{-x^2}}{2\ln 5} + C$

33. $\dfrac{1}{2\pi + 1}x^{2\pi + 1} + C$ 35. $\dfrac{3}{4\ln 2} - \dfrac{1}{2}$

37. $(\ln 2)/2$ 39. approximately 1.46497

41. approximately 0.873569

45. concave upward for $a > 1$; concave downward for $0 < a < 1$

51. 0.1 millimeters   53. approximately 0.794328

55. approximately 3.16228×10^{-5}

57. 30 decibels

59. approximately 1.15

Section 7.5

1. a. 4 days   b. $2\dfrac{\ln 3}{\ln 2} \approx 3.16993$ (days)

3. $\dfrac{41 \ln 1.25}{6 \ln 2} \approx 2.19984$ (days ago)

5. $\dfrac{20 \ln 2.5}{\ln 2} \approx 26.4386$ (years)

9. approximately $-\dfrac{1590}{\ln 2}\ln 0.9 \approx 241.685$ (years)

11. approximately 1.4116 percent

13. $100e^{(\ln 2)/4.07} \approx 118.567$ (milligrams)

15. $e^{-0.2} \approx 0.818731$

17. a. approximately 29.92 (inches of mercury)

     b. approximately $29.92e^{-0.1} \approx 11.0070$ (inches of mercury)

     c. approximately $29.92e^{-0.2} \approx 4.04923$ (inches of mercury)

19. $200e^{(\ln 2)/10} \approx 214.355$ (milligrams)

21. $10 \ln 2 \approx 6.93147$ (percent)

23. after $\dfrac{\ln \frac{1}{2}}{\ln \frac{3}{4}} \approx 2.40942$ minutes

29. a. approximately 1.15499×10^6 years

     b. approximately 0.907946

Section 7.6

1. $\pi/3$ 3. $\pi/4$ 5. $-\pi/6$

7. $\pi/6$ 9. $5\pi/4$ 11. $\pi/3$

13. $5\pi/6$ 15. $-\frac{1}{2}$ 17. $\sqrt{2}/2$

19. 1 21. 2 23. $\pi/3$

25. $\sqrt{1 - x^2}$ 27. $\sqrt{x^2 + 1}$

29. $x^2/\sqrt{x^4 + 1}$ 31. $1 - 2x^2$

33. $3/\sqrt{1 - 9x^2}$ 35. $\dfrac{1}{2(1 + t)\sqrt{t}}$

37. $\dfrac{x}{(2 - x^2)\sqrt{1 - x^2}}$ 39. $\dfrac{1}{x \ln x \sqrt{(\ln x)^2 - 1}}$

41. $\frac{1}{4}\arctan \frac{1}{4}x + C$ 43. $\frac{1}{12}\arctan \frac{3}{4}x + C$

45. $\frac{1}{4}\sqrt{2}\arctan\left[(x + 1)/\sqrt{2}\right] + C$

47. $\frac{1}{2}\arcsin \frac{2}{3}x + C$ 49. $\frac{1}{5}\operatorname{arcsec} \frac{1}{5}x + C$

51. $\frac{1}{4}\arcsin x^4 + C$ 53. $-\arctan e^{-x} + C$

55. $\frac{1}{4}(\arctan 2x)^2 + C$ 57. $\frac{1}{3}\arctan\left(\frac{1}{3}\sin t\right) + C$

59. $\frac{1}{8}\operatorname{arcsec}(2\sin 4x) + C$ 61. $\pi/6$

63. $\pi/6$ 65. $\frac{1}{3} - \pi\sqrt{3}/18$

67. $\pi/2 - 1$

77. On $(-\infty, 0)$ the constant is $-\pi/2$; on $(0, \infty)$ the constant is $\pi/2$

79. Its base is 2.5 feet above the floor: yes

81. $\pi^2/6$ 83. $\frac{4}{3} + \frac{1}{16}\pi$

Section 7.7

1. 0 3. 0 5. $\dfrac{1 + e^2}{1 - e^2}$

7. $\frac{4}{3}$ 9. $\frac{17}{15}$ 11. $\frac{2}{3}\sqrt{2}$

13. $\dfrac{x^2 - 1}{2x}$ 15. $\dfrac{x^2 - 1}{x^2 + 1}$ 17. $\operatorname{sech}^2 x$

19. $-\operatorname{sech} x \tanh x$

21. $-\dfrac{1}{2\sqrt{x}}\operatorname{sech}\sqrt{x}\tanh\sqrt{x}$

23. $-\dfrac{2x}{\sqrt{1 - x^2}}\sinh\sqrt{1 - x^2}\cosh\sqrt{1 - x^2}$

25. $\dfrac{2e^{2x}}{1 + e^{4x}}\sinh(\arctan e^{2x})$

29. $\tanh x + C$ 31. $\ln|\operatorname{sech} x + \tanh x| + C$

33. $\ln\dfrac{10+\sqrt{101}}{5+\sqrt{26}}$　　　35. $16(e-e^{-1})$

39. $\left(\dfrac{\sqrt{2}-1}{\ln(1+\sqrt{2})},\dfrac{\pi}{8\ln(1+\sqrt{2})}\right)$

Section 7.8

1. $16a^{15}$　　　3. 0　　　5. $\frac{8}{5}$
7. 1　　　9. 1　　　11. ∞
13. 1　　　15. 0　　　17. 2
19. $\frac{4}{9}$　　　21. 1　　　23. 0
25. $\ln 5 - \ln 3$　　　27. 0　　　29. 1
31. $-\frac{1}{6}$　　　33. 0　　　35. 0
37. 1　　　39. 0　　　41. 1
43. 2　　　45. ∞　　　47. 1
49. 0　　　51. $2a$
53. relative maximum value: $f(1) = e^{-1}$; inflection point: $(2, 2e^{-2})$

59. 1　　　65. 1

Chapter 7 Review Exercises

1. yes　　　3. yes　　　5. no
7. $f^{-1}(x) = (x+2)/(x+3)$
11. $(\ln 2)2^x/(1+2^x)$　　　13. $1/\cosh x$
15. $-\dfrac{2x}{3(1-x^2)^{2/3}\sqrt{1-(1-x^2)^{2/3}}}$
17. $\dfrac{2x}{(\ln 4)(1+x^4)\arctan x^2}$　　　19. $2\sqrt{1+e^x}+C$
21. $\sinh^{-1}(e^x)+C$　　　23. $e^{\sec x}+C$
25. $\dfrac{4}{15\ln 5}$　　　27. $\frac{3}{2}\arctan 2t + C$
29. $\pi/6$　　　31. $\dfrac{1}{\ln 2}\cosh 2^x + C$

33. $\frac{1}{12}\sqrt{6}\arctan\left(\dfrac{x^2+2}{\sqrt{6}}\right)+C$

37. d. $\frac{1}{21} \le \ln 1.05 \le \frac{1}{20}$　　　39. c. $\frac{1}{21} \le \ln 1.05 \le \frac{1}{20}$
41. a　　　43. 1
45. 0　　　47. e^2
51. relative maximum value: $f(e^{-1}/2) = 1/e^{1/2e}$

53. relative maximum value: $f(\ln(3/2)) = 4/27$; inflection point: $(\ln\frac{9}{4}, \frac{80}{729})$

55. $\pi/12$　　　57. $\frac{1}{2}(e-e^{-1})$
59. a. approximately 7.59130　　　b. approximately 4.35649

Cumulative Review Exercises (Chapters 1–6)

1. $\frac{1}{8}\sqrt{2}$　　　3. $\dfrac{6x^2}{(1-x^3)^2}$
9. relative minimum value: $f(-2) = -16 = f(2)$;
relative maximum value: $f(0) = 0$;
inflection points: $\left(-\frac{2}{3}\sqrt{3}, -\frac{80}{9}\right), \left(\frac{2}{3}\sqrt{3}, -\frac{80}{9}\right)$

11. $25\sqrt{3}$ feet per second　　　13. $\frac{1}{2}\ln(4+x^2)+C$
15. $80+5\pi \approx 95.7080$ miles　　　17. $\frac{74}{9}$
19. a. 32π　　b. $(0,2)$　　　21. $\frac{1}{2}\sqrt{3}-\frac{1}{6}\pi$

CHAPTER 8

Section 8.1

1. $-x\cos x + \sin x + C$　　　3. $\frac{1}{2}x^2\ln x - \frac{1}{4}x^2 + C$
5. $x(\ln x)^2 - 2x\ln x + 2x + C$

7. $\frac{1}{4}x^4 \ln x - \frac{1}{16}x^4 + C$

9. $\frac{1}{4}x^2 e^{4x} - \frac{1}{8}xe^{4x} + \frac{1}{32}e^{4x} + C$

11. $x^3 \sin x + 3x^2 \cos x - 6x \sin x - 6 \cos x + C$

13. $\frac{1}{6}e^{3x} \cos 3x + \frac{1}{6}e^{3x} \sin 3x + C$

15. $\frac{t}{\ln 2} 2^t - \frac{1}{(\ln 2)^2} 2^t + C$

17. $\left(\frac{t^2}{\ln 4} - \frac{2t}{(\ln 4)^2} + \frac{2}{(\ln 4)^3} \right) 4^t + C$

19. $t \cosh t - \sinh t + C$

21. $x \arctan x - \frac{1}{2} \ln (1 + x^2) + C$

23. $x \arccos (-7x) + \frac{1}{7}\sqrt{1 - 49x^2} + C$

25. $\frac{m}{n + 1} x^{n+1} \ln x - \frac{m}{(n + 1)^2} x^{n+1} + C$

27. $\frac{1}{2}[x \cos (\ln x) + x \sin (\ln x)] + C$

29. $\frac{4}{25}e^5 + \frac{1}{25}$

31. -2π

33. $\pi/4 - \pi\sqrt{3}/3 + \frac{1}{2} \ln 2$

35. $2 \ln 2 - 1$

37. $-\frac{x}{a} \cos ax + \frac{1}{a^2} \sin ax + C$

39. $-\cos x \arctan (\cos x) + \frac{1}{2} \ln (1 + \cos^2 x) + C$

41. $2\sqrt{t} \sin \sqrt{t} + 2 \cos \sqrt{t} + C$

43. $\dfrac{5\sqrt{2}}{6}$

51. $x(\ln x)^3 - 3x(\ln x)^2 + 6x \ln x - 6x + C$

55. $\frac{1}{4}\pi - \frac{1}{2} \ln 2$

57. $\frac{1}{4}\pi(e^2 + 1)$

59. $\frac{4}{3}\pi$

61. $\dfrac{n(n - 2)\cdots 4 \cdot 2}{(n - 1)(n - 3)\cdots 5 \cdot 3}$ if n is even; $\dfrac{n(n - 2)\cdots 5 \cdot 3}{(n - 1)(n - 3)\cdots 4 \cdot 2} \dfrac{\pi}{2}$ if n is odd

63. $\frac{64}{3}\pi r^2$

65. $\left(\dfrac{8 \ln 2 - 3}{8 \ln 2 - 4}, 1 \right)$

45. $\frac{1}{2} \cos x - \frac{1}{10} \cos 5x + C$

47. $\frac{1}{4} \cos 2x + \frac{1}{12} \cos 6x + C$

49. $3 \cos \frac{1}{6}x - \frac{3}{7} \cos \frac{7}{6}x + C$

51. $-\frac{1}{10} \sin 5x + \frac{1}{2} \sin x + C$

53. $\frac{1}{4} \sin 2x + \frac{1}{16} \sin 8x + C$

55. $2 - \sqrt{2}$

57. $\ln |1 - \cos x| + C$

61. $\frac{2}{15}$

63. $\ln 2 - \frac{9}{16}$

65. $\pi[\sqrt{2} + \ln (\sqrt{2} + 1)]$

Section 8.3

1. $\pi/8$

3. π

5. $\frac{1}{54} \arctan t/3 + \frac{1}{18}t/(9 + t^2) + C$

7. $\dfrac{x}{\sqrt{x^2 + 1}} + C$

9. $\frac{1}{25}\sqrt{5}$

11. $\dfrac{-3x^3}{16(9x^2 - 4)^{3/2}} + \dfrac{x}{16(9x^2 - 4)^{1/2}} + C$

13. $\dfrac{1}{3} \dfrac{x}{\sqrt{3 - x^2}} + C$

15. $\pi/12$

17. $\frac{1}{2} \ln |\sqrt{4x^2 + 4x + 2} + (2x + 1)| + C$

19. $\dfrac{2w^3}{3(1 - 2w^2)^{3/2}} + \dfrac{w}{(1 - 2w^2)^{1/2}} + C$

21. $\pi/2$

23. $\frac{1}{8} \arcsin (2x - 1) + \frac{1}{4}(2x - 1)\sqrt{x - x^2} + C$

25. $\frac{1}{18}x\sqrt{9x^2 - 1} + \frac{1}{54} \ln |3x + \sqrt{9x^2 - 1}| + C$

27. $-\frac{1}{2} \ln \left| \dfrac{\sqrt{x^2 + 4}}{x} + \dfrac{2}{x} \right| + C$

29. $\dfrac{\sqrt{4x^2 - 9}}{9x} + C$

31. $\frac{1}{2}x\sqrt{1 + x^2} - \frac{1}{2} \ln |\sqrt{1 + x^2} + x| + C$

33. $2 - \pi/2$

35. $\dfrac{x\sqrt{4 + x^2}}{2} + 2 \ln \left| \dfrac{\sqrt{4 + x^2}}{2} + \dfrac{x}{2} \right| + C$

37. $\dfrac{1}{81}\left(\dfrac{3\sqrt{3}}{8} - \dfrac{5\sqrt{2}}{12} \right)$

39. $\frac{1}{2}\sqrt{2x^2 + 12x + 19} - \frac{3}{2}\sqrt{2} \ln |\sqrt{2x^2 + 12x + 19} + \sqrt{2}(x + 3)| + C$

41. $\frac{1}{2}(x + 3)\sqrt{x^2 + 6x + 5} - 2 \ln \left| \dfrac{x + 3}{2} + \dfrac{\sqrt{x^2 + 6x + 5}}{2} \right| + C$

43. $\frac{1}{2}e^w\sqrt{1 + e^{2w}} + \frac{1}{2} \ln |\sqrt{1 + e^{2w}} + e^w| + C$

45. $\frac{1}{2}x^2 \arcsin x - \frac{1}{4} \arcsin x + \frac{1}{4}x\sqrt{1 - x^2} + C$

47. $\pi/4$

49. $\frac{9}{2}\sqrt{2} + \frac{9}{2} \ln (\sqrt{2} + 1)$

51. $480{,}000 \pi$ square feet

53. a. $2\pi^2 br^2$

b. The doughnut having $b = 4$ and $r = 2$ should cost more.

55. $\frac{1}{4}[\sqrt{2} + \ln (\sqrt{2} + 1)]$

57. $8\pi + 2\sqrt{2}\pi \ln (3 + 2\sqrt{2})$

59. $13{,}440\pi - 17{,}920$ foot-pounds

Section 8.2

1. $-\frac{1}{3} \cos^3 x + \frac{1}{5} \cos^5 x + C$

3. $\frac{1}{12} \sin^4 3x + C$

5. $\dfrac{1}{3} \cos^3 \dfrac{1}{x} - \dfrac{2}{5} \cos^5 \dfrac{1}{x} + \dfrac{1}{7} \cos^7 \dfrac{1}{x} + C$

7. $\frac{1}{8}y - \frac{1}{32} \sin 4y + C$

9. $-\frac{1}{128} \sin^3 2x \cos 2x - \frac{3}{256} \sin 2x \cos 2x + \frac{3}{128}x + C$

11. $-\frac{1}{9} \sin^{-9} x + \frac{1}{7} \sin^{-7} x + C$

13. $\frac{17}{8}x - \frac{1}{32} \sin 4x + C$

15. $\frac{3}{2}\sqrt{2} - 2$

17. $\frac{1}{6} \tan^6 x + C$

19. $\frac{7}{24}$

21. $-\frac{2}{3} - \frac{1}{3}\sqrt{2}$

23. $\frac{2}{5} \sec^5 \sqrt{x} - \frac{2}{3} \sec^3 \sqrt{x} + C$

25. $\frac{1}{6} \tan^6 x + \frac{1}{4} \tan^4 x + C$

27. $\frac{1}{5} \sec^5 x + C$

29. $-\frac{1}{4} \cot^4 x + C$

31. $\frac{2}{15}(\sqrt{2} + 1)$

33. $\frac{1}{2} \sin^2 x + C$

35. $\dfrac{1}{3 \cos^3 x} + C$

37. $\frac{1}{3} \sin^3 x - \frac{1}{5} \sin^5 x + C$

39. $\frac{1}{2} \sin^2 x + C$

41. $\tan x - x + C$

43. $\frac{1}{3} \tan^3 x - \tan x + x + C$

Section 8.4

1. $x - \ln|x + 1| + C$

3. $x + \frac{1}{2}\ln\left|\frac{x-1}{x+1}\right| + C$

5. $\ln\left|\frac{x^4}{(x-1)^3}\right| - \frac{5}{x-1} + C$

7. $\ln\frac{12}{7}$

9. $\frac{15}{2}\ln|t - 5| - \frac{9}{2}\ln|t - 3| + C$

11. $1 - \frac{1}{2}\ln 2$

13. $x - \frac{1}{2}\ln|x + 1| + \frac{3}{2}\ln|x - 1| + C$

15. $\frac{1}{2}\ln 3 - \frac{1}{6}\sqrt{3}\pi$

17. $3\ln|x - 2| - \frac{6}{x-2} + C$

19. $\ln\left|\frac{x-1}{x-2}\right| + C$

21. $\frac{1}{2}u^2 - 2u + 3\ln|u + 1| + \frac{1}{u+1} + C$

23. $\frac{-1}{4(x+1)} + \frac{1}{4}\ln\left|\frac{x+1}{x-1}\right| - \frac{1}{4(x-1)} + C$

25. $\frac{2}{9}\ln\left|\frac{x-2}{x+1}\right| - \frac{1}{3(x+1)} + C$

27. $\ln|x + 1| - \frac{1}{2}\ln(x^2 + 1) + \arctan x + \frac{x}{x^2+1} + C$

29. $\frac{1}{2}\ln|x^2 - x + 4| - \frac{1}{\sqrt{15}}\arctan\frac{\sqrt{15}(2x-1)}{15} + C$

31. $\ln\left|\frac{\sqrt{x+1}-1}{\sqrt{x+1}+1}\right| + C$

33. $\frac{6}{7}x^{7/6} - \frac{6}{5}x^{5/6} + 2x^{1/2} - 6x^{1/6} + 6\arctan x^{1/6} + C$

35. $\frac{4}{3} - \ln 3$

37. $\sin x - \arctan(\sin x) + C$

39. $-\frac{1}{3}\ln|1 - e^x| + \frac{1}{6}\ln(1 + e^x + e^{2x})$
$+ \frac{1}{\sqrt{3}}\arctan\frac{2e^x + 1}{\sqrt{3}} + C$

41. $\frac{1}{2}x^2\arctan x - \frac{1}{2}x + \frac{1}{2}\arctan x + C$

43. $x\ln(x^2 + 1) - 2x + 2\arctan x + C$

45. $\frac{2}{\sqrt{3}}\arctan\frac{1}{\sqrt{3}}\left(2\tan\frac{x}{2} + 1\right) + C$

47. $\frac{\sqrt{5}}{5}\ln\left|\frac{\sqrt{5}-1+2\tan x/2}{\sqrt{5}+1-2\tan x/2}\right| + C$

49. $\frac{9}{2} - \frac{1}{2}\ln 10$

51. $-\frac{5}{2}\ln 2 + \frac{9}{10}\ln 5 + \frac{\pi}{10} + \frac{2}{5}\arctan 2$

53. $2\pi(\frac{5}{2} - 3\ln 2)$

55. $3 + \ln 2$

Section 8.5

1. $\frac{x}{2}\sqrt{x^2 + 9} + \frac{9}{2}\ln|x + \sqrt{x^2 + 9}| + C$

3. $\frac{4}{101}e^5[(5\sin\frac{1}{2} - \frac{1}{2}\cos\frac{1}{2}) + \frac{1}{2}]$

5. $\frac{1}{12}\ln\left|\frac{2x-3}{2x+3}\right| + C$

7. $\sqrt{10x - \frac{1}{4}x^2} + 10\arccos\left(1 - \frac{x}{20}\right) + C$

9. $\frac{2}{3}e^{\sqrt{x}}(2\cosh 2\sqrt{x} - \sinh 2\sqrt{x}) + C$

11. $e\sin e + \cos e - \sin 1 - \cos 1$

13. $\frac{1}{2}x[\sin(\ln x) - \cos(\ln x)] + C$

15. $\frac{2x - \sqrt{x} - 3}{3}\sqrt{2\sqrt{x} - x} + \arccos(1 - \sqrt{x}) + C$

17. $-\frac{4 + x}{2}\sqrt{4 - x^2} + 2\arcsin\frac{x}{2} + C$

Section 8.6

T and S are the approximations by the Trapezoidal Rule and Simpson's Rule, respectively.

1. $T = \frac{1}{2}(1 + 2(\frac{1}{2}) + \frac{1}{3}) = \frac{7}{6}$;
$S = \frac{1}{3}(1 + 4(\frac{1}{2}) + \frac{1}{3}) = \frac{10}{9}$

3. $T = \frac{1}{2}(1 + 2(\frac{1}{2}) + 2(\frac{1}{3}) + 2(\frac{1}{4}) + 2(\frac{1}{5}) + 2(\frac{1}{6}) + \frac{1}{7})$
$= \frac{283}{140} \approx 2.02143$;
$S = \frac{1}{3}(1 + 4(\frac{1}{2}) + 2(\frac{1}{3}) + 4(\frac{1}{4}) + 2(\frac{1}{5}) + 4(\frac{1}{6}) + \frac{1}{7})$
$= \frac{617}{315} \approx 1.95873$

5. $T = \frac{1}{2}(1 + 2\sqrt{2} + 3) \approx 3.41421$;
$S = \frac{1}{3}(1 + 4\sqrt{2} + 3) \approx 3.21895$

7. $T = \frac{1}{2}[-1 + 2(2\ln 2 - 2) + 2(3\ln 3 - 3)$
$+ 2(4\ln 4 - 4) + 5\ln 5 - 5] \approx 2.25090$;
$S = \frac{1}{3}[-1 + 4(2\ln 2 - 2) + 2(3\ln 3 - 3)$
$+ 4(4\ln 4 - 4) + 5\ln 5 - 5] \approx 2.12158$

9. $T = \frac{1}{2}\sqrt{2}; S = \frac{2}{3}\sqrt{2}$

11. $T = \frac{\pi}{8}\left[1 + 2\left(\frac{1}{1 + \sqrt{2}/2}\right) + 2\left(\frac{1}{2}\right)\right.$
$\left. + 2\left(\frac{1}{1 + \sqrt{2}/2}\right) + 1\right] \approx 2.09825$;
$S = \frac{\pi}{12}\left[1 + 4\left(\frac{1}{1 + \sqrt{2}/2}\right) + 2\left(\frac{1}{2}\right)\right.$
$\left. + 4\left(\frac{1}{1 + \sqrt{2}/2}\right) + 1\right] \approx 2.01227$

13. approximately 1.06412

15. approximately 3.14191

17. 3

19. 20

21. $E_4^T \leq \frac{1}{12}; \frac{31}{30} \leq \int_1^3 \frac{1}{x}dx \leq \frac{6}{5}$

23. $E_4^S \leq \frac{1}{60}; \frac{13}{12} \leq \int_1^3 \frac{1}{x}dx \leq \frac{67}{60}$

25. $\frac{2\pi}{36}\left[\frac{1}{1 + 1/2} + 4\frac{1}{1 + \sqrt{3}/2} + 2\left(\frac{1}{2}\right) +\right.$
$\left. 4\frac{1}{1 + \sqrt{3}/2} + \frac{1}{1 + 1/2}\right] \approx 1.15550$

27. a. i. 3.13696; ii. 3.14157
b. $E_6^T \approx 0.00463$; $E_4^S \approx 0.00002$
29. a. 70   b. 8   35. c. no
39. $\frac{\pi}{12}(20 + 8\sqrt{26}) \approx 15.9153$
41. $\frac{1}{3}\pi^2(2\sqrt{31} + 3\sqrt{7} + 8) \approx 89.0659$

Section 8.7

1. converges; 10   3. diverges
5. converges; $\frac{3}{2}$ 7. diverges
9. converges; $2\sqrt{2}$ 11. diverges
13. converges; 0   15. diverges
17. diverges   19. diverges
21. converges; 0   23. converges; $\pi/2$
25. diverges   27. converges; $2^{-\pi+1}/(\pi - 1)$
29. diverges   31. converges; $\frac{1}{2}$
33. diverges   35. diverges
37. converges; $1/[2(\ln 2)^2]$ 39. diverges
41. diverges   43. diverges
45. converges; 1   47. diverges
49. converges; $\pi/2$ 51. converges; $\pi/4$
53. diverges   55. diverges
57. converges; 0   59. converges; π
61. b. no   63. finite area: $\frac{1}{3}$
65. infinite area   67. infinite area
75. V is finite if $p > \dfrac{1}{2}$; $V = \dfrac{\pi}{2p - 1}$

Chapter 8 Review Exercises

1. $x \ln(x^2 + 9) - 2x + 6 \arctan x/3 + C$
3. $-x \cot x + \ln|\sin x| + C$
5. $x \sinh x - \cosh x + C$
7. $\frac{1}{4}x^2 + \frac{1}{4}x \sin 2x + \frac{1}{8}\cos 2x + C$
9. $\frac{1}{6}\sin^2 x^3 + C$
11. $\frac{1}{4}\tan^4 x - \frac{1}{2}\tan^2 x - \ln|\cos x| + C$
13. $\frac{1}{2}x^2 \sin x^2 + \frac{1}{2}\cos x^2 + C$
15. $-\frac{2}{81}(1 - 3t)^{3/2} + \frac{4}{135}(1 - 3t)^{5/2} - \frac{2}{189}(1 - 3t)^{7/2} + C$
17. $-\csc x + \cot x + x + C$
19. $\dfrac{8}{243} \cdot \dfrac{(x + 1)^3}{(x^2 + 2x + 10)^{3/2}} + \dfrac{2}{3} \cdot \dfrac{1}{(x^2 + 2x + 10)^{3/2}} + \dfrac{1}{81} \cdot \dfrac{x + 1}{(x^2 + 2x + 10)^{1/2}} + C$
21. $x - \dfrac{3}{2}\arctan x + \dfrac{1}{2} \cdot \dfrac{x}{x^2 + 1} + C$
23. $\frac{2}{3}\ln|x + 6| + \frac{1}{3}\ln|x - 3| + C$
25. improper; diverges
27. proper; converges; $\sqrt{2}\pi/4 - \ln(\sqrt{2} + 1)$
29. proper; $2 + \ln\frac{4}{9}$

31. proper; $\frac{9}{560}\sqrt{2}$
33. proper; $\frac{1}{4}$
35. proper; $\ln(\sqrt{2} + 1) - \frac{1}{2}\sqrt{2}$
37. proper; $\pi/9$
39. proper; $\sqrt{3} + \frac{1}{2}\ln(2 + \sqrt{3})$
41. improper; diverges   43. improper; diverges
45. improper; $-\frac{1}{4}$ 47. improper; diverges
49. improper; $\frac{1}{2}$ 51. improper; $\frac{1}{2}\ln 2$
53. $\frac{1}{2}(\ln x)^2 + C$ 55. b. $\frac{1}{858}$
57. $3\pi/4 - \frac{3}{2}$
59. $\frac{1}{2}\ln|\sec^2 x + \sec x \tan x| - \frac{1}{2}\cos x + \frac{1}{2}x + C$
61. a. approximately 0.776130   b. approximately 0.781752
63. a. approximately 1.00709   b. approximately 1.00713
65. $9\pi/4$ 67. infinite area
69. infinite area   71. finite area: $\frac{1}{4}$
73. a. $(\frac{1}{2}\pi, \frac{5}{8}\pi)$
75. $\dfrac{\pi}{9}(\sqrt{2} + 2\sqrt{7} + \sqrt{5} + 2) \approx 3.81940$
77. $\frac{2}{5}\sqrt{2}(2e^\pi + 1)$
79. a. $231,000\pi$ foot-pounds   b. $308,000\pi$ foot-pounds

Cumulative Review Exercises (Chapters 1–7)

1. $\frac{1}{6}$ 3. 1
5. $\dfrac{e^{3x} + 2e^{2x}}{(e^x + 1)^2}$ 9. approximately 1.15
11. relative minimum value: $f(-3) = -\frac{1}{12}$; inflection point: $(-6, -\frac{2}{27})$

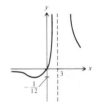

13. isosceles triangle   15. $\frac{\sqrt{3}}{18}\pi$
17. $\arcsin(x - 3) + C$ 19. $\frac{1}{6}\pi$
21. $\frac{2}{75}\pi(\sqrt{2} + 1)$

CHAPTER 9

Section 9.1

1.

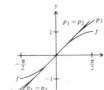